THE UNIVERSITY OF MAINE

RAYMOND H. FOGLER LIBRARY

# ENCYCLOPEDIA OF
# GEOLOGY

# ENCYCLOPEDIA OF GEOLOGY

EDITED BY

RICHARD C. SELLEY
L. ROBIN M. COCKS
IAN R. PLIMER

ELSEVIER
ACADEMIC
PRESS

Amsterdam  Boston  Heidelberg  London  New York  Oxford
Paris  San Diego  San Francisco  Singapore  Sydney  Tokyo

Elsevier Ltd., The Boulevard, Langford Lane, Kidlington, Oxford, OX5 1GB, UK

© 2005 Elsevier Ltd.

The following articles are © 2005, The Natural History Museum, London, UK:

FOSSIL VERTEBRATES/Hominids
Palaeontology
PALAEOZOIC/Silurian
PRECAMBRIAN/Overview
Terranes, Overview
Conservation of Geological Specimens
MINERALS/Olivines
MINERALS/Sulphates
TERTIARY TO PRESENT/Pleistocene and The Ice Age
Environmental Geochemistry
Biological Radiations and Speciation
PALAEOZOIC/Ordovician
TERTIARY TO PRESENT/Eocene
TERTIARY TO PRESENT/Paleocene
FOSSIL PLANTS/Angiosperms
FOSSIL PLANTS/Gymnosperms
Biozones
MESOZOIC/Cretaceous
MESOZOIC/End Cretaceous Extinctions
Stratigraphical Principles
FOSSIL INVERTEBRATES/Molluscs Overview
FOSSIL INVERTEBRATES/Trilobites
FOSSIL INVERTEBRATES/Echinoderms (Other Than Echinoids)
FOSSIL INVERTEBRATES/Echinoids
TERTIARY TO PRESENT/Pliocene
FOSSIL INVERTEBRATES/Bryozoans
MINERALS/Feldspathoids
Russia

The following article is a US Government work in the public domain and not subject to copyright:

NORTH AMERICA/Atlantic Margin

"Earth from Space" endpaper figure reproduced with permission from Reto Stockli, Nazmi El Saleous, and Marit Jentoft-Nilsen and NASA GSFC

*All rights reserved. No part of this publication may be reproduced or transmitted in any form or by any means, electronic or mechanical, including photocopy, recording, or any information storage and retrieval system, without permission in writing from the publishers.*

*Permissions may be sought directly from Elsevier's Rights Department in Oxford, UK: phone (+44) 1865 843830, fax (+44) 1865 853333, e-mail permissions@elsevier.com.*

*Requests may also be completed on-line via the homepage (http://www.elsevier.com/locate/permissions).*

First edition 2005

Library of Congress Control Number: 2004104445

A catalogue record for this book is available from the British Library

ISBN 0-12-636380-3 (set)

This book is printed on acid-free paper
Printed and bound in Spain

## Editors

### EDITORS

**Richard C. Selley**
Imperial College
London, UK

**L. Robin M. Cocks**
Natural History Museum
London, UK

**Ian R. Plimer**
University of Melbourne
Melbourne, VA
Australia

### CONSULTANT EDITOR

**Joe McCall**
Cirencester
Gloucestershire, UK

# Editorial Advisory Board

**Jaroslav Aichler**
Czech Geological Survey
Jeseník, Czech Republic

**Andrew R Armour**
Revus Energy A/S
Norway

**John Collinson**
Delos, Beech
Staffordshire, UK

**Alexander M Davis**
Infoscape Solutions Ltd.
Guildford, UK

**Peter Doyle**
University College London
London, UK

**Wolfgang Franke**
Institut für Geowissenschaften
Giessen, Germany

**Yves Fuchs**
Université Marne la Valle
France

**Paul Garrard**
Formerly Imperial College
London, UK

**R O Greiling**
Universität Heidelberg
Heidelberg, Germany

**Gwendy Hall**
Natural Resources Canada
Ottawa, ON, Canada

**Robert D Hatcher, Jr.**
University of Tennessee
Knoxville, TN, USA

**Georg Hoinkes**
Universität Graz
Universitätplatz 2
Graz, Austria

**R A Howie**
Royal Holloway, London University
London, UK

**Shunsho Ishihara**
Geological Survey of Japan
Tsukuba, Japan

**Gilbert Kelling**
Keele University
Keele, UK

**Ken Macdonald**
University of California Santa Barbara
Santa Barbara, CA, USA

**Norman MacLeod**
The Natural History Museum
London, UK

**Stuart Marsh**
British Geological Survey
Nottingham, UK

**Joe McCall**
Cirencester, Gloucestershire, UK

**David R Oldroyd**
University of New South Wales
Sydney, NSW, Australia

**Rong Jia-yu**
Nanjing Institute of Geology and Palaeontology
Nanjing, China

**Mike Rosenbaum**
Twickenham, UK

**Peter Styles**
Keele University
Keele, UK

**Hans D Sues**
Carnegie Museum of Natural History
Pittsburgh, PA, USA

**John Veevers**
Macquarie University
Sydney, NSW, Australia

**S H White**
Universiteit Utrecht
Utrecht, The Netherlands

# Foreword

Few areas of science can have changed as fast as geology has in the past forty years. In the first half of the last century geologists were divided, often bitterly, between the drifters and those who believed that the Earth and its continents were static. Neither side of this debate foresaw that the application of methods from physics, chemistry and mathematics to these speculations would revolutionize the study of all aspects of the Earth Sciences, and would lead to accurate and detailed reconstructions of world geography at former times, as well as to an understanding of the origin of the forces that maintain the continental movements. This change in world-view is no longer controversial, and is now embedded in every aspect of the Earth Sciences. It is a real pleasure to see this change, which has revitalized so many classic areas of research, reflected in the articles of this encyclopedia. Particularly affected are the articles on large-scale Earth processes, which discuss many of the new geological ideas that have come from geophysics and geochemistry. Forty years ago we had no understanding of these topics, which are fundamental to so many aspects of the Earth Sciences. The editors have decided, and in my view quite rightly, not to include detailed discussion of the present technology that is used to make geophysical and geochemical measurements. Such instrumental aspects are changing rapidly and become dated very quickly. They can easily be found in more technical publications. Instead the editors have concentrated on the influence such studies have had on our understanding of the Earth and its evolution, and in so doing have produced an excellent and accessible account of what is now known.

Any encyclopedia has to satisfy a wide variety of users, and in particular those who know that some subject like sedimentation or mineral exploration is part of geology, and go to an encyclopedia of geology to find out more. The editors have made a very thorough attempt to satisfy such users, and have included sections on such unexpected geological topics as the evolution of the Earth's atmosphere, the geology of Jupiter, Saturn, and their moons, aggregates, and creationism. I congratulate the editors and authors for producing such a fine summary of our present knowledge, and am particularly pleased that they intend to produce an online version of the encyclopedia. Though I have become addicted to using the Internet as my general encyclopedia, I will be delighted to be able to access something concerned with my own field that is as organized and scholarly as are these volumes.

**Dan McKenzie**
Royal Society Professor of Earth Sciences
Cambridge University, UK

# Introduction

*Civilization occurs by geological consent – subject to change without notice....*
Will Durant (1885–1981)

Richard de Bury, Bishop of Durham from 1333 to 1345, divided all knowledge into '*Geologia*', earthly knowledge, and '*Theologia*', heavenly knowledge. By the beginning of the last century, however, Geology was generally understood to be restricted to the study of rocks: according to the old dictum of the Geological Survey of Great Britain 'If you can hit it with a hammer, then it's geology.' Subsequently geology has been subsumed into Earth Science. This includes not only the study of rocks (the lithosphere), but also the atmosphere and hydrosphere and their relationship with the biosphere. Presently these relationships now form a nexus in Earth System Science.

The 'Encyclopedia of Geology' is what it says on the cover. What appealed to us when first approached to edit this work by Academic Press was a request that the encyclopedia should be rock-based. Readers are referred to the companion volumes, Encyclopedia of Atmospheric Sciences, Encyclopedia of the Solar System, Encyclopedia of Soils in the Environment and Encyclopedia of Ocean Sciences for knowledge on the other branches of Earth Science. Nonetheless we have extended our brief to include articles on the other planets and rocky detritus of our solar system, leaving others to argue, as no doubt Bishop Richard would have done, where the boundaries of earthly and heavenly knowledge might be. (His Grace would probably have charged the editors of the Encyclopedia of the Solar System with heresy.)

One of the first, and most difficult, tasks of editing this encyclopedia was to decide, not only which topics merited articles, but also how these articles should be grouped to facilitate the reader. This is easy for some branches of geology, but difficult for others. It is relatively easy to logically arrange articles on mineralogy and palaeontology, since they are defined by their chemistry and evolutionary biology. Articles that describe Earth history may be conveniently arranged in a chronological order, and articles on regional geology may be presented geographically. Other topics present problems, particularly in the area of sedimentology. There is, for example, a range of inter-related topics associated with deserts. This area could be described geomorphologically, and in terms of the aeolian and aqueous processes of deserts, aeolian sedimentary structures, and aeolian deposits. All of these aspects of deserts deserve mention, but there is no obvious logical way of arranging the discrete topics into articles. To help us in this task we relied heavily on our editorial board, whose individual members had more specialized knowledge of their field than we. To the Editorial Board Members, authors and anonymous referees of each article we give heartfelt thanks. We were also, of course, constrained by the willingness of expert authorities to contribute articles. To some degree therefore, the shape of the encyclopedia owes as much to the enthusiasm of experts to write for us, as for our 'wish list' of articles. To facilitate readers finding their way around the Encyclopedia of Geology great care has been taken in cross-referencing within and between articles, in providing 'See Also' lists at the end of articles, and in the index. No doubt it will be easier for readers to navigate around the online version of the work, than to manipulate the several hard copy volumes.

As geological knowledge expands there is always more to learn and understand. While preparing the 'Encyclopedia of Geology' we have ourselves learned a great deal about geology, both within and beyond our own specialties. We invite you to read this encyclopedia and join us in the field trip of a lifetime.

*Richard C. Selley*
*L. Robin M. Cocks*
*Ian R. Plimer*
*1 August 2004*

References to related encyclopedia published by Elsevier, Academic Press:

*Encyclopedia of the Solar System*, 1998
*Encyclopedia of Ocean Sciences*, 2001
*Encyclopedia of Atmospheric Sciences*, 2002
*Encyclopedia of Soils in the Environment*, 2005

# Guide to Use of the Encyclopedia

## Structure of the Encyclopedia

The material in the Encyclopedia is arranged as a series of entries in alphabetical order. Most entries consist of several articles that deal with various aspects of a topic and are arranged in a logical sequence within an entry. Some entries comprise a single article.

To help you realize the full potential of the material in the Encyclopedia we have provided three features to help you find the topic of your choice: a Contents List, Cross-References and an Index.

## 1. Contents List

Your first point of reference will probably be the contents list. The complete contents lists, which appears at the front of each volume will provide you with both the volume number and the page number of the entry. On the opening page of an entry a contents list is provided so that the full details of the articles within the entry are immediately available.

Alternatively you may choose to browse through a volume using the alphabetical order of the entries as your guide. To assist you in identifying your location within the Encyclopedia a running headline indicates the current entry and the current article within that entry.

You will find 'dummy entries' where obvious synonyms exist for entries or where we have grouped together related topics. Dummy entries appear in both the contents lists and the body of the text.

*Example*

If you were attempting to locate material on erosional sedimentary structures via the contents list:

**EROSION** *see* **SEDIMENTARY PROCESSES: Fluxes and Budgets; Aeolian Processes; Erosional Sedimentary Structures.**

The dummy entry directs you to the **Erosional Sedimentary Structures** article, in the SEDIMENTARY PROCESSES entry. At the appropriate location in the contents list, the page numbers for articles under Sedimentary Processes are given.

If you were trying to locate the material by browsing through the text and you looked up **Erosion** then the following information would be provided in the dummy entry:

# EROSION

See **SEDIMENTARY PROCESSES: Erosional Sedimentary Structures; Aeolian Processes; Fluxes and Budgets**

Alternatively, if you were looking up **Sedimentary Processes** the following information would be provided:

# SEDIMENTARY PROCESSES

Contents

**Erosional Sedimentary Structures**
**Depositional Sedimentary Structures**
**Post-Depositional Sedimentary Structures**
**Aeolian Processes**
**Catastrophic Floods**
**Deep Water Processes and Deposits**
**Fluvial Geomorphology**
**Glaciers**
**Karst and Palaeokarst**
**Landslides**
**Particle-Driven Subaqueous Gravity Processes**
**Deposition from Suspension**
**Fluxes and Budgets**

## 2. Cross-References

All of the articles in the Encyclopedia have been extensively cross-referenced.

The cross-references, which appear at the end of an article, serve three different functions. For example, at the end of the **PRECAMBRIAN: Overview** article, cross-references are used:

i. To indicate if a topic is discussed in greater detail elsewhere.

> **Africa:** Pan-African Orogeny. **Antarctic. Asia:** Central. **Australia:** Proterozoic. **Biosediments and Biofilms**. Earth Structure and Origins. Earth System Science. **Europe:** East European Craton; Timanides of Northern Russia. Gondwanaland and Gondwana. Grenvillian Orogeny. **Indian Subcontinent. North America:** Precambrian Continental Nucleus; Continental Interior. **Precambrian:** Eukaryote Fossils; Prokaryote Fossils; Vendian and Ediacaran. **Russia. Sedimentary Rocks:** Banded Iron Formations. Shields. Terranes, Overview.

ii. To draw the reader's attention to parallel discussions in other articles.

> **Africa:** Pan-African Orogeny. **Antarctic. Asia:** Central. **Australia:** Proterozoic. **Biosediments and Biofilms**. Earth Structure and Origins. Earth System Science. **Europe:** East European Craton; Timanides of Northern Russia. **Gondwanaland and Gondwana.** Grenvillian Orogeny. **Indian Subcontinent. North America:** Precambrian Continental Nucleus; Continental Interior. **Precambrian:** Eukaryote Fossils; Prokaryote Fossils; Vendian and Ediacaran. **Russia. Sedimentary Rocks:** Banded Iron Formations. Shields. Terranes, Overview.

iii. To indicate material that broadens the discussion.

> **Africa:** Pan-African Orogeny. **Antarctic. Asia:** Central. **Australia:** Proterozoic. **Biosediments and Biofilms. Earth Structure and Origins. Earth System Science. Europe:** East European Craton; Timanides of Northern Russia. **Gondwanaland and Gondwana. Grenvillian Orogeny. Indian Subcontinent. North America:** Precambrian Continental Nucleus; Continental Interior. **Precambrian:** Eukaryote Fossils; Prokaryote Fossils; Vendian and Ediacaran. **Russia. Sedimentary Rocks:** Banded Iron Formations. **Shields. Terranes, Overview.**

## 3. Index

The index will provide you with the page number where the material is located, and the index entries differentiate between material that is a whole article, is part of an article or is data presented in a figure or table. Detailed notes are provided on the opening page of the index.

## 4. Contributors

A full list of contributors appears at the beginning of each volume.

# Contributors

**Abart, R**
University of Basel, Basel, Switzerland

**Aldridge, R J**
University of Leicester, Leicester, UK

**Al-Jallal, I A**
Sandroses Est. for Geological, Geophysical Petroleum Engineering Consultancy and Petroleum Services, Khobar, Saudi Arabia

**Alkmim, F F**
Universidade Federal de Ouro Preto, Ouro Preto, Brazil

**Allen, P M**
Bingham, Nottingham, UK

**Allwood, A C**
Macquarie University, Sydney, NSW, Australia

**Al-Sharhan, A S**
United Arab Emirates University, Al-Ain, United Arab Emirates

**Anderson, L I**
National Museums of Scotland, Edinburgh, UK

**Arndt, N T**
LCEA, Grenoble, France

**Arnott, R**
Oxford Institute for Energy Studies, Oxford, UK

**Asimow, P D**
California Institute of Technology, Pasadena, CA, USA

**Atkinson, J**
City University, London, UK

**Bacon, M**
Petro-Canada, London, UK

**Bailey, J**
Anglo-Australian Observatory and Australian Centre for Astrobiology, Sydney, Australia

**Bani, P**
Institut de la Recherche pour le Développement, Nouméa, New Caledonia

**Bell, F G**
British Geological Survey, Keyworth, UK

**Bell, K**
Carleton University, Ottawa, ON, Canada

**Best, J**
University of Leeds, Leeds, UK

**Birch, W D**
Museum Victoria, Melbourne, VIC, Australia

**Bird, J F**
Imperial College London, London, UK

**Black, P**
Auckland University, Auckland, New Zealand

**Bleeker, W**
Geological Survey of Canada, Ottawa, ON, Canada

**Bogdanova, S V**
Lund University, Lund, Sweden

**Bommer, J J**
Imperial College London, London, UK

**Boore, D M**
United States Geological Survey, Menlo Park, CA, USA

**Bosence, D W J**
Royal Holloway, University of London, Egham, UK

**Boulanger, R W**
University of California, Davis, CA, USA

**Braga, J C**
University of Granada, Granada, Spain

**Branagan, D F**
University of Sydney, Sydney, NSW, Australia

**Brasier, M D**
University of Oxford, Oxford, UK

**Brewer, P A**
University of Wales, Aberystwyth, UK

**Bridge, M**
University College London, London, UK

**Brown, D**
Instituto de Ciencias de la Tierra 'Jaume Almera' CSIC, Barcelona, Spain

**Brown, A J**
Macquarie University, Sydney, NSW, Australia

**Brown, R J**
University of Bristol, Bristol, UK

## CONTRIBUTORS

**Bucher, K**
University of Freiburg, Freiburg, Germany

**Burns, S F**
Portland State University, Portland, OR, USA

**Byford, E**
Broken Hill, NSW, Australia

**Calder, E S**
Open University, Milton Keynes, UK

**Cameron, E M**
Eion Cameron Geochemical Inc., Ottawa, ON, Canada

**Carbotte, S M**
Columbia University, New York, NY, USA

**Carminati, E**
Università La Sapienza, Rome, Italy

**Chamberlain, S A**
Macquarie University, Sydney, NSW, Australia

**Charles, J A**
Formerly Building Research Establishment
Hertfordshire, UK

**Chiappe, L M**
Natural History Museum of Los Angeles County
Los Angeles, CA, USA

**Clack, J A**
University of Cambridge, Cambridge, UK

**Clayton, C**
Eardiston, Tenbury Wells, UK

**Clayton, G**
Trinity College, Dublin, Ireland

**Cocks, L R M**
The Natural History Museum, London, UK

**Coffin, M F**
University of Tokyo, Tokyo, Japan

**Collinson, J**
John Collinson Consulting, Beech, UK

**Comerford, G**
The Natural History Museum, London, UK

**Condie, K C**
New Mexico Tech, Socorro, NM, USA

**Cornford, C**
Integrated Geochemical Interpretation Ltd, Bideford, UK

**Cornish, L**
The Natural History Museum, London, UK

**Cosgrove, J W**
Imperial College London, London, UK

**Coxon, P**
Trinity College, Dublin, Ireland

**Cressey, G**
The Natural History Museum, London, UK

**Cribb, S J**
Carraig Associates, Inverness, UK

**Cronan, D S**
Imperial College London, London, UK

**Currant, A**
The Natural History Museum, London, UK

**Davies, H**
University of Papua New Guinea, Port Moresby
Papua New Guinea

**Davis, G R**
Imperial College London, London, UK

**DeCarli, P S**
SRI International, Menlo Park, CA, USA

**Dewey, J F**
University of California Davis
Davis, CA, USA, and University of Oxford, Oxford, UK

**Doglioni, C**
Università La Sapienza, Rome, Italy

**Dorning, K J**
University of Sheffield, Sheffield, UK

**Dott, Jr R H**
University of Wisconsin, Madison, WI, USA

**Doyle, P**
University College London, London, UK

**Dubbin, W E**
The Natural History Museum, London, UK

**Dyke, G J**
University College Dublin, Dublin, Ireland

**Echtler, H**
GeoForschungsZentrum Potsdam, Potsdam, Germany

**Eden, M A**
Geomaterials Research Services Ltd, Basildon, UK

**Eide, E A**
Geological Survey of Norway, Trondheim, Norway

**Eldholm, O**
University of Bergen, Bergen, Norway

**Elliott, D K**
Northern Arizona University, Flagstaff, AZ, USA

**Elliott, T**
University of Liverpool, Liverpool, UK

**Eriksen, A S**
Zetica, Witney, UK

**Fayers, S R**
University of Aberdeen, Aberdeen, UK

**Feenstra, A**
GeoForschungsZentrum Potsdam, Potsdam, Germany

**Felix, M**
University of Leeds, Leeds, UK

**Figueras, D**
BFI, Houston, TX, USA

**Fookes, P G**
Winchester, UK

**Forey, P L**
The Natural History Museum, London, UK

**Fortey, R A**
The Natural History Museum, London, UK

**Foster, D A**
University of Florida, Gainesville, FL, USA

**Frýda, J**
Czech Geological Survey, Prague, Czech Republic

**Franke, W**
Johann Wolfgang Goethe-Universität
Frankfurt am Main, Germany

**Franz, G**
Technische Universität Berlin, Berlin, Germany

**French, W J**
Geomaterials Research Services Ltd, Basildon, UK

**Fritscher, B**
Munich University, Munich, Germany

**Frostick, L**
University of Hull, Hull, UK

**Fuchs, Y**
Université Marne la Vallée, Marne la Vallée, France

**Gabbott, S E**
University of Leicester, Leicester, UK

**Garaebiti, E**
Department of Geology and Mines, Port Vila, Vanuatu

**Garetsky, R G**
Institute of Geological Sciences, Minsk, Belarus

**Garrard, P**
Imperial College London, London, UK

**Gascoyne, J K**
Zetica, Witney, UK

**Gee, D G**
University of Uppsala, Uppsala, Sweden

**Geshi, N**
Geological Survey of Japan, Ibaraki, Japan

**Giese, P**
Freie Universität Berlin, Berlin, Germany

**Giles, D P**
University of Portsmouth, Portsmouth, UK

**Glasser, N F**
University of Wales, Aberystwyth, UK

**Gluyas, J**
Acorn Oil and Gas Ltd., Staines, UK

**Gorbatschev, R**
Lund University, Lund, Sweden

**Gordon, J E**
Scottish Natural Heritage, Edinburgh, UK

**Gradstein, F M**
University of Oslo, Oslo, Norway

**Gray, D R**
University of Melbourne, Melbourne, VIC, Australia

**Greenwood, J R**
Nottingham Trent University, Nottingham, UK

**Grieve, R A F**
Natural Resources Canada, Ottawa, ON, Canada

**Griffiths, J S**
University of Plymouth, Plymouth, UK

**Hambrey, M J**
University of Wales, Aberystwyth, UK

**Hancock, J M**[†]
Formerly Imperial College London, London, UK

**Hansen, J M**
Danish Research Agency, Copenhagen, Denmark

**Harff, J**
Baltic Sea Research Institute Warnemünde, Rostock, Germany

[†]Deceased

**Harper, D A T**
Geologisk Museum, Copenhagen, Denmark

**Harper, E M**
University of Cambridge, Cambridge, UK

**Harrison, JP**
Imperial College London, London, UK

**Hatcher, Jr R D**
University of Tennessee, Knoxville, TN, USA

**Hatheway, A W**
Rolla, MO and Big Arm, MT, USA

**Hauzenberger, C A**
University of Graz, Graz, Austria

**Hawkins, A B**
Charlotte House, Bristol, UK

**Haymon, R M**
University of California–Santa Barbara
Santa Barbara, CA, USA

**He Guoqi**
Peking University, Beijing, China

**Head, J W**
Brown University, Providence, RI, USA

**Heim, N A**
University of Georgia, Athens, GA, USA

**Helvaci, C**
Dokuz Eylül Üniversitesi, İzmir, Turkey

**Hendriks, B W H**
Geological Survey of Norway, Trondheim, Norway

**Henk, A**
Universität Freiburg, Freiburg, Germany

**Herries Davies, G L**
University of Dublin, Dublin, Ireland

**Hey, R N**
University of Hawaii at Manoa, Honolulu, HI, USA

**Hoinkes, G**
University of Graz, Graz, Austria

**Hooker, J J**
The Natural History Museum, London, UK

**Horne, D J**
University of London, London, UK

**Hovland, M**
Statoil, Stavanger, Norway

**Howell, J**
University of Bergen, Bergen, Norway

**Howie, R A**
Royal Holloway, University of London, London, UK

**Hudson-Edwards, K**
University of London, London, UK

**Huggett, J M**
Petroclays, Ashtead, UK and The Natural History Museum, London, UK

**Hughes, N C**
University of California, Riverside, CA, USA

**Hutchinson, D R**
US Geological Survey, Woods Hole, MA, USA

**Idriss, I M**
University of California, Davis, CA, USA

**Ineson, J R**
Geological Survey of Denmark and Greenland
Geocenter Copenhagen, Copenhagen, Denmark

**Ivanov, M A**
Russian Academy of Sciences, Moscow, Russia

**Jäger, K D**
Martin Luther University, Halle, Germany

**Jarzembowski, E A**
University of Reading, Reading, UK and Maidstone Museum and Bentlif Art Gallery, Maidstone, UK

**Jones, B**
University of Alberta, Edmonton, AB, Canada

**Jones, G L**
Conodate Geology, Dublin, Ireland

**Joyner, L**
Cardiff University, Cardiff, UK

**Kaminski, M A**
University College London, London, UK

**Kay, S M**
Cornell University, Ithaca, NY, USA

**Kemp, A I S**
University of Bristol, Bristol, UK

**Kendall, A C**
University of East Anglia, Norwich, UK

**Kenrick, P**
The Natural History Museum, London, UK

# CONTRIBUTORS

**Kogiso, T**
Japan Marine Science and Technology Center, Yokosuka, Japan

**Krings, M**
Bayerische Staatssammlung für Paläontologie und Geologie, Geo-Bio Center, Munich, Germany

**Lancaster, N**
Desert Research Institute, Reno, NV, and United States Geological Survey, Reston, VA, USA

**Lang, K R**
Tufts University, Medford, MA, USA

**Laurent, G**
Brest, France

**Lee, E M**
York, UK

**Lemke, W**
Baltic Sea Research Institute Warnemünde, Rostock Germany

**Lesher, C M**
Laurentian University, ON, Canada

**Lewin, J**
University of Wales, Aberystwyth, UK

**Liu, J G**
Imperial College London, London, UK

**Long, J A**
The Western Australian Museum, Perth WA, Australia

**Loock, J C**
University of the Free State Bloemfontein, South Africa

**Lowell, R P**
Georgia Institute of Technology, Atlanta, GA, USA

**Lucas, S G**
New Mexico Museum of Natural History Albuquerque, NM, USA

**Lüning, S**
University of Bremen, Bremen, Germany

**Luo, Z-X**
Carnegie Museum of Natural History Pittsburgh, PA, USA

**Macdonald, K C**
University of California–Santa Barbara Santa Barbara, CA, USA

**Machel, H G**
University of Alberta, Edmonton, Alberta, Canada

**MacLeod, N**
The Natural History Museum, London, UK

**Maltman, A**
University of Wales, Aberystwyth, UK

**Martill, D M**
University of Portsmouth, Portsmouth, UK

**Martins-Neto, M A**
Universidade Federal de Ouro Preto, Ouro Preto, Brazil

**Marvin, U B**
Harvard-Smithsonian Center for Astrophysics Cambridge, MA, USA

**Mason, P J**
HME Partnership, Romford, UK

**Massonne, H-J**
Universität Stuttgart, Stuttgart, Germany

**Matte, P**
University of Montpellier II, Montpellier, France

**Mayor, A**
Princeton, USA

**McCaffrey, W**
University of Leeds, Leeds, UK

**McCall, G J H**
Cirencester, Gloucester, UK

**McCave, I N**
University of Cambridge, Cambridge, UK

**McGhee, G R**
Rutgers University, New Brunswick, NJ, USA

**McKibben, M A**
University of California, CA, USA

**McLaughlin, Jr P P**
Delaware Geological Society, Newark, DE, USA

**McManus, J**
University of St. Andrews, St. Andrews, UK

**McMenamin, M A S**
Mount Holyoke College, South Hadley, MA, USA

**Merriam, D F**
University of Kansas, Lawrence, KS, USA

**Metcalfe, I**
University of New England, Armidale, NSW, Australia

**Milke, R**
University of Basel, Basel, Switzerland

## CONTRIBUTORS

**Milner, A R**
Birkbeck College, London, UK

**Mojzsis, S J**
University of Colorado, Boulder, CO, USA

**Monger, J W H**
Geological Survey of Canada, Vancouver, BC, Canada
and Simon Fraser University Burnaby, BC, Canada

**Moore, P**
Selsey, UK

**Morris, N J**
The Natural History Museum, London, UK

**Mortimer, N**
Institute of Geological and Nuclear Sciences, Dunedin
New Zealand

**Mountney, N P**
Keele University, Keele, UK

**Mpodozis, C**
SIPETROL SA, Santiago, Chile

**Mungall, J E**
University of Toronto, Toronto, ON, Canada

**Myrow, P**
Colorado College, Colorado Springs, CO, USA

**Naish, D**
University of Portsmouth, Portsmouth, UK

**Nickel, E H**
CSIRO Exploration and Mining, Wembley, WA, Australia

**Nielsen, K C**
The University of Texas at Dallas, Richardson, TX, USA

**Nikishin, A M**
Lomonosov Moscow State University, Moscow, Russia

**Nokleberg, W J**
United States Geological Survey, Menlo Park, CA, USA

**Norbury, D**
CL Associates, Wokingham, UK

**O'Brien, P J**
Universität Potsdam, Potsdam, Germany

**Ogg, J G**
Purdue University, West Lafayette, IN, USA

**Oldershaw, C**
St. Albans, UK

**Oldroyd, D R**
University of New South Wales, Sydney, Australia

**Oneacre, J W**
BFI, Houston, TX, USA

**Orchard, M J**
Geological Survey of Canada
Vancouver, BC, Canada

**Orr, P J**
University College Dublin, Dublin, Ireland

**Owen, A W**
University of Glasgow, Glasgow, UK

**Pälike, H**
Stockholm University, Stockholm, Sweden

**Page, K N**
University of Plymouth, Plymouth, UK

**Paris, F**
University of Rennes 1, Rennes, France

**Parker, J R**
Formerly Shell EP International, London, UK

**Pfiffner, O A**
University of Bern, Bern, Switzerland

**Piper, D J W**
Geological Survey of Canada, Dartmouth, NS, Canada

**Price, R A**
Queens University Kingston, ON, Canada

**Prothero, D R**
Occidental College, Los Angeles, CA, USA

**Puche-Riart, O**
Polytechnic University of Madrid, Madrid, Spain

**Pye, K**
Royal Holloway, University of London, Egham, UK

**Rahn, P H**
South Dakota School of Mines and Technology
Rapid City, SD, USA

**Ramos, V A**
Universidad de Buenos Aires, Buenos Aires, Argentina

**Rankin, A H**
Kingston University, Kingston-upon-Thames, UK

**Rebesco, M**
Istituto Nazionale di Oceanografia e di Geofisica
Sperimentale (OGS), Italy

**Reedman, A J**
Mapperley, UK

**Reisz, R R**
University of Toronto at Mississauga
Mississauga, ON, Canada

**Retallack, G J**
University of Oregon, Eugene, OR, USA

**Rickards, R B**
University of Cambridge, Cambridge, UK

**Riding, R**
Cardiff University, Cardiff, UK

**Rigby, J K**
Brigham Young University, Provo, UT, USA

**Rigby, S**
University of Edinburgh, Edinburgh, UK

**Rodda, P**
Mineral Resources Department, Suva, Fiji

**Rona, P A**
Rutgers University, New Brunswick, NJ, USA

**Rose, E P F**
Royal Holloway, University of London, Egham, UK

**Rosenbaum, M S**
Twickenham, UK

**Rothwell, R G**
Southampton Oceanography Centre, Southampton, UK

**Roy, A B**
Presidency College, Kolkata, India

**Rushton, A W A**
The Natural History Museum, London, UK

**Russell, A J**
University of Newcastle upon Tyne, Newcastle upon Tyne, UK

**Schmid, R**
ETH-centre, Zurich, Switzerland

**Scott, E**
National Center for Science Education
Berkeley, CA, USA

**Scott, A C**
Royal Holloway, University of London, Egham, UK

**Scrutton, C T**
Formerly University of Durham, Durham, UK

**Searle, M**
University of Oxford, Oxford, UK

**Searle, R C**
University of Durham, Durham, UK

**Seibold, I**
University Library, Freiburg, Germany

**Selley, R C**
Imperial College London, London, UK

**Sellwood, B W**
University of Reading, Reading, UK

**Shields, G A**
James Cook University, Townsville, QLD, Australia

**Simms, M J**
Ulster Museum, Belfast, UK

**Slipper, I J**
University of Greenwich, Chatham Maritime, UK

**Smallwood, J R**
Amerada Hess plc, London, UK

**Smith, A B**
The Natural History Museum, London, UK

**Smith, I**
Auckland University, Auckland, New Zealand

**Snoke, A W**
University of Wyoming, Laramie, WY, USA

**Soligo, C**
The Natural History Museum, London, UK

**Stein, S**
Northwestern University, Evanston, IL, USA

**Steinberger, B**
Japan Marine Science and Technology Center
Yokosuka, Japan

**Stemmerik, L**
Geological Survey of Denmark and Greenland,
Geocenter Copenhagen, Copenhagen, Denmark

**Stern, R J**
The University of Texas at Dallas, Richardson, TX, USA

**Stewart, I**
University of Plymouth, Plymouth, UK

**Storey, B C**
University of Canterbury, Christchurch, New Zealand

**Storrs, G W**
Cincinnati Museum Center, Museum of Natural History
and Science, Cincinnati, OH, USA

**Strachan, R A**
University of Portsmouth, Portsmouth, UK

**Suetsugu, D**
Japan Marine Science and Technology Center, Yokosuka
Japan

**Surlyk, F**
University of Copenhagen, Geocenter Copenhagen,
Copenhagen, Denmark

**Tait, J**
Ludwig-Maximilians-Universität, München, Germany

**Talbot, M R**
University of Bergen, Bergen, Norway

**Taylor, P D**
The Natural History Museum, London, UK

**Taylor, T N**
University of Kansas, Lawrence, KS, USA

**Taylor, W E G**
University of Lancaster, Lancaster, UK

**Tazawa, J**
Niigata University, Niigata, Japan

**Theodor, J M**
Illinois State Museum, Springfield, IL, USA

**Timmerman, M J**
Universität Potsdam, Potsdam, Germany

**Tollo, R P**
George Washington University, Washington, DC, USA

**Torsvik, T H**
Geological Survey of Norway, Trondheim, Norway

**Trendall, A**
Curtin University of Technology, Perth, Australia

**Trewin, N H**
University of Aberdeen, Aberdeen, UK

**Turner, A K**
Colorado School of Mines, Colorado, USA

**Twitchett, R J**
University of Plymouth, Plymouth, UK

**Tyler, I M**
Geological Survey of Western Australia
East Perth, WA, Australia

**Valdes, P J**
University of Bristol, Bristol, UK

**van Geuns, L C**
Clingendael International Energy Programme
The Hague, The Netherlands

**van Staal, C R**
Geological Survey of Canada, Ottawa, ON, Canada

**Vaněček, M**
Charles University Prague, Prague, Czech Republic

**Vaughan, D J**
University of Manchester, Manchester, UK

**Veevers, J J**
Macquarie University, Sydney, NSW, Australia

**Verniers, J**
University of Ghent, Ghent, Belgium

**Wadge, G**
University of Reading, Reading, UK

**Walter, M R**
Macquarie University, Sydney, NSW, Australia

**Wang, H**
China University of Geosciences, Beijing, China

**Ware, N G**
Australian National University, Canberra, ACT, Australia

**Warke, P A**
Queen's University Belfast, Belfast, UK

**Weber, K J**
Technical University, Delft, The Netherlands

**Welch, M D**
The Natural History Museum, London, UK

**Westbrook, G K**
University of Birmingham, Birmingham, UK

**Westermann, G E G**
McMaster University, Hamilton, ON, Canada

**Whalley, W B**
Queen's University Belfast, Belfast, UK

**White, N C**
Brisbane, QLD, Australia

**White, S M**
University of South Carolina, Columbia, SC, USA

**Wignall, P B**
University of Leeds, Leeds, UK

**Williams, P A**
University of Western Sydney, Parramata, Australia

**Wise, W S**
University of California–Santa Barbara
Santa Barbara, CA, USA

**Worden, R H**
University of Liverpool, Liverpool, UK

**Wyatt, A R**
Sidmouth, UK

**Xiao, S**
Virginia Polytechnic Institute and State University
Blacksburg, VA, USA

**Yakubchuk, A S**
The Natural History Museum, London, UK

**Yates, A M**
University of the Witwatersrand, Johannesburg
South Africa

**Zhang Shihong**
China University of Geosciences, Beijing, China

**Ziegler, P A**
University of Basel, Basel, Switzerland

# Contents

## Volume 1

## A

AFRICA
    Pan-African Orogeny    *A Kröner, R J Stern*      1
    North African Phanerozoic    *S Lüning*      12
    Rift Valley    *L Frostick*      26

AGGREGATES    *M A Eden, W J French*      34

ALPS *See* EUROPE: The Alps

ANALYTICAL METHODS
    Fission Track Analysis    *B W H Hendriks*      43
    Geochemical Analysis (Including X-ray)    *R H Worden*      54
    Geochronological Techniques    *E A Eide*      77
    Gravity    *J R Smallwood*      92
    Mineral Analysis    *N G Ware*      107

ANDES    *S M Kay, C Mpodozis, V A Ramos*      118

ANTARCTIC    *B C Storey*      132

ARABIA AND THE GULF    *I A Al-Jallal, A S Al-Sharhan*      140

ARGENTINA    *V A Ramos*      153

ASIA
    Central    *S G Lucas*      164
    South-East    *I Metcalfe*      169

ASTEROIDS *See* SOLAR SYSTEM: Asteroids, Comets and Space Dust

ATMOSPHERE EVOLUTION    *S J Mojzsis*      197

AUSTRALIA
    Proterozoic    *I M Tyler*      208
    Phanerozoic    *J J Veevers*      222
    Tasman Orogenic Belt    *D R Gray, D A Foster*      237

## B

BIBLICAL GEOLOGY    *E Byford*      253

BIODIVERSITY    *A W Owen*      259

BIOLOGICAL RADIATIONS AND SPECIATION    *P L Forey*      266

BIOSEDIMENTS AND BIOFILMS    *M R Walter, A C Allwood*      279

BIOZONES    *N MacLeod*      294

BRAZIL    *F F Alkmim, M A Martins-Neto*      306

BUILDING STONE    *A W Hatheway*      328

## C

CALEDONIDE OROGENY *See* EUROPE: Caledonides Britain and Ireland; Scandinavian Caledonides (with Greenland)

| | |
|---|---|
| CARBON CYCLE    *G A Shields* | 335 |
| CHINA AND MONGOLIA    *H Wang, Shihong Zhang, Guoqi He* | 345 |
| CLAY MINERALS    *J M Huggett* | 358 |
| CLAYS, ECONOMIC USES    *Y Fuchs* | 366 |

COCCOLITHS *See* CALCAREOUS ALGAE

| | |
|---|---|
| COLONIAL SURVEYS    *A J Reedman* | 370 |

COMETS *See* SOLAR SYSTEM: Asteroids, Comets and Space Dust

| | |
|---|---|
| CONSERVATION OF GEOLOGICAL SPECIMENS    *L Cornish, G Comerford* | 373 |
| CREATIONISM    *E Scott* | 381 |

## D

DELTAS *See* SEDIMENTARY ENVIRONMENTS: Deltas

| | |
|---|---|
| DENDROCHRONOLOGY    *M Bridge* | 387 |

DESERTS *See* SEDIMENTARY ENVIRONMENTS: Deserts

| | |
|---|---|
| DIAGENESIS, OVERVIEW    *R C Selley* | 393 |

DINOSAURS *See* FOSSIL VERTEBRATES: Dinosaurs

## E

| | |
|---|---|
| EARTH | |
|     Mantle    *G J H McCall* | 397 |
|     Crust    *G J H McCall* | 403 |
|     Orbital Variation (Including Milankovitch Cycles)    *H Pälike* | 410 |
| EARTH STRUCTURE AND ORIGINS    *G J H McCall* | 421 |
| EARTH SYSTEM SCIENCE    *R C Selley* | 430 |

EARTHQUAKES *See* ENGINEERING GEOLOGY: Aspects of Earthquakes; TECTONICS: Earthquakes

| | |
|---|---|
| ECONOMIC GEOLOGY    *G R Davis* | 434 |
| ENGINEERING GEOLOGY | |
|     Overview    *M S Rosenbaum* | 444 |
|     Codes of Practice    *D Norbury* | 448 |
|     Aspects of Earthquakes    *A W Hatheway* | 456 |
|     Geological Maps    *J S Griffiths* | 463 |
|     Geomorphology    *E M Lee, J S Griffiths, P G Fookes* | 474 |
|     Geophysics    *J K Gascoyne, A S Eriksen* | 482 |
|     Seismology    *J J Bommer, D M Boore* | 499 |
|     Natural and Anthropogenic Geohazards    *G J H McCall* | 515 |
|     Liquefaction    *J F Bird, R W Boulanger, I M Idriss* | 525 |
|     Made Ground    *J A Charles* | 535 |

| | |
|---|---|
| Problematic Rocks    *F G Bell* | 543 |
| Problematic Soils    *F G Bell* | 554 |
| Rock Properties and Their Assessment    *F G Bell* | 566 |
| Site and Ground Investigation    *J R Greenwood* | 580 |

# Volume 2

| | |
|---|---|
| **ENGINEERING GEOLOGY** | |
| Site Classification    *A W Hatheway* | 1 |
| Subsidence    *A B Hawkins* | 9 |
| Ground Water Monitoring at Solid Waste Landfills    *J W Oneacre, D Figueras* | 14 |
| **ENVIRONMENTAL GEOCHEMISTRY**    *W E Dubbin* | 21 |
| **ENVIRONMENTAL GEOLOGY**    *P Doyle* | 25 |
| **EROSION** *See* **SEDIMENTARY PROCESSES:** Erosional Sedimentary Structures; Aeolian Processes; Fluxes and Budgets | |
| **EUROPE** | |
| East European Craton    *R G Garetsky, S V Bogdanova, R Gorbatschev* | 34 |
| Timanides of Northern Russia    *D G Gee* | 49 |
| Caledonides of Britain and Ireland    *R A Strachan , J F Dewey* | 56 |
| Scandinavian Caledonides  (with Greenland)    *D G Gee* | 64 |
| Variscan Orogeny    *W Franke, P Matte, J Tait* | 75 |
| The Urals    *D Brown, H Echtler* | 86 |
| Permian Basins    *A Henk, M J Timmerman* | 95 |
| Permian to Recent Evolution    *P A Ziegler* | 102 |
| The Alps    *O A Pfiffner* | 125 |
| Mediterranean Tectonics    *E Carminati, C Doglioni* | 135 |
| Holocene    *W Lemke, J Harff* | 147 |
| **EVOLUTION**    *S Rigby, E M Harper* | 160 |

# F

| | |
|---|---|
| **FAKE FOSSILS**    *D M Martill* | 169 |
| **FAMOUS GEOLOGISTS** | |
| Agassiz    *D R Oldroyd* | 174 |
| Cuvier    *G Laurent* | 179 |
| Darwin    *D R Oldroyd* | 184 |
| Du Toit    *J C Loock, D F Branagan* | 188 |
| Hall    *R H Dott, Jr* | 194 |
| Hutton    *D R Oldroyd* | 200 |
| Lyell    *D R Oldroyd* | 206 |
| Murchison    *D R Oldroyd* | 210 |
| Sedgwick    *D R Oldroyd* | 216 |
| Smith    *D R Oldroyd* | 221 |
| Steno    *J M Hansen* | 226 |
| Suess    *B Fritscher* | 233 |
| Walther    *I Seibold* | 242 |
| Wegener    *B Fritscher* | 246 |
| **FLUID INCLUSIONS**    *A H Rankin* | 253 |

| | |
|---|---|
| FORENSIC GEOLOGY    *K Pye* | 261 |
| **FOSSIL INVERTEBRATES** | |
|     Arthropods    *L I Anderson* | 274 |
|     Trilobites    *A W A Rushton* | 281 |
|     Insects    *E A Jarzembowski* | 295 |
|     Brachiopods    *D A T Harper* | 301 |
|     Bryozoans    *P D Taylor* | 310 |
|     Corals and Other Cnidaria    *C T Scrutton* | 321 |
|     Echinoderms (Other Than Echinoids)    *A B Smith* | 334 |
|     Crinoids    *M J Simms* | 342 |
|     Echinoids    *A B Smith* | 350 |
|     Graptolites    *R B Rickards* | 357 |
|     Molluscs Overview    *N J Morris* | 367 |
|     Bivalves    *E M Harper* | 369 |
|     Gastropods    *J Frýda* | 378 |
|     Cephalopods (Other Than Ammonites)    *P Doyle* | 389 |
|     Ammonites    *G E G Westermann* | 396 |
|     Porifera    *J K Rigby* | 408 |
| **FOSSIL PLANTS** | |
|     Angiosperms    *P Kenrick* | 418 |
|     Calcareous Algae    *J C Braga, R Riding* | 428 |
|     Fungi and Lichens    *T N Taylor, M Krings* | 436 |
|     Gymnosperms    *P Kenrick* | 443 |
| **FOSSIL VERTEBRATES** | |
|     Jawless Fish-Like Vertebrates    *D K Elliott* | 454 |
|     Fish    *J A Long* | 462 |
|     Palaeozoic Non-Amniote Tetrapods    *J A Clack* | 468 |
|     Reptiles Other Than Dinosaurs    *R R Reisz* | 479 |
|     Dinosaurs    *A M Yates* | 490 |
|     Birds    *G J Dyke, L M Chiappe* | 497 |
|     Swimming Reptiles    *G W Storrs* | 502 |
|     Flying Reptiles    *D Naish, D M Martill* | 508 |
|     Mesozoic Amphibians and Other Non-Amniote Tetrapods    *A R Milner* | 516 |
|     Cenozoic Amphibians    *A R Milner* | 523 |
|     Mesozoic Mammals    *Z-X Luo* | 527 |
|     Placental Mammals    *D R Prothero* | 535 |
|     Hominids    *L R M Cocks* | 541 |

# Volume 3

# G

| | |
|---|---|
| GAIA    *G J H McCall* | 1 |
| GEMSTONES    *C Oldershaw* | 6 |
| GEOARCHAEOLOGY    *L Joyner* | 14 |
| GEOCHEMICAL EXPLORATION    *E M Cameron* | 21 |
| GEOLOGICAL CONSERVATION    *J E Gordon* | 29 |
| GEOLOGICAL ENGINEERING    *A K Turner* | 35 |

| | |
|---|---:|
| GEOLOGICAL FIELD MAPPING    *P Garrard* | 43 |
| GEOLOGICAL MAPS AND THEIR INTERPRETATION    *A Maltman* | 53 |
| GEOLOGICAL SOCIETIES    *G L Herries Davies* | 60 |
| GEOLOGICAL SURVEYS    *P M Allen* | 65 |
| GEOLOGY, THE PROFESSION    *G L Jones* | 73 |
| GEOLOGY OF BEER    *S J Cribb* | 78 |
| GEOLOGY OF WHISKY    *S J Cribb* | 82 |
| GEOLOGY OF WINE    *J M Hancock*[†] | 85 |
| GEOMORPHOLOGY    *P H Rahn* | 90 |
| GEOMYTHOLOGY    *A Mayor* | 96 |
| GEOPHYSICS *See* EARTH: Orbital Variation (Including Milankovitch Cycles); EARTH SYSTEM SCIENCE; ENGINEERING GEOLOGY: Seismology; MAGNETOSTRATIGRAPHY; MOHO DISCONTINUITY; PALAEOMAGNETISM; PETROLEUM GEOLOGY: Exploration; REMOTE SENSING: Active Sensors; GIS; Passive Sensors; SEISMIC SURVEYS; TECTONICS: Seismic Structure at Mid-Ocean Ridges | |
| GEOTECHNICAL ENGINEERING    *D P Giles* | 100 |
| GEYSERS AND HOT SPRINGS    *G J H McCall* | 105 |
| GLACIERS *See* SEDIMENTARY PROCESSES: Glaciers | |
| GOLD    *M A McKibben* | 118 |
| GONDWANALAND AND GONDWANA    *J J Veevers* | 128 |
| GRANITE *See* IGNEOUS ROCKS: Granite | |
| GRENVILLIAN OROGENY    *R P Tollo* | 155 |

# H

| | |
|---|---:|
| HERCYNIAN OROGENY *See* EUROPE: Variscan Orogeny | |
| HIMALAYAS *See* INDIAN SUBCONTINENT | |
| HISTORY OF GEOLOGY UP TO 1780    *O Puche-Riart* | 167 |
| HISTORY OF GEOLOGY FROM 1780 TO 1835    *D R Oldroyd* | 173 |
| HISTORY OF GEOLOGY FROM 1835 TO 1900    *D R Oldroyd* | 179 |
| HISTORY OF GEOLOGY FROM 1900 TO 1962    *D F Branagan* | 185 |
| HISTORY OF GEOLOGY SINCE 1962    *U B Marvin* | 197 |

# I

| | |
|---|---:|
| IGNEOUS PROCESSES    *P D Asimow* | 209 |
| IGNEOUS ROCKS | |
|     Carbonatites    *K Bell* | 217 |
|     Granite    *A I S Kemp* | 233 |

[†]Deceased

| | |
|---|---|
| Kimberlite  *G J H McCall* | 247 |
| Komatiite  *N T Arndt, C M Lesher* | 260 |
| Obsidian  *G J H McCall* | 267 |

**IMPACT STRUCTURES**  *R A F Grieve* — 277

**INDIAN SUBCONTINENT**  *A B Roy* — 285

# J

**JAPAN**  *J Tazawa* — 297

**JUPITER** See SOLAR SYSTEM: Jupiter, Saturn and Their Moons

# L

**LAGERSTÄTTEN**  *S E Gabbott* — 307

**LARGE IGNEOUS PROVINCES**  *M F Coffin, O Eldholm* — 315

**LAVA**  *N Geshi* — 323

# M

**MAGNETOSTRATIGRAPHY**  *S G Lucas* — 331

**MANTLE PLUMES AND HOT SPOTS**  *D Suetsugu, T Kogiso, B Steinberger* — 335

**MARS** See SOLAR SYSTEM: Mars

**MERCURY** See SOLAR SYSTEM: Mercury

**MESOZOIC**
| | |
|---|---|
| Triassic  *S G Lucas, M J Orchard* | 344 |
| Jurassic  *K N Page* | 352 |
| Cretaceous  *N MacLeod* | 360 |
| End Cretaceous Extinctions  *N MacLeod* | 372 |

**METAMORPHIC ROCKS**
| | |
|---|---|
| Classification, Nomenclature and Formation  *G Hoinkes, C A Hauzenberger, R Schmid* | 386 |
| Facies and Zones  *K Bucher* | 402 |
| PTt-Paths  *P J O'Brien* | 409 |

**METEORITES** See SOLAR SYSTEM: Meteorites

**MICROFOSSILS**
| | |
|---|---|
| Acritarchs  *K J Dorning* | 418 |
| Chitinozoa  *F Paris, J Verniers* | 428 |
| Conodonts  *R J Aldridge* | 440 |
| Foraminifera  *M A Kaminski* | 448 |
| Ostracoda  *D J Horne* | 453 |
| Palynology  *P Coxon, G Clayton* | 464 |

**MICROPALAEONTOLOGICAL TECHNIQUES**  *I J Slipper* — 470

**MILANKOVITCH CYCLES** See EARTH: Orbital Variation (Including Milankovitch Cycles)

**MILITARY GEOLOGY**  *E P F Rose* — 475

**MINERAL DEPOSITS AND THEIR GENESIS**  *G R Davis* — 488

MINERALS
    Definition and Classification    E H Nickel    498
    Amphiboles    R A Howie    503
    Arsenates    K Hudson-Edwards    506
    Borates    C Helvaci    510
    Carbonates    B Jones    522
    Chromates    P A Williams    532
    Feldspars    R A Howie    534
    Feldspathoids    M D Welch    539
    Glauconites    J M Huggett    542
    Micas    R A Howie    548
    Molybdates    P A Williams    551
    Native Elements    P A Williams    553
    Nitrates    P A Williams    555
    Olivines    G Cressey, R A Howie    557
    Other Silicates    R A Howie    561
    Phosphates *See* SEDIMENTARY ROCKS: Phosphates
    Pyroxenes    R A Howie    567
    Quartz    R A Howie    569
    Sulphates    G Cressey    572
    Sulphides    D J Vaughan    574
    Tungstates    P A Williams    586
    Vanadates    P A Williams    588
    Zeolites    W S Wise    591
    Zircons    G J H McCall    601

MINING GEOLOGY
    Exploration Boreholes    M Vaněček    609
    Exploration    N C White    613
    Mineral Reserves    M Vaněček    623
    Hydrothermal Ores    M A McKibben    628
    Magmatic Ores    J E Mungall    637

MOHO DISCONTINUITY    P Giese    645

MOON *See* SOLAR SYSTEM: Moon

# Volume 4

# N

NEW ZEALAND    N Mortimer    1

NORTH AMERICA
    Precambrian Continental Nucleus    W Bleeker    8
    Continental Interior    D F Merriam    21
    Northern Cordillera    J W H Monger, R A Price, W J Nokleberg    36
    Southern Cordillera    A W Snoke    48
    Ouachitas    K C Nielsen    61
    Southern and Central Appalachians    R D Hatcher, Jr    72
    Northern Appalachians    C R van Staal    81
    Atlantic Margin    D R Hutchinson    92

## O

| | |
|---|---|
| OCEANIA (INCLUDING FIJI, PNG AND SOLOMONS)   *H Davies, P Bani, P Black, I Smith, E Garaebiti, P Rodda* | 109 |
| ORIGIN OF LIFE   *J Bailey* | 123 |

## P

| | |
|---|---|
| PALAEOCLIMATES   *B W Sellwood, P J Valdes* | 131 |
| PALAEOECOLOGY   *E M Harper, S Rigby* | 140 |
| PALAEOMAGNETISM   *T H Torsvik* | 147 |
| PALAEONTOLOGY   *L R M Cocks* | 156 |
| PALAEOPATHOLOGY   *S G Lucas* | 160 |
| PALAEOZOIC | |
|    Cambrian   *N C Hughes, N A Heim* | 163 |
|    Ordovician   *R A Fortey* | 175 |
|    Silurian   *L R M Cocks* | 184 |
|    Devonian   *G R McGhee* | 194 |
|    Carboniferous   *A C Scott* | 200 |
|    Permian   *P B Wignall* | 214 |
|    End Permian Extinctions   *R J Twitchett* | 219 |
| PANGAEA   *S G Lucas* | 225 |
| PETROLEUM GEOLOGY | |
|    Overview   *J Gluyas* | 229 |
|    Chemical and Physical Properties   *C Clayton* | 248 |
|    Gas Hydrates   *M Hovland* | 261 |
|    The Petroleum System   *C Cornford* | 268 |
|    Exploration   *J R Parker* | 295 |
|    Production   *K J Weber, L C van Geuns* | 308 |
|    Reserves   *R Arnott* | 331 |
| PLATE TECTONICS   *R C Searle* | 340 |
| PRECAMBRIAN | |
|    Overview   *L R M Cocks* | 350 |
|    Eukaryote Fossils   *S Xiao* | 354 |
|    Prokaryote Fossils   *M D Brasier* | 363 |
|    Vendian and Ediacaran   *M A S McMenamin* | 371 |
| PSEUDOFOSSILS   *D M Martill* | 382 |
| PYROCLASTICS   *R J Brown, E S Calder* | 386 |

## Q

| | |
|---|---|
| QUARRYING   *A W Hatheway* | 399 |

## R

| | |
|---|---|
| REEFS *See* SEDIMENTARY ENVIRONMENTS: Reefs ("Build-Ups") | |
| REGIONAL METAMORPHISM   *A Feenstra, G Franz* | 407 |

## REMOTE SENSING
    Active Sensors    *G Wadge* — 414
    GIS    *P J Mason* — 420
    Passive Sensors    *J G Liu* — 431

## RIFT VALLEYS *See* AFRICA: Rift Valley

## ROCK MECHANICS    *JP Harrison* — 440

## ROCKS AND THEIR CLASSIFICATION    *R C Selley* — 452

## RUSSIA    *A S Yakubchuk, A M Nikishin* — 456

# S

## SATURN *See* SOLAR SYSTEM: Jupiter, Saturn and Their Moons

## SEAMOUNTS    *S M White* — 475

## SEDIMENTARY ENVIRONMENTS
    Depositional Systems and Facies    *J Collinson* — 485
    Alluvial Fans, Alluvial Sediments and Settings    *K D Jäger* — 492
    Anoxic Environments    *P B Wignall* — 495
    Carbonate Shorelines and Shelves    *D W J Bosence* — 501
    Contourites    *M Rebesco* — 513
    Deltas    *T Elliott* — 528
    Deserts    *N P Mountney* — 539
    Lake Processes and Deposits    *M R Talbot* — 550
    Reefs ('Build-Ups')    *B W Sellwood* — 562
    Shoreline and Shoreface Deposits    *J Howell* — 570
    Storms and Storm Deposits    *P Myrow* — 580

## SEDIMENTARY PROCESSES
    Erosional Sedimentary Structures    *J Collinson* — 587
    Depositional Sedimentary Structures    *J Collinson* — 593
    Post-Depositional Sedimentary Structures    *J Collinson* — 602
    Aeolian Processes    *N Lancaster* — 612
    Catastrophic Floods    *A J Russell* — 628
    Deep Water Processes and Deposits    *D J W Piper* — 641
    Fluvial Geomorphology    *J Lewin, P A Brewer* — 650
    Glaciers    *M J Hambrey, N F Glasser* — 663
    Karst and Palaeokarst    *M J Simms* — 678
    Landslides    *S F Burns* — 687

# Volume 5

## SEDIMENTARY PROCESSES
    Particle-Driven Subaqueous Gravity Processes    *M Felix, W McCaffrey* — 1
    Deposition from Suspension    *I N McCave* — 8
    Fluxes and Budgets    *L Frostick* — 17

## SEDIMENTARY ROCKS
    Mineralogy and Classification    *R C Selley* — 25
    Banded Iron Formations    *A Trendall* — 37
    Chalk    *J R Ineson, L Stemmerik, F Surlyk* — 42
    Chert    *N H Trewin, S R Fayers* — 51

| | |
|---|---|
| Clays and Their Diagenesis    *J M Huggett* | 62 |
| Deep Ocean Pelagic Oozes    *R G Rothwell* | 70 |
| Dolomites    *H G Machel* | 79 |
| Evaporites    *A C Kendall* | 94 |
| Ironstones    *W E G Taylor* | 97 |
| Limestones    *R C Selley* | 107 |
| Oceanic Manganese Deposits    *D S Cronan* | 113 |
| Phosphates    *W D Birch* | 120 |
| Rudaceous Rocks    *J McManus* | 129 |
| Sandstones, Diagenesis and Porosity Evolution    *J Gluyas* | 141 |

SEISMIC SURVEYS    *M Bacon* — 151

SEQUENCE STRATIGRAPHY    *P P McLaughlin, Jr* — 159

SHIELDS    *K C Condie* — 173

SHOCK METAMORPHISM    *P S DeCarli* — 179

SOIL MECHANICS    *J Atkinson* — 184

SOILS
| | |
|---|---|
| Modern    *G J Retallack* | 194 |
| Palaeosols    *G J Retallack* | 203 |

SOLAR SYSTEM
| | |
|---|---|
| The Sun    *K R Lang* | 209 |
| Asteroids, Comets and Space Dust    *P Moore* | 220 |
| Meteorites    *G J H McCall* | 228 |
| Mercury    *G J H McCall* | 238 |
| Venus    *M A Ivanov, J W Head* | 244 |
| Moon    *P Moore* | 264 |
| Mars    *M R Walter, A J Brown, S A Chamberlain* | 272 |
| Jupiter, Saturn and Their Moons    *P Moore* | 282 |
| Neptune, Pluto and Uranus    *P Moore* | 289 |

SPACE DUST *See* SOLAR SYSTEM: Asteroids, Comets and Space Dust

STRATIGRAPHICAL PRINCIPLES    *N MacLeod* — 295

STROMATOLITES *See* BIOSEDIMENTS AND BIOFILMS

SUN *See* SOLAR SYSTEM: The Sun

# T

TECTONICS
| | |
|---|---|
| Convergent Plate Boundaries and Accretionary Wedges    *G K Westbrook* | 307 |
| Earthquakes    *G J H McCall* | 318 |
| Faults    *S Stein* | 330 |
| Folding    *J W Cosgrove* | 339 |
| Fractures (Including Joints)    *J W Cosgrove* | 352 |
| Hydrothermal Activity    *R P Lowell, P A Rona* | 362 |
| Mid-Ocean Ridges    *K C Macdonald* | 372 |
| Hydrothermal Vents At Mid-Ocean Ridges    *R M Haymon* | 388 |
| Propagating Rifts and Microplates At Mid-Ocean Ridges    *R N Hey* | 396 |
| Seismic Structure At Mid-Ocean Ridges    *S M Carbotte* | 405 |
| Mountain Building and Orogeny    *M Searle* | 417 |
| Neotectonics    *I Stewart* | 425 |

| | |
|---|---:|
| Ocean Trenches    *R J Stern* | 428 |
| Rift Valleys    *L Frostick* | 437 |
| **TEKTITES**    *G J H McCall* | 443 |
| **TERRANES OVERVIEW**    *L R M Cocks* | 455 |
| **TERTIARY TO PRESENT** | |
| Paleocene    *J J Hooker* | 459 |
| Eocene    *J J Hooker* | 466 |
| Oligocene    *D R Prothero* | 472 |
| Miocene    *J M Theodor* | 478 |
| Pliocene    *C Soligo* | 486 |
| Pleistocene and The Ice Age    *A Currant* | 493 |
| **THERMAL METAMORPHISM**    *R Abart, R Milke* | 499 |
| **TIME SCALE**    *F M Gradstein, J G Ogg* | 503 |
| **TRACE FOSSILS**    *P J Orr* | 520 |

# U

| | |
|---|---:|
| **ULTRA HIGH PRESSURE METAMORPHISM**    *H-J Massonne* | 533 |
| **UNCONFORMITIES**    *A R Wyatt* | 541 |
| **UNIDIRECTIONAL AQUEOUS FLOW**    *J Best* | 548 |
| **URALS** *See* **EUROPE:** The Urals | |
| **URBAN GEOLOGY**    *A W Hatheway* | 557 |

# V

| | |
|---|---:|
| **VENUS** *See* **SOLAR SYSTEM:** Venus | |
| **VOLCANOES**    *G J H McCall* | 565 |

# W

| | |
|---|---:|
| **WEATHERING**    *W B Whalley, P A Warke* | 581 |

**Index**    591

# AFRICA

Contents

**Pan-African Orogeny**
**North African Phanerozoic**
**Rift Valley**

## Pan-African Orogeny

**A Kröner**, Universität Mainz, Mainz, Germany
**R J Stern**, University of Texas-Dallas, Richardson TX, USA

© 2005, Elsevier Ltd. All Rights Reserved.

### Introduction

The term 'Pan-African' was coined by WQ Kennedy in 1964 on the basis of an assessment of available Rb–Sr and K–Ar ages in Africa. The Pan-African was interpreted as a tectono-thermal event, some 500 Ma ago, during which a number of mobile belts formed, surrounding older cratons. The concept was then extended to the Gondwana continents (**Figure 1**) although regional names were proposed such as Brasiliano for South America, Adelaidean for Australia, and Beardmore for Antarctica. This thermal event was later recognized to constitute the final part of an orogenic cycle, leading to orogenic belts which are currently interpreted to have resulted from the amalgamation of continental domains during the period ~870 to ~550 Ma. The term Pan-African is now used to describe tectonic, magmatic, and metamorphic activity of Neoproterozoic to earliest Palaeozoic age, especially for crust that was once part of Gondwana. Because of its tremendous geographical and temporal extent, the Pan-African cannot be a single orogeny but must be a protracted orogenic cycle reflecting the opening and closing of large oceanic realms as well as accretion and collision of buoyant crustal blocks. Pan-African events culminated in the formation of the Late Neoproterozoic supercontinent Gondwana (**Figure 1**). The Pan-African orogenic cycle is time-equivalent with the Cadomian Orogeny in western and central Europe and the Baikalian in Asia; in fact, these parts of Europe and Asia were probably part of Gondwana in pre-Palaeozoic times as were small Neoproterozoic crustal fragments identified in Turkey, Iran and Pakistan (**Figure 1**).

Within the Pan-African domains, two broad types of orogenic or mobile belts can be distinguished. One type consists predominantly of Neoproterozoic supracrustal and magmatic assemblages, many of juvenile (mantle-derived) origin, with structural and metamorphic histories that are similar to those in Phanerozoic collision and accretion belts. These belts expose upper to middle crustal levels and contain diagnostic features such as ophiolites, subduction- or collision-related granitoids, island-arc or passive continental margin assemblages as well as exotic terranes that permit reconstruction of their evolution in Phanerozoic-style plate tectonic scenarios. Such belts include the Arabian-Nubian shield of Arabia and north-east Africa (**Figure 2**), the Damara–Kaoko–Gariep Belt and Lufilian Arc of south-central and south-western Africa, the West Congo Belt of Angola and Congo Republic, the Trans-Sahara Belt of West Africa, and the Rokelide and Mauretanian belts along the western part of the West African Craton (**Figure 1**).

The other type of mobile belt generally contains polydeformed high-grade metamorphic assemblages, exposing middle to lower crustal levels, whose origin, environment of formation and structural evolution are more difficult to reconstruct. The protoliths of these assemblages consist predominantly of much older Mesoproterozoic to Archaean continental crust that was strongly reworked during the Neoproterozoic. Well studied examples are the Mozambique Belt of East Africa, including Madagascar (**Figure 2**) with extensions into western Antarctica, the Zambezi Belt of northern Zimbabwe and Zambia and, possibly, the little known migmatitic terranes of Chad, the Central African Republic, the Tibesti Massif in Libya and the western parts of Sudan and Egypt (**Figure 1**). It has been proposed that the latter type of belt represents the deeply eroded part of a collisional orogen and that the two types of Pan-African belts are not fundamentally different but constitute different crustal levels of collisional and/or accretional systems. For this reason, the term East African Orogen has been proposed for the combined upper crustal Arabian-Nubian Shield and lower crustal Mozambique Belt (**Figure 2**).

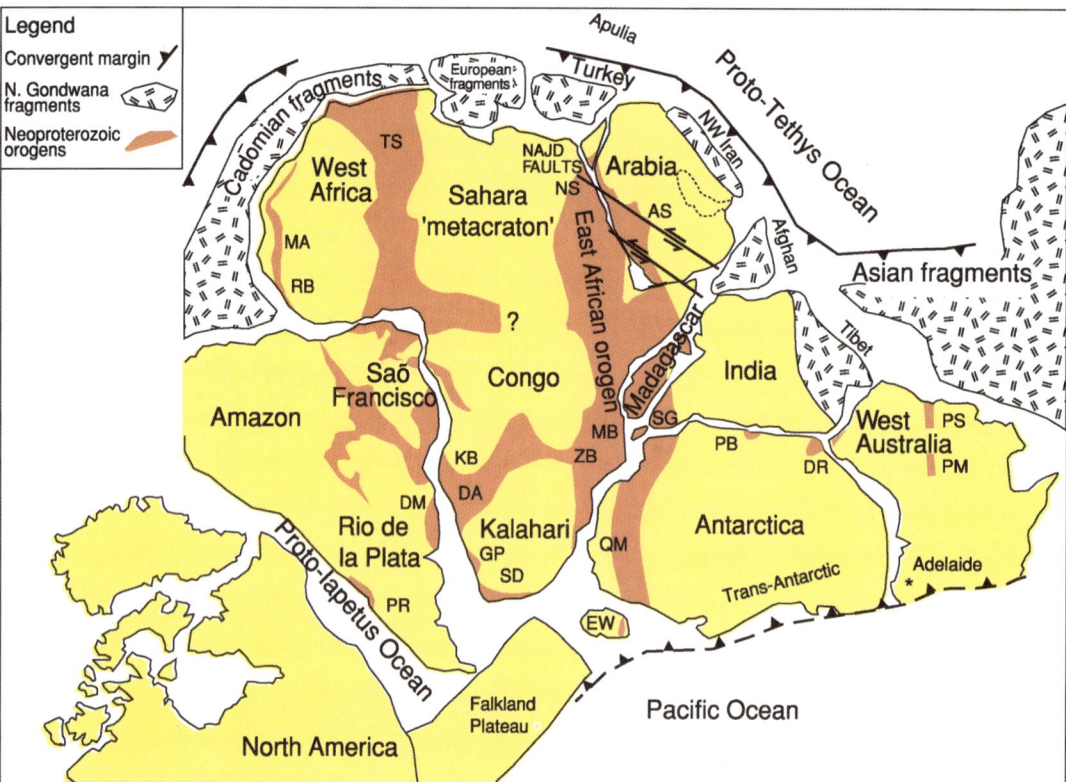

**Figure 1** Map of Gondwana at the end of Neoproterozoic time (~540 Ma) showing the general arrangement of Pan-African belts. AS, Arabian Shield; BR, Brasiliano; DA, Damara; DM, Dom Feliciano; DR, Denman Darling; EW, Ellsworth-Whitmore Mountains; GP, Gariep; KB, Kaoko; MA, Mauretanides; MB, Mozambique Belt; NS, Nubian Shield; PM, Peterman Ranges; PB, Pryolz Bay; PR, Pampean Ranges; PS, Paterson; QM, Queen Maud Land; RB, Rokelides; SD, Saldania; SG, Southern Granulite Terrane; TS, Trans-Sahara Belt; WB, West Congo; ZB, Zambezi. (Reproduced with permission from Kusky et al., 2003.)

The Pan-African system of orogenic belts in Africa, Brazil and eastern Antarctica has been interpreted as a network surrounding older cratons (**Figure 1**) and essentially resulting from closure of several major Neoproterozoic oceans. These are the Mozambique Ocean between East Gondwana (Australia, Antarctica, southern India) and West Gondwana (Africa, South America), the Adamastor Ocean between Africa and South America, the Damara Ocean between the Kalahari and Congo cratons, and the Trans-Sahara Ocean between the West African Craton and a poorly known pre-Pan-African terrane in north-central Africa variously known as the Nile or Sahara Craton (**Figure 1**).

## Arabian-Nubian Shield (ANS)

A broad region was uplifted in association with Cenozoic rifting to form the Red Sea, exposing a large tract of mostly juvenile Neoproterozoic crust. These exposures comprise the Arabian-Nubian Shield (ANS). The ANS makes up the northern half of the East African orogen and stretches from southern Israel and Jordan south as far as Ethiopia and Yemen, where the ANS transitions into the Mozambique Belt (**Figure 2**). The ANS is distinguished from the Mozambique Belt by its dominantly juvenile nature, relatively low grade of metamorphism, and abundance of island-arc rocks and ophiolites. The ANS, thus defined, extends about 3000 km north to south and >500 km on either side of the Red Sea (**Figure 3**). It is flanked to the west by a broad tract of older crust that was remobilized during Neoproterozoic time along with a significant amount of juvenile Neoproterozoic crust, known as the Nile Craton or 'Saharan Metacraton'. The extent of juvenile Neoproterozoic crust to the east in the subsurface of Arabia is not well defined, but it appears that Pan-African crust underlies most of this region. Scattered outcrops in Oman yielded mostly Neoproterozoic radiometric ages for igneous rocks, and there is no evidence that a significant body of pre-Pan-African crust underlies this region. The ANS is truncated to the north as a result of rifting at about the time of the Precambrian–Cambrian boundary, which generated crustal fragments now preserved in south-east Europe, Turkey and Iran.

The ANS is by far the largest tract of mostly juvenile Neoproterozoic crust among the regions of Africa that were affected by the Pan-African orogenic cycle. It

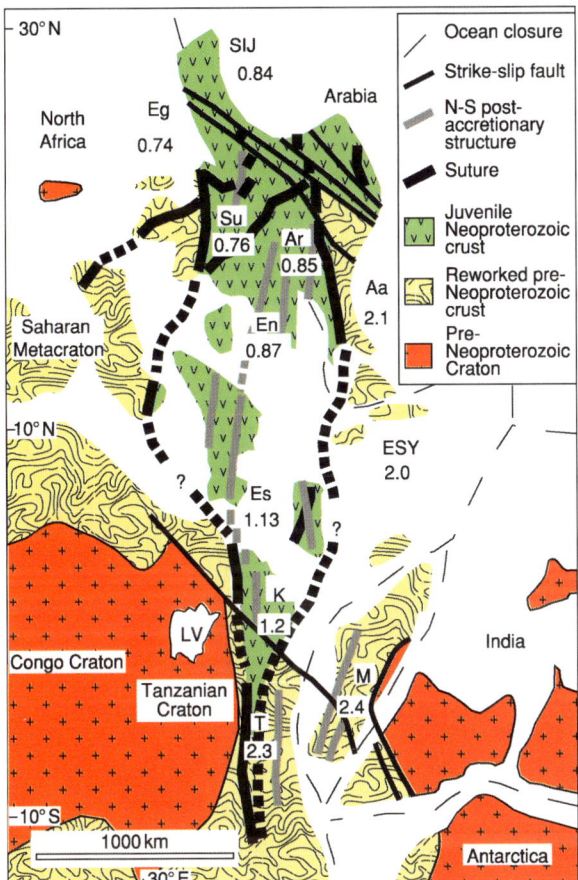

**Figure 2** Pre-Jurassic configuration of elements of the East African Orogen in Africa and surrounding regions. Regions include Egypt (Eg), Sudan (Su), Sinai–Israel–Jordan (SIJ), Afif terrane, Arabia (Aa), rest of Arabian Shield (Ar), Eritrea and northern Ethiopia (En), southern Ethiopia (Es), eastern Ethiopia, Somalia, and Yemen (ESY), Kenya (K), Tanzania (T), and Madagascar (M). Numbers in italics beneath each region label are mean Nd-model ages in Gy.

formed as a result of a multistage process, whereby juvenile crust was produced above intra-oceanic convergent plate boundaries (juvenile arcs) and perhaps oceanic plateaux (ca. 870–630 Ma), and these juvenile terranes collided and coalesced to form larger composite terranes (**Figure 4**). There is also a significant amount of older continental crust (Mesoproterozoic age crust of the Afif terrane in Arabia; Palaeoproterozoic and Archaean crust in Yemen, **Figure 2**) that was overprinted by Pan-African tectonomagmatic events. ANS terrane boundaries (**Figure 3**) are frequently defined by suture zones that are marked by ophiolites, and the terranes are stitched together by abundant tonalitic to granodioritic plutons. Most ANS ophiolites have trace element chemical compositions suggesting formation above a convergent plate margin, either as part of a back-arc basin or in a fore-arc setting. Boninites have been identified in Sudan and Eritrea and suggest a forearc setting for at least some ANS sequences. Sediments are mostly immature sandstones and wackes derived from nearby arc volcanoes. Deposits that are diagnostic of Neoproterozoic 'snowball Earth' episodes have been recognized in parts of the ANS, and banded iron formations in the northern ANS may be deep-water expressions of snowball Earth events. Because it mostly lies in the Sahara and Arabian deserts, the ANS has almost no vegetation or soil and is excellently exposed. This makes it very amenable to study using imagery from remote sensing satellites.

Juvenile crust of the ANS was sandwiched between continental tracts of East and West Gondwana (**Figure 4**). The precise timing of the collision is still being resolved, but appears to have occurred after ∼630 Ma when high-magnesium andesite 'schistose dykes' were emplaced in southern Israel but before the ∼610 Ma post-tectonic 'Mereb' granites were emplaced in northern Ethiopia. By analogy with the continuing collision between India and Asia, the terminal collision between East and West Gondwana may have continued for a few tens of millions of years. Deformation in the ANS ended by the beginning of Cambrian time, although it has locally continued into Cambrian and Ordovician time farther south in Africa. The most intense collision (i.e. greatest shortening, highest relief, and greatest erosion) occurred south of the ANS, in the Mozambique belt. Compared to the strong deformation and metamorphism experienced during collision in the Mozambique belt, the ANS was considerably less affected by the collision. North-west trending leftlateral faults of the Najd fault system of Arabia and Egypt (**Figures 1** and **2**) formed as a result of escape tectonics associated with the collision and were active between about 630 and 560 Ma. Deformation associated with terminal collision is more intense in the southern ANS, with tight, upright folds, steep thrusts, and strike-slip shear zones controlling basement fabrics in Eritrea, Ethiopia, and southern Arabia. These north–south trending, collision-related structures obscure the earlier structures in the southern ANS that are related to arc accretion, and the intensity of this deformation has made it difficult to identify ophiolitic assemblages in southern Arabia, Ethiopia, and Eritrea. Thus, the transition between the ANS and the Mozambique Belt is marked by a change from less deformed and less metamorphosed, juvenile crust in the north to more deformed and more metamorphosed, remobilized older crust in the south, with the structural transition occurring farther north than the lithological transition.

The final stages in the evolution of the ANS witnessed the emplacement of post-tectonic 'A-type'

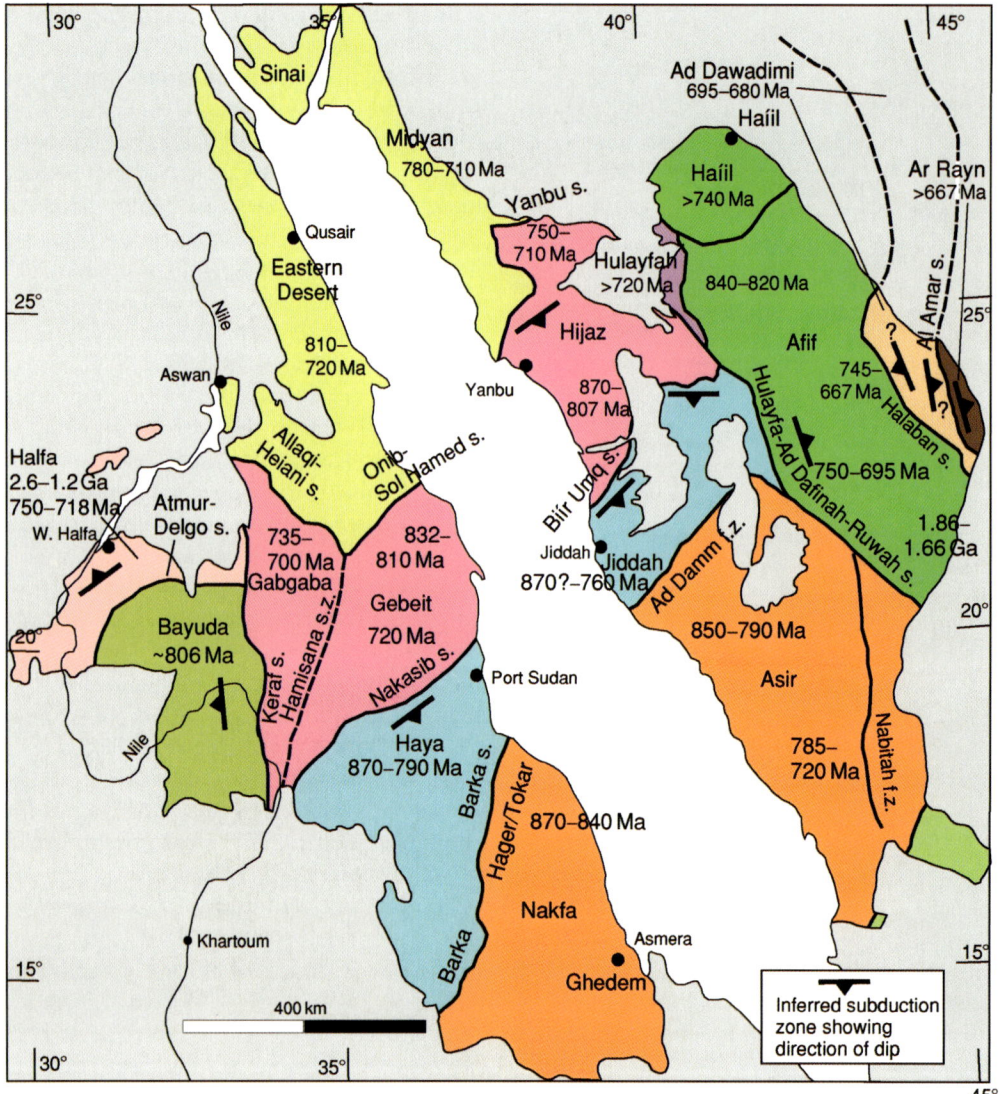

**Figure 3** Terrane map of the Arabian-Nubian Shield. (Reproduced with permission from Johnson PR and Woldehaimanot B (2003) Development of the Arabian-Nubian Shield: perspectives on accretion and deformation in the northern East African Orogen and the assembly of Gondwana. In: Yoshida M, Windley BF and Dasgupta S (eds.) *Proterozoic East Gondwana: Supercontinent Assembly and Breakup*. Geological Society, London, Special Publications 206, pp. 289–325.)

granites, bimodal volcanics, and molassic sediments. These testify to strong extension caused by orogenic collapse at the end of the Neoproterozoic. Extension-related metamorphic and magmatic core complexes are recognized in the northern ANS but are even more likely to be found in the more deformed regions of the southern ANS and the Mozambique Belt. A well developed peneplain developed on top of the ANS crust before basal Cambrian sediments were deposited, possibly cut by a continental ice-sheet.

The ANS has been the source of gold since Pharaonic Egypt. There is now a resurgence of mining and exploration activity, especially in Sudan, Arabia, Eritrea, and Ethiopia.

## Mozambique Belt (MB)

This broad belt defines the southern part of the East African Orogen and essentially consists of medium- to high-grade gneisses and voluminous granitoids. It extends south from the Arabian-Nubian Shield into southern Ethiopia, Kenya and Somalia via Tanzania to Malawi and Mozambique and also includes Madagascar (**Figure 2**). Southward continuation of the belt into Dronning Maud Land of East Antarctica (**Figure 1**) has been proposed on the basis of geophysical patterns, structural features and geochronology. Most parts of the belt are not covered by detailed mapping, making regional correlations difficult. There is no

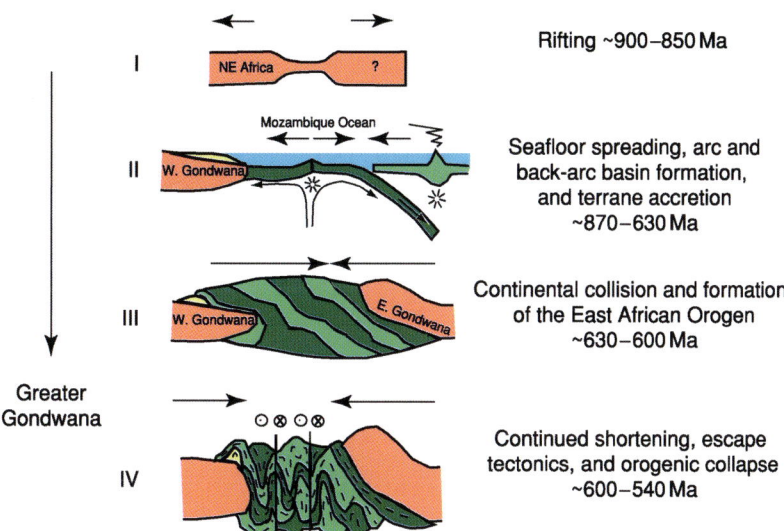

**Figure 4** A diagram of the suggested evolution of the Arabian-Nubian Shield.

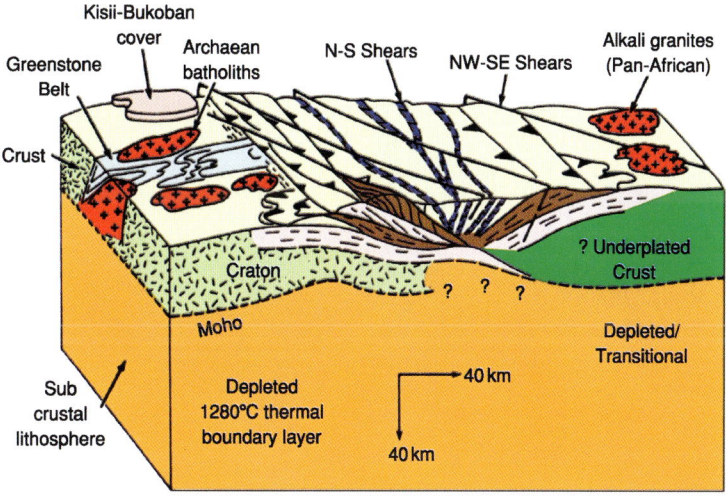

**Figure 5** A schematic block diagram showing tectonic interdigitation of basement and cover rocks in the Mozambique Belt of Kenya. (Reproduced with permission from Mosley PN (1993) Geological evolution of the Late Proterozoic 'Mozambique Belt' of Kenya. *Tectonophysics* 221: 223–250.)

overall model for the evolution of the MB although most workers agree that it resulted from collision between East and West Gondwana. Significant differences in rock type, structural style, age and metamorphic evolution suggest that the belt as a whole constitutes a Pan-African Collage of terranes accreted to the eastern margin of the combined Congo and Tanzania cratons and that significant volumes of older crust of these cratons were reconstituted during this event.

Mapping and geochronology in Kenya have recognized undated Neoproterozoic supracrustal sequences that are structurally sandwiched between basement gneisses of Archaean and younger age (**Figure 5**). A ~700 Ma dismembered ophiolite complex at the Kenyan/Ethiopian border testifies to the consumption and obduction of marginal basin oceanic crust. Major deformation and high-grade metamorphism is ascribed to two major events at ~830 and ~620 Ma, based on Rb–Sr dating, but the older of these appears questionable.

A similar situation prevails in Tanzania where the metamorphic grade is generally high and many granulite-facies rocks of Neoproterozoic age show evidence of retrogression. Unquestionable Neoproterozoic supracrustal sequences are rare, whereas Late Archaean to Palaeoproterozoic granitoid gneisses volumetrically greatly dominate over juvenile Pan-African intrusives. These older rocks, strongly reworked during the Pan-African orogenic cycle and locally migmatized and/or mylonitized, either represent eastward extensions of the Tanzania Craton that were structurally reworked during Pan-African events or are separate crustal entities (exotic blocks) of unknown origin. The significance of rare granitoid gneisses with protolith ages of ~1000–1100 Ma in southern Tanzania and Malawi is unknown. From these, some workers have postulated a major Kibaran (Grenvillian) event in the MB, but there is no geological evidence to relate these rocks to an orogeny. A layered gabbro-anorthosite complex was emplaced at ~695 Ma in Tanzania. The peak of granulite-facies metamorphism was dated at 620–640 Ma over wide areas of the MB in Tanzania, suggesting that this was the major collision and crustal-thickening event in this part of the belt.

In northern Mozambique the high-grade gneisses, granulites and migmatites of the MB were interpreted to have been deformed and metamorphosed during two distinct events, namely the Mozambican cycle at 1100–850 Ma, also known as Lurian Orogeny, and the Pan-African cycle at 800–550 Ma. Recent high-precision zircon geochronology has confirmed the older event to represent a major phase of granitoid plutonism, including emplacement of a large layered gabbro-anorthosite massif near Tete at ~1025 Ma, but there is as yet no conclusive evidence for deformation and granulite-facies metamorphism in these rocks during this time. The available evidence points to only one severe event of ductile deformation and high-grade metamorphism, with a peak some 615–540 Ma ago. A similar situation prevails in southern Malawi where high-grade granitoid gneisses with protolith ages of 1040–555 Ma were ductilely deformed together with supracrustal rocks and the peak of granulite-facies metamorphism was reached 550–570 Ma ago.

The Pan-African terrane of central and southern Madagascar primarily consists of high-grade ortho- and paragneisses as well as granitoids. Recent high-precision geochronology has shown that these rocks are either Archaean or Neoproterozoic in age and were probably structurally juxtaposed during Pan-African deformation. Several tectonic provinces have been recognized (**Figure 6**), including a domain consisting of low-grade Mesoproterozoic to Early Neoproterozoic metasediments known as the Itremo group which was thrust eastwards over high-grade gneisses. A Pan-African suture zone has been postulated in eastern Madagascar, the Betsimisaraka Belt (**Figure 6**), consisting of highly strained paragneisses decorated with lenses of mafic–ultramafic bodies containing podiform chromite and constituting a lithological and isotopic boundary with the Archaean gneisses and granites of the Antongil block east of this postulated suture which may correlate with similar rocks in southern India.

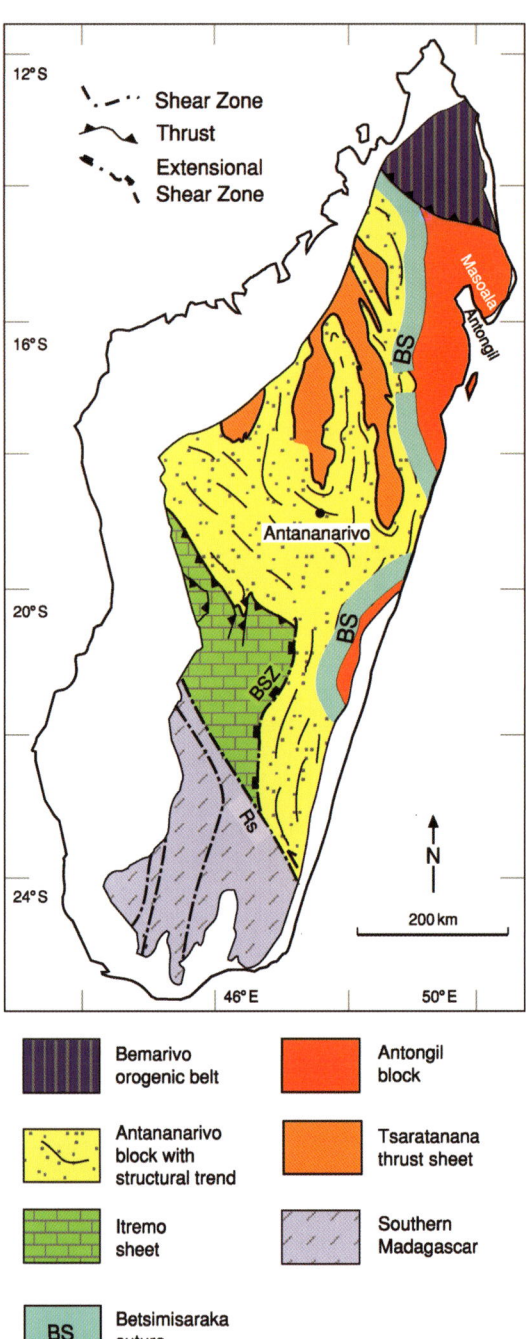

**Figure 6** A simplified geological map showing the major tectonic units of the Precambrian basement in Madagascar. Rs, Ranotsara Shear Zone; BSZ, Betsileo Shear Zone. (Reproduced with permission from Collins and Windley 2002.)

Central and northern Madagascar are separated from southern Madagascar by the Ranotsara Shear Zone (**Figure 6**), showing sinistral displacement of >100 km and correlated with one of the major shear zones in southern India. Southern Madagascar consists of several north–south trending shear-bounded

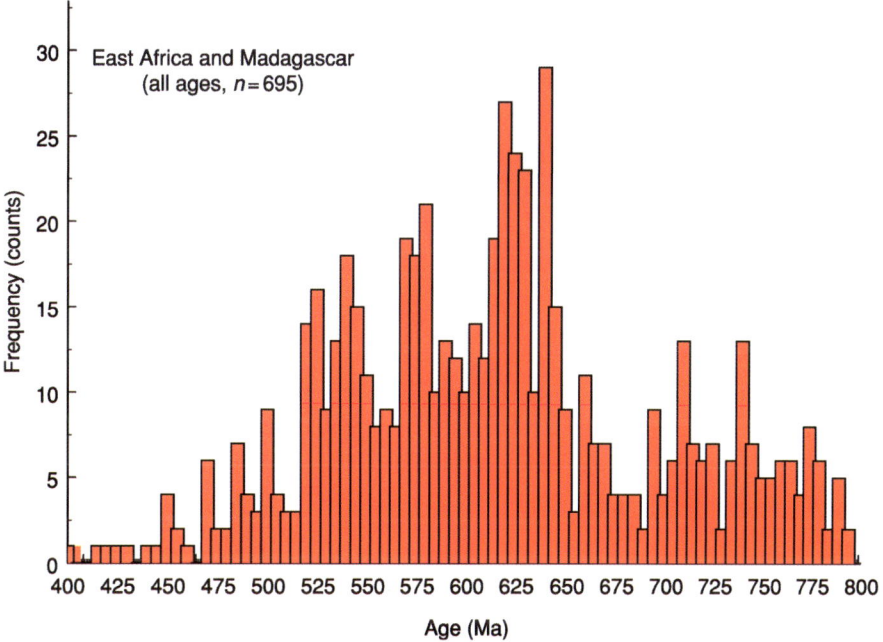

**Figure 7** Histogram of radiometric ages for the Mozambique Belt of East Africa and Madagascar. Data from Meert JG (2003) A synopsis of events related to the assembly of eastern Gondwana. *Tectonophysics* 362: 1–40, with updates.

tectonic units consisting of upper amphibolite to granulite-facies para- and orthogneisses, partly of pre-Neoproterozoic age. The peak of granulite-facies metamorphism in central and southern Madagascar, including widespread formation of charnockites, was dated at 550–560 Ma.

The distribution of zircon radiometric ages in the MB suggests two distinct peaks at 610–660 and 530–570 Ma (**Figure 7**) from which two orogenic events have been postulated, the older East African Orogeny (∼660–610 Ma) and the younger Kuunga Orogeny (∼570–530 Ma). However, the are no reliable field criteria to distinguish between these postulated phases, and it is likely that the older age group characterizes syntectonic magmatism whereas the younger age group reflects post-tectonic granites and pegmatites which are widespread in the entire MB.

## Zambezi Belt

The Zambezi Belt branches off to the west from the Mozambique Belt in northernmost Zimbabwe along what has been described as a triple junction and extends into Zambia (**Figures 1** and **8**). It consists predominantly of strongly deformed amphibolite- to granulite-facies, early Neoproterozoic ortho- and paragneisses which were locally intruded by ∼860 Ma, layered gabbro-anorthosite bodies and generally displays south-verging thrusting and transpressional shearing. Lenses of eclogite record pressures up to 23 kbar. Although most of the above gneisses seem to be 850–870 Ma in age, there are tectonically interlayered granitoid gneisses with zircon ages around 1100 Ma. The peak of Pan-African metamorphism occurred at 540–535 Ma. The Zambezi Belt is in tectonic contact with lower-grade rocks of the Lufilian Arc in Zambia along the transcurrent Mwembeshi shear zone.

## Lufilian Arc

The Lufilian Arc (**Figure 8**) has long been interpreted to be a continuation of the Damara Belt of Namibia, connected through isolated outcrops in northern Botswana (**Figure 1**). The outer part of this broad arc in the Congo Republic and Zambia is a north-east-verging thin-skinned, low-grade fold and thrust belt, whereas the higher-grade southern part is characterized by basement-involved thrusts. The main lithostratigraphic unit is the Neoproterozoic, copper-bearing Katanga succession which contains volcanic rocks dated between 765 and 735 Ma. Thrusting probably began shortly after deposition, and the main phase of thrusting and associated metamorphism occurred at 566–550 Ma.

## Damara Belt

This broad belt exposed in central and northern Namibia branches north-west and south-east near

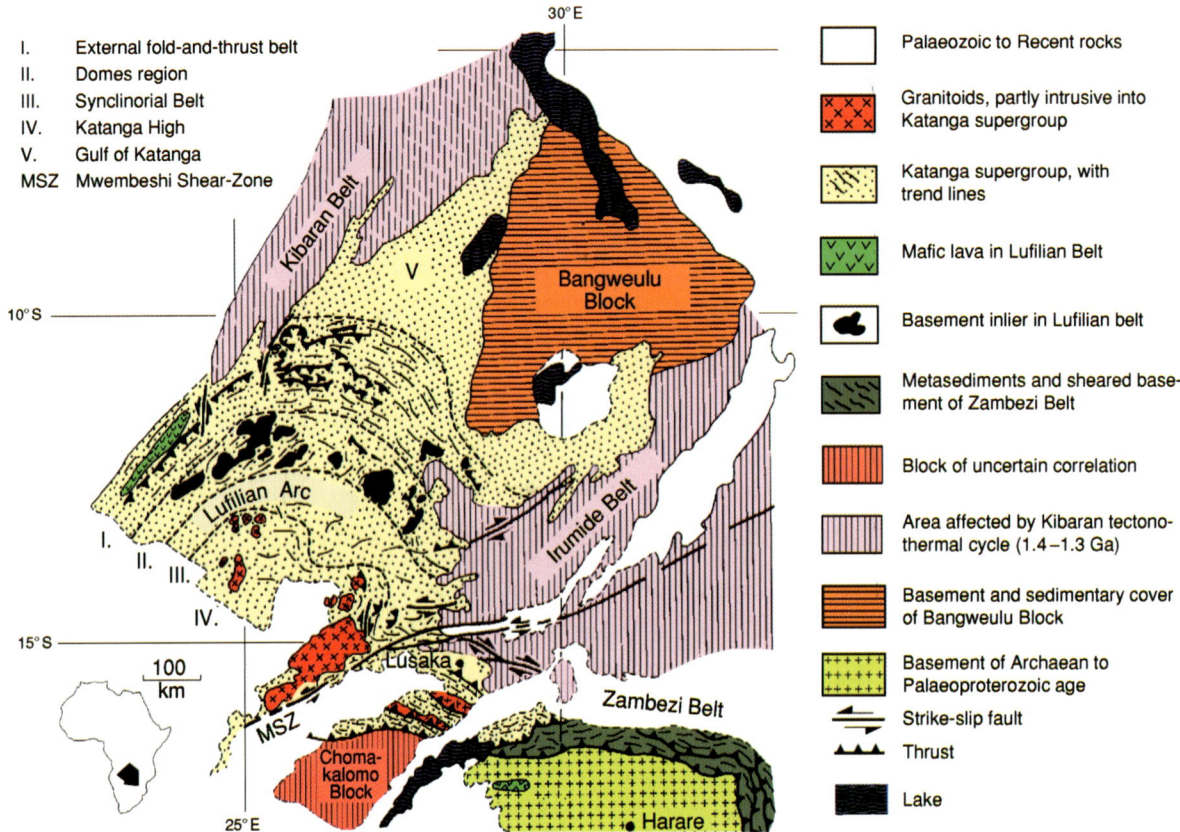

**Figure 8** A simplified geological map of the Lufilian Arc and Zambezi Belt. (Reproduced with permission from Porada H and Berhorst V (2000) Towards a new understanding of the Neoproterozoic-early Palaeozoic Lufilian and northern Zambezi belts in Zambia and the Democratic Republic of Congo.)

the Atlantic coast and continues southwards into the Gariep and Saldania belts and northwards into the Kaoko Belt (**Figure 1**). The triple junction so produced may have resulted from closure of the Adamastor Ocean, followed by closure of the Damara Ocean. The main lithostratigraphic unit is the Damara supergroup which records basin formation and rift-related magmatism at ∼760 Ma, followed by the formation of a broad carbonate shelf in the north and a turbidite basin in the south. The turbidite sequence contains interlayered, locally pillowed, amphibolites and metagabbros which have been interpreted as remnants of a dismembered ophiolite. Of particular interest are two distinct horizons of glaciogenic rocks which can probably be correlated with similar strata in the Katanga sequence of south central Africa and reflect a severe glaciation currently explained by the snowball Earth hypothesis.

The Damara Belt underwent north- and south-verging thrusting along its respective margins, whereas the deeply eroded central zone exposes medium- to high-grade ductilely deformed rocks, widespread migmatization and anatexis in which both the Damara supracrustal sequence and a 1.0–2.0 Ga old basement are involved. Sinistral transpression is seen as the cause for this orogenic event which reached its peak at ∼550–520 Ma. Voluminous pre-, syn- and post-tectonic granitoid plutons intruded the central part of the belt between ∼650 and ∼488 Ma, and highly differentiated granites, hosting one of the largest opencast uranium mines in the world (Rössing), were dated at 460 Ma.

Uplift of the belt during the Damaran Orogeny led to erosion and deposition of two Late Neoproterozoic to Early Palaeozoic clastic molasse sequences, the Mulden group in the north and the Nama group in the south. The latter contains spectacular examples of the Late Neoproterozoic Ediacara fauna.

## Gariep and Saldania Belts

These belts fringe the high-grade basement along the south-western and southern margin of the Kalahari craton (**Figure 1**) and are interpreted to result from oblique closure of the Adamastor Ocean. Deep marine fan and accretionary prism deposits, oceanic seamounts and ophiolitic assemblages were thrust over Neoproterozoic shelf sequences on the craton margin containing a major Zn mineralization just north of the Orange River in Namibia. The main deformation and metamorphism occurred at 570–540 Ma, and post-tectonic granites were emplaced 536–507 Ma ago. The famous granite at Sea Point, Cape Town, which was described by Charles Darwin, belongs to this episode of Pan-African igneous activity.

## Kaoko Belt

This little known Pan-African Belt branches off to the north-west from the Damara Belt and extends into south-western Angola. Here again a well developed Neoproterozoic continental margin sequence of the Congo Craton, including glacial deposits, was overthrust, eastwards, by a tectonic mixture of pre-Pan-African basement and Neoproterozoic rocks during an oblique transpressional event following closure of the Adamastor Ocean. A spectacular shear zone, the mylonite-decorated Puros lineament, exemplifies this event and can be followed into southern Angola. High-grade metamorphism and migmatization dated between 650 and 550 Ma affected both basement and cover rocks, and granitoids were emplaced between 733 and 550 Ma. Some of the strongly deformed basement rocks have ages between ∼1450 and ∼2030 Ma and may represent reworked material of the Congo Craton, whereas a small area of Late Archaean granitoid gneisses may constitute an exotic terrane. The western part of the belt consists of large volumes of *ca.* 550 Ma crustal melt granites and is poorly exposed below the Namib sand dunes. No island-arc, ophiolite or high-pressure assemblages have been described from the Kaoko Belt, and current tectonic models involving collision between the Congo and Rio de la Plata cratons are rather speculative.

## West Congo Belt

This belt resulted from rifting between 999 and 912 Ma along the western margin of the Congo Craton (**Figure 1**), followed by subsidence and formation of a carbonate-rich foreland basin, in which the West Congolian group was deposited between *ca.* 900 and 570 Ma, including two glaciogenic horizons similar to those in the Katangan sequence of the Lufilian Arc. The structures are dominated by east-verging deformation and thrusting onto the Congo Craton, associated with dextral and sinistral transcurrent shearing, and metamorphism is low to medium grade. In the west, an allochthonous thrust-and-fold stack of Palaeo- to Mesoproterozoic basement rocks overrides the West Congolian foreland sequence. The West Congo Belt may only constitute the eastern part of an orogenic system with the western part, including an 800 Ma ophiolite, exposed in the Aracuaí Belt of Brazil.

## Trans-Saharan Belt

This orogenic Belt is more than 3000 km long and occurs to the north and east of the >2 Ga West African Craton within the Anti-Atlas and bordering the Tuareg and Nigerian shields (**Figure 1**). It consists of pre-Neoproterozoic basement strongly reworked during the Pan-African event and of Neoproterozoic oceanic assemblages. The presence of ophiolites, accretionary prisms, island-arc magmatic suites and high-pressure metamorphic assemblages makes this one of the best documented Pan-African belts, revealing ocean opening, followed by a subduction- and collision-related evolution between 900 and 520 Ma (**Figure 9**). In southern Morocco, the ∼740–720 Ma Sirwa-Bou Azzer ophiolitic mélange was thrust southwards, at ∼660 Ma, over a Neoproterozoic continental margin sequence of the West African Craton, following northward subduction of oceanic lithosphere and preceding oblique collision with the Saghro Arc.

Farther south, in the Tuareg Shield of Algeria, Mali and Niger, several terranes with contrasting lithologies and origins have been recognized, and ocean closure during westward subduction produced a collision belt with Pan-African rocks, including oceanic terranes tectonically interlayered with older basement. The latter were thrust westwards over the West African Craton and to the east over the so-called LATEA (Laouni, Azrou-n-Fad, Tefedest, and Egéré-Aleksod, parts of a single passive margin in central Hoggar) Superterrane, a completely deformed composite crustal segment consisting of Archaean to Neoproterozoic assemblages (**Figure 9**). In Mali, the 730–710 Ma Tilemsi magmatic arc records ocean-floor and intra-oceanic island-arc formation, ending in collision at 620–600 Ma.

The southern part of the Trans-Saharan Belt is exposed in Benin, Togo and Ghana where it is known as the Dahomeyan Belt. The western part of this belt consists of a passive margin sedimentary sequence in the Volta basin which was overthrust, from the east, along a well delineated suture zone by an ophiolitic mélange and by a 613 My old high-pressure metamorphic assemblage (up to 14 kbar, ∼700°C), including granulites and eclogites. The eastern part of the belt consists of a high-grade granitoid–gneiss terrane of the Nigerian province, partly consisting of Palaeoproterozoic rocks which were migmatized at ∼600 Ma. This deformation and metamorphism is considered to have resulted from oblique collision of

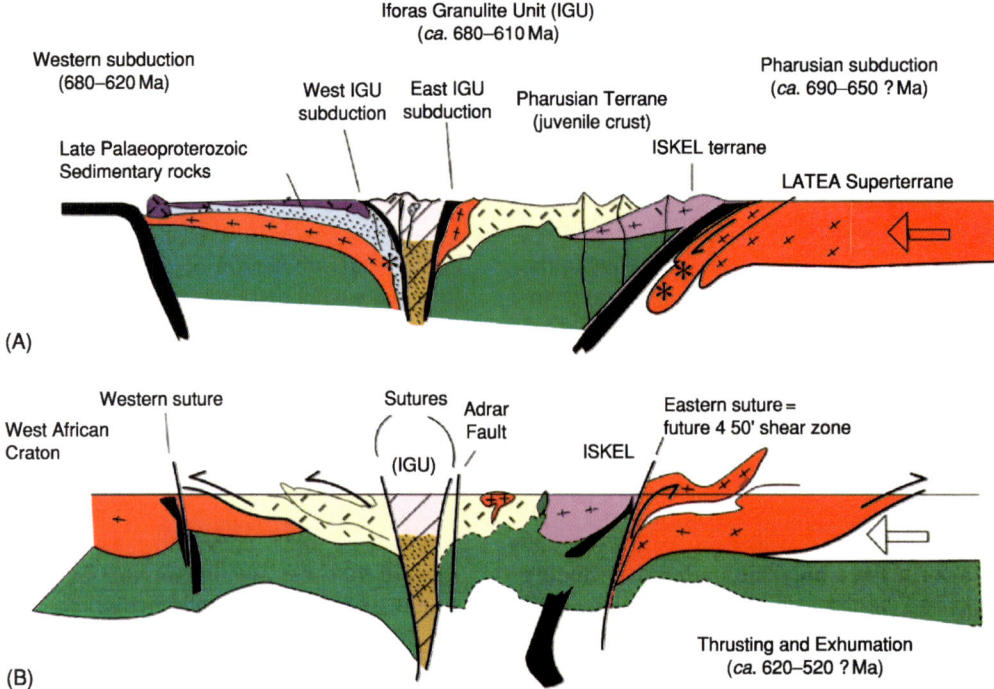

**Figure 9** Diagrams showing the geodynamic evolution of western central Hoggar (Trans-Sahara Belt) between ~900 and ~520 Ma. Stars denote high-pressure rocks now exposed. (Reproduced with permission from Caby R (2003) Terrane assembly and geodynamic evolution of central-western Hoggar: a synthesis.)

the Nigerian shield with the West African Craton, followed by anatectic doming and wrench faulting.

## Pan-African Belt in Central Africa (Cameroon, Chad and Central African Republic)

The Pan-African Belt between the Congo Craton in the south and the Nigerian basement in the north-west consists of Neoproterozoic supracrustal assemblages and variously deformed granitoids with tectonically interlayered wedges of Palaeoproterozoic basement (**Figure 10**). The southern part displays medium- to high-grade Neoproterozoic rocks, including 620 Ma granulites, which are interpreted to have formed in a continental collision zone and were thrust over the Congo Craton, whereas the central and northern parts expose a giant shear belt characterized by thrust and shear zones which have been correlated with similar structures in north-eastern Brazil and which are late collisional features. The Pan-African Belt continues eastward into the little known Oubanguide Belt of the Central African Republic.

## Pan-African Reworking of Older Crust in North-Eastern Africa

A large area between the western Hoggar and the river Nile largely consists of Archaean to Palaeoproterozoic basement, much of which was structurally and thermally overprinted during the Pan-African event and intruded by granitoids. The terrane is variously known in the literature as 'Nile Craton', 'East Sahara Craton' or 'Central Sahara Ghost Craton' and is geologically poorly known. Extensive reworking is ascribed by some to crustal instability following delamination of the subcrustal mantle lithosphere, and the term 'Sahara Metacraton' has been coined to characterize this region. A 'metacraton' refers to a craton that has been remobilized during an orogenic event but is still recognizable through its rheological, geochronological and isotopic characteristics.

## Rokelide Belt

This belt occurs along the south-western margin of the Archaean Man Craton of West Africa (**Figure 1**) and is made up of high-grade gneisses, including granulites (Kasila group), lower-grade supracrustal sequences (Marampa group) and volcano-sedimentary rocks with calc-alkaline affinity (Rokel River group). Pan-African deformation was intense and culminated in extensive thrusting and sinistral strike-slip deformation. The peak of metamorphism reached 7 kb and 800°C and was dated at ~560 Ma. Late Pan-African emplacement ages for the protoliths of some of the granitoid gneisses contradict earlier hypotheses arguing for extensive overprinting of

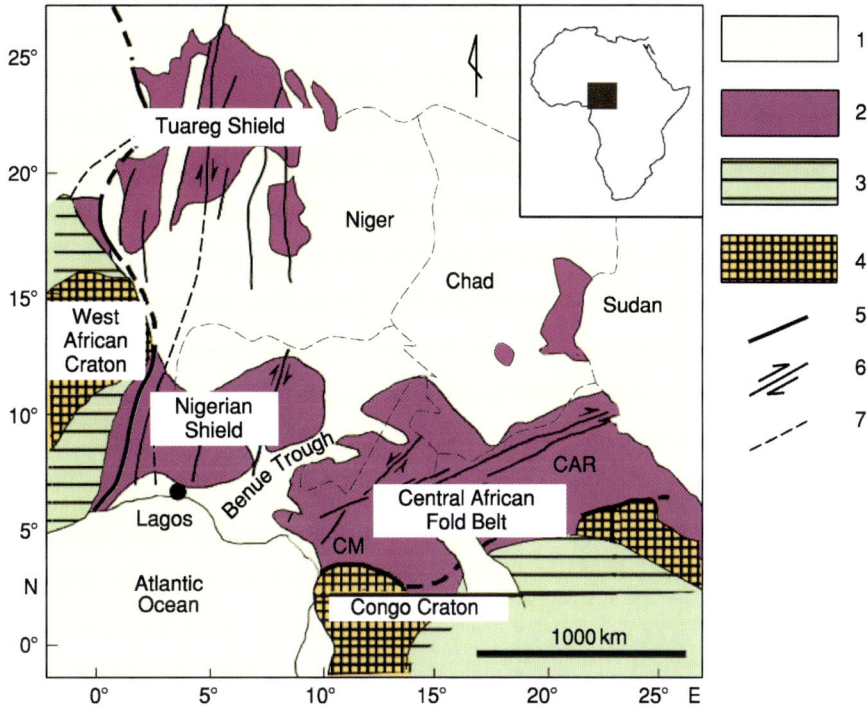

**Figure 10** A sketch map showing Pan-African domains in west central Africa. 1, Post-Pan-African cover; 2, Pan-African domains; 3, pre-Mesozoic platform deposits; 4, Archaean to Palaeoproterozoic cratons; 5, craton limits; 6, major strike-slip faults; 7, state boundaries. CAR, Central African Republic; CM, Cameroun. (Reproduced with permission from Toteu SF, Penaye J and Djomani YP (2004).)

Archaean rocks. The Rokelides may be an accretionary belt, but there are no modern structural data and only speculative geodynamic interpretations.

## Gondwana Correlations

The Pan-African orogenic cycle was the result of ocean closure, arc and microcontinent accretion and final suturing of continental fragments to form the supercontinent Gondwana. It has been suggested that the opening of large Neoproterozoic oceans between the Brazilian and African cratons (Adamastor Ocean), the West African and Sahara–Congo cratons (Pharusian Ocean) and the African cratons and India/Antarctica (Mozambique Ocean) (**Figure 1**) resulted from breakup of the Rodinia supercontinent some 800–850 Ma, but current data indicate that the African and South American cratons were never part of Rodinia. Although arc accretion and continent formation in the Arabian-Nubian shield are reasonably well understood, this process is still very speculative in the Mozambique Belt. It seems clear that Madagascar, Sri Lanka, southern India and parts of East Antarctica were part of this process (**Figure 1**), although the exact correlations between these fragments are not known. The Southern Granulite Terrane of India (**Figure 1**) consists predominantly of Late Archaean to Palaeoproterozoicc gneisses and granulites, deformed and metamorphosed during the Pan-African event and sutured against the Dharwar Craton. Areas in East Antarctica such as Lützow-Holm Bay, Central Dronning Maud Land and the Shackleton Range, previously considered to be Mesoproterozoic in age, are now interpreted to be part of the Pan-African Belt system (**Figure 1**). Correlations between the Pan-African belts in south-western Africa (Gariep–Damara–Kaoko) and the Brasiliano belts of south-eastern Brazil (Ribeira and Dom Feliciano) are equally uncertain, and typical hallmarks of continental collision such as ophiolite-decorated sutures or high-pressure metamorphic assemblages have not been found. The most convincing correlations exist between the southern end of the Trans-Saharan Belt in West Africa and Pan-African terranes in north-eastern Brazil (**Figure 1**). Following consolidation of the Gondwana supercontinent at the end of the Precambrian, rifting processes at the northern margin of Gondwana led to the formation of continental fragments (**Figure 1**) which drifted northwards and are now found as exotic terranes in Europe (Cadomian and Armorican terrane assemblages), in the Appalachian Belt of North

America (Avalonian Terrane assemblage) and in various parts of central and eastern Asia.

## See Also

**Arabia and The Gulf**. **Australia:** Proterozoic. **Brazil**. **Gondwanaland and Gondwana**. **Palaeomagnetism**. **Tectonics:** Mountain Building and Orogeny. **Tertiary To Present:** Pleistocene and The Ice Age.

## Further Reading

Abdelsalam MG and Stern RJ (1997) Sutures and shear zones in the Arabian-Nubian Shield. *Journal of African Earth Sciences* 23: 289–310.

Caby R (2003) Terrane assembly and geodynamic evolution of central-western Hoggar: a synthesis. *Journal of African Earth Sciences* 37: 133–159.

Cahen L, Snelling NJ, Delhal J, and Vail JR (1984) *The Geochronology and Evolution of Africa*. Oxford: Clarendon Press.

Clifford TN (1968) Radiometric dating and the pre-Silurian geology of Africa. In: Hamilton EI and Farquhar RM (eds.) *Radiometric Dating for Geologists*, pp. 299–416. London: Interscience.

Collins AS and Windley BF (2002) The tectonic evolution of central and northern Madagascar and its place in the final assembly of Gondwana. *Journal of Geology* 110: 325–339.

Fitzsimons ICW (2000) A review of tectonic events in the East Antarctic shield and their implications for Gondwana and earlier supercontinents. *Journal of African Earth Sciences* 31: 3–23.

Hanson RE (2003) Proterozoic geochronology and tectonic evolution of southern Africa. In: Yoshida M, Windley BF, and Dasgupta S (eds.) *Proterozoic East Gondwana: Supercontinent Assembly and Breakup*. Geological Society, London, Special Publications 206, pp. 427–463.

Hoffman PF and Schrag DP (2002) The snowball Earth hypothesis: testing the limits of global change. *Terra Nova* 14: 129–155.

Johnson PR and Woldehaimanot B (2003) Development of the Arabian-Nubian Shield: perspectives on accretion and deformation in the northern East African orogen and the assembly of Gondwana. In: Yoshida M, Windley BF, and Dasgupta S (eds.) *Proterozoic East Gondwana: Supercontinent Assembly and Breakup*. Geological Society, London, Special Publications 206, pp. 289–325.

Kröner A (2001) The Mozambique belt of East Africa and Madagascar; significance of zircon and Nd model ages for Rodinia and Gondwana supercontinent formation and dispersal. *South African Journal of Geology* 104: 151–166.

Kusky TM, Abdelsalam M, Stern RJ, and Tucker RD (eds.) (2003) Evolution of the East African and related orogens, and the assembly of Gondwana. *Precambrian Res.* 123: 82–85.

Meert JG (2003) A synopsis of events related to the assembly of eastern Gondwana. *Tectonophysics* 362: 1–40.

Miller RMcG (ed.) (1983) *Evolution of the Damara Orogen of South West Africa/Namiba*. Geological Society of South Africa, Special Publications, 11.

Mosley PN (1993) Geological evolution of the late Proterozoic 'Mozambique Belt' of Kenya. *Tectonophysics* 221: 223–250.

Porada H and Berhorst V (2000) Towards a new understanding of the Neoproterozoic-early Palaeozoic Lufilian and northern Zambezi belts in Zambia and the Democratic Republic of Congo. *Journal of African Earth Sciences* 30: 727–771.

Stern RJ (1994) Arc assembly and continental collision in the Neoproterozoic East African Orogen: implications for the consolidation of Gondwanaland. *Annual Reviews Earth Planetary Sciences* 22: 319–351.

Toteu SF, Penaye J, and Djomani YP (2004) Geodynamic evolution of the Pan-African belt in central Africa with special reference to Cameroon. *Canadian Journal of Earth Science* 41: 73–85.

Veevers JJ (2003) Pan-African is Pan-Gondwanaland: oblique convergence drives rotation during 650–500 Ma assembly. *Geology* 31: 501–504.

# North African Phanerozoic

**S Lüning**, University of Bremen, Bremen, Germany

© 2005, Elsevier Ltd. All Rights Reserved.

## Introduction

North Africa forms the northern margin of the African Plate and comprises the countries Morocco, Algeria, Tunisia, Libya, and Egypt (**Figure 1**). The region discussed here is bounded to the west by the Atlantic, to the north by the Mediterranean Sea, to the east by the Arabian Plate and to the south by political boundaries. Much of the geology across North Africa is remarkably uniform because many geological events affected the whole region (**Figure 2**). The geological study of North Africa benefits from large-scale desert exposures and an extensive subsurface database from hydrocarbon exploration. The region contains some 4% of the world's remaining oil (*see* **Petroleum Geology:** Overview) and gas reserves with fields mainly in Algeria, Libya, and

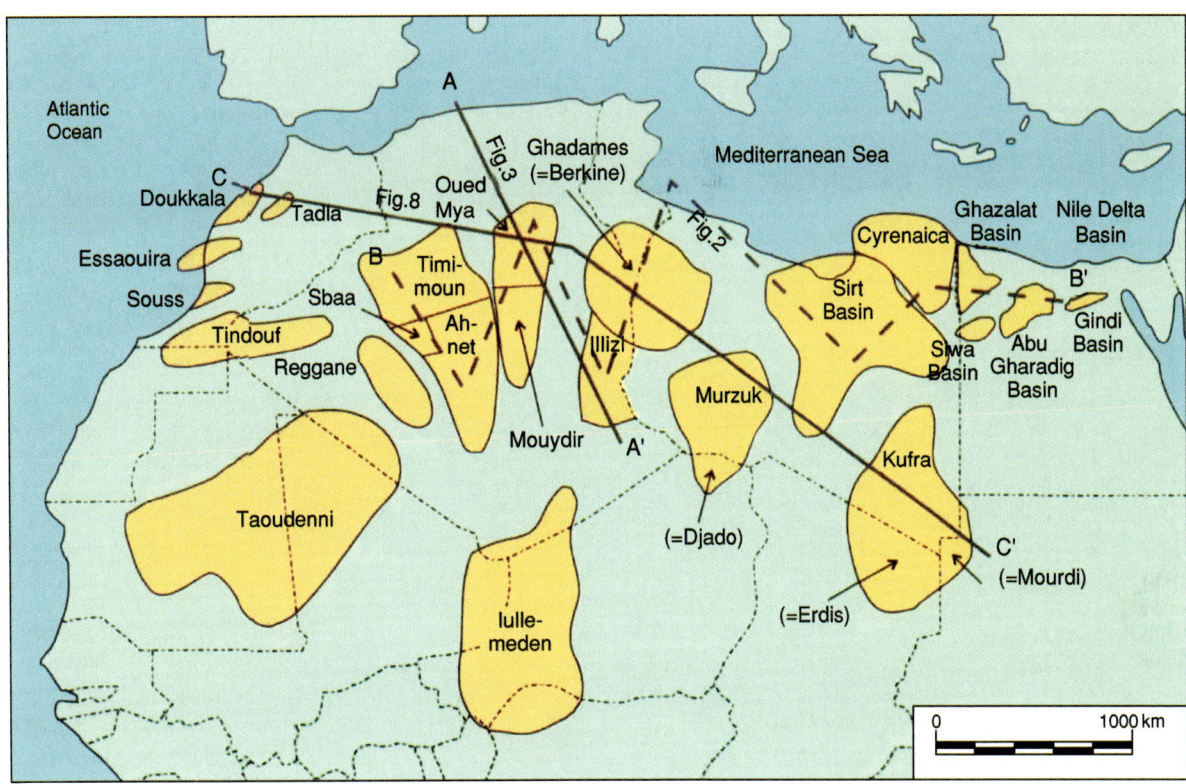

Figure 1  Location of major North African sedimentary basins. Lines indicate locations of cross-sections in **Figures 2, 3** and **8**.

Egypt. Other natural resources that are exploited include Saharan fossil groundwater, phosphate, (see **Sedimentary Rocks: Phosphates**) and mineral ores.

## Structural Evolution

Most of North Africa has formed part of a single plate throughout the Phanerozoic with the exception of the Atlas Mountains which became accreted during Late Carboniferous and Tertiary collisional events. North Africa can be structurally subdivided into a northern Mesozoic to Alpine deformed, mobile belt and the stable Saharan Platform (**Figure 3**). The latter became consolidated during the Proterozoic Pan-African Orogeny (see **Africa: Pan-African Orogeny**), a collisional amalgamation between the West African Craton and numerous island arcs, Andean-type magmatic arcs, and various microplates. The Late Neoproterozoic to Phanerozoic structural development of North Africa can be divided into six major tectonic (see **Plate Tectonics**) phases: (i) Infracambrian extension and wrenching; (ii) Cambrian to Carboniferous alternating extension and compression; (iii) mainly Late Carboniferous 'Hercynian' intraplate uplift; (iv) Late Triassic–Early Jurassic and Early Cretaceous rifting; (v) mid-Cretaceous 'Austrian' and Late Cretaceous–Tertiary 'Alpine' compression, and (vi) Oligo-Miocene rifting (see **Tectonics: Rift Valleys**).

### Infracambrian Extension and Wrenching

The Late Neoproterozoic to Early Cambrian ('Infracambrian') in North Africa and Arabia was characterized by major extensional and strike-slip movements. Halfgrabens and pull-apart basins developed, for example, in the Taoudenni Basin (SW Algeria) and in the Kufra Basin (SE Libya). These features are considered to be a westward continuation of an Infracambrian system of salt basins extending across Gondwana from Australia, through Pakistan, Iran and Oman, to North Africa.

### Post-Infracambrian – Pre-Hercynian

The structural evolution of North Africa between the Infracambrian extensional/wrenching phase and the Late Carboniferous 'Hercynian Orogeny' is complex. Local transpressional and transtensional reactivation processes dominated as a result of the interaction of intraplate stress fields with pre-existing fault systems of varying orientation and geometry. In some areas, such as the Murzuq Basin in SW Libya, these tectonic processes played an important role in the formation of hydrocarbon traps.

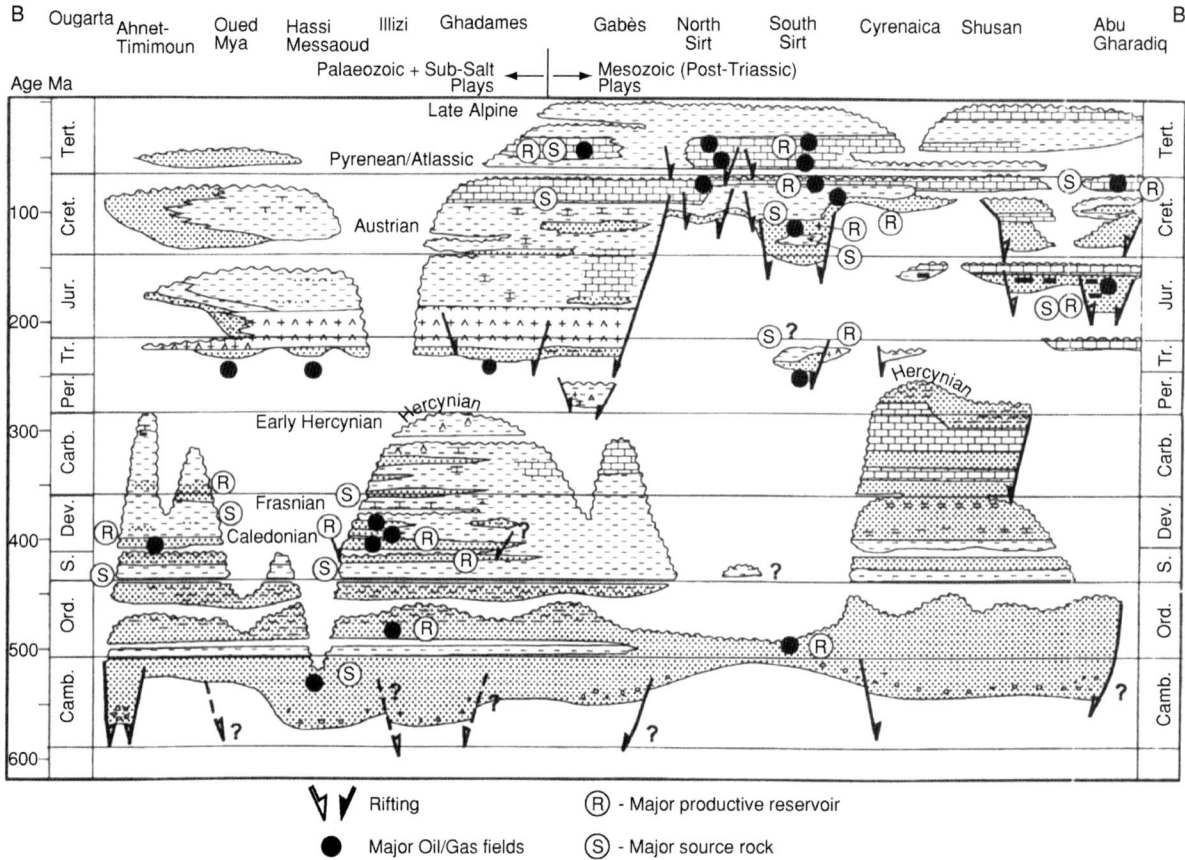

**Figure 2** Phanerozoic chronostratigraphy of petroliferous provinces in North Africa. (From MacGregor (1998).)

During most of the Early Palaeozoic the Saharan Palaeozoic basins were part of a large, interconnected North African shelf system that was in a sagging phase. Some relief, however, was locally already created associated with local uplift and increased subsidence, including, for example, late Cambrian uplift in the Hoggar and increased sagging in the SE Libyan Kufra Basin, the latter leading to thinning of Cambro-Ordovician strata towards the present-day basin margins. The Saharan basins differentiated mainly from the Late Silurian/Early Devonian onwards when ridges were uplifted, associated with a basal unconformity, that in the regional literature has often been referred to as 'Caledonian unconformity'. This term, however, is inappropriate as tectonic events during the Silurian in North Africa were independent of those in the 'Caledonian' collisional zone, located many thousands of kilometres to the north, involving the continents of Laurentia, Baltica, Armorica, and Avalonia.

### Hercynian Orogeny

Collision of Gondwana and Laurasia during the Late Carboniferous resulted in the compressional movements of the Hercynian Orogeny (**Figure 4**). In North Africa, the collisional zone was located in the north-west, leading to substantial thrusting and uplift in Morocco and western Algeria. Strong uplift associated with transpression on old faults occurred in the Algerian Hassi Massaoud region, leading to erosion into stratigraphic levels as deep as the Cambrian. The intensity of Hercynian deformation decreases eastwards across North Africa such that strong folding and erosion of anticlinal crests in the Algerian Sbaa and Ahnet basins is replaced towards the plate interior by low-angle unconformities and disconformities in the Murzuq Basin in south-west Libya. Notably, the present-day maturity levels of the main Palaeozoic hydrocarbon source rocks have a decreasing trend eastwards across North Africa (once present-day burial effects are removed) in parallel with the decrease in the intensity of the Hercynian deformation.

The gravitational collapse of the Hercynian Orogenic Belt in north-west Africa was accompanied by widespread Permo-Carboniferous volcanism in Morocco. The magmatism acted here as an 'exhaust valve' releasing the heat accumulated beneath the

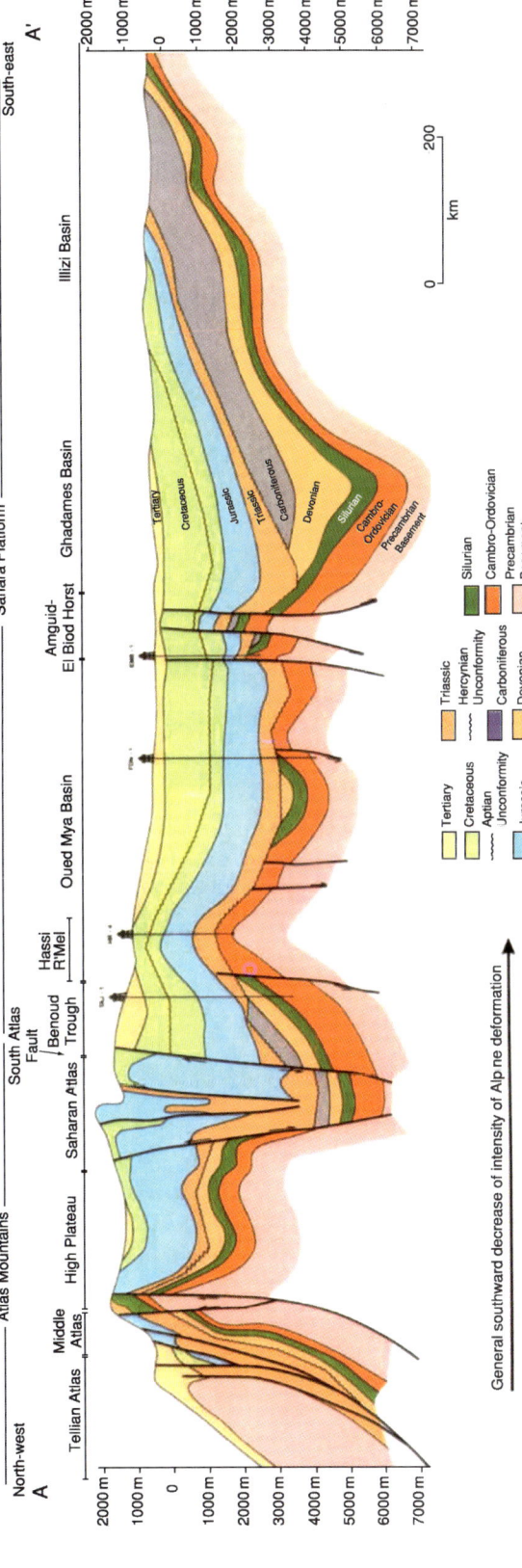

**Figure 3** Cross-section through Algeria illustrating typical the structural styles of North Africa. The Alpine-deformed Atlas Mountains are separated from the Saharan Platform to the south by the South Atlas Fault. Location of section in **Figure 1**. (Courtesy E. Zanella.)

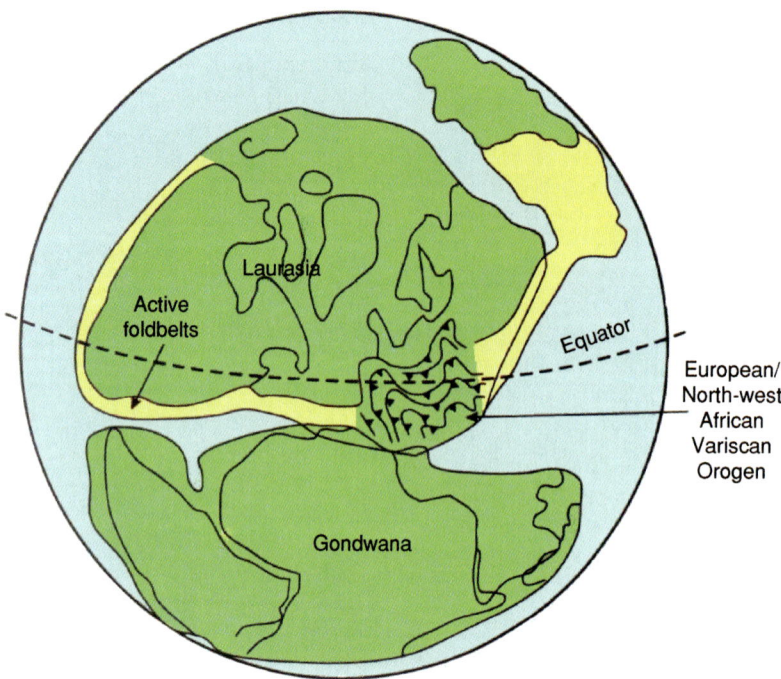

**Figure 4** Hercynian compression as result of a Late Carboniferous plate collision between Laurasia and Gondwana. (After Doblas et al. (1998).)

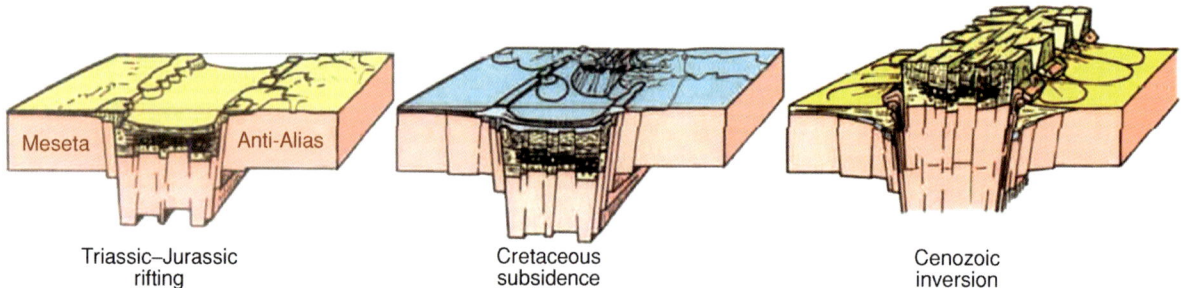

**Figure 5** Block diagrams illustrating the geological evolution of the High Atlas, including Triassic–Jurassic rifting, Cretaceous and Cenozoic inversion. (After Stets and Wurster (1981).)

Pangaean Supercontinent by insulation and blanketing processes which triggered large-scale mantle-wide upward convection and general instability of the supercontinent.

### Mesozoic Extension

The opening of the Central Atlantic in the Triassic–Early Jurassic and contemporaneous separation of the Turkish–Apulian Terrane from north-east Africa initiated a significant extensional phase in North Africa which included graben formation in the Atlas region (**Figure 5**), rifting from Syria to Cyrenaica (NE Libya) and extension in offshore Libya and in the Oued Mya and Ghadames (=Berkine) basins in central and eastern Algeria. Rift-related Triassic volcanism occured in the northern Ghadames and Oued Maya basins.

A second important Mesozoic extensional phase in North Africa occurred during the Early Cretaceous, related to the opening of the South and Equatorial Atlantic Ocean. As a result, a complex of failed rift systems originated across North and Central Africa with the formation of half-grabens in, for example, the Egyptian Abu Gharadig Basin and in the Libyan Sirte Basin.

The Mesozoic extensional phase also triggered increased subsidence in several Saharan Palaeozoic basins, leading to deposition of thick, continental deposits, for example, in the south-east Libyan Kufra Basin.

### Alpine Orogeny

The onset of rifting in the northern North Atlantic during the Late Cretaceous led to an abrupt change in the motion of the European Plate which began to move eastwards with respect to Africa. The previous sinistral transtensional movements were quickly replaced by a prolonged phase of dextral transpression resulting in the collision of Africa and Europe. The 'Alpine Orogeny' led to an overall compressional regime in North Africa from the mid-Cretaceous through to the Recent. Changes in the collisional process, such as subduction of oceanic crust after accretion of a seamount in the Eastern Mediterranean, produced localized stress-neutral or even extensional pulses within the overall compressive regime.

An Aptian compressional event may be considered as a precursor to the 'Alpine Orogeny', in the narrow sense. It affected parts of North and Central Africa, inverting Early Cretaceous rift systems and reactivating older structures. Large Aptian-age anticlines occur in the Berkine Basin in Algeria and result from sinistral transpression along the N–S trending Transaharian fracture system.

The post-Cenomanian 'Alpine' compression in North Africa resulted in folding and thrusting within the north-west African collisional zones, as well as in intraplate inversion and uplift of Late Triassic-Early Jurassic grabens. Major orogens formed during this phase include the Atlas Mountains (Morocco, Algeria, Tunisia; **Figure 5**) and the 'Syrian Arc' Fold Belt in north-east Egypt and north-west Arabia. The Cyrenaica Platform (Jebel Akhdar) in north-east Libya also is an 'Alpine' deformed region.

The structural boundary between the Atlas Mountains and the Saharan Platform is the South Atlas Front (South Atlas Fault), a continuous structure from Agadir (Morocco) to Tunis (Tunisia). The fault separates a zone where the Mesozoic-Cenozoic cover is shortened and mostly detached from its basement from a zone where the cover remains horizontal and attached to its basement. Thrust-belt rocks north of the fault are structurally elevated by about 1.5 km above the Saharan Platform.

Apatite fission track data (*see* **Analytical Methods: Fission Track Analysis**) suggests that large parts of Libya and Algeria were uplifted by 1–2 km during the 'Alpine' deformational phase. As a consequence, Palaeozoic hydrocarbon source rocks were lifted out of the oil window in some parts of the Saharan Palaeozoic basins, resulting in termination of hydrocarbon generation.

### Oligo-Miocene Rifting

Another major rifting phase in North Africa during the Oligo-Miocene was associated with the development of the Red Sea, Gulf of Suez, Gulf of Aqaba Rift system, which is the northern continuation of the Gulf of Aden, and East African rifts. Along the north-eastern margin of the Red Sea/Gulf of Suez axis, extension was associated with intrusion of a widespread network of dykes and other small intrusions. Rifting and separation of Arabia from Africa commenced in the southern Red Sea at about 30 Ma (Oligocene) and in the northern Red Sea and Gulf of Suez at about 20 Ma (Early Miocene). Subsequently, tectonic processes in the Arabian–Eurasian collisional zone changed the regional stress field in the northern Red Sea region, causing the rifting activity to switch from the Gulf of Suez to the Gulf of Aqaba. As a consequence the Gulf of Suez became a failed rift and was in part inverted.

Intense volcanic activity occurred in central and eastern North Africa during the Late Miocene to Late Quaternary. In places this had already commenced in the Late Eocene. Volcanic features include the plateau basalts in northern Libya, the volcanic field of Jebel Haruj in central Libya, the Tibesti volcanoes in south-east Libya and north-east Chad and the volcanism in the Hoggar (S Algeria, NE Mali, NW Niger). Some authors interpret this continental volcanism as related to a hot spot overlying a deep-seated mantle plume while others see the cause in intraplate stresses originating from the Africa–Europe collision that led to melting of rocks at the lithosphere/asthenosphere interface by adiabatic pressure release.

## Depositional History

### Infracambrian

The Infracambrian in North Africa is represented by carbonates, sandstones, siltstones, and shales, often infilling halfgrabens. In Morocco and Algeria, the unit includes stromatolitic carbonates as well as red and black shales, a facies similar to the Huqf Supergroup in Oman that represents an important hydrocarbon source rock there. Infracambrian siliciclastics are also known from several boreholes in the central Algerian Ahnet Basin and southern Cyrenaica (NE Libya). Infracambrian conglomeratic and shaly sandstones and siltstones occur at outcrop underneath Cambrian strata along the eastern margin of the Murzuq Basin and in some boreholes in the basin centre. In the Kufra Basin, the presence of some 1500 m of Infracambrian sedimentary rocks (of unknown lithology) is inferred for the southern basin centre, while strata of similar age, including dolomites, have been reported from the eastern and western margins of this basin. Notably, salt deposits like those in Oman have not yet been confirmed from North Africa, although some features from seismic studies in the Kufra Basin may represent salt diapirs.

## Cambro-Ordovician

The Cambro-Ordovician in North Africa is mostly represented by continental and shallow marine siliciclastics, dominated by sandstones with minor siltstone and shale intervals (**Figure 6**). Deposition occurred on the wide North African shelf in a generally low accommodation setting. The sediment source was the large Gondwanan hinterland to the south, with SE-NW directed palaeocurrents prevailing. The five reservoir horizons of the giant Hassi Messaoud oilfield are located in Upper Cambrian to Arenig quartzitic sandstones, including the Lower Ordovician Hamra Quartzite.

A major, shortlived ($\frac{1}{2}$–1 my) glaciation occurred in western Gondwana during the latest Ordovician, with the centre of the ice sheet located in central Africa. Features commonly attributed to pro- and sub-glacial processes reported from North Africa, Mauritania, Mali, the Arabian Peninsula, and Turkey include glacial striations, glacial pre-lithification tectonics, diamictites, microconglomeratic shales, and systems of km-scale channels. Several of these features, however, may also occur in deltaic systems unrelated to glaciation, complicating detailed reconstructions of the latest Ordovician glaciation in the region. The uppermost Ordovician in North Africa represents an important hydrocarbon reservoir horizon in Algeria (Unit IV) and Libya (Memouniyat Formation) (**Figure 7**).

## Silurian

Melting of the Late Ordovician icecap caused the Early Silurian sea-level to rise by more than 100 m, leading to a major transgression that flooded the North African Shelf to as far south as the northern parts of Mali, Niger, and Chad (**Figure 8**). Graptolitic, hemipelagic shales represent the dominant facies, while sandstone or non-deposition prevailed in palaeohigh areas, such as most of Egypt, which formed a peninsula at that time. In Libya, the total thickness of the shales (termed 'Tanezzuft Formation', **Figure 7**) increases north-westwards from 50 m in the proximal Kufra Basin, through 500 m in the Murzuq Basin to 700 m in the distal Ghadames Basin, reflecting the north-westward progradation of the overstepping sandy deltaic system ('Akakus Formation', **Figure 7**) during the mid-Llandovery to Ludlow/Přidolí (**Figure 8**).

The Silurian shales are generally organically lean, except for the Lower Llandovery (Rhuddanian) and Upper Llandovery/Lower Wenlock when anoxic phases occurred. During these phases, organically rich, black shales (often referred to as 'hot shale') with total organic carbon values of up to 16% were deposited. The older of the two black shale horizons is developed only in palaeodepressions that were already flooded in the Early Llandovery, while the upper black shale unit is restricted to areas that during the Late Llandovery/Early Wenlock had not yet been reached by the prograding sandy delta (**Figure 8**).

Silurian organic-rich shales are estimated to be the origin of 80–90% of all Palaeozoic-sourced hydrocarbons in North Africa. The same depositional system is also developed on the Arabian Peninsula, where age-equivalent black shales exist, for example, in Saudi Arabia, Syria, Jordan, and Iraq.

Characteristic limestone beds rich in '*Orthoceras*' are interbedded with the Ludlow-Přidolí shales in Morocco and western Algeria, the most distal parts of the North African shelf (**Figure 8**). In more

**Figure 6** Cambro-Ordovician Skolithos ('Tigillites') in Jebel Dalma (Kufra Basin, SE Libya).

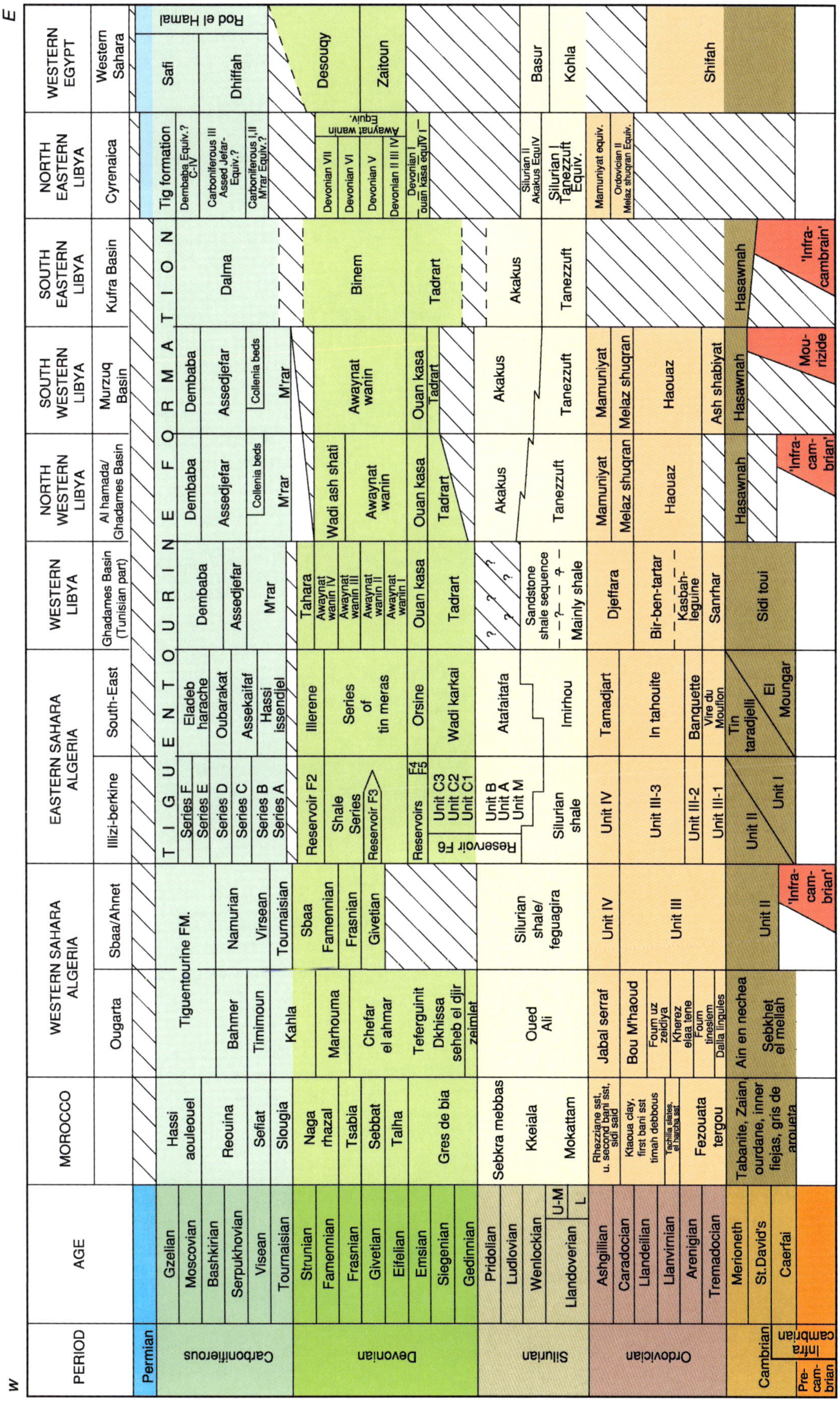

**Figure 7** Correlation chart of Palaeozoic formations in North Africa.

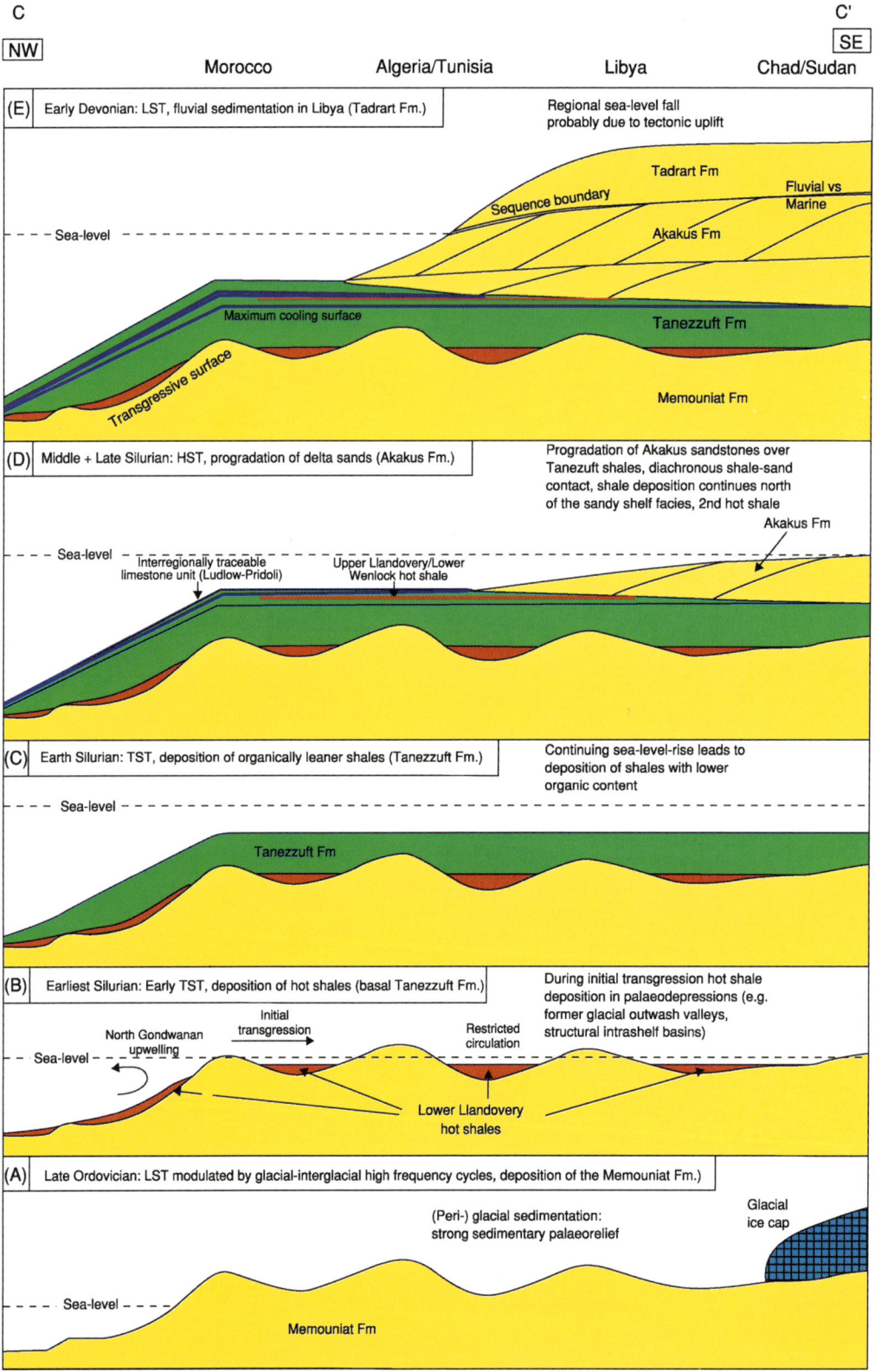

**Figure 8** Depositional model for Late Ordovician to Early Devonian sediments in North Africa. (Modified after Lüning *et al.* (2000).)

proximal shelfal locations, sand influx was already too great for limestones to develop. The '*Orthoceras* Limestone' in some areas is organic-rich. Similar age-equivalent limestones also occur in some peri-Gondwana terranes, such as in Saxo-Thuringia where the unit is termed 'Ockerkalk'.

### Devonian

A major eustatic sea-level fall occurred during the latest Silurian/Early Devonian, resulting in a change to a shallow marine/continental facies in eastern and central North Africa. Coastal sand bar, tidal, and fluvial deposits form important hydrocarbon reservoir horizons, for example, in the Algerian Illizi Basin (unit F6, **Figure 7**) and the Ghadames (=Berkine) Basin ('Tadrart Formation') in north-west Libya (**Figure 8**). On the distal side of the North African shelf towards Morocco fully marine conditions still prevailed. The Lower Devonian of Morocco is well-known for its rich trilobite horizons. A sea-level rise during the later part of the Early Devonian led to deposition of shelfal shales and sandstones in central North Africa. In Algeria significant hydrocarbon reservoirs exist in sandstones of the Emsian (units F4, F5). In western Algeria the base of the Emsian lies under a limestone bed termed 'Muraille de Chine' ('Chinese Wall'), because at exposure it commonly forms a characteristic, long ridge.

Due to their distal position on the North African shelf and a minimum of siliciclastic dilution Morocco and western Algeria were dominated by carbonate sedimentation during the mid-Devonian. The facies here includes prominent mud mounds, for example, in the southern Moroccan area of Erfoud and in the central Algerian Azel Matti area. Further to the east, the facies becomes more siliciclastic. Eifelian-Givetian tidal bar sandstones form the main reservoir (unit F3) in the Alrar/Al Wafa gas-condensate fields in the eastern Illizi Basin.

The beginning of the Late Devonian was characterized by a major eustatic sea-level rise which resulted in deposition of hemipelagic shales, marls, and limestones over wide areas of North Africa. The Moroccan Middle to Upper Devonian typically contains rich cephalopods faunas (goniatites, clymeniids).

The 'Frasnian Event', an important goniatite extinction event and a phase of anoxia, occurred during the Early Frasnian and led to deposition of organic-rich shales and limestones in various places across North Africa. In the Algerian, Tunisian, and Libyan Berkine (=Ghadames) Basin, Frasnian black shales contain up to 16% organic carbon and represent an important hydrocarbon source rock (**Figure 9**). The organic-rich unit also occurs in South Morocco and north-west Eygpt. In parts of north-west Africa, a second organically enriched horizon exists around the Frasnian–Famennian boundary, associated with the worldwide Kellwasser biotic crisis. The deposits in southern Morocco include black limestones.

A major fall in sea-level occurred during the latest Devonian, triggering progradation of a Strunian (latest Devonian–earliest Carboniferous) delta in central North Africa. These clastics form an important hydrocarbon reservoir unit (F2) in Algeria.

### Carboniferous

Sea-level rise during the Early Carboniferous resulted in the development of a widespread shallow marine to deltaic facies across large parts of North Africa. A carbonate platform was established in the Bechar Basin in western Algeria at this time. Early Carboniferous dolomites of the Um Bogma Formation in south-west Sinai host important Mn-Fe ores. Non-deposition and continental sandstone sedimentation occurred in southern and elevated areas, for example, in most of Egypt.

In the Late Carboniferous, deposition of marine siliciclastics was restricted to north-west Africa and the northernmost parts of north-east Africa, for example, Cyrenaica and the Gulf of Suez area. Paralic coal in the Westphalian of the Jerada Basin (NE Morocco) forms the only sizable Late Carboniferous coal deposit in North Africa. In the course of the latest Carboniferous Hercynian folding and thrusting, most of north-west Africa was uplifted, resulting in a change to a fully continental environment. Only Tunisia, north-west Libya and the Sinai Peninsula were still under marine influence at this time.

### Permo–Triassic

Marine Permo-Triassic sedimentary rocks are restricted to the northernmost margin of central and eastern North Africa. For example, Permian marine carbonates and siliciclastics crop out in southern Tunisia representing the only exposed Palaeozoic unit in this country. Most of North Africa, however, remained subaerially exposed during the Permian to mid-Triassic. Continental red clastics (sandstones, shales, conglomerates) represent the most important lithologies. The Permian of Morocco is restricted to a series of intramontane basins located around the margin of the central Moroccan Hercynian massif. The main facies associations in the Triassic TAGI (Trias Argilo-Gréseux Inférieur) in the eastern Algerian Berkine (=Ghadames) Basin are fluvial channel sandstones, floodplain silts and palaeosols, crevasse splay deposits, lacustrine sediments, and shallow marine transgressive deposits. Fluvial sandstones of the TAGI are the main oil and gas reservoirs in the

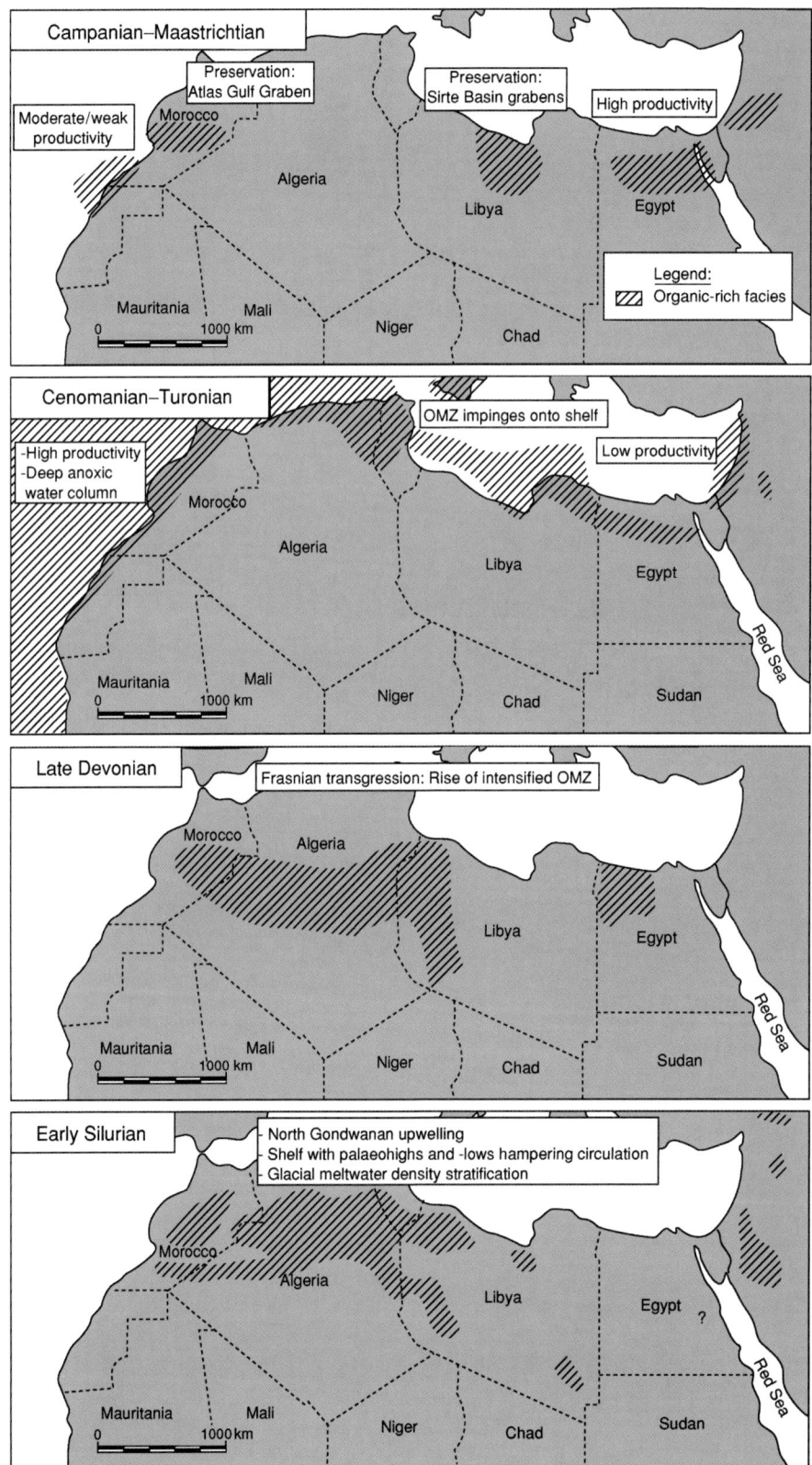

**Figure 9** Known distribution of organic-rich strata of Early Silurian, Late Devonian, Cenomanian–Turonian, and Campanian–Maastrichtian age in North Africa.

Algerian Berkine and Oued Mya basins, including the super-giant gas field in Hassi'R Mel. Similar Triassic sandstones also serve as a relatively minor hydrocarbon reservoir in the Sirt Basin, sourced from Cretaceous source rocks.

During the Late Triassic/Early Jurassic, evaporites were deposited in rift grabens associated with the opening of the Atlantic, and of the Atlas Gulf and with the separation of the Turkish-Apulian terrane from North Africa. Characteristic 'salt provinces' are located offshore along the Moroccan Atlantic coast, northern Algeria/Tunisia and offshore east-Tunisia/north-west Libya. In most areas the diapiric rise commenced in the Jurassic–Cretaceous.

The Late Triassic/Early Jurassic evaporites and shales in the north-east part of the Algerian Saharan Platform are up to 2 km thick and form a hydrocarbon caprock for the Triassic reservoir. In some cases, because of the Hercynian unconformity, they also form the caprock for Palaeozoic reservoirs such as at the super-giant Hassi Messaoud field in Algeria.

### Jurassic

Marine sedimentation during the Jurassic was restricted to the northern and western rims of North Africa, including, for example, northernmost Egypt, the Atlas region, and the Tarfaya Platform in southern Morocco. Carbonate platforms and intraplatform basins were widespread, including development of reefal limestones and oolites. In the Gebel Maghara area in northern Sinai, paralic coal was deposited during the Middle Jurassic. Locally the Lower and Upper Jurassic of North Africa contain organically enriched horizons, corresponding in age to the prominent Jurassic black shales of central Europe (e.g., Posidonia Shale in Germany and Kimmeridge Clay in England). Such Jurassic bituminous pelites occur, for example, in the Atlantic Basin, Atlas Rift of Morocco, and the Egyptian Abu Gharadig Basin. South of the North African Jurassic marine facies belt, continental redbeds were deposited (**Figure 10**). In the Egyptian Western Desert the Jurassic–Cretaceous contains several prolific hydrocarbon reservoir horizons.

### Cretaceous

Due to low eustatic sea level the Lower Cretaceous of North Africa is dominated by terrestrial clastics, termed the 'Nubian Sandstone' in Egypt and Libya ('Sarir Sandstone' in the Sirt Basin) (**Figure 10**). Once again, marine conditions existed only in a marine coastal belt in the north. During the Aptian to Maastrichtian, a series of transgressions gradually flooded the areas to the south. On the Sinai Peninsula, the transition phase is characterised by deltaic influenced,

**Figure 10** Cross-bedded fluvial 'Nubian Sandstone', Jurassic–Cretaceous, 'Coloured Canyon', central East Sinai (Egypt).

mixed siliciclastic-carbonate systems that during the Albian evolved into carbonate-dominated environments. During the latest Cenomanian, large parts of North Africa became submerged following a prominent eustatic sea-level rise that is thought to be one of the most intense Phanerozoic flooding event. As a consequence, the 'Transsaharan Seaway' was created, connecting the Tethys in central North Africa with the Atlantic in West Africa. Similar seaways and gulfs existed in north-west Africa into the Eocene. A seaway located within the Atlas rift system, the 'Atlas Gulf', was restricted temporally to the Cenomanian–Turonian.

The strong latest Cenomanian sea-level rise in combination with high productivity conditions in the southern North Atlantic are thought to form the basis for the Late Cenomanian–Early Turonian Oceanic Anoxic Event (OAE2) during which organic-rich strata were deposited in rift shelf basins and slopes across North Africa and in deep sea basins of the adjacent oceans. Characteristic sediments associated with this anoxia include oil shales in the Tarfaya Basin (southern Morocco), organic-rich limestones in north-west Algeria and northern Tunisia (Bahloul Formation), and black shales in offshore Cyrenaica, and the Egyptian Abu Gharadig Basin (Abu Roash Formation) (**Figure 9**). The unit represents a potential oil-prone hydrocarbon source rock in the region. A general decrease in peak organic richness and black shale thickness occurs in North Africa from west to east, which possibly is a result of upwelling along the Moroccan Atlantic coast and the absence of upwelling in the Eastern Mediterranean area.

The organic-rich Cenomanian-Turonian deposits also play an important role in the genesis of Zn/Pb ore deposits in northern Tunisia and eastern Algeria. The origin of these Zn/Pb ores is related to hypersaline basinal brines, made of ground water and dissolved Triassic evaporites, that leached metals from the Triassic-Cretaceous sediments. Ore deposition occurred when these metal-bearing solutions mixed with microbially reduced sulphate solutions that were associated with the organic carbon of the Cenomanian-Turonian strata.

Due to the generally high sea-level, the marine Upper Cretaceous in North Africa is dominated by calcareous lithologies, namely dolomites/limestones, chalks, and marls (**Figure 11**). Lateral and vertical facies distributions are strongly related to sea-level changes of various orders as well as to the changing structural relief associated with Late Cretaceous syn-depositional compression. Great variations in thickness and facies as well as onlap features, for example, are developed around the domal anticlines of the Syrian Arc Foldbelt in Sinai and within rift grabens of the Sirt Basin (N. Libya).

The Campanian–Maastrichtian was characterised by very high sea-level, resulting in a widespread distribution of hemipelagic deposits, such as chalks and marls. These deposits often contain abundant foraminiferal faunas and calcareous nannofossil floras, which allow high-resolution biostratigraphic and palaeoecological studies in these horizons. As on the Arabian Peninsula, the Santonian–Maastrichtian interval in North Africa contains significant amounts of phosphorites, which are mined in, for example, Morocco/Western Sahara and Abu Tartour (Western Desert), making North Africa one of the world's largest producers of phosphate (*see* **Sedimentary Rocks: Phosphates**).

In places, the Campanian–Maastrichtian contains organic-rich intervals with total organic carbon contents of up to 16%, for example, in the Moroccan Tarfaya Basin and Atlas Gulf area, the Libyan Sirt Basin and the Egyptian southern Western Desert, Red Sea Coast and Gulf of Suez (**Figure 9**). Notably, Algeria, Tunisia, and West Libya are dominated by organically lean deposition during this time. Campanian–Maastrichtian black shales form important hydrocarbon source rocks in the Sirt Basin and the Gulf of Suez.

### Palaeogene

Sea-level during most of the Paleocene–Eocene remained high resulting in deposition over wide areas (Egypt: Dakhla and Esna Shale) of hemipelagic marls and chalks that are rich in planktonic foraminifera. A sea-level fall occurred during the mid-Paleocene, resulting in the formation of a short-lived carbonate interbed ('Tarawan Chalk') in parts of Egypt. Within the Eocene, the facies typically changes here to hard dolomitic limestones with abundant chert nodules ('Thebes Limestone'). A similar Palaeogene facies development can also be found in parts of northern Libya and Tunisia.

The Eocene in Egypt, Libya, Tunisia, and Algeria includes nummulitic limestones up to several 100 metres thick, which were deposited in carbonate ramp settings. The unit forms major hydrocarbon reservoirs in offshore Libya and Tunisia. Well-exposed and continuous exposures occur in Jabal al Akhdar (Cyrenaica), where the nummulite body's geometry can best be studied (**Figure 12**). Notably, the Giza pyramids in Cairo are built from Eocene nummulite limestone.

The Eocene hydrocarbon play in the offshore of Tunisia is sourced by dark-brown marl and mudstone of the lower Eocene Bou Dabbous Formation. The unit contains type I and II kerogen and ranges in thickness from 50 to 300 m.

### Neogene and Quaternary

Marine conditions during the Miocene were again restricted to the northernmost margin of North Africa

**Figure 11** Contact between chalky limestones of the Early Eocene Bou Dabbous Formation (reddish) and the underlying Campanian–Maastrichtian Abiod Formation (bluish) (Ain Rahma Quarry, Gulf of Hammamet area, Tunisia).

**Figure 12** High energy nummulitic bank facies, Darnah Formation, Middle to Late Eocene, West Darnah Roadcut, Jebel Akhdar (Cyrenaica, Libya).

including the Atlas, Sirte Basin, Cyrenaica, and Red Sea. Carbonate platforms and ramps were developed in northern Morocco. The Miocene Gulf of Suez in Egypt is rich in hydrocarbons, containing more than 80 oilfields. Oils in the Gulf of Suez were mostly sourced from source rocks in the pre-rift succession, including the Campanian–Maastrichtian Brown Limestone. Hydrocarbon reservoir horizons include various Miocene syn-rift sandstones and carbonates as well as pre-rift reservoirs, including fractured Precambrian granites, Palaeozoic–Cretaceous sandstones, and fractured Eocene Thebes Limestone. The thickness distribution and facies of the syn-rift strata are strongly controlled by fault block tectonics. Shales and dense limestones of the pre-rift and the syn-rift units are the primary seals, while overlying Miocene evaporites form the ultimate hydrocarbon seals.

During the latest Miocene, more than 2 km thick evaporites were deposited in a deep and desiccated Mediterranean basin that had been repeatedly isolated from the Atlantic Ocean. In the near-offshore only a few tens to hundreds of metres of evaporites exist, whilst they are almost absent from the onshore area. As a consequence of the 'Messinian Salinity Crisis', a large fall in Mediterranean sea-level occurred, followed by erosion and deposition of non-marine sediments in a large 'Lago Mare' ('lake Sea') basin. Cyclic evaporite deposition is thought to be almost entirely related to circum-Mediterranean climate changes.

The Nile Delta system represents a major natural gas province. It was initiated during the Late Miocene with deep canyon incision into pre-existing Cenozoic/Mesozoic substrate, allowing transportation of huge amounts of sediments into the Mediterranean. The proximal infill of these canyons is thick, coarse alluvium becoming sandier with greater marine influence northwards. The far reaches of these canyon systems have proven to be a good Plio-Pleistocene hydrocarbon reservoir linked mainly to the lowstands, when sands were conveyed to the outer belts through incised canyons in the upper slopes which led to submarine fans farther northwards.

The Early Holocene (~9–7 kyr BP) was a relatively humid period in North Africa. During this phase, the African Humid Period, grasslands covered the Sahara/Sahel region, and many lakes and wetlands existed here. The humid conditions at this time were associated with a strengthening of the summer monsoon circulation due to an increase in the land–sea thermal contrast under the influence of relatively high summer insolation.

## See Also

**Africa:** Pan-African Orogeny. **Analytical Methods:** Fission Track Analysis. **Petroleum Geology:** Overview. **Plate Tectonics. Sedimentary Rocks:** Phosphates. **Tectonics:** Rift Valleys.

## Further Reading

Ben Ferjani A, Burollet PF, and Mejri F (1990) *Petroleum Geology of Tunisia*. Tunis: Entreprise Tunisienne d'Activités Pétrolières.

Beuf S, Biju-Duval B, de Charpal O, Rognon P, Gariel O, and Bennacef F (1971) Les grès du Paléozoïque inférieur au Sahara, Sédimentation et discontinuités, évolution d'un craton. *Publications de l'Institut français du Pétrole* 18: 464.

Coward MP and Ries AC (2003) Tectonic development of North African basins. In: Arthur TJ, MacGregor DS, and Cameron NR (eds.) *Petroleum Geology of Africa: New Themes and Developing Technologies*. Geological Society London, Special Publication 207: 61–83.

Dercourt JM, Gaetani B Vrielynck E, *et al.* (eds.) (2000) *Atlas Peri-Tethys, Palaeogeographical maps*. CCGM/CGMW, Paris.

Doblas M, Oyarzun R, Lopez-Ruiz J, Cebria JM, Youbi N, Mahecha V, Lago M, Pocovi A, and Cabanis B (1998) Permo-Carboniferous volcanism in Europe and northwest Africa: a superplume exhaust valve in the centre of Pangaea? *J. Afr. Earth Sciences* 26: 89–99.

Hallett D (2002) *Petroleum Geology of Libya*. Amsterdam: Elsevier.

Lüning S, Craig J, Loydell DK, Štorch P, and Fitches B (2000) Lower Silurian 'Hot Shales' in North Africa and Arabia: Regional Distribution and Depositional Model. *Earth Science Reviews* 49: 121–200.

Macgregor DS, Moody RTJ, and Clark-Lowes DD (eds.) (1998) Petroleum Geology of North Africa. *Geological Society London Special Publication* 132: 7–68.

Maurin J-C and Guiraud R (1993) Basement control in the development of the Early Cretaceous West and Central African Rift System. *Tectonophysics* 228: 81–95.

Piqué A (2002) *Geology of Northwest Africa*. Stuttgart: Gebr. Borntraeger.

Said R (1990) *The Geology of Egypt*. Rotterdam, Netherlands: Balkema Publishers.

Schandelmeier H and Reynolds PO (eds.) (1997) *Palaeogeographic-Palaeotectonic Atlas of North-eastern Africa and adjacent areas*. Rotterdam: Balkema.

Selley RC (1997) Sedimentary basins of the World: Africa. Amsterdam: Elsevier.

Stampfli GM, Borel G, Cavazza W, Mosar J, and Ziegler PA (2001) *The Paleotectonic Atlas of the Peritethyan Domain*. Strasburg European Geophysical Society.

Stets J and Wurster P (1981) Zur Strukturgeschichte des Hohen Atlas in Marokko. *Geologische Rundschau* vol. 70(3): 801–841.

Tawadros EE (2001) *Geology of Egypt and Libya*. Rotterdam: Balkema.

# Rift Valley

**L Frostick**, University of Hull, Hull, UK

© 2005, Elsevier Ltd. All Rights Reserved.

## Introduction

The East African and Dead Sea rifts are famous examples of rifts that have played prominent parts in human evolution and history. They are both areas where the Earth's crust has been put under tension and ripped apart to give deep valleys that snake across the landscape. They are linked tectonically, via the Red Sea–Gulf of Aden, which is an incipient ocean separating the African and Arabian plates. The differences between the two rifts are caused by differences in the relative movement of the crust. In the East African rift the tension that formed the rift is close to 90° to the rift axis, whereas the movement of Jordan relative to Israel is northwards, almost parallel to the Dead Sea, which is a small section pulled apart as a result of splaying and bending of the faulted plate boundary.

The ancient crust of Africa has been subjected to rifting many times in its very long geological history. Recognizable rift basins can be identified in many locations around the continent, and they range in age from Palaeozoic to Quaternary, a time-span of over 500 Ma. In some areas there is evidence of repeated activity, and it appears that there have been at least seven phases of rifting over the past 300 Ma. The older rifts, for example the Benue trough in West Africa, have been inactive for many millions of years, but the most spectacular rift features are to be found in East Africa, where recent rifting has left a scar on the landscape that is visible from space (**Figure 1**). North of the zone where Africa touches Europe at the eastern end of the Mediterranean there is another famous rift, which is linked tectonically to East Africa. The Dead Sea Rift straddles the border between Israel and Jordan and is the lowest point on the surface of the Earth, reaching more than 800 m below sea-level.

## Plate Tectonic Setting

Rifting occurs when the crust of the Earth is placed under tension, pulling it apart and causing faulting. The general term for the basins so produced is 'extensional' but they can occur in situations where the regional sense of movement is compressional or is tearing the crust, e.g. the Baikal and Dead Sea rifts, respectively. However, the main African rift basins were formed in a plate-tectonic setting that is dominated by extension, particularly during the Tertiary–Quaternary period. During this time the Great or East African Rift was formed as part of a larger plate-tectonic feature that stretches from south of Lake Malawi in Africa to the flanks of the Zagros mountains and the Persian Gulf in the north (**Figure 2**). It changes its nature along its length, resulting in a range of geological basins and geomorphological features. In Africa it is a volcanically active continental rift hundreds of kilometres wide that contains a range of river and lake sediments. As it quits Africa it passes into an incipient ocean with a newly formed seafloor spreading centre along the length of the Red Sea and the Gulf of Aden. The deposits in these basins include thick sequences of salt, which form effective traps for hydrocarbons generated from associated organic-rich shales. North of the Red Sea the type of plate margin alters as the boundary passes through the Gulf of Aqaba/Elat and

**Figure 1** Satellite remote-sensing image of the Horn of Africa and Arabia, showing the East African Rift system, the incipient ocean of the Red Sea–Gulf of Aden, and the conservative plate boundary that runs through the Dead Sea. Images collected by the TERRA satellite using the MODIS instrument (moderate resolution imaging spectroradiometer) and enhanced with SRTM30 (Shuttle Radar Topography Mission-1km resolution) shaded relief.

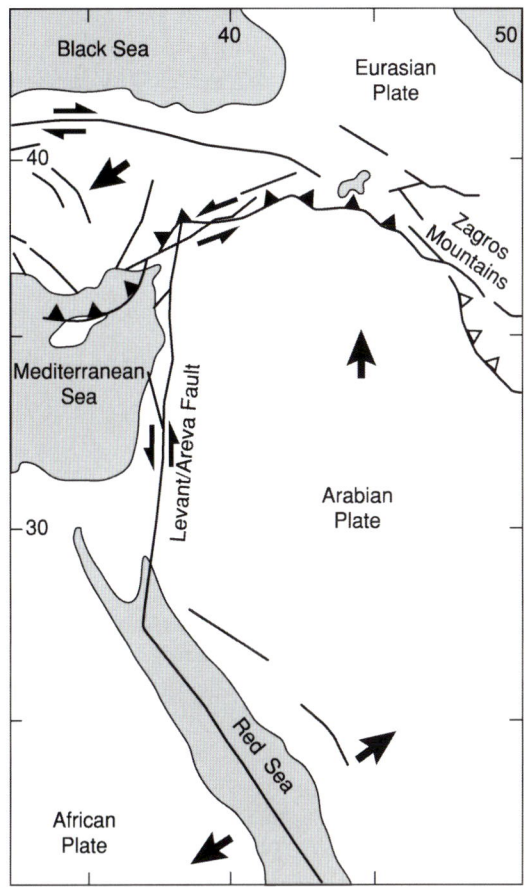

**Figure 2** Diagrammatic representation of the plate-tectonic setting of the area between the northern end of the East African Rift (Afar triangle) and the Zagros Mountains.

**Figure 3** Satellite remote-sensing image of the Sinai–Arabian plate boundary, showing the Dead Sea and Sea of Galilee (Lake Kinneret). Image collected by the TERRA satellite using the MODIS instrument (moderate resolution imaging spectroradiometer).

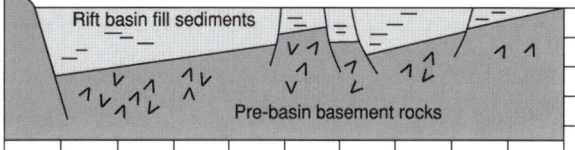

**Figure 4** Stylized half-graben structure typical of the basins in the East African Rift.

into the Levant/Areva valley between Israel and Jordan. Here, the Arabian plate is moving past the European plate without significant extension or compression. However, localized tension associated with fault bends and splays has resulted in the formation of two very well-known biblical lake basins, the Dead Sea and Lake Kinneret (otherwise known as the Sea of Galilee), both of which can be seen in **Figure 3**. These are also termed 'rifts' although the setting and geological history are different from those of their larger East African contemporary. The system terminates in the Zagros mountains, where the crust created in the new Red Sea–Gulf of Aden ocean is compensated for by the crustal shortening inherent in mountain-building processes.

## The East African Rift

### Topography and Structure

Within Africa certain features of the topography and structure are common to all the basins. Topographically they comprise a central valley, often referred to as a 'graben', flanked by uplifted shoulders that are stepped down towards the rift axis by more or less parallel faults. Often, one flank is more faulted than the other, so that the rift valley is in fact asymmetrical and should be referred to as a 'half graben' (**Figure 4**). The width of the structure varies from 30 km to over 200 km, with the widest section at the northern extremity where the rift links to the Red Sea in the Afar region of Ethiopia. The main faulted margin alternates from one side of the rift to the other along its length, producing a series of relatively separated basins, many of which contain lakes of varying depth and character (e.g. Lakes Tanganyika, Naivasha, and Malawi). These are separated into hydrologically distinct basins by topographical barriers crossing the rift axis where the border faults switch polarity. This surface separation reflects an underlying structure, the nature of which varies from basin to basin but often includes faulting with a tearing or scissor type of movement and flexing. Geologists are not agreed on the processes going on in these areas and have given these zones different names according to their assumptions about the mechanism of formation. These include transfer, relay, and accommodation zones, as well as ramps or just segment boundaries. The distance between adjacent boundaries varies from tens to hundreds of kilometres (**Figure 5**).

At each end of the individual border faults the displacement of the rift floor relative to rocks outside the valley reduces to zero. Displacement is greatest at the centre of the fault, and this leads to a subtle rise and fall of the rift floor along its length even without the intervention of major new cross-rift structures and processes.

The rift in Kenya is characterized by numerous caldera volcanoes and at least 3 to 4 phases of faulting, the most recent forming a narrow linear axial zone. the faulting ranges in age from Miocene to Recent.

In the southern half of the rift's 35 000 km length it divides into two distinct branches around Lake Victoria. The eastern branch contains only small, largely saline, lakes, while the western branch contains some of the largest and deepest lakes in the region, including Lake Tanganyika.

## Doming and Volcanicity

The East African Rift contains two large domes centred on Robit in Ethiopia and Nakuru in Kenya. These domes are over 1000 km in diameter and extend far beyond the structural margins of the rift valley. Geophysical studies of these domes have shown that they are underlain by zones of hot low-density mantle rocks and that the surface crust is thinned significantly relative to adjacent areas. The domes are centres of volcanic activity that began more than 25 Ma ago and continues to the present day (**Figure 6**). Volcanic features are widespread in Ethiopia and extend southwards into Kenya along the eastern branch of the rift. It is estimated that there are more than 500 000 km$^3$ of volcanic rocks in this area, over a third of which occur in Kenya. In the branch to the west of Lake Victoria volcanism is spatially more

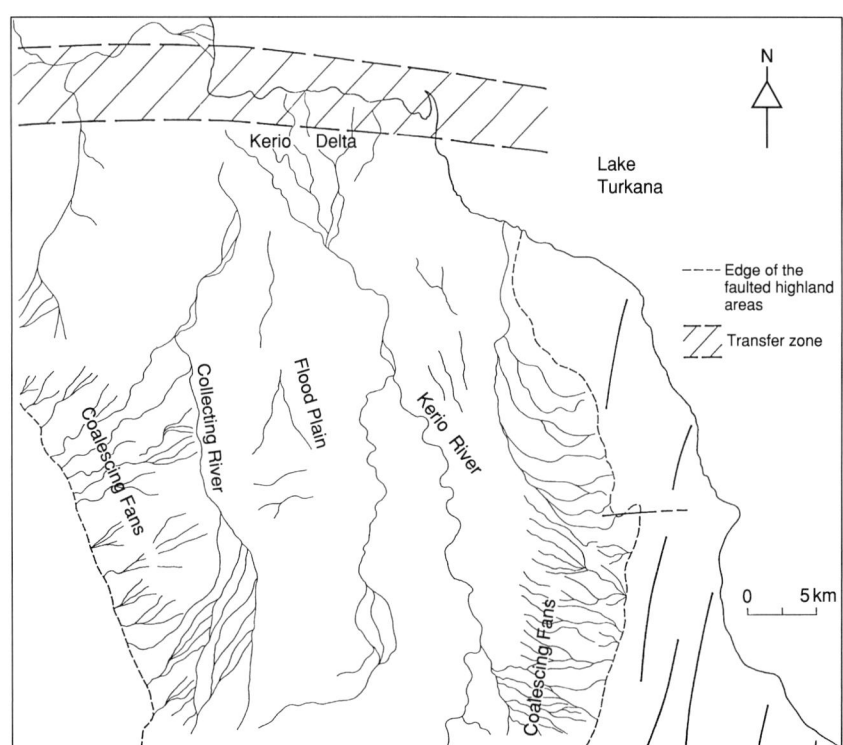

**Figure 5** Diagrammatic representation of the river drainage close to the west shore of Lake Turkana, northern Kenya, showing the Kerio River flowing into the Lake at a transfer zone and the alluvial fans issuing from the fault scarps.

**Figure 6** Geyser activity in the volcanically active area around Lake Bogoria, Kenya.

limited, occurring only to the north and south of Lake Tanganyika. This contributes to the different characters of the lakes in the two branches as not only can the volcanic rocks fill the basins, leaving less space for large lakes, but also many of the rock types are rich in salts, which contribute to the salinity of the lakes once they are released by weathering.

Large and active volcanoes that sit outside the rift structure are a striking feature of the landscape. Mounts Kilimanjaro and Kenya, for example, are favourite targets for climbers, and both sit on the flanks of the rift (**Figure 7**).

## Hydrology and Climate

The East African Rift system sits astride the equator, extending from 12° N to 15° S, and this dictates the overall character of the climate. Superimposed on this are the effects of the rift topography, with its uplifted domes, faulted flanks, and depressed central valleys. Rainfall is lowest in the northern parts of Ethiopia and increases southwards into northern Kenya. The region is generally desert or semi-desert with vegetation limited to sparse grasses and scrub. South of where the rift branches the rainfall is higher, with the western branch being wetter than the eastern one. The uplifted mountains that make up the margins of the rift are wetter and cooler than the valley bottom; for example, an annual figure of over 2000 mm of rainfall has been recorded in the Ruwenzori Mountains near Lake Mobutu.

The doming that accompanied the rifting in East Africa has had a major impact on the present river systems. The development of the rift disrupted a pre-existing continental drainage system in which a few large rivers with vast integrated drainage basins dominated the landscape. As the area was domed and faulted and the new valley formed, the rivers adjusted to the new landscape: some lost their headwaters, others were created, some gained new areas to drain. The overall effect was to divert much of the drainage north into the Nile system and west into the Congo drainage, with only a few small rivers now reaching the Indian Ocean. Inside the valley, the rivers are generally short and small, ending in a lake not far from the river source, but a few rivers run along the rift, often caught between faulted hills, and discharge into lakes far from their original sources, e.g. the Kerio River in Kenya has its source near Lake Baringo but discharges into Lake Turkana more than 200 km to the north (**Figure 8**).

The segregation of the underlying structure into topographically distinct sections exerts an overriding control on the character and distribution of lakes throughout the rift. It provides the framework within which the balance between movement of water into the basin, from rainfall and rivers, and evaporation from the surface will work. The largest and deepest lake, Lake Tanganyika, is in the wetter western branch of the rift in a particularly deep section. It covers an area of over 40 000 km$^2$ and is more than 1400 m deep at its deepest point. Lakes in the eastern branch are smaller and shallower; for example Lake Bogoria is an average of less than 10 m deep, and if the climate changes and rainfall decreases they soon become ephemeral, drying out completely during periods of drought.

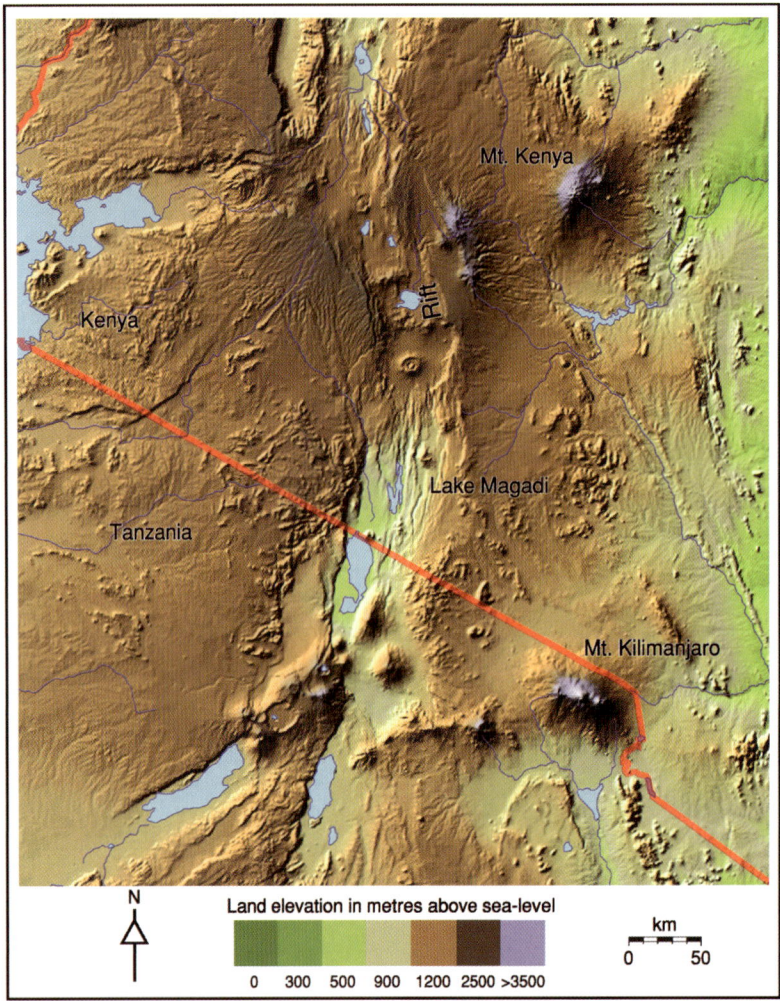

**Figure 7** Satellite image of Mount Kilimanjaro and Mount Kenya, showing how they sit outside the main East African Rift structure. This is a shaded relief map produced from SRTM30 data with colour added to indicate land elevations.

### Sedimentation and Basin Fills

As water flows into the rift basins it brings with it material dislodged and dissolved from the surrounding rocks, which is then deposited within the basin. How, where, and what is deposited depends on the shape of the basin and how surface processes work on and disperse the material. The overall shape of the basin fill is controlled by the pattern of faults and subsidence: deposits are thicker close to areas of the faults with greatest displacement (**Figure 4**). The geometry of the fill is therefore almost always asymmetric, thickening towards the main border fault and thinning in all other directions, giving a characteristic wedge shape.

There are no marine sediments in the rift: all the deposits are terrestrial and comprise river, delta, lake-coast, and lake sediments. Wind-blown sands and dunes are rare and of only local importance. The rivers vary in character from ephemeral, flowing only in response to seasonal rain storms, to perennial. The rivers carry and deposit sands and gravels in their beds, sweeping finer silts and clays into overbank lagoons and lake-shore deltas. The character of the lake deposits themselves depends on a variety of factors including the timing and character of river supplies, salinity, evaporation, water stratification, and animal and plant growth. In deep lakes such as Lake Tanganyika there are layered muds, which can be hundreds of metres thick and contain enough algal remains to generate oil. Shallower lakes can contain high numbers of diatoms, which leave deposits of a silica-rich rock called diatomite. Some lakes in volcanic areas of the rift have sufficiently high salt concentrations for precipitation and the development of exploitable salt deposits. One example is the trona, a complex carbonate of sodium, which is extracted seasonally from Lakes Magadi and Natron (**Figure 9**).

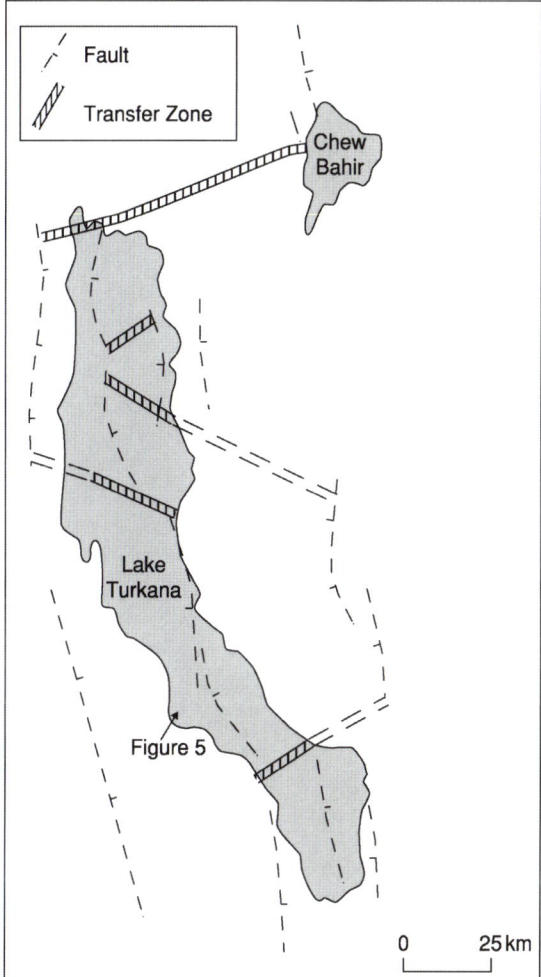

**Figure 8** Stylized diagram of the Lake Turkana area at 3° N in the East African Rift, showing the main faults and transfer zones crossing the rift axis.

## Hominid Finds and Evolution

The rift forms a striking geomorphological feature cutting across the African craton, segmenting the landscape, and controlling the local geology. Along most of its length it achieves a depth of in excess of 1 km and at its deepest, in Ethiopia, it is over 3 km deep. Its striking topography generates its own set of microclimatic and hydrological conditions, which have had a major impact on plant and animal distributions and evolution. It acts as a north–south corridor for the migration of animals and birds, but equally inhibits east–west movements. During periods of climatic stress at higher latitudes, when glaciers dominated much of the European and Asian continents, the lake basins of the rift were havens for animals, including early humans. Finds of early humans (hominids) in the rift are more numerous and more complete than in almost any other part of the world, and it has been postulated that all present-day humans are derived from ancestors that migrated out of the East African Rift (*see* **Fossil Vertebrates:** Hominids).

## Dead Sea Rift

### Topography and Structure

The Dead Sea Rift is superficially very similar to some of the individual lake basins in the East African Rift. It is a narrow depression in the surface of the Earth over 100 km long and only 25 km wide, reaching over 800 m deep at its lowest point (**Figure 10**). The Dead Sea is not, in reality, a sea at all but an enclosed salty lake, which occupies more than 80% of the surface area of the basin. It sits on the plate boundary that

**Figure 9** Lake Magadi, Kenya, during the dry season, showing the surface of the lake completely encrusted with salt.

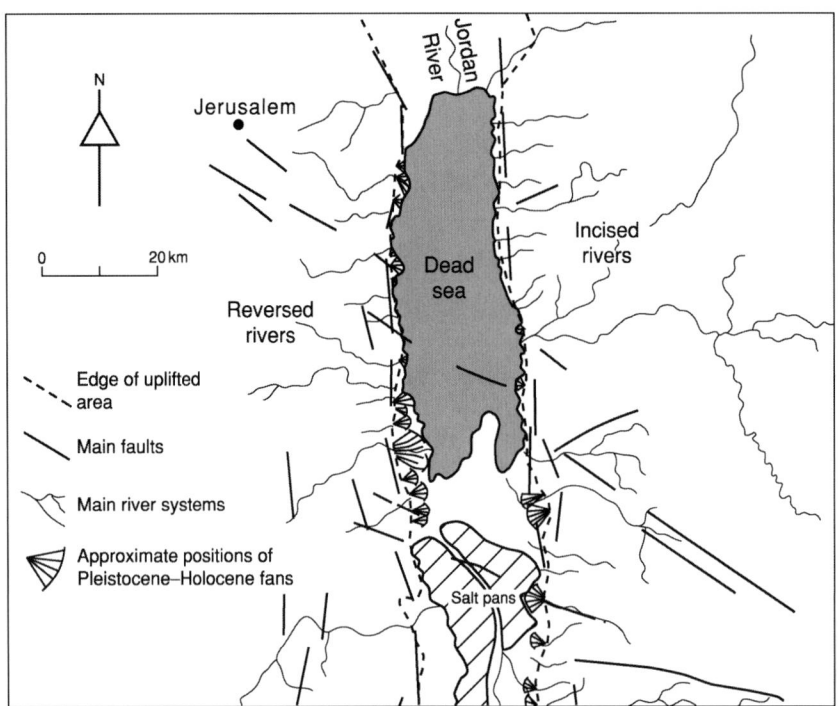

**Figure 10** Patterns of faulting and their influence on the development of river systems around the Dead Sea pull-apart basin.

spans the 1100 km between the Gulf of Elat/Aqaba and Turkey and separates the Arabian Plate to the east from the African Plate to the west (**Figure 2**). Since the Miocene, a period of about 20 Ma, Arabia is thought to have moved more than 105 km northwards, a type of movement that is termed strike-slip.

The Dead Sea Basin is a zone where the movement has resulted in local tension, producing faulting and leading to the sinking of a section of the crust. Such basins are termed 'pull-apart' basins and are characterized by very rapid subsidence and thick basin-fill sequences. The overall structure of the Dead Sea Rift is asymmetrical, not dissimilar to that of the East African Rift. The largest fault is in the eastern margin and forms the Jordanian shore of the lake. Here, the faulting exposes a spectacular rock sequence more than 1 km thick, which ranges in age from Precambrian (more than 544 Ma) to Pleistocene (less than 1 Ma). On the opposite side of the basin are a number of smaller subparallel faults, which cut the Cretaceous limestones of this margin into a series of structural steps (**Figure 10**). At either end of the basin are cross-rift structures that link movement along the Areva fault to the south with movement on the Jordan fault to the north.

A major feature of the southern part of the basin is the development of salt diapirs. These result from subsurface movements of thick deposits of rock salt, which can push up and punch through the overlying sediments and penetrate to the surface. One famous example of such a diapir is Mount Sedom, famous for its biblical links with the doomed and 'sinful' cities of Sodom and Gomorrah.

**Rivers and Hydrology**

The development of the Dead Sea Rift system disrupted a pre-existing drainage system that crossed from east to west across the Jordan plateau and drained into the Mediterranean. The headwaters of this system now run across the eastern scarp of the Dead Sea and have cut gorges over a kilometre deep to reach the lake shore (**Figure 10**). On the western shore a new set of rivers have evolved, which no longer drain into the Mediterranean Sea to the west but instead have been reversed and now drain from west to east. These rivers have also cut down into pre-rift rocks and run in gorges that are less deep than those of their eastern equivalents.

The present climate of the area is desert to semi-desert with rainfall of 50–200 mm year$^{-1}$. Because of this all rivers except the Jordan, which has headwaters in an area of higher rainfall to the north, are ephemeral and flow only in response to winter rain storms.

The lake water is renowned for its high salinity, which is 10 times that of normal seawater. Tourists are attracted to 'swim' in the waters, which are so buoyant that individuals can sit unsupported and read a paper. The high salinity is a result of a combination of evaporation in a closed basin and the influence of brines coming from the solution of subsurface rock salt. The lake brines are particularly rich in chlorine and bromine, which are extracted in salt ponds and exported worldwide.

### Climate Change and the Basin Fill

Lake levels in closed basins are very susceptible to the effects of climate change. Any increase in rainfall will upset the hydrological balance and cause lake levels to rise and salinity to fall. If rainfall decreases, lake levels will drop and evaporation will dominate, resulting in an increase in salinity. The surface of the Dead Sea shows evidence of having fluctuated between 180 m and 700 m below sea-level over the past 60 Ka in response to well-documented changes in climate. The rising and falling lake levels have a profound effect on the sedimentary deposits of the rift. High lake levels, such as those that prevailed during the deposition of the Pleistocene Lisan Formation, result in thick sequences of interlaminated chalk and silty clay (**Figure 11**). During periods of lower lake levels the river and fan deposits penetrate far into the basin and dominate the sequences. One surprising consequence of depressed lake levels is a change in the balance between saline and fresh groundwaters, with the latter penetrating further towards the axis of the basin. Since much of the basin axis is underlain by thick salt deposits, the fresh groundwater dissolves the preserved layers of salt, generating subsurface caverns and solution holes. This is currently happening in response to lake levels falling as a result of over abstraction of water from the Jordan River.

### Earthquakes, Archaeology, and Sodom and Gomorrah

Earthquakes have been a feature of the Dead Sea Rift throughout its history. The earthquakes are generated by movement along the main fault zone and are often accompanied by the release of asphalt, gases, and tars, which are trapped in the layers of rock beneath the surface. The asphalt in particular is well documented and is found in layers within the older lake deposits. Fault movements tend to happen sporadically: long periods of quiescence are succeeded by times when earthquakes are regular events.

The Dead Sea Basin has been inhabited by local peoples for many thousands of years. The alluvial plains of the valley were rendered fertile by irrigation, and trading routes to the south, east, and west allowed early settlers to exploit the mineral wealth of the area, including gathering and trading materials from oil seeps and asphalt, which have been found as far away as Egypt in the tombs of the Pharaohs. The early Bronze Age was a time when the basin was well populated and was also a tectonically quiet period when few earthquakes occurred. Towards the end of this period there was a large earthquake, which may have resulted in the destruction of two major cities, Sodom and Gomorrah. There has been speculation about precisely how and why these cities were so

**Figure 11** A section through the Lisan Formation of the Dead Sea, showing layers of chalk and silt (horizontal layers at the top and bottom of the section), some of which have been disturbed by earthquake activity (folded layers in the centre of the section).

comprehensively demolished that they were never rebuilt. One theory is that they were built on soft sediments that became liquid (liquefaction) as they were shaken, maximizing the instability of the ground (*see* **Engineering Geology: Liquefaction**). Interestingly, the occurrence of 'sulphurous' fires reported in the bible corresponds well with the release of the light fractions of oil from underground reservoirs as the ground moves and slides in response to shaking. It seems likely that the myths surrounding the destruction of Sodom and Gomorrah are based in fact and are a direct consequence of the unique geology of the area.

## See Also

**Biblical Geology**. **Engineering Geology:** Liquefaction. **Fossil Vertebrates:** Hominids. **Geomorphology**. **Sedimentary Environments:** Lake Processes and Deposits. **Tectonics:** Earthquakes; Faults; Mid-Ocean Ridges; Rift Valleys.

## Further Reading

Allen PA and Allen JR (1990) *Basin Analysis: Principles and Applications*. Oxford: Blackwells.

Enzel Y, Kadan G, and Eyal Y (2000) Holocene earthquakes inferred from a fan delta sequence in the Dead Sea graben. *Quaternary Research* 53: 34–48.

Frostick LE and Reid I (1989) Is structure the main control on river drainage and sedimentation in rifts? *Journal of African Earth Sciences* 8: 165–182.

Frostick LE and Steel RJ (eds.) (1993) *Tectonic Controls and Signatures in Sedimentary Successions*. International Association of Sedimentologists Special Publication 20. Oxford: Blackwells.

Frostick LE, Renaut RW, Reid I, and Tiercelin JJ (1986) *Sedimentation in the African Rifts*. Special Publication 25. London: Geological Society.

Girdler RW (1991) The Afro-Arabian Rift System: an overview. *Tectonophysics* 197: 139–153.

Gupta S and Cowie P (2000) Processes and controls on the stratigraphic development of extensional basins. *Basin Research* 12: 185–194.

Neev D and Emery KO (1995) *The Destruction of Sodom, Gomorrah and Jericho*. Oxford: Oxford University Press.

Selley RC (ed.) (1997) *African Basins*. Sedimentary Basins of the World 3. Amsterdam: Elsevier.

Summerfield MA (1991) *Global Geomorphology: An Introduction to the Study of Landforms*. Harlow: Longman.

# AGGREGATES

**M A Eden and W J French**, Geomaterials Research Services Ltd, Basildon, UK

© 2005, Elsevier Ltd. All Rights Reserved.

## Introduction

Aggregates are composed of particles of robust rock derived from natural sands and gravels or from the crushing of quarried rock. The strength and the elastic modulus of the rock should ideally match the anticipated properties of the final product.

Aggregates are used in concrete, mortar, road materials with a bituminous binder, and unbound construction (including railway-track ballast). They are also used as fill and as drainage filter media.

In England alone some 250 million tonnes of aggregate are consumed each year, representing the extraction of about $0.1 \text{ km}^3$ of rock, if necessary wastage is taken into account. Aggregates may be derived from rocks extracted from quarries and pits, or from less robust materials. For example, slate and clay can be turned, by heating, into useful expanded aggregates of low bulk density.

The principal sources of aggregate are sand and gravel pits, marine deposits extracted by dredging, and crushed rock from hard-rock quarries. As extracted, these materials would rarely make satisfactory aggregate. They need to be carefully prepared and cleaned to make them suitable for their intended purpose. The sources may also be rather variable in their composition and in the rock types present, so it is essential that potential sources are carefully evaluated. At the very least, the preparation of the aggregate involves washing to remove dust and riffling to separate specific size ranges.

The classification of aggregates varies greatly. An early classification involved the recognition of Trade Groups, which were aggregates consisting of rocks thought to have like properties and which could be used for a particular purpose. A fairly wide range of rock types was therefore included in a given Group. More recent classifications have been based on petrography. Again, these groups tend to be broad, and they focus on the macroscopic properties of the materials for use as aggregate rather than on detailed petrographic variation.

Because aggregates consist of particulate materials, whether crushed or obtained from naturally occurring sands and gravels, their properties are normally measured on the bulk prepared material. There are therefore numerous standard tests that relate to the intended use of the material. Standard tests vary from country to country, and, in particular, collections of standard tests and expected test results are given in specific British and American Standards.

Many defective materials can occur within an aggregate. It is therefore essential that detailed petrographic evaluation is carried out, with particular reference to the intended use. An example of failure to do this was seen in the refurbishment of a small housing estate: white render was applied to face degraded brickwork. At first the result was splendid, but within 2–3 years brown rust spots appeared all over the white render because of the presence of very small amounts of iron sulphide (pyrite) in the sand used in the render.

## Aggregate sources

Sands and gravels can be obtained from river or glacial deposits, many of which are relatively young unconsolidated superficial deposits of Quaternary age. They may also be derived from older geological deposits, such as Triassic and Devonian conglomerates (to take English examples). Flood plain and terrace gravels are particularly important sources of aggregate because nature has already sorted them and destroyed or removed much of the potentially deleterious material; however, they may still vary in composition and particle size. Glacial deposits tend to be less predictable than fluvial deposits and are most useful where they have been clearly sorted by fluvial processes.

Among the quarried rocks, limestones – particularly the Carboniferous limestones of the British Isles – have been widely used as aggregate. Similarly, many sandstones have suitable properties and are used as sources of aggregate, particularly where they have been thoroughly cemented. Compact greywackes have been widely used, notably the Palaeozoic greywackes of the South West and Wales.

Igneous rocks are also a very useful source of quarried stone when crushed to yield aggregates; their character depends on their mineralogy and texture. Coarsely crystalline rocks such as granite, syenite, diorite, and gabbro are widely used, as are their medium-grained equivalents. Some finer-grained igneous rocks are also used, but the very finest-grained rocks are liable to be unsatisfactory for a wide range of purposes. Reserves of rocks such as dolerite, microgranite, and basalt tend to be small in comparison with the coarse-grained intrusive plutons. Conversely, some of the high-quality granite sources lie within very large igneous bodies, which sustain large quarries and provide a considerable resource.

Regional metamorphic rock fabrics generally make poor aggregate sources. On crushing they develop an unsatisfactory flaky shape. Schists and gneisses can provide strong material, but of poor shape. On the other hand, metamorphism of some greywackes and sandstones can provide material of high quality, especially when it has involved contact metamorphism associated with the intrusion of igneous rocks, producing hornfels or marble. Such thermally metamorphosed rocks often have a good fabric and provide useful resources.

## Investigation of Sources

There are three levels of investigation of the potential aggregate source. The first is the field investigation, in which the characteristics and distributions of the rocks present in the source can be established by mapping, geophysics, and borehole drilling. The second concerns the specific petrography of the materials. The third involves testing the physical and chemical properties of the materials. The material being extracted from the source must also be tested on a regular basis to ensure that there is no departure from the original test results and specification. Because sources are inevitably variable from place to place, there is always the risk that certain potentially deleterious components may appear in undesirable abundance.

A number of features may make the aggregate unsuitable for certain purposes; these include the presence of iron sulphide (pyrite, pyrrhotite, and marcasite). Iron sulphide minerals are unacceptable because they become oxidized on exposure to air in the presence of moisture, producing iron oxides (rust) and sulphate. This can result in spalling of material from the surface of concrete and rendering. The presence of gypsum in the aggregate is also highly undesirable from the point of view of concrete durability. Gypsum is commonly found in aggregates from arid regions. The presence of gypsum in concrete leads to medium- to long-term expansion and cracking. Other substances can create both durability and cosmetic problems.

## Extraction of Aggregates

The development of aggregate quarries requires the removal of overburden and its disposal, the fragmentation of rock (usually by a scheme of blasting), and the collection and crushing of the blast product (*see* **Quarrying**). Critical to the success of the operation is the stability of the size of the feed material to the primary crusher. Screening is usually necessary to ensure that the particles are suitable for the crusher regime. At this stage it is also necessary to remove degraded and waste material that is not required as part of the aggregate.

In sand and gravel workings, the source material is excavated in either dry or wet pit working. In marine environments, the process is based on suction and dredging using two techniques. In the first, the dredger is anchored and a pit is created in the seabed; production continues as consolidated materials fall into the excavation. In contrast, trail dredging is performed by a moving vessel, which excavates the deposit by cutting trenches in the seabed.

Extracted crushed rock, sand, and gravel are then prepared as aggregates through the use of jaw, gyratory, impact, and cone crushers. The type of crusher is selected according to the individual sizes of the feed material. Grading by screening is an adjunct to comminution and is also necessary in the production and preparation of the finished aggregate in cases where the particle-size distribution of the aggregate is important. The product is also washed and cleaned. The process of cleaning often uses density separation, with weak porous rock types of low density being removed from the more satisfactory gravel materials.

## Classification

The classification of aggregates has changed significantly over the years but has always suffered from the need to satisfy many different interests. Most commonly aggregates are divided into natural and artificial and, if natural, into crushed rock, sand, and gravel. If the aggregate is a sand or gravel, it is further subdivided according to whether it is crushed, partly crushed, or uncrushed. It may then be important to state whether the material was derived from the land or from marine sources.

Once produced, the aggregate is identified by its particle size, particle shape, particle surface texture, colour, the presence of impurities (such as dust, silt, or clay), and the presence of surface coatings or encrustations on the individual particles.

Detailed petrographic examination is employed so that specific rock names can be included in the description. This also helps in the recognition of potentially deleterious substances. However, the diversity of rock names means that considerable simplification is required before this classification can be used to describe aggregates. Following recognition of the main category of rock from the field data, more specific names can be applied according to texture and mineral composition. Because aggregates are used for particular purposes, they are sometimes grouped according to their potential use. This means that they may be incorrectly named from a geological point of view. The most obvious example of this is where limestone is referred to as 'marble'. In 1913 a list of petrographically determined rock types was assembled, with the rocks being arranged in Trade Groups. This was thought to help the classification of road stone in particular. It was presumed that each Trade Group was composed of rocks with common properties. However, the range of properties in any one Group is so large as to make a nonsense of any expectation that the members of the Group will perform similarly, either in tests or in service. The Trade Groups were therefore replaced by a petrological group classification.

However, even rocks within a single petrographic group can vary substantially in their properties. For example, the basalt group includes rocks that are not basalt, such as andesite, epidiorite, lamprophyre, and spilite. Hence a wide range of properties are to be expected from among these diverse lithologies.

In the first place a classification describes the nature of the aggregate in a broad sense: quarried rock, sand, or gravel; crushed or otherwise. Second, the physical characteristics of the material are considered. Third, the petrography of the possibly diverse materials present must be established. This may require the examination of large and numerous samples. While it may be reasonable to describe as 'granite' the aggregate produced from a quarry in a mass of granite, that aggregate will inevitably contain a wide range of lithologies, including hydrothermally altered and weathered rocks. Whether a rock is geologically a granite, a granodiorite, or an adamellite may be less significant for the description of the aggregate than the recognition of the presence of strain within the quartz, alteration of the feldspar, or the presence of shear zones or veins.

## Aggregate Grading

Aggregate grading is determined by sieve analyses. Material passing through the 5 mm sieve is termed fine aggregate, while coarse aggregate is wholly retained on this sieve (**Figure 1**). The fine aggregate is often divided into three (formerly four) subsets – coarse, medium, and fine – which fall within specified and partly overlapping particle-size envelopes. The size range is sometimes recorded as the ratio of the sieve sizes at which 60% passes and at which 10% passes. The shapes of the particles greatly affect the masses falling in given size ranges. For example, an aggregate with a high proportion of elongate grains of a given grain size would be coarser than an aggregate with flaky particles. This can affect the properties of materials made using the aggregate for, say, concrete, road materials, and filter design. Commonly materials needed for particular purposes have standard

**Figure 1** Aggregate grades. (A) Fine sand suitable for mortars or render (width of image: 10 mm). (B) Coarse sharp sand or 'concreting' sand (width of image: 10 mm). (C) Coarse natural sand (width of image: 10 mm). (D) Flint gravel 5–10 mm (width of image: 100 mm). (E) Crushed granite 5–10 mm (width of image: 100 mm). (F) Crushed granite 10–20 mm (width of image: 100 mm).

aggregate gradings. These include, for example, mortars, concrete, and road-surface aggregates. It is sometimes useful to have rock particles that are much larger than the normal maximum, for example where large masses of concrete are to be placed. Commonly, however, the maximum particle size used in structural concrete is around 20 mm. An important parameter is the proportion of dust, which is often taken as the amount passing the 75 $\mu$m sieve. In blending aggregates for particular purposes, it is usually necessary to combine at least two and possibly more size ranges; for example, in a concrete the aggregate may be a mixture of suitable material in the size ranges 0–5 mm, 5–10 mm, and 10–20 mm.

The grading curve – a plot of the mass of material passing each sieve size – also determines the potential workability of mixtures and the space to be filled by binder and can be adjusted to suit particular purposes. The grading curve can be designed to reduce the volume of space to less than 10% of the total volume, but at this level the aggregate becomes almost completely unworkable.

## Particle Shape

Particle shape is important in controlling the ability of the aggregate to compact, with or without a binder, and affects the adhesion of the binder to the aggregate surface. Shapes are described as rounded, irregular, angular, flaky, or elongate, and can be combinations of these (**Figure 2**). The first three are essentially equidimensional. The shape is assessed by measuring the longest, shortest, and intermediate axial diameters of the fragments. In the ideal equidimensional fragment, the three diameters are the same. Particles with ratios of the shortest to the intermediate and the intermediate to the longest diameters of above about 0.6 are normally regarded as equidimensional.

For many purposes, it is important that the aggregate particles have equant shape: their maximum and minimum dimensions must be very similar. Spherical and equant particles of a given uniform size placed together have the lowest space between the particles. Highly angular particles and flaky particles with high aspect ratios of the same grading can have much more space between the particles. The shape of the particles can significantly affect the properties and composition of a mixture. The overall space is also determined by the grading curve. Sometimes highly flaky particles such as slate can be used in a mixture if they are accompanied by suitably graded and highly spherical particles.

### Flakiness Index (British Standard 812)

The flakiness index is measured on particles larger than 6.5 mm and is the weight percentage of particles that have a least dimension of less than 0.6 times the mean dimension. The sample must be greater than 200 pieces. The test is carried out using a standard plate that has elongate holes of a given size; the proportion passing through the appropriate hole gives a measure of the flakiness index.

### Elongation Index (BS 812)

The elongation index is the percentage of particles by mass having a long dimension that is more than 1.8 times the mean dimension. This measurement is made with a standard gauge in which pegs are placed an appropriate distance apart.

## Petrography

The petrography of the aggregate is mainly assessed on the basis of hand picking particles from a bulk sample. Thin-section analysis either of selected pieces or of a crush or sand mounted in a resin is also employed. The petrographic analysis is essential to determine the rock types present and hence to identify potential difficulties in the use of the material. It allows recognition of potentially deleterious components and estimation of physical parameters. The experienced petrographer, for example, can estimate the parameters relevant to the use of a material for road surfacing.

Published standards provide procedures for petrographic description, including the standards published by the American Society for Testing and

**Figure 2** Examples of particular particle shapes. (A) Well-rounded spherical metaquartzite. (B) Elongate angular quartzite. (C) Rounded flaky limestone.

Materials and the Rilem procedures. These standards list the minimum amounts of material to be examined in the petrographic examination. In BS 812, for example, it is specified that for an aggregate with a maximum particle size of 20 mm the laboratory sample should consist of 30 kg. The minimum mass of the test portion to be examined particle by particle is 6 kg. Normally the analysis would be carried out on duplicate portions. The samples are examined particle by particle, using a binocular stereoscopic microscope if necessary. Unfortunately, this procedure does not cover all eventualities, and some seriously deleterious constituents within the material may be missed. A rock particle passing a 20 mm sieve may have within it structures that give it potentially deleterious properties (**Figure 3**). It is therefore essential that the aggregate is examined in thin section as well as in the hand specimen. It is helpful if the aggregate sample is crushed and resampled to provide a representative portion for observation in thin section. A large thin section carrying several hundred particles is required. Some of the potentially deleterious ingredients may be present at relatively low abundance. For example, the presence of 1–2% of opaline vein silica would be likely to cause significant problems.

Where a sand or fine gravel is to be sorted by hand it is first divided into sieve fractions, typically using the size ranges <1.18 mm, 1.18–2.36 mm, 2.36–5 mm, and >5 mm. These size fractions are analysed quantitatively by hand sorting in the same way as for coarse aggregate. The stereoscopic microscope is used to help with identification. Thin sections are also prepared from the sample using either the fraction passing the 1.18 mm sieve or the whole fine aggregate. The sample is embedded in resin and a thin section is made of the briquette so produced.

## Specific Tests Measuring Strength, Elasticity, and Durability

For quarried rocks it is possible to take cores of the original source material and to measure the compressive and tensile strengths of that material directly. It may be necessary to take a large number of samples in order to obtain a reliable representative result. However, for sands and gravels the strength of the material can rarely be tested in this way, and so a series of tests has been developed that simulate the conditions in which the aggregate is to be used.

There is often a simple relationship between the flakiness index of the aggregate and its aggregate impact value (AIV) and aggregate crushing value (ACV). In general, the lower the flakiness index, the higher the AIV and ACV. Hence, comparing the AIV and ACV values with specifications requires knowledge of the flakiness index. Consideration also needs to be given to the shape of the aggregate following the test.

### Density and Water Absorption

Some of the most important quantities measured for an aggregate are various density values. These include the bulk density, which is the total mass of material in a given volume, including the space between the aggregate particles. The saturated surface-dry density is the density of the actual rock material when fully saturated with water but having been dried at the surface. The dry density is the rock density after drying. In making these measurements, the water absorption is also recorded. These provide data that are essential for the design of composite mixes.

### Aggregate Impact Value (BS 812)

The aggregate impact value provides an indirect measurement of strength and involves the impaction of a standard mass on a previously well-sorted sample. The result is obtained by measuring the amount of material of less than 2.36 mm produced from an aggregate of 10–14 mm. The lower the result, the greater the resistance of the rock to impaction. It is also useful to examine the material that does not pass the 2.36 mm sieve, and it is common to sieve the total

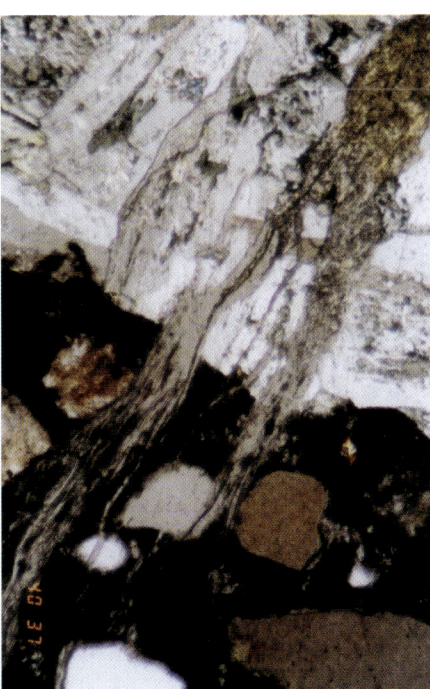

**Figure 3** An alkali-reactive granite coarse aggregate particle (top) with cracks filled with alkali-silicate gel. The cracks run into the surrounding binder, which appears dark and contains quartz-rich sand as a fine aggregate.

product at 9.5 mm to establish whether there is an overall general reduction in particle size.

### Aggregate Crushing Value (BS 812)

The aggregate crushing value provides an indirect assessment of strength and elasticity in which a well-sorted sample is slowly compressed. The lower the degradation of the sample, the greater the resistance to crushing. The size ranges used are the same as for the AIV test.

### 10% Fines Value (BS 812)

The 10% fines value is the crushing load required to produce degradation such that 10% of the original mass of the material passes a 2.36 mm sieve, the original test sample being 10–14 mm. The samples are subjected to two different loads, and the amount passing the 2.36 mm sieve in each test is measured. Typically the two results should fall between 7.5% and 12.5% of the initial weight. The force required to produce 10% fines is then calculated.

### Aggregate Abrasion Value (BS 812)

In determining the aggregate abrasion value, fixed aggregate particles are abraded with standard sand, and the mass of the aggregate is recorded before and after abrasion. The reduction in mass indicates the hardness, brittleness, and integrity of the rock.

### The Los Angeles Abrasion Value (ASTM C131 and C535)

To determine the Los Angeles abrasion value, a sample charge is mixed with six to twelve steel balls, and together these are rotated in a steel cylinder for 500 or 1000 revolutions at 33 rpm. This causes attrition through tumbling and the mutual impact of the particles and the steel balls. The sample is screened after the rotations are completed using a 1.68 mm sieve. The coarser fraction is washed, oven dried, and weighed. The loss in mass as a percentage of the original mass is the Los Angeles abrasion value.

### Micro Deval test

The Micro Deval test is widely used to determine the resistance of an aggregate to abrasion. Steel balls and the aggregate are placed in a rotating cylinder. The test may be carried out either wet or dry. The Micro Deval value is calculated from the mass of material that passes the 1.6 mm test sieve, as a percentage of the original aggregate mass.

### Polished Stone Value (BS 812, Part 114)

To determine the polished stone value, the aggregate is mounted in resin and the exposed surface is polished using a wheel and standard abrasive. The result is measured using a standard pendulum, with the ability of the rock to reduce the motion of the pendulum giving an indication of the potential resistance of the aggregate to skidding. The sample is small and the result can vary according to the proportions of rock that are present. This test is difficult to perform reliably, and considerable practice is required to obtain a consistent result. In practice it is found that good skid resistance is derived from a varied texture in the rock with some variation in particle quality. Well-cemented sandstones and some dolerites tend to have high polished stone values, while rocks such as limestones and chert have very low polished stone values.

### Franklin Point Load Strength

The Franklin point load strength can be directly assessed for large pieces of rough rock. A load is applied through conical platens. The specimen fails in tension at a fraction of the load required in the standard laboratory compressive-strength test. However, the values obtained in the test correlate reasonably well with those obtained from the laboratory-based uniaxial compressive test, so an estimated value for this can be obtained, if necessary, in the field.

### Schmidt Rebound Hammer Value

The Schmidt Rebound Hammer test is a simple quantitative test in which a spring-loaded hammer travelling through a fixed distance strikes the rock in a given orientation. The rebound of the hammer from the rock is influenced by the elasticity of the rock and is recorded as a percentage of the initial forward travel. A sound rock will generally give a rebound value in excess of 50%, while weathered and altered rock will tend to give a much lower value.

### Magnesium Sulphate Soundness Test (BS 812)

In the magnesium sulphate soundness test the degradation of the aggregate is measured following alternate wetting and drying in a solution of magnesium sulphate. The test provides a measure of the tendency of the rock to degrade through the crystallization of salts or ice formation. The result is influenced by the porosity and particularly by planes of weakness in the aggregate.

### Freeze–Thaw Test

In the freeze–thaw test the aggregate is subjected to cycles of freezing and thawing in water. Each cycle lasts approximately 24 h. The temperature is reduced over a period of several hours and then maintained at $-15°C$ to $-20°C$ for at least 4 h. The sample is then maintained in water at $20°C$ for 5 h. The cycle is repeated 10 times, and then the sample is dried and sieved, and the percentage loss in mass is determined.

### Slake Durability Index

A number of small samples of known mass are placed in a wire-mesh drum. The drum is immersed in water and rotated for 10 min. The specimens are dried and weighed, and any loss in weight is expressed as a percentage of the initial weight. This is the slake durability index.

### Methylene Blue Absorption Test

Methylene blue dye is dissolved in water to give a blue solution. It is absorbed from the solution by swelling clay minerals, such as montmorillonite. The quantity of potentially swelling clay minerals in a sample of rock is assessed by measuring the amount of methylene blue absorbed.

### Chemical Tests

Aggregates are commonly tested by chemical analysis for a variety of constituents, including their organic, chloride, and sulphate contents. Organic material is readily separated from the aggregate by, for example, the alkalinity of cement paste. Its presence leads to severe staining of concrete and mortar surfaces. Sulphate causes long-term chemical changes in cement paste, leading to cracking and degradation. Chloride affects the durability of steel reinforcement in concrete, accelerating corrosion and the consequent reduction in strength.

### Mortar Bar and Concrete Prism Tests

The durability of concrete made with a given aggregate is evaluated by measuring the dimensional change in bars made of mortar or larger prisms of concrete containing the specific aggregate. The mortar-bar test results can be obtained in a few weeks, but the prism test needs to run for many months or even years. The tests allow the recognition of components in the rocks or contaminants (e.g. artificial glass) that take part in expansive alkali–aggregate reactions.

## Aggregates for Specific Purposes

### Railway Track Ballasts

Railway track is normally placed on a bed of coarse aggregate. A lack of fines is required: the desirable particle size is generally 20–60 mm. The bed requires a free-draining base that is stable and able to maintain the track alignment with minimum maintenance. The aggregate is sometimes placed on a blanket of sand to prevent fines entering the coarse aggregate layer. The aggregate layer may be up to 400 mm thick.

The favoured rock types are medium-grained igneous rocks such as aplite and microgranite. Sometimes hornfels is used. Some of the more durable limestones and sandstones are also used. Weaker limestones and many sandstones are generally regarded as unsatisfactory because of their low durability and ready abrasion. The desirable qualities for an aggregate used for ballast are that it must be a strong rock, angular in shape, tending to be equidimensional, and free from dust and fines.

### Aggregates for Use in Bituminous Construction Materials

Aggregates for use with a bitumen binder in building construction (as used in bridge decks and in the decks and ramps of multistorey car parks) require a high skid resistance. They must also be highly impermeable, protecting the underlying construction from water and frost attack and from the effects of de-icing salts. The mix design is important: there should be a high bitumen content and a high content of fine aggregate and filler in the aggregate grading.

A wide range of rocks of diverse origin and a number of artificial materials are used in the bituminous mixes. The rocks must be durable, strong, and resistant to polishing. The aggregate must show good adhesion to the binder and have good shape. Skid resistance is also dependent on traffic density and, in some instances, a reduction in traffic has improved skid resistance. Visual aggregates have been developed where high skid resistance is required, and these include calcined bauxite, calcined flint, ballotini, and sinopal. Blast furnace slags yield moderately high polished stone values. The light-reflecting qualities are also important, and artificial aggregates such as sinopal, with their very high light reflectivity, are valued. Resistance to stripping, i.e., the breakdown of the bond between the aggregate and the bituminous binder, is also important. Stripping is likely to result in the failure of the wearing course and not necessarily in failure of the base course. The stripping tends to be most conspicuous in coarse-grained aggregates that contain quartz and feldspar. Basic rocks show little or no detachment. The aggregate has considerable strength, particularly in the wearing course. As an example, the aggregate crushing value for surface chasing and dense wearing courses will typically be 16 to 23, while for the base course it may be as high as 30. Similarly, the aggregate impact value might be 23 in the wearing course and 30 in the base course.

### Aggregates in Unbound Pavement Construction

Aggregate is sometimes used in construction without cement or a bitumin binder. Examples are a working platform in advance of construction, structural layers beneath a road system, a drainage layer, and a replacement of unsuitable foundation material. Aggregates for these purposes must be resistant to crushing and impact effects during compaction and in use, and when in place they must resist breakdown by weathering or by chemical and physical processes and must be able to resist freeze–thaw processes.

It is likely that recycled aggregates will become increasingly important in these situations, although levels of potentially deleterious components, such as sulphate, may point to a need for caution in the use of such material. Aggregates for unbound construction often need to resist the ingress of moisture, since moisture rise and capillary transfer can cause progressive degradation.

### Mortar

Mortar consists of a fine aggregate with a binding agent. It is used as a jointing or surface-rendering material. Sands for mortar production are excavated from sand and gravel pits in unconsolidated clastic deposits and are typically dominated by quartz. They are used in their natural form or processed by screening and washing. Rock fines of similar grade can also be used.

The most important feature of sand for mortar manufacture is that the space between the aggregate particles must generally be about 30% by volume. The volume of binder needs to be slightly greater than this volume, and hence a relatively high proportion of cement or lime may be required. Should the space be such that voids occur in the mix, the material will commonly show early signs of degradation and will be readily damaged by penetration of moisture. The space also appears to reduce the capacity of the mortar to bond with the substrate.

The workability and ease of use of the mixture also depends on the shape of the particles and the grading curve. Very uniform sand tends to have a high void space and therefore requires a high cementitious or water content and tends to develop a high voidage. On the other hand, the grading may be such that the space between the particles is too small and the mixture becomes stiff. The strength and elastic modulus of the rocks are also important because the resultant mixture of paste and aggregate must match the strength and elasticity of the material to which the mortar is applied. If it is not, then partings are liable to develop between the binder and the substrate. Similarly, the material must exhibit minimal shrinkage because again it might become detached from the substrate.

### Concrete

This very widely used material has a very diverse structure and composition and serves many purposes. It is composed of aggregate graded for the specific purpose and a binder containing cement. In general, the properties of the aggregate must match the intended strength and elasticity of the product, and it must be highly durable. For many purposes a combination of coarse and fine aggregate with a maximum particle size of 20 mm is used. The grading curve is designed such that an appropriate amount of space occurs between the particles – typically around 25% by volume of the mixture. There are numerous components of aggregate that perform adversely in the medium and long term, so careful study of the material is required before use. The defective components are described in several standards, along with procedures for measuring their effects on the concrete. Some of these are described below.

In the 1940s it was recognized in the USA that certain siliceous aggregates could react with alkalis derived from Portland Cement. This led to spalling of concrete surfaces and cracking, sometimes in a spectacular manner. The phenomenon occurs throughout the world, and few rock sources are immune. An enormous amount of work has been carried out to evaluate the reaction, both in the laboratory and in structures. Major international conferences on the subject have been held. The alkalis for the reaction derive from the cement and are extracted into the pore fluid in the setting concrete. The concentration of alkali in the pore fluid can be affected by external factors as well as by the internal composition of the cement matrix. The rock reacting with the alkalis is typically extremely fine grained or has extremely small strain domains. Hence, fine-grained rocks, such as opaline silica within limestone, some cherts, volcanic glass, slate, and similar fine-grained metamorphic rocks, may exhibit a high degree of strain and so be able to take part in the reaction. More recently it has been found that certain dolomitic siliceous limestones are also to be avoided, again because they react with alkalis to cause significant expansion of the concrete and severe cracking.

## See Also

**Building Stone**. **Geotechnical Engineering**. **Quarrying**. **Rock Mechanics**. **Sedimentary Environments:** Alluvial Fans, Alluvial Sediments and Settings. **Sedimentary Processes:** Glaciers. **Sedimentary Rocks:** Limestones; Sandstones, Diagenesis and Porosity Evolution.

## Further Reading

American Society for Testing and Materials (1994) *Annual Book of ASTM Standards (1994), Section 4, Construction, Volume 04.02, Concrete and Aggregates*. West Conshohocken: American Society for Testing and Materials.

Bérubé MA, Fournier B, and Durand B (eds.) (2000) *Alkali–Aggregate Reaction in Concrete*. Proceedings of the 11th International Conference, Quebec, Canada.

British Standards Institution (1990) *BS812 Parts 1 to 3: Methods for Sampling and Testing of Mineral Aggregates, Sands and Fillers, Parts 100 Series Testing Aggregates*. British Standards Institution.

Dolor-Mantuani L (1983) *Handbook of Concrete Aggregates: A Petrographic and Technological Evaluation*. New Jersey: Noyes Publications.

(1983) *FIP Manual of Leightweight Aggregate Concrete*, 2nd edn. Surrey University Press (Halsted Press).

Hobbs DW (1988) *Alkali–Silica Reaction in Concrete*. Thomas Telford.

Latham J-P (1998) *Advances in Aggregates and Armourstone Evaluation*. Engineering Geology Special Publication 13. London: Geological Society.

Popovics S (1979) *Concrete-Making Materials*. Hemisphere Publishing Corporation, McGraw-Hill Book Company.

Smith MR and Collis L (2001) *Aggregates, Sand, Gravel, and Crushed Rock for Construction Purposes*, 3rd edn. Engineering Geology Special Publication 17. London: Geological Society.

West G (1996) *Alkali–Aggregate Reaction in Concrete Roads and Bridges*. Thomas Telford.

# ALPS

*See* **EUROPE: The Alps**

# ANALYTICAL METHODS

Contents

**Fission Track Analysis**
**Geochemical Analysis (Including X-Ray)**
**Geochronological Techniques**
**Gravity**
**Mineral Analysis**

## Fission Track Analysis

**B W H Hendriks**, Geological Survey of Norway, Trondheim, Norway

© 2005, Elsevier Ltd. All Rights Reserved.

### Introduction

Ages obtained from isotopic dating methods are based on the ratio of parent and daughter isotopes. Radioactive decay of parent isotopes causes daughter isotopes to accumulate over time, unless they decay further or are lost by diffusion or emission. In the case of the fission track method, the daughter product is not another isotope, but a trail of physical damage to the crystal lattice resulting from spontaneous fission of the parent nucleus. When the rate at which spontaneous fission occurs is known, the accumulation of such trails, known as fission tracks, can be used as a dating tool. Analogous to diffusional loss of daughter isotopes, the damage trails in the crystal lattice disappear above a threshold temperature by the fission track annealing process. Although the physics behind the annealing process are poorly understood, the outcome is empirically well known. Annealing initially causes the length of fission tracks to decrease and may eventually completely repair the damage to the crystal lattice. The latter is known as total annealing. The rate at which annealing takes place is a function of both mineral properties and temperature history.

Fission tracks in geological samples have been well-studied in mica (see **Minerals: Micas**), volcanic glass, tektite glass; (see **Tektites**) titanite, and zircon (see **Minerals: Zircons**). However, most research has been done on fission tracks in apatite, a widely disseminated accessory mineral in all classes of rocks. Retention of fission tracks in natural minerals takes place only at temperatures well below that of their crystallization temperature. Fission track dating will, therefore, document the crystallization age of a crystal only when it has cooled rapidly to surface temperatures immediatley after crystallization (see **Analytical Methods: Mineral Analysis**). Fission track dating of volcanic rocks can provide an age of crystallization, while fission track dating of more slowly cooled rocks will always yield an age that is younger than the age of crystallization. The amount of fission tracks per volume and their length will be a sensitive function of the annealing process and of the cooling history of the sample being studied. A cooling history can be constrained by thermal history modelling of fission track data (fission track age and fission track length distribution). Fission track analysis and thermal history modelling of apatite fission track data provide powerful tools with which to assess regional cooling and denudation histories.

Following the rejuvenation of (U-Th)/He dating in the 1990s, the technique has become an important addition to the fission track method. (U-Th)/He dating can be applied to the same minerals as those commonly used in fission track analysis. (U-Th)/He dating is unique in its capability to constrain the very low temperature part of cooling histories of rock samples; the nominal closure temperature for apatite (U-Th)/He ages may be as low as $\sim 50°C$. Apatite (U-Th)/He dating today is a well-established technique in itself, but in most studies it is used in combination with apatite fission track analysis. Many fission track research groups now routinely apply (U-Th)/He dating in parallel with fission track analysis. An introduction to (U-Th)/He dating is, therefore, included here.

## Fission Tracks

Fission tracks are linear damage trails in the crystal lattice. Natural fission tracks in geological samples are formed almost exclusively by the spontaneous fission of $^{238}U$. Other naturally occurring isotopes, such as $^{235}U$ and $^{232}Th$, also fission spontaneously, but the respective isotopes have such low fission decay rates that it is generally assumed that all spontaneous fission tracks in naturally occurring crystals are derived from $^{238}U$. The frequency of fission events is low compared to $\alpha$-particle decay events, about 1 fission event for every $2 \times 10^6$ $\alpha$-particle decay events.

During spontaneous fission an unstable nucleus splits into two highly charged daughter nuclides (**Figure 1**). The two fission fragments are propelled in opposite directions, at random orientation with respect to the crystal lattice. The passage of the positively charged fission fragments through the host mineral damages the crystal lattice by ionization or electron stripping, causing electrostatic displacement. The end result is a cylindrical zone of atomic disorder with a diameter of a few nanometers – known as a fission track. Detailed information on the length of fission tracks is available for apatite only. Newly created apatite fission tracks have a length of $\sim 16.3 \pm 0.5\,\mu m$. Fission tracks can be observed directly through transmission electron microscopy, but with

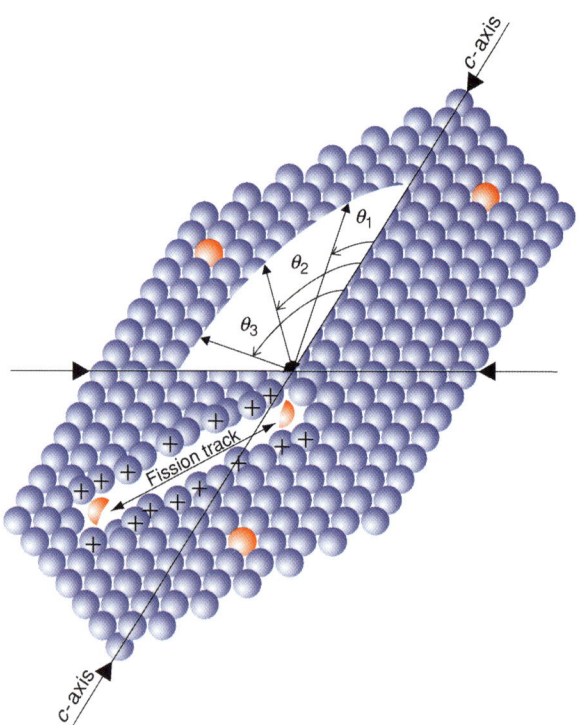

**Figure 1** Spontaneous fission of $^{238}U$ (red spheres) produces two highly charged fission fragments (red half spheres) that recoil as a result of Coulomb repulsion. They interact with other atoms in the crystal lattice by electron stripping or ionization. This leads to further deformation of the crystal lattice as the ionized lattice atoms (blue spheres with plus sign) repel each other. After the fission fragments come to rest, a damage trail ('fission track') is left, which can be observed with an optical microscope after chemical etching. In apatite, the fission track annealing rate is higher for tracks at greater angle ($\theta$) to the crystallographic $c$-axis. Therefore, tracks perpendicular to the $c$-axis are on average shorter than tracks that are parallel to the $c$-axis.

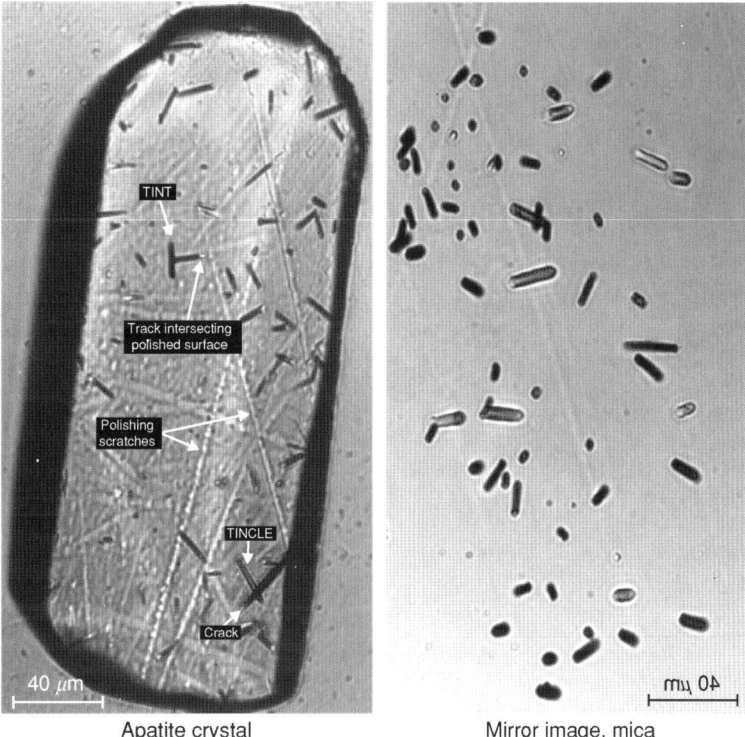

**Figure 2** Fission tracks in apatite (left) resulting from the spontaneous fission of $^{238}$U and induced fission tracks in mica (right) produced by irradiation in a nuclear reactor. Fission tracks in the mica outline a mirror image of the polished apatite crystal with which it was in close contact during irradiation. Fission tracks are revealed by chemical etching with $HNO_3$ (apatite) and HF (mica). Only fission tracks that intersect the polished surface, cracks (track-in-cleavage, TINCLE) or other tracks (track-in-track, TINT) can be reached and enlarged by the etchant.

an optical microscope they can only be observed after revelation by chemical etching. Seen through an optical microscope, chemically etched fission tracks appear as randomly oriented cigar-shaped features (**Figure 2**).

## Fission Track Annealing

Laboratory experiments show that residence at elevated temperatures induces shortening of fission tracks. This process of track shortening by solid state diffusion is called fission track annealing. The rate of the annealing process is dependent on mineral properties and thermal history. Pressure and stress dependency have been suggested, but the evidence is ambiguous and highly controversial.

When a sample cools below the total annealing temperature, it enters the Partial Annealing Zone (PAZ; APAZ in the case of apatite, ZPAZ for zircon). As the sample cools within the PAZ, tracks shorten by lesser amounts until becoming relatively stable at low (<60°C) temperatures. Fission track ages are based on counts of tracks in a polished cross-section through crystal. Shorter tracks have a smaller probability of intersecting the polished surface, and track length shortening in the PAZ consequently also leads to an apparent age reduction. This produces a characteristic pattern of fission track ages and mean track lengths within a PAZ (**Figure 3**). Such a pattern may be (partly) preserved in the case of very rapid cooling. This is referred to as a 'fossil PAZ' (**Figure 4**). The APAZ is sometimes referred to simply as the temperature interval between 60°C and 120°C, but this is an oversimplification because it neglects the impact of variations in chemical composition and cooling rate on the rate of the annealing process. The ZPAZ is more loosely constrained than the APAZ and probably lies somewhere between ∼200°C and ∼350°C.

A well-known problem with the interpretation of fission track data is that annealing can take place even below the temperatures normally associated with the PAZ. Apatite fission tracks that are formed in a nuclear reactor by irradiation with slow neutrons ('induced tracks') have an initial track length ($l_0$) immediately after irradiation of ∼16.3 ± 0.5 μm. However, natural fission tracks ('spontaneous tracks') in apatite crystals separated from rapidly cooled volcanic rocks have mean track lengths in the order of 14.5–15 μm. Since this track length reduction can only have taken place at surface temperatures after eruption, it is

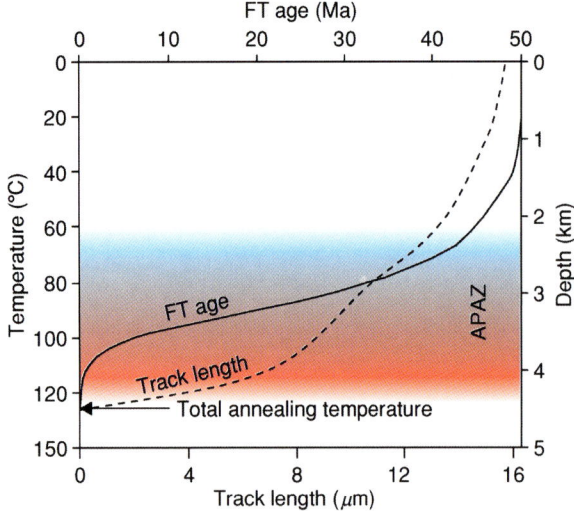

**Figure 3** Pattern of apatite fission track ages and mean track lengths in a borehole. Example based on 5 km of rapid denudation at 50 Ma (30°C/km geothermal gradient) followed by 50 Ma of stable thermal conditions (no denudation). Samples close to the surface (<2 km depth) have fission track ages that approximate the timing of rapid denudation and have very long mean track lengths. Samples inside the apatite partial annealing zone (APAZ) have much younger fission track ages and shorter mean track lengths. Above the total annealing temperature fission tracks will be erased almost immediately after formation.

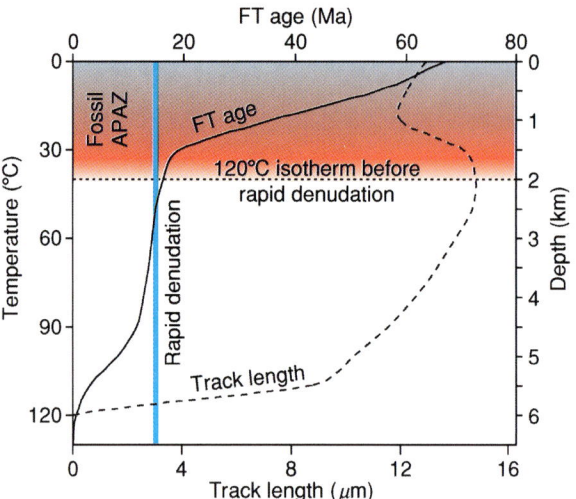

**Figure 4** Pattern of apatite fission track ages and mean track lengths in a borehole after a late phase of rapid cooling. Example based on stable thermal conditions from 80 Ma to 15 Ma and a geothermal gradient of 20°C/km. At 15 Ma, a phase of rapid denudation removes 4 km of overburden. As a result of this rapid cooling, the pattern of fission track ages and mean track lengths in the APAZ as it existed before the onset of rapid cooling ('fossil APAZ') is largely preserved. The break-in-slope in the fission track age profile coincides with the onset of rapid denudation.

apparent that low-rate annealing in apatite takes place even at these low temperatures. However, it is very difficult to determine the rate at which fission track annealing takes place at such low temperatures. The annealing rate becomes very low and extrapolation of results from lab experiments to the geological timescale would introduce large uncertainties. Calibration with data from boreholes is also difficult, because the low temperature history (<60°C) of the borehole will almost never be accurately known in detail.

Fission track annealing behaviour in apatite is correlated with its unit-cell parameters and thus with crystallographic structure and chemical composition. The apatite group, with its simplified molecular formula $Ca_5(PO_4)_3F$, has a wide range of possible chemical compositions. Of all possible substitutions, that of chlorine (Cl) in place of fluorine (F) has the largest impact on the fission track annealing process. Apatite crystals with high chlorine contents are very resistant to annealing. Fission tracks in such crystals will survive higher temperatures than fission tracks in apatite crystals with lower chlorine contents. Even though chlorine content has a dominant control on annealing behaviour, significant variation in unit-cell parameters, and therefore in annealing behaviour, exists between chlorine-poor apatites as a result of a variety of possible substitutions, other than that of chlorine in place of fluorine. Because chlorine-poor apatites are the most common in nature, relying solely on chlorine content as a proxy for annealing behaviour does not account for the variation in annealing behaviour between most natural samples. The most obvious way to assess variation in chemical composition between apatite samples is to analyse every individual grain in which fission tracks are counted and measured, for example with a microprobe. Alternatively, etch pit size (**Figure 5**) the size of the intersection of a chemically etched fission track with the polished apatite surface, can be used as a measure for solubility and thereby as a proxy for bulk chemical composition. This method has been applied successfully and has the advantage that etch pit size is easily and inexpensively measured under the microscope at the same time that the actual fission track counting and measuring is done. No matter which approach is taken, it is essential to account for variations in chemical composition, because they exert such a strong influence on the annealing behaviour. Moreover, variation in chemical composition does not only occur between apatites from different rock samples, but also between apatite grains from the same rock sample. This is not only the case for (meta-)sedimentary rock samples, but also for igneous rocks.

Constraints on how the annealing process affects fission tracks can be combined into so-called annealing

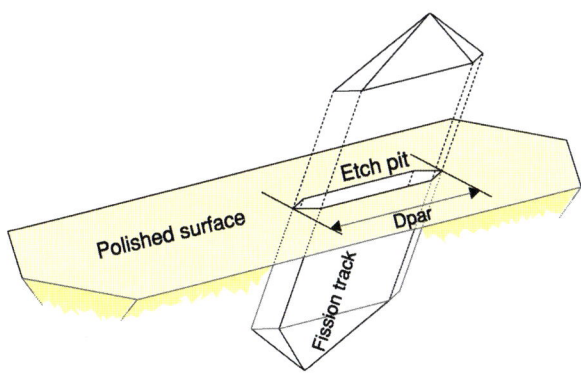

**Figure 5** Chemically etched fission track intersecting a polished crystal surface. An etch pit is the intersection of the track with this surface. Dpar is the maximum diameter of the etch pit. Please note that the size of the etch pit in this example is exaggerated; real etch pits are orders of magnitude smaller relative to the grain sizes of natural crystals. Dpar can be used as a proxy for annealing behaviour.

models. Annealing models are mathematical approximations of how the annealing process relates to temperature history and mineral properties. Modern annealing models take into account variation in chemical composition and other parameters. They are calibrated with lab experiments and data from vertical arrays of samples in drill holes of which the thermal history is well known from independent data.

## Fission Track Ages

A fission track age does not indicate the timing of cooling through a specific temperature boundary, but instead represents an integrated signal of the thermal history of a sample. Fission track ages, therefore, usually do not correspond directly to geological events. Additional fission track length data is needed to reconstruct a thermal history. Only in the case of very rapid cooling, may fission track ages correspond to a specific geological event.

The amount of spontaneous fission tracks in a crystal is dependent on its thermal history, its annealing characteristics, and the concentration of $^{238}$U, the parent isotope. With the External Detector Method (EDM), which is the preferred analytical method in most fission track studies, it is possible to determine the $^{238}$U concentration of every individual crystal in which tracks are counted. This is achieved by irradiating polished crystals that are mounted in epoxy (**Figure 6A,B**), together with a piece of low-uranium mica (the 'external detector') in direct contact with the polished surface. Spontaneous fission tracks in the crystals are revealed by chemical etching prior to irradiation (**Figure 6C**). Irradiation will induce the fission of $^{235}$U and thereby produce new fission fragments in the crystals. A proportion of them will be emitted from the etched crystals into the covering mica detector (**Figure 6D**). After irradiation, the mica is etched (**Figure 6E**). Because $^{235}$U/$^{238}$U has a constant ratio in normal rocks, the amount of fission tracks in the part of the mica that was in contact with a crystal ('induced tracks') can be used as a measure for the $^{238}$U concentration in that particular crystal.

To monitor the neutron fluence during irradiation, samples are irradiated together with pieces of uranium-bearing glass that are also covered with mica detectors. Irradiation of the glass will produce fission tracks in the mica in the same way as natural crystals do and the amount of tracks per area (fission track density) will be proportional to the neutron fluence. After determining the density of spontaneous fission tracks in the crystals, of induced tracks in the matching areas in the mica and of the tracks in the mica covering the uranium-bearing glass, single crystal ages can be calculated. From the single crystal ages, a combined sample age can then be determined.

Statistical tests can be performed on the distribution of single crystal ages. Significant spread of single crystal ages can be a result of variation in annealing characteristics between crystals from the same sample, or, in the case of a (meta-)sedimentary sample, mixing of crystals from different source areas with different thermal histories. To test whether or not single crystal ages are from the same statistical population, a $\chi^2$-test can be performed. Significant spread of single crystal fission track ages can also easily be detected in a radial plot (**Figure 7**). Zircon fission track dating is regularly used in provenance studies and the distribution of single crystal fission track ages is then used to tie age-populations to different source areas. Because of their greater resistance to annealing, zircon fission tracks will be preserved up to much higher temperatures than apatite fission tracks during burial in a sedimentary basin. Zircon crystals will, therefore, retain the source area signatures much better than apatite crystals and are also more resistant to abrasion during physical transport.

Fission track ages are normally calibrated against one or more age standards with the Zeta ($\zeta$) method. Standards of known age are irradiated and analysed in the same manner as regular samples in order to establish a calibration factor $\zeta$. Every analyst has to establish a personal $\zeta$ value in order to correct for observational bias. When a constant personal $\zeta$ value is obtained, every unknown fission track grain age can be calculated. The most commonly used apatite age standard is from Durango, Mexico, and has a $^{40}$Ar/$^{39}$Ar age of 31.4 ± 0.5 Ma.

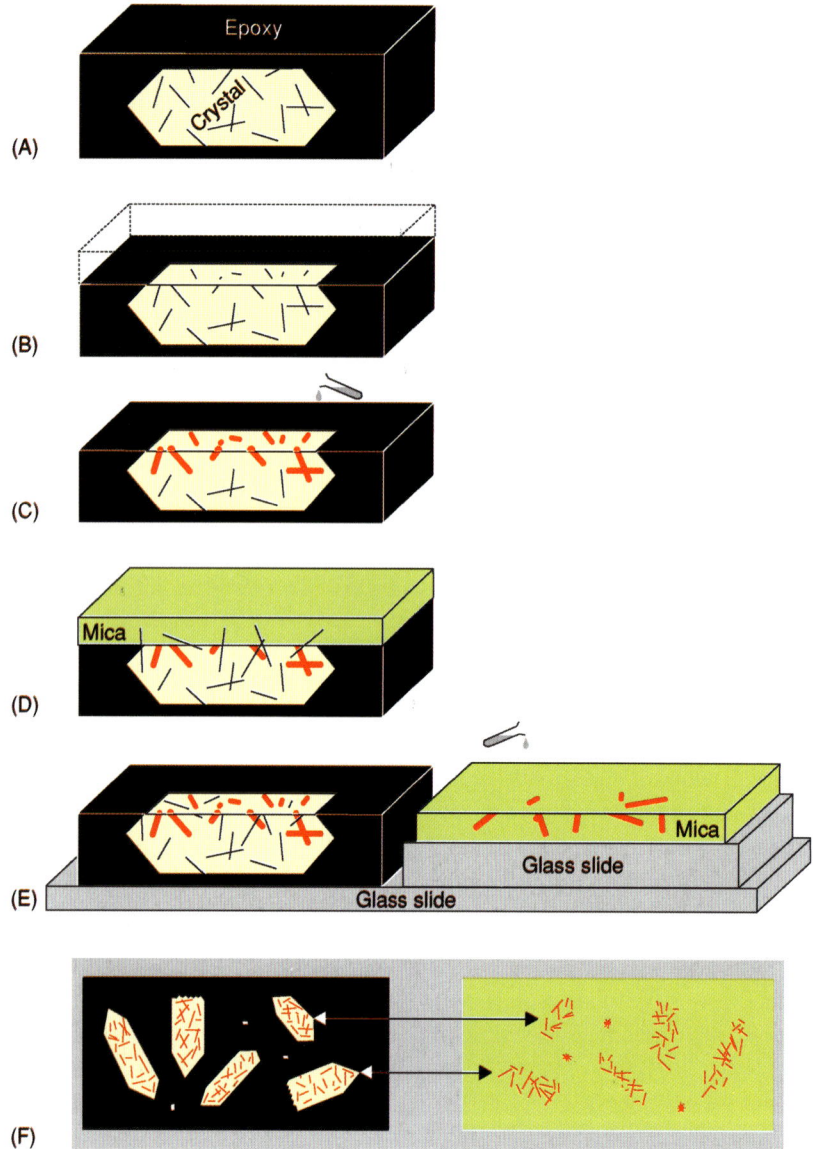

**Figure 6** Schematic overview of the laboratory procedures involved in the External Detector Method. (A) Crystals with latent tracks are mounted in epoxy. These latent tracks cannot be observed with an optical microscope. (B) The mount is ground and polished until a sufficiently large internal surface in the crystal is exposed. (C) Chemical etching. Tracks that intersect the polished surface, or that can be reached by the etchant through other tracks or cracks, become enlarged (red) and can now be observed with an optical microscope. (D) External Detector Mica is mounted in close contact with the polished mount. A stack of mounts is placed in a cylinder and irradiated. A proportion of the fission fragments that are created inside the crystal by irradiation with neutrons, will be ejected into the mica. On top and bottom of the stack is a piece of uranium bearing glass, also covered with mica, to monitor the neutron fluence during irradiation. (E) After irradiation the mica is removed and chemically etched with HF. (E,F) Mounts and micas are glued onto a glass slide as mirror images. Track densities in crystals and the matching areas in the mica can now be determined under the microscope. (F) Plan view of several mounted crystals. Induced tracks in mica define outline of the crystals they were in contact with during irradiation.

## Fission Track Length

Because detailed information on the length of fission tracks is available for apatite only, this section is solely applicable to apatite fission tracks.

Fission track length measurements are made on polished and chemically etched crystals (**Figure 6A,B,C**). To make sure that the full length of a track is measured, only tracks that are oriented parallel to the polished surface should be measured ('horizontal confined tracks'). This can easily be verified by focusing and defocusing the microscope; only tracks parallel to the polished surface will come into focus over their entire length at once.

The apatite crystal structure is anisotropic and so is the effect of the annealing process. The annealing rate

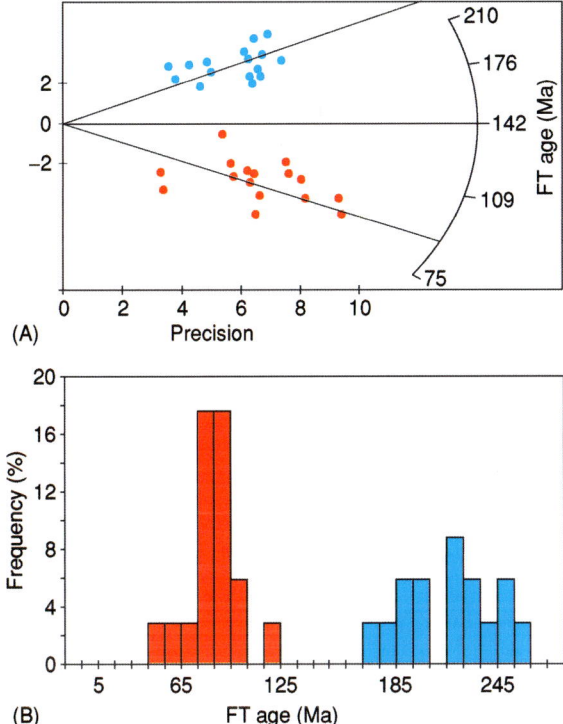

**Figure 7** (A) Radial plot for a sample with two clearly defined age populations (red and blue). More precise fission track ages plot further from the origin of the x-axis (precision). The y-axis gives the fission track age and its error. The two populations have ages of ∼90 Ma (red) and ∼225 Ma (blue) and a combined age of 142 Ma. (B) Histogram of the individual crystal ages in the radial plot.

of fission tracks depends on the orientation of the fission tracks with respect to the crystallographic orientation of the apatite crystal (**Figure 1**). Tracks perpendicular to the *c*-axis anneal slightly faster than those parallel to it. For weakly annealed samples, this effect will be very small and usually it is ignored. For samples that experienced higher degrees of annealing, ignoring the orientation dependency of the track length will introduce some error. Anisotropy of annealing can, however, easily be taken into account by measuring θ, the angle of the fission track with the crystallographic c-axis, when the track length is measured.

Tracks that are revealed by etching because they intersect cracks (track-in-cleavage, TINCLE **Figure 2**), tend to be longer than tracks that are revealed because they intersect other tracks (track-in-track, TINT; **Figure 2**). Several explanations have been proposed, such as widening of cracks during polishing, increased etch rates along cracks, and the smaller likelihood for shorter tracks to intersect a crack. Because of their bias towards longer track lenghts, TINCLEs tend to conceal the anisotropy of annealing. It is recommended to measure only TINTs.

## Thermal History Modelling

Fission tracks are produced continuously throughout the thermal history of a sample. Older fission tracks will, therefore, always have experienced a longer and older part of a sample's thermal history than younger fission tracks. This results in variation of fission track lengths within a sample and within single crystals. In the case of a purely cooling history, the older tracks will have experienced higher temperatures than younger tracks and so the older tracks will be annealed more, and thus will be shorter, than younger tracks. Track length reduction within the PAZ results in a mean length shorter than the initial length ($l_0$) and a skewed track length distribution. Different cooling histories and their resulting apatite fission track age and track length distributions are displayed in (**Figure 8**).

Thermal history modelling makes use of annealing models to calculate the apatite fission track age and track length distribution that would result from a particular thermal history. This synthetic modelling result can be compared to the fission track age and track length distribution obtained from a sample and the degree of fit can be assessed. By doing this for many different thermal histories, a range of thermal histories with a good degree of fit can be obtained. A geological interpretation can then be based on these thermal histories.

Independent geological observations and also additional age data such as those from (U-Th)/He dating can be used to constrain the range of thermal histories that a modelling program has to go through. For example, a time–temperature point with surface temperature at 100 Ma can be used as a constraint when modelling fission track data from a sedimentary sample of 100 Ma old (or data from a sample of the basement just below the sediment), because at the time of deposition the sediment (and the basement immediately below) must have been experiencing surface temperatures.

Thermal history modelling is a popular tool to retrieve a time–temperature path from apatite fission track data. However, in the absence of independent geological constraints, it cannot constrain anything other than cooling. The fission track age and fission track length distribution resulting from a thermal history that includes a reheating event will be virtually identical to that resulting from the same thermal history without the reheating event (**Figure 9**). Fission tracks essentially record cooling, and even in a thermal history that has included reheating, most of the record will come from the cooling segments.

A very common characteristic of thermal histories obtained from modelling of fission track data is that they include a late cooling event. The annealing

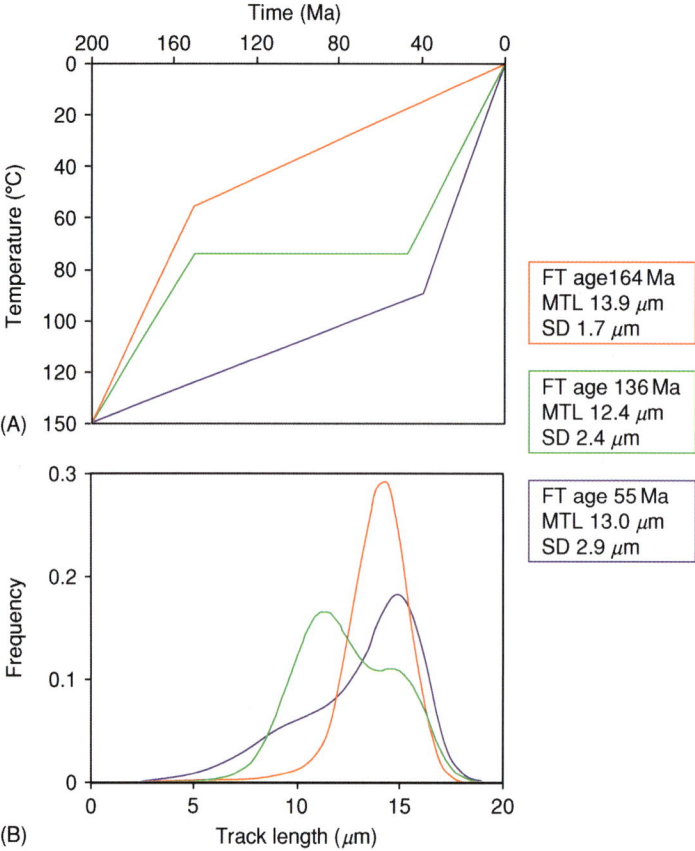

**Figure 8** (A) Three cooling histories and (B) the resulting apatite fission track ages and track length distributions. Rapid early cooling (red) produces a narrow track length distribution with a long mean track length and an old age. Rapid initial cooling followed by a long period of time in the APAZ (green) results in a skewed track length distribution and a younger apatite fission track age. Slow cooling in the APAZ followed by rapid late cooling (blue) produces a negatively skewed track length distribution and a very young apatite fission track age. FT age, fission track age. MTL, mean track length. SD, standard deviation of the mean track length.

process is poorly understood for temperatures below ~60°C and, in general, annealing at these low temperatures is underestimated. Modelling of fission track data, therefore, often results in thermal histories that include a rapid late cooling event, which in many cases is merely a modelling artefact. Therefore, one should always be very critical about any geological interpretations based on this part of the modelled thermal history. The (U-Th)/He method gives much better time–temperature constraints for such low temperatures.

## (U-Th)/He Dating

(U-Th)/He dating, like fission track analysis, provides information on a sample's low-temperature thermal history and not on its original, high-temperature igneous or metamorphic history. As is the case for fission track analysis, the (U-Th)/He technique is best established for apatite. (U-Th)/He dating of zircon and titanite has been explored, but at present this application is still in its infancy. At least for apatite, zircon, and titanite, the (U-Th)/He method is sensitive to lower temperatures than the fission track technique.

In nearly all minerals radiogenic helium is predominantly derived from the decay of uranium and thorium. Accumulation of α-particles ($^4$He nuclei) starts upon cooling into the Helium Partial Retention Zone (HePRZ). The HePRZ concept is similar to the Partial Annealing Zone concept of fission track analysis. At temperatures higher than that of the HePRZ, helium diffuses out of grains almost immediately after production. This means that there can be no accumulation of helium and the age will thus remain 0. At temperatures below that of the HePRZ, diffusion ceases and (U-Th)/He ages track calendar time. The nominal helium closure temperature ($T_c$) is dependent on mineral properties, cooling rate, and grain size. For apatite crystals in the size range of ~140–180 μm, $T_c$ will be ~70°C at a cooling rate of 10°C/Ma. For smaller grain sizes and at lower cooling rates, it may be as low as ~50°C. Unlike fission track data,

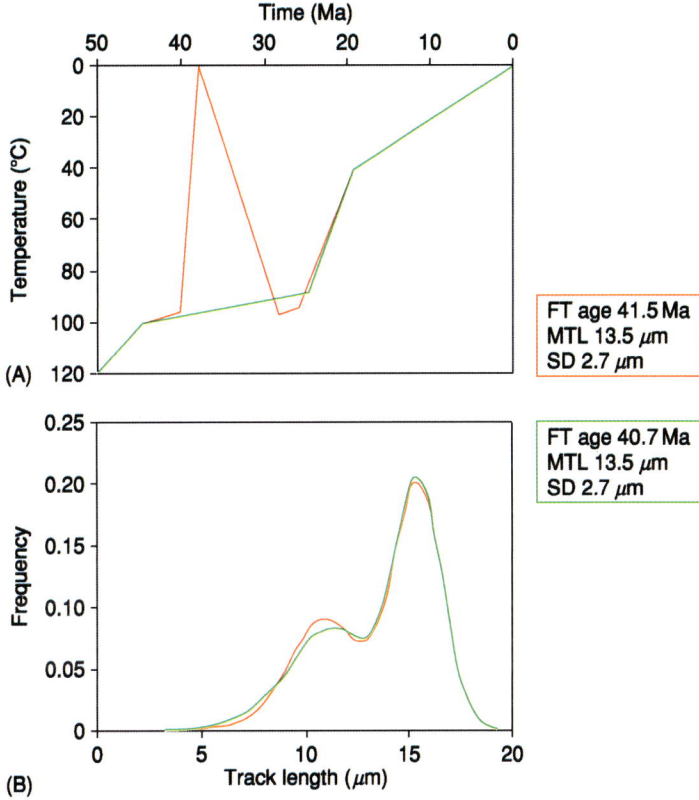

**Figure 9** (A) Two thermal histories and (B) the resulting apatite fission track ages and track length distributions. Despite the difference in thermal history before the onset of final cooling, both scenarios produce essentially the same apatite fission track age and track length distribution. The difference in age and track length distribution is well within the uncertainty normally associated with fission track data.

apatite (U-Th)/He data do not seem to be influenced by chemical composition.

Because of the high kinetic energy with which α-particles are emitted from their parent nuclides, they may be lost from grains as a result of α-ejection. For apatite, the so-called α-stopping distance is ∼19–23 μm, depending on the parent nuclide. Conditional on the distance of the parent nuclide from the physical grain boundary, α-particles may be either ejected from, or retained within grains (**Figure 10A**).

Because α-particles are emitted in random direction from their parent nuclei, they have a 50% chance to be ejected if their parent nuclei are located on the physical grain boundary itself. α-particles emitted from parent nuclei at more than the α-stopping distance from the grain boundary will never be ejected by α-emission. In principle, α-particles can also be implanted into grains when they are ejected from neighbouring grains. α-implantation can be ignored in most cases however, because the concentration of parent nuclei normally is much higher in minerals that are used for dating compared to the concentration in the host rock they were separated from. α-ejection must be corrected for by the so-called α-emission correction to obtain a geologically meaningful 'α-ejection corrected' (U-Th)/He age. The α-emission correction is dependent on the mineral and the appropriate α-stopping distances, grain size, and grain geometry. Smaller grains require bigger corrections because a larger percentage of parent nuclei will be within α-stopping distance from the grain boundary. A complicating factor for making the appropriate α-emission correction is that helium loss by ejection is interwoven with diffusional helium loss. The helium diffusion domain in some minerals (e.g., apatite) coincides with the physical grain. Within the HePRZ, helium will, therefore, be lost by diffusion at the grain boundary. Diffusion will tend to smooth the α-retention profile resulting from α-ejection, which is illustrated in a grain cross-section (**Figure 10B**). Because they are interlinked, the diffusion process and α-ejection cannot be accurately modelled separately and instead must be incorporated simultaneously in a numerical model. The diffusion process and its effect on the α-retention profile becomes more important with a longer residence time in the HePRZ, and application of such a model, therefore, is particularly important when working

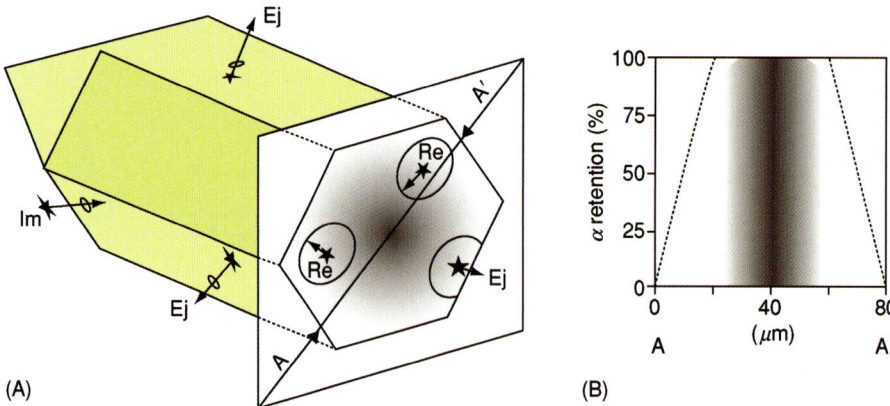

**Figure 10** (A) Three possible outcomes of α-emission in an apatite crystal: (Im) implantation, (Re) retention, and (Ej) ejection of α-particles. Radius of circles is equivalent to the α-stopping distance. (B) Diagram showing how α-retention changes from rim to core to rim along a cross-section perpendicular to the c-axis of the crystal.

with samples that spend considerable time in the HePRZ.

Successful application of (U-Th)/He dating is very much dependent on the quality of the material being analysed. Several criteria must be satisfied before mineral grains are studied by (U-Th)/He dating. This somewhat restricts the range of rock types to which this technique can be applied. Crystals must be sufficiently large to keep the α-emission correction and its potential error within reasonable bounds. The minimum diameter of apatites that can be accurately dated is about ~75 μm. Helium diffusion and α-emission correction models are based on euhedral crystal morphology (or, for example, a finite cylinder approximation) and a homogeneous distribution of parent nuclei. Only intact, minimally abraded crystals lacking uranium and thorium zonation should therefore be used. Zonation of uranium and thorium is fairly common in apatite and ignoring this can introduce large errors. Crystals must be free of cracks and inclusions and should, therefore, be thoroughly inspected under the microscope before analysis. Small inclusions may be overlooked however, and uranium and thorium zonation is not easily detected in the grains that are being dated. Even very carefully selected grains may thus not fulfil all the selection criteria. (U-Th)/He analysis should, therefore, always be done in duplicate and preferably triplicate on separate batches of crystals from the same sample, so that reproducibility of the data can be verified. Although few rock types are explicitly excluded from (U-Th)/He dating as a result of the above-mentioned criteria, it is obvious that, for example, sedimentary rocks are less likely to yield unabraded crystals and that volcanic rocks may not contain sufficiently large crystals.

The analytical procedure for (U-Th)/He analysis consists of two stages. First, helium is extracted from crystals in vacuum by heating in a furnace or with a laser and then measured on a noble gas mass spectrometer. Subsequently, the sample is removed from the vacuum system, dissolved in acid and the uranium and thorium concentrations are determined, usually by Inductively Coupled Plasma Mass Spectrometry (ICP-MS). The entire procedure is then repeated for the sample duplicates. Depending on the expected helium, uranium, and thorium concentrations, a batch of apatites typically consists of ~1–20 crystals. When the helium, uranium, and thorium concentrations have been determined, a 'raw' (U-Th)/He age can be calculated. This must then be corrected for loss of α-particles to obtain a geologically meaningful age. A standard of known age should always be included in a series of age determinations to verify data quality. For apatite (U-Th)/He analysis, the Durango apatite is the most common monitor.

## Applications of Fission Track Analysis and (U-Th)/He Dating

Fission track analysis can be applied to a large variety of geological problems and settings. In basins analysis, for example, apatite fission track analysis can help to estimate maximum palaeotemperatures and the postdepositional denudation history of the sedimentary basin, and zircon fission track data can be used to investigate sediment provenance. In orogenic belts (*see* **Tectonics:** Mountain Building and Orogeny), fission track analysis can provide estimates on erosion and denudation and, therefore, is important for mass balance studies. Non-orogenic settings, such as passive continental margins, are probably best suited to the capabilities of fission track analysis, because this type of setting in many cases will be dominated by cooling and denudation. As a conventional dating method, fission track analysis can be applied to date, for example, volcanic glass or tuff. Because of its unique ability to constrain temperature histories in the uppermost part of the crust (upper ~3 km, depending on the geothermal gradient), apatite (U-Th)/He dating can be used to constrain the development of topography. It is, therefore, a valuable tool in geomorphological studies (*see* **Geomorphology**). (U-Th)/He dating also is an important technique to constrain the last part of thermal histories obtained from thermal history modelling of fission track data.

## List of Units

| | |
|---|---|
| $\mu m$ | One millionth of a metre |
| $^{40}Ar/^{39}Ar$ | Argon – argon dating method |
| Ca | Calcium |
| Cl | Chlorine |
| F | Fluorine |
| Fission | Splitting of an atomic nucleus |
| $^4He$ | Helium – mass 4 |
| $HNO_3$ | Nitric acid |
| $l_0$ | Length of a fission track immediately after formation |
| O | Oxygen |
| P | Phosphorus |
| $T_c$ | Closure temperature |
| $^{232}Th$ | Thorium – mass 232 |
| $^{235}U$ | Uranium – mass 235 |
| $^{238}U$ | Uranium – mass 238 |

## Glossary

**α-ejection** Loss of α-particles from a crystal as a result of α-emission.
**α-emission** Emission of α-particles from parent nuclides.
**α-implantation** Gain of α-particles by a crystal resulting from α-ejection from neighbouring crystals.
**α-particles** $^4He$ nuclei.
**α-retention** Retention of α-particles in a crystal after α-emission.
**APAZ** Apatite Partial Annealing Zone.
**EDM** External Detector Method.
**HePRZ** Helium Partial Retention Zone.
**HF** Hydrofluoric acid.
**ICP-MS** Inductively Coupled Plasma Mass Spectrometry.
**Induced tracks** Fission tracks produced by irradiation in a nuclear reactor.
**PAZ** Partial Annealing Zone.
**Spontaneous tracks** Natural fission tracks, formed by spontaneous fission.
**TINCLE** Track-in-cleavage, fission track intersecting a crack.
**TINT** Track-in-track, fission track intersected by another track.
**ZPAZ** Zircon Partial Annealing Zone.

## See Also

**Analytical Methods:** Geochronological Techniques; Mineral Analysis. **Geomorphology**. **Minerals:** Micas; Zircons. **Petroleum Geology:** Exploration. **Tectonics:** Mountain Building and Orogeny. **Tektites**.

## Further Reading

Barbarand J, Carter A, Wood I, and Hurford T (2003) Compositional and structural control of fission-track annealing in apatite. *Chemical Geology* 198: 107–137.

Deer WA, Howie RA, and Zussman J (2004) *Rock Forming Minerals*, vol. 4B, *Silica Minerals, Feldspathoids and Zeolites*. London: The Geological Society.

Donelick RA, Ketcham RA, and Carlson WD (1999) Variability of apatite fission-track annealing kinetics: II. Crystallographic orientation effects. *American Mineralogist* 84: 1224–1234.

Farley KA (2002) (U-Th)/He Dating: Techniques, Calibrations and Applications. In: Porcelli DP, Ballentine CJ, and Wieler R (eds.) *Noble gases in geochemistry and cosmochemistry*, pp. 819–843. Washington: Mineralogical Society of America and Geochemical Society.

Fleischer RL, Price PB, and Walker RM (1975) *Nuclear tracks in solids*. Berkeley: University of California Press.

Gallagher K, Brown RW, and Johnson C (1998) Fission track analysis and its applications to geological problems. *Annual Review of Earth and Planetary Sciences* 26: 519–572.

Gleadow AJW and Brown RW (2000) Fission-track thermochronology and the long-term denudational response to tectonics. In: Summerfield MA (ed.) *Geomorphology and Global Tectonics*, pp. 57–75. Chichester: John Wiley and Sons Ltd.

Green PF, Duddy IR, and Hegarty KA (2002) Quantifying exhumation from apatite fission-track analysis and vitrinite reflectance data: precision, accuracy and latest results from the Atlantic margin of NW Europe. In: Doré AG, Cartwright JA, Stoker MS, Turner JP, and White N (eds.) *Exhumation of the North Atlantic Margin: Timing, mechanisms and implications for petroleum exploration*. London: Geological Society of London.

Palmer DC (1994) Stuffed derivatives of the silica polymorphs. *Reviews in Mineralogy* 29: 83–118.

Van den Haute P and De Corte F (1998) *Advances in Fission-Track Geochronology*. Dordrecht: Kluwer Academic Publishers.

Wagner G and Van den Haute P (1992) *Fission-Track Dating*. Dordrecht: Kluwer Academic Publishers.

Zentilli M and Reynolds PH (1992) *Low temperature thermochronology; short course handbook*. Mineralogical Association of Canada: Toronto.

# Geochemical Analysis (Including X-ray)

**R H Worden**, University of Liverpool, Liverpool, UK

© 2005, Elsevier Ltd. All Rights Reserved.

## Introduction

Geochemistry is the study of the occurrence and distribution of elements, isotopes, minerals, and compounds in the natural environment. Geochemical analysis, the measurement of the quantities of elements, isotopes, minerals, and compounds in a rock, natural liquid, or naturally occurring gas, is a colossal topic.

The applications of geochemistry stretch from the core to the outer atmosphere and beyond. The objects of study include minerals, metals, and salts, organic biomolecules, coal, bitumen, kerogen, and petroleum, atmospheric-, river-, ground-, and formation-water, and gases in the crust, sediments, and atmosphere. Analysis is also a huge topic, with subjects ranging from rocks, minerals, compounds, and species to isotopes, elements, and atoms.

### The Role of Analysis: Hypotheses, Questions, Problems, and Theories

In the context of Earth sciences, geochemical analysis commonly implies quantitative work with a numerical output. Geochemical analysis is thus the quantitative examination of the composition of natural Earth materials. Most geochemical work is undertaken to address a hypothesis, answer a question, or solve a problem. Even routine environmental geochemical monitoring is done in order to address the question of the continued safety of natural materials over time. It is typically important to formulate a hypothesis, question, or problem carefully before embarking on an analytical programme. Such an approach is best since the outcome should be cost-effective while being statistically rigorous and ultimately defensible.

### Producing Geochemical Data

When any analytical geochemical device is used to quantify the composition of a material, the detector output is usually in the form of some sort of electrical signal. The electrical signal must be converted into a meaningful geochemical parameter, such as a unit of concentration. This conversion is usually achieved via a calibration curve. For any technique, a series of previously characterized standards must be analysed, in exactly the same way as an unknown sample, and the output signals measured. In essence, the calibration is a plot of known concentration versus the output signal over a range of concentrations (**Figure 1**). The detector outputs from the standards must be converted into some sort of equation that permits the eventual back-conversion of output signals into concentrations for unknown samples. In the example in **Figure 1**, the output must be determined as a function of concentration (equation 1) and then inverted to allow concentration to be determined from the output signal for unknown samples (using equation 2).

The quality of the analytical output from a device is fundamentally a function of the quality of the calibration curve. No calibration is perfect, and the degree of imperfection can be described in terms of accuracy and precision. These discrete characteristics are best described for a dataset composed of repeated analyses of the same standard sample (**Figure 2**). When the results are widely dispersed about the known answer, the output is said to be imprecise. When the results are tightly clustered about the known answer, the output is said to be precise. If the results cluster around the correct figure, they are said to be accurate, whereas if they cluster about a figure other than the correct output, they are said to be inaccurate (**Figure 2**). Accuracy and precision are the combined result of the quality of calibration, the sensitivity of the device,

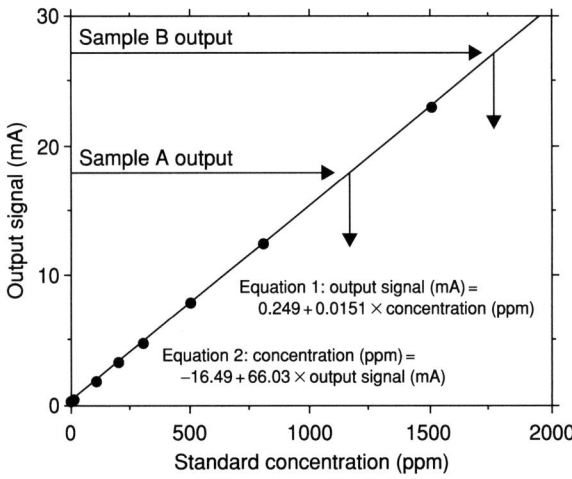

**Figure 1** Example of a calibration curve using standards of known concentration and their output signal strengths. The small black dots represent the calibration. Equation 1 describes how concentration could be converted to signal strength. The more useful equation 2 describes how signal strength from an unknown sample can be converted into concentration. Unknown sample A yields a valid result since it falls within the calibration range. The analysis of sample B is not valid since the output signal falls outside the calibration range.

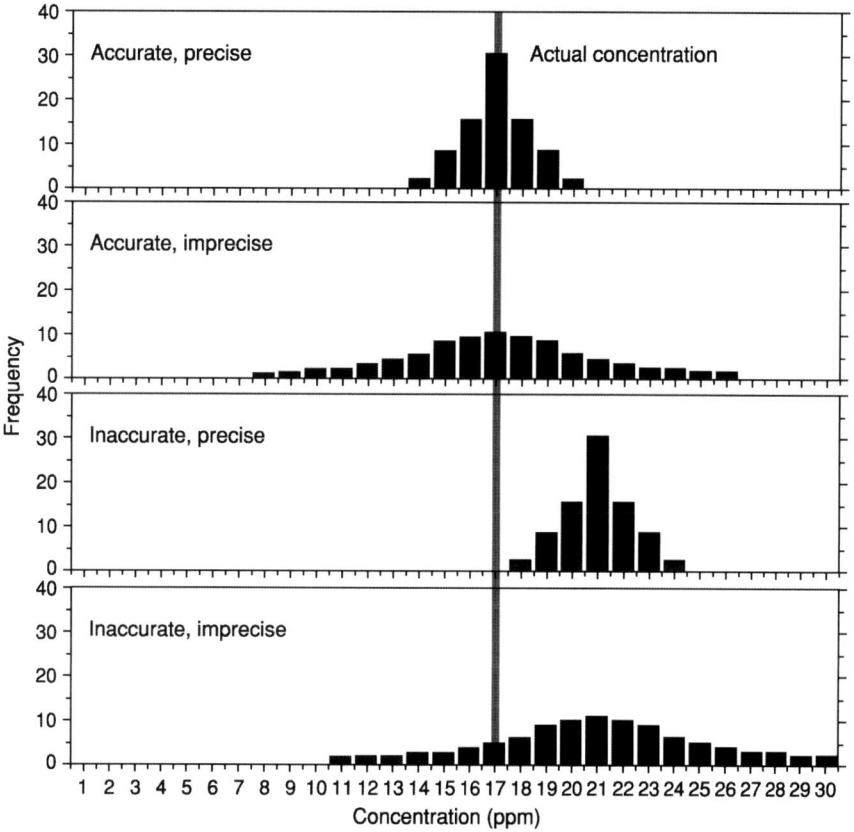

**Figure 2** Representation of analytical precision and accuracy. Precision is a measure of how tightly the results cluster about the mean result. Accuracy measures the proximity of the mean result to the expected result. Accuracy can be improved by recalibrating the device.

the attainment of constant operating conditions, and the skill of the device operator.

### Geochemical Analytical Protocol (Blanks, Standards, Repeats)

One of the most important aspects of any geochemical analytical protocol is to ascertain the credibility of the data. Every geochemical analysis has some degree of associated uncertainty (see **Figure 2**, where uncertainty is expressed in terms of precision). Every technique also has a minimum detection limit, below which the concentration cannot be determined. The theoretical lower detection limit is an intrinsic function of the technique, although the quality of the calibration and the cleanliness of the equipment used in sample and standard preparation also limit detection. It is seldom correct to report a figure of zero concentration. It is more correct to report the detection limit and state that the sample had a concentration below that value ('below detection' or 'none detected'). In any analytical run it is worthwhile running blanks to check the lower detection limit. It is also worth discretely running repeat samples to check the reproducibility of the analytical technique. Finally, it is worth running standards throughout an analytical run at regular intervals.

Developing and checking the calibration lines are the most crucial steps in geochemical analysis, and being aware of the upper and lower limits of the calibration is important. Many calibration lines do not vary linearly with concentration. Thus, if any geochemical analytical result falls outside the calibrated range of the detector and its output, then the result is theoretically invalid since it is not certain how to convert the instrumental signal (e.g. amps, volts etc) into geochemically significant units.

### The Range of Geochemical Analytical Techniques

As stated earlier, geochemical analysis is a large topic, and many analytical techniques have been employed in geochemical research and studies over the last four or five decades. Techniques may be grouped according to the attribute being measured (**Table 1**). Many important techniques use X-rays in a variety of ways for analytical purposes. Other techniques use the optical effects of samples, typically taking advantage of electromagnetic radiation emitted or absorbed by the material at elevated temperatures.

**Table 1** Summary of some of the main techniques of geochemical analysis and their sample types

| Technique | Output types | Abbreviation | Solid (rock, mineral, sediment, soil, filtrate) | Artificially dissolved solid | Water (natural solution) | Fluid organic | Solid organic | Gas | Fluids trapped in solids |
|---|---|---|---|---|---|---|---|---|---|
| **X-ray techniques** | | | | | | | | | |
| X-ray analysis (electron microprobe) | Point mineral composition | MPA | Yes | | | | | | Yes |
| X-ray fluorescence | Rock composition | XRF | Yes | | | Yes | Yes | | |
| X-ray diffraction | Mineralogy | XRD | Yes | | | | Yes | | |
| **Spectroscopic techniques** | | | | | | | | | |
| Atomic absorption spectroscopy | Solution composition | AAS | | Yes | Yes | | | | Yes |
| Inductively coupled plasma optical emission spectroscopy | Solution composition | ICP-OES | | Yes | Yes | | | | Yes |
| UV spectroscopy | Composition | UVS | Yes | | | Yes | | | Yes |
| **Chromatography** | | | | | | | | | |
| Gas chromatography | Gas composition | GC | | | Yes | Yes | | Yes | Yes |
| Ion-exchange chromatography | Aqueous-liquid composition | IC | | Yes | Yes | | | Yes | Yes |
| **Mass spectroscopy** | | | | | | | | | |
| Mass spectroscopy | Composition, isotope content | MS | Yes | Yes | Yes | Yes | Yes | Yes | Yes |
| Isotope dilution mass spectroscopy | Isotope ratio | IRMS | Yes | Yes | Yes | Yes | Yes | Yes | |
| Inductively coupled plasma mass spectroscopy | Composition | ICP-MS | Yes | Yes | Yes | | Yes | | Yes |
| Gas chromatography mass spectrocopy | Composition | GC-MS | | | Yes | Yes | | Yes | Yes |
| **Thermal analysis techniques** | | | | | | | | | |
| Pyrolysis | Elemental analysis (C,H,O,S,N), organic type | | | | | Yes | Yes | | |
| Evolved water analysis, thermogravimetry | Clay content, clay mineralogy | EWA-TG | Yes | | | | | | |
| Fluid inclusion microthermometry | Temperature of trapping, composition | FI | Yes | | | | | | Yes |
| **Related techniques** | | | | | | | | | |
| Wet chemistry techniques | Composition | | Yes | Yes | Yes | | | | |
| Microscopy (light and electron optics) | Mineralogy, fabric, crystallography | SEM,TEM | Yes | | | | Yes | | Yes |

Other techniques use chromatography, the time taken for one substance to move through another or through a capillary under a given gradient. A wide family of geochemical analytical techniques use mass spectrometry to split propelled material, converted into charged particles, using electromagnets. Other techniques involve examining the products of heating Earth materials under controlled conditions and studying either the evolved fluids or the changes in the properties of the residual solids.

## X-ray Techniques

### Origin of X-rays

X-ray technologies have proved to be useful in geochemical analysis (**Table 2**). X-rays are part of the electromagnetic spectrum (**Figure 3**) and have wavelengths ranging between 0.01 nm and 10 nm (0.1–100 Å). They are waveforms that are part of a family that includes light, infrared, and radio waves. Since X-rays have no mass and no electrical charge, they are not influenced by electrical or magnetic fields and travel in straight lines. X-rays, like all parts of the electromagnetic spectrum, possess a dual character, being both particles and waves. The name that has been given to the small packets of energy with these characteristics is photon.

The simple model of the atom, proposed by Niels Bohr in 1915, is not completely correct, but it has many features that are approximately correct. The modern theory of the atom is called quantum mechanics; the Bohr model is an approximation to quantum mechanics that has the virtue of being much simpler than the full theory. In the Bohr model neutrons and protons occupy a dense central region (the nucleus), and electrons orbit the nucleus. The basic feature of quantum mechanics that is incorporated in the Bohr model is that the energies of the electrons in the Bohr atom are restricted to certain discrete values (the energy is quantized) – only certain electron orbits with certain radii are allowed.

X-rays are generated when free electrons from an electron gun give up some of their energy when they interact with the orbital electrons or the nucleus of an atom (**Figure 4**). The energy given up by the electron during this interaction reappears as emitted electromagnetic energy, known as X-radiation. Two different atomic processes can produce X-ray photons. One is called bremsstrahlung and the other is called electron-shell emission (**Figure 5**). Bremsstrahlung means 'braking rays'. When an electron approaches an atom, it is affected by the negative force from the electrons of the atom, and it may be slowed or completely stopped. The energy absorbed by the atom during the slowing of the electrons is excessive to the atom and will be radiated as X-radiation of equal energy to that absorbed. Bremsstrahlung X-rays tend to have a broad range of energies since the degree of slowing can be variable and materials composed of mixtures have atoms with different properties (**Figure 6**). Bremsstrahlung tend not to be used for geochemical analysis; that is the preserve of electron-shell emission.

### Analysis of X-rays: Electron-Shell Emission

A common geochemical application of X-ray analysis is to direct a focussed electron beam at a polished rock or mineral surface and then collect and quantify the resulting secondary characteristic X-rays (**Figure 7**). The secondary X-rays help to reveal the elements present in that part of the sample that is directly under the electron beam. This technique is known as electron-beam microanalysis, or microprobe analysis, and gives spatially resolved major- and trace-element geochemical data from solid samples, including rocks, minerals, sediments, soils, and glass. Many ordinary electron microscopes are fitted with a secondary X-ray detector, making them suitable for geochemical analysis. All of these devices rely on electron optics, using electromagnetic lenses to focus and direct a stream of electrons, generated by an electron gun, onto a polished mineral or rock surface (**Figure 7**). The focused electron beam has a variable radius, but can typically be maintained at slightly greater than about 1 $\mu$m. The spatial resolution of a microprobe is actually somewhat greater than 1 $\mu$m. The impinging electron stream interacts with the polished surface and produces a wide range of signals, including secondary and backscattered electron and cathodoluminescence (light) as well as the secondary X-rays of concern here. There is an activation volume from which X-rays are generated, below the polished surface, which is several times larger than the primary beam. Samples must be highly polished (flat) to avoid scattering.

When a sample is bombarded by an electron beam, some electrons are knocked out of their quantum shells in a process called inner-shell ionization (**Figure 5**). Outer-shell electrons fall in to fill a vacancy in a process of self-neutralization. The shells are termed K, L, M, and N starting from the innermost most strongly bound shell.

Electrons moving from one shell to another produce characteristic X-rays. K-shell ionizations are commonly filled by electrons from the L shell (K$\alpha$ radiation) or M shell (K$\beta$ radiation). There are two K$\alpha$ peaks (K$\alpha$1 and K$\beta$2) corresponding to two discrete states of the in-falling electron. When outer-shell electrons drop into inner shells, they emit a quantized

**Table 2** X-ray techniques commonly used in geochemical analysis

| Technique | Output-1 | Output-2 | Output-3 | Sample type | Advantages | Disadvantages |
| --- | --- | --- | --- | --- | --- | --- |
| X-ray analysis | Quantitative elemental composition of small volume (several cubic mm) | Major elements detected using energy dispersive spectrometer (SEM or microprobe) | Trace elements detected using wavelength dispersive spectrometer (microprobe) | Polished rock samples, or grains set in resin and polished | High spatial resolution; gives quantitative data with estimate of uncertainty; well-established technology; wide range of elements | Sample preparation can be slow; errors can be large for some elements; problems of analysis statistics for heterogeneous minerals |
| X-ray fluorescence spectroscopy | Quantitative major and trace elemental composition of a crushed homogenized sample | | | Crushed and homogenized sample then compressed or melted and quenched into small disks | Gives major and trace elements simultaneously; produces data for Al and Si in rocks as well as a range of metals; well-established technology | Sample preparation can be slow; relies on sample type-specific calibration curve; matrix effects can be large; instrumental set-up can be arduous |
| X-ray diffraction | Qualitative presence or absence of minerals in crushed sample | Relative proportion of minerals in a rock or sediment mixture – for suitably prepared samples | Mineral composition | | Fairly quick; well-established technology; large database of minerals for comparison | Semi-quantitative in most cases; difficult for minerals with solid solutions; difficult for poorly crystalline materials; difficult with very complex mixtures |

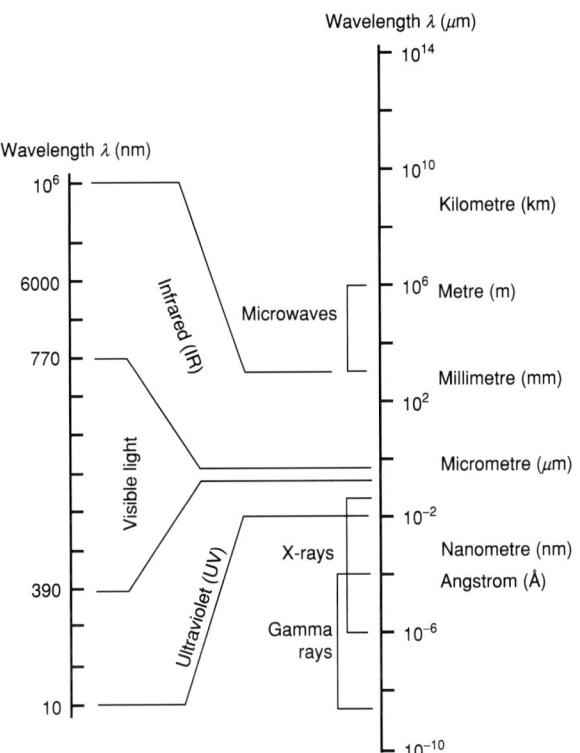

**Figure 3** Representation of the electromagnetic spectrum with the various wavelengths employed in geochemical studies. Visible light is not discussed here but forms the basis of optical microscopy. Gamma rays are used to differentiate radioactive fission reactions and thus identify and quantify radioactive elements.

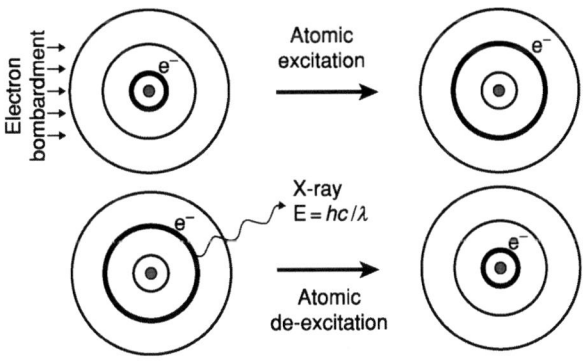

**Figure 4** General set of processes involved in the generation of X-rays by electron bombardment of atoms. The small grey filled circle represents the nucleus. The outer rings represent the quantized electron orbitals of the Bohr atomic model. The thicker black circle represents the location of a given electron (e$^-$). With electron bombardment, the highlighted electron jumps to a higher orbital. The energized electron quickly falls back to its original state, releasing an X-ray.

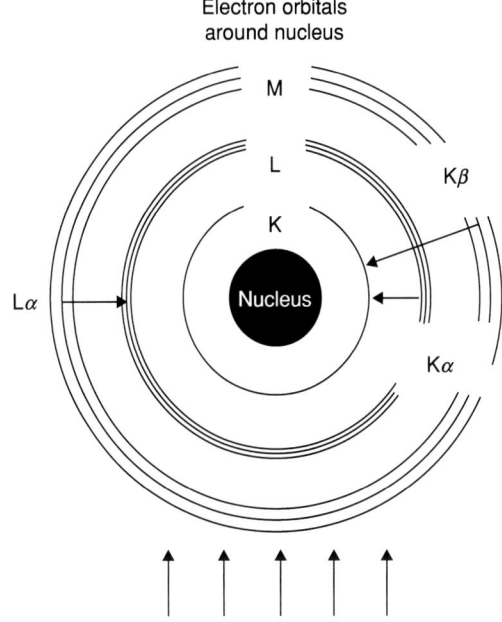

**Figure 5** Incident electrons cause electrons in metal atoms to become excited and jump to outer orbitals. K$\alpha$ and K$\beta$ X-rays (etc.) are emitted as electrons fall back to their original orbitals. The X-ray energy is characteristic of the element and of the starting and final orbitals. Background 'white' radiation (bremsstrahlung) is due to collisions between incident and orbital electrons.

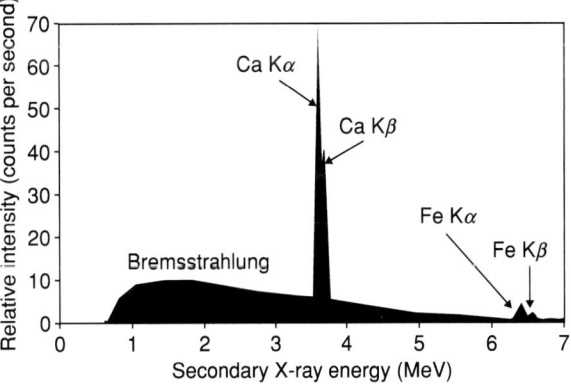

**Figure 6** Example of a secondary X-ray trace from a sample of calcite with a minor amount of iron substituting for the calcium (known as ferroan calcite). The X-ray energy values are characteristic of the elements. Note that the energy of the K lines increases with increasing atomic number. Note also the increasing separation of the K$\alpha$ and K$\beta$ lines with increasing atomic number. The bremsstrahlung can cause problems with quantification, especially at low secondary X-ray energies.

photon characteristic of the element. The resulting characteristic spectrum is superimposed on the bremsstrahlung (**Figure 6**). An atom remains ionized for a very short time (about $10^{-14}$ s) and thus the incident electrons, which arrive about every $10^{-12}$ s, can repeatedly ionize an atom.

It is common practice to measure X-rays in units of thousand electron volts (KeV). One electron volt is

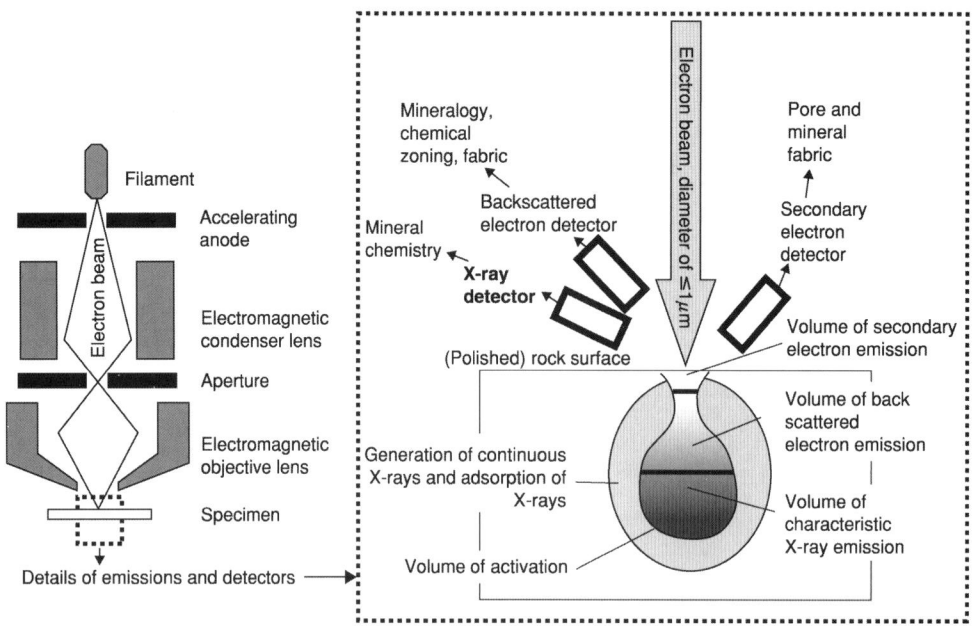

**Figure 7** Basic equipment used for secondary X-ray analysis following electron-beam bombardment of Earth materials. The various types of signals generated in the samples and the various detectors are highlighted in the enlarged diagram. Characteristic X-ray emission occurs several micrometres below the sample surface in a volume with a radius of a few times the electron-beam radius (spot size). Bremsstrahlung are generated in the outlying volume where adsorption of X-rays also occurs.

the same as $1.6021 \times 10^{-19}$ J. An electron volt is the amount of energy gained by one electron when it is accelerated by one volt. The output characteristic X-rays have an energy and wavelength controlled by the element that is present and the specific nature of the electron orbital transition. Elements with low atomic numbers produce low-energy X-rays, and many X-ray detectors have difficulty quantifying the output from elements with atomic numbers of less than 11 (sodium). The energy and wavelength of the characteristic X-rays are related by the equation $E = hc/\lambda$, where $E$ is the energy (in KeV), $h$ is the Plank constant ($6.626 \times 10^{-34}$ J s), $c$ is the speed of light ($2.99782 \times 10^8$ m s$^{-1}$), and $\lambda$ is the element and electron orbital-specific wavelength of the X-rays. Conversion of X-ray energy into wavelength is thus achieved using $E$ (eV) $= 123985/\lambda$ (in nm). X-ray spectra are typically plotted in terms of energy. The outputs of characteristic X-rays have intensities that are a complex function of the quantity of the element in the analysed volume. In general, the intensity is approximately proportional to the relative concentration. Several matrix (mineral) dependent processes alter the primary intensity of the secondary X-rays. The atomic mass of the element influences the efficiency of X-ray generation (Z-correction), some secondary X-rays are absorbed by the material (A-correction), and some of the absorbed X-rays result in localized X-ray fluorescence (F correction).

Secondary X-rays can be detected and quantified in two ways. One, which is best for major elements (>0.1 wt%), involves measuring the energy of the emitted X-rays using a scintillation counter (known as energy dispersive analysis). The other, best for trace elements (>1 ppm at best), measures the wavelength (and intensity) of the emitted X-rays (known as wavelength dispersive analysis). Energy dispersive analysis is typically faster than wavelength dispersive analysis but has much lower sensitivity.

Most modern devices are computerized and have inbuilt quantification and correction systems. The geochemical output from electron microprobes is in the form of oxide or element weight percentages.

**X-ray Fluorescence**

Although X-ray beams cannot be focussed or bent, they can be directed by a series of diffraction gratings, collimators, and slits. Such X-ray beams have been used for geochemical analysis by a technique known as X-ray fluorescence spectroscopy (XRF). The output from this technique gives the concentrations of most major elements (with atomic numbers of 11 (sodium) and above) and many useful trace elements.

It was stated earlier that some secondary X-rays can be absorbed by material and that some of this absorption can result in localized X-ray fluorescence (**Figure 8**). Just as with primary electron beams, when a primary X-ray excitation source (e.g. from an X-ray

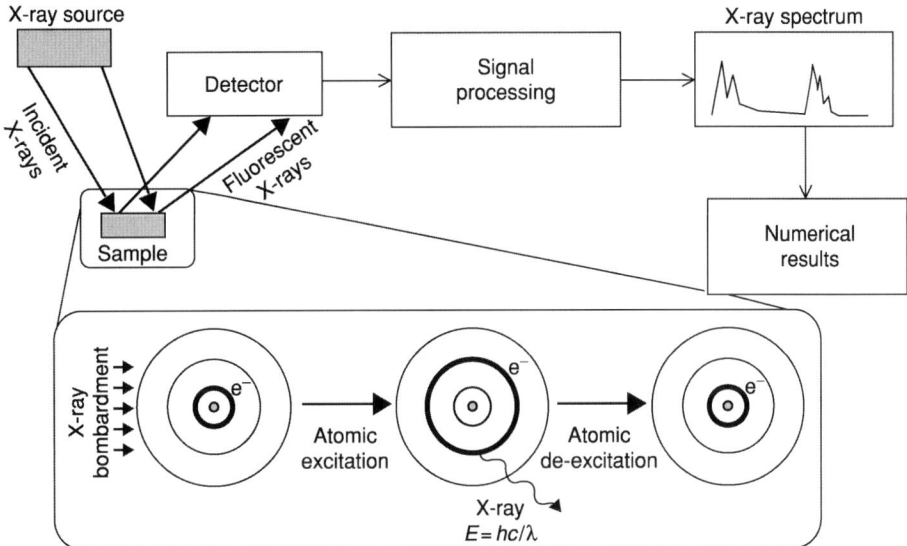

**Figure 8** General set of processes involved in the generation of fluorescent X-rays by X-ray bombardment of atoms. The small grey filled circle represents the nucleus; the outer rings represent electron orbitals. The thicker black circle represents the location of a given electron ($e^-$). With X-ray bombardment, the highlighted electron jumps to a higher orbital. The energized electron quickly falls back to its original state, releasing a secondary X-ray. The range of elements in a sample leads to a range of characteristic fluorescent X-rays with peak heights that are functions of the amounts of the elements in the sample.

tube or even a radioactive source) strikes a sample, the X-ray can be either absorbed by the atom or scattered through the material. Some incident X-rays are absorbed by the atom, and electrons are ejected from the inner shells to outer shells, creating vacancies. As the atom returns to its stable condition, electrons from the outer shells are transferred to the inner shells and in the process give off a characteristic X-ray with an energy equal to the difference between the two binding energies of the corresponding shells.

XRF is widely used to measure the elemental composition of natural materials, since it is fast and nondestructive to the sample. It can be used to measure many elements simultaneously. XRF can be used directly on rock surfaces, although there is a danger of natural heterogeneity resulting in variable results. Rock, soil, and sediment samples are typically crushed and made into pellets by compressing them or by melting the whole sample and then quenching to make a glass disk. XRF is useful for the geochemical analysis of a wide range of metals and refractory and amphoteric compounds (such as $SiO_2$ and $Al_2O_3$) and even some non-metals (chloride and bromide). XRF is also routinely used to measure the natural metal content of liquid petroleum samples. The quality of XRF data is a function of the selection of appropriate standards. It is considered to be best practice to use standards that are similar to the samples in question to minimize matrix effects. XRF can measure down to parts-per-million concentrations and lower, depending on the element and the material.

**X-ray Diffraction**

X-rays have a similar wavelength to the lattice spacing of common rock-forming minerals and have been used to characterize the crystal structure and mineralogy of Earth materials by using X-ray diffraction (XRD) analysis. This is most commonly used to define the presence of minerals, mineral proportions, mineral composition (in favourable circumstances), and other subtle mineralogical features of rocks, sediments, and soils.

X-rays, even from a pure elemental source bombarded with electrons, have a collection of peaks – X-rays characteristic of the quantized electron energy levels – and bremsstrahlung. X-ray beams of a tightly defined energy (and thus wavelength) have been used to investigate and characterize the minerals present in rocks, sediments, and soils by removing all but one X-ray peak from the spectrum of wavelengths generated by a source element. X-rays are useful in investigating mineral structure since they can be selected to have a wavelength that is only just smaller than the interlattice spacing (d-spacing) of common rock-forming minerals. A number of X-ray sources have historically been selected, but copper is the most commonly employed, and the copper K$\alpha$ peak is the one that is directed onto samples. This has a characteristic wavelength of 1.5418 Å (0.15418 nm). This is ideal for many minerals since they have high-order (dominant, most obvious) lattice spacings of this size or up to 10–15 times greater than this wavelength.

The melee of X-radiation from copper can be reduced to the Kα peak and then directed by a series of diffraction gratings, collimators, and slits.

The main features of an XRD analyser include an X-ray source with collimators, slits, etc, a sample, and an X-ray detector (**Figure 9**). The source and detector are both rotated about the sample, an arbitrarily fixed point, and define the same angle ($\theta$) relative to the sample. The angle between the source and the detector is thus $2\theta$ relative to the sample.

Diffraction occurs when X-rays, light, or any other type of radiation passes into, but is then bounced back out of, a material with a regular series of layers. Diffraction occurs within the body of the material rather than from the surface (and so is quite different from reflection). Regular layers are a characteristic of all crystalline materials (minerals, metals, etc). Each rock-forming mineral has a well-defined set of these layers, which constitute the crystal lattice. No two minerals have exactly the same crystal structure, so fingerprinting a mineral by its characteristic set of lattice spacings helps to identify minerals. A radiation beam from a pure source has a defined wavelength, and the rays from such a pure source will be 'in phase'. Constructive interference occurs only when all the outgoing (diffracted) X-rays are also in phase. Destructive interference, the norm, occurs when the diffracted X-rays are no longer in phase. Constructive interference occurs when the extra distance that X-rays travel within the body of the material is an integer (whole number) multiple of the characteristic wavelength of the incident X-ray (**Figure 10**). The geometry of the XRD equipment, the wavelength of the incident radiation, and the lattice spacing are all important in defining whether constructive interference occurs. The key equation is known as the Bragg Law, which must be satisfied for constructive interference ('diffraction') to occur: $2d\sin\theta = n\lambda$, where $d$ is the lattice spacing, $\lambda$ is the wavelength of the incident X-ray source, and $n$ is an integer (typically one in many cases). The value of $\theta$, defined in **Figure 9**, is a function of the variable geometry of the XRD equipment.

X-ray diffraction is most commonly used on crushed (powered) rock samples to ensure homogeneity of the sample and randomness of the orientations of all the crystal lattices represented by different minerals. This is known as X-ray powder diffraction.

XRD works by rotating the X-ray source and the detector about the sample from small angles (e.g. 4°) through to angles of up to 70°. The low angles can detect large interlattice spacings (large values of $d$) while the high angles detect smaller interlattice spacings. For CuKα radiation these angles equate to $d$-spacings from about 30 Å down to about 1.5 Å, covering the dominant $d$-spacings of practically all rock-forming minerals.

For a pure mineral sample, the diffraction peaks from different lattice planes with discrete $d$-spacings have different relative intensities. This is a function of the details of the crystal structure of a particular mineral, but the maximum-intensity trace (peak) for many minerals has a low Miller Index value (a simple notation for describing the orientation of a crystal). For example, many clay minerals dominated by sheet-like crystal structures have (001) as the maximum-intensity peak. All other XRD traces have intensities that are fixed fractions of the intensity of the maximum-intensity trace. The result for each pure mineral is a fingerprint of XRD peaks on a chart of intensity on the $y$-axis and $2\theta$ on the $x$-axis (**Figure 11**). This can be compared with collections of standards to identify the mineral.

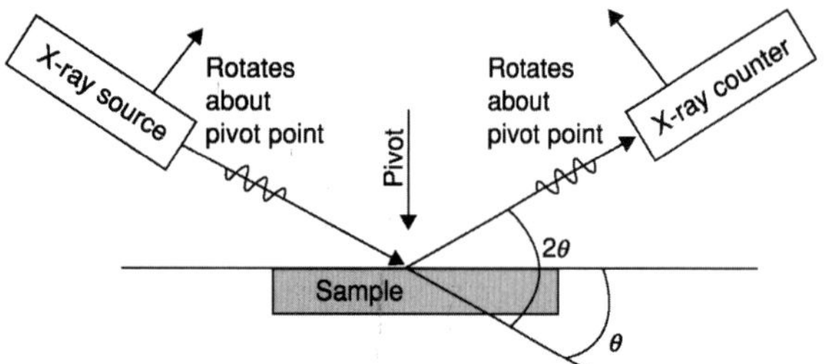

**Figure 9** Basic elements of an X-ray diffraction device. An X-ray source is directed at a sample at a controlled and variable angle ($\theta$). The X-ray detector is at the equivalent angle on the opposite side of the pivot point. The source and detector are at an angle of $180° - 2\theta$ to one another. The source and detector are thus simultaneously rotated about the pivot point. When diffraction occurs the X-ray detector records a signal above the background. The sample is usually a powder and preferably randomly orientated.

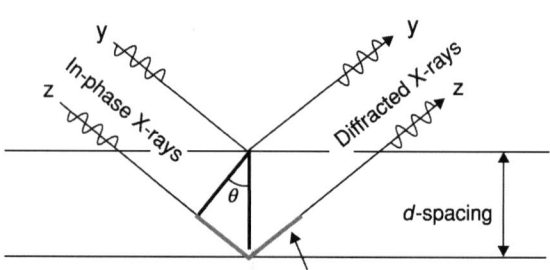

**Figure 10** Diffraction from a crystal. The incident X-rays are in phase as they hit the mineral surface. The grey line shows the path-length difference between the two X-rays. Constructive interference occurs when the extra path length ($2d\sin\theta$) is an integral multiple (typically one) of the wavelength of the X-rays. Constructive interference leads to an X-ray diffraction peak set against a low level of background noise.

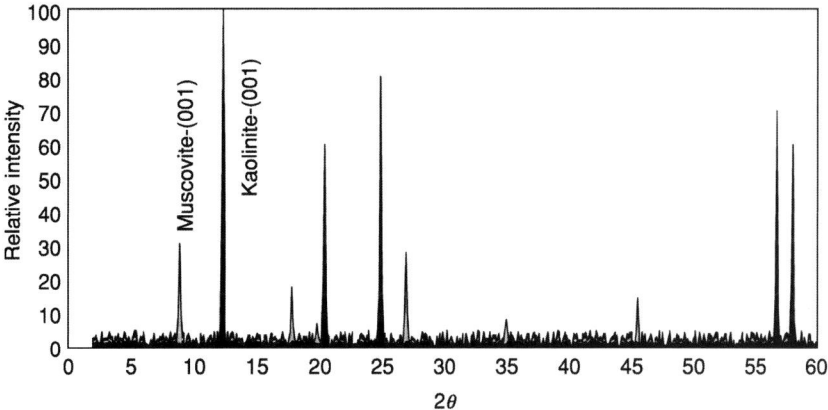

**Figure 11** The X-ray diffraction output from a rock composed of the clay minerals muscovite (pale grey peaks) and kaolinite (black peaks). The two minerals have discrete $d$ spacings, which are reflected in their discrete peaks on the chart. The peaks at the low $2\theta$ end have high $d$-spacings, and vice versa. The dominant basal spacings of these two sheet silicates (the (001) planes) are labelled. The maximum-intensity diffraction peaks from these minerals are produced by the (001) planes. From this diagram it would appear that there is much more kaolinite than muscovite in the mixture (approximately three times as much) since its maximum-intensity peak is much more intense. The mixture therefore has about 25% muscovite and 75% kaolinite.

The intensity of the collection of diffraction peaks from a given mineral in a mixture (e.g. rock, soil, or sediment) is broadly a function of the proportion of the mineral in the mixture. This is a subject of ongoing research since the issue is complicated by different minerals having different efficiencies at diffracting X-rays. In simple terms, the intensities of the maximum-intensity peaks from a mixture of minerals give a guide to the relative proportions of the different minerals (**Figure 11**).

There is a range of problems for XRD when dealing with natural Earth materials. The most obvious is the simple identification of minerals when each one has its own collection of diffraction peaks. These must be carefully deconvoluted by a process of elimination. Computer-based programs can be of great help in this task. Many rock-forming minerals do not conform to a perfectly defined composition. The set of interlattice $d$-spacings in minerals is a function of the way atoms are packed together in the lattice, so that compositional variation affects the precise $2\theta$ position of a given crystal orientation. In some cases, this variability can be put to good use since it can be used to identify the composition of a given mineral (**Figure 12**).

Another issue with XRD analysis, especially of sediments, soils, and sedimentary rocks is that many minerals have poorly defined crystal structure. This problem is typical of clay minerals, pedogenic oxides, etc. Poorly defined crystal structure translates into wider XRD peaks. This presents problems for quantification and results in a need to measure the area under an entire peak rather than the height of the maximum-intensity peak. However, this too can be put to good use by enabling the quantification of the transformation of a poorly defined crystal structure into a well-defined structure (typically as a function of time and temperature) during diagenesis or metamorphism (**Figure 13**).

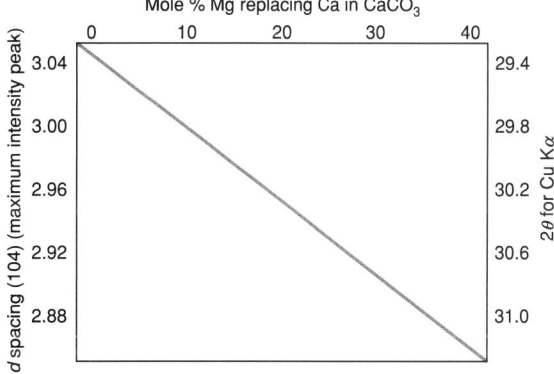

**Figure 12** X-ray diffraction for the same peak from the same mineral may occur at a different $2\theta$ value if there is solid solution and the substituted element has a different ionic radius. For example, in calcite the dominant (104) diffraction peak moves systematically as magnesium replaces calcium. Magnesium ions are smaller than calcium ions so the structure collapses slightly. In favourable circumstances, this approach can be used to help determine mineral composition.

## Optical Techniques

Spectroscopy is the use of the absorption, emission, or scattering of electromagnetic radiation by atoms or molecules (or atomic or molecular ions) to study the atoms or molecules qualitatively or quantitatively.

Isolated atoms can absorb and emit packets of electromagnetic radiation with discrete energies that

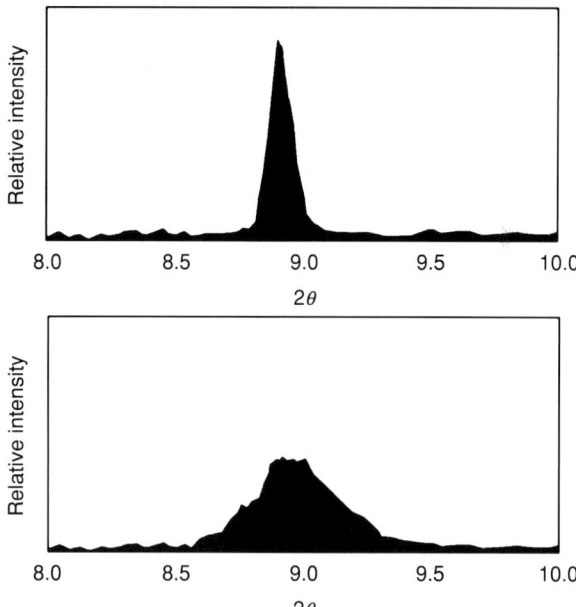

**Figure 13** X-ray diffraction outputs from a mineral (illite) with variable degrees of crystallinity. The top image represents a well-crystalline sample, e.g. a slate. The lower image represents a poorly crystalline material, e.g. from a juvenile sedimentary rock, soil, or weathering horizon. The degree of crystallinity can be quantified by measuring the peak width at half the peak height.

are dictated by the detailed electronic structure of the atoms. When the resulting altered light is passed through a prism or spectrograph it is separated according to wavelength. Emission spectra are produced by high-temperature gases. The emission lines correspond to photons of discrete energies, which are emitted when excited atomic states in the gas make transitions back to lower-lying levels. Generally, solids, liquids, and dense gases emit light at all wavelengths when heated. An absorption spectrum occurs when light passes through a cold dilute gas and atoms in the gas absorb light at characteristic frequencies. Since the re-emitted light is unlikely to be emitted in the same direction as the absorbed photon was travelling, this gives rise to dark lines (absence of light) in the spectrum.

Emission and absorption spectroscopy have been used widely in geochemical analysis for many years (**Table 3**). They are commonly used to analyse waters but can also be employed to analyse rock and other solid samples that have been quantitatively dissolved (e.g. in acid solutions).

### Atomic Absorption and Atomic Emission Spectroscopy

Atomic-absorption (AA) spectroscopy uses the absorption of light to measure the concentration of gas-phase atoms. Since samples are usually liquids or solids, the sample atoms or ions must be vaporized in a flame or graphite furnace (**Figure 14**). The atoms absorb ultraviolet or visible light. Concentrations are usually determined from a working curve after calibrating the instrument with standards of known concentration (**Figure 15**). The Lambert–Beer Law is the relationship between the change in light intensity for a given wavelength and the relative incident light energy: $\log I_o/I = aLc$, where $I$ is the light intensity after the metal is added, $I_o$ is the initial light intensity, $a$ is a machine-dependent constant, $L$ is the path length of light through the torch, and $c$ is the concentration.

The light source is usually a hollow-cathode lamp of the element that is being measured. A major disadvantage of these narrow-band light sources is that only one element can be measured at a time. AA spectroscopy requires that the target atoms be in the gas phase. Ions or atoms in a sample must be desolvated and vaporized in a high-temperature source such as a flame or graphite furnace. Flame AA can analyse only solutions. Sample solutions are usually aspirated with the gas flow into a nebulizing and mixing chamber to form small droplets before entering the flame. AA spectrometers use monochromators and detectors for UV and visible light. The main purpose of the monochromator is to isolate the absorption line from background light due to interferences. Photomultiplier tubes are the most common detectors for AA spectroscopy.

This technique can be used to analyse aqueous samples with negligible sample preparation. It can also be used for rock and mineral samples if they are quantitatively dissolved (typically using acids of various strengths). Flame AA spectroscopy can typically detect concentrations as low as $mg\,l^{-1}$ (ppm), although graphite-furnace AA spectroscopy has been shown to be able to detect concentrations that are orders of magnitude below this.

### Inductively Coupled Plasma Optical Emission Spectroscopy

ICP-OES is short for optical (or atomic) emission spectrometry with inductively coupled plasma. Plasma is a luminous volume of atoms and gas at extremely high temperature in an ionized state. The plasma is formed by argon flowing through a radio frequency field, where it is kept in a state of partial ionization, i.e. the gas consists partly of electrically charged particles. This allows it to reach very high temperatures of up to approximately 10 000°C. At these high temperatures, most elements emit light of characteristic wavelengths, which can be measured and used to determine the concentration of the elements in the solution.

ANALYTICAL METHODS/Geochemical Analysis (Including X-ray) 65

Table 3  Optical spectroscopic techniques commonly used in geochemical analysis

| Technique | Output-1 | Output-2 | Output-3 | Sample type | Advantages | Disadvantages |
|---|---|---|---|---|---|---|
| Atomic absorption spectroscopy (AAS) | Element concentrations in water | Element concentration in quantitatively dissolved rock samples | | Water sample from the natural or altered environment (or solid rocks and minerals dissolved in acid) | Well-established technique; relatively low costs | Relatively high detection limit; relatively slow – one element at a time; limited range of elements |
| Inductively coupled plasma optical emission spectroscopy (ICP-OES) | Element concentrations in water | Element concentration in quantitatively dissolved rock samples | | Water sample from the natural or altered environment (or solid rocks and minerals dissolved in acid) | Many elements analysed simultaneously; relatively fast; linear calibration over wide concentration range; wide concentration range; one calibration suitable for most material types; good for water samples; excellent (ppb) resolution for some trace elements | Expensive equipment; technique not for the novice analyst; carrier gases cause interference with some elements; minerals and rocks must be dissolved prior to analysis; relatively new technique still undergoing development |
| Infrared spectroscopy | Mineral proportions | Water content in quartz | | Clay and other minerals | Ultra-small samples; quantitative; relatively low cost; simple sample preparation | Complex to interpret absorption spectra; adsorbed water can interfere with diagnostic peaks |
| Ultraviolet spectroscopy | Presence of organic inclusions in minerals and rocks | Presence of liquid organics in porous materials (e.g. pollution or oil) | Approximate indication of the density and maturity of petroleum | Organic-bearing rock, sediment, or soil | Simple to make qualitative observations; easily repeated; can be used on bulk and microscopic samples | Some minerals have masking fluorescence; difficult to quantify |

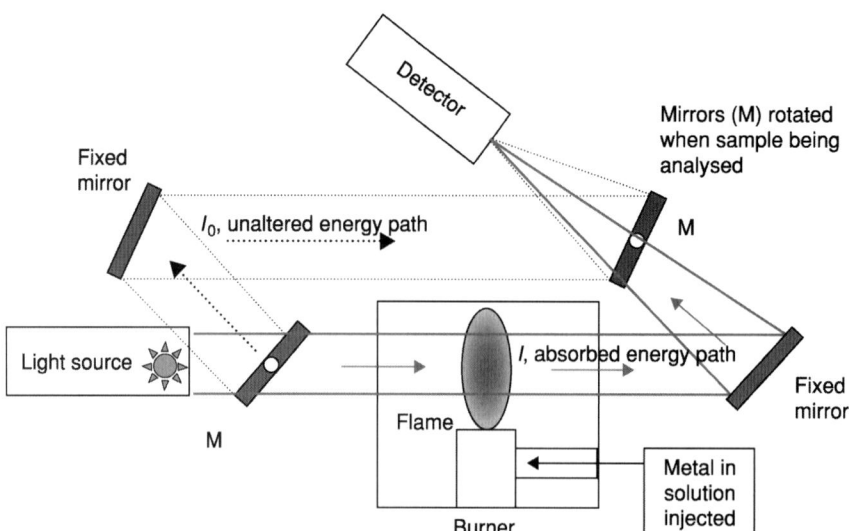

**Figure 14** Atomic absorption spectroscopy equipment. The aqueous sample for analysis is injected into the burner and light is shone through the flame. The mirrors (labelled M) are rotated to measure original and the absorbed light characteristics. The path that is deflected away from the flame (labelled as unaltered energy path) measures the unaffected light intensity. The path that goes through the flame (labelled as absorbed energy path) measures the intensity of the light after it has passed through a burner containing the co-injected dissolved sample.

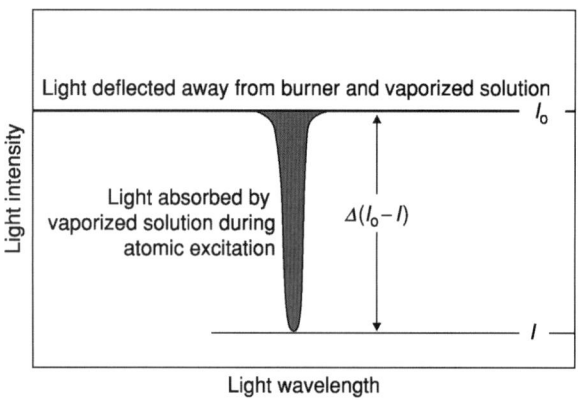

**Figure 15** Light absorption due to atomic excitation with baseline and post-excitation light intensities indicated. The frequency of the absorbed light depends on the element. Concentration is proportional to $I_0/I$, all other things being equal.

The sample being analysed is introduced into the plasma as an aerosol of fine droplets (**Figure 16**). Light from the different elements is separated into different wavelengths by means of a grating and is captured simultaneously by light-sensitive detectors. This permits the simultaneous analysis of up to 40 elements, and ICP-OES is consequently a multi-element technique. In terms of sensitivity, ICP-OES is generally comparable with graphite furnace AA, i.e. detection limits are typically at the $\mu g\, l^{-1}$ level in aqueous solutions.

## Ultraviolet Spectroscopy

Many organic-derived materials fluoresce under ultraviolet (UV) light – that is, they absorb light from the incident ultraviolet beam and release light, often in the visible range, that has a different wavelength from the primary UV light. UV spectroscopy can be used qualitatively to determine whether complex organic molecules are present in sediment or a rock. UV spectroscopy is commonly used to determine the presence of diffuse oil-shows in petroleum-reservoir cores.

There is a loose correspondence between the precise wavelength of the fluorescent light and the nature of the organic material. This relationship has been developed into an analytical technique to determine the thermal maturity of petroleum in rock samples. This technique has reached its apotheosis in its application to petroleum trapped in inclusions in mineral cements and healed fractures. These fluid inclusions have been used to track the petroleum generation of source rocks and the petroleum migration history into reservoirs (*see* **Fluid Inclusions**).

## Infrared Spectroscopy

The electrical bonds between molecules continually vibrate as a function of interaction between neighbouring molecules. These bonds can be excited by infrared radiation, resulting in higher amplitudes of vibration. Only discrete (quantile) increases in vibration energy are possible, and this results in an

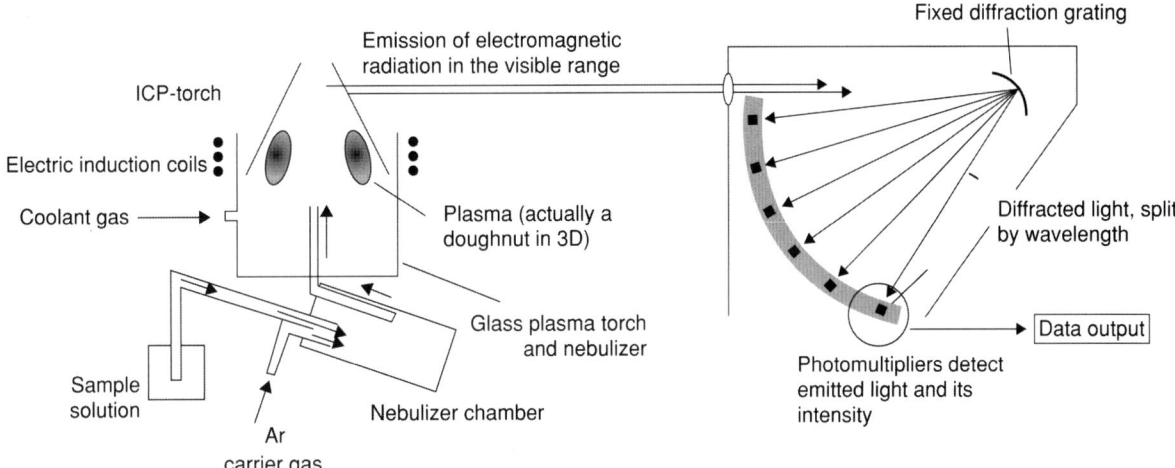

**Figure 16** Essential components of an inductively coupled plasma optical emission spectrometer. The induction coils produce a ring of plasma at temperatures of *ca.* 10 000°C. The water sample, drawn in by the flowing carrier gas, is dispersed and drawn into the plasma. Elements emit light due to thermal excitation. A wide range of light frequencies can be analysed simultaneously using a diffraction grating to disperse the light.

absorption spectrum. The molecular environment is different in all minerals, even in minerals of similar structure (e.g. the clay family), so that infrared spectroscopy can be used to identify minerals in rock and sediment samples. The strength of the absorption spectrum depends on concentration, so the technique can be used for quantitative analysis of minerals under favourable circumstances. Infrared spectroscopy has been used to identify clay minerals in sandstones but is most useful for identifying organic inclusions and non-hydrocarbon gases trapped in fluid inclusions.

## Chromatography

Chromatography has proved invaluable in geochemical analysis (**Table 4**). It is an analytical technique used to quantify and separate mixtures of fluid chemical compounds. There are many different kinds of chromatography, among them gas chromatography, organic liquid chromatography, and ion-exchange chromatography. All chromatographic methods share the same basic principles and mode of operation. In every case, a sample of the mixture to be analysed is applied to some stationary fixed material (the adsorbent or stationary phase) and then a second material (the eluent or mobile phase) is passed through or over the stationary phase. The compounds contained in the sample are then partitioned between the stationary phase and the mobile phase. The success of the approach depends on the fact that different materials adhere to the adsorbent with different forces. Those that adhere more strongly are moved through the adsorbent more slowly as the mobile phase flows over them. Other components of the sample that are less strongly adsorbed on the stationary phase and moved along more quickly by the moving phase. Thus as the mobile phase flows through the column, components in the sample move down the column at different rates and therefore separate from one another. At the end of the column, molecules or ions of the fastest-moving substance (least tightly bound to the stationary phase) emerge first, usually with each compound emerging over a well-defined time interval. A suitable detector analyses the output at the end of the column. Each time molecules or ions of the sample emerge from the chromatography column the detector generates a measurable signal, which is recorded as a peak on the chromatogram. The chromatogram is thus a record of detector output as a function of time and consists of a series of peaks corresponding to the different times at which components of the sample mixture emerge from the column.

By running standards and mixtures of known concentration, it is possible to relate the arrival time to species type and the size of the peak to concentration, making chromatography a valuable quantitative technique. There are three main types of chromatography employed in geochemical laboratories: liquid chromatography, gas chromatography, and ion chromatography. Gas chromatography is used to separate mixtures of gases or vaporized liquids. Ion chromatography is used to separate and analyse ions, typically but not exclusively anions, in aqueous solutions. Liquid chromatography is included in **Table 4** but is mainly used as a sample-preparation procedure prior to gas chromatography to split petroleum, e.g. into

**Table 4** Chromatographic techniques commonly used in geochemical analysis

| Technique | Output-1 | Output-2 | Output-3 | Sample type | Advantages | Disadvantages |
|---|---|---|---|---|---|---|
| Liquid chromatography | Physical separation of different components of complex liquid mixture (petroleum) | | | Whole petroleum or extracted bitumen samples | Excellent pre-separation technique for GC and GC-MS analyses; Gives quantities of groups of petroleum compounds | Limited separation capability |
| Gas chromatography | Physical separation of different components of complex gas-phase mixture (petroleum gas or heated volatilized liquid) | Quantitative measure of proportions of different compounds in gas and liquid petroleum | | Either whole petroleum or separate parts (achieved using liquid chromatography) | Splits gas and liquid range compounds; easily quantified; well-established technology; good for samples with dominant alkanes | Co-elution of different compounds; requires sample preparation; unknown GC-peaks can give ambiguous interpretation |
| Ion chromatography | Physical separation of different charged (aqueous) components in complex natural solutions | Quantitative measure of proportions of different anions in water (common application) | Quantitative measure of proportions of different cations in water (less common application) | Water sample or solid sample quantitatively dissolved in water | Splits a range of anions in water; high resolution; relatively fast and simultaneously analyses all anions; no real alternative | Unsuitable for bicarbonate analysis |

saturated, aromatic, and resin groups, and will not be covered further here.

### Gas Chromatography

Gas chromatography (GC) – specifically gas–liquid chromatography – involves a sample being vaporized and injected onto the head of the chromatographic column (**Table 4**). The sample is transported through the column by the flow of an inert gaseous mobile phase. The column itself contains a liquid stationary phase that is adsorbed onto the surface of an inert solid (**Figure 17**).

The carrier gas is chemically inert (e.g. helium). A sample of gas or petroleum is injected into the column quickly as a slug to prevent peak broadening and loss of resolution. The temperature of the sample port is somewhat higher than the boiling point of the least-volatile component of the sample. Sample sizes typically range from tenths of a microlitre to $20\,\mu l$. The carrier gas enters a mixing chamber, where the sample vaporizes to form a mixture of carrier gas, vaporized solvent, and vaporized solutes. A proportion of this mixture passes onto the chromatography column. Chromatography columns have an internal diameter of a few tenths of a millimetre and walls coated with a liquid but stationary phase. The optimum column temperature depends on the boiling point of the sample; typically a temperature slightly above the average boiling point of the sample results in an elution time of 2–30 min. As the carrier gas containing the chromatographically separated sample passes out of the end of the column, it is passed into one of a number of detectors such as a flame ionization detector, which has high sensitivity, a large linear response range, and low noise. An example of a flame-ionization-detector signal from a whole-petroleum sample injected onto a GC column is given in **Figure 18**.

One of the problems of GC analysis of geochemical samples is that different compounds can have similar elution times, rendering identification and quantification difficult. However, the output stream from a gas chromatograph can be passed into other types of analytical instrument (e.g. a mass spectrometer) for further analysis of the separated compounds over and above simple quantification of compounds with a common elution time. GC is useful for analysing organic compounds and can be used to quantify mixtures if suitable standards have been employed.

### Ion Chromatography

Ion chromatography can be used for both cations and anions. However, it is in the analysis of non-metal ions that the technique has proved most useful mainly because there are no real alternatives for the simultaneous quantitative analysis of these important species in waters or synthetic solutions.

Ion chromatography is used to analyse aqueous samples containing ppm quantities of common anions (such as fluoride, chloride, nitrite, nitrate, and sulphate). Ion chromatography is a form of liquid chromatography that uses ion-exchange resins to separate atomic or molecular ions based on their interaction with the resin (**Figure 19**). Its greatest utility is for the

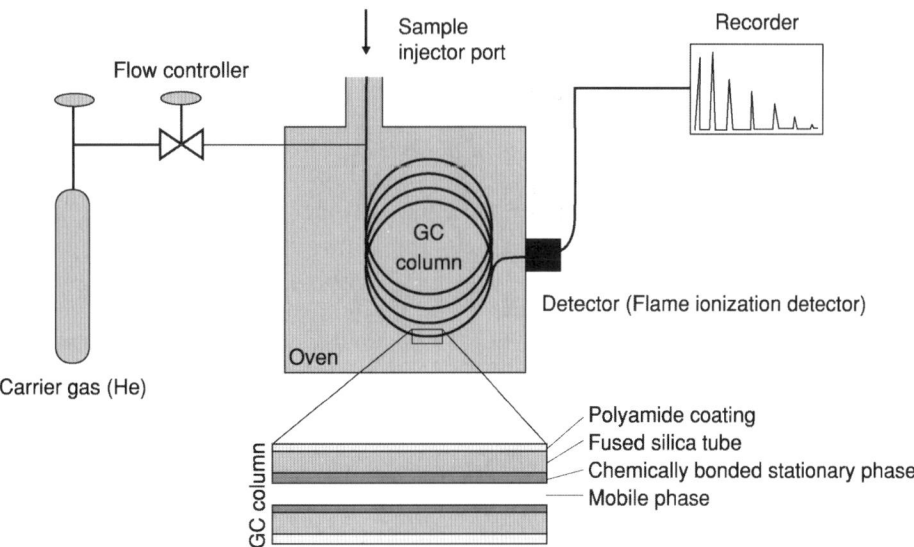

**Figure 17** Essential components of a gas chromatograph. The sample injection port is heated to volatilize liquid-phase organics. The GC column is held in an oven at a temperature above the boiling point of the compounds of interest. The inert carrier gas drives the sample through the capillary with its stationary phase. The stationary phase retards larger molecules more efficiently than small molecules. Smaller molecules thus pass out of the column more rapidly than large molecules.

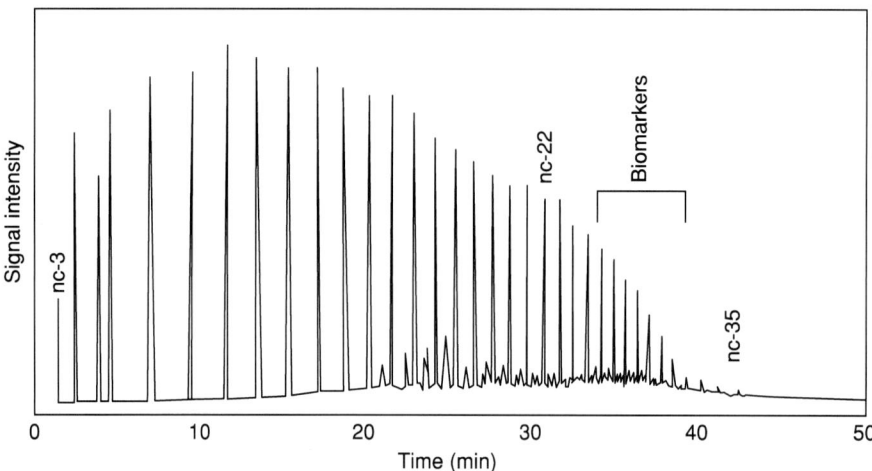

**Figure 18** Typical GC output from a black oil sample. Most of the peaks correspond to normal alkanes. The longest-chain alkane detected is $C_{35}H_{72}$. The biomarkers, molecules of clear biological origin, form an area of low-level noise that is difficult to discern with this technique. On the figure nc-3, nc-22, nc-35 refer to normal alkanes with 3, 22 and 35 carbon atoms. The three labelled peaks are thus from $C_3H_8$, $C_{22}H_{46}$ and $C_{35}H_{72}$.

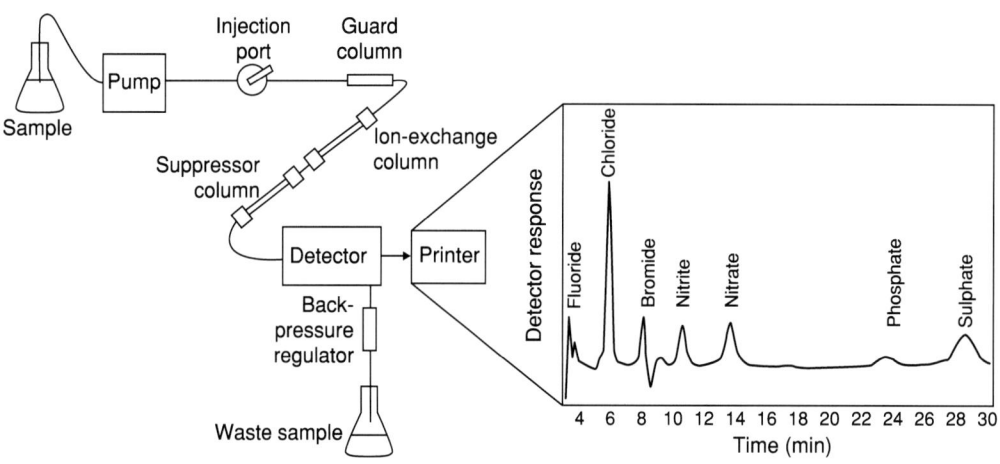

**Figure 19** Essential components of an ion-exchange chromatograph, and a typical output trace for a low-concentration standard. The suppressor column removes the bicarbonate and carbonate that are released from the ion-exchange chromatography column to avoid the real sample signal being swamped. The trace has 0.02 ppm fluoride and 0.1 ppm of all the other anions. Note that the transit time through the column increases with atomic number for the halogens and is greatest for the multivalent anions phosphate and sulphate.

analysis of anions for which there are no other rapid analytical methods. For anion chromatography the mobile phase is a dilute aqueous solution of sodium bicarbonate and sodium carbonate prepared with pure water.

The ion-exchange column is tightly packed with the stationary adsorbent. This adsorbent is usually composed of tiny polymer beads that have positively charged centres. These become coated with bicarbonate and carbonate anions if no sample is passing through the column. As anions in the sample enter the column, they are attracted to the positive centres on the polymer surface and may replace (exchange with) the bicarbonate and carbonate ions stuck to the surface. Usually, the greater the charge on the anion the more strongly it is attracted to the surface of the polymer bead. Also, larger anions generally move more slowly through the column than smaller anions. The result is that the sample separates into bands of different kinds of ion as it travels through the column.

The detector, usually a conductivity cell, measures the conductance of the solution passing through it. The conductance is proportional to the concentration of ions dissolved in the solution. It is essential to pass the sample–mobile-phase mixture through a suppresser column – another ion-exchange column – to remove the bicarbonate and carbonate ions and avoid the sample signal being masked before entering the detector. Anions can be qualitatively identified by analysing standards and standard mixtures. The concentrations of anions are quantified by the usual geochemical technique of calibration (**Figure 1**).

## Mass Spectroscopy

### Principles of Mass Spectroscopy

If a moving charged molecule or atom is subjected to a sideways electromagnetic force, it will move in a curve. The amount of deflection in a given electromagnetic field depends on the velocity and mass of the ion and the number of charges on it. The latter two factors are combined in the mass–charge ratio (given the symbol m/z or sometimes m/e). Most of the ions passing through the mass spectrometer will have a charge of +1, so the mass–charge ratio will be the same as the mass of the ion. Atoms and molecules are ionized by removing one or more electron in an ionization chamber to leave a positive ion. The beam of ions passing through the machine is detected electrically. It is important that the ions produced in the ionization chamber have a free run through the device without hitting air molecules so that equipment is operated under a high vacuum. The vaporized sample passes into the ionization chamber. The electrically heated metal coil gives off electrons that are attracted to the electron trap, which is a positively charged plate. The particles in the sample (atoms or molecules) are therefore bombarded with a stream of electrons, and some of the collisions are energetic enough to knock one or more electrons out of the sample particles to make positive ions. Most of the positive ions formed will carry a charge of +1 because it is much more difficult to remove further electrons from an already positive ion. These positive ions are persuaded out into the rest of the machine by the ion repeller, which is another metal plate carrying a slight positive charge.

When an ion hits the detector, its charge is neutralized by an electron jumping from the metal to the ion. That leaves a space amongst the electrons in the metal, and the electrons in the wire shuffle along to fill it. A flow of electrons in the wire is detected as an electric current, which can be amplified and recorded.

If the magnetic field is varied, the ion stream can be deflected on to the detector to produce a current that is proportional to the number of ions arriving. The mass of the ion being detected is related to the size of the magnetic field used to bring it to the detector. The

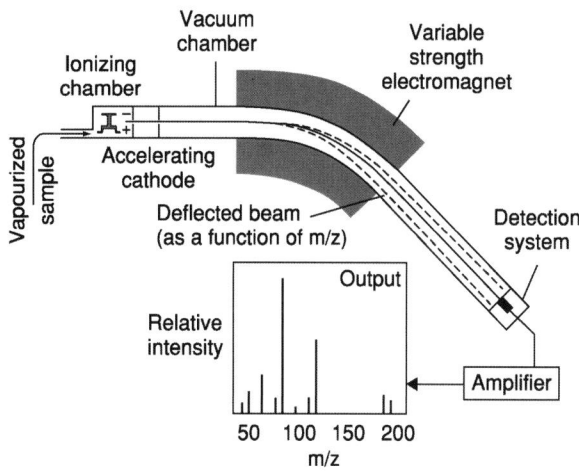

**Figure 20** Essential components of a mass spectrometer. A source (from thermal ionization, inductively coupled plasma, a GC column, etc) feeds vaporized sample into an ionization chamber. The vapour is ionized to produce positively charged ions. These are drawn into the mass spectrometer via an accelerating electrode. The ions are passed into an electromagnet, where they are deflected as a function of the mass–charge (m/z) ratio, their velocity, and the strength of the magnetic field. The electromagnets have their field strength varied to cause variable deflection. The signal strength is recorded as a function of magnetic-field strength, and the signal output is given in terms of the relative intensity of the peak versus m/z.

device can be calibrated to record current (which is a measure of the number of ions) directly against m/z. The output from the chart recorder is usually simplified into a 'stick diagram'. This shows the relative current produced by ions of varying mass–charge ratio (**Figure 20**). There are a wide range of mass-spectroscopy techniques; three have been summarized in **Table 5**.

### Thermal-Ionization Isotope-Ratio Mass Spectroscopy

The technique of isotope dilution is being used increasingly to improve precision and accuracy by reducing the problems of calibration and sample-preparation effects. Some materials for analysis and some analytical methods result in a large degree of uncertainty. Variability caused by such problems is usually partly compensated for or monitored by using internal standards and surrogate samples. An isotope-dilution standard is the 'perfect' internal standard or surrogate.

An internal standard or surrogate is a compound similar to the sample of interest. An isotope-dilution standard is an isotope of an element or a molecular compound labelled with an isotope. A good example of this is $^{204}$Pb, a minor isotope of lead. The natural stable lead isotopes are 204 (1.4%), 206 (24.1%), 207 (22.1%), and 208 (52.4%). By adding a known

**Table 5** Mass spectrometry techniques commonly used in geochemical analysis

| Technique | Output-1 | Sample type | Advantages | Disadvantages |
|---|---|---|---|---|
| Thermal ionization mass spectroscopy | Concentrations of individual elements and isotopes, typically heavier elements | Solid salt of the element in question deposited on a filament | Relatively simple; gives isotope ratios; can be quantitative if sample diluted with known concentration of isotope; high temperature of filament causes ionization directly | Only for limited range of elements; only for solid samples |
| Inductively coupled plasma mass spectroscopy | Concentrations of trace and minor elements | Water and quantitatively dissolved solids | High analytical resolution; small sample sizes; wide range of elements; rapid analysis; no matrix effects due to sample dissolution | Expensive equipment; difficult technique requiring expert operator; solid samples require quantitative dissolution |
| Gas chromatography mass spectroscopy | Concentrations of trace organic compounds | Volatile or dissolved organic compounds | High analytical resolution; easy to quantify; large range of compounds can be determined from one sample | Problems with *ab initio* determination of unknown compounds; expensive and tricky technique |

amount of $^{204}$Pb to a sample before testing for total lead and by testing for each of the lead isotopes, it is possible to determine accurately total lead and individual lead-isotope ratios. This approach leads to much smaller errors than simply calibrating signal strength for individual m/z values.

### Gas Chromatography Mass Spectroscopy

One of the main problems with analysis of GC column output is the co-elution (same rate of passage) of groups of compounds. This problem has been tackled by using a mass spectrometer as the detection system (rather than, for example, a flame ionization detector). Placed at the end of a chromatographic column in a similar manner to other GC detectors, a mass spectrometer detector is more complicated than other GC detector systems (**Figure 21**). A capillary column must be used in the chromatograph because the entire mass spectroscopy process must be carried out at very low pressures and in order to meet this requirement a vacuum is maintained via constant pumping using a vacuum pump.

The major components over and above the GC column are an ionization source, a mass separator, and an ion detector. There are two common mass analysers or separators commercially available for GC-MS: the quadrapole and the ion trap.

The power of this technique lies in its ability to produce mass spectra from each time-controlled GC peak (**Figure 18**). Thus, for each GC peak, coeluted molecules are ionized and separated into m/z fractions in the mass spectrometer. Complex organic molecules tend to fragment in predictable ways in the ion source, so a group of related molecules can be traced using the same m/z fraction for the different time-controlled GC fractions. A time plot of the same ionized molecular fragment can then be reconstructed from the individual intensities of the mass spectra. The data can be used to determine the identity and quantity of an unknown chromatographic component with an assuredness that is simply unavailable by other techniques. This approach can be quantified if the equipment is calibrated or if the sample is mixed with a known quantity of a standard. GC-MS analysis allows the quantification of trace organic components including biomarkers, thermally controlled optical isomers (e.g. steranes), and geological age-dependent molecules (e.g. olearane, derived from post-Cretaceous flowering plants).

### Inductively Coupled Plasma Mass Spectroscopy

Inductively coupled plasma mass spectroscopy (ICP-MS) uses plasma of the same type as in ICP-OES (**Figure 16**), but here it is used to convert elements to ions that are then separated by mass in a mass spectrometer. This allows the different elements in a sample (and their natural isotopes) to be separated and their concentrations determined. The core of the ICP-MS system is the interface through which ions from the inductively coupled plasma source enter the high-vacuum chamber of the mass spectrometer.

ICP-MS combines the advantages of inductively coupled plasma (simple and rapid sample handling) and mass spectrometry (high sensitivity, isotope measurement) in a multielement technique. Detection limits can be much lower than in ICP-AES: certain

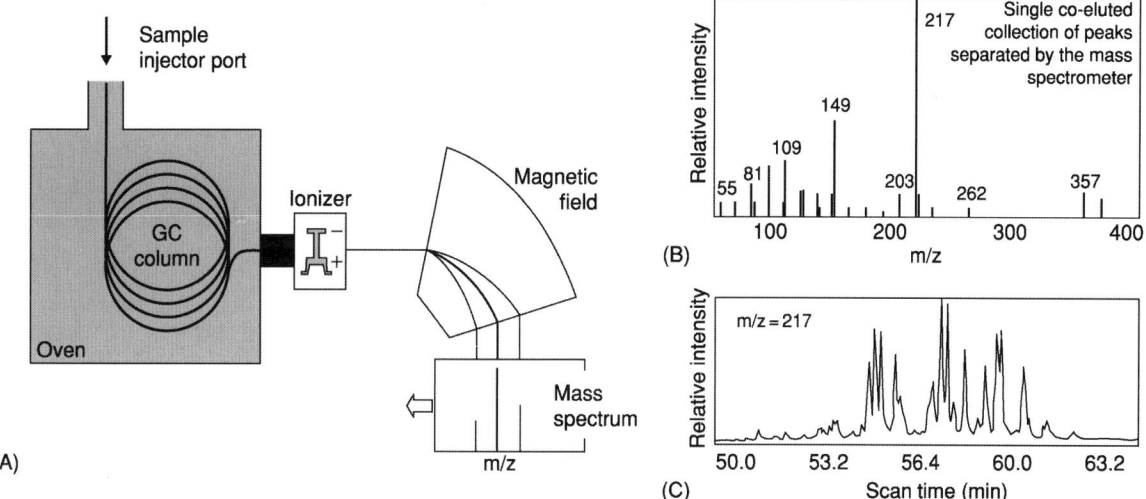

**Figure 21** Essential components of a gas chromatography mass spectrometer and its output. The sample injection port is heated to volatilize liquid-phase organics. The capillary GC column is held in an oven at a temperature above the boiling point of the compounds of interest. The GC column separates the sample into groups of molecules according to their elution rates. The GC-separated sample is fed into the ionization chamber, where the organic molecules are fragmented, drawn into the mass spectrometer, and separated into different m/z fragments. (B) Each peak on a GC trace (**Figure 18**) has its own mass spectrum, permitting high-resolution analysis of trace organic molecules. (C) The individual m/z fragments (e.g. 217) can be reconstructed as a plot of fragment concentration versus time. The high resolution of this technique permits the resolution of the biomarkers shown on the GC trace in **Figure 18**.

elements can be detected at the $ng\,l^{-1}$ (parts per trillion) level in aqueous solutions.

## Pyrolysis and Other Heating Techniques)

It is possible to characterize materials by heating them and either by studying the resulting change in optical and physical properties or by analysing the fluid evolved. This approach has been applied to organic materials, minerals, rocks, and fluid trapped as inclusions within mineral grains (**Table 6**).

### Thermogravimetry and Evolved Water Analysis

If a small sample of rock is subjected to a controlled heating cycle and simultaneously weighed, the temperature at which volatiles are driven off from the sample can be accurately monitored. If the volatiles are carried to a detector using an inert gas (e.g. nitrogen), then it is also possible to analyse the evolved gases using an infrared water-vapour analyser to determine independently the exact quantity of water driven off at each stage in the heating cycle. The first technique is known as thermogravimetry; the second technique is known as evolved water analysis. Carbonate minerals also undergo volatile loss during heating, so it is important to pair the weight loss with the identification of the mineral undergoing volatile loss. The combined approach allows accurate assessment of volatile loss from clay minerals. These techniques can be used to determine the quantity of clay minerals in a rock under ideal conditions the exact types and quantities of clays can be determined since different clay minerals dehydrate at different temperatures.

### Pyrolysis

Pyrolysis has been used to study organic material for a number of purposes. Heating organic matter is used to study the resulting total quantity of evolved $CO_2$, $SO_2$ etc. (when done in an oxygen atmosphere), which can be used to help determine the elemental composition of the organic matter. The elements in organic matter of all sorts, determined in this way, include carbon, hydrogen, sulphur, and nitrogen. The resulting evolved gas phases can be split using a GC column (see above) or analysed using various optical (e.g. infrared) techniques specific to the expected gas products. The output from this approach can be the total organic carbon content (e.g. of a rock or sediment) or the elemental analysis of pre-separated samples (**Figure 22**).

Another approach is to heat solid samples (e.g. of organic-rich sediment, coal, separated kerogen, or asphaltene exsolved from petroleum) in an oxygen-free environment and analyse the resulting fluid-phase products during a programmed heating cycle (known as rock eval pyrolysis). During heating the

**Table 6** Pyrolysis and other thermal techniques commonly used in geochemical analysis

| Technique | Output-1 | Output-2 | Output-3 | Sample type | Advantages | Disadvantages |
|---|---|---|---|---|---|---|
| Thermogravimetry–evolved water analysis | Estimate of total clay mineral content of rocks | | | Small samples of rocks, sediments, or minerals | Sub percentage level resolution; useful ally to XRD analysis; rapid | Difficult to resolve individual clay minerals in clastic rocks; difficult to repeat; destroys sample |
| Pyrolysis | Total organic carbon (on decarbonated samples) | Quantities of existing petroleum, petroleum-generation potential, $CO_2$-generating potential (rock-eval) | General character of petroleum that would be generated by organic matter with further heating (pyrolysis-GC) | Organic-bearing sediment or rock, solid asphaltene exolved from petroleum, bitumen | Rapid; approach allied to existing technology; quantitative; can reveal kinetic data about source rocks | Destroys sample; geologically unrealistic rates of heating (for rock-eval); results sensitive to geological history of sample |
| Fluid-inclusion microthermometry | Salinity of water trapped in minerals | Growth temperature of mineral | | Polished wafers of rocks and minerals | High level of precision; easily repeated; high spatial resolution; reveals geological evolution of samples | Difficult sample preparation; requires assumption that salinity is due to NaCl; requires assumption that vapour-saturated fluid was trapped |

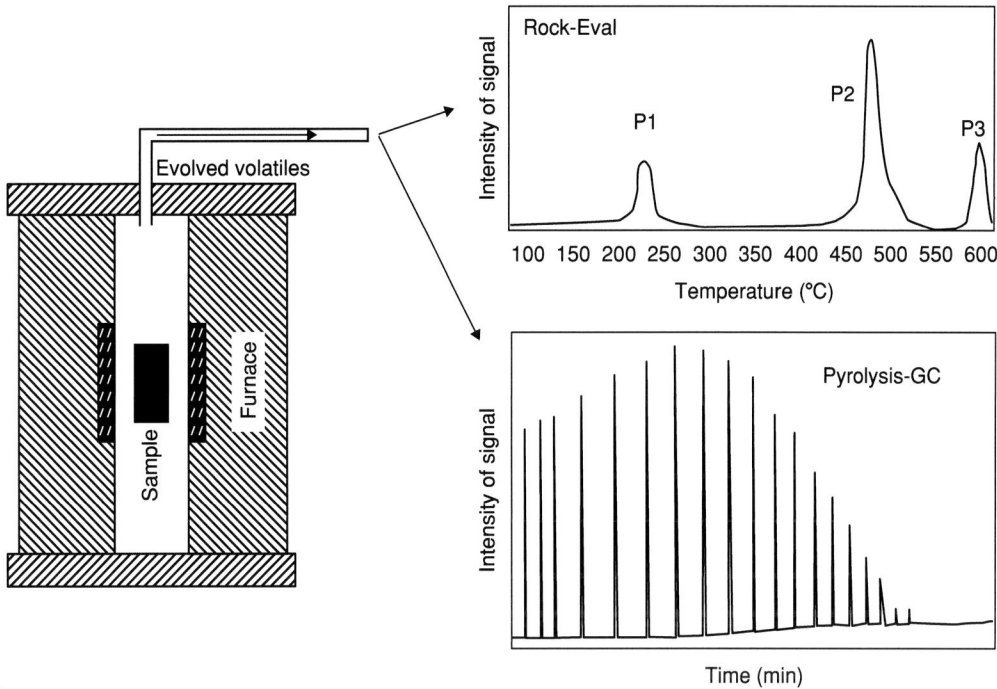

**Figure 22** Essential components of pyrolysis equipment, with either the volatiles sequentially emitted during a programmed heating cycle or GC analysis of the pyrolysate. P1 represents evaporated pre-existing petroleum. P2 represents the petroleum generated from kerogen. P3 represents the generation of oxygen-bearing species ($CO_2$). $T_{max}$ is the peak temperature of the P2 trace. The pyrolysis-GC trace is typically used to determine the gas (short alkanes, small elution time) versus oil (longer alkanes, longer elution time) generation capability of immature kerogen.

free hydrocarbons contained in the sample are driven off first (the P1 peak) followed by experimentally generated petroleum (the P2 peak) and then oxygen-containing compounds ($CO_2$; the P3 peak). Detection in simple instruments is performed using a flame ionization detector (for free and laboratory-generated hydrocarbons) and a thermal conductivity detector for carbon dioxide. Three peaks are thus usually produced and quantified during heating and analysis. These peaks are expressed in terms of $mg\,g^{-1}$. P1 indicates how much petroleum exists in an organic-rich rock now. P2 indicates the petroleum-generating capability of the rock and is used to calculate the hydrogen index of the source rock. P3 is an indication of the amount of oxygen in the kerogen and is used to calculate the oxygen index. The temperature at which the maximum release of hydrocarbons from cracking of kerogen occurs during pyrolysis (P2 peak) is termed $T_{max}$ and is an indication of the stage of maturation of the organic matter. The hydrogen index ($HI = (100 \times P2)$/total organic carbon) and the oxygen index ($OI = (100 \times P3)$/total organic carbon) correlate with the ratios of hydrogen to carbon and oxygen to carbon, respectively. These parameters have been usefully employed to study the type and thermal evolution of sedimentary organic matter.

The products of pyrolysis in the absence of oxygen can also be passed into a GC column to study their composition. This is known as pyrolysis-GC and can usefully simulate the type of petroleum that would be expected from the organic-rich source rock during thermal evolution.

### Fluid Inclusion Microthermometry

Small samples of the fluid from which a mineral grew are commonly trapped in inclusions. When minerals are fractured, they sometimes re-heal, trapping the ambient fluid. Petroleum, which is immiscible with the aqueous mineral growth medium, can also get trapped in these inclusions. These fluid inclusions have proved to be very valuable to geochemists since they provide a snapshot of fluid evolution and rock geohistory. The fluid itself can be analysed if it is released by crushing. This approach is especially useful for petroleum inclusions. Analysis is by GC, GC-MS, or simply mass spectroscopy. UV spectroscopy can also be used to analyse petroleum trapped in inclusions to help reveal the broad characteristics of the petroleum.

Aqueous inclusions can be analysed either by freezing the sample and using electron micoprobe (secondary X-ray) analysis, or by crushing the sample and collecting the fluid for conventional water analysis (inductively coupled plasma techniques, ion chromatography). Fluid inclusions are listed here since they are most commonly analysed by using a high powered microscope and heating–cooling stage. Most fluid inclusions are composed of discrete liquid and vapour phases even though they would have been trapped as a single-phase liquid, which is typically assumed to have been saturated with vapour at the time of trapping. During a heating cycle, the two phases homogenize; the precise temperature of homogenization reveals the temperature at which the mineral grew. The salinity of the water can be assessed by monitoring the temperature at which it starts freezing, since this temperature can be related to salt content (assuming that the water is dominated by dissolved NaCl). A combination of thorough petrography and thermometric studies of aqueous inclusions can help to reveal details of the thermal and mineral-growth history as well as the fluid evolution history.

## Related Geochemical Techniques

The techniques listed and briefly discussed here are only some of the vast panoply of techniques that have been employed during geochemical studies over the last 50 years or so. Some have now fallen out of favour. For example, a technique called neutron activation analysis was used for a long time to measure trace elements in solids. It has fallen out of favour mainly owing to developments in ICP-OES and ICP-MS.

The vast range of light and electron optical techniques are routinely used in conjunction with a wide range of solid-state and even organic geochemical studies. Scanning electron microscopy has recently been extensively developed and can now give fabric, mineralogy, mineral chemistry, and high-resolution crystallographic information. Transmission electron microscopy can provide ultra-high spatial resolution (of the order of tens of nm) geochemical data (using secondary X-rays) as well as fabric and crystallographic data.

A wide range of wet geochemical techniques have been employed routinely in studies of natural waters from all near-surface and surface environments. Titration, electrochemical techniques, and colorimetry are essential techniques that are used routinely in many geochemical studies.

## See Also

**Clay Minerals**. **Fluid Inclusions**. **Minerals:** Definition and Classification; Native Elements. **Petroleum Geology:** Chemical and Physical Properties; The Petroleum System. **Rocks and Their Classification**.

## Further Reading

Emery D and Robinson AC (1993) *Inorganic Geochemistry: Applications to Petroleum Geology*. Oxford: Blackwells.

Farmer VC (1974) *The Infrared Spectra of Minerals*. London: Mineralogical Society.

Faure G (1986) *Principle of Isotope Geology*. New York: John Wiley and Sons.

Goldstein RH and Reynolds TJ (1994) *Systems of Fluid Inclusions*. Tulsa: Society of Sedimentary Geology.

Hagemann HW and Hollerbach A (1986) The fluorescence behaviour of crude oils with respect to their thermal maturation and degradation. *Organic Geochemistry* 10: 473–480.

Jarvis I and Jarvis KE (1992) Plasma spectrometry in the Earth sciences: techniques, applications and future trends. *Chemical Geology* 95: 1–33.

Jenkins R (1999) *X-ray Fluorescence Spectrometry*. New York: Wiley Interscience.

Lico MS, Kharaka YK, Carothers WM, and Wright VA (1982) *Methods for the Collection and Analysis of Geopressured Geothermal and Oil Field Waters*. Water-Supply Paper 2194. United States Geological Survey.

Moore DM and Reynolds RC (1997) *X-ray Diffraction and the Identification and Analysis of Clay Minerals*. Oxford: Oxford University Press.

Rollinson HR (1993) *Using Geochemical Data: Evaluation, Presentation, Interpretation*. New York: Longman Scientific and Technical.

Tissot B and Welte D (1984) *Petroleum Formation and Occurrence*. Berlin: Springer Verlag.

Tucker ME (1988) *Techniques in Sedimentology*. Oxford: Blackwell Scientific.

Weiss J (2000) *Ion Chromatography*. New York: John Wiley and Sons.

Zussman J (1967) *Physical Methods in Determinative Mineralogy*. London: Academic Press.

# Geochronological Techniques

**E A Eide**, Geological Survey of Norway, Trondheim, Norway

© 2005, Elsevier Ltd. All Rights Reserved.

## Introduction

Geochronology is the study of time as it relates to Earth history. As a distinct discipline within the natural sciences, geochronology emerged fully during the late nineteenth and early twentieth centuries with the discovery of radioactivity and the advent of radiometric dating methods. Importantly, the appearance of modern geochronology was the result of a strong interest in Earth history and the development of relative methods to estimate the age of Earth, both of which had been aspects of natural science research since at least the seventeenth century.

The human fascination with studying time and marking its passage can be traced to ancient cultures, exemplified through the precise astronomical calendars produced by numerous early civilizations (**Figure 1**). These calendars were based on calculations of the movements of celestial bodies relative to Earth and helped to raise speculations about the position and motion of Earth within this celestial system. These speculations led to efforts to understand Earth's origin and calculate its age, which today is generally agreed to be 4.5–4.6 billion years (By), and is the starting point for the geological time-scale (GTS) (**Figure 2**). The GTS is an iterative solution between 'absolute' and 'relative' ages determined by absolute and relative geochronological techniques. The formal distinction between absolute and relative ages has its roots in ancient calendars for which the passage of time was calculated from astronomical events linked to the solar year. Broadly, an absolute age is one that is based on processes affected only by the passage of time and which may thus be valid worldwide. In a strict sense, an absolute age should have direct correspondence to the absolute time-scale, determined on the basis of the solar year (**Table 1**). Relative ages are applicable to a restricted geographic area and usually pertain to a limited geological time period. Relative ages place the formation of different rock units or their physical features (faults, unconformities, etc.) in a relative chronological order. Though knowledge of the exact formation ages of different rock units is useful, numerical (absolute) ages are not prerequisite for establishing their relative chronology. Nonetheless, relative ages must eventually be calibrated against independently established (absolute) time-scales if they are to be extrapolated globally.

The framework for the GTS is based on relative ages, represented by the established, sequential subdivisions of geological time (**Figure 2**). The nomenclature of this framework was developed largely through the studies of natural scientists in the eighteenth and nineteenth centuries (*see* **Famous Geologists:** Sedgwick; Murchison; Darwin; Smith; Cuvier; Hutton). During the twentieth century, absolute age determinations for rocks around the globe allowed refinement of the GTS and adjustments were made to the initially imprecise or disputed boundaries between the geological systems. The absolute ages were derived using radiogenic isotope geochronological techniques. Calculating an age for a rock or mineral using these techniques combines precise measurement of naturally occurring, radioactive isotopes and their stable decay products with the physical principle that the radioactive decay of the isotopes occurred at a constant, known rate. Because radiogenic ages are 'absolute' in the sense that the decay of a radioactive isotope primarily depends only on the passage of time, radiogenic ages for rocks found in one area of the world should, in principle, be directly comparable 'in time' to other rocks dated with similar methods in other areas of the world. Regardless of the geochronological technique used, the combination of relative and absolute ages has yielded the opportunity not only to generate geological time-scales, but also to determine the ages of rocks and geological structures, the timing of geological 'events', and, importantly, the rates at which geological processes occur.

Today, the primary techniques for relative dating of geological materials include biostratigraphy, palaeomagnetism and magnetostratigraphy, and chemostratigraphy (*see* **Palaeomagnetism, Magnetostratigraphy, Analytical Methods:** Fission Track Analysis). Of the absolute dating methods, radiogenic isotope geochronology, astronomical time calibrations, and dendrochronology (*see* **Dendrochronology**) are the most widely used. However, it is the rock type that usually dictates the geochronological technique appropriate for obtaining the rock's age. Thus, basic knowledge of the relative and absolute geochronological techniques is useful not only to select the appropriate method to date the rock, but also to interpret the age(s) produced, and to give a higher degree of confidence to comparisons made between geological ages and the processes to which they are linked.

# ANALYTICAL METHODS/Geochronological Techniques

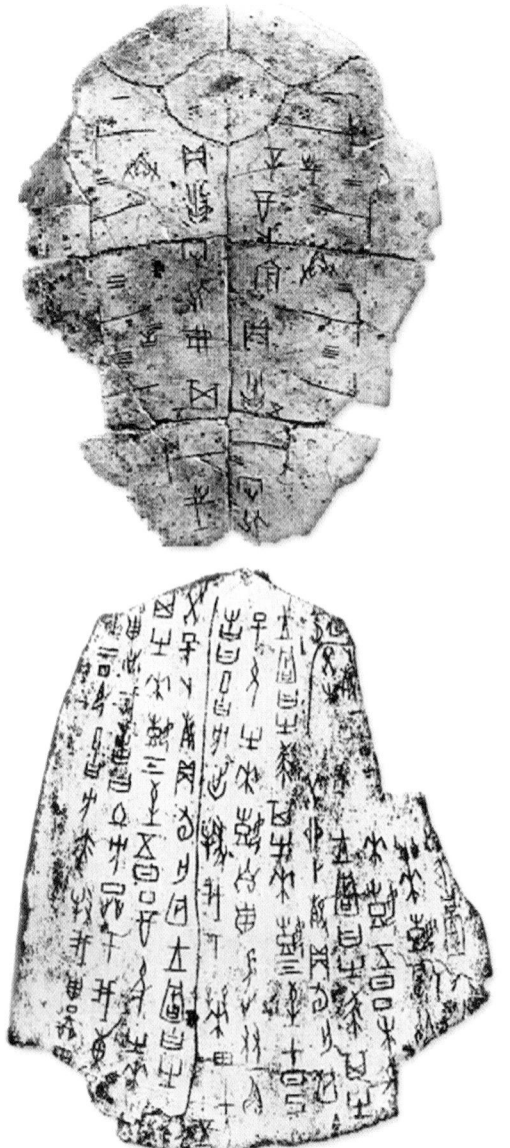

**Figure 1** Shang oracle bones. Precise calendars developed by early civilizations were based on calculations combining the rotation of Earth on its axis (day), Earth's revolution about the Sun (year), and the Moon's revolution about Earth (month). These individual astronomical cycles are neither constant nor synchronous with one another, and ancient peoples had to determine the appropriate lengths for days, months, and years so as to allow the seasonal cycles of the sun to coincide with the monthly cycles of the moon. The two oracle bones from the Shang Dynasty, dating back to the fourteenth century BCE in China, show that the Chinese had established the solar year at $365\frac{1}{4}$ days and a lunar cycle at $29\frac{1}{2}$ days; in this way, they had recognized and accounted for the differences in astronomical cycles between the Sun, Moon, and Earth in a consistent manner. The causes for these shifts in the astronomical cycles are now known to be primarily the gravitational forces acting between the different celestial bodies (see **Figure 6** for modern use of astronomical calibrations). Figure used with permission from M Douma. (see http://webexhibits.org for additional information on calendars).

## Historical Perspective

### Relative Ages

The conceptual background for modern geochronological techniques was established largely through the investigation of sedimentary rocks and the development of relative geochronology tools to study them. Studies of fossils and of the depositional order of sedimentary layers led to the principles of fossil succession and correlation and of superposition and relative chronology. These principles, accepted today as basic aspects in geoscience training, were crucial to the development of all relative geochronological techniques and can be attributed to the work of Steno in the seventeenth century, and of Smith, Cuvier, Brongniart, Lehmann, Füchsel, Pallas, and Arduino in the eighteenth century (see **Famous Geologists:** Smith; Cuvier).

Studies in the late eighteenth and early nineteenth centuries by Hutton, Lyell, Darwin, Murchison, and Sedgwick, among others, built on these early observations and the concept of relative geochronology. Murchison and Sedgwick named the Cambrian and Silurian systems in western Wales based on the systematic definition of sedimentary rock units using distinctive fossils; this procedure was fundamental for establishing all of the system subdivisions in the GTS. Hutton, Lyell, and Darwin each promoted the idea that Earth was very old. Darwin, in his first edition of the *Origin of Species*, estimated that about 300 million years (My) had passed since the end of the Mesozoic. Though today we know this estimate to be too high, Darwin's suggestion of this order of magnitude for the passage of geological time was important for carrying forward concepts such as evolution, and also for encouraging efforts to calculate absolute ages of rocks and of Earth (see **Famous Geologists:** Hutton; Lyell; Darwin; Murchison; Sedgwick).

### Absolute Ages

In the latter half of the nineteenth century, calculations of Earth's age incorporated physical measurements in the field and laboratory and were, in this sense, quantitative; however, the methodologies initially employed were based on flawed assumptions that precluded their yielding accurate ages. Examples of these early attempts included calculating the rate of salinity increase over time for the world's oceans, and determining the age of the oldest sediment on Earth by estimating the total thickness and deposition rates for the sedimentary rock record. The salinity method assumed (incorrectly) that the world's oceans had initially been fresh and that no net exchange of sodium had occurred between seawater and rock.

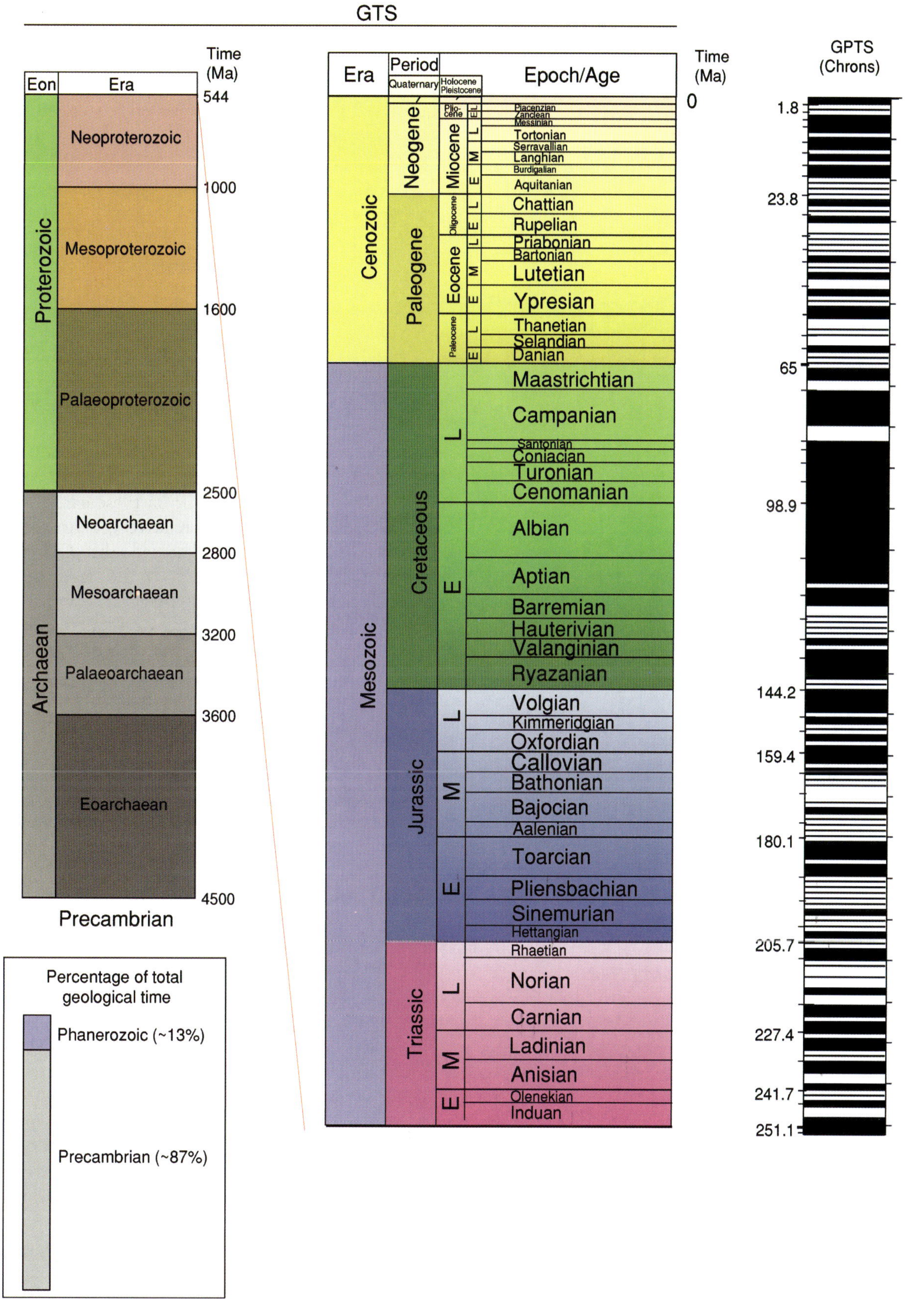

ANALYTICAL METHODS/Geochronological Techniques 79

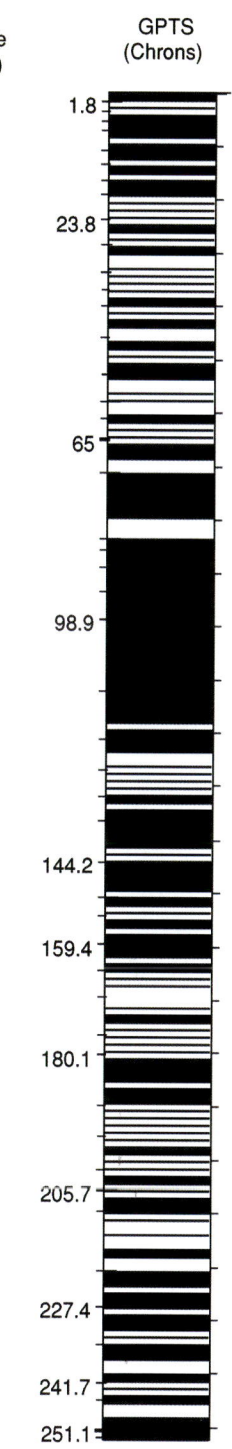

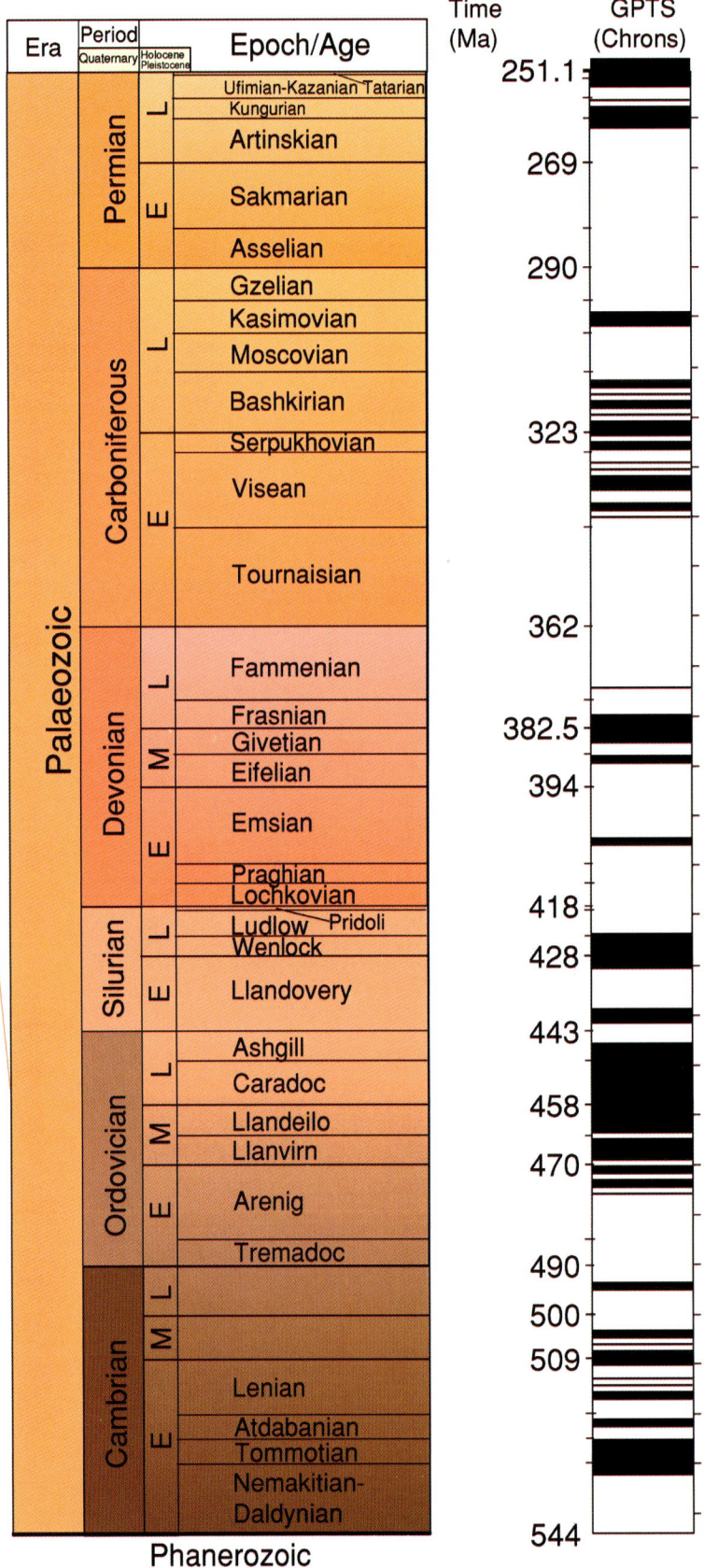

The calculations using the sedimentary rock record and deposition rates were inhibited by items such as missing sections and different rates of sedimentary deposition around the globe and throughout time. Both the salinity and sedimentation rate calculations yielded very low estimates for Earth's age (**Table 2**).

The first truly quantitative and influential effort to calculate an absolute age for Earth was made by the renowned physicist William Thomson (Lord Kelvin) during the middle to late nineteenth century. His concept was based on the idea that Earth had cooled from an originally molten state and was continuously losing heat from its surface through this cooling process. He made calculations for the length of time this process should have taken based on physical measurements of the rates of heat flow through a cooling body and of radiation of heat from the body's surface. Kelvin's calculations involved measurement of physical processes that were dependent only on the passage of time, so his conclusion that Earth was 20–40 My old fell technically within the realm of 'absolute' age determination and was widely accepted in the science community. This young age for Earth was at odds with the concepts put forward by Lyell and Darwin and produced an intense debate between Kelvin and promoters of evolution theory. However, a fundamental feature was missing from Kelvin's calculations – that of radioactive heat generation within Earth. At the time of Kelvin's initial calculations, radioactivity had not yet been discovered, so his equations greatly underestimated the amount of continuous heat generation within the crust and resulted in large underestimates of Earth's age (**Table 2**).

Henri Becquerel's discovery of radioactivity in 1896 launched the development of modern, radiogenic geochronological techniques. Radioactivity accounted for constant heat production from Earth's crust, as well as the production of heat from the sun, and eroded the premises of Kelvin's calculation. Soon after Becquerel had discovered that uranium (U) was radioactive, the radioactive properties of the elements radium, thorium, rubidium, and potassium (Ra, Th, Rb, and K) were also identified. The production of the isotopes helium (He), Th, and lead (Pb)

**Table 1** Major time periods and definitions used for astronomical calendars[a]

| Term | Definition | Comment |
| --- | --- | --- |
| Solar or tropical year | Equal to 365.24219 days; the mean interval between two successive vernal equinoxes | The interval from one vernal equinox to the next may vary from this mean value by several minutes; this is because Earth's position in its orbit shifts slightly at the time of the equinoxes every year |
| Sidereal year | Equal to 365.25636 days; the time for Earth to make one revolution around the Sun, measured according to consecutive observations from Earth of the positions of stars | The precession of the equinoxes causes the sidereal year to be slightly variable and longer than the tropical year |
| Lunar or synodic month | Equal to 29.5305889 days; the mean period of time between new moons (or between exact conjunctions of the Sun and Moon); the lunar year contains 12 lunar months and is equal to 354.3671 days | The synchronization of calendar months with the lunar phases requires a combined sequence of months of 29 and 30 days in length; alternatively, as in **Figure 1**, the length of a month in days can be designated to be a non-integer |

[a]Time reference frames for astronomical calendars show the difficulties faced by early civilizations as they attempted to synchronize the movements of celestial bodies in a consistent calendar for measuring the passage of time. The cycles of the Moon and Sun relative to Earth change slowly with time, and a calendar year with an integral number of days cannot be perfectly synchronized to any of the astronomical reference frames. The astronomical formulas developed in the twentieth century to describe the changes in the orbital cycles of these celestial bodies yield the best approximations available for the length of any type of year (solar, sidereal, or synodic); however, the solutions to these formulas are descriptions of a constantly changing system and cannot be considered exact solutions. Thus the term 'absolute' age, in practice, when referring to astronomically calibrated time-scales, is not strictly correct. A rather more general definition of absolute age is used herein.

**Figure 2** The geological time-scale (GTS; two coloured columns) and geomagnetic polarity time-scale (GPTS; column with alternating black and white pattern) are often used together in geochronological studies. On the left side of the GPTS, the linear time-scale hachures correspond to the epoch/age boundaries in the GTS; on the right side of the GPTS, the linear time-scale hachures are placed at 10-My intervals. Note that different linear scales are used for denoting the Phanerozoic and Precambrian divisions. The true scale relationship between Precambrian and Phanerozoic times as a percentage of total geological time is shown on the lower left. Reproduced with permission from Eide EA (2002) Introduction – plate reconstructions and integrated datasets. In: Eide EA (coord.) *BATLAS – Mid Norway Plate Reconstruction Atlas with Global and Atlantic Perspectives*, pp. 8–17. Trondheim: Geological Survey of Norway.

**Table 2** Selected historical review of estimates for the age of Earth[a]

| Age of Earth (million years) | Method | Year/author |
|---|---|---|
| ≃1973 | Hindu chronology | ca. 120–150 BCE/priests |
| >300 | Time for natural selection | 1859/Darwin |
| 100 | Sediment thickness/deposition rate | 1869/Huxley |
| <100 | Cooling of Earth | 1871/Kelvin |
| 90 | Sediment thickness/deposition rate | 1890/de Lapparent |
| 20–40 | Cooling of Earth | 1897/Kelvin |
| 90 | Salinity accumulation | 1899/Joly |
| >1640 | U–Pb age of a Precambrian rock | 1907/Boltwood |
| 80 | Sediment thickness/deposition rate | 1908/Joly |
| >1300 | Cooling of Earth | 1917/Holmes |
| 1600–3000 | Decay of U to Pb in crust | 1927/Holmes |
| 3350 | Terrestrial Pb isotope evolution | 1947/Holmes |
| 4000–5000 | Radioactive isotope abundances | 1949/Suess |
| 4500 ± 300 | Terrestrial Pb isotope evolution | 1953/Houtermans |
| 4540 | Terrestrial Pb isotope evolution | 1981/Tera |

[a]In addition to these estimates, Jewish and Christian Biblical scholars from the second through seventeenth centuries suggested that the age of Earth ranged between –5000 and 7500 years, based on Julian, Gregorian, or Hebrew calendars. Some of the most well-known sources for these age estimates include James Ussher, John Lightfoot, and St. Augustine. Regardless of the source, most ages of Earth published prior to the twentieth century were greatly underestimated. Research on the decay rates and processes for radioactive elements in Earth's crust finally led to more accurate calculations for Earth's age by the middle the 1900s. These calculations were based on the reconstruction of terrestrial Pb isotopic compositions from a primordial Pb reservoir, of composition similar to meteorites. The meteorite reference for these calculations has been the Canyon Diablo troilite.

from the radioactive decay of U was discovered at the start of the twentieth century by physicists Rutherford, Soddy, Strutt, Thomson, and Boltwood. Boltwood measured Pb–U ratios in unaltered minerals using a very rough estimate of the rate for the radioactive decay of U to Pb; he noted that the older the mineral, the greater the ratio (greater amount of the decay product, Pb). Rutherford applied the decay of U to He in a similar way to attempt to obtain ages for rock samples. At this important watershed for geochronological techniques, the realms of physics and geology became linked in a quantitative tool for measuring geological time. Through the first half of the twentieth century, great advances were made in understanding and applying radiogenic isotope geochronology to determine the ages of rocks and the age of Earth. Arthur Holmes was among those who made important contributions to the development of radiogenic geochronological techniques in this period (**Table 2**). Despite the progress through the middle of the twentieth century in producing absolute age constraints on Earth and its rocks, scientists lacked a cohesive Earth model in which to place the geological processes they were dating. In the 1950s and 1960s, the fundamental step was made in this regard through development of the plate tectonic paradigm and magnetic stratigraphy; plate tectonics and magnetostratigraphy also contributed significantly to development of high-fidelity time-scales and geochronological tools (*see* **History of Geology Since 1962**).

Oceanographic cruises in the 1950s identified the presence of alternating 'stripes' of high and low magnetic intensity on the ocean floor. This pattern was clarified in the 1960s marine geophysical work of Hess and Dietz, who proposed the theory of seafloor spreading, and Vine and Matthews, who suggested that new oceanic crust was generated at ocean ridges and became magnetized in the direction of Earth's magnetic field. The ocean-floor stripes revealed alternating periods in Earth's history during which the magnetic field had changed from normal to reversed polarity. When these theories were combined with new results from palaeomagnetic studies conducted on sedimentary and volcanic rocks onshore, a globally applicable pattern of periods of normal and reversed magnetic polarities was gradually defined (**Figure 3**). This magnetic 'stratigraphy' was a relative time-scale useful for global 'pattern matching' of magnetic anomalies and for relative geochronology. The potassium-argon (K–Ar) radiogenic isotope geochronological technique, employed since the 1950s, was used to determine ages for fine-grained basalts used in the palaeomagnetic studies and thus placed absolute age constraints on points in the magnetic anomaly stratigraphy. Through combination of palaeomagnetic and K–Ar dating methods, the magnetic stratigraphy became better defined and, eventually, globally correlatable in terms of geological time. From the 1970s to the present, ties between palaeomagnetism, radiogenic isotope geochronology,

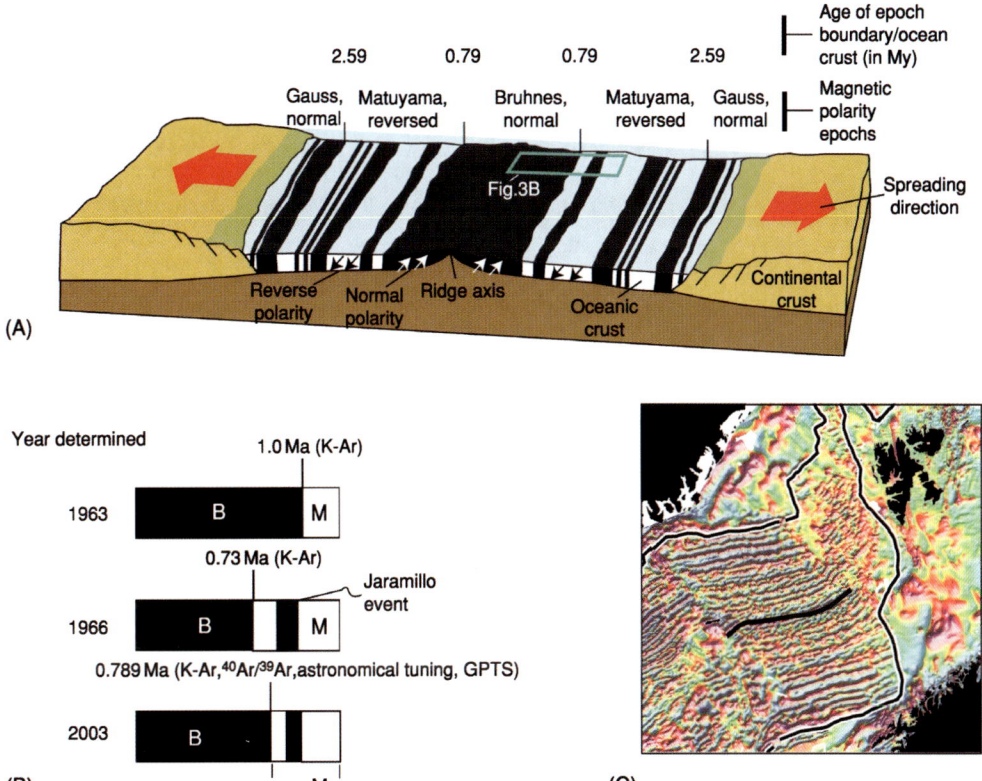

**Figure 3** Seafloor spreading. (A) Genesis of mirror-image, normal, and reversed magnetic polarity patterns in new oceanic crust, on either side of an oceanic ridge axis. The rifted continental margins yielded to new oceanic crust as seafloor spreading commenced. Alternating black (normal) and white (reversed) polarity patterns would normally be recorded by shipborne or satellite surveys. Historically, magnetic reversals were subdivided into major epochs (Bruhnes, normal; Matuyama, reversed; etc.); smaller normal and reversed 'events' were identified within these overall periods of normal or reversed polarity. Precise ages for these reversal epochs and, importantly, the boundaries between epochs were initially obtained with potassium–argon (K–Ar) geochronology. Refinements since the 1960s of the number and duration of magnetic reversals as well as their absolute ages have been accomplished by detailed comparison to biostratigraphy, the astronomically calibrated time-scale, and ages from radiogenic isotope dating methods. (B)Historical refinement of the Bruhnes (B)–Matuyama (M) boundary, where, in 1963, K–Ar dating indicated the epoch boundary to be at ~1 Ma. The Jaramillo 'event' close to the Bruhnes–Matuyama boundary had been discovered by 1966, and more precise K–Ar dating placed the age of the epoch boundary at 0.73 Ma. By 2003, the combination of several dating methods, including K–Ar and $^{40}Ar/^{39}Ar$ calibrations, astronomically calibrated time-scales, and geomagnetic polarity time-scales (GPTS), further refined the age of the boundary to a precise 0.789 Ma. (C) The magnetic anomaly map of the northern Atlantic Ocean between northern Norway, East Greenland, and Svalbard shows a real example of the alternating striped pattern of magnetic anomaly highs (red, normal polarity) and lows (blue, reversed polarity) on either side of the mid-ocean ridge axis. The mid-ocean ridge axis (trace identified with the single black line) separates a relatively symmetric, mirror-image anomaly pattern in this part of the seafloor. Continent–ocean boundaries are schematically indicated by thick black-on-white lines on the Norway and Greenland margins. (C) Reproduced with permission from Eide EA (coord.) *BATLAS – Mid-Norway Plate Reconstruction Atlas with Global and Atlantic Perspectives*, pp. 8–17. Trondheim: Geological Survey of Norway.

astronomically calibrated time-scales (ATSs), and biostratigraphy have facilitated definition of the geomagnetic polarity time-scale (GPTS) (**Figure 2**). Because of its tight calibration with these other methods, the GPTS provides the framework for most of the integrated time-scales presently in use for Jurassic and younger times (*see* **Plate Tectonics, Magnetostratigraphy**).

Today, the GTS, the GPTS, and the ATS have been intercalibrated for some geological time periods. Continued refinement and intercalibration of these time-scales will increase the possibility to make accurate age correlations for rocks and the geological events they represent. Important to recall is the fact that different geochronological techniques have been used to generate specific features of each time-scale, and that many techniques have particular geological time periods to which they are best suited; thus, complete intercalibration of these time-scales remains a challenging objective.

## Relative Geochronological Techniques

### Biostratigraphy

**Methodology** Biostratigraphy refers to correlation and age determination of rocks through use of fossils. Determining the environment in which the fossil species lived is inherent in this type of analysis. Theoretically, any fossil can be used to make physical correlations between stratigraphic horizons, but fossils that are best suited for making precise age correlations (time-stratigraphic correlations) represent organisms that (1) had wide geographic dispersal, (2) were short-lived, and/or (3) had distinct and rapidly developed evolutionary features by which they can now be identified. Fossils fulfilling these criteria are termed 'index' fossils. Both evolution and changes in local environment can cause the appearance or disappearance of a species, thus the time-significance of a particular index fossil must be demonstrated regionally through distinctions made between local environmental effects and time-significant events. Environmental effects may bring about the appearance/disappearance of a species because of local conditions, whereas time-significant effects may bring about the appearance/disappearance of a species because of evolution, extinction, or regional migration. Local environmental effects are not necessarily time significant and cannot be used in time correlations between different sedimentary units.

**Application** Fossils from the marine sedimentary record indicate existence of primitive life perhaps as early as 2.1 By ago, although the explosion of abundant life in the seas is usually tied to the start of the Palaeozoic era 544 million years ago (Ma). The continental sedimentary record indicates existence of plants and animals by Early Palaeozoic times, with recent indications of animals making forays from the seas onto land perhaps 530 Ma. Palaeozoic biostratigraphy, especially for the marine sedimentary record, is tied to precise, absolute ages for most period and stage boundaries, but gaps in the fossil record and/or the lack of isotopically datable rocks at key boundaries leave some discrepancies yet to be resolved. Biostratigraphy and fossil zone correlation are most precisely defined for the Mesozoic and Cenozoic eras; this is largely due to the ability to calibrate biostratigraphy not only with radiogenic isotope ages, but also with the GPTS and the ATS for these time periods.

### Palaeomagnetism and Magnetostratigraphy

**Methodology** Earth's magnetic field, generated in the liquid outer core, undergoes periodic reversals, with magnetic reversal frequencies typically between 1 and 5 My. Some rock minerals (such as hematite or magnetite) may become magnetized in the same direction as Earth's magnetic field (normal or reversed), either when a magmatic rock cools or when sedimentary rocks are deposited. As geochronological tools, palaeomagnetism and magnetostratigraphy rely on determining the magnetic polarity, including magnetic declination and inclination, of the sample's remanent magnetic component. Palaeomagnetism uses these parameters to calculate a palaeomagnetic pole for the sampling site. An age for the pole is determined by matching the pole to a part of the apparent polar wander path (APWP) for that continent (**Figure 4**). Instead of using poles, magnetostratigraphy, as outlined previously, identifies a sequence of magnetic reversals in a sedimentary or volcanic section (**Figure 2**). The magnetostratigraphic profile is compared and matched to similar patterns in the GPTS and a chronology for the sampled interval is established. The absolute chronology of the GPTS is tied by radiogenic isotope methods, by calibration against the ATS, and/or by calibration with a well-defined biostratigraphic zone (*see* **Magnetostratigraphy, Palaeomagnetism**).

**Application** Palaeomagnetism and magnetostratigraphy are most successfully applied to fine-grained volcanic and sedimentary rocks; the latter include red beds, siltstones, mudstones, and limestones. Matching of palaeomagnetic poles to established APWPs yields imprecise ages for rocks, but is useful for reasonable, first-order age estimates, probably within about ±10 My for Phanerozoic through Late Proterozoic rocks. The GPTS is most accurately refined through about 175 Ma because of the availability of marine magnetic anomaly profiles to which onshore data can be referenced; nonetheless, magnetic stratigraphy and the GPTS extend through the Palaeozoic to the earliest datable Cambrian sedimentary rocks (**Figure 2**). Well-constrained magnetostratigraphy yields very precise ages for the following reasons: (1) geomagnetic polarity reversals are rapid, globally synchronous events, and lend themselves well to global, time-significant correlations; (2) polarity reversals are not predictable and yield unique reversal patterns; (3) significant parts of the GPTS have been astronomically tuned, intercalibrated with detailed biostratigraphy, and/or constrained with absolute radiometric ages.

### Chemostratigraphy

**Methodology** Non-radiogenic chemical geochronological tools for sedimentary rocks fall into one of

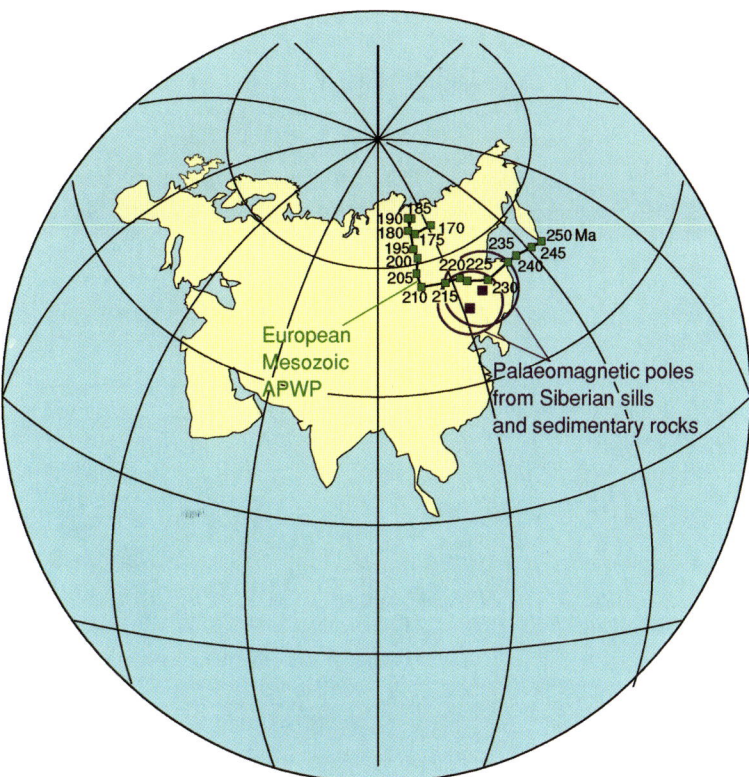

**Figure 4** Palaeomagnetic poles from gabbroic sills and interleaved sedimentary rocks of initially unknown ages were obtained from a study in northern Siberia. The poles for these rocks were compared to the apparent polar wander path (APWP) for Europe in the Mesozoic. Well-known ages are indicated in millions of years (Ma) for different segments of the APWP (designated with green squares). Within the uncertainty ellipses for the poles from the Siberian samples, the ages of the rocks were suggested to be between ~215 and 235 My. Subsequent radiogenic isotope age determinations on the sills confirmed this suggestion and refined the ages for the rocks to lie between 220 and 234 My.

three categories: pattern matching of time-stratigraphic shifts in stable isotope (O, C, or S) values and $^{87}Sr/^{86}Sr$ ratios, identification of siderophile element anomalies (Ir, Au, Pd, Pt, etc.), and chemical dating using amino acids. The principles for stable isotope methods are based on the fractionation of heavy and light isotopes of the stable elements O, C, and S. The heavy isotopes, $^{18}O$, $^{13}C$, and $^{34}S$, are compared, respectively, to the lighter isotopes $^{16}O$, $^{12}C$, and $^{32}S$. Stable isotopic compositions are reported as ratios (for example, $^{18}O/^{16}O$) relative to a standard for the same isotopic ratios. Processes causing fractionation of these isotopes depend primarily on temperature, isotope exchange reactions, and, in the case of S, change in oxidation state of sulphur compounds from action of anaerobic bacteria. The isotopic composition of Sr in sedimentary rocks is characterized by the $^{87}Sr/^{86}Sr$ ratio of the water from which the sediment precipitated; the water in the catchment area or in the ocean, in turn, will have an $^{87}Sr/^{86}Sr$ ratio that represents contributions from chemical weathering of rocks. Rocks of varying ages and different mineralogies have distinct $^{87}Sr/^{86}Sr$ ratios that will make different contributions of Sr to the water cycle. These contributions have been shown to vary over geological time in response to changes in the exposure and weathering of different landmasses.

For purposes of geochronology, the principle of 'pattern-matching' is also used with these isotopic methods (**Figure 5**). Measured isotopic ratios in a stratigraphic sample suite representing some interval of geological time yield a curve (or excursion pattern) that is compared to a global reference or supraregional curve for the same isotopes. The global reference curve must, in turn, be calibrated to an absolute timescale by some independent means, usually matching the stratigraphic section in question to another section that is tied either to the GPTS or to absolute ages.

Anomalously high concentrations of siderophile elements have been identified globally at three precisely determined time intervals: the Cretaceous–Tertiary boundary (65 Ma), the Eocene–Oligocene

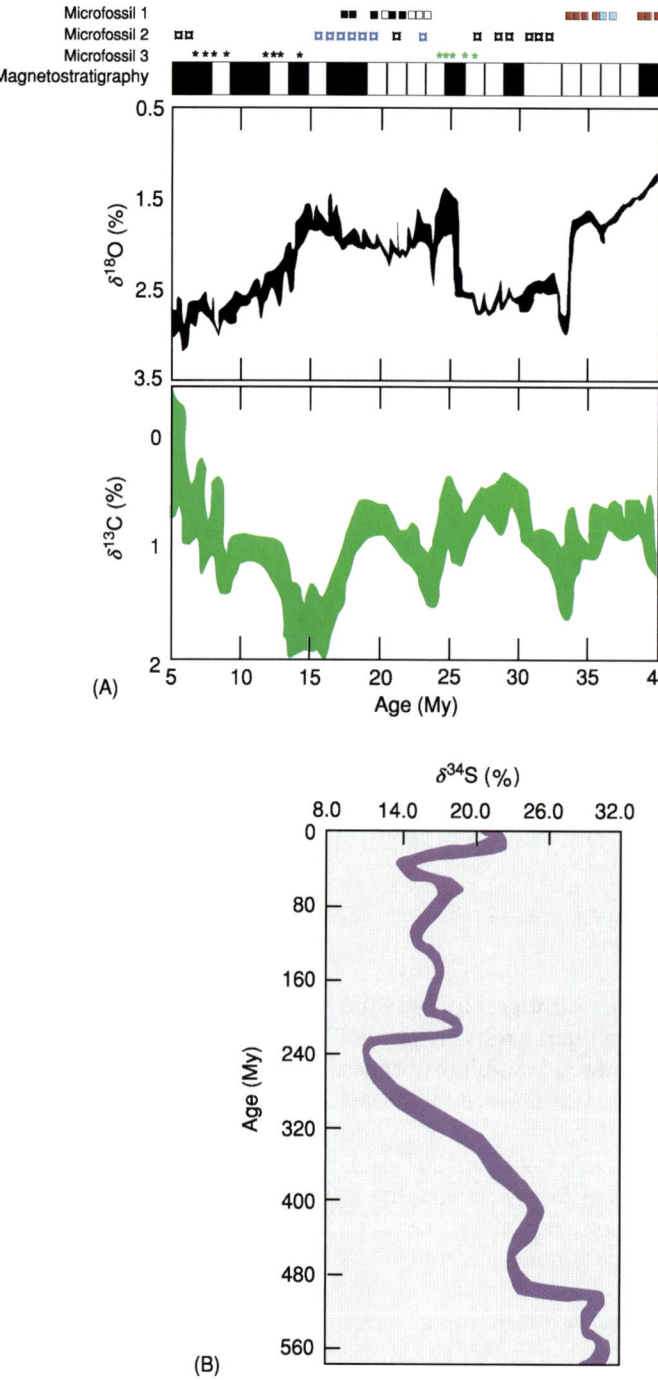

**Figure 5** (A) Stable isotopes used in chemostratigraphy are commonly coupled with magnetostratigraphic and biostratigraphic information. In this fictive example, the stable isotope values for O and C were acquired for an entire sedimentary sequence of Cenozoic age. Magnetostratigraphy over the same zone may have revealed a pattern similar to that shown on the bar above the stable isotope curves, and this stratigraphy could then be correlated to the geomagnetic polarity time-scale and used to calibrate the ages for the sedimentary column, which in this case spanned Pliocene through latest Eocene time. Biostratigraphy over the same stratigraphic column may have revealed a predominance of three types of microfossils, with different species within each microfossil group identified (designated here with different coloured symbols). Biostratigraphy might also be used to tie together and calibrate the stable isotope curves and make fine adjustments to ages determined with the magnetostratigraphic profile. Especially interesting would be to attempt to link any significant excursions in the isotope curves, either to changes observed in the microfossil distribution or to a specific time boundary. (B) Stable isotope stratigraphy can also be used over a larger time-span for more regional or global correlations. This isotope curve for sulphur shows a marked change at about 240 Ma following a steady decrease through the Palaeozoic.

boundary (33.7 Ma), and 2.3 Ma. Other anomalies – specifically, spikes in iridium concentrations in sedimentary sequences – have been suggested at the Triassic–Jurassic boundary and at the Devonian–Carboniferous boundary. These anomalous concentrations have been associated with catastrophic events, usually meteor impacts or massive volcanic eruptions, and faunal crises or mass extinctions. Because of their global nature, limited duration, and precisely defined ages, anomalous siderophile concentrations can serve as indirect dating tools in sedimentary sequences (*see* **Impact Structures**).

The amino acid racemization (AAR) method uses the asymmetry of isomeric forms of several amino acids in fossil skeletal material to determine the time since the start of racemization. Racemization is the reversible conversion of one set of amino acid isomers to another set of isomers and begins with death of the organism. Sample materials are chemically treated and the amino acid types and isomer ratios are determined through chromatography methods. These ratios are used to calculate the time since the start of racemization through a formula containing a sample-site constant for the racemization rate. Because the racemization rate depends on external factors such as temperature, pH, and moisture, the rate varies between one sample site and another and must be calibrated for each site and each sample. This usually involves calibration against other samples (from the same sites) that have been dated by other methods.

**Application** Oxygen isotope stratigraphy may be applied to planktonic foraminiferal tests in pelagic sediments that are at least 1 My old. Sulphur isotopes are most commonly used to date marine evaporites with ages of deposition extending through ∼650 Ma. Carbon isotopes may be used to date marine evaporites, marine carbonates, and (metamorphosed) marbles through Neoproterozoic age. Similarly, strontium, which substitutes readily for calcium, can also be used to date marine carbonates, apatite in marine sediments, and marbles through the Neoproterozoic. All of the isotope methods generally require samples that have been relatively unaltered by postdepositional events such as erosion, bioturbation, metamorphism, or recrystallization during diagenesis. Notably, work with metamorphosed marbles has indicated that C and Sr isotopes may maintain their original sedimentary deposition ratios despite having undergone extreme changes in pressure, temperature, and deformation subsequent to deposition.

Siderophile element anomalies are confined to the sedimentary rock record; the most well-documented anomaly is at the Cretaceous–Tertiary boundary (*see* **Mesozoic: End Cretaceous Extinctions**). The AAR method is restricted primarily to dating Holocene foraminifers extracted from pelagic sediments, although ages have also been determined for coprolites and mollusc shells.

## Absolute Geochronological Techniques

### Radiogenic Isotope Techniques

**Methodology** The natural decay of a radioactive isotope to a stable isotope occurs at a regular rate that is described by the decay constant ($\lambda$). The decay process is defined by an exponential function represented by the decay 'half-life' ($t_{1/2}$); the half-life is equivalent to the amount of time necessary for one-half of the radioactive nuclide to decay to a stable nuclide form. Radiogenic isotope techniques use this principle to calculate the age of a rock or mineral through measurement of the amount of radioactive 'parent' isotope and stable 'daughter' isotope in the sample material. The parent/daughter ratio and the decay constant for that isotope series are used to calculate how much time had to elapse for all of the stable daughter isotope to have been produced from an initial reservoir of radioactive parent isotope in the material (**Table 3**). This calculation presumes (1) no net transfer of radiogenic parent, stable daughter, and/or intermediate radioactive isotopes in or out of the sample material (mineral or rock) since time zero, (2) no unknown quantity of daughter isotope in the sample at time zero, and (3) that decay constants have not changed over the history of Earth. Many radiogenic isotope techniques are presently used to determine the ages of geological materials; the choice of appropriate isotopic system to determine an age of a sample depends primarily on the composition of the sample material, the geological 'event' or 'process' to be dated, and the sample's age. The latter is directly linked to the half-life of the isotope system: radionuclides with long half-lives can be used to date very old samples, whereas those with shorter half-lives are restricted to dating younger rocks. In addition to the naturally occurring radioactive isotopes, a number of nuclear reactions of cosmic rays with gas molecules will produce radionuclides, the so-called cosmogenic radionuclides. The most long-lived of these can be used for age determinations based on principles similar to those outlined for the other radioactive isotopes.

**Applications** The methods routinely used to date terrestrial metamorphic or igneous rocks and their minerals include techniques utilizing U/Th/Pb, Pb/Pb, Sm/Nd, Lu/Hf, Re/Os, Rb/Sr, K–Ar, and Ar/Ar (**Table 3**). All of these isotopes have half-lives >1 By,

**Table 3** Common radiogenic isotope geochronological techniques

| Method | Radioactive parent | Stable daughter | Intermediate products[a] | Decay scheme | Half-life (years) | Sample material | Typical geological 'events' dated | Comments |
|---|---|---|---|---|---|---|---|---|
| U/Th/Pb, Pb/Pb | $^{238}$U | $^{206}$Pb | From $^{238}$U: $^{234}$Th, $^{234}$Pa, $^{234}$U, $^{230}$Th, $^{226}$Ra, $^{222}$Rn, $^{218}$Po, $^{218}$At, $^{218}$Rn, $^{214}$Po, $^{210}$Pb, $^{210}$Bi, $^{210}$Po | Chain: $^{238}$U → $^{206}$Pb, $^{235}$U → $^{207}$Pb, $^{232}$Th → $^{208}$Pb | $^{238}$U = 4.468 × 10$^9$, $^{235}$U = 0.7038 × 10$^9$, $^{232}$Th = 14.01 × 10$^9$ | Zircon, thorite, monazite, apatite, xenotime, titanite, uraninite, thorianite | Crystallization age (from melt or from medium to high metamorphic grade); age of Earth | U and Th are concentrated in the liquid phase and are typically incorporated in more silica-rich fractions; half-lives of the parent isotopes are much longer than those of intermediate products; Pb isotopes alone in rocks without U or Th can be used to calculate 'model ages' (with information on crustal growth) |
| | $^{235}$U | $^{207}$Pb | From $^{235}$U: $^{231}$Th, $^{231}$Pa, $^{227}$Ac, $^{227}$Th, $^{223}$Ra, $^{219}$Rn, $^{215}$Po, $^{214}$At, $^{211}$Bi, $^{211}$Po | | | | | |
| | $^{232}$Th | $^{208}$Pb | From $^{232}$Th: $^{228}$Ra, $^{228}$Ac, $^{228}$Th, $^{224}$Ra, $^{220}$Rn, $^{216}$Po, $^{212}$Pb, $^{212}$Bi, $^{212}$Po, $^{208}$Pb | Decay schemes produce alpha ($^4$He) particles; used for (U/Th)/He dating | | | | |
| Sm/Nd | $^{147}$Sm | $^{143}$Nd | None | Simple: $^{147}$Sm → $^{143}$Nd (alpha decay) | 1.06 × 10$^{11}$ | Garnet, pyroxene, amphibole, plagioclase; mafic and ultramafic igneous and metamorphic whole rocks; lunar rocks | Crystallization age (from melt or from medium to high metamorphic grade) | Ages calculated from analysis of isotopes in separated minerals or cogenetic rocks; Sm and Nd are rare earth elements that tend to be less mobile during metamorphism and weathering |
| Lu/Hf | $^{176}$Lu | $^{176}$Hf | None | Branched: $^{176}$Lu → $^{176}$Hf (gamma ray emission); $^{176}$Lu → $^{176}$Yb (electron capture) | 3.54 × 10$^{10}$ | Apatite, garnet, monazite, zircon, xenotime, meteorites, lunar rocks | Meteorite formation; high-grade metamorphism; igneous crystallization | Can also be used for information on differentiation of the mantle and crustal growth; $^{176}$Yb branch of decay can be ignored for purpose of geochronology |

*Continued*

ANALYTICAL METHODS/Geochronological Techniques 89

**Table 3** Continued

| Method | Radioactive parent | Stable daughter | Intermediate products[a] | Decay scheme | Half-life (years) | Sample material | Typical geological 'events' dated | Comments |
|---|---|---|---|---|---|---|---|---|
| Re/Os | $^{187}$Re | $^{187}$Os | None | Simple: $^{187}$Re $\rightarrow$ $^{187}$Os (beta particle emission) | $4.56 \times 10^{10}$ | molybdenite, osmiridium, laurite, columbite, tantalite, Cu-sulphides; ores, meteorites | Ore deposit formation; iron-meteorite formation | Enriched in metallic and sulphide phases; relatively depleted in silicates |
| Rb/Sr | $^{87}$Rb | $^{86}$Sr | None | Simple: $^{87}$Rb $\rightarrow$ $^{86}$Sr (beta particle emission) | $4.88 \times 10^{10}$ | Mica, feldspar, leucite, apatite, epidote, garnet, ilmenite, hornblende, pyroxene, clay minerals, some salts; felsic whole rocks, meteorites | Crystallization age (from melt or metamorphism); cooling (after high-grade 'event'); diagenesis | Because Rb and Sr have close relationships to K and Ca, respectively, the method is especially useful for study of granitic rocks |
| K–Ar, $^{40}$Ar/$^{39}$Ar | $^{40}$K | $^{40}$Ar | None | Branched: $^{40}$K $\rightarrow$ $^{40}$Ca (beta emission); $^{40}$K $\rightarrow$ $^{40}$Ar (beta emission and electron capture) | $1.25 \times 10^{10}$ | Mica, feldspar, feldspathoids, amphibole, illite, volcanic rocks, lunar rocks, low-grade metamorphic rocks, glass, salts, clay minerals, evaporites | Crystallization of quickly cooled igneous rocks; cooling of metamorphic and plutonic rocks | K–Ar method involves splitting the sample to measure K and Ar; $^{40}$Ar/$^{39}$Ar uses $^{39}$Ar as a proxy for K and measures only Ar isotopes, with no sample splitting; the $^{40}$Ar/$^{39}$Ar method is commonly used today |
| Carbon-14 | $^{14}$C | $^{14}$N | $^{14}$C produced in atmosphere by collision of thermal neutrons (from cosmic rays) with $^{14}$N; $^{14}$C is oxidized rapidly and radioactive $CO_2$ enters the carbon cycle; radioactive $^{14}$C decays | $^{14}$C $\rightarrow$ $^{14}$N | ~5700 | Organic matter: wood, charcoal, seeds, leaves, peat, bone, tissue, mollusc shells | Time since the organic material ceased to take up carbon | Dendrochronology and varve chronology are often used in carbon-14 dating to account for secular variation in the $^{14}$C content in the atmosphere |

[a]Note that the U-Th-Pb decay series involves numerous intermediate radioactive isotopes with short half-lives ('chain' decay); only the direct intermediate products are listed here (products from branched decay have not been listed).

so the samples can be used to date Earth's oldest geological materials and events. Lunar and cosmogenic materials have also been dated with some of the same methods. The relatively shorter half-life of the K–Ar decay series, as well as the very short half-lives of the intermediate nuclides in the U and Th decay series, allow these isotope systems to be used for dating certain geological materials of Pleistocene (the U-series nuclides) and Holocene (the K–Ar and Ar/Ar methods) ages. Of the cosmogenic radionuclides, the most well known is probably carbon-14. The carbon-14 method is used to date organic materials; $^{14}$C has a half-life of ~5700 years and is restricted to materials less than about 100 000 years old (**Table 3**). Aside from $^{14}$C, other cosmogenic radionuclides include $^{10}$Be, $^{26}$Al, $^{36}$Cl, $^{41}$Ca, $^{53}$Mn, $^{81}$Kr, and $^{129}$I; these can be used for dating relatively young materials (on the order of several 100 000 years for Ca and Kr and up to 1 My or more for Be, Al, Cl, Mn, and I). Though not treated in detail here, these isotopes can be applied to date a range of materials, including Quaternary sediments, ice, manganese nodules, groundwater, and soils, and to determine the age of exposure of terrestrial land surfaces and meteorites (*see* **Analytical Methods: Fission Track Analysis**).

### Astronomically Calibrated Time-Scales

**Methodology** Perturbations in the orbit of Earth about the sun are generated by gravitational interactions between Earth and the sun, moon, and other celestial bodies. These orbital perturbations cause cyclical climatic changes that are recorded in some sedimentary rocks. This principle was recognized by G K Gilbert in the nineteenth century, and he noted the potential to use this climatically driven, sedimentary cyclicity to place age constraints on certain parts of the rock record. Since Gilbert's time, astronomically calibrated time-scales have generated astronomical solutions for these perturbations in Earth's orbit that match sedimentary cycles recognized in nature, such as glacial varve sequences (**Figure 6**). These gravity-induced perturbations apply specifically to the obliquity of Earth's orbit, Earth's axial precession, and the eccentricity of Earth's orbit about the sun. Obliquity refers to the angle between Earth's axis of rotation and the orbital plane, whereas precession is the movement ('wobble') of the rotation axis about a circular path that describes a cone. Eccentricity is the elongation of Earth's orbit about the sun; this varies between a circular and an elliptical shape. The main periods of eccentricity of Earth's orbit are 100 000 and 413 000 years. The obliquity of Earth's axis has a main period of 41 000 years and precession of the axis has a main period of 21 000 years. Because the astronomically calibrated

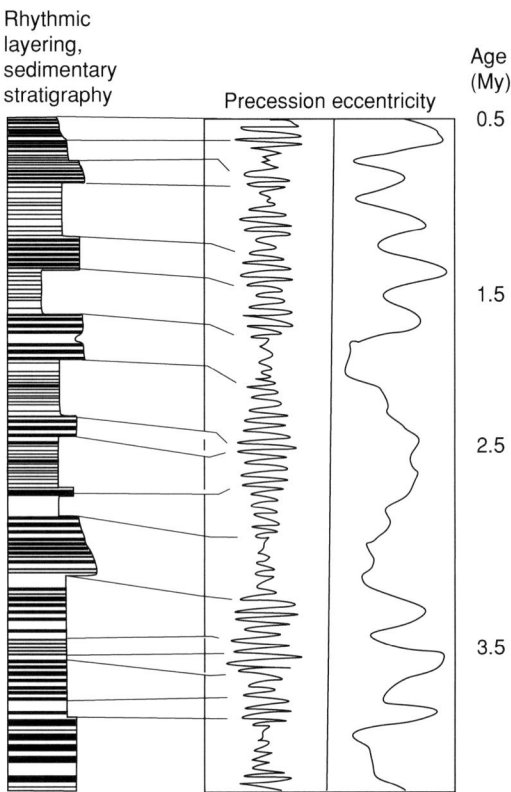

**Figure 6** Astronomically calibrated time-scales attempt to resolve the long-term gravitational perturbations in Earth's orbit about the Sun. The mathematical solutions for the cyclicity of these perturbations are projected backward in time to determine the geological age of seasonal (solar) cycles preserved in the sedimentary rock record. Most astronomical calibrations define solutions for the precession and eccentricity of Earth's orbit. In this example, cyclical sedimentation patterns (alternating dark and light sedimentary layers) in a fictitious marine sequence were carefully logged, as on the left-hand column. The log is matched to the calculated solutions for orbital precession and eccentricity that are tied to absolute time. Where possible, the stratigraphic column may also be tied to magnetostratigraphic, biostratigraphic, and/or radiogenic isotope geochronology data.

time-scales are based only on factors related to Earth's orbit about the sun, they are the only truly 'absolute' time-scales, following the strict definition of this word, and are mainstays for tying together or intercalibrating the other time-scales (*see* **Earth: Orbital Variation (Including Milankovitch Cycles)**).

**Applications** The geologically short periodicity of Earth's orbital perturbations has allowed calibration of precise astronomical time-scales for the past 15 My. Climate changes associated with ice ages have been the most easily recognized events in the rock record and the astronomical calibration of the Plio-Pleistocene time-scale remains one of the best. Although the Miocene-and-younger time-scales have been based primarily on the marine rock record, continental sedimentary sections have increasingly been incorporated in these calibrations. Work with astronomically calibrated lacustrine sections of Triassic–Jurassic age has demonstrated that older rocks can also be anchored to the astronomical time-scale.

### Dendrochronology

**Methodology and Applications** Dendrochronology applies the nonsystematic, climate-dependent variations in the thickness of annual tree rings of particular tree species to determine very exact dates for young events. Although restricted to use on Holocene samples, the high precision of the method (trees produce one ring per year, and uncertainties in ages determined with the method are usually ±1 year) has also been used to calibrate carbon-14 ages (see also Table 3).

## Future Considerations

Geochronology furnishes the temporal framework for the study of geologic processes, giving data necessary to evaluate the rates, quantity, and significance of different rocks and geological 'events'. Both relative and absolute ages are important in this regard and should be viewed as complementary methods through which different rock types may be correlated in time. Today, a big challenge facing geochronologists is the intercalibration of the various timescales. As part of this work, geologists working with radiogenic isotopes are attempting to refine the decay constants for a number of the commonly used radiogenic isotope dating methods. Inaccurate decay constants would clearly affect the accuracy of an age for a rock determined with a particular isotope system, and would have corresponding spin-off effects for ties made to magnetostratigraphic, biostratigraphic, chemostratigraphic, and astronomically calibrated datasets. Intercalibration of the various time-scales back through Mesozoic and Palaeozoic times will probably incorporate all of these methods, with extension of astronomical calibrations to the Palaeozoic probably involving 'floating' astronomical time-scales intercalibrated with the continually updated and refined GPTS and GTS.

## Glossary

**decay constant** A number describing the probability that a radioactive atom will decay in a unit time.
**half-life** The time required for half of a quantity of radioactive atoms to decay.
**isotopes** Atoms with the same number of protons (= the same element), but a different number of neutrons (= different mass).
**radioactive decay** The spontaneous disintegration of certain atoms whereby energy is emitted in the form of radiation; a new, stable atom is the result.
**siderophile** An element preferring a metallic phase, with a weak affinity for oxygen or sulphur.

## See Also

**Analytical Methods:** Fission Track Analysis. **Conservation of Geological Specimens. Creationism. Earth:** Orbital Variation (Including Milankovitch Cycles). **Dendrochronology. Famous Geologists:** Cuvier; Darwin; Hutton; Lyell; Murchison; Sedgwick; Smith; Steno. **Magnetostratigraphy. Mesozoic:** End Cretaceous Extinctions. **Palaeomagnetism. Palaeozoic:** Cambrian; End Permian Extinctions. **Plate Tectonics. Time Scale**.

## Further Reading

Butler RF (1992) *Palaeomagnetism: Magnetic Domains and Geologic Terranes.* Cambridge, MA: Blackwell Scientific Publications.

Cox A (ed.) (1973) *Plate Tectonics and Geomagnetic Reversals.* San Francisco, CA: WH Freeman and Company.

Dalrymple BG (1991) *The Age of the Earth.* Palo Alto, CA: Stanford University Press.

Dickin AP (1995) *Radiogenic Isotope Geology.* Cambridge: Cambridge University Press.

Doyle P, Bennett MR, and Baxter AN (1994) *The Key to Earth History: An Introduction to Stratigraphy.* Chichester: John Wiley and Sons.

Eicher DL (1976) *Geologic Time*, 2nd edn. Englewood Cliffs, NJ: Prentice-Hall.

Eide EA (2002) Introduction – plate reconstructions and integrated datasets. In: Eide EA (coord.) *BATLAS – Mid Norway Plate Reconstruction Atlas with Global and Atlantic Perspectives*, pp. 8–17. Trondheim: Geological Survey of Norway.

Faure G (1986) *Principles of Isotope Geology*, 2nd edn. New York: John Wiley and Sons.

Geyh MA and Schleicher H (1990) *Absolute Age Determination: Physical and Chemical Dating Methods and Their Application.* Berlin: Springer-Verlag.

Hilgen FJ, Krijgsman W, Langereis CG, and Lourens LJ (1997) Breakthrough made in dating of the geological record. *EOS* 78(28): 285, 288–289.

Lewis C (2000) *The Dating Game – One Man's Search for the Age of the Earth.* Cambridge: Cambridge University Press.

Renne PR, Deino AL, Walter RC, *et al.* (1994) Intercalibration of astronomical and radioisotopic time. *Geology* 22: 783–786.

# Gravity

J R Smallwood, Amerada Hess plc, London, UK

© 2005, Elsevier Ltd. All Rights Reserved.

## Introduction

The law of gravitational attraction between objects was deduced by Isaac Newton in the late seventeenth century. His 'inverse square' law stated that the force attracting two objects was proportional to the masses of the two objects and inversely proportional to the square of the distance between them (**Table 1**). Since the mass of the Earth is so great relative to the mass of objects on its surface, attraction of objects towards the Earth, i.e., their response to the Earth's gravity field, is often an important factor affecting geological processes. Measurement of the gravity field of the Earth is in itself a useful tool for investigating the sub-surface, as mass variations below the surface cause variations in the gravity field. The measurement of the shape of the Earth and its mass distribution have been important to defining the baseline gravity field from which deviations can be measured, as usually the anomaly rather than the overall field strength is useful for geological applications. There are now many ways of acquiring gravity data on land, sea, air, and from space, appropriate to the many scales on which gravity studies can be applied. Gravity variations over thousands of kilometres can be used for studies of mantle convection, variations over hundreds and tens of kilometres are relevant for studies such as lithospheric flexure, plate tectonics (*see* **Plate Tectonics**), crustal structure, and sedimentary basin development, hydrocarbon (*see* **Petroleum Geology:** Exploration) and mineral exploration (*see* **Mining Geology:** Exploration), while gravity variations over tens of metres can be used in civil engineering applications.

## The Earth's Shape and its Gravity Field

The gravitational potential of a perfectly uniform sphere would be equal at all points on its surface. However, the Earth is not a perfect sphere; it is an oblate spheroid, and has a smaller radius at the poles than at the equator. Surveys in the early eighteenth century, under the direction of Ch-M de La Condamine and M de Maupertius found that a meridian degree measured at Quito, Equador, near the equator, was about 1500 m longer than a meridian degree near Tornio, Finland, near the Arctic circle.

Subsequently, various standard reference spheroids or ellipsoids have been proposed as first-order approximations to the shape of the Earth, such as the World Geodetic System 1984 (**Table 1**). Given such an ellipsoid, a gravity field can be calculated analytically as a function of latitude. For example, a reference gravity formula was adopted by the International Association of Geodesy in 1967 (IGF67, **Table 1**), and another introduced in 1984 (WGS84, **Table 1**).

The mean density of the Earth, which is fundamental to the calculation of gravitational attraction, was first estimated following an experiment in 1775 by the Rev. Neville Maskelyne, using a technique suggested by Newton. If the Earth was perfectly spherical and of uniform density, then a plumbline would point down towards the centre of the Earth because of the force of gravity on the bob. However, any nearby mass would deflect the plumbline off this 'vertical'. Maskelyne and his co-workers measured plumb-bob deflections on the Scottish mountain, Schiehallion (**Figure 1**). They discovered that the mountain's gravitational pull deflected the plumb line by 11.7 seconds of arc. This allowed Charles Hutton to report in 1778 that the mean density of the Earth was approximately $4500 \, \text{kg m}^{-3}$. This density value leads to an estimate of the mass of the Earth of about $5 \times 10^{24}$ kg, not far from the currently accepted value of $5.97 \times 10^{24}$ kg. The Schiehallion experiment had another distinction, in that in order to calculate the mass and centre of gravity of the mountain a detailed survey was carried out, and the contour map was invented by Hutton to present the data.

Since the mass of the Earth is not distributed uniformly, the real gravity field does not correspond to that calculated for an ellipsoid of uniform density. The 'geoid' is a surface which is defined by points of equal gravitational potential or equipotential (**Table 1**), which is chosen to coincide, on average, with mean sea-level. The geoid is not a perfect ellipsoid, because local and regional mass anomalies perturb the gravitational potential surface in their vicinity by several tens of metres. For example, a seamount on the ocean floor, which is denser than the surrounding seawater, will deflect the geoid downwards above it. 'Geoid anomalies' are defined as displacements of the geoid above or below a selected ellipsoid. The concept of the geoid as the global mean sea-level surface can be extended across areas occupied by land. This provides both a horizontal reference datum and a definition of the direction of the vertical, as a plumbline will hang perpendicular to the geoid.

**Table 1** Gravity formulae

| Quantity | Formula | Constants and variables | |
|---|---|---|---|
| Gravitational Force between two masses, F | $F = \dfrac{GMm}{r^2}$ | G | Gravitational or Newtonian constant, $6.67 \times 10^{-11} \, m^3 kg^{-1} s^{-1}$ |
| | | M | Mass of body (Mass of earth approx. $5.97 \times 10^{24}$ kg) |
| | | m | Mass of second body |
| | | r | Distance |
| Gravitational Acceleration, a | $a = \dfrac{GM}{r^2}$ | | As above |
| Gravitational Potential, V | $V = \dfrac{GM}{r}$ | | As above |
| (Vertical) Gravity anomaly above a buried sphere, $\delta g_z$ See **Figure 6** | $\delta g_z = \dfrac{4G\Delta\rho b^3 h}{3(x^2 + h^2)^{3/2}}$ | $\Delta\rho$ | Density contrast |
| | | b | Radius of sphere |
| | | h | Depth of sphere |
| | | x | Horizontal distance |
| International Gravity Formula 1967 Gravitational acceleration, $g_t$ | $g_t = g_0(1 + \alpha \cdot \sin^2\lambda + \beta \cdot \sin^4\lambda)$ | $g_0$ | Mean gravitational acceleration at equator, $9.7803185 \, ms^{-2}$ |
| | | $\alpha$ | $5.278895 \times 10^{-3}$ |
| | | $\beta$ | $2.3462 \times 10^{-5}$ |
| | | $\lambda$ | Latitude |
| WGS84 Ellipsoidal Gravity Formula Gravitational acceleration, $g_t$ | $g_t = \dfrac{g_0(1 + d \cdot \sin^2\lambda)}{\sqrt{(1 - e \cdot \sin^2\lambda)}}$ | $g_0$ | Mean gravitational acceleration at equator, $9.7803267714 \, ms^{-2}$ |
| | | d | $1.93185138639 \times 10^{-3}$ |
| | | e | $6.6943999103 \times 10^{-3}$ |
| | | $\lambda$ | Latitude |
| WGS Formula atmospheric correction, $\delta g_t$ | $\delta g_t = 0.87 \times 10^{-5} \cdot \exp(-0.116 h^{1.047})$ | h | Elevation |
| Latitude correction for relative gravity measurements, $\delta g_L$ | $\delta g_L = 8.12 \times 10^{-5} \cdot \sin 2\lambda \cdot \delta l$ | $\delta l$ | Distance in N–S direction between readings |
| | | $\lambda$ | Latitude |
| Bouguer plate correction, $\delta g_B$ | $\delta g_B = 2\pi\rho \, Gh$ | $\rho$ | Bouguer correction density |
| | | h | Elevation |
| Free air correction, $\delta g_{FA}$ | $\delta g_{FA} = 308.6 \cdot h$ | h | Elevation |
| Free air anomaly, $g_{FA}$ | $g_{FA} = g_{obs} - g_t + (\delta g_L + \delta g_{FA})$ | $g_{obs}$ | Observed gravity |
| Bouguer anomaly, $g_B$ | $g_B = g_{obs} - g_t + (\delta g_L + \delta g_{FA} - \delta g_B + \delta g_T)$ | $\delta g_T$ | Terrain correction |
| Flattening factor for ellipsoid, f | $f = \dfrac{a - c}{c} = \dfrac{1}{298.26}$ | a | Equatorial radius of Earth, 6378.14 km |
| | | c | Polar radius of Earth, 6356.75 km |

## Measurement of Gravity

The first measurements of Earth's gravity, by timing the sliding of objects down inclined planes, were made by Galileo, after whom gravitational units were named. 1 Gal is $10^{-2} \, m \, s^{-2}$, and the gravitational acceleration at the Earth's surface is about 981 Gal. For convenience in geophysical studies of gravity anomalies, the mGal is usually used, or for local surveys 'gravity units' (g.u.) where 1 mGal = 10 g.u. Gravity may be measured as an absolute or relative quantity.

Classically, absolute gravity has been measured with a pendulum consisting of a heavy weight suspended by a thin fibre. The period of the oscillation is a function of gravitational acceleration and the length of the pendulum. H Kater designed a compound, or reverse, pendulum in 1815, that allowed some instrument-dependent factors to be cancelled out. The instrument was superceded by methods based on observations of falling objects. In a development of the free-fall method, a projectile is

94 ANALYTICAL METHODS/Gravity

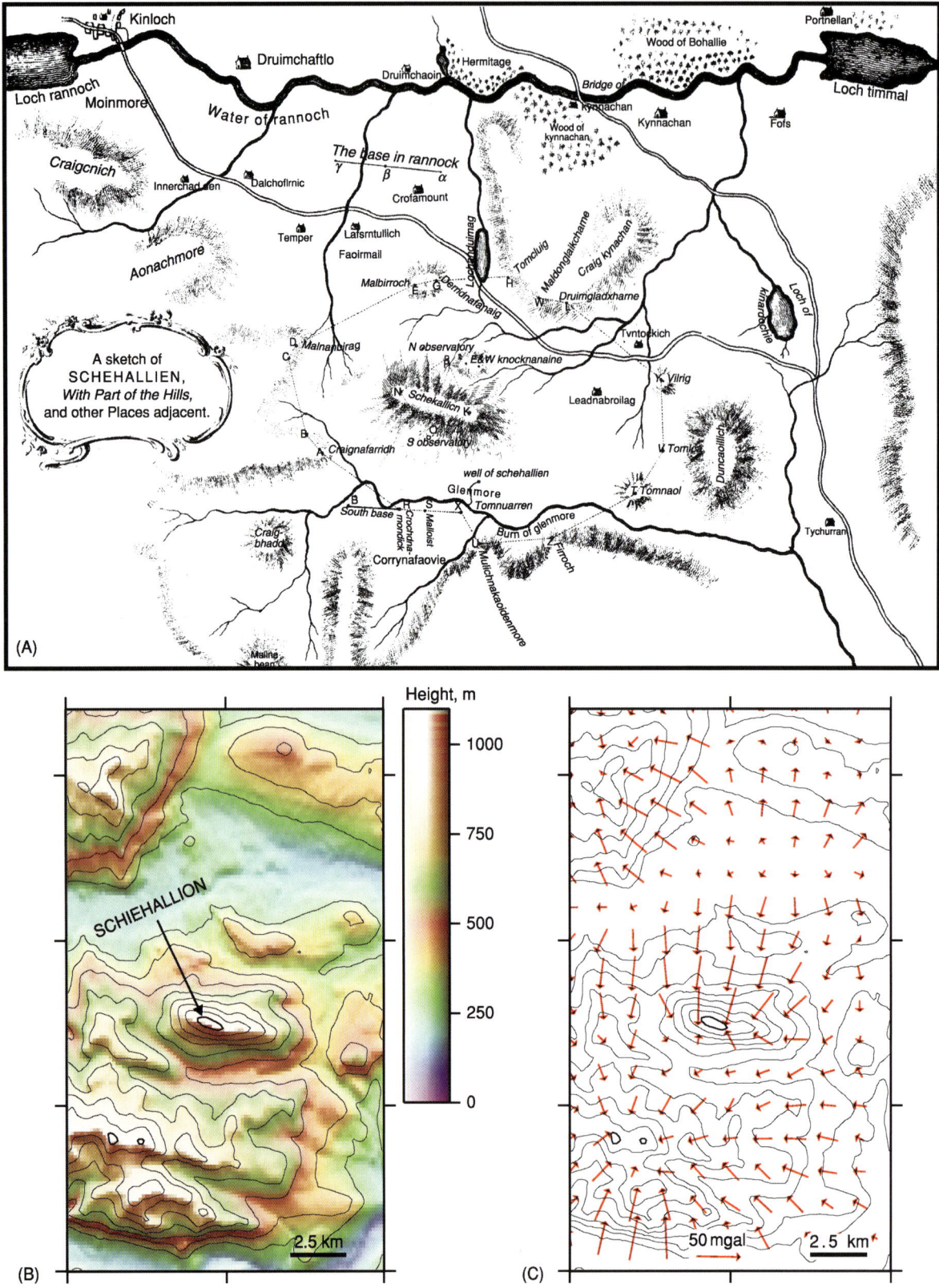

**Figure 1** Sketch map of the area around Schiehallion, Scotland, by Charles Hutton. Plumb-line deflections measured at stations north and south of the mountain allowed the first estimate of the density of the earth. (After R.M. Sillito with permission from Hutton (1778) © The Royal Society.)

fired vertically upwards and allowed to fall back along the same path. The gravity measurement depends on timing the upward and downward paths, which may be by light beam-controlled timers or interferometry.

Gravity differences can be measured on land with a stable gravity meter or gravimeter based on Hooke's law. A mass extends a spring under the influence of gravity and changes in extension are proportional to changes in the gravitational acceleration. More sensitive are 'astatic' gravity meters, which contain a mass supported by a 'zero-length' spring for which tension is proportional to extension. When the meter is in position, a measurement is made of an additional force needed to restore the mass to a standard position, supplied by an auxiliary spring or springs, an electrostatic system, or an adjustment of the zero-length spring itself. Gravity meters working on this principle measure differences in gravity between stations and surveys may be tied to one or more base stations at which repeated measurement can be made. Astatic gravity meters can have a sensitivity of about 0.01 mGal.

For applications where the gravity meter is subject to tilting and vibration, such as on board a ship or in an aircraft, isolation of the instrument is required such as providing a moving stabilised platform for the gravity meter and damping vibrations with appropriate shock absorption. When the gravity meter is moving, accurate data on the location and trajectory of the platform is required along with the gravity measurement. For airborne application, this requirement has been greatly assisted by the advent of the global positioning system (GPS) which allows rapid, precise, and accurate positioning (see **Remote Sensing: GIS**). Airborne gravity surveys, whether flown using fixed wing or helicopters, can provide economic, rapid, and non-invasive geophysical reconnaissance ideal for difficult terrain such as tundra, jungles, and wildlife reserves.

Deviations in artificial satellite orbits can be used to determine the long-wavelength components of the Earth's gravity field. Altimetry tools mounted on satellites have allowed much more detailed gravity mapping over the oceans, as sea surface height data can be processed to give the marine geoid. Geoid data can then be converted to gravity data with a series of numerical operations (**Figure 2**). Since the mean sea-level surface is the geoid, an equipotential surface, variations in sea surface height from the reference ellipsoid reflect density changes below the sea surface, largely from the density contrast at the seabed, but also from sub-seabed changes, such as crustal thickness changes.

## Adjustments to Measured Gravity Signals

The first correction that can be applied to measured gravity values is the correction for latitude, to account for the centrifugal acceleration which is maximum at the equator and zero at the poles (**Table 1**). For gravity measurements made on land, several further corrections must be made (**Table 1**). The 'free-air correction' is made to adjust for difference in height between the measurement point and sea-level. This does not make any assumptions about the material between the sea-level datum level and the observation point and uses the inverse square law and the assumption of a spherical Earth. The 'Bouguer correction', named after the French mathematician and astronomer, is used to account for the gravitational effect of the mass of material between measurement point and sea-level. This requires assumptions to be made about the density of material, and the Bouguer plate or slab formula is applied (**Table 1**), which further assumes that this material is a uniform infinite plate. Historically a 'density correction' value of $2670 \text{ kg m}^{-3}$ has been used as a standard density for crustal material, and this corresponds to a Bouguer correction of 1.112 g.u./m, negative above sea-level. A 'terrain correction' may be applied to compensate for the effect of topography, again requiring assumptions about densities. Nearby mass above the gravity measurement station will decrease the reading and any nearby topographic lows will have been be artificially 'filled in' by the Bouguer correction so the correction is always positive. An additional correction to gravity measurements made on a moving vehicle such as an aeroplane or boat is the Eötvös correction, which depends on horizontal speed vector, latitude, and flight altitude.

## Gravity Anomalies and Derivatives

Since for most geological applications the perturbations in the gravity field across an area or feature of interest are more important than the absolute gravity values, it is standard to compute gravity anomalies by subtracting the theoretical gravity value from the observed. The Bouguer gravity anomaly is the observed value of gravity minus the theoretical gravity value for a particular latitude and altitude, as outlined in **Table 1**. The Bouguer gravity is commonly used on land where maps of gravity anomalies can be used to view gravity data in plane view and it is convenient to have topographic effects (approximately) removed. Offshore, the free-air gravity anomaly is most useful, as the measurements are straightforward to correct to the sea-level datum.

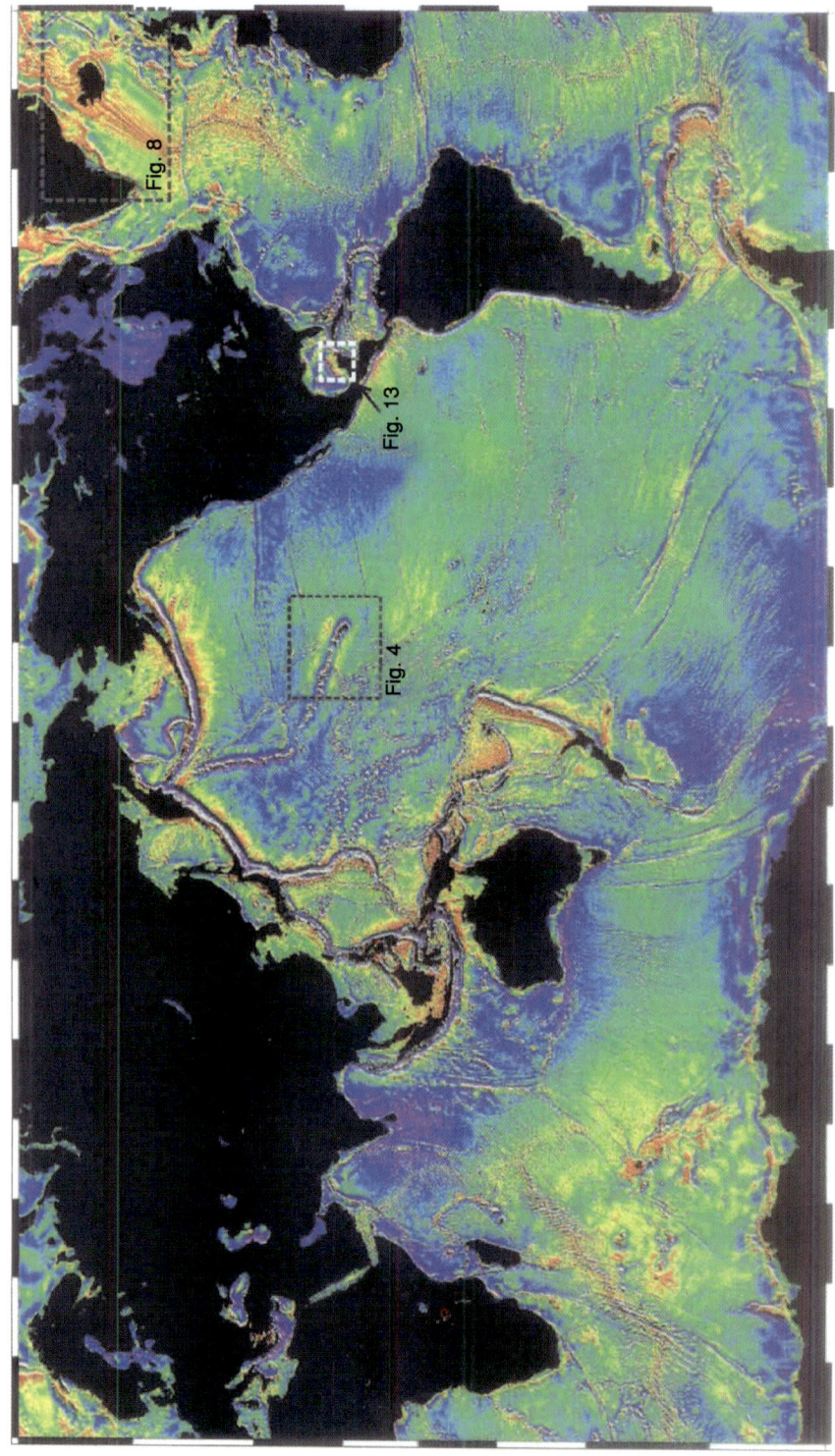

**Figure 2** Marine free-air gravity anomaly map derived from satellite altimetry (Sandwell and Smith (1997)). Warm colours indicate positive gravity anomalies. The gravity anomaly primarily indicates the shape of the seafloor, due to the strong density contrast from seawater to oceanic crust. Oceanic island chains, subduction zone trenches, and mid-ocean ridges form features visible on this world map. Locations of **Figures 4, 8** and **13** indicated. (Image courtesy of NGDC.)

High-pass filtering or subtraction of a planar function from gravity anomaly data may be undertaken to remove a 'residual' or background trend if the feature of interest is known to be shallow or a subtle perturbation to a strong regional gradient. Other treatments of gravity data include upward and downward continuation, by which different observation levels can be simulated, and computation of vertical or horizontal derivatives, which may emphasise structural trends in the data.

## Applications and Examples

### Submarine Topography

The satellite-derived free-air gravity anomaly map over the oceans (**Figure 2**) strongly reflects the nearest significant density change, the seabed. There are positive gravity anomalies over seabed topographic highs such as submarine seamounts and mid-ocean ridges and negative anomalies over bathymetric deeps such as the trenches associated with subduction zones, although long-wavelength isostatically compensated structures have no gravity anomaly above them.

The coverage of the marine free-air gravity anomaly data can be exploited to produce sea-floor topography data (**Figure 2**). For this purpose, shipboard depth surveys, usually made with sonar equipment, are used to supply the long-wavelength part of the transfer function from gravity to topography. The shipboard data is usually considered accurate but limited in global coverage due to the spacing and orientation of survey ship tracks. Bathymetry interpolation using the satellite-derived gravity data highlights isostatically

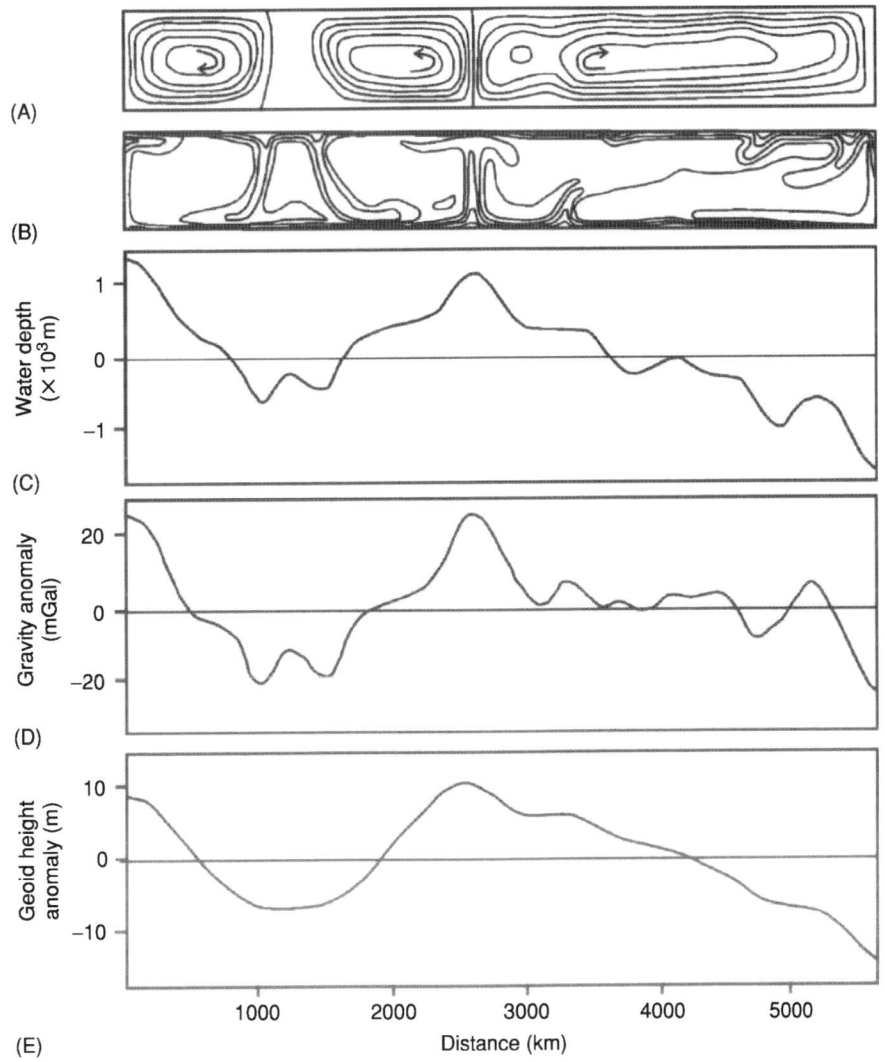

**Figure 3** Model of the gravity effect of convection in the Earth's mantle. (A) Stream function of computer modelled mantle flow (B) 100°C temperature contours (C) Variation in seafloor depth given a 30 km-thick elastic lithosphere above the convecting mantle (D) Modelled free-air gravity anomaly (E) Modelled geoid (sea-surface height) anomaly over the convecting mantle. (Reproduced with kind permission from McKenzie et al. (1980) © Nature Publishing Group. http:www.nature.com)

compensated topography which has no long-wavelength expression in the gravity data alone.

### Mantle Convection

It is commonly accepted that the Earth's mantle convects, and the flow of mantle material gives rise to gravity anomalies. Where mantle material is anomalously hot, it has a lower density than surrounding cooler mantle, and it will give rise to a negative gravity anomaly at the surface. This effect is, however, overprinted by the positive gravity anomaly caused by the upward deflection of the lithosphere above the rising anomalously hot mantle column or sheet (*see* **Mantle Plumes and Hot Spots**). There will, therefore, be a positive gravity anomaly over rising mantle material and a negative gravity anomaly where mantle is cool and sinking (**Figure 3**).

### Isostasy and Lithospheric Strength

Not all mountains would cause a gravitational plumbline deflection such as that observed at Schiehallion. Bouguer had observed that a plumb-line was only deflected by 8 seconds of arc towards the mountains during Condamine's Quito survey, while his calculations suggested that it should have been deflected as much as 1′ 43″. This anomalous lack of deflection was attributed by R. Boscovich in 1755 to 'compensation' for the mass excess of the mountain by underlying mass deficiency at depth. This fed into the development of 'isostasy', which addresses the issue of support for topography on the Earth's surface. Two alternative early views of isostatic theory were put forward in the 1850s. John Henry Pratt suggested that the amount of matter in a vertical column from the surface to some reference level in the Earth was always equal, and that this was achieved by the material in the column having lower density material below mountains than below topographic lows. George Biddell Airy advanced the alternative view using the analogy of icebergs, that elevated surface topography was underlain by low-density crustal roots which effectively displaced denser underlying material. Subsequent studies have used gravity data to investigate these alternative models in different tectonic settings and included the additional factor of the strength of the lithosphere to support loads.

At wavelengths shorter than about 500 km, the relationship between the gravity anomaly and topography

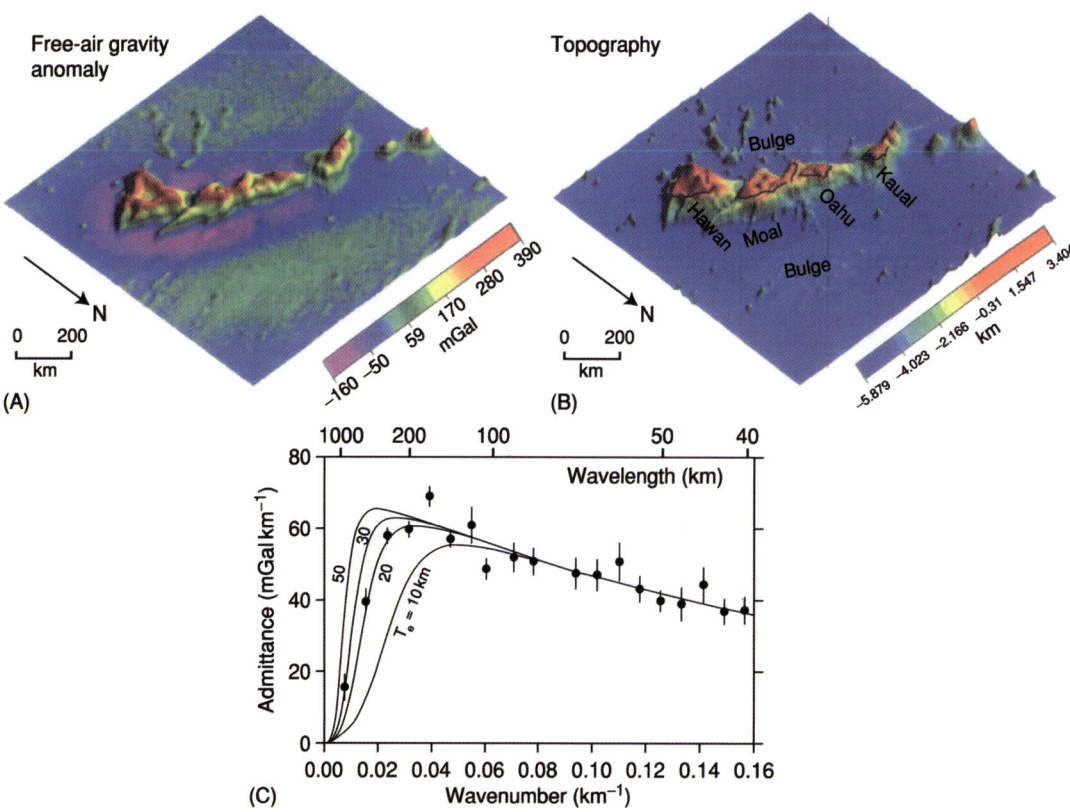

**Figure 4** Free-air gravity anomaly (A) and topography (B) in the region of the Hawaiian islands, Pacific Ocean (see **Figure 2** for location). (C) A comparison of the observed admittance along the Hawaiian-Emperor seamount chain (dots) with the predictions of a simple flexure model of isostasy, with varying elastic thickness, Te (lines). The observed admittance can be best explained with an elastic thickness for the lithosphere of 20–30 km. See Watts (2001) for more details. (Reproduced with kind permission from Watts (2001) © Cambridge University Press.)

is controlled by the mechanical properties of the lithosphere, which may be strong enough to support short wavelength loads, for example, isolated mountains. At longer wavelengths, the flexural strength of the lithosphere is commonly insufficient to support loads. The relationship between the gravity anomaly and topography can described by the wavelength-dependent 'admittance' function. The rate of change from flexurally-supported topography at short wavelength to topographic support by base-lithospheric pressure variations and regional density variation at long-wavelength depends on the effective 'elastic thickness' of the lithosphere.

**Figure 4** shows the topography and gravity anomaly of some of the Hawaiian island chain and the calculated admittance. For these islands, a modelled elastic thickness of about 25 km matches the admittance data. Recently, methods have been developed to also include the effect of lithospheric loads both with and without topographic expression in estimation of the elastic thickness.

## Density Contrasts, Analytical Models, and Non-Uniqueness

On a smaller scale, gravity anomaly maps provide the opportunity to identify and delineate sub-surface structures, as long as there are lateral density changes associated with the structure. Rocks at and near the surface of the Earth are much less dense than the Earth's average density of approximately $5155\,\text{kg m}^{-3}$, and crustal rocks are almost universally less dense than mantle rocks. An approximate density value of $2670\,\text{kg m}^{-3}$ is often taken as an average value for upper crustal rocks while values of $2850\,\text{kg m}^{-3}$ and $3300\,\text{kg m}^{-3}$ have been used for overall crustal rocks and uppermost mantle, respectively, although these values vary with composition and temperature. Many sedimentary rocks are less dense than metamorphic and igneous rocks. Coal ($1200-1500\,\text{kg m}^{-3}$) is one of the least dense rocks, while chalks and siliciclastic sedimentary rocks ($1900-2100\,\text{kg m}^{-3}$) are generally less dense than massive carbonates ($2600-2700\,\text{kg m}^{-3}$). With the exception of porous extrusive examples, crustal igneous rocks have densities approximately ranging from 2700 to $3000\,\text{kg m}^{-3}$. Density is not a diagnostic for lithology and variation in parameters such as porosity, temperature, and mineralogy can give significant density variability. Rocks with the lowest densities are those with very high porosities such as volcanic pumice, and in sub-aqueous environments recently deposited sediments. Density of sediments in a sedimentary basin tends to increase with depth as grains are compacted together (**Figure 5**). Igneous and metamorphic rocks tend to have higher densities than sediments as they frequently have negligible porosity and consist of relatively dense minerals.

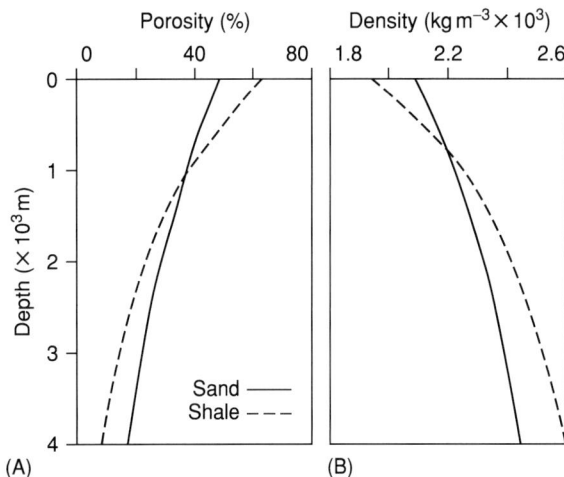

**Figure 5** Typical variation of (A) porosity and (B) density with depth below seafloor for sands and shales in a sedimentary basin. Increasing vertical effective stress with depth causes compaction of the rock, reducing porosity and correspondingly increasing density. Deeply buried sedimentary rocks, therefore, have higher densities than shallower rocks of similar lithology.

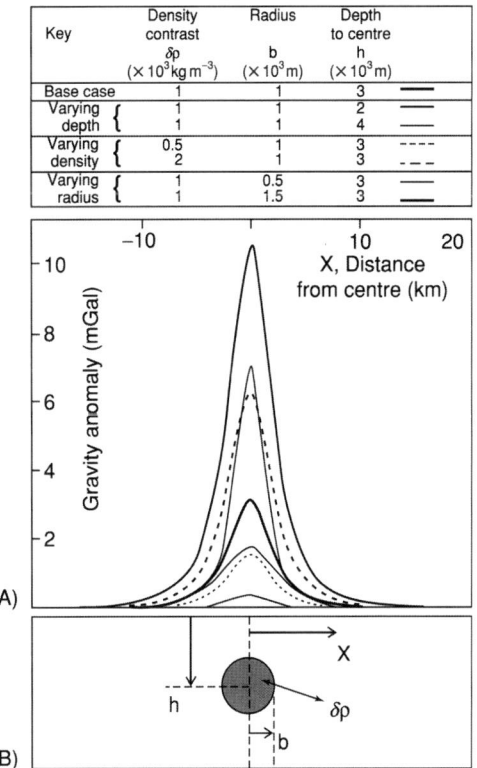

**Figure 6** Modelled gravity anomaly (A) along a transect through the centre of a buried sphere (B) of varying radius b, density contrast $\delta\rho$, and depth of burial h. The similarity in shape between the various cases shown highlights the difficulties of interpretation of gravity anomalies, as there are no unique solutions to explain a particular gravity anomaly.

Any non-uniformity in mass distribution results in lateral variability of the gravity field. For some simple geometrical shapes, a gravity anomaly can be calculated analytically (**Figure 6**). The buried sphere example illustrates the observation that deep density anomalies give rise to an anomaly over a wider surface distance than otherwise similar shallow anomalies, while greater density contrasts give larger anomalies than small density contrasts. The similarity in the gravity anomaly curves for the example of a buried sphere (**Figure 6**) illustrates one of the problems that arises in interpreting gravity data: there is no unique density distribution that produces a particular gravity anomaly. Gravity models tend to be constructed using additional geological or geophysical data such as seismic refraction or reflection profiles, surface geology (**Figure 7**), borehole density measurements, magnetic, magneto-telluric, or

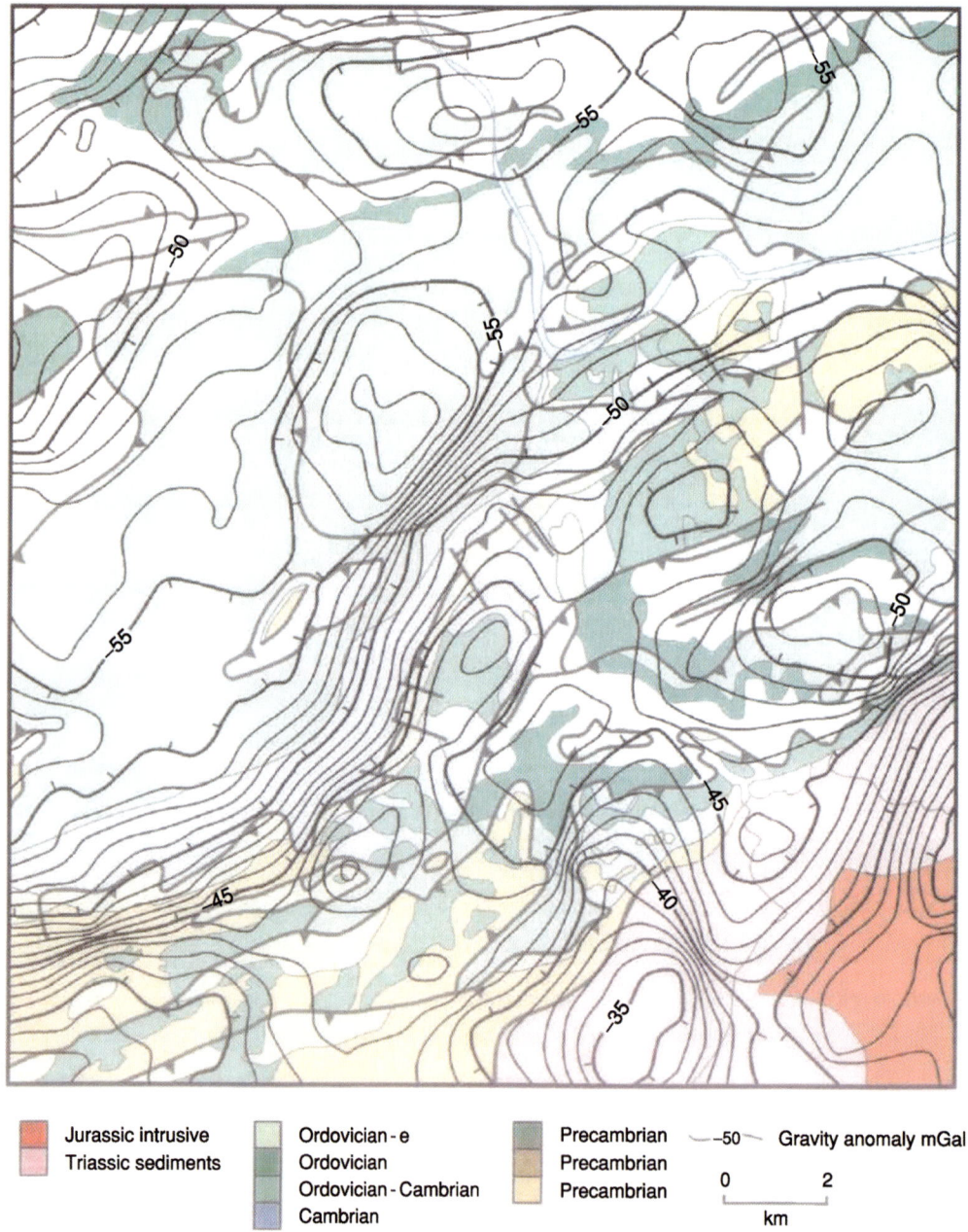

**Figure 7** Bouguer gravity anomaly contours overlain on geological map of part of Eastern Pennsylvania, USA. Bouguer gravity anomaly highs occur over the horst blocks of dense Precambrian material and other lows and highs in the gravity field are associated with formations of varying densities. (Geological map courtesy of the Bureau of Topographic and Geologic Survey, Pennsylvania Department of Conservation and Natural Resources, gravity data courtesy of W. Gumert and Carson Services Inc. Aerogravity Division, PA, USA.)

electromagnetic surveys as appropriate to arrive at a plausible and consistent interpretation.

### Crustal Observations from Satellite Gravity

Circular gravity anomalies similar in shape to that calculated for a buried sphere are observed over discrete igneous centres (**Figure 8**), where dense igneous rocks are inferred to have intruded less dense crustal rocks at a point of weakness in the lithosphere. Some of these 'bulls-eyes' in the gravity field have topographic expression but others may not have been identified without the satellite-derived gravity map. This map also allows identification of other large-scale crust-mantle interactions. One example is the set of south-ward pointing 'V-shaped' gravity anomalies flanking the mid-ocean ridge south of Iceland (**Figure 8**), which are caused by ridges and troughs in the top of the igneous crust. Although partially buried by sediment, these ridges have an expression in the gravity anomaly map because there is a significant density contrast between the igneous upper crustal rocks and the young pelagic sediments draping them.

### Modelling in Conjunction with Other Data

A combination of gravity data and other data types is often productive. For example, oceanic fracture zones identified in the satellite-derived gravity anomaly map are useful in conjunction with 'sea-floor stripes' in magnetic anomaly data to determine the relative movement between tectonic plates (**Figure 8**).

Gravity data is commonly used to verify interpreted seismic models. Empirical relationships between seismic velocity and density can be used to convert a seismic (*see* **Seismic Surveys**) velocity model into a density model and the predicted gravity anomaly compared with observations. The example shown in **Figure 9** shows a crustal velocity model along a 400 km line in the North Atlantic that has been

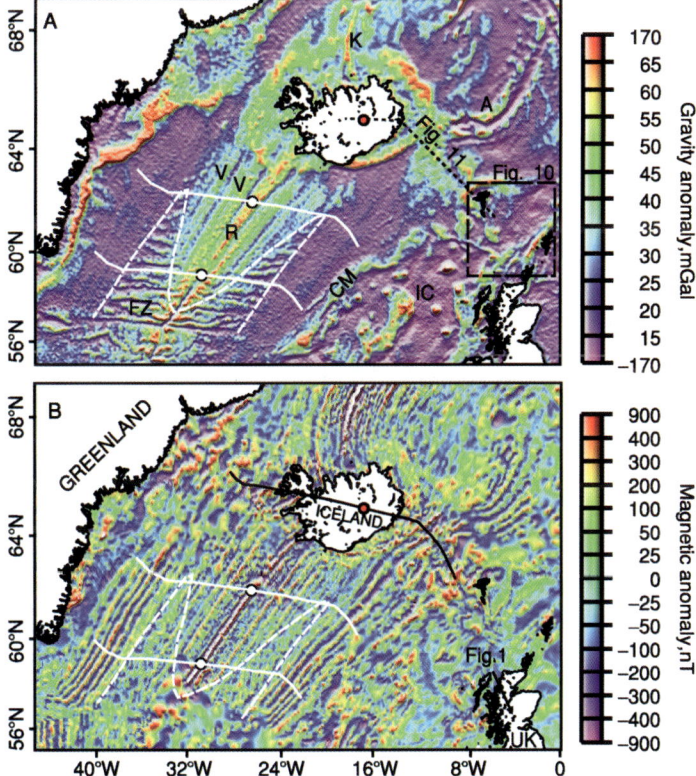

**Figure 8** Free-air gravity anomaly (A) and Magnetic anomaly (B) over the area surrounding Iceland (see **Figure 2** for location). The magnetic stripes form the record of magnetic field reversals during production of oceanic crust at the spreading centre. There are gravity anomaly highs over topographic highs such as the Reykjanes Ridge (R) and Kolbeinsey Ridge (K) spreading centres, and the extinct Aegir Ridge spreading centre (A). There are also circular gravity highs over igneous centres (IC) and linear anomalies along the continental margins (CM) and 'V-shaped' ridges (V) which flank the Reykjanes Ridge and reflect propagating pulses of anomalously hot mantle beneath the spreading centre. Red and white circles show the position of the present-day spreading centre plate boundary. Solid white line shows flowlines from present spreading centre indicating direction of paleo-seafloor spreading. These are determined from reconstruction of the magnetic stripes parallel to the fracture zones seen in the gravity data. Dashed white line shows area of oceanic crust disrupted by fracture zones (FZ); outside this area, oceanic crust was formed at a spreading centre without fracture zones on this scale. Dotted black line indicates approximate line of **Figure 12**. Location of **Figure 9** indicated. (After Smallwood and White (2002) with permission Geological Society of London.)

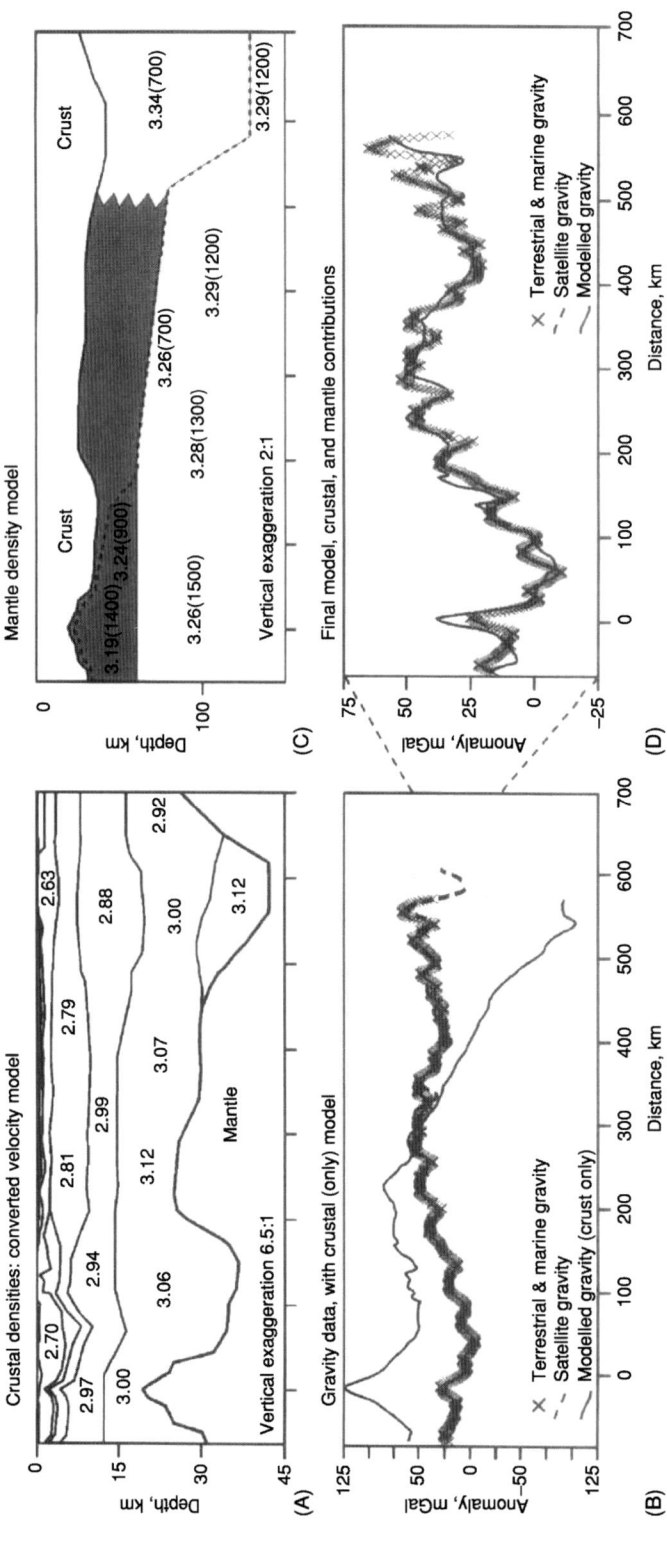

**Figure 9** Gravity model along a profile from Iceland to the Faroe Islands (see **Figure 8** for location). The crustal density model (A) was constructed by converting a wide-angle seismic velocity model to density using published empirical relationships. The gravity signature of the crustal model (B) did not match the satellite-derived (crosses) or ship-board/land-based gravity anomaly measurements. When effects of varying lithosphere thickness and mantle density variations were included (C), a good match between model and data could be achieved (D). (After Smallwood et al. (1999) by permission © American Geophysical Union.)

converted to density. While the gravity anomaly signal expected from the crust alone does not match the observed gravity, when reasonable mantle temperature and compositional variations are included, a good match to the data can be obtained.

Gravity data is increasingly being incorporated into multivariable mathematical inversion projects in which multiple datasets are simultaneously modelled in order to increase confidence in a particular interpretation of the subsurface.

## Modelling Over Sedimentary Basins

Since there is often a significant density contrast between crustal and mantle rocks, gravity data may provide useful constraints on crustal thickness variations, which can occur in continental as well as oceanic settings. Lithospheric extension, for example, may thin the crust along with the rest of the lithosphere. As the relatively low density crust is thinned, it may isostatically subside and the resulting topographic low may form a sedimentary basin (**Figure 10**). If assumptions are made about rock densities, gravity anomaly data can be modelled to infer the extent of crustal thinning. Simplified models of the subsurface can be constructed and adjusted until a match or matches can be made to gravity observations. Mathematical inversion may assist by identifying a model which produces a gravity field that has a minimum misfit to observations.

In the example of this, shown in **Figures 11** and **12**, from the UK/Faroe-Shetland Basin, the gravity data is particularly valuable as flood basalts to the west of the basin make seismic imaging difficult. Although the top of the relatively dense mantle is elevated in the position where the crystalline crust is modelled to be most highly extended, there is a free-air gravity low caused by the dominance of the relatively low density water column and sedimentary fill which are constrained by seismic data, and the long wavelength effect of the thicker continental crust on the basin

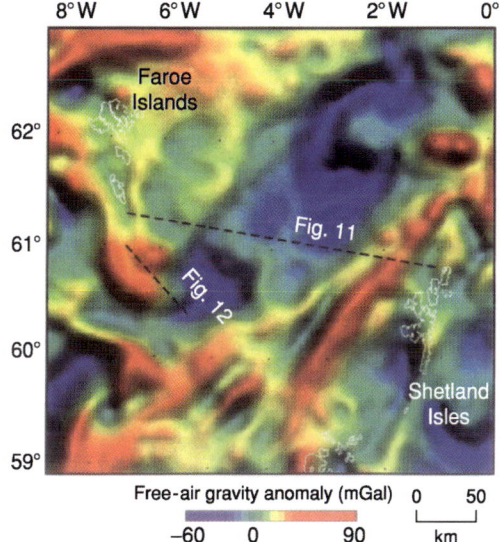

**Figure 10** Free-air gravity anomaly over Faroe–Shetland area derived from satellite altimetry (Sandwell and Smith 1997) and shiptrack data (see **Figure 8** for location). The dominant signal is the NW–SE gravity reflecting the area of deepest water between the Faroe Islands and the Shetland Isles. Shorter wavelength features arise from geological structures (see **Figures 11** and **12**).

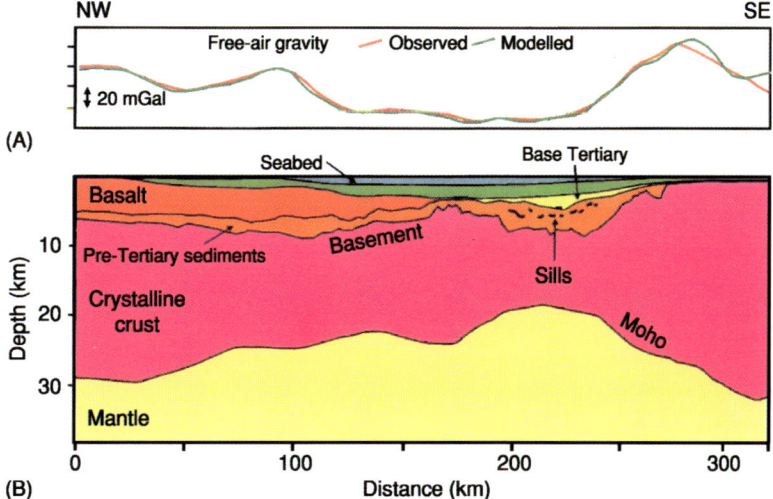

**Figure 11** Modelled and observed free-air gravity (A) along a profile between the Faroe Islands and the Shetland Isles (see **Figure 10** for location). The seafloor and other horizons (B) were partly constrained by seismic reflection data but beneath the basalt wedge reflections were not easy to interpret and the gravity modelling along this and other intersecting lines constrains a possible crustal model. (After Smallwood *et al.* (2001) with permission, Geological Society of London.)

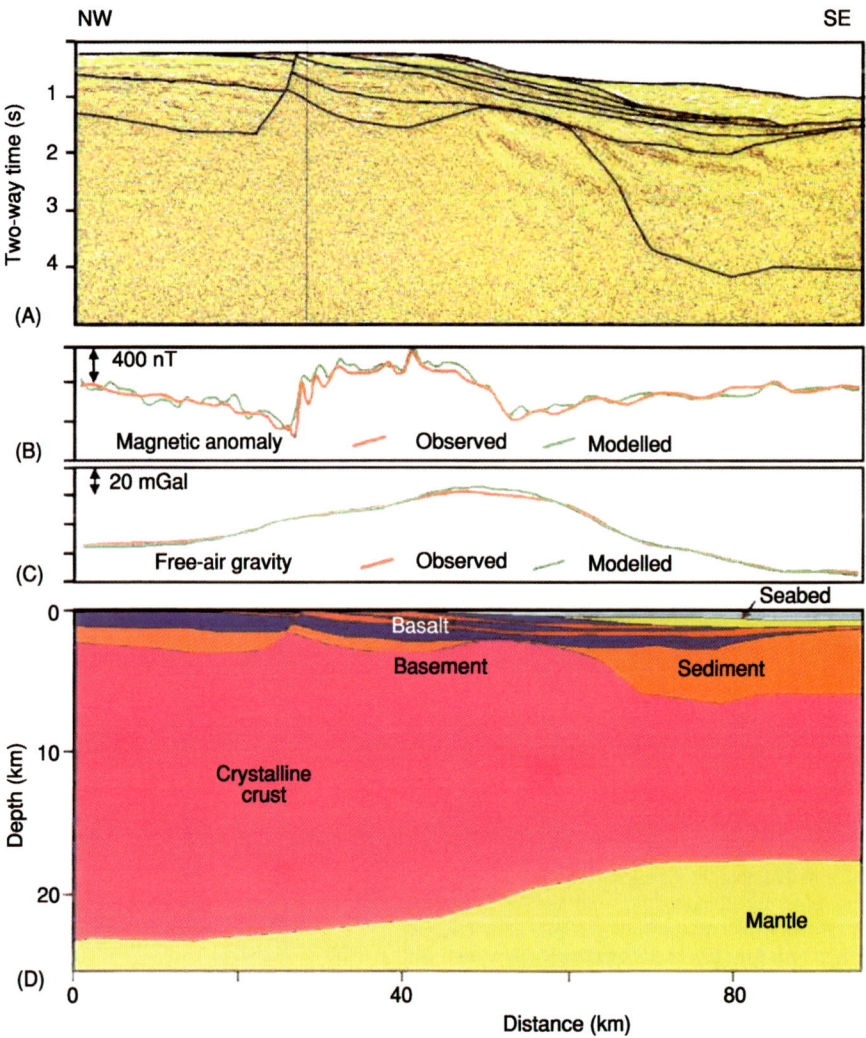

**Figure 12** Modelled and observed free-air gravity along a profile southeast of the Faroe Islands (see **Figure 10** for location). The seafloor and other horizons were partly constrained by seismic reflection data (A) but beneath the basalt reflections were not easy to interpret and magnetic anomaly (B) and gravity anomaly (C) modelling along this and other intersecting lines constrains a possible crustal model (D). (After Smallwood et al. (2001) with permission, Geological Society of London.)

margins. Another benefit added by gravity data to the understanding of this sedimentary basin was the requirement to add a unit with elevated density approximately 1 km thick in the centre of the basin to represent an interval intruded by igneous sills. The top of this unit was imaged well by seismic data but the thickness could not be estimated without the gravity model. **Figure 12** shows the value of modelling magnetic anomaly data along with the gravity to constrain basalt thickness and internal structure.

Another geological structure for which gravity data provides a useful tool of investigation is the Chicxulub impact crater in the northern Yukatan peninsular of Mexico. There is no dramatic surface expression of the site, but there are concentric circular rings apparent in the gravity anomaly (**Figure 13**). The gravity anomaly arises as the crater has been infilled with relatively low-density breccias and Tertiary sediments. The double humped central gravity high is thought to correspond to a central uplift buried deep within the crater. The Chicxulub crater is one example where 3D gravity modelling has proved useful to constrain crustal structures in three dimensions.

## Smaller Scale Surveys

Spatial deviation of gravity measurements is often used to infer lateral variations in density. If sufficiently accurate measurements can be made, then small-scale lateral variations in density can be inferred. Gravity surveying may be the best tool to identify mineral deposits if the target ores have densities contrasting with their host rocks. Massive

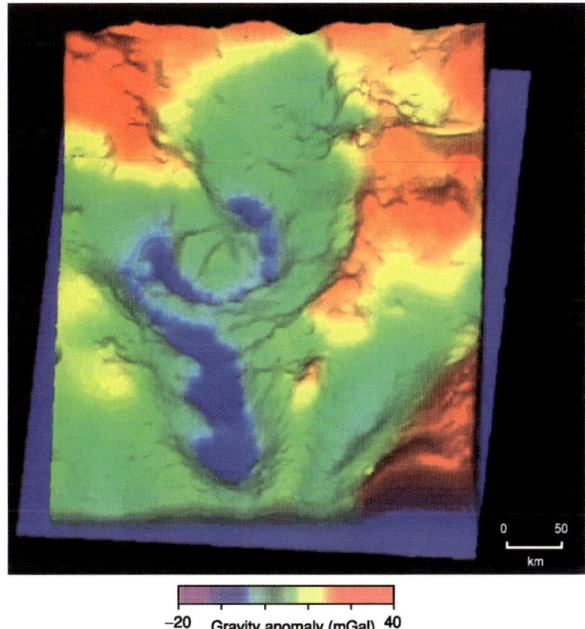

**Figure 13** Merged free-air gravity (offshore) and Bouguer (onshore) gravity anomalies across Chicxulub impact crater, Yukatan peninsular, Mexico (see **Figure 2** for location). Bouguer anomaly calculated with a reduction density of 2670 kg m$^{-3}$. Gravity anomaly over Chicxulub is a 30 mGal circular low with a 180 km diameter, with a central 20 mGal high. (Courtesy of Mark Pilkington, Natural Resources Canada.)

sulphides have densities ranging up to 4240 kg m$^{-3}$, and within host rocks of densities around 2750 kg m$^{-3}$, a sulphide body having a width of 50 m, a strike length of 500 m, and a depth extent of 300 m would give a gravity anomaly of about 3 mGal.

On a smaller scale, 'micro-gravity' surveys typically involve a large number of closely spaced gravity measurements aiming to detect gravity variations at levels below 1 mGal. These surveys may be designed for civil engineering projects where underground natural cavities in limestone or disused mine workings need to be detected, or depth to bedrock needs to be established. As with any gravity interpretation, any additional available information such as outcrop geological boundaries, density values of samples, or depths to important horizons may be incorporated in order to give a more realistic model.

## Gravity Gradiometry

Sometimes knowledge of the magnitude of the gravity field is not sufficient to resolve between competing geological or structural models. In the example shown in **Figure 14**, the conventional gravity data is rather insensitive to the geometry of the salt diapir as a dominant long-wavelength gravity signal originates

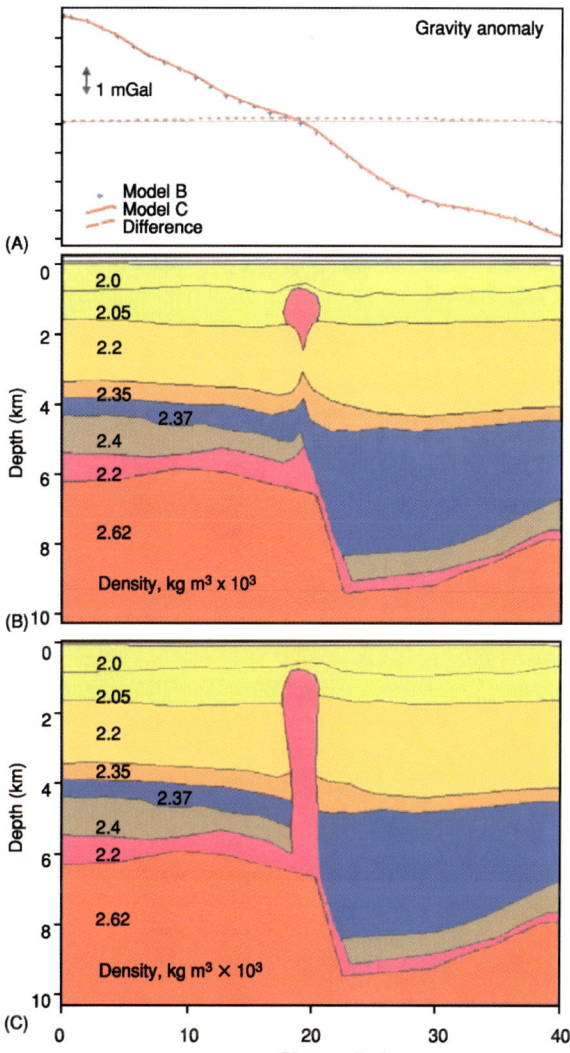

**Figure 14** Cross-sections across two gravity models. The modelled gravity response (A) for a cross-section over a fault block and small detached salt pillow (B) is very similar to the response over a bigger salt diapir (C) offset by some other changes to the model layers. Since uncertainty and noise in marine gravity data may be at a 1 mGal level, gravity modelling of the total field may not be able to distinguish between these models. Seismic data is often poor below the top of the salt. Courtesy of A. Cunningham.

from an underlying fault block. In this case, the gradients of the gravity field may provide additional assistance. An instrument to measure the gradient of the gravity field was developed by Baron von Eötvös in 1886, and a unit of gravity gradient was named after him (1 Eötvös = 0.1 mGal km$^{-1}$). The concept of his torsion balance was that two weights were suspended from a beam at different heights from a single torsion fibre, and the different forces experienced by the two weights would deflect the beam. The torsion balance was accurate but somewhat cumbersome and slow, and it was superceded by the more

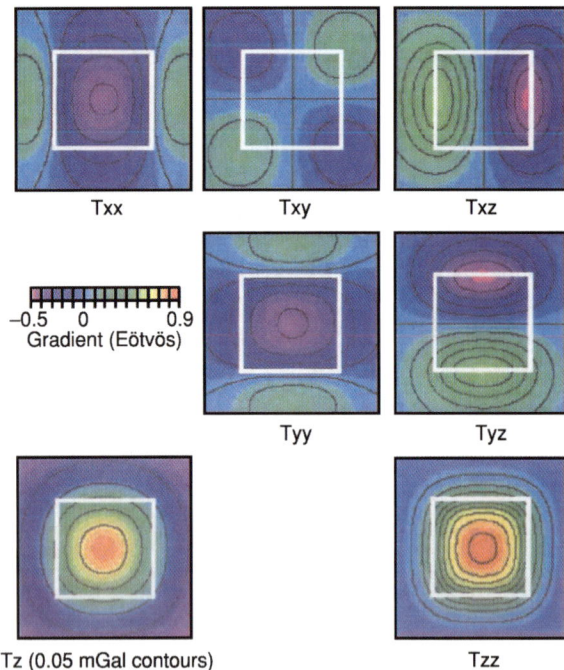

**Figure 15** Modelled components of the gradients of the gravity field over a cube. The different tensors represent changes in the gravity field gradient in different directions, for example, Tzz is the vertical gradient of the field measured in a vertical direction. It highlights edges of subsurface density contrasts. Tz is the (directionless) full gravity anomaly field. (Courtesy of C. Murphy and Bell Geospace Inc.)

convenient astatic gravity meters. However, recently declassified military technology has seen a renaissance of gravity gradiometry, with the availability of a full tensor gradiometer consisting of 12 separate accelerometers arranged in orthogonal pairs on three separate rotating disks. This instrument has a quoted instrumental accuracy of $10^{-11}$ gal. Five independent tensor measurements are made and can be modelled, each sensitive to different aspects of subsurface density variation (**Figure 15**). In addition to the valuable insights gained from this multicomponent data, the components can be recombined to give a high resolution 'conventional' gravity map which benefits from the precision of the instrumentation. In cases similar to the salt diapir example, inversion of such precise gravity data has been used together with a correlation between seismic velocity and density to produce a modelled seismic velocity field which can be used in seismic depth processing.

## Extra-terrestrial Gravity Fields

Gravity fields have been computed for the moon (see **Solar System: Moon**), and for Venus (see **Solar System: Venus**) and Mars (see **Solar System: Mars**), from observations of variation in artificial satellite orbits. The Doppler shift of spacecraft signals is observed, giving the spacecraft velocity in the 'line-of-sight' direction. The gravity field is calculated using a combination of many observations of line-of-sight acceleration.

In a similar method to that outline for terrestrial studies, the wavelength-dependent relationship between gravity anomalies and topography can be used to study the internal dynamics and the support of surface loads by the lithosphere. Gravity studies on Venus show that it has a similar lithospheric rigidity to the continents on Earth, despite its higher surface temperature, and that active mantle convection is responsible for the observed volcanic rises. In contrast, large gravity anomalies on Mars for example, a maximum anomaly of 344 mGal (from a spacecraft altitude of 275 km) over the crest of the Olympus Mons volcano, have led to the suggestion that the Martian lithosphere is extremely rigid. On the moon, circular positive gravity anomalies of up to 300 mGal have been identified, associated with basaltic lava flows infilling giant impact craters. These 'mascons' (mass concentrations) have provided a focus for debate on isostatic lunar history.

## Conclusion

Gravity is a versatile tool for investigation and can provide constraints on sub-surface structure on a wide variety of scales from man-made structures to the size of an entire planet. To unlock the information contained within the gravity field, gravity observations are best used in conjunction with other types of data such as surface topography, geological mapping, borehole information, and seismic data.

## See Also

**Mantle Plumes and Hot Spots**. **Mining Geology:** Exploration. **Petroleum Geology:** Exploration. **Plate Tectonics**. **Seismic Surveys**. **Solar System:** Venus; Moon; Mars.

## Further Reading

Bott MHP (1982) *The Interior of the Earth*, 2nd edn. Amsterdam: Elsevier.

Fowler CMR (1990) *The Solid Earth: An Introduction to Global Geophysics*. Cambridge, UK: Cambridge University Press.

Gibson RI and Millegan PS (eds.) (1998) *Geologic Applications of Gravity and Magnetics: Case Histories*. SEG Geophysical Reference Series 8/AAPG Studies in Geology 43. Tulsa, OK: Society of Exploration Geophysicists and the American Association of Petroleum Geologists.

Gumert WR (1998) A historical review of airborne gravity. *The Leading Edge* 17: 113–117. http://www.aerogravity.com/carson2.htm.

Hansen R (1999) The gravity gradiometer: basic concepts and tradeoffs. *The Leading Edge* 18: 478, 480.

Hildebrand AR, Pilkington M, Connors M, Ortiz-Aleman C, and Chavez RE (1995) Size and structure of the Chicxulub crater revealed by horizontal gravity gradients and cenotes. *Nature* 376: 415–417.

Hutton C (1778) An account of the calculations made from the survey and measures taken at Schiehallion, in order to ascertain the mean density of the Earth. *Phil. Trans. Royal Soc.* LXVIII: 689–788.

McKenzie DP, Watts AB, Parsons B, and Roufosse M (1980) Planform of mantle convection beneath the Pacific. *Nature* 288: 442–446.

McKenzie DP and Nimmo F (1997) Elastic thickness estimates for Venus from line of sight accelerations. *Icarus* 130: 198–216.

Milsom J (2002) *Field Geophysics*, 3rd edn. Chichester, UK: John Wiley and Sons.

Sandwell DT and Smith WHF (1997) Marine gravity anomaly from Geosat and ERS 1 satellite altimetry. *Journal of Geophysical Research* 105: 10039–10054. (www.ngdc.noaa.gov)

Smallwood JR, Staples RK, Richardson KR, White RS, and the FIRE working group (1999) Crust formed above the Iceland mantle plume: from continental rift to oceanic spreading center. *Journal of Geophysics Research* 104(B10): 22885–22902.

Smallwood JR, Towns MJ, and White RS (2001) The structure of the Faeroe-Shetland Trough from integrated deep seismic and potential field modelling. *Journal of the Geological Society of London* 158: 409–412.

Smallwood JR and White RS (2002) Ridge-plume interaction in the North Atlantic and its influence on continental breakup and seafloor spreading. In: Jolley DW and Bell BR (eds.) *The North Atlantic Igneous Province: Stratigraphy, Tectonic, Volcanic and Magmatic Processes*, pp. 15–37. London: Geological Society of London, Spec. Publ. 197.

Smith WH and Sandwell DT (1997) Global Sea Floor Topography from Satellite Altimetry and Ship Depth Soundings. *Science* 277: 1956–1962.

Telford WM, Geldart LP, and Sheriff RE (1990) *Applied Geophysics*, 2nd edn. Cambridge, UK: Cambridge University Press.

Watts AB (2001) *Isostasy and Flexure of the Lithosphere*. Cambridge, UK: Cambridge University Press.

# Mineral Analysis

**N G Ware**, Australian National University, Canberra, ACT, Australia

© 2005, Elsevier Ltd. All Rights Reserved.

## Mineral Analysis

Mineral analysis involves determining the chemical relationships between and within mineral grains. Microanalytical techniques are essential, and methods include X-ray spectrometry and mass spectrometry. Electron probe and laser ablation procedures are commonly used techniques for major and trace element analysis, respectively (*see* **Analytical Methods: Geochemical Analysis (Including X-Ray)**).

A chemical analysis of a mineral is expressed as a table of weight percent (wt.%) of its component elements or oxides. Concentrations lower than about 0.5 wt.% are often expressed as parts per million (ppm) by weight of element. These mineral analyses are easily converted into atomic formulas and thence into percentages of the end-member 'molecules' within the mineral group (see **Table 1**). Mineral analyses are used in descriptive petrology, geothermometry, and geobarometry, and in the understanding of petrogenesis. Sometimes thousands of analyses are collected in the completion of a single research project. Large amounts of data are presented graphically, plotting concentrations of elements or ratios of elements against each other, thus illustrating chemical trends or chemical equilibrium (see **Figure 1**).

In addition to the chemical analysis, a complete description of a mineral requires a knowledge of its crystallography. Both chemical composition and crystallography are required to predict the behaviour of minerals, and hence rocks, in geological processes. The discovery of each new mineral involves the determination of its crystal structure as a matter of routine using X-ray and electron diffraction techniques. Thus, when a monomorphic mineral is identified from its composition, its crystallography follows. Polymorphs may be identified by optical microscopy. Whereas it is sometimes convenient to identify an unknown mineral from its diffraction pattern, and although cell parameters can be used as a rough measure of end-member composition, crystallography no longer plays a major role in quantitative mineral analysis.

It was once necessary to separate a mineral from its parent rock by crushing, followed by use of heavy liquids and magnetic/isodynamic separators. Up to a

**Table 1** Analysis of garnet by EMPA/WDS for major elements and by LA-ICP-MS for trace elements[a]

| Oxides (wt.%) | | Atoms (Oxygen = 12) | | Trace (ppm) | | End member (%) | |
|---|---|---|---|---|---|---|---|
| $SiO_2$ | 37.12 | Si | 2.9985 | Ti | 172 | Almandine | 76.4 |
| $TiO_2$ | 0.03 | Ti | 1.9798 | Cr | 37 | Spessartine | 7.9 |
| $Al_2O_3$ | 20.80 | Cr | 0.0000 | Mn | 2716 | Grossularite | 2.7 |
| $Cr_2O_3$ | <0.01 | Fe | 2.3149 | Ni | 12 | Pyrope | 13.0 |
| FeO | 34.27 | Mn | 0.2386 | Cu | 32 | | |
| MnO | 3.49 | Zn | 0.0024 | Zn | 331 | | |
| ZnO | 0.04 | Mg | 0.3931 | Pb | 3.2 | | |
| MgO | 3.27 | Ca | 0.0809 | Y | 12 | | |
| CaO | 0.94 | Na | 0.0000 | La | 0.09 | | |
| $Na_2O$ | <0.02 | K | 0.0000 | Ce | 0.11 | | |
| $K_2O$ | <0.01 | | | Yb | 07.1 | | |
| Total | 99.96 | Total | 8.0098 | | | Total | 100.0 |

[a]EMPA/WDS, Electron microprobe analysis/wavelength-dispersive spectrometry; LA-ICP-MS, laser ablation inductively coupled mass spectrometry.

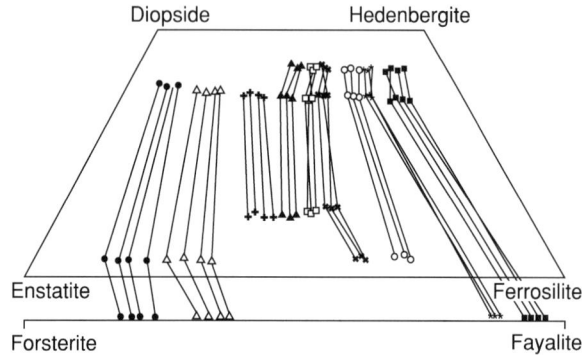

**Figure 1** Depiction of plots of multiple mineral analyses: coexisting olivine, orthopyroxene, pigeonite, and two augites at various levels (different symbols) in a differentiated tholeiitic sill. Including duplicates, 246 analyses were used.

gram of a mineral was obtained in this laborious way and then analysed gravimetrically, destroying the aliquot in the process. With the advent of the electron microprobe X-ray analyser in the 1950s and reliable matrix correction methods in the 1960s, non-destructive microanalysis of minerals *in situ* in a polished thin section became possible. To the mineralogist, microanalysis refers not to microgram quantities, but to analysis on the micrometre scale. Today, a variety of microanalytical techniques are available and used almost to the exclusion of bulk techniques, which are reserved for petrological analysis and control of mining operations. Microanalytical techniques available to the mineralogist are listed with their acronyms in **Table 2** and are discussed in the following sections. Nearly all major element mineral analysis is now carried out by electron microprobe analysis (EMPA), and laser ablation inductively coupled plasma mass spectrometry (LA-ICP-MS) is fast becoming the technique of preference for trace element work.

## Sample Preparation

The sample must be sectioned and have a polished surface and be of a form so that selection of any point on the surface is rapid and exact. Samples may be mounted in epoxy resin, often using a 25 mm diameter mould, then are well polished, typically finishing with 0.25 μm particle-size diamond paste. The sample is usually top-loaded in a specimen holder: this requires the sides of the mount to be orthogonal, and the polished surface should have minimal bevelling. Alternatively, rock sections may be prepared by gluing them to a glass slide, then grinding them to a 30 μm thickness and polishing well to create a section suitable for both optical microscopy and microanalysis. The thickness of the section must conform to the requirements of the analytical techniques used. In microanalysis, the term 'thin' is usually reserved for a section thin enough for the exciting beam to pass through it. This is essential in analytical electron microscopy (AEM), in which a <500 nm film prepared by ion beam milling is supported on a metallic grid. In proton-induced X-ray emission (PIXE) spectroscopy, it is sometimes advantageous to use sections 50–80 μm thick so that the proton beam can transit the section without causing damage. For micro-X-ray fluorescence (XRF) techniques, the sections should be no thicker than the grain size of the minerals.

When beams of charged particles are used, the sample must be an electrical conductor, and samples are coated with an evaporated conductive film. For

**Table 2** Microanalytical techniques used in mineral analysis

| Technique | Acronym | Probe | Product | Best practical concentration range | Analytical volume Diameter | Depth |
|---|---|---|---|---|---|---|
| Electron microprobe analysis | EMPA | Electrons | X-rays | 5 ppm-100% | 0.5–4 μm | 0.3–8 μm |
| Scanning electron microscopy | SEM | Electrons | X-rays | 200 ppm-100% | 0.5–4 μm | 0.3–8 μm |
| Transmission electron microscopy (analytical electron microscopy) | TEM (AEM) | Electrons | X-rays | 100 ppm-100% | 5–20 nm | 50–100 nm |
| Proton-induced X-ray emission | PIXE | Protons | X-rays | 2 ppm-5% | 1–5 μm | 40–80 μm |
| Micro-X-ray fluorescence | XRF | X-rays | X-rays | 10 ppm-1% | 10 μm–1 mm | 100 μm–2 mm |
| Synchrotron X-ray fluorescence | SXRF | X-rays | X-rays | 500 ppb-2% | 2 μm–1 mm | 100 μm–2 mm |
| Laser ablation inductively coupled plasma mass spectrometry | LA-ICP-MS | Laser | Ions | 1 ppb-0.5% | 10–200 μm | 10–100 μm |
| Secondary ion mass spectrometry (sensitive high-resolution ion microprobe analysis) | SIMS (SHRIMP) | Ions | Ions | 0.1 ppb-1% | 5–50 μm | 100 nm–2 μm |

electron excitation, the resistivity of the film should be less than $10 \, k\Omega \, mm^{-1}$ and a carbon coat about 25 nm thick is usually sufficient. Passing a current of 50 to 150 A through sharpened electrodes in a $10^{-5}$ torr vacuum or carbon thread at $10^{-2}$ torr creates a homogeneous coating at a distance of 5 to 10 cm from the carbon. Secondary ion mass spectrometry and PIXE analysts mostly use 5 nm gold films, which are readily produced by heating a weighed quantity of gold in a tungsten wire basket.

## Electron Microprobe Analysis

An electron probe is a beam of electrons accelerated through a high voltage and focused at the surface of the sample. The beam is commonly produced by passing an electric current through a hairpin tungsten filament that is maintained at up to 30 kV in the scanning electron microscope (SEM), 50 kV in the EMPA, and 400 kV in the AEM. The electrons are channeled through a hole in an anode maintained at zero potential and then through a series of magnetic lenses that shape the beam into a circular cross-section and focus it to a size less than the scattering expected in the sample. The energy of the electrons at the surface of the sample (depth 0) is denoted $E_0$ and is measured in thousands of electron volts (keV).

### Electron Scattering

When an electron in the beam interacts with an atom in the sample, the most common result is for the electron to be scattered elastically, with a resulting change in direction and insignificant loss of energy. This is a simple Coulomb attraction between the approaching electron and the multiply charged nucleus. By comparison, the diffuse electron cloud in the atom, repelling in all directions, has a much smaller effect. The probability of elastic scattering increases with atomic number and decreases with higher $E_0$. It is possible for the electron to reverse direction, usually after several scattering events, and to leave the sample; this is known as back scattering and, at high atomic number and low $E_0$, more than half the electrons may be back scattered. In the AEM, the very high accelerating voltage ensures that the electrons pass through the thin sample with minimal scattering.

Electrons are also prone to inelastic scattering, which again involves a Coulomb reaction with the nucleus, but here the force of attraction by the opposite charges is actually translated into a reduction of momentum and the loss of kinetic energy is released in the form of an X-ray. These X-rays are known as *bremsstrahlung* ('braking radiation') and may have any energy up to the value of $E_0$. Inelastic scattering eventually causes the electron to come to rest and so defines the size of the interaction volume: the shape of the volume is determined by elastic scattering. Figure 2 illustrates such volumes.

### Characteristic X-Ray Generation

Electrons may also cause ionization by ejecting an electron from its atomic orbital, provided $E_0$ is greater than the critical excitation energy ($E_c$). The probability of ionization is much lower than that of inelastic deceleration or inelastic scattering. If an inner orbital electron is ejected, an outer electron may fill the

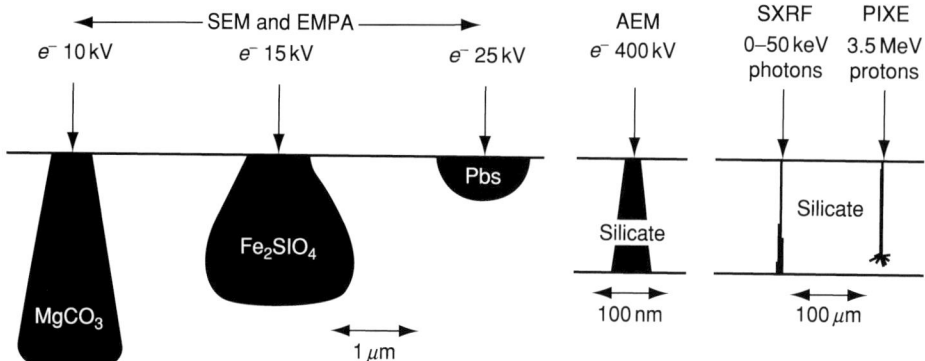

**Figure 2** Reaction volumes; the shape of the volume is determined by elastic scattering. SEM, Scanning electron microscopy; EMPA, electron microprobe analysis; AEM, analytical electron microscopy; SXRF, synchrotron X-ray fluorescence; PIXE, proton-induced X-ray emission.

vacancy and an X-ray photon having an energy equal to the difference in the energy levels of the two orbitals involved is emitted. The energy levels are a function of the number of protons in the nucleus, hence the energy of the X-ray is characteristic of the element. Thus, an electron-induced X-ray spectrum consists of characteristic lines superimposed on a continuum of *bremsstrahlung* of significant intensity.

Nearly all electron transitions occur by an electric dipole mechanism that is possible between only a limited number of orbitals; thus X-ray spectra are not complicated, containing only a few lines in any given energy range. The 1920s-era Siegbahn notation for X-ray lines is still in common use: for example, $L\beta_3$ denotes the third most intense line in the second most intense band in the L-shell spectrum; modern nomenclature would describe this line as $L_1-M_3$ which is the actual electron transition.

### X-Ray Spectrometry

X-Rays may be measured using three types of spectrometer: (1) the Bragg-law crystal monochromator, usually (mis)named the wavelength-dispersive spectrometer (WDS), (2) the energy-dispersive spectrometer (EDS), which uses semiconductors such as lithium-drifted silicon or intrinsic germanium, and (3) the new X-ray bolometers or microcalorimeters, which operate at temperatures close to absolute zero.

**Wavelength-dispersive spectrometer technology** The linear focusing spectrometer has proved the most efficient WDS design for the point source of X-rays generated in the electron probe (**Figure 3**). The analysing crystal is curved in a barrel shape so that the diffracted X-rays fall on the entrance slit of the detector. The crystal is mounted on a worm gear and moves linearly with changing inclination. The detector

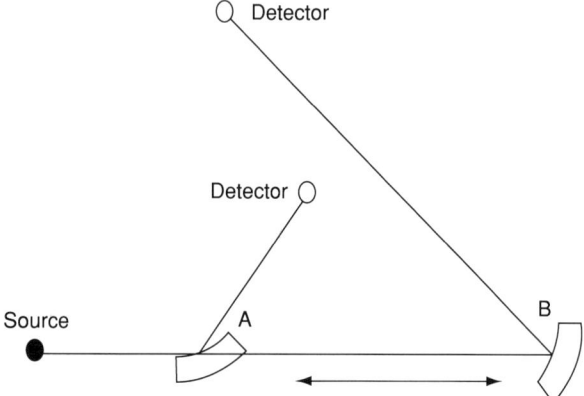

**Figure 3** Linear focusing spectrometer, showing the low (A) and high (B) angles.

moves in a parabola so that the angles subtended by the source and detector at the crystal are equal and so that the source-crystal and detector-crystal distances are equal. Most EMPA instruments are fitted with three to five spectrometers, each mounting a selection of two to four crystals that cover a full range of wavelengths. Lithium fluoride (LiF) is commonly used for short wavelengths, pentaerythritol (PET; $C(CH_2OH)_4$) is used for intermediate wavelengths, thallium acid phthalate (TAP; $C_8H_5O_4Tl$) is used for long wavelengths, and various vacuum-deposited multilayers (e.g., tungsten silicides, $WSi_x$) are used for ultralong wavelengths. Sealed xenon and gas-flow counters are used, the voltage on the wire being adjusted for operation in the proportional region with an internal gain of $10^3$ to $10^5$. A low-gain preamplifier is positioned as close as possible to the detector and a separate main amplifier shapes the counting pulses for digitization. The X-ray intensities are measured by collecting counts sequentially at characteristic X-ray peak positions and at predetermined background positions, if possible, on either side of the peak.

**The energy-dispersive spectrometer system** The detector consisting of lithium-drifted silicon, Si(Li), is a wafer made from a single crystal of silicon having a surface area of 10–30 mm$^2$. The electrons in the outer atomic orbitals are shared by several neighbouring atoms, forming an essentially covalent bond. The energies of these electrons in the 'valence band' orbitals are about 1.1 eV lower than the semiconductor's 'conduction band levels'. However, it takes about 3.8 eV to promote an electron from the valence to the conduction band. When an X-ray enters the crystal, it may be absorbed in an interaction with an electron of one of the silicon atoms, producing a high-energy photoelectron. This photoelectron dissipates its energy in interactions that promote valence band electrons to the conduction band, leaving holes in the once-filled valence band. The number of electron-hole pairs formed is proportional to the energy of the X-ray: for example, Ca K$\alpha$ (3.691 keV) would yield on average some 970 electron-hole pairs. A bias of 300–1500 V, depending on the thickness of the Si(Li) chip, is applied and the electrons are swept to the rear, where they enter the preamplifier as a weak pulse, the amplitude of which is proportional to the energy of the X-ray photon. Photo-optic feedback around a field-effect transistor is used in the preamplifier. Room temperature thermal excitation is sufficient to promote electrons from the valence to conduction band, so both the detector and the preamplifier are operated at liquid nitrogen temperatures. Sophisticated electronics are necessary to process the weak pulses in order that they may be stored accurately in a multichannel analyser. The whole X-ray spectrum is stored simultaneously, resulting in an X-ray histogram. Examples of EDS spectra are given in **Figure 4**; using these spectra as fingerprints, identification of minerals is possible with counting times as short as 100 ms. Within the EDS histogram, it is not always possible to measure background on either side of the peaks, and for quantitative analysis, spectrum deconvolution methods are necessary. Commercial software uses either background modelling or filter fitting.

*Energy resolution* The resolution of an EDS is quoted as the full-width at half-maximum (FWHM)

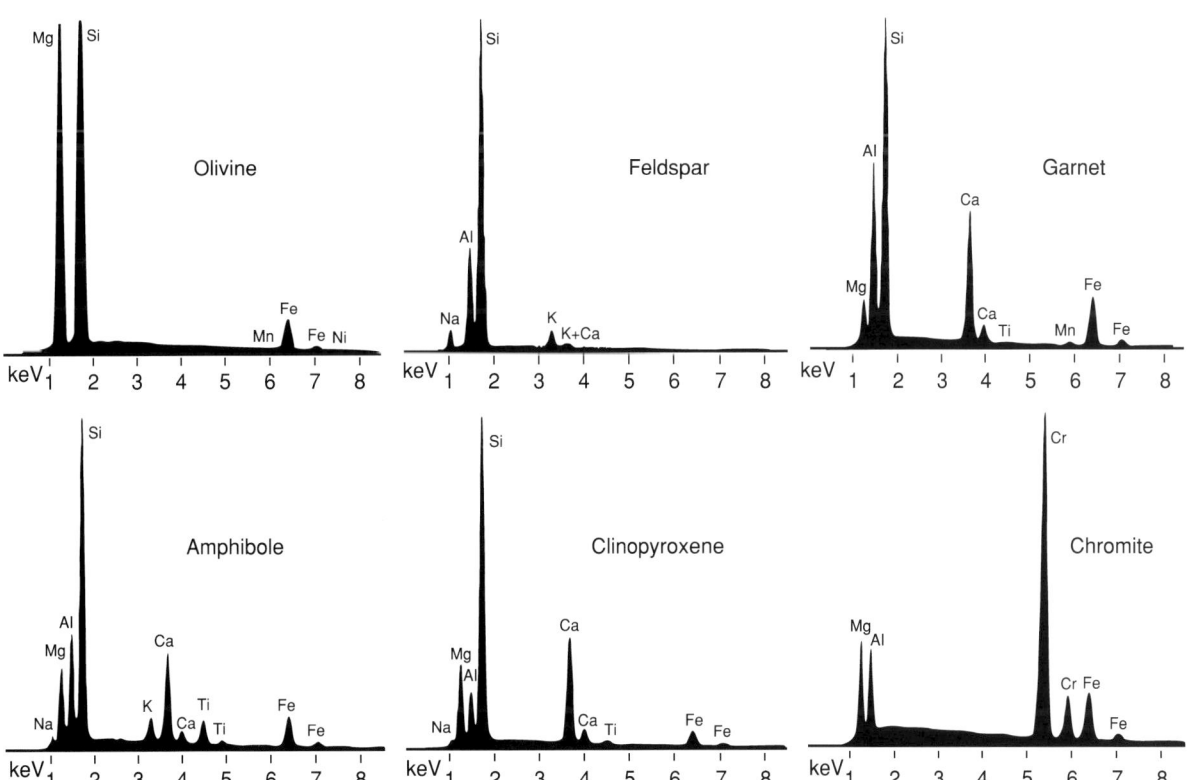

**Figure 4** Scanning electron microscope/wavelength-dispersive spectrometer spectra of minerals (15 kV, 1 nA, 20 s, 14 000 counts s$^{-1}$).

**Table 3** X-ray spectrometer devices

| Crystal/device[a] | Dispersion method[b] | Resolution FWHM (eV)[c] | Max practical count rate (kHZ) | Energy range (keV) | Collection area (mm$^2$) |
| --- | --- | --- | --- | --- | --- |
| LiF | WDS | 5–10 | 50 | 4–12 | 300–1200 |
| PET | WDS | 4–8 | 50 | 1.5–6 | 300–1200 |
| TAP | WDS | 3–5 | 50 | 0.8–2.2 | 300–1200 |
| WSi$_x$ | WDS | 4–6 | 40 | 0.2–0.9 | 300–1200 |
| Si(Li) | EDS | 100–200 | 20 | (0.3) 1–20 | 10–30 |
| Ge | EDS | 90–190 | 20 | 1–40 | 10–30 |
| SDD | EDS | 110–250 | 500 | 1–20 | 100–400 |
| Bolometer | EDS | 2–10 | 1 | 0.2–8 | 0.5–1 |

[a]LiF, Lithium fluoride; PET, pentaerythritol; TAP, thallium acid phthalate; WSi$_x$, tungsten silicides; Ge, germanium; SDD, silicon drift detector.
[b]WDS, Wavelength-dispersive spectrometry; EDS, energy-dispersive spectrometry.
[c]FWHM, Full-width half-maximum.

of the Mn K$\alpha$ peak. This measure is chosen because readily available $^{55}$Fe is a source of this X-ray line. For the Si(Li) detector, the FWHM at energy $E$ is given by $\text{FWHM}_E = (\text{FWHM}_0^2 + 21.1 FE)^{0.5}$, where $\text{FWHM}_0$ is the resolution at energy 0 and $F$ is the Fano factor, which is a measure of the statistical fluctuations in the ionization and charge collection processes. Table 3 lists the performance of a range of X-ray spectrometers.

**Other energy-dispersive spectrometer technology** Germanium detectors have properties similar to those of Si(Li) detectors and are preferred for use at higher energies. They are found on AEM, PIXE, and synchrotron X-ray fluorescence (SXRF) instruments. The silicon drift detector (SDD) is based on charge-coupled semiconductor technology and can provide energy resolution similar to that of the monolithic Si-crystal EDS, but at a count rate of 500 kHz. Furthermore, an energy resolution of 140 eV can be achieved at only $-13\,^\circ$C. The detector area can be made as large as 400 mm$^2$ so that low currents can be used for high count rates. It is possible to count at more than 1 MHz, but the resolution degrades as the count rate increases. This makes the detector unsuitable for quantitative analysis but ideal for mapping of mineral grains.

X-Ray bolometry has been developed using thin-film Ag microcalorimeters, transition edge sensors, and superconducting quantum interference devices. Such detectors have energy resolutions down to 2 eV and count rates of only 1 kHz. In theory, arrays of these millimetre-sized devices could be constructed giving a high overall count rate. The operating temperature is 70–100 mK and it is possible to achieve this using multistage Peltier cooling and an adiabatic demagnetization refrigerator.

**Matrix Corrections**

X-Ray intensities are measured in units of counts per second per nanoampere of beam current. The weight percent concentration of an element in a sample, $C_{\text{samp}}$, is related to the characteristic X-ray intensity, $I_{\text{samp}}$, by the equation $C_{\text{samp}} = C_{\text{std}}(I_{\text{samp}}/I_{\text{std}})([\text{MATRIX}]_{\text{samp}}/[\text{MATRIX}]_{\text{std}})$, where [MATRIX] denotes the effect of the chemical composition of the matrix on the X-ray intensity and 'std' refers to a standard of known composition.

There are four approaches to matrix corrections:

1. Empirical methods assume that each element linearly influences the X-ray intensity of each other element. A table of coefficients, analysed element against matrix element, is drawn up using extrapolations from measurements of binary alloys and solid solution series. These are known as alpha coefficients.

2. The ZAF corrections separately compute the effects of atomic number ($Z$), absorption ($A$), and secondary fluorescence ($F$): $ZAF = R/S f(\chi)(1+\gamma)$, where $R$ is the back-scattering fraction and $S$ is the X-ray generation factor due to stopping power; both of these are functions of atomic number. The function of the mass attenuation coefficient, $f(\chi)$, corrects for the absorption of the X-rays as they pass through the sample towards the detector. The additional contribution when a matrix X-ray fluoresces an analysed element ($E_m > E_{c,a}$) is represented by $\gamma$.

3. The $\phi(\rho z)$ methods: $\phi$ is defined as the ratio of the X-ray intensity from a thin layer, $\delta z$, of sample at a mass depth ($\rho z$) to the X-ray intensity of a similar layer isolated in space. The $\phi(\rho z)$ procedures integrate this X-ray intensity ratio function, corrected for multicomponent systems, from the surface to a

depth where $\phi$ becomes zero. Work by several groups spanning 30 years has established $\phi(\rho z)$ methods that give reliable matrix corrections for nearly all mineral analysis.

4. The quantitative microanalysis procedure developed by J L Pouchou and F Pichoir, and so named the PAP procedure, has affinities with both the *ZAF* and the $\phi(\rho z)$ methods. The depth distribution of X-ray generation is modelled using parts of two adjoining parabolas, functions that are easy to integrate. Stopping power and absorption are carried out together. Fluorescence and back scattering are corrected separately.

The EMPA instrument is designed specifically for X-ray analysis and current designs provide up to five WDSs, one EDS, and an optical microscope built round the electron optical column. Scanning-beam imaging by back-scattered electrons, secondary electrons, and cathodoluminescence is also possible. The specimen stage and spectrometers are automated and software has been developed that has transformed the electron microprobe analyser into a turnkey instrument.

## The Scanning Electron Microscope

Good quality mineral analyses may by obtained by scanning electron microscopy (SEM) and energy-dispersive spectrometry. SEM does not incorporate an optical microscope and the minerals must be located using back-scattered electron imaging. The energy-dispersive spectrometer must be robust and properly calibrated, requires stable electronics, and an appropriate spectrum deconvolution method must be employed. Many systems offer 'standardless analyses'; these methods work well over a restricted energy range but should be used with caution in mineral analysis, in which the energy range extends from Na (1.0 keV) to Fe (6.4 keV). Notwithstanding the high peak-to-background ratio, EDS analysis, properly executed, should have a better precision than WDS has for concentrations greater than 5 wt.%.

## The Analytical Transmission Electron Microscope

In analytical electron microscopy, sufficient current may be focused on a region as small as 5 nm in diameter, so that an X-ray flux greater than 1000 counts s$^{-1}$ measured by an EDS system may be generated in a 150-nm-thick film. Thus, extremely small domains of mineral grains may be analysed. The high value of $E_0$ helps to improve the peak-to-background ratios and enables the analysis of X-rays at the upper energy limits of the EDS detector. The intensity of the X-ray spectrum is a function of the indeterminate thickness of the sample. Quantification is attained by using ratios of peak intensities to that of a common element (Si for silicates, Fe for opaques) and assuming a value (usually 100%) for the sum of the analysed components: $C_x/C_{Si} = k_{x,Si}(I_x/I_{Si})$, $\sum C_x = 1.0$, where $C$ is concentration and $I$ is intensity. The calibration constants, $k_{x,Si}$, are evaluated empirically and correct for the difference in EDS efficiency and resolution at different energies. Matrix corrections are slight and are often ignored but can be applied using a rough estimation of the film thickness. Beryllium-window Si or Ge detectors that do not detect elements lower than sodium are used, but the technique of electron energy loss spectrometry (EELS) can be used in the analytical transmission electron microscope to determine elements down to beryllium.

## Proton Induced X-Ray Emission

In the proton probe, the protons are typically accelerated through 2.5–3.5 MeV. This is below the threshold of the 8 MeV required for atom smashing yet high enough to generate a reasonable flux of X-rays. The mass of the proton is 1837 times that of the electron, and this large mass, combined with the higher energy, predicates very low scattering, either elastic or inelastic. The X-ray spectrum has a very low background and X-rays are generated to high $E_c$. The limiting factor for the use of high-energy characteristic X-rays is the capability of the X-ray detector. The protons may be focused down to submicrometre beams using collimation and a series of quadrupole magnetic lenses. The penetration of the proton beam is considerably longer than it is in the electron microprobe analyser: 3.5 MeV protons have a range of 100 $\mu$m in aluminium. Most of the X-rays are detected from depths of 20–50 $\mu$m, where the protons penetrate, causing very little damage. However, just before coming to rest, much of the kinetic energy is absorbed by the sample and considerable damage to the crystal lattice results (see **Figure 2**). The protons end up forming hydroxyl ions, free hydrogen, and hydrides.

The proton trajectory in the sample is essentially linear, with a smooth deceleration of about 100 eV per collision. The mechanisms of energy loss and ionization are well understood, and the algorithm of integration of PIXE X-ray yields along the path of the beam provides the foundation for a standardless microanalytical method. However, unavoidable uncertainties in mass attenuation coefficients, when applied over the relatively long distances, can give rise to unacceptable errors in the analysis of high concentrations of elements such as Na, Mg, Al, and Si.

## X-Ray Fluorescence

A beam of X-rays is not much scattered by solid matter and is absorbed only when ionization occurs. Ionization is caused when the energy of the X-ray waveform resonates with the energy of the electron orbital, increasing the amplitude of vibration so that the electron leaves the atom. The probability of ionization increases exponentially as the energy of the photon approaches the critical excitation energy. Thus, a primary X-ray beam is an efficient producer of characteristic X-rays in a sample. There is little background and the peak-to-background ratios are even better than is obtained in PIXE.

Commercial X-ray microprobes are available. A low-voltage X-ray tube and a waveguide focuses the X-ray beam down to a $2\,\mu m$ spot. In common with PIXE, the secondary X-ray intensity is too low for WDS and the spectrometers used are conventional Be-window Si(Li) detectors.

The intense 'white' radiation in a synchrotron beam line has been used in mineral analysis since the mid-1980s; X-ray photon fluxes have since increased by factors of $10^5$. Simple collimation with fine apertures can create a $<10\,\mu m$ beam that may be used to detect X-rays up to 40 keV in energy. Focusing mirrors (the Kirkpatrick–Baez method) produce a $<2\,\mu m$ spot, but the maximum useful energy is about 10 keV. Phase zone plates can obtain a nanometre focus and are used for X-ray mapping and microtomography.

## Laser Ablation

Laser probing started in the 1970s with ultraviolet (UV; e.g., 266 nm) lasers focused with multiple-lens UV optics. The ablated material from a single laser pulse was ionized by a second laser beam that horizontally flooded the space above the sample, and the ions were then extracted through a tube in the optics and into a time-of-flight (TOF) mass spectrometer (MS). This technology has improved so that spatial resolution is $<0.5\,\mu m$ and mass resolution is $>5000$ $M/\Delta M$. Resonant postionization techniques have made great improvements in sensitivity. These instruments are dedicated mass microprobes and are useful in sensitive qualitative applications, but there are problems in attempting quantitative analysis of refractory elements. Unfortunately, so many of the trace elements of interest to the mineralogist are refractory.

Since the mid-1980s, lasers have been used in conjunction with inductively coupled plasma mass spectrometers to form very successful laser microprobes. Today, the technique of laser ablation in conjunction with ICP-MS is used for the majority of published trace element mineral analyses. The polished sample is inserted in a cell and an inert gas is passed over the surface. The cell in **Figure 5** is of a sophisticated design; most ablation cells use a cylindrical box and only one gas, which serves both as an ablating and a carrier medium. A UV laser, collimated to 20–200 $\mu m$ and pulsing typically at 5 Hz, ablates the sample and the material is carried by the gas, usually argon, into a plasma torch, where most of the material is converted into monatomic cations. These ions are usually analysed by a quadrupole mass spectrometer.

The laser ablates the sample to a depth approximating the radius of the beam. Several hundred pulses are used and a stream of material enters the plasma torch over a period of up to several minutes. The ionized product of the ICP enters the high vacuum of the MS through a series of metal cones having small apertures in their tips. As little as 0.01% of the sample may enter the MS. The MS is cycled to detect the required isotopes in sequence; a cycle takes of the order of a second, and usually 10–30 isotopes are counted. Each cycle is deemed to analyse a 'slice' of the sample and the counts of each isotope in each slice are recorded by the on-board computer. Notwithstanding the inefficiency of transporting material into the MS, counts of $10^3$ to $10^5$ ppm$^{-1}$ are obtained. As analysis proceeds, material can build up on the surfaces within the equipment and isotopes may be detected when the laser is switched off. The MS is cycled for a period before starting ablation so that the background levels may be determined. Some operators are concerned that the act of running the laser

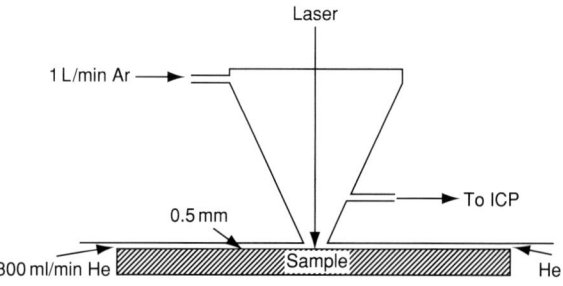

**Figure 5** Laser ablation cell (the HelEx system). ICP, Inductively coupled plasma.

may of itself dislodge material in the system, thus adding to the background. As a precaution, an ultrapure silicon standard may be analysed to quantify this effect.

The National Institute of Standards and Technology (US Department of Commerce) prepares and issues a range of glass standards that contain 61 elements at trace concentration, at approximately 50 and 500 ppm. These standards are commonly used in LA-ICP-MS and are measured repeatedly throughout an analytical session, perhaps once to every 10 or 20 sample measurements. Corrections for drift in the performance of the equipment with each sample analysis are applied linearly between standard measurements.

Quantification of the isotope counts requires knowledge of the concentration of at least one element, used as an internal standard; concentrations of other elements may be determined from the ratios of isotopes to that of the internal standard. A convenient element to use is calcium, which is present as a major element in most silicates and has five stable isotopes of widely ranging abundance. It is usually possible to select a calcium isotope giving a signal similar to that of the trace elements. Other elements can be used for example, nickel in olivine, vanadium in oxides, and titanium in micas. With effort, the EMPA laboratory can provide such internal standard concentrations to $\pm 10$ ppm at the 1000 ppm level, but isotopes of major elements such as Si and Mg can give good results. In sulphides, the sulphur concentration is usually known and, being an electronegative element, the cation signal is weak enough to be comparable with those of the trace elements. During ablation, much material condenses in and around the ablation pit, and the more refractory an element, the less likely it will be carried away by the gas. Thus fractionation processes occur even when the laser couples well with the mineral, and there is always a crater rim to the ablation pit (e.g., **Figure 6A**). In general the worse the coupling (**Figure 6D**), the greater the fractionation.

The first stage in quantification is to obtain isotope ratios corrected for background and fractionation. The background signal obtained with the laser off is averaged to give intensity values per slice, and these are subtracted (together with the values from the 'null' pure silicon standard if any) from the isotope signals measured with the laser on. Then ratios are calculated for each slice. These ratios, if plotted against slice number, will have a positive slope if the unknown undergoes less fractionation than the internal standard does; if not, then the slope will be negative. Either way, the plots are regressed to the point at which the laser is switched on and the value there is adopted for further calculation. Linear regression is often adequate; some operators prefer a polynomial. Anomalous slices, such as those containing inclusions in the mineral, may be excluded from the regression. Editing the data is facilitated by a good graphics computer program, but operations with a simple spreadsheet are adequate.

Quantification of the isotope ratios is continued by adjusting them with reference to the changes in ratios in the glass standard taken before and after the sample. Finally, the concentration of element x in the sample, $C_{x,samp}$, is given by the following equation:

$$C_{x,samp} = C_{int,samp}(I_{x,samp}/I_{int,samp})(I_{int,std}/I_{x,std})(C_{x,std}/C_{int,std})$$

where 'std' denotes the glass standard and 'int' is the internal element.

Differences in the coupling of the laser and hence the process of ablation between the glass standard and the sample are responsible for the major source of error in LA-ICP-MS. Another error is in the failure to predict overlaps on the analysed isotope. Overlap of isotopes of different elements and equal mass is either avoidable or readily quantified, but overlap from argon-sample dimers and from doubly charged ions may not be so obvious.

## The Ion Microprobe

In secondary ion mass spectrometry (SIMS) and in sensitive high-resolution ion microprobe (SHRIMP) analysis (the 'big brother' of SIMS), beams of $O^-$, $O^+$, $O_2^-$, or $Cs^+$ at 10–20 keV sputter the surface of the sample, yielding a mixture of ions, molecules, and plasma. Three types of mass spectrometers are used: magnetic sector, quadrupole, and time-of-flight. Although few useful ions are produced, unlike the LA-ICP-MS, nearly all the ions can be analysed by the mass spectrometer, and SIMS is a more sensitive technique. Erosion of the sample is usually 1–10 nm s$^{-1}$, which is much slower than laser ablation and much less sample is required. In **Figure 6**, the volume of material excavated in the SIMS pit is 0.3% that of the laser pits. SIMS is primarily an isotope ratio technique, but quantitative elemental analysis at very low levels is possible.

In contrast to EMPA, a general theory for matrix corrections in SIMS may never eventuate. Several specialized methods have been applied; for example, in the infinite velocity method, emission velocities, calculated from experimentally measured energy distributions, are extrapolated to infinite velocity, where there are no matrix effects. This approach works well

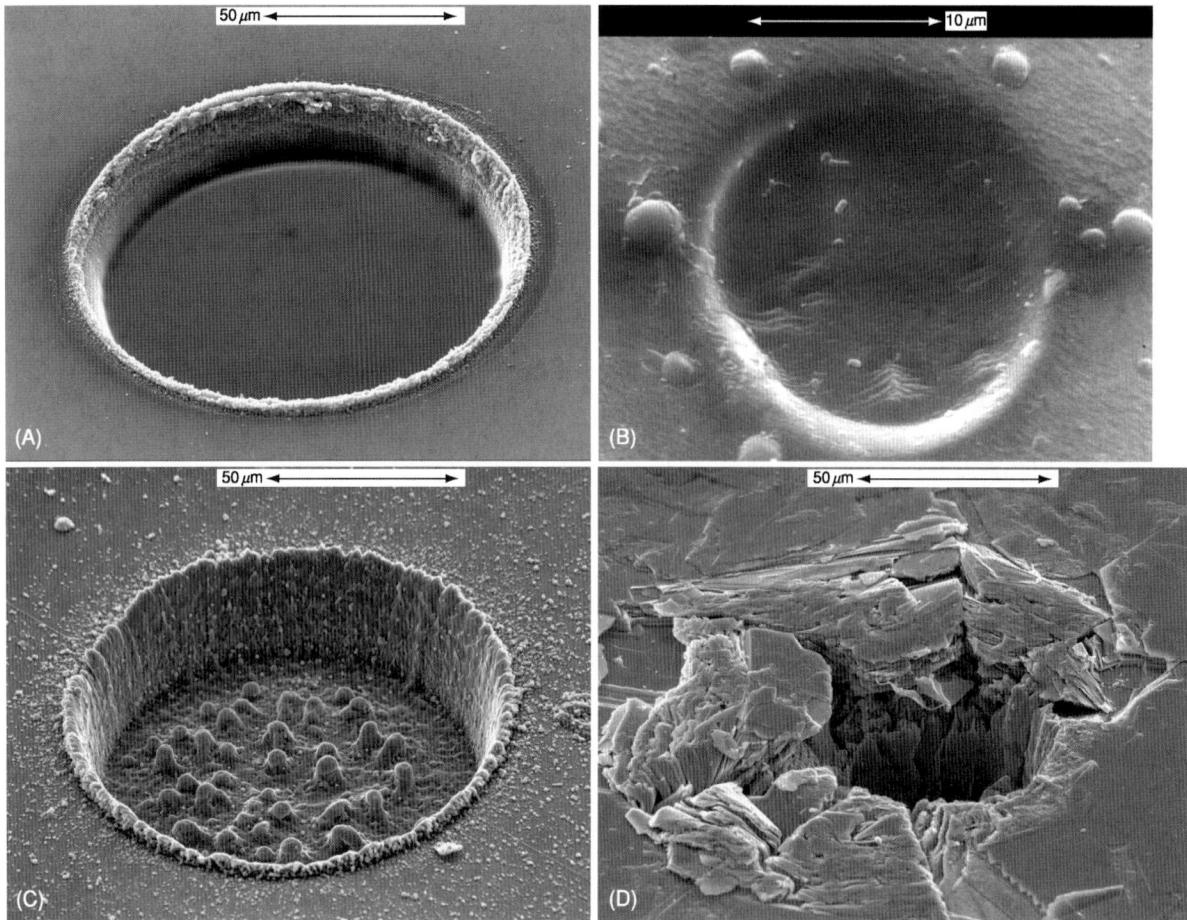

**Figure 6** Scanning electron microscope images of ablation pits, using (A) a 193-nm laser on calcite, (B) secondary ion mass spectrometry, with $O_2^-$ on zircon, and (C) 193-nm and (D) 226-nm lasers on molybdenite.

for trace elements implanted in simple matrices such as high-purity silicon. In mineral analysis, standards with the same crystal lattice as the unknown are required. Nevertheless, various laboratories have set up routine procedures for SIMS analysis in applications involving rare-earth elements, platinum group elements, and light elements, including hydrogen.

## Compositional Mapping

By moving the automated stage under the beam in any microanalytical instrument, it is possible to build up an array of data that may be transformed into false-colour maps of the sort shown in **Figure 7**. This has become a routine overnight procedure in many EMPA laboratories: the X-ray peak intensities may be recorded for each position together with the backscattered electron signal, which furnishes both an image of the area scanned and a template for the *bremsstrahlung* background. For some applications, it is possible to obtain a full quantitative analysis at each point (i.e., pixel) with acceptable precision. Usually the colour scale is calibrated roughly (as in **Figure 7**) from the software's calibration file and matrix effects are ignored. In addition to EMPA, compositional maps have been obtained using PIXE, SXRF, and SIMS. LA-ICP-MS does not have high spatial resolution but line scans are attempted with useful results.

## Other Mineral Analysis Methods

Analysis of $OH^-$, $CO_3^{2-}$, B, Be, and Li and the allocation of iron between $Fe^{2+}$ and $Fe^{3+}$ pose problems in mineral analysis by the methods outlined in the preceding sections. Of the light elements, only F may be analysed by EMPA with an accuracy comparable with heavier elements. However, B, Be, and Li may

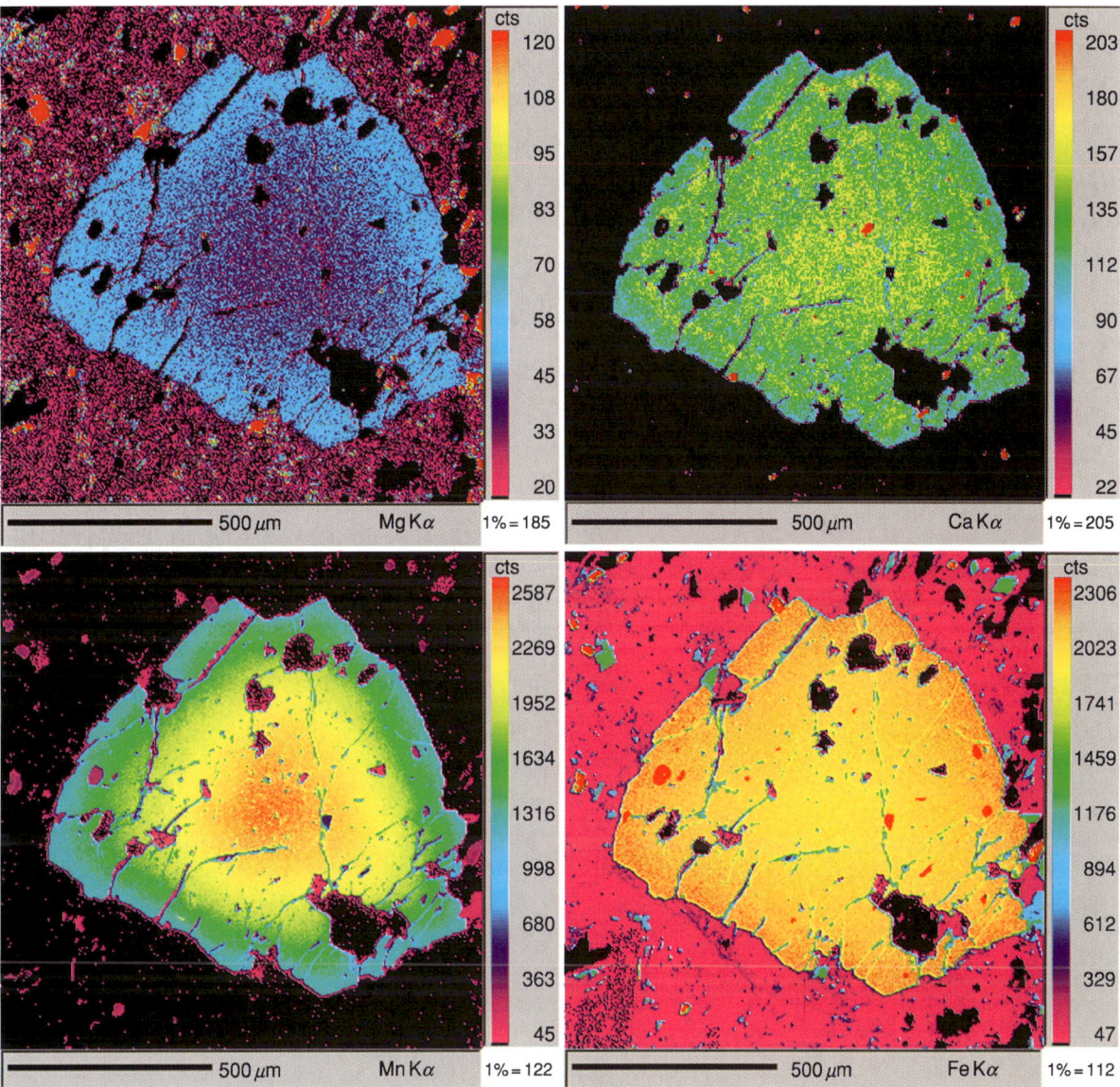

**Figure 7** Electron microprobe analysis/wavelength-dispersive spectrometry maps of garnet $(Mg,Ca,Mn,Fe)_3Al_2Si_3O_{12}$ ($600 \times 600$ pixels, 50 ms pixel$^{-1}$, with a 25-kV, 100-nA, 1-$\mu$m beam).

be determined by LA-ICP-MS, though at a limit of detection >50 ppm and with indeterminate accuracy at concentrations corresponding to borates and beryllates. Light elements are readily detected by SIMS but quantification is beset with uncertainties.

The local atomic environment around the nucleus, including the electronic, chemical, and magnetic state, may be studied using Mössbauer spectroscopy. The Mössbauer effect is the recoilless absorption and emission of $\gamma$-rays by specific nuclei in a solid, and the spectroscopy of $^{57}$Fe has been much studied with respect to mineral analysis.

Fourier transform infrared spectroscopy using an optical microscope can give quantitative information about anions such as $OH^-$ and $CO_3^{2-}$, but the thickness of the slide must be measured accurately. Multiple valency may be determined by X-ray absorption near-edge spectroscopy (XANES), which is performed on the synchrotron, and by X-ray photoelectric spectroscopy (XPS), which requires an ultrahigh vacuum and analyses the outer 10 nm of the sample.

For most silicates, the $FeO/Fe_2O_3$ ratio may be estimated by allocation of Fe to FeO and $Fe_2O_3$ so that the cation total equals the theoretical amount. A general equation can be used:

$$\text{wt.\% Fe(trivalent)} = (\text{total} - \text{theoretical})/\text{theoretical} \times \text{wt.\% O} \times 6.98125$$

In the garnet analysis in **Table 1**, the cation total is 8.0098 and the theoretical total is 8.0000; the wt.% oxygen is 39.54%, which is the sum of the oxides, 99.96%, less the sum of the elements. Application of this formula gives 0.48% $Fe_2O_3$ and 33.83% FeO, and recalculation of the mineral formula gives exactly 8.0000 cations. It is possible to analyse directly for oxygen with the EMPA, and if this is done with care, a similar result can be obtained but at the cost of extra instrument time.

## See Also

**Analytical Methods:** Fission Track Analysis; Geochemical Analysis (Including X-Ray). **Minerals:** Definition and Classification; Micas; Olivines; Sulphides.

## Further Reading

Cabri LJ and Vaughan DJ (eds.) (1998) *Modern Approaches to Ore and Environmental Mineralogy, Short Course Series*, vol. 27. Ottawa: Mineralogical Association of Canada.

Deer WA, Howie RA, and Zussman J (1997) *Rock Forming Minerals* (5 vols.). Bath, UK: Geological Society Publ. House.

Henderson G and Baker D (eds.) (2002) *Synchrotron Radiation: Earth, Environmental and Material Science Applications, Short Course Series*, vol. 30 Ottawa: Mineralogical Association of Canada.

Hurlbut CS and Sharp WE (1998) *Dana's Minerals and How to Study Them*, 4th edn. New York: Wiley.

Ireland TR (1995) Ion microprobe mass spectrometry: techniques and applications in cosmochemistry, geochemistry and goechronology. In: Hyman M and Rowe M (eds.) *Advances in Analytical Geochemistry*, vol. 2, pp. 1–118. Greenwich, CT: JAI Press.

Johansson SAE, Campbell JL, and Malmqvist KG (1995) *Particle Induced X-ray Emission Spectrometry (PIXE)*. New York: Wiley.

McCammon C (1995) Mossbauer spectroscopy of minerals. In: Ahrens TJ (ed.) *A Handbook of Physical Constants: Mineral Physics and Crystallography*, vol. 2, pp. 332–347. Washington, DC: American Geophysical Union.

Potts PJ (1992) *A Handbook of Silicate Rock Analysis*. London: Blackie Academic & Professional.

Reed SBJ (1993) *Electron Microprobe Analysis*, 2nd edn. Cambridge, UK: Cambridge University Press.

Ryan CG (1995) The nuclear microprobe as a probe of Earth structure and geological processes. *Nuclear Instruments and Methods* B104: 377–394.

Schulze D, Bertsch P, and Stucki J (eds.) (1999) *Synchrotron X-ray Methods in Clay Science*. Aurora, CO: Clay Minerals Society of America.

Sylvester P (ed.) (2001) *Laser Ablation-ICPMS in the Earth Sciences. Principles and Applications, Short Course Series*, vol. 29. Ottawa: Mineralogical Association of Canada.

# ANDES

**S M Kay**, Cornell University, Ithaca, NY, USA
**C Mpodozis**, SIPETROL SA, Santiago, Chile
**V A Ramos**, Universidad de Buenos Aires, Buenos Aires, Argentina

© 2005, Elsevier Ltd. All Rights Reserved.

## Introduction

The Andean mountains are the type example of an 'Andean'-type subduction zone characterized by subduction of an oceanic plate beneath a continental margin and uplift of a mountain range without continental collision. They extend some 8000 km from Venezuela to Tierra del Fuego and are the major morphological feature of South America. On the Earth's continents, they include the highest active volcanoes (>6800 m), the highest peaks outside of the Himalayas (*ca.* 7000 m), the thickest crust (>70 km), the second greatest plateau in height and area (after Tibet), the most important volcanic plateau with the largest Tertiary ignimbrite calderas, and among the most shortened continental crust, deepest foreland sedimentary basins, and largest and richest precious metal (Cu, Au, Ag) and oil deposits. The central Andes are also the type example of the effects of shallowly subducting oceanic plates and of non-accreting margins where continental lithosphere has been removed by the subduction erosion process. The evolution of the Andes began in the Jurassic coincident with the arc system that developed above subducting oceanic plates along the western margin of South America during and after the breakup of the

Mesozoic Pangaean supercontinent. The history of the Andes is predominantly a story of magmatism, uplift related to contractional deformation, intervening episodes of oblique extension, collision of oceanic terranes in the north, formation of sedimentary basins, mineralization, loss of continental crust by fore-arc subduction erosion, and removal of the base of overthickened crust by delamination. The mechanisms of uplift and crustal thickening along with the amount, timing and fate of removed continental lithosphere are hotly debated topics.

## Principal Geological Features of the Modern Andes

### Subducting Oceanic Crust and Distribution and Character of Andean Magmatism

The morphology and geology of the modern Andes are strongly influenced by the age, geometry, and morphology of the subducting oceanic plates (**Figures 1 and 2**). A first-order feature related to these subducting plates is the division of the active volcanic arc into the Northern (NVZ), Central (CVZ), Southern (SVZ) and Austral (AVZ) Volcanic zones (**Figures 3–6**). The NVZ, CVZ and SVZ are underlain by segments of the Nazca Plate that are subducting at an angle of ∼20–30°, and in which the magmas are principally generated by hydrous fluxing and melting of the mantle wedge. CVZ and northern SVZ magmas, erupted through the thick crust of the Central Andes, are primarily andesites and dacites, whereas central and southern SVZ and NVZ magmas, erupted through thinner crust, are primarily basalts and mafic andesites. Between these segments are the Peruvian and Chilean (or Pampean) amagmatic flat slab segments under which the subducting Nazca Plate forms a flat bench at ∼100 km that can extend up to ∼300 km east of the high Andes. The near absence of an asthenospheric wedge accounts for the lack of volcanism. The origin of the shallowly subducting segments of the Nazca Plate has been debated and variously attributed to subduction of the Juan Fernandez and Nazca ridges near their southern margins or complex interactions between the underriding and overriding plates. All of the models call for a 'collision' between a shallowly dipping Nazca Plate and the overriding South American Plate.

Other factors come into play at the northern and southern ends of the Andes where the geometry of the subducting plate is less well known. To the north, the NVZ is flanked by the amagmatic Bucaramanga segment under which the weakly defined subducting Caribbean Plate appears to dip at ∼20°. In the south,

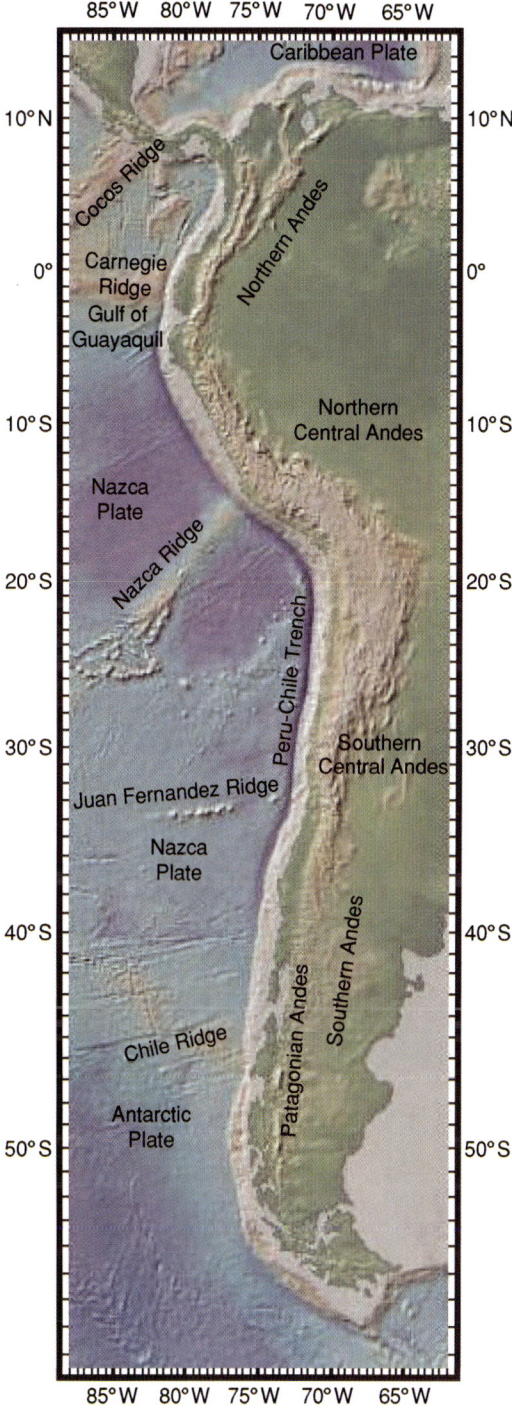

**Figure 1** Digital elevation model (DEM) of western South America and surrounding oceans based on global bathymetry database at the Lamont Doherty Observatory of Columbia University. The figure shows major features on the subducting oceanic plate and the correspondence with the division of the Andes into the Northern Andes bounded to the south by the Gulf of Guayaquil, the Central Andes bounded to the south by the Juan Fernandez Ridge, and the Southern Andes.

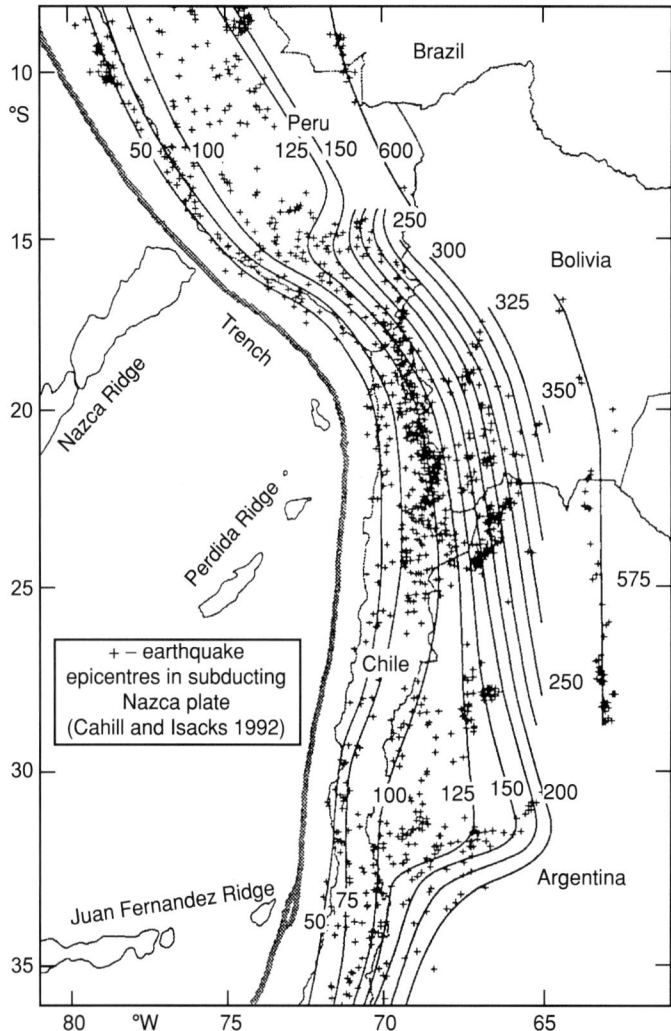

**Figure 2** Map showing the Benioff zone geometry of the portion of the Nazca oceanic Plate subducting beneath the Central Andes. Major north to south changes in distribution and style of volcanism, basin development, and deformational styles can be correlated to a first order with the shape of the Nazca Plate. (Reproduced with permission from Cahill TA and Isacks BL (1992) Seismicity and shape of the subducted Nazca Plate. *Journal of Geophysical Research* 97: 17 503–17 529.)

the SVZ is separated from the AVZ by a volcanic gap that coincides with the Chile Triple Junction where the Chile Rise is colliding with the Chile Trench (**Figure 7**). The net convergence rate of the South American Plate with the Nazca Plate is ∼9 cm year$^{-1}$ whereas that with the Antarctic Plate is ∼2 cm year$^{-1}$. Magmatism is absent in this region as the subducting slab is too young and hot to provide the volatiles to flux melting of the mantle wedge. The andesitic to dacitic 'adakitic' magmas of the AVZ are distinctive in that they are attributed to melting of the young hot subducting Antarctic Plate. The most convincing slab melt 'adakites' erupted anywhere in continental crust are the ∼14–12 Ma Patagonian adakites (e.g. Cerro Pampa) that are attributed to melting of the trailing edge of the subducted Nazca Plate near the time of ridge collision.

## Character of the Ranges, Basins and Faults of the Northern, Central and Southern Andes

The principal ranges and basins of the Andes reflect both the geometry of the subducting plate and the pre-existing continental crust and mantle lithosphere. The Andes are generally discussed in terms of a northern, a central, and a southern sector. Here the limit between the Northern and Central Andes is near the northern boundary of the Peruvian flat slab at ∼4° S, and the limit between the Central and Southern Andes is at the southern margin of the Chilean Flat Slab near 33° S (**Figure 1**).

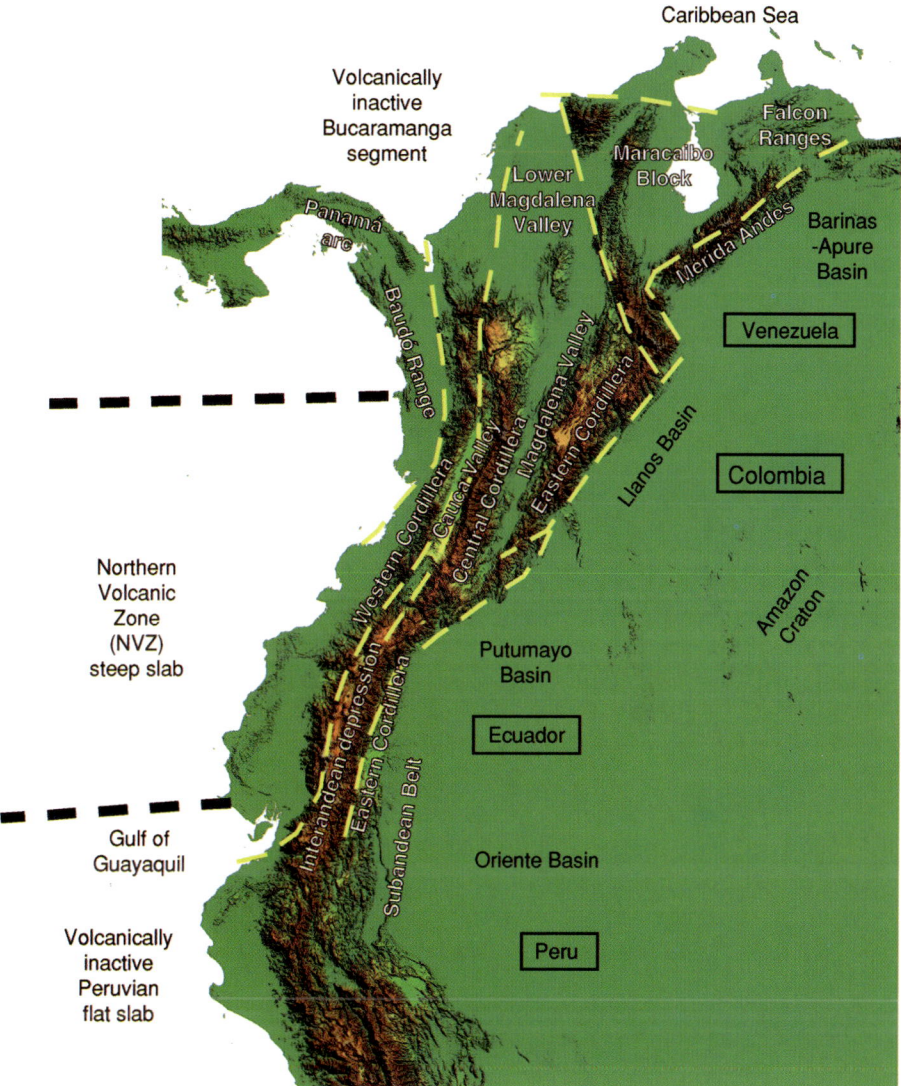

**Figure 3** Digital elevation map (DEM) based on 1 km grid SRTM (Shuttle Radar Topographic Mission) satellite data (available through the United States Geological Survey) assembled by the Cornell University Andes group showing the major features of the northern Andes (Venezuelan, Colombian, and Ecuadorian Andes). Elevations are indicated by colours (green is lowest and white is highest). Dark regions are areas of rapid elevation change. Dashed yellow lines are principal faults. They include the fault zone east of the Baudó Block that corresponds with a Late Miocene suture, and the Romeral Fault that runs through the western part of the Central Cordillera and marks the eastern limit of oceanic terranes accreted in the Cretaceous. Faults surrounding the Maracaibo Block are the Bucaramanga Fault on the west, the Boconó Fault in the Mérida Andes on the east, and the Oca Fault on the north. Regions of active and inactive volcanism are bounded by dashed lines that intersect the coast at the boundaries.

*Northern Andes* The Northern Andes include the Andes of Venezuela, Colombia, and Ecuador, and are bounded to the south by the Gulf of Guayaquil at $\sim 3°$ S (**Figure 3**). A distinctive feature of this region, well recognized by Gansser in the 1970s, is that the western part of the Northern Andes has been built on oceanic terranes accreted to South America as the Andes evolved, whereas, except for the southernmost part, the rest of the Andes has been built on continental basement that was already part of South America. Pervasive strike-slip deformation has played a major role in the evolution of the Northern Andes and is important in accounting for their extreme width. The Northern Andes are discussed relative to three segments that basically coincide with the political boundaries.

*Venezuelan Andes* The Venezuelan Andes face the Caribbean Plate and are geomorphologically the north-eastern extension of the Eastern Cordillera of

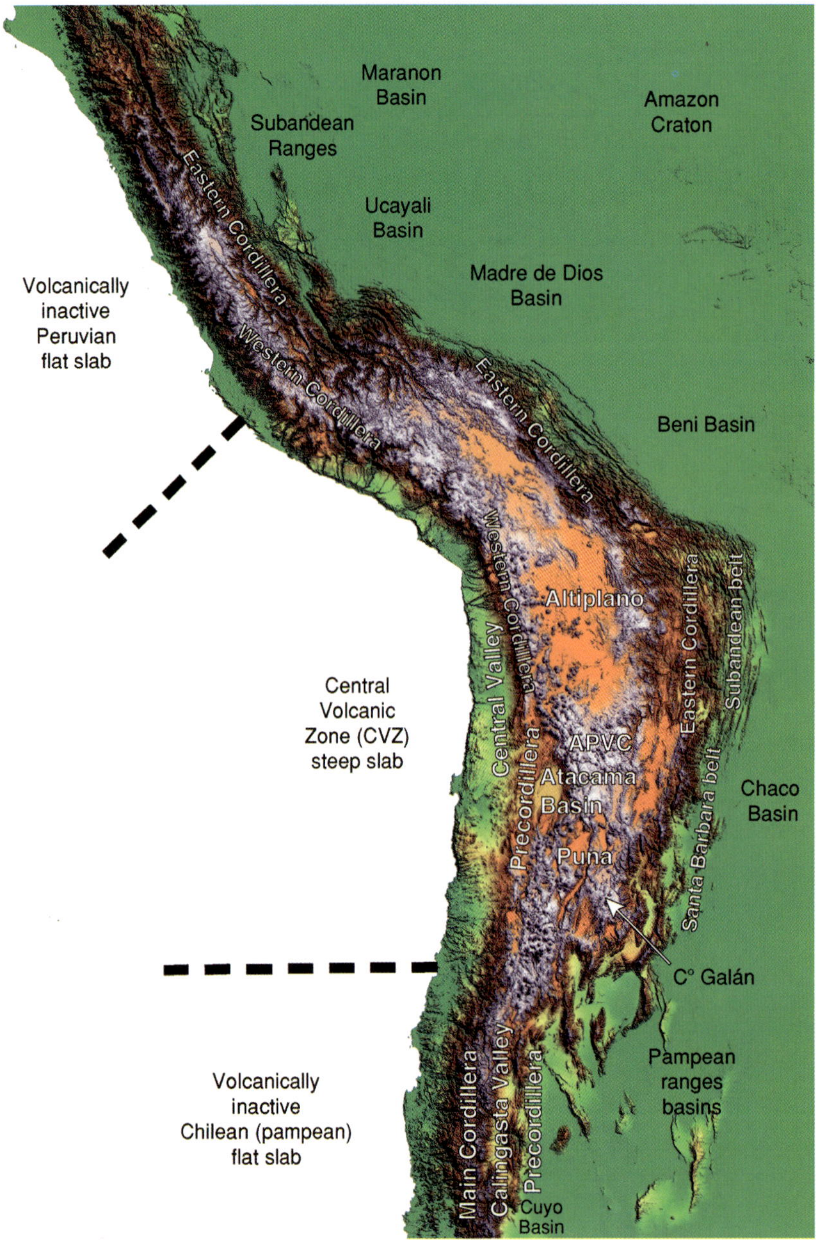

**Figure 4** Continuation of digital elevation map (DEM) in **Figure 3** showing the principal features of the Central Andes of Peru, Bolivia, northern Chile and northern Argentina. The presence of white and blue colours attests to the higher elevation of the Central Andes. Individual faults are not indicated. Cerro Galán is the ~2 Ma ignimbrite caldera shown on the image on **Figure 5**. The highest average elevations and the greatest amounts of crustal shortening occur in this part of the Andes.

Colombia. They consist primarily of the north-east-trending Mérida Andes that merges northward into the Falcón fold–thrust belt and north-eastward into the Caribbean ranges. The Mérida Andes are largely composed of the Palaeozoic rocks of the Mérida Terrane that was accreted to South America in the Palaeozoic. The Caribbean ranges are a south-verging Cenozoic fold–thrust belt north of the oil-rich Venezuela Basin. The Mérida Andes are bounded to the north-west by the Maracaibo Block and to the south-east by the complex, multicycle Barinas–Apure foreland basin whose evolution initiated in the Late Palaeozoic. The Maracaibo Block is a triangular lithospheric scale wedge that is being squeezed northward between the right-lateral Boconó Fault to the east and the left-lateral Bucaramanga Fault to the west. The Oca-El Pilar right-lateral Fault lies to the north. The traces of the ~500 km long Boconó Fault system

**Figure 5** Thematic Mapper satellite image of the southern Puna plateau of the Central Andes showing the development of salars (closed evaporite basins), Neogene andesitic to dacitic volcanic centres, and ignimbrite flows and caldera complexes.

project north-east through the Mérida Andes into the Caribbean ranges. The Cretaceous to Cenozoic Maracaibo Basin is within the Maracaibo Block and is an important petroleum-producing basin.

*Colombian Andes* The Colombian Andes consist of the distinct Eastern, Central and Western Cordilleras. They are bounded on the east by the Borde Llanero thrust system that marks the boundary with the Putumayo Basin in the south and the Llanos Basin that connects with the Barinas–Apure foreland basin system to the north. The active volcanic arc over the Nazca Plate runs through the Central Cordillera. The Central and Eastern Cordilleras are largely composed of Precambrian continental crust covered by Palaeozoic to Tertiary sedimentary and volcanic sequences. The Central Cordillera was uplifted and subjected to pervasive Late Cretaceous to Early Tertiary deformation, whereas the Eastern Cordillera was most affected by Late Miocene to Recent compression. The Eastern and Central Cordilleras are separated by the upper and middle Magdalena Valley Neogene successor foreland basins. The asymmetrical shape of the basin results from post-Miocene west-verging thrusting on the east and the east-dipping Central Cordillera Block that partially underlies the basin. The Western Cordillera consists primarily of Cretaceous oceanic magmatic and deep-water sedimentary rocks. Along with the part of the Central Cordillera west of the Romeral Fault, the Western Cordillera is considered to be an amalgamation of terranes accreted to South America during the Cretaceous. The Central and Western Cordilleras are separated by the Cauca Valley whose western side is largely bounded by the Romeral Fault. This fault is considered to be a terrane suture that continues northward through the Lower Magdalena Valley Basin. On the west, the Western Cordillera is bounded by the San Juan Atrato fore-arc valley which is underlain by oceanic terranes. Part of this valley along with the Seranía de Baudó Block and the Panamá Arc were accreted in the Miocene (~12–13 Ma) along the Istmina deformed zone and the Uramita Fault.

*Ecuadorian Andes* The central Colombian Cordillera continues southward to form the Cordillera Real or Eastern Cordillera of Ecuador, which is composed

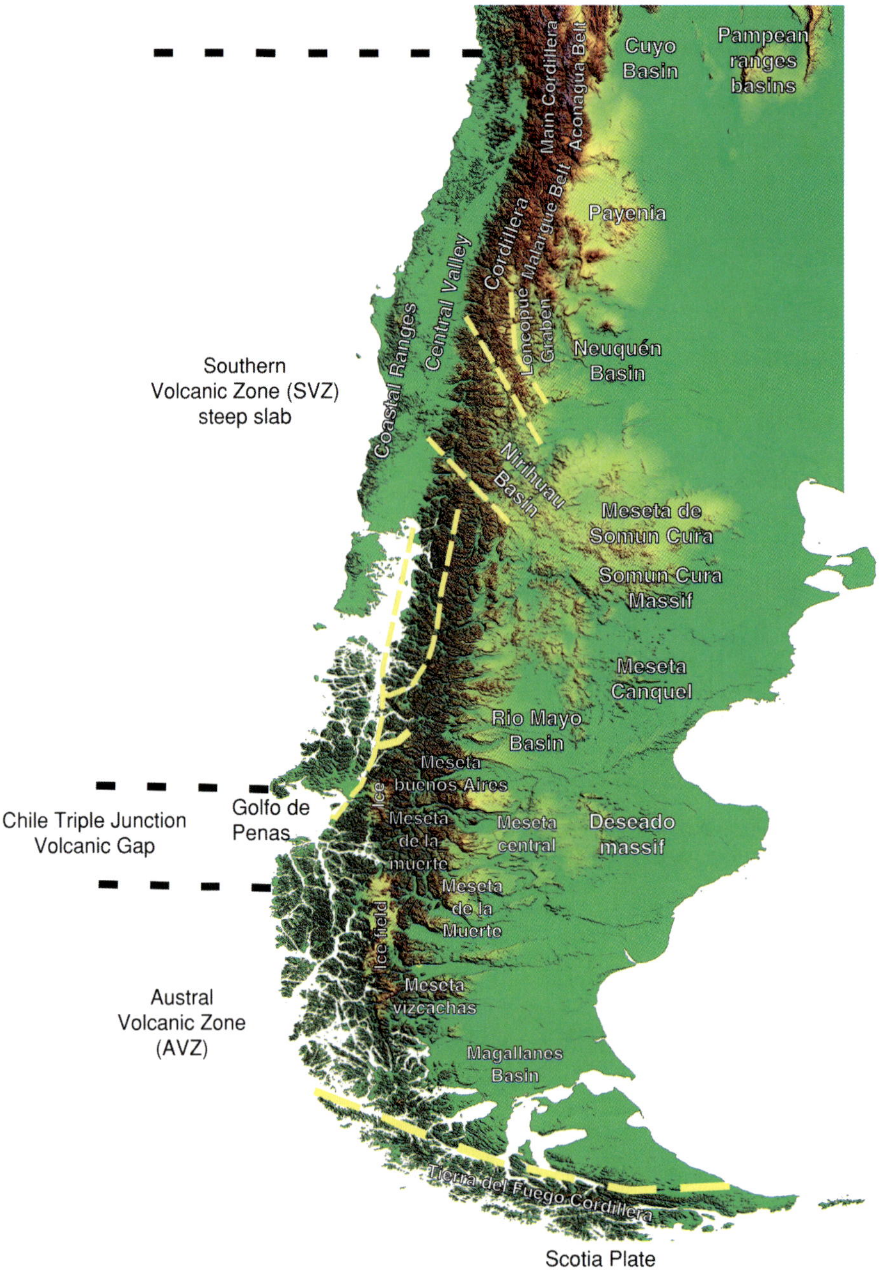

**Figure 6** Continuation of digital elevation map (DEM) in **Figure 4** showing the principal features of the Southern Andes of Chile and Argentina. The evolution of the southern part of the region has been strongly influenced by successive collisions of segments of the Chile Rise with the trench as shown in more detail in **Figure 7**. Yellow lines are faults. Those along the Pacific coast belong to the Liquiñe–Ofqui strike-slip fault system.

of multiply deformed, variably metamorphosed Palaeozoic to Mesozoic rocks covered by Cenozoic volcanic and sedimentary units. The Putumayo and Oriente Basins to the east are part of a large composite Mesozoic to Cenozoic foreland basin that overlies older cratonic basement. The western 50–80 km of the basin, known as the Subandean Zone, has been uplifted to elevations of 500–1000 m by late Cenozoic thrusts. The Cordillera Real is separated from the Ecuadorian Western Cordillera by the Interandean Valley that is filled by Tertiary to Quaternary volcanic and volcaniclastic deposits. The valley is bounded to the west by the Pujili Fault and to the east by the Peltetec Fault, and is probably floored by Cordillera

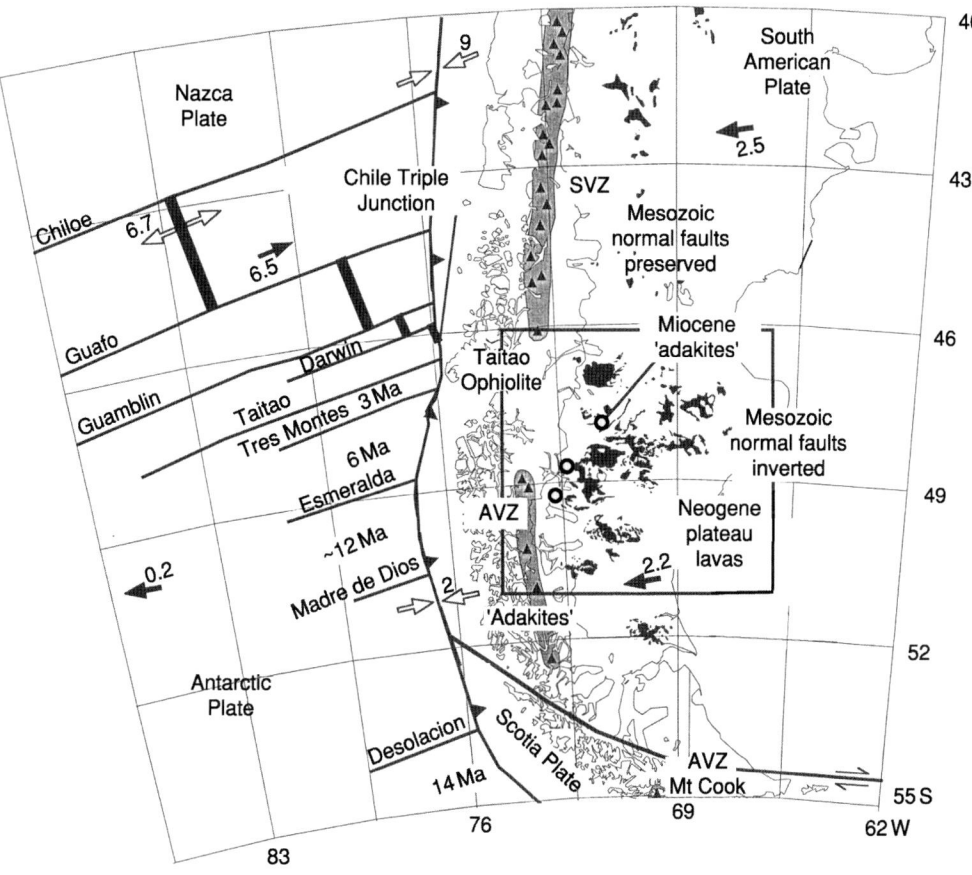

**Figure 7** Map of southern Patagonia and surrounding ocean basins showing distribution of Southern Volcanic Zone (SVZ), Austral Volcanic Zone (AVZ adakites), and Miocene slab melt 'adakitic' volcanic centres, deformational styles, plate convergence velocities, and oceanic fracture zones and ridges. Active volcanism is absent east of where the Chile Ridge has collided in the last ~6 My. The boxed region is where features related to ridge collision are best developed. (Modified from Gorring ML, Kay SM, Zeitler PK, et al. (1997) Neogene Patagonian plateau lavas: continental magmas associated with ridge collision at the Chile Triple Junction. *Tectonics* 16: 1–17.)

Real-like basement. The axis of the Recent NVZ arc runs through the valley with volcanic centres occurring from the eastern part of the Western Cordillera into the Subandean Zone. The Western Cordillera consists largely of Cretaceous to Eocene oceanic crust, turbidites, and oceanic arc sequences covered by Late Eocene–Oligocene continental sediments and intruded by Eocene and younger magmatic rocks. The low-lying, Pacific coastal region is comprised of Cretaceous to Cenozoic basins underlain by oceanic basement.

**Central Andes (~3° to 33° S)** The best known part of the Andes is the Central Andes (**Figure 4**) which can be separated into the Peruvian flat slab segment, the Altiplano–Puna plateau segment that includes the Central Volcanic Zone Arc, and the Chilean flat slab segment. A surprising degree of bilateral symmetry exists in the shape of the subducting slab (**Figure 2**) and the topography of the land surface across the region (**Figure 4**).

*Peruvian flat slab segment (~3° S to 15° S)* The Peruvian flat slab segment largely coincides with the inactive part of the magmatic arc where active volcanism has been absent from 2.5° to 16° S for the last ~3–4 My. The Western and Eastern Cordilleras of Ecuador narrow into this region as the Interandean Valley virtually disappears and the Subandean belt broadens to a width of 120–250 km across into the composite Marañon–Ucayali foreland basin. A west to east profile shows a topographically low, narrow coastal region west of the Western Cordillera, a narrow Interandean depression, the Eastern Cordillera, the Subandean Zone, and the basins (e.g. Madre de Dios) of the eastern lowlands. A series of partly submerged Tertiary sediment-filled fore-arc basins (Trujillo, Lima and Pisco) along the narrow continental shelf reflect the effects of crustal thinning due to fore-arc tectonic erosion. The basement of the southern coastal region consists of the high-grade metamorphic rocks of the Precambrian (~1 Ga) Arequipa block. To the east, the Western Cordillera is mainly composed of deformed metamorphic rocks and Mesozoic sediments that are cut and covered by Mesozoic to Tertiary magmatic rocks. The extensive Middle Cretaceous to Palaeogene Peruvian batholith that extends into the coastal region forms the western part. The highest peaks occur in the ~8 Ma Cordillera Blanca pluton whose western boundary, the Cordillera Blanca shear zone, has been interpreted as a low-angle detachment fault. The Eastern Cordillera is made up of pre-Mesozoic metamorphic rocks along with subordinate Palaeozoic and Tertiary intrusives. It hosts the Early Tertiary 'inner arc' and was the locus of intense Late Paleocene and Middle Eocene east-verging thrusting. The Subandean belt further east consists of Mesozoic and Tertiary sedimentary rocks and Palaeozoic basement cut by Late Miocene to Recent reverse faults that extend almost to the Brazilian border.

*Central Volcanic Zone/Altiplano–Puna Plateau segment (~15° to 28° S)* This segment includes the active Central Volcanic Zone arc and the widest part of the high Andean Cordillera that includes the Altiplano–Puna Plateau. The whole region is built on a variable age Precambrian and Palaeozoic basement. Uplift of the plateau is principally attributed to ductile thickening of the lower crust in response to compressional shortening, and subordinately to magmatic addition. Compressional shortening is reflected in the fold–thrust belts of the upper crust. Total amounts of shortening are widely debated with values in various transects ranging from 300 to 500 km in the north to <150 km in the south. The principal uplift is considered to be Eocene to Late Miocene in age in the north and Late Miocene in the south. A generalized west to east section across the central plateau crosses the coastal Cordillera, the fore-arc Central Valley, the Chilean Precordillera, the Western Cordillera, the Altiplano–Puna Plateau, the Eastern Cordillera, the Subandean Belt, and the Madre de Dios–Beni–Chaco composite foreland basin. These features all follow the bend in the trench (**Figures 1** and **2**) and coast (**Figure 4**) known as the Bolivian Orocline.

The Andes of northernmost Chile include the fore-arc region from the coast to the CVZ volcanic front. The Neogene sediment-filled Central Valley is largely flanked to the west by Jurassic to Middle Cretaceous arc sequences cut by the Cretaceous Atacama strike-slip Fault system, and to the east by the Cretaceous to Eocene volcanic and sedimentary rocks in the Chilean Precordillera that are cut by the Late Palaeozoic to Early Mesozoic Palaeogene Domeyko strike-slip Fault system. The Chilean Precordillera hosts the giant Chuquicamata and Escondita Cu deposits. The coastal Cordillera and Atacama Fault system run offshore at the bend in the coast in northernmost Chile where the western margin appears to have been removed by fore-arc subduction erosion. The Western Cordillera marks the western limit of the high Andes and contains the CVZ arc. Both diverge to the east around the pre-Andean Atacama Basin near 23° S to 24° S. The highest peaks reach ~5000–6550 m in the north and over 6800 m in the south. Underlying crustal thicknesses range from 70 km in the north to ~55 km in the south.

Flanking the Western Cordillera to the east is the most prominent feature of the segment: the ~700 km long and ~200 km wide Altiplano-Puna Plateau with an average elevation of 3700 m. This internally drained plateau hosts extensive Late Cenozoic sedimentary evaporate-filled basins (salars) and chains of Neogene volcanic centres (**Figure 5**) that extend to its eastern edge. The highest regions (>6300 m) are Late Miocene to Pliocene andesitic to dacitic stratovolcanic and giant dacitic ignimbrite complexes. Only sequences older than ~10 My are significantly deformed. The Altiplano section in Bolivia north of ~22° S is mostly comprised of a relatively flat sediment-filled basin covering a largely Palaeozoic sedimentary sequence over a Brazilian shield-like Precambrian basement. Tertiary volcanic rocks are concentrated in the Western and Eastern Cordilleras where they can host important tin and silver (Potosi) deposits. One of the most famous is the silver deposit at Potosi in the Eastern Cordillera. Crustal thicknesses under the Altiplano range from ~66 to ~78 km. The Puna section in Argentina to the south is distinctive in being broken into ranges with high peaks and basins and east–west chains of Neogene volcanic rocks (**Figure 5**). This region also differs in being underlain by a basement that includes Palaeozoic mafic to silicic magmatic sequences formed in complex arc and back-arc settings. The greatest average elevation is in the southern Puna where crustal thicknesses appear to be <58 km.

The distribution of large <10–12 Ma ignimbrite calderas is uneven across the plateau. Late Miocene centres occur in the Altiplano–Puna Volcanic Complex (APVC) in the central plateau and along the eastern margin. Early Pliocene centres occur in the western APVC and near the southern Western Cordillera. The youngest major eruption, the ~2 Ma Cerro

Galán ignimbrite (see **Figure 5**), occurred in the southern Puna where latest Miocene to Recent monogenetic calc-alkaline and intraplate mafic flows not seen elsewhere on the plateau are also found. Small young northern Puna and Altiplano mafic flows have shoshonitic chemistry.

The plateau is bounded to the east by the Eastern Cordillera and adjoining Subandean Ranges. East of the Altiplano, the Eastern Cordillera of Bolivia consists mainly of deformed Palaeozoic sedimentary rocks overlain by Eocene to Miocene magmatic rocks that have been subjected to Middle Eocene to Late Miocene contractional deformation. Peaks can reach up to 6000 m. To the east, elevations drop off abruptly into the Subandean Belt where compressional deformation has been concentrated for the last ~10 My since the active thrust front moved eastward. The Subandean Belt of Bolivia and northernmost Argentina is largely comprised of deformed Palaeozoic sedimentary rocks cut by Late Miocene to Recent thin-skinned thrusts. The picture changes along the Puna where the principal Tertiary deformation of the Eastern Cordillera is Miocene in age and the Subandean Belt is characterized by the 400 km long Santa Barbara Fault system in which Neogene compressional deformation is characterized by inversion of normal faults related to the complex Cretaceous to Palaeogene Salta Group Rift system. A sharp contrast in structural style with the thin-skinned Subandean belt to the north and the Pampean basement uplifts that characterize the Chilean flat slab to the south corresponds with the boundaries of the Salta Group Rift basins.

*Chilean flat slab segment (28° to 33° S)* The Chilean (or Pampean) flat slab corresponds with the current gap in the active volcanic arc between 28° and 33° S over the shallow Benioff zone (**Figure 2**). In accord with the geometry of the subducting plate, the northern boundary is geologically transitional, whereas the southern boundary is relatively abrupt. A west to east traverse crosses the Coastal Cordillera, the high Andes consisting of the Main and Frontal Cordilleras, the Uspallata–Calingasta Valley, the Argentine Precordillera, and the Pampean Ranges that are flanked by a series of Neogene foreland basins. The fore-arc differs from the adjacent segments in that the Chilean Central Valley is missing. The Main Cordillera hosts Miocene volcanic arc rocks that along with Mesozoic and Palaeogene sedimentary and magmatic rocks cap major Late Palaeozoic/Triassic magmatic sequences similar to those that comprise the Frontal Cordillera. The 6962 m Aconcagua peak in the Main Cordillera near 32° S, the highest in the world outside of the Himalayas, is comprised of Miocene (~15–9 Ma) arc volcanic rocks thrust over Mesozoic and Tertiary sediments. The Main Cordillera hosts rich Miocene Au (El Indio–Veladero system) and Cu (e.g. Rio Blanco) deposits. The Uspallata–Calingasta Valley to the east is a piggy-back foreland basin filled with Miocene sedimentary and volcanic rocks. It coincides with a major Palaeozoic terrane suture and fills the volcanic gap between the presently active CVZ and SVZ arc segments. The Argentine Precordillera primarily consists of deformed Palaeozoic sedimentary sequences associated with Miocene sedimentary and minor dacitic volcanic rocks overlying ~1 Ga metamorphic rocks. These sequences are deformed by Middle to Late Miocene east-verging thrusts in the west and younger west-verging thrusts in the east. The west-verging thrusts are attributed to reactivation of basement structures. Further east, the Pampean ranges are thick-skinned blocks of Precambrian to early Palaeozoic crystalline rocks uplifted by Late Miocene to Pliocene high-angle reverse faults. Elevations in the north (Sierra de Aconquija) reach over 6000 m. The easternmost range, which is ~700 km from the trench, hosts small ~7–5 Ma volcanic rocks whose geochemical signatures indicate slab-derived components. The westernmost range (Pie de Palo) has been uplifted in the last 3 My. Total Neogene shortening estimates across the Chilean flat slab segment near 30° S are near 200 km with ~120 km of that occurring on thrust faults in the Precordillera.

**Southern Andes (33° to 56° S)** The Southern Andes begin south of the Chilean flat slab and the subducting Juan Fernandez Ridge on the Nazca Plate (**Figures 1, 2** and **6**). The northern part includes the Southern Volcanic Zone that overlies a moderately steep subduction zone in which the Nazca Plate decreases in age until reaching the Chile Triple Junction near 47° S. The southern segment is south of the Chile Triple Junction. *See* **Argentina** for more information on these regions.

*Southern Volcanic Zone Segment (33° to 46° S)* The Southern Volcanic Zone comprises the region of active volcanism from 33° to 46° S. From west to east, a section crosses the Late Palaeozoic metamorphic rocks of the Coastal Cordillera, the largely Tertiary sediment-filled Chilean Central Valley, Miocene arc rocks and uplifted and deformed Mesozoic sequences of the Main Cordillera upon which the active SVZ volcanic arc is built, and a southwardly narrowing retro-arc fold–thrust belt called the Aconcagua Belt in the northernmost part (~32° to 34° S), and the Malargue Belt to the south (~34° to 37° S).

The height of the Main Cordillera decreases from ~6600 to <2000 m southward as the fore-arc Central Valley widens and back-arc shortening related to both thin- and thick-skinned thrusting decreases from ~100 to <35 km at 37° S. Between 38° S and 40° S, the arc front has migrated westward since the Pliocene in conjunction with mild extension in the region of the Loncopué Graben. From ~38° S southward, the modern volcanic arc overlaps with the >900 km long Liquiñe–Ofqui Fault system (**Figure 6**), which has had dextral displacement since at least the Early Miocene. The Mesozoic and Tertiary arc fronts are in the modern fore-arc north of 36° S and near the modern arc to the south. The giant Late Miocene El Teniente Cu deposit occurs in the westernmost Cordillera near 34° S.

The foreland has important basins with both rift and foreland histories that surround the pre-Mesozoic Somuncura and Deseado massifs. They include the Jurassic through Tertiary Neuquén Basin north of ~38° S, the Early Cretaceous Rio Mayo Embayment near 45° S and the Palaeogene to Early Neogene Nirihuau Basin near 41° S. A series of smaller segmented basins occur in the fore-arc.

Andean magmatism is well developed in the foreland in two regions. The first is in the Payenia area between ~35° S and 38° S where extensive volcanic sequences cover the northern half of the Neuquén Basin. These volcanic sequences, which can occur more than 500 km east of the modern trench, consist of Miocene andesitic/dacitic volcanic centres with arc geochemical signatures, and Early Miocene and Late Pliocene to Recent mafic alkaline flows. The second region is between ~41° S and 44° S where Palaeogene and Miocene andesitic to rhyolitic sequences, Eocene alkaline basalts, and the extensive Late Oligocene to Miocene mafic plateau flows associated with the Mesetas de Somun Cura and Canquel occur.

*Chile Triple Junction and Austral Volcanic Zone Segment (47° to 56° S)* Important changes occur south of the Chile Triple Junction near ~47° S where the Chile Ridge is colliding with the trench (**Figure 7**). The modern volcanic arc disappears between ~47° S and 49° S and resumes as the Austral Volcanic Zone in response to subduction of the Antarctic Plate. The high Andes at this latitude are dominantly composed of post-Triassic magmatic rocks with the principal part of the Jurassic to Miocene Patagonian Batholith forming the backbone of the Cordillera. Late Palaeozoic/Early Mesozoic fore-arc accretionary complexes occur in the fore-arc. The Patagonian Cordillera reaches a maximum elevation of ~4000 m near ~47° S east of where the Chile Ridge is currently colliding and decreases in elevation to the south. The 30 000 km² Patagonian ice-field, the world's third largest, occurs at the higher elevations.

The importance of back-arc crustal shortening abruptly increases at ~47° S as does the amount of Tertiary foreland sedimentary deposits. To the south, Mesozoic normal faults have largely been inverted by Tertiary compression whereas similar age normal faults are preserved to the north. The Patagonian region east of where the Chile Ridge has collided is notable for Neogene fore-arc volcanism (Taitao Ophiolite), abundant Late Neogene mafic plateau flows east of where ridge collision occurred at ~12 and 6 Ma, and widespread Pleistocene to Recent mafic flows. Extensive Eocene mafic retroarc plateau lavas also occur in this area.

The southernmost part of Patagonia includes the Jurassic to Tertiary Magallanes (austral) Basin whose axis coincides with the Early Cretaceous Tortuga and Sarmiento ophiolite complexes. The Andes of Tierra de Fuego contain the Cretaceous metamorphic complex of the Cordillera Darwin. They include an east to west trending fold–thrust belt cut by a major northwest to south-east trending left-lateral fault system that delimits the northern boundary of the Scotia Plate. The SVZ Mount Cook volcanic centre also occurs in this region.

## Jurassic to Recent Evolution of the Andean Chain

The evolution of the Andes that began with the Mesozoic breakup of Pangaea can be divided into a Late Triassic to Early Cretaceous stage dominated by rifting processes and extensional arc systems, a Late Cretaceous to Early Oligocene stage in which the Andes evolved from an extensional dominated to a compressional regime, and a latest Oligocene to Recent stage in which most of the main Andean range was uplifted.

### Stage 1: Rifting and Extensional Arc Systems

The Late Triassic to Early Cretaceous stage began with rifting associated with the breakup of Pangaea. The geometries of the rifts that began all across western South America in the Triassic and Jurassic reflect extensional directions, basement fabrics and old terrane boundaries. The north-east to south-west rifting *en echelon* pattern in the Northern Andes matches the counterclockwise rotation associated with rifting from the conjugate North America Yucatan Block. North-west trending rifts in the rest of the Northern, Central and Southern Andes are aligned with Gondwana Palaeozoic sutures. The north-west trending dextral shear pattern of rifts in Patagonia could have been influenced by clockwise rotation of the Atlantic Peninsula initiating spreading of the Weddell Sea between 175 and 155 Ma.

This rifting was well underway when a subduction regime was diachronically established along the western margin of South America in the Early to Middle Jurassic. The Jurassic to Early Cretaceous history of this system is dominated by a complex series of fore-arc, intra-arc, and retro-arc basins in which extension was associated with subduction zone rollback and oblique convergence. Middle Jurassic to Early Cretaceous thermal subsidence in the back-arc began at different times in different places. The expansion of marine sedimentation in the Early Cretaceous reflects the thermal subsidence that followed this rifting and a long-term rise in global sea-level.

In the Southern Andes, extensional faulting in the Neuguen Basin was interrupted by inversion in the Jurassic (Araucanian Event). This inversion was followed by rifting along the east coast of northern Argentina, Uruguay and Africa at about the same time that the Late Jurassic magmatic arc in central Chile migrated ~30–40 km to the east and the Atacama intra-arc strike-slip fault system became active. Extension reached a height in the Southern Andes with the Late Jurassic formation of the oceanic Rocas Verdes Basin in southernmost Patagonia that incorporates the Sarmiento and Tortuga ophiolites. In the Northern Andes, Jurassic extension precipitated thermal sag that led to extensive shallow marine embayment by the end of the middle of the Early Cretaceous (Neocomian).

The Early Cretaceous (130–110 Ma) marks the opening of the South Atlantic Ocean off the shore of Brazil and rapid westward drift of South America relative to the underlying mantle. Active extension was most pronounced in Bolivia, northern Argentina and Chile as the central South Atlantic Ocean began opening at ~130 Ma at the time of the eruption of the Parana flood basalts in Brazil. Intracratonic extension occurred in the southern Altiplano and in the Salta Rift system and Pampean ranges where it was associated with basaltic volcanism. To the west, the Atacama intra-arc strike-slip fault system became transtensional at ~132–125 Ma and the Aptian period (~121–112 Ma) brought increased negative trench rollback velocity causing the intra-arc and back-arc extension that produced the marine sediment-filled aborted marginal basin in Peru associated with Casma volcanism (after ~112 Ma), the Sierra de Fraga low-angle detachment faults near 27° S, and the aborted marginal basin linked with the eruption of the extensive ~119–110 Ma Veta Negra and related volcanic groups in central and south central Chile (29 to 33° S).

In the Southern Andes, intracratonic rifting had essentially ceased by the Aptian as shown by the terminal stages of the Rocas Verdes Basin in southern Patagonia, a peak in the emplacement of the Patagonian batholith, and the end of marine sedimentation in the Neuquén Basin. In the Northern Andes, the presence of blueschist and high-pressure metamorphic rocks (~130 Ma) along the Romeral and Peltetec faults can be associated with the obduction of small island-arc systems to the Colombian Central Cordillera and western side of the Ecuadorian Cordillera Real. The Cretaceous (112–90 Ma) drowning of the Barinas–Apure Basin can be attributed to thermal subsidence.

### Stage 2: Basin Inversion and Formation of the Early Andes

The main compressional stage that built the Andes began after 10 Ma as active spreading in the South Atlantic accelerated the separation of South America from Africa, and South America began to actively override the trench. A series of compressional events of variable intensity occurred all along the margin as foreland basins responded to flexural loading. Compressional events took place in the Late Cretaceous near ~105–95 Ma (Mochica Phase), near 85–75 Ma (Peruvian Phase), at the beginning of the Paleocene near 65–50 Ma, and during the Eocene (Incaic Phase). These are all times of changes in plate convergence rates and directions along the margin.

Late Cretaceous compression in Peru and Bolivia is marked by tectonic inversion in the Mochica Phase at ~105 Ma. In Chile, the final phase of pluton emplacement in the Coastal Cordillera took place at 106 Ma, after which the intra-arc Atacama Fault system was largely abandoned. In Patagonia, the Late Cretaceous (98–85 Ma) marks the closure of the Rocas Verde Basin a peak in production, of the Patagonian Batholith, and the formation of the thrust-loaded Magallanes foreland basin. An inversion at ~99 Ma marks the change from an extensional to a foreland setting for the Neuquén Basin.

The Peruvian phase corresponds with the final emplacement of the Late Cretaceous Coastal Batholith in Peru and the end of marine sedimentation in the large back-arc marine basin to the east in Peru and Bolivia. In Chile, a new arc and fault system was established at ~86 Ma in what is now the Central Valley, some 50 km east of the old Atacama system. The proto-Cordillera de Domeyko was uplifted at this time. In the Northern Andes, Late Cretaceous closure of an ocean basin led to accretion of the Cretaceous oceanic Piñon–Dagua Terrane to the Colombian and Ecuadorian Western Cordilleras between ~80 and 60 Ma (Calima Orogeny). The collision is generally associated with an east-dipping subduction zone along the continental margin. The progressive southward drowning of the basin to the east that had been occurring through the Cretaceous ended in association with the deformation and uplift of the Central and Eastern Cordillera. Partial obduction of the Caribbean Plate at ~85–80 Ma caused the Latest Cretaceous to Early Eocene Caribbean orogeny and the southward migration of the Venezuelan fore-deep.

By the Palaeogene, marine deposition had ended everywhere but the northernmost Andes and the large basins in Patagonia; uplift had built a barrier to Pacific marine ingressions; and Andean detritus was being spread into foreland basins. In central Chile, the Paleocene was characterized by explosive silicic calderas. Progressive Palaeogene to Eocene southward collision of the Aluk–Farallon Ridge with the trench leading to a gap (slab window) between the subducting plates has been suggested to explain a pattern of arc volcanism at ~56–50 Ma near ~37° S, bimodal arc volcanism near 40° S, paucity of arc magmatism south of ~43° S and widespread basaltic volcanism further south.

Middle to Late Eocene deformation (47–32 Ma, Incaic Orogeny) occurred during a phase of fast oblique convergence. Compression produced the fold–thrust belts of the Western Cordillera of Peru, deformation associated with contractional and strike-slip faulting along the Cordillera de Domeyko in northern Chile, and the west-directed thrusting that resulted in uplift of the Bolivian Eastern Cordillera. This deformation, along with back-arc magmatism and deformation in the Zongo–San–Gabán Zone on the Altiplano/Eastern Cordillera, has been related to southward migration of a shallow subduction zone as well as to a more normal convergence direction here than to the north or south. These events were followed by a relative Oligocene magmatic lull along most of the Andean margin during a period of slow plate convergence.

## Stage 3: Formation of the Modern Andes (~27–0 Ma)

The Neogene history of the Andes begins with the breakup of the Farallon Plate into the Nazca and Cocos Plates at ~28–26 Ma. This event is linked to the change in convergence direction and increase in relative convergence velocity that led to the major uplift of the Andes that culminated in the Late Miocene and Pliocene. Although the continuity of Miocene deformational phases in the broad Quechua Orogeny associated with this uplift has been widely debated, there is some agreement that general margin-wide changes occurred at ~19–16 Ma and 8–4 Ma. The formation of major Neogene Cu and Au deposits during this period correlates with periods of shallowing and steepening subduction zones, migrating arc fronts, and crustal thickening.

Events in the Central Andes can be related with shallowing and steepening of the subducting Nazca Plate. Shallowing beneath the Chilean flat slab can be correlated with Miocene uplift and cessation of arc volcanism in the Main Cordillera, Late Miocene thrusting in the Argentine Precordillera, Late Miocene to Pliocene uplift of the Sierras Pampeanas, and Pliocene termination of volcanism across the region. Deformational and magmatic events associated with shallowing of the subducting slab under the Peruvian flat slab region are superimposed on older Palaeogene features. Steepening of a formerly shallow subduction zone beneath the Central Altiplano–Puna Segment can be correlated with a Late Oligocene to Early Miocene magmatic gap associated with widespread compressional deformation that was followed by large volume Late Miocene to Pliocene ignimbrite eruptions. The large eruptions can be explained by hydrated mantle melts from above the slab intruding the crust and causing massive melting. The progressive westward narrowing of the magmatic arc fits with steepening of the subducting slab. Ductile deformation of the hot crust can explain Miocene plateau uplift as the thrust front migrated eastward into the Subandean belt. Magmatic and geophysical data support delamination of a thickened crustal root beneath the Altiplano–northern Puna during this time. The southern Puna and Altiplano regions were in transitional positions between the steepening and shallowing segments of the subducting slab. The presence of Late Pliocene ignimbrites and mafic magmas, a relatively thin crust and mantle lithosphere, a high average elevation, and a change in Late Miocene/Pliocene faulting patterns in the southern Puna have been related to Pliocene delamination of a dense, thickened crustal root.

The Neogene magmatic and deformational history of the Southern Volcanic Zone region can be tied to eastward migration of the Neogene arc front at ~20–16 Ma north of 36° S, and again at ~8–4 Ma north of ~34.5° S. The back-arc andesitic/dacitic centres and large retro-arc volcanic fields in the Puyenia region may be due to melting of hydrated mantle that formed as the result of transient Miocene shallowing of the subducting Nazca Plate. The Neogene evolution of the Andes south of 47° S can be tied to the northward propagation of the Chile Triple Point as segments of the Chile spreading centre collided progressively northward with the Chile Ridge. The effects of ridge collision include uplift and compressional deformation in the fore-arc and arc region before collision, emplacement of fore-arc lavas, eruption of

Late Miocene to Recent plateau lavas above the gap (slab window) that opened between the trailing edge of the Nazca and the leading edge of the Antarctic Plate, and eruption of arc magmas generated by melting of the hot young subducting plate. The presence of Paleocene to Recent Patagonian mafic back-arc magmatism at different times and different places (Payenia, Somun Cura region, and east of where the Chile Ridge has collided) attests to the readiness of the Patagonian mantle to melt given provocation.

Most of the present configuration of the Northern Andes is also due to Neogene tectonics. The Miocene opening of the Gulf of Guayaquil in Ecuador due to strike-slip motion on the Pujili–Cauca and Peltetec-Romeral faults led to the generation of pull-apart coastal and intermontane basins. The Late Miocene collision of the Carnegie Ridge against the coast of Ecuador resulted in compressional inversion of extensional structures and subsequent strike-slip displacement of the Northern Andes. The accretion of the Choco Terrane, which includes the Panama–Baudó Block, starting at ~13–12 Ma, caused deformation and uplift of the Colombian Eastern Cordillera and the Mérida Andes. The most intense phases of deformation in the Venezuelan Andes and the Maracaibo Block are linked to Caribbean oblique convergence and underthrusting in the last 10 My. The present convergence vectors of the Nazca and Caribbean plates and ongoing collision of the Panama–Baudó Arc have generated a state of intracontinental compressional stress partitioned into reverse and strike-slip faulting.

## Concluding Remarks

This short synthesis presents the state of knowledge of the nature of the Andean mountain range. What is presented here should be viewed only as one stage in the understanding of the evolving picture of this magnificent and complex range which is a natural laboratory for investigating the origin and evolution of continental mountain belts and the formation and destruction of continental crust.

## See Also

**Antarctic**. **Argentina**. **Brazil**. **Tectonics:** Mountain Building and Orogeny. **Volcanoes**.

## Further Reading

Allmendinger RW, Jordan TE, Kay SM, and Isacks BL (1997) The evolution of the Altiplano-Puna Plateau of the Central Andes. *Annual Reviews of Earth and Planetary Sciences* 25: 139–174.

ANCORP Working Group (2003) Seismic imaging of a convergent continental margin and plateau in the Central Andes (Andean Continental Research Project 1996 ANCORP'96). *Journal of Geophysical Research* 108: 2328 doi:10.1029/2002JB001771.

Cahill TA and Isacks BL (1992) Seismicity and shape of the subducted Nazca Plate. *Journal of Geophysical Research* 97: 17 503–17 529.

Cordani UJ, Milani EJ, Thomaz F, Ilho A, and Campos DA (eds.) (2000) *Tectonic Evolution of South America*. 31st International Geological Congress, Río de Janeiro. Available through the Geological Society of America.

Gansser A (1973) Facts and theories on the Andes. *Journal of the Geological Society, London,* 129: 93–131.

Gorring ML, Kay SM, Zeitler PK, *et al.* (1997) Neogene Patagonian plateau lavas: continental magmas associated with ridge collision at the Chile Triple Junction. *Tectonics* 16: 1–17.

Gutscher MA (2002) Andean subduction styles and their effect on thermal structure and interplate coupling. *Journal of South American Earth Sciences* 15: 3–10.

Isacks BL (1988) Uplift of the central Andean plateau and bending of the Bolivian orocline. *Journal of Geophysical Research* 93: 3211–3231.

Jordan TE, Isacks BL, Allmendinger RW, *et al.* (1983) Andean tectonics related to geometry of subducted Nazca plate. *Bulletin of the Geological Society of America* 94: 341–361.

Kay SM, Godoy E, and Kurtz AJ (2004) Episodic arc migration, crustal thickening, subduction erosion, and Miocene to Recent magmatism along the Andean Southern Volcanic Zone Margin. *Bulletin of the Geological Society of America* (in press).

Kley J, Monaldi CR, and Salfity JA (1999) Along-strike segmentation of the Andean foreland: causes and consequences. *Tectonophysics* 301: 75–96.

Skinner BJ (ed.) (1999) *Geology and Ore Deposits of the Central Andes*. Society of Economic Geology Special Publication No. 7.

Taboada A, Rivera LA, Fuenzalida A, *et al.* (2000) Geodynamics of the northern Andes: subduction and intracontinental deformation (Colombia). *Tectonics* 19: 787–813.

Tankard AJ, Suarez Soruco R, and Welsink HJ (eds.) (1995) *Petroleum Basins of South America*. American Association of Petroleum Geologists Memoir 62.

# ANTARCTIC

B C Storey, University of Canterbury, Christchurch, New Zealand

© 2005, Elsevier Ltd. All Rights Reserved.

## Introduction

Antarctica has not always been the cold, isolated, polar continent that it is today, with all but a minor fraction covered by ice. Throughout geological time, Antarctica has amalgamated and combined with other continental fragments and has rifted to form new ocean seaways. For a period of nearly 300 million years (My), during much of the Phanerozoic, from 450 to 180 million years ago (Ma), Antarctica formed the keystone of the Gondwana supercontinent (**Figure 1A**). Prior to Gondwana, Antarctica may have formed part of a supercontinent called Rodinia (**Figure 1B**), whereby East Antarctica was joined to the western side of North America; this possibility has been developed within the framework of the Southwest US–East Antarctic (SWEAT) connection hypothesis. Today, Antarctica is isolated in a south polar position and, with the exception of one small segment, is completely surrounded by spreading ridges. Consequently, the continent has low seismicity and few active volcanoes.

The continent (**Figure 2**) is divided physiographically by the Transantarctic Mountains, one of Earth's major mountain ranges. These mountains, which extend for some 3500 km across the continent between the Ross and Weddell seas, are typically 100 to 200 km wide and reach elevations locally in excess of 4500 m. This spectacular topographic feature defines a fundamental lithospheric boundary that has profound crustal anisotrophy due to repeated cycles of tectonism, and marks a boundary between two regions, East and West Antarctica, with quite different geological histories. The main continental features are shown in **Figure 3A** and the main geological events are summarized in **Figure 3B**. East Antarctica is the old crystalline core of the continent (East Antarctic Shield) and is largely covered by the 4-km-thick East Antarctic ice-sheet. With the exception of the central Gamburtsev Mountains, a high sub glacial mountain range of unknown origin, East Antarctica has a subdued topography. West Antarctica, on the other hand, has a horst and graben topography, with the spectacular Ellsworth Mountains hosting the highest mountain in Antarctica, Mt Vinson (4820 m). A high Andean mountain chain occupies the spine of the Antarctic Peninsula.

## The East Antarctic Shield

The East Antarctic Shield comprises a Precambrian to Ordovician basement of igneous and sedimentary rocks deformed and metamorphosed to varying degrees and intruded by syn- to posttectonic granites. This basement is locally overlain by undeformed Devonian to Jurassic sediments (Beacon Supergroup) and intruded by Jurassic tholeiitic plutonic and volcanic rocks. Although much of the East Antarctic Shield is covered by the thick East Antarctic ice-sheet, the application of traditional and modern U/Pb zircon and other dating techniques has shown that the shield has a three-stage tectonic history, the details of which are only just becoming evident. This history involves the following events:

1. The stabilization of various Archaean to Palaeoproterozoic cratons (dating from 3.0 to 1.6 Ga). These areas of ancient crust can be divided into an extensive central craton, the Mawson Continent, inferred to occupy much of the continental interior of the East Antarctic shield, and various marginal cratons exposed along the coast. The marginal Archaean cratons are correlated with similar cratons that rifted from Antarctica during the Mesozoic breakup of Gondwana (for example, the Kaapvaal–Zimbabwe Craton of southern Africa and the Dharwar Craton of southern India). One of these, the Napier Complex, contains the oldest rock, dated at $3930 \pm 10$ My, currently known from Antarctica.

2. The development of three high-grade Late Mesoproterozoic to Early Neoproterozoic mobile belts. These were previously assumed to form one single Grenville-age orogen around the coastline of East Antarctica, but now appear to represent distinct crustal fragments juxtaposed in the Cambrian. One of these belts, according to the SWEAT hypothesis, constituted a piercing point, linking North America and East Antarctica in the Rodinia Supercontinent between 1100 and 750 Ma.

3. Two Late Neoproterozoic to Cambrian 'Pan-African' mobile belts that rework, truncate, and offset the preceding mobile belts, indicating that the East Antarctic segment of Gondwana underwent significant reorganization during the assembly of Gondwana at the Precambrian–Cambrian boundary. One of these belts is a continuation of the East African Orogen and developed during closure of the Mozambique Ocean and the ultimate amalgamation of Gondwana.

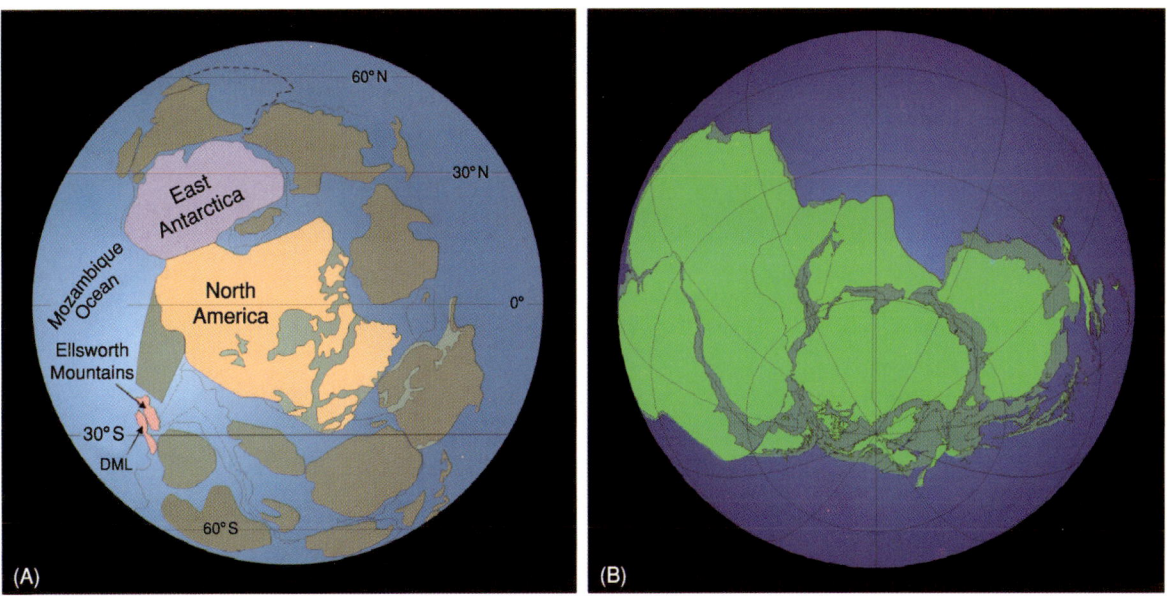

**Figure 1** (A) A proposed Rodinia Supercontinent linking East Antarctica and North America, 1000 Ma, according to the 'SWEAT' hypothesis. (B) Antarctica within Gondwana 260 Ma (courtesy of R Livermore, British Antarctic Survey).

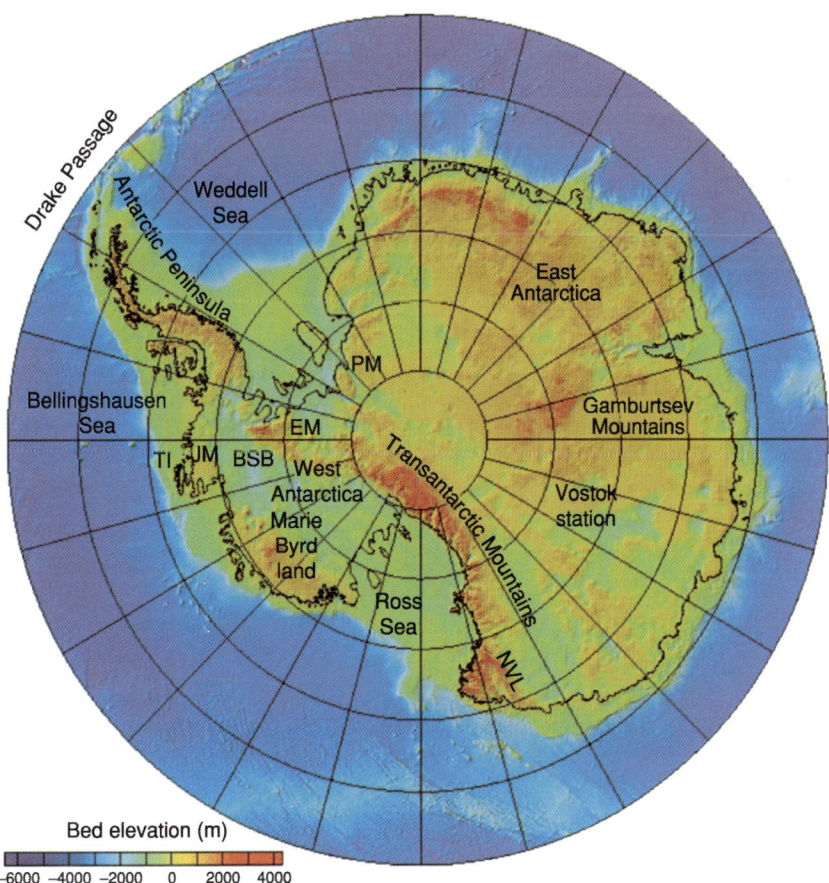

**Figure 2** Subglacial topographic map of Antarctica. BSB, Byrd Subglacial Basin; EM, Ellsworth Mountains; JM, Jones Mountains; PM, Pensacola Mountains; NVL, North Victoria Land; TI, Thurston Island.

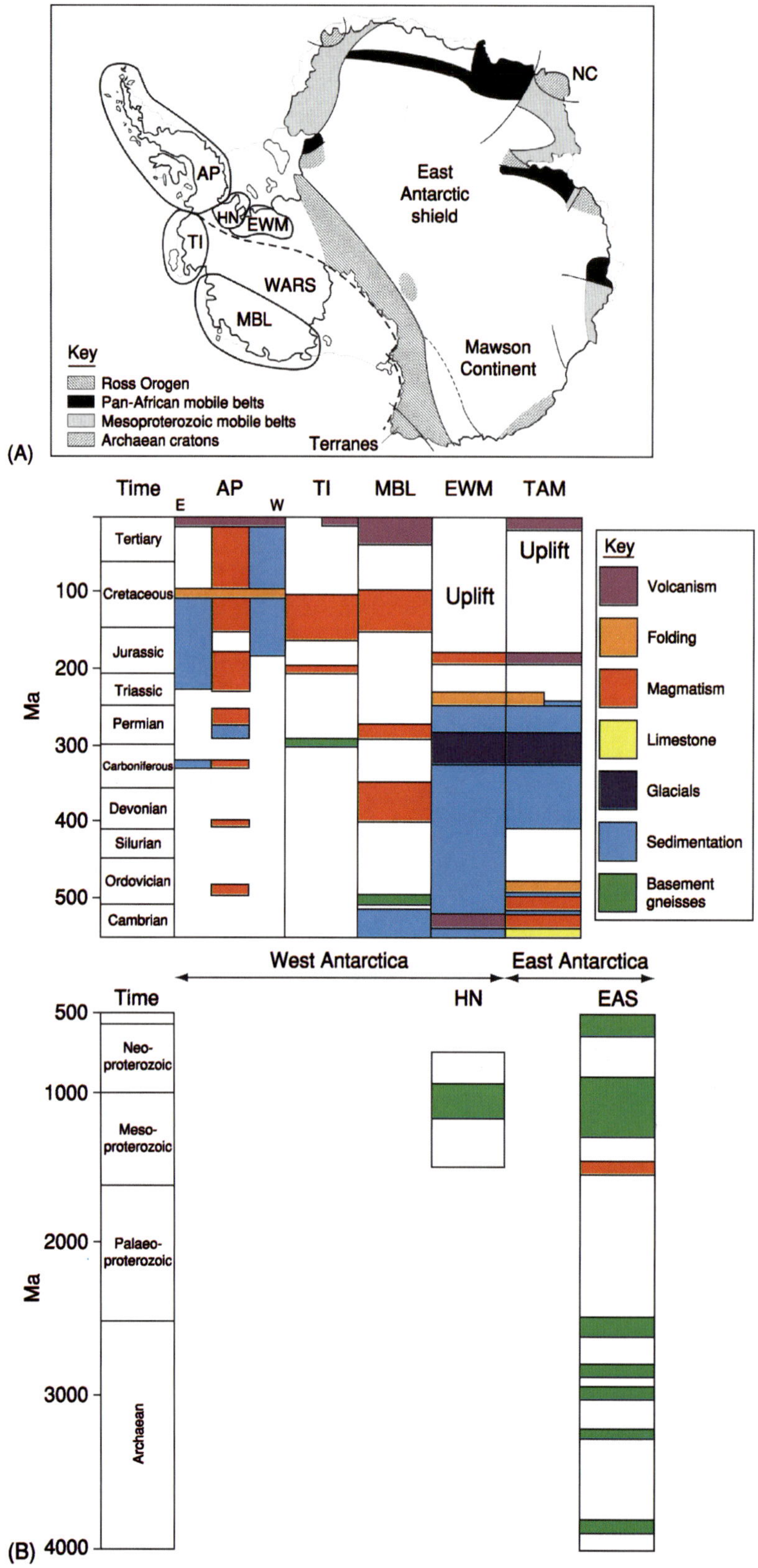

**Figure 3** (A) Simplified geological map of Antarctica. AP, Antarctic Peninsula; EWM, Ellsworth–Whitmore Mountains; HN, Haag Nunataks; MBL, Marie Byrd Land; TI, Thurston Island; WARS, West Antarctic Rift System. (B) Summary of the main geological events within the different crustal segments of Antarctica. E, East; W, West; EAS, East Antarctic Shield; TAM, Transantarctic Mountains; other abbreviations as in (A).

It is unclear as to what extent the other belt involved ocean closure or regional-scale transcurrent tectonics.

It is not known if the centre of the East Antarctic Shield is a single Palaeoproterozoic craton (Mawson Continent), as commonly assumed, or whether it is cut by Grenville-age or pan-African mobile belts, which would require modification of the boundaries of the Mawson Continent. This uncertainty means that the various terranes exposed on the coast can be joined in any number of ways and it follows that current models for Phanerozoic tectonics in East Antarctica and for global tectonics, given the crucial location of Antarctica in proposed supercontinents, will remain poorly constrained until some understanding of the continental structure beneath the ice-cap is achieved.

## The Transantarctic Mountains

Marking the boundary between East and West Antarctica, the present-day intracratonic mountain chain has undergone episodic uplift since the Early Cretaceous and has been modelled as a major rift shoulder. The unifying geological feature of the mountains is a Middle Palaeozoic erosion surface (Kukri Peneplain) that separates gently tilted Devonian to Triassic sedimentary rocks (Gondwana cover sequence) and Jurassic continental tholeiites (Ferrar Supergroup) from a Proterozoic to Early Palaeozoic orogenic belt known as the Ross Orogen.

### The Ross Orogen: The Palaeo-Pacific Margin of Gondwana

The recently formulated SWEAT hypothesis linking the North American Laurentian continent with East Antarctica has provided a powerful tectonic framework for interpreting the Late Proterozoic and Early Palaeozoic siliciclastic turbidites and volcanic rocks exposed along the Transantarctic Mountains as being deposited in a rift margin setting following the separation of Laurentia from Antarctica ~750 Ma. Following the Beardmore folding event, carbonates were deposited along the margin in Early Cambrian times. Outboard of the Early Cambrian limestone, Middle Cambrian carbonates, sedimentary rocks, and a bimodal volcanic sequence formed. The margin was subsequently transformed to an active Early Palaeozoic orogenic setting following the initiation of subduction of newly created proto-Pacific oceanic lithosphere beneath the rifted margin. Active deformation, volcanism, metamorphism, and emplacement of subduction-related igneous rocks (the Granite Harbour Intrusives) occurred along the mountain front from about 520 to 480 Ma. Large volumes of molasse sediment were shed into fore-arc marginal basins in the Middle Cambrian and Ordovician, primarily by erosion of volcanic rocks of the early Ross magmatic arc. The fore-arc deposits were intruded by late orogenic plutons as the locus of magmatism shifted offshore. Deposition of individual molasse sequences continued until 490–485 Ma. Three exotic terranes were emplaced along the North Victoria Land margin during the late stage of the Ross Orogeny.

### Gondwana Cover Sequences: A Stable Continent

Unconformably overlying the Ross Orogen in the Transantarctic Mountains and in the once-neighbouring Gondwana continents is a flat-lying cover sequence up to 2.5 km thick. Known in Antarctica as the Beacon Supergroup, this represents a period of tectonic stability within the Gondwana continent that spans the Devonian to the Triassic. It is capped by basalt flows and is intruded by thick dolerite dykes and sills of Middle Jurassic age (~180 My old) in the Ferrar province, a large igneous province that heralded the breakup of Gondwana (**Figure 4**).

There are four phases of sedimentation within the Beacon Supergroup and its scattered equivalents around the periphery of the East Antarctic shield: the Devonian strata, Permo-Carboniferous glacial deposits, Permian sediments, and Triassic cross-bedded sandstones and mudstones. The Devonian strata, which were deposited on an extensive undulating surface known as the Kukri Peneplain, are largely mature quartzose sandstones with red and green siltstones deposited by a mix of shallow marine and alluvial sedimentation during warm and semiarid climatic conditions. Some of the strata contain fossil fish and characteristic trace fossils. The Permo-Carboniferous glacial deposits derive from Late Carboniferous times, when a thick ice-sheet covered a large part of Gondwana, including Antarctica, depositing a thick diamictite unit. The diamictite deposited from the continental ice-sheet ranges from 5 to 50 m thick and records at least four advance and retreat cycles. A thick glacial section in the northern part of the Ellsworth Mountains is considered to be glacial marine in origin. Glacial striae and associated features have been measured and provide a relatively simple picture of ice flow away from the central Transantarctic Mountains.

Permian sediments, the most widely distributed Beacon strata, are cross-bedded fine- to medium-grained sandstones, shale, and coal, the products of meandering braided rivers. Coal beds average about 1 m thick but range up to 10 m. The finer overbank deposits commonly contain the famous glossopterid

**Figure 4** Middle Jurassic mafic sills related to the initial breakup of Gondwana, emplaced within a Triassic sedimentary sequence within the Theron Mountains, which are located in the Transantarctic Mountains.

leaves and fossilized stems and roots of the now-extinct glossopterid tree. The rapid and virtually complete disappearance of Late Palaeozoic ice is recorded in the earliest Permian strata. The *Glossopteris* leaves, dropped from the widespread woody deciduous tree or shrub, indicate a cool and wet rather than cold climate.

Following the end-Permian extinction event and the disappearance of the *Glossopteris* flora, the Triassic cross-bedded sandstones and mudstones were deposited by low-sinuosity rivers on a north-west-sloping plain. The reversal in palaeoslope is attributed to uplift associated with a Late Permian–Early Triassic Gondwanian folding that deformed the Cape Fold Belt and also the Permian and older strata of the Ellsworth and Pensacola mountains in Antarctica. The climate was mild and arid with a varied reptilian and amphibian fauna, and a flora characterized by the fossil fern known as *Dicroidium*.

Beacon sedimentation culminated in the Early Jurassic with explosive rhyolitic volcanism and volcanic debris flows, and ultimately by basaltic flows and associated intrusive rocks of the Ferrar Supergroup.

## West Antarctica: A Collage of Crustal Blocks

The basin and range topography of West Antarctica can be used to delineate five physiographically defined crustal blocks (**Figure 3**) that have distinctive geological features and that may have existed as distinct microplates during the Mesozoic breakup of Gondwana.

### Haag Nunataks: Part of the East Antarctic Shield

This small crustal block is situated between the southern tip of the Antarctic Peninsula and the Ellsworth Mountains. It is formed entirely of Proterozoic basement amphibolites and orthogneiss of Grenvillian ages (dating to between 1176 and 1003 Ma). Although the gneisses are exposed only on three small nunataks, aeromagnetic surveys show the full extent of the block and suggest that similar Proterozoic basement may underlie part of the Weddell Sea embayment region. Isotopic studies also suggest that this basement may be present beneath the neighbouring Antarctic Peninsula region. The gneisses correlate with Proterozoic basement gneisses within the East Antarctic Shield and may represent a fragment of the East Antarctic Shield displaced during the breakup of Gondwana.

### Ellsworth Whitmore Mountains: A Displaced Fragment of the Gondwanian Fold Belt

The Ellsworth Mountains form a 415-km-long NNE–SSW-trending mountain range that contains the highest mountain in Antarctica. The mountains are situated along the northern periphery of the crustal block of the Ellsworth–Whitmore Mountains, which represents part of a displaced terrane once situated along the palaeo-Pacific margin of Gondwana, prior to supercontinent breakup, adjacent to South Africa and the Weddell Sea coast of East Antarctica. It was assembled in its present position by Late Cretaceous times, following the Middle to Late Jurassic breakup of Gondwana. Some 13 km of sedimentary succession are exposed within the Ellsworth Mountains, representing a continuous Middle Cambrian to Permian succession that was deformed during the Late Permian to Early Triassic Gondwanian folding. Such a complete succession is unusual in Antarctica and is part of the Gondwana cover sequence that has similarities to the Transantarctic Mountains, including the Pensacola Mountains. The Early Palaeozoic succession of the Ellsworth Mountains comprises the Heritage Group (Early to Late Cambrian) and the Crashsite Group (Late Cambrian to Devonian). The Heritage Group is composed of clastic sedimentary and volcanic rocks that make up half the entire stratigraphic thickness within the Ellsworth Mountains. The volcanic rocks formed in a mid-Cambrian continental rift environment with mid-ocean ridge-type basalts erupted near the rift axis. The Heritage Group is overlain by 3000 m of quartzites of the Crashsite Group, Permo-Carboniferous glacial diamictites of the Whiteout conglomerate, and the Permian Polar Star Formation. Middle Jurassic granites related to the Gondwana breakup intrude the Ellsworth Mountain succession in scattered nunataks throughout the remainder of the Ellsworth-Whitmore Mountains crustal block.

### Thurston Island: Pacific Margin Magmatic Arc

The Thurston Island block, which includes Thurston Island on the adjacent Eights Coast and Jones Mountains, records Pacific margin magmatism from Carboniferous to late Cretaceous times. The igneous rocks form a uniform calc-alkaline suite typical of subduction settings. On Thurston Island, the observable history began with Late Carboniferous (~300 Ma) emplacement of granitic protoliths of orthogneiss. Cumulate gabbros were emplaced soon after gneiss formation, followed by diorites that have Triassic ages. In the nearby Jones Mountains, the oldest exposed rock is a muscovite-bearing granite with an Early Jurassic age of 198 My. The subsequent evolution of the Thurston Island area was dominated by Late Jurassic and Early Cretaceous bimodal suites. Between 100 and 90 Ma, volcanism in the Jones Mountains became predominantly silicic prior to cessation of subduction along this part of the margin by collision or interaction of a spreading ridge with the trench. In the Jones Mountains, the basement rocks are unconformably overlain by postsubduction Miocene alkalic basalts.

### Marie Byrd Land: Pacific Margin Magmatic Arc

Small scattered exposures throughout the coastal parts of Marie Byrd Land suggest that the block may contain two geological provinces or superterranes. In the western part, the oldest Palaeozoic rocks are a thick uniform sequence of folded sandstone-dominated quartzose turbidites of the Swanson Formation. In the eastern part, the older basement rocks include biotite paragneiss, calc-silicate gneiss, marble, amphibolite, and granitic orthogneiss with protolith ages of 504 My. In western Marie Byrd Land, the Swanson Formation was intruded by the Devonian–Carboniferous Ford Granodiorite and Cretaceous granitoids, whereas in eastern Marie Byrd Land, magmatic rocks are predominantly Permian and Cretaceous in age, indicating a long-lived magmatic history for Marie Byrd Land. The Cretaceous magmatic rocks include mafic dyke suites and anorogenic silicic rocks, including syenites and peralkaline granites that record an important change in the tectonic regime, from a subducting to an extensional margin, prior to separation of New Zealand from Marie Byrd Land and seafloor spreading from 84 Ma.

### The Antarctic Peninsula: Long-Lived Andean-Type Margin

The Antarctic Peninsula is a long-lived magmatic arc built at least in part on continental crust with a record of magmatism and metamorphism that stretches back at least to Cambrian times. Pre-Mesozoic rocks are only sparsely exposed and include orthogneisses with protolith ages dating to ~450–550 Ma, paragneisses that form the basement to Triassic granitoids no older than Late Cambrian, and a few small granitic bodies ~400 My old. Locally, the basement underwent metamorphism, migmatization, and granite emplacement during Carboniferous (~325 Ma) and Permian (~260 Ma) times.

The Mesozoic geology of the Antarctic Peninsula has traditionally been interpreted in terms of a near-complete arc/trench system with Mesozoic accretion subduction complexes on the western Pacific margin of the Peninsula, a central magmatic arc active from 240 to 10 Ma (represented by the Antarctic Peninsula Batholith), and thick back-arc and retro-arc basin sequences on the eastern Weddell Sea side. The polarity of the system is consistent with east-directed subduction of proto-Pacific Ocean floor. However, the discovery of major ductile shear zones along the spine of the peninsula has led to the identification of separate domains and the possibility of the collision and accretion of separate terranes along the margin. Triassic and Early Jurassic plutons were emplaced along the palaeo-Pacific margin of Gondwana prior to the breakup of the supercontinent. The earliest plutons were peraluminous granites with S-type characteristics. By 205 Ma, metaluminous, type-I granitoids were emplaced. Magmatism associated with the Jurassic breakup of Gondwana is represented by extensive silicic volcanism and associated subvolcanic plutonism that is part of the large Gondwana wide volcanic igneous province. Cretaceous and younger plutons were emplaced as a result of east-directed subduction of proto-Pacific ocean crust ranging in composition from gabbro to granodiorite, with a peak of activity between 125 and 100 Ma. The Tertiary part of the batholith is restricted to the west coast of the northern Antarctic Peninsula, signifying a major westward jump in the locus of the arc. With the exception of one small segment, subduction and its associated magmatism ceased in the Antarctic Peninsula between ~50 Ma and the present day, following a series of northward-younging (northward-facing) ridge trench collisions.

## Gondwana Breakup: The Isolation of Antarctica

Four main episodes in the disintegration of the Gondwana continent led eventually to the isolation of Antarctica, to the development of the circumpolar current, and to an Antarctic continent covered in ice (**Figure 5**). The initial rifting stage started in Early Jurassic times (180 Ma) and led to formation of a seaway between West Gondwana (South America and

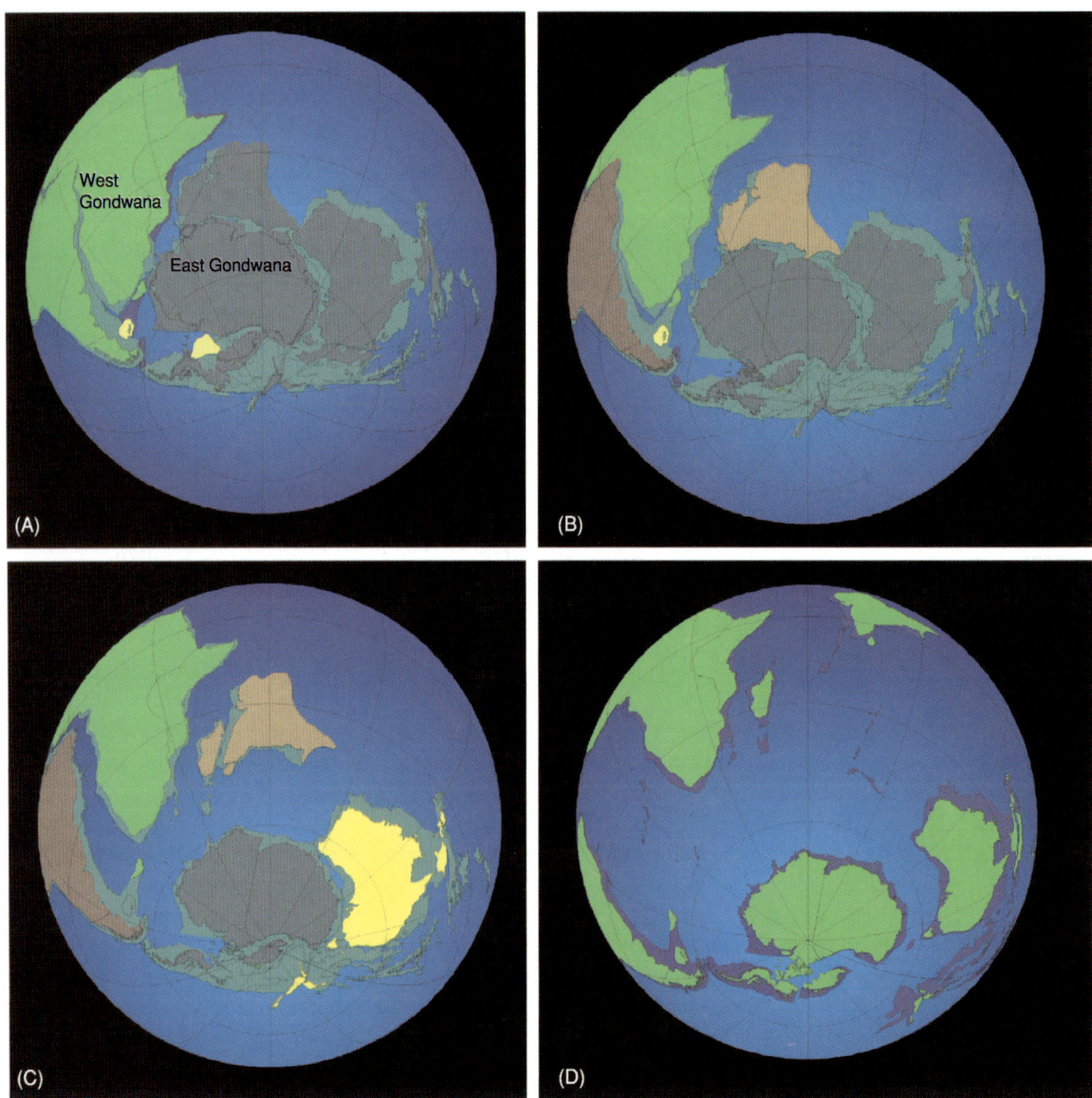

**Figure 5** Snapshots of the distribution of the continents at (A) 150, (B) 130, (C) 100, and (D) 35 Ma during the breakup of Gondwana, leading to the development of the circumpolar current and the polar isolation of Antarctica. Courtesy of Roy Livermore, British Antarctic Survey.

Africa) and East Gondwana (Antarctica, Australia, India, and New Zealand), and to seafloor spreading in the Somali, Mozambique, and possibly Weddell Sea basins. The second stage occurred in Early Cretaceous times (~130 Ma) when this two-plate system was replaced by three plates, with South America separating from an African–Indian plate, and the African–Indian plate separating from Antarctica. In Late Cretaceous times (90–100 Ma), New Zealand and South America started to separate from Antarctica until finally, approximately 32 Ma, the breakup of that once large continent was complete, when the tip of South America separated from the Antarctic Peninsula by opening of the Drake Passage, allowing the formation of the circumpolar current and thermal isolation of Antarctica.

### The West Antarctic Rift System

Although the final isolation of Antarctica did not occur until opening of the Drake Passage ~32 Ma, a rift system formed within West Antarctica during Tertiary times. The rift system extends over a largely ice-covered area extending 3000 × 750 km, from the Ross Sea to the Bellinghausen Sea, comparable in area to the Basin and Range and the East African rift systems. A spectacular rift shoulder scarp, along which peaks reach 4 to 5 km maximum elevation, extends from northern Victoria Land to the Ellsworth Mountains. The rift shoulder has a maximum present-day physiographical relief of 7 km in the Ellsworth Mountains–Byrd Subglacial Basin area. The Transantarctic Mountains part of the rift shoulder has been rising episodically since Late Cretaceous times. The rift system is characterized by bimodal, mainly basaltic alkali volcanic rocks ranging in age from Oligocene or earlier to the present day. The large Cenozoic volcanic centres in Marie Byrd Land have been related to a mantle plume. Sedimentary basins within the rift system in the Ross Sea embayment contain several kilometres of Tertiary sediments that preserve a record of climate change from a greenhouse environment to an ice-covered world. There are 18 large central vent volcanoes in Marie Byrd Land that rise to elevations between 2000 and 4200 m above sea-level.

## Antarctic Climate History: The Past 100 Million Years

The large-scale palaeogeography of the Antarctic region varied little from Late Cretaceous to Eocene times (~35 Ma), with the Antarctic continent still connected to South America and Australia and in a polar position. The West Antarctic rift system had begun to develop in the Late Mesozoic, with uplift in the Paleocene (45–50 Ma) as the Transantarctic Mountains began to rise and steadily erode. The earliest evidence of glaciers forming on the Antarctic continent comes as sand grains in fine-grained uppermost Lower Eocene and younger deep-sea sediments from the South Pacific, with isolated sand grains interpreted to record ice-rafting events centred on 51, 48, and 41 Ma. In the Ross Sea area, close to the Transantarctic Mountains, glaciers were calving at sea-level during the Eocene, becoming more extensive and spreading in earliest to Late Oligocene (~25 Ma), this being characterized by a number of thin till sheets separated by thin mudstone intervals. One of the mudstones contains a *Notofagus* leaf, which, along with contemporaneous beech palynomorphs, indicates a cool temperate climate on land during interglacial episodes, with many episodes of temperate ice-sheet growth and collapse. There was progressive disappearance of the *Notofagus*-dominated flora by the Late Oligocene (~24 Ma).

By the Early Miocene (~15 Ma), the Antarctic was completely isolated; there was development of the vigorous circumpolar currents and the present topography of the continent was in place. There was a large increase in ice cover beginning around 15 Ma. The majority view now is that since the Middle Miocene, Antarctic temperatures have persisted close to the present levels and that the East Antarctic ice-sheet has been a semipermanent feature during the past 15 My. This view is supported by work on well-preserved ice-desert landforms and deposits in mountains at the head of the Dry Valleys. These are dated from fresh volcanic ash deposits ranging between 4 and 15 My old, indicating that mean annual temperatures have been no more than 3°C above present at any time in the Pliocene. These and the geomorphic data suggest an enduring polar ice-sheet since Middle Miocene times. However, an alternative view of the post-Middle Miocene behaviour of the East Antarctic ice-sheet was presented as a consequence of finding a diverse biota of diatoms, sponge spicules, radiolarians, palynomorphs, and foraminifera in glacial diamictites or till deposits (the Sirius Group) at a number of locations high in the Transantarctic Mountains. These biota include Pliocene-age marine diatoms that may have been deposited in seas in the East Antarctic interior, subsequently to be glacially eroded and transported to their present sites by an enlarged East Antarctic ice-sheet. Although the transport processes and the depositional setting for the tills are well established, the origin of the age-diagnostic microfossils found in them has been in dispute. It is likely that some of the Pliocene-age diagnostic diatoms were deposited from the atmosphere and thus cannot be used to date the associated deposit. The Sirius Group also includes terrestrial sequences that record many advances and retreats of inland ice through the Transantarctic Mountains. The uppermost strata include a shrubby vegetation that indicates a mean annual temperature 20° warmer than the present-day mean. Because the flora cover a long time range, the precise age of these deposits is not known.

The cyclical pattern of ice-volume change through the Quaternary is well established from the deep-sea isotopic records and is inferred from ice-core studies representing the past 400 000 years; the ice cores were taken from near the middle of the East Antarctic ice-sheet at Vostok Station. The data show a cyclical advance and retreat at 100 000-year intervals and that temperatures in the Antarctic interior, which were the same as present-day temperatures during the last interglacial, fell episodically to about 10° cooler during the Last Glacial Maximum, then rose rapidly to Holocene temperatures around 10 000 years ago.

## See Also

**Africa:** Pan-African Orogeny; North African Phanerozoic; Rift Valley. **Andes. Argentina. Australia:** Proterozoic; Phanerozoic; Tasman Orogenic Belt. **Brazil. Gondwanaland and Gondwana. Indian Subcontinent. New Zealand. Oceania (Including Fiji, PNG and Solomons)**.

## Further Reading

Barrett P (1999) Antarctic climate history over the last 100 million years. *Terra Antarctica Reports* 3: 53–72.

Collinson JW, Isbell JL, Elliot DH, Miller MF, and Miller JMG (1994) *Permian–Triassic Transantarctic Basin. Geological Society of America Memoir 184*, pp. 173–222. Boulder, CO: Geological Society of America.

Dalziel IWD (1992) Antarctica; a tale of two supercontinents? *Annual Revue of Earth and Planetary Sciences* 20: 501–526.

Fitzsimons ICW (2000) A review of tectonic events in the East Antarctic Shield and their implications for Gondwana and earlier supercontinents. *Journal of African Earth Sciences* 31: 3–23.

Gamble JA, Skinner DNB, and Henrys S (eds.) (2002) *Antarctica at the Close of a Millennium*. Wellington: The Royal Society of New Zealand.

Lawver LA (1992) The development of paleoseaways around Antarctica. *Antarctic Research Series* 56: 7–30.

Leat PT, Scarrow JH, and Millar IL (1995) On the Antarctic Peninsula batholith. *Geological Magazine* 132: 399–412.

Macdonald DIM and Butterworth PJ (1990) The stratigraphy, setting and hydrocarbon potential of the Mesozoic sedimentary basins of the Antarctic Peninsula. *Antarctica as an Exploration Frontier* 139: 100–105.

Miller IL, Willan RCR, Wareham CD, and Boyce AJ (2001) The role of crustal and mantle sources in the genesis of granitoids of the Antarctic Peninsula and adjacent crustal blocks. *Journal of the Geological Society, London* 158: 855–867.

Moores EM (1991) The Southwest U.S–East Antarctic (SWEAT) connection: A hypothesis. *Geology* 19: 425–428.

Storey BC (1996) Microplates and mantle plumes in Antarctica. *Terra Antarctica* 3: 91–102.

Stump E (1995) *The Ross Orogen of the Transantarctic Mountains*. Cambridge, UK: Cambridge University Press.

Tingey RJ (1991) *The Geology of Antarctica*. Oxford, UK: Oxford Science Publications.

Vaugham APM and Storey BC (2000) The eastern Palmer Land shear zone: a new terrane accretion model for the Mesozoic development of the Antarctic Peninsula. *Journal of the Geological Society, London* 157: 1243–1256.

# ARABIA AND THE GULF

**I A Al-Jallal**, Sandroses Est. for Geological, Geophysical Petroleum Engineering Consultancy and Petroleum Services, Khobar, Saudi Arabia
**A S Al-Sharhan**, United Arab Emirates University, Al-Ain, United Arab Emirates

© 2005, Elsevier Ltd. All Rights Reserved.

## Introduction

Arabia consists of two main geological features: the Arabian shield and the Arabian shelf. Charles Doughty, who in 1888 produced the first geological map of Arabia, wrote "the Geology of the Peninsula of the Arabs consists of a stack of plutonic rock, whereupon lie sandstones, and upon the sand-rocks limestones. There are besides great land-breadths of lava and spent volcanoes". These two sentences encapsulate the geology of Arabia, and indeed of the whole of the southern shoreline of Palaeo-Tethys, from the modern Atlantic Ocean to the Arabian Gulf.

The Arabian shield is a Precambrian complex of igneous and metamorphic rocks that occupies roughly one-third of the western part of the Arabian Peninsula. The Arabian shield is a continuation of the adjacent African shield from which it is separated by the Red

Sea rift (*see* **Africa: Rift Valley**). The rocks in West Arabia, Yemen, Aden, and Oman are dated as Precambrian by radioisotopic dating and by the presence of Cambrian fossils in sediments above. Radiometric dates show that the Arabian shield was involved in the Pan-African Orogeny (*see* **Africa: Pan-African Orogeny**). The shield crops out along the east coast of the Red Sea rift, south to Yemen, and extend eastward into central Arabia with varied degrees of exposures.

The Arabian shield dips eastwards beneath some 6000 m of sedimentary rocks of Infracambrian to Recent age. These rocks crop out as belts surrounding the shield that dip gently east and north-east into the subsurface before they crop out again in Oman, United Arab Emirates, and Iran, mostly during the Mesozoic time that brought to surface the famous ophiolites of Oman. The geological formations are generally well exposed, due to a lack of vegetation, and can be traced along their outcrops for hundreds of kilometres. Geologists from Charles Doughty to the present day have noted that the uniform stratigraphy of Arabia can be traced westwards across North Africa to the Atlantic Ocean time (*see* **Africa: North African Phanerozoic**). This uniformity is most marked in the Lower Palaeozoic, and degrades thereafter through time.

Subsequent to the work of Doughty, geological mapping and fieldwork in Arabia in the 1930s was tied to oil exploration and covered more than 1 300 000 km$^2$. **Figure 1** is the first published geological column of Arabia compiled by Powers and Ramirez in 1963.

During the early days of Aramco, geologists identified, measured, and mapped nearly 6000 m of sediments ranging in age from presumed Infracambrian to Recent. The main rock units have been identified and mapped since 1966 by Aramco and the US Geological Survey (USGS). Through the years much additional work has been performed by Saudi Aramco, USGS, and Bureau de Recherches Geologiques et Minieres (BRGM). Some formation names have changed and some new ones have been introduced in Arabia and the Gulf states, the result being a wealth of information both from outcrops and from the subsurface that has affected our knowledge of the formations, their contacts, ages, and nomenclatures; however, the original stratigraphic framework remains largely intact. This may be seen by comparing **Figure 1** with **Figure 2**, which is the stratigraphic column for Saudi Arabia, recently produced by Aramco.

In central Arabia, the Palaeozoic and Mesozoic rocks crop out as curved belts bordering the Arabian shield, dominated by parallel west-facing escarpments capped by resistant limestone. In eastern Arabia the older sediments are mostly covered by a gently dipping belt of low relief Tertiary and younger deposits that include the Rub al Khali and north-eastern Arabia (Nafud) of Quaternary sands.

In north-western Arabia some 2000 m of largely lower Palaeozoic sandstone is exposed. The lower units of this Palaeozoic sandstone can be correlated into Jordan and across the Arabo-Nubian shield into North Africa. To the east lies a basin of Upper Cretaceous to Tertiary sediments. Volcanic rocks of Tertiary to Recent age cover substantial parts of the shield and adjacent cover. These result from deep crustal tension associated with the development of the Red Sea rift system.

## The Stratigraphy of Arabia and the Gulf

The geological section above the Precambrian basement Arabia falls into eight major divisions separated by unconformities (**Figures 1** and **2**). These may be summarized as follows from base to top:

1. Infracambrian and Lower Palaeozoic clastic rocks (Cambrian through Lower Devonian);
2. Carboniferous, Permian, and Triassic carbonate/clastic rocks (Upper Permian through Upper Triassic);
3. Lower and Middle Jurassic clastic and carbonate rocks (Toarcian to Callovian?);
4. Upper Jurassic and early Lower Cretaceous carbonate rocks (Callovian through Valanginian);
5. Late Lower Cretaceous clastic rocks (Hauterivian through Aptian);
6. Middle Cretaceous clastic rocks (Cenomanian through Turonian?);
7. Upper Cretaceous to Eocene carbonate rocks (Campanian through Lutetian); and
8. Neogene clastic rocks (Miocene and Pliocene).

### Infracambrian and Lower Palaeozoic Clastic Rocks (Cambrian through Lower Devonian)

Above the igneous and metamorphic basement of the Arabo-Nubian shield the Infracambrian cover consists of sandstone, carbonates, shale, and salts. The Infracambrian shows the oldest fossils. The Infracambrian Huqf Group of Oman contains potential petroleum source rocks. The Huqf Group includes the Mahara, Khufal, Shuram, Buah, and Ara Salt formations, ranging in age from Precambrian to Lower Cambrian. The evaporites are usually referred to as the Hormoz Salts, and have formed many diapiric structures in Oman and throughout the Gulf, many of which are petroliferous. In central Arabia the Precambrian shield is overlain by the Infracambrian

| Age | | | | Formation | Generalized lithologic description | Thickness (Type or reference section) | Major stratigraphic divisions |
|---|---|---|---|---|---|---|---|
| Cenozoic | Quaternary and Tertiary | | | Surficial deposits | Gravel, sand, and silt | | |
| | Tertiary | Miocene and Pliocene | | Kharj | Limestone, lacustrine limestone, gypsum, and gravel | 28 m | Miocene and Pliocene clastic rocks |
| | | | | Hofuf | Sandy marl and sandy limestone; subordinate calcareous standstone. Local gravel beds in lower part | 95 m | |
| | | | | Dam | Marl and shale; subordinate sandstone, chalky limestone, and coquina | 91 m | |
| | | | | Hadrukh | Calcareous, silty sandstone, sandy limestone; local chert | 84 m | |
| | | Eocene | Lutetian | Dammam | Limestone, dolomite, marl, and shale | 33 m | Upper Cretaceous to Eocene carbonate rocks |
| | | | Ypresian | Rus | Marl, chalky limestone, and gypsum; common chert and geodal quartz in lower part. Dominantly anhydrite in subsurface | 56 m | |
| | | Paleocene | Thanetian Montian(?) | Umm er Radhuma | Limestone, dolomitc limestone, and dolomite | 243 m | |
| Mesozoic | Cretaceous | | Maestrichtian Campanian | Aruma | Limestone; subordinate dolomite and shale. Lower part grades to sandstone in northwestern and southern areas of outcrop | 142 m | |
| | | | Turonian(?) Cenomanian | Wasia (Sakaka Sandstone of north-west Arabia) | Sandstone; subordinate shale, rare dolomite lenses | 42 m | Middle Cretaceous clastic rocks |
| | | | Aptian Barremian | Biyadh | Sandstone; subordinate shale | 425 m | Late Lower Cretaceous clastic rocks |
| | | | Hauterivian | Buwaib | Biogenic calcarenite and calcarenitic limestone interbedded with fine sandstone in upper part | 18 m | |
| | | | Valanginian | Yamama | Biogenic-pellet calcarenite; subordinate aphanitic limestone and biogenic calcarenitic limestone | 46 m | |
| | | | Berriasian | Sulaiy | Chalky aphanitic limestone; rare biogenic calcarenite and calcarenite limestone | 170 m | |
| | Jurassic | | Tithonian | Hith | Anhydrite | 90 m | Upper Jurassic and Early Lower Cretaceous carbonate rocks |
| | | | | Arab | Calcarenite, calcarenitic and aphanitic limestone, dolomite and some anhydrite. Solution-collapse carbonate breccia on outcrop due to loss of interbedded anhydrite | 124 m | |
| | | | Kimmeridgian | Jubaila | Aphanitic limestone and dolomite; subordinate calcarenite and calcarenitic limestone. Lower part sandstone between 20° N and 22° N | ±118 m | |
| | | | | Hanifa | Aphanitic limestone, calcarenitic limestone, and calcarenite | 113 m | |
| | | | Oxfordian Callovian | Tuwaiq Mountain | Aphanitic limestone; subordinate calcarenitic limestone and calcarenite. Abundant corals and stromatoporoids in upper part | 203 m | |
| | | | Callovian(?) Bathonian Bajocian | Dhruma | Aphanitic limestone and shale; subordinate calcarenite. Dominantly sandstone south of 22°N, and north of 26°N | 375 m | Lower and Middle Jurassic Clastic and carbonate rocks |
| | | | Toarcian | Marrat | Shale and aphanitic limestone; subordinate sandstone | 103 m | |
| | Triassic | | Upper | Minjur | Sandstone, some shale | 315 m | Permian and Triassic clastic rocks |
| | | | Middle | Jilh | Sandstone, aphanitic limestone, and shale; subordinate gypsum | ±316 m | |
| | | | Lower | Sudair | Red and green shale | 116 m | |
| Palaeozoic | Permian | | Upper | Khuff | Limestone and shale; dominantly sandstone south of 21°N | 171 m | |
| | | | Lower | Wajid | Sandstone, gravel, and basement erratics (Recognized only in southwestern Saudi Arabia and northern Yemen) | 950 m Calculated | |
| | | ? | Undated | | | | |
| | Devonian | | Lower | Jauf | Limestone, shale, and sandstone | 299 m | Lower Palaeozoic clastic rocks |
| | Ordovician and Silurian | | | Tabuk | Sandstone and shale | 1,072 m | |
| | Cambrian | | | Saq (Umm Sahm, Ram, Quweira, Siq) | Sandstone | +600 m | |

Precambrian basement complex

Compiled by RW Powers and LF Ramirez, June 3, 1963

**Figure 1** The first geological column of Saudi Arabia, compiled by RW Powers and LF Ramirez in 1963. After Powers *et al.* (1966).

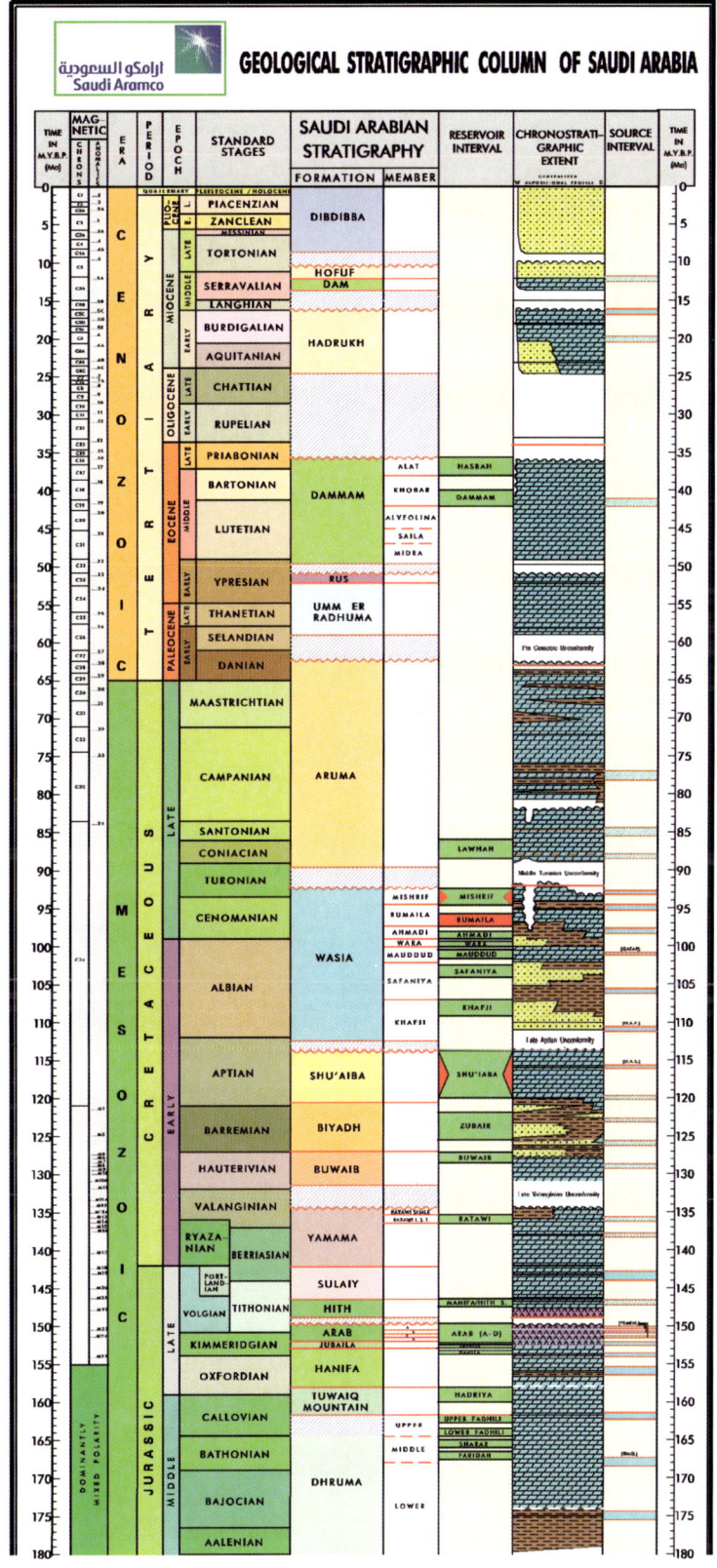

**Figure 2** Modern geological stratigraphic column of Saudi Arabia. After Qahtani et al. (2004).

Continued

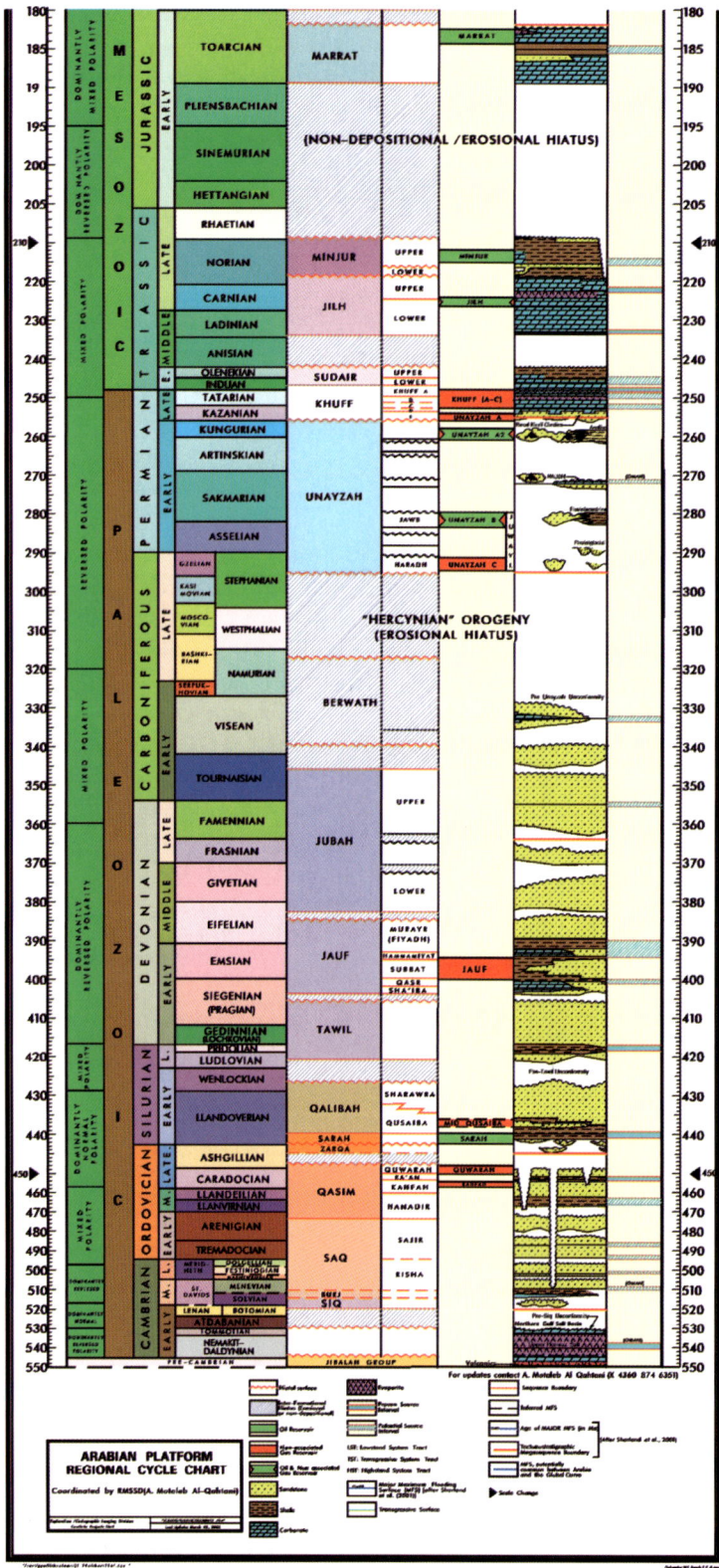

**Figure 2** Continued

Jubailah Group, whose biostratigraphy is poorly known, but can be correlated with Huqf Group of Oman. The Lower Palaeozoic rocks are mainly quartz, sandstone, and shale, with some thin carbonate beds in the upper part. They range in age from Cambrian to Lower Devonian. These include the Saq, Tabauk, Tawil, and Jauf formations. Some of these old names have been replaced by newer names. For example, the Tabuk has been replaced by the Qasim. Additional formations have been added, such as the Kahfah and Juba. The 'old' Wajid Sandstone (**Figure 1**) in south-western Arabia has been divided into four new formations. The Cambro-Ordovician Saq and Qasim formations can be correlated with the Haima Group in Oman, which includes the Amin, Andam, Ghudun, and Safiq formations. In Jordan the Saq Sandstone can be divided into four units: the Siq, Quweira, Ram, and Um Sahm. The Siq is probably equivalent to the Yatib Formation of the BRGM in central Arabia. The Burj Formation of Jordan (Middle Cambrian carbonates and shale) was introduced to the Aramco chart in 1992, and Laboun suggested that it be renamed the Farwan Formation in al-Jauf near Jordan. The Burj carbonates overlie the Siq or Yatib in north-western Saudi Arabia. The Saq Sandstone is of mainly fluvial origin, though the presence of *Cruziana* tracks in abandoned shale channels towards the top suggest marine influence. The Saq Formation has two members, the Resha and the Sajir.

The Saq Formation is uncomfortably overlain by the Middle Ordovician Qasim Formation of sandstone with alternating Shale members. These include the Hanadir Shale, the Kahfah Sandstone, the Raan Shale, and the Qawarah Sandstone. The shales contain rare graptolites, of great biostratigraphic significance, and the sandstones locally contain abundant burrows variously termed *'Tigillites'*, *'Sabellarifex'*, and *'Scolithos'*, which are indicative of intertidal conditions.

The Qasim is unconformably overlain by the Zarga and Sarah formations of Late Ordovician and Early Silurian ages; these two formations are famous for their glacial features. The Mid- to Late Silurian age Qalibah Formation follows with its two members, Sharwrah and Qusaiba. Formerly the Qalibah was called the Tayyarat Formation developed by BRGM studies. The Qusaiba Shale is the main source rock for the Palaeozoic gas in the Jauf (Devonian), Unayzah (Permian), and Khuff (Late Permian–Early Triassic) reservoirs.

Sedimentological studies reveal that this Lower Palaeozoic sequence resulted from deposition on alluvial braid plains that passed down-slope into intertidal and shallow marine shelves, which in turn passed into basinal mud. Graptolites in the latter can be used to demonstrate the diachronous progradation of these facies across the Arabian shield. These events were coeval with similar progradational episodes northwards across the Saharan shelf.

The Tawil Formation (Late Silurian and Early Devonian) unconformably overlies the Sharwrah member of the Qalibah Formation. This sandstone is very different from other Palaeozoic sandstones because it contains an abundance of heavy minerals. These give a spiky character to the spectral gamma ray log, which aids subsurface correlation because the Palaeozoic sands commonly lack palynomorphs that can be used for age dating. The Tawil Formation is conformably overlain by the Jauf Formation. The Jauf Formation is remarkably different in character from where it crops out in the north to where it is found in the subsurface of eastern Arabia. Where it crops out near the town of Jauf it is mostly marine limestone, whereas in the subsurface in eastern Arabia it consists of deltaic sandstone. The Jauf Formation is one of the main Palaeozoic gas reservoirs in Arabia. In the Jauf Formation reservoir drilling problems are usually encountered due to considerable amounts of permeability-inhibiting illite. The palynology of the Jauf Formation is more useful for environmental than for stratigraphic analysis (**Figures 1** and **2**).

The Jauf Formation is overlain by the Jubah Formation. This consists of Middle Devonian to Lower Carboniferous cross-bedded sandstone that used to be considered part of the Jauf Formation, but is now separated from it. It is unconformably overlain by the Wasia Formation in the Sakaka area of northern Saudi Arabia. The Jubah Formation can be correlated with the Sakaka Sandstone, whose age was controversial, having once been considered Cretaceous, and equivalent to the Wasia Formation.

The Wajid Sandstone of south-western Arabia has been divided into five members: the Dibsiyah (Cambrian, equivalent to Saq), the Sanamah (Late Ordovician, equivalent to Sarah/Zarga), the Silurian Qusaiba, the Khusyyayn (Devonian–Carboniferous, equivalent to Tawil, Jauf, Jubah, and Berwath) and the Juwayl (Permo-Carboniferous) equivalent to Unayzah and its units. The lower parts of Unayzah has recently been interpreted by Aramco as the equivalent of the Juwayl or Wajid Sandstone and to the Al Khlata Formation of Oman.

## Carboniferous, Permian, and Triassic Carbonate/Clastic Rocks (Upper Permian through Upper Triassic)

Lower Palaeozoic strata are succeeded by some 1000 m of interbedded carbonate and clastic sediment of Upper Permian to Triassic ages, with some basal Permo-Carboniferous rocks. The old stratigraphic terminology included the Berwath (Carboniferous), Khuff, Sudair, Jilh, and Manjur Formations. Of these, only the Berwath has been renamed (**Figures 1** and **2**). The Unayzah Formation was introduced to include the Pre-Khuff clastics. The new formation includes the clastics below the Khuff together with the basal Khuff clastics. The Unayzah is one of main sweet gas reservoirs in the Palaeozoic rocks of Arabia. It is divided into the A, B, and C reservoirs. The age of this formation is still ill-defined. Currently it is believed to span the Carboniferous/Early Permian to Late Permian, usually referred as a Post-Hercynian orogeny event. Many intervals of the Unayzah reservoirs, however, differ from each other and from the lower unit (the Unayzah C). This is more cemented than the overlying B and A units, suggesting a very wide gap in age within the Hercynian period. This in turn led geologists to generate other nomenclatures to separate these sections. The Unayzah rocks are mostly alternating red beds with three sandstone reservoirs. Several depositional environments have been suggested for the Unayzah Formation. These include eolian dunes, meandering streams, incised valleys, deltas and parabolic and coastal plain deposits, and variations of the above. The Berwath Formation, however, is only known in the subsurface, and it has been suggested that the name be discarded. The Berwath rocks are similar to those of the Unayzah Formation. The Unayzah Formation in Saudi Arabia can be correlated with the Gharif and Al Khlata (Houshi group) in Oman and Faragan in Iran, while the Unayzah upper reservoirs (partly A and B) can be correlated with the Gharif Formation and the Unayzah C with the Al Khlata Formation. Debates on the nomenclature and correlation of the stratigraphic units of the Unayzah Formation continue. The Unayzah is one of the main Palaeozoic sweet gas reservoirs in Arabia.

The shallow marine sabkha carbonates and evaporites of the Khuff Formation unconformably overly the Unayzah Formation. The sequence starts with a basal shallow marine clastic unit. The Khuff reservoirs hold large volumes of gas in both Saudi and the Gulf states. The Khuff gas is usually sour due to its sulphur content, which increases northwards with increasing anhydrite. The Khuff reservoirs are normally dolomitized, with lenses of limestone. The reservoirs are heterogeneous, even though the main units are correlatable for long distances. The formation has undergone extensive diagenesis, including leaching, anhydrite cementation, and dissolution, that made it very hard to predict reservoir character. The Khuff Formation is equivalent to the Akhdar Formation of Oman. In the subsurface the Khuff has been divided into seven members: Khuff A, Khuff B, Khuff C, Khuff D, Khuff E, Northern Sandstone/Evaporate Member, and Southern Sandstone/Shale Member.

The Khuff Formation is overlain with local unconformities by the Sudair shale. This marks a change to a more restricted depositional environment from sabkha evaporites to terrestrial red beds. The Sudair Shale is followed by the Jilh Formation, which consists of interbedded sandstone, shale, limestone, anhydrite, and salt. It is often overpressured and hazardous to drill through. The formation has few oil shows in Arabia. The Jilh Formation is conformably overlain by the Minjur Formation (Upper Triassic), which consists mainly of sandstone and shale and is a very good aquifer for central Arabia. These rocks correlate with the lower part of the Sahtan unit in Oman.

### Lower and Middle Jurassic Clastic/Carbonate Rocks (Toarcian to Callovian?)

Jurassic rocks are mainly marine shale interbedded with carbonates in central Arabia, grading to sandstone in northern and southern areas. These include the Marrat, Dhruma, and Tuwaiq Mountain Formations (**Figures 1** and **2**). The Marrat Formation unconformably overlies the Minjur Formation. The Jurassic system in central Arabia is dominated by the Tuwaiq Mountains escarpment in the outcrop with coral heads. Marrat and Dhruma are exposed in lower relief structures, marked by the red and green shales that alternate with resistant caps of yellowish limestone. These formations contain few oil reservoirs in Arabia.

The Lower and Upper Jurassic formation names have remained unchanged in Saudi Arabia, but other names have been introduced in other Gulf countries. For example, the Marrat Formation in Kuwait carries the same name, but the Tuwaiq Mountain Formation has been replaced by the Sargelo Formation.

### Upper Jurassic and Early Lower Cretaceous Carbonate Rocks (Callovian through Valanginian)

The Upper Jurassic to Lower Cretaceous rocks are mostly cyclic carbonate sands and evaporites that close several stages of the Jurassic. These include the most important oil-bearing Arab formation in Arabia. The formation names are largely unchanged from the earliest days of research (**Figures 1** and **2**). The upper parts of the Tuwaiq Mountain Formation consist of mainly carbonate grainstone and packstone with corals and stromatoporoids, followed by the Hanifa Formation, which is composed of carbonate mudstone, wackestone, and grainstone. The Hanifa is the main source rock of the Jurassic oil of Arabia. This is overlain by the Jubaila Formation, which is composed of mainly mudstone and wacke-to packstone; the famous Arab D reservoir can extend to include the upper parts of the Jubaila.

The Jubaila is succeeded by the Arab Formation (**Figures 1** and **2**), the most famous oil reservoir in Arabia, especially the Arab D unit. The Arab Formation includes four members A, B, C, and D, each of which consists of a carbonate reservoir with an anhydrite cap rock. The reservoirs include alternating ooidal, skeletal, and peloidal grainstone, wackestone, and packstone containing *Cladocropsis*, stromatoporoids, and foraminiferans. They also include some mudstone. The facies indicate shallowing upward sequences from high-energy shoal, through intertidal flat, to supratidal sabkha (salt marsh) and subaqeous anhydrites (*see* **Sedimentary Environments:** Carbonate Shorelines and Shelves). The Arab D Member of the Arab Formation contains most of the oil reserves of the Ghawar Field, the largest oil field in the world, more than 250 km long and 50 km wide.

The Arab Formation is well known in the subsurface by cores and logs, but the outcrop is poorly known from small hilly exposures that show signs of anhydrite karst terrain near Riyadh (*see* **Sedimentary Processes:** Karst and Palaeokarst). The Arab Formation extends into the Gulf states with minor modifications. For example, in Qatar, it is divided into the Fahahil and Qatar formations. The Arab Formation extends into Abu Dhabi where the amount of anhydrite increases offshore. Major facies changes of the Arab Formation occur in Kuwait where the Jubaila and Arab formations of the Gotnia Basin are largely evaporites.

The Jurassic rocks of Yemen are very different yet again, consisting of sandstone and shale with minor carbonates. The Naifa and Hajar formations (Kimmeridgian-Tithonian) are similar to the Arab Formation further north.

In Saudi the Arab Formation is overlain by the Hith Formation, which consists mainly of anhydrite with minor carbonate reservoirs and crops out at Dahl Al-Hith in the Al-Kharj area near Riyadh; elsewhere it has generated karst terrane. The Hith is overlain by the Early Cretaceous Sulaiy Formation, which is composed of fossiliferous chalky limestone with wackestone and packstone, and is overlain in turn by the Yamama Formation (Berriasian-Valanginian) composed of bioclastic and pelletal grainstone, wacke to grainstone and mudstone (**Figures 1** and **2**).

### Late Lower Cretaceous Clastic Rocks (Hauterivian through Aptian)

The Late Cretaceous marks a change from predominantly carbonate to terrigenous sedimentation. The sequence commences with the Buwaib Formation, which is a thin basal carbonate composed of mainly bioclastic grainstone and packstone that pass up into sandstone, overlain by sandstone and shale of the Biyadh Formation. This is truncated by an unconformity between the Aptian and Cenomanian stages. In Kuwait, however, the Buwaib and Biyadh Formations are grouped together as the Zubair Formation, while in Abu Dhabi and Oman the Yamama, Buwaib, and parts of the Biyadh Formations are termed the Lekhwair Formation. The Biyadh Formation is overlain by the Shu'aiba Formation (Aptian). This does not crop out at the surface, and like most formations, thickens eastwards from the shield to the shelf. The Shuaiba Formation contains many rudist 'reefs' which are important oil fields, such as the Shaybah field in the Rub Al-Khali (empty quarter) of Arabia and the Bu Hasa field in the United Arab Emirates (*see* **Sedimentary Environments:** Reefs ('Build-Ups')).

### Middle Cretaceous Clastic Rocks (Cenomanian through Turonian?)

The Shuiaba Formation is overlain by deltaic sandstone and shale of the Albian–Turonian Wasia Formation (**Figures 1** and **2**). This crops out occasionally as low-lying hills. The Wasia Formation includes the Ahmadi, Moudoud, Khafji, and Safaniyah members in the northern parts of Arabia. The Safaniyah and Khafji members contain huge oil reservoirs in northern Saudi Arabia and Kuwait. In some of the Gulf states the Wasia Formation and its constituent members have been raised to group and formation status, respectively. In Kuwait its equivalent is the Burgan Formation. In Oman its equivalent is the Nahr Umr Formation, a predominantly shaley unit that serves as a source rock and seal to the underlying Shuaiba reservoirs.

### Upper Cretaceous to Eocene Carbonate Rocks (Campanian through Lutetian)

Wasia sedimentation culminated in an important regional unconformity of Turonian age, overlain by some 500 m of limestone of the Aruma Formation (**Figures 1** and **2**) that crops out in an escarpment extending northwards from Arabia into Iraq. The Aruma Formation ranges in age from Turonian to Danian, and is composed mainly of carbonate rocks with some shale towards the base and increasing amounts of sand towards the north. It is equivalent to the Fiqa Formation in Oman.

The Aruma Formation is overlain unconformably by the Um er Radhuma Formation (Selandian–Ypresian), which is a highly porous and permeable dolomitic carbonate and an important aquifer in eastern Arabia. This formation is overlain by the Ypresian Rus Formation, which consists of interbedded marls, chalky limestone, gypsum, and anhydrite with quartz geodes. The gypsum and anhydrite usually dissolve to form collapse breccias that create drilling problems and also increase the salinity of adjacent aquifers. The

Rus Formation is overlain by the Dammam Formation, whose type locality is the famous Dammam Dome, on which was drilled the first well to discover oil in Saudi Arabia. This was Dammam well 7, which found oil in the Jurassic Arab Formation in 1933. The Dammam Formation consists of a number of members that include shale and carbonates, are locally reefal, and serve as local aquifers. The formation is truncated by an Eocene–Miocene unconformity.

### Neogene Clastic Rocks (Miocene through Pliocene)

Sediments of Neogene age (Miocene–Pliocene) include the Hadrukh, Dam, Hofuf, Kharj, and Dibdibah formations (**Figures 1** and **2**). These comprise alternating limestone, chalky limestone, marly sandstone, gravel, and gypsum. The upper part consists of about 200–600 m of nonmarine marly sand and sandy limestone that crop out across the Rub Al-Khali (Empty Quarter) and northeastern Arabia. The Tertiary in Oman is represented by the Hadhramout and Fars groups.

All of the sediment above is locally covered by unconsolidated Quaternary sand and gravel, which is the major contributor to the Rub Al-Khali sand in southern Arabia, an area of about $600 \text{ km}^2$. These include both eolian dunes in the sand seas and vast plains of fluvial sand and gravel.

## The Structural Geology of Arabia and the Gulf

As mentioned in the Introduction, Arabia consists of two main structural elements, the Precambrian shield of igneous and metamorphic rocks in the west, and the shelf whose sediment thickens eastwards towards the great mobile belt of the Taurus, Zagros, and Oman Mountains. The Arabian Plate (**Figure 3**) extends from the eastern Mediterranean to the greater part of Arabia (Arabian shield, Arabian platform, and Arabian Gulf), and the western Zagros Thrust Zone—an area enclosed by latitudes $13°$ and $38°$ N and longitudes $35°$ and $60°$ E. The natural boundaries of the Arabian Plate are most easily defined to the north and north-east, where the Taurus Mountains pass eastwards into the Zagros Fold Belt, which passes in turn eastwards into the Makran ranges. The structures of the Zagros can be traced into the northern Oman Mountains. The region is bounded to the south by the Owen Fracture Zone in the Indian Ocean and the Gulf of Aden Rift, and to the west by the rift system of the Red Sea and the Gulf of Aqaba and Dead Sea.

The area of the Arabian Plate is more than $3\,000\,000 \text{ km}^2$. Geologically nearly one-third of the area is composed of Neoproterozoic igneous and metamorphic rocks of the Arabian shield, of which, by far, the greatest part lies within western Saudi Arabia with minor inliers in Yemen. This shield was formed by the accretion of a series of Precambrian volcanic island arcs that can be traced into north-east Africa. The Arabian shield represents a fragment of Gondwana that separated as a result of the Phanerozoic tectonic events that were involved with the demise of the Palaeo-Tethys Ocean. Prior to breakup, Gondwana was an important source of widespread clastic sedimentation in the Palaeozoic (from Cambrian to Mid-Permian) and its spread over the platform/interior homocline. The main structural elements of the Arabian Plate are shown in **Figure 4**.

Sedimentary rocks were deposited over the Arabian Plate from Late Precambrian to Late Cenozoic as a result of a series of major tectonic phases. These began with an intracratonic phase (Late Precambrian to Mid-Permian), followed by a passive margin phase (Mesozoic), and concluded with an active margin phase (Cenozoic). During the Palaeozoic era much of the Arabian Plate lay south of the tropics, and was affected by glaciation in the Late Proterozoic, in the Late Ordovician and in the Carboniferous–Permian. It was dominated by the deposition of clastics, but interrupted by episodes of warm-water carbonate deposition in the Middle Cambrian, the Devonian, and the Upper Permian.

In contrast, during the Late Permian to the Holocene the area lay in subtropical and equatorial latitudes and was dominated by carbonate deposition.

The Arabian Plate experienced a number of events, including the rifting and sea-floor spreading of the Red Sea and Gulf of Aden, collision along the Zagros and Bitlis sutures, subduction along the Makran zone, and transform movement along the Dead Sea and Owen-Sheba fault zones. The Makran and Zagros convergence zones separate the Arabian Plate from the microplates of interior Iran.

The main structural elements in the Arabian platform contain several inherited, mechanically weak trends. These include:

1. North-trending highs as exemplified by the En Nala (Ghawar) anticline and the Qatar Arch;
2. The north-west-trending Mesozoic grabens of Azraq (Wadi Sirhan and Jauf) and Ma'rib; and
3. North-east-trending systems like the south Syria Platform and the Khleissia and Mosul trends.

These trends suggest that rejuvenation of basement discontinuities played an important role in the evolution of Arabia.

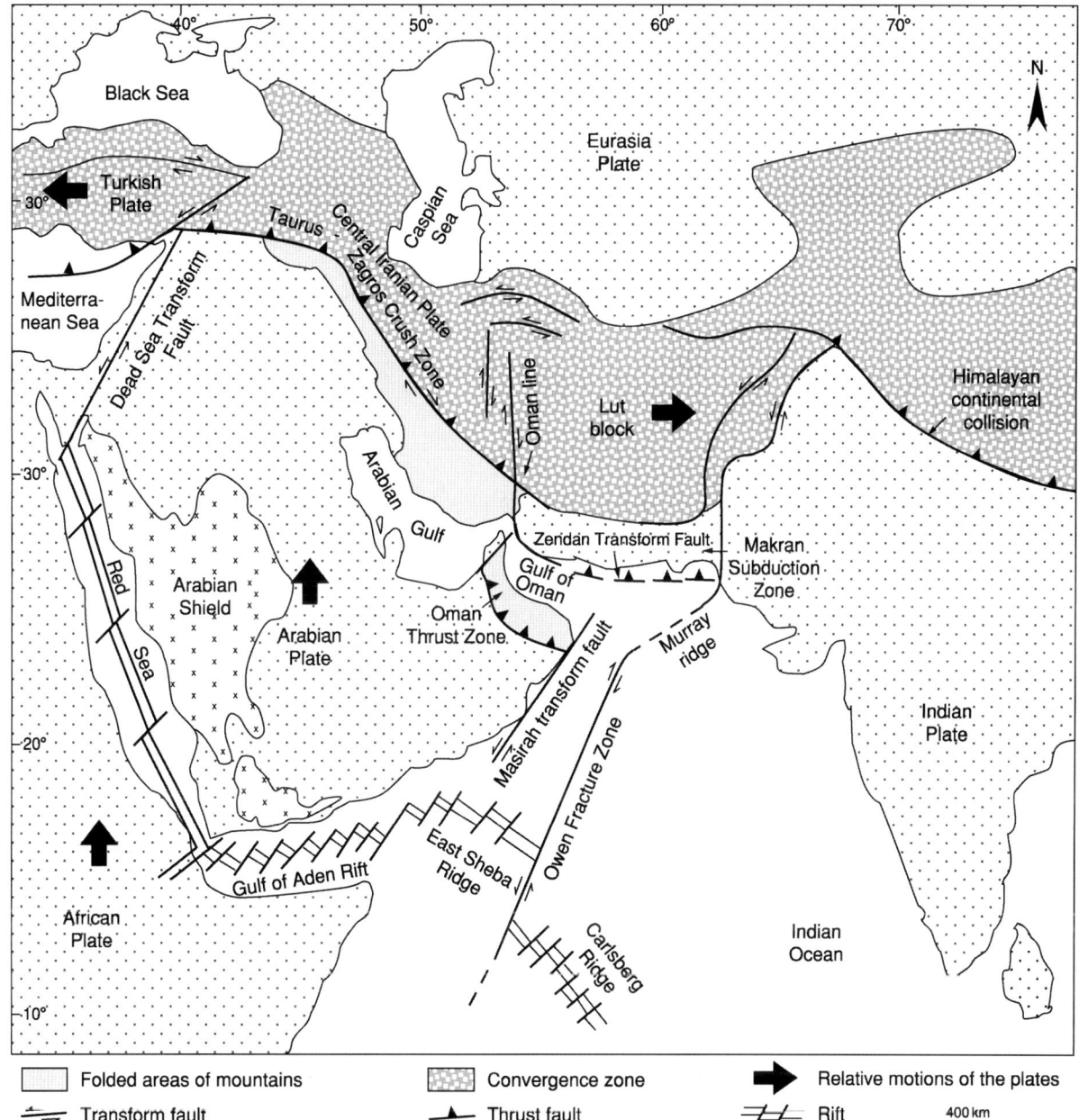

**Figure 3** Tectonic setting of the Arabian Plate in relation to adjacent plates.

The Late Precambrian rocks of the Arabian Plate result from the accretion of a mosaic of terranes and ophiolitic sutures (dating to about 870–650 Ma) with later Neoproterozoic intrusions and depositional basins that together formed the basement of Arabia. The basement evolved and consolidated through the coalescence of several island-arc terranes over a long time span in the Proterozoic. Each closure and arc collision resulted in deformation and ophiolite obduction preserved as cryptic sutures.

Faulting in the Najd and the development of rift basins with thick salt sequences, including the salt basins of Oman and the Hormuz in the Arabian Gulf, occurred in the Late Precambrian to Early Cambrian (dated to about 610 to 520 Ma).

A Late Palaeozoic Orogeny is believed to have caused uplift and erosion over most of the Arabian Gulf and some parts of the Middle East Craton. Erosion related to this event tentatively dated as Early Carboniferous cut deeply into Cambrian and Precambrian strata. Following the earlier development of a carbonate shelf along the north and northwest margins in the Early to Middle Cambrian, the plate was covered by continental, deltaic, and

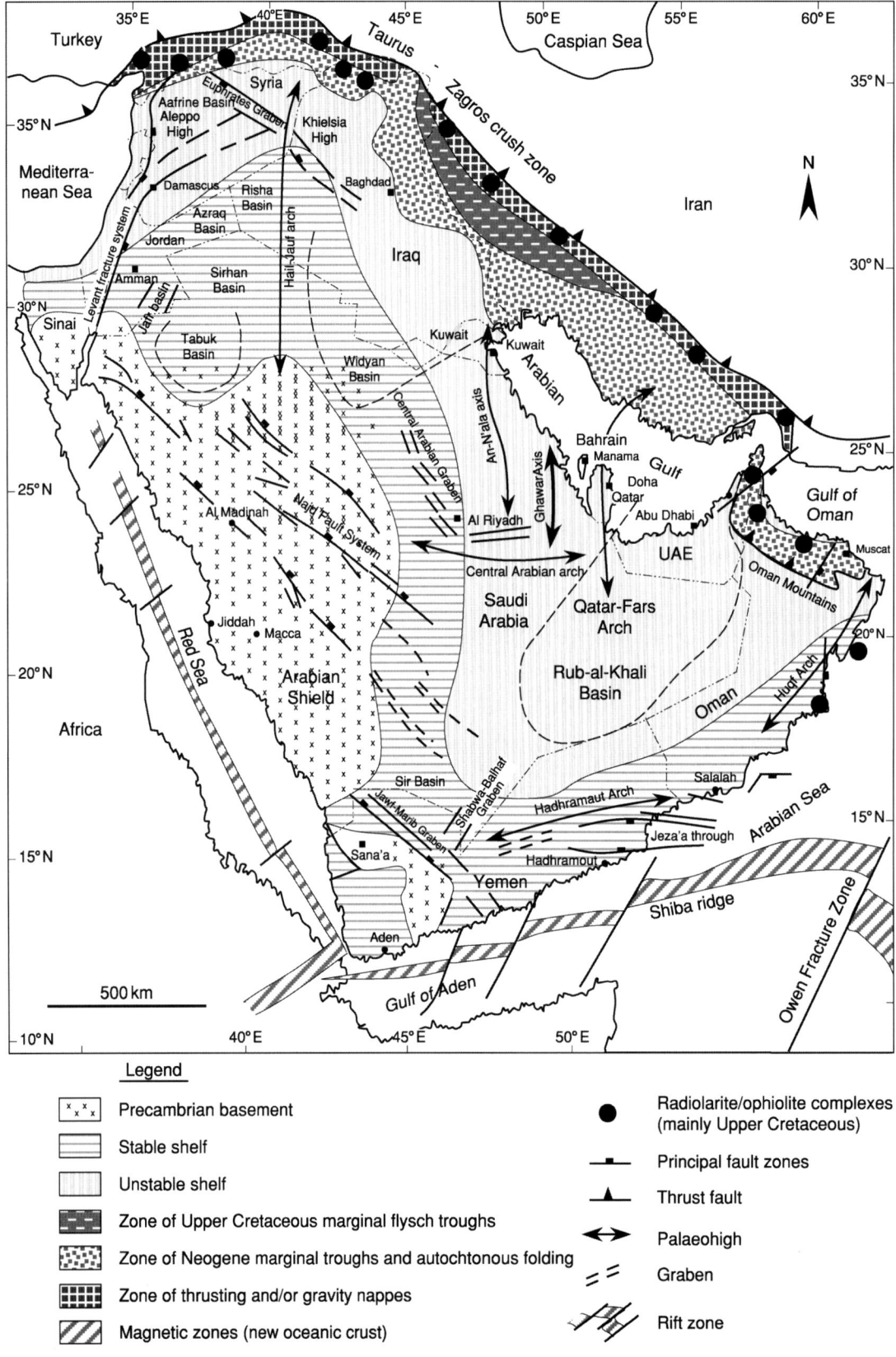

**Figure 4** Main structural elements in the Arabian Plate. Reprinted with permission from Alsharhan and Nairn (1997) *Sedimentary Basins and Petroleum Geology of the Middle East*.

shallow-marine clastics. Carbonate deposition continued into the Early Ordovician on the Iranian terraces (**Figure 4**).

During the Late Ordovician and Early Silurian, the central and western parts of the Arabian Peninsula were covered by Saharan glaciers that advanced from the south pole, which was then located in African Gondwana. During this period, nondeposition, erosion, or marginal marine conditions prevailed in eastern and northern Arabia. Deglaciation in the Early Silurian led to a sharp sea-level rise, and the Palaeo-Tethys Ocean transgressed the Arabian and adjoining plates, depositing a thick, widespread, organic-rich shale directly over the glaciogenic and periglacial rocks of Arabia.

There is a general absence of Devonian deposits over the north-eastern Arabian shelf region with the exception of parts of north-eastern Iraq and locally in Oman, and the carbonate deposition reflects a return to lower latitudes. In the Late Devonian to Early Carboniferous the onset of south-west-directed subduction along the former passive margin initiated a phase of back-arc rifting and volcanism. Lower Carboniferous carbonates occur in northern Iraq, but elsewhere the Carboniferous is largely missing, reflecting regional emergence, nondeposition, or erosion.

Following the Hercynian orogeny in Late Carboniferous to Late Permian times the Central Arabian Arch developed as a nondepositional uplift, and the Rub Al Khali became a large nonmarine intracratonic basin. In south and south-east Arabia uplifted areas developed in the Hadhramaut–Huqf and in the vicinity of the present Oman Mountains.

Glaciation occurred in Oman, southern Saudi Arabia, and Yemen, and periglacial and fluviatile conditions existed in central Arabia. In Oman and Yemen tillites rest directly on a glacially striated Precambrian basement.

During the deposition of the Permo-Triassic sequence, back-arc rifting continued at the northern end of the Arabian Plate, and the new north-east passive margin was transgressed by a shallow Permian sea from which was deposited carbonates and evaporates, thickness variations of which indicate syndepositional tectonic activity. Over Arabia the thickness is almost uniform, ranging between 300 and 600 m, but thickens dramatically to more than 1200 m east- and southwards. The main depocentres for the Late Permian carbonates trend approximately north-west–south-east, parallel to the axes of the opening Neo-Tethyan and Hawasina oceans in Oman. There is a general thickening towards the north-east Arabian Gulf and Gulf of Oman and Iran.

A major period of Late Triassic uplift and erosion affected the southern part of the Arabian Gulf and led to the progradation of continental clastic sediment across the southern Arabian Gulf region.

In the Early Jurassic progressive back-arc rifting in the eastern Mediterranean led to the development of a new northern passive margin. During the Jurassic era rift basins in Syria and south-east Yemen were active, and intrashelf basins in the south-western Arabian Gulf, eastern Saudi Arabia, and southern Iraq–Kuwait were well developed. These intrashelf basins formed the main source and reservoir rocks for the large reserves of oil in Arabia.

The Neo-Tethys spreading ridge continued migrating north-eastwards and progressively subducted under Eurasia. This Early Cretaceous sedimentation is dominated by a carbonate sequence related to major flooding of the Arabian Peninsula.

The onset of Late Cretaceous thrusting in the Oman Mountains marks a distinctive change in the pattern of the basin subsidence, and represents the main phase of thrust tectonics in south-east Arabia. The Late Cretaceous thrusting during the closure of the Neo-Tethys is directly related to the change in plate translation (from a south-west to north-east direction) in response to the opening of the South Atlantic Ocean.

Significant and widespread breaks in sedimentation occurred across the Arabian Gulf region in Late Cenomanian and Turonian times. These stratigraphic breaks correspond to major tectonic events in eastern Arabia. In the Late Cretaceous the obduction of a series of ophiolites along the Neo-Tethys margin led to reactivation of some of the basement features in Arabia and localized basin inversions in Syria.

The end of the Cretaceous was marked by a regional unconformity which resulted in the Late Maastrichtian and Danian sediments being absent over most of the Arabian Plate. During the Early Cenozoic subduction of the Neo-Tethys beneath the Sanandaj–Sirjan Terrane along the northern margin of the Neo-Tethys caused the ocean basin to close, a process assisted by the initiation of rifting and opening of the Red Sea. The Late Paleocene–Eocene consists of predominantly shallow marine carbonates and evaporites. The onset of collision between the Arabian and Eurasian continents, which commenced in the Late Eocene, initiated the Zagros orogeny by suturing the Arabian and Eurasian plates. The collision created the Zagros foreland basin on the outer edge of the north-eastern Arabian shelf margin during the final closure of the Neo-Tethys.

Coeval with the Late Alpine Orogeny in Europe, the Neogene was a time of maximum compression between Arabia and Asia. During this period, the Arabian

Plate began to separate from Africa, the Gulf of Aden opened, and the Dead Sea transform fault acted as a complex sinistral strike-slip fault.

## Economic Geology

The Precambrian basement rocks of Arabia contain a range of mineral deposits, notably gold, anciently mined in the Yemen. Important phosphate deposits occur in a belt that extends from Syria, through Jordan, and along the Palaeo-Tethyan shoreline into Egypt and across the Sahara to the Atlantic Ocean. These are found in Late Cretaceous and Eocene limestone (see **Sedimentary Rocks: Phosphates**). A range of evaporite minerals occurs in the Cambrian Hormuz salt, and in modern sabkhas of the Gulf coast, where magnesite occurs in significant quantities (see **Sedimentary Rocks: Evaporites**).

It is, of course, for its petroleum reserves that Arabia and the Gulf are best known. This basin is the largest petroleum province in the world. Various figures have been calculated for its reserves, but it is generally agreed (figures vary in time and with author) that it contains some $700 \times 10^9$ BOE (barrels of oil equivalent). This is about 40% of the world's petroleum reserves (see **Petroleum Geology: Reserves**). Petroleum is produced from reservoirs throughout the stratigraphic column. As a generalizations deep gas comes from Palaeozoic sandstones, and oil from Mesozoic carbonates. Petroleum source beds range from Infracambrian to Cretaceous. Petroleum entrapment is largely stratigraphic, occurring in truncated or onlapped sandstone, and in reefal or shoal carbonate, within which diagenesis in general, and dolomitization in particular, has often played an important part in controlling reservoir quality.

## See Also

**Africa:** Pan-African Orogeny; North African Phanerozoic; Rift Valley. **Petroleum Geology:** Reserves. **Sedimentary Environments:** Carbonate Shorelines and Shelves; Deserts; Reefs ('Build-Ups'). **Sedimentary Rocks:** Dolomites; Limestones; Phosphates. **Shields**.

## Further Reading

Al-Hajri S and Owens B (2000) Subsurface palynostratigraphy of the Palaeozoic of Saudi Arabia, joint study between Saudi Aramco and CIMP, in stratigraphic palynology of the paleozoic of Saudi Arabia, Special Publication 1. *GeoArabia*.

Al-Jallal IA (1994) *The Khuff Formation, its regional reservoir potential in Saudi Arabia and other Gulf countries; depositional and stratigraphic approach*, GEO 1994, Middle East Petroleum Geosciences Conference, vol. 1, pp. 103–119. Bahrain: Gulf Petrolink.

Al-Laboun A (1993) *Lexicon of the Paleozoic and Lower Mesozoic of Saudi Arabia*. Riyadh, Saudi Arabia: Al-Hudhud Publishers.

Alshahran AS and Kendall CGStC (1986) Paleozoic to Early Mesozoic facies, depositional setting and hydrocarbon habitat in the Middle East: an overview and some play concepts. *American Association Petroleum Geologists Bulletin* 70: 977–1002.

Alsharhan AS and Nairn AEM (eds.) (1997) *Sedimentary Basins and Petroleum Geology of the Middle East*. Elsevier.

Alsharhan AS, Nairn AEM, and Mohammed AA (1993) Late Palaeozoic glacial sediments of the southern Arabian Peninsula: Their lithofacies and hydrocarbon potential. *Marine and Petroleum Geology* 10: 1–78.

Alsharhan AS and Scott RW (2000) Hydrocarbon potential of Mesozoic carbonate platform-basin systems. In: Alsharhan AS and Scott RW (eds.) *Middle East Models of Jurassic/Cretaceous Carbonate Systems*, Special Publication 69, pp. 335–358. Society for Sedimentary Geology.

Beydoun ZR (1991) Arabian Plate hydrocarbon, geology and potential—A plate tectonic approach. *American Association of Petroleum Geologists, Studies in Geology* 33: 77.

Doughty C (1888) *Travels in Arabia Deserta* (includes geological map and paper). London: Jonathan Cape.

Glennie KW, Boeuf MGA, Hughes-Clarke MW, et al. (1974) Geology of the Oman Mountain. *Royal Geology and Mining Society (Netherlands) Transactions* 31: 423.

Konert G, Afifi AM, Al-Hajri SA, and Droste HJ (2001) Paleozoic stratigraphy and hydrocarbon habitat of the Arabian Plate. *GeoArabia* 6(3): 407–441.

Murris RJ (1980) Middle East: Stratigraphic evolution and oil habitat. *American Association of Petroleum Geologists Bulletin* 64: 597–618.

Murris RJ (1981) Middle East: Stratigraphic evolution and oil habitat. *Geologie en Mijnbouw* 60: 467–480.

Sharland PR, Archer R, Casey DM, et al. (2001) Arabian Plate sequence stratigraphy, Special Publication 2. *GeoArabia*.

Stump TE and Van der Eem JG (1994) *Overview of the stratigraphy, depositional environments and periods of deformation of the Wajid outcrop belt, southwestern Saudi Arabia*, GEO 1994, Middle East Petroleum Geosciences Conference, pp. 867–876. Bahrain: Gulf Petrolink.

Vaslet D, Manivit J, Le Nindre YM, Brosse JM, Fourniquet J, and Delfour J. (1983). Explanatory notes to the geological maps of Wadi ar Rayn quadrangle.

Ziegler MA (2001) Late Permian to Holocene paleofacies evolution of the Arabian Plate and its hydrocarbon occurrences. *GeoArabia* 6(3): 445–503.

# ARGENTINA

**V A Ramos**, Universidad de Buenos Aires, Buenos Aires, Argentina

© 2005, Elsevier Ltd. All Rights Reserved.

## Introduction

Argentina has a complete stratigraphic record of the Late Precambrian and Phanerozoic history of the southern sector of South America, and a large variety of present and past tectonic settings. As a whole, it can be divided into an Andean orogenic region and a stable basement area that is associated with one of the largest offshore continental platforms of the world. In order to understand its geology, a brief description of the present geological setting provides the key to the comprehensive record of its geological history. The Andes (see **Figure 1**) have controlled the distribution of volcanic and plutonic terrains, the sedimentary basins, and the structure of the different fold-and-thrust belts since the early Mesozoic. The opening of the Weddell Sea and the South Atlantic Ocean dominated the Mesozoic tectonics with a large period of extension. The Palaeozoic rocks were the result of an intricate interaction of sialic terranes or microcontinents that collided against the western Gondwana margin: ophiolitic belts, magmatic rocks, and different basins are testimony of these processes. Isolated patches of Late Precambrian rocks are widespread throughout most of Argentina, indicating the complex relationships of cratons, mobile belts, and shear zones.

## Geological Setting

A large variety of geological provinces exist, from the active margin related to subduction of the oceanic Nazca and Antarctic plates beneath South America in the west, to the extra-Andean stable regions and large continental platform of the passive Atlantic margin in the east. The Andean chain in Argentina has the highest peaks of the western hemisphere; the Aconcagua, for example, a Miocene volcanic massif, rises 6965 m above sea-level (a.s.l.) (**Figure 2**). One of the main geological features of Argentina is the segmented nature of the Andes and the associated foreland regions, as depicted in **Figures 3** and **4**. This segmentation is the result of subhorizontal subduction in the central segment, as represented by the Pampean flat-slab region in the foreland. This contrasts with normal subduction in the northern and southern segments, which have Wadati–Benioff zones with a dip of about 30°.

### Northern Segment

Four distinctive geological provinces characterize the northern segment. The first is the Cordillera Occidental, located along the international border with Chile; it is an active volcanic belt with large stratovolcanoes, several of them active, including the Lascar volcano, and has large ignimbritic fields of Late Cenozoic age. The second is the Puna, a high plain at 3750 m elevation; this southern extension of the Altiplano plateau is mainly composed of Miocene volcanic rocks and partially cannibalized foreland basins filled with Palaeogene to Miocene synorogenic deposits unconformably lying on a low-grade Palaeozoic basement. The third is the Cordillera Oriental; this thrust belt rises up to 6200 m high and has tectonic slices of Early Palaeozoic sedimentary and plutonic rocks. The fourth is the Sub-Andean system, consisting of middle- and low-elevation ranges ranging from 1500 to less than 1000 m high; the Sub-Andean province includes a series of large folds that expose a complete sequence of Silurian to Permian sedimentary rocks thrust on Cenozoic synorogenic deposits. The active orogenic front is located along the eastern foothills, 750 km east of the present oceanic trench. The Sub-Andean belt changes towards the south into the Santa Bárbara System, which is a range characterized by tectonic inversion of previous Cretaceous rifts (**Figure 3**).

### Central Segment

The main feature of the central segment is the lack of active volcanoes along the Andean chain. The central segment includes the Cordillera Principal, the highest mountains in the Andes; the mountains are composed of Miocene uplifted volcanic rocks unconformably overlying marine Cretaceous and Jurassic carbonates, sandstones, and shales, interfingered with andesitic and dacitic lavas and tuffs. These rocks are in tectonic contact with the Cordillera Frontal, a series of north-trending mountains composed of Late Palaeozoic 5000- to >6000-m-high volcanics and granitoids that are overlain by continental Miocene to Pliocene synorogenic deposits. A wide tectonic valley, the Calingasta–Uspallata tectonic trench, separates the Cordillera Frontal from the Precordillera. The Precordillera fold-and-thrust belt, a 5000- to 3000-m-high range, is composed of Early Palaeozoic carbonate platform

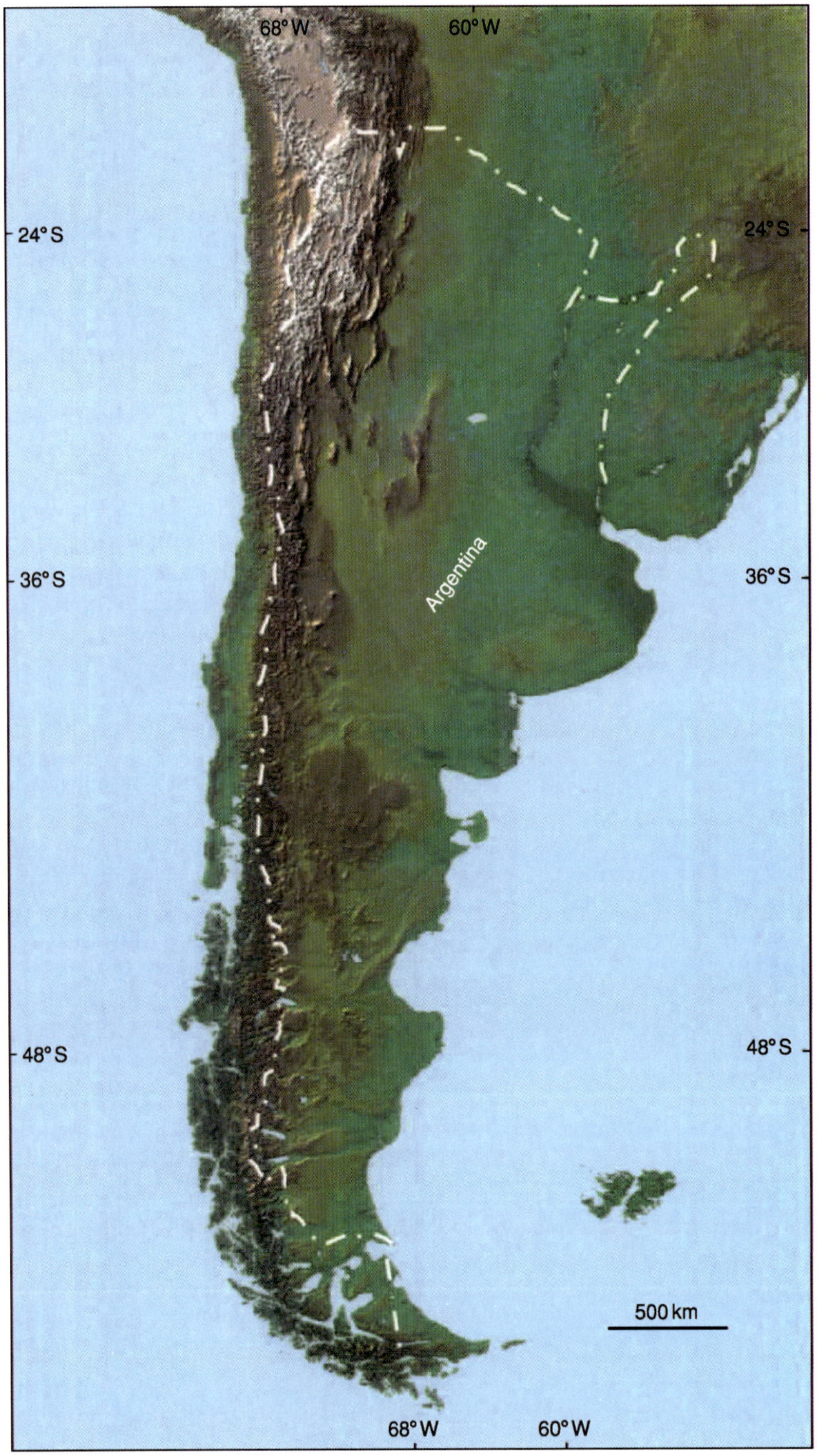

**Figure 1** Digital elevation model of Argentina. The extension and importance of the distinct cordilleras can be seen in the different segments of the Andean chain; note also the presence of plains, volcanic fields, basement uplifts, and large topographic massifs in the extra-Andean regions. Image courtesy of United States Geological Survey.

deposits, clastic foreland basin Silurian and Devonian deposits, and Late Palaeozoic sedimentary rocks, including glacial deposits. The broken foreland is characterized by the crystalline basement uplifts of the Sierras Pampeanas, a block-mountain system formed during the shallowing of the subducted slab during the past 12 million to 9 million years. Isolated patches of Miocene to Pleistocene stratovolcanoes and volcanic

**Figure 2** View of Mount Aconcagua, the highest peak in the western hemisphere (6965 m), characterized by Miocene volcanic rocks, unconformably overlying folded Mesozoic rocks.

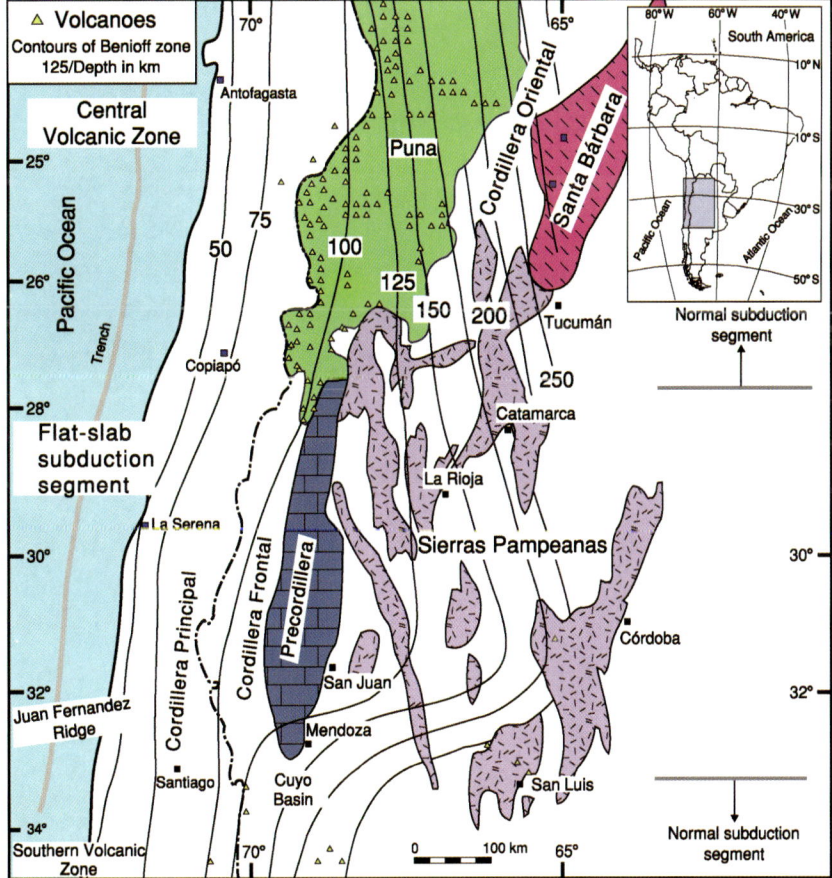

**Figure 3** The different geological provinces of Argentina are controlled by the segmented nature of the Nazca oceanic slab. The northern and southern segments have a normal 30° dipping subduction zone that contrasts with the Pampean flat slab. Contours indicate depth (in kilometres) relative to the oceanic slab; triangles are Late Cenozoic volcanoes. The dash-dot line represents the border between Argentina and Chile. From Ramos, V.A., Cristalliniy, E., Pérez, D.J. 2002. The Pampean flat-slab of the Central Andes. Journal of South American Earth Sciences 15(1): 59–78.

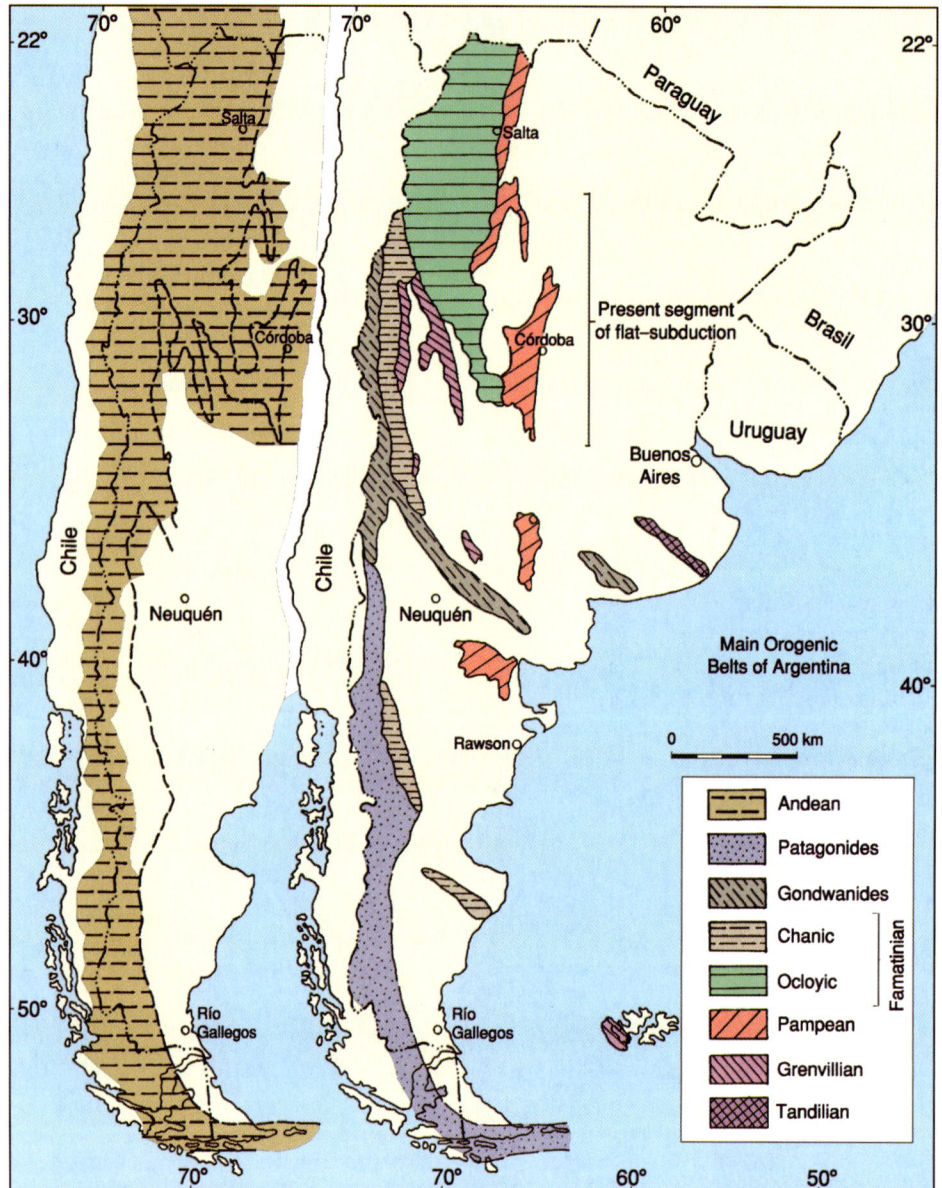

**Figure 4** Main orogenic belts in the Andes of Argentina and Chile and adjacent regions. From the Tandilian to the Andean, the corresponding geological ages are Early Proterozoic, Middle Proterozoic, Late Proterozoic–Early Cambrian, Early Palaeozoic (for both Chanic and Ocloyic), Late Palaeozoic, Late Mesozoic, and Cenozoic.

domes, with lavas and ignimbritic flows, spread over the Precordillera and the western part of the Sierras Pampeanas as a result of this flattening subduction process.

**Southern Segment**

South of the Tupungato stratovolcano, at about 33° 30′ S latitude, active volcanism resumes along the Cordillera Principal and continues down to the southernmost Andes. The main Cordillera is composed of a thin to thick-skinned fold-and-thrust belt formed by Jurassic and Cretaceous marine sediments derived from the Pacific Ocean. The altitudes, with the only exception being the active volcanic edifices, do not exceed 4000–3000 m and are even lower further to the south. The Cordillera Frontal, the Precordillera, and the Sierras Pampeanas disappear a few kilometres south of the flat subduction segment. The Palaeozoic rocks in the foothills are mildly uplifted in a series of basement blocks, including the San Rafael Block, yet still preserve the Late Palaeozoic peneplain. The foreland region has a Late Cenozoic large volcanic plateau that was formed by a few stratovolcanoes of andesitic to dacitic composition, and large basaltic fields. This retroarc volcanism, developed between 35° and 38° S latitudes, is related to the late Cenozoic steepening of the subduction zone. Between 36° and 39° S, a large Mesozoic embayment exposes the Early Cretaceous and Jurassic marine clastic and carbonate sediments of the Neuquén retroarc basin, which are unconformably overlain by continental molasse of Late Cretaceous age.

### Southernmost Segment

South of the Aysén Triple Junction between the Nazca, Antarctic, and South America plates, at about 46° 30′ S, the Patagonian Cordillera developed along the border between Argentina and Chile. The main characteristic of this segment, the continuous elongated batholith along the axis of the Patagonian Cordillera, is exposed for 2000 km and is thrust on Mesozoic volcanic and sedimentary rocks. The foothills are characterized by Late Cretaceous molasse deposits and synorogenic deposits of the Palaeogene and Neogene foreland basins. South of 52° S latitude, an oroclinal bend is characteristic of the Fueguian Cordillera in the southernmost tip of the Andes; it is composed of Jurassic volcanics and Cretaceous deposits that were heavily deformed during the Late Cretaceous closure of the Rocas Verdes marginal basin.

## Andean Region

### Volcanic Rocks

The Andes of Chile and Argentina are characterized by thick volcanic sequences developed in the past 26 million to 28 million years. These volcanic rocks are composed of andesite and dacite lavas interbedded with pyroclastic rocks in stratovolcanoes and volcanic domes, interfingered with large ignimbritic fields. The peaks of volcanic activity in the main arc are represented by thick (up to several kilometres) packages of highly differentiated rocks that constitute the Central Volcanic Zone of the Andes, as defined many years ago in southern Peru, Bolivia, and northern Chile and Argentina. The Ojos del Salado, the highest active volcano in the world (6800 m a.s.l.), is an example of a thick pile of andesites, dacites, and basalts that developed during Late Cenozoic times. These highly evolved rocks were the result of large eruptions through a thick crust (**Figure 5**); the eruptions contaminated the magmas by assimilation and differentiation through crystal fractionation. The eruptions alternated in Late Miocene and Pliocene times with rhyolitic tuffs and ignimbrites, which together make up large volcanic plateaux. These acidic plateaux are interpreted as the result of the lithospheric and crustal delamination that followed a period of steepening of the subduction zone after subhorizontal subduction north of 26° S.

The volcanic rocks south of this latitude record an opposite trend. Large stratovolcanoes developed during most of the Palaeogene along the border with Chile and they extended into Argentina in Early Neogene times. The large volcanic fields are younger towards the east and they gradually migrated to the foreland along corridors. The volcanic activity decreased in the Late Miocene and ceased along the main axis between 8 and 6 million years ago. Isolated high-K volcanoes, such as the Sierra del Morro in San Luis, are located 750 km away from the trench. South of 38° S

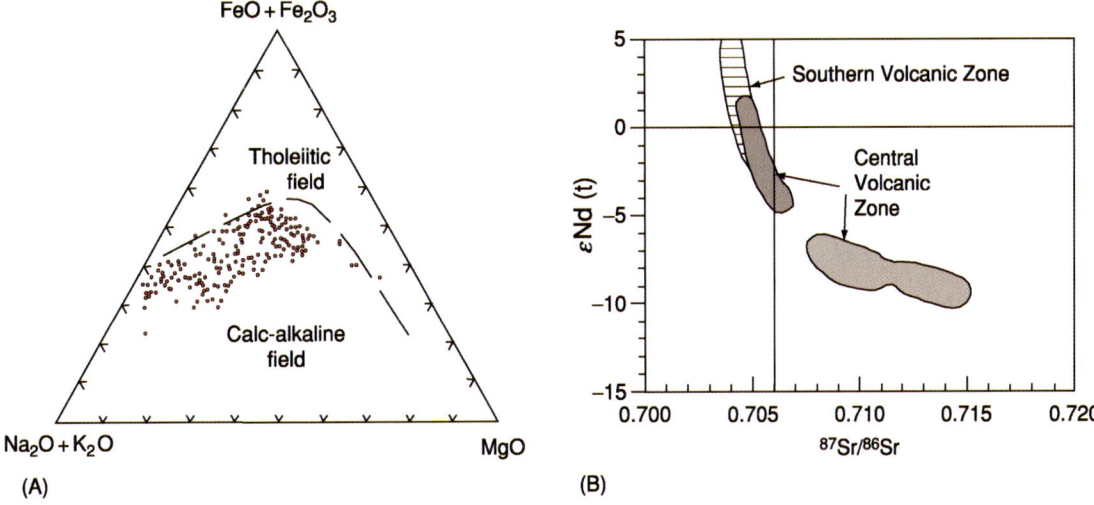

**Figure 5** (A) Main geochemical characteristics of the subduction-related volcanic rocks of Argentina, with a continuous calc-alkaline trend. (B) Contrasting neodymium (Nd) and strontium (Sr) isotopic compositions of the central and southern volcanic rocks. Note the more primitive composition of the southern volcanic zone rocks, in comparison with rocks from the central volcanic zone.

and down to 46° 30′ S, large isolated patches of volcanic rocks represented the main arc along the Andes. These volcanic rocks, termed the Southern Volcanic Zone of the Andes, have two distinct segments. The northern one corresponds to andesites, dacites, and minor basalts, which erupted through large stratovolcanoes, such as the Tupungato, Marmolejo, and San José. The southern segment is represented by basalts, basandesites, and scarce dacites and rhyolites, which erupted through several Late Cenozoic volcanoes. The difference in composition is closely related to the thickness of the continental crust along the Andean chain; this varies from over 60 km north of 33° S down to 42 km at 46° S. South of 46° 30′ S there is a volcanic gap in arc volcanism; this is indicated by isolated adakite outcrops such as Cerro Pampa and Puesto Nuevo, which are related to exceptional oceanic slab melting of young oceanic crust prior to the subduction of an oceanic seismic ridge. Further south, from 48° S to 52° S, there are five small isolated volcanoes formed of basalts and basandesites; these (including the Lautaro, Aguilera, and Monte Cook volcanoes) constitute the Austral Volcanic Zone of the Andes. All of these rocks have an adakitic signature that indicates small partial melting of the oceanic slab superimposed onto the asthenospheric wedge magmas.

**Fold-and-Thrust Belts and Their Synorogenic Deposits**

The eastward migration of volcanic activity was coeval with the deformation, uplift, and development of synorogenic deposits. The eastern slope of the Andean chain records a series of discontinuous foreland basins, with thick sequences of continental deposits up to more than 10 000 m thick. The fold-and-thrust belts vary in style and kinematics from north to south and were controlled by the segmented nature of the subduction zones and the previous Palaeozoic and Mesozoic geological history.

**Sub-Andean fold-and-thrust Belt** The thin-skinned sub-Andean Belt, which moved on Late Ordovician, Silurian, and Devonian shales, has open folds and thrusts. The area was covered by marine deposits about 13.5 million years ago, showing that deformation in the sub-Andean Belt is later than Middle Miocene. The foreland basin deposits reach over 6000 metres. The syngrowth strata of Late Pliocene and Pleistocene age, as well as Global Positioning System (GPS) data and earthquake locations, indicate active tectonics in the thrust front. The Cenozoic orogenic shortening rate was of the order of 6.7–6.9 mm year$^{-1}$, but probably with periods of higher activity during Late Miocene and Plio-Quaternary times.

**Santa Bárbara fold-and-thrust Belt** The Santa Bárbara Belt is controlled by tectonic inversion of basement faults, and therefore shortening is less important than in the Sub-Andean Belt. Fault vergence is towards the west, controlled by the polarity of Mesozoic extension. Sag and synrift sequences of Cretaceous and Palaeogene age have marine deposits of Maastrichtian–Palaeogene age with a Pacific provenance. Most of the synorogenic deposits have been preserved in intermontaneous basins, up to 3-4 km thick.

**Sierras Pampeanas Belt** The Sierras Pampeanas Belt is a region of uplifted basement in which crystalline basement of Precambrian–Early Palaeozoic age is widely exposed. The reverse faults that bound the basement blocks have a dominant west vergence and are controlled by previous crustal sutures inherited from the Early Palaeozoic tectonics. This sector coincides with the subhorizontal subduction segment. The synorogenic deposits record two different stages. The older stage was an open-to-the-east large foreland basin associated with uplift and shortening in the main Andes. The second stage, which is characterized by a broken foreland, has the largest subsidence. Consequently, basins with more than 10 000 m of Late Miocene–Pliocene deposits have accumulated in a continental environment.

**Cordillera Principal Belt** The Cordillera Principal Belt comprises (1) a northern sector (30°–32° S latitude) characterized by the tectonic inversion of basement blocks where Late Palaeozoic–Triassic volcanic and plutonic rocks are exposed, (2) a central sector (32°–33° 30′ S latitude) where thin-skinned deformations of Jurassic and Early Cretaceous marine deposits are widely exposed, and (3) a southern sector (33° 30′–37° S latitude) where inversion tectonics occurred again. All of these belts were deformed in Late Cenozoic times, resulting in accumulation in the foothills of up to 3000-m-thick synorogenic deposits of continental sequences. The southern sector is linked to the Neuquén Embayment, a wide retroarc basin that developed during Early Mesozoic times and which covered most of the adjacent extra-Andean platform.

**Patagonia fold-and-thrust Belt** The Patagonia Belt also has two different segments. The northern segment (37°–46° S latitude) has a mild basement uplift with only local Cenozoic depocentres, because the main phase of uplift was produced in Cretaceous times. South of 46° 30′ S latitude, thick foreland basin sequences constitute the Austral or Magallanes retroarc Basin. The uplift and subsequent deposition of the synorogenic deposits were controlled by collision of several segments of the Chile active spreading ridge from south to north in the past 14 million years, a collision that is still taking place at 46° 30′ S latitude.

**Fueguian fold-and-thrust Belt** The Fueguian Belt was controlled by the Late Cretaceous closure of the Rocas Verdes marginal basin, a back-arc basin of Late Jurassic–Early Cretaceous age. This segment is the only one that records the metamorphism and obduction of ophiolitic rocks during the Andean deformation. The basin development is linked to the formation of the Austral Basin; as a result of the northern vergence of the Fueguian fold-and-thrust Belt, the synorogenic deposits fill a series of east- and west-trending successor basins.

## Stable Platform

The extra-Andean region consists of large Late Cenozoic plains, known as the Pampas, where thin Tertiary and Quaternary sediments cover the basement. These sequences preserve the first Atlantic transgressions, which occurred in Maastrichtian–Paleocene times, and several others, which were produced during the different episodes of high sea-level in the South Atlantic. Most of these sediments are capped by thick sequences of loessic deposits. The stable platform is characterized by several geological provinces that encompass uplifts and basinal areas.

### Uplift Areas

Several old mountain systems emerge from the extra-Andean plains; from north to south, these regions consist of the Tandilia and Ventania systems and the Somuncurá and Deseado massifs.

The Tandilia System is a basement uplift produced during the Cretaceous at the time of opening of the South Atlantic Ocean, which exposed remnants of an Archaean basement (**Figure 4**), deformed by the Early Proterozoic Tandilian Orogen of Transamazonian age (~2000 Ma). The basement is covered by thin sequences of quartzites and carbonates of Late Proterozoic–Early Palaeozoic age. The Ventania System, a thin-skinned fold belt 600 km south of Buenos Aires, developed as a consequence of the Patagonia collision against Gondwana during Permian times. Thick sequences of clastic platform marine deposits, capped by the Sauce Grande Tillites in the Late Carboniferous, were thrust with a north vergence over the Gondwana margin. These rocks are a western extension of the Cape Fold Belt of South Africa, and the correlation between its glacial deposits with the Dwyka Tillite was one of the first geological arguments to support the existence of continental drift in the early twentieth century.

The Somuncurá basement massif is preserved in the northern sector of Patagonia (**Figure 6**). A thermally uplifted region during the development of a transient hotspot 27 Ma, the Somuncurá Massif is bounded by two east- and west-trending aulacogenic basins. The resulting thick alkaline basalts constitute the large Somuncurá shield volcano. These basalts are associated with other alkaline plugs and volcanic flows of Miocene age. This volcanic cover is unconformably overlying a Late Triassic–Early Jurassic rhyolitic plateau, Late Palaeozoic arc granitoids, and a metamorphic basement. The Deseado Massif is another uplifted basement; it is located in south central Patagonia (**Figure 6**), where metamorphic rocks of Early Palaeozoic age are covered by a thick pile of rhyolitic lavas and ignimbrites of Middle Jurassic age. The plateau basalts that characterize this sector of Patagonia, which was formed in the past 14 million years, were controlled by the thin sequences of alkaline basalts that are related to asthenospheric slab-windows generated during the Chile seismic ridge collision.

### Basinal Areas

The platform area of the extra-Andean region has several basins, most of them with an orthogonal trend to the continental margin. The main basins are from north to south.

**Chaco-Paraná Basin** The 5-km-thick Chaco-Paraná Basin records a complex history. This region first consisted of Early Palaeozoic redbeds that were part of an early foreland basin associated with the Pampean deformation in Late Proterozoic–Early Cambrian times; this was followed by Silurian and Devonian marine and continental deposits related to a reactivation of the deformation in the western sector. Late Palaeozoic glacial and marine sequences subsequently formed during a period of generalized extension. Cretaceous redbeds and tholeiitic basalts then formed during the development of the Parana plume, about 130 Ma, and, finally, a thin sequence of distal foreland deposits of Cenozoic age was associated with dynamic subsidence of the stable platform during Andean deformation.

**Salado and Colorado Basins** The Salado Basin is an aulacogenic basin that formed during the opening of the South Atlantic Ocean. Synrift deposits of Late Jurassic–Early Cretaceous age are covered by marine Late Cretaceous and Cenozoic deposits. The sedimentary sequence is coeval with the development of the continental platform in the Atlantic margin. The

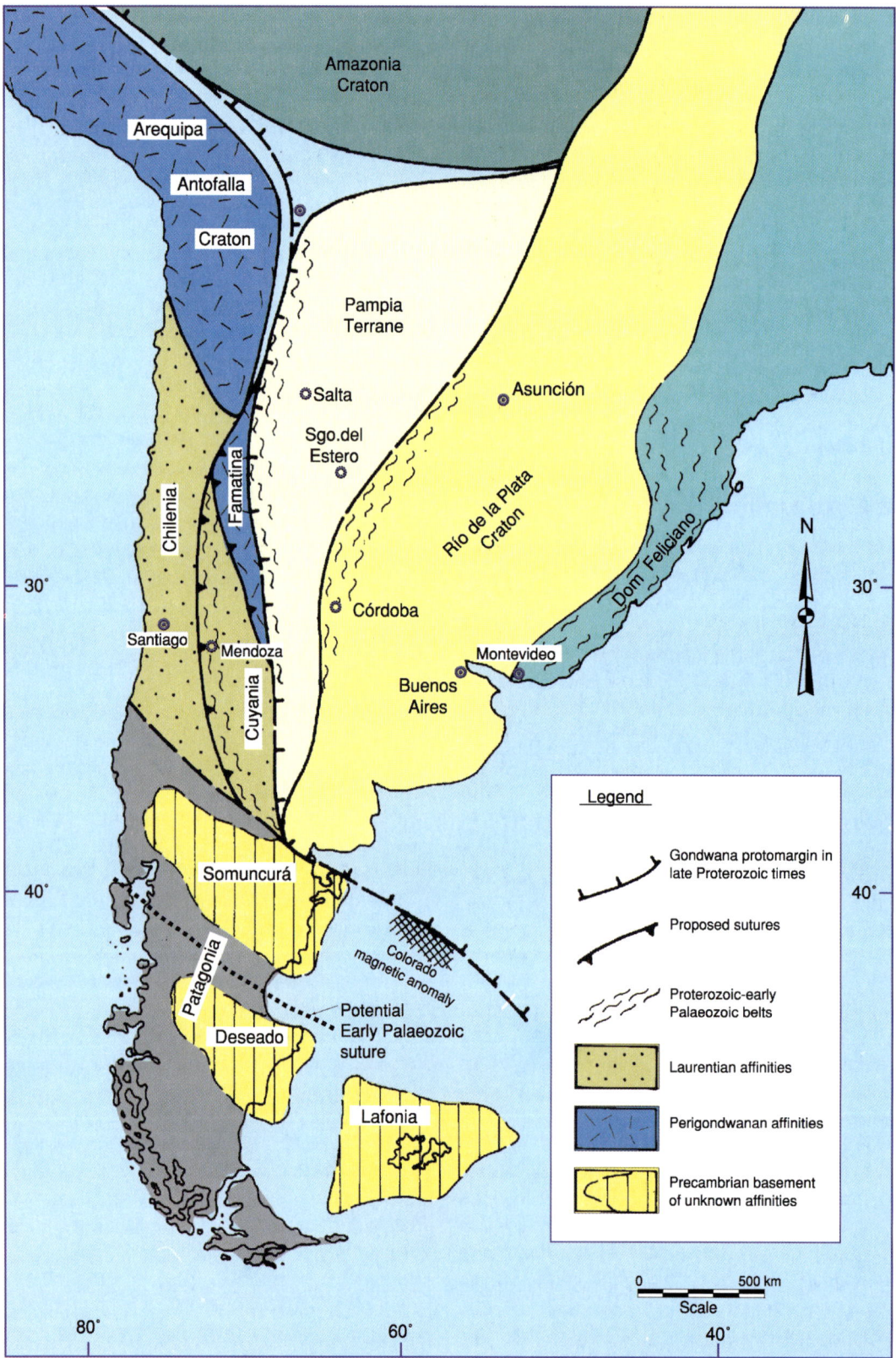

**Figure 6** Generalized map of the Early Palaeozoic terranes and cratonic blocks of Argentina.

basin is over 5–6 km thick and opens towards the east. The Colorado Basin, another aulacogenic basin, developed along the suture between Patagonia and the south-western Gondwana continent. The timing and geometry of this basin are similar to those of the Salado Basin. Towards the west, the Colorado Basin is linked with the retroarc basin of the Neuquén embayment.

**San Jorge Basin** The intracratonic San Jorge Basin developed with an east–west trend during Middle Jurassic to Early Cretaceous times, during the opening of the Weddell Sea and the South Atlantic Ocean. During the Late Cretaceous, thick sequences of molasse deposits unconformably overlaid the previous rocks. Thick sequences of clastic deposits and tuff layers record the volcanic activity in the Andes during Cretaceous and Palaeogene times. Most of the basin was filled by continental deposits, with the first Atlantic marine transgression in Maastrichtian–Paleocene times.

## Mesozoic Provinces

Important rift deposits that accumulated in north-western-trending depocentres are recorded in most of central and southern Argentina. The synrift deposits are of Late Triassic–Early Jurassic age, and consist of redbeds and alkaline basalts. A younger reactivation of these rift systems occurred in the Early Cretaceous, the time at which (~125 Ma) the first oceanic crust formed in the South Atlantic Ocean at these latitudes. The Cretaceous rifts are subparallel to the continental margin (**Figure 7**) and were controlled by the sutures of the Palaeozoic accreted terranes.

Two important rhyolitic geological provinces, the Choiyoi and the Chon Aike, were developed in the Andean and extra-Andean regions in Mesozoic times. Together these provinces constitute one of the largest rhyolitic provinces in the world, with an extension exceeding 1 000 000 km$^2$.

### The Choiyoi Province

The Choiyoi province is characterized by an extensive rhyolitic plateau that developed between northern Chile, at 24° S latitude, and the northern part of Patagonia, at 40° S latitude. The region consists of thick piles of rhyolitic lavas, ignimbrites, and volcaniclastic deposits, occasionally interbedded with lacustrine limestones. These rocks were deposited in half-graben systems, with frequent thickness changes that indicate a synextensional deposition. The age of formation varies from 280 Ma in the north to 226 Ma in the southern sector. Volcanic activity had a climax 265 Ma, when ash-flow tuffs fell over most of the extra-Andean part of Argentina and southern Brazil, even reaching South Africa. This generalized extension followed terrane accretion and deformation in Late Palaeozoic times, which was associated with the cessation of subduction along the Pacific margin.

### The Chon Aike Province

The rhyolitic Chon Aike Province covers most of Patagonia south of 40° S and also extends into the Antarctic Peninsula and further south. It is composed of abundant rhyolitic lavas, ignimbrites, and tuffs, with minor amounts of alkaline basalts interbedded in the lower part of the succession. Seismic lines through the Austral Basin show the half-graben geometry of these deposits. The volcanic sequences exposed in the Somuncurá Massif are between 185 and 170 My old, whereas the rhyolites exposed in the Deseado Massif are younger, between 165 and 155 My old. The equivalent rocks in the Tierra del Fuego Island are about 140 My old. All together, these rocks again show a decrease in age to south of the acidic volcanism, and are associated with the early rifting that led to the opening of the Weddell Sea and the South Atlantic Ocean.

## Palaeozoic Terranes

Most of the basement of Argentina was the result of collision and amalgamation of continental fragments formed during Palaeozoic times. Ophiolitic assemblages, magmatic arcs, and older orogens cover most of the present stable extra-Andean platform. Among the main continental fragments are the Cuyania, Pampia, Chilenia, and Patagonia terranes (**Figure 6**).

### Cuyania

Previously known as the Precordillera Terrane, Cuyania is one of the best known allochthonous pieces that docked against the western Gondwana margin during Middle Ordovician times. It consists of a middle Proterozoic metamorphic basement, with amphibolites, marbles, and granitoids with a typical Grenvillian signature. Cambrian limestones bearing *Olenellus* trilobites denote a Laurentian provenance; this has been confirmed by palaeomagnetic studies. The Early Ordovician carbonate rocks are followed by black shales that are unconformably overlaid by conglomerates and sandstones, interbedded with typical Gondwanian glacial deposits of latest Ordovician age. The arrival of this terrane is documented by an active calc-alkaline series of granitoids that developed along the continental margin of Gondwana from Late Cambrian to Early Ordovician times. These rocks,

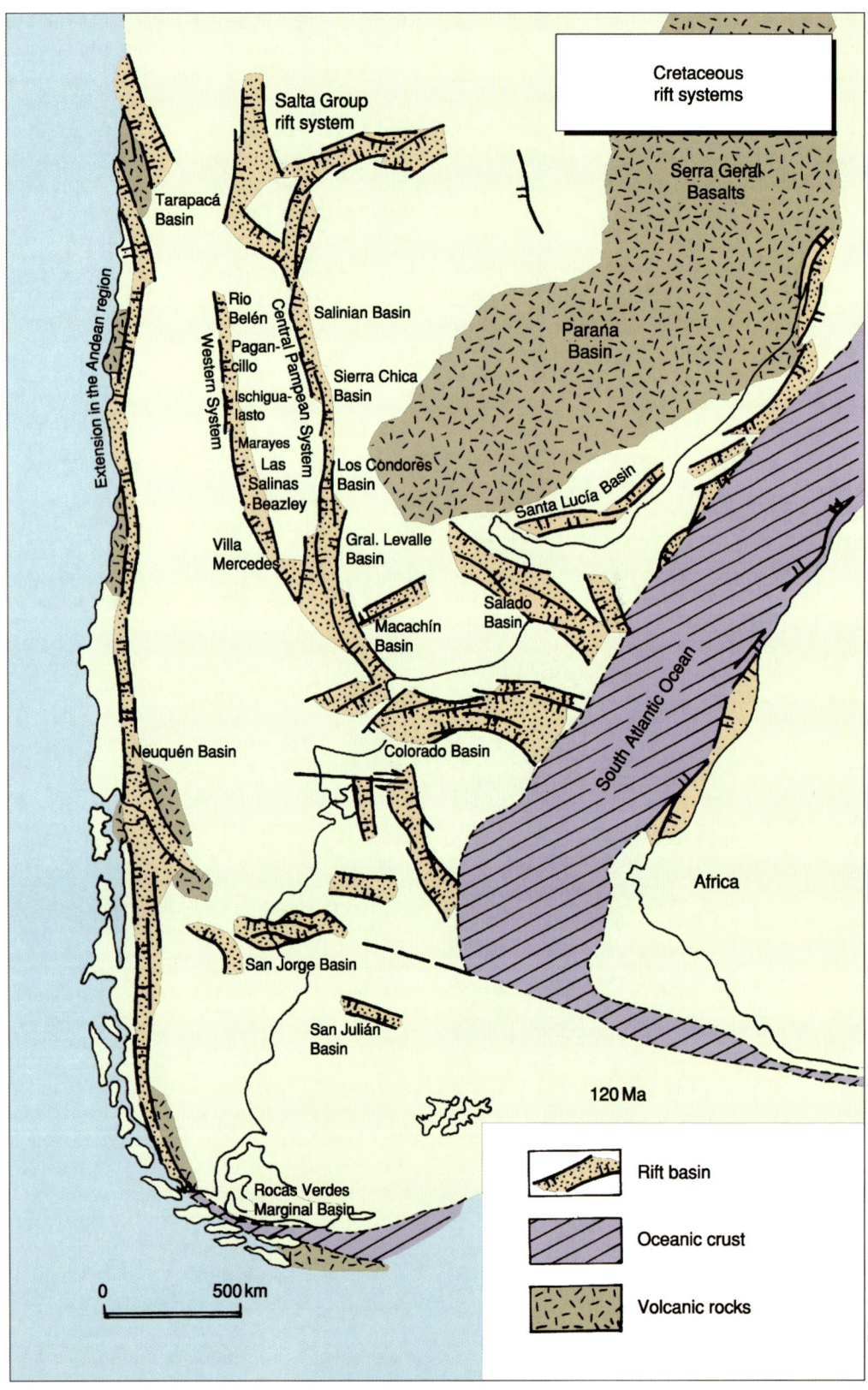

**Figure 7** Mesozoic rift systems associated with the opening of the Atlantic margin. The intracratonic rifts were partially controlled by the sutures of the Palaeozoic accreted terranes. Along the main Andes, extensional basins were controlled by negative trench roll-back velocity in the retroarc basins.

together with subsequent peraluminous granites, show a magmatic arc and postcollisional plutons along 1000 km of central Argentina. The Cuyania Terrane is bounded by two series of ophiolites, one in the western margin of the present Precordillera, and the other along the western margin of the Pampia Terrane.

### Pampia

The Pampia cratonic block hosted an Early Palaeozoic magmatic arc, which consists of metamorphic and igneous rocks penetratively deformed during Late Proterozoic–Early Cambrian times, at the time of the Pampean Orogeny. A Late Proterozoic ophiolite bounds the eastern margin of the terrane. The protomargin of western Gondwana is exposed along the eastern Sierras Pampeanas, where relics of Late Proterozoic–Early Cambrian tonalites, granodiorites, and granites constituted the magmatic arc. The collision between the Pampia Terrane and the Gondwana protomargin is constrained by a series of rhyolites and other volcanic rocks of Early to Middle Cambrian age that unconformably overlie the deformed basement rocks. Low-grade metamorphic rocks, up to 5 km thick, constituted the synorogenic sequence preserved in the Cambrian foreland basin.

### Chilenia

The amphibolites and schists with Grenvillian-age zircons, poorly exposed along the main Andes between Chile and Argentina, formed the basement of the Chilenia Terrane. Although the basement is well preserved along the Cordones del Plata and El Portillo, most of the metamorphic rocks are covered by several kilometres of sedimentary and volcanic rocks on the Principal and Frontal Cordilleras. An ophiolitic belt of Ordovician to Devonian age separates this basement from the Cuyania Terrane. The docking of the rocks is constrained by the Early Carboniferous marine deposits that finally amalgamated the allochthonous terranes of central Argentina.

### Patagonia

The Patagonian basement records a Precambrian history that started in the Middle Proterozoic, as detected in detrital zircon fractions from the Deseado Massif and from the western part of the Somuncurá Massif. The northern sector of the Somuncurá Massif has evidence of a superimposed Brasiliano deformation.

Crustal growth of Patagonia began in the Early Palaeozoic, when a magmatic arc was developed in the northern margin of the Somuncurá Massif. Age constraints for this magmatic belt indicate Middle to Late Ordovician to Late Carboniferous and Early Permian ages, showing that an active margin lasted until Early Permian times. The Deseado Massif was also the locus of Early Palaeozoic magmatism. Granitoids of Ordovician to Silurian age show a north-western-trending arc, parallel to penetrative deformation preserved in slates and schists of Late Precambrian age. It has been proposed that both basement blocks were independent terranes that amalgamated during the Early Palaeozoic to form a microcontinent that collided with Gondwana in Permian times, to form the Ventania fold-and-thrust belt.

## See Also

Andes. Antarctic. Brazil. Gondwanaland and Gondwana. Grenvillian Orogeny.

## Further Reading

Mpodozis C and Ramos VA (1990) The Andes of Chile and Argentina. In: Ericksen GE, Cañas, Pinochet MT, and Reinemud JA (eds.) *Geology of the Andes and Its Relation to Hydrocarbon and Mineral Resources, Earth Sciences Series*, vol. 11, pp. 59–90. Menlo Park, CA: Circum-Pacific Council for Energy and Mineral Resources.

Ramos VA (1994) Geology of South America. *Encyclopaedia Britannica* 27: 580–583.

Ramos VA and Alemán A (2000) Tectonic evolution of the Andes. In: Cordani UJ, Milani EJ, Thomaz Filho A, and Campos DA (eds.) *Tectonic Evolution of South America*, pp. 635–685. Río de Janeiro: 31st International Geological Congress.

Ramos VA and Keppie D (eds.) (1999) *Laurentia–Gondwana Connections before Pangea*. Special Paper 336. Boulder, CO: Geological Society of America.

Ramos VA and McNulty B (eds.) (2002) Flat-slab subduction in the Andes. *Journal of South American Earth Sciences* 15(1): Special Issue.

Tankard AJ, Suárez SR, and Welsink HJ (eds.) (1995) *Petroleum Basins of South America*. American Association of Petroleum Geologists Memoir 62. Tulsa, OK: American Association of Petroleum Geologists.

# ASIA

Contents

**Central**

**South-East**

## Central

**S G Lucas**, New Mexico Museum of Natural History, Albuquerque, NM, USA

© 2005, Elsevier Ltd. All Rights Reserved.

### Introduction

Central Asia encompasses a land area of 5.9 million km$^2$ and includes the countries of Kazakstan, Turkmenistan, Uzbekistan, Kyrgyzstan, and Tajikistan (**Figure 1**), which became independent of the Soviet Union in 1991. Some geographers also include portions of neighbouring Afghanistan, Pakistan, western China, and Mongolia in Central Asia, but this review focuses on the regional geology of the five former Soviet republics. Most of Central Asia is steppe (arid grassland) and desert, which contrasts sharply with its mountain ranges, which are some of the world's longest (the Tien Shan is 2500 km long) and tallest (more than 7000 m).

### Kazakstan

The largest state in Central Asia, Kazakstan, is half the size of the USA and has a land area of 2 717 300 km$^2$. It thus occupies about half the total area of Central Asia. Kazakstan is mostly steppe and desert, but in the south and southeast the Tien Shan, Zailiski Alatau, and Dzhungarian mountain ranges have peaks that exceed 7000 m in altitude (**Figure 2**). In contrast, the Karagie Depression in western Kazakstan in the Caspian Sea basin is 132 m below sea level, the lowest point on land after the Dead Sea. Large Kazak lakes include one of the largest in the world, Lake Balkash, with a surface area in excess of 17 000 km$^2$.

Physiographically, Kazakstan is a vast country south of the western Siberian lowlands, north of the Tien Shan Mountains, east of the Caspian Sea and west of the Altai ranges that run along its eastern borders with China and Mongolia. Kazakstan includes part of the South Caspian Basin (Depression), with a sedimentary fill of Jurassic–Neogene age as much as 20 km thick, one of the most important oil-producing regions in the world. It also encompasses the southern tip of the Ural Mountains, most of which are in Russia. The extensive Kazak deserts and uplands cover most of the country. Internally-drained basins with saline lakes extend across the northern part of Kazakstan–the Tengiz, Balkash and Alakol basins. A chain of deserts (including the Betpak Dala) north of the Tien Shan Mountains and south of Lake Balkash grade northward into steppes and forested uplands.

Precambrian and Palaeozoic rocks are exposed in all of the major mountain ranges of Kazakstan, as well as in the Mangyshlak Peninsula (which juts into the Caspian Sea) and in the North Caspian basin. Mesozoic and Cenozoic rocks are the widespread fill of various sedimentary basins, including the South Caspian, North Caspian, Turan, Turgay, Balkash, Alakol, Ily (Chu), and Zaysan basins.

The Precambrian rocks exposed in the Kazak mountain ranges consist of mostly schists, gneisses, marbles, and igneous rocks. Palaeozoic rocks are mostly of sedimentary origin, being Cambrian–Ordovician marine strata and volcanic rocks, Silurian–Early Carboniferous clastic sediments, volcanic rocks, cherts and ophiolites and Late Pennsylvanian–Permian nonmarine and marine sedimentary rocks. Notably, in the North Caspian Basin, salt domes have formed in a thick (4–5 km) succession of Permian salt. From the Tien Shan to north of Lake Balkash, the Kazak mountain ranges also expose remnants of a huge magmatic belt of mid-Carboniferous to Permian age, composed of lavas, tuffs, and ignimbrites, with an aggregate thickness of 5 to 6 km.

In Kazakstan, Early Triassic rocks are mostly tuffs and basalts, but, younger Mesozoic rocks are mostly nonmarine sedimentary rocks, though some marine strata of Mesozoic age are present.

The large lakes of Kazakstan have long geological histories. A good example is Lake Zaysan in northeastern Kazakstan. Today, Lake Zaysan is located

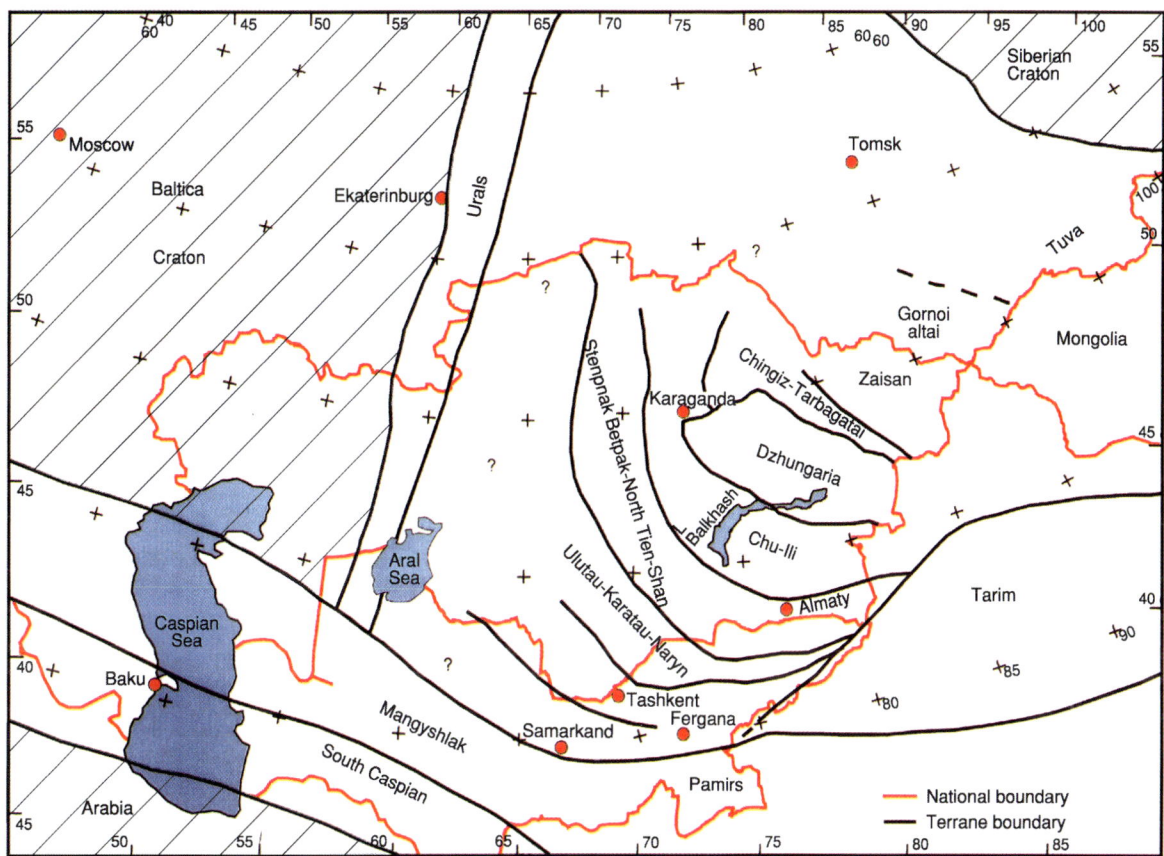

**Figure 1** Map of Central Asia (modern topography) showing the boundaries of the Palaeozoic terranes. (With permission from Fortey and Cocks (2003).)

**Figure 2** The Zaily Alatau in eastern Kazakstan is typical of the high mountain ranges of the Tien Shan Mountains that characterise much of Central Asia.

between the Altai and Targabatay ranges, is fed by the Cherny Itrysh River to the east and historically was about 100 km long, 32 km wide and 8 m deep. Beginning in the Late Cretaceous, a lake basin formed in the Zaysan basin and still exists today. The contracting and expanding lake basin thus has endured for more than 65 million years, and it has left nearly a kilometer of sedimentary basin fill, which is well exposed along the uplifted basin flanks (**Figure 3A**). These sedimentary rocks contain one of the most remarkable and continuous Cenozoic fossil records in Asia, including fossils of charophytes, land plants, ostracods, molluscs, fishes, turtles, crocodiles, birds, and mammals. Other Cenozoic lake basins in eastern Kazakstan also contain thick and fossiliferous sedimentary basin fills (**Figure 3B**).

Kazakstan is a country rich in oil, gas, and mineral resources. Oil and gas in Mesozoic and Palaeogene strata in the South Caspian basin are one of the world's great oil reserves. More than 300 coal deposits have been identified in Kazakstan, constituting proven reserves in excess of 170 billion tons. These coal deposits are mostly in Carboniferous rocks in the Kazak mountain ranges.

Kazakstan has a diverse and extensive mineral wealth, including important deposits of lead, zinc, iron, manganese, nickel, copper, tin, molybdenum, aluminum, and gold. Palaeozoic rocks are generally host to these metals.

**Figure 3** (A) Uplifted Palaeogene strata along the Kalmakpay River at the foot of the Targabatay Range in the Zaysan basin. (B) Miocene lake sediments near Aktau Mountain on the northern flank of the Ily basin in eastern Kazakstan.

## Turkmenistan

Turkmenistan encompasses an area of 448 100 km$^2$, most of which (about 80%) is the vast Kara Kum (Black Sand) Desert. Elevations across this desert drop from 100–200 m in the east, to 28 m below sea level at the coastline of the Caspian Sea. Along the southwestern frontier of Turkmenistan, the Kopetdag and Balkhan Mountains reach elevations of up to 2912 m, and the western edge of the Tien Shan Mountains just reaches the eastern border of Turkmenistan. The vast Kara Kum is mostly blanketed by Quaternary deposits, and the mountain ranges in the southwestern part of Turkmenistan mostly expose Mesozoic (especially Triassic) rocks, as is true of the mountainous southeastern part of the country. The only significant outcrops of Palaeozoic strata, and of metamorphic and igneous rocks, are in the western Tien Shan Mountains.

Turkmenistan can thus be divided into two geologic regions, the Kara Kum platform and the southern mountainous (orogenic) belt. In the subsurface of the Kara Kum, extensive drilling has revealed Precambrian metamorphic rocks overlain by a thick, much deformed succession of Palaeozoic strata (mostly of Silurian, Devonian, and Carboniferous age), Palaeozoic granites and less deformed, mostly nonmarine, strata of Late Carboniferous, Permian, and Triassic age, with an aggregate thickness of 5 to 6 km. These strata are overlain by essentially flat-lying Jurassic, Cretaceous, and Cenozoic sedimentary rocks, that typically have an aggregate thickness of 1 to 2 km, though they reach thicknesses of 8 to 10 km in local sedimentary basins.

The mountainous belt of Turkmenistan is mostly made up of strongly deformed Mesozoic and Paleogene rocks that represent a wide array of environments, from nonmarine Triassic red beds, to Jurassic marine limestones, to Cretaceous shallow marine glauconitic claystones. The Tien Shan Mountains in Turkmenistan are a continuation of the same rocks better exposed in Uzbekistan.

**Figure 4** This 10-metre-long fishing boat was abandoned along the arid and receding shoreline of the Aral Sea.

Most of the topography in Turkmenistan is the result of Late Cenozoic tectonism. This tectonic activity continues today and is most evident in numerous earthquakes, some with Richter values greater than 7, especially in the southwestern part of the country. Pliocene–Pleistocene oscillations of the water level in the Caspian Sea drove water all the way out to the main part of the Turkmenistan plains. During the Soviet era, between the 1950s and 1980s, the Great Kara Kum Canal was built across southern Turkmenistan, from its eastern border to the Caspian Sea, a distance of some 1400 km. This canal, major irrigation projects along the Amu-Darya and Syr-Darya rivers, and the damming of other rivers helped aridify the Aral Sea region, turning it into a rapidly shrinking waterbody (**Figure 4**).

Turkmenistan is a major oil and gas producer, particularly in the area that adjoins the Caspian Sea. Deposits of potassium salt, halite, and sulphate in the Kara Kum platform have yielded important deposits of iodine and bromine. All economic production of these chemical elements is from rocks of Jurassic–Eocene age.

## Uzbekistan

Uzbekistan encompasses an area of 447 400 km$^2$ and has a varied landscape from the western portion of the Tien Shan Mountains in the east and the Pamir-Altai ranges to the south-east, to the Kyzyl Kum ('Red Sand') Desert and shores of the Aral Sea to the northwest. The country is located between two of the great rivers of Central Asia, the Amu-Darya and the Syr-Darya, both of which have headwaters in the Tien Shan Mountains and flow to the Aral Sea.

The Tien Shan Mountains in Uzbekistan, as elsewhere, have a core of Palaeozoic (especially Silurian, Devonian, and Carboniferous age) sedimentary rocks and Precambrian–Early Palaeozoic metamorphic rocks that are intensely deformed. The metamorphic rocks include ophiolites, which are evidence of the ancient collapse of ocean basins as island arcs and other microcontinents were amalgamated to form what is now Central Asia. Granitic intrusions, that range in age from Cambrian to Triassic, are also present in the Uzbek portion of the Tien Shan. North of the Tien Shan, the bedrock of Uzbekistan is primarily Mesozoic and Cenozoic sedimentary rocks.

The Mesozoic rocks were mostly deposited in small, tectonically active basins in the southern and western parts of Uzbekistan. On the Ustyurt Plateau (the portion of Uzbekistan southwest of the Aral Sea) and in southern Uzbekistan, the Triassic rocks are mostly nonmarine sediments. Jurassic rocks, however, are much more widespread and are a mixture of marine and nonmarine sediments. The Cretaceous rocks are similarly widespread. The Lower Cretaceous strata are mostly of nonmarine origin, but some marine strata are present and are the host rocks for petroleum. The Upper Cretaceous rocks are mostly marine strata of limestone and gypsum; they contain important uranium deposits.

Palaeogene rocks in Uzbekistan are of both marine and nonmarine origin and contain reserves of oil and gas. The thickest cover of the low-lying regions of the country, however, is the Neogene strata, which are up to 6 km thick. Much of this thickness is alluvial sediments that were shed from the rising Tien Shan. Quaternary sediments are also products of the Tien Shan uplift and provide evidence of four pulses of uplift in the form of four prominent river terrace levels. To the north of the mountains, they are covered by a veneer of reddish orange sand dunes, 5 to 60 m thick, that make up the Kyzyl Kum Desert of northern Uzbekistan.

Oil, gas, and uranium deposits of Uzbekistan have just been mentioned. Mineral deposits are diverse and include copper, zinc, gold, and mercury, mostly in hydrothermal concentrations. Some Uzbek basaltic diatremes of Triassic age even yield diamonds.

## Kyrgyzstan

A relatively small country, with an area of 198 500 km$^2$, Kyrgyztan is a mountainous land dominated by the Tien Shan Mountains, with peaks as high as 7439 m. Exposed rocks range in age from Archaean to Cenozoic. Much of the older bedrock of Kyrgyzstan (the bedrock core of the Tien Shan) is constructed primarily from a few Palaeozoic (primarily of Late Ordovician and Carboniferous age) island arcs that collided during the Late Palaeozoic–Early Mesozoic.

In Kyrgyzstan, Precambrian rocks have small outcrop areas in the mountains, which mostly expose Palaeozoic sedimentary rocks (especially of Ordovician, Silurian, Carboniferous, and Permian ages) as well as many Palaeozoic granitic intrusive rocks. Basin floors are covered with Cenozoic rocks, and Mesozoic strata are primarily exposed along uplifted basin margins.

The principal and largest basin in Kyrgyzstan is the Fergana Basin, which occupies much of the western part of the country and extends into eastern Uzbekistan. In the various Kyrgyz basins, Triassic and Jurassic strata are intensely folded, coal-bearing rocks that reach an impressive 5 km thick in the Fergana Basin. These rocks yield an extensive record of fossil plants (**Figure 5**). Overlying Cretaceous and Paleocene–Eocene rocks are a mixture of shallow marine, shoreline, and terrestrial sediments that are more than 2 km thick in the Fergana basin. Subsequent Cenozoic sediments are wholly nonmarine in origin, and are more than 4 km thick in the Fergana Basin. At various times, the Fergana Basin was filled with a lake surrounded by mountains several hundred meters high. Today, however, the largest lake basin in Kyrgyzstan is Issyk-Kul in the Tien Shan along the northern frontier with Kazakstan.

Tectonic activity in Kyrgyzstan during the Plio-Pleistocene produced the final uplift of the mountains and the deposition of thick continental sediments in the intermontane basins under cold and arid Pleistocene glacial conditions. This is when the current Tien Shan Mountains developed, and there was intense folding of the basinal Mesozoic and Cenozoic rocks. Kyrgyzstan is a highly seismic country where numerous large earthquakes occur, especially in the eastern Fergana Basin and in the Kyrgyz and Kungey mountain ranges along the northern frontier.

In Kyrgyzstan, oil and gas is produced in the Fergana Basin from Jurassic, Cretaceous, and Cenozoic

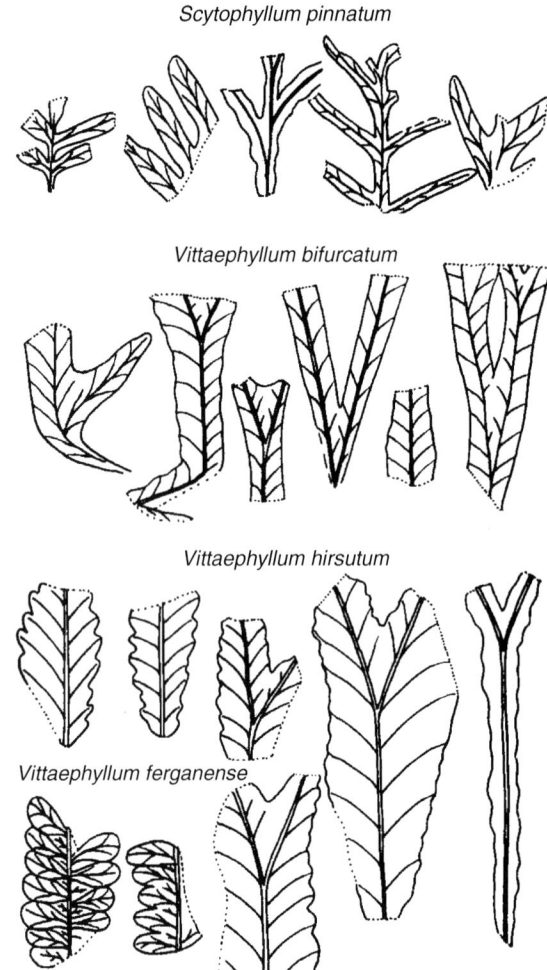

**Figure 5** Triassic fern-like plants from the Fergana Basin in Kyrgyzstan.

strata, and coal is also mined there. The Jurassic coal-bearing rocks in the Uzgen basin of western Kyrgyzstan are up to 5 km thick, and they are the most extensive exposures of Jurassic rocks in Central Asia. Minerals of economic importance are mercury, antimony, gold, and tin, mostly found in hydrothermal or skarn deposits.

## Tajikistan

The smallest Central Asian country, with a land area of 143 100 km$^2$, more than 90% of Tajikistan is covered by mountains. The southern edge of the Tien Shan Mountains (also called the Trans-Altai Mountains) is along the northern edge of Tajikistan. Most notable are the Pamir Mountains, with glaciated peaks as high as 7495 m. The Pamirs (the name is from a Farsi word meaning 'high level valleys') are actually a high plateau with an average elevation of about 6000 m.

The eastern two-thirds of Tajikistan is the Trans-Altai and Pamir Mountains. The Pamirs are faulted belts of rocks that range from Late Palaeozoic in the north to Early Cenozoic in the south. In the western part of Tajikistan, where its capital city of Dushanbe is located, the Tajik depression is a basin with a sedimentary basin fill 6 km thick that consists of Jurassic salt deposits, Cretaceous–Oligocene marine strata, and younger Cenozoic nonmarine strata. These rocks are intensely folded, and the Jurassic salt includes a detachment surface that is being underthrust. Indeed, Tadjikistan (particularly under the Pamirs) is a place where the subduction of continental lithosphere is taking place. Because of this, it is a region of high seismicity, with earthquakes concentrated along the Pamirs and the southern margin of the Trans-Altai Mountains.

## Geologic History

Central Asia is composed of a mosaic of ancient continental blocks which were microplates or independent terranes at various times in the Palaeozoic. **Figure 1** shows a number of these terranes which are positioned between the old cratons of Baltica (most of northern and eastern Europe), Siberia (only part of modern Siberia) and the southern peri-Gondwanan terranes which include Mangyshlak, the South Caspian Terrane, and the various terranes which make up Arabia today. The many central Asian terranes shown together make up the tectonically complex 'Altaids': some have some Precambrian slivers in their cores, but most originated within island arc settings at various times in the Palaeozoic. The Lower Palaeozoic faunas contained within many of them are abundant and often distinctive, indicating liaisons and/or oceanic separations between their Lower Palaeozoic positions. These relatively small terranes progressively collided and amalgamated with each other, mostly during the Upper Palaeozoic during the formation of Pangaea. The results of the collisions are most evident in the bedrock cores.

In Central Asia, Late Palaeozoic compression, collision, and subduction was followed by Mesozoic extension, with widespread emplacement of granite batholiths and diabase dike intrusions, as well as the development of large sedimentary basins. During the Late Palaeozoic–Mesozoic, Central Asia was part of the Pangaean supercontinent, and as Pangaea split apart it remained part of southern Eurasia.

The Cenozoic geological history of Central Asia has been dominated by the collision of India with Eurasia during the Paleogene. This collision reactivated old structures along the Pangaean collisional zones. The northward propagation of deformation from the collision saw India thrust under Tibet, subsidence of many of the Cenozoic sedimentary basins of Central Asia, rise of the southern and then the northern Tien Shan, and also the deformation of more northerly uplifts. This deformation thus produced the dramatic basins and ranges that characterize the Central Asian landscape today.

## See Also

**China and Mongolia**. **Europe:** The Urals. **Pangaea**. **Plate Tectonics**. **Russia**.

## Further Reading

Burtman VS (1980) Faults of Middle Asia. *American Journal of Science* 280: 725–744.

Burtman VS (1997) Kyrgyz Republic. In: Moores EM and Fairbridge RW (eds.) *Encyclopedia of European and Asian Regional Geology*, pp. 483–492. London: Chapman and Hall.

Dobruskina IA (1995) Keuper (Triassic) flora from Middle Asia (Madygen, southern Fergana). *New Mexico Museum of Natural History & Science*, Bulletin 5: 49p.

Fortey RA and Cocks LRM (2003) Palaeontological evidence bearing on global Ordovician–Silurian reconstructions. *Earth Science Reviews* 61: 245–307.

Hendrix MS and Davis GA (eds.) (2001) Paleozoic and Mesozoic Tectonic Evolution of Central and Eastern Asia. *Geological Society of America Memoir* 194: 447p.

Leith RW (1982) Rock assemblages in Central Asia and the evolution of the southern Asian margin. *Tectonics* 1: 303–318.

Lucas SG, Emry RJ, Chkhikvadze V, *et al.* (2000) Upper Cretaceous-Cenozoic lacustrine deposits of the Zaysan basin, eastern Kazakstan. In: Gierlowski-Kordesch EH and Kelts KR (eds.) *Lake Basins Through Time and Space*, 46, pp. 335–340. AAPG Studies in Geology.

Moores EN (1997) Tajikistan. In: Moores EM and Fairbridge RW (eds.) *Encyclopedia of European and Asian Regional Geology*, pp. 71–80. London: Chapman and Hall.

Mukhin P (1997) Uzbekistan. In: Moores EM and Fairbridge RW (eds.) *Encyclopedia of European and Asian Regional Geology*, pp. 766–773. London: Chapman and Hall.

Rastsvetaev L (1997) Turkmenistan. In: Moores EM and Fairbridge RW (eds.) *Encyclopedia of European and Asian Regional Geology*, pp. 743–759. London: Chapman and Hall.

Yakubchuk A (1997) Kazakhstan. In: Moores EM and Fairbridge RW (eds.) *Encyclopedia of European and Asian Regional Geology*, pp. 450–464. London: Chapman and Hall.

Zonenshain LP, Kuzmin MI, and Natapov LM (1990) Geology of the USSR: A plate tectonic synthesis. Washington, DC, *AGU Geodynamic Series* 21: 242.

# South-East

**I Metcalfe**, University of New England, Armidale, NSW, Australia

© 2005, Elsevier Ltd. All Rights Reserved.

## Introduction

East and South-east Asia is a giant 'jigsaw puzzle' of allochthonous continental lithospheric blocks and fragments (terranes) that are bounded by suture zones (representing the remnants of closed ocean basins) or by geological discontinuities such as major strike-slip faults (**Figure 1**).

The complex assemblage of South-east Asian continental terranes, accretionary complexes, ophiolites, volcanic arcs, and marginal ocean basins occurs in the zone of convergence between the Eurasian, Indo-Australian, and Pacific plates (**Figures 2** and **3**). In this region, two important biogeographical boundaries are recognized: the extant Wallace's Line and the Late Palaeozoic Gondwana–Cathaysia Divide (see **Figures 1, 2,** and **3**). Wallace's Line marks the boundary between Eurasian faunas and floras to the north-west (including tigers, deer, and woodpeckers) and Australasian faunas and floras to the south-east (including wallabies, possums, and cockatoos) and was recognized by and named after Alfred Russel Wallace, now regarded as the Father of Biogeography. The Late Palaeozoic Gondwana–Cathaysia Divide forms the boundary between high-latitude cold-climate southern-hemisphere Gondwanan faunas and floras and equatorial or northern-hemisphere warm-climate sub-tropical or tropical Cathaysian faunas and floras. Both of these biogeographical boundaries are the result of convergent plate-tectonic processes bringing together allochthonous continental lithospheric terranes on which had developed contrasting faunas and floras owing to their prior geographical separation, different palaeoclimates, and biogeographical isolation. The regional geology of South-east Asia can be understood only in the context of the kinematic history and framework of the allochthonous continental terranes of the region and the evolution of the various ocean basins that once separated them.

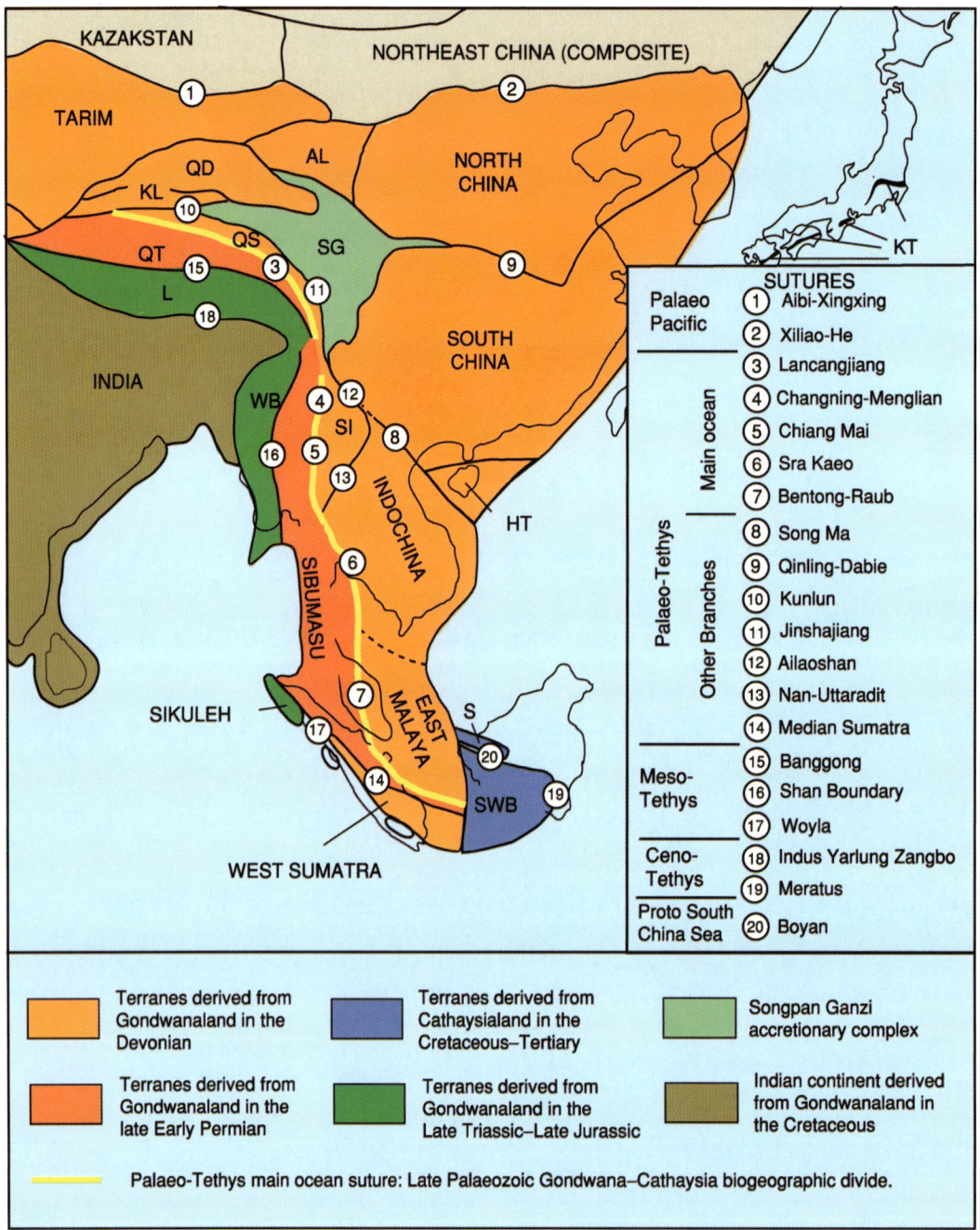

**Figure 1** Distribution of the principal continental terranes and sutures of East and South-east Asia. WB, West Burma; SWB, South West Borneo; S, Semitau Terrane; HT, Hainan Island terranes; L, Lhasa Terrane; QT, Qiangtang Terrane; QS, Qamdo-Simao Terrane; SI, Simao Terrane; SG, Songpan Ganzi accretionary complex; KL, Kunlun Terrane; QD, Qaidam Terrane; AL; Ala Shan Terrane; KT, Kurosegawa Terrane.

Multidisciplinary data, including stratigraphic, biostratigraphic, biogeographical, palaeoclimatological, palaeomagnetic, and structural or tectonic information, indicate that in the Early Palaeozoic (545–410 Ma) all of the principal East and South-east Asian continental terranes were located on the margin of eastern Gondwana, where they formed an Indo-Australian 'Greater' Gondwana (see **Gondwanaland and Gondwana**)

(**Figure 4**). They also probably formed part of the Indo-Australian element of the ancient supercontinent of Rodinia at 1000 Ma, prior to its breakup around 700 Ma and the subsequent formation of Gondwana at about 500 Ma (**Figure 5**).

Multidisciplinary data also suggest that the East and South-east Asian terranes rifted and separated from Gondwana as three continental slivers around

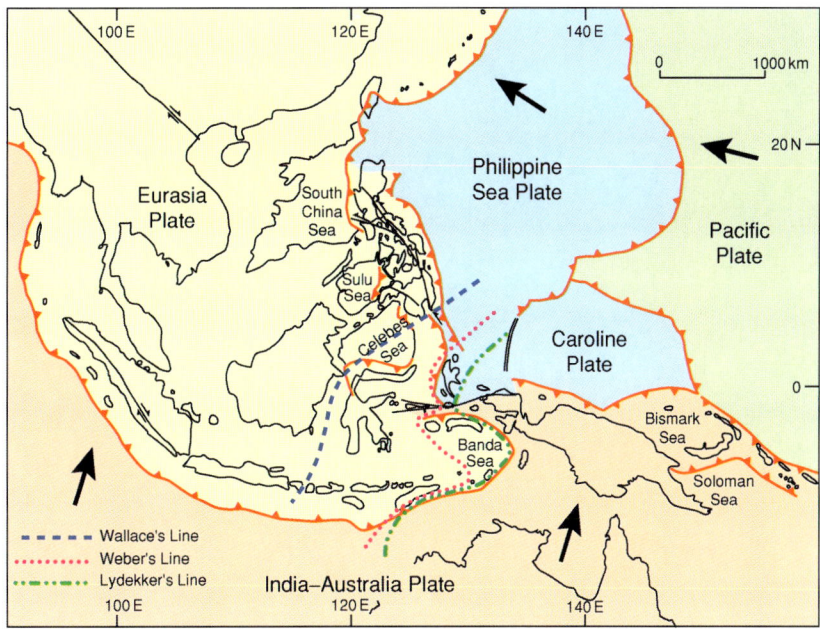

**Figure 2** Principal tectonic plates of South-east Asia. Arrows show relative plate motions.

350 Ma, 270 Ma, and 200–140 Ma, in the Devonian (North China, South China, Tarim, Indochina, East Malaya, West Sumatra), Lower Permian (Sibumasu, Qiangtang), and Late Triassic–Late Jurassic (Lhasa, West Burma, possibly Sikuleh, possibly West Sulawesi etc.), respectively. As these three continental slivers separated from Gondwana and drifted northwards, successive ocean basins opened between each sliver and Gondwana: the Palaeo-Tethys, Meso-Tethys and Ceno-Tethys, respectively (**Figure 6**). The Meso-Tethys and Ceno-Tethys are broadly equivalent to the Neo-Tethys of some workers. Destruction and closure of these ocean basins by subduction during Carboniferous to Cenozoic times led to the juxtaposition, by amalgamation and accretion (continental collisions), of once widely separated continental fragments, and the remnants of the ocean basins are now preserved in the suture zones of the region.

Smaller continental fragments, distributed in eastern South-east Asia, were derived from Indochina and South China during the opening of the South China Sea and southwards subduction and destruction of the Proto-South China Sea, or were transported westwards along major strike-slip faults from the northern Australian margin during its collision with the westwards-moving Philippine Sea, Caroline, and Pacific plates in a kind of 'bacon-slicer' tectonic mechanism.

The regional geology of South-east Asia is thus characterized by Gondwanan dispersion and Asian accretion of terranes and the subsequent collisions of India and Australia with these terranes following the breakup of Gondwana and their northwards drift. The geological evolution of South-east Asia is thus essentially the combined and cumulative history of these terranes, the ocean basins that once separated them, and the plate-tectonic processes that have shaped the region. A variety of multidisciplinary data (**Table 1**) is used to constrain the origins of the terranes, their times of rifting and separation from the parent cratons, the timing, directions, and amount of drift, and the timing of suturing (collision and welding) of the terranes to each other. Some terranes sutured to each other (amalgamated) within a major ocean before, as an amalgamated composite terrane, they sutured (accreted) to proto-Asia.

## Origins of the South-East Asian Terranes

Multidisciplinary data (**Table 1**) suggest that all the East and South-east Asian continental terranes originated on the Indian or northern or north-western Australian margin of Gondwana. Cambrian and Ordovician shallow-marine faunas of the North China, South China, and Sibumasu terranes have close affinities with those of eastern Gondwana, especially Australian Gondwana. This is observed in trilobites, brachiopods, corals and stromatoporoids, nautiloids, gastropods, and conodonts (**Figure 4**). More recently, the Gondwanan acritarch *Dicrodiacroium ancoriforme* Burmann has been reported from the Lower Ordovician of South China. The Cambrian–Ordovician faunas of Indochina are poorly known, but the Silurian brachiopods of Indochina

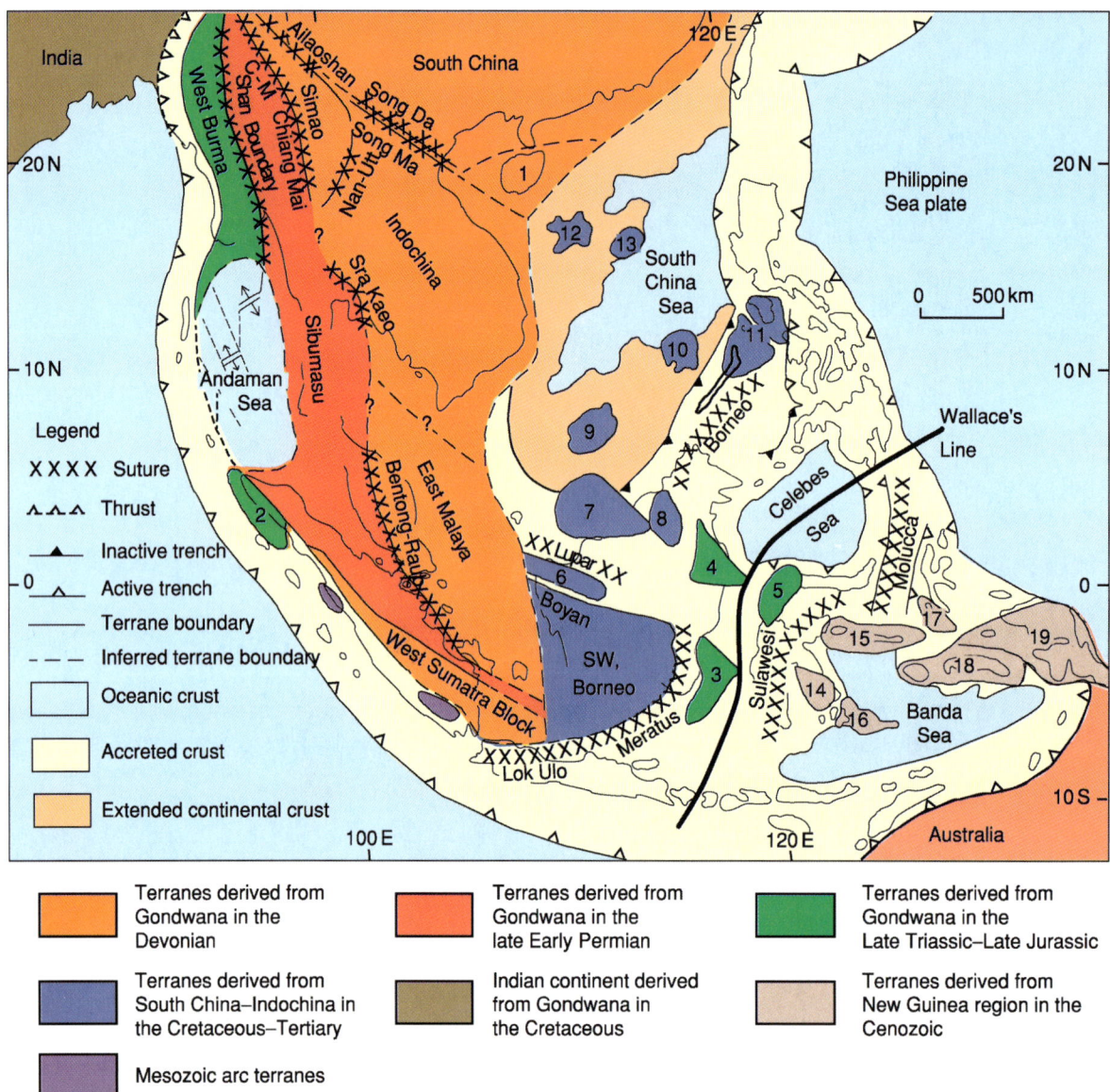

**Figure 3** Distribution of continental blocks, fragments, and terranes, together with the principal sutures of South-east Asia. Numbered microcontinental blocks: 1, Hainan Island terranes; 2, Sikuleh; 3, Paternoster; 4, Mangkalihat; 5, West Sulawesi; 6, Semitau; 7, Luconia; 8, Kelabit-Longbowan; 9, Spratley Islands-Dangerous Ground; 10, Reed Bank; 11, North Palawan; 12, Paracel Islands; 13, Macclesfield Bank; 14, East Sulawesi; 15, Bangai-Sula; 16, Buton; 17, Obi-Bacan; 18, Buru-Seram; 19, West Irian Jaya. C–M, Changning-Menglian Suture.

along with those of South China, North China, Eastern Australia, and the Tarim terrane define a Sino-Australian province characterized by the *Retziella* fauna (**Figures 4** and **7**). Lower Palaeozoic sequences and faunas of the Qaidam, Kunlun, and Ala Shan blocks are similar to those of the Tarim block and South and North China, and these blocks are regarded as disrupted fragments of a larger Tarim terrane. These biogeographical data suggest that North China, South China, Tarim (here taken to include the Qaidam, Kunlun, and Ala Shan blocks), Sibumasu (with the contiguous Lhasa and Qiangtang blocks), Indochina, East Malaya, and West Sumatra formed the outer margin of northern Gondwana in the Early Palaeozoic. The close faunal affinities, at both lower and higher taxonomic levels, suggest continental contiguity of these blocks with each other and with Gondwana at this time rather than mere close proximity.

Early Palaeozoic palaeomagnetic data for the various East and South-east Asian terranes are often equivocal, varying in both quantity and quality. This makes reconstructions based purely on palaeomagnetic data difficult and suspect. However, in some cases, reasonable constraints on palaeolatitudes (but

ASIA/South-East 173

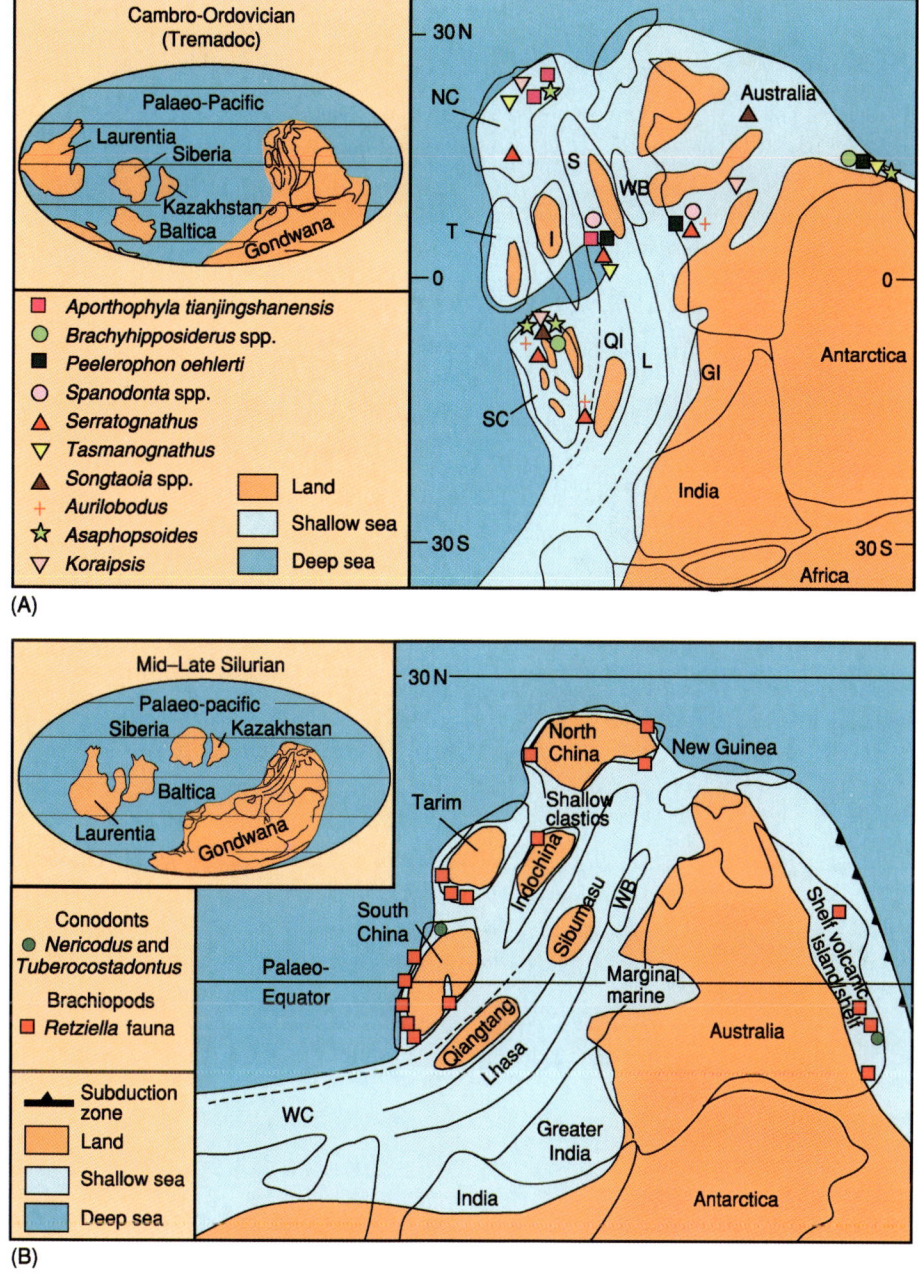

**Figure 4** Reconstructions of eastern Gondwana for (A) the Cambro-Ordovician (Tremadoc) and (B) the Mid–Late Silurian, showing the postulated positions of the East and South-east Asian terranes, the distribution of land and sea, and the distributions of shallow-marine fossils that illustrate Asia–Australia connections at these times. NC, North China; SC, South China; T, Tarim; I, Indochina–East Malaya–West Sumatra; QI, Qiangtang; L, Lhasa; S, Sibumasu; WB, West Burma; WC, Western Cimmerian Continent; GI, Greater India.

not always the hemisphere) and in some cases the actual position of attachment to Gondwana can be made. Data from North China provide a Cambrian–Late Devonian pole-path segment that, when rotated about an Euler pole to a position of fit with the Australian Cambrian–Late Devonian pole path, positions North China adjacent to North Australia. This position is consistent with reconstructions presented here. The gross stratigraphies of North China and the Arafura Basin show a remarkable similarity in the Early Palaeozoic, also supporting this position for North China. Positions for South China, Tarim, and Indochina are more equivocal, but latitudes of between 1° and 15° are indicated for South China in the Late Cambrian–Early Ordovician. The Tarim Block is placed between 6° S and 20° S for the same time period, consistent with a position on the Gondwanan margin between the North and South

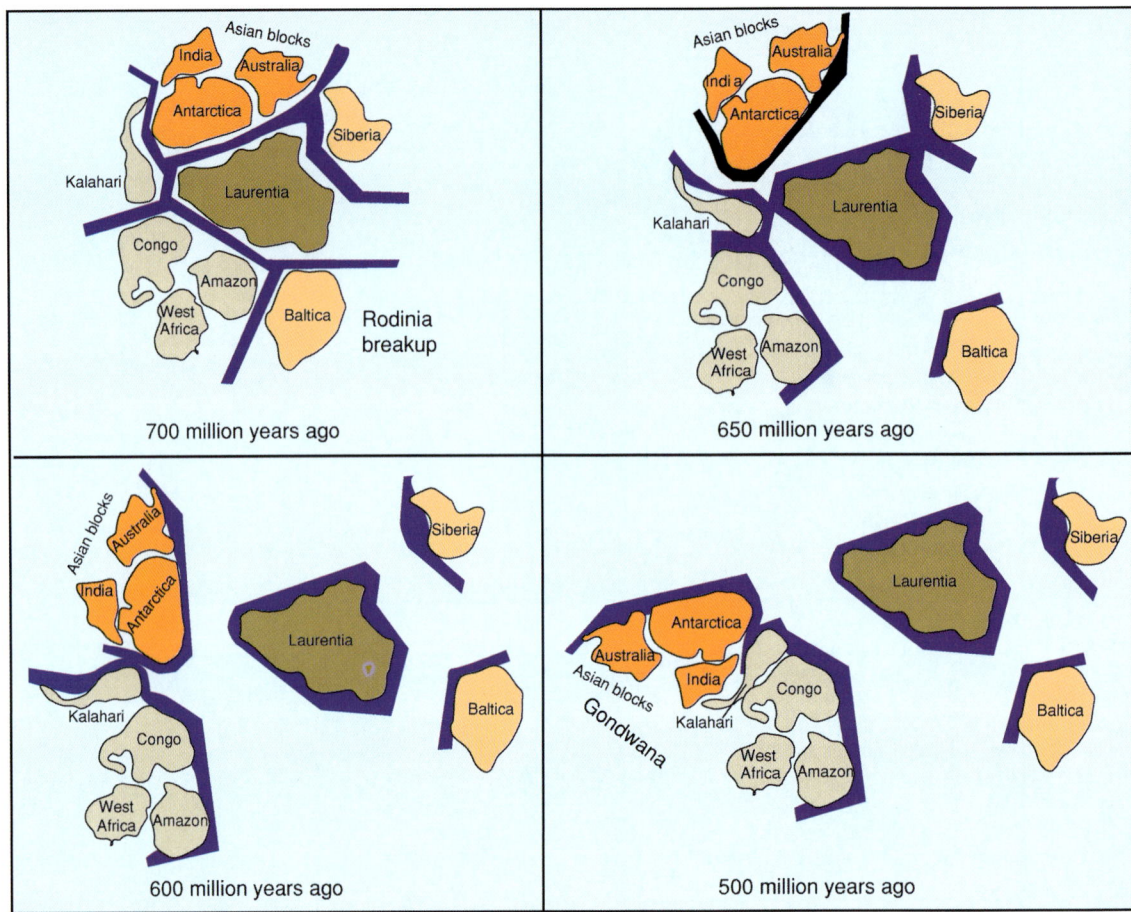

**Figure 5** The supercontinent Rodinia at 700 Ma and its subsequent breakup, showing the formation of Gondwana at about 500 Ma.

China blocks. An origin for the Tarim Block outboard of the Kimberley region of Australia is suggested by comparisons of Precambrian sequences. The Cambrian–Early Permian faunas of the Sibumasu Terrane have unequivocal Gondwanan affinities and in particular show close relationships with western Australian faunas. Gondwanan plants and spores are also reported from this terrane. Glacial–marine diamictites, with associated cold-water faunas and sediments, of possibly Late Carboniferous and (mainly) Early Permian age are also distributed along the entire length of Sibumasu (**Figure 8**) and indicate attachment to the margin of Gondwana, where substantial ice reached the sea during glaciation. The most likely region for attachment of this terrane is north-western Australia, and a Late Carboniferous palaeolatitude of 42° S supports this placement. Striking similarities in the Cambrian–Early Permian gross stratigraphies of Sibumasu and the Canning Basin of north-western Australia also support a position outboard of the Canning Basin during this time. Both the Qiangtang and Lhasa blocks of Tibet exhibit Gondwanan faunas and floras up to the Early Permian and have glacial–marine diamictites, till, and associated cold-water faunas and sediments in the Late Carboniferous–Early Permian. Thus, all the East and South-east Asian continental terranes appear to have originated on the margin of Gondwana.

## Rifting and Separation of South-East Asian Terranes from Gondwana

### Devonian Rifting and Separation

South China, North China, Tarim, Indochina, East Malaya, and West Sumatra were attached to Gondwana in the Cambrian–Silurian but by Carboniferous times were separated from the parent craton, suggesting a Devonian rifting and separation of these blocks. This timing is supported by the presence of a conspicuous Devonian unconformity in South China and a subsequent Devonian–Triassic passive-margin sequence along its southern margin. Devonian basin formation in South China has also been shown to be related to rifting. The splitting of the Silurian Sino-Australian brachiopod province into two sub-provinces and the apparent loss of links

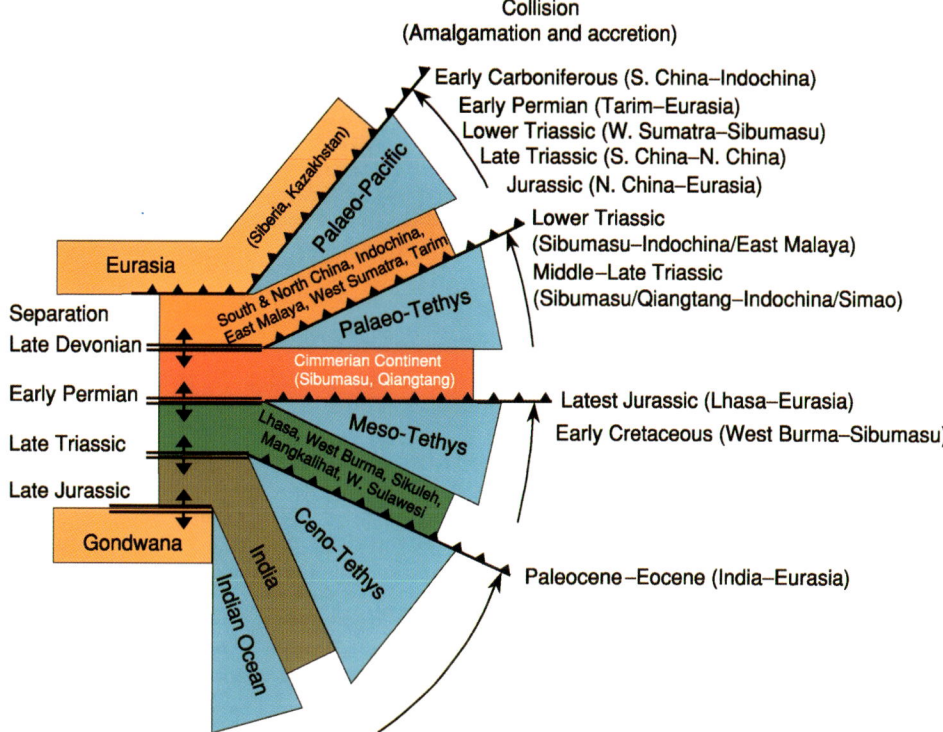

**Figure 6** Times of separation and subsequent collision of the three continental slivers or collages of terranes that rifted from Gondwana and were translated northwards by the opening and closing of three successive oceans: the Palaeo-Tethys, Meso-Tethys, and Ceno-Tethys.

between the Asian terranes and Australia in the Early Devonian may be consequences of the northwards movement and separation of these terranes from Gondwana. A counterclockwise rotation of Gondwana in the Devonian about an Euler pole in Australia is also consistent with clockwise rotation of the separating Asian terranes and spreading of the Palaeo-Tethys at this time.

### Carboniferous–Permian Rifting and Early Permian Separation

There is now substantial evidence for rifting along the northern margin of Gondwana in the Carboniferous–Permian accompanied by rift-related magmatism. This rifting episode led to the late Early Permian (Late Sakmarian) separation of the Sibumasu and Qiangtang terranes, as part of the Cimmerian continent, from the Indo-Australian margin of Gondwana. The late Early Permian separation and Middle–Upper Permian rapid northwards drift of the Sibumasu Terrane are supported by palaeolatitude data, which indicate a change of latitude from 42° S in the Late Carboniferous to low northern latitudes by the Early Triassic. In addition, the faunas of the Sibumasu Terrane show a progressive change in marine provinciality from peri-Gondwanan Indoralian Province faunas in the Early Permian (Asselian–Early Sakmarian), to endemic Sibumasu province faunas in the Middle Permian, before being absorbed into the equatorial Cathaysian province in the Late Permian.

### Late Triassic to Late Jurassic Rifting and Separation

The separation of the Lhasa Block from Gondwana has been proposed by different authors to have occurred in either the Permian or the Triassic. A Permian separation is advocated, either as part of the Cimmerian continent or as a 'Mega-Lhasa' Block that included Iran and Afghanistan. Permian rifting on the North Indian margin and in Tibet is here regarded as being related to separation of the Cimmerian continental strip, which included Iran, Afghanistan, and the Qiangtang Block of Tibet, but not the Lhasa Block. Sedimentological and stratigraphic studies in the Tibetan Himalayas and Nepal have documented the Triassic rifting and Late Triassic (Norian) separation of the Lhasa Block from northern Gondwana. This Late Triassic episode of rifting is also recognized along the north-western shelf of Australia, where it continued into the Late Jurassic, resulting in the separation of West Burma.

**Table 1** Multidisciplinary constraining data for the origins and the rift–drift–suturing of terranes

| Origin of terrane | Age of rifting and separation | Drifting (palaeoposition of terrane) | Age of suturing (amalgamation/ accretion) |
| --- | --- | --- | --- |
| Palaeobiogeographical constraints (fossil affinities with proposed parent craton) | Ocean floor ages and magnetic-stripe data | Palaeomagnetism (palaeolatitude, orientation) | Ages of ophiolite and ophiolite obduction ages (pre-suturing) |
| Tectonostratigraphic constraints (similarity of gross stratigraphy with that of parent craton, presence of distinctive lithologies characteristic of parent craton, e.g. glacials) | Divergence of apparent polar wander paths indicates separation | Palaeobiogeography (shifting from one biogeographical province to another with drift) | Melange ages (pre-suturing) |
| Palaeolatitude and orientation from palaeomagnetism consistent with proposed origin | Divergence of palaeolatitudes (indicates separation) | Palaeoclimatology (indicates palaeolatitudinal zone) | Age of 'stitching' plutons (post-suturing) |
| | Age of associated rift volcanism and intrusives | | Age of collisional or post-collisional plutons (syn- to post-suturing) |
| | Regional unconformities (formed during pre-rift uplift and during block faulting) | | Age of volcanic arc (pre-suturing) |
| | Major block-faulting episodes and slumping | | Major changes in arc chemistry (syn-collisional) |
| | Palaeobiogeography (development of separate biogeographical provinces after separation) | | Convergence of apparent polar wander paths |
| | Stratigraphy–rift sequences in grabens and half grabens | | Loops or disruptions in apparent polar wander paths (indicates rapid rotations during collisions) |
| | | | Convergence of palaeolatitudes (may indicate suturing but no control on longitudinal separation) |
| | | | Age of blanketing strata (post-suturing) |
| | | | Palaeobiogeography (migration of continental animals and plants from one terrane to another indicates terranes have sutured) |
| | | | Stratigraphy/sedimentology (e.g. provenence of sedimentary detritus from one terrane on to another) |
| | | | Structural geology (age of deformation associated with collision) |

Reproduced with permission from Metcalfe I (1998) Palaeozoic and Mesozoic geological evolution of the SE Asian region: multidisciplinary constraints and implications for biogeography. In: Hall R and Holloway JD (eds.) (1998) *Biogeography and Geological Evolution of SE Asia*, pp. 25–41. Amsterdam: Backhuys Publishers.

## Amalgamation and Accretion of Terranes

The continental terranes of East and South-east Asia progressively amalgamated during the Palaeozoic to Cenozoic. Most of the major terranes had coalesced by the end of the Cretaceous and proto South-east Asia had formed. The time of welding of one terrane to another can be determined using the various criteria given in **Table 1**. **Table 2** lists the East and South-east Asian suture zones, colliding lithospheric blocks, interpreted times of suturing, and constraints on the times of suturing. The tectonostratigraphic record of each continental terrane in the region documents the geological history of that terrane, including variations in sedimentary environment, climate, faunal and

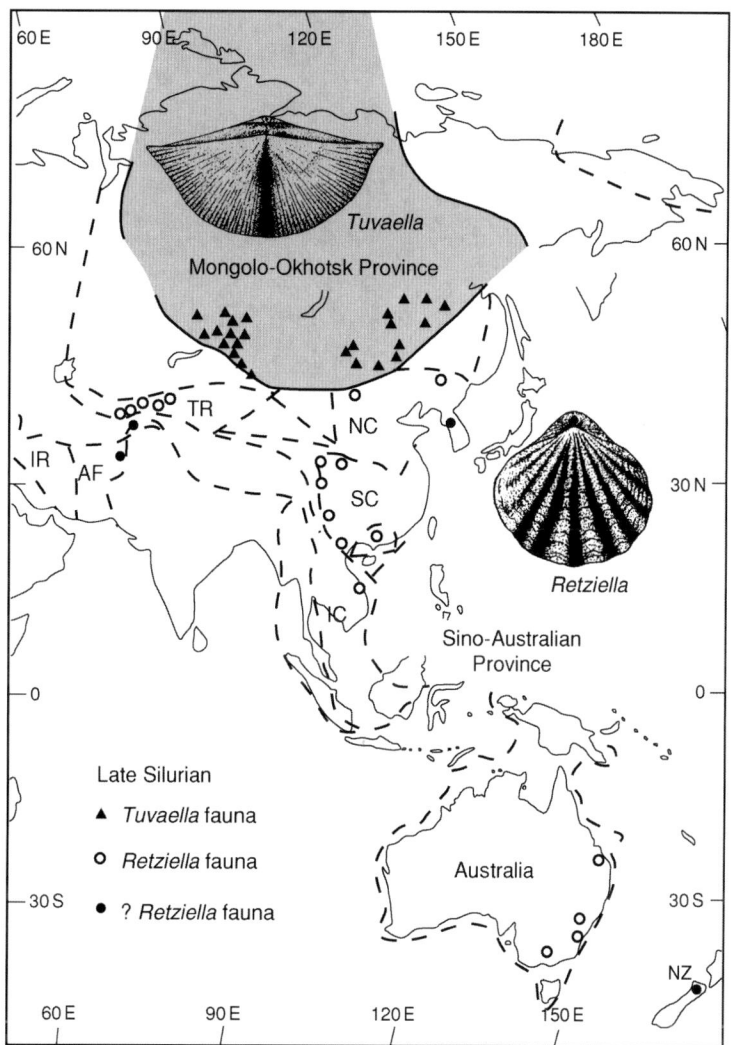

**Figure 7** Late Silurian brachiopod provinces of Asia–Australasia. IR, Iran; AF, Afghanistan; TR, Tarim; NC, North China; SC, South China; IC, Indochina; NZ, New Zealand.

floral affinities (changes in biogeographical regime) and latitudinal shifts, rifting events, episodes of deformation, and plutono-volcanic igneous activity. The regional geological history of South-east Asia is discussed below chronologically and in terms of the evolution of the various tectonic elements of the region.

## Geological and Tectonic Evolution of South-East Asia

### Proterozoic (2500–545 Ma)

The East and South-east Asian terranes, together with India and Australia, formed an integral part of the ancient supercontinent of Rodinia around 1000 Ma (see **Precambrian: Overview**). Fragmentation of this ancient supercontinent about 700 Ma ago led to Australia, India, Antarctica, and elements that now constitute South Africa and South America colliding and coalescing to form Gondwana about 500 Ma (**Figure 5**). Proterozoic basements of the mainland South-east Asian terranes are indicated by limited outcrops of schists and gneisses with Proterozoic radio-isotopic ages in Thailand and Vietnam, by more substantial relatively unmetamorphosed Proterozoic sedimentary sequences in South China, and by Proterozoic radioisotopic ages of inherited zircons in granitoids in Peninsular Malaysia.

### Phanerozoic (545–0 Ma)

Table 3 summarizes the principal geological events, and their timings, that have affected the South-east Asian region during the Palaeozoic, Mesozoic, and Cenozoic Eras. Continental collisions (amalgamation

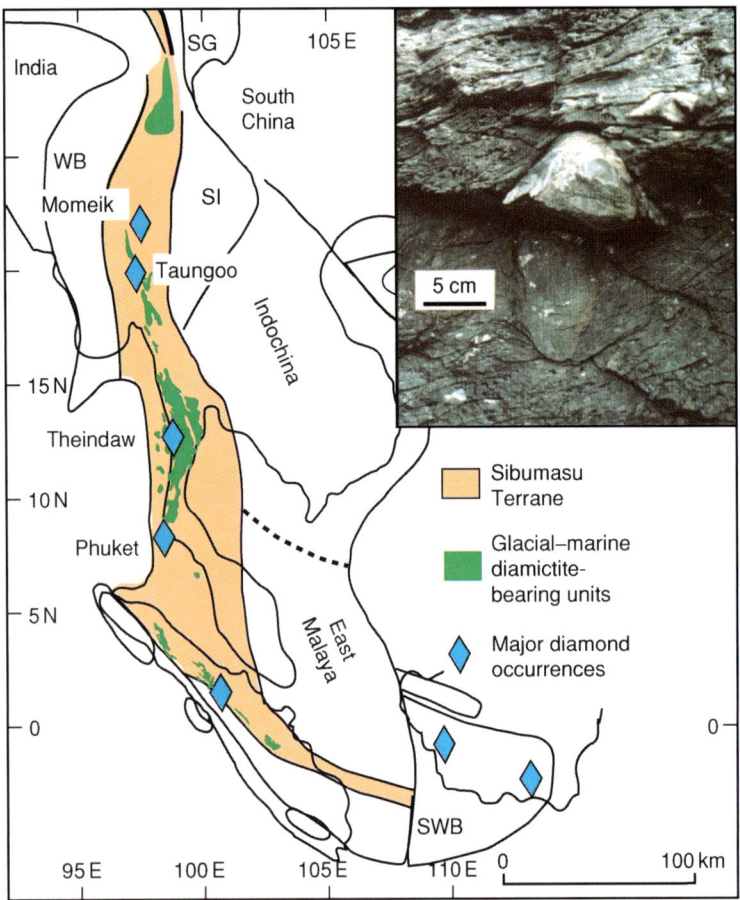

**Figure 8** Map of mainland South-east Asia, showing the distributions of Carboniferous–Early Permian glacial–marine sediments and major alluvial diamond deposits. Inset photo: dropstone in glacial–marine diamictite orientated vertical to bedding; Singa Formation, Langkawi Islands, Peninsular Malaysia.

or accretion) are dated by the various suture zones (**Table 2**). The principal events that have affected the South-east Asian region during the Phanerozoic are outlined below chronologically.

**Cambrian–Ordovician–Silurian (545–410 Ma)** The East and South-east Asian terranes formed part of Indian–Australian 'Greater Gondwana'. Faunas of this age on the Asian blocks and in Australasia define Asian–Australian palaeo-equatorial warm-climate 'provinces', for example the Sino-Australian brachiopod province in the Silurian – see **Figures 4** and **7**.

**Devonian (410–354 Ma)** Australian eastern Gondwana continued to reside in low southern latitudes during the Devonian but rotated counterclockwise. This counterclockwise rotation mirrors a clockwise rotation of the North and South China, Tarim, Indochina, East Malaya, and West Sumatra terranes as they separated from Gondwana as an elongate continental sliver. Separation of this sliver from Gondwana opened the Palaeo-Tethys Ocean (**Figure 9**). Devonian faunas on the Chinese terranes still have some Australian connections. Early Devonian endemic faunas of South China, including some fish faunas and the distinctive *Chuiella* brachiopod fauna (**Figure 9**), are interpreted to be a result of the rifting process and isolation of South China on the rifting continental promontory, and do not necessarily imply continental separation of South China from the other Chinese blocks and Australia at this time.

**Carboniferous (354–298 Ma)** During the Carboniferous, Gondwana rotated clockwise and collided with Laurentia in the west to form Pangaea (*see* **Pangaea**). Australia, Sibumasu, Qiangtang, Lhasa, and West Burma were still attached to north-eastern Gondwana and drifted from low southern latitudes in Tournaisian–Visean times to high southern latitudes in Middle–Late Namurian times. The major Gondwanan glaciation commenced in the Namurian and extended through to the Early Permian. There were

**Table 2** East and South-east Asian sutures and their interpreted ages and age constraints. For location of sutures see **Figures** 1 and 2

| No. on Fig. 1 | Suture name | Colliding lithospheric blocks | Suture age | Age constraints |
|---|---|---|---|---|
| 1 | Aibi-Xingxing | Tarim, Kazakhstan | Permian | Lower Carboniferous ophiolites. Major arc magmatism ceased in the Late Carboniferous. Late Permian post-orogenic subsidence and continental sedimentation in Junggar Basin. Palaeomagnetic data indicate convergence of Tarim and Kazakhstan by the Permian. Upper Permian continental clastics blanket the suture |
| 2 | Xiliao-He | North China, Altaid terranes | Jurassic | Late Jurassic–Early Cretaceous deformation and thrust faulting. Widespread Jurassic–Cretaceous granites. Triassic–Middle Jurassic deep-marine cherts and clastics. Upper Jurassic–Lower Cretaceous continental deposits blanket suture |
| 3 | Lancangjiang Suture | Qiangtang, Qamdo-Simao | Early Triassic | Suture zone rocks include Devonian and Carboniferous turbiditic 'flysch'. Ocean-floor basic extrusives of Permian age and Carboniferous–Permian mélange. Carboniferous–Permian island arc rocks are developed along the west side of the suture. Upper Triassic collisional granitoids are associated with the suture. Suture zone rocks are blanketed by Middle Triassic continental clastics |
| 4 | Changning-Menglian Suture | Sibumasu, Simao | Late Permian–Late Triassic | Oceanic ribbon-bedded chert–shale sequences have yielded graptolites, conodonts, and radiolarians indicating ages ranging from Lower Devonian to Middle Triassic. Limestone blocks and lenses dominantly found within the basalt sequence of the suture and interpreted as seamount caps, have yielded fusulinids indicative of Lower Carboniferous to Upper Permian ages |
| 5 | Chiang Mai Suture | Sibumasu, Simao | Middle Triassic | Basic volcanics (including pillow basalts) are dated as Carboniferous and Permian. Ages of oceanic deep-marine bedded cherts within the suture zone range from Devonian to Middle Triassic. Seamount limestone caps are dated as Lower Carboniferous (Visean) to Upper Permian (Changxingian) in age |
| 6 | Sra Kaeo Suture | Sibumasu, Indochina | Late Triassic | Suture zone rocks include melange and chert-clastic sequences, which include ultrabasics, serpentinites, Carboniferous pillow basalts, and Early Permian to Late Triassic oceanic sediments and associated pillow basalts. Limestone blocks in the mélange range from Upper Lower Permian to Middle Permian and a granitic lens has yielded a zircon U-Pb age of $486 \pm 5$ Ma. Imbricate thrust slices dated as Middle Triassic by radiolarians. Jurassic continental sandstones blanket the suture zone |
| 7 | Bentong-Raub Suture | Sibumasu, East Malaya | Early Triassic | Upper Devonian to Upper Permian oceanic ribbon-bedded cherts. Mélange includes chert and limestone clasts with Lower Carboniferous to Upper Permian ages. The Main Range 'collisional' 'S' type granites of Peninsular Malaysia range from Late Triassic ($230 \pm 9$ Ma) to earliest Jurassic ($207 \pm 14$ Ma) in age, with a peak of around 210 Ma. Suture zone is covered by latest Triassic, Jurassic, and Cretaceous, mainly continental, overlap sequence |
| 8 | Song Ma | Indochina, South China | Late Devonian–Early Carboniferous | Large-scale folding and thrusting and nappe formation in the Early to Middle Carboniferous. Middle Carboniferous shallow-marine carbonates are reported to blanket the Song Ma suture in North Vietnam. Pre-middle Carboniferous faunas on each side of the Song Ma zone are different whilst the Middle Carboniferous floras on the Indochina block in north-eastern Thailand indicate continental connection between Indochina and South China in the Carboniferous |
| 9 | Qinling-Dabie | South China, North China | Triassic–Jurassic | Late Triassic subduction-related granite. Late Triassic U-Pb dates of zircons from ultrahigh-pressure eclogites. Late Triassic–Early Jurassic convergence of apparent polar wander-paths and palaeolatitudes of South China and North China. Initial contact between South and North China is indicated by isotopic data in Shandong and sedimentological records along the suture. Widespread Triassic to Early Jurassic deformation in the North China block north of the suture |

*Continued*

**Table 2** Continued

| No. on Fig. 1 | Suture name | Colliding lithospheric blocks | Suture age | Age constraints |
|---|---|---|---|---|
| 10 | Kunlun | Qamdo-Simao, Kunlun | Triassic | Permo-Triassic ophiolites and subduction-zone mélange. Upper Permian calc-alkaline volcanics, strongly deformed Triassic flysch and Late Triassic granitoids |
| 11 | Jinshajiang | Simao, South China | Late Permian–Late Triassic | Ophiolites are regarded as Upper Permian to Lower Triassic in age. Mélange comprises Devonian, Carboniferous, and Permian exotics in a Triassic matrix. Upper Permian to Jurassic sediments unconformably overlie Lower Permian ophiolites in the Hoh Xil Range |
| 12 | Ailaoshan Suture | Simao, South China | Middle Triassic | Plagiogranite U-Pb ages of $340 \pm 3$ Ma and $294 \pm 3$ Ma indicate that the oceanic lithosphere formed in the latest Devonian to earliest Carboniferous. Ophiolitic rocks are associated with deep-marine sedimentary rocks including ribbon-bedded cherts that have yielded some Lower Carboniferous and Lower Permian radiolarians. Upper Triassic sediments (Carnian conglomerates and sandstones, Norian limestones, and Rhaetian sandstones) blanket the suture |
| 13 | Nan-Uttaradit Suture | Simao, Indochiona | Middle Triassic | Pre-Permian ophiolitic mafic and ultramafic rocks with associated blueschists. Mafic and ultramafic blocks in the mélange comprise ocean-island basalts, back-arc basin basalts and andesites, island-arc basalts and andesites and supra-subduction cumulates generated in Carboniferous to Permo-Triassic times. Permo-Triassic dacites and rhyolites associated with relatively unmetamorphosed Lower Triassic sandstone–shale turbidite sequence. Suture zone rocks are overlain unconformably by Jurassic redbeds and post-Triassic intraplate continental basalts |
| 14 | Median Sumatra | West Burma, Sibumasu | Early Triassic | No remnants of the ocean basin that once separated West Sumatra and Sibumasu so far known. It is likely that West Sumatra was slid into juxtaposition with Sibumasu from east to west along a major strike-slip fault associated with oblique subduction |
| 15 | Banggong Suture | Lhasa, Qiangtang | Late Jurassic–Earliest Cretaceous | Suture is blanketed in Tibet by Cretaceous and Palaeogene rocks. Structural data indicate collision around the Jurassic–Cretaceous boundary |
| 16 | Shan Boundary Suture | West Burma, Sibumasu | Early Cretaceous | Cretaceous thrusts in the back-arc belt. Late Cretaceous age for the Western Belt tin-bearing granites |
| 17 | Woyla Suture | Sikuleh, Sibumasu/W Sumatra | Late Cretaceous | Cretaceous ophiolites and accretionary complex material |
| 18 | Indus-Yarlung-Zangbo Suture | India, Eurasia | Late Cretaceous–Eocene | Jurassic–Cretaceous ophiolites and ophiolitic mélange with Jurassic–Lower Cretaceous radiolarian cherts. Late Cretaceous blueschists. Eocene collision-related plutons. Palaeomagnetic data indicates initial collision around 60 Ma. Palaeogene strata blanket the suture |
| 19 | Lok Ulo-Meratus Suture | Paternoster, SW Borneo | Late Cretaceous | Subduction mélange of Middle–Late Cretaceous age. Ophiolite with Jurassic ultramafic rocks. Pillow basalts of Jurassic and Early Cretaceous ages. Oceanic cherts of Late Jurassic to Late Cretaceous ages. Turbiditic flysche of Early–Late Cretaceous age. Ophiolite obducted in Cenomanian. Suture overlain by Eocene strata |
| 20 | Boyan Suture Lupar-Adio-Borneo Suture | Semitau, SW Borneo South China, Mangkalihat | Late Cretaceous Early to middle Miocene | Upper Cretaceous mélange. Ophiolites of probable Cretaceous age. Subduction-related melanges of Early Miocene and older ages. Melanges include tectonic melanges, olistostromes and mud diapirs. Blueschist metamorphism reported. Accretionary deformation in South Sabah until Early Miocene, continued until Middle Miocene in the north |
| | Sulawesi Suture | W. Sulawesi, Australian fragments | Late Oligocene–Early Miocene | Complex accretion of ophiolites and arc and continental fragments (Australia-derived). At least three phases of accretion. Cretaceous to Early Oligocene marine sediments associated with ophiolites. Oligo-Miocene blueschist metamorphism. Large granite plutons in West Sulawesi dated at 12–4 Ma |
| | Molucca Suture | Arc–arc collision | Pliocene and ongoing | Arc–arc collision. Double subduction system with complete subduction of Molucca Sea Plate. Melange wedge/collision complex in Molucca Sea |

Table 3  Palaeozoic, Mesozoic, and Cenozoic events and their ages in South-east Asia

| Process | Age |
|---|---|
| *Palaeozoic evolution* | |
| Rifting of South China, North China, Indochina, East Malaya, West Sumatra, Tarim, and Qaidam from Gondwana | Early Devonian |
| Initial spreading of the Palaeo-Tethys ocean | Middle–Late Devonian |
| Amalgamation of South China, Indochina, and East Malaya to form Cathaysialand | Late Devonian to Early Carboniferous |
| Development of the Ailaoshan-Nan-Uttaradit back-arc basin and separation of the Simao Terrane by back-arc spreading | Late Early Carboniferous |
| Rifting of Sibumasu and Qiangtang from Gondwana as part of the Cimmerian continent | Late Early Permian (Sakmarian) |
| Initial spreading of Meso-Tethys ocean | Middle Permian |
| Collision and suturing of Sibumasu to Indochina and East Malaya | Latest Permian to Early Triassic |
| Initial collision of South and North China and development of Tanlu Fault | Late Permian to Triassic |
| *Mesozoic evolution* | |
| Suturing of South China with North China and final consolidation of Sundaland | Late Triassic to Early Jurassic |
| Rifting of Lhasa, West Burma and other small terranes | Late Triassic to Late Jurassic |
| Initial spreading of Ceno-Tethys ocean | Late Triassic (Norian) in west (northern India) and Late Jurassic in east (NW Australia) |
| Northward drift of Lhasa, West Burma, and small terranes | Jurassic to Cretaceous |
| Collision of the Lhasa Block with Eurasia | Late Jurassic–Earliest Cretaceous |
| Accretion of West Burma and Sikuleh? terranes to Sibumasu | Late Early Cretaceous |
| Suturing of Semitau to SW Borneo | Late Cretaceous |
| *Cenozoic evolution* | |
| Collision of India with Eurasia | Initial contact around 60 Ma, major indentation from around 45 Ma onwards |
| Final separation of Australia from Antarctica and establishment of the circum-Antarctic ocean current | 45 Ma, Late Eocene |
| Northwards drift of the isolated Australian continent over 27° of latitude. Gradual drying of the continent and evolution of the distinctive Australian fauna and flora from Gondwana ancestry | 45–0 Ma, Late Eocene to present |
| Clockwise rotation and extrusion of Indochina and parts of northern Sibumasu | During the early Cenozoic but precise age not known |
| Major strike-slip faulting | 30–15 Ma, Middle Oligocene to Middle Miocene |
| Anti-clockwise rotation of Borneo and the Malay Peninsula | Progressively during the Cenozoic but mostly during the Miocene |
| Initial spreading of the South China Sea basin | 30 Ma, Middle Oligocene |
| Proto-South China Sea destroyed by southwards subduction | 40–15 Ma, Middle Eocene to Middle Miocene |
| Collision of the Australian continent with the Philippine Sea Plate (initiating clockwise rotation of Philippine Sea Plate and causing structural inversion in many Cenozoic basins of the region) | 25 Ma, Late Oligocene |
| Clockwise rotation of the Philippine Sea Plate | From about 20 Ma onwards |
| Opening of the Sulu Sea | 20 Ma, Early Miocene |
| Molucca Sea double subduction established | 10 Ma, Late Miocene |
| Collision of the Philippine Arc and Eurasian continental margin in Taiwan | 5 Ma, End Miocene |

major global shifts in both plate configurations and climate during this time, and there was a change from warm to cold conditions in Australasia. This is reflected in the change from a high to a low diversity of faunas in Australasia, especially eastern Australia, where low-diversity endemic faunas developed in the Upper Carboniferous. The faunas and floras of North and South China, Indochina, East Malaya, West Sumatra, and Tarim are tropical or sub-tropical Cathaysian or Tethyan types during the Carboniferous and show no Gondwanan affinities (**Figure 10**). These terranes had already separated from Gondwana and were located in low-latitude or equatorial positions during the Carboniferous (**Figure 11**). Indochina and South China collided and amalgamated within Tethys during the Early Carboniferous along the Song Ma suture zone, which is now located in Laos and Vietnam. Ice-sheets and glaciers extended across much of eastern Gondwana during the Late Carboniferous: ice reached the marine environment of the

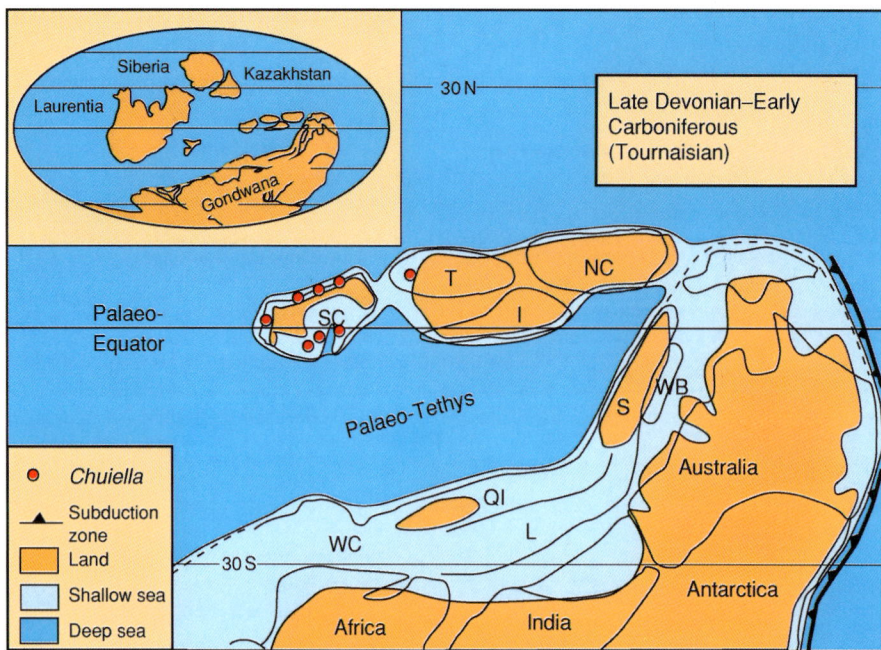

**Figure 9** Reconstruction of eastern Gondwana for the Late Devonian to Lower Carboniferous (Tournaisian), showing the postulated positions of the East and South-east Asian terranes, the distribution of land and sea, and the opening of the Palaeo-Tethys ocean at this time. Also shown is the distribution of the endemic Tournaisian brachiopod genus *Chuiella*. NC, North China; SC, South China; T, Tarim; I, Indochina–East Malaya–West Sumatra; QI, Qiangtang; L, Lhasa; S, Sibumasu; WB, West Burma; WC, Western Cimmerian Continent.

India–Australian continental shelf of Gondwana, and glacial–marine sediments (diamictites; pebbly mudstones interbedded with normal marine shales and sands) were deposited on the continental shelf of eastern Gondwana. Subduction beneath South China–Indochina in the Carboniferous led to the development of the Ailaoshan-Nan-Uttaradit back-arc basin (now represented by the Ailaoshan and Nan-Uttaradit Suture Zones) and separation of the Simao Terrane by back-arc spreading.

**Early Permian (298–270 Ma)** During the Permian, Australia remained in high southern latitudes. Glacial ice continued to reach the marine environment of the north-eastern Gondwanan margin, and glacial–marine sediments continued to be deposited on the Sibumasu, Qiangtang, and Lhasa terranes (**Figures 8, 11, and 12**). Gondwanan cold-climate faunas and floras characterized the Sibumasu, Qiangtang, and Lhasa terranes at this time. In addition, the distinctive cool-water-tolerant conodont genus *Vjalovognathus* defines an eastern peri-Gondwanan cold-water province (**Figure 11**). Floral provinces are particularly marked at this time, and the distinctive Cathaysian (*Gigantopteris*) flora developed on North China, South China, Indochina, East Malaya, and West Sumatra, which were isolated within Tethys and located equatorially (**Figure 13**). During the late Early Permian (End-Sakmarian), the Cimmerian continental sliver separated from the north-eastern margin of Gondwana, and the Meso-Tethys Ocean opened by seafloor spreading between it and mainland Gondwana (**Figure 11C**).

**Late Permian (270–252 Ma)** By early Late Permian times the Sibumasu and Qiangtang terranes had separated from Gondwana, and the Meso-Tethys had opened between this continental sliver and Gondwana (**Figure 11C**). The Palaeo-Tethys continued to be destroyed by northwards subduction beneath Laurasia, North China, and the amalgamated South China–Indochina–East Malaya terranes. Following separation, and during their northwards drift, the Sibumasu and Qiangtang terranes initially developed a Cimmerian Province fauna and were then absorbed into the Cathaysian Province. North and South China begin to collide during the Late Permian, and a connection between mainland Pangaea and Indochina, via South and North China or via the western Cimmerian continent, is indicated by the occurrence of the genus *Dicynodon* in the Upper Permian of Indochina (**Figure 11C**). Sibumasu began to collide with Indochina and East Malaya in the Late Permian, and collision continued into the Early Triassic. Deformation associated with this event, and with the collision of North and south China, is known as the Indosinian Orogeny in South-east Asia and China.

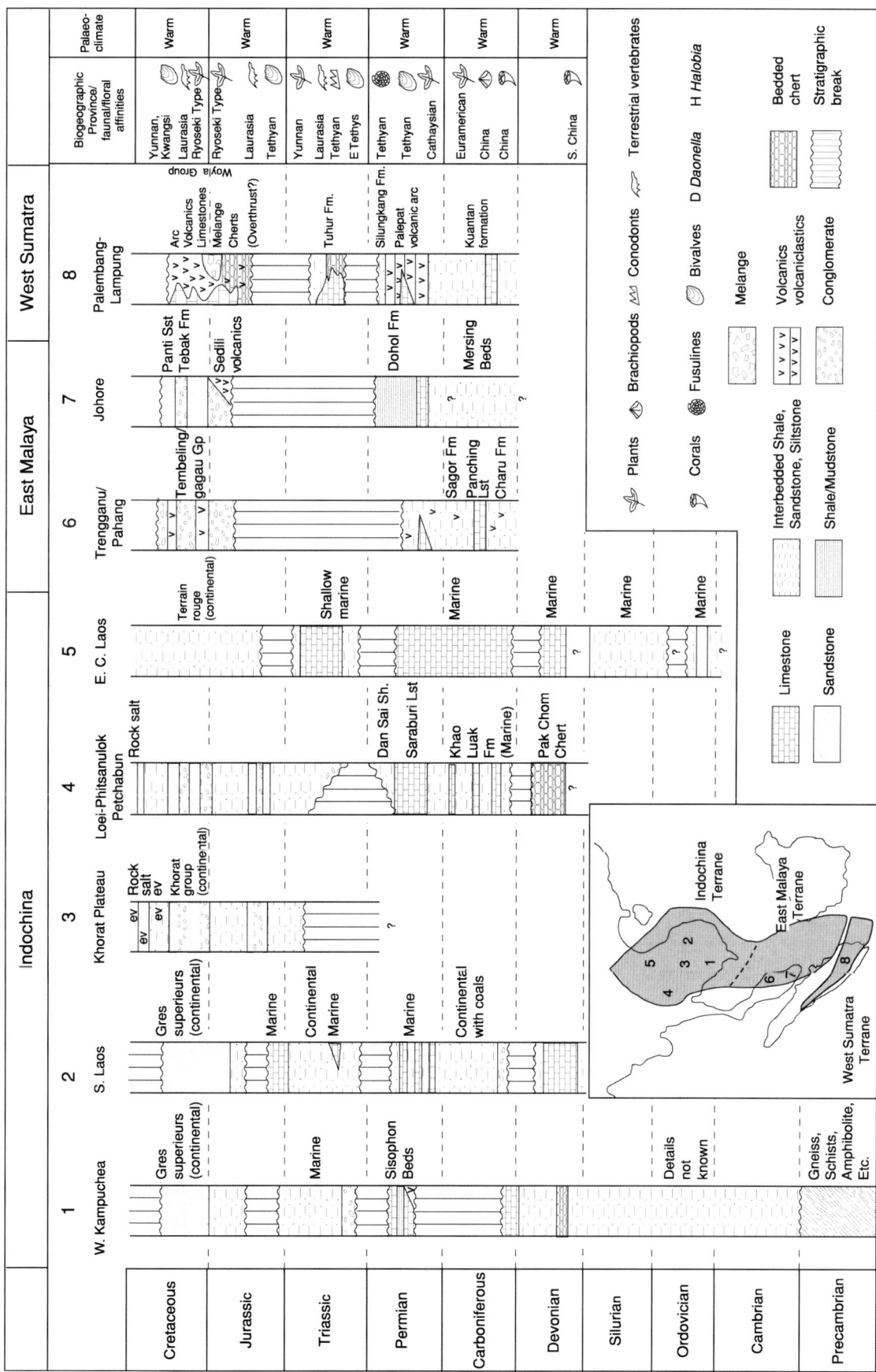

**Figure 10** Representative stratigraphic columns for the Indochina, East Malaya, and West Sumatra terranes, and faunal provinces and affinities of faunas and floras. ev = evaporites.

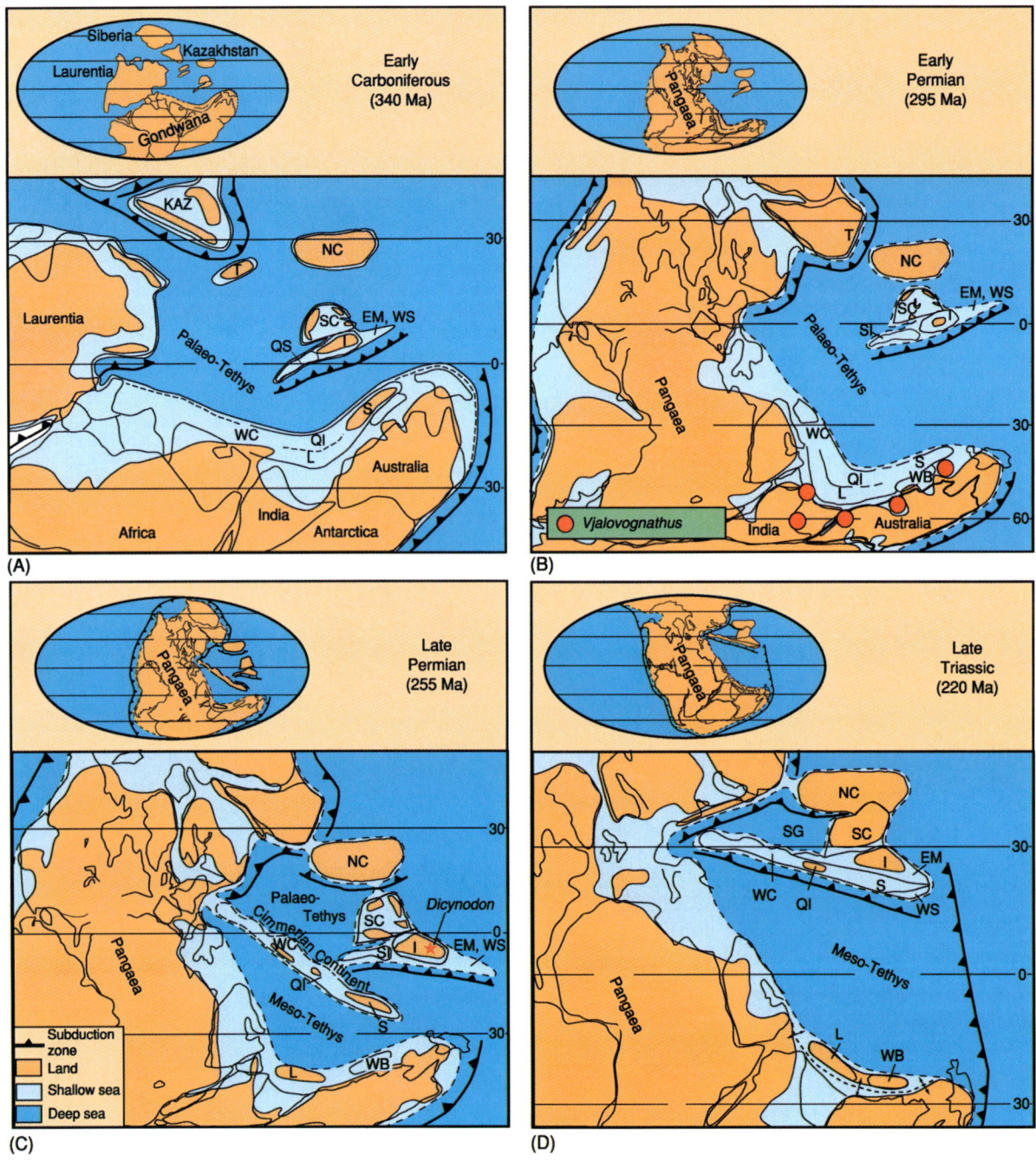

**Figure 11** Palaeogeographical reconstructions of the Tethyan region for (A) the Early Carboniferous, (B) the Early Permian, (C) the Late Permian, and (D) the Late Triassic, showing the relative positions of the East and South-east Asian terranes and the distribution of land and sea. Also shown are the distribution of the Early Permian cold-water–tolerant conodont, *Vjalovognathus*, and the Late Permian *Dicynodon* locality in Indochina. SC, South China; T, Tarim; I, Indochina; EM, East Malaya; WS, West Sumatra; NC, North China; SI, Simao; S, Sibumasu; WB, West Burma; QI, Qiangtang; L, Lhasa; WC, Western Cimmerian Continent; KAZ, Kazakhstan; QS, Qamdo-Simao; SG, Songpan Ganzi.

**Triassic (253–205 Ma)** Australia was in low-to-moderate southern latitudes during the Triassic. The Sibumasu and Qiangtang terranes collided and sutured to the Indochina–South China amalgamated terrane. The West Sumatra Block was pushed westwards by the interaction of the westwards-subducting Palaeopacific Plate with the northwards-subducting Palaeo-Tethys during the Sibumasu–Indochina–East Malaya collisional process and was translated along major strike-slip faults to a position outboard of Sibumasu in the Early Triassic. The North–South China collision was nearly complete, and ultrahigh-pressure

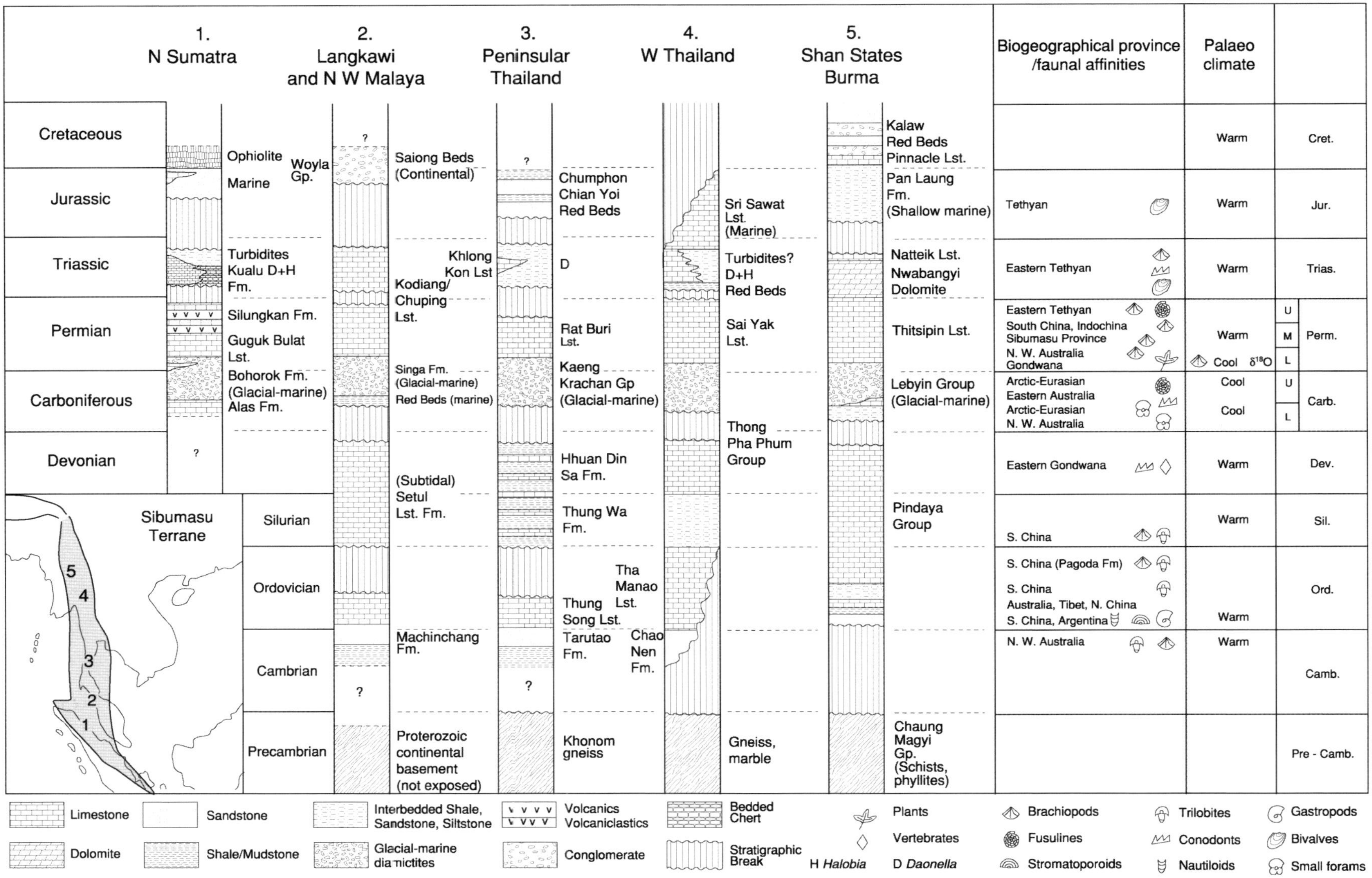

ASIA/South-East 185

**Figure 12** Representative stratigraphic columns for the Sibumasu Terrane, and faunal provinces and affinities of faunas and floras.

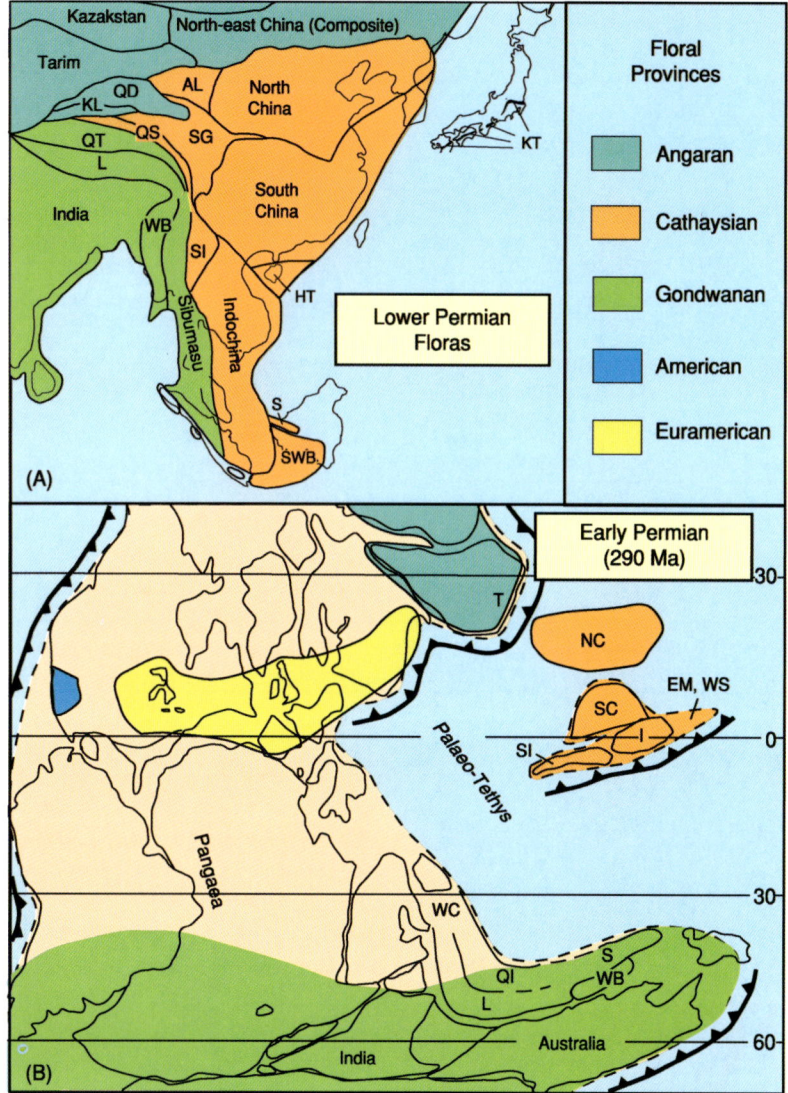

**Figure 13** The distribution of Early Permian floral provinces plotted on (A) a present-day geographical map, and (B) an Early Permian palaeogeographical map. (A) KL, Kunlun; QD, Qaidam; AL, Ala Shan; QT, Qiantang; L, Lhasa; QS, Qamdo-Simao; SG, Songpan Ganzi; WB, West Burma; SI, Simao; HT, Hainan Island Terranes; S, Semitau; SWB, South West Borneo. (B) T, Tarim; NC, North China; SI, Simao; SC, South China; I, Indochina; EM, East Malaya; WS, West Sumatra; WC, Western Cimmerian Continent; QI, Qiangtang; S, Sibumasu; L, Lhasa; WB, West Burma.

metamorphics were exhumed along the Qinling-Dabie suture zone. Sediment derived from the North–South China collisional orogen poured into the Songpan Ganzi accretionary-complex basin producing huge thicknesses of flysch turbidites. The Ailaoshan-Nan-Uttaradit back-arc basin was closed when the Simao Terrane collided with South China–Indochina in the Middle to early Late Triassic. By Late Triassic (Norian) times, the North China, South China, Sibumasu, Indochina, East Malaya, West Sumatra, and Simao terranes had coalesced to form proto-East and South-east Asia (**Figure 11D**). During the final collisional consolidation of these terranes, the major economically important Late Triassic–Early Jurassic collisional tin-bearing Main Range granitoids were formed in South-east Asia (**Figure 14**).

**Jurassic (205–141 Ma)** Australia remained in low-to-moderate southern latitudes in the Jurassic. Rifting and separation of the Lhasa, West Burma, Sikuleh, Mangkalihat, and West Sulawesi terranes from north-western Australia occurred progressively from west to east during the Late Triassic to Late Jurassic. The Ceno-Tethys Ocean opened behind these terranes as they separated from Gondwana (**Figure 15**). Final welding of North China to Eurasia (the Yanshanian deformational orogeny) took place in the Jurassic with the closure of the Mongol-Okhotsk Ocean. Initial pre-breakup rifting of the main Gondwanan supercontinent also began in the Jurassic.

**Cretaceous (141–65 Ma)** The Lhasa Block collided and amalgamated with Eurasia in latest Jurassic–earliest Cretaceous times. Gondwana broke up and India drifted north, making initial contact with Eurasia at the end of the Cretaceous (**Figure 15**). This early contact between India and Eurasia is indicated by palaeomagnetic data from the Ninetyeast Ridge and also by Late Cretaceous biogeographical links, including frogs and other vertebrates. The small West Burma and Sikuleh terranes also accreted to Sibumasu during the Cretaceous. Australia began to separate from Antarctica and drift northwards, but a connection with Antarctica via Tasmania remained.

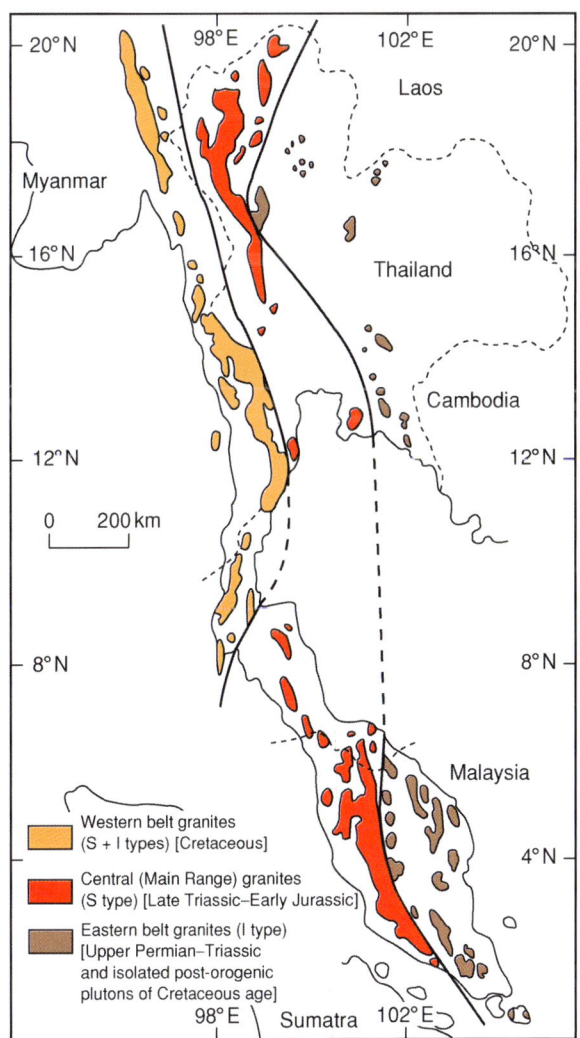

**Figure 14** The three granitoid belts (provinces) of South-east Asia.

**Cenozoic (65–0 Ma)** The Cenozoic evolution of East and South-east Asia involved substantial movements along and rotations of strike-slip faults, rotations of continental blocks and oceanic plates, the development and spreading of 'marginal' seas, and the formation of important hydrocarbon-bearing sedimentary basins. This evolution was essentially due to the combined effects of the interactions of the Eurasian, Pacific, and Indo-Australian plates and the collisions of India with Eurasia and of Australia with South-east Asia.

Various tectonic models have been proposed for the Cenozoic evolution of the region, which invoke different mechanisms for the India–Eurasia collision and different interpretations of rotations of continental blocks and oceanic plates. Three basic mechanisms have been proposed for the northwards motion of India into Eurasia. The first involves underthrusting of greater India beneath Eurasia; the second involves crustal shortening and thickening; and the third involves major eastwards lateral extrusion and progressive clockwise rotations of China, Indochina, and Sundaland.

Cenozoic clockwise rotations of crustal blocks are observed in Indochina and western Thailand, but major progressive counterclockwise rotations are seen in Borneo and the Malay Peninsula. The counter-clockwise rotations observed in Borneo and Malaya are at variance with the extrusion model. Most models for the region neglect the clockwise rotation of the Philippine Sea Plate and/or the counterclockwise rotation of Borneo. The Cenozoic reconstruction model proposed by Robert Hall and illustrated here (**Figures 16, 17,** and **18**), which takes into account both the clockwise rotation of the Philippine Sea Plate and the counterclockwise rotation of Borneo and Peninsular Malaysia, is preferred. Knowledge of the Cenozoic geodynamics of South-east Asia, the movements of microcontinents, and the shifting distribution of land and sea in the area (**Figure 19**) underpins and enhances our understanding of the biogeography of the region.

## South-East Asian Geological Resources

### Oil and Gas

Significant oil and gas accumulations occur widely in East and South-east Asia, generally in Cenozoic sedimentary basins (**Figure 20**), and these have contributed markedly to the economies of South-east Asian countries. The oil and gas accumulations are commonly associated with rocks of Middle and Upper Miocene age, with locally significant Oligocene and

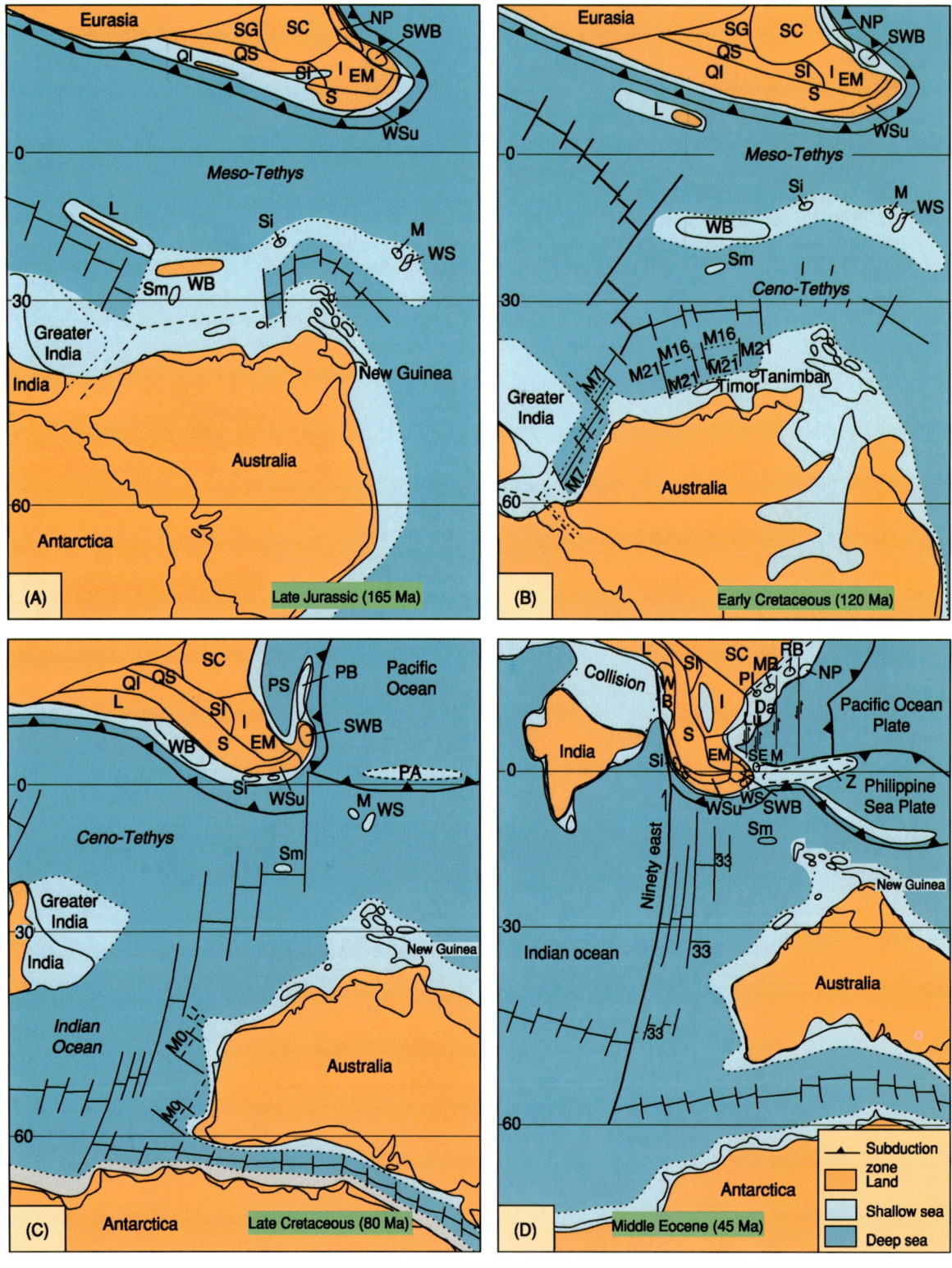

**Figure 15** Palaeogeographical reconstructions for Eastern Tethys in (A) the Late Jurassic, (B) the Early Cretaceous, (C) the Late Cretaceous, and (D) the Middle Eocene, showing the distribution of continental blocks and fragments of South-east Asia–Australasia and the areas of land and sea. SG, Songpan Ganzi accretionary complex; SC, South China; QS, Qamdo-Simao; SI, Simao; QI, Qiangtang; S, Sibumasu; I, Indochina; EM, East Malaya; WSu, West Sumatra; L, Lhasa; WB, West Burma; SWB, South West Borneo; SE, Semitau; NP, North Palawan and other small continental fragments now forming part of the Philippines basement; Si, Sikuleh; M, Mangkalihat; WS, West Sulawesi; PB, Philippine Basement; PA, Incipient East Philippine Arc; PS, Proto-South China Sea; Z, Zambales Ophiolite; RB, Reed Bank; MB, Macclesfield Bank; PI, Paracel Islands; Da, Dangerous Ground; Lu, Luconia; Sm, Sumba. M numbers represent Indian Ocean magnetic anomalies.

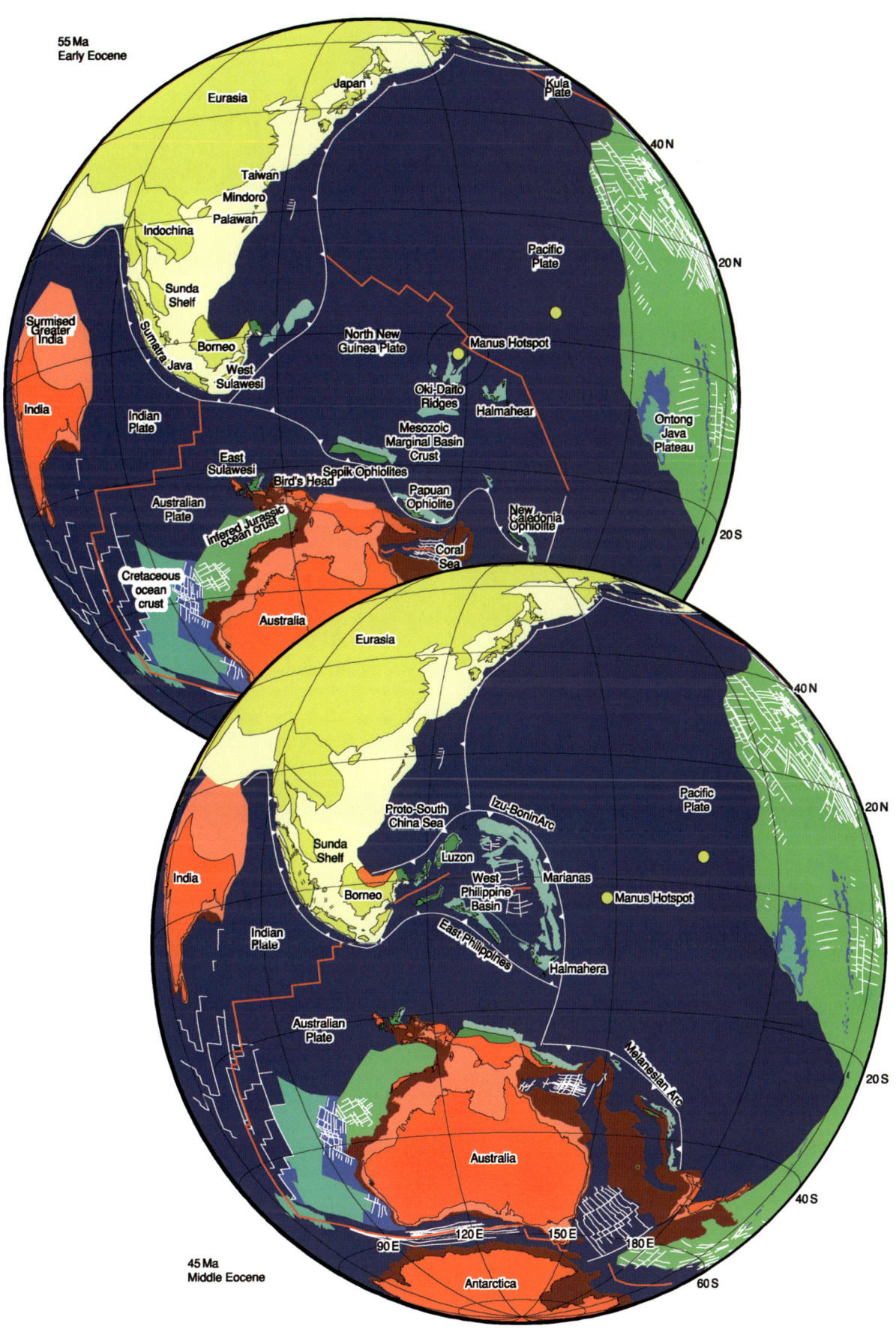

Pliocene occurrences. Some older Mesozoic accumulations do occur (e.g. in North Thailand), but, in general, the petroleum industry of the region is almost exclusively concerned with Tertiary sedimentary basins, and the pre-Tertiary is regarded as economic basement. However, in some cases, oil has migrated laterally and accumulated in fractured granitoids and other basement rocks, including Triassic limestones in the Gulf of Thailand.

**Oil- and gas-bearing basins** Space constraints preclude a full detailed description of the basins and hydrocarbon fields. A number of attempts have been made to classify South-east Asian hydrocarbon-bearing basins – particularly the Cenozoic basins of the region – genetically, but these attempts have failed owing to the complex and changing pattern of compressional and extensional tectonics in South-east Asia. Basins that were previously regarded as 'typical' back-arc basins are now being interpreted by some workers as the result of major strike-slip faults. Most basins were initiated in the Eocene or Oligocene, following a major Eocene break in sedimentation. Interpretation of the genesis of the South-east Asian basins is somewhat model-dependent, and there are competing models. Whichever model one favours, there has certainly been major strike-slip faulting in the region during the Cenozoic. This strike-slip faulting can clearly be related to basin formation, for example in North and Central Thailand and the Gulf of Thailand. However, the sense and displacement of many (if not most) of these strike-slip faults are poorly known. Other basins in the region are clearly related to plate convergence and subduction; for example, the North, Central, and South Sumatra basins and the Barito Basin of Borneo appear to be back-arc basins behind the arc developed on Sundaland as the India–Australian Plate was subducted northwards. The Palawan and Sabah basins could be classified as fore-arc basins. Other basins in the region, not related genetically to strike-slip faulting or to subduction processes, have been classified variously as continental failed rifts (aulacogens), cratonic basins, or basins that have formed on or between continental fragments.

## Minerals

The distribution of the principal mineral deposits of South-east Asia is shown in **Figure 21**. Mineralization in South-east Asia is primarily associated with the ophiolites, volcanic arcs, and granitoid plutons of the region.

**Mineralization associated with ophiolites** Chromites are found in economic concentrations in ophiolites derived from the marginal-basin lithosphere of the Celebes, Sulu, and South China seas. Nickel sulphides and platinum also occur in dunite and serpentinite, and remobilized nickel also occurs as sulphides in veins in andesitic rocks. Deep tropical weathering of ophiolite ultramafics has resulted in nickel-bearing laterites that can be valuable ores (with a nickel oxide content of around 2.7%). Other important mineral deposits associated with ophiolites include manganese and Cyprus-type copper on pillow lavas and Besshi-type copper–iron massive sulphide deposits.

**Mineralization associated with volcanic arcs** Volcanic arcs are characterized by a wide range of mineralization types, ranging from epithermal vein deposits associated with near-surface fracture systems to higher-temperature vein and dissemination deposits associated with epizonal plutons. Ore deposits include porphyry copper, Kuroko-type copper–lead–zinc, and gold and gold–silver epithermal deposits.

**Non-volcanic epithermal deposits** The non-volcanic epithermal deposits of the region comprise stibnite, stibnite–gold, and stibnite–gold–scheelite mineralization, which is confined to 'Sundaland', i.e. the continental cratonic core of South-east Asia. The gold association is quite different from that of the Cenozoic volcanic arcs of the region and is confined to areas characterized by high-level felsic-to-intermediate plutons. There is commonly an important mercury association. This mineralization is younger than, and unrelated to, the tin-bearing granites of the region. Important deposits occur in western Borneo, Palawan, Sulawesi, the Malay Peninsula, Thailand, and Myanmar.

**Tungsten deposits** South-east Asia is a major tin–tungsten metallogenic province. The tungsten mineralization in South-east Asia is spatially and genetically related to the three main granite provinces of the region (**Figure 14**), and four types of deposit occur: hydrothermal quartz veins, scheelite skarn deposits, wolframite–scheelite–sulphide veins, and placer deposits (eluvial or marginally alluvial).

---

**Figure 16** Plate reconstructions for South-east Asia–Australasia at 55 Ma (Early Eocene) and 45 Ma (Middle Eocene). The reconstruction at 55 Ma precedes the collision of India with Eurasia. The reconstruction at 45 Ma coincides with a major period of plate reorganization. (Reproduced with permission from Hall R (2002) Cenozoic geological and plate tectonic evolution of SE Asia and the SW Pacific: computer-based reconstructions and animations. *Journal of Asian Earth Sciences* 20: 353–434.)

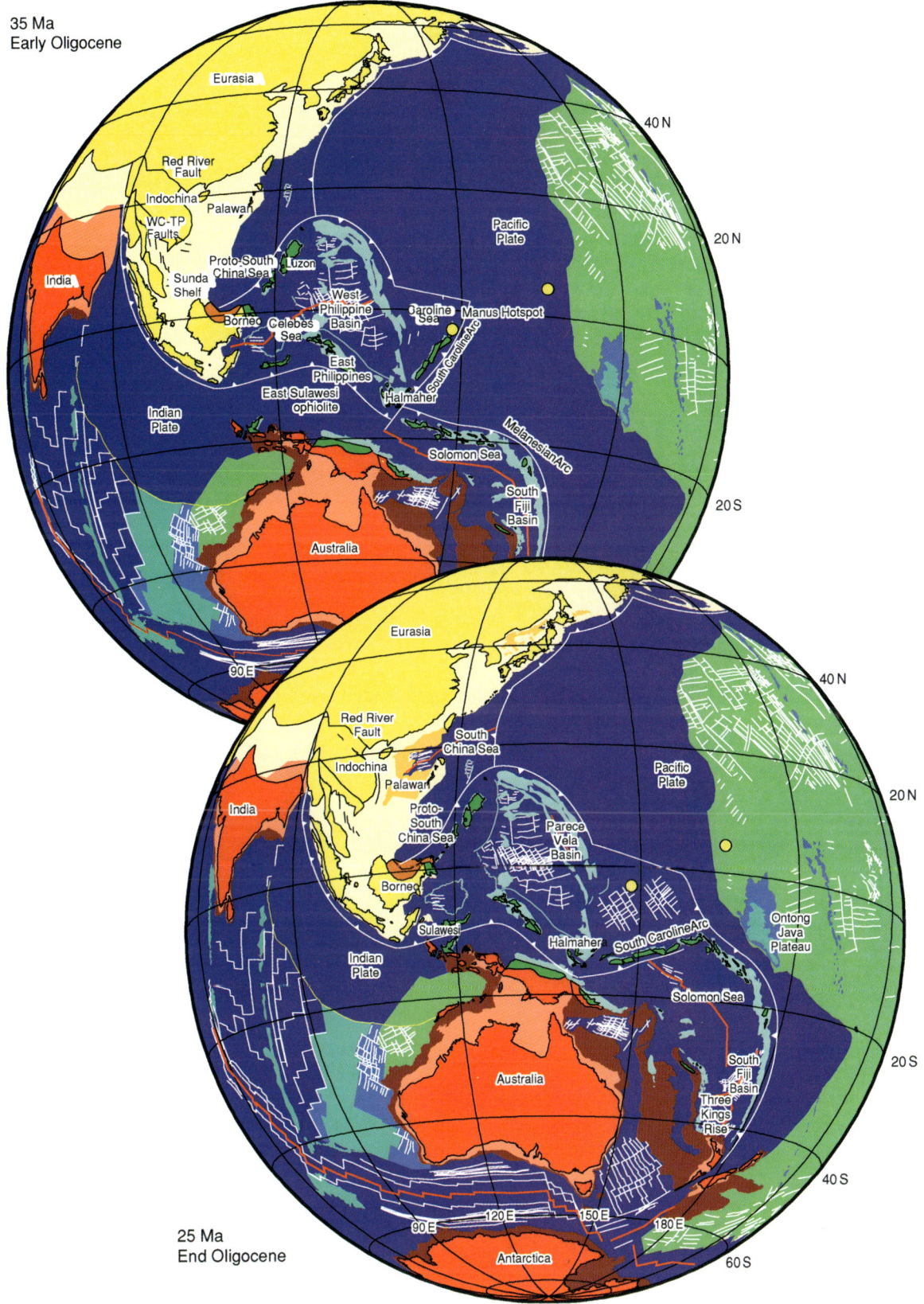

**Figure 17** Plate reconstructions for South-east Asia–Australasia at 35 Ma (Early Oligocene) and 25 Ma (End Oligocene). (Reproduced with permission from Hall R (2002) Cenozoic geological and plate tectonic evolution of SE Asia and the SW Pacific: computer-based reconstructions and animations. *Journal of Asian Earth Sciences* 20: 353–434.)

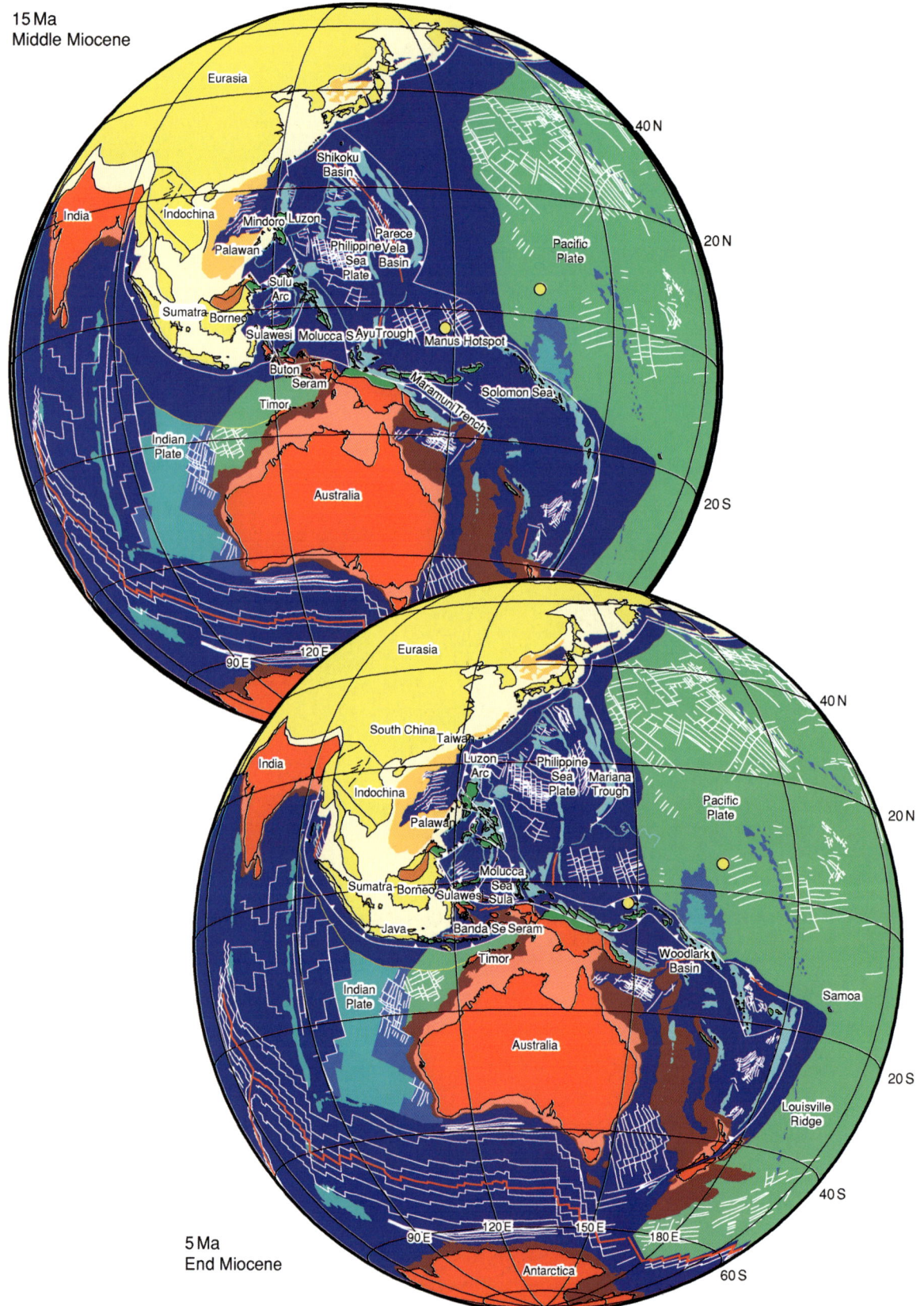

**Figure 18** Plate reconstructions for South-east Asia–Australasia at 15 Ma (Middle Miocene) and 5 Ma (End Miocene). (Reproduced with permission from Hall R (2002) Cenozoic geological and plate tectonic evolution of SE Asia and the SW Pacific: computer-based reconstructions and animations. *Journal of Asian Earth Sciences* 20: 353–434.)

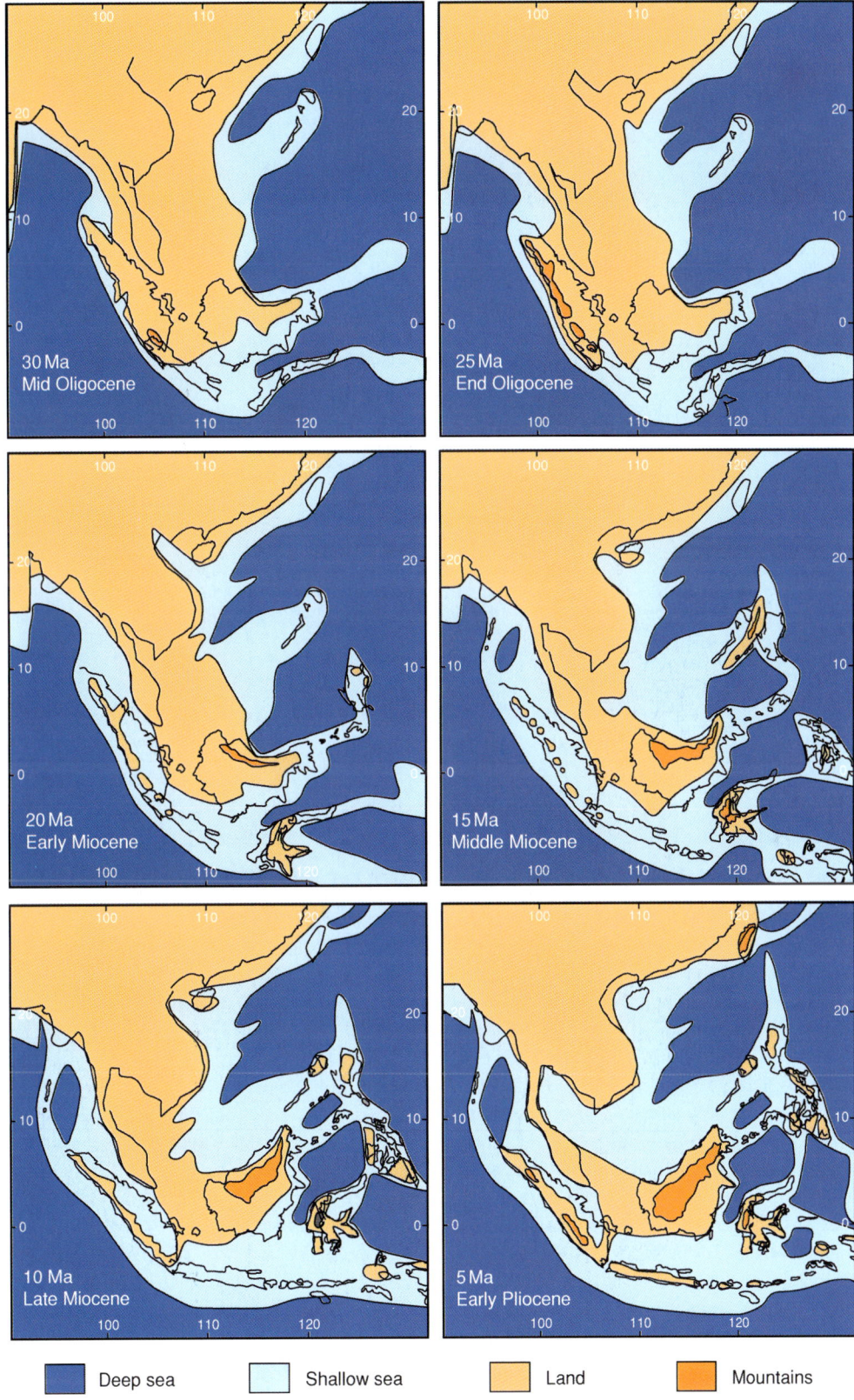

**Figure 19** Reconstructions of East and South-east Asia for the last 30 Ma, showing the variation in the distributions of deep sea, shallow sea, land, and mountains.

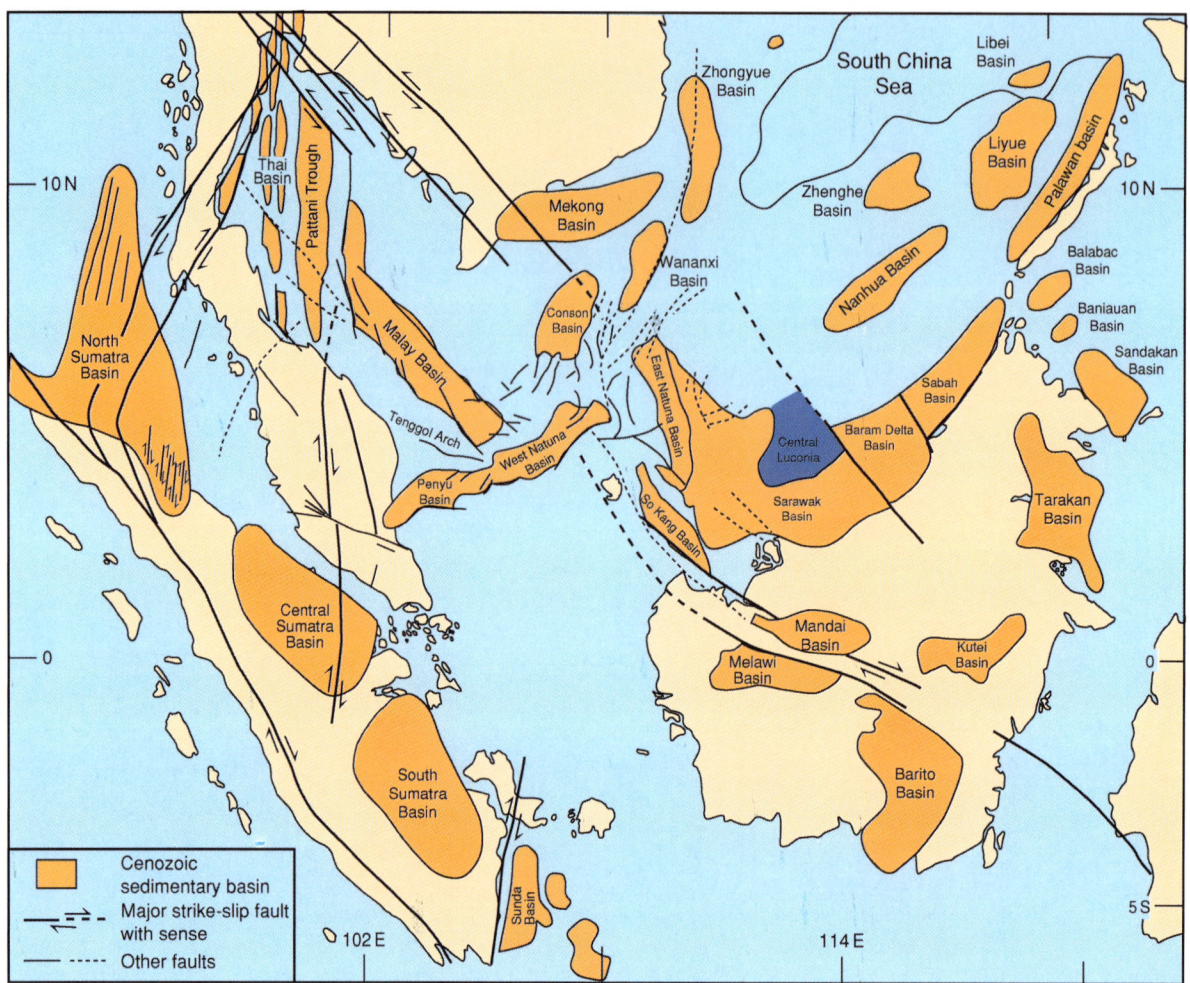

**Figure 20** Oil- and gas-bearing Cenozoic sedimentary basins and major fault zones of South-east Asia.

**Tin deposits** During the twentieth century, 70% of the tin mined in the world came from the South-east Asian tin belt. The tin mainly occurs as cassiterite ore in Late Cenozoic alluvial placer deposits that form part of the 400 km wide South-east Asian tin belt, which extends 2800 km from Burma and Thailand through Peninsular Malaysia to the Indonesian tin islands of Bangka and Belitung (**Figure 21**). Historically, these deposits have been extremely important to the economic development of Malaysia and South-east Asia, and, altogether, 9.6 million tonnes of tin (54% of the world's total tin production) has come from this tin belt. The tin deposits of the region are associated with the major granite/granitoid bodies of South-east Asia and formed during the emplacement of these intrusive igneous bodies, which were produced by major plate-tectonic processes as the continental blocks of the region converged and collided with each other. The South-east Asian granites form three major north–south belts or provinces: a Western Granitoid Province of Cretaceous–Cenozoic (149–22 Ma) age, a Central Granitoid Province of mainly Late Triassic and Early Jurassic (230–184 Ma) age, and an Eastern Granitoid Province of Permian, Triassic, and Early Jurassic (257–197 Ma) age (**Figure 14**). The Eastern Province was formed by subduction on the western margins of the Indochina and East Malaya terranes during convergence with the Sibumasu terrane. When these two continental terranes collided, the Central Province granites were generated. The younger Western Province granites were principally formed when the West Burma terrane collided with Sibumasu and by subsequent subduction processes. The Main Range and Eastern Belt granites of Peninsular Malaysia form parts of the Central and Eastern Granitoid Provinces, respectively. The tin mineralization occurred within the granites themselves and more especially in the country rocks that immediately surrounded them. Subsequent to mineralization, the granites and the country rocks into which they were intruded were uplifted and eroded, and the tin ore was deposited and concentrated in river alluvium.

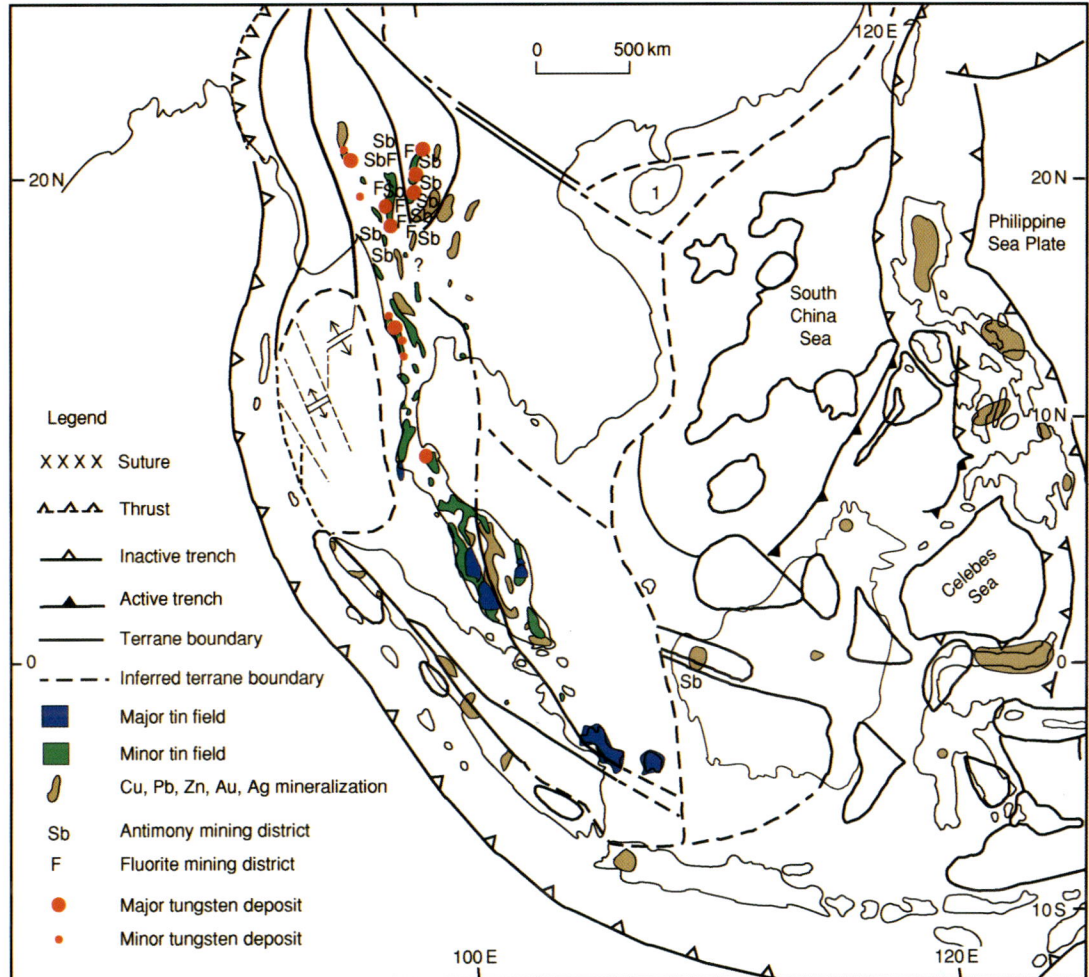

**Figure 21** Distribution of the principal mineral deposits of South-east Asia.

In addition to placer tin deposits there are also a wide variety of primary tin deposits. However, only two major primary tin fields have led to significant deep-mine production: the Sungei Lembing Mine near Kuantan, Pahang, Peninsular Malaysia, and Kelapa Kampit on Billiton Island. The lodes of these mines are developed in contact-metamorphosed Carboniferous–Permian argillaceous sedimentary and andesitic tuffaceous rocks within aureoles around high-level Permian–Triassic granitoids.

**Mississippi Valley-type epigenetic deposits** Stratiform deposits of predominantly galena and sphalerite are most commonly hosted in shallow-marine carbonate-dominated sequences on the rims of large sedimentary basins where changes in salinity caused deposition. The continental regions of South-east Asia, with their extensive Palaeozoic carbonate platforms, are suited to Mississippi Valley-type epigenetic deposits, and some have been identified. One of the best-known deposits in the region, now considered to be of this type, is at the Bawdwin lead–zinc mine north-north-east of Rangoon, Myanmar. Mining here can be traced back to 1412.

**Iron ore deposits** Iron ore deposits are widespread in the region, but significant production has been restricted to three areas: South-east China (Hainan, Kwangtung, Fukien), the Malay Peninsula, and the Philippines. The deposits are predominantly of contact pyrometasomatic type and are associated with plutons emplaced at a wide range of depths. Peninsular Malaysia was a major producer of iron ore, with production reaching 7 million tonnes in 1964. However, all major deposits have now been worked out.

**Coal and lignite deposits** South-east Asian coal deposits are almost entirely confined to deposits of sub-bituminous and lignitic coal in relatively small Tertiary terrestrial basins. There are minor occurrences of bituminous coal and anthracite in Palaeozoic and Mesozoic formations of the older Sundaland continental core of the region, but these are generally uneconomic except for deposits of Triassic anthracite mined in northern Vietnam.

**Diamonds, sapphires, and other gems** Alluvial diamonds with no obvious sources ('headless placers') are found in several areas of South-east Asia, including Myanmar, southern Thailand (Phuket), Sumatra, and Kalimantan (**Figure** 8). These deposits occur in relatively young geological terranes, in contrast to the Archaean or Proterozoic terranes that host most primary diamond deposits and their associated alluvial workings.

Significant quantities of diamond have been recovered from two areas in Myanmar: Momeik in the northern part of the country and Theindaw in the southern part. The Momeik diamonds are recovered during mining of gemstone gravels; the Theindaw and Phuket diamonds are by-products of tin dredging. These sites, and those in Sumatra, lie within the Sibumasu terrane, which was detached from north-western Australia in Palaeozoic times and drifted northwards to become part of South-east Asia. The Myanmar–Thailand diamonds may be derived either from lamproitic intrusives within this terrane or secondarily from Permian glacial–marine sediments. The Kalimantan diamonds lie in another terrane, with a different origin, and, like eastern Australian diamonds, they may be directly or indirectly related to subduction processes.

Sapphire, ruby, zircon, and spinel gemstones and jade are worked at a number of localities in South-east Asia from alluvial and eluvial gravels derived from nearby outcrops of extensive Pliocene–Pleistocene high-alkali basalt fields. The basalts are usually basinite and nephelinite. The main centres for the mining of these gemstones are at Pailin, Bokeo, and Rovieng in Cambodia, Chanthaburi, Trat, and Kancharaburi in Thailand, south of Haikou on Hainan Island, in the Mogok and Lonkin areas of Myanmar, and in Taiwan. Most of the South-east Asian rubies come from the Trat area of Thailand, which supplies 70% of the world's production. Ruby, sapphire, and spinel are mined in the Mogok area of Myanmar from placers and eluvial deposits derived from marbles and calc-silicate migmatites associated with alaskite (Mogok Gneiss). Jadeite is mined in the Lonkin area of Myanmar, which is the world's foremost source of high-grade jadeite.

## See Also

**Asia:** Central. **Australia:** Phanerozoic. **China and Mongolia**. **Gondwanaland and Gondwana**. **Palaeoclimates**. **Pangaea**. **Plate Tectonics**. **Precambrian:** Overview. **Tectonics:** Convergent Plate Boundaries and Accretionary Wedges; Mountain Building and Orogeny.

## Further Reading

Fraser AJ, Matthews SJ, and Murphy RW (eds.) (1997) *Petroleum Geology of Southeast Asia*. Special Publication 126. London: Geological Society.

Hall R and Blundell D (eds.) (1996) *Tectonic Evolution of Southeast Asia*. Special Publication 106. London: Geological Society.

Hall R and Holloway JD (eds.) (1998) *Biogeography and Geological Evolution of South-East Asia*. Amsterdam: Backhuys Publishers.

Hamilton W (1979) *Tectonics of the Indonesian Region*. USGS Professional Paper 1078. Washington: US Geological Survey.

Hutchison CS (1989) *Geological Evolution of South-East Asia*. Oxford: Clarendon Press.

Hutchison CS (1996) *South-East Asian Oil, Gas, Coal and Mineral Deposits*. Oxford: Clarendon Press.

Meng LK (ed.) The Petroleum Geology and Resources of Malaysia. Kuala Lumpur. Petronal. P.

Metcalfe I (1988) Origin and assembly of South-east Asian continental terranes. In: Audley-Charles MG and Hallam A (eds.) *Gondwana and Tethys*, pp. 101–118. Special Publication 37. London and Oxford: Geological Society and Oxford University Press.

Metcalfe I (ed.) (1994) *Gondwana dispersion and Asian accretion* Special Issue of *Journal of Southeast Asian Earth Sciences* 9: 303–461.

Metcalfe I (ed.) (1999) *Gondwana Dispersion and Asian Accretion*. IGCP 321 Final Results Volume. Rotterdam: A A Balkema.

Metcalfe I and Allen MB (eds.) (2000) Suture zones of East and Southeast Asia Special Issue of *Journal of Asian Earth Sciences* 18: 635–808.

Metcalfe I, Smith JMB, Morwood M, and Davidson I (eds.) (2001) Faunal and floral migrations and evolution in SE Asia–Australasia Lisse: A A Balkema.

Shi GR and Metcalfe I (eds.) (2002) Permian of Southeast Asia. Special Issue of *Journal of Asian Earth Sciences*, 20: 549–774.

# ASTEROIDS

*See* **SOLAR SYSTEM: Asteroids, Comets and Space Dust**

# ATMOSPHERE EVOLUTION

**S J Mojzsis**, University of Colorado, Boulder, CO, USA

© 2005, Elsevier Ltd. All Rights Reserved.

## Introduction

Earth is unique among the planets of the Solar System in having sustained conditions of temperature and pressure that permit stable liquid water at the surface for more than 4 Ga. The long-term stability of liquid water is fundamental to the origin and propagation of life, and Earth is exceptional among the known planets in having a biosphere. Abundant free oxygen accumulated in the atmosphere as a metabolic waste product of oxygenic photosynthesis; the high free-oxygen content contrasts markedly with the atmospheres of the neighbouring planets of the inner Solar System (**Table 1**). This globally oxidative condition was at first deleterious to microbial life, but led to the emergence of aerobic metabolisms, sexual reproduction, and multicellularity.

Appreciating how the current physical and chemical state of the atmosphere came to be requires knowledge of the initial conditions. The elemental ingredients that go into making a habitable world are ultimately derived from cosmologic nucleosynthesis – the formation of primarily hydrogen ($^1$H) and helium ($^4$He), with minor amounts of $^2$D, $^3$He, and $^6$Li, at the Big Bang approximately $13 \pm 1$ Ga ago. All the other naturally occurring elements in the periodic table were created by stellar nucleosynthetic and supernova reactions. These events contributed gas and dust to the interstellar medium, and a cloud of such supernova ashes collapsed to form the solar nebula and, ultimately, the Sun and planets about 4.56 Ga ago (*see* **Solar System: The Sun**).

## Origin of a Habitable World

### Solar System and Planetary Formation: Relevance to Atmospheric Evolution

Following Big Bang nucleosynthesis, local concentrations of gas collapsed to form the first galaxies and the nebulae and stars that comprise them. Gravity binds these immense structures together, and it is the gravitational collapse when stars form that creates the pressures needed to initiate thermonuclear fusion reactions, which convert four $^1$H nuclei to one $^4$He nucleus and release energy. This energy provides an outward force that counters the inward gravitational collapse. Although seemingly violent and catastrophic in our Earth-bound (solid–liquid–gas) view of matter in the universe, such systems are remarkably stable and are sustained over many millions of years throughout the lifetime of a star on the main sequence. In smaller stars, such as the Sun, luminosity changes over the stellar lifetime, beginning with a short (less than 100 Ma) early phase of variable brightness, which lasts until the so-called main sequence of hydrogen fusion commences (**Figure 1**). Time in the main sequence is long for a star such as the Sun, of the order of 8 Ga, but it eventually draws to a close when much of the hydrogen fuel is exhausted. At that stage, the star expands and enters a

**Table 1** Atmospheric constituents of the inner planets

|  | Earth | Venus | Mars |
| --- | --- | --- | --- |
| $N_2^*$ | 78.084% | 3.5% | 2.7% |
| $O_2$ | 20.946% | tr. | 0.13% |
| $H_2O$ | <4% | 30 ppm | 0.03% |
| Ar | 9340 ppm | 70 ppm | 1.6% |
| $CO_2$ | 350 ppm | 96.5% | 95.32% |
| Ne | 18.18 ppm | 7 ppm | 2.5 ppm |
| $^4$He | 5.24 ppm | ~12 ppm | nd |
| $CH_4$ | 1.7 ppm | nd | nd |
| Kr | 1.14 ppm | ~25 ppb | 0.3 ppm |
| $H_2$ | 0.55 ppm | nd | nd |
| $N_2O$ | 320 ppb | nd | ? |
| CO | 125 ppb | 17–45 ppm | 0.08% |
| Xe | 87 ppb | ~1.9 ppb | 0.08 ppm |
| $O_3$ | 10–100 ppb | nd | >0.04 ppm |
| HCl | 1 ppb | ~0.6 ppm | nd |
| $SO_2$ | 20–90 ppt | 25–150 ppm | nd |
| $H_2S$ | 30–100 ppt | 3 ppm | nd |

*Abundance values are given for dry air. Abbreviations: tr., trace; ppm, $10^{-6}$ g g$^{-1}$; ppb, $10^{-9}$ g g$^{-1}$; ppt, $10^{-12}$ g g$^{-1}$; nd, not detected.

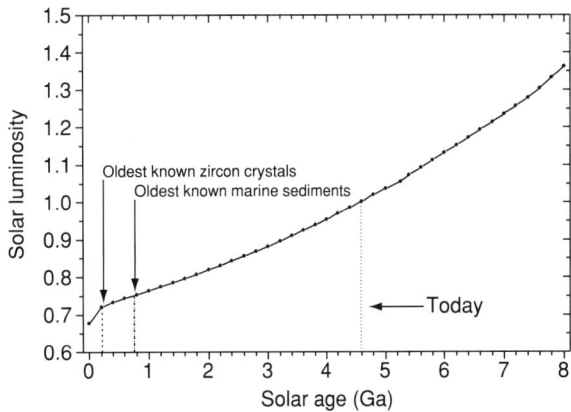

**Figure 1** Secular changes in solar luminosity with time. Solar luminosity is expressed as fraction of current solar luminosity.

red giant phase, when helium is consumed to synthesize the isotopes $^{12}C$ and $^{16}O$ (via $^{8}Be$). This is a rapid process in comparison to the length of time spent in the main sequence. When helium becomes depleted, the star collapses from a red giant into a white dwarf, which is dominantly composed of oxygen and carbon. Some of these stellar objects can continue to accrete mass from companion stars and become increasingly unstable, resulting in what are termed type-I supernovae. In the absence of a massive stellar companion, white dwarf stars slowly cool and darken over eons. The solar nebula of approximately 4.6 Ga ago was derived from the violent remains of perhaps several supernovae. There is evidence that more than one event occurred around the time of the collapse of the solar nebula, injecting it with short-lived nuclides such as $^{244}Pu$, $^{60}Fe$, and $^{26}Al$, which were important heat-producing elements in the larger differentiated asteroids and proto-planets. During nebular collapse, the central region of the disk concentrated hydrogen and helium and eventually formed the Sun. The Sun entered its main sequence about 50 Ma after the initiation of cloud collapse. Solar radius and luminosity increased steadily as hydrogen was converted to helium; this is because, as the average density of the Sun increased, the rate of hydrogen fusion also increased.

For stars greater than ten times the solar mass, hydrogen fusion proceeds at higher temperatures and pressures. In such stars, time in the main sequence may be brief, as short as 10 Ma, before they enter the red giant phase. Since these stars burn hotter and at higher pressures, fusion of helium continues beyond $^{12}C$ to yield $^{20}Ne$ and $^{24}Mg$, as well as beyond $^{16}O$ to create $^{32}S$, $^{28}Si$, and, finally, $^{56}Fe$. Iron has the greatest binding energy per nucleon (**Figure 2**) and represents the limit of the energy-yielding thermonuclear-fusion products. When even this elemental fuel becomes exhausted, a super-massive star will rapidly collapse and can explode as a type-II supernova, one of the most exergonic events in the universe. Such an event was recently observed as SN 1987A, a supernova approximately 150 000 light-years distant, which proceeded with fusion beyond $^{56}Fe$ to create gold, lead, isotopes of thorium ($^{232}Th$) and uranium ($^{235}U$, $^{238}U$), and transuranic elements up to $^{244}Pu$ and beyond. Supernova events randomly fertilize the galaxy with heavy nuclides, including the unstable radioactive heat-producing elements (potassium, uranium, and thorium) that are responsible for keeping planetary interiors warm and dynamic over geological timescales. It is estimated that about 50% of all interstellar dust is created in supernova explosions. Cosmic dust is the crucial raw material in the creation of a habitable world.

**Formation of the Earth: Primary Atmosphere**

A rapid accretion by asteroid-like bodies (termed 'planetesimals') formed the planets of the Solar System (*see* **Earth Structure and Origins**). A volatile envelope – captured from nebular gases and perhaps containing significant amounts of hydrogen, helium, methane, and ammonia – would have already started to accumulate when the growing proto-Earth reached a critical mass

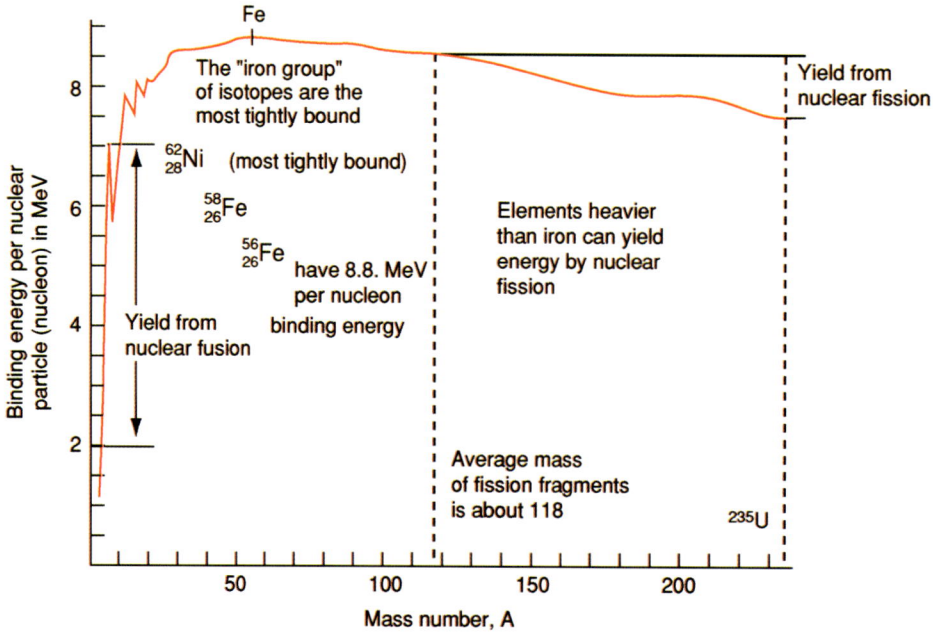

**Figure 2** Binding-energy curve of the nuclides.

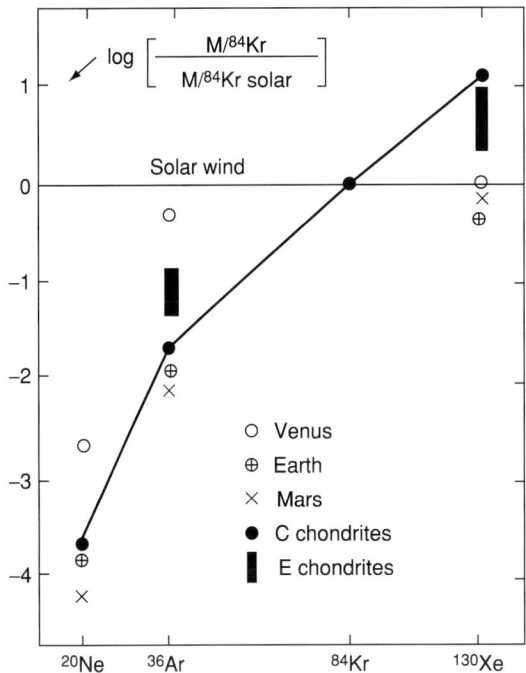

**Figure 3** Noble-gas abundances on Earth in comparison with cosmic abundances (adapted with permission from D. Hunten et al. in *Meteorites and the Early Solar System*, eds. John F. Kerridge and Mildred Shapley Matthews. © 1988 The Arizona Board of Regents).

of 10–15% of its present value. Whatever this primary atmosphere was, intense solar wind from the young Sun and later energetic impacts, culminating in the Moon-forming event, would have led to significant gas loss. Evidence for this can be seen in the concentrations of the noble gases (helium, neon, argon, krypton, xenon). Abundances of these chemically unreactive gases in Earth's atmosphere are far less than their solar or cosmic equivalents (**Figure 3**). This could be due to either solar wind or impact-induced erosion of planetary atmospheres (or both).

### Formation of the Moon: Loss of the Primary Atmosphere

Evidence from a combination of lunar samples, meteoritics, interplanetary probes, isotopic studies, and physical models shows that the accretionary phase of the planets was intense and short-lived, lasting for less than 100 Ma after the initial collapse of the solar nebula. Under conditions of intense early bombardment, the conversion of the kinetic energy of millions of impacting planetesimals to thermal energy would have heated the proto-Earth to melting, further facilitating differentiation into core, mantle, crust, and atmosphere (*see* **Earth: Mantle; Crust**).

Sometime between 4.5 Ga and 4.45 Ga an object the mass of Mars entered an Earth-crossing orbit. The consequences of this event were profound. The giant impact hypothesis envisages total Earth melting, blow-out of the tattered remnants of primary atmosphere captured during accretion, and devolatilization of the material that coalesced to form the Moon (*see* **Solar System: Moon**).

### Abatement and Cooling

As planetary accretion slowed, heat rapidly radiated from the surface into space, and the Earth's surface rapidly cooled, forming a solid chill crust. Outgassing due to the internal differentiation of the planet into core and mantle resulted in the formation of a dense atmosphere, probably composed of carbon dioxide, steam, and other gases. Condensation of oceans onto the primordial surface could then have occurred as a consequence of either intrinsic water from planetary outgassing or later contributions from comets and meteorites. Comets and meteorites are primitive undifferentiated objects rich in volatiles and reduced organic compounds – attractive procreative ingredients for the origin of life. Some postulate that all the water on the Earth's surface could have been delivered from comets without the need for outgassing of water vapour from the mantle. However, if the $^2D/^1H$ composition of the three comets measured so far is typical of these objects, the oceans could not have been produced solely by cometary water (**Figure 4**). To explain the data whilst still accepting the cometary origin of the hydrosphere, terrestrial water requires an additional component with a $^2D/^1H$ much less than that of average seawater. The weight of present evidence suggests that the likeliest source of this water is intrinsic planetary outgassing. However, it could be that prebiotic reactions on the early Earth leading to the origin of life required a significant contribution from comets, and several models predict that these contributions could have been large.

Liquid water, a source of energy, and organic raw materials are presumed to be necessary to initiate biological processes. Depending on the compositions of the earliest retained (secondary) atmosphere and hydrosphere, prebiotic molecules could have been synthesized in the atmosphere, formed at hydrothermal vents, or delivered by comets. The oceans were an early feature of the Earth. The oldest known rocks of sedimentary origin are >3.8 Ga ferruginous quartzites and banded iron formations from southern West Greenland and unequivocally support the presence of surface seas at that time. Rare zircons from the Jack Hills region of Western Australia are as old as 4.38 Ga and contain oxygen isotopic compositions that are consistent with the presence of substantial liquid water at or near the Earth's surface and interacting with the crust within 150 Ma of the origin of the Moon.

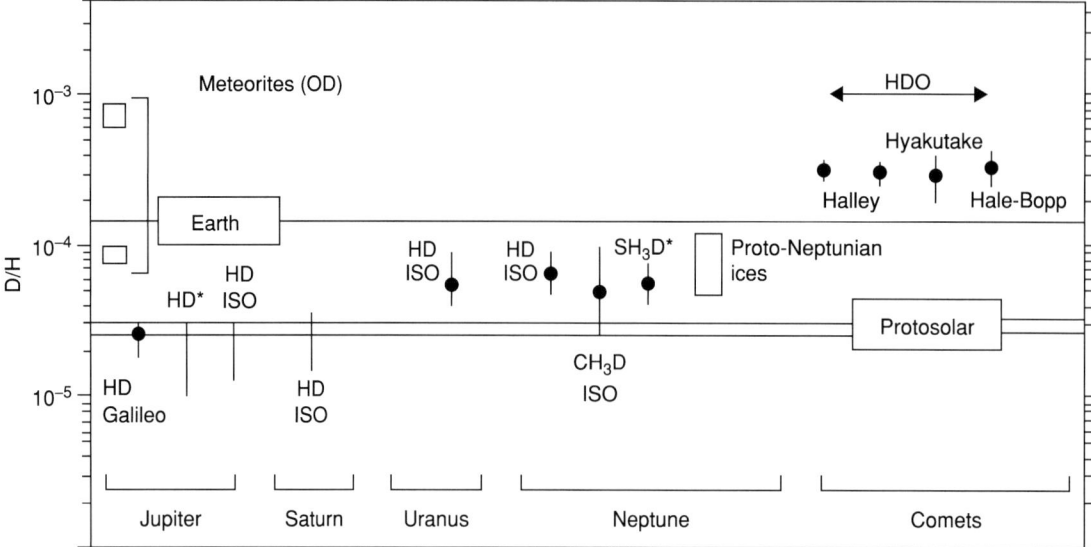

**Figure 4** Hydrogen isotopic composition of various Solar System objects (adapted from A. Drouart et al. 1999).

## Origin of the Secondary Atmosphere

Following the loss of the primary planetary atmosphere as a result of the energetic solar wind, bombardments, and the Moon-forming event, the secondary atmosphere began to accumulate. The large masses of the Earth ($5.97 \times 10^{24}$ g) and Venus ($4.87 \times 10^{24}$ g) and their planetary inventories of heat-producing elements allow the long-term persistence of high geothermal gradients and associated volcanism. In the case of the Earth, plate tectonics is possible in part because of a steep geothermal gradient that allows for the ultimate recycling of sediments and volatiles. On smaller bodies, such as the asteroids, the Moon, Mercury, and the icy moons of the outer planets, geothermal gradients are shallow; these bodies cooled quickly and early in Solar System history and are effectively cold dead worlds. Mars is an intermediate case between the warmer Earth and Venus and 'dead' worlds such as Mercury and the Moon. Comparisons of terrestrial volcanic-gas compositions with the atmospheres of the neighbouring planets show broad similarities (**Tables 1** and **2**). Water appears to have been lost on Venus due to thermal escape and was trapped in the crust of Mars by weathering and cold temperatures.

There is a growing consensus that the early (secondary) atmosphere of Earth was dominated by carbon dioxide, nitrogen, and water vapour, with minor components of noble gases, hydrogen, methane, and sulphurous compounds. To explain the early presence of liquid water when solar output was considerably less than today (**Figure 1**), it is necessary to assume that a dense carbon dioxide greenhouse atmosphere (possibly including methane) increased insolation on the early Earth. Such an atmosphere must have had a partial pressure of carbon dioxide ($pCO_2$) of at least 1 bar (and probably more) to keep the planet from freezing over (*see* **Solar System**: Asteroids, Comets and Space Dust; Meteorites; Mercury; Venus; Mars; Jupiter, Saturn and Their Moons; Neptune, Pluto and Uranus).

Weathering and hydrothermal reactions of crustal materials immediately commenced once the atmosphere and hydrosphere were established. Hydrothermal vents would have been ubiquitous on the early Earth, and it is estimated that the total volume of Earth's oceans could have cycled through the crust in less than 1 Ma. Carbon dioxide and water (as carbonic acid) react with silicate minerals to add carbonate and silica to the oceans, as represented in the Urey equation:

$$CO_2 + H_2O + CaSiO_2 = H_2CO_3 + SiO_2 + Ca^{2+}$$

**Table 2** Terrestrial volcanic-gas compositions

|  | St Helens | Etna | Kilauea |
| --- | --- | --- | --- |
| $H_2O$* | 91.58 | 53.69 | 78.7 |
| $H_2$ | 0.269 | 0.57 | 1.065 |
| $CO_2$ | 0.913 | 20.00 | 3.17 |
| CO | 0.0013 | 0.42 | 0.0584 |
| $SO_2$ | 0.073 | 24.85 | 11.5 |
| $H_2S$ | 0.137 | 0.22 | 3.21 |
| OCS | $2 \times 10^{-5}$ | nd | 0.0054 |
| $S_2$ | 0.0003 | 0.21 | 1.89 |
| SO | nd | 0.03 | nd |
| HCl | 0.089 | nd | 0.167 |
| HF | nd | nd | 0.20 |

*Values are given in volume % (=mol %); nd, not detected.

This reaction was the major control on the amount of carbon dioxide in the atmosphere before the onset of photosynthesis, respiration, and organic-matter sequestration. During subduction, carbonate minerals carried in the descending oceanic crust are heated, releasing carbon dioxide and water as volatile components in arc magmas, completing the cycle back to the atmosphere. In hydrothermal vents within basaltic crust, water oxidizes $Fe^{2+}$ in olivine, yielding hydrogen, magnetite, and serpentine. In gas-phase high-temperature reactions, hydrogen can reduce carbon dioxide to methane, but it is not likely that much hydrogen was maintained in the atmosphere, owing to its high rate of escape from the top of the atmosphere. The currently held view is that the early secondary atmosphere was, at most, weakly reducing, with the major components – carbon dioxide, nitrogen, water vapour, and carbon monoxide – supplemented by minor mixing fractions of hydrogen and methane. Estimates have been made of the rate of decline of carbon dioxide over geological time that provided the major greenhouse forcing to the atmosphere, while tracking secular changes in solar luminosity. Initially high partial pressures of carbon dioxide, possibly supplemented by biogenic methane, would have kept surface temperatures warm enough to avoid freezing of the planet until the crisis conditions of the snowball Earth in the Late Proterozoic. Since the era of severe Proterozoic glaciations, there has been a generally gentle decline in levels of carbon dioxide, maintaining the stability of the biosphere in the face of rising solar luminosity (of the order of about 6% per billion years).

## Atmospheric Evolution

### Early Anoxic Atmospheres

Free oxygen was rare or absent on the early Earth, so that the atmosphere was effectively anoxic. Constraints on ancient levels of atmospheric oxygen have been sought for many years, and most models of the early Earth consider oxygen levels to have been very low (less than $10^{-2}$ of present atmospheric level (PAL)) before about 2.3 Ga. Evidence to support this comes from detrital gold, uraninite, and pyrite grains in sediments older than 2.3 Ga, which it has been argued could not have been preserved in a high-oxygen atmosphere. Other geochemical indications, such as $Fe^{2+}/Fe^{3+}$ in palaeosols and discussions of the possible origins of banded iron formations (*see* **Sedimentary Rocks: Banded Iron Formations**) and of carbon isotopic signatures in Archaean and Palaeoproterozoic organic matter and carbonates, have also been used to support the contention that the early atmosphere was essentially devoid of oxygen. In the absence of a ready source of free oxygen and, therefore, no effective ozone ($O_3$) screen from solar ultraviolet, atmospheric water vapour would have photodissociated into hydrogen and oxygen. Since the oxygen formed by this reaction produces just enough ozone to inhibit the reaction, accumulation of oxygen by this means is limited. Oxygen combines readily with hydrogen to form water vapour, so little oxygen is lost from the atmosphere to space. The bottom of the stratosphere acts as a cold trap and maintains a constant temperature of $-56°C$, keeping water from escaping from the top of the atmosphere (**Figure 5**).

**Sulphur isotopes: a unique window onto early atmospheric chemistry** Further evidence in support of an early anoxic atmosphere comes from the unusual chemistry of sulphur isotopes. Most physical processes fractionate the isotopes of an element because of the relative mass differences between the isotopic species. For instance, because the mass difference between $^{33}S$ and $^{32}S$ is half that between $^{34}S$ and $^{32}S$, any mass-dependent (defined by $\Delta^{33}S = 0‰$; where $\Delta^{33}S = 0.515\delta^{34}S - \delta^{33}S$) physical effects on the ratio of $^{33}S$ to $^{32}S$ are expected to be about half those on the ratio of $^{34}S$ to $^{32}S$. However, certain unusual gas-phase chemical reactions in the absence of free oxygen are capable of producing anomalous sulphur isotopic values that do not follow the mass-dependent relationship described above. The recent discovery of the signal of the mass-independent fractionation effect (where $\Delta^{33}S \neq 0‰$) in geological materials has important implications for understanding the chemical evolution of

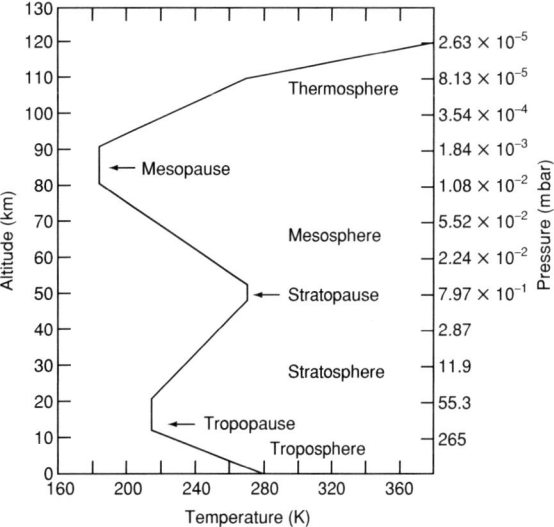

**Figure 5** Temperature and pressure profile of the Earth's atmosphere.

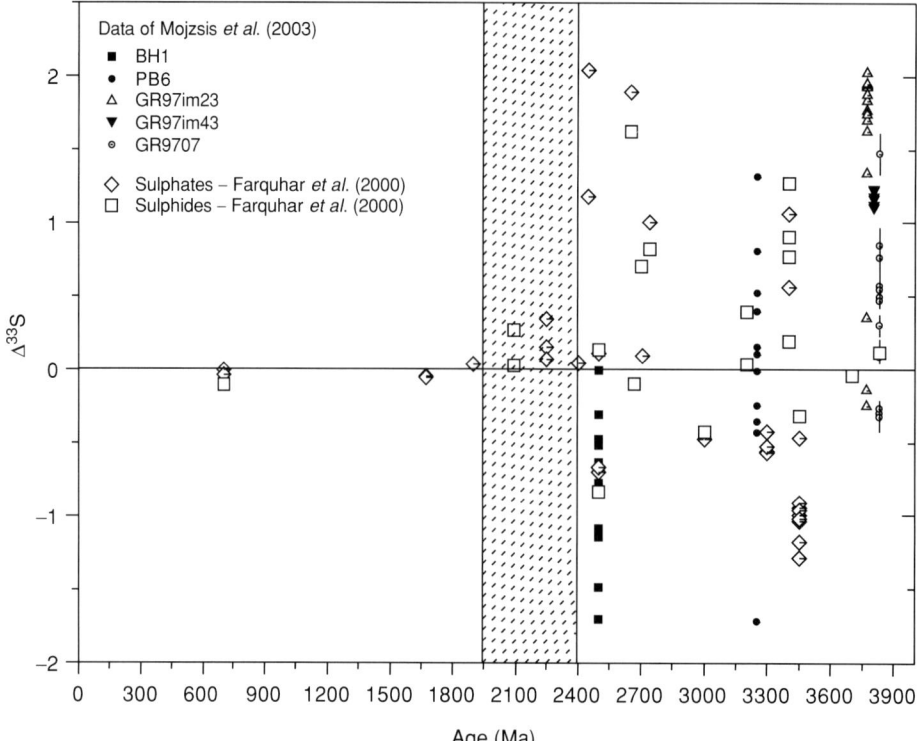

**Figure 6** Mass-independent sulphur isotopic compositions track changes in atmospheric composition with time. (Reproduced from Mojzsis et al. (2003).)

the atmosphere. It has been found that sedimentary sulphide and sulphate minerals in Archaean rocks contain sulphur where the levels of the minor isotope ($^{33}S$) deviate from the mass-fractionation line with other post-2.3 Ga terrestrial sulphur minerals. It appears that the signal of this chemistry can be transferred from the atmosphere to surface materials and thereafter to the geological record. The mass-independent fractionation effect of sulphur in ancient sulphides records chemical reactions in early anoxic terrestrial atmospheres, which were destroyed in the Early Proterozoic (2.4–2.3 Ga), and tracks the rise in levels of atmospheric oxygen (**Figure 6**).

## Metabolic Energy and the Rise of Oxygen

Changes in atmospheric composition have had a profound effect on the evolution of life. Sequence analyses used to compare genomes have been used to infer the phylogenetic relatedness of living organisms and their evolutionary history in terms of molecular biology and metabolic style. It has been found that three great domains exist within the ribosomal RNA tree of life: Bacteria, Archaea, and Eukarya (**Figure 7**) (*see* **Precambrian: Prokaryote Fossils**). The organisms with the deepest branches in the tree, corresponding to the most ancient pedigrees, close to the root in **Figure 7**, are hyperthermophilic (heat-loving) chemoautotrophs (organisms that obtain their metabolic energy from chemical disequilibria). It appears that at least some of the earliest organisms lived off chemical energy, in much the same way as contemporary hot-spring microbial communities. Such environments are widespread in volcanic centres on land and on the seafloor, and they would have been even more widespread on a geologically restive early Earth. However, hyperthermophilic organisms are restricted to zones around hydrothermal vents where optimal chemical and temperature conditions persist. Photosynthesizers, on the other hand, are less restricted and have evolved to inhabit environments away from those of their chemosynthetic ancestors into the open ocean.

Light from the Sun provides a readily accessible, stable, and inexhaustible energy source to drive metabolic reactions. Photosynthesis uses light energy to capture electrons and move low-energy-state ions to higher energy states. The energy released can then be used to power metabolic reactions such as growth and reproduction. Photosystem I was the earliest photosynthetic pathway and used light-activated chlorophylls and electron donors such as $H_2S$ and $Fe^{2+}$ to yield $S^0$ and $Fe^{3+}$ as oxidative products. At a later stage, photosystem II developed; this uses water as the electron

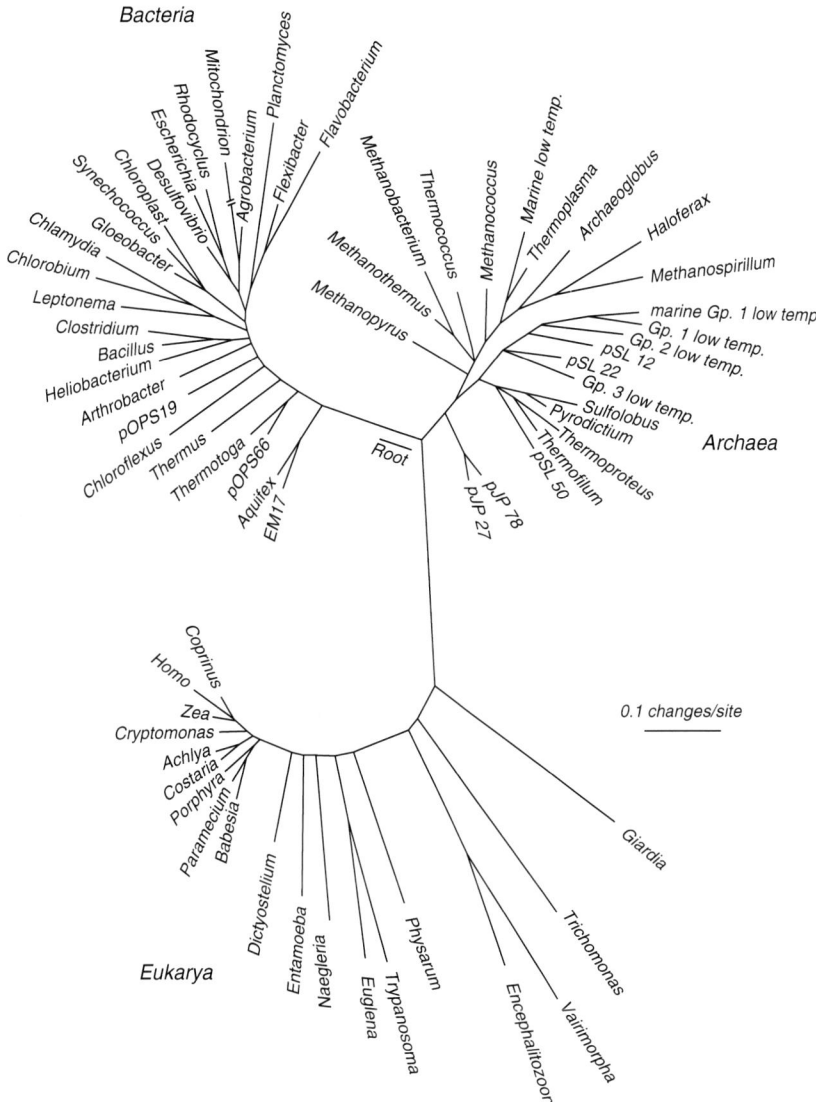

**Figure 7** The phylogenetic tree of life based on comparative 16s rRNA sequences. (Reproduced from Pace NR (2001) The universal nature of biochemistry. *Proceedings of the National Academy of Sciences USA* 98: 805–808.)

donor to reduce carbon dioxide to biologically useful organic compounds such as carbohydrates. Liquid water and carbon dioxide are available to life in practically unlimited quantities; sunlight drives the reaction, and cyanobacteria that use it, although relative late-comers in phylogenetic terms, are arguably the most successful organisms in the history of life on Earth. It is photosystem II that releases free oxygen as a by-product in the reversible reaction:

$$6CO_2 + 6H_2O + \text{light energy} \leftrightarrow C_6H_{12}O_6 + 6O_2$$

The gradual build-up of oxygen in the atmosphere was accomplished by the slow toilsome efforts of cyanobacteria over many hundreds of millions of years. The immense importance of this metabolic reaction to the history of all life beyond the microbial level cannot be overstated (*see* **Earth System Science**).

**Rise in atmospheric oxygen** Initially, the vast amounts of reduced iron ($Fe^{2+}$) and other chemical species dissolved in the oceans, and perhaps some reduced gases in the atmosphere (such as methane), provided a sink for the oxygen produced by photosynthesis. As these sinks were exhausted, local concentrations of oxygen would have appeared, creating a dilemma for the early anaerobic microbial biosphere. High levels of oxygen produce toxic radicals in the environment, which cause extensive damage to cell components. This would have been a strong driving force for natural selection leading to adaptation: microbes that did not immediately go extinct survived by retreating, for example, to anaerobic environments, such as deep in crustal rocks, organic-rich muds, and animal digestive tracts, where they continue to thrive today. New forms of life emerged with superoxide dysmutase and catalase enzymes, which catalyze the reduction of oxic free radicals to water. Finally, another group (the Eukarya; **Figure 7**) developed aerobic metabolism, providing energy yields approximately twenty times greater than those of anaerobic metabolism (*see* **Precambrian:** Eukaryote Fossils). The mitochondria – cell organelles contained in most Eukarya, which facilitate oxidative metabolism – arose from the endosymbiosis of proteobacteria. These organelles actually contain their own subunits of DNA that further implicate a eubacterial heritage.

The increased energy yield of aerobic metabolism set the stage for the evolution of all life above the unicellular level. The energy source that supports the aerobic biosphere is the Sun – thus established oxygenic photosynthesis and aerobic respiration governed the flow of carbon and oxygen through the atmosphere–hydrosphere system. During the Archaean the atmosphere and hydrosphere initially contained small amounts of free oxygen from photolysis of water vapour and then increasing amounts from oxygenic photosynthesis. By the end of the Archaean–Early Proterozoic, oxygen levels had begun to creep upwards, balanced by oxygen-consuming reactions such as weathering, hydrothermal activity, respiration, oxidation of organic matter, and differential rates of organic-matter burial. By about 2 Ga, levels of oxygen were about 1% of PAL. A transition era ensued, with oxygen levels fluctuating around the 1% PAL value and the oceans remaining reducing and sulphidic. After about 1.8 Ga, oxygen-consuming reactions were generally exhausted and free-oxygen concentrations reached about 10% of PAL. Over time, these levels stabilized at near 'normal' Phanerozoic atmospheric concentrations. It remains unquantified how rapidly oxygen levels increased during the Proterozoic.

The rise in atmospheric oxygen had an acute effect on the surface environment, not only because of its toxicity to many microbial organisms (providing the motive force to drive the evolution of more efficient aerobic metabolisms) but also by establishing an effective ultraviolet screen. Ozone is far more effective than diatomic oxygen at absorbing ultraviolet. The ozone screen formed in the stratosphere from accumulating $O_2$ that photodissociated to produce free oxygen radicals ($O°$), which then recombined with $O_2$ to make ozone. This ozone screen effectively made dry land habitable for plants and animals by the Palaeozoic. Industrial pollutants such as chlorofluorocarbons are now severely damaging the ozone layer.

## A Neoproterozoic Snowball Earth

The increased oxidation of the surface zone in the Middle Proterozoic was probably a consequence of the sequestration of large quantities of organic carbon in sediments. From the oxygenic photosynthesis reaction, it can be seen how this scenario leads to a net loss of carbon dioxide and a net increase of oxygen in the atmospheric reservoir. The boosted greenhouse effect of heightened levels of carbon dioxide before about 2.5 Ga was lost to the Earth at a most inopportune time. Levels of solar luminosity were about 10%–15% lower at 2.2 Ga than at present (**Figure 1**). Loss of insolation from the carbon dioxide greenhouse spelled disaster for the Proterozoic Earth. The planet froze over, locking the oceans in ice and creating a high-albedo feedback loop that kept the planet frozen for extended periods, until levels of carbon dioxide increased due to passive volcanism and outgassing. Heightened levels of carbon dioxide warmed the atmosphere and dark dusty ice reduced the albedo, causing catastrophic melting and massive planetary warming. Subsequent weathering in a carbon dioxide-rich atmosphere, combined with massive algal blooms, led to enhanced burial of organic carbon, drawdown of carbon dioxide levels, and a renewed snowball Earth. The cycle is thought to have been broken by the secular increase in solar luminosity and a steady redistribution of the continents via plate motions. The survivors of these repeated 'icehouse' and 'hothouse' Earths were multicellular organisms that inherited a more stable environment high in oxygen (*see* **Palaeoclimates**).

## The Phanerozoic Atmosphere

### Phanerozoic Atmospheric Changes

Geochemical evidence from the study of palaeosols, coupled with data from carbonate and organic carbon in sediments as well as sedimentary pyrite and the chemistry of sedimentary silicate minerals, has been used to improve models of the carbon cycle of palaeoenvironments. These models have been used to document and explain fluctuations in levels of oxygen and carbon dioxide over Phanerozoic time. Long-term changes in the carbon dioxide and oxygen concentrations in the Phanerozoic atmosphere are summarized in **Figure 8**.

In the first part of the Phanerozoic, the Early Palaeozoic (Cambrian–Ordovician), evidence indicates that levels of carbon dioxide were about fifteen times PAL. These declined to within a few percent of

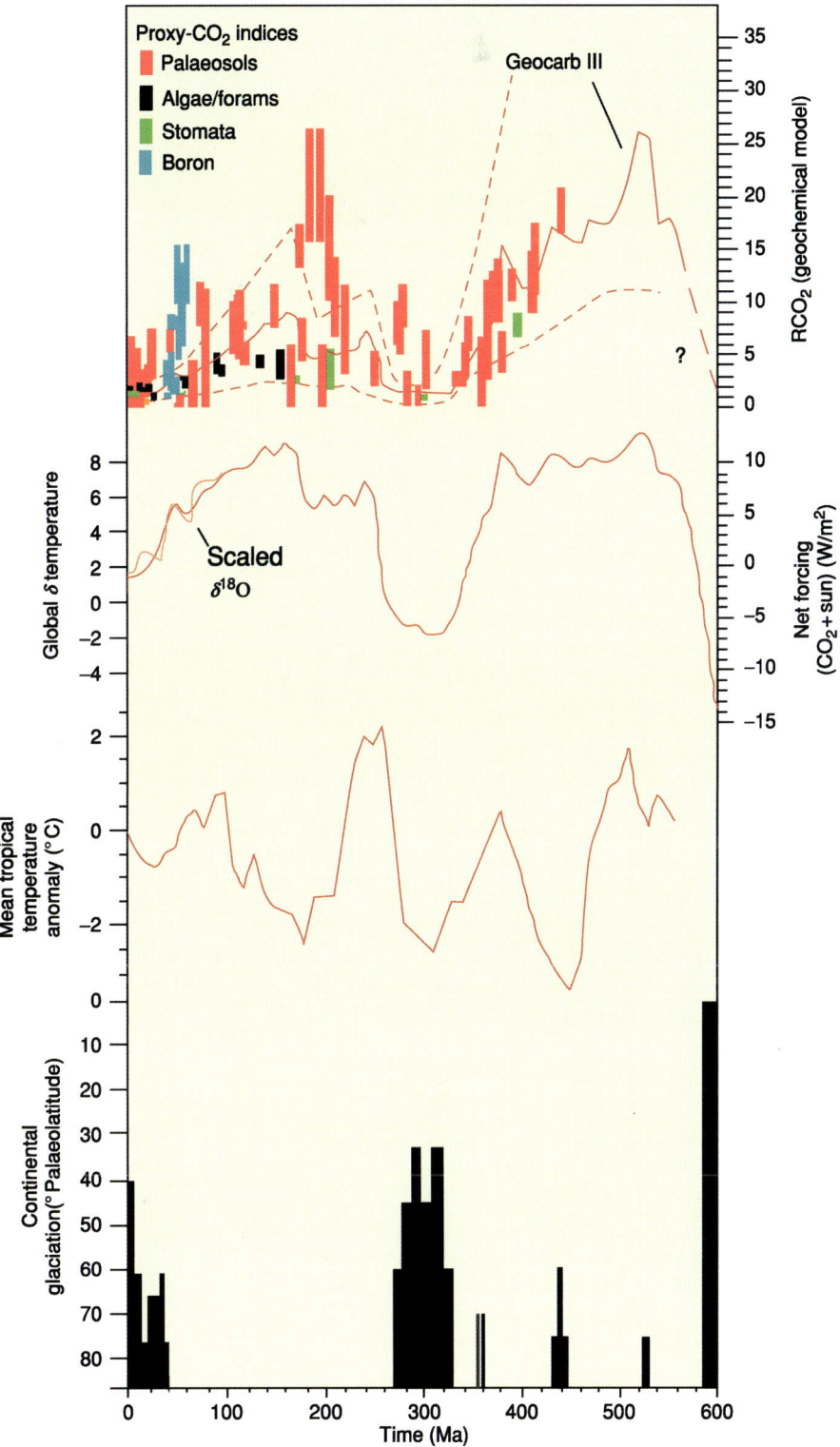

**Figure 8** Phanerozoic temperature history. (Reproduced from with permission from Berner RA. Geocarb II: A revised model of atmospheric CO$_2$ over phanerozoic time, *Am. Jour. Sci.*, 294: 56–91)

PAL at the end of the Carboniferous. Massive carbon burial and increased solar radiation by the Carboniferous meant that atmospheric oxygen concentrations were greater than at any other time in Earth history (*see* **Palaeozoic:** Carboniferous). This may have allowed the existence of the huge Carboniferous insects observed in the fossil record (*see* **Fossil Invertebrates:** Insects). Insects rely on diffusion-limited flow of oxygen through their exoskeletons: to be big they need to live in conditions of high atmospheric oxygen, so that sufficient oxygen passively diffuses into their blood to power muscles for flight.

Reduced greenhouse forcing and glaciation at the end of the Palaeozoic with the assembly of Pangaea reduced organic-matter burial, and there was a slow rise in carbon dioxide levels to around six times PAL in the Permian and Triassic. Levels of carbon dioxide gradually decreased, and stabilized near present-day levels in the Mesozoic. Oxygen tends to track carbon dioxide inversely; geological evidence and palaeoclimate models suggest a maximum of near 35% oxygen in the atmosphere at the beginning of the Permian. During the Permian, the oceans were highly stratified, with carbonate-rich water at depth that was depleted in oxygen. This system was unstable: ocean hypoxia could occur if ocean circulation intensified enough to mix deep-water carbon dioxide and hydrogen sulphide into surface waters. A protracted (20 Ma) whole-ocean hypoxia event is considered to be a major mechanism in the Permian–Triassic extinction event, which wiped out 90%–95% of all marine species (*see* **Palaeozoic: End Permian Extinctions**).

### Carbon Dioxide and Climate Changes

High-resolution information about changes in atmospheric chemistry over the past 160 000 years can be obtained by studying the record of trapped gases in ice cores from the Greenland and Antarctic ice-caps. Furthermore, oxygen isotopic values from marine sediments, marine planktonic and benthonic fossils, cave deposits, and other sources can be used to estimate marine palaeotemperatures. Direct measurements of carbon dioxide and methane concentrations in ice cores permit assessment of past atmospheric levels of these gases, providing factors to incorporate into models of past air temperatures and sea-levels.

Data from deep ice cores taken in polar regions, coupled with complex palaeoclimate models (**Figure 9**), show large fluctuations in atmospheric carbon dioxide, oxygen and methane levels, leading to long-term temperature changes of the order of ±6°C or more. There is a strong correlation between levels of atmospheric greenhouse gases and palaeotemperature. The periodicities in these data provide clear evidence of the role of Milankovitch forcing by changes in the Earth's orbital parameters. The two strongest Milankovitch cycles observed correspond to the 26 000 year precession of the equinoxes and the 100 000 year period of rotation of the Earth's orbital axis (*see* **Earth: Orbital Variation (Including Milankovitch Cycles)**). Changes in greenhouse-gas concentrations appear to follow rather than guide long-term climate, suggesting that Milankovitch

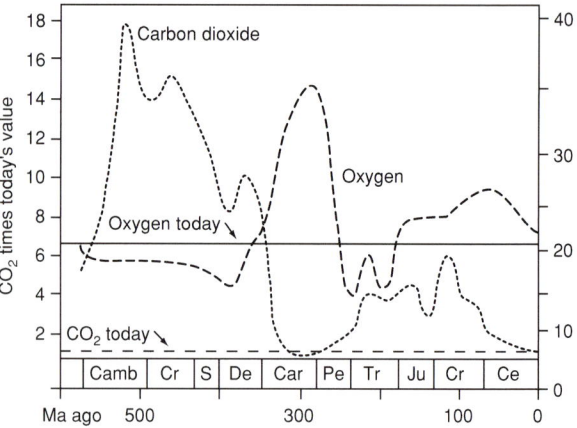

**Figure 9** Phanerozoic carbon dioxide and oxygen concentrations.

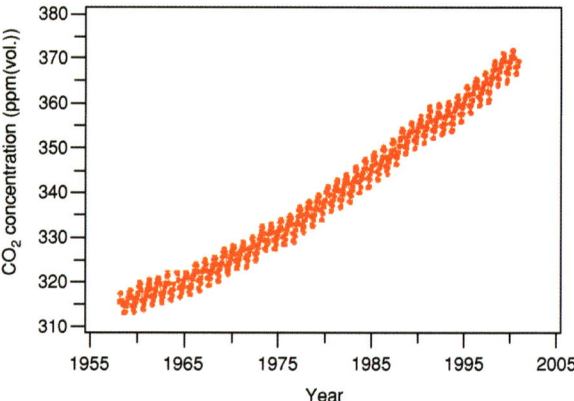

**Figure 10** Changes in the concentration of atmospheric carbon dioxide over the last 60 years as measured at Mauna Loa, Hawaii. Data provided by D. Keeling and T. Whorf.

cycles are the prime mechanism bringing the Earth into a greenhouse or icehouse condition. Greenhouse gases do provide positive feedback at the beginning of temperature changes by boosting insolation. Anthropogenic emissions of greenhouse gases are not governed by Milankovitch cycles and represent a separate and increasingly important climate-forcing mechanism. **Figure 9** shows that carbon dioxide concentrations changed by almost 100 ppm(vol.) towards the end of the last glaciation. Modern levels of carbon dioxide are near 370 ppm(vol.) and rising. Almost all of this change has occurred since the Industrial Revolution, and high-resolution monitoring at the Mauna Loa observatory shows clear diurnal and annual cycles in carbon dioxide levels (**Figure 10**), with a mean annual increase of 1.16 ppm(vol.) year$^{-1}$. This is over one hundred times the rate of increase of carbon dioxide levels inferred from all available geological data, with no clear stabilization of these levels in sight.

**Other greenhouse gases** Neglecting the small contribution from internal heating driven by radioactive decay, planetary surface temperatures are governed by the amount of solar radiation received and its interaction with the atmosphere. Re-radiation to the surface of infrared radiation by greenhouse gases, chiefly carbon dioxide but also water vapour, methane, nitrous oxide species ($NO_x$), and chlorofluorocarbons (CFCs), keeps the present average surface temperature some 33 K above the black-body temperature of 255 K. Although they are present in small amounts relative to carbon dioxide and water vapour, methane, nitrous oxide species, and CFCs (which can form only from industrial processes) are very effective greenhouse gases. Methane in the atmosphere (concentration of 1.7 ppm(vol.)) is dominantly formed by biological processes and absorbs infrared radiation approximately 21 times more effectively than carbon dioxide; its levels have increased rapidly in the last several centuries. Gaseous nitrous oxide compounds are present at low concentrations (*ca.* 0.3 ppm(vol.)), are produced by biological nitrification, and have a long residence time in the atmosphere. The concentration of CFCs is very low (less than 0.003 ppm(vol.)), but they have a greenhouse effect ten thousand times greater than that of carbon dioxide and likewise have long residence times in the atmosphere.

## Conclusions

The splendour of contemporary life on Earth, as revealed by the geochemical, palaeontological, and molecular phylogenetic records, took billions of years to achieve, and its evolution occurred for the most part within the envelope of the atmosphere and hydrosphere. In the context of astrophysical changes to the Sun and geophysical changes to the solid Earth, the atmosphere has evolved through a complex set of interrelated cycles, within which biology has been of central importance. Life has affected the planetary atmosphere, and biological evolution was in turn affected by it. The gaseous envelope of the planet evolved in response to changing geophysical conditions, such as mantle heat flow and solar luminosity. Early in Solar System history, Venus and Mars apparently had geochemical paths that paralleled that of the Earth, probably including liquid water at their surfaces. However, long ago they diverged from a physicochemical regime that promoted habitability. The nascent biosphere on these planets, if it ever existed, was either burned to a crisp (on Venus) or freeze-dried (on Mars). In the far future, as the luminosity of the Sun continues to increase, Earth will go the way of Venus.

## See Also

**Earth:** Mantle; Crust; Orbital Variation (Including Milankovitch Cycles). **Earth Structure and Origins**. **Earth System Science**. **Fossil Invertebrates:** Insects. **Palaeoclimates**. **Palaeozoic:** Carboniferous; End Permian Extinctions. **Precambrian:** Overview; Eukaryote Fossils; Prokaryote Fossils. **Sedimentary Rocks:** Banded Iron Formations. **Solar System:** The Sun; Asteroids, Comets and Space Dust; Meteorites; Mercury; Venus; Moon; Mars; Jupiter, Saturn and Their Moons; Neptune, Pluto and Uranus.

## Further Reading

Hoffman PF and Schrag DP (2000) Snowball Earth. *Scientific American* 282: 68–75.

Holland HD (1984) *The Chemical Evolution of the Atmosphere and Oceans*. Princeton: Princeton University Press.

Margulis L (1984) *Early Life*. Boston: Jones and Bartlett.

Mojzsis SJ and Harrison TM (2000) Vestiges of a beginning: clues to the emergent biosphere recorded in the oldest known sedimentary rocks. *GSA Today* 10: 1–6.

Pace NR (2001) The universal nature of biochemistry. *Proceedings of the National Academy of Sciences USA* 98: 805–808.

Royer DL, Berner RA, Montañez IP, Tabor NJ, and Beerling DJ (2004) $CO_2$ as a primary driver of Phanerozoic climate. *GSA Today* 14: 4–10.

Sagan C and Mullen G (1972) Earth and Mars: evolution of atmospheres and surface temperatures. *Science* 177: 52–56.

# AUSTRALIA

Contents

**Proterozoic**
**Phanerozoic**
**Tasman Orogenic Belt**

## Proterozoic

**I M Tyler**, Geological Survey of Western Australia, East Perth, WA, Australia

© 2005, Elsevier Ltd. All Rights Reserved.

## Introduction

The Proterozoic is the period of geological time extending from the end of the Archaean, 2500 million years ago (Ma), to the start of the Phanerozoic (the base of the Cambrian System), 545 Ma. The Proterozoic is divided into the Palaeoproterozoic (2500–1600 Ma), the Mesoproterozoic (1600–1000 Ma), and the Neoproterozoic (1000–545 Ma). In Australia, Proterozoic rocks are present to the west of the 'Tasman Line' that separates 'Proterozoic Australia', where geophysical datasets show that Precambrian basement is continuous beneath Phanerozoic sedimentary basins, from the Tasmanides, where predominantly Palaeozoic basement is overlain by Mesozoic and younger sedimentary basins. Extensive exposures of Proterozoic rocks are present in western Australia, in northern, central, and north-eastern Australia, and in southern Australia and Tasmania. Proterozoic Australia is made up of three distinct cratons, the West Australian Craton, the North Australian Craton, and the South Australian Craton (**Figure 1**). These probably formed originally as parts of larger cratons or continental blocks (the South Australian Craton together with the previously adjacent part of Antarctica formed the Mawson Craton) and are dominated by Archaean and Palaeoproterozoic to Mesoproterozoic crust. The three cratons are separated by two predominantly Mesoproterozoic to Neoproterozoic orogenic belts, the Paterson Orogen and the Albany–Fraser Orogen. The Palaeoproterozoic to Neoproterozoic Pinjarra Orogen is present along the western margin of Australia.

Plate tectonic models can be applied to Proterozoic Australia. The increasing availability of high-quality geochronological data has highlighted the presence of distinct tectonostratigraphic terranes with differing geological histories within orogenic belts, and geophysical datasets reveal the heterogeneous nature of the crust throughout Proterozoic Australia. However, Proterozoic plate-tectonic processes may differ from modern processes, and the real lack of accretionary complexes and ophiolites, and significant differences in the geochemical and isotopic compositions of igneous rocks, may reflect changes in the nature and composition of the oceanic lithosphere through time. Palaeomagnetic evidence is placing greater controls on the movement and relative positions of the constituent crustal components. Diverse cratons and continental blocks aggregated during the Palaeoproterozoic and Early Mesoproterozoic to form Proterozoic Australia, which then played an integral part in the formation and breakup of the Meso- to Neoproterozoic supercontinent, Rodinia (**Figures 2–7**; summarized in **Table 1**). Proterozoic Australia is host to a wide variety of minerals, including world-class deposits of iron, uranium, gold, copper–gold, lead–zinc–silver, and diamond orebodies (**Figure 8**).

## Neoarchaean to Palaeoproterozoic Assembling Proterozoic Australia: (2770–1600 Ma)

### West Australian Craton

Within the West Australian Craton, large areas of Archaean rocks are exposed in the geologically distinct Pilbara and Yilgarn cratons (**Figures 1** and **2**). The Hamersley Basin was initiated on the southern part of the Pilbara Craton at the start of the Neoarchaean. West- to south-westerly directed rifting within cratonized granite–greenstone basement represents a distinct change to a Proterozoic and Phanerozoic tectonic style. The flood basalts (2770–2690 Ma) of the Fortescue Group dominate the rift-related lower part of the basin. These were buried beneath a breakup unconformity overlain by a passive margin sequence characterized by cherts and banded iron formations

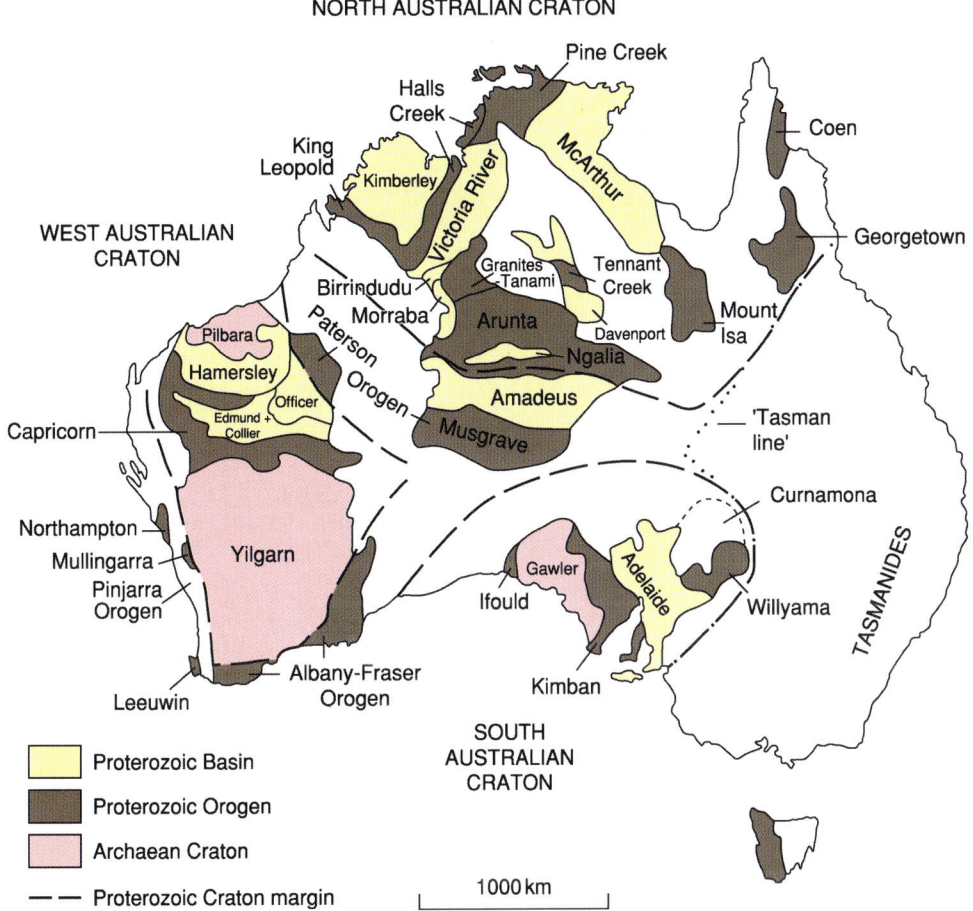

**Figure 1** North, South, and West Australian cratons, showing the main outcrops of Precambrian rocks in Australia.

(BIFs) of the uppermost Fortescue Group and the Hamersley Group (2600–2440 Ma), which were deposited on a shelf or platform open to an ocean (*see* **Sedimentary Rocks: Banded Iron Formations**). A collisional, intracontinental back-arc setting has been proposed for the BIFs and mafic and felsic magmatic rocks of the upper part (2470–2440 Ma) of the Hamersley Group. The overlying Turee Creek Group, which includes probable glacial deposits, and the lower Wyloo Group were deposited in the McGrath Trough, a foreland basin developed in front of a northward-verging fold-and-thrust belt during the Ophthalmian Orogeny (~2200 Ma) (**Figure 2**). The Glenburgh Terrane in the southern part of the Gascoyne Complex within the Capricorn Orogen to the south contains basement from 2550–2450 Ma and may represent a remnant of the colliding continent that drove this orogeny.

In the Yilgarn Craton, tectonic processes typical of the Archaean developed the Eastern Goldfields granite–greenstone terrane 2700–2600 Ma, at the same time as the upper Fortescue Group on the Pilbara Craton. In the southern Capricorn Orogen, rifting and continental breakup took place in the gneiss terranes and granite–greenstones along the northern margin of the Yilgarn Craton. The ~2150-My-old rocks of the basal Yerrida Basin (**Figure 2**) formed initially in a sag basin followed by an abrupt change to a rift setting. The development of the Bryah and Padbury basins 2020–1900 Ma at the north-western margin of the Yilgarn Craton may reflect the development of a back-arc basin, which was then overlain by a foreland basin during the Glenburgh Orogeny 2005–1960 Ma (**Figure 2**). Geochemical and isotopic data indicate that suprasubduction zone magmatism formed the Dalgaringa Supersuite at an Andean-type margin during this event; this may represent the coming together of the combined Pilbara Craton and Glenburgh Terrane with the Yilgarn Craton.

The Capricorn Orogeny from 1830–1780 Ma (**Figure 3**) has previously been regarded as marking the collision between the Pilbara and Yilgarn Cratons. Neodymium isotopes show that melting of pre-existing, Early Palaeoproterozoic crust without the introduction of mantle-derived material formed

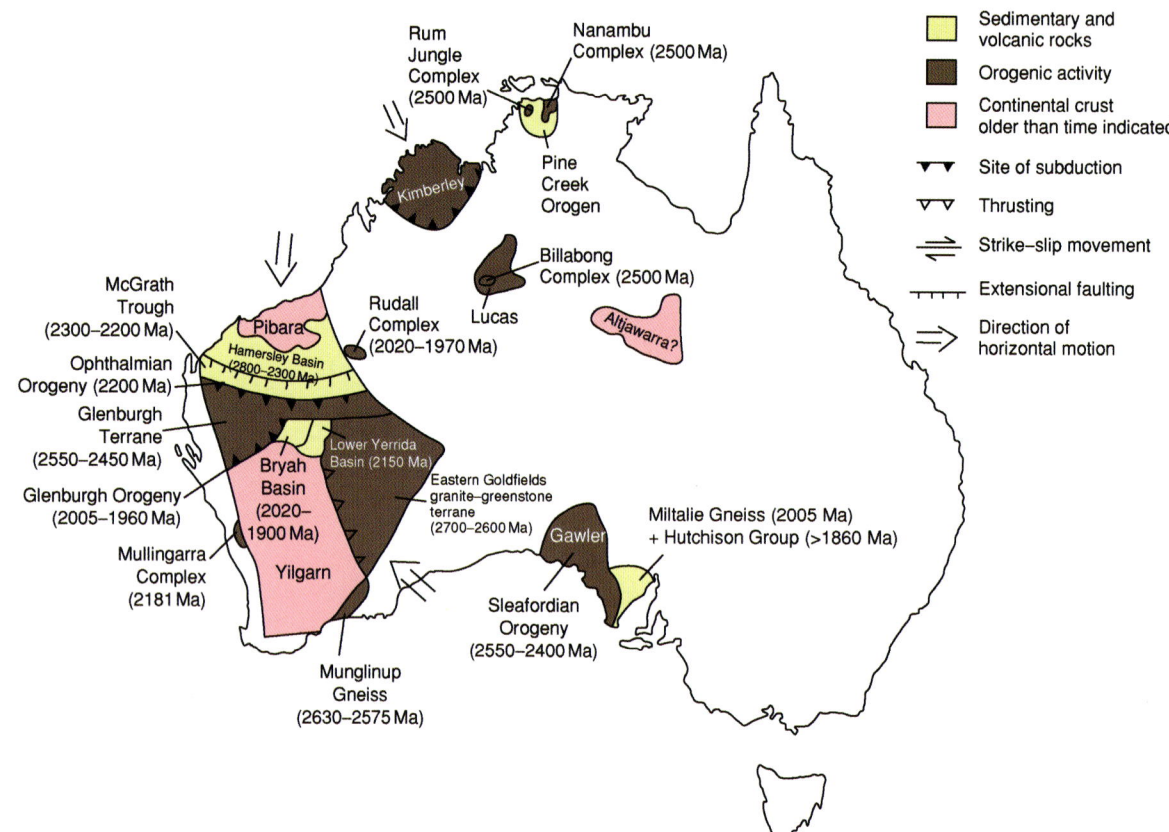

**Figure 2** Assembly of the West Australian Craton, 2800–1900 Ma.

voluminous granitic magmatism, which, together with associated deformational and metamorphic events, extended across the entire orogen. The Capricorn Orogeny may represent an intracratonic event resulting from reactivation, possibly during amalgamation of the West Australian Craton with the North Australian Craton. The 1840- to 1805-My-old upper part of the Ashburton Basin and the ∼1805-My-old Blair Basin developed as a foreland basin along the northern margin of the orogen at this time. Uplift of the Gascoyne Complex and the Yilgarn Craton, together with a presently unexposed Early Palaeoproterozoic terrane, provided the sediment deposited in the upper Ashburton Basin and the Blair Basin, the ∼1840-My-old upper part of the Yerrida Basin and the 1840- to 1800-My-old Earaheedy Basin (**Figure 3**).

In the Rudall Complex along the eastern margin of the Pilbara Craton (**Figure 3**), the Talbot Terrane contains quartzite, amphibolite, and serpentinite as inclusions within complex orthogneisses that contain components that crystallized, respectively, ∼2015, ∼1970, and ∼1800 Ma. A younger sequence (from <1790 Ma) of clastic metasedimentary rocks was possibly deposited in a foreland basin. In the adjacent Connaughton Terrane, a sequence of deformed and metamorphosed mafic volcanic rocks, and chemical and clastic sedimentary rocks, may have been deposited in a rift prior to ∼1780 Ma. Deformation, metamorphism, and further granite intrusion took place in the Talbot and Connaughton terranes during the Yapungku Orogeny (1790–1760 Ma). West-verging thrusting and high-P metamorphism may have accompanied collision of the West Australian Craton with the North Australian Craton, which palaeomagnetic evidence indicates were unified by ∼1700 Ma.

The Mount Barren Group and equivalent sedimentary rocks were deposited on the south-eastern part of the Yilgarn Craton ∼1700 Ma (**Figure 4**). Reactivation of the Capricorn Orogen took place between 1670 and 1620 Ma with the intrusion of granitic rocks and the occurrence of medium- to high-grade metamorphism, which was synchronous with localized shear zones and hydrothermal alteration. In the Albany–Fraser Orogen, the Biranup Complex consists of granitic gneisses that include both Archaean (2630–2575 Ma) (**Figure 2**) and Palaeoproterozoic (1700–1600 Ma) protoliths (**Figure 3**). In the Pinjarra Orogen, a monzogranite in fault contact with

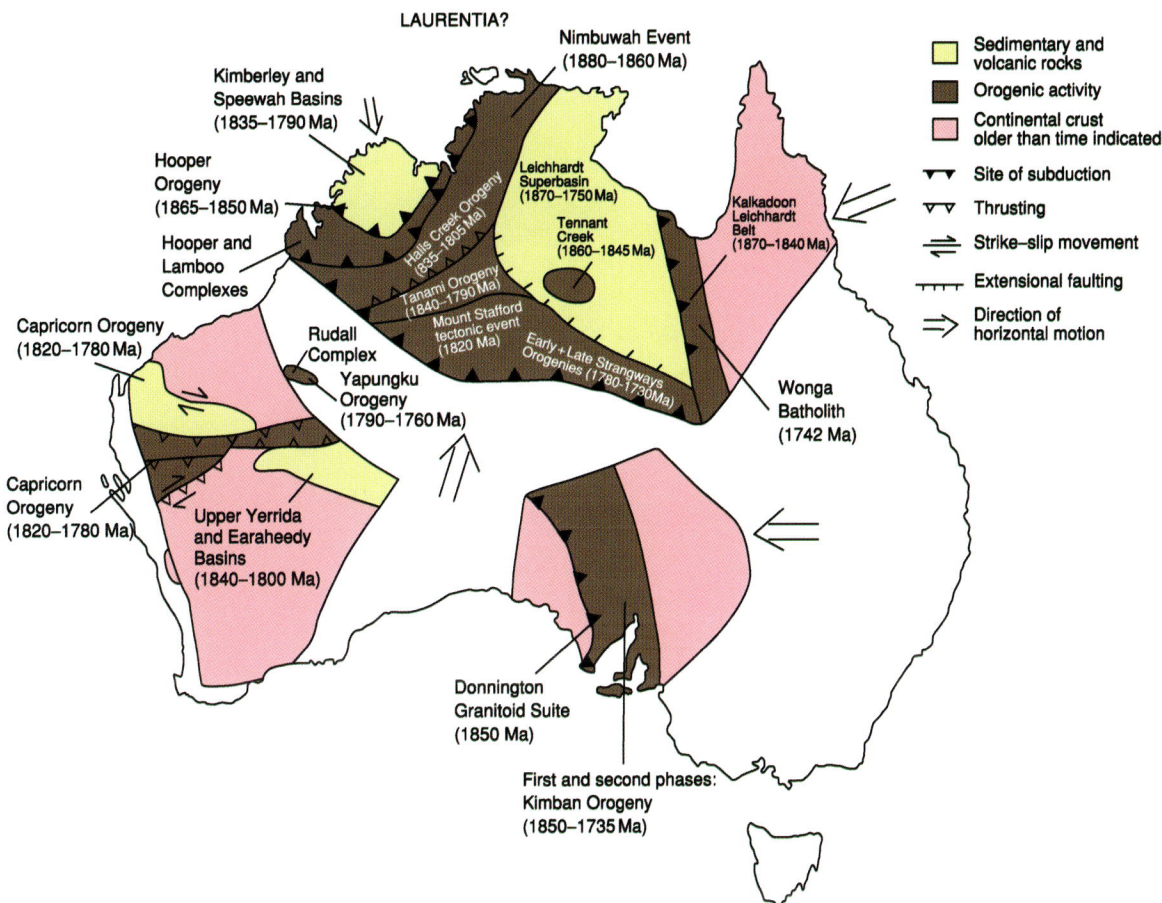

**Figure 3** Assembly of the North Australian and South Australian cratons, 1900–1730 Ma.

metasedimentary gneiss in the Mullingarra Complex has been dated to ~2181 Ma (**Figure 2**). This is consistent with Sm–Nd model ages of 2280–2030 My from the Northampton and Mullingarra complexes, and boreholes in the Perth Basin, indicating the involvement of Early Palaeoproterozoic crust in the evolution of the orogen.

**North Australian Craton**

In the North Australian Craton, the presence of probable Archaean to Early Palaeoproterozoic cratons and continental blocks within the basement is inferred from geophysical datasets. Latest Neoarchaean basement is exposed as granite and granite gneiss within the Rum Jungle and Nanambu complexes in the Pine Creek Orogen (**Figure 2**). These basement inliers are overlain by arkosic rocks of the Namoona Group and the shallow-marine carbonates and mudstones and siltstones of the Mount Partridge Group deposited before ~1885 Ma. Archaean crust of the Lucas Craton is also present as inliers partly covered by arkosic and conglomeratic rocks within the Browns Range Dome and the Billabong Complex in the Granites–Tanami Complex (**Figure 2**). The presence of further concealed Archaean crust that may be as old as ~3600 My is indicated by Sm–Nd and Pb isotopic data, together with a U–Pb detrital zircon studies.

A widespread orogenic event took place throughout the North Australian Craton between 1910 and 1790 Ma. This event, originally defined with a shorter duration (between 1870 and 1840 Ma), has been termed the 'Barramundi Orogeny'. It was interpreted as an intracratonic event involving a linked polygonal system of rifts and sag basins, which evolved rapidly to became the sites of crustal shortening, voluminous magmatism, and high-temperature/low-pressure metamorphism. More recent interpretations have suggested that the 'Barramundi Orogeny' represents a series of linked collisional events that assembled the North Australian Craton. Based on similarities that can be recognized with the evolution of the Trans-Hudson Orogen, the North Australian Craton may have been originally contiguous with Laurentia, which would have lain to the north.

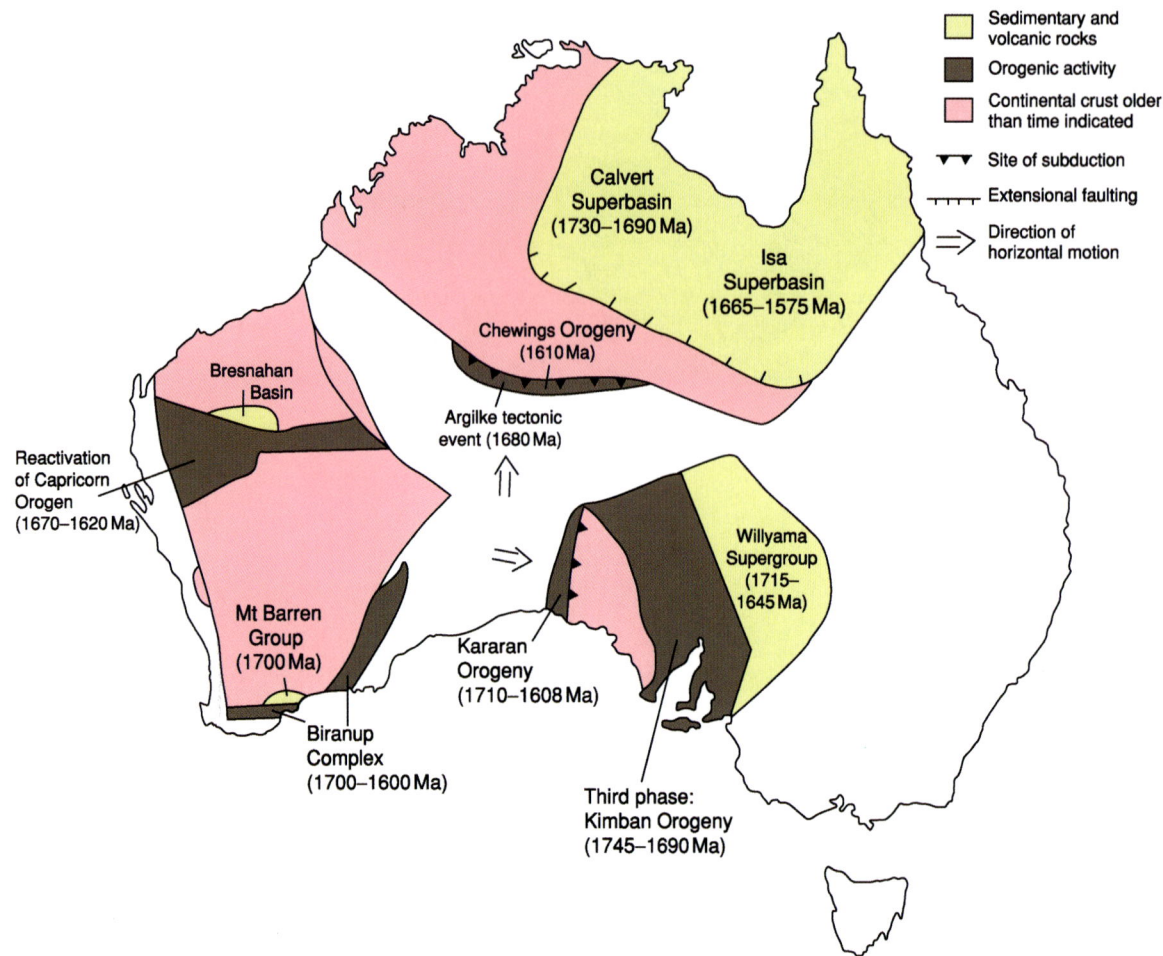

**Figure 4** Final assembly of the cratonic elements of Proterozoic Australia, 1730–1600 Ma.

In the north-western part of the craton, the King Leopold and Halls Creek orogens were initiated by rifting along the western margin of the Lucas Craton ~1910 Ma. Renewed rifting is represented by deposition of the lower Halls Creek Group ~1880 Ma. Accretion of Neoarchaean to early Palaeoproterozoic continental fragments to the opposing eastern edge of the Kimberley Craton occurred before ~1910 Ma. Turbidites derived by erosion from these fragments were deposited ~1870 Ma. Deformation, metamorphism, and extensive felsic and mafic magmatism (Whitewater Volcanics, Paperbark Supersuite) occurred during the Hooper Orogeny (1865–1850 Ma) (**Figure 3**). The central part of the orogen (Tickalara Metamorphics) formed ~1865 Ma as an oceanic island arc above a north-west-dipping subduction zone. Layered mafic–ultramafic intrusions were emplaced ~1855 Ma. Deformation and metamorphism to high grade ~1845 Ma followed intrusion of numerous felsic and basic to intermediate sheetlike bodies during convergence and collision of the arc with the Kimberley Craton. Alkaline volcanism between 1870 and 1850 Ma marked further rifting along the Lucas Craton margin. A submarine fan deposited turbiditic rocks of the upper Halls Creek Group parallel to the craton margin.

Eruption of felsic and mafic volcanic rocks during rifting of the accreted arc ~1840 Ma was accompanied by the emplacement of further layered mafic–ultramafic intrusions. Continued subduction of oceanic crust to the north-west led to collision and suturing of the Kimberley Craton with the rest of the North Australian Craton by ~1820 Ma, during the Halls Creek Orogeny (**Figure 3**). Folding and thrusting accompanied metamorphism. During and immediately following collision, plutons of granite and gabbro were intruded to form the Sally Downs Supersuite (1835–1805 Ma) at the same time as the intrusion of additional large, layered mafic–ultramafic bodies. As the Sally Downs Supersuite was being intruded, the Speewah Group was deposited ~1835 Ma in a foreland basin on the Kimberley

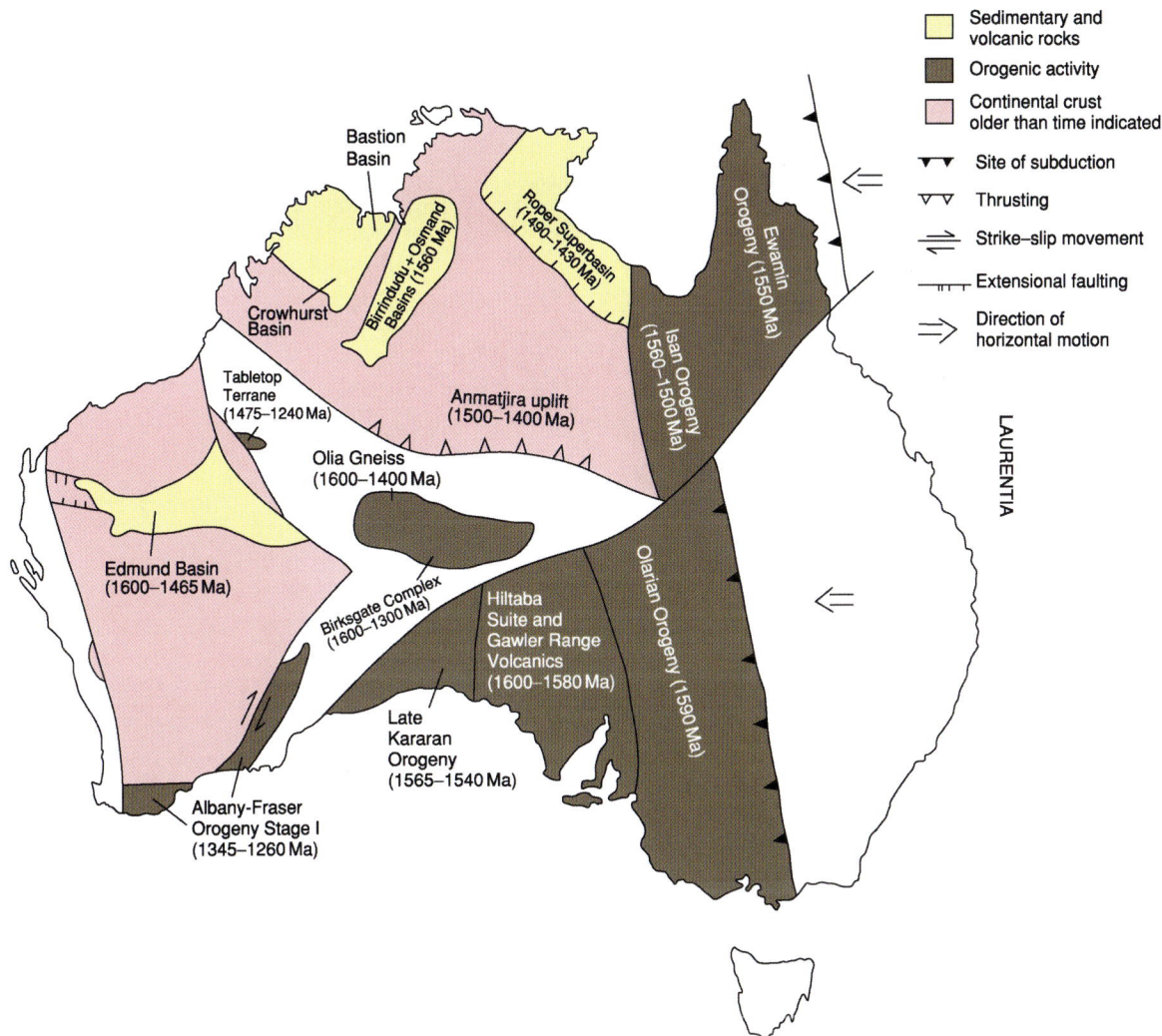

**Figure 5** Docking with Laurentia, 1600–1200 Ma.

Craton. The overlying ~1800-My-old Kimberley Group was derived from the north. The intrusion of the voluminous Hart Dolerite ~1790 Ma may be related to a continental breakup, possibly from Laurentia, centred to the north (**Figure 3**).

To the south-east of the Halls Creek Orogen, the Granites–Tanami Complex and the northern part of the Arunta Inlier developed on the thinned Archaean to Early Palaeoproterozoic basement of the Lucas Craton (**Figure 3**). The <1877-My-old MacFarlane Peak Group and Dead Bullock Formation represent a rift, followed by a sag basin, with the overlying <1840-My-old Killi Killi Formation and Lander Rock beds turbidites being deposited during the Tanami Orogeny, which probably represents a within-plate response to the collisional Halls Creek Orogeny to the north-west. The orogenic event was followed by voluminous bimodal and granitic magmatism between 1825 and 1790 Ma in both the Granites–Tanami Complex and the northern Arunta Inlier, including the high-temperature/low-pressure metamorphism, deformation, and granite magmatism of the Mount Stafford tectonic event ~1820 Ma (**Figure 3**). Over the same time period, granite was intruded into the southern part of the Halls Creek Orogen. The Lichfield Complex is the north-eastward continuation of the Halls Creek Orogen, and the Pine Creek Orogen may represent a within-plate tectonic setting similar to that of the Granites–Tanami Complex (**Figure 3**). The ~1885-My-old South Alligator Group and the overlying Finniss River Group represent a transition from shallow-marine sediments to deep-marine turbidites, possibly related to the onset of the 'Barramundi Orogeny' in the form of the Nimbuwah event 1880–1860 Ma. Further deformation, equivalent to the Halls Creek and Tanami orogenies, accompanied extensive intrusion of granites between ~1840 and ~1820 Ma.

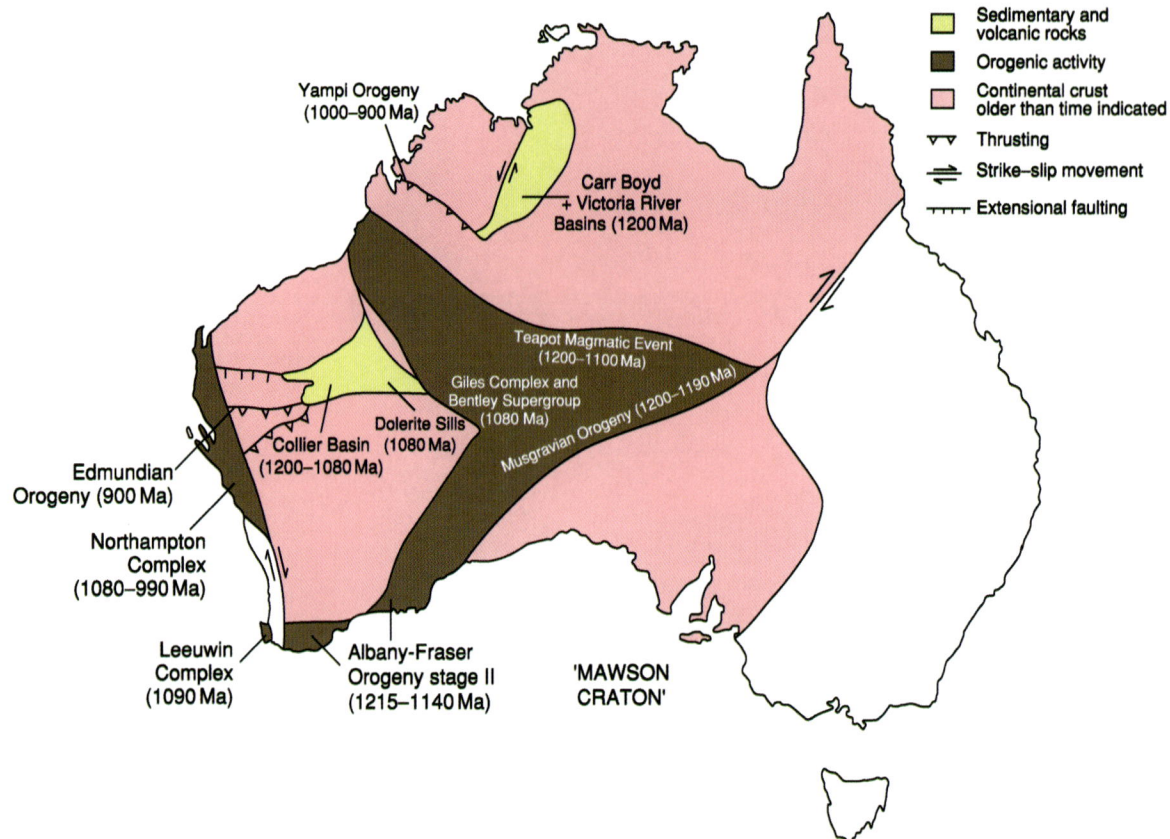

**Figure 6** Intracratonic reactivation, 1200–900 Ma.

In the Mount Isa Inlier, the Kalkadoon–Leichhardt Belt from 1870–1840 Ma, now regarded as the remnants of a magmatic arc, represents the 'Barramundi Orogeny' (**Figure 3**). Along the southern edge of the northern Arunta Inlier, the Atnarpa igneous complex from 1880–1860 Ma has been interpreted as being generated by subduction at a convergent plate margin. Rocks of 'Barramundi Orogeny' age are also present in the Tennant Creek Inlier, where turbidites of the <1860-My-old Warramunga Formation are intruded by ~1850-My-old granites and are overlain by ~1845-My-old felsic volcanic rocks (**Figure 3**).

Between 1780 and 1730 Ma, a magmatic arc developed along the southern margin of the northern Arunta Inlier, above a northward-dipping subduction zone. The high-temperature/low-pressure metamorphism associated with the Early Strangways Orogeny 1780–1770 Ma overlapped the development of the magmatic arc. This event is contemporaneous with the Yapungku Orogeny (1790–1760 Ma) in the opposing margin of the West Australian Craton (**Figure 3**), and together they may record the suturing of those cratons, although probably not in their current configuration. The Late Strangways Orogeny (1740–1730 Ma) was accompanied by moderate-pressure metamorphism. Subduction of oceanic crust associated with the Early and Late Strangways orogenies was at a low angle to the north. A back-arc setting resulted, causing southward tilting within the crust to the north of the Arunta Inlier, together with high heat flows and extension followed by thermal subsidence. Between 1800 and 1670 Ma, two cycles of extensive volcano-sedimentary basin formation took place across the North Australian Craton, separated by a period of uplift and erosion. The first cycle (1800–1750 Ma) is represented by the development of the Leichhardt Superbasin (**Figure 3**). Bimodal volcanism, dominated by flood basalts, and fluvial to shallow-marine sedimentation occurred within the lower parts of the McArthur Basin successions (lower Katherine River Group, lower Donydji Group, Groote Eylandt Group, and lower Tawallah Group), the upper Hatches Creek Group in the Tennant Creek Inlier, and the lower parts of the successions within the Mount Isa Inlier (Haslingden Group, Quilalar Formation, Magna Lyn Metabasalt, Argylla Volcanics, Malbon Group, and Corella Formation). In the

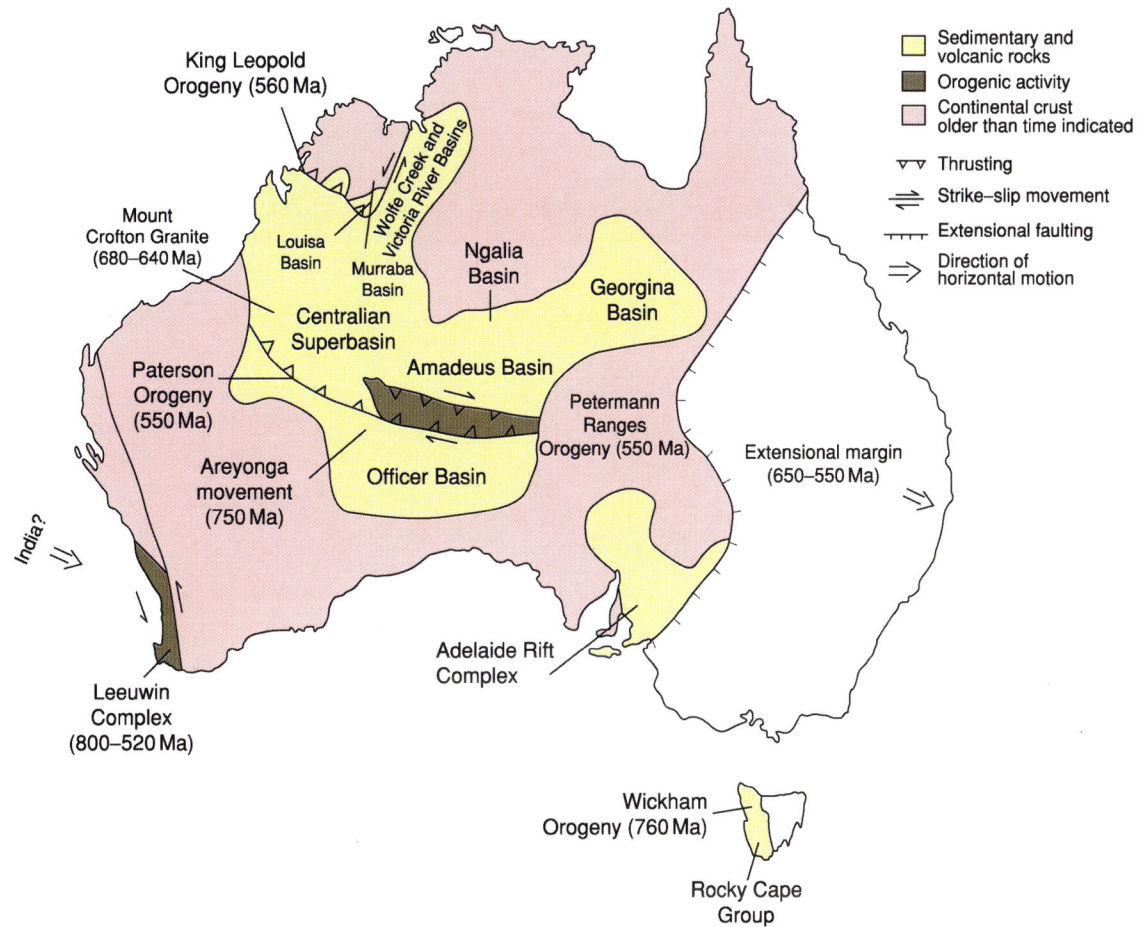

**Figure 7** Rifting from Laurentia, collision with India, 900–520 Ma.

Mount Isa Inlier, rifting in the upper crust was accompanied by the development of a regional-scale extensional shear zone, high-temperature/low-pressure metamorphism, and igneous intrusion (the ~1742-My-old Wonga batholith) in the mid-crust.

A period of basin inversion ~1730 Ma produced a regionally extensive unconformity before the onset of the second depositional cycle (1730–1690 Ma), which formed the Calvert Superbasin, and the third depositional cycle (1665–1575 Ma), which formed the Isa Superbasin (**Figure 4**). The Sybella batholith was intruded in the Mount Isa Inlier ~1665 Ma. Again the successions are characterized by bimodal volcanics and rift-related sediments in the McArthur Basin (upper Katherine River Group, upper Donydji Group, Parsons Range Group, Spencer Creek Group, and upper Tawallah Group), the Mount Isa Inlier (Fiery Creek Volcanics, lower McNamara Group, lower Mount Isa Group, Kuridala Formation, and lower Soldiers Cap Group), and the Georgetown Inlier (lower Etheridge Group). The Redbank Thrust separates the southern Arunta Inlier from the northern Arunta Inlier. The Argilke tectonic event represents a significant magmatic and metamorphic event ~1680 Ma. A major north-directed thrust system, associated with granite intrusion, took place ~1610 Ma during the Chewings Orogeny (**Figure 4**).

### South Australian Craton

In the Gawler Craton, Archaean gneissic rocks of the Sleaford and Mulgathing complexes were deformed, metamorphosed to granulite facies, and intruded by granites during the Sleafordian Orogeny 2550–2400 Ma (**Figure 2**). At the eastern margin of the Gawler Craton, orthogneiss of the Miltalie Gneiss (~2003 Ma) is unconformably overlain by shallow-water continental shelf deposits. These include metamorphosed amphibolite facies mafic volcanics, carbonates, and iron formation of the Hutchison Group >1860 Ma (**Figure 2**). The Kimban Orogeny has affected much of the Gawler Craton. It has been regarded as consisting of three tectonic events and dates from ~1850 to ~1690 Ma. It is contemporaneous with the Capricorn Orogeny and

**Table 1** Summary of Proterozoic tectonic events in Australia

| Pinjarra Orogen | West Australian Craton | | | Albany-Fraser Orogen | Paterson Orogen | | South Australian Craton | |
|---|---|---|---|---|---|---|---|---|
| | | | | | *Paterson Orogeny (550 Ma)* | *Petermann Ranges Orogeny (550 Ma)* | | |
| Leeuwin Complex (800–520 Ma) | Western Officer Basin (Supersequences 3+4) | | | | Mount Crofton granite (680–640 Ma) | Amadeus Basin (supersequences 3+4) | Eastern Officer Basin (Supersequences 3+4) | |
| | | | | | | Souths Range movement | | |
| | | | | | | Amadeus Basin (Supersequence 2) | Eastern Officer Basin (Supersequence 2) | |
| | | | | | Aeryonga movement (750 Ma) | | | |
| | Officer + Amadeus Basins (Supersequence 1) | | | | | | | |
| Northampton + Leeuwin complexes (1090–990 Ma) | *Edmundian Orogeny (900 Ma)* | | | | | | | |
| | Dolerite sills (1080 Ma) | | | | | Giles Complex + Bentley Supergroup (1080 Ma) | | |
| | Proterozoic Australia (Rodinia) | | | | | | | |
| | Collier Basin (1200–1080 Ma) | | | *Stage II Albany-Fraser Orogeny (1215–1140) Ma* | *Musgravian Orogeny (1200–1190 Ma)* | Teapot Magmatic Event (1200–1100 Ma) | | |
| | | | | *Stage I Albany-Fraser Orogeny (1345–1260) Ma* | Tabletop Terrane (1475–1290 Ma) | | | |
| | Edmund Basin (1600–1465 Ma) | | | | | *Anmatjira Uplift (1500–1400 Ma)* | | |
| | | | | | Musgrave Complex: Olia gneiss + Birksgate Complex (1600–1300 Ma) | | *Late Kararan Orogeny (1565–1540 Ma)* | |
| | | | | | | | | Hiltaba Suite + Gawler Range volcanics (1600–1580 Ma) |
| | Bresnahan Basin | | | | | | South Australian Craton | |
| | | | | Biranup Complex (1700–1600 Ma) | | *Chewings Orogeny (1610 Ma)* | *Second Episode Kararan Orogeny (1630–1610 Ma)* | |
| | *Reactivation of Capricorn Orogen (1670–1620 Ma)* | | | | | | St Peter's Suite (1630–1608 Ma) | |
| | | | Mount Barren Group (1700 Ma) | | | *Argilke Tectonic Event (1680 Ma)* | *First episode Kararan Orogeny (1710–1670 Ma)* | |
| | | | | | | Southern Arunta Inlier | Tunkillia Suite | |
| | | | | | *Yapungku Orogeny (1790–1760 Ma)* | | | |
| | *Capricorn Orogeny (1830–1780 Ma)* | | | | | | | |
| | Ashburton and Blair Basins (1840–1805 Ma) | Gascoyne Complex | Upper Yerrida and Earaheedy basins (1840–1800 Ma) | | Rudall Complex (1800 Ma) | | | |
| | West Australian Craton | | | | | | | |
| | | *Glenburgh Orogeny 2005–1960 Ma* | Bryah and Padbury Basins (2020–1900 Ma) | | Rudall Complex (2020 1970 Ma) | | | |
| Mullingarra Complex (2181 Ma) | *Ophthalmian Orogeny (2200 Ma)* | | Lower Yerrida Basin (2150 Ma) | | | | | |
| | McGrath Trough (2300–2200 Ma) | | | | | | | Gawler Craton |
| | Hamersley Basin (2800–2440 Ma) | Glenburgh Terrane (2550–2450 Ma) | | | | | | *Sleafordian Orogeny (2550–2400) Ma* |
| | Pilbara Craton | | Yilgarn Craton | Munglinup Gneiss (2630–2575 Ma) | | | | Sleaford and Mulgathing complexes |
| | Granite-greenstones (3560–2800 Ma) | | Granite-greenstones (3730–2600 Ma) | | | | | |

the Yapungku Orogeny in the West Australian Craton, and the Early and Late Strangways orogenies in the North Australian Craton, and may represent a series of intraplate and plate margin events reflecting convergence and accretion of the Gawler Craton to what was then the south-eastern North Australian Craton.

Intrusive rocks of the Donnington Granitoid Suite (~1850 Ma) were emplaced synchronously with a deformation event that has been regarded as an early tectonic phase of the Kimban Orogeny (**Figure 3**). The second phase of the orogeny produced medium- to high-pressure/high-temperature metamorphism and associated deformation. Felsic and

**Table 1** Continued

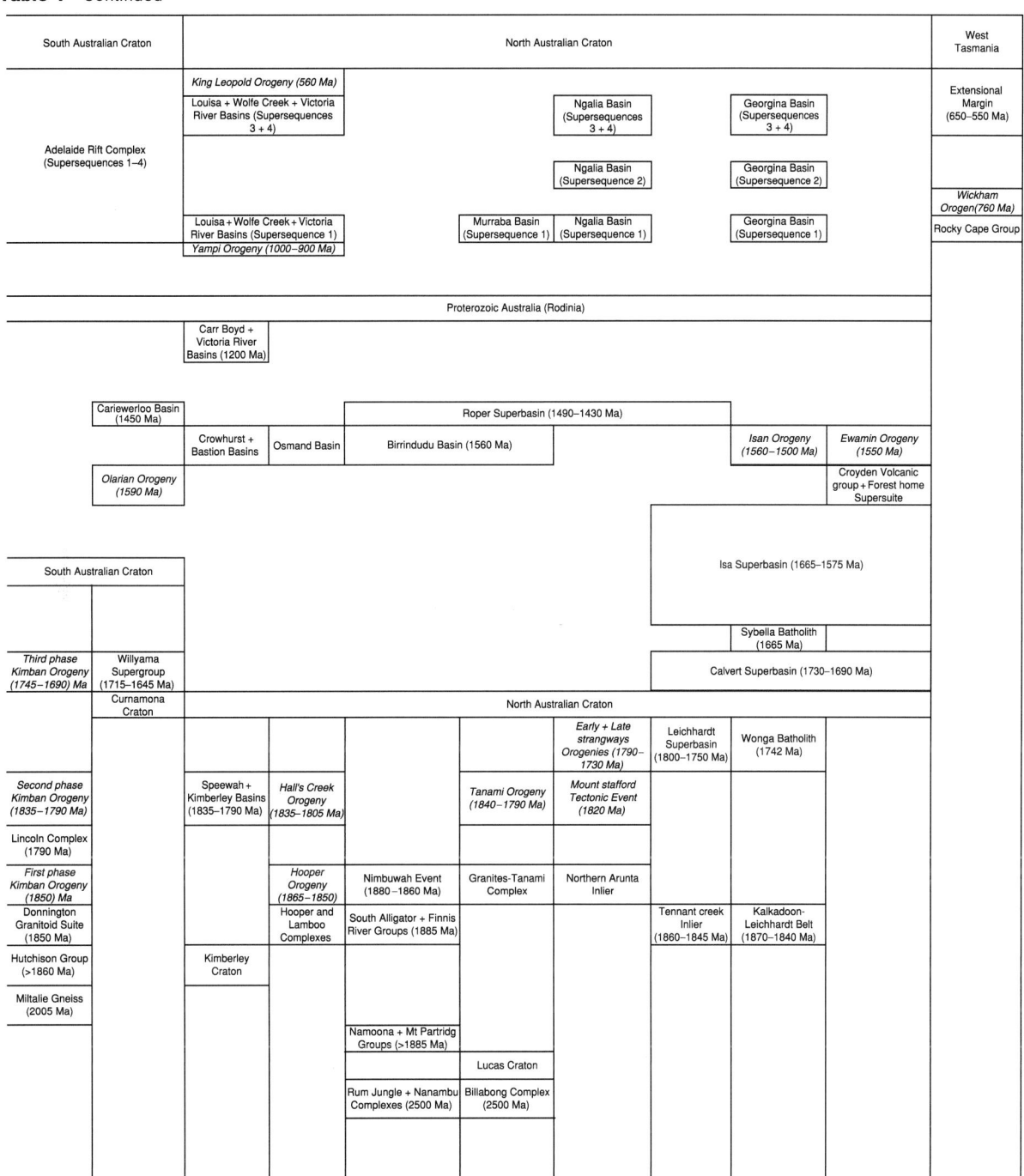

bimodal volcanics and interlayered metasedimentary rocks of the Myola Volcanics and the Broadview Schist, and the Peake Metamorphics were deposited on the Gawler Craton ~1790 Ma, contemporaneous with syntectonic granitic rocks of the Lincoln Complex. Further volcanic rocks and metasedimentary rocks were deposited between ~1765 and ~1735 Ma (McGregor Volcanics, Moonabie Formation, and Wallaroo Group). The third phase of the orogeny took place between 1745 and 1690 Ma and involved deformation, medium-grade metamorphism, and further granite intrusion (**Figure 4**).

Following accretion to the North Australian Craton, the site of subduction may have moved to

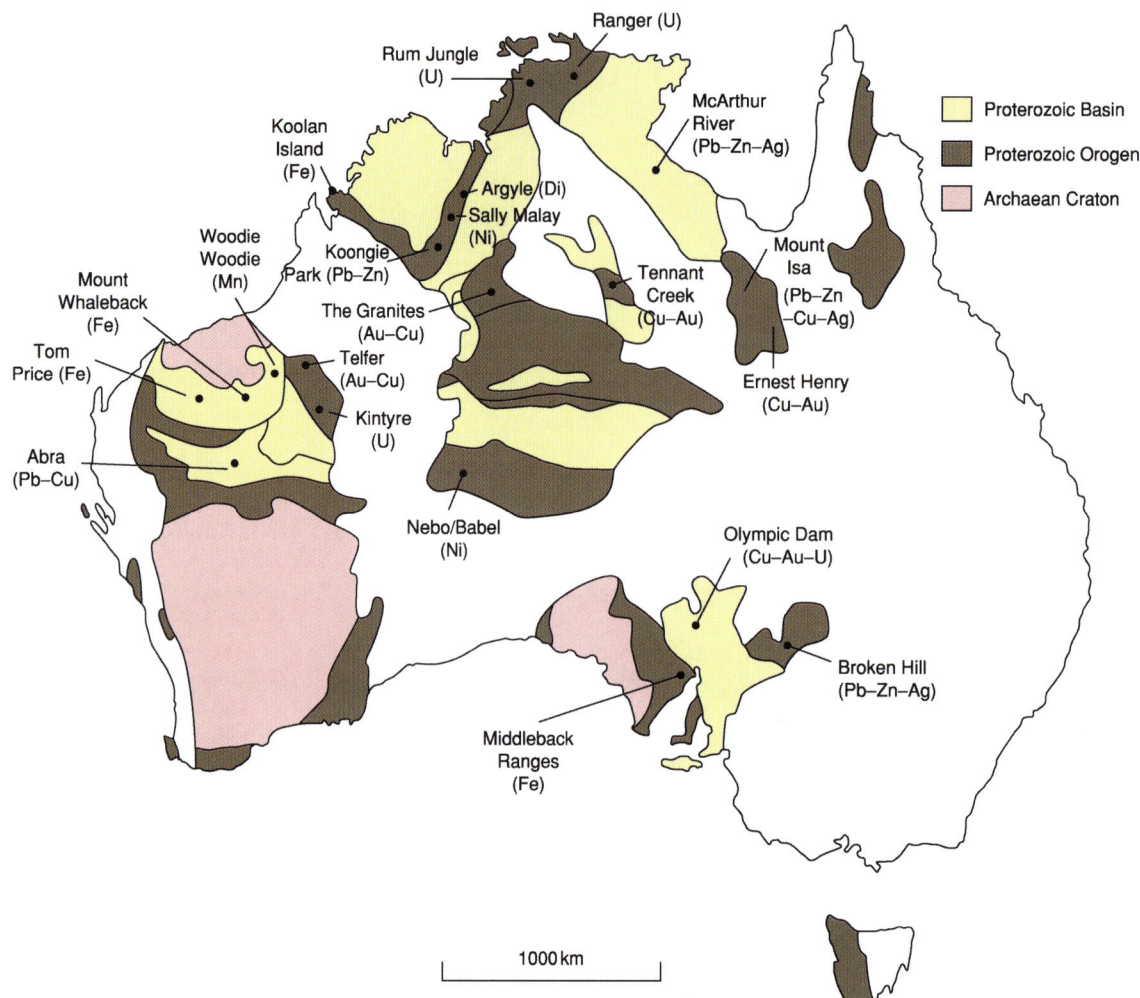

**Figure 8** Significant mineral and diamond (Di) deposits in the Proterozoic of Australia.

the south-western Gawler Craton (**Figure 4**) with the development of the first tectonic episode of the Kararan Orogeny (1710–1670 Ma) and the intrusion of voluminous granitic rocks of the Tunkillia Suite in a possible back-arc setting. The second tectonic phase involved the intrusion of the St Peter Suite (1630–1608 Ma) as a continental magmatic arc, and probably reflects accretion of continental blocks from the west, now buried beneath the Phanerozoic Eucla Basin. The Kararan Orogeny is coincident with the Argilke and Chewings events in the southern Arunta Inlier, with the deposition of the Mount Barren Group and the reactivation of the Capricorn Orogen in the West Australian Craton, and with the development of the Calvert and Isa superbasins on the North Australian Craton (**Figure 4**). In the Curnamona Craton, the shallow-marine, immature clastic sedimentary rocks and bimodal volcanics of the Willyama Supergroup were deposited between ∼1715 and ∼1645 Ma. They may represent a continuation of the Calvert and Isa superbasins that has subsequently been displaced dextrally during the Mesoproterozoic (**Figure 3**).

## Mesoproterozoic – The Assembly of Rodinia (1600–1000 Ma)

Along the eastern edge of the North Australian Craton and in the north-eastern part of the South Australian Craton, a major orogenic event took place between 1600 and 1500 Ma (**Figure 5**). This event included the Isan Orogeny (1560–1500 Ma) in the Mount Isa Inlier, the Ewamin Orogeny (1550 Ma) in the Georgetown Inlier, the Olarian Orogeny (1590 Ma) in the Curnamona Craton, and the late event (1565–1540 Ma) of the Kararan Orogeny. Subduction was to the west, culminating in collision with a continent to the east that may have been

Laurentia. In the Georgetown Inlier, the felsic rocks of the Croydon Volcanic Group and the Forest Home Supersuite represent a magmatic arc. Further to the west, the eruption of the voluminous Gawler Range Volcanics, and the intrusion of the Hiltaba Suite in the Gawler Craton, occurred between ~1600 and ~1580 Ma, contemporaneously with the early stages of this orogenic event.

Elsewhere in central and western Australia, Mesoproterozoic rocks of poorly understood tectonic setting include medium- to high-grade ~1600-My-old quartzofeldspathic gneisses (Olia Gneiss) to the north of the Woodroffe Thrust in the Musgrave Complex. These were intruded by granitic rocks ~1500 Ma, before being metamorphosed ~1400 Ma. South of the Woodroffe Thrust, metasedimentary gneisses and felsic and mafic orthogneisses (~1550 and 1330 Ma) represent a separate terrane. Granulites in the Birksgate Complex at the northern margin of the South Australian Craton were derived from metamorphism of felsic volcanics and sedimentary rocks (1600–1300 Ma). Uplift along the Redbank Thrust Zone at the northern margin of the southern Arunta Inlier took place 1500–1400 Ma (Anmatjira uplift phase). The eastern Tabletop Terrane of the Rudall Complex is formed by metasedimentary and metavolcanic rocks intruded by granitoids that date to ~1475 to ~1290 Ma.

On the North Australian Craton, clastic and carbonate sedimentary rocks of the Birrindudu Group were deposited in the Birrindudu Basin ~1560 Ma, unconformably overlying the Granites–Tanami Complex. In the Kimberley region, the Mount Parker Sandstone and the Bungle Bungle Dolomite in the Osmand Basin, and the Crowhurst Group and the Bastion Group, may be equivalent to the rocks of the Birrindudu Basin. The Roper Group and the South Nicholson Group were deposited in the intracratonic Roper Superbasin between ~1490 and ~1430 Ma, and subsidence may be related to late stages of the Isan Orogeny, and to the Anmatjira uplift phase in the Arunta Inlier (**Figure 5**). The intracratonic Cariewerloo Basin (~1450 Ma) developed on the South Australian Craton. On the West Australian Craton, the Edmund Group, the lower part of the Bangemall Supergroup, was deposited in the intracratonic Edmund Basin after ~1600 Ma. Initial deposition was restricted to narrow rift basins before the development of a broad, marine basin. Extensive intrusion of dolerite sills into the Edmund Group took place ~1465 Ma.

The Albany–Fraser Orogen developed between the South Australian Craton and the West Australian Craton (**Figures 5** and **6**). It has been interpreted as part of a Grenvillian collisional orogeny (1345–1140 Ma) that originally extended through the Musgrave Complex and possibly the now displaced eastern Rudall Complex, separating the South Australian Craton from the combined West and North Australian cratons. This event has been interpreted as marking the final assembly of Proterozoic Australia as part of the Rodinia Supercontinent. Two distinct tectonic stages took place in the Albany–Fraser Orogen, with the first producing deformation, high-grade metamorphism, and granitic magmatism between 1345 and 1260 Ma. The granulite facies layered mafic intrusions of the Fraser Complex crystallized ~1300 Ma. The second phase produced further granulite facies metamorphism and deformation ~1214 Ma and subsequent granite intrusion between 1190 and 1140 Ma. In the Tabletop Terrane of the Rudall Complex, granitic magmatism occurred ~1300 Ma, whereas in the southern part of the Musgrave Complex, granulite facies metamorphism occurred ~1200 Ma, with the intrusion of granitic rocks ~1190 Ma. Granitic rocks were also intruded into the southern Arunta Inlier during the Teapot magmatic event between 1200 and 1100 Ma. In northern Australia, the ~1200-My-old Auvergne Group and the Carr Boyd Group were deposited in the Victoria River and Carr Boyd Basins, respectively (**Figure 6**).

The Collier Group, the upper part of the Bangemall Supergroup, was deposited in the Collier Basin on the West Australian Craton and, together with the underlying Edmund Group, was intruded by a suite of syn-depositional dolerite sills ~1080 Ma. Large, layered mafic–ultramafic intrusions forming the Giles Complex (~1080 Ma) were intruded into the Musgrave Complex, followed by the eruption of the mafic and felsic volcanic rocks of the Bentley Supergroup 1080–1060 Ma. Together with the intrusion of the contemporaneous Stuart and Kulgera dyke swarms, the extensive mafic magmatism of 1080–1060 Ma has been described as a Large Igneous Province, and may represent the influence of a mantle plume centred beneath the Musgrave Complex (*see* **Large Igneous Provinces**).

In the Pinjarra Orogen, metasedimentary rocks in the Northampton Complex and Mullingarra Complex that may have had their source in the Albany–Fraser Orogen were metamorphosed, deformed, and intruded by granites between ~1080 and ~990 Ma. They are probably allochthonous and may have been transported and accreted in their present position by dextral movement along a proto-Darling Fault prior to 755 Ma (**Figure 6**). The protolith to ~1090-My-old granitic orthogneiss in the Leeuwin Complex may have a syn-collisional origin. Intracratonic reactivation of the King Leopold, Halls

Creek, and Capricorn orogenic belts took place between 1000 and 900 Ma during the Yampi and Edmundian orogenies (**Figure 6**).

## Neoproterozoic-Proterozoic Australia in Rodinia (1000–545 Ma)

The Centralian Superbasin developed throughout much of Proterozoic Australia from ~830 Ma, either as an extensive intracratonic sag basin or as a series of interconnected basins that lapped onto intervening, emergent basement highs (**Figure 7**). The Adelaide Rift Complex developed to the south-east at the same time, possibly centred over a mantle plume. The lower part of the depositional succession in the superbasin, Supersequence 1, is contemporaneous with the intrusion of the Amata and Gairdner mafic dykes in the Gawler Craton and eastern Musgrave Complex, and the volcanics of the Callana Group in the Adelaide Rift Complex. Supersequence 1 is characterized by a basal sand sheet, overlain by stromatolitic carbonates and evaporates that are correlated throughout the component basins of the superbasin (Officer Basin, Amadeus Basin, Ngalia Basin, Georgina Basin, Wolfe Creek Basin, Louisa Basin, and Murraba Basin). Siliciclastic rocks of the Rocky Cape Group and the Burnie Formation in west Tasmania may be equivalent to Supersequence 1, deposited prior to the Wickham Orogeny ~760 Ma.

Deposition of Supersequence 1 was followed by a period of tectonic activity, the Areyonga movement, which may represent the breakup of Rodinia as Laurentia rifted away from the eastern margin of Proterozoic Australia to form the proto-Pacific Ocean. The northern part of the Adelaide Rift Complex between the Gawler and Curnamona Cratons has been interpreted as a failed arm, but there is no evidence of the development of an adjacent passive margin at this time. The main locus of Rodinia breakup may have developed much further to the east, now buried within the Tasmanides. Tilting and flexuring of basement fault blocks to the west within the Centralian Superbasin was controlled by the reactivation of major structures such as the Redbank Thrust, and a proto-Woodroffe Thrust. Widespread folding, uplift, and erosion took place within the basins. The 755-My-old Mundine Well dyke swarm was intruded into the West Australian Craton and the adjacent Northampton Complex.

Supersequece 2 is restricted to northern and central Australia and is marked by glacigene deposits correlated with the Sturtian glaciation ~700 Ma in the Adelaide Rift Complex. As sea-level rose, the glacial deposits were overlain by silt and mud deposited in a shallow epeiric sea. This was followed by a further period of uplift and erosion, the South Range Movement, and by the emplacement of the ~680- to 640-My-old Mount Crofton Granite and associated intrusions into the Lamil Group of the Yeneena Basin in the north-western part of the Paterson Orogen. Supersequence 3 is also marked by glacigene deposits, which extend into the Kimberley region and the northwest Officer Basin. These are correlated with the ~600-My-old Marinoan glaciation in the Adelaide Rift Complex. Again, rising sea-levels and the establishment of a shallow epeiric sea followed glaciation.

Supersequence 4 is dominated by sandstone and conglomerate deposited in submarine fan, deltaic, fluvial, and alluvial fan settings in local foreland basins. These are associated with the development of intracratonic orogenies, which resulted from the reactivation of the ancient sutures between the North Australian Craton, the West Australian Craton, and the South Australian Craton during the Paterson and Petermann Ranges orogenies ~550 Ma (**Figure 7**). In the Kimberley region, the King Leopold Orogeny reactivated the suture between the Kimberley Craton and the rest of the North Australian Craton at about the same time. Large dextral strike–slip movements produced folding and thrusting in the Musgrave Complex, with the exhumation of eclogite facies rocks as part of a crustal-scale flower-type structure. The Musgrave Complex now truncates the Albany–Fraser Orogen, and may have been displaced to the west at this time, extending as far as the Mesoproterozoic Tabletop Terrane of the Rudall Complex, with which the Warri–Anketell Gravity Ridge connects it beneath the Centralian Superbasin.

In the Pinjarra Orogen at the western margin of Proterozoic Australia, Leeuwin Complex orthogneisses from ~1090 and 800–650 Ma were metamorphosed at upper amphibolite to granulite facies conditions ~540 Ma. Further granitic rocks were intruded 540–520 Ma. These events were coincident with sinistral tectonic transport of the Leeuwin Complex along the Darling Fault during an oblique collision of Australia with India (**Figure 7**). This collision may be responsible for the contemporaneous intracratonic reactivations that produced major unconformities at the base of the Phanerozoic basins overlying Proterozoic Australia to the west of the Adelaide Rift Complex. An extensional margin developed to the east of the Adelaide Rift Complex between ~650 and ~550 Ma, and is now exposed in Tasmania and western Victoria (**Figure 7**). It is marked by the intrusion of a dolerite dyke swarm, the eruption of tholeiitic basalts, and the deposition of associated volcanogenic sediments, carbonates, and shallow-water siliciclastics. This margin was involved in an arc–continent collision ~505 Ma during the Delamarian Orogeny, initiating the development of the Palaeozoic Lachlan Orogen. Extensive eruption of continental flood basalts (Kalkarinji large igneous province) of the ~510-My-old Antrim Plateau Volcanics and the Table Hill Volcanics in northern and western Australia may reflect further mantle plume activity coincident with this collision.

## Mineral Deposits

The Proterozoic rocks of Australia are endowed with substantial mineral resources, containing significant deposits of iron ore, manganese, uranium, gold, copper, lead, zinc, silver, nickel, and diamond (**Figure 8**) (*see* **Mineral Deposits and Their Genesis**). World-class hematite deposits (Mount Whaleback, Tom Price) have been produced by enrichment of the banded iron formations of the Hamersley Basin during the Palaeoproterozoic. The highest grade low-P deposits (62–69% Fe), have been interpreted as the product of burial metamorphism of original supergene enrichment prior to ~1840 Ma. More recently, they have been reinterpreted as the product of hypogene processes related to the expulsion of fluids into the foreland of the Ophthalmian Orogeny (~2200 Ma). Proterozoic iron ore deposits have also been mined in the Palaeoproterozoic iron formations of the Hutchison Group in the Gawler Craton, and from enrichment of ferruginous placer deposits in the upper Kimberley Group in the Kimberley Basin (Koolan Island). Manganese has been mined from supergene deposits (Woodie Woodie) associated with Palaeoproterozoic karst developed on dolomites in the eastern Hamersley Basin.

'Unconformity-related' uranium ($\pm$ Au $\pm$ platinum group elements) deposits (Rum Jungle, Alligator River) are found in Palaeoproterozoic sedimentary successions overlying predominantly Archaean basement rocks in northern Australia, and may be related to interaction of highly oxidized, acidic and Ca-rich meteoric brine with reduced basement fluids. Similar deposits have been found elsewhere in the Australian Proterozoic, including Kintyre, where Neoproterozoic metasedimentary rocks overlie Palaeoproterozoic basement. Metasomatic U deposits and the metamorphic-related Mary Kathleen deposit occur in Palaeoproterozoic rocks of the Mount Isa Inlier.

Gold is found in a number of settings throughout the Proterozoic, and historic production is related to generally small lode-gold occurrences and associated alluvial deposits. Major gold (Cu) mines have been developed at Telfer in the Yeneena Basin, and at The Granites in the Granites–Tanami Complex. Mineralization is centred on metasedimentary rocks in structural domes and is hypogene, related to the emplacement of granitic pluton(s) near a periodically reactivated, regional-scale, fluid-focusing structure, often a strike–slip fault. 'Proterozoic Cu–Au deposits' are found associated with iron oxide in the Tennant Creek Inlier, at Olympic Dam in the northern Gawler Craton, and in the Mount Isa Inlier (Ernest Henry). There is usually a spatial and temporal relationship between this style of deposit and granite intrusion (e.g., Olympic Dam and the ~1590-My-old Hiltaba Suite). Cu mineralization at Mount Isa is regarded as syn-deformational, late metamorphic, and is a separate event from Pb–Zn–Ag mineralization. The Mount Isa orebody is a world-class example of stratiform sediment-hosted Pb–Zn–Ag mineralization, produced by oxidized fluids moving through the sediment pile and being deposited either by seafloor exhalative processes or within the sediments. A similar orebody is present at McArthur River. The Broken Hill orebody is generally regarded as a metamorphosed example of stratiform sediment-hosted mineralization, although syn-metamorphic, skarn-type processes may have modified it. Volcanic-hosted massive sulphide (VHMS) Pb–Zn mineralization has been found at Koongie Park in the Halls Creek Orogen.

Nickel mineralization in the Proterozoic, together with platinum group elements, Cu, and V, is generally associated with layered mafic–ultramafic intrusions such as those in the Halls Creek Orogen (Sally Malay). Large, layered intrusions are also present in the Musgrave Complex (Giles Complex–Nebo and Babel deposits), the Albany–Fraser Orogen (Fraser Complex), and the Arunta Inlier.

The Argyle diamond mine, which is the world's largest, is developed on the ~1200-My-old AK1 lamproite pipe intruded into the Halls Creek Orogen. Proterozoic diamondiferous kimberlites from ~815 Ma are present in the Kimberley Basin (*see* **Igneous Rocks: Kimberlite**).

## See Also

**Igneous Processes**. **Igneous Rocks:** Kimberlite. **Large Igneous Provinces**. **Mantle Plumes and Hot Spots**. **Metamorphic Rocks:** Classification, Nomenclature and Formation. **Mineral Deposits and Their Genesis**. **Mining Geology:** Exploration; Hydrothermal Ores. **Plate Tectonics**. **Sedimentary Environments:** Depositional Systems and Facies. **Sedimentary Rocks:** Mineralogy and Classification; Banded Iron Formations. **Tectonics:** Mountain Building and Orogeny. **Time Scale**.

## Further Reading

Australian Geological Survey Organisation (1998) Geology and mineral potential of major Australian mineral provinces. *AGSO Journal of Australian Geology and Geophysics* 17(3): 1–260.

Australian Geological Survey Organisation (1998) Exploration models for major Australian mineral deposit types. *AGSO Journal of Australian Geology and Geophysics* 17(4): 1–313.

Cawood PA and Tyler IM (2004) Assembling and reactivating the Proterozoic Capricorn Orogen: lithotectonic elements, orogenies, and significance. *Precambrian Research* 128: 201–218.

Collins WJ and Shaw RD (eds.) (1995) *Time limits on tectonic events and crustal evolution using geochronology: some Australian examples Precambrian Research* 71: 1–346.

Direen NG and Crawford AJ (2003) The Tasman Line: where is it, what is it, and is it Australia's Rodinian breakup boundary? *Australian Journal of Earth Sciences* 50: 491–502.

Drexel JF and Parker AJ (1993) *The Geology of South Australia: Volume 1. The Precambrian. Geological Survey of South Australia, Bulletin 54.* Adelaide: Geological Survey of South Australia.

Fitzsimons ICW (2003) Proterozoic basement provinces of southern and south-western Australia, and their correlation with Antarctica. In: Yoshida M, Windley BF, and Dasgupta S (eds.) *Proterozoic East Gondwana: Supercontinent Assembly and Breakup. Geological Society of London, Special Publication 206*, pp. 93–130. London: Geological Society of London.

Geological Survey of Western Australia (1990) *Geology and Mineral Resources of Western Australia. Western Australia Geological Survey, Memoir 3.* East Perth: Geological Survey of Western Australia.

Glikson AY, Stewart AJ, Ballhaus CG, Clarke GL, Feeken EHJ, Leven JH, *et al.* (1996) *Geology of the Western Musgrave Block, Central Australia, with Particular Reference to the Mafic–Ultramafic Giles Complex. Australian Geological Survey Organisation, Bulletin 239.* Canberra: Australian Geological Survey Organisation.

Hoatson DM and Blake DH (2000) *Geology and Economic Potential of the Palaeoproterozoic Layered Mafic–Ultramafic Intrusions in the East Kimberley, Western Australia. Australian Geological Survey Organisation, Bulletin 246.* Canberra: Australian Geological Survey Organisation.

Hunter DR (1981) *Precambrian of the Southern Hemisphere. Developments in Precambrian Geology 2.* Amsterdam: Elsevier.

Korsch RJ (ed.) (2002) Thematic issue: geodynamics of Australia and its mineral systems: technologies, syntheses and regional studies. *Australian Journal of Earth Sciences* 49(4): 593–771.

Myers JS, Shaw RD, and Tyler IM (1996) Tectonic evolution of Proterozoic Australia. *Tectonics* 15: 1431–1446.

Powell C McA and Meert JG (eds.) (2001) Assembly and breakup of Rodinia. *Precambrian Research* 110(1-4): 1–386.

Southgate PN (ed.) (2000) Thematic issue: Carpentaria–Mt Isa zinc belt: basement framework, chronostratigraphy and geodynamic evolution of Proterozoic successions. *Australian Journal of Earth Sciences* 47(3): 337–657.

Walter MR (ed.) (2000) Neoproterozoic of Australia. *Precambrian Research* 100(1-3): 1–433.

# Phanerozoic

**J J Veevers**, Macquarie University, Sydney, NSW, Australia

© 2005, Elsevier Ltd. All Rights Reserved.

## Introduction

This account of Phanerozoic (544–0 Ma) Australia is told through a timetable of events (**Figure 1**), a set of fold-belt cross-sections (**Figure 2**), and a set of maps, which stretch from the initial Pan-Gondwanaland phase to the tectonics of today (**Figures 3–27**).

Phanerozoic Australia developed through three stratitectonic regimes (uniform plate tectonics and climate at similar or slowly changing latitude), which generated depositional successions of distinct facies that are bounded by unconformities at the margins and by stratigraphical lacunas in the interior (**Figure 1**). These stratitectonic regimes are summarized in **Table 1**.

Terranes successively broke off the west and northwest along divergent margins with rim basins. The eastern convergent margin grew by accretions of arc-subduction complexes during successive rollback of the trench. Changes of regime were driven by global events (*see* **Gondwanaland and Gondwana; Australia: Tasman Orogenic Belt**): the 544 Ma aftermath of the Pan-Gondwanaland collisional rotations and the onset of an active Pacific margin; the 320 Ma collisional merger of Gondwanaland and Laurussia to form Pangaea and the subsequent far-field contractional stress in central Australia, with ensuing tectonics driven by Pangaean-induced heat; and the 99 Ma change in the azimuth of subduction of the Pacific Plate from head-on to side-swipe.

The 650–580 Ma collision of East Gondwanaland and West Gondwanaland resulting from the closure of the Mozambique Ocean, together with continued Pan-Gondwanaland contraction until 500 Ma, generated the Mozambique Orogenic Belt and an

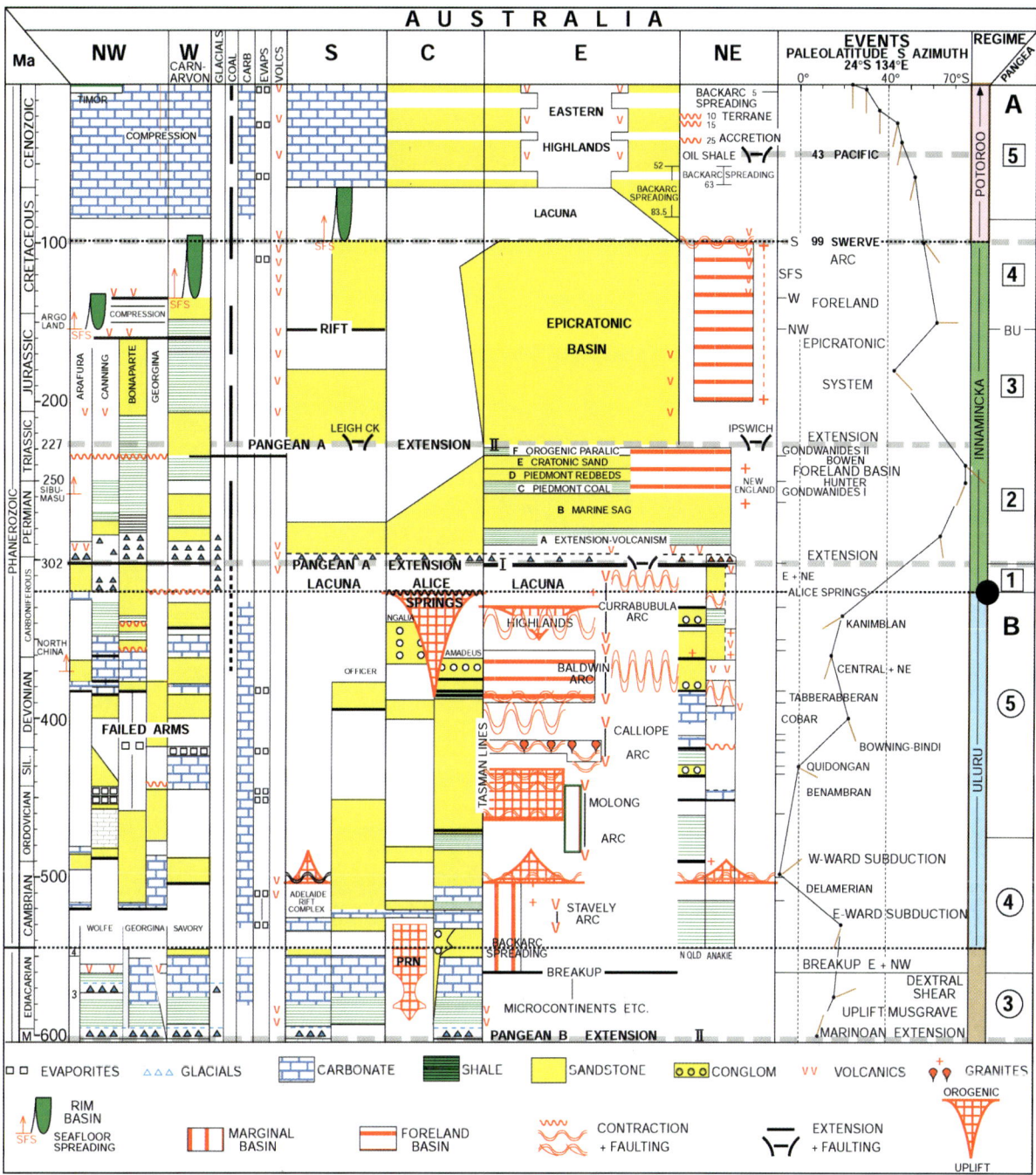

**Figure 1** Timetable of events during the past 600 Ma. Columns from left to right are divergent margins (northwest (NW) with break-off of terranes and west (W) represented by the Carnarvon Basin), depositional facies (glacial, coal, carbonate, evaporites, volcanics), and southern region (S), the centre (C), across the Tasman Lines to the Tasman Fold Belt System in the east (E) and north-east (NE). Under 'events', the palaeolatitude of Alice Springs is plotted and the chief deformations are listed. Under 'regime', the last Neoproterozoic supersequences and the three Phanerozoic regimes are indicated. Under 'Pangaea', the stages of Pangaea B and A are indicated and the 320 Ma merger is marked by a filled circle. BU, breakup; carb, carbonate; evaps, evaporites; PRN, Petermann Ranges Nappe; QLD, Queensland; SFS, seafloor spreading; volcs, volcanics. Reproduced with permission from Veevers JJ (ed.) (2000) *Billion-Year Earth History of Australia and Neighbours in Gondwanaland*. Sydney: GEMOC Press.

orthogonal belt through Prydz Bay to Cape Leeuwin, with distributed thrusting and associated metamorphism in the rest of Australia. These events, together with the inception of subduction of the Palaeo-Pacific Plate beneath Antarctica–Australia at about 550 Ma, inaugurated Phanerozoic Australia.

The events along the Pacific margin during the Palaeozoic are shown in the set of cross-sections in

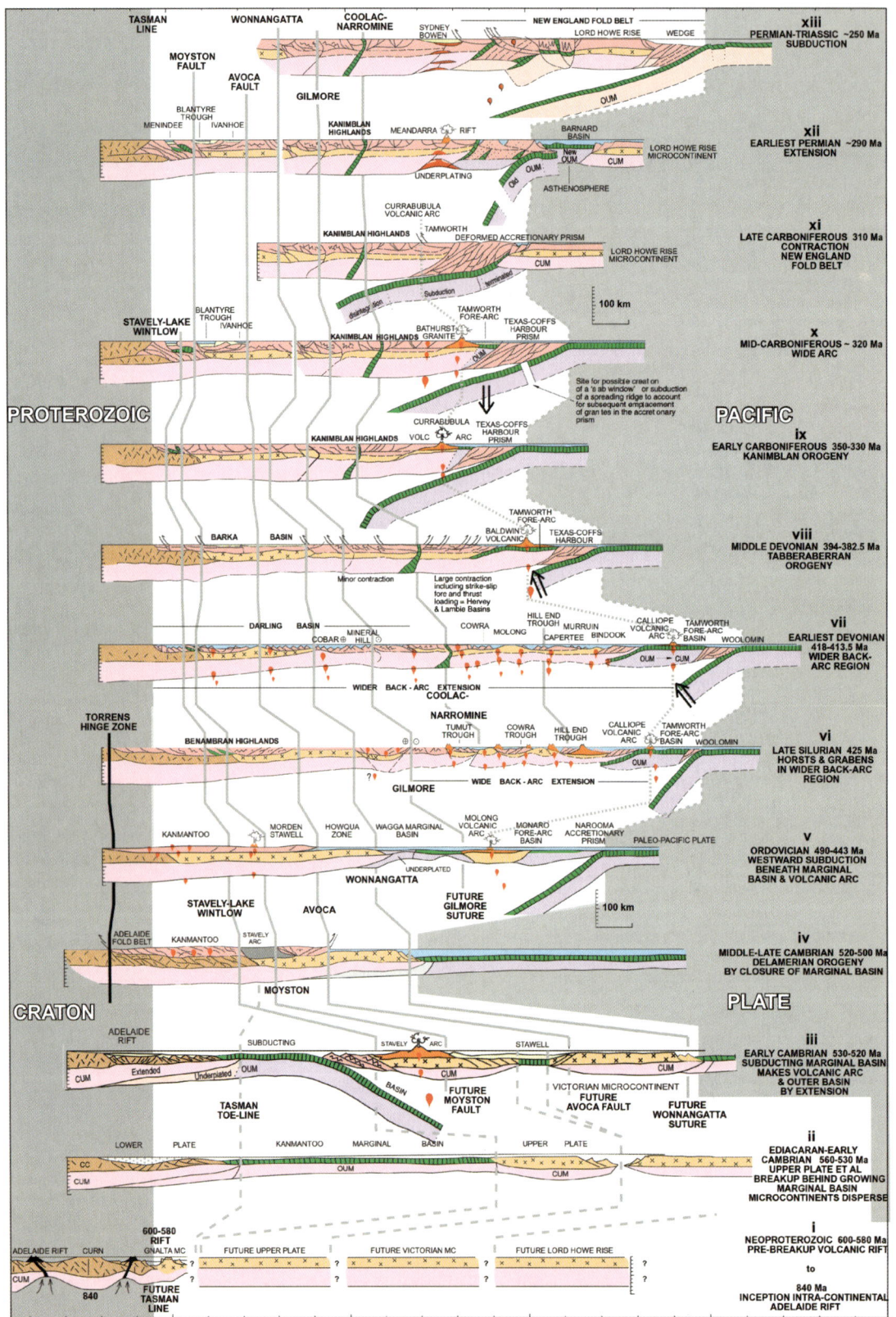

**Figure 2** Cross-sections of the lithosphere of the Tasman Fold Belt System developing between the fixed Proterozoic Craton and the active Pacific Plate. Ratio of vertical to horizontal scales is 1:1; the vertical scale has ticks every 10 km; the horizontal scale has ticks every 100 km. CC, continental crust; CUM, continental upper mantle; OUM, oceanic upper mantle; CURN, Curnamona Craton; MC, microcontinent. Reproduced with permission from Veevers JJ (ed.) (2000) *Billion-Year Earth History of Australia and Neighbours in Gondwanaland*. Sydney: GEMOC Press.

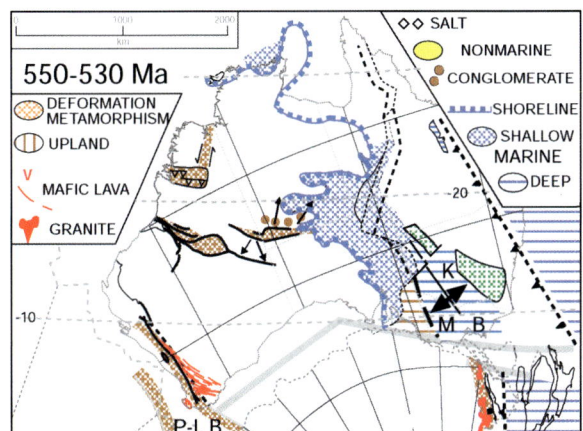

**Figure 3** Latest Neoproterozoic–earliest Cambrian (550–530 Ma) palaeogeography. The parallel lines crossing central-eastern Australia are the Tasman Line (heavy broken line) and the Tasman Toe-Line (light broken line), which mark the eastern outcrops of the Proterozoic craton and its wedge-out, respectively. KMB, Kanmantoo Marginal Basin; P-LB, Prydz-Leeuwin Belt. Reproduced with permission from Veevers JJ (ed.) (2000) *Billion-Year Earth History of Australia and Neighbours in Gondwanaland*. Sydney: GEMOC Press.

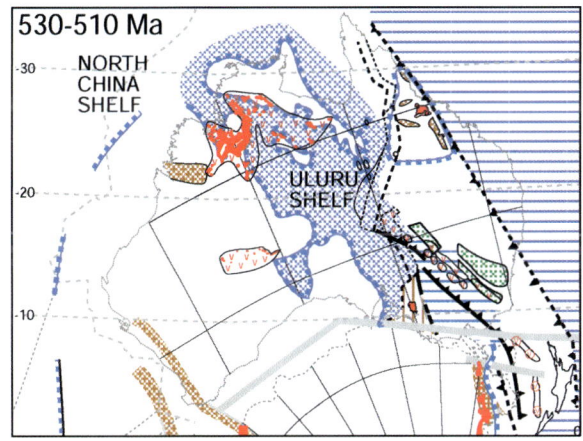

**Figure 4** Early and Middle Cambrian (530–510 Ma) palaeogeography. Legend as in **Figure 3**. Reproduced with permission from Veevers JJ (ed.) (2000) *Billion-Year Earth History of Australia and Neighbours in Gondwanaland*. Sydney: GEMOC Press.

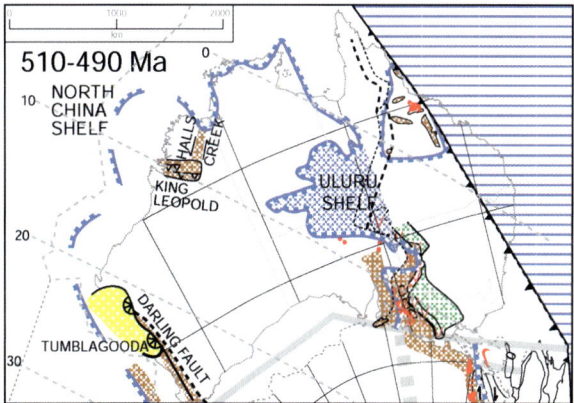

**Figure 5** Late Cambrian (510–490 Ma) palaeogeography. Legend as in **Figure 3**. Reproduced with permission from Veevers JJ (ed.) (2000) *Billion-Year Earth History of Australia and Neighbours in Gondwanaland*. Sydney: GEMOC Press.

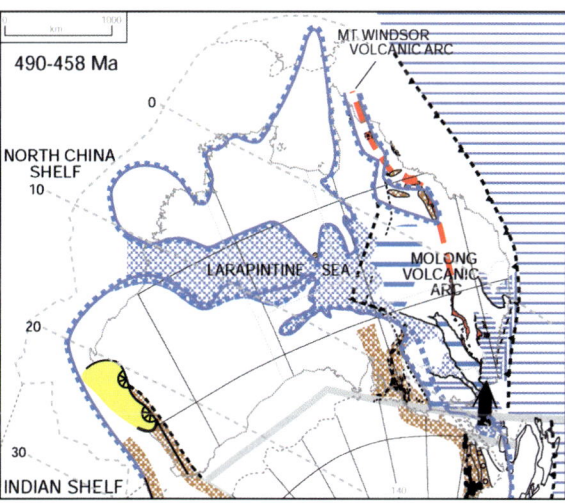

**Figure 6** Early Ordovician (490–458 Ma) palaeogeography. The cooling Ross and Delamerian granites are shown in black. Legend as in **Figure 3**. Reproduced with permission from Veevers JJ (ed.) (2000) *Billion-Year Earth History of Australia and Neighbours in Gondwanaland*. Sydney: GEMOC Press.

Figure 2. The initial dispersal of microcontinents by growth of a marginal basin and their return by closure of the basin were followed by subduction, generating a magmatic arc behind which granites were emplaced in an extended region. When the subducted slab shallowed, the crust was shortened; when it steepened, the trench rolled back in front of a growing marginal basin.

The events across Australia and neighbouring Antarctica are shown in plan view in the set of palaeogeographical figures.

## Latest Neoproterozoic–Earliest Cambrian (550–530 Ma)

The Uluru regime was inaugurated by collisions in West Gondwanaland and subduction in Antarctica, which deformed central and north-western Australia to produce mountain ranges and intervening basins where coarse arkose was deposited at the foots of the mountains, as at Uluru (**Figure 3**). In the south-west, the Prydz-Leeuwin Belt was terminally metamorphosed, and its periphery was intruded by a swarm of dykes.

In the south-east, the Kanmantoo Marginal Basin grew by back-arc spreading behind the subducting

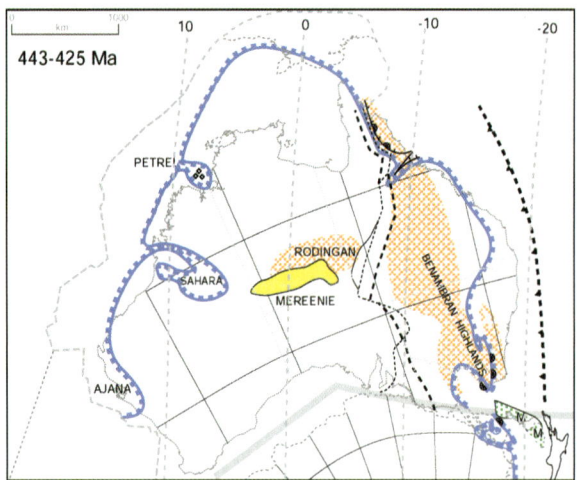

**Figure 7** Early Silurian (443–425 Ma) palaeogeography. MM, Melbourne-Mathinna Terrane. Legend as in **Figure 3**. Reproduced with permission from Veevers JJ (ed.) (2000) *Billion-Year Earth History of Australia and Neighbours in Gondwanaland*. Sydney: GEMOC Press.

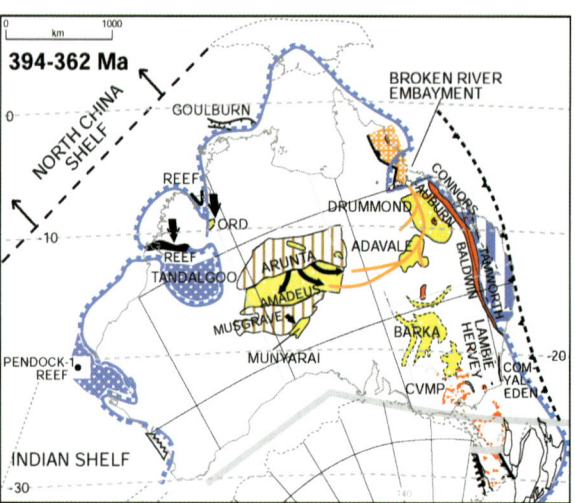

**Figure 9** Middle and Late Devonian (394–362 Ma) palaeogeography. COM–YAL–EDEN, Comerong–Yalwal–Eden Rift Zone; ORD, Ord Basin; CVMP, Central Victorian Magmatic Province. Legend as in **Figure 3**. Reproduced with permission from Veevers JJ (ed.) (2000) *Billion-Year Earth History of Australia and Neighbours in Gondwanaland*. Sydney: GEMOC Press.

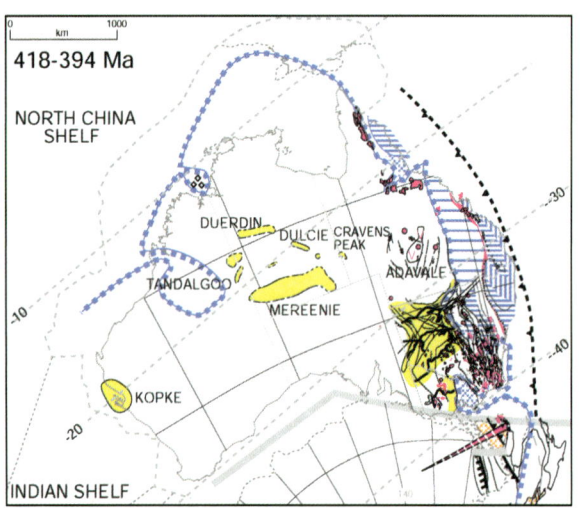

**Figure 8** Early Devonian (418–394 Ma) palaeogeography. Legend as in **Figure 3**. Reproduced with permission from Veevers JJ (ed.) (2000) *Billion-Year Earth History of Australia and Neighbours in Gondwanaland*. Sydney: GEMOC Press.

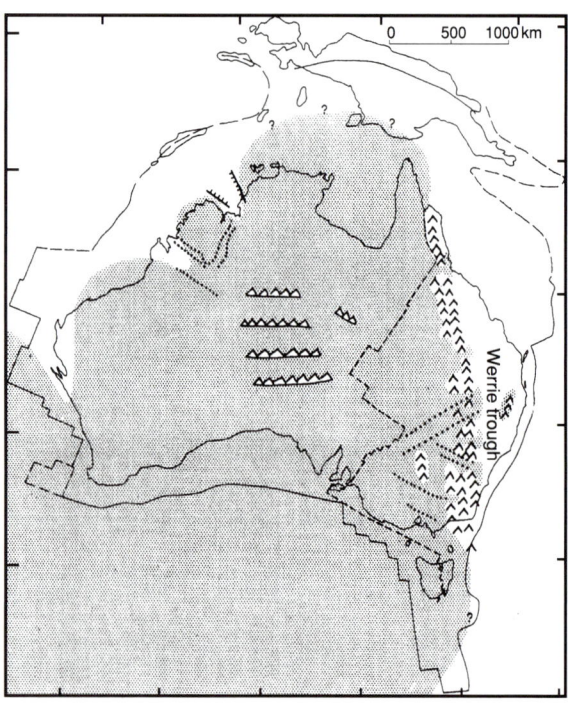

**Figure 10** Postulated extent of the mid-Carboniferous ice-sheet. Reproduced with permission from Veevers JJ (ed.) (2000) *Billion-Year Earth History of Australia and Neighbours in Gondwanaland*. Sydney: GEMOC Press.

Palaeo-Pacific Plate and wedged microcontinents off the mainland in the first episode of the Tasman Fold Belt System (**Figure 2ii**). In the north-east, an accretionary wedge grew in front of the trench. Central Australia was flooded by a shallow sea that lapped over the Uluru arkose, and another arm of the sea crossed the north. On the Antarctic side of a transform fault, another accretionary wedge grew in front of the trench and the first Ross granites were emplaced along the Transantarctic Mountains.

## Early–Middle Cambrian 530–510 Ma

Basalt flows over much of the north and centre (**Figure 4**) were crossed by the Uluru Shelf, which connected with the North China Shelf. An island in

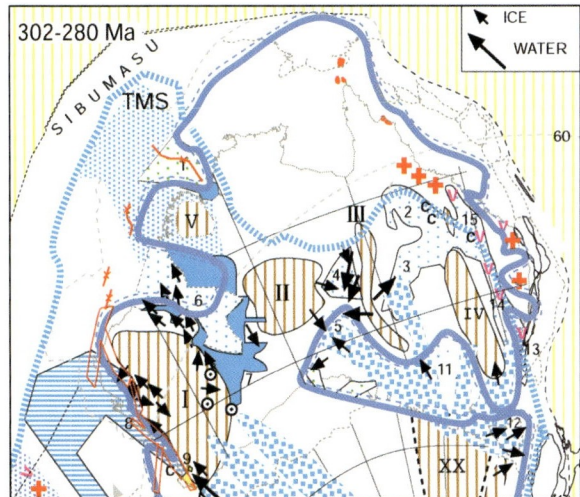

**Figure 11** Latest Carboniferous–earliest Permian (302-280 Ma) palaeogeography. Uplands are: I, ancestral Great Western Plateau; II, Central; III, Central-Eastern; IV, ancestral Eastern Australian foreswell; V, ancestral Kimberley Plateau; XX, Ross. Basins are 1, Bonaparte; 2, Galilee; 3, Cooper; 4, Pedirka; 5, Arckaringa; 6, Canning; 7, Officer; 8, Carnarvon–Perth; 9, Collie; 11, Oaklands; 12, Tasmania; 13, Sydney; 14, Gunnedah; 15, Bowen. Bullseyes indicate glacigenic sediment in mines. Coarse pattern, ice flow; solid blue, outcrops of glacigenic sediment; coarse blue stipple, subsurface glacigenic sediment; fine stipple, inferred glacigenic sediment. Red lines (west) indicate structure formed during Extension I. Legend as in **Figure 3**. Reproduced with permission from Veevers JJ (ed.) (2000) *Billion-Year Earth History of Australia and Neighbours in Gondwanaland*. Sydney: GEMOC Press.

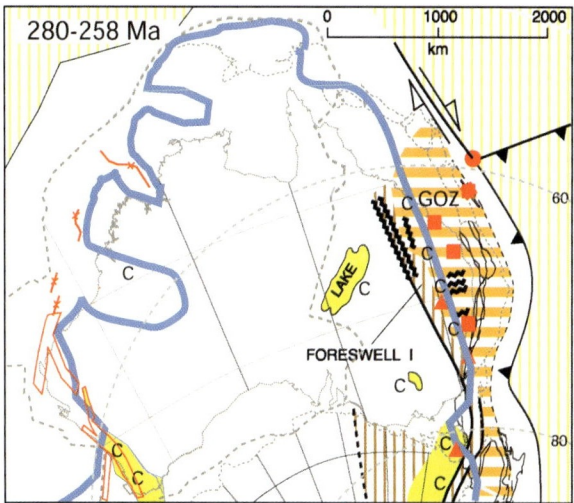

**Figure 12** Permian (280–258 Ma) palaeogeography. C, coal; GOZ, Gogango Overfolded Zone; red squares, granite; red triangles, tuff. Legend as in **Figure 3**. Reproduced with permission from Veevers JJ (ed.) (2000) *Billion-Year Earth History of Australia and Neighbours in Gondwanaland*. Sydney: GEMOC Press.

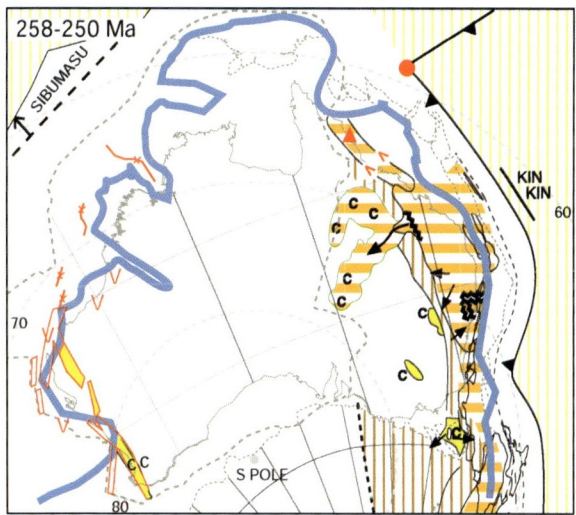

**Figure 13** Late Permian (258–250 Ma) palaeogeography. Red triangle, tuff. Legend as in **Figure 3**. Reproduced with permission from Veevers JJ (ed.) (2000) *Billion-Year Earth History of Australia and Neighbours in Gondwanaland*. Sydney: GEMOC Press.

the north-east was intruded by granite and deformed behind the trench. The Kanmantoo Marginal Basin started to close by north-eastward-directed subduction beneath an overlying volcanic arc (**Figure 2iii**). In Antarctica, an accretionary wedge continued to grow above the westward-subducting limb, and related Ross granites and marine volcanics were erupted.

## Late Cambrian (510–490 Ma)

At the tectonothermal climax of Gondwanaland, granites stretched from the Antarctic Peninsula through the Transantarctic Mountains to Tasmania (**Figure 5**), and the Delamerian granites were generated during closure of the marginal basin to form the Kanmantoo Fold Belt (**Figure 2iv**). The trench continued past north Queensland along a margin with a wide accretionary prism that was metamorphosed and intruded by plutons at 500 Ma.

In the west, uplift along the Darling Fault and within the Prydz-Leeuwin Belt shed fans of Tumblagooda sand into the nascent Perth–Carnarvon Basin. In the north-east, the King Leopold and Halls Creek Fold Belts were terminally metamorphosed and thrusted, and the enclosed block was uplifted and stripped.

## Early Ordovician (490–458 Ma)

The Delamerian–Ross Fold Belt, the zone of deformation with cooling granite (black dots), was intensely denuded in a tropical setting in which labile minerals and rock fragments were weathered away to provide a copious amount of quartzose sediment downslope to

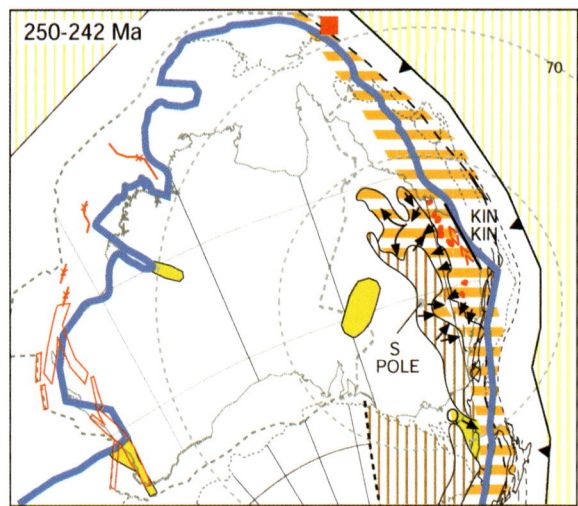

**Figure 14** Early Triassic (250–242 Ma) palaeogeography. Legend as in **Figure 3**. Reproduced with permission from Veevers JJ (ed.) (2000) *Billion-Year Earth History of Australia and Neighbours in Gondwanaland*. Sydney: GEMOC Press.

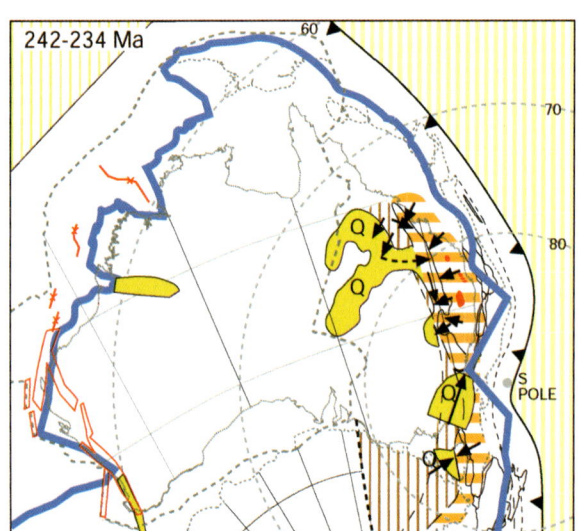

**Figure 15** Early–Middle Triassic (242–234 Ma) palaeogeography. Red, rejuvenated and initiated faults and depositional axes; Q, quartzose sandstone. Legend as in **Figure 3**. Reproduced with permission from Veevers JJ (ed.) (2000) *Billion-Year Earth History of Australia and Neighbours in Gondwanaland*. Sydney: GEMOC Press.

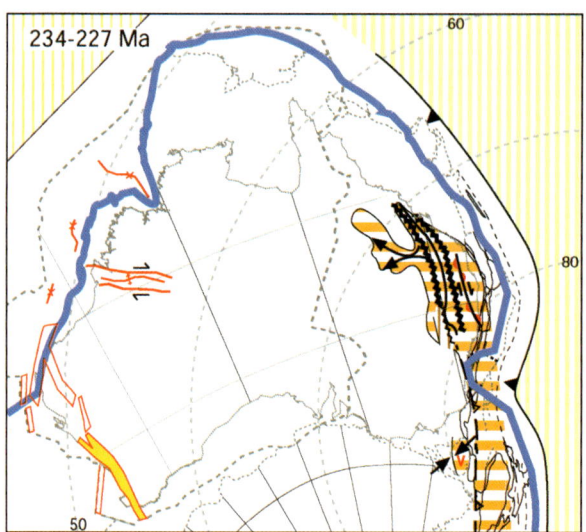

**Figure 16** Mid-Triassic (234–227 Ma) palaeogeography. Legend as in **Figure 3**. In the northern Canning Basin, an anticlinal axis (red line crossed by double-headed arrow) formed between transcurrent faults (red lines) with black arrows. Reproduced with permission from Veevers JJ (ed.) (2000) *Billion-Year Earth History of Australia and Neighbours in Gondwanaland*. Sydney: GEMOC Press.

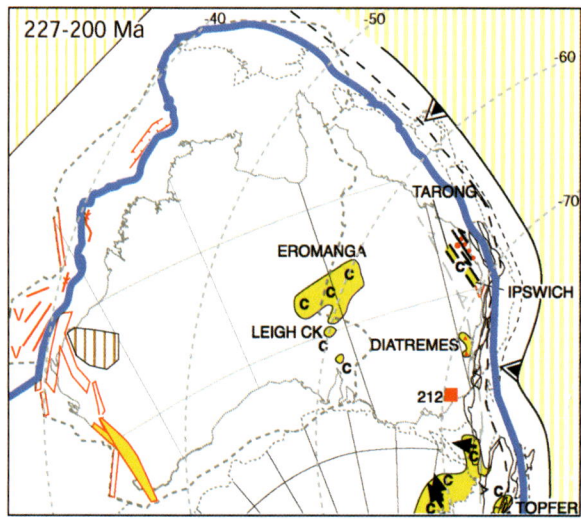

**Figure 17** Late Triassic (227–200 Ma) palaeogeography. The double-headed barbs on the trench indicate the change from shallowly dipping to steeply dipping subduction. Legend as in **Figure 3**. The red lines in the west represent faults. Reproduced with permission from Veevers JJ (ed.) (2000) *Billion-Year Earth History of Australia and Neighbours in Gondwanaland*. Sydney: GEMOC Press.

Australia (**Figure 6**). Westward-directed subduction of the Palaeo-Pacific Plate beneath Eastern Australia generated a volcanic arc that stretched 3000 km northwards in front of a marginal basin (**Figure 2v**). Quartzose turbidites were shed eastwards (transversely) from the Delamerian upland, and volcanogenic sediment was shed from the volcanic arc. Turbidites were shed northwards (longitudinally) from Antarctica, as shown by 700–500 Ma zircons, along the fore-arc basin and abyssal plain in vast submarine fans, comparable to the modern Bengal fan. Behind the shelf, the Larapintine Sea crossed the interior towards the western sea, which intermittently advanced across the Tumblagooda sand.

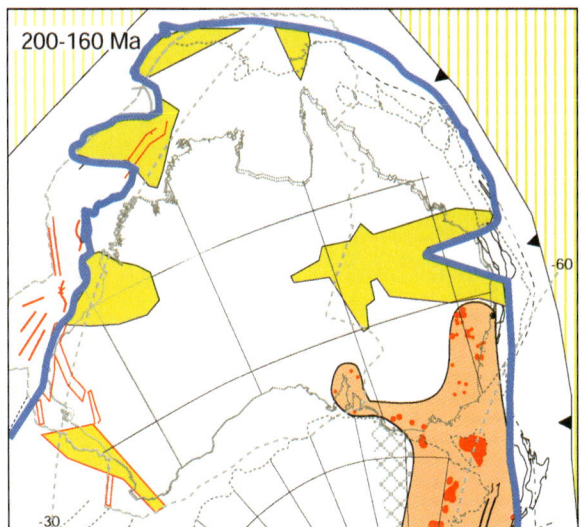

**Figure 18** Early and Middle Jurassic (200–160 Ma) palaeogeography. The deep red shading represents basalt or dolerite within the large igneous field (pink). Legend as in **Figure 3**. Reproduced with permission from Veevers JJ (ed.) (2000) *Billion-Year Earth History of Australia and Neighbours in Gondwanaland*. Sydney: GEMOC Press.

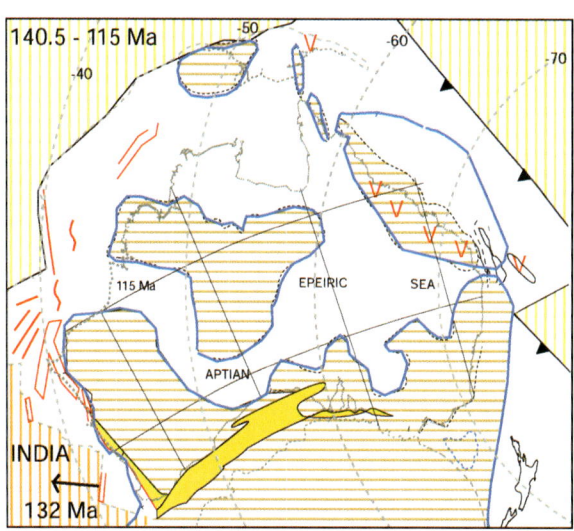

**Figure 20** Neocomian to Aptian (140.5–115 Ma) palaeogeography. Legend as in **Figure 3**. Reproduced with permission from Veevers JJ (ed.) (2000) *Billion-Year Earth History of Australia and Neighbours in Gondwanaland*. Sydney: GEMOC Press.

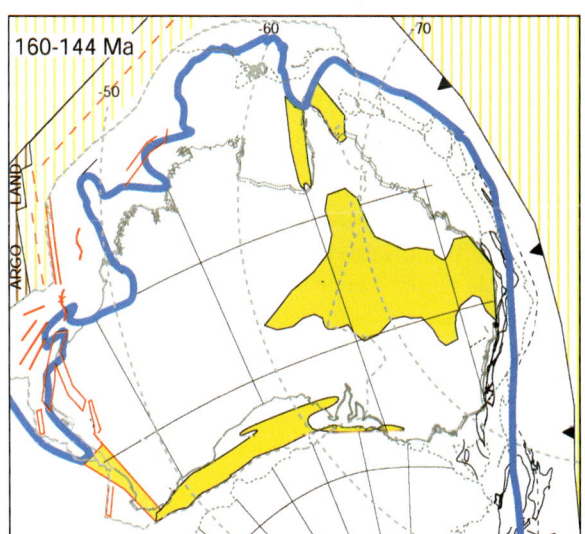

**Figure 19** Late Jurassic (160–144 Ma) palaeogeography. Legend as in **Figure 3**. Reproduced with permission from Veevers JJ (ed.) (2000) *Billion-Year Earth History of Australia and Neighbours in Gondwanaland*. Sydney: GEMOC Press.

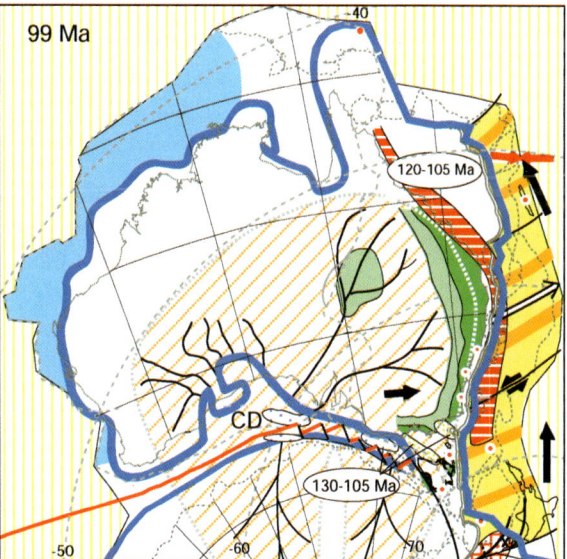

**Figure 21** Cenomanian (99–93 Ma) palaeogeography. Red arrows, previous velocities; black arrows, prospective velocities (which are unchanged in Australia but swerve through 70° in the Pacific Plate); double line, detachment fault; yellow and tan shading, lower-plate margin (Lord Howe Rise), white dotted line, division of drainage; CD, Ceduna depocentre; green, eroding uplands; oblique orange lines, south-western drainage into Australian–Antarctic depression; red horizontal shading, Whitsunday Volcanic Province; red dots, 99 Ma alkaline volcanics; light blue, carbonate deposition. Legend as in **Figure 3**. Reproduced with permission from Veevers JJ (ed.) (2000) *Billion-Year Earth History of Australia and Neighbours in Gondwanaland*. Sydney: GEMOC Press.

## Early Silurian (443–425 Ma)

Closure of the marginal basins in eastern Australia generated the Benambran and related highlands (**Figure 7** and **Figure 2vi**). The Rodingan uplift in the centre shed the Mereenie Sandstone. In the Canning Basin, the Sahara Formation was deposited in a

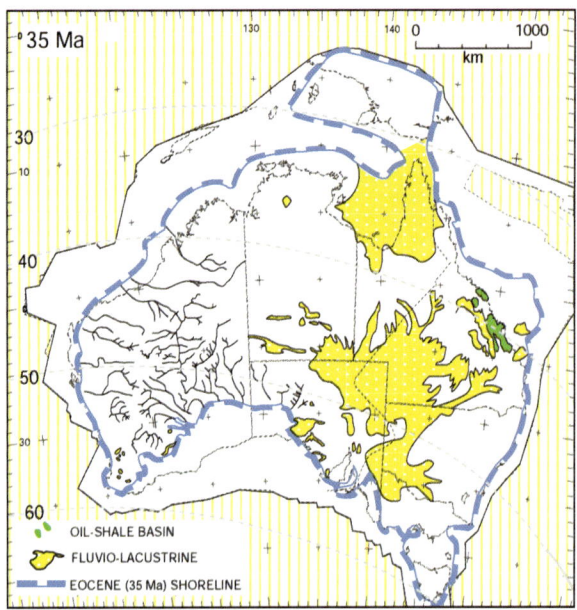

**Figure 22** Eocene (35 Ma) palaeogeography. Legend as in **Figure 3**. Reproduced with permission from Veevers JJ (ed.) (2000) *Billion-Year Earth History of Australia and Neighbours in Gondwanaland*. Sydney: GEMOC Press.

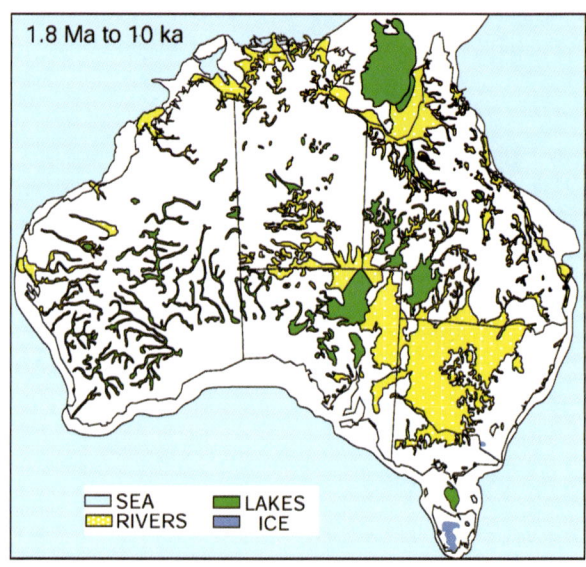

**Figure 24** Pleistocene (1.8 Ma to 10 ka) palaeogeography. Reproduced with permission from Veevers JJ (ed.) (2000) *Billion-Year Earth History of Australia and Neighbours in Gondwanaland*. Sydney: GEMOC Press.

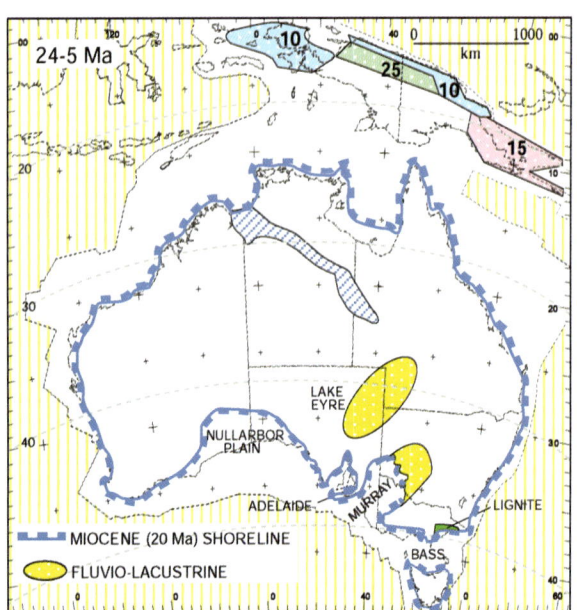

**Figure 23** Miocene (24–5 Ma) palaeogeography. Numerals indicate the times (Ma) at which the terranes docked. The blue oblique shading represents marine to nonmarine sediment. Legend as in **Figure 3**. Reproduced with permission from Veevers JJ (ed.) (2000) *Billion-Year Earth History of Australia and Neighbours in Gondwanaland*. Sydney: GEMOC Press.

restricted sea. In the Carnarvon Basin, the Ajana Formation succeeded the Tumblagooda Sandstone. In the Petrel Sub-Basin, salt was probably deposited during the Silurian and Devonian. Western Tasmania was covered by a shallow sea, while the Melbourne–Mathinna terrane slid past on its way to docking at 400 Ma.

## Early Devonian (418–394 Ma)

The New Zealand terranes amalgamated at 415 Ma, and north-east and West Tasmania amalgamated at 400 Ma (**Figure 8**). In south-east Australia, relocation of the trench was accompanied by wide back-arc extension and emplacement of granite (**Figure 2vii**). The Adavale Basin was initiated with flows of andesite. Farther north, inner shallow-marine and outer deep-marine deposits were flanked by granitic plutons. A group of nonmarine deposits (Cravens Peak, Dulcie, Duerdin, and Mereenie) was deposited in the centre, salt in the Petrel Sub-Basin, aeolian and playa deposits of the Tandalgoo Formation in the Canning Basin, and the Kopke redbeds of the Carnarvon Basin.

## Middle and Late Devonian (394–362 Ma)

Granodiorite was erupted in North Victoria Land during continuing westward-directed subduction (**Figure 9**). Following the contractional Tabberabberan Orogeny, plutonic rocks were emplaced and volcanic rocks were erupted in a broad zone in Tasmania, Victoria, and New South Wales, which included the Comerong–Yalwal–Eden volcanic rift. Fluvial sediment was deposited in the Barka, Lambie,

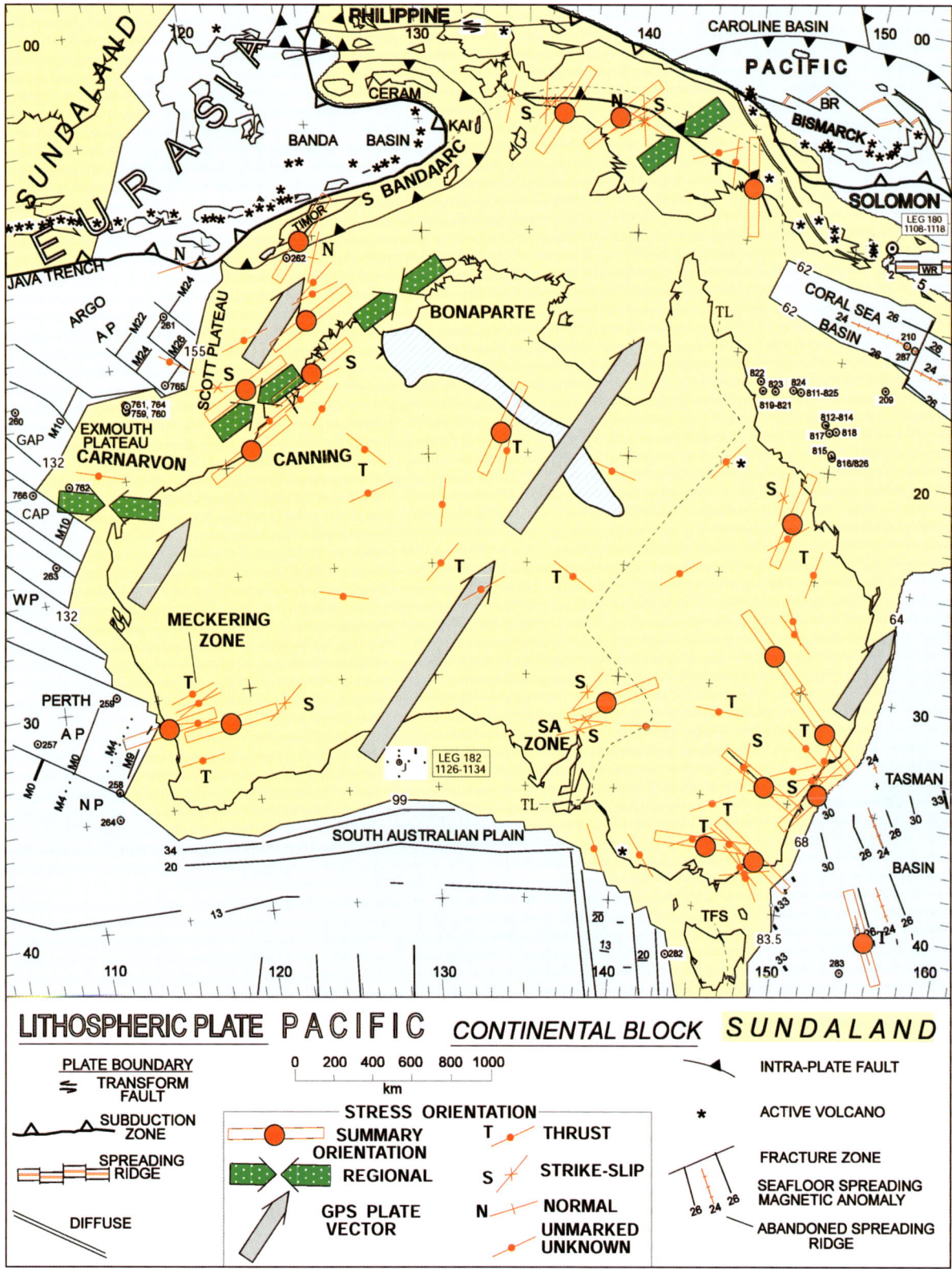

**Figure 25** Tectonic map of the continental lithosphere of Australia–New Guinea (yellow) and Sundaland (part of the Eurasian Plate; yellow) and the surrounding oceanic lithosphere (blue), including the Pacific Plate and the Solomon sub-Plate to the north-east. The ocean is marked by Deep Sea Drilling Project and Ocean Drilling Program sites (bullseyes) and selected seafloor-spreading magnetic anomalies. The age of the continent–ocean boundary is given in Ma. The oblique blue band across northern Australia signifies Miocene or younger sediment. Also shown are the orientation of maximum horizontal stress and the plate vector. AP, Abyssal Plain;

*Continued*

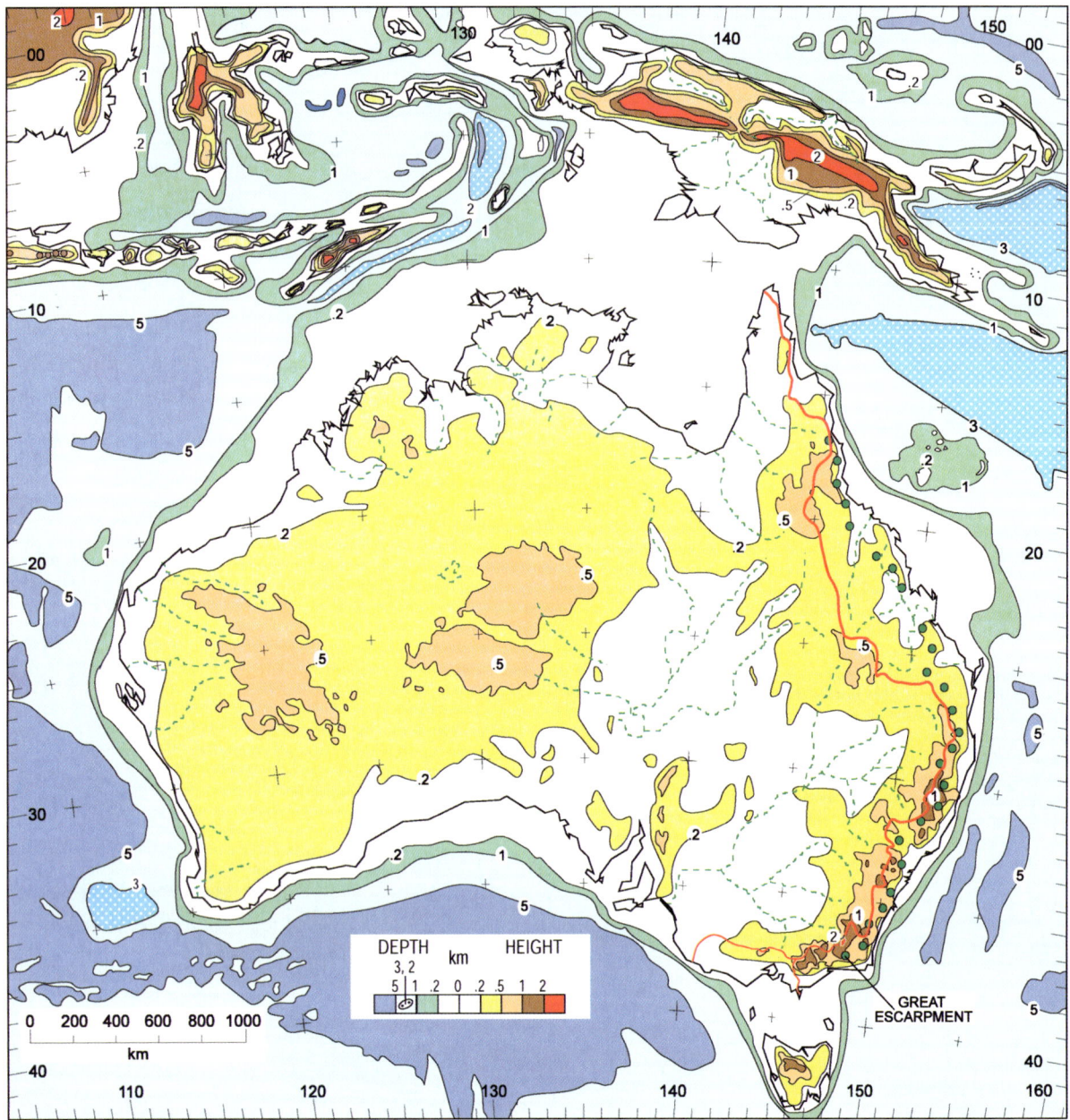

**Figure 26** Morphology of Australia and the neighbouring land, sea, and ocean. Shown in eastern Australia are the Great Escarpment (green dots) and the drainage divide (red line). Reproduced with permission from Veevers JJ (ed.) (2000) *Billion-Year Earth History of Australia and Neighbours in Gondwanaland.* Sydney: GEMOC Press.

and Hervey Basins behind the Baldwin arc. The fore-arc basin and accretionary prism (**Figure 2viii**) extended north as the Connors–Auburn arc.

In central Australia, the uplifted Arunta Block shed alluvial fans to the south, which were themselves folded and eroded. Fans in the Munyarai Trough were likewise shed from the overthrusting Musgrave Block. The eastward flow from central Australia through the Adavale Basin swung northwards (parallel to the Connors–Auburn arc) in the Drummond Basin and debouched in the Broken River Embayment, which was deformed in the latest Devonian.

BR, Bismarck Rift; C, Cuvier; G, Gascoyne; J, Jerboa-1; NP, Naturaliste Plateau; SA, South Australia; TL, Tasman Line; TFS, Tamar Fracture System; WP, Wallaby Plateau; WR, Woodlark Rift. Reproduced with permission from Veevers JJ (ed.) (2000) *Billion-Year Earth History of Australia and Neighbours in Gondwanaland.* Sydney: GEMOC Press.

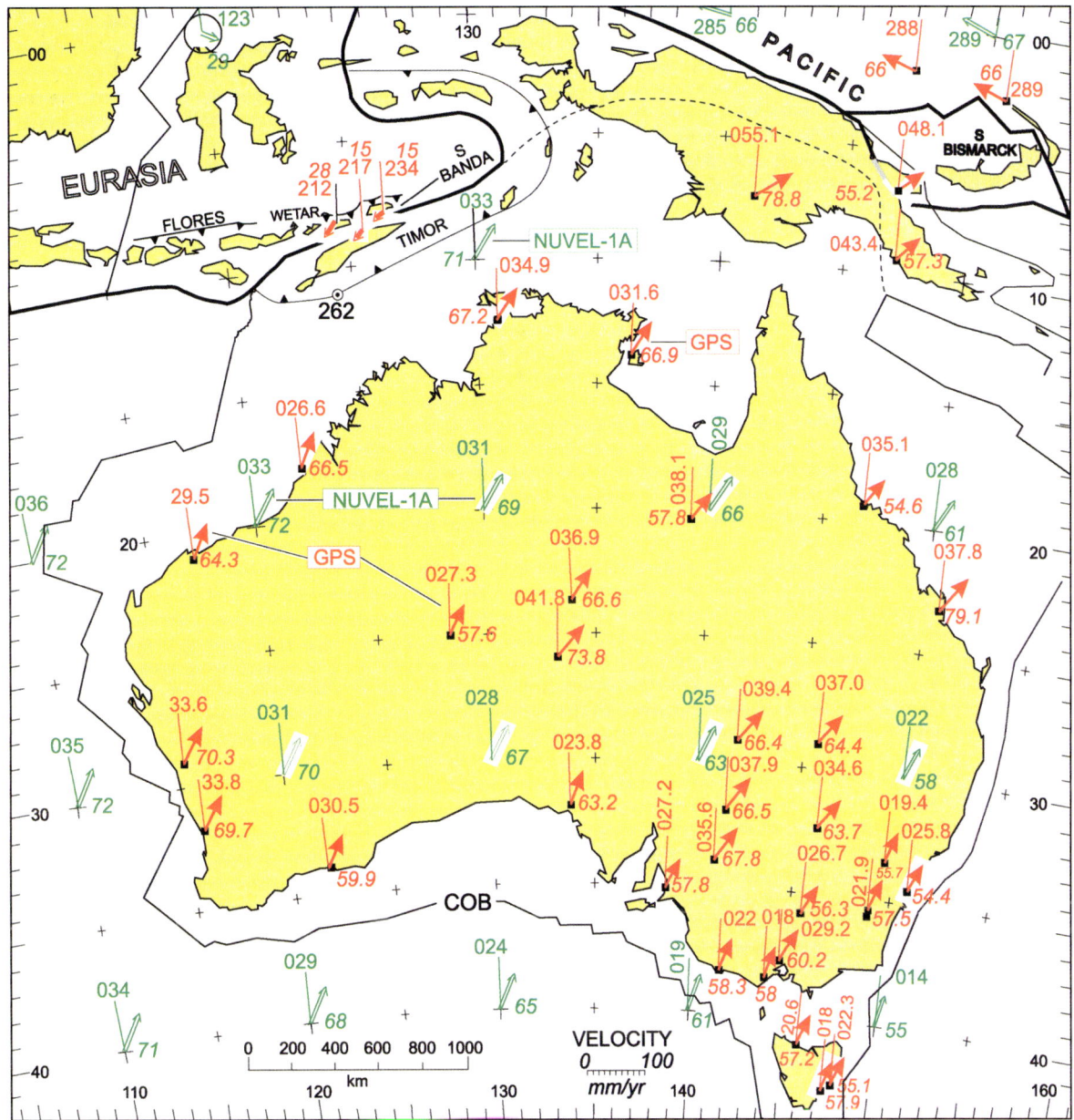

**Figure 27** Velocities of sites on the Australian Plate and adjacent Pacific and Eurasian plates. Red arrows indicate velocities measured by GPS; double-shafted green arrows indicate velocities calculated from NUVEL-1A at selected 10° crosses. For each arrow, the fine line is the north azimuth, the three-digit numeral the bearing, and the italicized numeral the velocity in mm year$^{-1}$. The thrust line of Timor passes through DSDP (Deep Sea Drilling Project) site 262. Reproduced with permission from Veevers JJ (ed.) (2000) *Billion-Year Earth History of Australia and Neighbours in Gondwanaland.* Sydney: GEMOC Press.

**Table 1** The three stratitectonic regimes in the development of Phanerozoic Australia

| Regime | Age (Ma) | Platform stratigraphy | Tectonics | Latitude |
|---|---|---|---|---|
| Potoroo | 99–0 | Quartz, carbonate, and evaporite | Post-Pangaean | High to tropical |
| Innamincka | 320–99 | Glacigenic and coal measures | Pangaean | Polar |
| Uluru | 544–320 | Quartz and carbonate | Gondwanaland | Tropical |

The North China terrane broke off the Palaeo-Tethyan margin. In the north, the Goulburn Graben was initiated; the Bonaparte Basin was rejuvenated and Frasnian alluvial fans were deposited at the foot of a fault and succeeded by a Famennian reef complex. In the Ord Basin, sand was deposited southwards from the same fault block. In the northern Canning Basin, gravel fans from the Kimberley Block mingled with a reef complex on a platform in front of a deep axis; to the south, the paralic Tandalgoo Formation was deposited in an arm of the sea. In the Carnarvon Basin, shelf limestone and inshore quartz sand give way in Pendock-1 to reef carbonate. Reworked Late Devonian conodonts in the Perth Basin suggest that the shoreline extended southwards.

## Carboniferous (350–300 Ma)

South-eastern Australia was subjected to an Early Carboniferous (350–330 Ma) east–west contraction, followed by Mid-Carboniferous emplacement of granites (**Figure 2x**) and finally north–south contraction in megakinks. Similar events took place in North Queensland. In between, nonmarine deposition in the Drummond Basin ended at 330 Ma when gentle folding began. Central Australia was terminally folded and thrusted in the Mid-Carboniferous (320 Ma) Alice Springs Orogeny, which was echoed by transpression in the Canning Basin, by folding and faulting in the Carnarvon Basin, and by overthrusting in the Officer Basin. These movements, represented by the Mid-Carboniferous lacuna, mark the termination of the Uluru regime (when the Uluru Arkose was upended) and the inception of the Innamincka regime.

This rapid uplift and concomitant downwearing must have produced copious amounts of sediment, yet the depositional record over the Australian platform, except the eastern and north-western margins, is blank. Where could the sediment have gone? The paradox is resolved by postulating that uplift combined with rapid polar movement triggered not only an alpine glaciation in the east but also a continent-wide dry-based ice-sheet (**Figure 10**), which carried away sediment shed from the nunataks of the central uplifts and from the eastern cordillera and deposited nonmarine glacial sediment in the Werrie Trough. Only during its retreat in the earliest Permian did the ice release its load of sediment.

The Mid-Carboniferous marked a turning point in Australia, from the Uluru regime, with warm-water low-latitude carbonate facies, to the Innamincka regime, with cold-water high-latitude detrital facies.

## Latest Carboniferous–Earliest Permian (302–280 Ma)

The Mid-Carboniferous lacuna ended with the latest Carboniferous (306 Ma) onset of Pangaean Extension I, which was accompanied by the eruption of S-type granites and felsic volcanics in north-eastern Australia and followed at 302 Ma by deposition of glacigenic sediment in the space provided by the extension (**Figure 11**). Eastern Australia was subjected to continuing dextral transtension, which produced an orocline, related pull-apart basins, and widespread magmatism (**Figure 2xii**). At polar latitudes ice flowed into depressions between uplands, and on the periphery into marine basins.

The west was extended into grabens and synclines. Ice flowed northwards over and around the ancestral Great Western Plateau and across the Canning Basin. Farther north, the ice possibly flowed through the Bonaparte Basin and through the restored position of the Tangchong–Malay–Sumatra Block in Sibumasu to the shore of Palaeo-Tethys. In central and eastern Australia, ice flowed from the central and central-eastern uplands into the Pedirka and Arckaringa basins, and from the central-eastern upland eastwards into the Cooper Basin, which also received ice from the ancestral eastern Australian foreswell to the east. From the Ross upland in Antarctica, ice flowed northwards into areas that were later, at 288 Ma, covered by an ephemeral shallow sea that followed the melting of the ice.

## Early Permian (280–258 Ma)

By 265 Ma, the magmatic arc, marked by granite and tuff, had migrated northwards opposite central Queensland (**Figure 12**), and the following tracts, from craton to margin, can be identified: first, an epicontinental basin with nonmarine sediment; second, foreswell I (vertical bands), which shed quartzose sediment into, third, the initial foreland basin of the Bowen–Gunnedah–Sydney Basin, which is marked by coal alternating with shallow-marine deposits; and, fourth, an orogen and magmatic arc that were intensely deformed in the Gogango Overfolded Zone, which shed sediment back into the foreland basin. In adjoining Antarctica, foreswell I separated coal measures on the craton from the foreland basin.

In the west, marine sediment alternated with nonmarine sediment including coal.

## Late Permian (258–250 Ma)

The magmatic arc propagated north-east to a position opposite 250 Ma tuff (**Figure 13**). Coal measures reached a climax. The New England Fold Belt and adjacent basins were deformed (**Figure 2xiii**), and copious sediment flowed into the foreland basin and the distal Galilee Basin. The last ice melted. The Kin Kin terrane approached the trench.

In the west, Sibumasu broke off, the shoreline regressed, and rhyolitic to undersaturated volcanics were erupted.

## Early Triassic (250–242 Ma)

The arc advanced to New Guinea, and plutons saturated the suture zone of the amalgamated Kin Kin terrane (**Figure 14**). In the foreland basin, cratonic and orogenic drainage joined in axial flow from a saddle just north of 30° S (present coordinates), and volcanolithic sediment poured westwards into the Galilee Basin. Tectonics was as before, but the surface environment was radically different. Coal measures and tuff were replaced by measures with redbeds devoid of both coal and tuff. The coal gap reflects the global extinction of peat-forming plants at the Permian–Triassic boundary, and tuff was eliminated by weathering in the warmer climate, even at polar latitudes.

In the west, shale was deposited behind a transgressive shoreline.

## Early–Middle Triassic (242–234 Ma)

Volcanolithic sandstone was deposited in the foreland basin (**Figure 15**). Sediment flow in the Bowen Basin was now wholly southerly, and this was joined later in the stage by cratonic (quartzose) sediment flowing in an easterly direction from the Cooper–Galilee Basin. In the Sydney Basin the Hawkesbury Sandstone was derived from the Ross craton to the south-west, which also fed a quartz sandstone in Tasmania. The pole was nearby but ice is not indicated.

## Mid-Triassic (234–227 Ma)

The complex of events called Gondwanides II included thrusting of the fold belt over the final deposits of the foreland basin and the spread of volcanolithic sediment to the west (**Figure 16**). The fold belt itself was intruded by granite along a transcurrent fault. At the same time, the northern Canning Basin was broadly folded during wrenching.

## Late Triassic (227–200 Ma)

The Late Triassic saw the onset of Pangaean Extension II. In the east the change from contraction to extension reflected a change from shallowly to steeply dipping subduction (**Figure 17**). Maar diatremes were erupted in the Sydney Basin, and a 212 Ma syenite complex was emplaced in Victoria.

Deposition of coal resumed after a gap of 23 Ma. New intramontane basins included the Ipswich and Tarong coal basins; the Eromanga Basin with carbonaceous sediment subsided in the sump between the fold belt and the craton; the Leigh Creek and associated coal basins subsided on Proterozoic crust; and deposition of coal resumed in Tasmania and Antarctica.

In the west, the margins were rejuvenated by wrenching, downfaulting, and folding. Uplands in the Pilbara Block and Yilgarn Craton shed coarse detritus into the rifted Perth and Carnarvon Basins. Volcanics accompanied rifting in the Exmouth Plateau.

## Early and Middle Jurassic (200–160 Ma)

The Permian–Triassic sedimentary successions were capped and intruded by a flood of 184–179 Ma tholeiitic basalts in southern Africa, Antarctica, and Tasmania, and by scattered volcanics in south-eastern Australia (**Figure 18**). The Eromanga Basin expanded across central-eastern Australia, with quartzose sediment derived from the craton and volcanolithic sediment from a presumed arc. In Tasmania, the dolerite rose into the Permo-Triassic sedimentary rocks of the Tasmania Basin as sills up to 400 m thick and dykes up to 1 km wide. Extrusive tholeiite is known near the southern tip only. In south-eastern Australia, the magma forms flows and shallow intrusions.

Thick sands accumulated in the western rifts, and sand sheets lay behind transgressive shorelines in the Canning and Bonaparte Basins.

## Late Jurassic (160–144 Ma)

Argo Land split off north-western Australia at 156 Ma as a result of seafloor spreading in the Argo Abyssal Plain (**Figure 19**). The rifts along the western and southern margins continued to fill with sand and silt. The Eromanga Basin expanded northwards to New Guinea.

## Neocomian to Aptian (140.5–115 Ma)

In a radical change, the shoreline advanced into the interior of Australia (**Figure 20**). The epeiric sea isolated large islands in the north-west and north and was confined by the rising rifted arch to the south and by the magmatic arc of the Pacific margin to the north-east. Volcanolithic sediment from the arc overran sediment from the craton.

The Southeast Indian Ocean opened at 132 Ma between India and Australia–Antarctica.

## Cenomanian (99–93 Ma)

The change from the Innamincka regime to the Potoroo regime was marked by breakup along the southern margin, cessation of subduction in the east, and intense uplift of the eastern Highlands and epeirogenic uplift elsewhere, all occasioned by the 99 Ma change in the azimuth of subduction of the Pacific Plate from head-on (red arrows in **Figure 21**) to sideswipe (black arrows). A detachment fault along the continent–ocean boundary separated passive-margin mountains on the upper-plate margin, buoyed by thick underplating of mantle-derived melts, from the conjugate extremely attenuated lower-plate margin of the Lord Howe Rise. Drainage was divided near the present coast into a long south-western slope feeding the Ceduna depocentre, which accumulated 8 km of Late Cretaceous sediment, and a short eastern slope feeding the low-lying eastern borderland.

The emergence of the land, in particular the rising (and eroding) of the eastern highlands, forced a major Cenomanian (99 Ma) regression at a time of marine flooding elsewhere. The inception of south-western drainage into the Australian–Antarctic depression coincided with the inception of the juvenile ocean that grew out of the rifted arch and was preceded by 130–105 Ma deposition of volcanogenic sediment in the Otway, Bass, and Gippsland basins. In the east, the 120–105 Ma Whitsunday Volcanic Province was succeeded by *ca.* 99 Ma alkaline volcanics, and the previous volcanogenic sediment from the arc and in the south-eastern rift basins was succeeded by quartzose sediment. In the north and west, the shoreline regressed to and beyond its present position, and carbonate replaced detrital sediment on the margin of the widening Indian Ocean.

## Eocene (35 Ma)

Following the opening of the Tasman Sea and Coral Sea, the Potoroo regime can be shown on a modern geographical base except for the higher palaeolatitudes (**Figure 22**). The sea transgressed the western half of the southern margin, and an ancestral Arafura Sea lay between northern Australia and New Guinea. Relaxation of the platform saw sediment accumulate in the central-eastern lowlands and in grabens in central Australia and south-eastern Queensland, including deposits of oil shale and, in isolated pockets, fluviolacustrine gravel, sand, silt, clay, and lignite, which were overlain by basalt in the eastern Highlands, too small to show in **Figure 22**. In the west, the uplift was etched by rivers except in an area of lakes in the south-west.

## Miocene (24–5 Ma)

At 20 Ma, another transgression in the south crossed the Nullarbor Plain, the Adelaide region, the Murray Basin, and Bass Strait, and thick lignite accumulated in the onshore Gippsland Basin (**Figure 23**). Sheets of sand accumulated in the Lake Eyre drainage system and in the Murray Basin.

In New Guinea, the Sepik terrane had docked by 25 Ma and thrusted south-westwards to form a foreland basin; it was joined at 15 Ma by a composite terrane in the south-east, and at 10 Ma by terranes in the north and north-west. Miocene and younger marine to nonmarine sediment was deposited behind a foreswell across northern Australia. By the end of the Miocene the geography of the region was as it is today except that the palaeolatitude (at 16 Ma) was 10° farther south than today.

## Pleistocene (1.8 Ma to 10 ka)

By the end of the Pleistocene, at 10 ka, sea-level was halfway between the level at the Last Glacial Maximum and its present level (**Figure 24**). Lakes occupied the Gulf of Carpentaria and the Bass Strait, and, as today, dry lakes covered much of interior Australia The central-eastern lowlands in New South Wales were covered by sheets of fluvial sediment. Glacial deposits were confined to Tasmania and the Mount Kosciusko area. Aeolian dunes covered much of the arid interior, including the relict drainage of the south-west.

## Present Tectonics and Morphology

The continental lithosphere of Australia–New Guinea is surrounded by the oceanic lithosphere of divergent oceans on its western, southern, and eastern margins and of convergent oceans on its northern margin from Timor to the Papuan Peninsula (**Figure 25**). The convergent boundary is marked by high relief, 3 km above sea-level in Timor and 5 km in New Guinea (**Figure 26**), intense Earth movement and seismicity, and scattered volcanicity. South of this boundary, as befits its intraplate position, Australia is low-lying, with a relief of 2.2 km in the south-east and little more than 1 km elsewhere, relatively quiescent, except for diffuse seismicity in the north-west sector, in zones about Meckering in the south-west, in South Australia, and in the south-east, and has only two areas of volcanicity, one in north-east Queensland, the other on the southern margin at 141° E.

The submarine features of the divergent margins have a certain bilateral symmetry about the meridian halfway across Australia: complexes of marginal plateaus and abyssal plains are backed by a broad shelf in the north, and narrow margins expand into long appendages (Naturaliste Plateau and Tasmania–South Tasman Rise) in the south. This pattern extends to the arrangement of seafloor-spreading magnetic anomalies in that the azimuth of the anomalies off the west coast is reflected in that off the east coast, while the set to the south is crossed by the line of symmetry. The age of onset of spreading follows another pattern, with anticlockwise propagation from 156 Ma in the north-west through 132 Ma in the west, and 99 Ma in the south, and as back-arc spreading at 83.5 Ma, 68 Ma, and 64 Ma in the south-east, 62 Ma in the north-east, and 5 Ma off eastern New Guinea.

## Present Motion

Figure 27 shows the velocities (azimuth and magnitude) from a few years of Global Positioning System (GPS) measurements and from the NUVEL-1A global-plate-motion model averaged over several million years. For example, the GPS estimate in western Victoria of 58.3 mm year$^{-1}$ at 022° approximates the NUVEL-1A estimate of 61 mm year$^{-1}$ at 019°, and that near Darwin of 67.2 mm year$^{-1}$ at 034.9° matches the NUVEL-1A estimate of 71 mm year$^{-1}$ at 033°. Most of this motion comes from seafloor spreading in the Southeast Indian Ocean at a full rate of 60 mm year$^{-1}$ from an almost stationary Antarctica.

The stations on mainland New Guinea, including one near the plate boundary, record an azimuth of ca. 050°, which locates them on the south-western (Australian) side of the plate boundary. On the other side, the Pacific Plate is moving at a velocity of 66 mm year$^{-1}$ at 289°, almost at a right angle to Australia's motion, confirming the idea expressed by Alfred Wegener that the Pacific's westerly motion has planed off the obtruding anvil of New Guinea.

According to NUVEL-1A, the Eurasian Plate is moving slowly south-eastwards. GPS measurements on either side of the Flores and Wetar thrusts show that the southern Banda arc is virtually accreted to the Australian plate margin and that most of the convergence seems to be accommodated by left-lateral slip at the plate boundary, suggesting that the Timor Trough is now inactive as a thrust.

## See Also

**Antarctic. Australia:** Proterozoic; Tasman Orogenic Belt. **Gondwanaland and Gondwana. New Zealand. Oceania (Including Fiji, PNG and Solomons). Palaeoclimates. Pangaea. Plate Tectonics. Tectonics:** Mountain Building and Orogeny.

## Further Reading

BMR Palaeogeographic Group (1990) *Evolution of a Continent.* Canberra: Bureau of Mineral Resources, Geology and Geophysics.

Veevers JJ (ed.) (1984) *Phanerozoic Earth History of Australia.* Oxford: Clarendon Press.

Veevers JJ (ed.) (2000) *Billion-Year Earth History of Australia and Neighbours in Gondwanaland.* Sydney: GEMOC Press.

Veevers JJ (2001) *Atlas of Billion-Year Earth History of Australia and Neighbours in Gondwanaland.* Sydney: GEMOC Press.

# Tasman Orogenic Belt

**D R Gray**, University of Melbourne, Melbourne, VIC, Australia
**D A Foster**, University of Florida, Gainesville, FL, USA

© 2005, Elsevier Ltd. All Rights Reserved.

## Introduction

The eastern part of Australia formed along the margin of Gondwana during the Palaeozoic due to the accretion of oceanic platform and basinal sequences (**Figure 1**). A region of eastern Australia over 1000 km wide now consists of three distinct orogenic belts (the Delamerian, Lachlan–Thomson, and New England orogens), which are collectively referred to as the Tasman Orogenic Belt (**Figures 1 and 2**). This deformed and metamorphosed tract of rocks was once part of a major orogenic system that extended some 20 000 km along the Gondwanan margin, incorporating parts of the Andes in South America (see **Andes**), the Cape Fold Belt in southern Africa, and the Ross Orogen in Antarctica (see **Antarctic**) (**Figure 1**). In eastern Australia accretion occurred in a stepwise fashion, with an eastward younging from the Cambrian to the Triassic, reflected

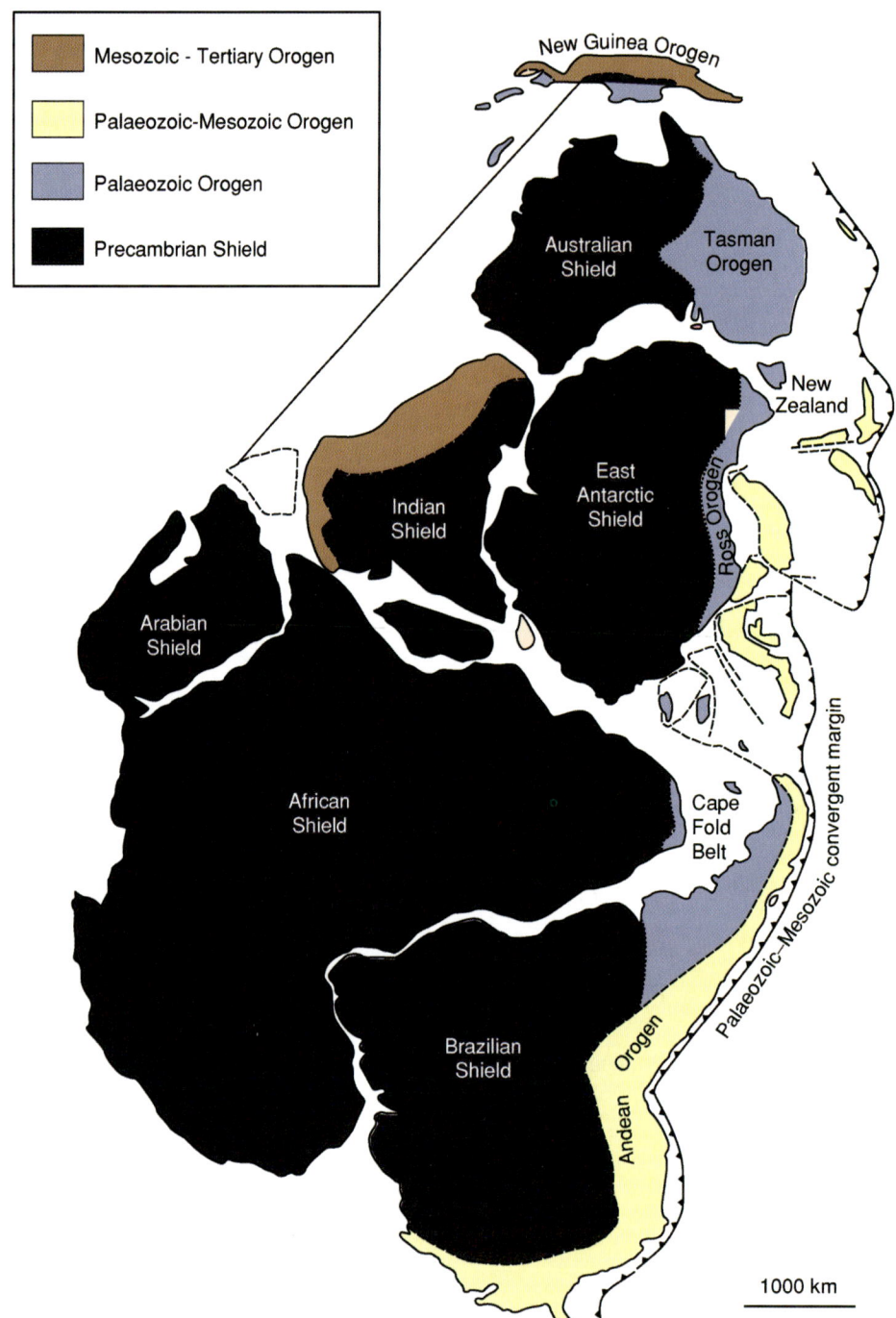

**Figure 1** Simplified geological map of Gondwana with modern-day continental outlines, showing regions of Precambrian shield and the younger orogenic belts. IGCP, Gondwana map base.

by peaks of deformation in the Early to Middle Cambrian, Late Ordovician–Silurian, and Permian–Triassic in the respective orogenic belts (**Table 1**).

Crustal growth occurred largely by the addition of turbidites (ocean-floor submarine-fan deposits), cherts (ocean-basin deposits), mid-ocean ridge basalts (oceanic crust), andesites (island arc), and granites. These younger, largely Palaeozoic, rocks occur to the east of the exposed Precambrian cratonic crystalline basement; the boundary between the two is known as the Tasman Line (**Figure 2**). They represent a significant addition (approximately 30% by surface area) to the Australian continent whilst it was part of Gondwana. Younger basinal sedimentary sequences

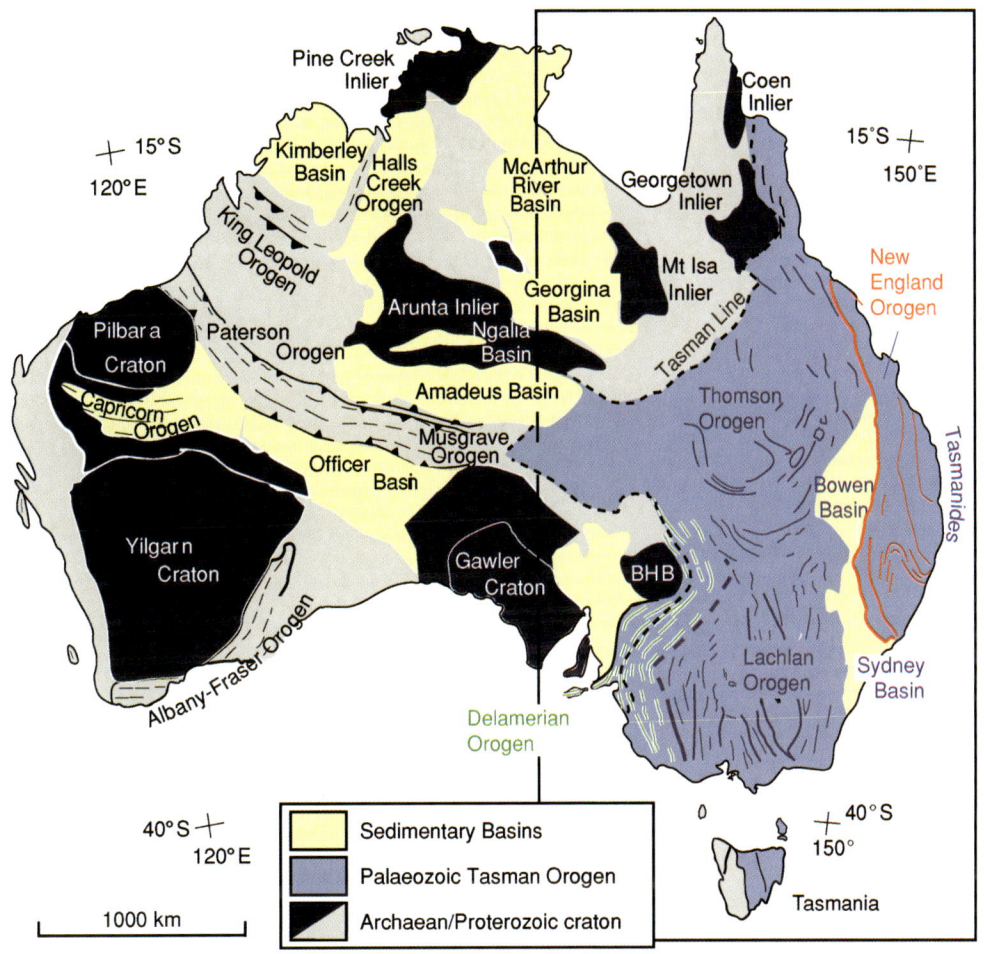

**Figure 2** The major geological elements of Australia, showing the Tasman Orogen along the eastern margin. Location of the more detailed map area of **Figure 3** is shown.

(e.g. Great Artesian, Murray, and Sydney–Bowen Basins) now cover much of belt (**Figure 3**).

Accretion was accompanied by marked structural thickening and shortening (by about 75%) of Neoproterozoic platform and rift basinal sequences (Delamerian part), a Cambrian–Orodovician composite turbidite submarine fan overlying Cambrian oceanic crust (Lachlan part), and Carboniferous–Permian arc, fore-arc, and trench (subduction complex) and Permian foreland basin sequences (New England part) (**Figure 3**). For a large part of the history (Silurian–Early Carboniferous), deformation occurred by massive telescoping, with thrusting largely directed away from the cratonic core, together with components of strike-slip translation within a progressively deforming and prograding sediment fan along the Gondwanan margin. The margin was intruded by voluminous granitic plutons (accounting for up to 35% of the area), derived from subduction-generated mantle magmas and lower- to middle-crustal melting, between the Late Ordovician and the Triassic (**Figure 3**).

The spatial and temporal variations in deformation, metamorphism, and magmatism across the Tasman Orogenic Zone show how subduction–accretion complexes are monotonous turbidite sequences evolve through time and eventually form crust of continental thickness and character.

## Tasman Orogen Make-up

The Delamerian, Lachlan–Thomson, and New England Orogens, which form the composite Palaeozoic and Mesozoic Tasman Orogen (**Figure 2**), are distinguished by their lithofacies (**Figure 3**), differing tectonic settings, and timing of orogenesis and eventual consolidation with the Australian craton (**Table 1**). The Delamerian Orogen involved an inverted intracratonic rift, the Lachlan Orogen involved closure of a south-west-Pacific style marginal basin incorporating Bengal Fan-sized turbidite fans, and the New England Orogen involved a deformed arc–subduction complex belt.

**Table 1** Orogens and subprovinces of the Tasman Orogenic Belt

| Orogen | Delamerian (West) | Western | Lachlan Central | Eastern | New England (East) |
|---|---|---|---|---|---|
| Main lithofacies | Platform to deep-water clastic margin sequence | Quartz-rich turbidite sequence on oceanic crust | Quartz-rich turbidite sequence on oceanic crust | Platform carbonates and clastics with rhyolites and dacitic tuffs | Volcanogenic clastic |
| Initial setting | Late Proterozoic rifted margin | Cambro-Ordovician submarine fan in back-arc basin | Ordovician submarine fan in back-arc basin | Ordovician rifted arc | Late Palaeozoic fore-arc–arc–trench |
| Oldest lithology | Basic volcanics | Cambrian basic volcanics (ca. 505 Ma) | Tremadocian chert | Ordovician andesitic volcanics (ca. 485 Ma) | Ordovician basic volcanics |
| Main deformation | Early to Middle Cambrian (ca. 520–500 Ma) | Late Ordovician–Silurian (ca. 450–420 Ma) | Late Ordovician–Early Devonian (ca. 450–410 Ma) | Silurian–Early Carboniferous (ca. 430–360 Ma) | Late Permian–Triassic (ca. 265–230 Ma) |
| Tectonic vergence | Westwards-directed thrusting | Eastwards-directed thrusting | Overall strike-slip (ca. 410–400 Ma) with south-westwards-directed thrusting (ca. 440–420 Ma) | Eastwards-directed thrusting | Westwards-directed thrusting |
| Main plutonism | Cambrian (520–490 Ma) | Late Devonian (380–360 Ma) | Late Silurian (ca. 430–400 Ma) | Late Silurian–Devonian (ca. 430–380 Ma) | Late Permian–Triassic (ca. 270–210 Ma) |

## Delamerian Orogen

The Delamerian Orogen is the oldest and innermost part, immediately adjacent to the Precambrian cratonic core. The Delamerian Orogen was tectonically active from the Late Precambrian (ca. 650–600 Ma) to the Late Cambrian (ca. 500 Ma). It is now an arcuate craton-verging thrust belt with foreland-style folds and detachment-style thrusts. Duplexes in deformed Cambrian deep-water sandstones and mudstones (Kanmantoo Group) overlie less deformed and metamorphosed Neoproterozoic shelf sequences (Adelaidean). Metamorphic grade increases to the east, where polyphase deformation and amphibolite-grade metamorphism with local development of kyanite–sillimanite assemblages are spatially and temporally confined to the aureoles of synkinematic granites that are conformably aligned with the structural grain.

## Lachlan Orogen

The Lachlan Orogen is the largest and the centrally located orogenic belt (**Figure 2**) with a marked similarity of sedimentary facies and structural style. It was active from the Ordovician and has a complex amalgamational and deformational history. The western and central parts of the Lachlan Orogen are dominated by a turbidite succession consisting of quartz-rich sandstones and black shales. These are laterally extensive over a present width of 800 km and have a current thickness exceeding 10 km. The eastern Lachlan Orogen consists of shoshonitic volcanics, volcaniclastic rocks, and limestone, as well as quartz-rich turbidites and extensive black shales in the easternmost part (**Figure 3**).

The western Lachlan Orogen consists of an eastwards-vergent thrust system with alternating zones of north-westwards- and northwards-trending structures. The central Lachlan Orogen is dominated by north-westwards-trending structures and consists of a south-westwards-vergent thrust belt linked to a fault-bounded metamorphic complex (Wagga-Omeo Metamorphic Belt). The eastern Lachlan Orogen is dominated by a northwards-trending structural grain and eastwards-directed thrusting, which caused inversion of extensional basins in the west. Tight-to-open chevron folds (accommodating between 50% and 70% shortening) cut by predominantly westwards-dipping high-angle reverse faults are part of different thrust systems within the Lachlan Orogen. The chevron folds are upright and gently plunging, but become

AUSTRALIA/Tasman Orogenic Belt 241

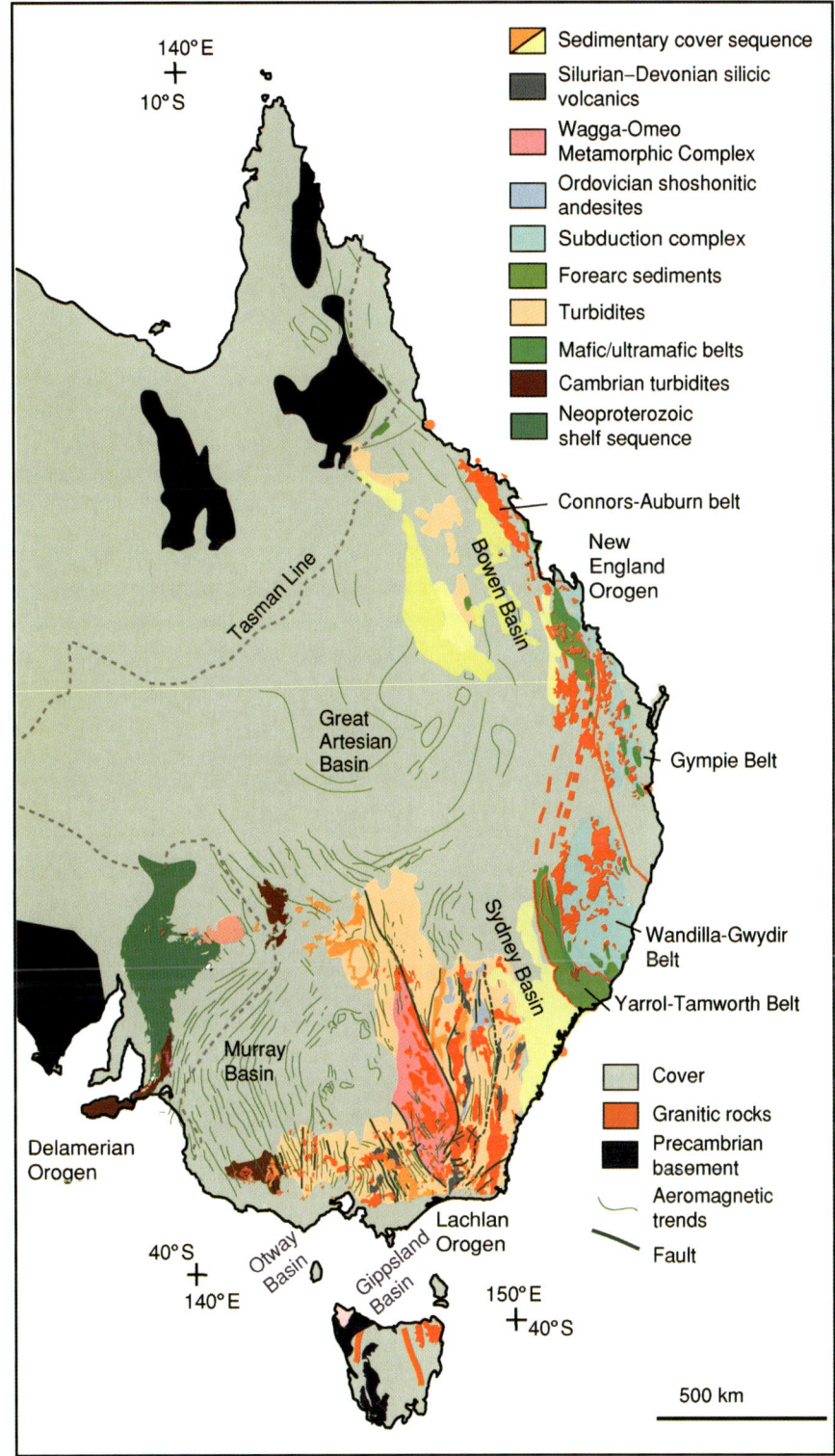

**Figure 3** Geological map of the Tasman Orogenic Belt, showing the dominant lithologies, granite intrusives, major fault traces, and trend lines from regional aeromagnetic datasets.

inclined and polydeformed as they approach major faults. Metamorphism is greenschist facies or lower across the Lachlan Orogen, except in the fault-bounded Wagga-Omeo and several smaller (Cooma, Cambalong, Jerangle, and Kuark) metamorphic complexes, where high-temperature–low-pressure metamorphism is characterized by andalusite–sillimanite assemblages.

### Thomson Orogen

The Thomson Orogen is meant to be coeval with, as well as the northern continuation of, the Lachlan Orogen (**Figure 2**). The orogen is hidden under a cover of younger sedimentary basins that make up the Great Artesian Basin, but geophysical aeromagnetic imagery shows that it has eastwards-trending structures that truncate the trends of the Lachlan Orogen. Drill samples and limited surface exposures have shown that it consists of low-metamorphic-grade turbiditic sandstone and mudstone lithologies similar to those of the Lachlan Orogen, and like the Lachlan Orogen it is intruded primarily by Silurian and Devonian granites.

### New England Orogen

The New England Orogen is the youngest and most easterly part of the Tasman Orogen (**Figure 2**) and incorporates arc, fore-arc, and accretionary complexes. The New England Fold Belt was tectonically active from the Early Carboniferous to the mid-Triassic (a period of *ca*. 130 Ma). Westwards-directed Permian–Triassic thrusting caused interleaving and imbrication of the arc magmatic belt (Connors–Auburn Belt) and fore-arc (Yarrol–Tamworth Belt) and oceanic assemblages, including subduction complexes (Wandilla–Gwydir Belt) and ophiolite (Gympie Belt). Subduction-complex assemblages show a strong thrust-related fabric, polyphase deformation, and greenschist to amphibolite facies metamorphism. The Hunter–Bowen Orogeny (*ca*. 265–230 Ma) consolidated the terranes with Australia and resulted in the development of a Permo-Triassic foreland basin (the Sydney–Bowen Basin) (**Figure 3**).

## Lithofacies

The Tasman Orogen is dominated by turbidites that were once part of large submarine fans overlying oceanic crust (**Figure 3**). Dismembered ophiolitic rocks of Neoproterozoic to Cambrian age and mafic to ultramafic affinities are preserved as slivers along major faults (**Figure 4**). The mafic rocks dominantly include tholeiitic pillow basalts and dolerites with some andesitic and boninitic volcanics, and are generally associated with hemipelagic black shales and cherts. These are interpreted as remnants of the oceanic crust that regionally underlies the Cambro-Ordovician turbidite fans. The ophiolitic slivers were incorporated into the turbidites as offscraped slices, as imbricated fault–duplex slices, and as blocks in mélange. In the eastern Lachlan Orogen Ordovician shoshonitic andesites are structurally interleaved with the Ordovician turbidites (**Figure 3**).

Inboard of the turbidites are Upper Proterozoic (Adelaidean) intracratonic rift sequences of marine to deltaic sandstones and shales, lagoonal evaporites, dolomites, and limestone, overridden by the Delamerian Orogen (**Figure 3**). These sediments were transgressed by Lower Cambrian shelf sediments that are transitional into deep-water sandstones and mudstones of the Kanmantoo Group. Outboard of the turbidites the New England Orogen consists of a collage of deformed and imbricated terranes of largely Middle to Upper Palaeozoic and Lower Mesozoic marine to terrestrial sedimentary and volcanic rocks, as well as strongly deformed flysch, argillite, chert, pillow basalts, ultramafics, and serpentinites.

## Deformation

Based on unconformities, sedimentary facies changes, and, most recently, geochronology (**Figure 5**), deformation across the Tasman Orogenic Belt occurred in the Late Cambrian (Delamerian Orogen), Late Ordovician–Silurian and Early Devonian (Lachlan Orogen), and Late Permian–Middle Triassic (New England Orogen). The deformation was partitioned into regional-scale migrating thrust systems. Tectonic vergence, recorded largely by the dips of the major faults, is craton-directed for the Delamerian and New England Orogens, whereas the Lachlan Orogen shows mixed vergence, but with thrusting largely directed away from the craton (**Figure 6**).

The deformation style is chevron folding cut by high-angle reverse faults. Fault zones are characterized by higher than average strain and intense mica fabrics, transposition foliation, isoclinal folds, and polydeformation with overprinting crenulation cleavages.

## Metamorphism

Metamorphic grade is generally low across the Tasman Orogen (**Figure 7**), with greenschist facies (epizonal) and subgreenschist facies (anchizonal) metamorphism of the turbidite sequences. Most of the turbidites are within the chlorite zone, with localized development of biotite in contact aureoles around granites. High-temperature–low-pressure metamorphism is indicated by localized migmatites and alkali feldspar–cordierite–andalusite–sillimanite gneisses in the Mount Lofty

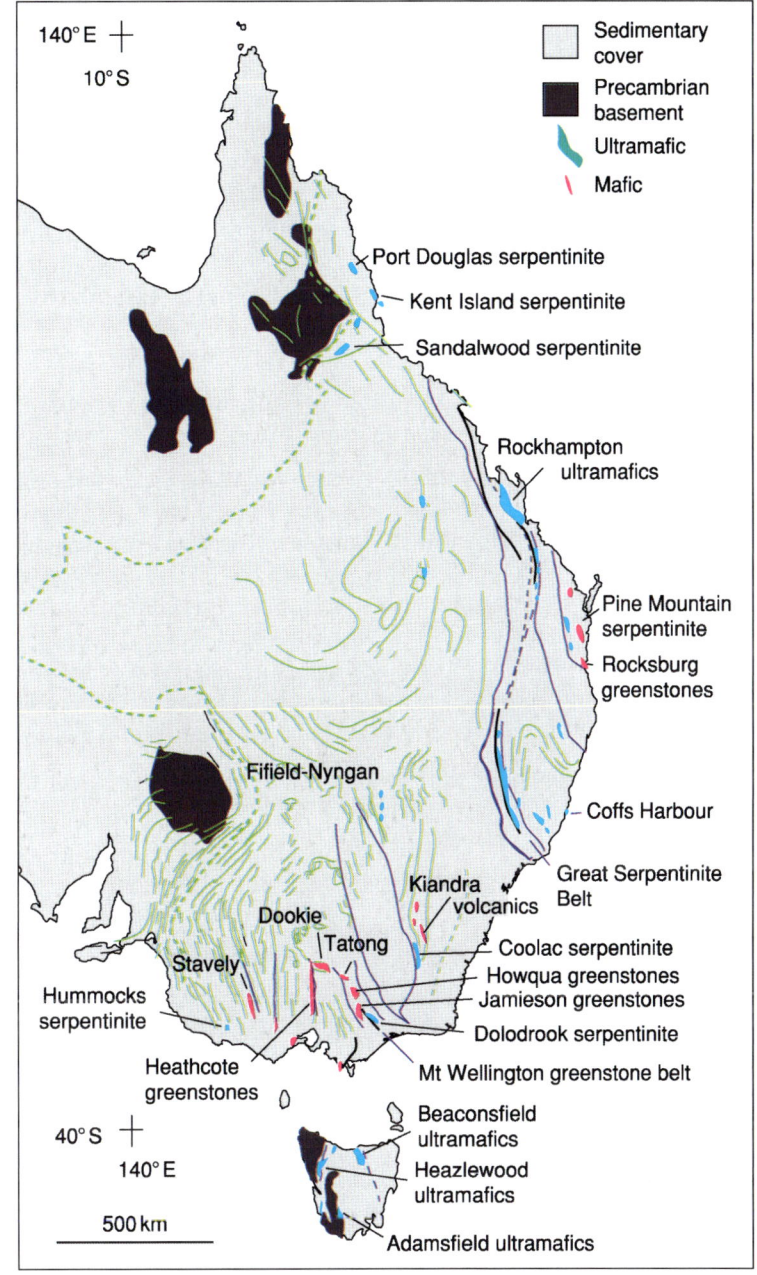

**Figure 4** Geological map of the Tasman Orogenic Belt, showing the distribution of rocks with mafic and ultramafic affinities (dismembered ophiolites), major fault traces, and trend lines from regional aeromagnetic datasets.

Ranges, Glenelg Zone, and Moornambool Complex (Delamerian Orogen), Wagga-Omeo Complex and Cooma, Cambalong, and Kuark belts (Lachlan Orogen), and Wongawibinda Complex (New England Orogen) (**Figure 7**). Delamerian metamorphism is characterized by peak conditions of 650–700°C and 4–5 kbar in the Glenelg Zone, and 540–590°C and approximately 7–8 kbar in the Moornambool Complex. Peak metamorphic conditions in the Lachlan Orogen are 700°C and 3–4 kbar in the Wagga-Omeo Metamorphic Belt and Eastern Metamorphic complexes (e.g. Cooma Complex), but these belts are intimately associated with S-type granitic bodies.

Turbidites from the low-grade parts of the Lachlan Orogen show intermediate pressure series metamorphism, based on $b_0$ or x-ray diffraction lattice spacing measurements of phengitic micas, and are inconsistent with previous interpretations of widespread low-pressure 'regional-aureole' metamorphism. Moderately high-pressure metamorphism has been inferred from co-existing chlorite and actinolite in metabasites from one of the major fault zones. Furthermore,

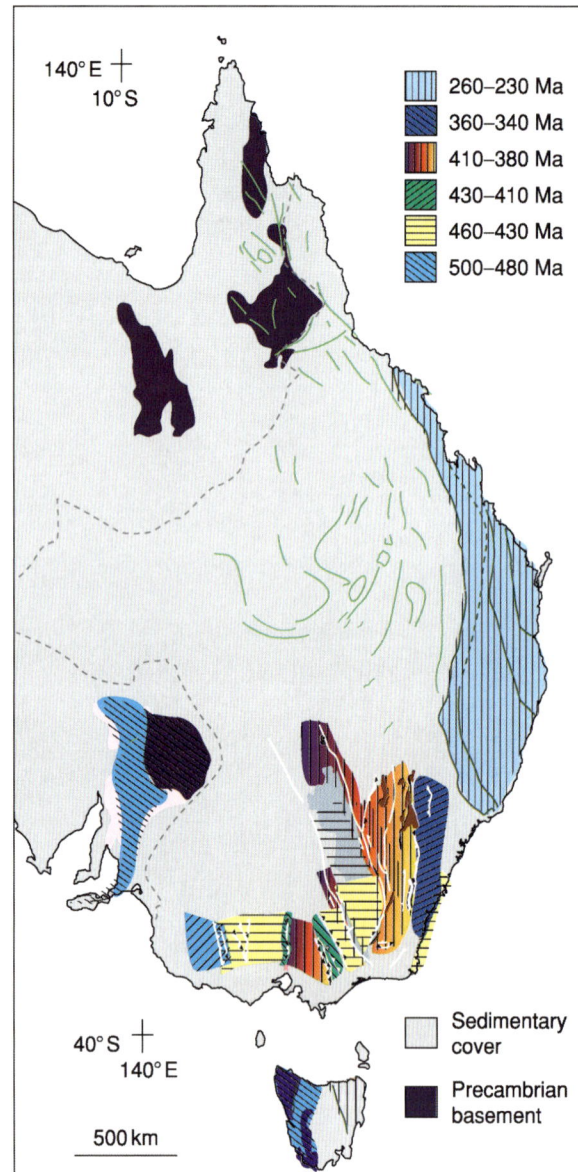

**Figure 5** Geological map of the Tasman Orogenic Belt showing geochronological data.

## Magmatism

Distinct pulses of magmatism are recorded by the granitic rocks that occupy up to 35% of the exposed area of the Tasman Orogenic Belt (**Figure 8**). Granitic bodies tend to be large (up to 10 000 km$^2$) and commonly elongated subparallel to the structural grain. Regional aureole, contact aureole, and subvolcanic field associations, as well as S and I types based on geochemistry and mineralogy, have been recognized. They are largely post-tectonic with undeformed and narrow (1–2 km wide) contact aureoles, although some (e.g. the Wagga-Omeo Metamorphic Complex and the Eastern metamorphic complexes of the Lachlan Orogen; **Figure 7**) are coeval with regional deformation. In the eastern Lachlan Orogen some granites coincident with the Eastern Metamorphic Belt were emplaced along major westwards-dipping shear zones. Some (e.g. the Central Victorian Magmatic Province; **Figure 8**) are subvolcanic granites associated with rhyolites and ash flows of similar composition, and reflect intrusion to shallow crustal levels (ca. 1–4 km). The regional-aureole types are less common and are associated with the high-temperature–low-pressure metamorphism, migmatites, and alkali feldspar–cordierite–andalusite–sillimanite gneisses (e.g. Mount Lofty Ranges, Glenelg Zone, Moornambool Complex, Cooma, Cambalong, and Kuark belts, Wongawibinda Complex; **Figure 7**).

Granite belts with distinctive shapes (**Figure 8**) reflect distinct tectonic regimes. The north-north-westwards-trending granites of the Central Lachlan Orogen (1 in **Figure 8**), the more northwards-trending granites of the eastern Lachlan Orogen (2 in **Figure 8**), and the northwards- to north-north-westwards-trending mid-Permian to Triassic granites of the New England Orogen (3 in **Figure 8**) are representative of arc and continental-margin–arc associations comparable to the Cordilleran batholiths of the western USA. Mid- to lower-crustal melting is due to crustal thickening and/or subduction, with some granitic activity – in particular the Silurian–Early Devonian granites of the eastern Lachlan Orogen – being linked to subduction rollback. Major magmatic activity occurred in the Cambrian–Ordovician (Delamerian part), Silurian–Early Devonian, Late Devonian, and Carboniferous (Lachlan part), and Carboniferous and mid-Permian–Triassic (New England part) (**Figure 8**).

## Eastern Australian Plate Tectonic Evolution in the Gondwanan Context

During the Palaeozoic the Tasman Orogen was part of a greater Gondwanan oceanic accretionary system (**Figures 9** and **10**). The cycle of extension, sedimentation, and orogeny that formed eastern

low-temperature–intermediate-pressure blueschist facies metamorphism is recorded by blueschist blocks (knockers) in serpentinite–talc-matrix mélanges within other major fault zones of the Lachlan Orogen. Winchite (blue sodium–calcium amphibole) and glaucophane (blue sodium amphibole)–actinolite assemblages give estimated pressures of 6–7 kbar and 5–6.5 kbar, respectively, with temperatures of less than 450°C. These conditions match those inferred from metabasite blueschists from the Franciscan Complex, California and the Sanbagawa Belt of Japan. This low-temperature–intermediate-pressure metamorphism is coincident with the regional deformation at around 450–445 Ma (Lachlan Orogen).

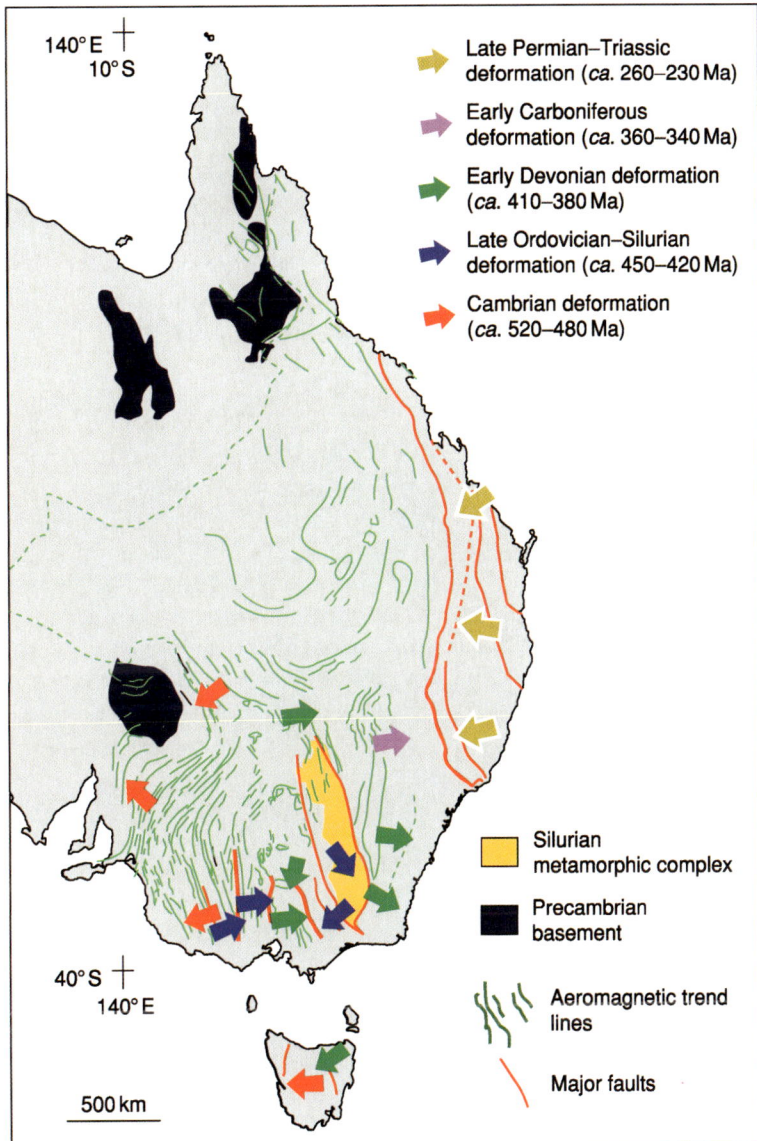

**Figure 6** Geological map of the Tasman Orogenic Belt, showing tectonic vergence based on major fault dip. The major fault traces and trend lines from regional aeromagnetic datasets are also shown.

Australia was preceded during the Late Neoproterozoic by initial rifting between cratonic Australia and another large continental plate, probably Laurentia (North America) within the supercontinent of Rodinia (**Figure 9A**).

Tasman Orogen tectonic evolution has been broken down into a number of significant phases described below and is also presented as a series of Gondwanan-based maps (*see* **Figures 9** and **10**).

### Rodinia Breakup (750–650 Ma): Rifting of Eastern Australia in the Late Neoproterozoic

Rifting began at about 800 Ma but did not lead to separation of the continents until about 700 Ma (**Figure 9A**). The early phases of rifting were epicontinental and are expressed as rift sequences in the Adelaide Rift and other graben structures. After separation, this sequence takes on a character that is more like that of a passive margin. In the Early Cambrian back-arc spreading in the south-east formed the deep-water Kanmantoo Basin, which rapidly subsided and was filled with clastic sediments. Very shortly after deposition of the Kanmantoo sediments the basin was inverted in the first stage of the Delamerian Orogeny.

### Basin Inversion along the Gondwanan Margin (520–500 Ma): Ross–Delamerian Orogeny

The Delamerian orogenic event involved closure of the Kanmantoo Basin by folding and thrusting, metamorphism, partial melting, and intrusion of granitic

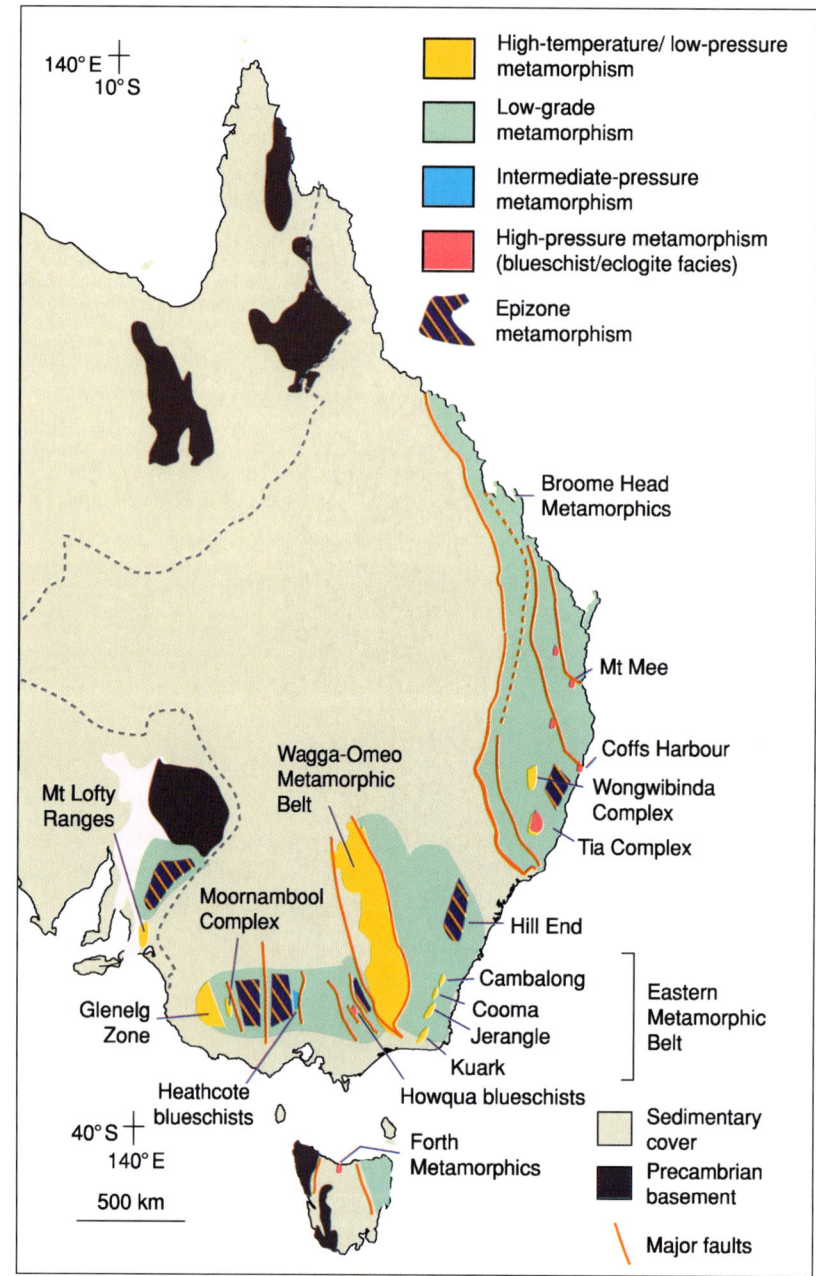

**Figure 7** Simplified metamorphic map of the Tasman Orogenic Belt, showing the distribution of the main regional metamorphic facies and the locations of the high-grade metamorphic complexes.

plutons (**Figure 9B**). The driving force for this orogenic event was presumably related to subduction that resulted in basin inversion. Subduction on the Pacific margin of the orogen is suggested by the presence of arc volcanics and plutons. The ultimate closure of the Kanmantoo Basin may have resulted from the collision of the margin with formerly rifted continental fragments that form the basement of the Glenelg Terrane or Tasmania or from ophiolite obduction. The Delamerian event began at about 520 Ma, continued until about 490 Ma, and involved westwards-directed thrusting. It is uncertain whether any oceanic crust was consumed between cratonic Australia and rifted continental ribbons in western Victoria, or whether the Kanmantoo Basin was developed only on thinned continental crust. The early phases of deformation were dominated by folding, metamorphism, and thrusting, followed shortly by syntectonic plutonism. The 'orogeny' culminated in local high-temperature–low-pressure metamorphism involving partial melting in

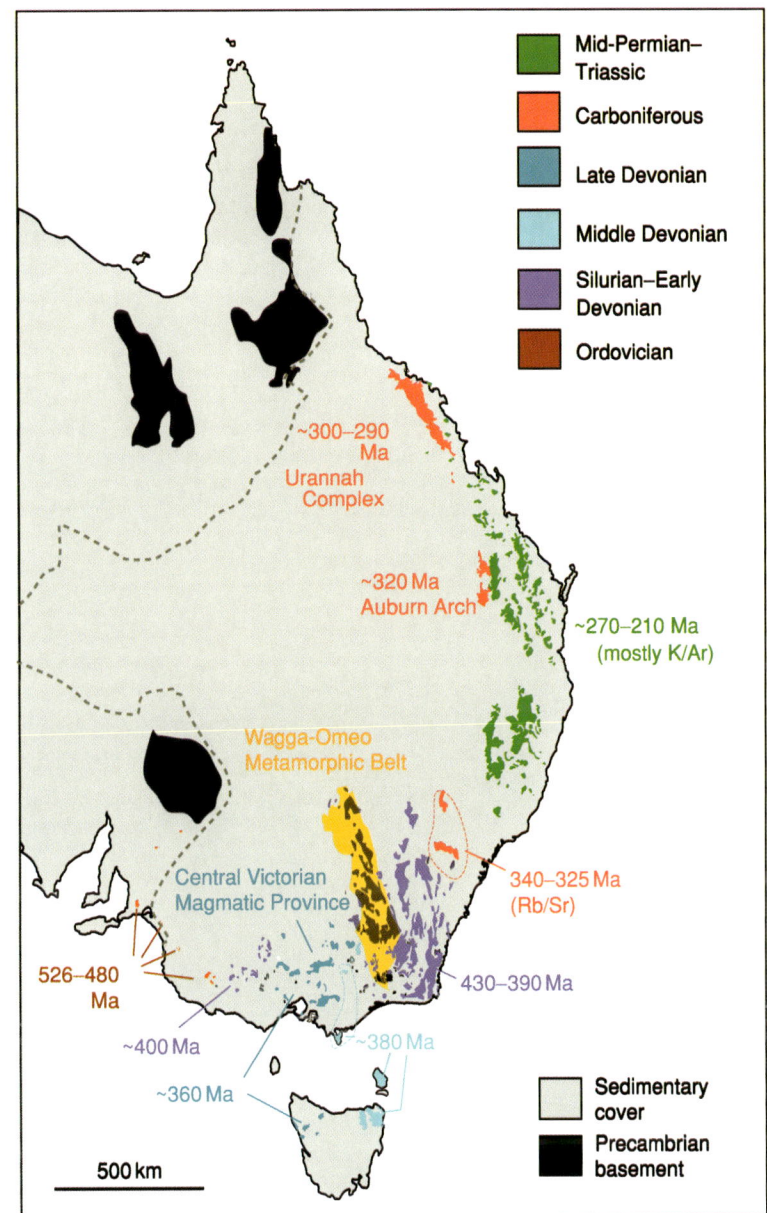

**Figure 8** Ages and distribution of granite intrusives associated with the Tasman Orogen.

the highest-grade areas (at about 506 Ma) and was followed by post-tectonic plutonism (*ca*. 490–480 Ma).

### Back-arc Basin Formation (520–500 Ma): Evolution of the Lachlan Orogen

A magmatic arc grew on the eastern side of the Delamerian Orogen, starting at about 510–500 Ma, and included plutons and volcanic rocks of the Stavely Belt in Victoria (**Figure 4**). Shortly after this major crustal thickening the Delamerian belt began to extend rapidly. Extension is suggested by, first, the dominantly alkaline and extension-related nature of the post-tectonic plutons (intrusion ages younger than about 490 Ma), second, the inferred rift setting of the Mount Stavely volcanic complex, and, third, thermochronological data indicating relatively rapid cooling of the higher-grade areas. Post-orogenic extension may have been caused by subduction roll-back after the Delamerian event, and this probably formed most of the oceanic (back-arc) basement for the Lachlan turbidites that were deposited during the Late Cambrian and Ordovician. Most of the mafic metavolcanic rocks and plutons of the central Victorian basement give ages of about 500 Ma. Extension induced by slab roll-back may also have rifted small continental fragments away from the cratonized Delamerian Orogen and distributed them within the developing back-arc basin.

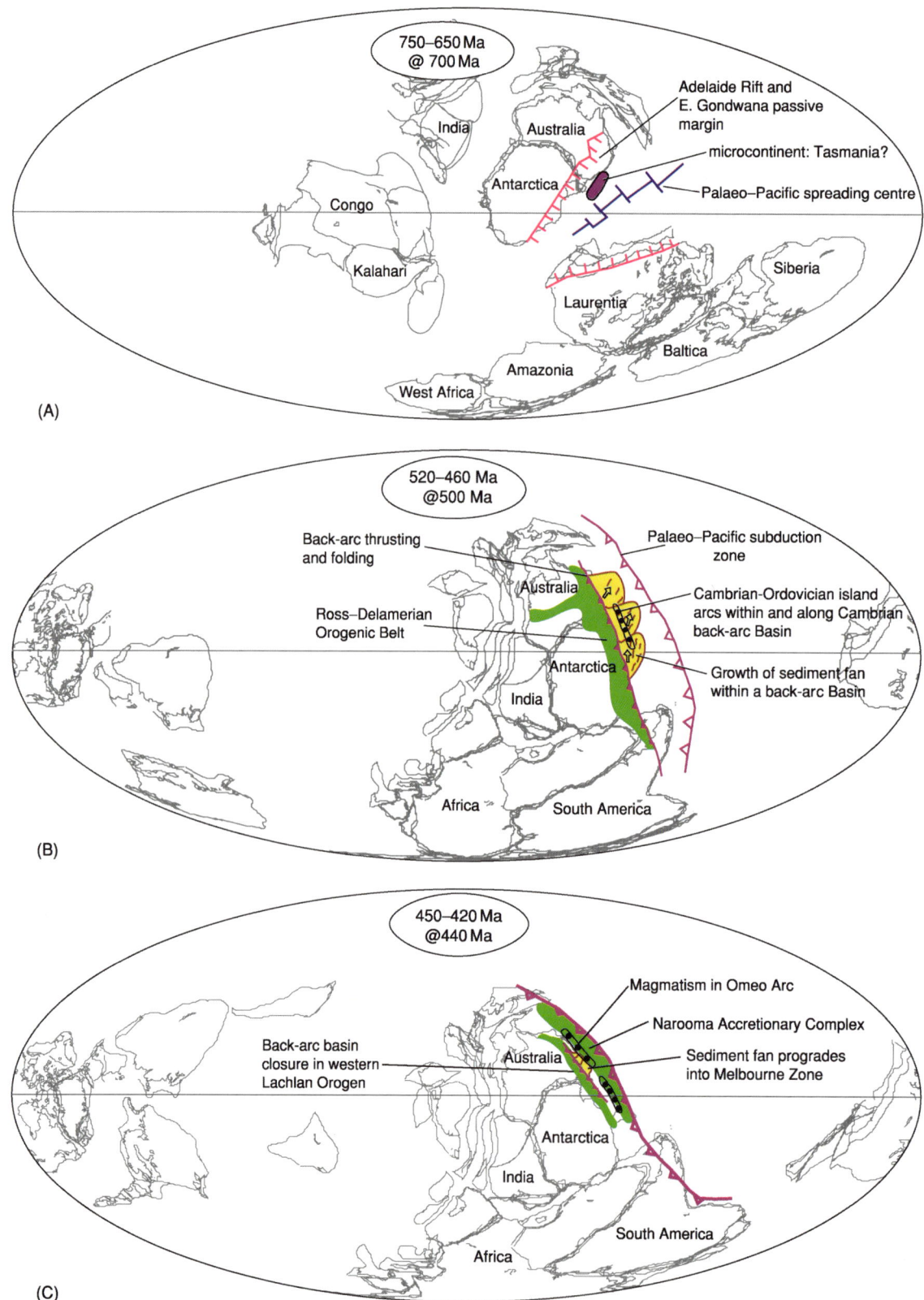

**Figure 9** Plate-tectonic reconstructions at (A) 750–650 Ma, (B) 520–460 Ma, and (C) 450–420 Ma. Adapted and reprinted, with permission, from the Annual Review of Earth and Planetary Sciences, Volume 28, © 2000 by Annual Reviews, www.annualreviews.org. The plate positions and continental outlines were calculated and drawn using data in Paleogeographic Information System/MacTM verison June 7, 1997 by MI Ross & CR Scotese (PALEOMAP Project, Arlington Texas).

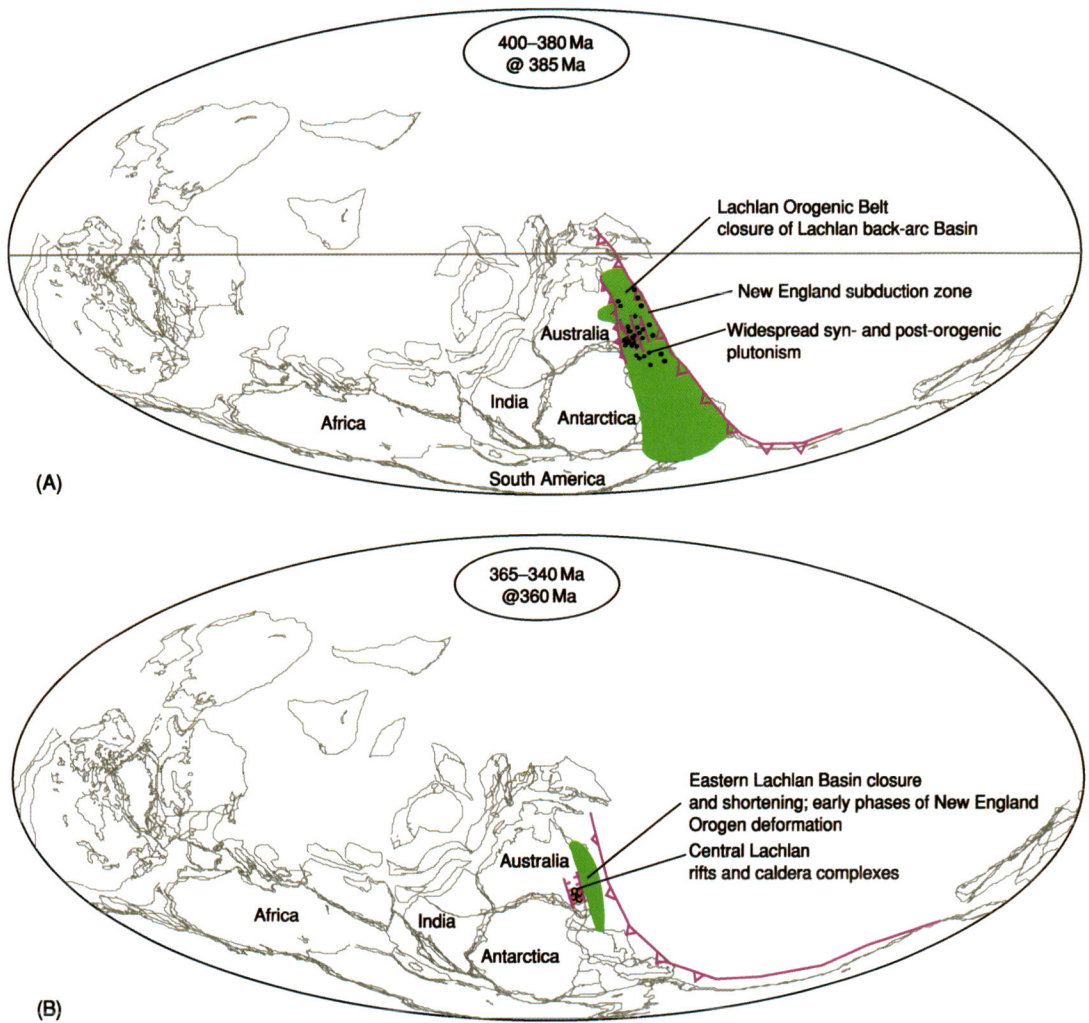

**Figure 10** Plate-tectonic reconstructions at (A) 400–380 Ma and (B) 365–340 Ma. Adapted and reprinted, with permission, from the Annual Review of Earth and Planetary Sciences, Volume 28, © 2000 by Annual Reviews, www.annualreviews.org. The plate positions and continental outlines were calculated and drawn using data in Paleogeographic Information System/MacTM verison June 7, 1997 by MI Ross & CR Scotese (PALEOMAP Project, Arlington Texas).

During this time large turbidite fans grew off the shore of the Ross–Delamerian mountain chain and spread onto the mafic crust of the developing marginal ocean (**Figure 9B**). Some thousands of kilometres off shore subduction had initiated within the oceanic plate, leading to the development of a volcanic arc complex at about 485 Ma, now represented by the Ordovician shoshonites in the eastern Lachlan Orogen, and an associated accretionary complex at about 445 Ma (Narooma Accretionary Complex), exposed in south-eastern New South Wales.

### Back-arc Basin Closure (450–420 Ma): Evolution of the Lachlan Orogen

The Lachlan Orogen formed by amalgamation of a series of thick turbidite-dominated thrust wedges and volcanic-arc terranes in a tectonic setting similar to that of the Philippine, Mulucca, and New Guinea sectors of the western Pacific today (**Figure 9C**). The Lachlan Orogen resulted from the closure of a small ocean-basin–arc system situated along the Pacific margin of Gondwana, inboard of a larger major long-lived Palaeozoic subduction system that is now exposed in the New England Orogen (**Figure 10A**). The total amount of subducted oceanic lithosphere was relatively small (less than 1000 km).

Basin closure involved Woodlark Basin-style double divergent subduction (**Figure 9C**), inferred from the multiple subduction-related thrust systems (**Figure 6**) and the presence of the blueschist blocks in the serpentinite-matrix mélange along major faults (**Figure 7**). The thrust systems developed by duplexing in the oceanic crust and imbrication in the overlying turbidite wedge. Shallow-angle subduction had probably initiated in the west, under the backstop of the Delamerian Orogen, by about 460 Ma, when major faults began forming in the western Lachlan Orogen. Sedimentation was continuing outboard in the basin as the western Lachlan sediment wedge thickened and the basal décollement propagated eastwards. A magmatic arc associated with this westwards-dipping subduction zone did not develop until Late Silurian times, probably because the subduction was of very shallow dip and the ocean basin that closed was relatively small, so the amount of subduction was limited. Shallow-dipping subduction is consistent with the moderately high metamorphic pressure–temperature ratios. The eastwards-dipping subduction system of the central Lachlan Orogen resulted in the high-temperature–low-pressure metamorphism and elongated north-westwards-trending Late Silurian to Early Devonian granitoids of the Wagga-Omeo Metamorphic Belt (**Figures 7** and **8**) as well as the emplacement of ophiolite slices (e.g. Howqua, Dookie, and Tatong greenstones; **Figure 4**) and the Howqua blueschists (**Figure 7**).

### Andean-type Margin (400–380 Ma): Evolution of the Lachlan Orogen

During the Early Devonian, at around 400 Ma, the last undeformed part of the marginal basin in the west collided with the arc–metamorphic complexes of the central Lachlan Orogen (**Figure 10A**). This caused some crustal thickening and deformation of the younger sediments (former Melbourne 'trough'). Combined structural thickening and the removal of the oceanic lithosphere of the marginal basin formed crust of continental thickness and character, and led to a switch to continental sedimentation and the development of a major regional fold-belt-wide angular unconformity. These effects are related to the Tabberabberan orogenic event.

Collision followed by attainment of freeboard over most of the Lachlan Orogen led to the development of an Andean-type margin (**Figure 10B**). Continued westwards-dipping or craton-directed subduction of proto-Pacific lithosphere beneath this part of the Gondwanan margin along the major outboard long-lived subduction zone caused the magmatism that was responsible for the extensive northwards-trending elongated granitoids in the eastern Lachlan Orogen (**Figure 8**). An easterly younging in these granitoids suggests associated slab rollback during the mid-Late Silurian to Early Devonian, with rollback rates of the order of 6 mm year$^{-1}$. Thrusting was also taking place in the eastern Lachlan Orogen at this time.

### Roll-back and Gondwanan Margin Post-Orogenic Extension (365–340 Ma)

Marked crustal-scale extension at around 380 Ma caused rifts, basin-and-range faulting, voluminous explosive volcanism and caldera development, and high-level plutonism in the western and central Lachlan Orogen (e.g. the Central Victorian magmatic province; **Figure 8**). By 340 Ma subduction in the eastern Lachlan Orogen had probably stopped, and reefs were growing around the volcanic edifices. The major subduction zone at this time probably stepped east more than 1000 km to form the New England Orogen (**Figure 10B**).

### Andean Margin (350–280 Ma) and Arc–Continent Collision (260–230 Ma): Evolution of the New England Orogen

During the Carboniferous, westwards-dipping subduction beneath the Gondwanan margin led to silicic and intermediate volcanism from 350 Ma to 310 Ma, coincident with an inboard belt of Late Carboniferous rocks (330–325 Ma). At around 300 Ma magmatic activity migrated outboard into the developing associated Devonian–Carboniferous accretionary prism. Migration of the volcanic arc was associated with widespread extension in the Late Carboniferous to Early Permian due to slab retreat along the long-lived subduction zone, with the development of an outboard intraoceanic arc (Gympie Belt). Collision of this arc with the Gondwanan margin in the Late Permian led to the Hunter–Bowen Orogeny (260–230 Ma), which juxtaposed and accreted the Carboniferous flank–fore-arc sequence (Yarrol–Tamworth Belt), the Devonian arc complex, and the Devonian–Carboniferous subduction complex (Wandilla–Gwydir Belt) by craton-directed overthrusting. Margin overthrusting led to molasse sedimentation in a developing foreland basin (Sydney–Bowen Basin). Cordilleran-style granites (e.g. New England batholith) reflect renewed subduction magmatism at this time, with the development of a Late Permian–Early Triassic magmatic arc. This was followed by Late Triassic extension, which is recorded by silicic caldera-related volcanism and granitic plutonism as well as the development of small extensional basins with bimodal volcanism.

## See Also

**Andes**. **Antarctic**. **Australia:** Proterozoic; Phanerozoic. **Gondwanaland and Gondwana**. **Tectonics:** Convergent Plate Boundaries and Accretionary Wedges; Mountain Building and Orogeny.

## Further Reading

Ashley PM and Flood PG (eds.) (1997) *Tectonics and Metallogenesis of the New England Orogen.* Special Publication 19. Sydney: Geological Society of Australia.

Birch WD (ed.) (2003) *Geology of Victoria.* Special Publication 23. Sydney: Geological Society of Australia.

Burrett CF and Martin EL (eds.) (1989) *Geology and Mineral Resources of Tasmania.* Special Publication 15. Sydney: Geological Society of Australia.

Coney PJ, Edwards A, Hine R, Morrison F, and Windrum D (1990) The regional tectonics of the Tasman orogenic system, eastern Australia. *Journal of Structural Geology* 125: 19–43.

Flöttmann T, Klinschmidt D, and Funk T (1993) Thrust patterns of the Ross/Delamerian orogens in northern Victoria Land (Antarctica) and southeastern Australia and their implications for Gondwana reconstructions. In: Findlay RH, Unrug R, Banks HR, and Veevers JJ (eds.) *Gondwana Eight, Assembly, Evolution and Dispersal,* pp. 131–139. Rotterdam: Balkema.

Foster DA and Gray DR (2000) The structure and evolution of the Lachlan Fold Belt (Orogen) of eastern Australia. *Annual Reviews of Earth and Planetary Sciences* 28: 47–80.

Foster DA, Gray DR, and Bucher M (1999) Chronology of deformation within the turbidite-dominated Lachlan orogen: implications for the tectonic evolution of eastern Australia and Gondwana. *Tectonics* 18: 452–485.

Gray DR and Foster DA (1997) Orogenic concepts – application and definition: Lachlan fold belt, eastern Australia. *American Journal of Science* 297: 859–891.

Scheibner E and Basden H (eds.) (1998) *Geology of New South Wales—Synthesis. Volume 2: Geological Evolution.* Memoir Geology 13(2). p. 666. Sydney: Geological Survey of New South Wales.

Veevers JJ (ed.) (1984) *Phanerozoic Earth History of Australia.* Oxford Monographs on Geology and Geophysics 2. Oxford: Oxford University Press.

Veevers JJ (ed.) (2000) *Billion-year Earth History of Australia and Neighbours in Gondwanaland.* Sydney: GEMOC Press.

# BIBLICAL GEOLOGY

**E Byford**, Broken Hill, NSW, Australia

© 2005, Elsevier Ltd. All Rights Reserved.

## World-View of the Hebrew Scriptures

The world-view of those who wrote or edited the Hebrew Scriptures reflects the commonly held contemporary understandings of the structure of the universe. The editors of the Hebrew Scriptures (including those who wrote the first chapter of the Book of Genesis) did their work in the period immediately following the end of Babylonian exile. The return from exile followed the conquest of Babylon by Cyrus, King of Persia, in 538 BCE. The world-view in the scriptures was that of the desert people who lived in the Fertile Crescent of modern Iran, Iraq, Syria, Israel, and Jordan. The clearest, detailed articulation of the structure of the world and the universe is to be found in the Book of Job, Chapters 38–40.

What is described in Genesis 1:1 to 2:3 was the commonly accepted structure of the universe from at least late in the second millennium BCE to the fourth or third century BCE. It represents a coherent model for the experiences of the people of Mesopotamia through that period. It reflects a world-view that made sense of water coming from the sky and the ground as well as the regular apparent movements of the stars, sun, moon, and planets. There is a clear understanding of the restrictions on breeding between different species of animals and of the way in which human beings had gained control over what were, by then, domestic animals. There is also recognition of the ability of humans to change the environment in which they lived. This same understanding occurred also in the great creation stories of Mesopotamia; these stories formed the basis for the Jewish theological reflections of the Hebrew Scriptures concerning the creation of the world. The Jewish priests and theologians who constructed the narrative took accepted ideas about the structure of the world and reflected theologically on them in the light of their experience and faith. There was never any clash between Jewish and Babylonian people about the structure of the world, but only about who was responsible for it and its ultimate theological meaning.

The envisaged structure is simple: Earth was seen as being situated in the middle of a great volume of water, with water both above and below Earth. A great dome was thought to be set above Earth (like an inverted glass bowl), maintaining the water above Earth in its place. Earth was pictured as resting on foundations that go down into the deep. These foundations secured the stability of the land as something that is not floating on the water and so could not be tossed about by wind and wave. The waters surrounding Earth were thought to have been gathered together in their place. The stars, sun, moon, and planets moved in their allotted paths across the great dome above Earth, with their movements defining the months, seasons, and year.

In the world of the Jewish scribes and priests who assembled the Hebrew Scriptures, the security of the world (the universe) was 'guaranteed' by God. It was God who made the world secure. It was God who made Earth move or not move, as the case might be. When this is understood, the ancient sources can be understood, and many of the geological events that produced the theological reflections can be deciphered.

## Geological Events with References in the Hebrew Scriptures

### The Angel with the Flaming Sword (Genesis 3:24)

The location of the Garden of Eden was described as being "in the east". Four rivers were said to have flowed out of Eden. The naming of the Tigris, the Euphrates, and two other streams that have names that translate literally from the Hebrew to 'gusher' and 'bubbler' suggests that the garden was located at the top of the Persian Gulf. This area is well watered and would have been lush compared to the desert area that was the habitual location of the Jewish people from the beginning of the first millennium BCE.

The creatures with flaming sword are described as cherubim. In the ancient world, these were supposed awesome (hybrid) creatures and were the steeds that carried the high god of the Canaanites through the air. What is interesting is that the reference to the sword was made using the definite article, as if 'it' (whatever it was) were well known. This is a region well known for its petroleum deposits and oil and gas fires. A gas release or petroleum seep that caught fire could last for many years and extend hundreds of metres above the ground. It would have released much light, heat, and sound and would have been of the very nature described in the Biblical text as keeping people away from the locality.

### The Flood: Genesis 6–9

Since, at latest, the middle of the nineteenth century, scholars, both scientific and theological, have rejected as impossible the idea of a flood that inundated all of Earth's landmasses. The flood story has been treated as ancient myth or as a theological metaphor for the return to supposed primeval chaos as a consequence of human sin. It can also be regarded as an expansion of local flood stories from the flood plains of great rivers and coastal areas to some sort of universal status. But since the work of William Ryan and Walter Pitman, there has been renewed interest in the occurrence of a catastrophic flood.

At the time of maximum glaciation at the height of the last Ice Age (about 16 000 BCE), sea-levels were about 120 m lower than present levels are. There was no longer a connection between the Black Sea Basin and the Mediterranean Sea. With the first great melting of the Eurasian Ice Sheet (12 500 BCE), the Black Sea filled with fresh water, which flowed through the Sakarya Outlet into the Sea of Marmara and down the Dardanelles Outlet into the Aegean. Later, with only seasonal flows down the rivers feeding the Black Sea, the level of this body of water fell far below its great melt maximum. But the level of the Mediterranean, connected to the waters of the world's oceans, continued to rise over the next 6000 years while at the same time the former channels from the Black Sea to the Sea of Marmara filled with rubble, thus creating a dam between the Mediterranean and the Black Sea Basin. By 6000 BCE, the sea-level in the Mediterranean was at least 100 m higher than the surface of the great Black Sea freshwater lake.

Ryan and Pitman have proposed a single geological event, namely, the breaching of that dam through what is now the Bosphorus Straight, from which followed the sudden and catastrophic flooding of the Black Sea Basin with salt water. Ryan and Pitman date the disappearance of freshwater molluscs and the introduction of saltwater molluscs to 5600 BCE. This happened suddenly, not gradually. Their submarine survey work has indicated that the Bosphorus was eroded by immense flows from the Mediterranean to the Black Sea. The inclination of the channel towards the Black Sea also accounts for the surface current of relatively fresh water towards the Mediterranean Sea and the lower more saline current directed towards the Black Sea. The dating of this geological event to early agricultural times links it to the Noah/Gilgamesh story.

There is some controversy in the early twenty-first-century geology literature concerning Ryan and Pitman's hypothesis of a single catastrophic breaching of the Bosphorus. At the time of publication, that controversy continues. However, Ryan and Pitman's hypothesis is not based simply on the dating of marine molluscs, but significantly on the erosion profiles of the Bosphorus Channel. These profiles reveal the continuation of the erosion channel beyond the escarpment, well below present sea-levels.

Most flood stories came from areas of flood plains and coastal districts and can be related to the sea-level rises associated with the end of the last Ice Age (10 000 to 12 000 years ago). The Noah legend is significantly different from these in that it is associated with the Anatolian highlands, well away from the coast and well above sea-level. The Black Sea had been a saltwater sea, but had become a large freshwater lake. Following the last Ice Age, it then shrank to become a lake with a surface about 100 m lower than the present level, but suddenly filled and became salty at about 5600 BCE. More recent discoveries by Robert Ballard indicate that there were quite sophisticated settlements on the shore of the freshwater lake before the reintroduction of salt water. These settlements seem to have been built on well-established agricultural and animal husbandry practices. Ryan and Pitman estimate that, after the breach of the Bosphorus, water would have flowed into the Black Sea Basin at about $40\,km^3\,day^{-1}$ and that the water level would have risen by about $15\,cm\,day^{-1}$. They estimate that the sound of the great inrush could have been heard up to nearly 500 km away. The water would have surged down the escarpment accompanied by large volumes of spray and deafening noise.

According to the Genesis account of the flood (Genesis 6–8), Noah was warned of the impending disaster and was told to make preparations for the preservation of animal and plant stock, which were to be preserved and transported to safety by means of a vessel that Noah was to build. The flood is described as coming from the sky and from the deep (Genesis 7:11). Noah had time to make the necessary preparations to save his domestic breeding and seed stock, for it would have taken nearly two years to fill the Black Sea Basin. In the flat country on the northern shores of the old freshwater lake, the advance of shoreline would have been about $1\,km\,day^{-1}$. The legend could well refer to an actual historical event, because people could have known of the rising water level on the Mediterranean side of the Bosphorus barrier, prior to its rupture.

It is suggested that those who fled the Black Sea Basin went in all directions, into northern and western Europe, and, significantly, across the Anatolian Plateau (adjacent to the location of the mountains of Ararat), into the Fertile Crescent, where the newcomers introduced their more sophisticated agricultural practices and preserved the story of the destruction of their world in the great flood.

### The Destruction of Sodom and Gomorrah (Genesis 19:24–26)

The description of the destruction of the Sodom and Gomorrah, two cities on the Plain, contains significant elements that correspond to massive earthquakes and the opening of sulphurous springs. Traditionally, Sodom and Gomorrah have been thought to have been located in the valley of the Dead Sea, perhaps even where the Dead Sea is now located. In spite of clear suggestions of a catastrophic geological event, no particular event has been satisfactorily identified with this story, nor have any identifiable ruins of early second-millennium-BCE towns or cities been found. Thus the description stands, but a particular event or location remains a matter of conjecture.

### The Exodus

**The Crossing of the Red Sea—Two Stories in Exodus 14** There are two accounts of the crossing of the Red Sea (the Hebrew name of which can also be translated as 'Sea of Reeds' or 'Distant Sea') in Chapter 14 of the Book of Exodus. The accounts are intricately intertwined but refer to two readily identifiable and distinguishable phenomena. It is generally agreed that the route of the Exodus from Egypt into the Sinai was close to the northern end of the Red Sea.

The first of these phenomena concerned the action of steady, strong winds on shallow expanses of water. Water can be removed a long way from the normal shoreline of one side of a lake or inlet, for the duration of the wind, and can then return to normal levels when the wind subsides. (This is a commonly observed phenomenon at Lake George, Australia, near Canberra.) At the beginning of the first account of the crossing (Exodus 14:21), we are told "the Lord drove the sea back by a strong east wind all night, and made the sea dry land". This account then tells of the attempt of the Egyptians to follow through after the Hebrews and that their chariot wheels became clogged and they fled. When it is recognized that the Hebrews had herds and were mostly on foot and would have further softened ground that was usually under water, then the clogging of the wheels of chariots would be expected. The dropping of the wind and the return of the water could have been a perfectly natural phenomenon.

The second account is much more dramatic, with references to great walls of water (Exodus 14:22). The destruction of the Egyptians was said to have been accomplished by the catastrophic return of the water (Exodus 14:28). This second account is of a far more destructive phenomenon, compared to the first, and probably more familiar one. In the second account, the sea pulls back and then surges over those who have ventured onto the ground from which the sea receded. This is a classic description of the destruction wrought by a tsunami.

The dating of the events of the Exodus has never been exactly agreed by scholars. The generally agreed range of dates is from the fifteenth century BCE to the middle of the thirteenth century BCE. Near the beginning of this period, there was, in fact, an event that could have produced a catastrophic tsunami of the type described in Exodus. In 1470 BCE, the volcano Santorini erupted with about the same force as that of Krakatau in 1883. These eruptions are the largest in historical times. The tsunami resulting from the Santorini eruption wrought havoc in northern Crete and on Milos and in the Peloponnese. This same tsunami would have swept south, causing immense damage along the Egyptian coast and especially in the low-lying agricultural areas of the Nile Delta. The sea would have withdrawn, possibly for 15 or 20 min, and then would have come the first return wave. The wave that hit Milos had a height of about 100 m and travelled at about 300 km h$^{-1}$. Even at half this speed and height, the destruction in Egypt would have been immense. Such an event would have been associated with the gods, and, for the Hebrews, with their own God. Thus, in Exodus 14, the accounts are of a real catastrophic event and of another event, of more common experience, combined into a single story of the power of the Hebrew God to save the Hebrew people.

**The Plagues (Exodus 7–11)** Evidence from the Greenland ice sheets implies that the Santorini eruption generated high levels of sulphuric acid. There would have been sustained acid rain in the eastern Mediterranean and cooling for many years as result of the dust blasted into the atmosphere. There would have been initial crop failures as a result of the acid rain and continued low yields as a result of lower temperatures. The dust blasted into the high atmosphere would have darkened the sky, making the sun and moon appear red, perhaps for several years.

Acid rain would also have contaminated water, causing destruction of aquatic plants and fishes. If this were combined with significant ash-falls, the waterways could have been rendered anoxic for a considerable length of time. Moreover, contaminated water and acid rain, with sudden loss of pasture, would be associated with a rise in stock disease and invasion of settled areas by insects and other vermin in search of food sources.

In the plague stories, descriptions of agricultural phenomena are consistent with the effects that would have followed the Santorini eruption. The sequence of plagues in Exodus were as follows: (1) blood (the Nile turns to 'blood' and everything in it died), (2) frogs, (3) gnats, (4) flies, (5) livestock disease, (6) boils, (7) hail and thunderstorms, (8) locusts, (9) darkness, and (10) death of the first-born child. Except for the last (which is clearly related to Jewish religious and cultic practice), each of the plagues could have been caused by the eruption of Santorini. Whatever else is concluded, it is clear that phenomena that would have been consequences of the eruption of Santorini were associated directly, in the stories and literature of the Hebrews, with the supposed divine intervention that led to the Exodus.

**The Crossing of the River Jordan (Joshua 3–4)** The crossing of the Jordan on dry land, by Joshua and descendants of those who had fled Egypt, is said to have been accomplished as a result of the waters "rising up in a single heap far off at Adam, the city that is beside Zarethan" (Joshua 3:16). Adam is at the junction of the Jordan and Jabbok rivers, about 25 km north (upstream) of the closest crossing point of the Jordan to Jericho. The Dead Sea/Jordan Valley is part of geologically active rift system where the Eurasian and African plates interact. Movements along the fault system could have produced a rubble dam that blocked the flow of the Jordan, as described in Joshua, thus allowing a crossing on dry ground or through very shallow water.

**The Tablets of Stone with the Law Inscribed on Them** The Ark of the Covenant, containing the tablets of stone that Moses brought down from the mountain, was the significant agent in the story of the Jordan crossing. The tablets were understood to have been given to Moses after God inscribed the law on them. The story of the initial giving of the tablets is in a series of events described in Exodus 24:12 and 31:18. The breaking of the stones is described in Exodus 32:19 and the giving of new stones is told in Exodus 34.

There is Precambrian graphic granite in the southern Sinai, the rock being formed by an intergrowth of quartz in feldspar derived from a quartz–feldspar melt, which gives the appearance of writing in stone (see **Figure 1**). Tablets of this material can be up to 20 by 20 cm. They are brittle, and when dropped, they cleave along the plane of crystallographic weakness.

**Moses Strikes the Stone to Produce Water: Exodus 17:1–7** The regions associated with the wanderings

**Figure 1** Example of graphic granite (from Thackaringa Hills, west of Broken Hill, NSW, Australia). Photograph by Edwin Byford.

of the Exodus are in the Sinai Peninsula, south of the Dead Sea and in the approach to the Holy Land on the eastern side of the Dead Sea and the Jordan. The river was crossed from the east and the siege of Jericho began. To the south and east of the Dead Sea, there are permeable sandstones overlaying impermeable granites. Water seeps down through the sandstones until it reaches the impermeable granite. Where there are vertical fractures, springs occur where water emerges from the rock face to produce a rock pool. A sequence of such springs occurs along the edge of the formations. One such pool, Wadi Musa (in modern Jordan), is identified by the local Bedouin with the story of Musa (Moses) striking the rock to get water.

### References to Earth Movements

The Hebrew and Christian scriptures originated at the eastern end of the Mediterranean, a region of geological activity throughout historical times. God was described as both securing the stability of the earth (Psalms 93 and 96) or as shaking the earth (Isaiah 29). God thus both protected people from earthquakes and caused them.

Earthquakes were used to indicate dates (e.g., Amos 1:1; Zechariah 14:5). Zechariah 14:5 referred back to the Amos earthquake (middle of the eighth century BCE), but the whole passage (Zechariah 14:4, 5) described what may now be construed as the lateral shift caused by an earthquake near Jerusalem. The Mount of Olives tear/wrench fault is part of the complex of tear faults that form the Dead Sea/Jordan Valley. In these regions, the rocks on the western side of the valley are displaced southwards compared to those on the east.

## The Scientific Revolution Beginning in the Sixteenth Century, and Christian Responses Thereto

For a convenient point of departure, the beginning of the scientific revolution can be taken to be with Nicolas Copernicus (1473–1543), who, while a canon of Frauenburg (1497–1543), formulated a model of the solar system with the sun, not Earth, at its centre. It is commonly accepted that Copernicus was responsible for the seminal ideas that laid the foundations of the work of Kepler, Galileo, and Newton. In spite of great reservations about the new learning (note the condemnations of Galileo and the fact that the work of Copernicus remained on the Index from 1616 to 1757), Pope Gregory XIII revised and corrected the calendar in 1582, following advice from the Vatican Observatory. Although Protestant countries did not adopt the revisions for some centuries (in England and the American colonies not until 1752, and in the eastern Orthodox countries not until 1924), the Gregorian calendar is now the accepted norm.

Seventeenth-century England saw the theoretical foundations for modern science established in the work of Isaac Newton (1642–1727). But it was seventeenth-century Britain that also produced what has become the centrepiece of a perceived clash of science, and especially geology, and Christianity. In the politically and religiously charged atmosphere of the Protectorate, James Us(s)her (1581–1656), who was a graduate of Trinity College, Dublin, and Archbishop of Armagh, calculated the age of Earth, using the ages of the Patriarchs as they were recorded in the Old Testament, and interlocking these dates with those available from the great civilisations of the ancient world. His calculations were given in his *Annales Veteris et Novi Testamenti*, written between 1650 and 1654. Ussher dated the creation of Earth at 23 October 4004 BC. From 1679, Ussher's dates for various events were included as marginal notes in the 'Oxford Bibles' and in modified form in subsequent editions of the Authorised Version of the Bible (the 'Lloyd Bible', 1701), thereby establishing Ussher's dates in the minds of those who read the Bible in English.

By the eighteenth century, a mechanistic understanding of the universe was accepted among theologians and church leaders, especially in Britain and the Protestant maritime nations. Studies in botany, zoology, and anatomy, as well as the developments in mechanics and astronomy, provided data for arguments for the existence of God based on ideas of divine design. William Paley (1743–1805) published his *Natural Theology* (1802), which remained a basic text in apologetics until well into the twentieth century. The newly emerging sciences were understood to provide further proof concerning the existence and benevolent nature of God. The convergence of science and Christian faith was further encouraged by endowments such as that by Robert Boyle (1627–91), of Boyle's Law fame, for annual lectures on Christian apologetics. As yet, there was little to bring Ussher's date of the creation of Earth into question.

All this began to change from the middle of the eighteenth and early nineteenth centuries, with the beginnings of the classifications of fossils and sedimentary strata. For some decades, accommodation of Christian faith and the emerging pool of geological data and interpretation was achieved (more or less) through what has been called 'flood geology', based chiefly on the 'catastrophist' ideas of the Frenchman Georges Cuvier (*see* **Famous Geologists: Cuvier**). But by the middle of the nineteenth century, this was no longer accepted in learned scientific or theological circles. Following on the work of James Hutton (*see* **Famous Geologists: Hutton**), the influential geologist Charles Lyell (1830) (*see* **Famous Geologists: Lyell**) argued for noncatastrophist geology and, using evidence from his work at Mount Etna and elsewhere, for the idea that Earth was immensely old.

Many clergy, especially in England and Scotland, had been in the forefront of the new biological and geological sciences. But this did not prepare them for the publication of Charles Darwin's *Origin of Species* (1859) (*see* **Famous Geologists: Darwin**). The creation of the world was pushed back thousands of millions of years, and human beings could no longer be understood as a unique creation, qualitatively different from all other life forms. Though many scientists, including William Thomson (Lord Kelvin), questioned the extreme age of Earth advocated by Lyell and Darwin, it was certain clergy, most notably Samuel Wilberforce, Bishop of Oxford, who attacked Darwin's theory. (Thomson's objections were based on his understanding of Earth's cooling. He calculated that it would have 'died' were it as old as geological theory indicated. Radioactivity had not then been discovered, and so he had no mechanism by which Earth could be both very old and warm.)

Wilberforce's objections (which successfully drew attention to weaknesses in Darwin's original theory) were not long shared by the theological faculties of the universities of England and Scotland. On the contrary, by the early 1880s, Darwin's theory was being embraced as evidence of the immense providence of God. Frederick Temple's Bampton Lectures of 1884 and the *Lux Mundi* collection of essays edited by Charles Gore in 1889 embraced the new scientific developments.

Through the twentieth century, there has been a perceived opposition between science and Christianity, in particular, driven by those who adhere to Ussher's dating of the creation of the heavens and Earth. Generally known as Creationists, opponents of the 'new' geology and biology have portrayed Darwinian evolutionary theories as anti-Christian. But, interestingly, they have tried to conscript science to the Creationist cause. So-called Creation science (*see* **Creationism**) has tried to adapt scientific learning to prove that Earth is very young (only several thousand years old). In no way does Creation science satisfy the methodological rubrics of the scientific community, and Creationist interpretations of the data are universally rejected. It is interesting that Creation scientists want to appeal to science to establish their case. Even for religious literalists, it would appear that the best way to describe the physical universe is what is generally understood as scientifically. But this is understandable in that they operate chiefly in the 'scientistic' culture of the United States.

In the United States in the late twentieth century, objections to Creation science being taught in school science curricula have been led by mainstream Christians and Jews. (The plaintiffs in the 1981 action to set aside as unconstitutional Act 590 of the Arkansas Legislature, "Balanced Treatment for Creation-Science and Evolution-Science Act", included the resident Arkansas bishops of the United Methodist, Episcopal, Roman Catholic, and African Methodist Episcopal Churches, the Principal Officer of the Presbyterian churches in Arkansas, and other United Methodist, Southern Baptist, and Presbyterian clergy. Organisational plaintiffs included the American Jewish Congress, the Union of American Hebrew Congregations, and the American Jewish Committee.) For most Christians and Jews there is no perceived clash between their religious faith and the scientific enterprise.

The Bible is a collection of religious texts gathered and edited over a period in excess of a 1000 years, concluding in the early decades of the second century CE (anno domini). The texts are products of the times and places in which people have struggled to understand the ultimate significance and meaning of life. They reflect religious conviction and searching and have never been scientific texts, in the way that scientific texts have been understood since the late Middle Ages when modern science began to develop. However, events that may have appeared miraculous to the authors of the Old Testament can be understood satisfactorily in terms of modern geological knowledge. Religious knowledge and scientific knowledge deal with the same world and universe, but they operate with fundamentally different presuppositions of causality. This means that they can overlap and interact, but the causalities for science are proximate and immediate, and, for religion, are ultimate and perhaps eternal.

## See Also

**Creationism**. **Famous Geologists:** Cuvier; Darwin; Hutton; Lyell. **Geomythology**. **Tectonics:** Earthquakes. **Volcanoes**.

## Further Reading

Alter R (1996) *Genesis*. New York: WW Norton.

Görür N, Cagatay MN, Emre O, *et al.* (2001) Is the abrupt drowning of the Black Sea shelf at 7150 yr BP a myth? *Marine Geology* 176: 65–73.

Lewis CLE and Knell SJ (eds.) (2001) *The Age of the Earth: From 4004 BC to AD 2002*. London: The Geological Society.

Numbers RL (1992) *The Creationists*. New York and Toronto: Alfred A. Knopf.

Plimer I (2001) *A Short History of Planet Earth*. Sydney: ABC Books.

Ruse M (1999) *Mystery of Mysteries*. Cambridge: Harvard University Press.

Ryan W and Pitman W (2000) *Noah's Flood*. New York: Touchstone.

Wilson I (2001) *Before the Flood*. London: Orion Books.

# BIODIVERSITY

**A W Owen**, University of Glasgow, Glasgow, UK

© 2005, Elsevier Ltd. All Rights Reserved.

## Introduction

In its broadest sense, biodiversity encompasses all hereditarily based variation at all levels of organization from genes through populations and species to communities and ecosystems. The term is a modern one, stemming from the early 1980s, and is a contraction of 'biological diversity', which has a long pedigree as an area of scientific investigation. Concerns over the rate of extinction of modern species have brought the study of biodiversity (and indeed the word itself) to the forefront of scientific, political, and popular concern. The fossil record both establishes the ecological and palaeobiogeographical origins of present-day biodiversity and provides an understanding of the longer-term influences on biodiversity of changes in climate, eustasy, and many other processes operating at the Earth's surface.

Biodiversity is measured in several ways, and there are considerable uncertainties in determining the number of living species, let alone the assessment of diversity through geological time. Databases compiled at a variety of geographical, taxonomic, and temporal scales have become powerful tools for assessing Phanerozoic biodiversity change, but there are significant concerns about the biases that might be inherent in the fossil data. The patterns emerging for the marine and terrestrial realms are very different, and there is a lively debate about how the global curves should be interpreted. Nonetheless, they have an important role to play in understanding the patterns and processes of biodiversity change in the modern world and therefore in the development of appropriate conservation strategies.

## The Measurement of Biodiversity

### Types of Biodiversity

Biodiversity is variously defined in terms of genes, species (or higher taxa), or ecosystems. The genetic and organism components have been combined by many authors and, in many palaeontological studies, a clear distinction is now commonly made between taxonomic diversity and disparity – the degree of morphological variability exhibited by a group of taxa. Disparity can be measured either by phylogenetic distance or by various phenetic indices. Within the history of a major clade, disparity may reach a peak before the peak in species richness is attained.

Setting aside disparity, two basic categories of biodiversity measurement have become widely used: inventory diversity, which records the number of taxa per unit area, and differentiation diversity, which provides a measure of the difference (or similarity) between levels of inventory diversity. Alpha (or within-habitat) diversity is the most common form of inventory diversity and records the number of taxa per area of homogenous habitat, thus reflecting species packing within a community. At its simplest, alpha diversity is species richness (i.e. the number of species in one area), but, especially in the study of modern diversity, its measurement also includes some assessment of abundance. Beta diversity, a differentiation metric, measures the variation in taxonomic composition between areas of alpha diversity. For larger areas, in many studies of terrestrial environments, the term gamma diversity has been used to reflect the number of taxa on an island or in a distinctive landscape, and epsilon diversity has been used for the inventory diversity of a large biogeographical region; the term delta diversity is occasionally used for the variation between areas of gamma diversity within such a region. However, in palaeontological analyses of marine faunas, many workers have used the term gamma diversity for differentiation diversity measuring taxonomic differentiation between geographical regions and therefore reflecting provinciality or the degree of endemicity.

Assessment of the numbers of higher taxa has been an important facet of the analysis of biodiversity – one point sample or community may contain more species than another while having fewer higher taxa and possibly, therefore, a lower genetic diversity and morphological disparity. Moreover, higher taxa are widely used as surrogates for species in biodiversity analyses, especially but not exclusively in the fossil record, where this approach reduces the effects of preservational biases, uncertainties in species identification, and the worst excesses of over-divisive species-level taxonomy. It also facilitates the analysis of biodiversity in biotas within which species-level identifications have not been undertaken because of lack of time or expertise. However, there is a need for caution: the correlation between number of species and number of higher taxa is at best a crude one and diminishes with increasing taxonomic rank. As a result, there is an increasing discordance between the diversity curves calculated from progressively higher taxa and those calculated from species.

## Modern Biodiversity

About 1.8 million modern species have been formally described, but there are enormous gaps in our knowledge of the true number of living species, and the data are patchy in terms of both taxonomic group and geographical coverage. Estimates of modern biodiversity vary considerably depending on the methods used and range from 3.5 to 111.5 million species, with about 13.5 million being considered a reasonable working figure. Less than 15% of the described living species are marine, but over 90% of all classes and virtually all phyla are represented in that environment, with two-thirds of all phyla known only from there. However, conservative estimates of the actual numbers of species of plants and animals (excluding micro-organisms) suggest that marine species may comprise as little as 4.1% of the total.

There is a crude inverse relationship between body size and species diversity, and the problems of quantifying prokaryotic diversity are the most acute. The development of molecular methods over the last few decades has revealed levels of species diversity that are orders of magnitude greater than those recorded previously using culture techniques and has more than trebled the number of major prokaryote divisions now recognized. Though still fraught with uncertainties, not least the definition of what constitutes a species, the estimation of prokaryote diversities from species-abundance curves holds considerable potential; recent calculations suggest that the oceans may have a total bacterial diversity of less than $2 \times 10^6$ species, whereas a ton of soil could contain twice this figure.

## Ancient Biodiversity

The quantification of ancient biodiversity has a long history, with the 1860 compilation by John Phillips of the changing number of marine species through geological time being widely regarded as including the first published biodiversity curve. This was based on fossil faunas from the British Isles and was calibrated to take account of the total thickness of each stratigraphical interval. Its overall shape is remarkably similar to many of the biodiversity curves published over the last four decades based on global datasets. Since the 1960s, there has been a lively debate surrounding the compilation and interpretation of data at a range of taxonomic levels to show the changing patterns of biodiversity through geological time. This scientific endeavour has been greatly facilitated by the development of increasingly sophisticated large computer databases. These vary greatly in both the scope and refinement of their geographical, temporal, and taxonomic coverage and in the origins of the data included in them. The data range from first-hand sample data through assessments of the primary literature to secondary compilations. Similarly there is considerable variation in the metrics used to plot and analyse the data.

Global marine diversity curves for the Phanerozoic compiled at the family level from different databases have proved to be remarkably similar. Moreover, the most widely cited family- and genus-level curves, developed by JJ Sepkoski (e.g., **Figure 1**), have proved to be robust when recalculated to take account of substantial revisions of the underlying database in terms of the taxonomy and stratigraphy of many of the entries. This reflects the random distribution of such errors and inaccuracies and is a common phenomenon when analysing large palaeontological databases. There has been some debate over the likelihood of an artificial steepening of the Cenozoic

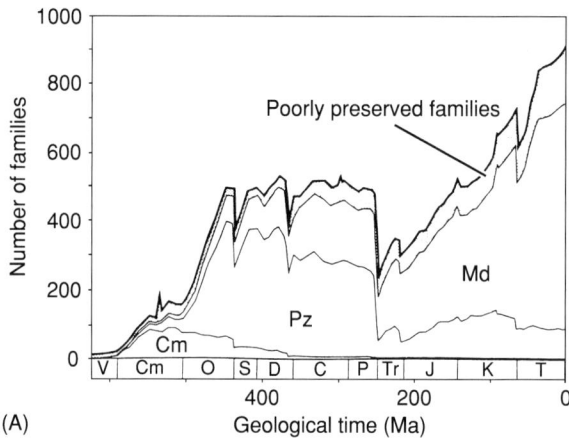

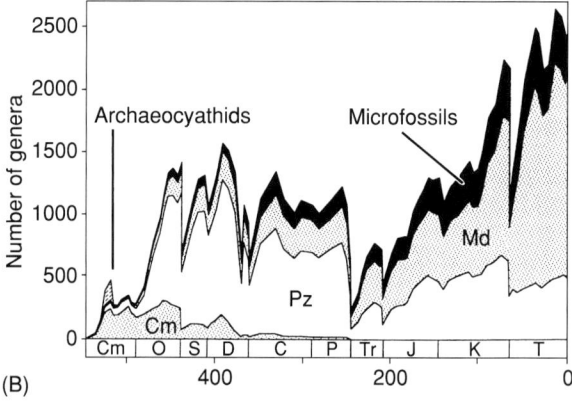

**Figure 1** (A) Family-level and (B) genus-level marine diversity curves, showing the changing proportions of the Cambrian (Cm), Palaeozoic (Pz), and Modern (Md) evolutionary faunas through the Phanerozoic. (Adapted from Sepkoski JJ (1997) Biodiversity: past, present, and future. *Journal of Paleontology* 71: 533–539.)

parts of the diversity curves by the 'pull of the recent' – the possibility that the relatively well-sampled living fauna may extend the stratigraphical ranges of those taxa that also occur in the fossil record, whereas more ancient fossil taxa without modern representatives are less likely to have such complete range records. However, detailed analysis of those modern bivalve genera that have a fossil record has revealed that 95% occur in the Pliocene or Pleistocene, indicating that, for this major group at least, the modern occurrences have only a very small influence on the shape of the Cenozoic diversity curve.

Setting aside the 'pull of the recent', there have been increasing concerns that the global diversity curves may be strongly influenced by taxonomic practice, the temporal sampling pattern used, major geographical gaps in sampling, the absence of information on abundance, and, perhaps most significantly, major biases in the rock record from which the fossil data are derived.

Biases in the marine rock record principally reflect changes in sea-level and can be understood in terms of sequence stratigraphy (see **Sequence Stratigraphy**). They result both from variations in the total outcrop area of rock available for study and from the periodic restriction of the available record to particular environments. Thus, for example, in the post-Palaeozoic succession in western Europe, patterns of standing diversity and (apparent) origination and extinction on time-scales of tens of millions of years show a strong correlation with surface outcrop area. Perhaps most tellingly, the 'mass extinction' peaks are strongly correlated with second-order depositional cycles, coinciding with either sequence bases or maximum flooding surfaces. In the former case, the amount of fossiliferous marine rock is very limited (and biased towards shallow-water facies), whereas, in the latter case, there is a strong bias towards deep-water environments and hence a taphonomic bias against shallower-water faunas being available for sampling. In contrast, evidence from fossil tetrapods indicates that, despite earlier assumptions to the contrary, there is no correlation between sea-level change and the quality of the continental fossil record.

## Biodiversity Change

### Precambrian Biodiversity

From its appearance in the Archaean, life had a profound effect on the chemistry of the oceans and atmosphere during much of Precambrian time, but the overall pattern of diversity change during most of this interval is still weakly constrained. Given the difficulties of assessing diversity in living bacteria and the importance of molecular techniques therein, any understanding of their ancient diversity as represented by fossil material is severely limited. In addition, there is still debate about links between organism diversity and disparity of stromatolites and even about the biotic origin of some such structures (see **Biosediments and Biofilms**). The development of eukaryotic cells in the Early Proterozoic marked a major change in organism complexity, and the widely distributed fossil record of protists suggests that their diversity (probably, more strictly, disparity) remained low until about 1000 Ma, the age of the Mesoproterozoic–Neoproterozoic boundary, after which there were major increases in diversity and turnover rates. A trough in diversity during the interval of the Late Neoproterozoic that was characterized by major glacial events was immediately followed by another peak. This was a short-lived event, which overlapped only slightly with the flourishing of the enigmatic soft-bodied Ediacara biota. The dip in protist diversity during the latest Precambrian was followed by a major diversification that was coincident with that of marine invertebrates.

The Ediacara biota flourished for about 30 Ma and represents the first unequivocal record of metazoans in the body-fossil record. The oldest fossil representatives of this soft-bodied biota are recorded from a horizon in Newfoundland that is about 10–15 Ma younger than diamictites of the last of the Neoproterozoic glaciations, which have been dated at about 595 Ma, but the size (up to about 2 m) and complexity of these organisms indicate an earlier history. Whether this indicates an origin prior to the latest glacial event or very rapid evolution thereafter remains uncertain. Either way, the Late Neoproterozoic glaciations (the so-called 'Snowball Earth') probably had a profound effect on the early evolution of metazoans.

A few weakly mineralized metazoans are known from the latest Neoproterozoic, and a large fully mineralized metazoan, *Namapoikia*, of probable cnidarian or poriferan affinities has been described from rocks dated at about 549 Ma. Molecular data suggest that most benthic metazoan groups had a Proterozoic history prior to their acquisition of mineralized skeletons and their appearance in the fossil record during the Cambrian 'explosion'. Some of this earlier history may be indicated by trace fossils, but many trace fossils below the uppermost Neoproterozoic are controversial. Recent estimates using molecular data suggest that the metazoans originated at some time between 1000 Ma and 700 Ma, but the origins and early history of metazoan groups and hence their diversity during the Proterozoic remain unknown.

## Phanerozoic Diversity Change

**Marine biodiversity change** The patterns of marine biodiversity change at the family level during the Phanerozoic emerging from the current databases consistently show a rise from a handful of taxa at the beginning of the Cambrian to some 1900 families at the present day (e.g. **Figure 1A**). This rise included steep increases in the Early to mid-Cambrian, the Early and mid-Ordovician, and from the Triassic to the Holocene. The diversity level attained in the Ordovician established a plateau, which was essentially maintained until the end Permian mass extinction event (*see* **Palaeozoic: End Permian Extinctions**). Other mass extinctions in the latest Ordovician and Late Devonian and at the ends of the Triassic and Cretaceous periods (*see* **Mesozoic: End Cretaceous Extinctions**) punctuated the overall pattern. The genus-level curve (**Figure 1B**) shows a similar overall pattern but, unsurprisingly, is much more saw-toothed, and major extinction events (not just the 'big five' listed above) had a much more profound effect at lower taxonomic levels.

The overall diversity curves have been resolved into three so-called evolutionary faunas (**Figure 1**), each characterized by the dominance of a set of major clades and together representing an increase in ecological complexity through time (**Table 1**). The Cambrian Fauna dominated its eponymous period but declined thereafter. The Palaeozoic Fauna had its origins in the Cambrian but rose during the Ordovician biodiversification event to a dominance that was largely maintained for the rest of the Palaeozoic. The Palaeozoic Fauna on the low-latitude Laurentian plate showed an onshore–offshore pattern of innovation and expansion of communities during the Ordovician. The Modern Fauna has been an important component of the marine biota, initially in nearshore habitats, since the Ordovician but did not become dominant until after the end Permian extinction event.

Mass extinctions produced the major troughs in the diversity curves but represent one extreme of a continuum of rates of species extinction per million years. The end-Palaeozoic, end-Mesozoic, and, to a lesser extent, Late Devonian mass extinctions did not simply reduce overall biodiversity; they disrupted the ecological patterns so severely that major changes in overall community and even ecosystem structure could occur. However, whilst some significant changes in the proportions of marine taxa grouped by fundamental differences in autecology and physiology took place after mass extinctions, many of the episodes of change or of relative stasis of such features did not mirror the concurrent trajectory in global diversity. Thus, for example, while there were stepped increases in the proportion of predator taxa following the end-Permian and end-Cretaceous extinction events, this proportion then fluctuated only within a fairly narrow band during the subsequent rises in overall diversity.

**Non-marine biodiversity change** The shape of the non-marine biodiversity curve is markedly different from that of the marine biodiversity curve (**Figure 2**), and its principal components are the plants, insects, and tetrapod vertebrates.

Whilst the land probably had a microbial flora extending back into the Precambrian, spores and phytodebris indicate that land plants were present

**Table 1** The major components and ecological structures of the three marine evolutionary faunas

| Evolutionary fauna | Community diversity | Food webs | Tiering | Suspension feeders | Detritivores and carnivores | Planktonic food | Animals in water column |
|---|---|---|---|---|---|---|---|
| Modern evolutionary fauna (demosponges, gastropods, bivalves, gymnolaemate bryozoans, malacostracans, echinoids, vertebrates) | High | Highly complex | Complex | Epifaunal and infaunal | Common | Abundant | Many |
| Palaeozoic evolutionary fauna (anthozoans, cephalopods, stenolaemate bryozoans, 'articulated' brachiopods, ostracods, stelleroids, crinoids, graptolites) | Intermediate | Intermediate | Develops | Epifaunal common | Common | Common | Many |
| Cambrian evolutionary fauna (hyolithids, monoplacophorans, 'inarticulate' brachiopods, trilobites, eocrinoids) | Low | Simple | Limited | Few | Dominant | Limited | Few |

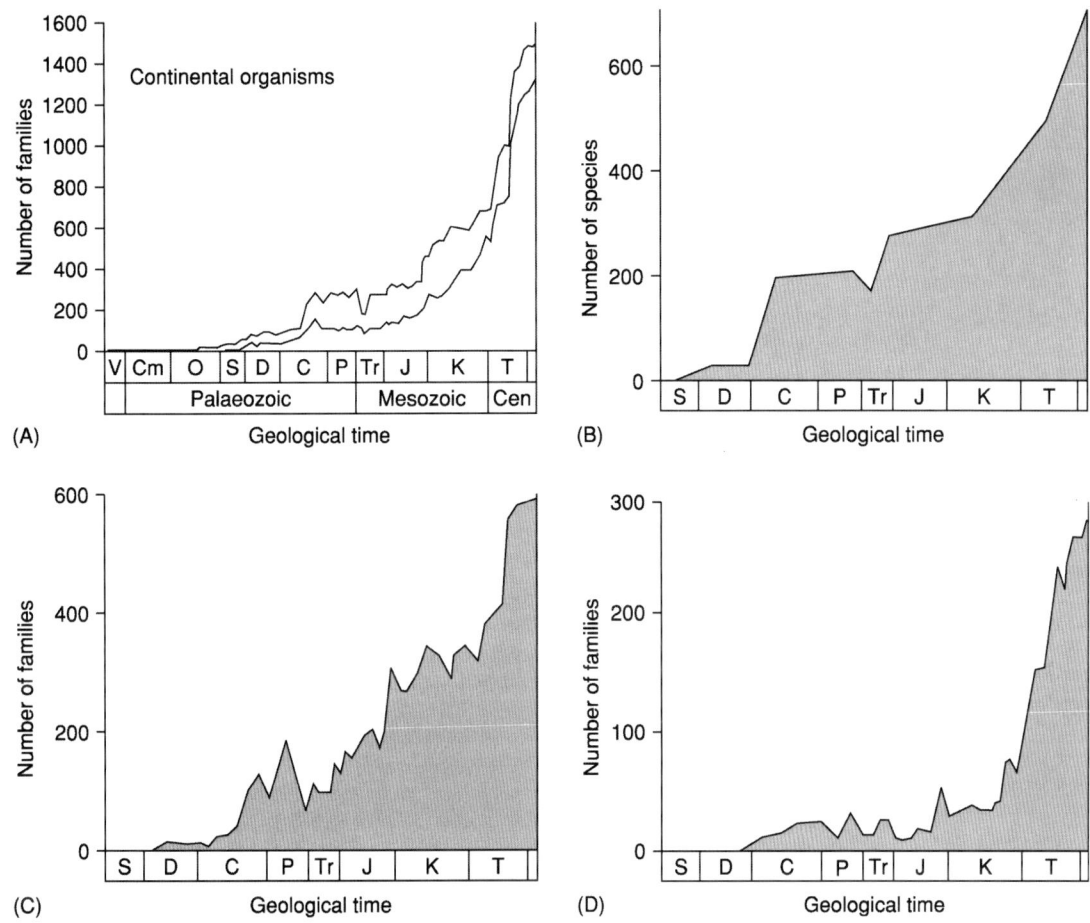

**Figure 2** Terrestrial biodiversity through the Phanerozoic. (A) All multicellular families (lower curve represents families undoubtedly present within each stage and restricted to the non-marine environment; upper curve includes less stratigraphically well-attributed families and those that include environmentally equivocal that include marine or environmentally equivocal taxa); (B) vascular land plants; (C) insects; and (D) non-marine tetrapods. (Adapted from Benton MJ (2001) Biodiversity through time. In: Briggs DEG and Crowther P (eds.) (2001) *Palaeobiology II*, pp. 211–220. Oxford: Blackwell Publishing.)

from the mid-Ordovician onwards. These earliest plants were probably bryophytes; spore evidence suggests that vascular plants or their immediate ancestors appeared in the Early Silurian. The first unequivocal land-plant megafossils are known from the mid-Silurian (Wenlock). Following the Silurian and Devonian development and radiation of early vascular-plant groups, the plant biodiversity curve (**Figure 2B**) reflects the successive acmes of the pteridophytes and early gymnosperms in the Carboniferous–Permian and the gymnosperms (*see* **Fossil Plants:** Gymnosperms) in the Triassic–Jurassic and the rise of the angiosperms from the Cretaceous onwards.

The diversity curve for the insects (**Figure 2C**) shows a rather saw-toothed exponential rise from the first appearance of the group in the Devonian. The tetrapods (**Figures 2D** and **3A**) also show an essentially exponential pattern, with a long (Devonian to mid-Cretaceous) period of weak diversity increase followed by a steep rise. The curve primarily reflects the radiations and acmes of three major sets of groups: the basal tetrapods and synapsids in the Palaeozoic, archosaurs in the Mesozoic, and lissamphibians, lepidosaurs, birds, and mammals in the Cenozoic.

The colonization of freshwater and the increase in diversity therein was achieved both by air-breathing groups and by taxa from several major marine clades that independently became adapted to reduced salinity by movement upstream through estuarine environments. However, the occupation by invertebrates of freshwater habitats, especially those within the substrate, took a remarkably long time, and it was not until the Mesozoic that a significant range of originally marine groups developed the necessary osmoregulatory, reproductive, and dispersal capacities to occupy freshwater environments.

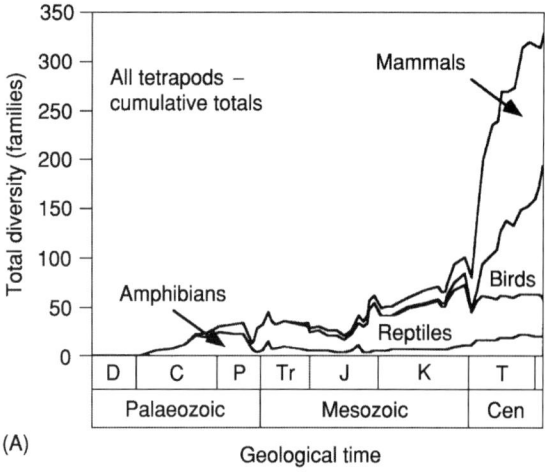

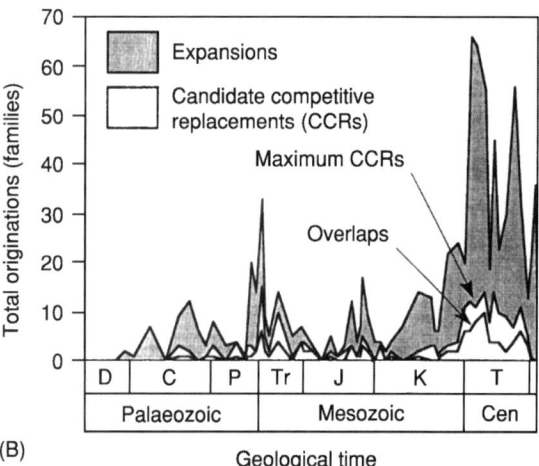

**Figure 3** Biodiversity change in 840 non-singleton tetrapod families, broken down into (A) the component groups and (B) the styles of origination of new families. Overlap candidate competitive replacements are those in which the range of an originating family overlapped with that of another family rather than replacing it in the fossil record. (Adapted from Benton MJ (1999) The history of life: large databases in palaeontology. In: Harper DAT (ed.) *Numerical Palaeobiology*, pp. 249–283. Chichester: Wiley.)

## Understanding Biodiversity Curves

If the existing curves are even an approximation of the true pattern, the great challenge is to interpret them in terms of the processes leading to biodiversity change through geological time. At the simplest level, the curves reflect the complex interplay of species originations, extinctions, and stasis, with the major changes in trajectory representing episodes of radiation punctuated by mass extinctions. Although the traditional assumption has been that radiations were 'adaptive', representing the acquisition of new characters by the radiating group that gave them superiority over their competitors, there is increasing evidence for the expansion of many clades into previously unoccupied ecospace. This is perhaps most strikingly illustrated by the tetrapods, for which it has been calculated that a maximum of only 13% of families could have originated by competitive replacement of earlier taxa (**Figure 3**). It has also been argued that clade radiations could arise randomly, without being driven by any deterministic cause such as competition or expansion associated with a new adaptation or opportunity.

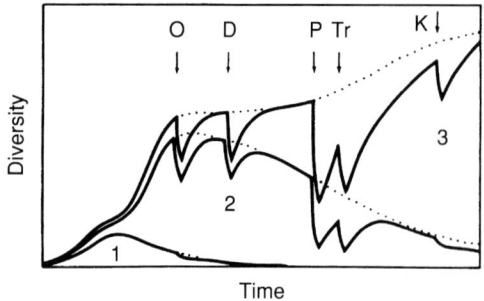

**Figure 4** Three-phase coupled logistic model of changes in family-level marine biodiversity during the Phanerozoic, including perturbations simulating the mass extinctions at the end of the Ordovician (O), in the late Devonian (D), and at the ends of the Permian (P), Triassic (Tr), and Cretaceous (K). The numbers 1, 2 and 3 refer to the Cambrian, Palaeozoic and Modern evolutionary faunas shown on **Figure 1**. Dotted lines represent the trajectories of the unperturbed three-phase system. (Adapted from Sepkoski JJ (1984) A kinetic model of Phanerozoic taxonomic diversity. III. Post-Paleozoic families and mass extinctions. *Paleobiology* 10: 246–267.)

**The shapes of biodiversity curves** Several attempts have been made to model mathematically the shapes of global Phanerozoic diversity curves and therefore to determine whether they reflect overarching global processes and patterns of evolution.

Continental biodiversity, either as a whole or of individual large clades (**Figure 2**), seems to show an exponential rise without a sustained levelling off. In contrast, the family-level marine curve has been modelled fairly closely in terms of logistic curves representing the Cambrian, Palaeozoic, and Modern Faunas perturbed by the major mass extinction events (**Figure 4**). In this coupled logistic model, each evolutionary fauna showed a slow initial increase in diversity, followed by a steep increase to a plateau and then a slow decline corresponding to the rise of the following evolutionary fauna (which had a lower initial rate of diversification and a higher maximum diversity level). Within such a model, the trajectory of the curve for the Modern Evolutionary Fauna is currently still in its exponential phase, but its roughly convex shape suggests that an asymptotic stage could be projected in the future. The effects of the mass extinction events enhance the fit of the theoretical curves to the measured global patterns.

By extending the theory of island biogeography developed in the 1960s to the global scale, the logistic curves applied to marine familial diversity have been interpreted as reflecting evolution into new habitats and empty niches until a level of dynamic equilibrium is reached at the carrying capacity of the environment. This theory has received widespread, but not universal, acceptance. It has been argued, for example, that the shapes of the curves may represent not global-scale biotic interactions but rather the total effect of physical perturbations operating at a wide range of geographical and temporal scales. Moreover, the evidence for equilibrium, even on an ecological rather than a geological time-scale, and the upper limit on diversity that it would impose have also been widely contested. The Palaeozoic diversity plateau, taken to be strong evidence for equilibrium, may have been maintained by factors other than biotic interactions and may even be an artefact of the taxonomic level of the data. Importantly, it is clear that different clades or parts thereof may show very different patterns of biodiversity change.

**The causes of biodiversity change** Despite considerable research effort, there has been only limited success in identifying 'rules' governing species and community diversity and the relative abundances of species within communities. It is highly unlikely, even for mass extinction events, that a single factor can be invoked to explain global changes in biodiversity. In addition to factors intrinsic to individual clades, a host of interrelated extrinsic factors undoubtedly influence the evolution, distribution, and diversity of organisms across the spectrum of spatial and temporal scales. Crucially, the factors operating at one scale may be very different from those operating at another. Intriguingly though, there is some evidence to suggest that the family-level global diversity curve has properties of self-organized criticality, and, thus, Phanerozoic diversity change may be driven by the internal dynamics of life itself, as well as responding to external factors.

The present-day concerns over biodiversity change stem from the recognition of the deleterious consequences of anthropogenic activities, including habitat destruction, pollution, and influences on global climate. The fossil record provides a time perspective not only on the patterns of biodiversity change but also on the natural physical factors that drive it, from local fluctuations in environmental conditions to plate tectonics, eustasy, ocean circulation patterns, and climate changes. Heightened awareness of the quality and detail of the data involved in the generation and analysis of diversity curves at all spatial and temporal scales should result in greater confidence in the conclusions drawn from them.

## See Also

**Biological Radiations and Speciation**. **Biosediments and Biofilms**. **Evolution**. **Fossil Plants:** Gymnosperms. **Mesozoic:** End Cretaceous Extinctions. **Palaeoecology**. **Palaeozoic:** End Permian Extinctions. **Sequence Stratigraphy**.

## Further Reading

Bambach RK, Knoll AH, and Sepkoski JJ (2002) Anatomical and ecological constraints on Phanerozoic animal diversity in the marine realm. *Proceedings of the National Academy of Sciences USA* 99: 6854–6859.

Benton MJ (1999) The history of life: large databases in palaeontology. In: Harper DAT (ed.) *Numerical Palaeobiology*, pp. 249–283. Chichester: Wiley.

Benton MJ (2001) Biodiversity on land and in the sea. *Geological Journal* 36: 211–230.

Briggs DEG and Crowther P (eds.) (2001) *Palaeobiology II*. Oxford: Blackwell Publishing.

Crame JA and Owen AW (eds.) (2002) *Palaeobiogeography and Biodiversity Change: The Ordovician and Mesozoic–Cenozoic Radiations*. Special Publication 194. London: Geological Society.

Gaston KJ and Spicer JI (1998) *Biodiversity: An Introduction*. Oxford: Blackwell Science.

Hewzulla D, Boulter MC, Benton MJ, and Halley JM (1999) Evolutionary patterns from mass originations and mass extinctions. *Philosophical Transactions of the Royal Society of London Series B* 354: 463–469.

Knoll AH (1994) Proterozoic and early Cambrian protists: evidence for accelerating evolutionary tempo. *Proceedings of the National Academy of Sciences USA* 91: 6743–6750.

Levin SA (ed.) (2001) *Encyclopedia of Biodiversity*, 5 vols. San Diego: Academic Press.

May RM (1992) How many species inhabit the Earth? *Scientific American* October 1992: 18–24.

Miller AI (2000) Conversations about Phanerozoic global diversity. *Paleobiology* 26(4 Suppl.): 53–73.

Sepkoski JJ (1984) A kinetic model of Phanerozoic taxonomic diversity. III. Post-Paleozoic families and mass extinctions. *Paleobiology* 10: 246–267.

Sepkoski JJ (1997) Biodiversity: past, present, and future. *Journal of Paleontology* 71: 533–539.

Smith AB (2001) Large-scale heterogeneity of the fossil record: implications for Phanerozoic biodiversity studies. *Philosophical Transactions of the Royal Society of London, Series B* 356: 351–367.

Ward BB (2002) How many species of prokaryotes are there? *Proceedings of the National Academy of Sciences USA* 99: 10234–10236.

# BIOLOGICAL RADIATIONS AND SPECIATION

**P L Forey**, The Natural History Museum, London, UK

Copyright 2005, Natural History Museum. All Rights Reserved.

## Introduction

This entry discusses the evidence for speciation and the patterns of species multiplication and associated morphological change as revealed in the fossil record. Speciation is regarded as the key stage in the generation of biological diversity as it is the point at which reproductive isolation is achieved – the hallmark of the biological species (*see* **Evolution**). Many studies of modern species concern themselves with the short-term genetic fluctuations between populations of a single species, how those fluctuations may come about and how they are maintained. Such studies are not possible in the fossil record. On the other hand, the fossil record documents extended timescales and may, therefore, identify patterns of long-term changes relevant to different theories of speciation that are unavailable in the Recent world.

Irrespective of the mode of speciation, the fossil record demonstrates repeated instances, geologically speaking, of sudden increases in the numbers of species and these are usually associated with rapid morphological divergence. Such events are commonly referred to as radiations, more usually adaptive radiations, and are inferred to have come about by causes intrinsic to the organisms and/or extrinsic causes. Although studies of radiations involving populations and closely related species are studied in the Recent world, broad-scale patterns can only be studied through the time dimension supplied by the fossil record.

## Species, Species Recognition and Speciation in the Fossil Record

There is a plethora of species definitions and criteria for recognizing species in the Recent world, but few are applicable to the fossil record where only a limited amount of morphological data, usually the skeleton, is available for study. Three such definitions that rely on some underlying process of speciation are shown in **Figure 1**. The biological species concept of Ernst Mayr (**Figure 1A**) is probably that most favoured by people studying the Recent world. Species status is achieved when populations diverge genetically to such an extent that reproductive cohesion is broken. There may be several reasons for fragmentation of reproductive continuity (geography, ecology, and behaviour are most commonly cited), although Mayr stressed geographical isolation. The parental species may continue to live alongside the daughter species. In practice, most modern species are not recognized on reproductive criteria: instead some measure of morphological, genetic, behavioural or ecological difference is used as a surrogate for reproductive incompatibility. As such, palaeontological species, recognized almost exclusively on morphological differences, are equally as valid as modern species. In fact, there has been empirical justification for equating genetic differences with morphological differences between modern species of some groups such as bryozoans and Darwin's finches. It is, however, recognized that morphological difference may not always indicate species differences (e.g., sibling species or polymorphic species including mimetic species).

The evolutionary concept of George Gaylord Simpson (**Figure 1B**) defines a species as an ancestor-descendent sequence of populations changing through time with their own trends and tendencies.

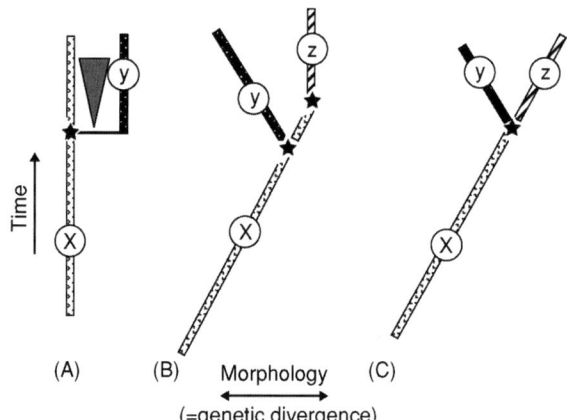

**Figure 1** Three concepts of species. (A) Mayr's biological species. Populations reach species distinction when reproductive cohesion is broken, usually by the imposition of some isolating mechanism such as geographic separation of populations (triangle). The original species (X) may or may not continue to live contemporaneously with the daughter species (Y). (B) Simpson's evolutionary species concept. Species are ancestor-descendent sequences of populations changing through time. A species may change gradually into another (X → Z) through a process of anagenesis or the lineage may split (X → Y) through cladogenesis. (C) The Hennigian species concept recognizes species as lineage segments. As soon as a cladogenetic event occurs there are automatically two new species (Y and Z), regardless of any perceived morphological change. Stars represent speciation events.

This definition embraces asexual species. Such a definition allows for one species to change gradually into another (anagenesis) as well as instances where a species is budded off or where one species splits into two or more (cladogenesis).

The Hennigian species concept (**Figure 1C**) (named after the German entomologist Willi Hennig), like the biological species concept, is based on the criterion of reproductive isolation. However, the species limits are recognized only at the point where one species splits to two or more and the interbreeding pattern of gene flow is disrupted, at which point the ancestral species cannot by definition live alongside the descendants. This species definition is strictly dependent upon the shape of the phylogenetic tree.

**Recognition of Species**

Species in the fossil record are nearly always recognized on morphological criteria. There are some exceptions, such as instances where species are distinguished from one another purely on the basis of stratigraphic occurrence or geographic location, but these are usually accepted as stopgap measures; to be revised when more information about morphology becomes available.

Monophyletic species are recognized on the basis of possessing one or more unique morphological characters and this corresponds most closely with the Hennigian species concept. Species viewed in this light have the same ontological status as higher taxa such as genera, families, orders, etc. In reality, very few species can be recognized in this way and logically it denies the existence of ancestors since ancestors have no unique features.

Typological species were once commonly recognized, whereby a single specimen (usually the first discovered or the most complete) was chosen as the Linnean type and this formed the nucleus of morphological variation considered worthy of species status. How much variation was to be allowed before a new species was recognized was never stated. Such species are usually frowned upon now as being steeped in the philosophy of pre-Darwinian essentialism. Yet, there remain many instances where such species are still used.

Phenetic species arose out of numerical taxonomy. Here species are measured for as many morphologically continuous variables as possible. The variables are then subjected to multivariate analysis, and clusters – (the species) – are then recognized. Although this may appear to be objective and theory free, it is heavily dependent upon the original samples analyzed and the multivariate analysis algorithms used. It is not always easy to identify which of the individual variables is contributing to the differences and, therefore, it is difficult to diagnose the species.

Phylogenetic species are the smallest diagnosable assemblages of specimens showing a unique combination of characters. Combination is the operative word here since there is no requirement for identifying unique characters (cf. monophyletic species recognition). Of all the recognition criteria this most faithfully agrees with all of the process-based definitions of species, as well as agreeing with the day-to-day practice of palaeontologists.

Estimates of species longevity may depend on which of the above ways the species have been recognized. Surveys across various fossil groups give average figures that range from 1 million years (for small mammals where generation time is very short) to 13 million years for dinoflagellates (which are largely asexual).

**Speciation in the Fossil Record**

It is generally accepted that most speciation events, with the possible exception of those caused by hybridization and polyploidy (chromosomal duplication), are too protracted to be detected in the Recent world and too brief to be seen in the fossil record. The lower limits of time that can be successively sampled in the fossil record appear to be in the order of 5000–10 000 years and such deposits are very rare, restricted to lake deposits of relatively short duration and small geographic range, and some deep-sea deposits. Nevertheless, there have been many studies seeking to establish the precise pattern of speciation.

There are two polarized theories on the tempo and pattern of speciation (**Figure 2**), punctuated equilibria and phyletic gradualism. Both have support from theoretical and empirical studies in the modern world (see legend to **Figure 2**) and it is likely that both modes of speciation can happen.

*Phyletic gradualism* Phyletic gradualism holds that morphological evolution is gradual and occurs independently of speciation. Most of the studies supporting this mode of evolution involve organisms that can be minutely sampled through continuous rock sequences.

**Figure 3** shows an example of one gastropod species (*see* **Fossil Invertebrates:** Gastropods) changing or replacing another in an ancestor-descendent sequence (gradual speciation by anagenesis) by a shift in the modal shell form as measured by morphometric discriminant analysis. In this example the gradual change has been correlated with increasing depth of water. The populations intermediate between the two species were presumed to have lasted for 73 000–250 000 years. In this pattern there is no

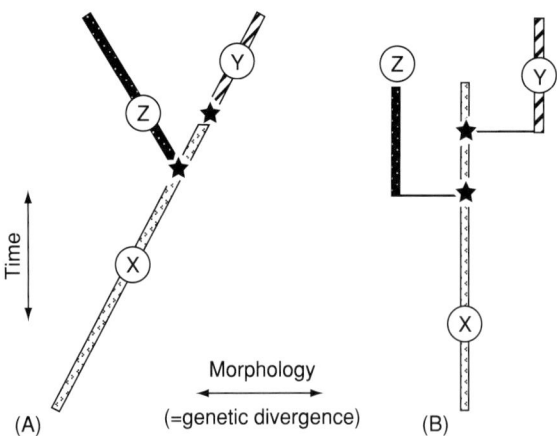

**Figure 2** Two theories of speciation modes in the fossil record. (A) Phyletic gradualism suggests that speciation is a by-product of changing ratios of gradual morphological variation (implied gene frequency) within populations through time and that speciation may happen through anagenesis (species X → Y) as well as cladogenesis (X → Z). Phyletic gradualism agrees with Simpson's evolutionary species concept (**Figure 1B**), coupled with Fisher's notion of 'directed' selection of gene frequencies within populations. (B) Punctuated equilibrium claims that individual species remain morphologically constant throughout time and that all or most of the morphological (and implied genetic) change is associated with cladogenesis: the speciation event. This theory agrees with Mayr's biological species concept (**Figure 1A**) coupled with Sewell Wright's idea of stabilizing selection. In these diagrams the species are denoted by letters and different shadings and the speciation events by stars.

rapid shift in morphology and, in studies of this kind, the species limits are usually recognized by comparing the morphological variation over a period of time with morphological variation of modern species and dividing up the continuum accordingly.

In another study (**Figure 4**) involving radiolarians recovered from Pacific Ocean deep-sea drilling cores, an instance of speciation was detected occurring during a period of about 500 000 years. In this case both the ancestral and daughter species showed marked deviation in many aspects of skeleton form. Further, it is interesting to note that morphological deviation, although slightly accelerated during the speciation event, continued for some time after the separation of morphotypes, suggesting strongly that morphological evolution is decoupled from speciation. In this instance both parental and daughter species showed morphological evolution and it may have been possible to suggest that a single ancestral species gave rise to two daughter species. The author of this study refrained from this because the ancestral species, which still lives today, shows a high degree of polymorphism and geographic variation, the scale of which is comparable to the variation seen in the lineage in this study.

It needs to be pointed out that many (but not all) of the studies that support phyletic gradualism, report information on continuously varying morphologies such as lengths and shapes. These parameters are inherently gradualist in their measurement, unlike many characters used in cladistic analysis such as presence/absence or four toes versus five toes in which there can be no intermediates.

**Punctuated equilibrium** Speciation following the punctuated equilibrium model (**Figure 2B**) has been claimed for many groups of organisms. This theory suggests that species themselves exhibit morphological stasis and that speciation is, geologically speaking, instantaneous and is accompanied by rapid morphological shifts. In order to demonstrate this at least three objections raised by advocates of phyletic gradualism must be overcome. The first objection is that what appear to be sudden changes only seem so because either the sampling intervals are too wide or the fossil record is lacking in the key years where intermediate populations might have lived. Therefore, it is necessary to demonstrate that the sampling is dense. A second objection is that it is possible that an isolated population, destined to become a new species, was accumulating genetic and morphological distinctiveness in a gradualist manner, but that the population was either so small that fossilization potential was virtually zero, or that it was living elsewhere. The sudden appearance is, therefore, explained not as any intrinsic property of speciation, but by the new species population growing large enough to leave a fossil record, or by immigration into the area of study. The third objection is a taxonomic argument. Many of the claims of punctuated equilibrium involve species recognized by qualitative characters where there are no intermediate states (see above). If this is true then the sudden appearance of morphologic changes must coincide with speciation and must also remain static until the next 'speciation'.

Despite the objections there are cases where punctuated equilibrium can be demonstrated. **Figure 5** illustrates the case of speciation in the bryozoan genus *Metrarhabdotus* during the Neogene and Quaternary of tropical America (see **Fossil Invertebrates: Bryozoans**). The species were distinguished on both continuously varying characters, as well as qualitative differences. A stratophenetic tree was constructed. There was particularly dense sampling of the fossil record in the Dominican Republic between 8 and 4 million years ago, within which time there was the successive origin of 12 species. Each of the species appeared suddenly, accompanied by substantial morphological differences. The species themselves exhibited little or no change throughout their individual

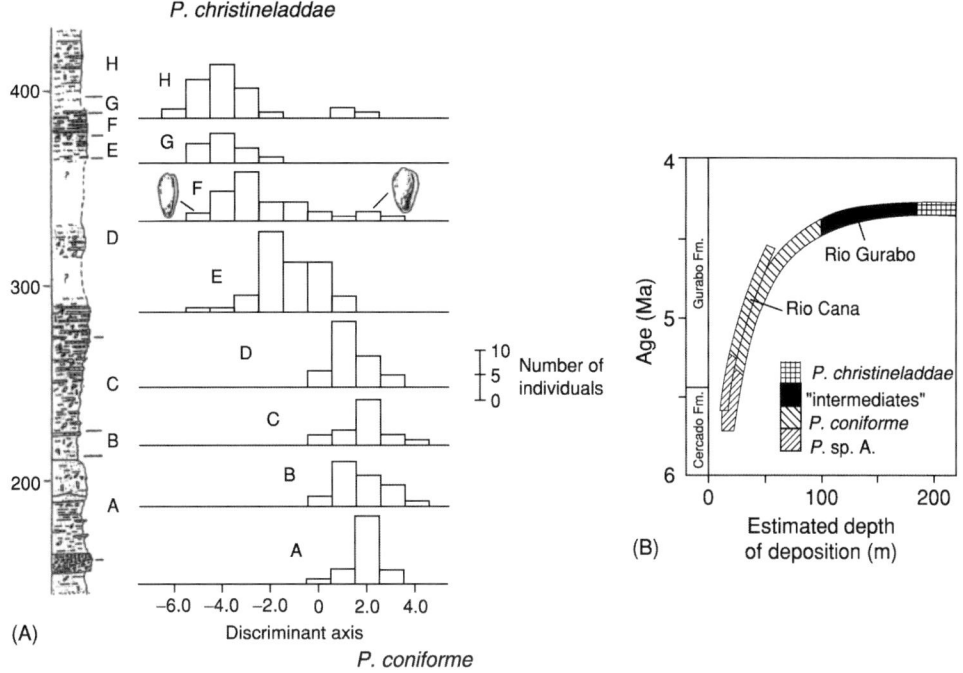

**Figure 3** Speciation through anagenesis. In this study a species of marginellid gastropod – (*Prunum coniforme*) – living in the Mio-Pliocene of the Dominican Republic was seen to show strong directional selection in shell morphology towards a new species – (*Prunum christineladdae*). (A) Histogram of individuals on the discriminant axis that best distinguishes the two species (endpoints of sampling) collected from successively higher stratigraphic levels within the section at Rio Gurabo, Dominican Republic (section thickness marked in meters at left). Notice that at level F there appears to be a mixed population which appears to have occupied 0.6–2.5% of the entire range of the ancestral species – (*P. coniforme*). (B) Over a longer time scale the species are distributed in sediments inferred to have been deposited in increasingly deeper water. Part of the history at Rio Gurabo is repeated at neighbouring Rio Cana. Reproduced from Nehm RH and Geary DH (1994) A gradual morphologic transition during a rapid speciation event in marginellid gastropods (Neogene; Dominican Republic). *Journal of Paleontology* 68: 787–795.

existence and this is a key feature of the punctuated equilibrium model. Although the possibility of immigration cannot be completely dismissed this is unlikely because the basal species of this lineage are morphologically very similar to species occurring earlier in the same area.

Therefore, it appears as though there is evidence for both patterns of speciation in the fossil record. Indeed, there may be instances where both kinds can be recognized within the same sequence, affecting closely related organisms. More commonly, species within the same lineage may show gradualism at some times and punctuated patterns at others (but here there is always the objection of differential sampling). Rather than trying to prove that all speciation complies with one or the other model it may be more productive to try to determine when and why one or the other is predominant.

## Radiations

Radiations are episodes of increased rates of speciation relative to extinctions resulting in rapid net increases in diversity. They may be accompanied by evolution of more diverse body forms than in normal periods when extinction and origination rates are approximately equal. Radiations are nearly always explained in terms of adaptation, hence 'adaptive radiations', the latter being defined by Dolph Schluter as the evolution of ecological and morphological diversity within a rapidly speciating lineage. There are difficulties with defining both ecological and phenotypic diversity and it is not always easy to describe precisely the adaptation and even more difficult to isolate the cause, but the following factors have been suggested as triggers for radiations (adaptive or otherwise):

1. Abiotic causes – such as changing levels in $O_2/CO_2$ concentrations or significant changes in ambient temperature, changes in ocean current circulation, fragmentation of landmasses and consequent changes in both the length and ecological differentiation of coastlines, and changes in latitudinal distributions.

2. Biotic factors – such as the evolution of a particular character or characters that may change aspects of life history leading to rapid speciation and/or ecological differentiation.

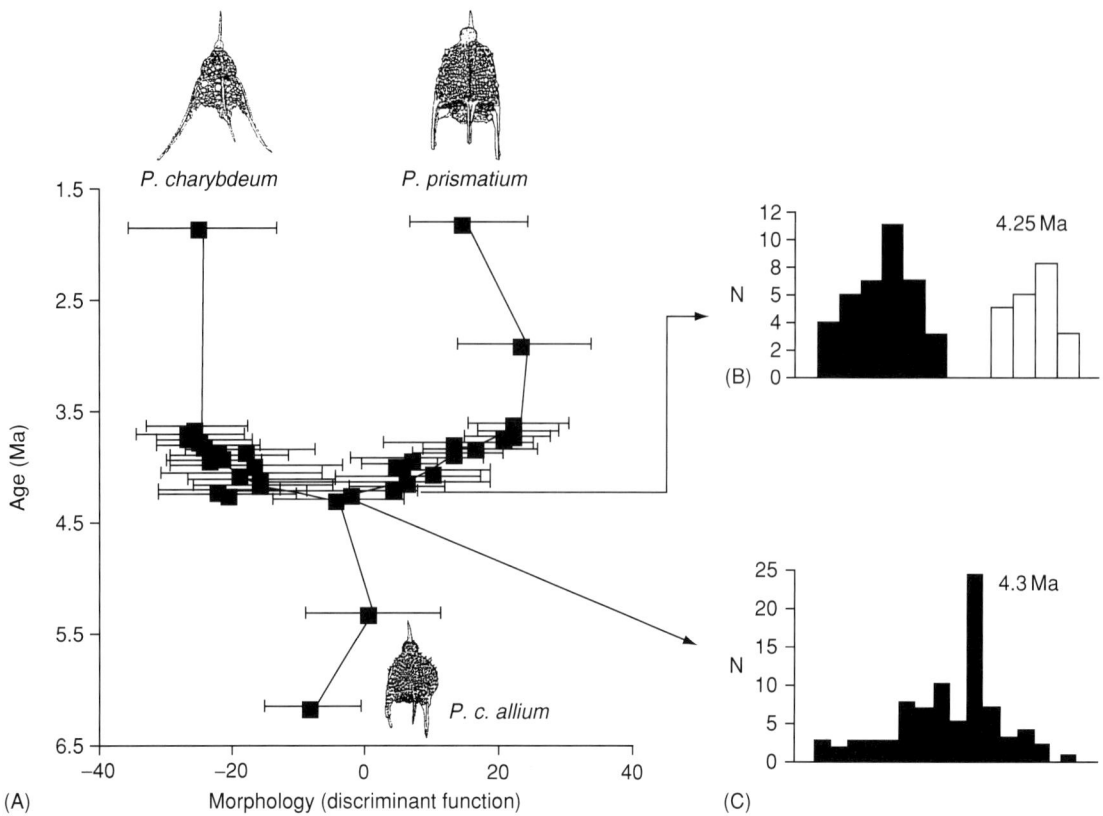

**Figure 4** Speciation and the gradualist model. The present day radiolarian species *Pterocanium charybdeum* extends well back into the Miocene where a subspecies (*P. c. allium* is recognized). At about 4 million years ago a new species, *P. prismatium* arose and subsequently went extinct about 1.8 million years ago. To compile this diagram samples representing populations at 50 000 year intervals were taken from piston cores during the key 4.5–3.5 million year interval (much more widely spaced outside of this time band). The radiolarian tests were measured for 32 morphological characters such as length, breadth, angles and outlines and subjected to a multivariate discriminant analysis. The x-axis of the diagram shows the value of the discriminant function that best separates the species. The boxes are the population means ± 1 standard error. The histograms on the right show discriminate scores at the point of the intermediate population (lower diagram) and the sampling level 50 000 years later (upper diagram), showing clear separation at this time; these two documenting the speciation event. Notice that both lineages continue to diverge morphologically in a gradualist manner long after the speciation event. Reproduced from Lazarus DB (2001) Speciation and morphological evolution. In: Briggs DE and Crowther PR (eds.) *Palaeobiology II*, pp. 133–137. Oxford: Blackwell Scientific.

3. Interactions between organisms – such as the extinction of one kind allowing ecological and phenotypic diversification in another. Alternatively it may be true that the morphological and diversification of one group allows similar phenomena in another (e.g., insects and plants). For example, in fishes, it has long been noted that the evolution of many durophagous (mollusc crushing) lineages coincides with the Mesozoic diversification of bivalve and gastropod molluscs. Many of these 'associations' are anecdotal and, while they may well be true, they need firm experimental evidence for their justification.

It is rarely possible to isolate a single cause and in all probability complex interactions between two or more factors are ultimately responsible for radiations. Therefore, the study of these evolutionary phenomena is both frustrating and challenging, demanding lengthy and detailed data collection.

Factors which may distort our view of radiations include the following:

1. Imperfections of the fossil record. Large hiatuses or differential preservation will inevitably distort our perception of radiations. For example, the sudden appearance and apparent radiation of many catfishes (most of which are freshwater) in the Early Tertiary must be judged against the knowledge that there are very few freshwater fish-bearing deposits in the underlying Upper Cretaceous.
2. Appearance of many diverse animals and plants associated with Lagerstätten deposits. Such deposits dramatically increase the numbers and morphological diversity of species that may, in reality, have had a long unrecorded history.

# BIOLOGICAL RADIATIONS AND SPECIATION 271

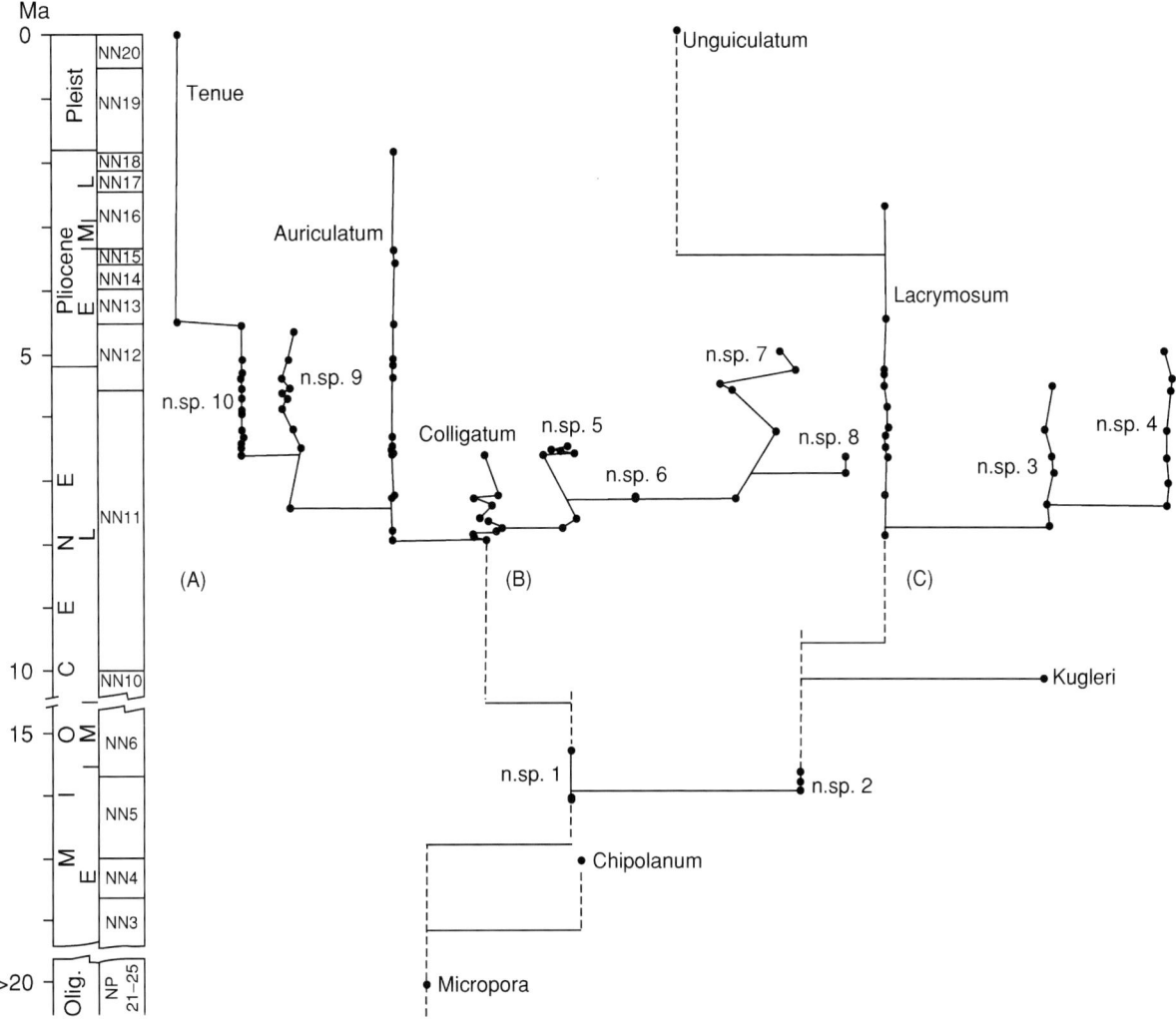

**Figure 5** Punctuated equilibrium in Neogene bryozoans. The cheilostome genus *Metrarhabdotos* is widely distributed in Miocene and Pliocene deposits of the Caribbean. Sections in the Dominican Republic are particularly rich and uninterrupted, such that is possible to sample extensively. This study used 46 measurements, counts or codings of colony morphology of the species (some were left unnamed). These were subjected to multivariate analysis and the stratophenetic tree shown here was produced. The dots are sampling points. The x-axis is a measure of the morphologic distance. Notice the sudden shift in morphological variation at points of cladogenesis. The within species variability is very small compared with the between species differences as predicted by the punctuated equilibrium model, as is also the prediction that the ancestral species continues to live alongside the descendant. The age, epoch and nannoplankton zones are shown on the y-axis. Reproduced from Cheetham AH (1986) Tempo of evolution in a Neogene bryozoan: rates of morphological change within and across species boundaries. *Paleobiology* 12: 190–202.

3. Taxonomic artifact. Many diversity studies are based on estimates of numbers of genera or families. Dependent on how those taxa are recognized in successive stratigraphic levels, the expansion (or decrease) in diversity may be exaggerated or distorted. For example, it has been pointed out that the apparent sudden increase in new trilobite families in the earliest Ordovician may be an artifact of taxonomy of Cambrian trilobites, since many phylogenetic lines can now be drawn across the boundary linking families that were previously thought to be quite distinct.

Despite these problems there are good examples of association of cause with radiations.

## Environmental Shift

One general pattern that has been identified after extensive collection of both taxic and geological data is that many invertebrate clades originated in nearshore marine environments and subsequently expanded in species numbers to occupy deeper water, at the same time becoming morphologically more diverse. **Figure 6** shows one example of the

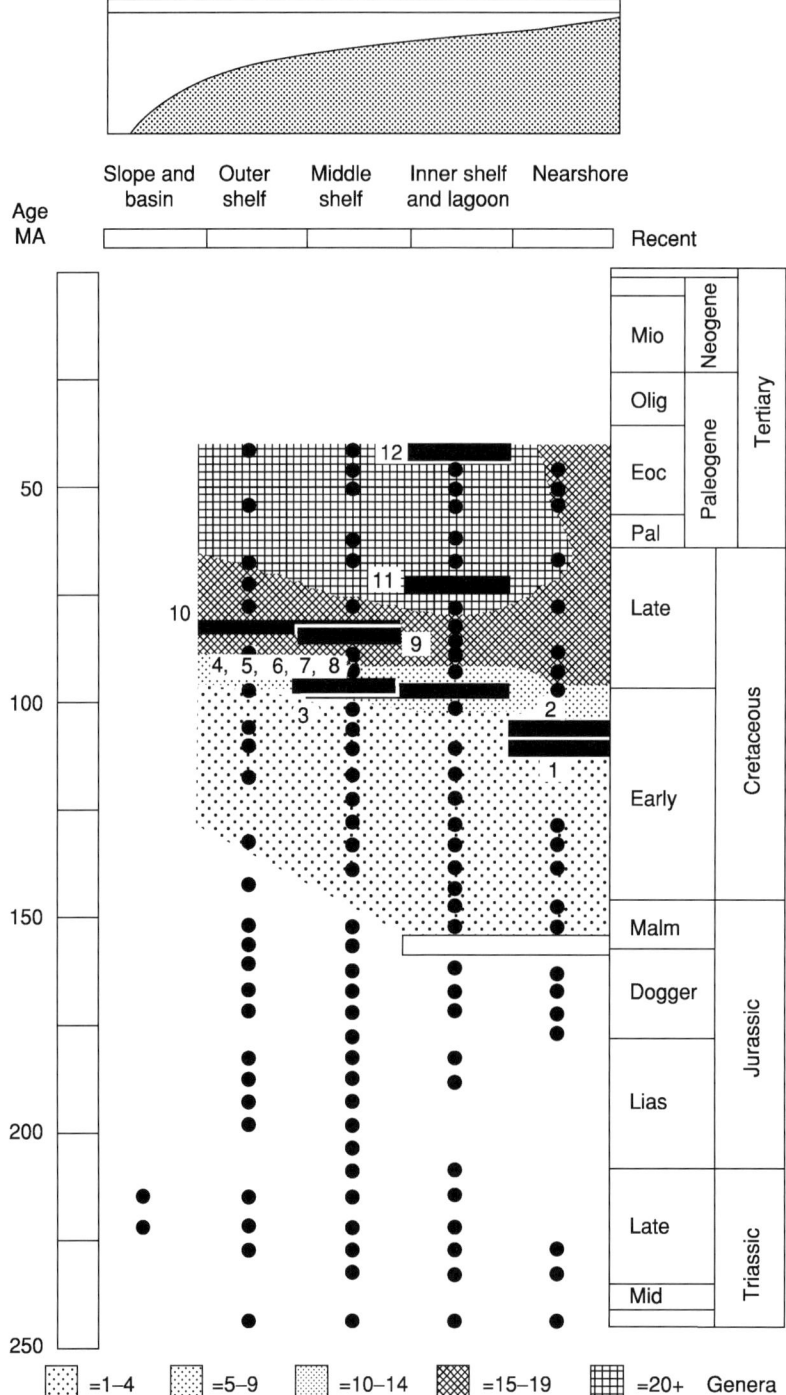

**Figure 6** Study exemplifying that the early phases of a history of marine invertebrate clades began in nearshore environments and subsequently expanded to the outer shelf. In this study the environmental history of the cheilostome bryozoan radiation suggests that the group originated in nearshore and inner shelf environments (white box). As the group diversified in both numbers of genera (contours denoted by the different shadings) with the acquisition of evolutionary novelties (numbers associated with black boxes) they spread to middle and outer shelf habitats. The black dots represent localities yielding bryozoans at different horizons and depositional environments. Those dots below the shaded areas represent localities at which only non-cheilostomes are found (i.e., cyclostomes and trepostomes) and these impose a taphonomic control documenting that the cheilostome pattern is not an artifact of preservation. The dots within the shaded area represent localities at which all kinds of bryozoan are found. Reproduced from Jablonski D, Lidgard S and Taylor PD (1997) Comparative ecology of bryozoan radiations: origin of novelties in cyclostomes and cheilostomes. *Palaios* 12: 505–523. SEPM (Society for Sedimentary Geology).

evolution of cheilostome bryozoans in the Cretaceous (see **Fossil Invertebrates:** Bryozoans). It needs to be emphasized that this is a general pattern; there is no implication that the ecological shift was the cause of speciation.

### Abiotic Causes – Fragmentation of Areas

Radiations coincident with fragmentation of continental masses can be documented for a variety of organisms, but usually only over long timescales.

One example, is the increased diversity of families of terrestrial amniotes between the Lower Jurassic and Paleocene following the breakup of Pangaea (**Figure 7**). It is very difficult to demonstrate a direct cause and effect, but there is a considerable shift in the latitudinal distribution of the continental masses such that substantial diversification in ecological conditions occurred leading to a tempting association with cladogenesis. The breakup of Pangaea also resulted in a considerable increase in available coastline,

|  | First | 1 (%) | 2 (%) | 3 (%) |
|---|---|---|---|---|
| Early Jurassic (Hettangian Toarcian) Total first appearances: three (sample too small) |  |  |  |  |
| Middle Jurassic (Aalenian-Callovian) | 12 | 1 (8) | 3 (25) | 8 (67) |
| Late Jurassic (Oxfordian-Portlandian) | 36 | 11 (30.5) | 14 (39) | 11 (30.5) |
| Early Cretaceous (Berriasian-Albian) | 17 | 4 (23.5) | 9 (53) | 4 (23.5) |
| Early Late Cretaceous (Cenomanian-Santonian) | 18 | 8 (44) | 9 (50) | 1 (6) |
| Latest Cretaceous (Campanian-Maastrichtian) | 46 | 26 (56) | 10 (22) | 10 (22) |
| Paleocene (Danian-Thanetian) | 74 | 49 (66) | 21 (28) | 4 (6) |

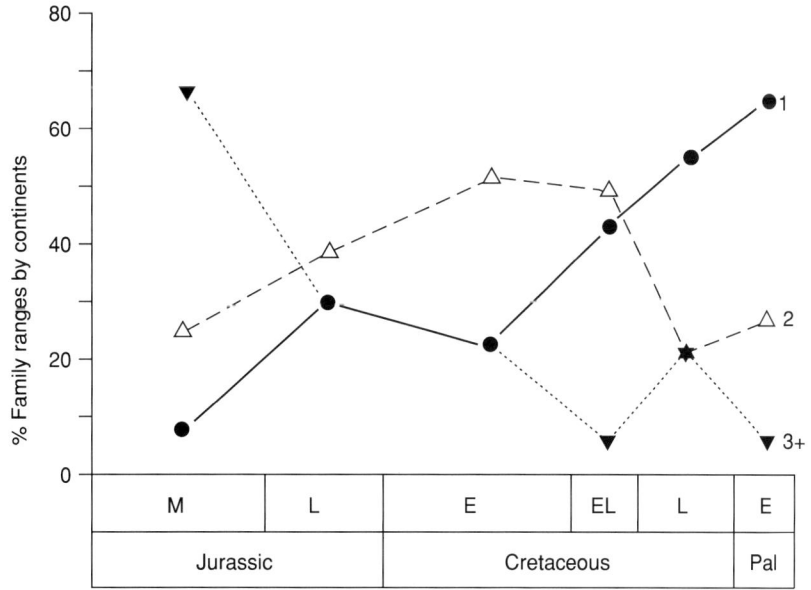

**Figure 7** Radiations associated with fragmentation of continents. The breakup of Pangaea was associated with increasing provinciality of terrestrial tetrapod families, plotted here from the Middle Jurassic to the Paleocene. This plot shows the percentage of families of terrestrial tetrapods that occupied one, two and three of the modern continental areas. Data given in table above. The plot shows that the percentage of families occupying only one of the modern continental areas rose from 8% of the families originating in the Middle Jurassic to 66% of those originating in the Paleocene. Conversely, 67% of families originating in the Middle Jurassic occupied three of the modern continental areas and this decreased to just 3% in the Paleocene. Reproduced from Benton MJ (1985) Patterns of diversification of Mesozoic non-marine tetrapods and problems of historical diversity analysis. *Special Papers in Palaeontology* 33: 185–202. Published by permission of the Palaeontological Association.

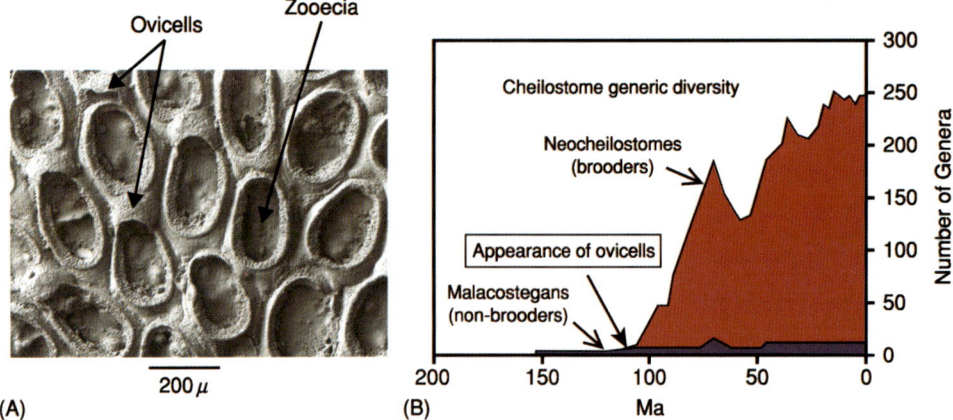

**Figure 8** The cause of a clade radiation has often been identified as the evolution of a key innovation. One such example is the appearance of ovicells (brood chambers) in some cheilostome bryozoans. (A) *Welbertopora mutabilis*, upper Albian, Fort Worth Formation, Denton County, Texas. Part of the colony of one of the earliest neocheilostomes with ovicells in which larva develop. (B) Generic plots of non-brooding and brooding cheilostomes showing the dramatic rise in the latter. The presence of Jurassic and Lower Cretaceous cheilostomes demonstrates that appearance of brooders is not a taphonomic artifact. Images supplied by Dr Paul Taylor, The Natural History Museum, from unpublished data.

along with environmental differentiation, and this almost certainly influenced the radiation of many marine organisms.

### Origination of an Evolutionary Novelty Leading to Taxic Diversity

Often the radiation of a particular group has been linked with the appearance of a particularly significant character (e.g., development of the cleidoic egg, enabling the organisms to become reproductively independent of water, or wings to fly). In many of the usual examples there is very little taphonomic control, so that it is difficult to separate the appearance of an evolutionary novelty and increase in species diversity from the preservation bias that may be present in the fossil record. But there are some examples that avoid this problem. Cheilostomes are the dominant bryozoans (*see* **Fossil Invertebrates:** Bryozoans) living today and they are distinguished from other Bryozoa by the possession of box-shaped zooecia. The fossil record of cheilostomes begins in the Upper Jurassic, with a rapid rise in numbers of species and genera in late Lower Cretaceous. There are two kinds of cheilostomes: malacostegans (about 150 Recent species) and their species descendents, the neocheilostomes (about 5000 Recent species). Modern malacostegans release planktotrophic larvae that spend up to about 1 month free living in the water column before settling and beginning colony growth. During this time there is potential for considerable dispersal as well as associated potential for continued gene flow. Neocheilostomes brood their larvae within a specialized brood chamber: – the ovicell. The larvae have a very short, free life span before settling to the substrate (often adjacent to the parent) and beginning colony formation. Thus, gene flow between populations may be very restricted and as a consequence there is the greater potential for genetic divergence

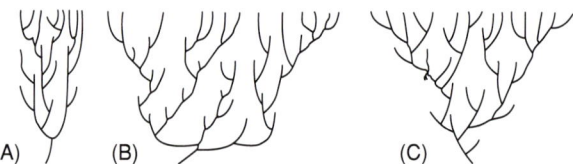

**Figure 9** The history of a clade during phylogenetic diversification may follow one of several patterns, idealized here. The horizontal axis represents morphological divergence, the vertical axis time. (A). Morphological evolution constrained, species origination high. (B) Morphological evolution rapid in early periods relative to speciation. (C) Morphological evolution and speciation approximately in step with one another. From Foote M (1993).

**Figure 10** Comparison of morphological and taxic evolution in coelacanth fishes using phylogenetic measures. (A) Phylogenetic tree showing 5 time bands selected as sample stages. Within each time band two types of comparison were made on a pair wise basis and then averaged. (B) The mean number of character changes occurring between taxa in each of the five time bands. This is a measure of morphological disparity. (C) The mean number of nodes (cladogenetic events) occurring between taxa in each of the five time bands. This is a measure of taxic diversity. In this case morphological diversity tracks taxic diversity except for a slight insignificant increase in taxic diversity relative to morphological diversity in the early stage (Namurian). Data from Forey PL (1998) *History of the coelacanth fishes.* London: Chapman & Hall.

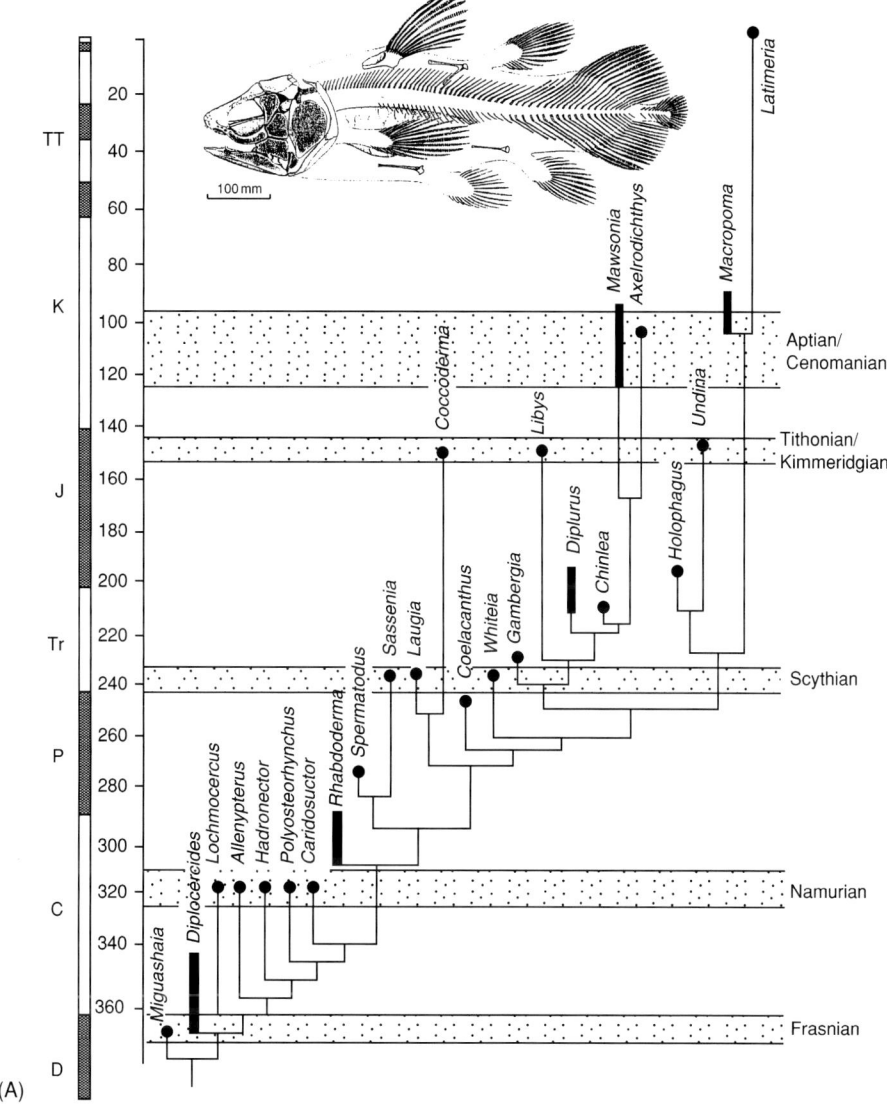
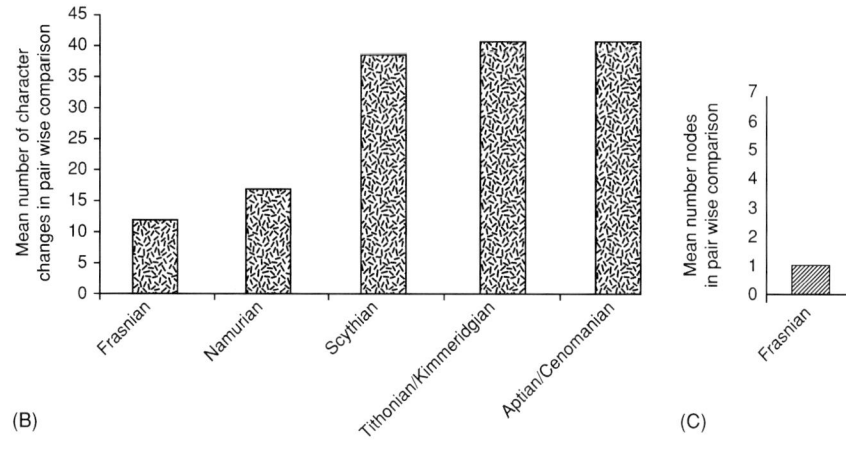
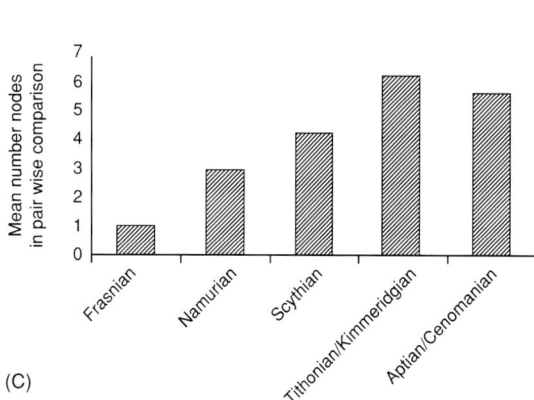

and, eventually, speciation. The ovicells and hence brooding of larvae can be identified in the skeleton of the neocheilostomes (**Figure 8A**). Plots at the generic-level of malacostegans and neocheilostomes show little change in the former, but a dramatic rise in the latter (**Figure 8B**), suggesting strongly that the acquisition of brooding is responsible for the radiation. It is pertinent to note that very similar patterns involving similar contrasting reproductive strategies have been observed in turritellid gastropod evolution.

### Morphological and Taxic Evolution During Radiations

Radiations involve particular patterns of morphological as well as taxic evolution. **Figure 9** shows three types of relationship between morphological and taxic evolution during clade history. Studies examining the relationship between these two aspects may be phylogeny dependent or independent, and several different methods have been used to analyze both. Much literature suggests that the pattern in **Figure 9B** is common.

One study of the radiation of coelacanth fishes (*see* **Fossil Vertebrates: Fish**) used a strictly phylogenetic approach for both measures (**Figure 10**). The phylogeny of 24 genera was sampled at five time intervals, which collectively spanned the Upper Devonian to Upper Cretaceous. Therefore, this study dealt with a radiation that was long term and involved few taxa and, while demonstrating a method,

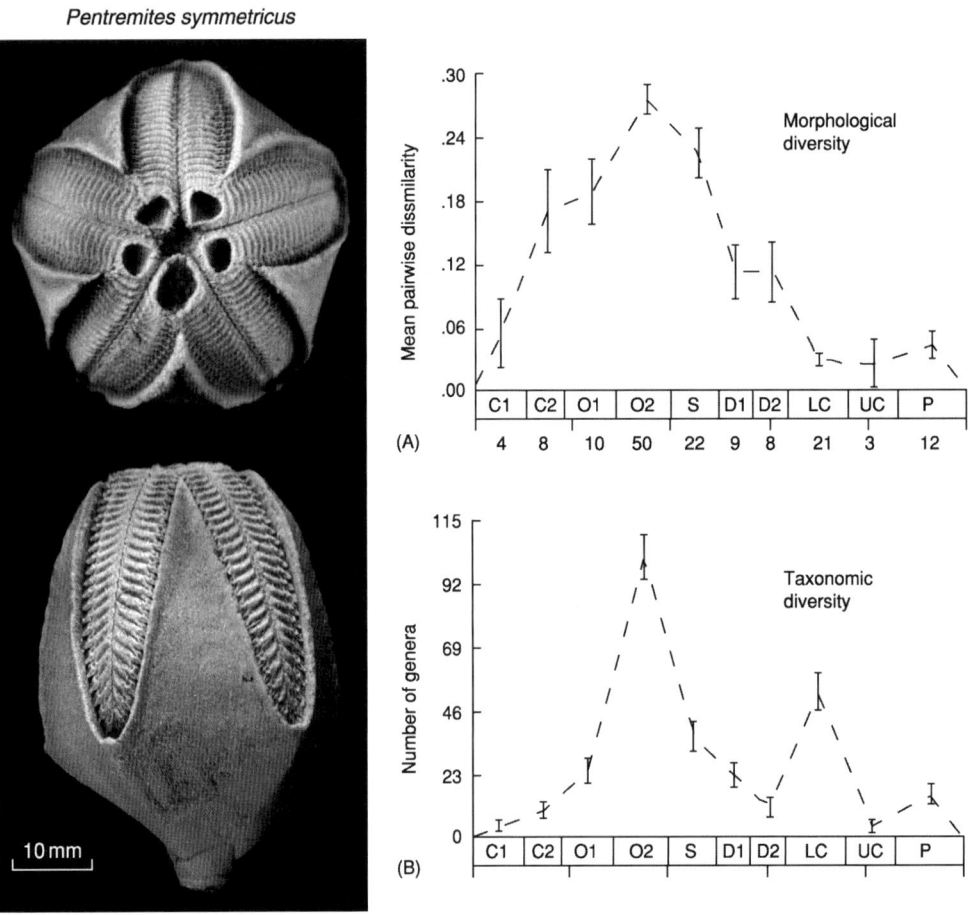

**Figure 11** Comparison of morphological and taxic evolution in Blastozoa echinoderms using non-phylogenetic measures taken at approximately equal time slices from Lower Cambrian to Permian. (A) Mean pair wise dissimilarity, calculated by comparing the number of dissimilar codes for 65 characters between each pair and averaging over the number of comparisons made. The number of genera (one species of each was examined) is given against each time slice examined. This gave an average measure of morphological disparity. (B) Number of genera at each time slice as a measure of taxic diversity. This study showed a rapid burst of morphological evolution that was not accompanied by taxic evolution in the early stages of clade history (C2 and O1). Note that quite the reverse trend was evident later in clade history (LC). Time slice abbreviations: C1 – Lower Cambrian, C2 – Middle and Upper Cambrian, O1 – Lower Ordovician, O2 – Middle and Upper Ordovician, S – Silurian, D1 – Lower Devonian, D2 – Middle and Upper Devonian, LC – Lower Carboniferous, UC – Upper Carboniferous, P – Permian. (After Foote 1992, with permission of author). Photograph supplied by Dr AB Smith.

# BIOLOGICAL RADIATIONS AND SPECIATION 277

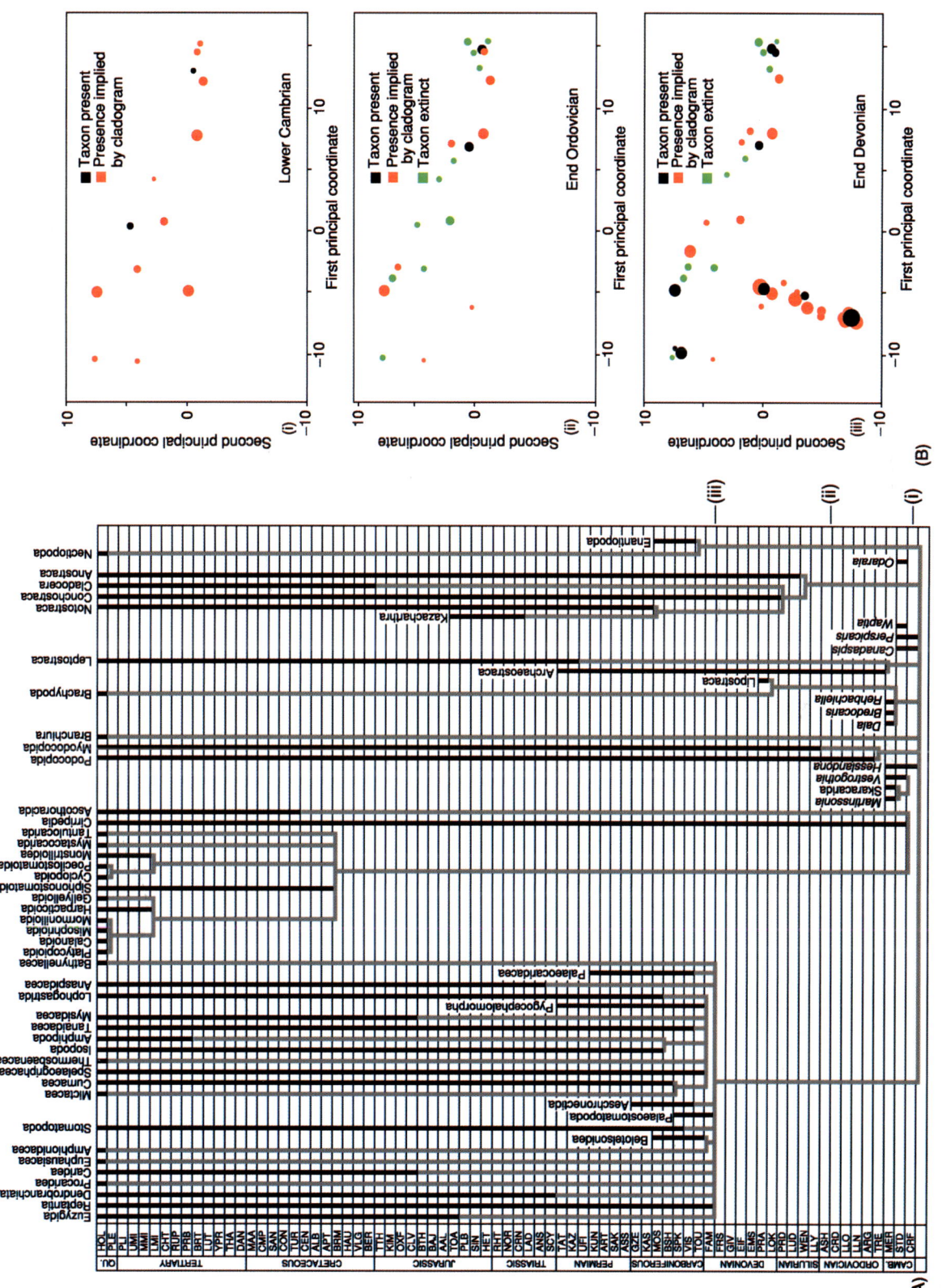

**Figure 12** (Continued)

may not be typical. There was very little taphonomic control in this study. At each interval the distance, as measured by number of character changes along the inferred lineage between terminal taxa, was averaged over all pair wise comparisons. This gave an average morphological distance for successive time periods in the clade's history. The values for successive time periods could then be compared to each other and, in turn, compared to the number of cladogenetic events within each time period (which provided a measurement of taxic diversity). In this case the taxic diversity tracked morphological diversity except for the early stages where taxic diversity slightly exceeded morphological diversity.

Some other studies have not considered the phylogenetic relationships within clades (although monophyly of the entire clade has been accepted). **Figure 11** shows the results of such a study of the radiation of members of the echinoderm subphylum Blastozoa (*see* **Fossil Invertebrates:** Echinoderms (Other Than Echinoids)). In this study, one species each from 147 genera, ranging in time from the Lower Cambrian to Permian, were examined for 65 morphological characters. Ten stratigraphic intervals were chosen and within each interval pair wise comparisons between characters coded for each of the genera were made and averaged. The phylogeny was not considered here. One common belief is that morphological diversity is disproportionately high compared to taxic diversity at the beginning of a radiation and that such a relationship is reversed later on. The blastozoan study appears to confirm this since the morphological diversity was far greater than the taxic diversity in the mid-Cambrian to Lower Ordovician; that is, in the early stages of clade history. The reverse relationship was true in the Lower Carboniferous.

The Cambrian explosion must be the most spectacular of all radiations that the planet has witnessed. Yet it remains one of the most enigmatic. Within 15 million years of the beginning of the Cambrian period, where tiny shelly faunas are first detected, a wide variety of skeletonized animals appeared, already well diversified and exquisitely adapted. Today they are classified amongst the major invertebrate phyla. Molecular evidence, alongside phylogenetic studies on the first appearing fossils all point to there being a long Precambrian history to many lineages. But exactly what form this history took is very uncertain. Steven Jay Gould, in his book *Wonderful Life*, – suggested that when skeletonized animals first appeared in abundance in the Cambrian there was a remarkably diverse array of body forms and that many of these were weeded out by extinctions leading to differential survival of a few body forms that eventually gave rise to our modern fauna. In other words, morphological disparity was greatest in the early phase of the animal lineage history and gradually became canalized. On a broad scale this theory echoed the radiation pattern shown by the blastozoan echinoderms mentioned above. Gould's thesis was grounded in a large part on the arthropods, and especially the crustacean-like animals, conspicuous in the famous Burgess Shale Lagerstätten, but was largely anecdotal, with little phylogenetic evidence or any precise measures of morphological disparity applied. More detailed studies comparing levels of disparity in Cambrian and Recent arthropods (*see* **Fossil Invertebrates:** Arthropods) showed that there was very little difference between then and now. More specifically detailed studies of the early evolution of the Crustacea have shown that morphological disparity grew in concert with taxic diversity (**Figure 12**).

The study of the pattern of radiations in the fossil record has implications for our understanding of post-extinction recovery and potentially for our efforts to put conservation efforts in place. If there are general patterns that emerge that can be associated with

**Figure 12** Morphological disparity of crustaceans tracked through time against a phylogeny. In this study one exemplar taxon from each of the crustacean orders or suborders was coded for 135 characters. (A) A cladistic analysis plotted against stratigraphy resulted in the consensus tree shown here. The actual stratigraphic ranges are shown in black and those which must be assumed because of the shape of the tree in grey (ghost ranges). Notice that most of the species of modern crustacea did not appear until late Palaeozoic. (B) Plots showing disparity of the clades present through three time slices. The disparity is measured by making pair wise comparisons between the coded characters for each of the taxa present in any one time interval (the ghost range taxa were coded according to reconstructed ancestral states). Some differential weights were applied according to the particular type of character and in this it differed from the Blastozoan study (**Figure 11**). A principal coordinate analysis was then carried out to illustrate the difference in morphospace between the taxa. In the diagrams the first three principal coordinates are shown, the third coordinate being at right angles to the paper (filled symbols positive and open symbols negative). The black symbols represent taxa with a range extending through the stage illustrated, red symbols represent taxa implied by ghost ranges and the green symbols are taxa extinct at stage illustrated. This diagram shows that both morphological disparity and taxic diversity increased from the Cambrian to Upper Devonian at which time a new area of morphospace (bottom left) was occupied. Reproduced from Wills MA (1998) Crustacean disparity through the Phanerozoic: comparing morphological and stratigraphic data. *Biological Journal of the Linnean Society of London* 65: 455–500.

specific ecological conditions, then some predictions may be made as to where and what to protect.

## See Also

**Evolution**. **Fossil Invertebrates:** Arthropods; Bryozoans; Echinoderms (Other Than Echinoids); Gastropods. **Fossil Vertebrates:** Fish.

## Further Reading

Benton MJ (1985) Patterns of diversification of Mesozoic non-marine tetrapods and problems of historical diversity analysis. *Special Papers in Palaeontology* 33: 185–202.

Cheetham AH (1986) Tempo of evolution in a Neogene bryozoan: rates of morphological change within and across species boundaries. *Paleobiology* 12: 190–202.

Erwin DH and Anstey RL (1995) Introduction. In: Erwin DH and Anstey RL (eds.) *New approaches to speciation in the fossil record*, pp. 1–10. New York: Columbia University Press.

Foote M (1992) Paleozoic record of morphological diversity in blastozoan echinoderms. *Proceedings of the Academy of Natural Sciences* 89: 7325–7329.

Foote M (1993) Discordance and correspondence between morphological and taxonomic diversity. *Paleobiology* 19: 185–204.

Forey PL (1998) *History of the coelacanth fishes*. London: Chapman & Hall.

Jablonski D, Lidgard S, and Taylor PD (1997) Comparative ecology of bryozoan radiations: origin of novelties in cyclostomes and cheilostomes. *Palaios* 12: 505–523.

Lazarus DB (2001) Speciation and morphological evolution. In: Briggs DE and Crowther PR (eds.) *Palaeobiology II*, pp. 133–137. Oxford: Blackwell Scientific.

Levington JS (2001) *Genetics, Palaeontology and Macroevolution*, 2nd edn. p. 617. Cambridge: Cambridge University Press.

Nehm RH and Geary DH (1994) A gradual morphologic transition during a rapid speciation event in marginellid gastropods (Neogene; Dominican Republic). *Journal of Paleontology* 68: 787–795.

Schluter D (2000) *The ecology of adaptive radiation*. Oxford: Oxford University Press.

Stanley SM (1979) *Macroevolution: pattern and process*. San Francisco: WH. Freeman.

Taylor PD and Larwood GP (eds.) (1990) *Major evolutionary radiations*. Oxford: Clarendon Press.

Wheeler QD and Platnick NI (eds.) (2000) *Species concepts and phylogenetic theory*. New York: Columbia University Press.

Wills MA (1998) Crustacean disparity through the Phanerozoic: comparing morphological and stratigraphic data. *Biological Journal of the Linnean Society of London* 65: 455–500.

# BIOSEDIMENTS AND BIOFILMS

**M R Walter and A C Allwood**, Macquarie University, Sydney, NSW, Australia

© 2005, Elsevier Ltd. All Rights Reserved.

## Introduction

Throughout most of the geological record we find traces of the past inhabitants of our planet. Bones, woody stems, and hard outer carapaces provide evidence of the passing of species, entombed by the sediments that were deposited around them. However, the organisms that yield these relatively familiar kinds of fossils represent only the evolutionary branch tips of the 'Tree of Life' (**Figure 1**), and have risen only recently in the history of life on Earth. The bulk of the 'Tree of Life', and most of the history of life on Earth (**Figure 2**), is represented by microscopic organisms. Unfortunately, these tiny, soft-bodied creatures are not readily fossilized. Moreover, where fossils are preserved, their tiny size, simple morphology, and imperfect preservation make them extremely difficult to recognize. This challenge increases in the search for the earliest fossils on Earth within rocks that have been subjected to over 3 billion years of geological processes: not only are the fossils likely to be heavily degraded, but the very nature of the host sedimentary deposits can be quite different to younger deposits, so that the misinterpretation of unfamiliar structures can be a risk.

The challenging search for the oldest traces of life is further limited by the availability of suitable rocks. There are just three occurrences of Early Archaean sedimentary rocks, from remnant Archaean cratons in Greenland (ca. 3.8 Ga), South Africa, and Western Australia (3.4–3.5 Ga; **Figure 2**). The rocks of Greenland, however, have been subjected to pressures and temperatures that have obliterated original depositional fabrics. Thus, there are just two potentially fossiliferous sedimentary sequences remaining from the era during which life arose on Earth. Over recent decades, these areas have become the focus of studies of the Earth's earliest biosphere and, despite the imagined impossibility of finding microbial remains in rocks of such antiquity, there have indeed been several 'fossil' discoveries. Evidence for life at *ca.* 3.5 Ga comprises three different types of 'fossil': (1) microscopic bacteria-like structures, interpreted as microfossils; (2) chemical signatures, interpreted as resulting from

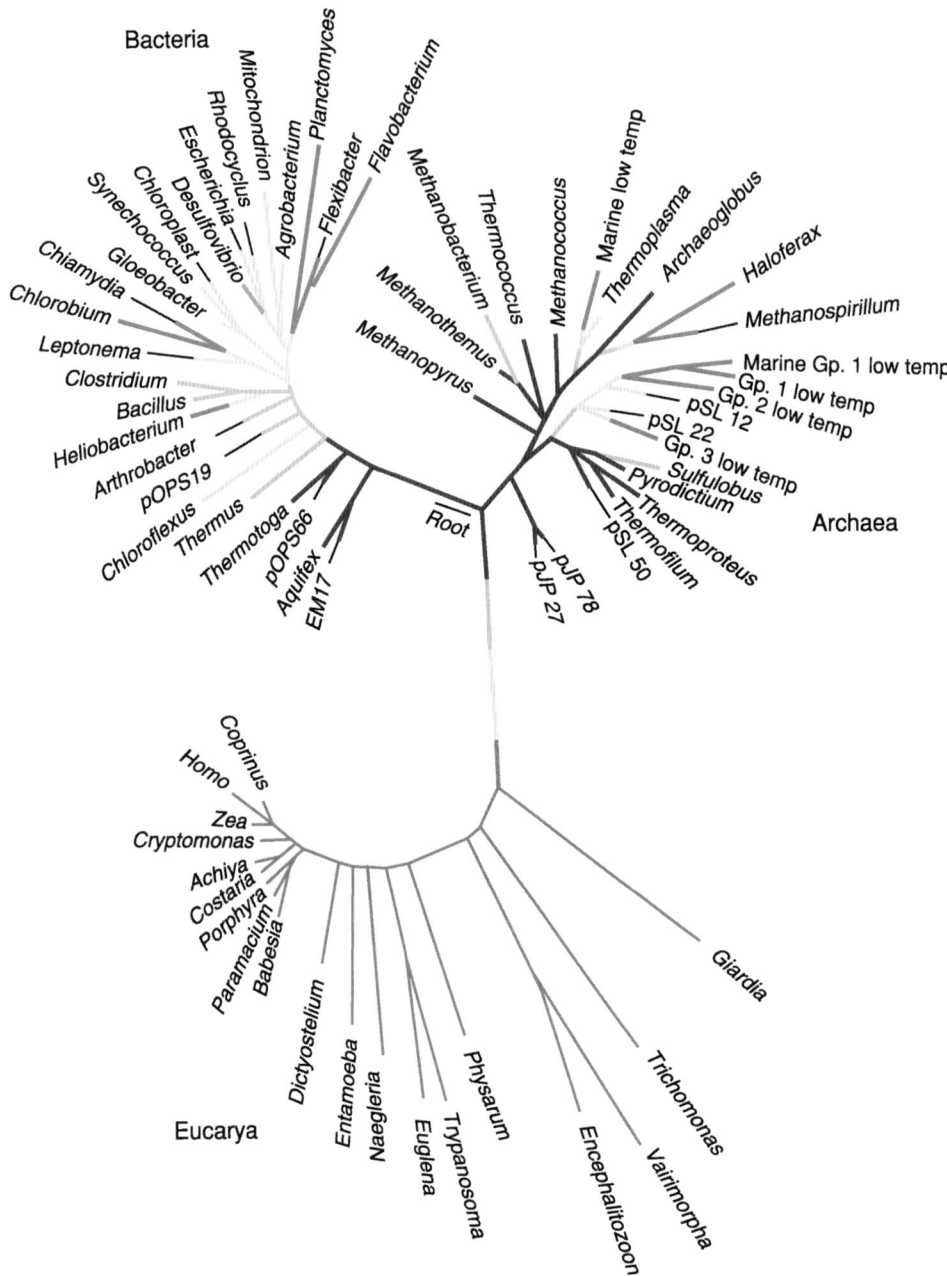

**Figure 1** The 'Tree of Life' showing the genetic relationships between all living organisms. All branches are microbial, except for *Coprinus*, *Homo*, and *Zea*, at the extreme tip of the lower branch.

**Figure 2** (A) Geographical distribution of Archaean stromatolite occurrences on the surface of the Earth today. All occurrences are clustered on remnants of Archaean crust in Africa, Australia, North America, Europe, and Asia. The oldest known stromatolites and microfossils occur in the Pilbara craton of Australia and the Kaapvaal craton in South Africa. (B) Temporal distribution of Archaean stromatolite and microfossil occurrences. Data are coloured by continent. The age of some occurrences is approximate, and some may extend into the Proterozoic, as indicated by dotted lines. (C) Geological time-scale highlighting the dominance of microbes throughout most of the Earth's history. Higher organisms represent only the most recent fraction of the history of life on Earth.

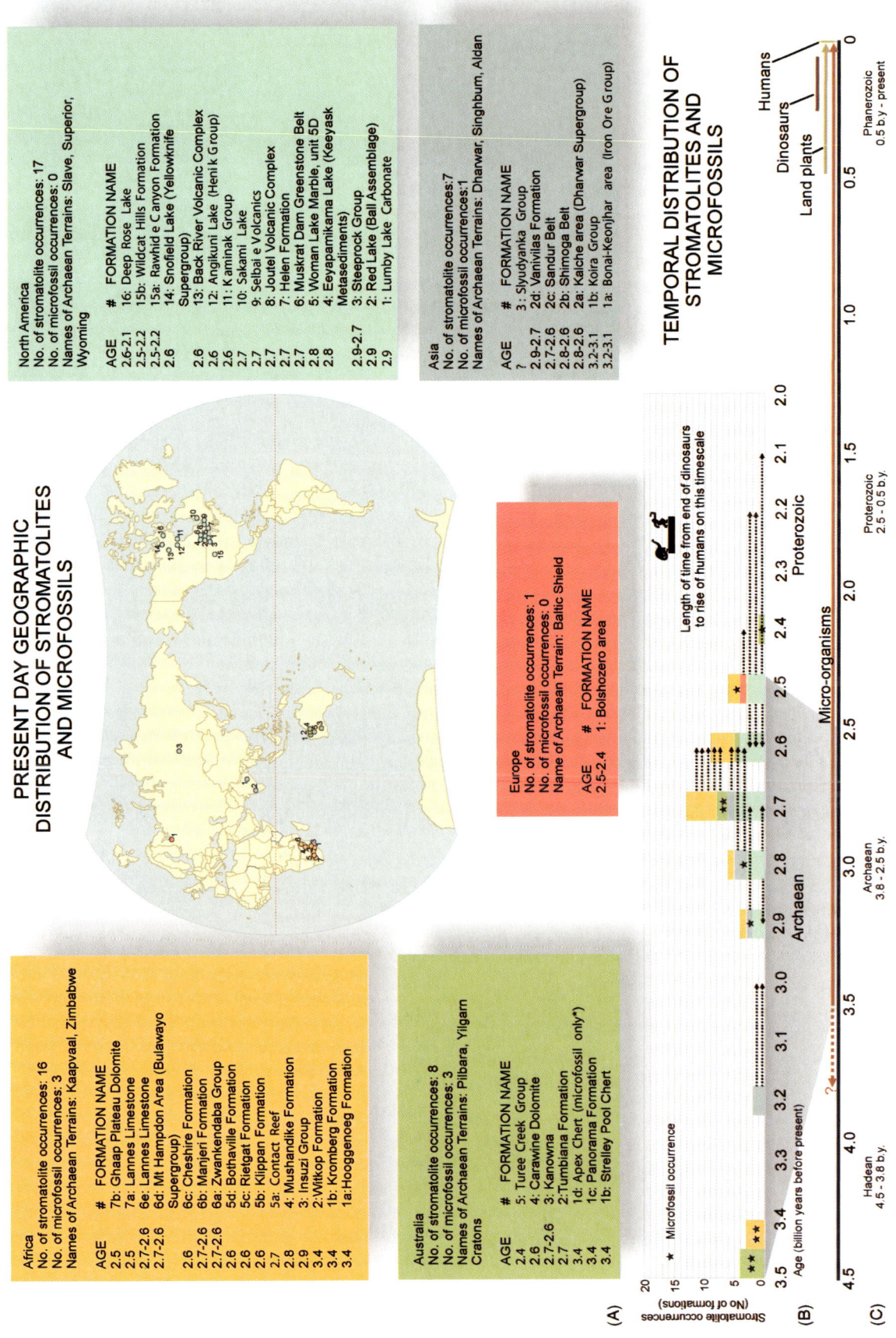

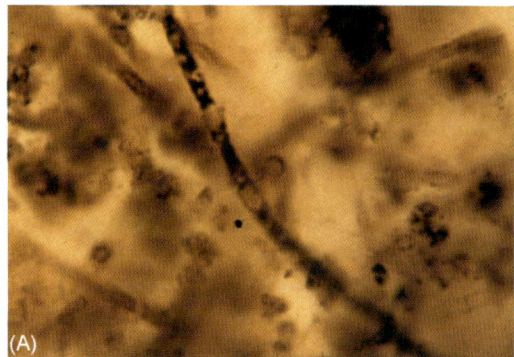 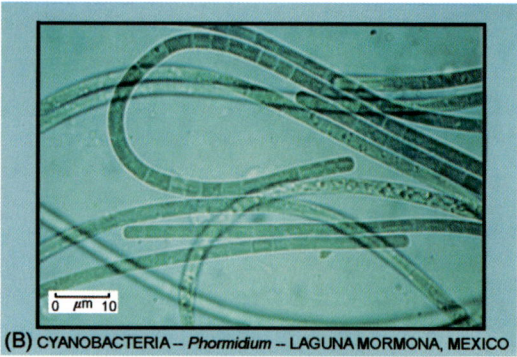

**Figure 3** (A) Filamentous microfossils from the 1.9 Ga Gunflint Formation, Ontario, Canada, arguably the oldest widely accepted microfossils, although older ones have been reported. Photographs supplied by J. W. Schopf. (B) Modern filamentous microbes. Part A and B based on data from: Hoffman H (2000) Archean Stromatolites as Microbial Archives. In: Riding R and Awramik S (eds.) *Microbial Sediments*, pp. 315–327. Berlin: Springer-Verlag.

microbial activity during formation of the rock; and (3) unique sedimentary structures, interpreted as stromatolites (laminated sedimentary structures formed by the activities of microbes) (**Figure 3**).

Arguably, these are the oldest traces of life on Earth. However, owing to the significance of determining when, where, and how life on our planet arose, a great deal of controversy surrounds the interpretation of the claimed fossils. As Carl Sagan once said, "extraordinary claims require extraordinary proof". Consequently, rigorous criteria have been developed for authenticating Early Archaean fossils. Moreover, early life on Earth is now recognized as an analogue for ancient microbial life on Mars, stimulating considerable research into the recognition of ancient microbial life using techniques that can be carried out by a rover or, eventually, a geologist on the surface of Mars.

Life is thought to have arisen on Earth sometime early during the 1.3 billion year period from 3.8 to 2.5 Ga, although the exact timing is still debated (*see* **Origin of Life**). In order to understand how we might interpret Archaean fossils, we will examine the way in which microbes interact with sediments in modern settings. Not only is this critical for interpreting the very oldest traces of life, but it is also important to hone our techniques in the familiar geological environments of Earth to aid us in the search for signs of life elsewhere.

## Nature of Modern and Ancient Biosedimentary Systems

### Microbial Sediments: Significance and Distribution

Today, almost no sedimentary environments are devoid of life. Through sheer numbers, microbes have played a fundamental role in shaping surface environments through time by their metabolic contributions to the global cycling of important elements, such as carbon, oxygen, sulphur, nitrogen, phosphorus, and iron. Microbes have also shaped the geological record of sedimentary environments, particularly through the activities of microbial communities living at the sediment–water interface. Although microbes live both as free-floating or swimming individuals (plankton), as well as structured, surface-adhered communities (biofilms), here we are particularly interested in the latter. This is because some of the activities undertaken by these film or mat-like communities can influence the way in which sediment is deposited and impart unique physical and chemical signatures that may be preserved in the geological record.

Although microbes are ubiquitous in all watery surface environments today, formation of microbial sedimentary structures, such as stromatolites, that could be recognized in the geological record is less common. Today, communities that form such structures occur mostly in environments that are inhospitable to higher, grazing organisms that feed upon the microbes. Such environments include those with extremes of temperature, pH, oxygen depletion (eg., hot springs), salinity (stranded lakes) or where the organisms are repeatedly exposed to drying conditions (intertidal zones). Sediment influx must also be minimal to prevent rapid burial of the mats. A famous modern example occurs in a hypersaline restricted coastal embayment at Hamelin Pool, Shark Bay, Western Australia (**Figure 4**). Salinity levels in Shark Bay are up to twice those of normal seawater, creating an inhospitable environment for most marine animals but hospitable for microorganisms associated with the stromatolites.

Although Shark Bay may not provide a perfect analogue for the first ecosystems, its lack of predators and consequent free reign of microbial activity certainly mimic one important aspect of the early biosphere: the dominance of micro-organisms. Microbes were the only form of life on Earth during the Archaean (3.8–2.5 Ga) and much of the Proterozoic (2.5–0.5 Ga) aeons (**Figure 2C**). Indeed, it was the rise of the first grazing organisms at the end of the Proterozoic (0.5 Ga) that is thought to have triggered the near disappearance of stromatolite-forming microbial activity from that point onwards.

## Nature of Mats and Biofilms

Biofilms are structured communities of micro-organisms adhering to a submersed surface. The species within a biofilm have planktonic equivalents (individual cells suspended in the water column), but only organisms in a biofilm are structured in a way that can significantly affect sedimentation. Several microbial species may be present within a biofilm, arranged in tiny mushroom- and pillar-type structures within an extracellular matrix of organic material (**Figure 5**). The

**Figure 4** (A) Modern stromatolites from the subtidal realm of Hamelin Pool, Shark Bay, Western Australia. A range of stromatolitic microbial communities also occur in the intertidal zone, forming different structures, including tabular stromatolites and microbially bound ripples. (B) Shark Bay stromatolite in cross-section, showing a clotted, slightly laminated texture.

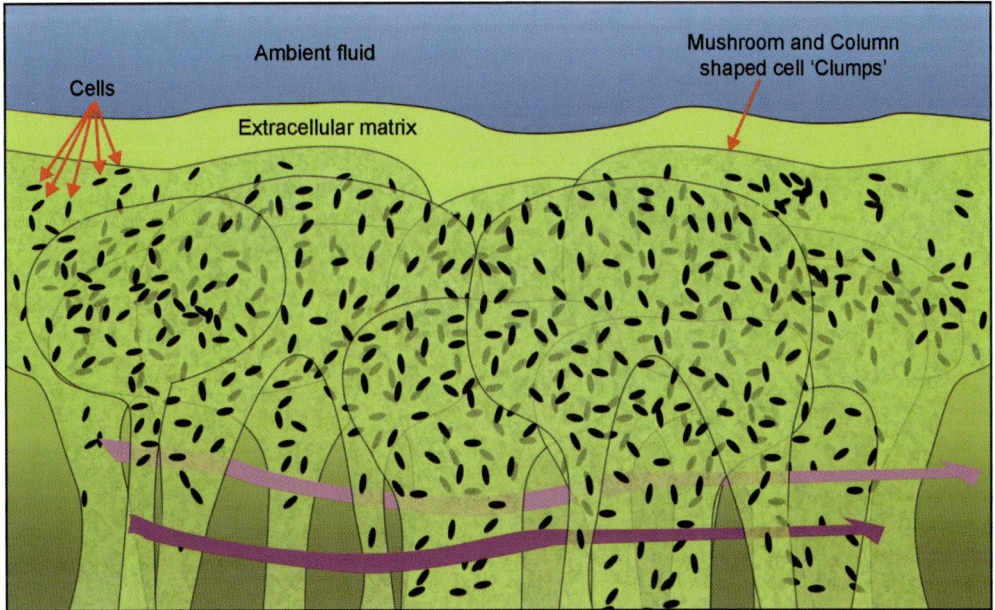

**Figure 5** Schematic cross-section of a biofilm. Cells are grouped together in mushroom- and pillar-type structures surrounded by a matrix of extracellular material. The intervening matrix provides a pathway for the diffusion of nutrients and wastes. Adapted from an illustration by Dirckx P (1999) Not Just Slime. In: Ben Ari E (ed.) *Bioscience* V. 49, no. 9. p. 691.

matrix provides a stable micro-environment that offers several advantages for cells, including:

1. Protection from harmful environmental fluxes and agents (e.g., temperature change, radiation damage, toxic materials).
2. Adhesive properties to anchor the community to the surface.
3. Efficient nutrient uptake from the surrounding environment.
4. Pretreatment of assimilated molecules, such as the breakdown of large organic molecules to smaller, more manageable components, which are then able to diffuse through the matrix and be taken up by the cells.
5. Conduits for the diffusion of metabolic by-products away from the cells.

Thus, cells in the biofilm are able to efficiently assimilate a range of organic and inorganic molecules from the environment through the matrix for use in reactions that provide energy and building blocks for cells. Moreover, the by-products from one species may provide the nutrients for another species within the biofilm. These processes are a significant component of the interaction between microbes and their environment.

A 'microbial mat' is a thicker and more complex type of biofilm, comprising many different species divided into specific layers. Different layers are favoured by different species because of their preference for certain light conditions, oxygen levels, sulphide concentrations, and so forth (**Figure 6**).

### Effects of Microbial Activity upon Sedimentation

**Mineral precipitation** Mineral precipitation is commonly associated with microbial mat growth. However, precipitation may or may not be actively caused by the metabolic activities of the organisms in the mat. The difference between active and passive precipitation processes is described in **Table 1**.

Biotic (active) precipitation occurs when the by-products of microbial reactions alter the chemical balance within the micro-environment and cause the precipitation of compounds such as carbonate, iron oxide, iron sulphide, or silica. Microbial photosynthesis, for example, is one of the most important mechanisms for biological calcium carbonate (limestone) precipitation on a global scale. In microbial mats, this process involves the photosynthetic removal of $CO_2$ from a bicarbonate-bearing solution, which increases the carbonate ($CO_3$) concentration in the residual fluid. The carbonate may then bond with a cation, such as calcium, and precipitate out of the fluid as calcium carbonate. Photosynthesis is not the only mechanism for $CO_2$ removal and carbonate precipitation; the same can be achieved through certain biological oxidation, reduction, or hydrolysis reactions. These reactions could have provided a mechanism for biological carbonate precipitation prior to the rise of photosynthetic organisms. Mineral

**Figure 6** A piece of modern microbial mat from the shallow subaqueous sediments at Shark Bay (cross-sectional view). The tufted, dark green layers at the top (photosynthetic cyanobacteria) are underlain by a light brown layer, a thin red–brown layer, and then a grey–black layer. The different colours in each layer reflect pigments, mineral matter, and organic material associated with different microbial species.

**Table 1** Comparison of non-biological and biological precipitation processes

| Passive (non-biological) precipitation | Active (biological) precipitation |
| --- | --- |
| Mechanism | |
| Fluid saturation levels are sufficiently high such that mineral precipitation would occur without biological influence | Fluid saturation levels are lower than those necessary for spontaneous precipitation, yet precipitation continues due to microbial mediation of chemical reactions in the environment, i.e., microbes are responsible for maintaining physicochemical parameters, such as minimum levels of carbonate concentration (via photosynthesis) necessary for precipitation |
| Microbes simply act as nucleation sites, much as a twig or fallen piece of rock | |
| Microbially influenced crystal fabrics may form and be preserved within such deposits, but precipitation is not caused by microbial activity | |
| Example environment | |
| Common in hot spring environments where supersaturated fluids emanate from springs or vents | May occur in marine environments where minerals will not spontaneously precipitate from seawater |

deposition can also occur when elements, such as iron, become concentrated by complexing (bonding) with microbially produced organic molecules.

Abiotic precipitation may occur in certain environments, such as hot springs, where fluids may be sufficiently saturated to spontaneously precipitate minerals even without microbial influence. These environments are commonly inhabited by microbes, perhaps simply because they are the only organisms that are able to survive under the prevailing conditions. Where fluid saturation approaches a level high enough for spontaneous precipitation, the microbes may simply act as templates for precipitation, a concept that is supported by the observation that fallen branches, insects, and other detritus in the same hot springs provide equally efficient substrates for mineral precipitation.

**Trapping and binding** In addition to causing the *in situ* precipitation of new minerals, thereby forming new, localized sediment deposits, microbial communities can also trap and bind detritus washed in from elsewhere. As material is trapped, the microbial community moves up through the sediment layer to form a new organic layer at the surface. Upward movement may be driven by, for example, a need to reach sunlight. The rate of sediment influx must be such so as to avoid either deeply burying the microbial mat or starving the mat of substrate material for growth.

## Ancient Microbial Sediments

Microbes of the Early Archaean left traces of their passing in much the same way as microbes do today, forming distinctive fabrics, structures, and chemical signatures in the sediments that fell and precipitated around them. There are four basic kinds of fossil evidence, or biosignatures, that we can study in the geological record to unravel the nature of the oldest biosedimentary systems. These are listed in **Table 2**.

The interpretation of ancient biosedimentary systems is aided by the study of younger analogues, but this is limited by the fact that surface processes and environments on the early Earth were very different to modern systems. Thus, direct comparisons cannot always be made between the Archaean Earth and the Phanerozoic or modern Earth in order to understand the types of processes that were responsible for the formation of certain fossil-like structures. Nonetheless, careful and selective study of modern and younger analogues provides an extremely useful starting point for the interpretation of ancient systems. A discussion of the formation, occurrence, and interpretation of different microbial sedimentary products and their fossils follows.

## Stromatolites

Stromatolites constitute by far the most abundant and important component of the early fossil record and thus provide the main body of evidence regarding the origins of life. The term 'stromatolite' has been widely used to refer to a diverse range of organic and/or laminated sedimentary structures. However, most researchers have used the definition by Awramik and Margulis, which defines a stromatolite as: "an organosedimentary structure produced by sediment trapping, binding, and/or precipitation resulting from the growth and metabolic activity of microorganisms".

Stromatolites accrete through the trapping and binding of sediment and/or by mineral precipitation that is influenced or caused by microbes at the sediment–water interface. Growth occurs because, as sediment accumulates on a stromatolite, either by deposition or precipitation, the microbes must move upwards to remain at the sediment–water interface. Thus, in living stromatolites, the constructing microbes generally live only in the top 1–10 mm of the structure. The physical shape of stromatolites may be columnar, domed, branching, tabular, or

**Table 2** Summary of the main types of fossil evidence, or biosignatures, that may be sought in the geological record

| Fossil type | Type of evidence provided | Description |
| --- | --- | --- |
| Microfossils | Direct, morphological | Occurs where mineral precipitation has entombed and preserved actual microbes. Should be made of biological organic matter, but fossil casts occur where organic matter has been replaced by other minerals |
| Stromatolites | Indirect, morphological | Sedimentary structure formed by the combination of sedimentation and the activities of microbes. Classed as trace fossils (similar to a dinosaur footprint where the remains of the organism itself may be absent) |
| Chemical fossils | Indirect, chemical | Remnant environmental effects and by-products of biochemical reactions that take place during microbial metabolism. Perhaps the subtlest of all fossils are the chemical fossils |
| Biomarkers | Indirect, biochemical | Remnant biological marker molecules that survive alteration processes that degrade morphological fossils and turn the original organic material to kerogen |

mound-shaped (**Figure 7**). The combined attributes of a stromatolite reflect the interaction of physical, chemical, and biological processes (**Figure 8**) in the environment. For example, currents may cause elongation of the stromatolite along the current direction, a feature observed in modern stromatolites in Shark Bay and in Archaean stromatolites of Western Australia. When interpreting the morphology of stromatolites, it is difficult to determine what proportion of the physical attributes of a structure can be attributed to environmental influences and how much to purely biological factors.

The hard, laminated interior of the stromatolite records the depositional history of the stromatolite-building process. We can look at the nature of the laminations in order to understand the behaviour and character of a stromatolite-building community. Lamination within any sedimentary deposit reflects some past cyclicity in the sedimentation processes. In the case of stromatolites, lamination reflects cyclic variations in non-biological sedimentation processes, coupled with the variation in microbial activity. Lamination may accrete rapidly or slowly, depending on the causes of accretion and cyclicity. **Table 3** describes four generalized mechanisms observed in modern stromatolites.

**Interpretation of stromatolite-like structures**
Stromatolite-like structures can form by a variety of abiological mechanisms. In order to verify a biologic origin for purported stromatolites, all possible non-biological origins ought to be considered, including:

1. Mechanical sedimentary processes such as those caused by waves or currents.
2. Post-burial deformation of laminar deposits by dewatering processes.
3. Tectonic deformation, i.e., folding.
4. Soft sediment deformation or slumping, i.e., cohesive downhill movement of a portion of sedimentary layers.
5. Chemical precipitation, e.g., from saturated fluids, such as those associated with hot spring environments.

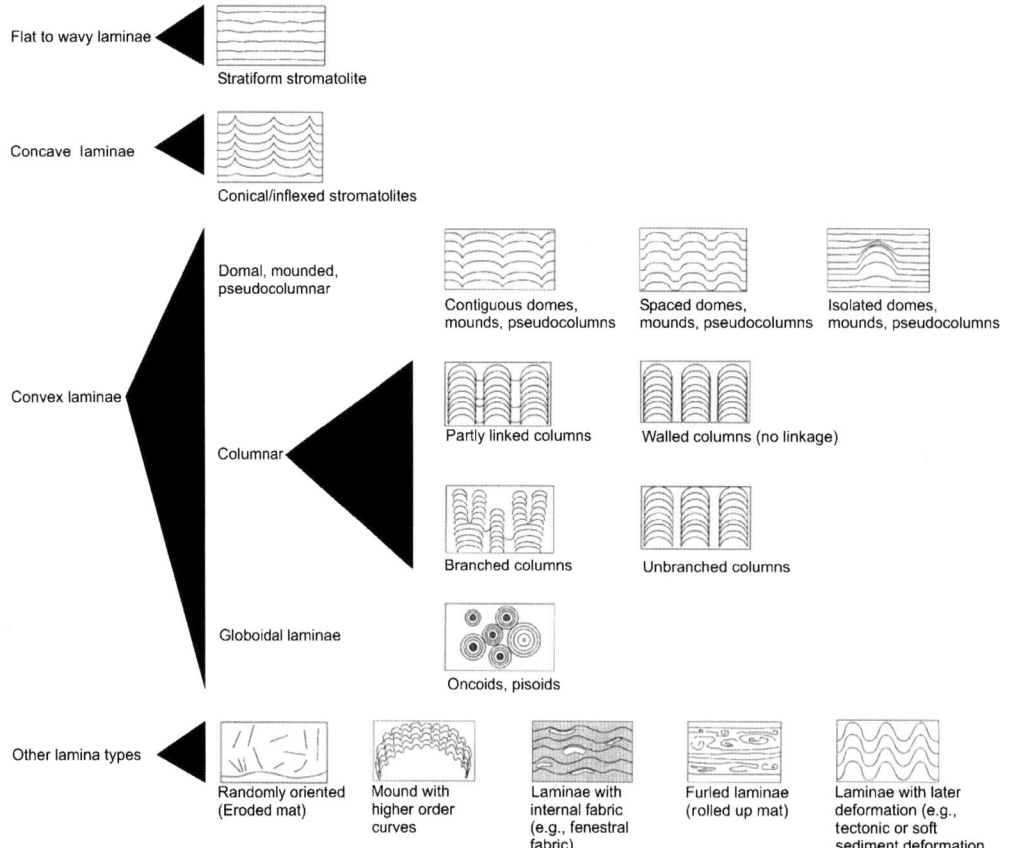

**Figure 7** Physical attributes of Archaean stromatolites. Different attributes can be combined; for example, one 'community' may comprise spaced domes with higher order curvature and later deformation. These basic characteristics can be used to describe and classify stromatolites observed in the field. Adapted from Hoffman HJ (2000) Archaean stromatolites as microbial archives. In: Riding R and Awramik S (eds.) *Microbial Sediments*, pp. 315–327. Berlin, Heidelberg: Springer-Verlag.

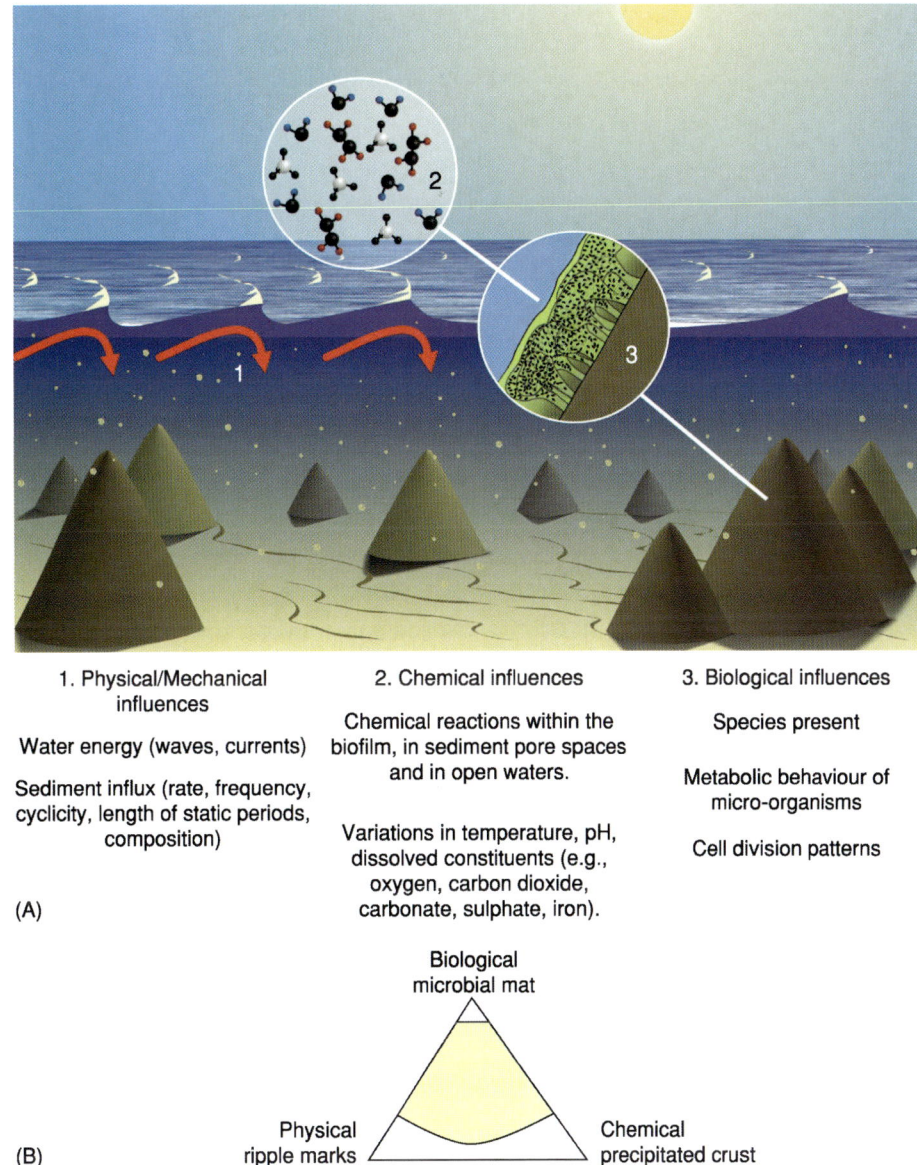

**Figure 8** (A) Schematic representation of the processes affecting stromatolite formation. Although stromatolites form as a result of interactions between microbes and sediments, the final characteristics of the stromatolite are also affected by environmental parameters, such as waves and currents, sediment influx, water chemistry, temperature, and pH. All of these parameters can affect the structure and development of the microbial communities and their interaction with the materials around them. (B) Stromatolites are influenced by physical, chemical, and microbial processes, which can be represented on a triangular diagram as shown. Stromatolites in the geological record are clustered in the shaded area. The apices represent end members; for example, in a system in which only physical and/or chemical processes dominate, microbial communities, and therefore stromatolites, will not form.

Ideally, microfossils, chemical fossils, and microscale fabrics should be used in the interpretation of stromatolite-like structures, as the macroscale physical shapes created by biological and non-biological processes may be the same for a given set of environmental parameters. Unfortunately, microfossils and microfabrics are commonly destroyed by diagenesis (pressure- and temperature-related alteration during burial) (*see* **Diagenesis, Overview**). This is the case in all known Early Archaean stromatolites.

**Archaean stromatolites** Except for the latest Archaean Transvaal (Africa), Carawine and Fortescue (Australia) sequences, Archaean stromatolites occur mainly in thin, discontinuous, localized carbonate units formed at times of transient stability between tectonism and volcanism. These stromatolitic deposits are now preserved in the volcano-sedimentary sequences (greenstone belts) of Archaean cratons (**Figure 2**). A summary of the attributes of Archaean stromatolites is presented in **Table 4**.

**Table 3** Comparison of four generalized mechanisms for the formation of stromatolite lamination

| Cause of accretion cycles | Microbial response | Example |
|---|---|---|
| Daily solar cycle. Independent of sedimentation | Phototactic (light-responsive). Filamentous cells lie flat at night, but stand upright during the day in order to reach sunlight | *Phormidium hendersonii, Conophyton* (Yellowstone National Park) |
| Episodic sediment influx. Independent of light conditions | Normally flat-lying filamentous organisms; glide vertically after new sediment influx in order to escape burial. Lamination is dependent on sediment supply | *Schizothrix* (subtidal stromatolites, Bahamas), *Microcoleus* |
| Seasonal changes in water chemistry | Low rainfall and warmer temperatures during the summer cause a change in organism dominance and behaviour | Lacustrine stromatolites, south-western Australia |
| Extended pauses in sedimentation | Profound changes in species composition in the mat. Laminae construction styles vary due to different behaviours of the new community | Multispecies mats at Great Sippewissett Salt Marsh, Massachusetts |

Based on data from Seong-Joo et al. (2000) On Stromatolite Lamination. In: Riding R and Awramik S (eds.) *Microbial Sediments*, pp. 16–24. Berlin: Springer-Verlag.

The geographical distribution of Early Archaean stromatolites today reflects the occurrence of preserved Early Archaean sedimentary rocks, which are restricted to narrow zones of sedimentary rocks, known as greenstone belts, of the Barberton (South Africa) and East Pilbara (Western Australia) regions. The oldest reputed stromatolites, from the 3.48 Ga Dresser Formation (Warrawoona Group, Pilbara), are domical laminated structures made of chert, barite, and iron oxide-rich laminae with minor carbonate (**Figure 9**; Dresser Formation stromatolites). The biological origin of these structures has been contested, with claims made that they could be artefacts of structural deformation (folding) and other post-burial processes. Moreover, the fact that these stromatolites occur in isolated clusters within structurally deformed rocks makes it difficult to acquire supporting contextual information.

Also in the Warrawoona Group lies the 3.46 Ga Strelley Pool Chert, a formation that contains abundant, well-preserved stromatolites (**Figure 10**; Strelley Pool Chert stromatolites). The stromatolites in the Strelley Pool Chert display a range of conical morphologies that grade into small ripple structures, suggesting that the stromatolites formed in a shallow environment intermittently washed by gentle waves. However, several abiological hypotheses have been proposed for the formation of these stromatolites, including non-biological chemical precipitation and/or wave/current deposition. Nonetheless, the Strelley Pool Chert stromatolites offer good prospects in the search for early evidence of life. Even better prospects are offered by the younger, but more complex, stromatolites in the 2.7 Ga Fortescue Group. These examples may provide a useful benchmark for the study of other Archaean stromatolites (**Figure 11**; Fortescue stromatolites).

### Microfossils

**Fossilization processes** Microbial fossilization occurs only under special circumstances, typically where mineral precipitation entombs the microbes as or soon after they die. This may occur in highly precipitative environments, such as hot springs, in which fluids laden with dissolved chemicals emanate at the Earth's surface. The mineral deposits associated with these environments are called sinters (silica precipitates) or travertine (carbonate precipitates). Modern examples occur at Yellowstone Park (USA), Taupo Volcanic Zone (New Zealand), and Lake Bogoria (Kenya). However, even in these environments, microbial fossilization is never complete and perfect. In multispecies systems, some species may be more readily fossilized, whereas others are destroyed. Furthermore, studies of modern sinters have shown that all cellular level information of colonies is lost on geologically short time-scales as opaline silica recrystallizes to a more stable quartz phase. In many sedimentary environments, microbial remains degrade or are completely obliterated almost immediately (in geological terms) through physical and chemical processes, including oxidation and thermal degradation (in which organic material breaks down to kerogen as sediments are heated through burial, igneous intrusion, or circulation of hydrothermal waters). Under most circumstances, the remains of microbes will not survive intact.

**Interpreting microfossil-like structures** Some researchers have suggested that, for the important question of the oldest evidence of life, the 'null hypothesis' should be adopted, which stipulates that a fossil cannot be authenticated until all abiological explanations (e.g., abiotic sedimentation, post-burial alteration, sample contamination) can be refuted. The

**Table 4** Summary of selected data on Archaean stromatolites. Entries in italics are questionably stromalites or questionably Archaean. Colours correspond to date in **Figure 2**

| Continent | Age (Ga) | Formation name | Map ref. (Figure 2A)[a] | Depositional association[b] | No. Different stromatolite types | Size range[c] | Morphological attributes[d] |
|---|---|---|---|---|---|---|---|
| North America | 2.6–2.1 | *Deep Rose Lake* | *16* | *Volcanic* | *1* | *M L* | *p, d* |
| | 2.5–2.2 | *Wildcat Hills Formation* | *15b* | *Volcanic* | *3* | *S M L* | *s, p, b* |
| | 2.5–2.2 | *Rawhide Canyon Formation* | *15a* | *Volcanic* | *1* | *S M L XL* | *p, d, m* |
| | 2.6 | Snofield Lake (Yellowknife Super group) | 14 | Volcanic | 7 | S M L | s, p, d, co, b, o, m |
| | 2.6 | Back River Volcanic Complex | 13 | Volcanic | 3 | M L | s, p, d |
| | 2.6 | *Angikuni Lake (Henik Group)* | *12* | *Volcanic* | *2* | *L* | *s, d* |
| | 2.6 | Kaminak Group | 11 | Volcanic | 1 | | s |
| | 2.7 | Sakami Lake | 10 | Volcanic | 1 | M | s? |
| | 2.7 | Selbaie Volcanics | 9 | Volcanic | 1 | | s? |
| | 2.7 | Joutel Volcanic Complex | 8 | Volcanic | 3 | M L | s, p, c |
| | 2.7 | Helen Formation | 7 | Volcanic | 1 | M | s? c |
| | 2.7 | Muskrat Dam Greenstone Belt | 6 | Volcanic | 1 | M | p |
| | 2.8 | Woman Lake Marble, unit 5D | 5 | Volcanic | 2 | M | s, d |
| | 2.8 | Eeyapamikama Lake (Keeyask Metasediments) | 4 | Volcanic | 3 | M | s, d, c |
| | 2.9–2.7 | Steeprock Group | 3 | Volcanic | 8 | S M L XL XXL | s, p, d, c, co, b |
| | 2.9 | Red Lake (Ball Assemblage) | 2 | Volcanic | 3 | M L XL | w, o, m |
| | 2.9 | *Lumby Lake Carbonate* | *1* | *Volcanic* | *1* | | *s, p, d* |
| Europe | 2.5–2.4 | *Bolshozero area* | *1* | | | *M* | *s?* |
| Asia | ? | *Slyudyanka Group* | *3* | | *1* | | *o?* |
| | 2.9–2.7 | Vanivilas Formation | 2d | Volcanic | 4 | M L | s, p, d, b, o, c |
| | 2.7–2.6 | Sandur Belt | 2c | Volcanic | 2 | M L | d, c? |
| | 2.8–2.6 | Shimoga Belt | 2b | Volcanic | 4 | M L | s, p, d, c, co, w |
| | 2.8–2.6 | Kalche area (Dharwar Supergroup) | 2a | Volcanic | 4 | M L | s, p, d, c, co, w |
| | 3.2–3.1 | Koira Group | 1b | Volcanic | 4 | M L | s, p, d, o |
| | 3.2–3.1 | Bonai–Keonjhar area (Iron Ore Group) | 1a | Volcanic | 4 | M | s, p, d, o |
| Australia | 2.4 | *Turee Creek Group* | *5* | *Volcanic* | *1* | *M* | *p, c* |
| | 2.6 | Carawine Dolomite | 4 | Volcanic | 7 | M L XL | s, p, d, c, co, o |
| | 2.7–2.6 | Kanowna | 3 | Volcanic | 1 | S M | c |
| | 2.7 | Tumbiana Formation | 2 | Volcanic lacustrine | 9 | S M L | s, p, d, c, co, b |
| | 3.5 | Strelley Pool Chert | 1c | Volcanic | 3 | M L | s, c, o |
| | 3.5 | Panorama Formation | 1b | Volcanic | 1 | M | s, c |
| | 3.5 | Dresser Formation | 1a | Volcanic | 4 | M L | s, p, d, o |
| Africa | 2.5 | Ghaap Plateau Dolomite | 7b | Volcanic | 14 | S M L XL XXL | s, p, d, c, co, b, w, r, m |
| | 2.5 | Malmani Dolomite | 7a | Volcanic | 10 | XS S M L XL XXL XXXL | s, p, d, c, co, b, o |
| | 2.7–2.6 | Lannes Limestone | 6e | Volcanic | 1 | | s? |
| | 2.7–2.6 | Mt Hampdon Area (Bulawayo Supergroup) | 6d | Volcanic | 1 | M | p, co |
| | 2.6 | Cheshire Formation | 6c | Volcanic | 4 | S M L | s, p, d, c, b, m |

*Continued*

**Table 4** Continued

| Continent | Age (Ga) | Formation name | Map ref. (Figure 2A)[a] | Depositional association[b] | No. Different stromatolite types | Size range[c] | Morphological attributes[d] |
|---|---|---|---|---|---|---|---|
| Africa | 2.7–2.6 | Manjeri Formation | 6b | Volcanic | 2 | M | s, p |
| | 2.7–2.6 | Zwankendaba Group | 6a | Volcanic | 4 | M L XL | p, d, c, o, m |
| | 2.6 | Bothaville Formation | 5d | Volcanic, fluvial | 2 | S M | s, c |
| | 2.6 | Rietgat Formation | 5c | Volcanic, lacustrine | 3 | S M L | p, d, b, o, m |
| | 2.6 | Klippan Formation | 5b | Volcanic, lacustrine | 1 | M | s, p |
| | 2.7 | Contact Reef | 5a | Volcanic | 1 | M | o? |
| | 2.8 | Mushandike Formation | 4 | Volcanic | 2 | S M | s, p, d, b |
| | 2.9 | Insuzi Group | 3 | Volcanic | 6 | S M L XL | s, p, d, c, co, b, m |
| | 3.4 | Witkop Formation | 2 | Volcanic | 2 | M L | p, d |
| | 3.4 | Kromberg Formation | 1b | Volcanic | 4 | M L | s, p, c, o |
| | 3.4 | Hooggenoeg Formation | 1a | Volcanic | 1 | | s? |

[a]The references in the fourth column correlate with the map in **Figure 2**.
[b]Depositional association refers to the type of overall geological setting in that the stromatolites occur in. Note that most stromatolites are found in volcanic settings; the actual deposits are typically hot spring-type sediments such as exist today at Yellowstone National Park (USA), Lake Bogoria (Kenya), and Taupo Volcanic Zone (New Zealand).
[c]Size ranges: XS = micrometres, S = millimetres, M = centimetres, L = tens of centimetres, XL = metres, XXL = tens of metres, XXXL = hundreds of metres.
[d]Morphology codes refer to: b = branching, c = conical, co = columnar, d = domed, m = ministromatolite, o = oncoids, p = Pseudocolumnar, r = rolled up, s = stratiform, w = walled.
Based on data from Hoffman H (2000) Archean Stromatolites as Microbial Archives. In: Riding R and Awramik S (eds.) *Microbial Sediments*, pp. 315–327. Berlin: Springer-Verlag.

**Figure 9** Domical stromatolite from the Dresser Formation, Pilbara region, Western Australia.

identification of microfossils is a topic of particular scrutiny in view of recent work demonstrating that microfossil-like objects can be reproduced abiotically in the laboratory. 'Biomorphs' made of silica and witherite (minerals that could occur in hydrothermal systems) were produced in the laboratory (**Figure 12**), displaying many features previously thought to be indicative of a biological origin, including spherical, segmented (septate), filamentous structures reminiscent of bacterial cells, and carbonaceous composition. Whereas kerogen is biological in origin, it should not be confused with

**Figure 10** Conical stromatolite from the 3.42 Ga Strelley Pool Chert, Pilbara region, Western Australia. The layering comprises alternating fine chert and carbonate laminations. These are relatively flat, except in the stromatolite itself, where the laminae form a distinct cone. This kind of structure may be attributed to the activity of microbes.

non-biological organic carbonaceous material, which can form by a number of different processes, and has been shown experimentally to adhere to abiological fossil-like structures, lending them additional biological character.

**The oldest microfossils** The microfossils that were, for a long time, cited as the earliest evidence for life come from the Apex Chert in the Warrawoona Group (Pilbara region, Western Australia). This discovery comprised septate, filamentous structures (**Figure 13**) in what was thought to be a sedimentary chert layer. However, this chert layer was later shown to be a black chert dyke, which led to the contention that the putative microfossils could be hydrothermal artefacts or biomorphs (see above). Other carbonaceous structures interpreted as microfossils have been found in hydrothermal black chert deposits in the 3.49 Ga Dresser Formation (Warrawoona Group). Spheroidal structures of similar age, but more poorly preserved, have been reported from the Onverwacht Group in the Swaziland Sequence of South Africa. Although a biological origin for these discoveries is plausible, every report of microfossils from rocks of Early Archaean age has been contested and alternative abiotic explanations have been offered for their formation. Less contentious filamentous structures, interpreted as fossilized bacteria, have been found in the Late Archaean (2.7 Ga) Tumbiana Formation (Fortescue Group, Pilbara), but the oldest microfossils of widely accepted microbial origin come

**Figure 11** Columnar stromatolites, exposed in cross-section, from the 2.7 Ga Tumbiana Formation (Fortescue Group) in the Pilbara region, Western Australia. Hammer is approximately 40 cm in length.

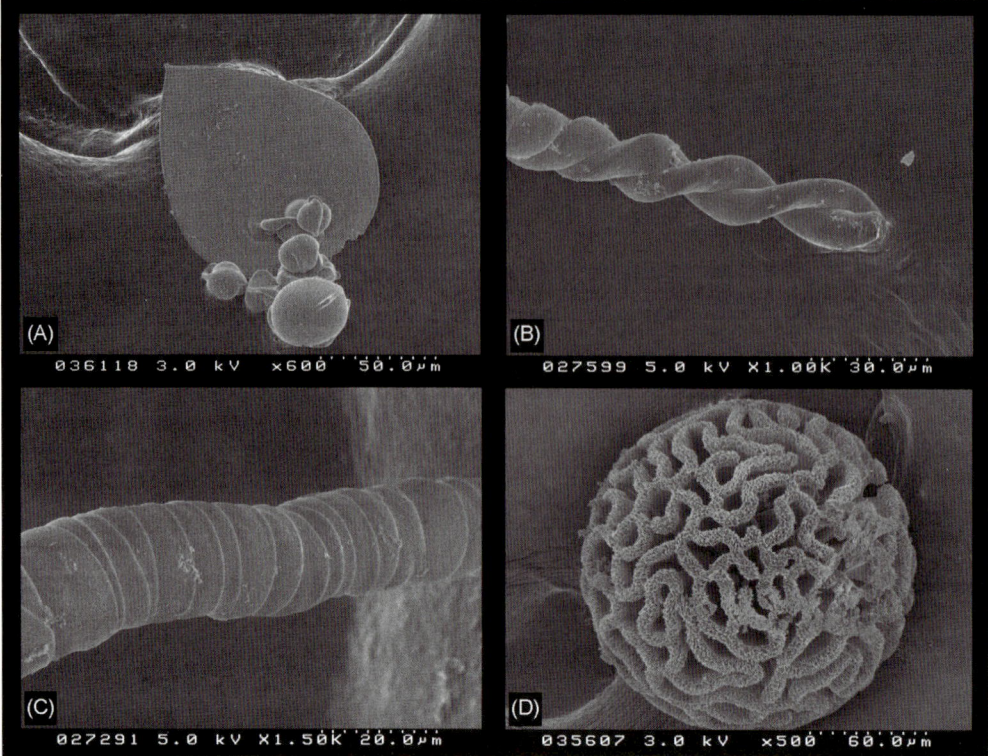

**Figure 12** Scanning electron microphotographs of microfossil-like structures that were produced abiologically in the laboratory. These 'biomorphs' display many complex characteristics that mimic structures of biological origin. Photographs courtesy of Stephen Hyde, Australian National University.

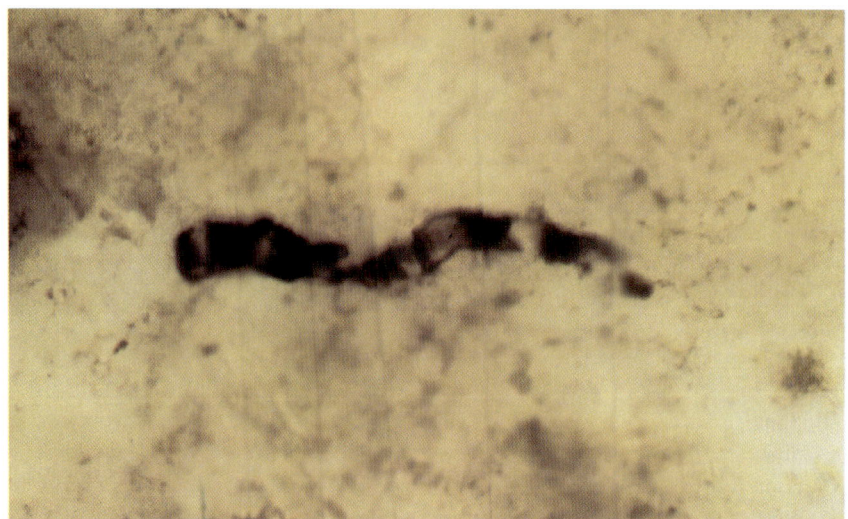

**Figure 13** Putative microfossil from the Apex Chert, Pilbara region, Western Australia. Such structures have been interpreted as the fossilized remains of bacteria that flourished around 3.5 billion years ago. Photograph courtesy of J. W. Schopf.

from the 2.5 Ga Ghaap Plateau Dolomite in South Africa and the 2.1–1.8 Ga Gunflint Formation in western Ontario, Canada. These microfossils display coccoid (roughly spherical), septate filamentous, unbranched tubular, and budding bacteria-like structures (**Figure 3**).

### Biomarkers

Biomarkers are biologically formed (biosynthesized) organic molecules that can be attributed to a specific biological origin. They possess molecular structures and isotopic characteristics that enable them to be distinguished from abiogenic organic compounds that

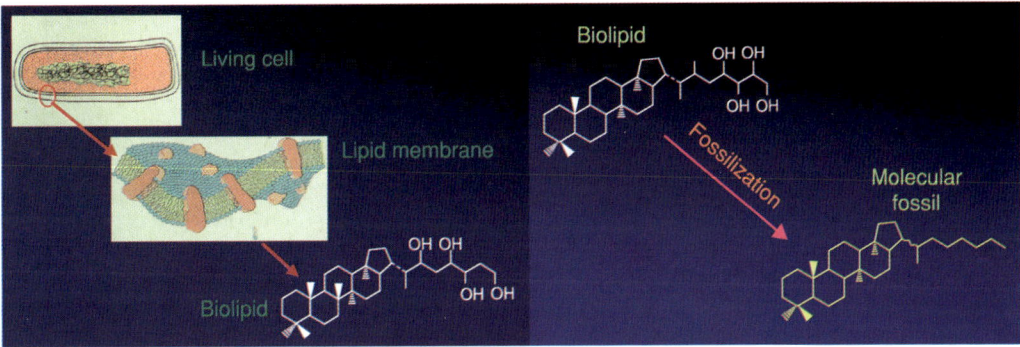

**Figure 14** Schematic diagram of biomarker formation from the degradation of biological organic material. As the organic material is heated over millions of years, functional groups are lost from the molecular structure. In some cases, the remaining structure may still provide clues to the organism from which the organic material was derived. Adapted from original unpublished illustration by J. Brocks, with permission.

are distributed throughout the cosmos. In other words, they are a kind of molecular fossil that can: (1) provide evidence of life; and (2) provide information about the nature of those life forms, even in the absence of recognizable fossils. **Figure 14** describes the formation of biomarker molecules as organisms degrade during burial and heating.

The oldest known biomarkers are from organic material in the 2.7 Ga Fortescue Group of the Pilbara region, Western Australia. The molecules were identified as originating from the group of organisms known as eukaryotes (*see* **Precambrian:** Eukaryote Fossils) (**Figure 1**).

### Chemical Fossils

Microbial activity can be traced, in some instances, by the chemical signatures left behind in rocks. However, at present, our ability to detect these signatures may exceed our ability to interpret them. We know that microbes act as agents of dispersion, concentration, or fractionation of different chemical components and, in doing so, may impart specific chemical signatures upon their environment. However, we also know that similar patterns may be produced by abiological mechanisms. In the following paragraphs, we investigate some of the chemical signatures that can be used to trace biological activity, and some of the proposed problems with their interpretation.

Biological isotopic fractionation signatures form when organisms discriminate between light and heavy isotopes of elements, such as carbon, hydrogen, nitrogen, and sulphur. Both the light and heavy isotopes are involved in abiotic reactions, whereas organisms that utilize carbon compounds in their metabolic reactions almost always prefer the lighter $^{12}C$ isotope. The carbonaceous remains of the organisms are enriched in $^{12}C$, whereas the source material from which they derived their carbon is enriched in $^{13}C$. Carbon-bearing minerals (e.g., calcium carbonate or limestone) that precipitate from the residual $^{13}C$-enriched material will acquire the heavy carbon signature. Other known biological fractionation processes involve the preferential assimilation of $^{14}N$ over $^{15}N$, $^{32}S$ over $^{34}S$, and hydrogen over deuterium. Thus, isotopic analyses of organic matter and their host mineral deposits can yield evidence of biological activity through fractionated C, N, S, and H signatures.

Carbon isotope fractionation patterns have been found in Early Archaean carbon-bearing rocks, such as the Strelley Pool Chert and Dresser Formation (Pilbara, Western Australia) and the Buck Reef Chert (Barberton, South Africa). Carbonaceous material (biogenic material?) with low values for $^{13}C/^{12}C$ from these formations may have derived from the biological fractionation of carbon isotopes. However, Fischer–Tropsch synthesis has been proposed as an alternative mechanism for carbon isotope fractionation. Fischer–Tropsch synthesis is a chemical reaction process involving carbon that is thought to occur in the mantle in the presence of certain catalytic compounds, such as Fe and Mn. Because Fischer–Tropsch synthesis is a high temperature process, and requires the presence of certain compounds, the geological setting of a fractionated C isotope signature may provide clues to differentiate Fischer–Tropsch-type occurrences from biological occurrences. However, the distinction is not always easily made and controversy persists with regard to the biological interpretation of light carbon signatures in Early Archaean rocks.

Beyond the morphological fossil record, which extends back to around 3.5 Ga, evidence for life may only be found via chemical fossils in *ca*. 3.8 Ga rocks from Greenland. These rocks are too metamorphosed to contain any morphological remains. However, carbon within the rocks is enriched in the light isotope, $^{12}C$. This enrichment may have resulted from the biological fractionation of carbon, and has thus been interpreted as the oldest evidence for life on Earth. However, in the absence of original sedimentary features and context, the interpretation of these signatures is contentious.

Clearly, the fossil that sets the irrefutable starting point for life on Earth has not been found. However, if the interpreted complexity of microbial communities seen in the fossil record at around 3.5 Ga is anything to go by, it seems likely that life on Earth began well before then. The remains of the very first microbes, or the rocks that they were part of, have been destroyed or metamorphosed beyond recognition.

## Glossary

**biogenic** Of biological origin.
**Ga** Billion years before present/billion years of age.
**lacustrine** Associated with lake environments.
**metamorphism** Pressure- and temperature-related alteration of rocks during burial (involving processes that result in new textural and mineralogical characteristics).

## See Also

**Diagenesis, Overview**. **Geysers and Hot Springs**. **Origin of Life**. **Precambrian:** Eukaryote Fossils; Vendian and Ediacaran. **Pseudofossils**. **Sedimentary Rocks:** Limestones.

## Further Reading

Banfield JF and Nealson KH (1997) *Geomicrobiology: Interactions Between Microbes and Minerals*. Washington: Mineralogical Society of America.

Brocks JJ and Summons RE (2003) Sedimentary hydrocarbons, biomarkers for early life. In: Holland HD and Turekian K (eds.) *Treatise in Geochemistry*, pp. 63–115. Amsterdam: Elsevier.

Buick R (1990) Microfossil recognition in Archaean rocks: an appraisal of spheroids and filaments from a 3500 M.Y. old chert–barite unit at North Pole, Western Australia *Palaios* 5: 441–459.

Ehrlich HL (2000) *Geomicrobiology*. New York: Marcel Dekker.

Grotzinger JP and Knoll AH (1999) Stromatolites in Precambrian carbonates: evolutionary mileposts or environmental dipsticks? *Annual Reviews, Earth and Planetary Science Letters* 27: 313–358.

Hoffman HJ (2000) Archaean stromatolites as microbial archives. In: Riding R and Awramik S (eds.) *Microbial Sediments*, pp. 315–327. Berlin, Heidelberg: Springer-Verlag.

Lowe DR (1994) Abiological origin of described stromatolites older than 3.2 Ga. *Geology* 22: 387–390.

Schopf JW (1983) *Earth's Earliest Biosphere*. Princeton, NJ: Princeton University Press.

Schopf JW and Packer BM (1987) Early Archaean (3.3 billion to 3.5 billion year old) microfossils from Warrawoona Group, Australia. *Science* 237: 70–73.

Walter MR (1976) *Stromatolites*. Amsterdam: Elsevier.

# BIOZONES

**N MacLeod**, The Natural History Museum, London, UK

Copyright 2005, Natural History Museum. All Rights Reserved.

## Introduction

The contemporary concept of biozone is that of a stratigraphical interval defined on the basis of its biotic content. This deceptively simple definition serves as the door to vast realms of complexity, access to which is important not only because those realms contain much of the early history of geology, but also because they encompasses the single most important set of tools yet devised for reconstructing earth history. The biozone concept is built on the work of William Smith, who first demonstrated the importance of fossils for establishing sequences of geological events. Smith's ideas were extended by many nineteenth-century geologists and were instrumental in creating the geological time-scale (*see* **Time Scale**): one of the greatest scientific achievements of that century. For the last 100 years, various biozone types – range biozones of diverse types, assemblage biozones, interval biozones, acme biozones – have been employed throughout the Earth sciences where they have proven their worth in fields as diverse as palaeoceanography, palaeogeography, palaeoecology, and palaeoclimate studies; wherever there is a need for relative time correlation. In the twenty-first century, use of biozones remains at the forefront of stratigraphical analysis, with methodological improvements being made through use of the concept in quantitative stratigraphy (*see* **Stratigraphical Principles**).

In order to understand the concept of a biozone, one must understand its development. As noted above, foundations for an understanding of biozones were laid by William Smith who, in 1796, realised that the rock layers cropping out around the southern English town of Bath always occurred in the same superposed order and that individual strata contained the same types of fossils. This insight provided Smith, who was engaged in surveying the routes of canals, with a practical tool whereby he could predict what rock types would be encountered during a canal's construction. Smith's observations of fossils were particularly important to his method because strata that superficially appeared similar (e.g., limestones, sandstones, shales), but that contained different sets of fossils, could be distinguished from one another and assigned to their correct positions within the overall sequence (*see* **Sequence Stratigraphy**). Using this approach, Smith was, by 1799, able to reconstruct the basic sequence of strata from the British Coal Measures (Carboniferous (*see* **Palaeozoic**: Carboniferous)) to the Chalk (Cretaceous (*see* **Mesozoic**: Cretaceous)). Extension of this method to other regions resulted in Smith producing the first geological map in 1815, the fossil evidence for which was publishing in his 1817 book *Stratigraphical System of Organized Fossils*.

Smith's work contained two further insights that proved crucial to development of the biozone concept. The first of these was that even a single stratum could be subdivided on the basis of its fossil content. This observation established variations in lithological type and fossil content as independent of one another, with fossil content often providing the more refined basis for subdivision. The second was that fossils from the same lithological types in different parts of the sequence often resembled one another. This observation established that there was a relation between the general types of fossils that occurred in different depositional environments.

In 1808 Smith showed his collection of fossils arranged stratigraphically to members of the Geological Society of London. By the 1820s Smith's methods had been accepted by most geologists and extensions had begun to appear, most notably by the work of Georges Cuvier and Alexander Brongniart (in 1822) in establishing the stratigraphical sequence of the Paris Basin. Later, Gerard Deshayes (in 1830), Heinrich Georg Bronn (in 1831), and especially Charles Lyell (in 1833) extended Smith's concepts by formulating subdivisions of Tertiary strata based on the sequence of fossils alone. This represented an important step toward the conceptualisation of the biozone in that it demonstrated the independence of palaeontological and lithological observations. Since changes in fossil morphologies were arranged in a recognisable and predictable sequence over both time and space, careful comparison of fossils with established sequences allowed stratigraphers to infer relative time correctly. Sequences of lithological changes failed to exhibit a similarly predictable pattern of variation over time.

The term 'zone' had been used informally by a number of geologists in the early and mid-1800s to denote a vertical interval of strata or assemblage of co-occurring fossil species. For example, Alcide d'Orbigny (1842–1851) used the term 'stage' and 'zone' interchangeably as subdivisions of Jurassic strata based on their ammonite content (*see* **Mesozoic**: Jurassic). The modern concept of the biostratigraphic zone, however, can be traced to the writings of Albert Oppel (e.g., *Die Juraforation Englands, Frankreichs und des südwestlichen Deutchlands*, 1856–1858) who used the term to denote 'the constant and exclusive occurrence of certain species [that] mark themselves off from their neighbours as distinct horizons' (translated in Arkell, *The Jurassic System of Great Britain*). Critical to Oppel's biozone concept was its abstract nature. Oppel-type zones existed independently of variations in the local lithological or palaeontological succession.

The success of Oppel's formulation for establishing long-range correlations within Europe and even between Europe and North America was beyond question by the late nineteenth century. Controversy remained though, regarding just what Oppel-type zones represented. In particular, Thomas Henry Huxley pointed out that, whereas Oppel's biozones could be construed to document the identity of, or homotaxic, arrangement in remote stratigraphical successions, the extent to which such zones represented equivalent intervals of time within such successions was unclear. In terms of practicality, such distinctions rarely mattered. Biostratigraphical analysis using Oppel-type zones had been established as an accurate method for reconstructing the local, regional, and (at least in principal) global sequence of strata, all of which were understood to represent some measure of geological time. Nevertheless, interest in the relation between biozones and measures of absolute time remained unabated, especially insofar as estimates of absolute time were critical to support for the concept of uniformitarianism that was held to underpin so much of late nineteenth-century geological theory.

The tools for radiometric – later radioisotopic – dating (*see* **Analytical Methods**: Geochronological Techniques), this volume) that were developed in the early 1900s offered an empirical way to resolve this controversy. Development of these methods was later augmented with other physio-chemical techniques, including magnetostratigraphy, chemostratigraphy, isotopic stratigraphy, and most recently, orbital stratigraphy. It should be noted though, that all of these latter, supplementary methods are, to a greater or lesser extent, dependent on accurate biostratigraphical analysis before their unique properties can be exploited with confidence. What has been made

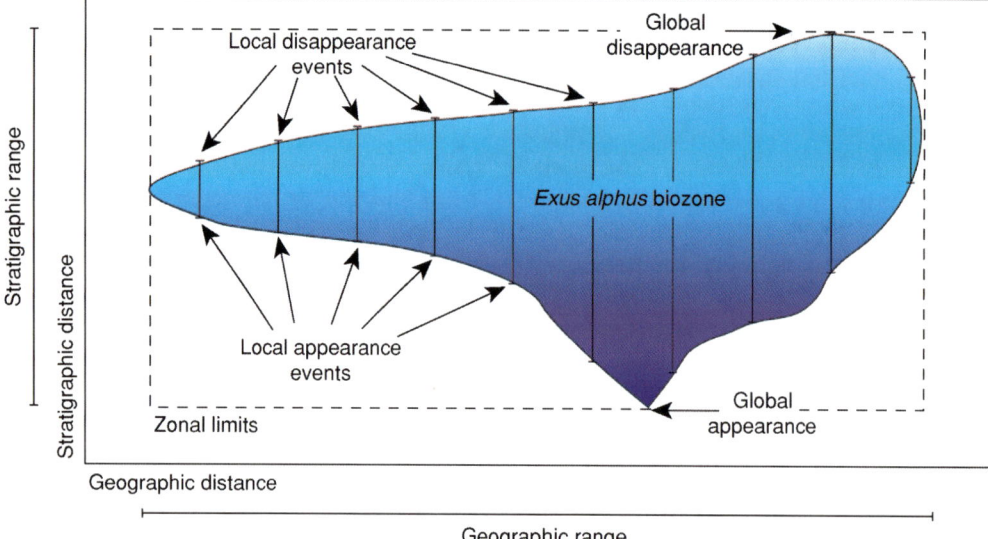

**Figure 1** Relation between the spatio-temporal concept of a biozone and various local and global events. A biozone is a three-dimensional concept (two dimensions represented here) that encompasses all strata delimited by the zone's boundary definitions. All local identifications of the zone span the interval from the first appearance (equivalent to the appearance or stratigraphically lowest juxtaposition of defining criteria) to the last appearance (equivalent to disappearance or stratigraphically highest juxtaposition of defining criteria). The global maxima and minima of these juxtapositions also define a combined chronostratigraphical-geographical interval within which the zone exists. Note that local first and last appearances of zone-defining criteria are all diachronous. Note also that, even though the stratigraphic level or chronostratigraphic time represented in any local succession can, in principle, be correlated outside the geographic limits of the zone, the zone itself is unrecognizable outside these limits.

abundantly clear by the various methods of absolute dating, however, is that the boundaries between most biozones are often measurably diachronous on regional and global scales. This diachrony is often modest and, in many cases, can be ignored without fatally compromising the results of a biostratigraphical analyses. Nevertheless, biozones (**Figure 1**) are now conceptually recognised to have no necessary chronological implication in the sense that identification of the same biozone or biozone boundary cannot be taken as evidence of contemporaneous sediment deposition. This, in turn, has led to a proliferation of methods that can be used to combine various types of stratigraphical data into synthetic and mutually reinforcing comparison systems that can be used to make accurate geochronological assessments of stratigraphical successions (see below).

## Types of Biozones

A wide variety of biozone types have been used to subdivide stratigraphical successions. Biozones are defined by single or combinations of biostratigraphical ranges for fossil species, genera, families, and so forth. Naturally, the ability to define and to recognise biozones is also strictly dependent on the existence of accurate systematic descriptions of fossil taxa such that individuals from locations remote from the primary study area can be quickly, easily, and correctly identified. A biostratigraphical range is the body of strata delineated by a fossil taxon's first appearance horizon or datum (FAD) and its last appearance horizon or datum (LAD, **Figure 2**). This 'first appearance' refers to the taxon's initial, or stratigraphically lowest appearance in a local succession while its 'last appearance' refers to the taxon's localised extinction at the stratigraphically highest level. Unfortunately, these first-lowest and last-highest conventions are usually reversed when working with wells that are being drilled, in which case the LAD is referred to as the 'first appearance' because it will be encountered before the FAD, and the FAD is referred to as the 'last appearance' because it will be encountered after the LAD.

### Range Zones

A biostratigraphical range zone is a body of strata delineated by the total range of occurrences of any selected element(s) of the total fossil assemblage recovered from the body of strata (**Figure 3**). Although the conventional representation of a range zone is that of a one-dimensional interval (**Figure 3A**), a better mental image of a range zone is that of a two-dimensional body of rock delineated by the correlation of fossil distributions between stratigraphic successions (**Figure 3B**) or a three-dimensional

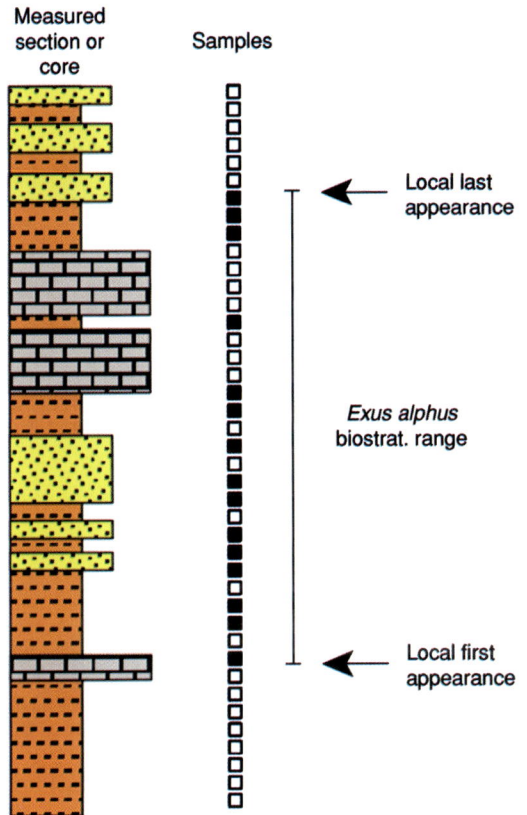

**Figure 2** The manner in which the limits of a local biozone are established. Samples are taken up through a stratigraphic succession (or down through a core) and analyzed for their fossil content. The stratigraphically lowest sample containing a fossil species marks that fossil's local first appearance and the stratigraphically highest sample containing the species marks the local last appearance. The interval between these two datums represents the species' local biostratigraphic range. Note that gaps in the appearance pattern can occur within the species' biostratigraphical range. These gaps can be the result of many factors (e.g., low absolute abundance, small sample size, variable preservation, environmental exclusion, structural complications). The existence of such gaps means that the observed end-points of each species' range represent minimum approximations of the true endpoints.

conceptualization that takes variation in occurrence and co-occurrence patterns across two geographical axes into account. This distinction between the conventional (one-dimensional) and actual (three-dimensional) conceptualization of biozone geometries holds true for all types of biozones. As better tools become available to reconstruct the three-dimensional geometries of stratigraphical bodies, the need for accurate three-dimensional concepts of biostratigraphical units will increase. There are four basic types of range zones: taxon range zones, concurrent range zones, Oppel zones, and lineage zones.

**Taxon range zone** A taxon range zone (also referred to as a teilzone, local zone, local-range zone, or topozone) is a body of strata delineated by the total occurrence range of any specimens belonging to a selected taxon (e.g., species, genus, family, **Figure 3**). Another way of thinking of this type of zone is as the region defined by all local appearances and extinctions of a selected fossil taxon (see **Figure 1**). While the geographic scope of such zones is set by the spatial occurrence pattern of their defining taxon, their boundaries are inherently diachronous insofar as the taxon's speciation and final extinction events will almost always vary from one locality to another (see **Mesozoic: End Cretaceous Extinctions; Palaeozoic: End Permian Extinctions**). Moreover, the boundaries of such zones are imprecise because taxon abundances usually diminish as both local and global speciation/appearance and extinction events are approached from within the zone. As a result, estimates of the local appearance and disappearance horizons of a taxon can be effected by sampling frequency and sample size near both ends of a taxon's stratigraphical range. Indeed, any abrupt appearance or disappearance of a taxon from a local succession is often taken as evidence of artificial truncation due to facies shift or the presence of a depositional hiatus.

Biostratigraphers estimate the boundaries of taxon range zones by assessing the boundaries of local range zones and correlating these boundaries between successions. By convention the names of taxon range zones refer to the defining taxon (e.g., *Dictyoha aculeata* Total Range Zone, *Parvularugoglobigerina* Total Range Zone).

**Concurrent range zone** A concurrent range zone (also referred to as an overlap zone or range-overlap zone) is a body of strata delineated by the those parts of the biostratigraphic ranges of two or more taxa that coincide in space and time (**Figure 4**). The term concurrent is a somewhat unfortunate choice for the name of this range-zone type since the colloquial definition of 'concurrent' is to 'happen at the same time'. This would, in principle, allow a concurrent range zone to be defined on the basis of two taxa that existed at the same time, but that never shared the same environment (e.g., a hypothetical *Tyrannosaurus rex–Abathomphalus mayaroensis* Concurrent Range Zone could be constructed as that interval of time represented by the overlap between these dinosaurian and planktonic foraminiferal species regardless of the fact that these two species have rarely, if ever, been found in the same stratigraphical succession). Such a zone would be of limited practical utility however, and the concept is understood to be restricted to those species whose stratigraphical ranges overlap in time and that occur together in the same samples.

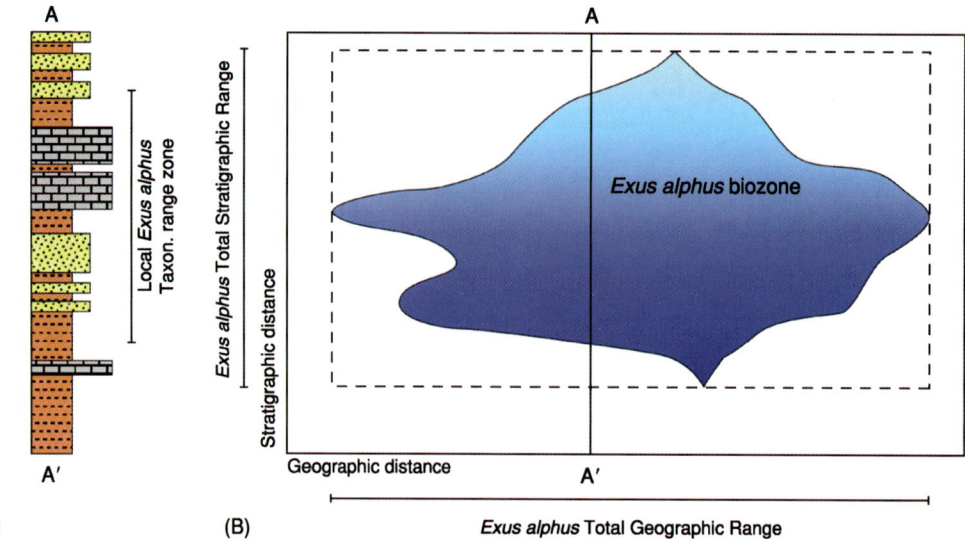

**Figure 3** One-dimensional (A) and two-dimensional (B) representations of a taxon range biozone. The zone begins at the speciation event that gave rise to the taxon, encompasses all strata containing fossils assignable to the taxon, and ends at the taxon's global extinction horizon. Globally, these speciation and extinction events, in additional to the taxon's geographic limits of distribution, define a spatio-temporal region within which the taxon's inferred Total Stratigraphic Range (vertical) and Total Geographic Range (horizontal) are defined (dashed box). In local sections however, the biozone will be represented by a one-dimensional interval whose boundaries are included within the taxon's global biozone. Note the strong diachrony of the biozone boundaries in this example. Intervals of strata that lie outside the biozones' geographical boundaries cannot be referred to the biozone *per se*, but can be placed within the taxon's range biochronozone (= time interval represented by the Total Stratigraphic Range).

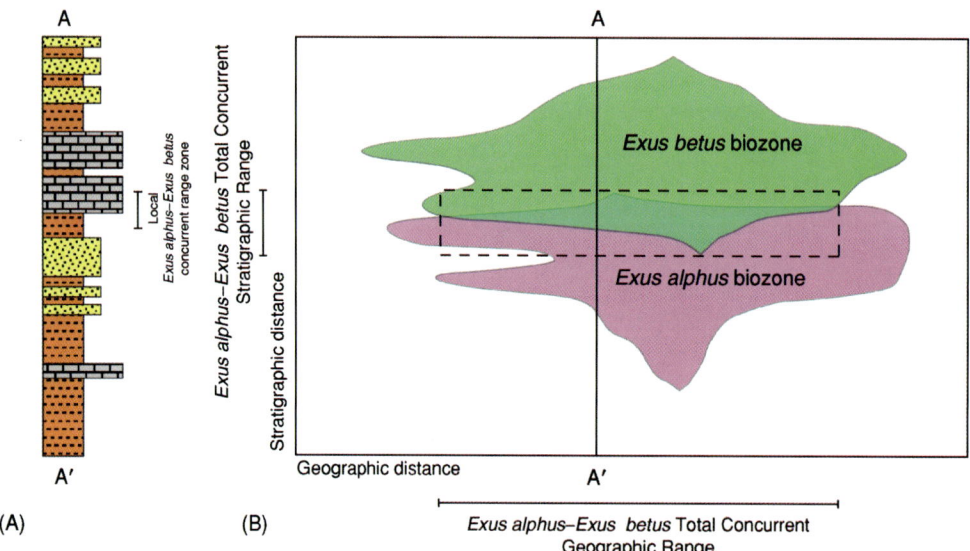

**Figure 4** One-dimensional (A) and two-dimensional (B) representations of a concurrent range biozone. In this example the zone begins at the speciation event of the hypothetical species *Exus betus* and encompasses all strata containing fossils assignable to the taxon range zones of *Exus alphus* and *Exus betus*. Globally, the speciation event of *Exus betus*, the extinction event of *Exus alphus*, and the geographic limits of the concurrent distribution of these two species' range biozones define a spatio-temporal region within which the taxon's inferred Total Stratigraphic Range (vertical) and Total Geographic Range (horizontal) are defined (dashed box). In local sections however, the biozone will be represented by a one-dimensional interval whose boundaries are included within this concurrent range biozone. Note the strong diachrony of the biozone boundaries. Intervals of strata that lie outside the biozones' geographical boundaries cannot be referred to the biozone *per se*, but can be placed within the taxons' concurrent range biochronozone (= time interval represented by the Total Concurrent Stratigraphic Range). Other definitional geometries are also possible (e.g., two taxa, one of whose biozone is completely enclosed by the stratigraphic and geographic ranges of another) so long as the concurrent range criterion is respected.

The concurrent range zone concept is distinguished from the taxon range zone concept by being based on more than a single taxon, and from the assemblage zone concept by not including all elements of a naturally occurring assemblage of fossil organisms. Ideal taxa to define a concurrent range zone are those whose range overlap represents a unit with advantageous recognizability, similar ecological tolerance, wide geographic scope, and limited temporal duration. In order to distinguish the concurrent range zone from an Oppel zone, the presence of all zone-defining taxa should be used in recognising the zone, though this can serve to limit the concurrent range zone's scope. In practice, many biostratigraphers adopt a flexible approach to the use of defining taxa when recognising concurrent range zones, in which case the concept intrudes inevitably into the realm of the Oppel zone.

Properly defined concurrent range zones can have boundaries that are less diachronous than taxon range zones and less facies-controlled that many assemblage range zones. They also typically delineate finer stratigraphic and temporal intervals than taxon range zones. By convention, current range zones are named for the taxa that define them (e.g., *Globigerina selli–Pseudohastigerina barbadoensis* Concurrent Range Zone, *Neodenticula koizumii–Neodenticula kamtschatica* Concurrent Range Zone).

**Oppel zone** An Oppel zone is a body of strata delineated by the ranges and range-limits of a group of fossil taxa selected in such a way as to minimize zonal boundary diachrony and maximize the geographic scope of the interval so defined (**Figure 5**). This type of zone was made popular by Albert Oppel in studies of Jurassic stratigraphy in the mid-1900s and served as a 'type' example of a biozone in the days when distinctions between biostratigraphy and chronostratigraphy were less clearly drawn. Oppel even made use of certain lithological beds (e.g., 'spongy limestones') that he felt had chronostratigraphic significance in defining his zones.

Many stratigraphic reference works still list the Oppel zone as the 'preferred' or 'most useful' type of biozone; particularly those written by biostratigraphers specializing in invertebrate macrofossil taxa. This author's experience, however, suggests that in modern, high-resolution stratigraphical applications, and especially in the context of microfossil zonations which represent the contemporary stratigraphic standard for Jurassic through Recent marine sediments, explicit Oppel zones are rarely established in

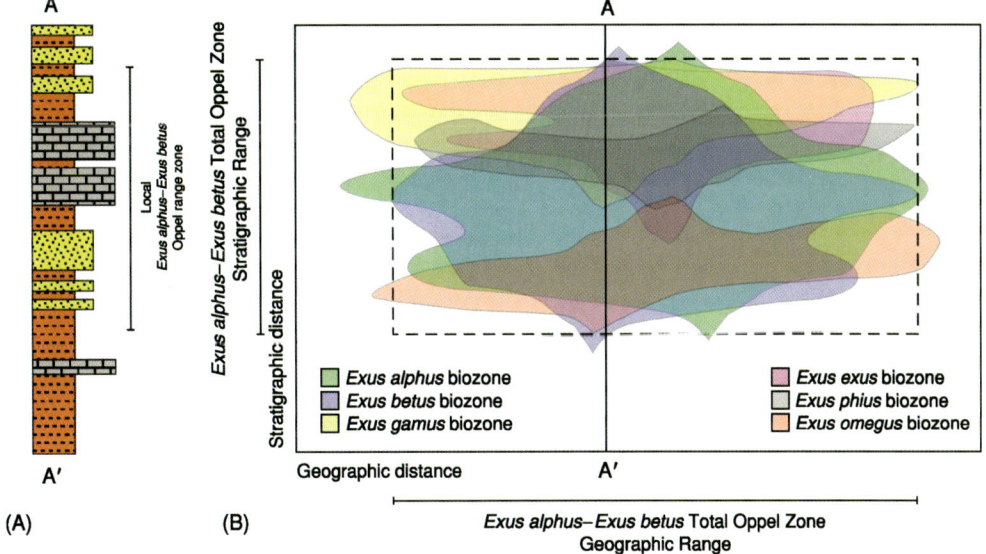

**Figure 5** One-dimensional (A) and two-dimensional (B) representations of an Oppel range biozone. In this example the zone begins at the speciation event of the hypothetical species *Exus omegus* (representing the lowest interval of a concurrent range based on *Exus alphus*, *Exus betus*, and *Exus omegus*), and encompasses all strata containing fossils assignable to the upper limit of the *Exus alphus-Exus beta-Exus gamus* concurrent range zone. As is always the case with Oppel-type biozones, these boundary definitions are chosen so as to enhance the ability of this zone to approximate a biochronozone. Globally, these two concurrent range biozones define a spatio-temporal region within which the Oppel biozone's inferred Total Stratigraphic Range (vertical) and Total Geographic Range (horizontal) are defined (dashed box). In local sections however, the biozone will be represented by a one-dimensional interval whose boundaries are included within this total Oppel biozone. Note the relatively modest diachrony of the biozone boundaries. Intervals of strata that lie outside the biozones' geographical boundaries cannot be referred to the biozone *per se*, but can be placed within the corresponding biochronozone ( = time interval represented by the Total Stratigraphic Range). Many other definitional geometries are also possible.

principle. Practice can be different though and, with the rise of neocatastrophism in the latter decades of the twentieth century, the common usage of certain standard microfossil zones has acquired a distinctly Oppelian character (e.g., the lowermost Paleocene planktonic foraminiferal Zone P0, also called the *Guembelitria cretacea* Zone) (*see* **Tertiary To Present: Paleocene**).

Oppel zones are more like abstract or Gestalt constructs than geometric juxtapositions of biostratigraphic ranges. They are largely constructed from sets of coincident range segments among taxa selected for the expressed purpose – or under the unexpressed assumption – of representing chronostratigraphic intervals. Unlike concurrent range zones (*sensu stricto*), Oppel zones do not rely on the necessary presence of all defining taxa to be recognised. Uniquely, this zonal concept embraces the idea that the biostratigrapher's judgement of the chronostratigraphical value of a particular taxon's presence (or absence) can be used in recognizing the zone. The rationale underlying this concept is that chronostratigraphical correlations are desirable and that, through long years of patient study, biostratigraphers can become sufficiently familiar with subtle morphological signals within their fossil faunas and floras that a higher level of time-based correlation than that afforded by any other zone concept can be achieved. The danger, of course, is that recognition of such a zone becomes an exercise in aesthetics with no obvious way to adjudicate judgemental disagreements between equally experienced stratigraphers. Interestingly, many – if not most – biostratigraphers approach the identification of fossil species in a similar manner. Employment of this Gestalt aesthetic in systematic palaeontology has led to the realisation that the reproducibility of many species lists is quite low. One cannot help but suspect that Oppel-zone recognisability is likely beset by similar problems.

By convention, Oppel zones are named for the dominant taxon used to recognise the zone's presence (e.g., *Subcolumbites-Prohungarites* Oppel Zone, *Globigerina selli–Pseudohastigerina barbadoensis* Oppel Zone).

**Lineage zone** A lineage zone (also referred to as a phylozone, evolutionary zone, lineage zone, morphogenic zone, phylogenetic zone, or morphoseries zone) is defined as a body of strata delineated by the combined biostratigraphic range of a series of ancestor–descendant species couplets (**Figure 6**). As such, a lineage zone is a type of taxon range zone in which

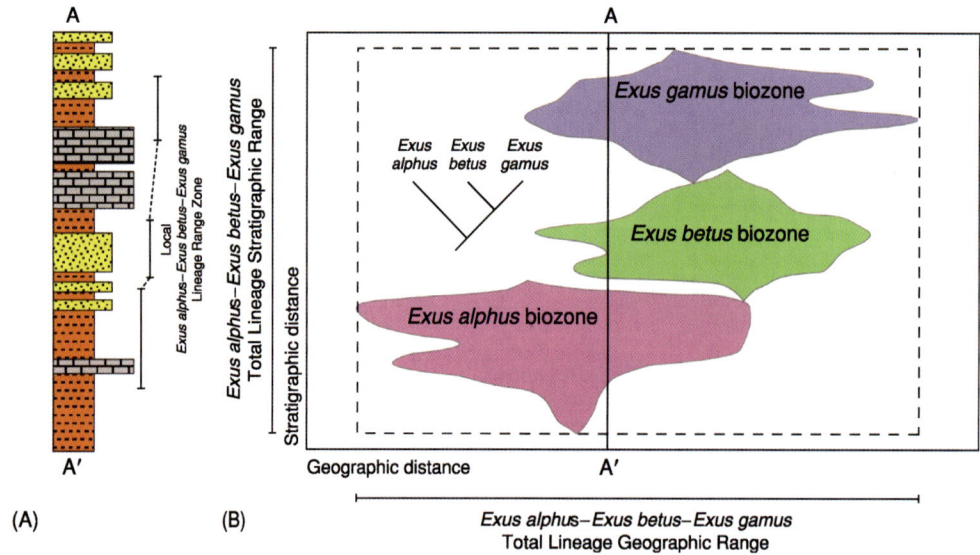

**Figure 6** One-dimensional (A) and two-dimensional (B) representations of a lineage range biozone. In this example the zone begins at the speciation event of the hypothetical species *Exus alphus* (representing the oldest species of the lineage) and encompasses all strata containing fossils assignable to the lineage. In this example, the stacked taxon range biozones for a three-species lineage define the overall lineage biozone. Globally, these taxon range biozones define a spatio-temporal region within which the lineage's inferred Total Stratigraphic Range (vertical) and Total Geographic Range (horizontal) are defined (dashed box). In local sections, however, the biozone will be represented by a one-dimensional interval whose boundaries are included within the range of the global biozone. Note the strong diachrony of the biozone boundaries. Intervals of strata that lie outside the biozones' geographical boundaries cannot be referred to the biozone *per se*, but can be placed within the corresponding biochronozone (= time interval represented by the lineage's Total Stratigraphic Range). Lineage range zones also have the added uncertainty that, since lineages are inferred—not observed, their composition may change with new evidence and/or analyses.

the 'taxon' concept has been replaced by the evolutionary concept of a lineage or lineage segment. Given the controversial nature of attempts to identify ancestors in the fossil record using modern methods of phylogenetic systematics, one suspects that, were these systematic methods applied to some lineage zone-defining taxa, the basic rationale for recognizing many lineage zones might be called into question. Nevertheless, there are a number of instances in which morphological transitions between stratigraphically successive taxa are so well structured as to pass even the most stringent morphological tests.

A number of previous writers have commented that lineage zones should be particularly useful for chronostratigraphical correlations. I do not believe this to be the case. Lineage zones are prone to all the problems of taxon range zones (see above), and to the problem of being uniquely susceptible to the discovery of new fossils, or new characters that, by altering our understanding of phylogenetic relations between fossil species, would serve to change the zone's concept or rationale. A strictly comparative and geometric approach to the determination of FAD/LAD sequences in local sections, along with efforts to estimate reliable chronostratigraphic ages for those events based on global composite standard sections (see below), are likely to be of more practical use in achieving a chronostratigraphically-based biostratigraphy than efforts to mix phylogenetic inferences with biostratigraphic observations. Notwithstanding the issues highlighted above, lineage-based zonations remain popular, particularly among students of Tertiary planktonic microfossils where the tradition of qualitative phylogenetic analysis remains strong.

By convention, lineage zones are named for the lineages on which they are defined (e.g., *Globorotalia foshi* Lineage Zone, *Allevium praegallowayi-galloway-superbum* Lineage Zone).

## Assemblage Zone

An assemblage zone (also referred to as a cenozone, ecozone, ecological zone, faunizone, biofacies zone, and association zone) is defined as that body of strata delineated by a natural assemblage of fossils (**Figure 7**). This assemblage can be defined in a variety of ways. For example, it may represent an ensemble of many groups (e.g., a coral–algal assemblage zone) or of a single group (e.g., a mollusc assemblage zone). It may also take a specific habitat into consideration (e.g., a planktonic foraminiferal assemblage zone), or cross habitat boundaries. The distinguishing character of this zone is that the assemblage be composed

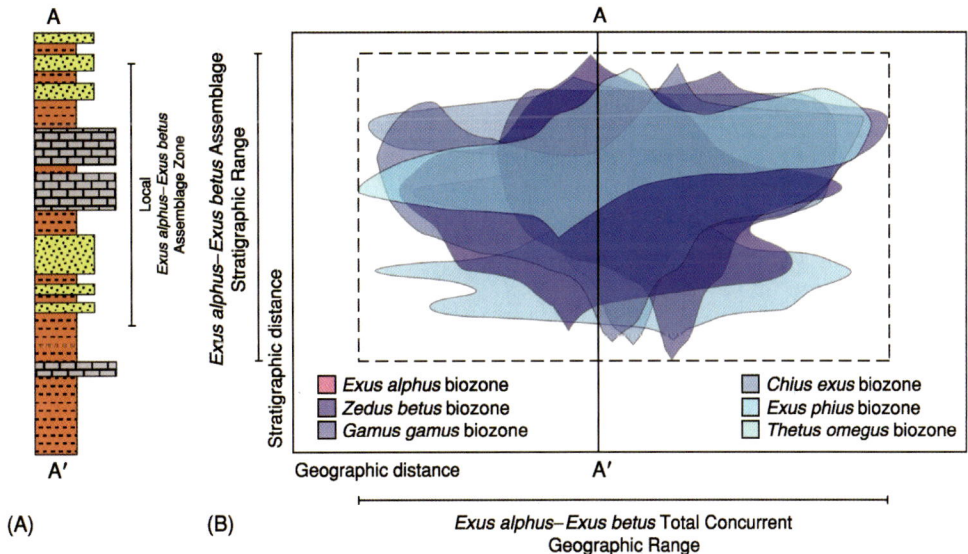

**Figure 7** One-dimensional (A) and two-dimensional (B) representations of a assemblage range biozone. In this example the zone begins at the speciation event of the hypothetical species *Exus alphus* (representing the oldest species in an assemblage of six fossil species usually found together in the same environment) and encompasses all strata containing fossils assignable to this assemblage. As is always the case with assemblage range biozones, the boundary definitions are chosen so as to base the zone on a series of ecologically related taxa. Globally, this assemblage of taxon range biozones defines a spatio-temporal region within which the assemblage biozone's inferred Total Stratigraphic Range (vertical) and Total Geographic Range (horizontal) are defined (dashed box). In local sections, however, the biozone will be represented by a one-dimensional interval whose boundaries are included within this global biozone. Note the strong diachrony of the biozone boundaries. Intervals of strata that lie outside the biozones' geographical boundaries cannot be referred to the biozone *per se*, but can be placed within the corresponding biochronozone (= time interval represented by the Total Stratigraphic Range).

of fossils that actually lived together and (ideally) interacted with one another to the extent that the grouping itself achieved a degree of stability – and therefore recognizability – as a result of these interactions.

Because properly defined assemblage zones reflect ecological relations, this type of zone tends to be restricted to particular facies. The ability of such zones to track spatial shifts in environments through time provides them with a distinctive utility in terms of palaeoenvironmental analyses. However, this utility comes at a price, and that price is a relatively reduced ability to achieve long-distance chronostratigraphical correlations. For this reason, assemblage zones of benthic organisms tend to be most useful on local and regional scales. Assemblage zones of planktonic organisms do perform well in terms of chronological correlations, but the degree to which such organismal groupings are maintained by close inter-specific interactions is debatable.

Like Oppel zones, the specific criteria used to recognize assemblage zone boundaries are flexible. Not all of the groups present in the zone's 'type area' need be present to recognise the zone in remote locations. Unlike Oppel zones though, there is an objective and independent rationale underlying this definitional latitude. In the case of assemblage zones, one seeks to recognise a set of dependent ecological relations among species and between organisms and their environment that transcend mere faunal and/or floral lists. The objective reality of such patterns in nature is well established by numerous studies of modern faunas and floras and is reasonably well understood from a theoretical point of view. Oppel zones, on the other hand, are unified only in the vague sense that the species used to recognize the zone are thought to be useful in chronostratigraphical analysis. Although there is certainly ample justification for suspecting that, in many cases, the biostratigraphic ranges of the species in different regions and habitats will coincide, there is much less justification for regarding these organisms as part of transcendent causal association than is the case with assemblage zones.

As a result of the flexible manner in which assemblage zones are defined, the same groups can be used to define different assemblage zones (e.g., a coral–bryozoan assemblage zone and a coral–foraminiferal assemblage zone can have zone-defining taxa in common) and different members of the same ecological association can be used to define different assemblage zones. Assemblage zones have been used frequently in areas where suitably short-ranging taxa are not present or have not been studied.

By convention, the name of an assemblage zone should be based on two or more taxa that figure prominently in the zone's definition (e.g., *Eponides–Planorbulinella* Assemblage Zone, *Eodicynodon* Assemblage Zone).

### Interval Zone

An interval zone (also referred to as an interbiohorizon zone, gap zone, or a partial-range zone) is defined as a body of strata delineated by the region between two distinctive biostratigraphic horizons, but that has no distinctive biostratigraphic identity of its own (**Figure 8**). The boundaries of interval zones can be marked by a wide variety of criteria. These zones typically represent the undefined regions between other types of zones; especially taxon range zones.

An interval zone's existence assumes a complementary relation with the underlying and overlying biostratigraphically defined horizons that serve as their inferior and superior boundaries. As with all other types of biozones, the traditional one-dimensional concept of biozone geometry (**Figure 8A**) can mask the more complex geometries evident in two and three-dimensional conceptualizations (**Figure 8B**). In particular, interval zones are confined geographically to only those regions in which the defining biozones overlap. So long as one's region of interest is confined to the geographical area encompassed jointly by the zone's defining taxa, recognition of the zone can be made with confidence. Outside this geographic envelope, though, recognition of an interval zone becomes problematic, if not impossible.

By convention, interval zones are either named for the taxa used to define their boundaries (e.g., *Globigerinoides sicanus–Orbulina suturalis* Interval Zone) or for a taxon that occurs in the interval, but is not itself used in the zone definition (e.g., *Globigerina ciperoensis* Zone).

### Acme Zone

An acme zone (also referred to as a peak zone, flood zone, or epibole) is defined as a body of strata delineated by the region of 'maximal development' of a taxon (e.g., species, genus, family), but not its total range (**Figure 9**). In this context, the term 'maximal development' is meant to be used flexibly. In some cases it might refer to an initial increase and subsequent decrease in the relative abundance of a taxon that takes place within the confines of its biostratigraphic-geographic range. In others, it might refer to an increase/decrease in body size, an increase or decrease in diversity, etc. Since these aspects of a taxon's evolutionary/ecological history tend to be strongly associated with local and regional conditions, it is on these spatial scales that acme zones have their

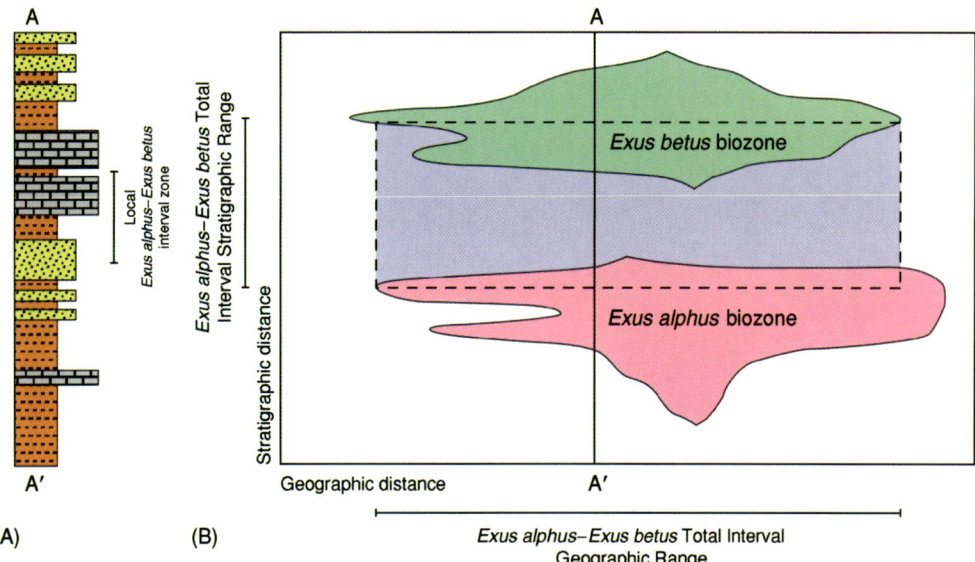

**Figure 8** One-dimensional (A) and two-dimensional (B) representations of an interval biozone. In this example the zone begins at the geographic acme of the hypothetical species *Exus alphus* and encompasses all strata between this extinction datum and the overlying *Exus betus* geographic expansion datum. Globally, this interval between taxon range biozones defines a spatio-temporal region within which the interval biozone's inferred Total Stratigraphic Range (vertical) and Total Geographic Range (horizontal) are defined (dashed box). In local sections however, the biozone will be represented by a one-dimensional interval whose boundaries are included within this global biozone. Note the strong diachrony of the biozone boundaries. Intervals of strata that lie outside the biozones' geographical boundaries cannot be referred to the biozone *per se*, but can be placed within the corresponding biochronozone (= time interval represented by the Total Stratigraphic Range). Many other definitional geometries are also possible (e.g., interval between two species' last appearances, interval between two species' first appearances).

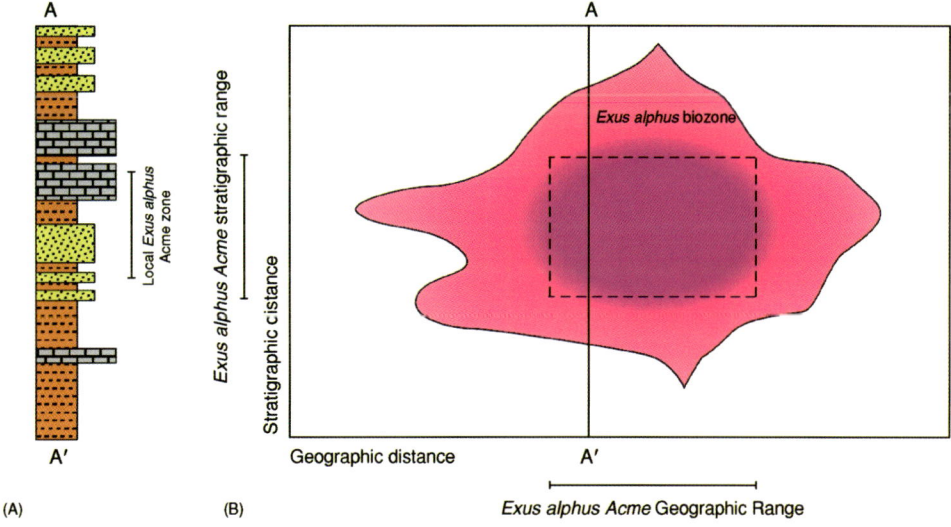

**Figure 9** One-dimensional (A) and two-dimensional (B) representations of an acme biozone. In this example the zone begins at the first substantial increase in the relative abundance of the species (symbolized by dark shading) of the hypothetical species *Exus alphus* and encompasses all strata in which the species is regarded as abundant. Globally, these abundance-based data define a spatio-temporal region within which the interval biozone's inferred Total Stratigraphic Range (vertical) and Total Geographic Range (horizontal) are defined (dashed box). In local sections, however, the biozone will be represented by a one-dimensional interval whose boundaries are included within this global biozone. Note the strong diachrony of the biozone boundaries. Intervals of strata that lie outside the biozones' geographical boundaries cannot be referred to the biozone *per se*, but can be placed within the corresponding biochronozone (= time interval represented by the Total Stratigraphic Range). Many other definitions of acme are also possible (e.g., increased size, increased ornamentation).

greatest utility. Care must be taken when defining such zones, that the recognition criteria employed are as explicit and objective as possible. Nevertheless, the popularity of such zones in regional basin analyses (particularly in commercial biozonations) stands as testimony to their practical utility. By convention, interval zones are named for the taxon used in its definition (e.g., *Emiliana huxleyi* Acme Zone, *Tylosaurus* Acme Zone).

### Other Types of Biozones

Although the foregoing descriptions represent the most frequently used biozone types, other types do exist. These include such exotica as barren zones, coiling-direction zones, negative association zones, species pre-lap zone, species post-lap zone, etc. In addition, biozone types exist that are not defined on fossils *per se*, but rather on the trace fossils left in the sediment as a result of animal and plant activity (e.g., track zones). All such zones are valid only to the extent that they are useful to the stratigrapher and geologist, and only to the extent that their definition is based on unambiguous observational evidence. As was noted by Hedberg in 1971, 'Time and usage will then be the surest means of determining whether it is desirable for such [zones] to persist'.

## Biozones and Biochronozones

Ever since the days of Smith, Lyell, and Oppel, the purpose of establishing biozones has been to achieve chronostratigraphical correlations. Other types of stratigraphic intervals (e.g., key beds, magnetochrons, isotope zones) may be based on boundaries whose emplacement is effectively synchronous over geological time-scale. None is uniquely identifiable in the same manner as biozones, however. Indeed, the use of biostratigraphic methods to place, in an approximate temporal framework, the observations upon which these other types of chronostratigraphic inference depend, is so common it is rarely even mentioned. This failure to give biostratigraphical data due credit for the basic contribution it makes would not be so unfortunate, were it not for the fact that biostratigraphy has come to be viewed by many as a lacklustre and unimaginative field of study. Nothing could be further from the truth. Biostratigraphic data underpin not only most of what we know about geological time over the last 600 million years, but are critical to studies of palaeogeography, palaeoecology, extinctions, diversifications, and patterns of morphological evolution. In order to remind oneself of the importance of biostratigraphical data, it is worth recalling that the lack of a detailed time-scale in which to place Proterozoic and Archaean events is not due to a lack of palaeomagnetic reversals, catastrophic depositional events, isotopic excursions, etc., but rather to the lack of fossil biotas on which to base a coherent, finely resolved, and widely accessible stratigraphy.

This having been said, the distinction between biozones and chronozones must always be borne in mind. All biozone boundaries are diachronous by definition and, without recourse to radioisotopic

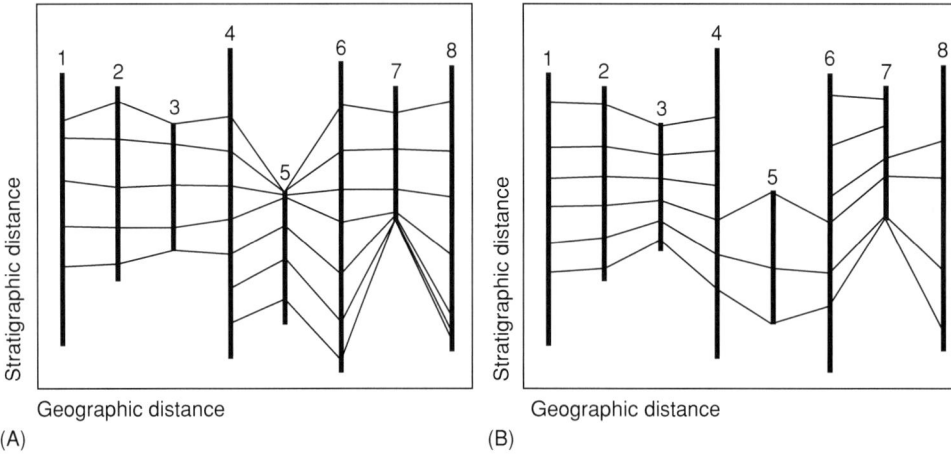

**Figure 10** Results of multivariate biostratigraphic analyses on simulated biostratigraphic data from various geological sections numbered 1 to 8 using the graphic correlation (A), as implemented by constrained optimization) and ranking and scaling (B) approaches. These methods focus on the estimation of global, composite, taxon range biozones from datasets containing a series of biostratigraphic first and last appearance datums for two or more stratigraphic successions or cores. Such approaches offer more consistency, objectivity, and higher reproducibility of results than either qualitative or pairwise quantitative approaches to biostratigraphic analysis. In these diagrams, connecting lines between the eight sections represent best-estimate lines of biochronostratigraphic correlation.

data, it is never correct to regard local biozones as being equivalent to chronozones. Even when the physical data necessary for absolute dating are available, it is no simple matter to accurately interpolate the ages of biozone boundaries from radioisotopically dated horizons.

Fortunately, several methods exist whereby stratigraphers can compare sets of biostratigraphical observations in different successions or cores and, through use of a few simple rules, create a summary of the chronostratigraphically relevant data contained in all such sections/cores. Graphic correlation (**Figure 10A**) and ranking and scaling (**Figure 10B**) are two of the more common quantitative methods used for bridging the gap between observational biostratigraphies and the ideal of a 'biochronostratigraphy'. The ongoing incorporation of raw biostratigraphical data into internally consistent and synthetic composite successions based on the application of these methods, along with improvements in consistent identifying fossil morphologies (e.g., via automated object recognition) promise to take the science of biostratigraphy, and its foundation concept – the biozone – into the twenty-first century. There it will, once again, serve as the single most useful contrivance in the practical stratigrapher's toolkit.

## Glossary

**Constrained optimization** A mathematical technique involving the search for the optimal condition or structure of a system of observations given one or more consistently applied external rules.

**Depositional hiatus** A horizon within a body of sedimentary rock that represents a gap in time due to the nondeposition of sediment, active erosion, or structural complications.

**Diachrony** The condition of taking place at different times.

**Facies** A stratigraphic body distinguished from other such bodies by a difference in appearance or composition.

**Homotaxis** The condition of occupying the same position in a sequence.

**Isochrony** The condition of being created at the same time.

**Microfossil** Fossil shells or other hard parts of tiny organisms or parts of organisms studied with the aid of a microscope.

**Phylogenetic systematics** A method of determining evolutionary relations between taxa based on the nested patterns of shared characteristics derived from the same pre-existing condition.

**Stratum (strata)** A tabular section of a rock body that consists of the same type of rock material.

**Taxon (taxa)** Generalized term for any level within the Linnean hierarchy of biological classification (e.g., order, family, genus).

## See Also

**Analytical Methods:** Geochronological Techniques. **Mesozoic:** Triassic; Jurassic; Cretaceous; End Cretaceous Extinctions. **Palaeozoic:** Cambrian; Ordovician; Silurian; Devonian; Carboniferous; Permian; End Permian Extinctions. **Precambrian:** Vendian and Ediacaran. **Sequence Stratigraphy. Stratigraphical Principles. Tertiary To Present:** Paleocene; Eocene; Oligocene; Miocene; Pliocene; Pleistocene and The Ice Age. **Time Scale.**

## Further Reading

Cubitt JM and Reyment RA (1982) *Quantitative stratigraphic correlation*. Chichester: Wiley.

Gradstein FM, Agterberg FP, Brower JC, *et al.* (1985) *Quantitative stratigraphy*. Dordrecht: D. Reidel.

Hedberg HD (1971) *Preliminary report on biostratigraphic units*. Montreal, Canada: International Subcommission on Stratigraphic Classification.

Hedberg HD (1976) *International stratigraphic guide: a guide to stratigraphic classification, terminology, and procedure*. New York: John Wiley & Sons.

Kauffman EG and Hazel JE (1977) *Concepts and methods in biostratigraphy*. Stroudsburg, Pennsylvania: Dowden, Hutchinson & Ross, Inc.

Mann K, Lane HR, and Stein J (1995) *Graphic correlation*. Tulsa: Society of Economic Paleontologists and Mineralogists, Special Publication 53.

Miall AD (1984) *Principles of sedimentary basin analysis*. New York: Springer-Verlag.

Rawson PF, Allen PM, Brenchley PJ, *et al.* (2002) *Stratigraphical procedure*. London: The Geological Society.

Reyment RA (1980) *Morphometric methods in biostratigraphy*. London: Academic Press.

Shaw A (1964) *Time in stratigraphy*. New York: McGraw-Hill.

# BRAZIL

**F F Alkmim and M A Martins-Neto**, Universidade Federal de Ouro Preto, Ouro Preto, Brazil

© 2005, Elsevier Ltd. All Rights Reserved.

## Brazil in the Geological Scenario of South America

The South American continent comprises five major tectonic units: the Pacific active margin, the Atlantic passive margin, the Andean Orogen, and the Patagonian and South American platforms (**Figure 1**). The continental margins and the Andean Orogen are the younger portions of the continent; the platforms, on the other hand, correspond to the mature and ancient parts of South America.

Brazil is located entirely on the South American platform, which is defined as the Precambrian core of the continent, not affected by the Andean orogenies (**Figure 1**). The exposures of the South American platform are collectively referred to as the Brazilian Shield, which in reality encompasses three distinct morphotectonic domains: the Guyanas, the Central Brazil (also called Guaporé) and the Atlantic shields (**Figure 1**). Sedimentary basins, including large Palaeozoic sags, Cretaceous passive and transform margins, and Tertiary rifts make up the Phanerozoic cover of the South American platform (**Figure 1**).

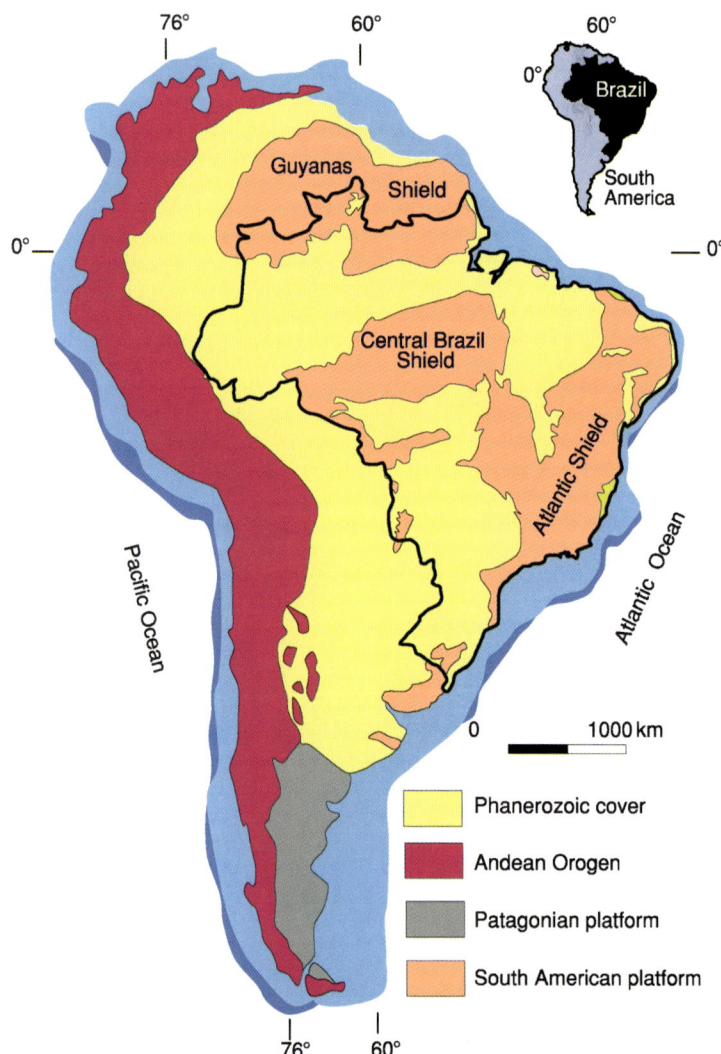

**Figure 1** Tectonic map of South America, showing the main subdivisions. Modified from Almeida FFM, Hasui Y, Brito Neves BB, and Fuck RA (1981) Brazilian structural provinces: an introduction. *Earth Science Reviews* 17: 1–29, with permission from Elsevier.

Two fundamentally distinct lithospheric components, namely cratons and Neoproterozoic orogens, form the Precambrian core of South America. The cratons correspond to the relatively stable pieces of the continent that escaped the Neoproterozoic orogenies recorded in all the remaining shield areas of Brazil. Four cratons have been delimited in the South American platform: the São Francisco, Amazon, São Luis, and Rio de la Plata cratons (**Figure 2**). The noncratonic segments of the platform are the Neoproterozoic Mantiqueira, Tocantins, and Borborema orogenic domains (**Figure 2**). (In the Brazilian geological literature each component of the South American platform and its cover (i.e. cratons, Brasiliano orogens, and Phanerozoic basins) is referred to as a tectonic province.)

Together with Africa, the South American platform once lay in the western portion of Gondwana, the supercontinent assembled by the end of the Neoproterozoic and split apart in the Cretaceous, which also encompassed Antarctica, India, and Australia (**Figure 3**). The assembly of western Gondwana resulted from a series of diachronic collisions, predominantly between 640 Ma and 520 Ma, the

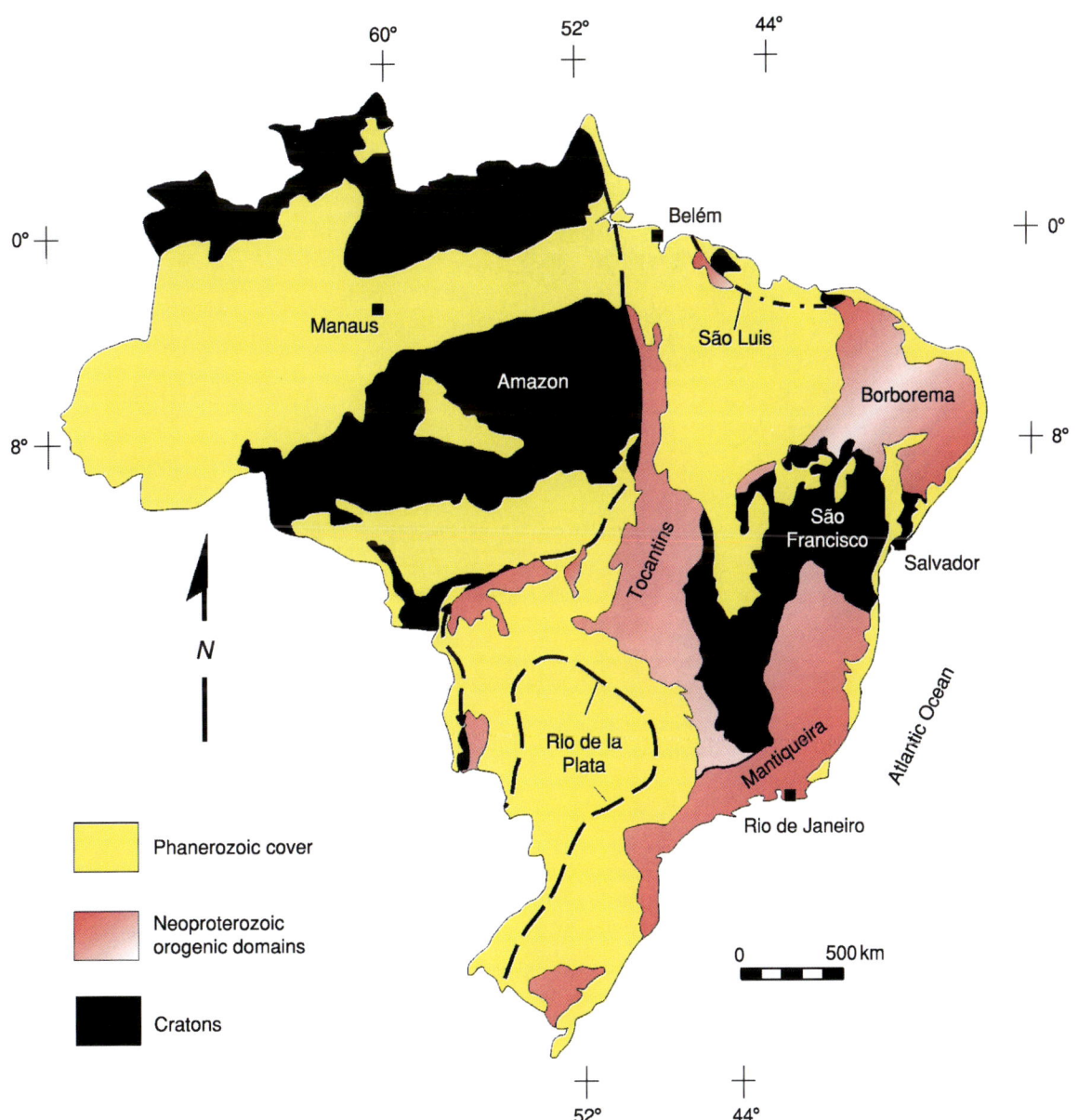

**Figure 2** Simplified tectonic map of Brazil, showing the subdivisions of the South American platform and its Phanerozoic cover. Modified from Almeida FFM, Hasui Y, Brito Neves BB, and Fuck RA (1981) Brazilian structural provinces: an introduction. *Earth Science Reviews* 17: 1–29, with permission from Elsevier.

so-called Brasiliano or Pan-African orogenies. In this context, the cratons of South America and Africa are the preserved and more internal portions of the plates that collided to build up western Gondwana; the Neoproterozoic orogenic domains encompass the margins of these plates and other lithospheric pieces also involved in the tectonic collage that made up western Gondwana. The dispersal of western Gondwana and the opening of the South Atlantic in the Lower Cretaceous broke apart Neoproterozoic orogens and cratons. Consequently, the Neoproterozoic orogens and cratons of eastern Brazil have African counterparts (**Figure 3**).

At first glance the geological panorama of Brazil reflects only the Neoproterozoic assembly, Early Palaeozoic to Jurassic permanence, and Cretaceous

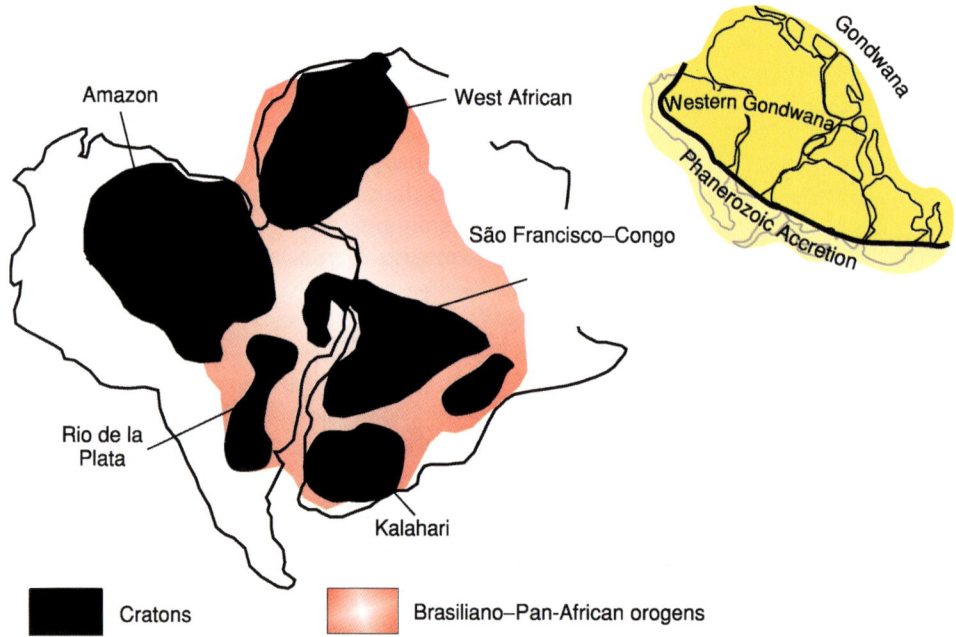

**Figure 3** Schematic map of western Gondwana, showing the cratons and Neoproterozoic Brasiliano–Pan-African belts.

**Table 1** The main thermotectonic events recorded in the South American platform

| Event | Age | Occurrence | Manifestation | Significance |
| --- | --- | --- | --- | --- |
| Tertiary reactivation | Eocene–Oligocene | Eastern and Northern Brazil | Rifting, alkaline magmatism | |
| South Atlantic | 130–78 Ma | Continental margin and interior | Flood basalts, followed by rifting and a late phase of alkaline magmatism | Gondwana breakup, opening of the South Atlantic |
| Brasiliano | 640–520 Ma | Non-cratonic areas | Deformation, metamorphism, magmatism, minor accretion | Western Gondwana assembly |
| Macaúbas rifting | 900–800 Ma | São Francisco Craton and adjacent belts | Rifting, bimodal magmatism, glaciation | Rodinia breakup |
| Rondonian/San Ignácio/Sunsás | 1500–1000 Ma | South-western Amazon Craton | Deformation, metamorphism, magmatism | Rodinia assembly |
| Staterian rifting | ~1750 Ma | São Francisco Craton and adjacent belts | Rifting, bimodal magmatism | Atlantica breakup |
| Uatumã | ~1800 Ma | Amazon Craton | Anorogenic magmatism | Atlantica breakup, plume |
| Transamazonian | 2200–1950 Ma | Amazon and São Francisco cratons, all Brasiliano orogenic domains | Deformation, metamorphism, magmatism, crustal accretion | Assembly of Atlantica supercontinent |
| Jequié, Rio Das Velhas, Aroense | 2900–2780 Ma | São Francisco and Amazon cratons, some Brasiliano orogenic domains | Deformation, metamorphism, magmatism, crustal accretion | |

dispersal of western Gondwana. A closer examination of the cratons and Neoproterozoic orogenic belts reveals, however, a long and diverse pre-Gondwana history, as well as a whole series of post-Gondwana features. Thus, in addition to the Brasiliano and South Atlantic events, other events of the same significance and extent are recorded in different ways in the South American platform and its cover. The most important of these events, together with their ages, areas of occurrence, and geotectonic significance, are shown in **Table 1**.

## Regional Structures and Topography of Brazil

The topography of Brazil to a large extent reflects the constitution of the South American platform discussed in the previous section. The cratons underlie the low areas, hosting the main river basins; the highlands, on the other hand, have the Neoproterozoic orogenic domains as their substrata (**Figure 4**).

Each of the large-scale topographical highs and lows of the Brazilian territory is the expression of a particular regional tectonic structure. Some of these structures had already nucleated by the Palaeozoic. The majority, however, were initiated in the Mesozoic and underwent significant reactivation during the Cenozoic. The most prominent structural and topographical lows correspond to the Paraná, São Francisco, Parnaíba, and Amazonas basins. The Borborema Plateau, the Serra do Mar Uplift, and the Alto Paranaíba Arch are the largest structural and topographical highs (**Figure 4**).

## Cratons

The cratons of the South American platform, consisting of Archaean crust with substantial

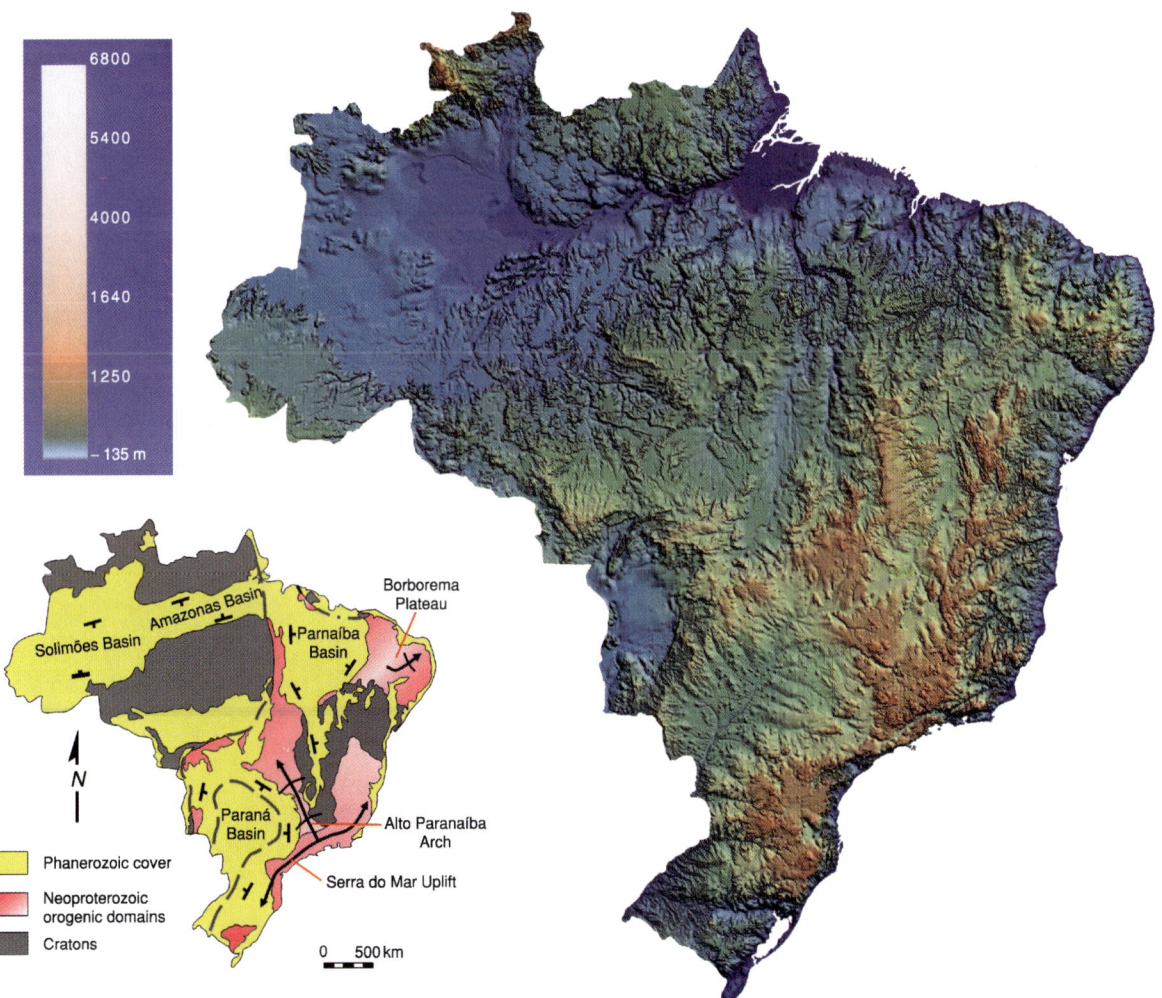

**Figure 4** Correlation between the large-scale tectonic structures and the topographical relief of Brazil. Relief map compiled by JBL Françolim based on EROS GTOPO 30 dataset; reproduced with the permission of the author.

Palaeoproterozoic accretions, are the portions of the Precambrian basement that were unaffected by the Brasiliano orogenies. Their boundaries are defined by the deformation styles of Neoproterozoic cover strata, patterns of geophysical anomalies, and geochronological data. Accordingly, the cratons of Brazil are bounded on all sides by Brasiliano basement-involved fold–thrust belts, but contain Brasiliano thin-skinned foreland fold–thrust belts. In this regard they differ from similar features delimited on other continents. The cratons of Brazil do, however, exhibit the typical attributes of their equivalents worldwide, such as mantle roots, low heat-flow values, and high lithospheric strength.

### São Francisco Craton

A map view of the São Francisco Craton shows a shape that resembles a horse's head, with the northern segment merging with the east coast of Brazil (**Figure 5**). Except for the Atlantic coast, the São

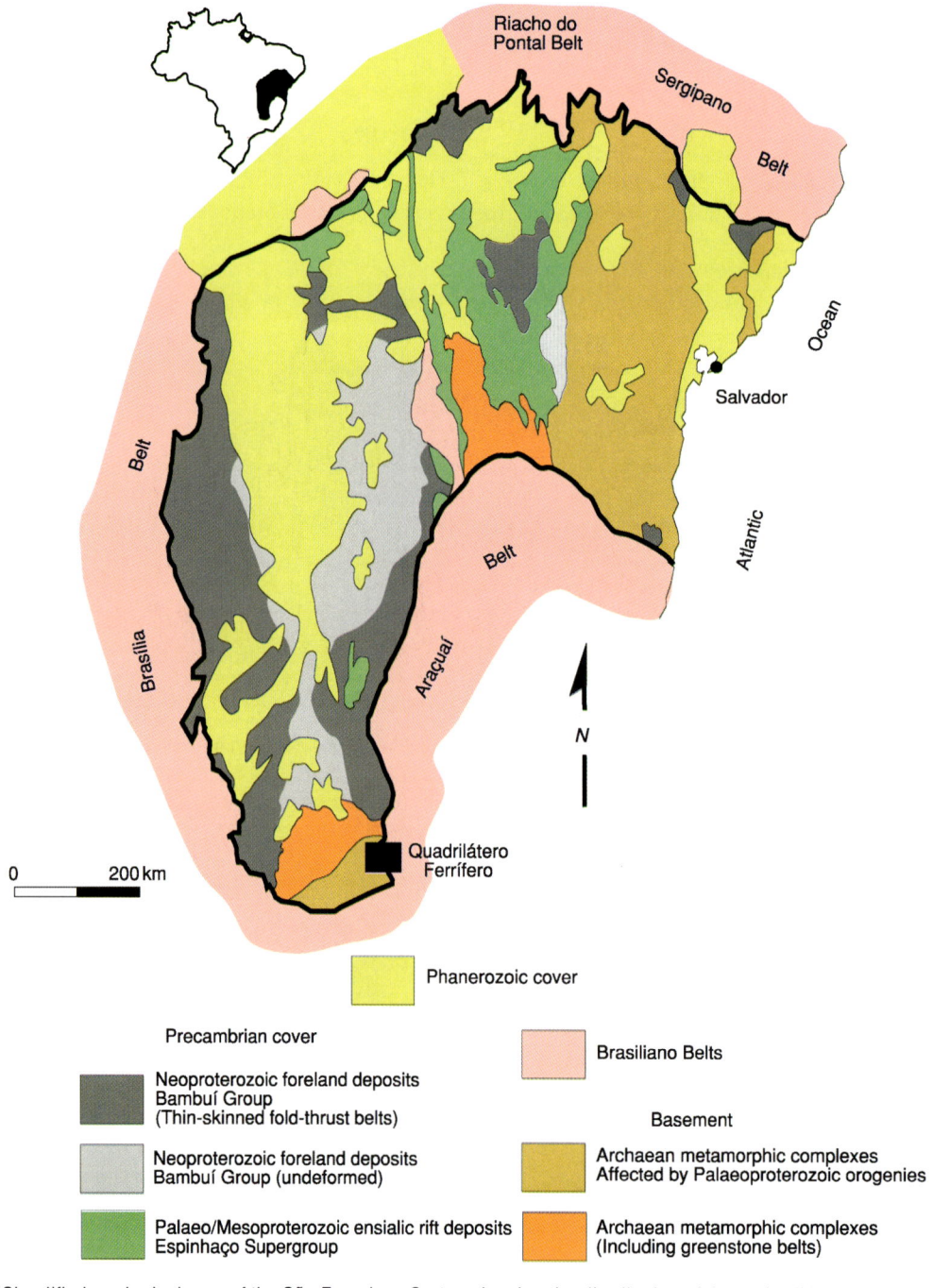

**Figure 5** Simplified geological map of the São Francisco Craton showing the distribution of the major lithostratigraphical units.

Francisco Craton is bounded on all sides by the external fold–thrust belts of the Mantiqueira, Borborema, and Tocantins orogenic domains (**Figures 2** and **5**). As indicated by reconstructions of western Gondwana, the São Francisco Craton and the Congo Craton of Africa formed a single lithospheric unit from the Palaeoproterozoic until the Cretaceous opening of the South Atlantic.

The basement, made up of Archaean metamorphic complexes (tonalite–trondhjemite–granodiorite association and voluminous calc-alkaline plutons), Archaean greenstone belts, and Palaeoproterozoic metasedimentary successions, is exposed in the northern lobe of the craton and in a smaller area close to the southern boundary (**Figure 5**). In both exposures, Archaean nuclei are bounded on the east and south-east by Palaeoproterozoic Transamazonian fold–thrust belts, which involve the Archaean basement and Palaeoproterozoic metasedimentary units. The second-largest and best-studied mineral province of Brazil, the Quadrilátero Ferrífero (Iron Quadrangle), lies partly in the southern portion of the craton and partly in the adjacent Brasiliano Araçuaí belt. In this province, gold is found in an Archaean greenstone-belt sequence, and high-grade iron-ore deposits occur in a Palaeoproterozoic banded iron formation.

Alluvial to marine sediments with 1750 Ma acid-volcanic intercalations at the base (representing the fill of an ensialic rift system), glacial-influenced Tonian (*ca.* 850 Ma) rift sediments, and Cryogenian (720–600 Ma) foreland strata form the Precambrian cover of the craton. The foreland-basin strata are deformed in the areas adjacent to the craton boundaries, thereby defining two thin-skinned foreland fold–thrust belts with opposite vergences. These belts represent the orogenic fronts of the Neoproterozoic Brasília and Araçuaí belts, which fringe the craton to the west and to the east, respectively.

### Amazon Craton

The Amazon Craton encompasses an area of approximately 430 000 km$^2$ in north-western South America, extending far beyond the Brazilian borders into Colombia, Venezuela, Guyana, Suriname, and French Guiana. The eastern and south-eastern limits of the craton are marked by the external portion of the Brasiliano Tocantins Orogen, represented by the Araguaia and Paraguay belts, respectively. The western and north-western boundaries are covered by the sub-Andean basins, and the northern limit is marked by the South American equatorial margin (**Figures 2** and **6**).

More than two-thirds of the Brazilian portion of the craton is covered by Phanerozoic sediments and the Amazon forest, so that the geological knowledge

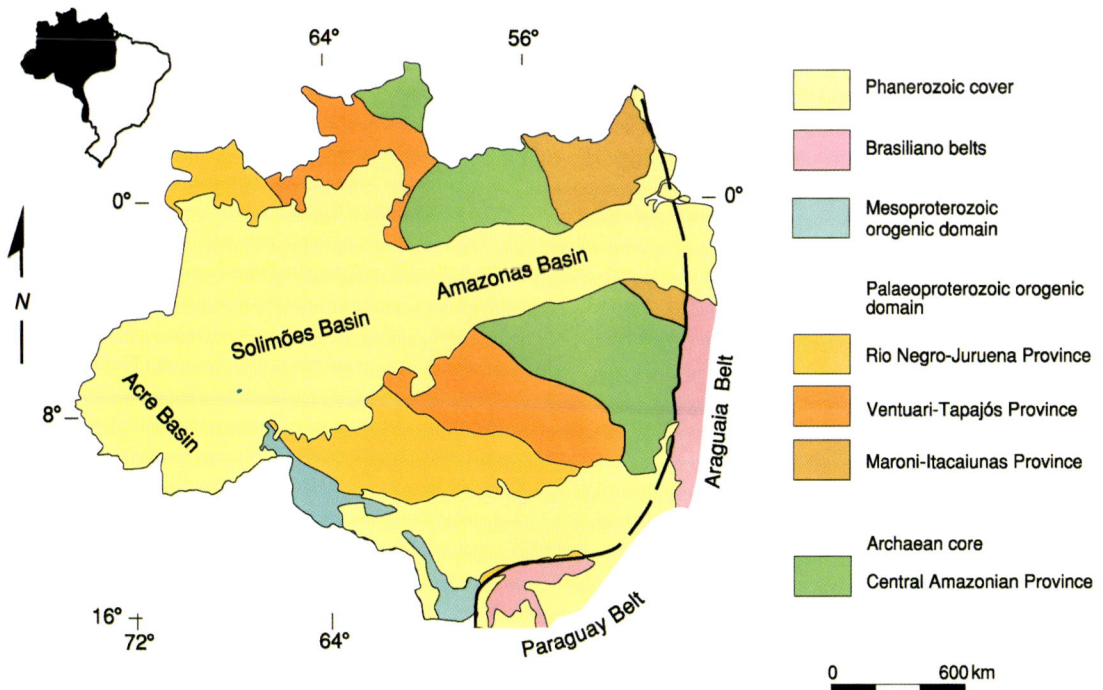

**Figure 6** Schematic map of the Amazon Craton, illustrating the main basement components and cover units. Modified from Tassinari CG, Bettencourt S, Geraldes M, Macambira JB, and Lafon JM (2000) The Amazonian Craton. In: Cordani UG, Milani EJ, Thomaz FA, and Campos DA (eds.) *Tectonic Evolution of South America*, pp. 41–95. 31st International Geological Congress, Rio de Janeiro.

of the region is limited. The cratonic basement is composed of an Archaean core, a large Palaeoproterozoic orogenic domain, and a Mesoproterozoic orogenic zone to the south-west.

The Archaean core, or Central Amazonian geochronological province, is the portion of the craton not affected by the Palaeoproterozoic orogenies. It consists of granite–greenstone and high-grade terranes, overlain by Archaean volcanosedimentary sequences and very thick packages of undeformed Palaeoproterozoic to Mesoproterozoic sedimentary and volcanic units. The Carajás mining district, located in the north-western portion of the province, contains the largest high-grade iron-ore reserves in Brazil as well as important copper, manganese, gold, and nickel deposits.

According to recent estimates, approximately 70% of the Amazon Craton consists of Palaeoproterozoic juvenile crust, which forms the basement of three geochronological provinces: Maroni-Itacaiunas, Ventuari-Tapajós, and Rio Negro-Juruena (**Figure 6**). The Maroni-Itacaiunas Province encompasses a segment of a Palaeoproterozoic accretionary Orogen, which is the main manifestation of the 2200–1950 Ma Transamazonian event. The Ventuari-Tapajós and Rio Negro-Juruena provinces are also collages of juvenile terranes generated in the intervals 1950–1800 Ma and 1800–1550 Ma, respectively.

The south-western part of the Amazon Craton, along the border with Bolivia, comprises parts of a Mesoproterozoic Orogen developed between 1500 Ma and 1300 Ma. It involves a Palaeoproterozoic basement, a juvenile terrane, and late to post-tectonic granites.

### São Luis and Rio de la Plata Cratons

The existence of the São Luis and Rio de la Plata cratons has been inferred from structural features of the various Brasiliano belts together with geochronological and geophysical data.

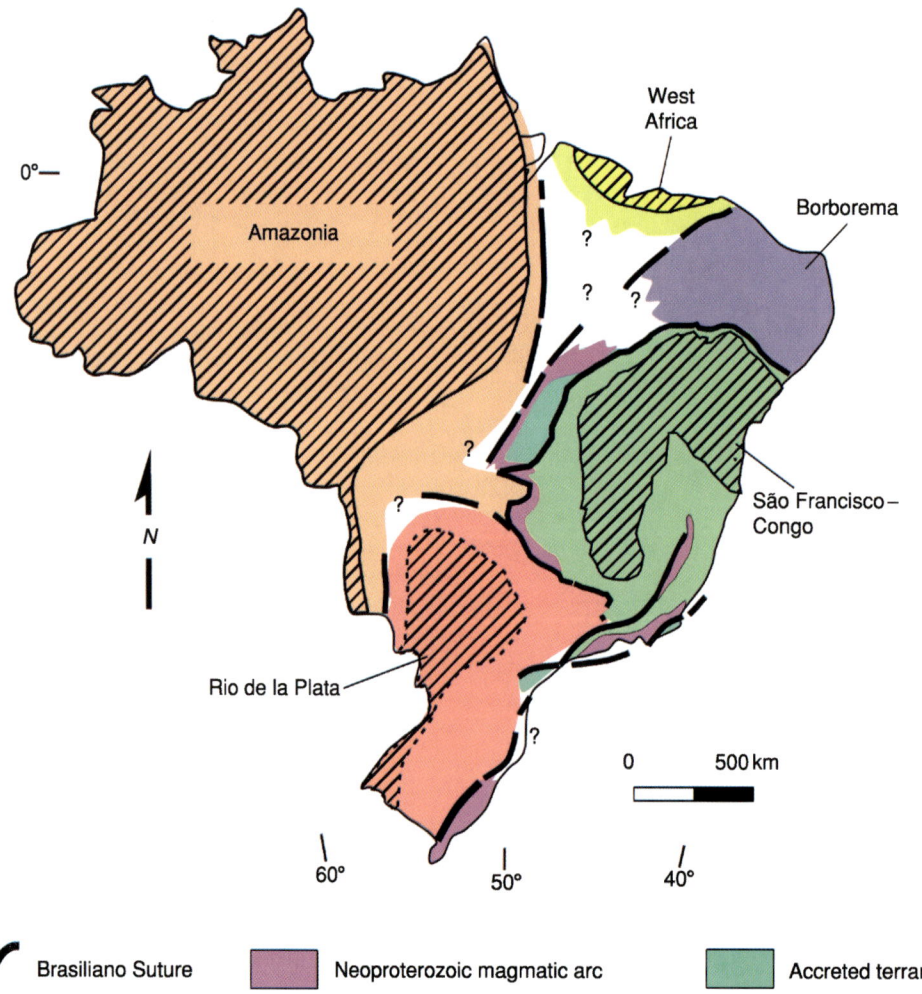

**Figure 7** Neoproterozoic suture zones of Brazil, and the plates that collided to form the South American portion of western Gondwana. Note that the internal parts of the plates correspond to the cratons.

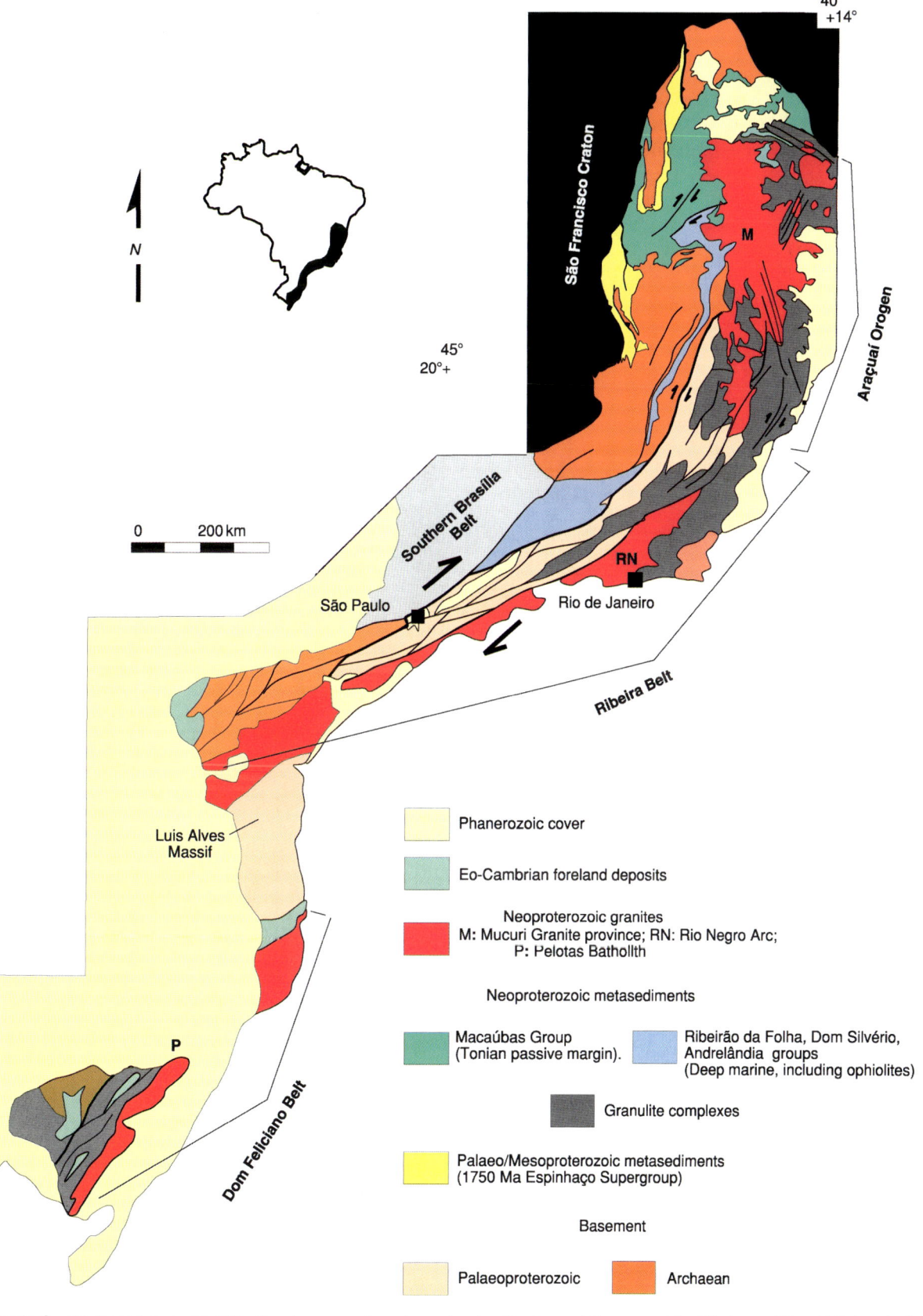

**Figure 8** Geological map of the Mantiqueira orogenic system, showing the Araçuaí Orogen and the Ribeira and Dom Feliciano belts.

The São Luis Craton on the northern coast has only small exposures of a granite–greenstone-belt association, which yield potassium–argon cooling ages of around 2100 Ma. Potassium–argon dates obtained from the exposures of adjacent areas fall in the interval between 660 Ma and 500 Ma, indicating that they were formed during the Brasiliano event. This led to the interpretation that a small piece of the West African Craton, represented by the São Luis Block, was left in South America during the opening of the Atlantic (**Figures 2** and **3**).

The Rio de la Plata Craton underlies the Palaeozoic Paraná Basin in southern Brazil (**Figure 2**). The presence of a cratonic block that underlies the Paraná Basin, originally inferred from the tectonic polarity of the Neoproterozoic orogenic belts exposed in the region, was confirmed by gravity and teleseismic travel-time studies. Furthermore, exposures of Palaeoproterozoic crust unaffected by the Brasiliano event in Uruguay and Argentina are now viewed as extensions of the cratonic block that underlies the Paraná Basin.

## Neoproterozoic Orogenic Domains

The Neoproterozoic orogenic domains form a complex network of interacting orogens and were generated by the collisions that led to the assembly of western Gondwana between 640 Ma and 520 Ma.

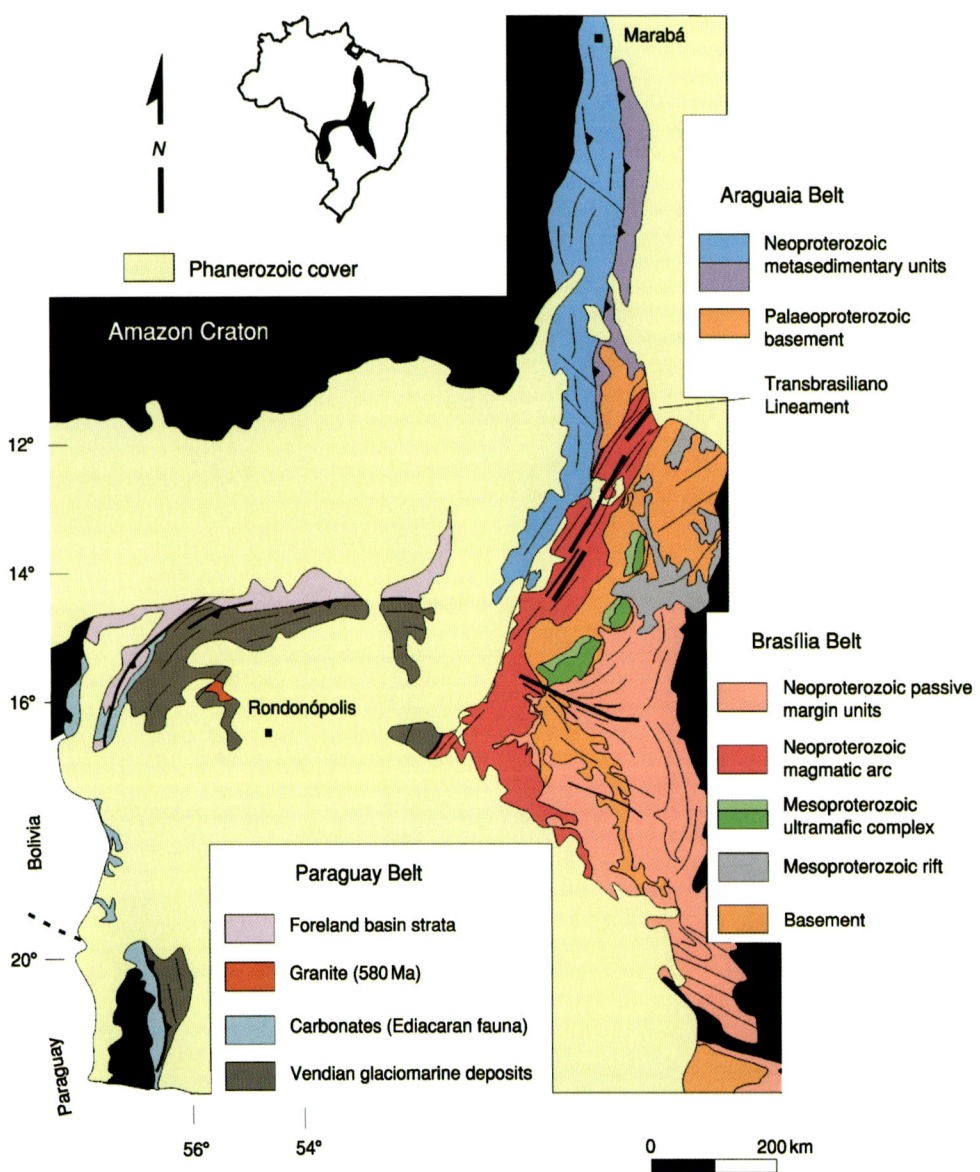

**Figure 9** Geological map of the Tocantins orogenic system, showing the Araguaia, Paraguay, and Brasília belts.

Accretion of juvenile crust played a minor role in the Brasiliano orogenies; recycling of the pre-existing continental lithosphere was the dominant process. Besides wide zones of deformation and metamorphism, the Brasiliano orogenic domains contain a relatively large volume of syn- and post-collisional granites. Furthermore, a substantial part of the orogenic zone is either transpressional in origin or has at least been modified by strike-slip movement in the late stages of development.

Neoproterozoic suture zones have been directly mapped or inferred with the help of geophysics in almost all the orogenic domains, so that the plates involved in the creation of the Brazilian portion of western Gondwana can be delineated as shown in **Figure 7**.

### Mantiqueira Orogenic System

The Mantiqueira orogenic system extends over a zone 2500 km long and 200 km wide along the south-east coast of Brazil, and comprises the Araçuaí Orogen and the Ribeira and Dom Feliciano belts (**Figure 8**); the West Congolian, Kaoko, and Gariep belts are their African counterparts. The Mantiqueira system overprints a branch of the Tocantins system in the area to the south of the São Francisco Craton and continues further south, reaching the Uruguayan shield beyond the Brazilian border.

**Araçuaí Orogen** The Araçuaí–West Congo Orogen (**Figure 8**) encompasses the terrane between the São Francisco Craton and the east coast of Brazil, as well as the West Congolian Belt of Africa. This orogen is the northern termination of the Mantiqueira system and its African counterpart, and it displays the unusual shape of a tongue, being surrounded on all sides, except the south, by cratonic areas. The Brazilian or Araçuaí portion of the orogen is made up of the Araçuaí fold–thrust belt to the west and a wide zone of high-grade and granitic rocks to the east, which represents its crystalline core. This internal zone is also known as the northern Ribeira, Atlantic, or Coastal Mobile Belt.

The Araçuaí fold–thrust belt verges towards the adjacent craton and involves a basement older than

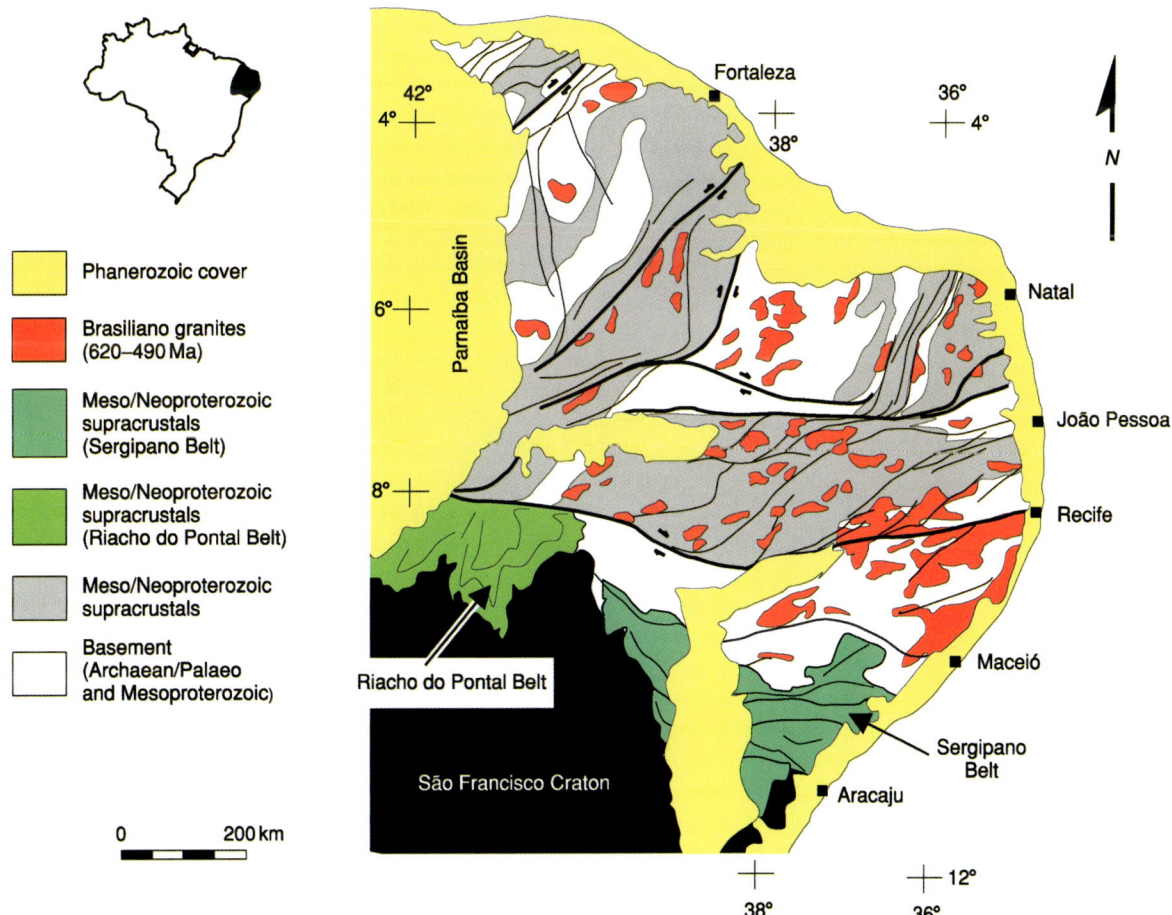

**Figure 10** Geological map of the Borborema Strike-slip System.

1800 Ma, a 1750 Ma ensialic rift assemblage, Neoproterozoic passive margin to foreland strata, and other correlative units. The crystalline core, dominated by an array of north-north-east–south-south-west-striking dextral transpressional zones, comprises a Palaeoproterozoic basement, amphibolite or granulite facies Neoproterozoic sedimentary units, and a batholithic body of Brasiliano granites (625–530 Ma).

The Araçuaí–West Congo Orogen is currently viewed as the product of the closure of a Red Sea type ocean, whose development started with rift formation in the Tonian between 900 Ma and 800 Ma.

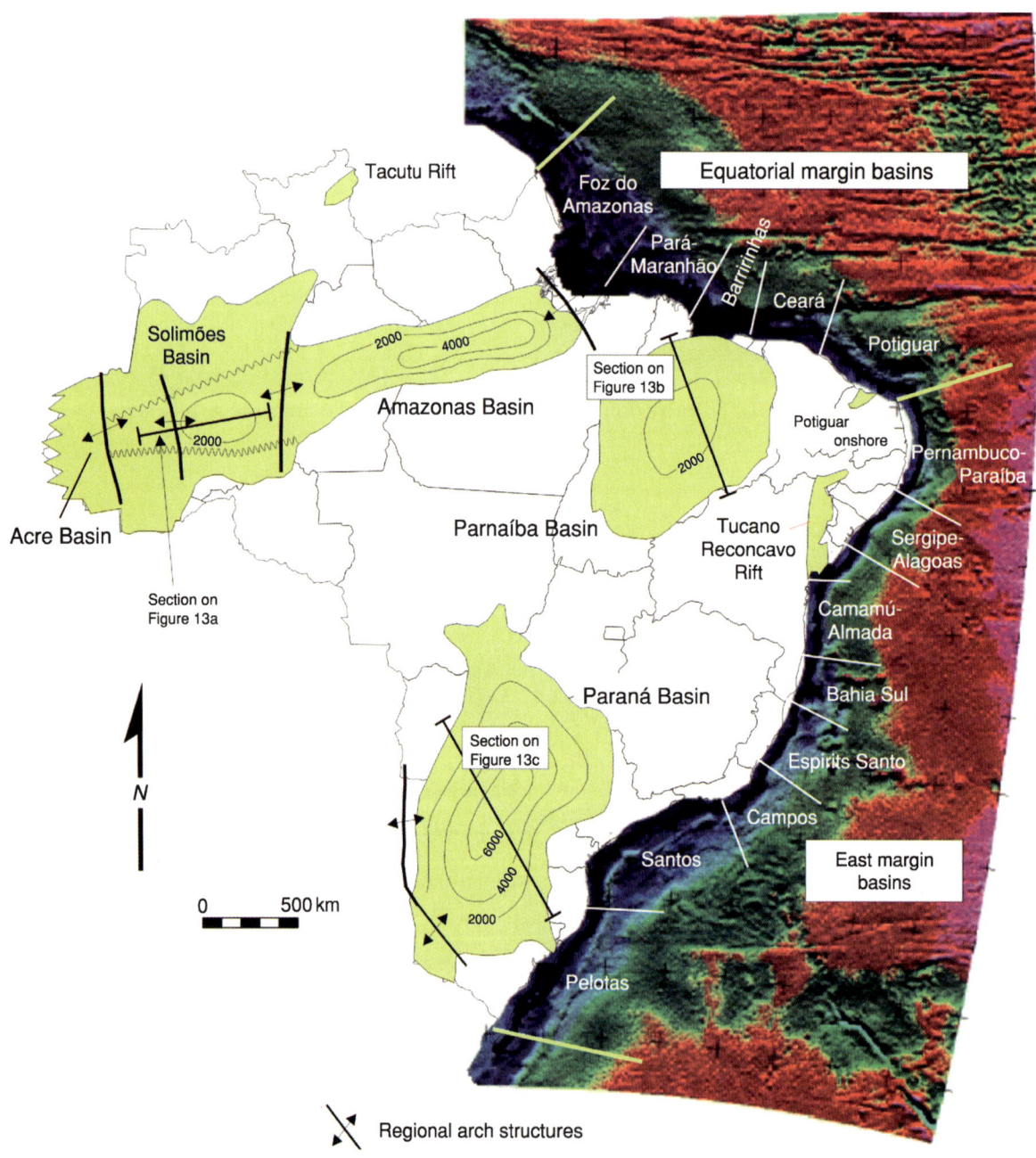

**Figure 11** Locations of the most important Brazilian Phanerozoic sedimentary basins. Maps and isopachytes (in metres) of Palaeozoic sags, modified from Milani E and Thomaz-Filho A (2000) Sedimentary basins of South America. In: Cordani UG, Milani EJ, Thomaz FA, Campos DA (eds.) *Tectonic Evolution of South America*, pp. 389–449. 31st International Geological Congress, Rio de Janeiro. Offshore Bouguer anomaly map from Mohriak (2003) Rift architecture and salt tectonics in South Atlantic sedimentary basins. In: Mohriak WU *Rifted Sedimentary Basins of South Atlantic: Turbidite Reservoirs, Sedimentary and Tectonic Processes*. 41st Brazilian Geological Congress and 1st International Conference of the Geophysical Society of Angola, Short Course Notes, CD-ROM, Socïedade Brasiliera de Geologìa, João Pessoa, Brazil.

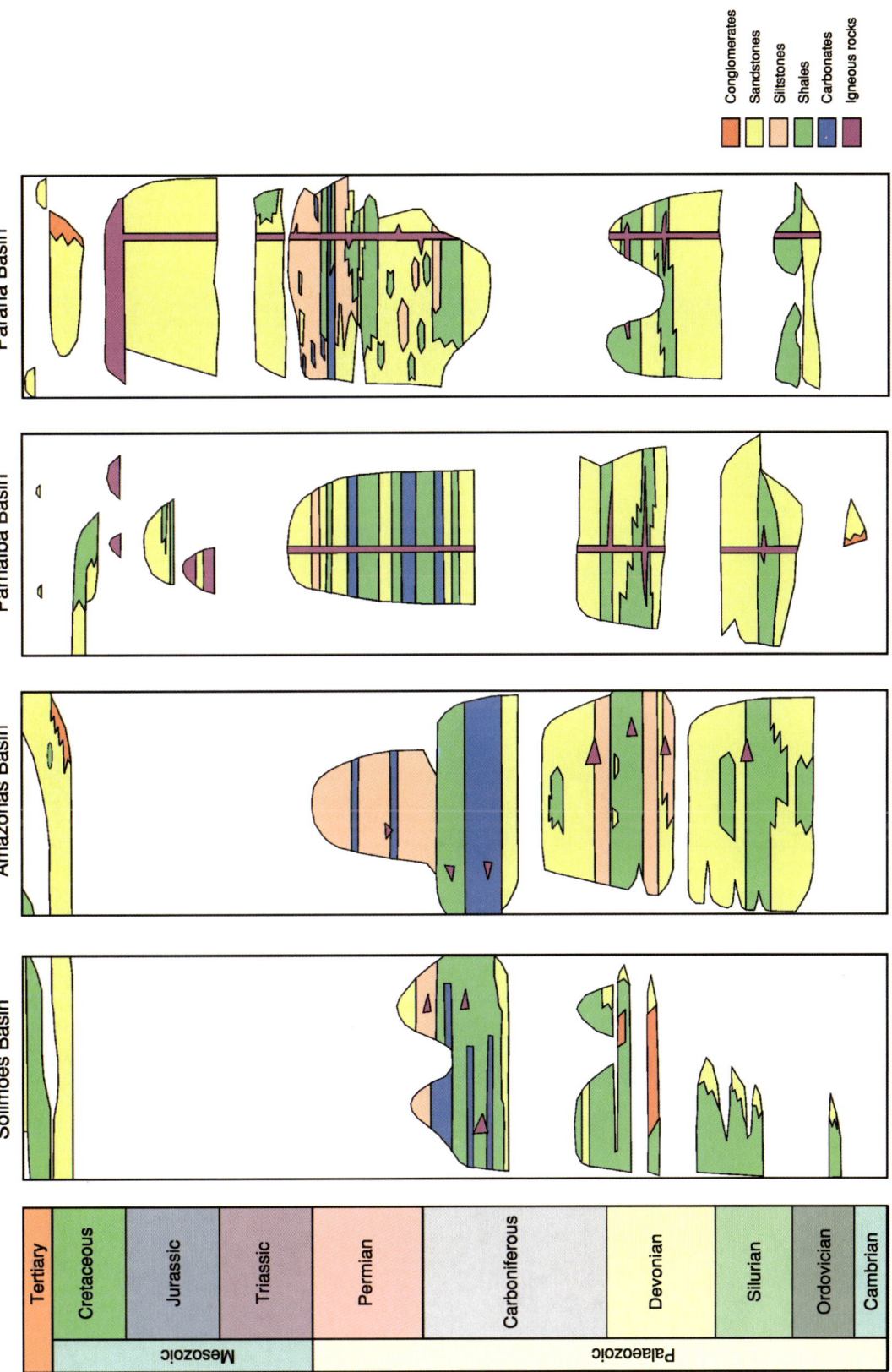

**Figure 12** Stratigraphical chart of Brazilian Palaeozoic sag basins. Modified from the Brazilian Petroleum Agency (Agência Nacional do Petróleo, ANP) home page at www.anp.gov.br

**Ribeira and Dom Feliciano belts** Together, the north-east–south-west orientated Ribeira and Dom Feliciano belts occupy a strip approximately 200 km wide and 1600 km long along the south-east coast of Brazil. The northern boundary with the Araçuaí Orogen is artificially taken to be where the fabric elements bend slightly to become north-north-east–south-south-west. The Ribeira Belt truncates the easternmost portion of the Tocantins orogenic system, represented by the southern Brasília Belt, and is overlain by the Paraná-Basin strata to the south, where the Luis Alves basement block separates it from the Dom Feliciano Belt.

A Palaeoproterozoic basement with some Archaean inliers, Mesoproterozoic to Neoproterozoic metasedimentary units, Eocambrian foreland deposits, and

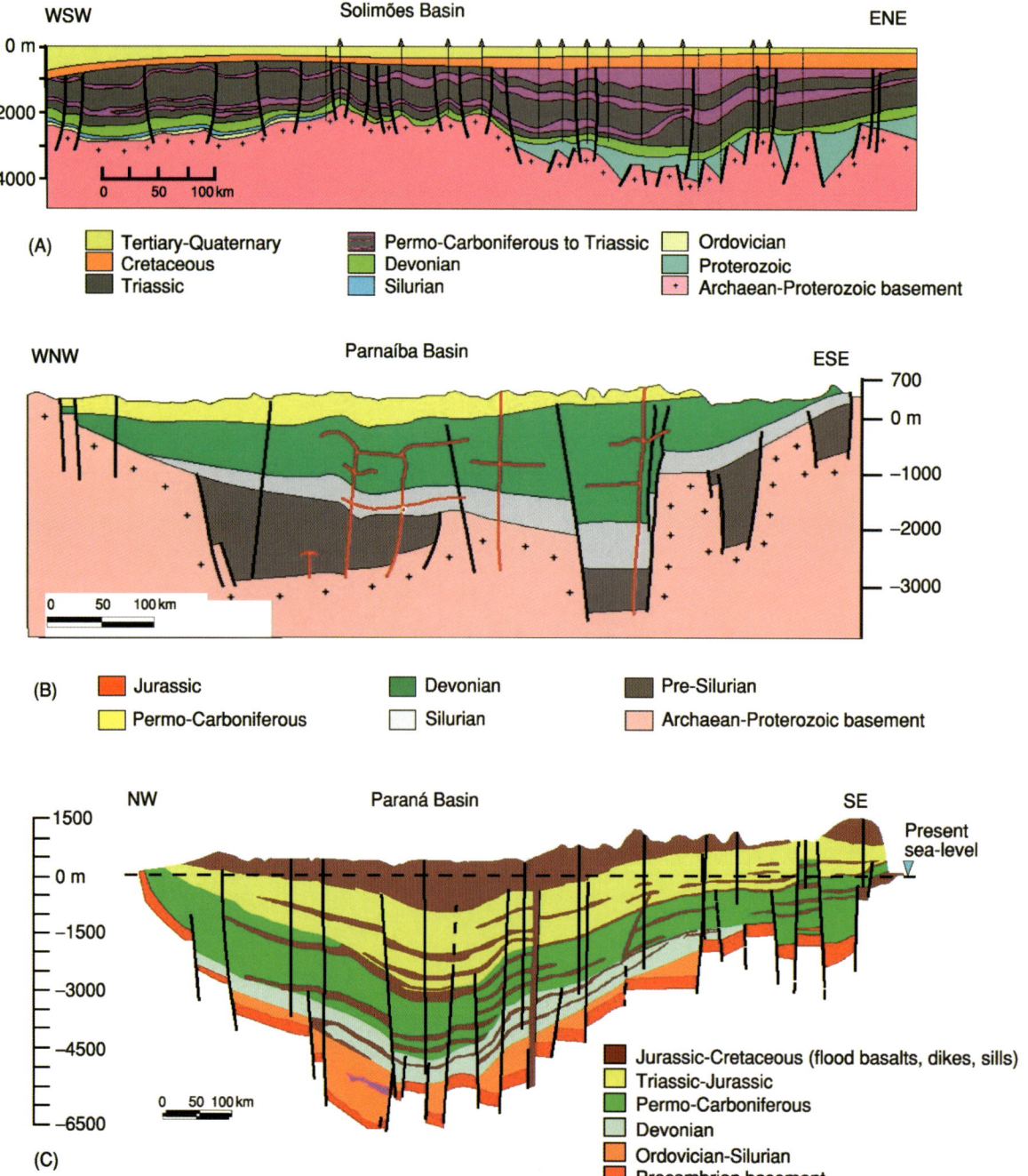

**Figure 13** Schematic geological sections of (A) Solimões, (B) Parnaíba, and (C) Paraná sag basins. Modified from the Brazilian Petroleum Agency (Agência Nacional do Petróleo, ANP) home page at www.anp.gov.br

voluminous granites emplaced in the pre-collisional to post-collisional stages are the main components of the Ribeira and Dom Feliciano belts. Various terranes and three magmatic arcs have been characterized along these belts.

The overall architecture of both the Ribeira and Dom Feliciano belts is a system of north-west-verging thrusts and nappes, which pass to the south-east and south into a transpressional zone. In the Ribeira Belt the internal transpressional zone is right lateral, whereas left lateral displacements dominate the Dom Feliciano Belt. This tectonic picture can be interpreted in two basic conflicting ways. It is portrayed either as a result of a late transpressional modification of a pre-existing frontal collisional orogen or as a product of an oblique plate convergence that led to the development of a transpressional orogen. Regardless of which hypothesis is correct, the evolution of the Ribeira and Dom Feliciano belts is associated with the interaction of three continental masses (the São Francisco–Congo, Rio de la Plata, and Kalahari cratons), which were separated by an ocean that contained magmatic arcs and microcontinents. The whole convergence stage lasted from 600 Ma to 520 Ma.

### Tocantins Orogenic System

The Tocantins system comprises the orogens developed between the Amazon, Rio de la Plata, and São Francisco cratons, and occupies a large area in northern and central Brazil. The Tocantins System, together with the Borborema strike-slip province of north-eastern Brazil (see next section), resembles the present-day Himalayan system in architecture. The Araguaia, Paraguay, and Brasília belts are the exposed components of the Tocantins System (**Figure 9**).

**Araguaia Belt** The north–south trending Araguaia Belt on the eastern margin of the Amazon Craton comprises a west-verging basement-involved fold–thrust belt, which is 1200 km long and 150 km

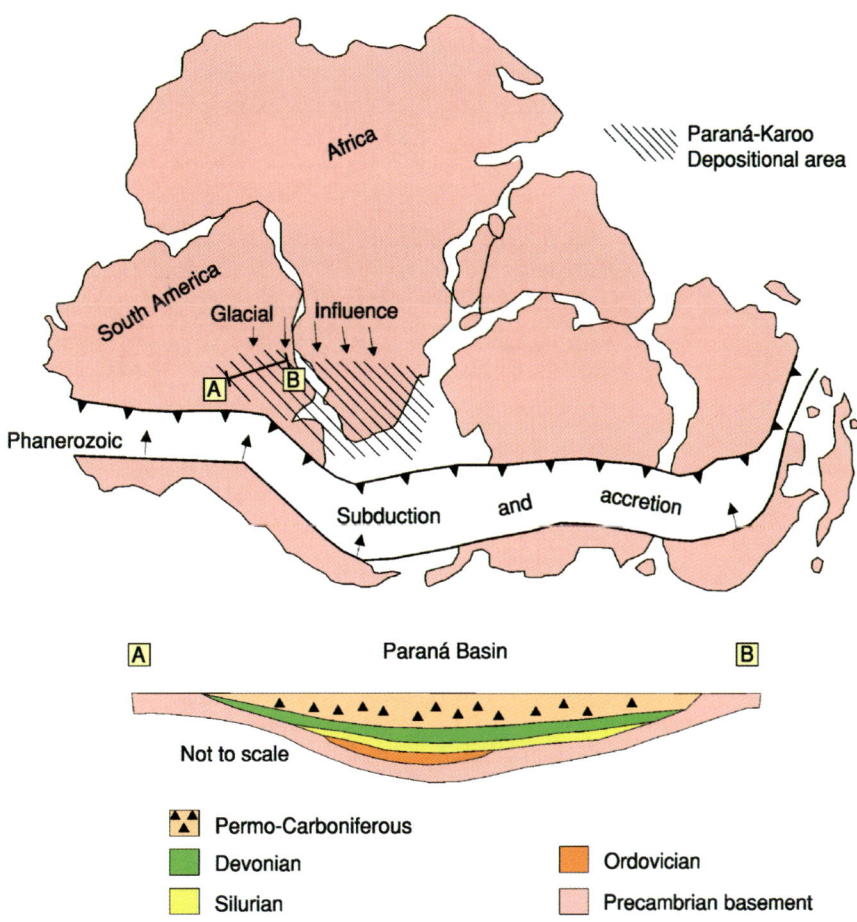

**Figure 14** Palaeogeographical reconstruction of western Gondwana in the Late Palaeozoic, showing the location of the Paraná Basin in the foreland domain of accretionary processes on the Gondwanan margin. Modified from Milani EJ (1992) Intraplate tectonics and the evolution of the Paraná Basin, SE Brazil. In: De Wit MJ, Ransome IGD (eds.) *Inversion tectonics of the Cape Fold Belt, Karoo and Cretaceous Basins of Southern Africa*. Rotterdam, Balkema, pp. 101–108.

wide. To the south it is separated from the Brasília Belt by a basement block, which is affected by the north-east–south-west-trending Transbrasiliano Fault Zone; to the east and to the north it is covered by the sediments of the Palaeozoic Parnaíba Basin (**Figure 9**). The Archaean and Palaeoproterozoic basement is overlain by Neoproterozoic metasediments, which are the best exposed in the belt. A 720 Ma ophiolite occurs within thrust sheets in the central portion of the belt.

**Paraguay Belt** The Paraguay Belt, the youngest member of the Tocantins orogenic system, is a pronounced salient on the south-eastern border of the Amazon Craton (**Figure 9**). Vendian glaciomarine deposits overlain by carbonates containing an Ediacaran fauna, together with continental foreland deposits, are the main units exposed in the belt. Its overall architecture is characterized by a system of open to tight upright folds in the culmination zone of the large salient, which grade into two cratonward-verging systems of folds and thrusts at its north-eastern and southern ends. The deformation and metamorphism in the Paraguay Belt is estimated to have occurred between 550 Ma and 500 Ma and probably represents the last steps towards the final assembly of western Gondwana.

**Brasília Belt** The Brasília Belt is an east-verging fold–thrust belt (1200 km long and up to 300 km wide) that fringes the São Francisco Craton to the west and south-west (**Figure 9**). The basement is an extension of the Archaean–Palaeoproterozoic substratum of the adjacent São Francisco Craton. In the internal zone, the Palaeoproterozoic basement received two Neoproterozoic accretions: an Archaean granite–greenstone terrane and a large magmatic arc (**Figure 9**). A suture zone juxtaposes the Rio de la Plata and São Francisco plates in the southern portion of the belt.

A 1770 Ma ensialic rift sequence, a thick Neoproterozoic passive-margin package, and deep-marine and probably ocean-floor assemblages are the main cover units. Mesoproterozoic layered mafic–ultramafic complexes and foreland-basin sediments are also involved.

From a structural standpoint the Brasília Belt consists of two distinct compartments that join along the pronounced east–west-trending Pireneus syntaxis: an east–south-east-verging basement-involved fold–thrust belt to the north, and a system of spoon-shaped east–south-east-verging nappes to the south. The Transbrasiliano Lineament, a north-east–south-west-orientated dextral strike-slip fault zone, overprints the internal portions of both the Brasília and

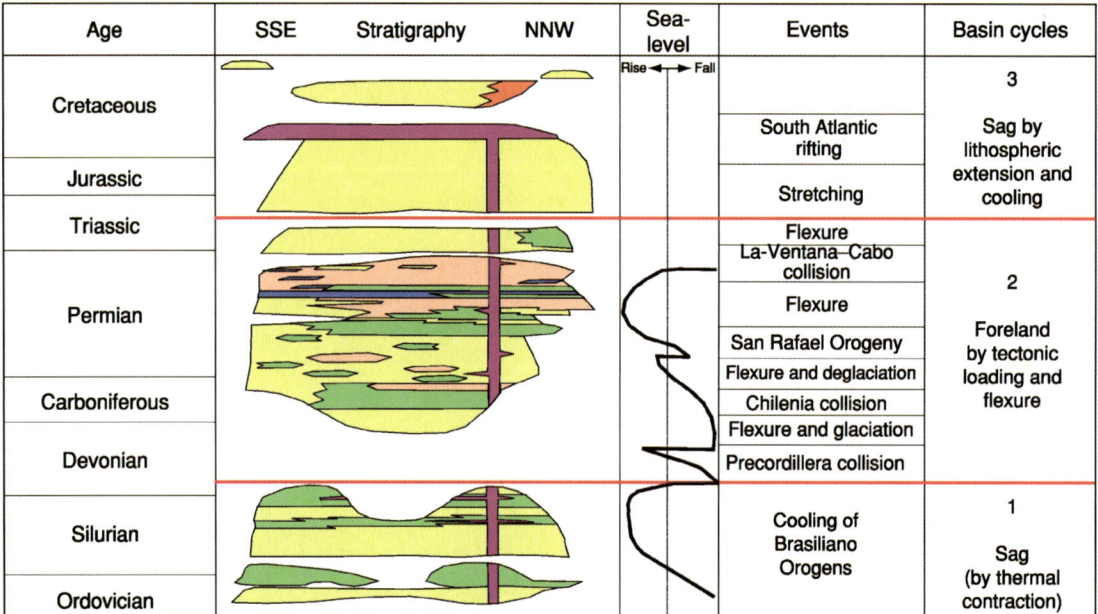

**Figure 15** Stratigraphical chart of the Paraná Basin showing the evolutionary stages and their genetic controls. Note the correspondence between depositional stages/unconformities and orogenic events in the Andean chain during the foreland stage. Based on: (1) Zalan PV, Wolff S, Astolfi MA, et al. (1990) The Paraná Basin, Brazil. In: Leighton MW, Kolata DR, Oltz DF, and Eidel JJ (eds.) *Interior Cratonic Basins*, pp. 681–708. AAPG Memoir 51. Tulsa: American Association of Petroleum Geologists; (2) Milani EJ (1992) Intraplate tectonics and the evolution of the Paraná Basin, SE Brazil. In: De Wit MJ, Ransome IGD (eds.) *Inversion tectonics of the Cape Fold Belt, Karoo and Cretaceous Basins of Southern Africa*. Rotterdam, Balkema, pp. 101–108.

Araguaia belts and seems to continue further north beneath the Phanerozoic cover, merging with the Borborema Strike-slip System (see next section).

The evolution of the Tocantins System probably started with an interaction between São Francisco–Congo and the Rio de la Plata at around 790 Ma. Arc magmatism took place between 900 Ma and 630 Ma, reflecting subduction and progressive closure of a large ocean separating São Francisco–Congo from Amazonia. After accretion of the arcs

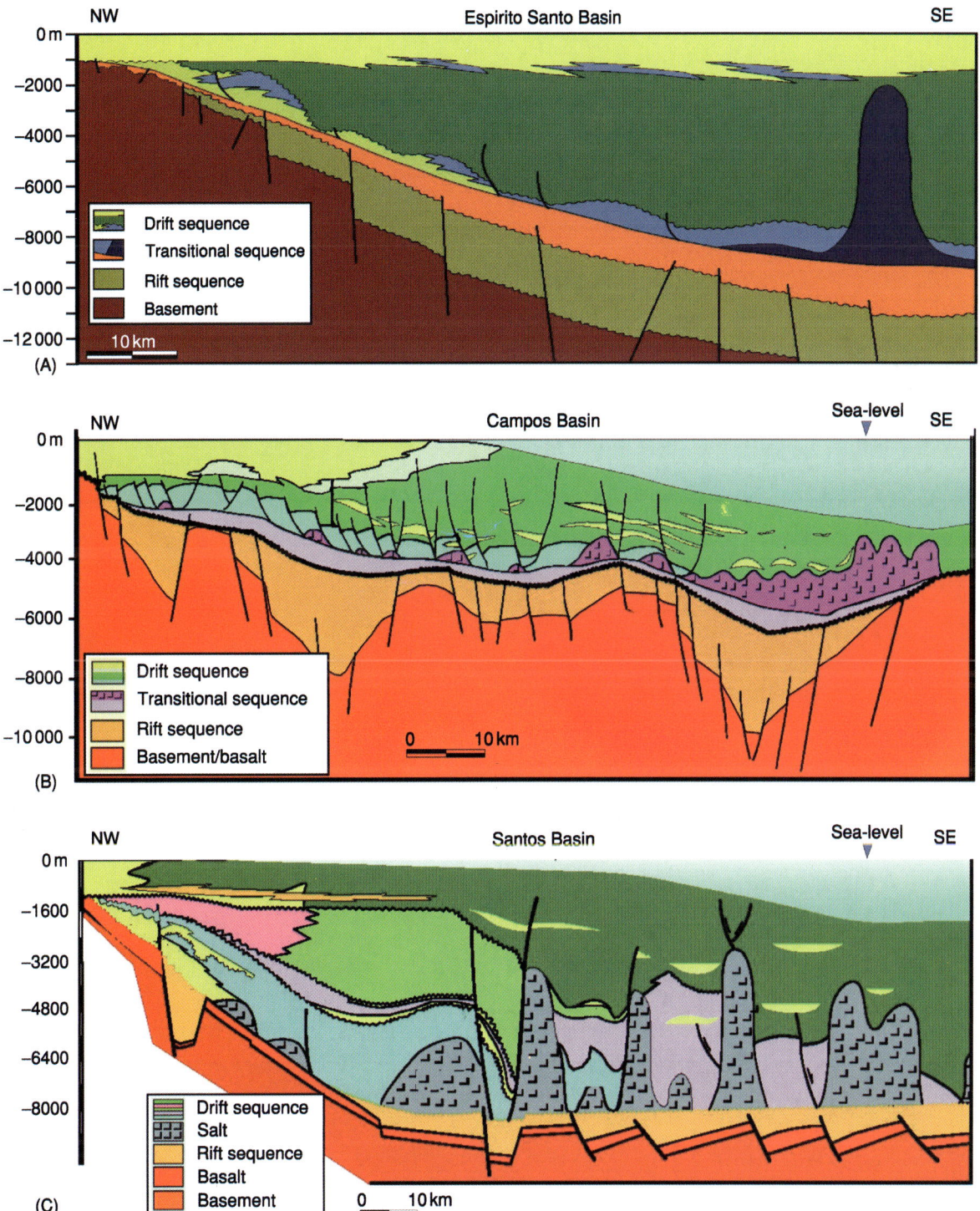

**Figure 16** Schematic geological sections of Brazilian eastern margin (A) Espirito Santo, (B) Campos, and (C) Santos basins. Modified from the Brazilian Petroleum Agency (Agência National do Petróleo, ANP) home page at www.anp.gov.br

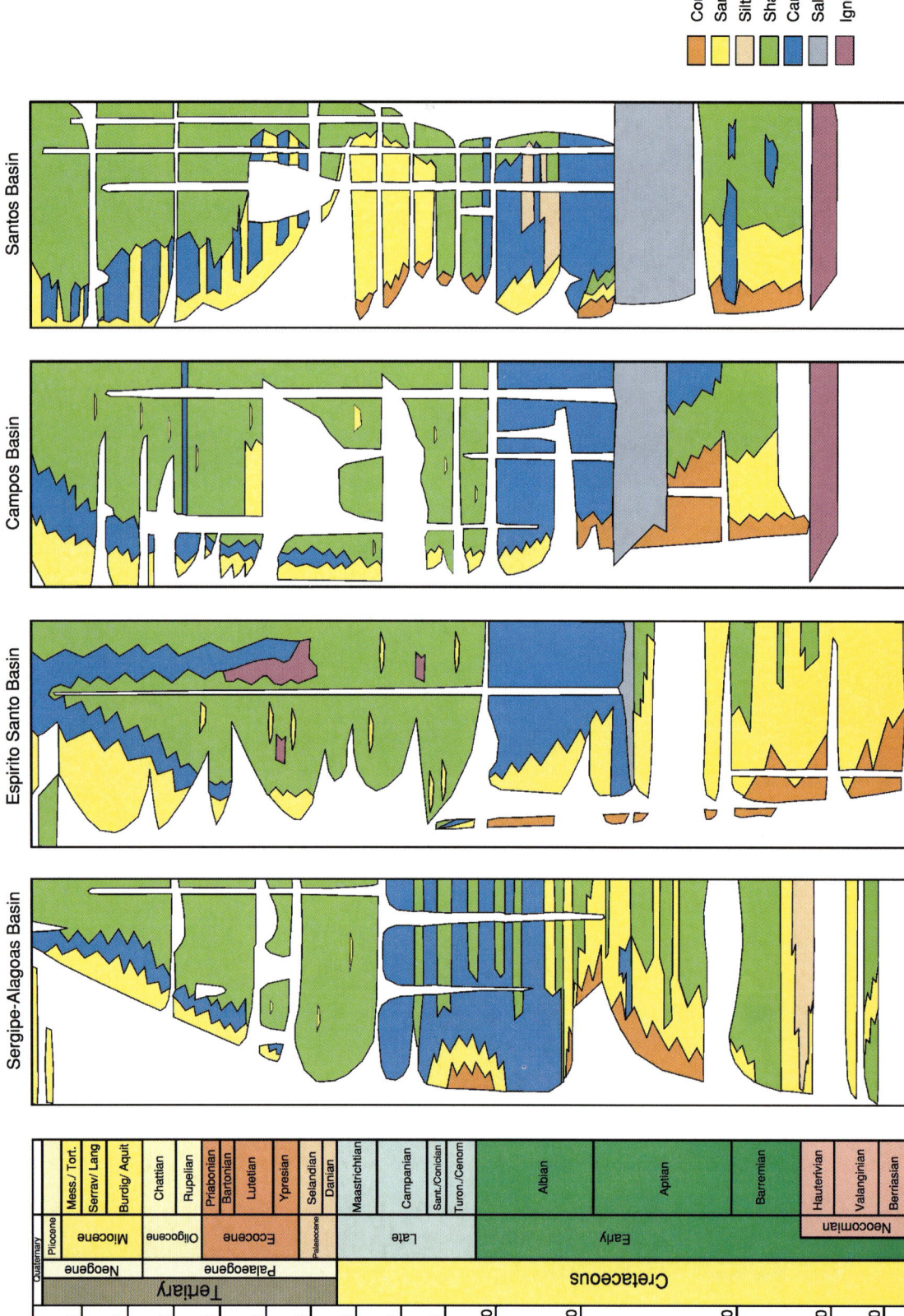

**Figure 17** Stratigraphical charts of the Sergipe-Alagoas, Espirito Santo, Campos, and Santos Basins of the Brazilian eastern margin.

and the microcontinent, the convergence of São Francisco–Congo, Amazonia, and West Africa between 630 Ma and 590 Ma led to the development of the central and northern segments of the Tocantins System. The final collision between Rio de la Plata and Amazonia led to the development of the Paraguay Belt in the Early Cambrian.

### Borborema Strike-Slip System and Associated Features

The Borborema Strike-slip System in north-eastern Brazil (**Figure 10**) comprises a gigantic Brasiliano strike-slip deformation zone associated with two south-verging fold–thrust belts, which were once continuous with the Nigerian, Hoggar, and Central African provinces. A fan-like array of dextral shear zones outlines elongated basement blocks, which correspond to the various massifs delimited in the region. These blocks are composed of Archaean, Palaeoproterozoic, and Mesoproterozoic rocks, covered by Mesoproterozoic to Neoproterozoic supracrustals, and intruded by voluminous Neoproterozoic granites. The metasedimentary units form schist belts, which wrap around the basement massifs.

The west-north-west–east-south-east-trending Sergipano Belt in the south-eastern portion of the Borborema System is made up of a system of south-south-west-verging folds and thrusts, involving an Archaean basement and a package of Neoproterozoic passive-margin to foreland strata. The thrusts of the Sergipano Belt are rooted in a dextral transpressional shear zone, which marks a Brasiliano suture, juxtaposing a continental mass represented by the remaining parts of the system to the north with the São Francisco–Congo Plate to the south. In Africa, the Oubanguides or Central African Belt on the northern

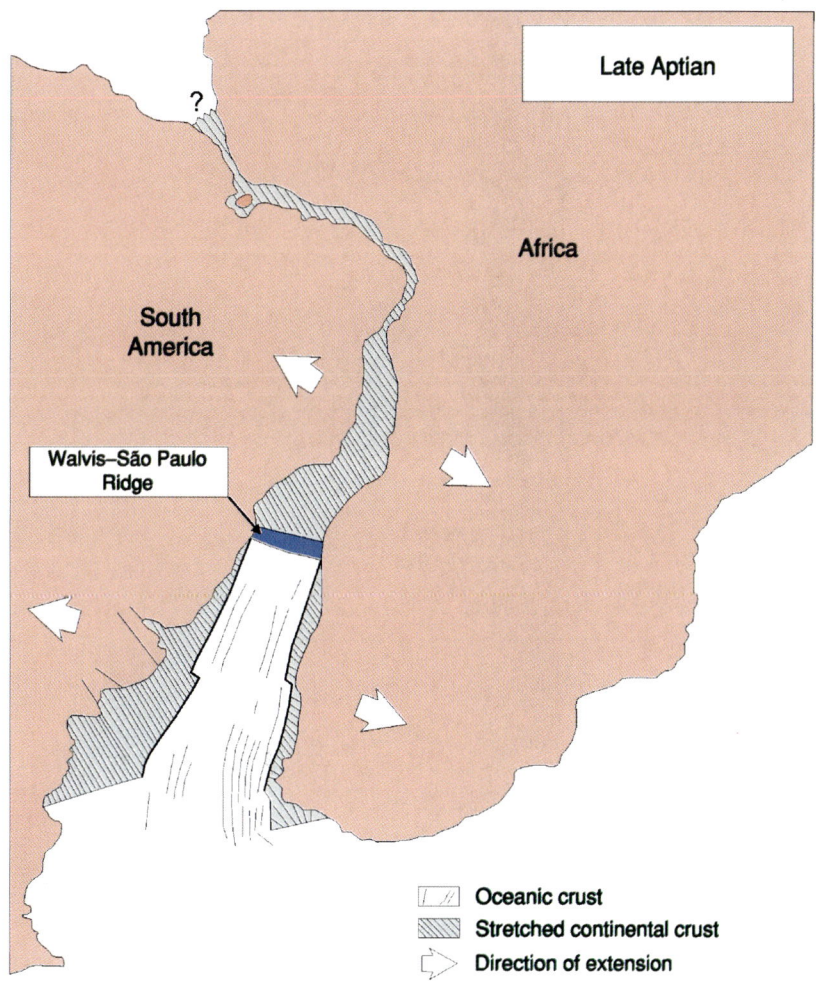

**Figure 18** Palaeogeographical reconstruction of the South Atlantic in the Late Aptian, showing the location of the Walvis–São Paulo Ridge, which acted as a barrier to free oceanic circulation, thereby creating conditions favourable to evaporitic (salt) deposition in the central segment of the eastern margin. Modified from Chang HK, Kowsmann RO, Figueiredo AMF, and Bender AA (1992) Tectonics and stratigraphy of the East Brazil rift system: an overview. *Tectonophysics* 213: 97–138, with permission from Elsevier.

border of the Congo Craton is an extension of the Sergipano Belt.

Separated from the Sergipano Belt by a basement high, the Riacho do Pontal Belt, comprises a pile of south-verging nappes involving Archaean basement and schists and quartzites of uncertain ages.

Peak metamorphism and the emplacement of syntectonic granites are dated by U-Pb and other methods at between 630 Ma and 580 Ma, whereas $^{39}$Ar-$^{40}$Ar dates obtained in some of the main shear zones indicate cooling ages of between 580 Ma and 500 Ma.

## Phanerozoic Sedimentary Basins

The Phanerozoic sedimentary record of Brazil (**Figure 11**) reflects three distinct plate-tectonic settings: the continental interior, where large Palaeozoic sag basins and Early Cretaceous intracontinental rifts are recognized; the passive margin, represented by the eastern border of South America; and the transform margin, represented by the Brazilian northern or equatorial margin. The continental margins developed from Mesozoic times onwards in response to the breakup of Gondwana and the spreading of the Atlantic Ocean.

### Palaeozoic Sag Basins

During the Ordovician, following the cooling of Brasiliano–Pan-African orogenic systems, sedimentation started in the newly assembled western Gondwana in large interior sag basins, namely the Solimões (600 000 km$^2$), Amazonas (500 000 km$^2$), Parnaíba (600 000 km$^2$), and Paraná (1 400 000 km$^2$) basins (**Figure 11**). The stratigraphical records of these long-lived basins are characterized by a layer-cake arrangement, reflecting mostly shallow-marine sandy–shaly deposition (**Figures 12 and 13**). The major stratigraphical sequences are bounded by inter-regional unconformities, which are coeval along the basins (**Figure 12**), suggesting global control mechanisms. The existence of precursor rifts in these

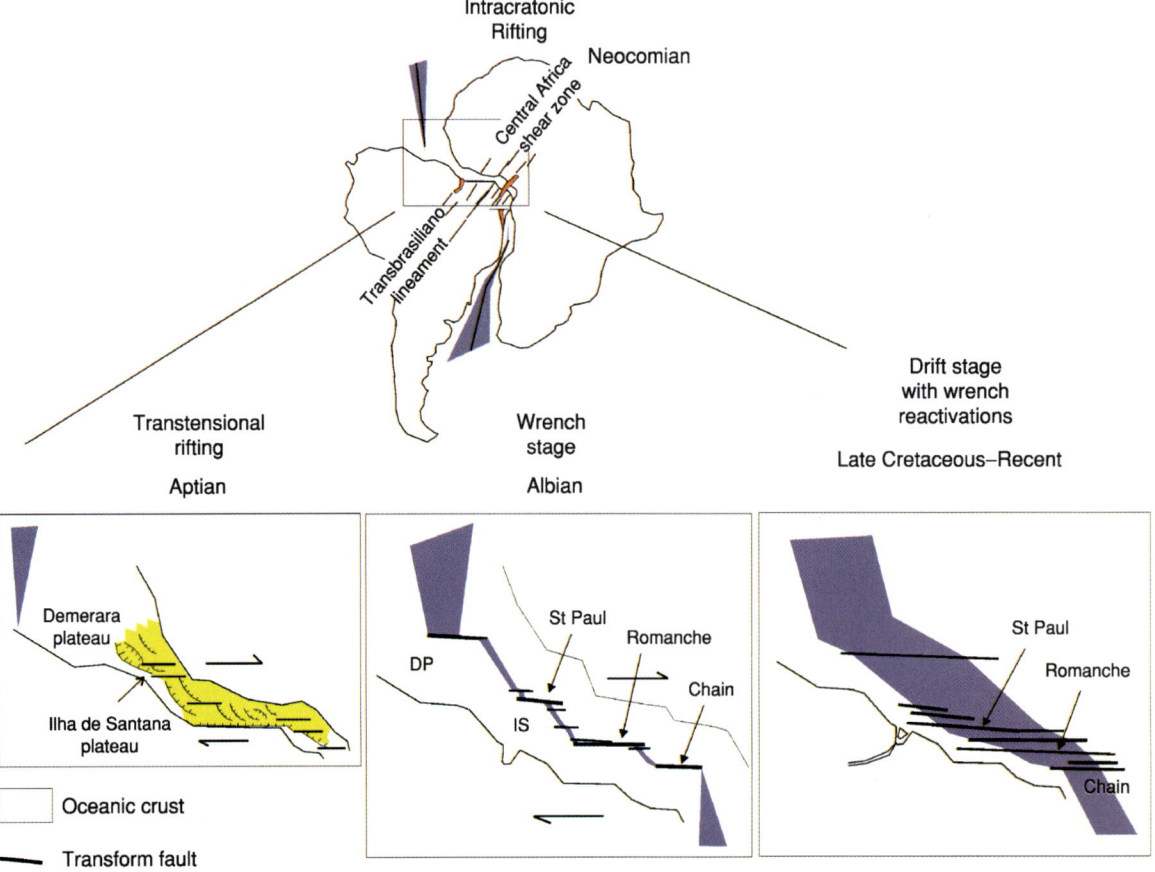

**Figure 19** Tectonic evolution of the Brazilian equatorial transform margin. Modified from Falkenhein K and Martins-Neto MA (eds.) (2003) *Equatorial Margin Project, Part 2: Sequence Stratigraphy, Turbidites, Tertiary Wrench, Petroleum Systems*. Multiclient project with Petrobras, Agip, Shell, Chevron Texaco, JNOC, ConocoPhillips, Encana, ElPaso, KerrMcGee, Statoil, Devon, Ouro Preto, Fundação Gorceix-NUPETRO. Final report (unpublished).

basins is still controversial. Some deeper-processed seismic lines and gravity–magnetometry maps suggest a steer-head geometry, with a rift basin beneath the main sag (**Figure 13**).

Some authors have considered the Paraná and Acre basins to be foreland basins, because of their possible relationships to accretionary processes along the margin of western Gondwana (**Figure 14**). The Acre Basin is the name for the eastern pinch-out of the Peruvian sub-Andean Marañon-Ucayali foreland wedge. The Paraná Basin comprises three basin cycles: an Ordovician–Silurian thermal sag; a Devonian–Triassic foreland basin; and a Jurassic–Cretaceous sag associated with the opening of the Atlantic Ocean (**Figure 15**). Permo-Carboniferous glacial deposits and an enormous volume of Early Cretaceous (137–128 Ma) flood basalts also characterize the Paraná Basin fill.

## Continental-Margin Basins and Associated Interior Rifts

The breakup of western Gondwana in the Early Cretaceous created the continental margins of South America and Africa. During the Neocomian, Barremian, and Aptian, rifting propagated from south to north, generating the eastern Brazilian margin basins. The north-west–south-east-orientated Brazilian equatorial margin opened later, between the Aptian and the Albian, as a transform margin. In the continental interior, Precambrian lineaments were reactivated to form interior rift basins.

**Eastern Brazilian margin basins** The eastern margin basins are classical passive-margin basins, comprising three sequences, reflecting precursor rift, transitional proto-oceanic, and thermal drift (**Figure 16**).

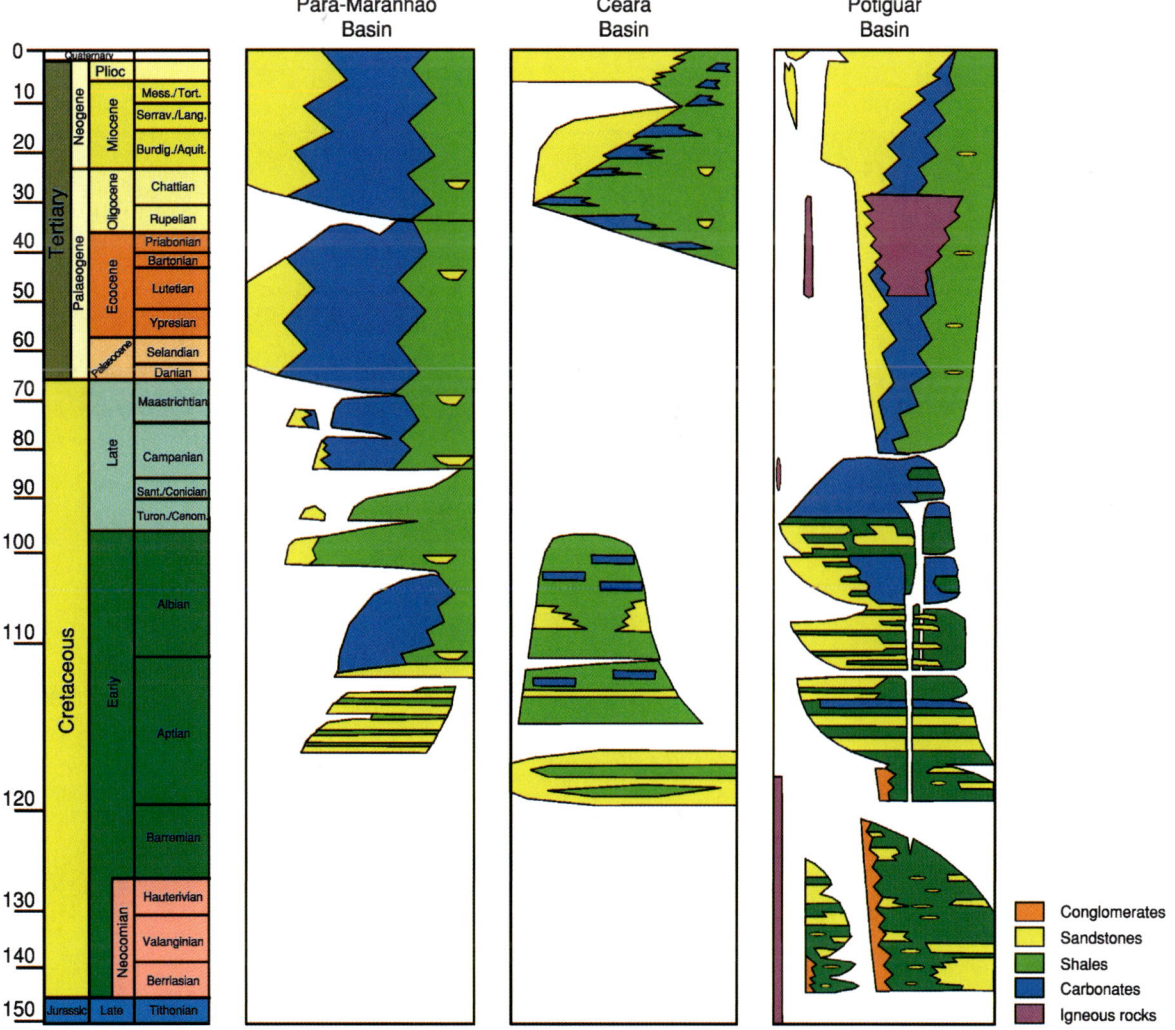

**Figure 20** Stratigraphical charts of the Pará-Maranhão, Ceará, and Potiguar Basins of the Brazilian equatorial margin. Modified from the Brazilian Petroleum Agency (Agência National do Petróleo, ANP) home page at www.anp.gov.br

Neocomian to Early Aptian rift-stage deposits consist of basalts and continental siliciclastics, deposited in depocentres controlled by normal faulting in alluvial, fluvial, deltaic, and lacustrine environments.

The Aptian transitional proto-oceanic supersequence, characterized by low subsidence rates, marks the incursion of marine waters into the basin system, and comprises coastal to shallow-marine carbonates and mudstones and thick evaporites (**Figures 16 and 17**). Salt was deposited in the central segment of the eastern margin (from the Santos to the Sergipe-Alagoas basins; **Figure 11**) in an evaporative basin created by a volcanic barrier, the Walvis–São Paulo Ridge (**Figure 18**).

The first-order drift package comprises an Albian–Cenomanian retrogradational transgressive supersequence, a Cenomanian–Turonian maximum flooding interval, and a Late Cretaceous–Holocene progradational regressive supersequence. Whereas the transgressive supersequence is mostly carbonate, the regressive supersequence comprises unconformity-bounded third-order depositional sequences of turbidite-prone lowstand systems tracts and siliciclastic or, subordinately, mixed-carbonate transgressive highstand systems tracts (**Figure 17**). Salt tectonics (deformational structures in response to gravity driving motion of salt layers) controlled the sediment bypass and the main depocentres, as well as the architecture of the drift package (**Figure 16**).

The Campos Basin in the southern segment of the eastern margin (**Figure 11**) contains approximately 85% of the recoverable hydrocarbon reserves in Brazil, forming together with the adjacent Santos and Espirito Santo basins the most prolific oil province of Brazil.

**Equatorial margin basins** The Brazilian equatorial margin evolved in four tectonosedimentary stages (**Figures 19, 20 and 21**): Neocomian interior rift; Aptian transtensional rift; Albian wrench; and Cenomanian–Holocene drift with wrench reactivations.

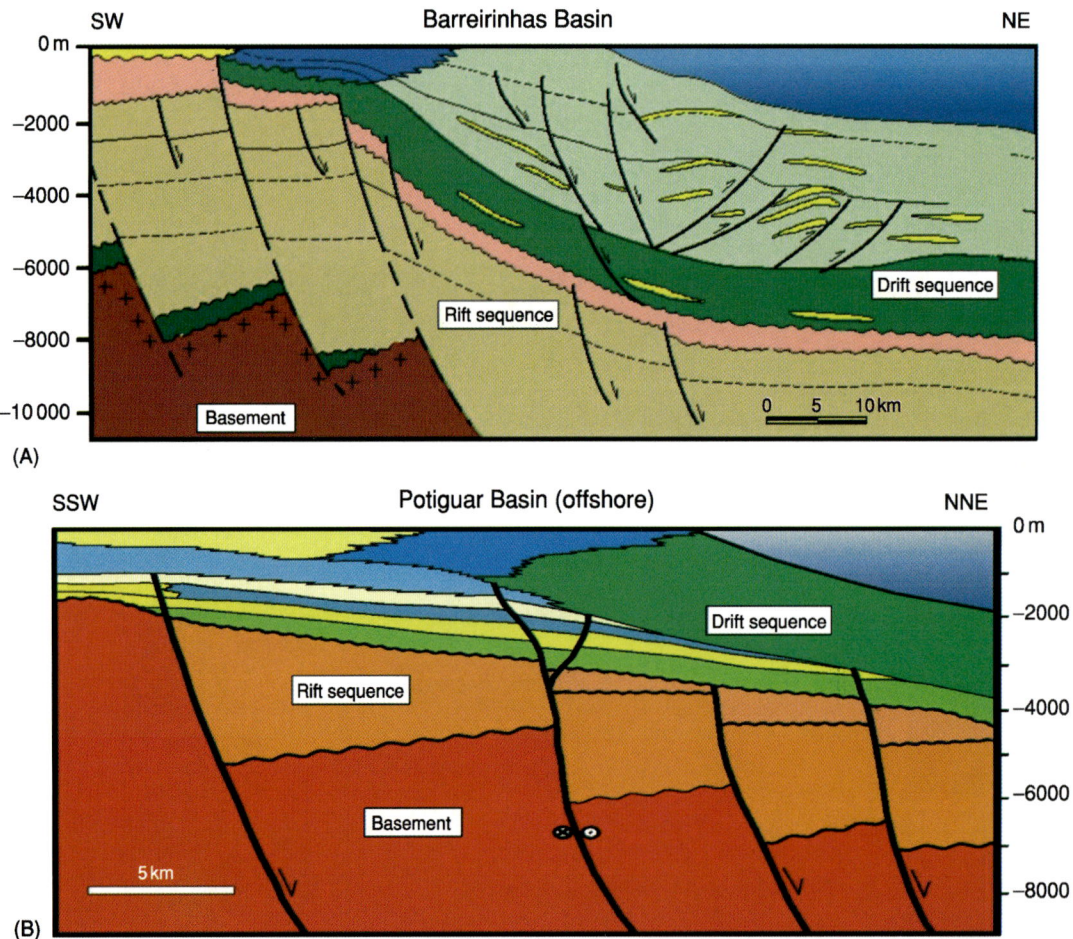

**Figure 21** Schematic geological sections through the Brazilian equatorial-margin (A) Barreirinhas and (B) Poriguar offshore basins. Modified from the Brazilian Petroleum Agency (Agência National do Petróleo, ANP) home page at www.anp.gov.br

During the Aptian, after the development of the Neocomian interior rifts, the connection between the Central and South Atlantic oceans was initiated by the nucleation of an east–west-trending megashear corridor, which controlled the evolution of the Brazilian and West African equatorial margins until the full breakup in the Early Cenomanian. This dextral transtensional corridor was made up of north-west–south-east-orientated releasing bends, alternating with east–west-orientated transfer segments (**Figure 19**). The releasing bends hosted a series of frontal rifts, whereas the transfer segments evolved as sets of en echelon oblique rifts and east–west-orientated transfer faults. During the Albian, with the establishment of oceanic spreading centres in the north-west–south-east-trending releasing bends, a series of east–west-trending transform faults nucleated, among them the St Paul, Romanche, and Chain transforms. The previously formed rifts were partially deformed in the vicinity of the fault zones. Carbonate sedimentation prevailed in the releasing bends (Pará-Maranhão, Barreirinhas, and Potiguar basins), whereas syn-wrench siliciclastic sediments locally accumulated along the east–west-trending transfer segments. During the Late Cretaceous and Tertiary, strike-slip displacements along the oceanic fracture zones induced wrench deformation in the distal portions of the basins. Tertiary wrenching also triggered gravitational gliding in slope areas (**Figure 21A**). Siliciclastic sedimentation prevailed until the Early Eocene. Carbonate and mixed systems dominated later, except in the Amazon Cone area (see **Figure 11**), where a sand–shale package several kilometres thick accumulated from the mid-Miocene.

**Early Cretaceous interior rifts**  Interior rifts formed during the Early Cretaceous in response to intraplate stresses related to the opening of the Central and South Atlantic oceans. The rift basins, the Potiguar Basin onshore, and the Tucano-Recôncavo-Jatobá and Tacutu rifts (**Figure 11**), display half-graben geometry and are filled with continental siliciclastic rocks (**Figure 22**).

## Tertiary Rifts and Related Features

An episode of rifting, whose driving mechanism is still unknown, affected the whole eastern margin of Brazil at the end of the Eocene and beginning of the Oligocene. Reactivation of the Early Cretaceous rift structures, basic to alkaline magmatism, and nucleation of a series of small rifts along the Serra do Mar on the south-eastern coast are the main manifestations of this event.

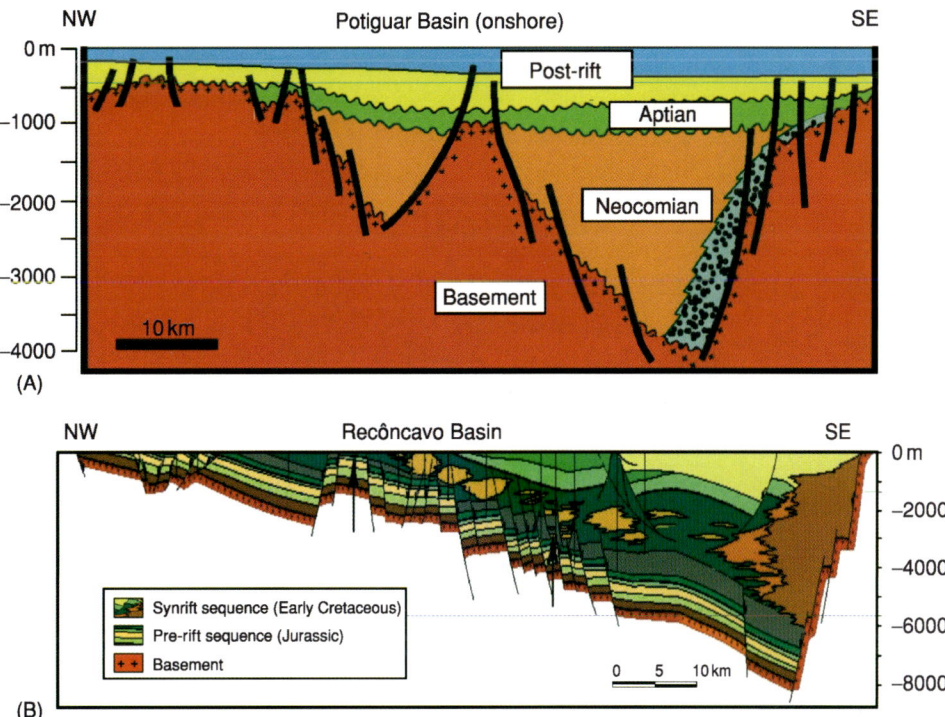

**Figure 22**  Schematic geological sections through the (A) Potiguar onshore and (B) Recôncavo interior rift basins. Modified from the Brazilian Petroleum Agency (Agência National do Petróleo, ANP) home page at www.anp.gov.br

## Glossary

**Craton** An ancient platform, i.e. a stable portion of a continent, not affected by Phanerozoic orogenies. Cratons also are differentiated pieces of continental lithosphere, characterized by a relatively thin crust coupled to a mantle root, whose thickness can reach up to 400 km.

**Geochronological province** Region in which a characteristic geochronological pattern predominates and dates obtained by distinct methods in different rock units are coherent.

**Platform** Relatively stable portion of a continent, not affected by a given orogenic event.

**Shields** Areas of the continental interior where Precambrian basement is exposed.

## See Also

**Africa:** Pan-African Orogeny. **Gondwanaland and Gondwana**. **Mesozoic:** Cretaceous. **Plate Tectonics**. **Precambrian:** Vendian and Ediacaran. **Shields**. **Tectonics:** Mountain Building and Orogeny; Rift Valleys.

## Further Reading

Almeida FFM, Hasui Y, Brito Neves BB, and Fuck RA (1981) Brazilian structural provinces: an introduction. *Earth Science Reviews* 17: 1–29.

Almeida FFM, Brito Neves BB, and Dal Ré Carneiro C (2000) The origin and evolution of the South American platform. *Earth Science Reviews* 50: 77–111.

Bizzi LA, Schobbenhaus C, Gonsalves JH, *et al.* (2001) *Geology, Tectonics and Mineral Resources of Brazil: Geographic Information System and Maps at the 1 : 2 500 000 scale*. Brasília: CPRM-Serviço Geológico do Brazil. [CD-ROM. Online at http://www.cprm.gov.br].

Brito Neves BB, Campos Neto MC, and Fuck RA (1999) From Rodinia to Western Gondwana: an approach to the Brasiliano–Pan African cycle and orogenic collage. *Episodes* 22: 155–199.

Chang HK, Kowsmann RO, Figueiredo AMF, and Bender AA (1992) Tectonics and stratigraphy of the East Brazil rift system: an overview. *Tectonophysics* 213: 97–138.

Cordani UG and Sato K (1999) Crustal evolution of the South American Platform, based on Sm–Nd isotopic systematics on granitoid rocks. *Episodes* 22: 167–173.

Cordani UG, Milani EJ, Thomaz FA, and Campos DA (eds.) (2000) *Tectonic Evolution of South America*. 31st International Geological Congress, Rio de Janeiro.

Heilbron M, Mohriak WU, Valeriano CM, *et al.* (2000) From collision to extension: the roots of the southeastern continental margin of Brazil. In: Mohriak WU and Talvani M (eds.) *Atlantic Rifts and Continental Margins*, pp. 1–32. Geophysical Monograph 115. Washington DC: American Geophysical Union.

Matos RMD (2000) Tectonic evolution of equatorial South Atlantic. In: Mohriak WU and Talvani M (eds.) *Atlantic Rifts and Continental Margins*, pp. 331–354. Geophysical Monograph 115. Washington DC: American Geophysical Union.

Mello MR and Katz BJ (2001) *Petroleum Systems of South Atlantic Margins*. AAPG Memoir 73. Tulsa: American Association of Petroleum Geologists.

Schobbenhaus C, Campos DA, Derze GR, and Asmus HE (1984) *Geologic Map of Brazil and Adjoining Ocean Floor Including Mineral Deposits Scale 1 : 2 500 000*. Brasília: Ministério das Minas e Energia.

Tankard AJ, Soruco RS, and Welsink HJ (1998) *Petroleum Basins of South America*. AAPG Memoir 62. Tulsa: American Association of Petroleum Geologists.

Trompette R (1994) *Geology of Western Gondwana (2000–500 Ma). Pan-African–Brasiliano Aggregation of South America and Africa*. Rotterdam: Balkema.

# BUILDING STONE

**A W Hatheway**, Rolla, MO and Big Arm, MT, USA

© 2005, Elsevier Ltd. All Rights Reserved.

## Introduction

Unlike most other materials available to early man, stone was the most durable and could be stacked in courses to make walls to support roofs and thereby afford protection from the elements. Building stone continues to be used, valued for its strength and for aesthetic beauty purposes. Hence there is now and will remain a market for the technical talents of geologists capable of locating and selecting deposits of durable and attractive building stone, and in scoping and devising the environmental protection measures now necessary in extraction of building stone.

## Historic Use of Building Stone

Historically, building stone has been employed for a wide variety of structural and load-bearing purposes (**Table 1**) Facade use of building stone was popular in the world's cities from the very advent of the high-rise building (about 1890) and this situation stimulated

**Table 1** Historic and modern uses for building stone

| Use | Historic | Modern |
|---|---|---|
| Housing | Generally structurally practical only to one or two storeys | Replaced since mid-nineteenth century by brick and timber frames with occasional stone facades |
| Fortifications | Durable material; employed in massive, protective wall | Replaced by reinforced concrete since 1900 |
| Road/street paving | Sole material for all-weather roads and streets | Infrequently used and then only for aesthetic purposes; replaced since 1940 by use of crushed stone in asphaltic concrete |
| Towers | Watch towers and lighthouses; thick-walls required for stability and resistance to attack | Replaced by reinforced concrete in order to meet cost of construction labour |
| Water supply | Traditional usage for aqueducts and for canal linings | Largely replaced by iron and steel conduits by 1900; lately by synthetic organics (plastics) |
| Sewers | Stone replaced by brick in mid-nineteenth century | Obsolete today due to leakage and high cost of emplacement labour |
| Commercial buildings | Mainly as facades | Used as facades on steel-framed buildings up to ten or more stories |
| Civil buildings | Limited to two storeys or requiring massive wall thickness | Restricted to facades, for aesthetic purposes |
| Mansions and institutions | Preferred construction material for wealthy patrons | Employed only for special esthetic reasons due to high costs of placement |
| Houses of worship | Traditional; thick-walled or with flying buttresses | Now replaced with reinforced concrete, often with stone as cladding |
| Monuments | The preferred material | Remains the preferred material |
| Fireplaces and furnaces | Replaced by firebrick by ca 1840 for furnaces | Remains popular for fireplace facing; now used in conjunction with steel or ceramic flues |
| Wall dressing, interior and exterior | Seldom used as was trivial and unnecessary | Now the prime use of building stone |
| Bounding walls | Commonplace where rock outcrops are exposed for quarry removal | Ongoing demand in conservation areas and national parks |

**Table 2** Attractions and qualities of natural building stone

| Quality | Nature | Geological consideration |
|---|---|---|
| Durability | When other than 'weak' rock, stone has been conditioned by the attack of geological agents and properties in preparation for its selection and use today | Most potential physical/chemical degradation has already occurred and the stone likely will be durable in its intended construction or architectural use |
| Resistance to abrasion | Necessary quality for stairs, walkways, and road and drive paving | Choice generally relegates to crystalline igneous or metamorphic rock type low in presence of micas |
| Variety of intact dimensions | Generally a result of geological depositional conditions of origin or of diagenetic or metamorphic changes that promote jointing and fissility as a result in natural separation into articulated blocks, preferably in a rectilinear pattern | Generally the result of either or both: 1. Joint spacing in igneous or crystalline metamorphic work 2. Foliation jointing in high-grade metamorphic rock 3. Fluvial rounding in high-energy stream regimes, often associated with paleglaciation |
| Planarity or flatness; when so selected | Considered 'flagstone' when length or breadth-to-thickness ratio is equal to or greater than about 10:1 | Influenced by fissility imparted by sedimentologic (bedding) character or by metamorphically-induced open foliation |
| Colour | Generally a wide variety, to please the client. Colour variability generally greatest in volcanic terrain, such as Italy | Directly related to lithology and geological origin. Locating geologist should examine for absence of reactive minerals such as iron pyrites |
| Texture | To please the client; but selected to avoid unsightly stains | Visible as a form of surface roughness. Directly related to lithology and geological origin |

**Table 3** Basic geological controls over nature of building stone

| Control | Defined | Geological considerations |
|---|---|---|
| Sedimentary rock | Most varieties are unsuitable for use as building stone when uniaxial compressive strength exceeds about 15 MPa | Rock younger than Cretaceous (as Tertiary era) aged tends to be 'weak' and unsuitable because of immature induration (density) |
| | | High clay mineral content generally promotes unsuitability by way of spalling and deterioration in exposure to the elements |
| Glacial ice-contact deposits | Often represents ground rich in durable and colour-lithologic diverse specimens | Only the most durable rock survives the glacial plucking and transport experience |
| | | Often excellent fireplace and pillar stone |
| | | Requires considerable experience and labour in selection of useful individual stones |
| Rock lying in tectonic shear zones | Tends to be mylonitinised or otherwise shear-damaged and therefore excessively fissile | Generally damaged beyond potential resistance to long-term exposure to the elements in exterior applications |
| Pluton-bounding zones of foliation in granitic rock | Lends itself well as a fall facing | Often has a pleasing and useful foliation character leading to slightly planar dimension ratio |
| Pluton-capping rhyolites in post-Jurassic terrains | Somewhat rare due to removal by erosion | Generally durable in climatically-temperate regions while exhibiting good planar dimensions for wall facing |
| Streams and rivers draining tertiary volcanic centres of moderately high rainfall or subject to cloudburst-type periodic events | Usually present as boulders of a variety of colours and textures | Fluvial action destroys weak varieties and delivers only the most durable specimens for collection |
| Foliated metamorphic rock in general | Foliated rock splits or rips well with minimum to no application of explosives | Tends to scar when removed and handled by machinery; requires care in recovery at quarries |
| High-grade metamorphic rock in general (Gneisses) | Gneisses and migmatites generally suitable only as having experienced glacial action | As grain size diminishes to that of schist, desirable building stone qualities decrease; especially in the presence of micas |
| Rock within the contact aureole of a few hundred meters of intrusive igneous bodies | Sometimes visually undetectable cation exchange effects generated by hydrothermal effects of intrusion | Best to place samples in the weather for observation; best to use application of wet-dry interludes of testing by laboratory cyclic wetting, and drying |
| Aggregate pits servicing great urban centers on the fringes of igneous-metamorphic complexes | Pits will be located along modern rivers and streams colocated with Plio-Pleistocene to recent fluvial deposits | Pits operating by suction dredging tend to have a 'bone pile' of vertebrate fossil bones |

great interest in the national geological surveys. Most geological surveys produced serial sets of geological technical monographs on building stone and these books today are the most valuable published references for those geological professionals who are interested in building stone.

## Geological Character of Building Stone

Stone possesses several strong qualities, valued by clients and architects and was so recognized in the early textbook on Engineering Geology in which Penning (1880) cautioned that:

"In the selection of a stone, it is not merely in the testing it by hardness, composition, and appearance, that judgment should be displayed, but also in ascertaining the conditions under which it lay before removal from its position in the quarry."

These qualities (**Table 2**) have geological origins, most effectively identified by professional geologists.

## Geological Controls on Nature of Building Stone

Geological conditions exert first-order control over the kinds of building stone available within a given economic radius for recovery. These controls are largely

**Table 4** Geologic tasks in prospecting for building stone

| Task | Defined | Considerations |
| --- | --- | --- |
| Determine general nature of stone desired | Nature of stone determines the likely source geological formation(s) to be prospected | Reduces the exploration task to identification of suitable terrain in lands available for recovery |
| Determine the economic radius for stone recovery | This defines the area of exploration | Acquire medium-scale geological maps of the area of economic consideration |
| Assess suitability of deposit to produce stone of desirable end-use qualities | 1. Dimensions on initial recovery, without treatment<br>2. Durability as exposed in place, on construction | Most stone undergoes physical/chemical adjustment to stress-removal on excavation and in the new environment of placement |
| Assess ultimate quantity of rock available | Must recognise costs involved in recovery site development | Quantity as affected by property bounds and stability of cuts made to recover the stone |
| Evaluate means of removal of stone | Cost of production of each cubic meter of stone prepared for shipment to market | Suitable rate of production must be balanced by lack of unacceptable damage to the produced stone |
| Evaluate cost of production of units of the stone (cubic meter) | Placement of accumulated stone produce between haul loads | Most stone today is palletised for unit-transport packages of one cubic meter; avoids labour costs in distribution to building sites |
| Evaluate cost of transport of units of stone (cubic meter) to market areas | Resting place for spoil accumulated from stone removal and selection for transport | Ideally removed periodically to the final disposal site, in accordance with quarry closure plans or for sale as off-site embankment or fill material |
| Assess unit-cost impact of providing environmental protection | Planning and permitting, and meeting environmental requirements of the permit | 1. Control of sedimentation and dust<br>2. Post-closure land use<br>3. Implementing approved plan of closure of the quarry |

**Table 5** Types of stone masonry

| Type | Description | Considerations |
| --- | --- | --- |
| Ashlar | Individual blocks of rock are cut or trimmed,<br>Broken: As placed, the joints are not continuous<br>Small: When individual stones are less than 300 mm thick<br>Rough: Squared-stone masonry laid as 'range' work | Masons attempt to make use of a high percentage of rock as delivered from the quarry, sized in accordance with provisional specifications of supplier's contract, not necessarily of comparable size or ratio |
| Squared-stone | Masonry in which individual stones are roughly squared in layers and roughly dressed on beds and joints | Distinction between Ashlar and Squared-stone is that the latter generally is laid dry or with mortar spacing being 150 mm or greater between surfaces of stones |
| Quarry-faced | Stone facing is left untreated as it comes from the quarry | Commonly jointed or bedded stone comes with iron or manganese-stained natural planar surfaces |
| Cut-and-polished stone | A relatively new market with durable stone being sold mainly for interior use as flooring and counter surfaces | Durable stone is recovered from 'the ends of the earth' and sold at high prices controlled by vendor-tradesmen |
| Rubble | Consists of stone unshaped by hand or machine and placed in its unaltered condition from the quarry | Applied mainly to river-run fluvial boulders previously shaped by the stone-to-stone contact in transport from outcrop to mined deposit<br>Also applies to informal recovery of naturally discontinuous rock recovered from small quarries by fairly intensive hand labour |
| Dressed rubble | Some attention to rough sizing by manual hammering | May be dressed by tensile-splitting of field-broken or split surfaces |
| Range-work | Exposed fact of individual stone is roughly dressed | Laid in rough courses |
| Broken range-work | Courses are not continuous throughout the wall or face | Facilitated with joint-bounded rock blocks |
| Random-work, or 'uncoursed rubble' | Unsquared stone without attempted regular courses | Laid without attempt to achieve horizontal or sub-horizontal levelling between individual courses of stone |
| Coursed rubble | Sometimes roughly shaped by hammer so as to fit approximately | Unsquared stone levelled off at specific heights to an approximately horizontal surface of each course |

**Table 6** Sequential environmental permitting planning for recovery of building stone

| Step | Considerations | Geologic detail |
| --- | --- | --- |
| Topographic layout | Defines where the recovery is to take place | Scale of 1:1000–1:2000 |
| | | Contour interval of one meter |
| | Enlarged ordnance map or plane-tabled topographic map of deposit bounds | Property bounds to 100 m lateral buffer space |
| | | Present hydrological features |
| | | Present cultural features |
| Exposed geological detail | Contact of soil with bedrock, if exposed or as predicted | Include notion of geological structure |
| Proposed product stone | Dimensions and character | 1. Note possible variations in the character of stone across the breadth and depth of the planned quarry |
| | How stored | |
| | Frequency and method of removal from the site | 2. Subdivide the quarry into separate "domains" if differences exceed the flexibility of marketing |
| Mining concept | Explains how the deposit will be opened up and how the stone will be accessed | 1. Cutting bounds, crown of cut and toe of slopes |
| | | 2. Sequential advance of the cutting with final proposed lateral bounds |
| | | 3. Nature of machines to be employed |
| | | 4. System of parting the stone from the outcrop or rock mass |
| | | 5. Use of explosives |
| Management of runoff | Describe how precipitation and snowmelt will be managed | Indicate presence of drainage gradient and channelization to be placed to remove runoff to suitable discharge |
| Management of spoil | Describe dust, fine particles and sediment likely to be generated | Show locations for interim spoil placement |
| | Rejection policy/procedure for discrimination of stone vs. spoil | |
| Determine general nature of stone desired | Nature of stone determines the likely source geological formation(s) to be prospected | Reduces the exploration task to Identification of suitable terrain in lands available for recovery |
| Determine the economic radius for stone recovery | This defines the area of exploration | Acquire medium-scale geologic maps of the area of economic consideration |
| Access and haul roads | Within about 8 km of centre of area of interest | Examine road cuts and outcrops to ascertain the general presence of suitable stone |
| Product storage area | Placement of accumulated stone produce between haul loads | Most stone today is palletised for unit-transport packages of one cubic meter; avoids labour costs in distribution to building sites |
| Spoil storage area | Resting place for spoil accumulated from stone removal and selection for transport | Ideally removed periodically to the final disposal site, in accordance with quarry closure plans or for sale as off-site embankment or fill material |
| Sequential quarry reclamation | Ongoing plan of converting incremental phase-areas of the overall stone removal site to suit the environmental permitting permit | Normally comes as a consequence of working within site ownership bounds and in consideration of stability of the highwall resulting from removal of stone from the bedrock outcrop |
| Management of incidental waters | 1. Runoff from rainfall and snowmelt | Requires temporary ditching and/or area sumps and pump-withdrawal to authorised discharge point |
| | 2. Accommodation of seepage from the highwall | |
| Management of sediment | Dust and small particles subject to wind and runoff transport from the quarry to the environment | Generally managed by placement of sedimentation sumps and bales of cut vegetation such as 'hay' |
| | | Requires ultimate on-site manipulation as part of the closure plan |
| Closure plan | Presumes that the quarry will be provided with an acceptable after-use | Should be worked into the long-term operational plan and be in conformance with the required end-use of the operating permit |

related to lithology, conditions of origin, and geological history (**Table 3**) and are reflected in mineralogy and petrography, and include fabric and texture. Large expenditure on the acquisition of building stone should include a petrographical analysis for the latter features.

## Locating Sources of Building Stone

Building stone has a relatively low economic value in the outcrop, its true value accruing from its placement work by stone mason tradesmen. Its recovery is expensive in terms of specialized skills and by the judicious use of explosives and of construction machinery for its recovery.

A number of factors related to prospecting for building stone deposits are well suited to engineering geological training (**Table 4**).

## Environmental Planning for Recovery of Building Stone

As with most earth materials, there are two levels of activity that apply to recovery of building stone, informal and formal. For all but the weekend mason, all other recovery activities quickly will be recognized by environmental regulatory authorities as a form of 'quarrying', for which even a small stone recovery enterprise will require submittal of a Permit Application and some scheme of site development and integrated measures for protection of the environment (**Tables 4** and **6**).

## Stone Masonry

Modern stone masonry nearly always serves the aesthetic purpose of the architect or landscape designer. No longer are we concerned with use of stone for its original structural role, but the manner in which the masonry is placed provides ample opportunity for meaningful artistic statement (**Table 5**).

## Petrography

Petrographical examination of thin-sectioned hand specimens recovered from prospecting will be helpful in some instances in which the shape, colour, and/or texture meets the client's approval, but the geologist remains wary of the ability of the stone to meet expectations when placed in exterior locations exposed to the elements. As with basic identification of the stone, such manuals as listed in the **Further Reading** Section below provide an excellent desk reference.

## Summary

Building stone provides a durable, aesthetically pleasing construction material where wear-resistant and climate-resistant natural stone can be used to enhance the qualities of modern construction. Significant costs must be paid for accessing and recovery machinery, palleting, and transportation to the vendor's yard. Application of geological knowledge is essential to providing appropriate attractive and durable stone at an economic production cost.

## See Also

**Aggregates**. **Engineering Geology:** Rock Properties and Their Assessment; Site and Ground Investigation.

## Further Reading

Bates RL (1987) *Stone, Clay, Glass; How Building Materials are Found and Used*: Enslow.

Bell FG (1993) Engineering Geology: Oxford: Blackwell Scientific Publications.

Blyth FGH and de Freitas MH (1984) *A Geology for Engineers*, 7th edn. New York: Elsevier.

Geological Society of London (1999) Stone; Building Stone, Rock Fill and Amourstone in Construction. *Engineering Geology Special Publications* 16: 480.

Parsons and David (eds.) (1990) *Stone; Quarrying and Building in England*. London: Phillmore, in Association with The Royal Architectural Institute.

Shadman A (1996) *Stone; An Introduction*. London: Intermediate Technical Publications.

Smith MR and Collis L (eds.) (1993) *Aggregates; Sand, Gravel and Crushed Rock Aggregates for Construction Purposes*, 2nd edn. London: Geological Society Special Publication no. 9, p. 339.

# CALEDONIDE OROGENY

See **EUROPE: Caledonides of Britain and Ireland; Scandinavian Caledonides (with Greenland)**

# CARBON CYCLE

**G A Shields**, James Cook University, Townsville, QLD, Australia

© 2005, Elsevier Ltd. All Rights Reserved.

## Introduction

The recycling of elements between Earth's interior, its sedimentary sheath, the oceans, and the atmosphere is vital to the continued functioning of Earth as a living planet. Of all the biogeochemical cycles operating on Earth, the carbon cycle (**Figure 1**) is probably the most fundamental process, without which life could not exist. Carbon, in the form of carbon dioxide ($CO_2$), represents the starting component in almost all food chains, while at the same time performing a complementary role as Earth's thermostat, providing an equable climate, suitable for the retention of liquid water on the surface of the planet. Carbon in the form of bicarbonate and carbonate ions helps to regulate ocean acidity and, in combination with calcium, provides a hard skeleton for many marine organisms. Without the biological sequestering of carbon and its eventual storage in the crust through geological processes over millions of years, there would be little or no free oxygen on Earth's surface. Similarly, without the deposition of organic carbon and calcium carbonate shells on the seafloor, Earth would inevitably become too hot for even the most radical of known extremophile metabolisms. All these important roles reflect different aspects of the global carbon cycle, which is in reality a hierarchy of subcycles, all operating on different time-scales.

From the moment a molecule of $CO_2$ first enters the surface environment, whether from a volcano or from the chimney stack of a power station, it enters a dynamically changing realm of continual recycling. Within a decade, the $CO_2$ molecule will very likely be consumed by photosynthesis and converted into organic carbohydrate, either as part of a land plant or in one of the trillions of single-celled algae in the world's oceans. In either case, death and decay will eventually release the $CO_2$ back into the surface environment, thus closing the cycle. This 'short-term' carbon cycle may replay hundreds of times before burial finally results in the addition of this carbon to the sedimentary pile, where some of it will remain for millions of years as part of the 'long-term' carbon cycle of sedimentation, tectonics, and weathering. On such huge time-scales, the sequestering of atmospheric $CO_2$ during chemical weathering and the eventual redeposition as carbonate rock are generally balanced by the release of $CO_2$ from volcanoes and through metamorphism. Although negative feedbacks tend to regulate $CO_2$ levels over the long term, shorter term perturbations to the carbon cycle may occur due to pulses of massive volcanism, carbon burial, or methane clathrate dissociation, all of which may have caused considerable climatic perturbation at different times in Earth history. The carbon cycle not only acts as a trigger for climate change but can react to such change, as happened during glacial–interglacial cycles, when regional warming trends appear to have preceded $CO_2$ increases. Interest in the short-term carbon cycle has increased greatly of late because of the observed elevation in atmospheric $CO_2$ concentrations due to fossil fuel combustion, cement manufacture, and land use changes. Fears that elevated $CO_2$ concentrations will change the future world climate have inspired research into the effects of an enhanced greenhouse effect on climate, sea-level, and biodiversity. This global research effort necessitates a good understanding of how Earth regulates climate through the carbon cycle, including the development of increasingly sophisticated models to describe the 'Earth system'.

## Short-Term Carbon Cycle

The short-term carbon cycle refers to the natural cyclical processes of organic matter production and organic decay (**Figure 2**). The key step in this cycle is photosynthesis, whereby $CO_2$ is consumed by reaction with water to form organic carbohydrates, releasing oxygen in the process (eqn [1]):

$$CO_2 + H_2O = CH_2O + O_2 \qquad [1]$$

Photosynthesis $\rightarrow$ $\leftarrow$ Respiration

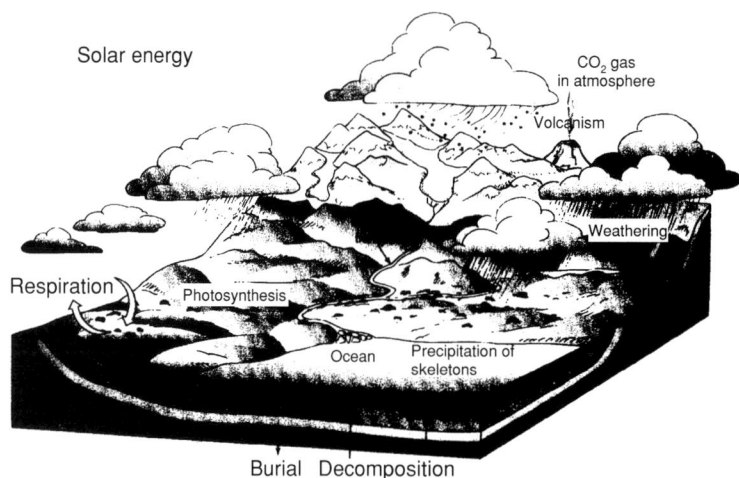

**Figure 1** The natural global carbon cycle. Reprinted from Davidson JP, Reed WE, and Davis PM (1997) *Exploring Earth: An Introduction to Physical Geography.* Upper Saddle River, NJ: Prentice Hall.

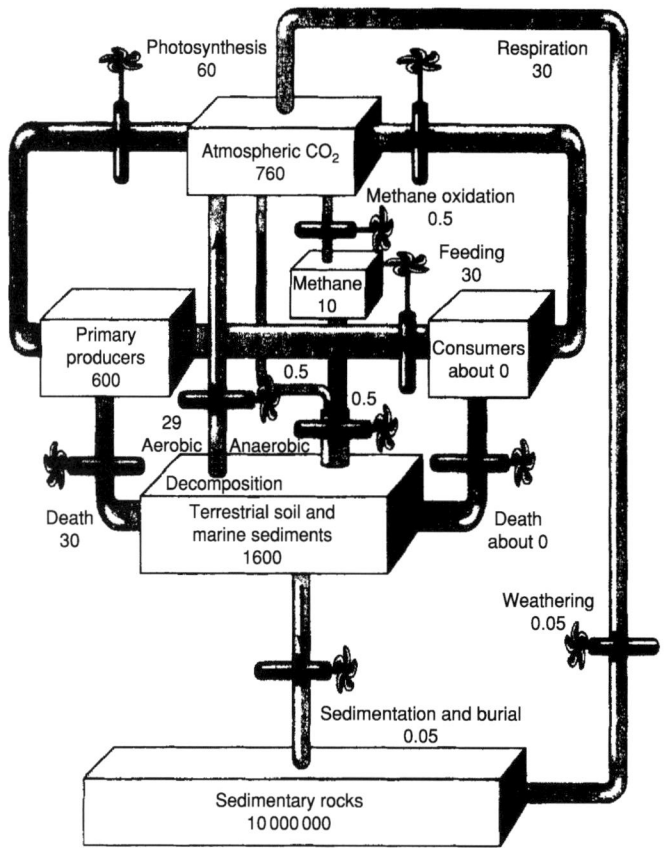

**Figure 2** The short-term and long-term organic carbon cycles of productivity, sedimentation, burial, and weathering. Numbers refer to carbon reservoirs (in gigatons) and carbon fluxes (in gigatons year$^{-1}$). Reprinted from Kump LR, Kasting JF, and Crane RG (1999) *The Earth System.* Upper Saddle River, NJ: Prentice Hall.

Photosynthesis does not occur spontaneously but requires the input of solar energy. Consequently, photosynthesis can take place only where sunlight can freely penetrate, on the land or within the uppermost 100 m or so of oceans and lakes. On land, photosynthesis is carried out mostly by plants, whereas in the sea it is almost exclusively the domain of algae in the form of minute, free-floating silica-shelled diatoms and calcite-shelled coccoliths. Just as photosynthesis always consumes $CO_2$, the reverse process, organic respiration or decay, returns that $CO_2$ after death of a plant or animal; this process is catalysed by the enzymatic action of chiefly anaerobic microbes that can be found in the digestive tracts of animals, in the soil, or beneath the seafloor. Because organic decay is essentially an oxidation process, oxygen is consumed instead of being released, which is why oceanic regions of high productivity are commonly underlain by 'oxygen minimum zones', i.e., zones of low oxygen concentration caused by the decomposition of organic matter in the water column. Organic raindown not only transfers $CO_2$ to the deep ocean but also acts as a biological pump for nutrients that, once stored in the deep ocean 'conveyor belt', generally reemerge only in regions of upwelling and high productivity. Reoxidation of organic matter that has reached the sediment may take place by a variety of mechanisms (oxic respiration, denitrification, or sulphate reduction). However, in addition, in the deepest, most reducing environments, methane may be produced by the actions of methanogenic bacteria using two main pathways (eqns [2a] and [2b]):

$$CH_3COOH \rightarrow CO_2 + CH_4 \qquad [2a]$$

or

$$4H_2 + CO_2 \rightarrow CH_4 + 2H_2O \qquad [2b]$$

Methane produced in this fashion may seep back into seawater to be reoxidized to $CO_2$ or may be stored temporarily for thousands to millions of years as the volatile methane clathrate.

Under normal circumstances, any particular ecosystem will be experiencing both photosynthesis and respiration, but in different proportions. If photosynthesis outweighs respiration, then the ecosystem will act as a sink for $CO_2$, whereas in the reverse case it will act as a source. This 'breathing' of the biosphere can best be seen in the annual fluctuations of atmospheric $CO_2$ in the northern hemisphere (**Figure 3**). During the northern hemisphere summer, when terrestrial photosynthesis and leaf growth surpass respiration and decomposition, $CO_2$ levels decrease by as much as 15 ppm in the boreal forest zone (55–65° N), a decrease almost completely balanced by the winter increase in $CO_2$ caused by the inevitable decomposition of fallen leaves. In the southern hemisphere, the cycle is reversed but the effect never attains more than 1 ppm. This hemisphere inequity confirms that such seasonal $CO_2$ fluctuations are driven by the terrestrial carbon cycle, rather than by the oceanic carbon cycle, there being far more terrestrial biomass in the northern than in the southern hemisphere. Because the fluxes involved in the short-term carbon cycle are so large, persistent imbalances would lead to intolerable fluctuations in atmospheric $CO_2$. Therefore, on timescales longer than about a century, this cycle of productivity and decay must be perfectly in balance. Nevertheless, not all $CO_2$ is eventually returned to the atmosphere. A small proportion of the carbon locked up in soils as organic matter may be washed away to be buried indefinitely in coastal and marine sedimentary basins. Similarly, a small proportion of marine organic matter will also escape the processes of decay by being swiftly buried in areas of high sedimentation rate. Organic burial constitutes a leak in the short-term carbon cycle, whereby a relatively small proportion of $CO_2$ is continually removed from the surface environment to be stored within Earth's crust. Some of this leaked $CO_2$ may eventually be

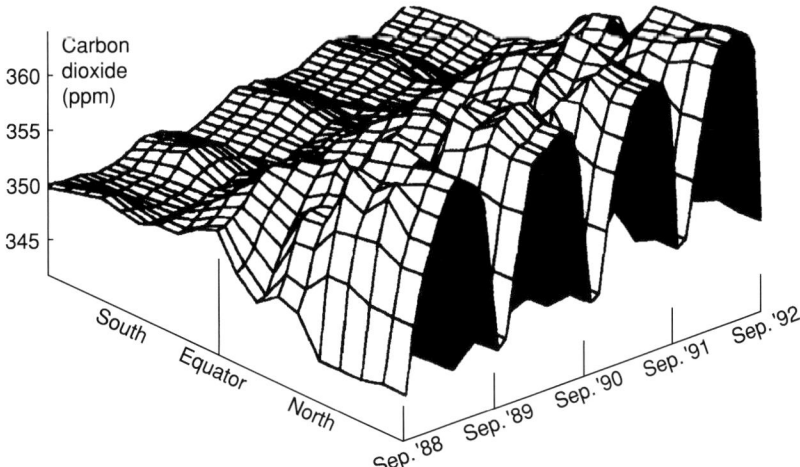

**Figure 3** A three-dimensional perspective of terrestrial biosphere breathing. Reprinted from Volk T (1998) *Gaia's Body: Toward a Physiology of the Earth.* New York: Springer-Verlag.

returned to the surface environment by tectonic processes at convergent margins (deep-sea trenches), but not for many millions of years, as part of the 'long-term' carbon cycle of carbon burial, subduction, uplift, weathering, and carbonate deposition.

## The Long-Term Carbon Cycle

Although some carbon reaches Earth from space in the form of impacting comets, most carbon first enters the surface environment as volcanic $CO_2$, originating from deep within Earth's interior at mid-ocean ridges, convergent margins, and the Great Rift Valley and other terrestrial volcanic provinces (e.g., Kamchatka). Some of this carbon is entirely new to the surface environment, being derived directly from the mantle, but some will be 'old' carbon, recycled from the sediment pile by tectonic processes such as metamorphism. At convergent plate margins, carbonate sediments can be carried to great depths, up to hundreds of kilometres, riding on subducting ocean lithosphere. The high pressures and temperatures at great depths encourage biogenic calcite (calcium carbonate) to react with biogenic and detrital silica (silicon dioxide) to form metamorphic calcium silicate minerals. This process releases $CO_2$ (eqn [3]) that may eventually find its way into the atmosphere via hot springs and seeps (**Figure 4**).

$$CaCO_3 + SiO_2 \rightarrow CaSiO_3 + CO_2 \qquad [3]$$

Much study is devoted to estimating the proportion of genuinely new $CO_2$ from the mantle relative to recycled $CO_2$ at convergent margins, but there is currently no consensus. If volcanic outgassing were allowed to proceed unchecked, then atmospheric partial pressure of $CO_2$ ($pCO_2$) would rise until an equilibrium concentration was reached, balanced only by escape of $CO_2$ into space. Exactly how hostile such an Earth would be can be appreciated by comparison with the furnace-like Venus, Earth's planetary neighbour and twin. Although Venus and Earth are of a similar size and are a similar distance from the sun, Venus has a much thicker atmosphere, made up almost entirely of $CO_2$, over 200 000 times more concentrated than in Earth's atmosphere. There is no reason to suspect that Venus overall is any richer in carbon than Earth is, so Earth must possess powerful mechanisms of $CO_2$ sequestration, unique in the solar system.

The first step in the permanent removal of $CO_2$ from the atmosphere is chemical weathering (**Figure 5**), whereby rainwater, made acidic by the dissolution of aqueous $CO_2$, corrodes rock, forming minerals such as silicates and carbonates. In the presence of soil, this process is accelerated by plants, which help to concentrate $CO_2$ in the soil to levels generally 10 to 100 times higher than in the atmosphere. $CO_2$ used up in this way may then be transported via rivers and groundwater to the oceans, chiefly as bicarbonate anions ($HCO_3^-$), along with the weathered-out major cations such as Si, Al, Ca, Mg, K, Na, and Fe. Once in the ocean, some $CO_2$ may be removed permanently from the exogenic system by chemical precipitation as calcium carbonate, mostly in the form of shells and skeletal elements that rain down to the seafloor. Because $CaCO_3$ precipitation also releases $CO_2$ back into the water column (eqn [4]), $CO_2$ taken up by the weathering of carbonate rocks does not result in any long-lasting net change to atmospheric $CO_2$ levels. In other words, the $CO_2$ taken up during carbonate dissolution is simply returned during carbonate precipitation in the ocean:

$$CaCO_3 + CO_2 + H_2O = 2HCO_3^- + Ca^{2+} \qquad [4]$$

Carbonate weathering $\rightarrow\leftarrow$ Carbonate precipitation

Similarly, $CO_2$ consumed during the weathering of Na and K silicates does not result in any net change in atmospheric $CO_2$ because Na and K carbonate minerals are very soluble and do not readily precipitate from seawater. Equation [5] shows the net result of Ca–Mg silicate weathering, which is the most important mechanism of permanently removing $CO_2$ from the atmosphere:

$$(CaMg)SiO_3 + CO_2 \rightarrow (CaMg)CO_3 + SiO_2 \qquad [5]$$

Much research effort has been expended to better understand the controls on chemical weathering rates

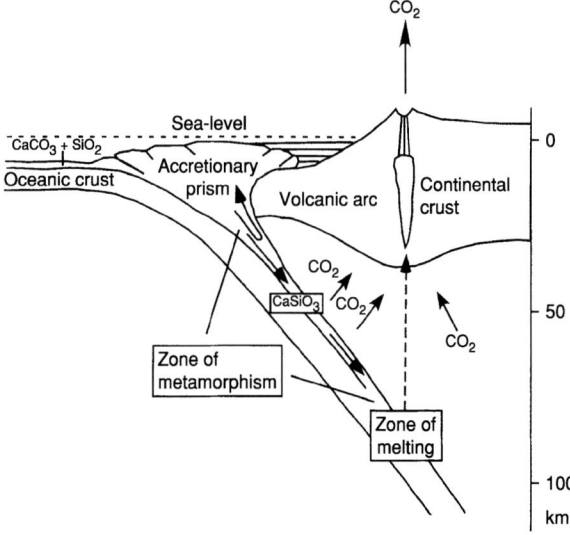

**Figure 4** The long-term carbon cycle of sedimentation, subduction, and outgassing.

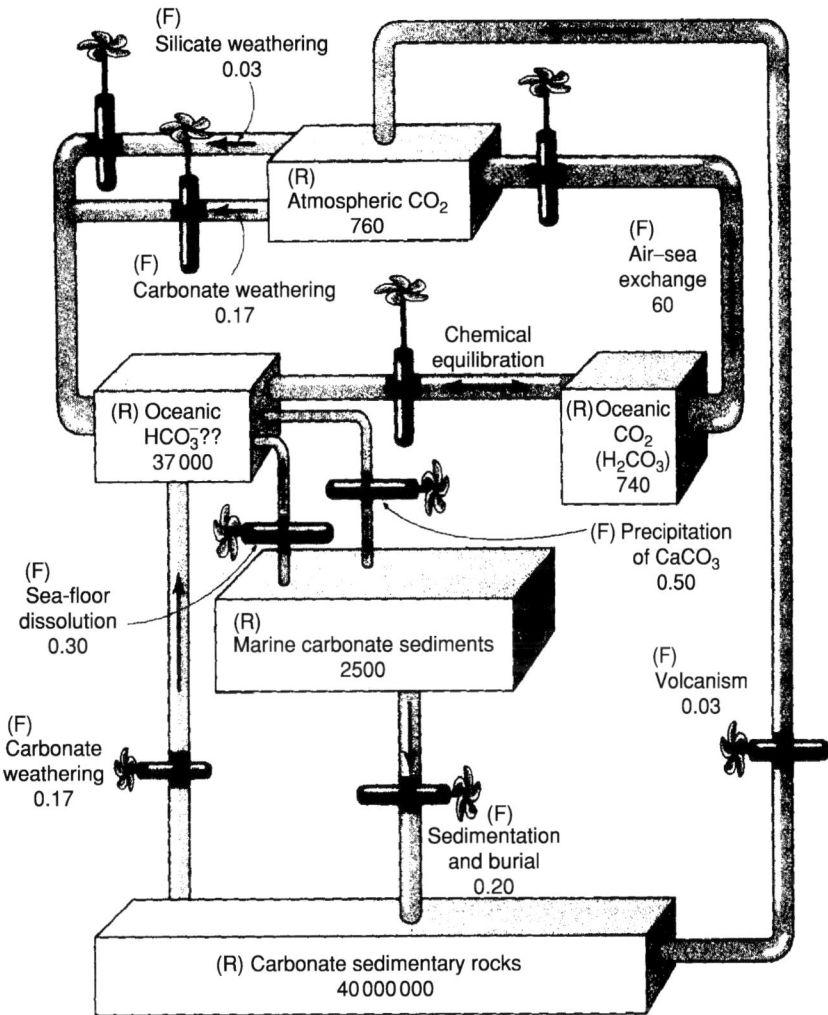

**Figure 5** The global inorganic carbon cycle of weathering, runoff, carbonate precipitation, and burial. Numbers refer to carbon reservoirs (R; in gigatons) and carbon fluxes (F; in gigatons year$^{-1}$). Reprinted from Kump LR, Kasting JF, and Crane RG (1999) *The Earth System*. Upper Saddle River, NJ: Prentice Hall.

and therefore the effects of these controls on the long-term global carbon cycle. There is no consensus on these issues because several unrelated factors are involved, such as climate (both rainfall and temperature), denudation rates, weathering history, and rock type. Through a combination of these factors, certain tropical, humid areas of the world with freshly exposed volcanic rock can exert an outsized influence over the global carbon cycle. This is because fresh basalts are particularly susceptible to chemical weathering and are made up predominantly of Ca–Mg silicates. The weathering of basalts today may contribute as much as a third of all the silicate-weathering $CO_2$ flux. Thus volcanism acts not only as a source for atmospheric $CO_2$, but also as one of the major long-term sinks.

The products of chemical weathering, including Ca, Mg, and bicarbonate ions, arrive in the oceans in solution; concentrations build up in the oceans until an equilibrium state is reached between input and carbonate precipitation. Bicarbonate ($HCO_3^-$) and carbonate ($CO_3^{2-}$) ions, which together make up 99% of all dissolved carbon in seawater, help to regulate seawater pH by transferring protons (or hydrogen ions) between carbonate species (eqn [6]) according to the rules of carbonate equilibria. In effect, the carbonate system in seawater is capable of neutralizing a large portion of any extra $CO_2$ added from, say, fossil fuel burning or volcanism.

$$CO_2 + H_2O = H_2CO_3 = H^+ + HCO_3^-$$
$$= 2H^+ + CO_3^{2-} \quad [6]$$

Carbon dioxide ↔ Carbonic acid ↔ Bicarbonate
↔ Carbonate

Many organisms catalyze the precipitation of $CaCO_3$ from seawater through active biomineralization and shell formation; for example, foraminifera, coccoliths, corals, and shellfish contribute to the permanent removal of atmospheric $CO_2$. Marine organisms generally form their shells at relatively shallow depths and sink to the shelf floor or abyssal depths after death. In regions where the water depth is no greater than about 1 km, the shells reach the seafloor more or less intact because these waters are saturated with respect to calcite and aragonite. However, at greater depths, where pressures are greater, temperatures are lower, and $CO_2$ concentrations are elevated, seawater is far more corrosive to $CaCO_3$. The level at which the rate of carbonate dissolution balances the downward flux of carbonate settling to the seafloor is called the carbonate compensation depth (CCD). This depth varies from ocean to ocean and has varied over time in response to carbonate productivity, temperature, atmospheric $CO_2$, and ocean circulation changes.

## Geological Evolution of the Global Carbon Cycle

Many scientists suspect that the greenhouse effect and, in particular, the level of $CO_2$ in the atmosphere were different in the geological past. One major reason for this suspicion derives from the 'Faint Young Sun Paradox'. Because the sun is probably heating up over time – owing to the increasingly exothermic nuclear fusion of lighter elements, forming heavier elements in the sun's central core – it is probable that the input of solar energy to Earth was significantly lower in the past, by as much as 1% for every 100 million years, according to some estimates. However, Earth's sedimentary record provides evidence for the existence of liquid water at least 4 billion years ago, indicating that surface temperatures were not much lower then than they are now. Therefore, it appears that, over time, Earth's atmosphere has become increasingly less efficient in retaining solar energy. Model calculations indicate that, were $CO_2$ the only relevant greenhouse gas, $CO_2$ levels must have been as much as 10 000 times higher during the Early Precambrian. However, because methane is likely to have been a major greenhouse gas during early Earth history, before the increase of atmospheric oxygen levels, such estimates are likely to represent maximum $CO_2$ concentrations only.

On geological time-scales, atmospheric $CO_2$ levels are thought to be regulated by negative feedbacks between climate and silicate weathering rates. First, increased temperatures due to higher $CO_2$ levels would accelerate chemical weathering rates, thus effectively slowing any $CO_2$ increase. Second, weathering requires rainfall, which would most likely increase as temperatures rise, due to the acceleration of the hydrological cycle. Third, enhanced $CO_2$ fertilization of plant growth would also help to increase weathering rates by stabilizing soils and encouraging chemical weathering over physical weathering. This does not mean, however, that $CO_2$ levels have remained constant over geological time. Chemical weathering rates are imperfectly tied to $CO_2$ levels, being related also to additional parameters, including tectonics, vegetation cover, and palaeogeography, whereas atmospheric $CO_2$ may be influenced by independent changes in the input and output of $CO_2$. In the geological past, higher $CO_2$ levels could be sustained because chemical weathering rates on the continents were lower in the absence of deep soils, before the introduction of vascular plants in the Devonian period. A higher $PCO_2$ was also made possible by higher volcanic outgassing rates of mantle $CO_2$ and the abiotic nature of much carbonate deposition. Before the introduction of pelagic carbonate producers, such as planktic foraminifera by the Jurassic, carbonate deposition would have been restricted to shallow shelf environments, thus increasing calcite saturation levels and rendering deposition rates vulnerable to changing sea-level (palaeogeography). Although the introduction of land plants in the Devonian period does not appear to have had an immediate effect on climate (by sequestering more atmospheric $CO_2$), the deposition of massive peat deposits in shallow coastal environments during the Carboniferous and Permian periods almost certainly did. Long-term ($10^7$ years) increases in organic carbon burial during this interval are thought to have reduced $CO_2$ levels by as much as 70% (**Figure 6**), causing a series of glaciations, while allowing oxygen to build up in the atmosphere. One consequence of higher $O_2$ is to increase air pressure; it seems plausible that the well-recorded insect gigantism during this interval was related to the added buoyancy provided by greater air pressures. Such long-term, but ultimately unsustainable, imbalances in the proportion of carbon buried as organic carbon relative to carbonate carbon may have caused changes to both the $CO_2$ and the $O_2$ atmospheric budgets at other times, too, but the carbon isotopic record suggests that this proportion has been more or less constant at 1:4, respectively, on $>10^7$ year time-scales since 3 Ga. Geochemical models show that $CO_2$ levels during the Phanerozoic were also higher than at present (**Figure 6**), and reached peaks during the Early Phanerozoic and more recently during the Cretaceous period when high $PCO_2$ is likely to have contributed to the equable and balmy 'Greenhouse' climates typical of that interval in Earth history.

Despite the existence of regulatory feedbacks on atmospheric $CO_2$, the Earth system is still susceptible

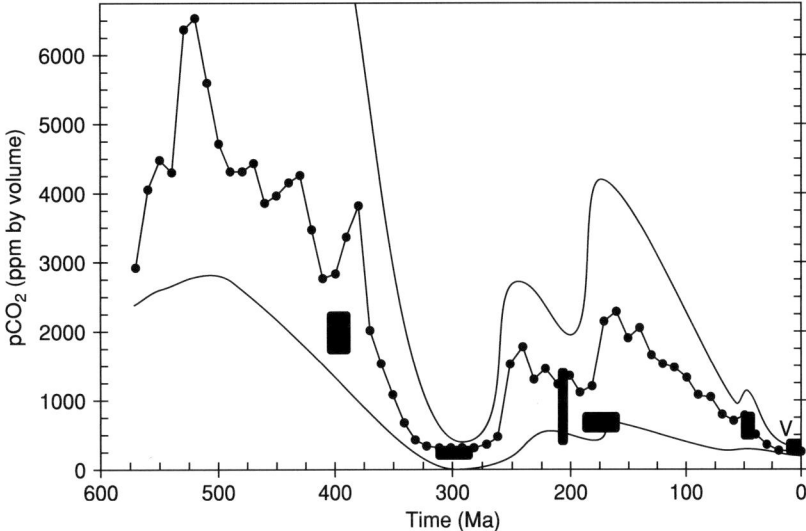

**Figure 6** Comparison of model predictions of atmospheric $CO_2$ over the Phanerozoic, and plant stomata-based data (black boxes). Reprinted from Royer DL, Berner RA, and Beerling DJ (2001) Phanerozoic atmospheric $CO_2$ change: evaluating geochemical and palaeobiological approaches. *Earth Science Reviews* 54: 349–392.

to shorter term ($<10^6$ years) perturbations caused by abrupt changes to the input or output fluxes of $CO_2$. Because volcanism is insensitive to surface negative feedbacks, short-term fluctuations in volcanic activity may have considerable, but short-lived, consequences for atmospheric composition and global climate due to the release of massive amounts of greenhouse $CO_2$. Such volcanic episodes, termed 'flood basalt events', are rare, but they have occurred several times over the course of the Phanerozoic; they resulted in Large Igneous Provinces, producing more than 1 million km$^3$ of volcanic rock, such as the Deccan and Siberian traps, in less than 1 million years (*see* **Large Igneous Provinces**). Correlations between the ages of rare igneous events and equally rare mass extinctions suggest that the resultant global warming may have caused mass extinctions, such as at the Permian–Triassic boundary (Siberian traps) (*see* **Palaeozoic: End Permian Extinctions**). Rapid episodes of global warming may also cause the sudden decomposition of volatile methane clathrate, which constitutes a massive reservoir of carbon, semi-permanently stored within sediments at high pressures and low temperatures, both on land and at sea. The abrupt release of methane (a greenhouse gas that quickly reverts to $CO_2$, another greenhouse gas) into the atmosphere would accelerate global warming. Precisely this may have happened at the Paleocene–Eocene boundary (**Figure 7**) as well as at other times in Earth history.

## Glacial–Interglacial Cycles

$CO_2$ and $CH_4$ concentrations can be reconstructed back more than 200 000 years using air bubbles locked up in the Antarctic ice-cap; past concentrations clearly parallel changes in temperature during the more recent glacial–interglacial cycles (**Figure 8**), although they appear to lag consistently behind temperature by as much as 1000 years. This time lag suggests that whatever mechanisms are involved, a finite amount of warming is required before $CO_2$ and $CH_4$ outgassing becomes significant. Because temperature changes in Antarctica generally preceded temperature changes in the northern hemisphere, it is also possible that $CO_2$ changes may have been caused initially by changes in climate, but that increased $CO_2$ and other trace gases acted to amplify those climatic changes. The 'coral reef hypothesis' was originally proposed to explain the observed increase in atmospheric $CO_2$ recorded in glacial ice during times of deglaciation. This hypothesis considers that shallow shelf (or neritic) carbonate deposition is significantly lower when shelf space is decreased during glacial sea-level lowstands, whereas carbonate deposition would increase during sea-level transgressions. Because carbonate precipitation returns $CO_2$ back into the atmosphere, an increase in shelf space during deglaciation would serve to increase atmospheric $pCO_2$ and decrease the alkalinity of surface seawater, at least in the short term (**Figure 9**). Calculations suggest that sea-level can account for at least half of the observed increase in $CO_2$ levels. On longer time-scales, the effects of sea-level on carbonate deposition would be offset by changes to the deep-sea preservation of carbonate shells, but the efficacy of this feedback depends on the link between the shallow environment and the deep, which is limited by the sluggish pace of ocean

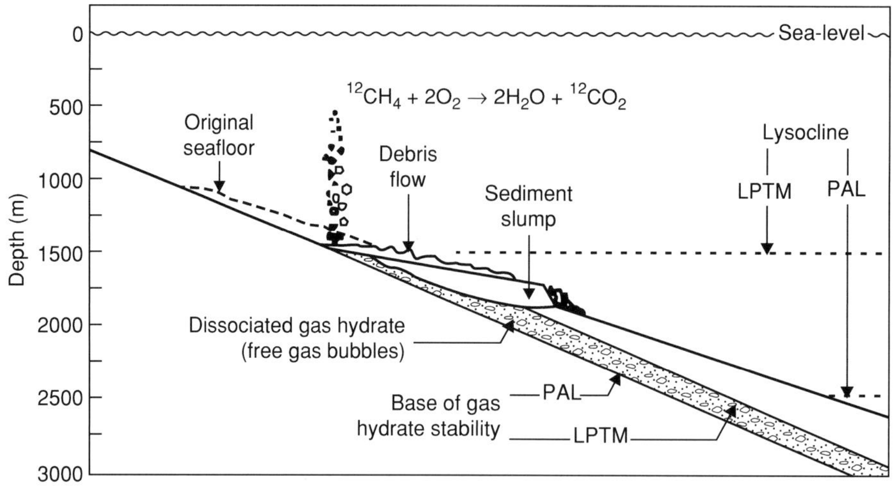

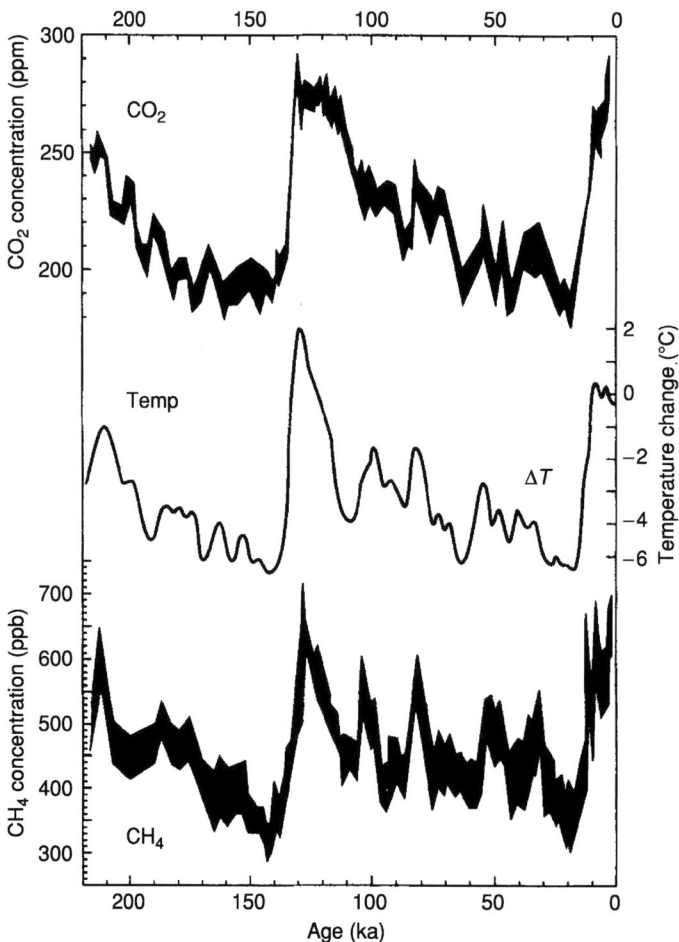

**Figure 7** Schematic diagram showing the Latest Paleocene Thermal Maximum (LPTM) dissociation hypothesis. Massive quantities of $^{12}C$-rich $CH_4$ are released into the water column via sediment failure, causing short-lived negative excursions in the marine carbonate $\delta^{13}C$ record and increases in atmospheric $CO_2$. PAL, Paleocene. Reprinted from Katz ME, Pak DK, Dickens GR, and Miller KG (1999) The source and fate of massive carbon input during the Latest Paleocene Thermal Maximum. *Science* 286: 1531–1533.

**Figure 8** Atmospheric $CO_2$, temperature, and $CH_4$ records determined from the Vostok ice-cores in Antarctica. Reprinted from Wigley TML and Schimel DS (eds.) (2000) *The Carbon Cycle*. Cambridge: Cambridge University Press.

circulation. Although the coral reef hypothesis seems unlikely to explain the entire $CO_2$ shift between glacials and interglacials, the same mechanism would have been more effective before the introduction of shelly plankton and before the establishment of a CCD.

The causes of $CH_4$ changes are unlikely to be the same as for $CO_2$ changes. The largest sources of $CH_4$ are the methane clathrate reservoirs that form beneath cold ocean floors at depths of several hundreds of metres (**Figure 7**); on land, the $CH_4$ reservoirs are in deep lakes or deep beneath the tundra floor in Siberia, Greenland, and North America (*see* **Petroleum Geology: Gas Hydrates**). Rising $CH_4$ levels could represent a response to global warming in the form of decomposition of these volatile methane reservoirs, but whether on land, beneath the seafloor, or beneath retreating glaciers is unclear. Because $CH_4$ is unstable in the presence of oxygen and has a very short half-life in the atmosphere (about 10 years), any increase in $CH_4$ levels would translate eventually into a more sustained increase in the $CO_2$ content of the atmosphere.

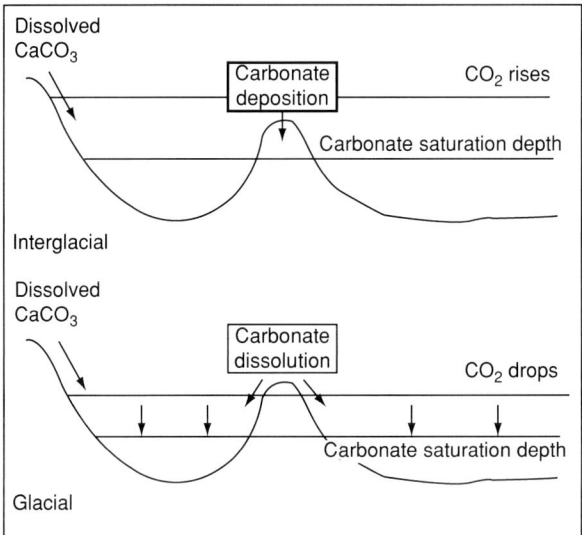

**Figure 9** A qualitative comparison of $CaCO_3$ deposition and erosion to and from carbonate reef platforms from interglacial to glacial episodes. Reprinted from Milliman JD (1974) Marine carbonates. In: Milliman JD, Mueller G, and Foerstner U (eds.) *Recent Sedimentary Carbonates*, vol. 1. Berlin: Springer-Verlag.

## Anthropogenically Induced $CO_2$ Increase and Future Predictions

Atmospheric $CO_2$ levels have been measured directly since 1957 at Mauna Loa, Hawaii; these and other data reveal that $PCO_2$ has risen by nearly 40% since 1800, most of half of this rise occurring during only the past half century (**Figure 10**). Ice-core

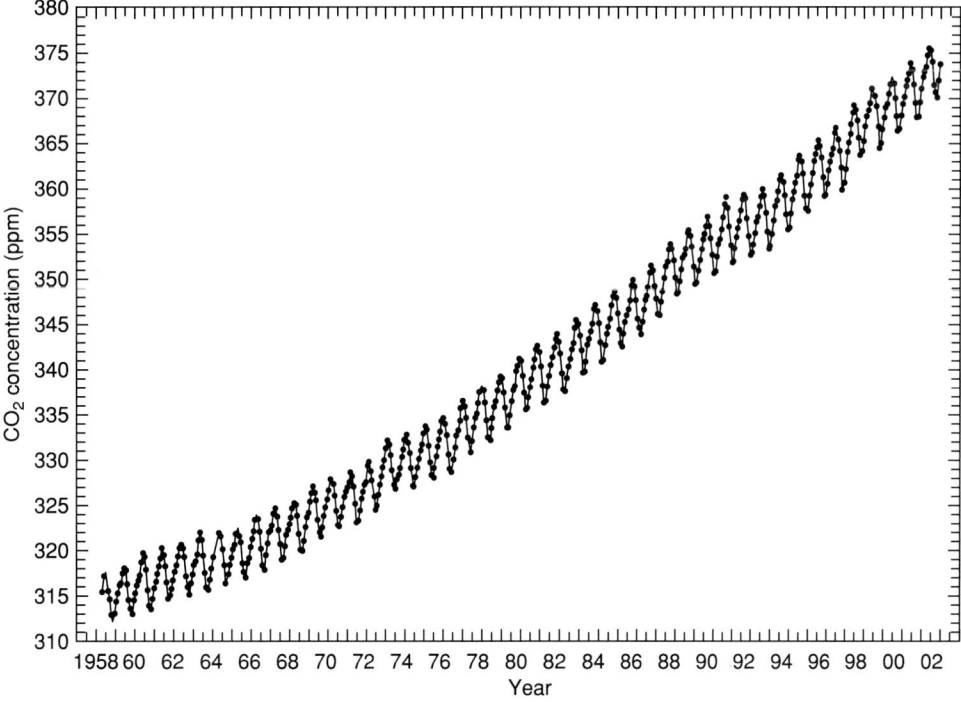

**Figure 10** The 'Keeling' curve. Measurements of monthly average atmospheric $CO_2$ concentrations at the top of Mauna Loa, Mauna Loa Observatory, Hawaii. Source: CD Keeling and TP Wharf, Scripps Institute of Oceanography, La Jolla, California, USA (available on the Internet at http://cdiac.esd.ornl.gov/trends/co2/graphics/mlo144e.pdf).

measurements of trapped air demonstrate that $PCO_2$ for the 18 000 years before 1800 was much lower than it is today, and consistently averaged 280 ppm. There is a widespread agreement among scientists that the almost 100-ppm rise since the beginning of industrialization is unprecedented, being more than the increase during the last interglacial warming, which took place over thousands of years rather than over a mere 200 years. Isotopic analyses of atmospheric $CO_2$ confirm beyond doubt that this extra $CO_2$ derives in large part from the incineration of fossil fuels such as coal, oil, and gas.

The major components of the anthropogenic perturbation to the atmospheric carbon budget are anthropogenic emissions, ocean and terrestrial exchanges, and their effects on the atmospheric $CO_2$ increase (**Figure 11**). Models suggest that each year more than half of all the anthropogenic emissions from fossil fuel combustion, cement production, and tropical land clearing are offset by uptake in the oceans, by forest regrowth largely in the northern hemisphere, and by enhanced growth caused by increased levels of nutrients, mostly $CO_2$ and nitrate from fertilizers. Although several potential feedbacks exist to further counteract $CO_2$ increase through the effects of climate on ocean and terrestrial biosphere uptake of $CO_2$, it is not thought that these are significant in the short term. However, the possible effects of climate change on ocean circulation could be very significant indeed.

Much current research is devoted to predicting how $CO_2$ levels are likely to rise in the future and what impact this will have on global climate, sea-level, and biodiversity. The greenhouse effect of $CO_2$ is roughly logarithmic, which means that each factor-of-two increase in $CO_2$ produces roughly the same amount of warming. Global circulation models predict a doubling of $CO_2$ levels by the end of the twenty-first century (**Figure 12**) associated with global warming of between 1.5 and 4.5°C. Although the effects of global warming will be felt worldwide, changes in temperature will be distributed unevenly, with by far the greatest impact felt in the high latitudes of the northern hemisphere. Because the cycling of carbon in the terrestrial and ocean biospheres occurs slowly, on time-scales of decades to millennia, the effect of additional $CO_2$ through industrialization inevitably represents a long-lasting disturbance to the carbon cycle. Model predictions of future atmospheric $CO_2$ levels indicate that they will continue to

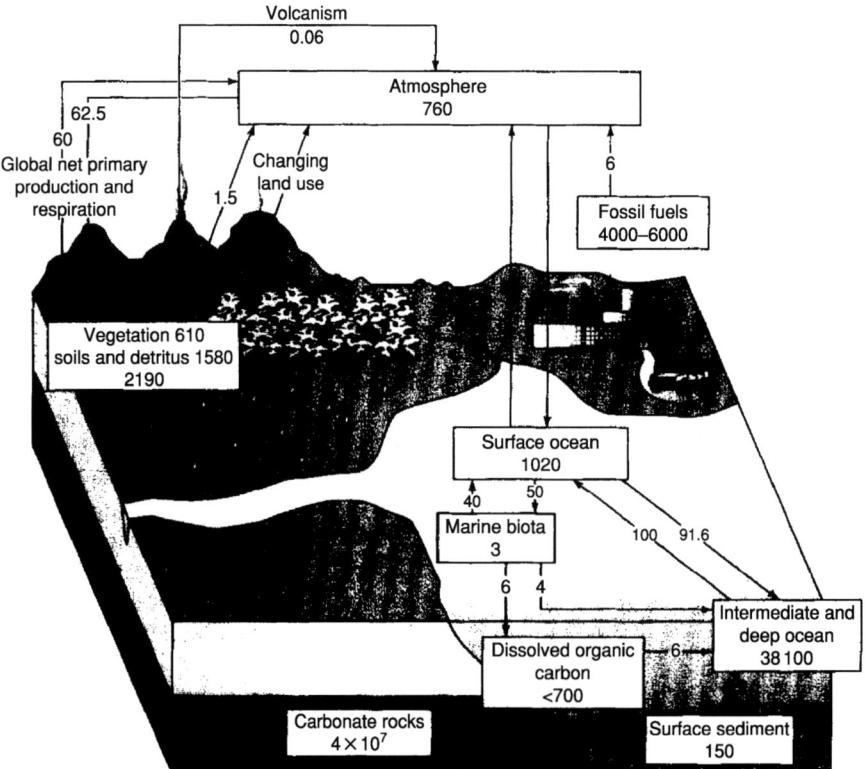

**Figure 11** Major carbon reservoirs and fluxes (in gigatons and gigatons year$^{-1}$, respectively) in the global anthropogenically influenced carbon cycle. Reprinted from Kump LR, Kasting JF, and Crane RG (1999) *The Earth System*. Upper Saddle River, NJ: Prentice Hall.

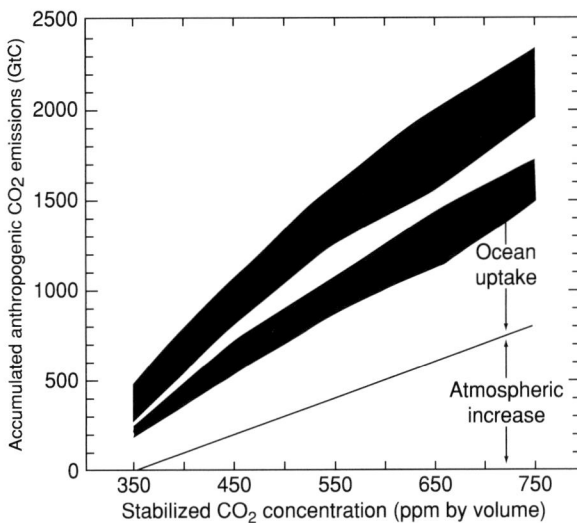

**Figure 12** Accumulated anthropogenic $CO_2$ emissions over a period beginning in 1990 and projected out to 2200 (in gigatons of carbon), plotted against the final stabilized concentration level. Reprinted from Wigley TML and Schimel DS (eds.) (2000) *The Carbon Cycle*. Cambridge: Cambridge University Press.

rise throughout the present century, even if emissions are held constant. If the level at which $CO_2$ stabilizes is kept below 750 ppm, almost three times preindustrial levels, emissions will still have to be cut relative to today's emission rates.

There is continual improvement in our understanding of the modern-day carbon cycle, and increasingly sophisticated models are being developed to describe the Earth system (*see* **Earth System Science**). However, it is already clear that we have begun an unprecedented experiment, the course of which we cannot control over the short-term, but which will shortly become clear.

## See Also

**Atmosphere Evolution**. **Earth System Science**. **Gaia**. **Palaeoclimates**. **Petroleum Geology:** Gas Hydrates; The Petroleum System. **Sedimentary Environments:** Carbonate Shorelines and Shelves; Reefs ('Build-Ups'). **Sedimentary Processes:** Fluxes and Budgets. **Sedimentary Rocks:** Limestones. **Solar System:** Venus. **Weathering**.

## Further Reading

Berger WH (1982) Increase of carbon dioxide in the atmosphere during deglaciation: the coral reef hypothesis. *Naturwissenschaften* 69: 87–88.

Berner RA (1992) Weathering, plants and the long-term carbon cycle. *Geochimica et Cosmochimica Acta* 56: 3225–3231.

Gaillardet J, Dupré B, Louvat P, and Allègre CJ (1999) Global silicate weathering and $CO_2$ consumption rates deduced from the chemistry of the large rivers. *Chemical Geology* 159: 3–30.

Katz ME, Pak DK, Dickens GR, and Miller KG (1999) The source and fate of massive carbon input during the Latest Paleocene Thermal Maximum. *Science* 286: 1531–1533.

Kump LR, Kasting JF, and Crane RG (1999) *The Earth System*. Upper Saddle River, NJ: Prentice Hall.

Pavlov AA, Kasting JF, Brown LL, Rages KA, and Freedman R (2000) Greenhouse warming by $CH_4$ in the atmosphere of early Earth. *Journal of Geophysical Research* 105: 11981–11990.

Royer DL, Berner RA, and Beerling DJ (2001) Phanerozoic atmospheric $CO_2$ change: evaluating geochemical and paleobiological approaches. *Earth Science Reviews* 54: 349–392.

Sunquist ET and Broecker WS (1985) *The Carbon Cycle and Atmospheric $CO_2$*. American Geophysical Monograph 32. Washington, DC: American Geophysical Union.

Wigley TML and Schimel DS (eds.) (2000) *The Carbon Cycle*. Cambridge: Cambridge University Press.

# CHINA AND MONGOLIA

**H Wang and Shihong Zhang**, China University of Geosciences, Beijing, China
**Guoqi He**, Peking University, Beijing, China

© 2005, Elsevier Ltd. All Rights Reserved.

## Introduction

China and Mongolia are among the most composite in geological structures in the Asian countries. The tectonic units of different ranks comprise tectonic domains, platforms, and massifs with Precambrian basement, and Phanerozoic orogenic belts and zones. A tectonic domain usually encloses a platform (craton) and its surrounding orogenic belts, which are composed of orogenic zones of various ages and interstitial Precambrian median massifs. In the outline tectonic map of Asia (**Figure 1**), are shown the main cratons, the major massifs, the orogenic belts, and the major sutures of various ages. The history of crustal evolution may be subdivided into tectonic megastages based on fundamental

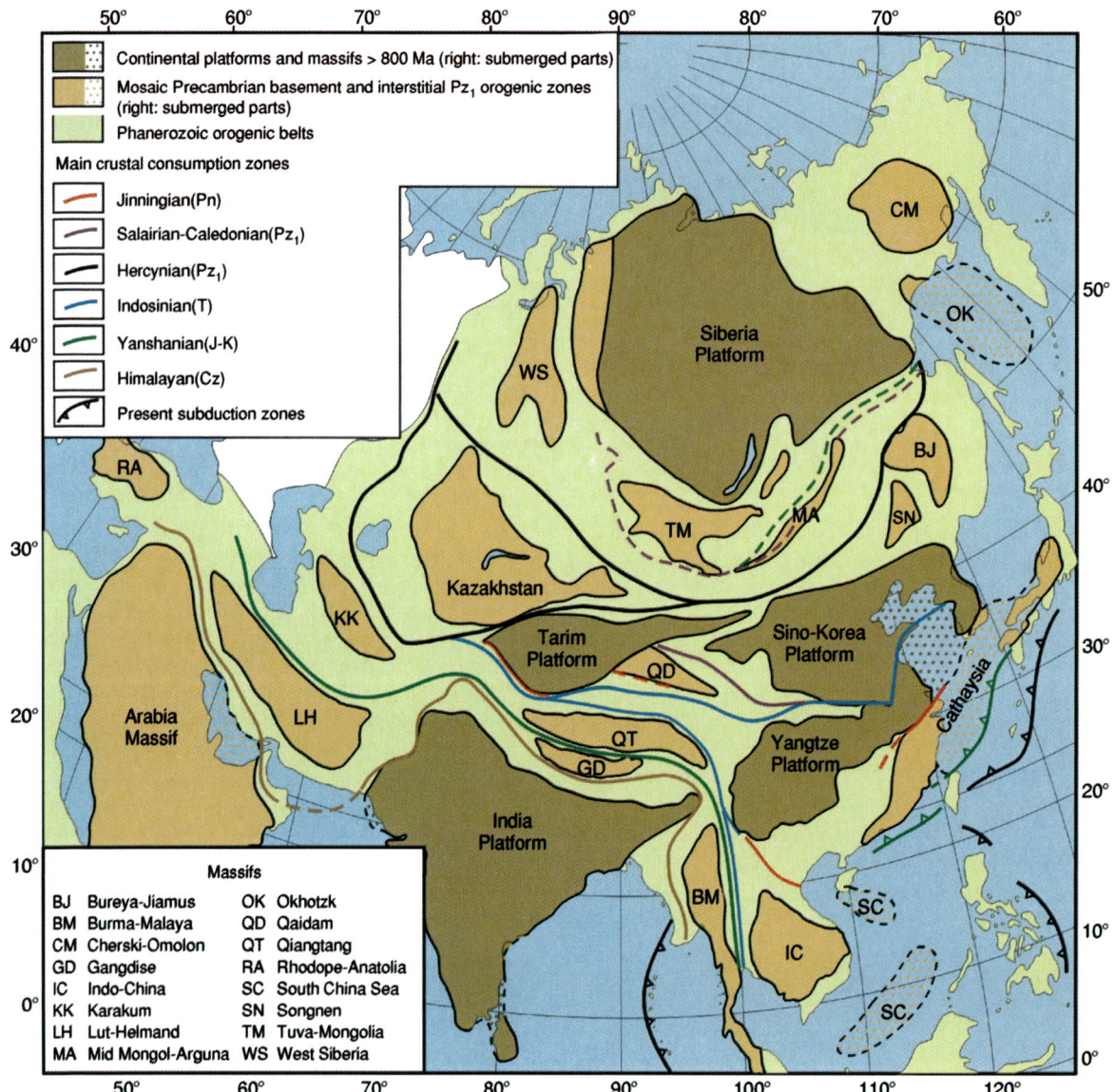

**Figure 1** Simplified tectonic map of Asia.

changes of tectonic frame of continental and even global scale, and tectonic stages marked by pronounced changes of widespread tectonic regimes, during which revolutionary orogenic epochs of short duration and rapid changes may be recognised (**Figure 2**). The global stratigraphic chart used follows the International Stratigraphic Chart (2000) with some changes in the Precambrian part. The tectonic development of China and Mongolia are discussed in terms of the megastages and the main geologic events that occurred in the various orogenic epochs. Emphasis are put on the Neoproterozoic Jinningian and the Upper Triassic Indosinian orogenies in China and on the Palaeozoic Salairian and Hercynian in Mongolia. The terms 'Caledonian' and 'Hercynian' for orogenies in China have different usages from their original locations in Europe. A world palaeocontinental reconstruction map for the Pangaea in mid-Permian time (**Figure 4**) is presented to show the possible position of the component parts of China and Mongolia and the world floral provinces at that time.

## The Geology of China

### The Main Tectonic Units and Crustal Evolution of China

The main tectonic units of China comprise three principal continental platforms (cratons), Sino-Korea, Yangtze, and Tarim, a deformed palaeocontinent (Cathaysia), and orogenic belts of various ages. The orogenic belts are situated between the platforms and

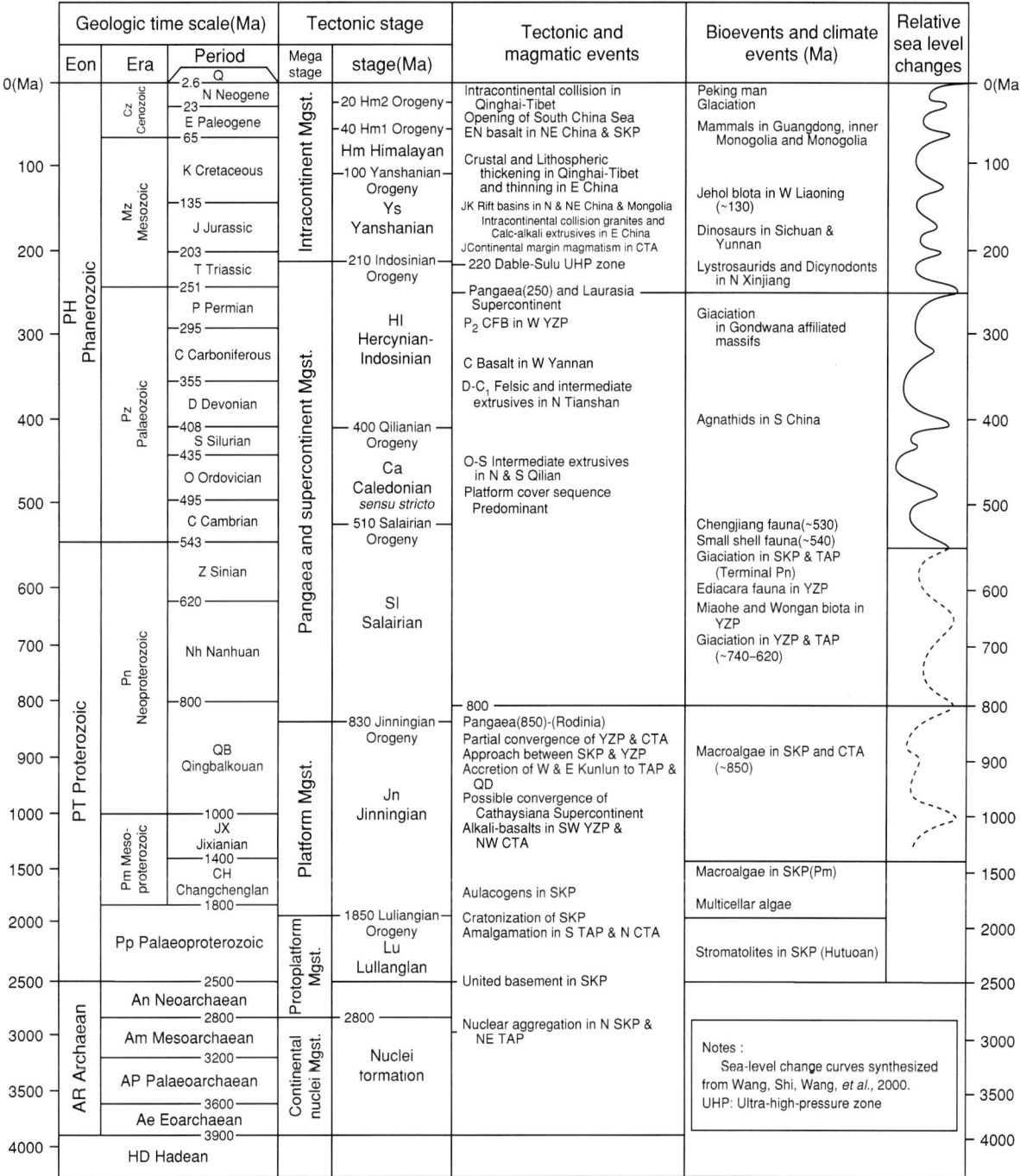

**Figure 2** Tectonic megastages and geoevents in China and Mongolia.

major massifs, and contain small massifs that were probably split from neighbouring palaeocontinents. The plate boundary does not lie at the platform border but in a demarcation zone between the once distantly opposite marginal tracts of the palaeocontinents. We have called the major crustal sutures, or the demarcation zones that represent the consumed open seas, convergent crustal consumption zones (CZ), and the sutures that represent accreted island arcs onto the palaeocontinents, accretional crustal consumption zones (AZ).

Geologically, three main regions may be recognised in China. The northern region comprises the narrow Altai-Arguna Belt representing the southernmost part of the wide southern continent marginal tract of Siberia Platform, and the Tianshan-Xingan Belt representing the northern continent marginal tract of Tarim (TAP) and Sino-Korea (SKP) platforms.

The middle region consists of TAP and SKP, the Qilian Caledonides between them, and the Kunlun-Qinling Belt, the central orogenic belt of China, which is mainly composed of the Indosinides. The southern boundary of this region is the Indosinian Muztagh-Maqen convergent zone (MMCZ) in the west and the superimposed Indosinian Fengxian-Shucheng convergent zone (FSCZ) in the east. The southern region includes two parts: the eastern part, South China, consists of the Yangtze Platform (YZP), the Cathaysia palaeocontinent (CTA), and the Caledonides and Indosinides between them, while the western part covers the Qinhai-Tibet Plateau and the broad Indosinides to the south of MMCZ.

The history of crustal evolution of China may be subdivided into five megastages: (i) continental nucleus formation (*ca.* 2.7–2.8 Ga); (ii) protoplatform formation (*ca.* 1.9–1.8 Ga); (iii) platform formation (*ca.* 0.85–0.8 Ga); (iv) Laurasia Supercontinent or Pangaea formation (*ca.* 230–210 Ma); and (v) intracontinental development (210 Ma-Present) (**Figure 2**). Two orogenic epochs of revolutionary changes occurred in the Jinningian (*ca.* 1000–830 Ma) and the Indosinian (*ca.* 230–210 Ma). The close of the Jinningian Orogeny probably witnessed a convergence of the Chinese platforms and massifs to form the supercontinent of Cathaysiana, which was a part of the Neoproterozoic Rodinia (850 Ma). After the close of the Indosinian Orogeny, the tectonic frame of China underwent a basic change from a contrast between the north and the south to that between the east and the west. Thus, the crustal evolution of China underwent three stages; (i) the Jinningian and pre-Jinningian (Archaean to Early Neoproterozoic >800 Ma); (ii) the post-Jinningian to the Indosinian (Nanhuan to Triassic); and (iii) the post-Indosinian (Jurassic to Quaternary) (**Figure 2**).

## China in the Pre-Jinningian and Jinningian (Archaean to Qingbaikouan)

The time span from Archaean to Early Neoproterozoic (*ca.* 800 Ma) may be subdivided into three megastages which resulted respectively in the formation of continental nuclei at the end of the Mesoarchaean (*ca.* 2.8 Ga), the formation of protoplatforms at the end of the Palaeoproterozoic (*ca.* 1.8 Ga) through the Luliangian Orogeny, and the formation of platforms at the end of the Early Neoproterozoic (*ca.* 800 Ma) through the Jinningian Orogeny.

**The continental nuclei** Several continental nuclei may be recognized in SKP, by the presence of Mesoarchaean and older metamorphic supracrustal and TTG rocks. The Jiliao Nucleus (Jl) encloses northern Shanxi, northern and eastern Hebei, northern Liaoning and Jilin provinces (**Figure 3**). U-Pb and Pb-Pb ages of 3850–3550 Ma, representing the Eoarchaean primordial sialic crust and supercrustals, were identified from Caozhuang, eastern Hebei, and Anshan, Liaoning. Mesoarchaean rocks older than 2.8 Ga are known from the Jiaodong Nucleus (Jd) in eastern Shandong, the Sanggan gneiss, the Jining Nucleus (Jn) in the southern Inner Mongolia, and northern Shanxi. The Ordos Nucleus (Or) is inferred to exist under the northern part of the Ordos basin based on geophysical as well as geological data. Further in the west is the small Alxa Nucleus (Ax) near the western border of SKP.

In the Yangtze Platform (YZP), Mesoarchaean TTG rocks older than 2.8 Ga are known from the Kongling Group in the Yangtze Gorges region, probably representing the uplifted eastern margin of the Chuanzhong Nucleus (Cz) beneath the Sichuan basin. In the eastern part of the Tarim Platform (TAP), Mesoarchaean rocks are reported from the Quruktagh Nucleus (Qr) and from the Dunhuang Nucleus (Dh), where a single grain zircon U-Pb age of 3.6 Ga of Eoarchaean age was recently obtained. Generally, these ancient continental nuclei are distributed in the northern part of SKP and the north-eastern part of TAP. It is noteworthy that isotopic model age studies for various Archaean rocks in the SKP show a cluster around 2.8–2.7 Ga, which denotes an epoch of rapid continental growth, as in many other cratons in the world.

### The Protoplatforms and the Luliangian Orogeny

The Neoarchaean is widely distributed in the platform regions of China, especially in SKP. The juvenile crust generated in the Neoarchaean may be mostly assigned to the granite-greenstone terrains with a high proportion of granitoid intrusives, as in the Taishan Complex in Shandong. The Neoarchaean usually occurs as granite and TTG belts around and within the reworked nuclear regions in SKP. On the other hand, Neoarchaean greenstone belts, represented by the Wutai, Dengfeng and Taihua groups in Shanxi and Henan, contain more khondalites and mafic volcanics, and are probably of extensional origin. In general, the Neoarchaean witnessed a profuse intrusive and extrusive magmatism of 2.6–2.4 Ga age throughout SKP, which is responsible for the formation of united Archaean basement.

The Palaeoproterozoic (Pp) in SKP, generally called the Hutuoan 'System' (2500–1800 Ma), often occurs in long narrow belts and represents ancient aulacogens formed on the Archaean basement. The aulacogen sequence is mainly composed of a lower fluvial, a middle immature volcanisedimentary, and an upper molasse deposits. This sequence occurs in the Hutuo

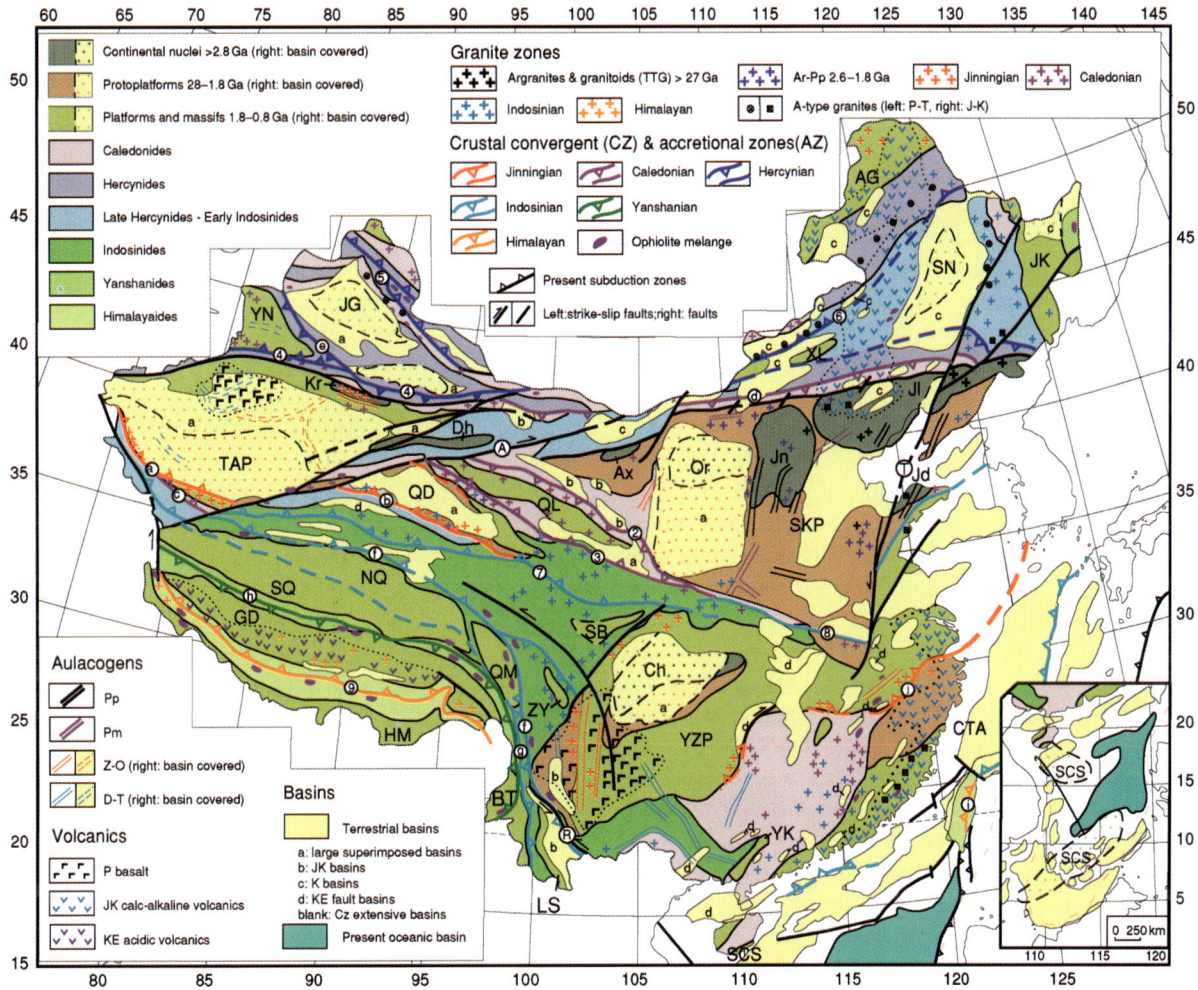

**Figure 3** Outline map showing the tectonic units and crustal evolution of China. Hercynian and Yanshanian granites are wide spread and planar in distribution, and are not shown on the map.
Tectonic units: AG, Arguna Massif; Ax, Alxa Nucleus; BT, Baoshan-Tengchong Massif; Ch, Chuanzhong Nucleus; CTA, Cathaysia palaeocontinent; Dh, Dunhuang Nucleus; GD, Gangdise Massif; HM, Himalaya Massif; Jd, Jiaodong Nucleus; JG, Junggar Massif; JK, Jamus-Xingkai Massif; Jl, Jiliao Nucleus; Jn, Jining Nucleus; Kr, Kuruktagh Nucleus; LS, Lanping-Simao Massif; NQ, North Qiangtang Massif; Or, Ordos Nucleus; QL, Qilian Massif; QM, Qamdo Massif; SB, Songpan-Bikou Massif; SCS, South China Sea Massif; SKP, Sino-Korea Platform; SN, Songnen Massif; SQ, South Qiangtang Massif; TAP, Tarim Platform; XL, Xilinhot Massif; YN, Yining Massif; YZP, Yangtze Platform; ZY, Zhongza-Yidun Massif.
Crustal consumption convergent zones (CZ): ① JSCZ, Jiangshan-Shaoxing (Jinningian); ② QQCZ, Qilian-Qinling (Caledonian); ③ SQCZ, South Qilian (Caledonian); ④ STCZ, South Tianshan (Hercynian); ⑤ EACZ, Ertix-Almantai (Hercynian); ⑥ HGCZ, Hegenshan (Hercynian); ⑦ MMCZ, Muztagh-Maqen (Indosinian); ⑧ FSCZ, Fengxian-Shuchang (Indosinian); ⑨ YZCZ, Yarlung-Zangbo (Himalayan).
Accretional zones (AZ): ⓐ WKAZ, West Kunlun (Jinningian); ⓑ EKAZ, East Kunlun (Jinningian); ⓒ KDAZ, Kudi (Indosinian); ⓓ OSAZ, Ondur Sum (Caledonian); ⓔ NTAZ, North Tianshan (Hercynian); ⓕ JSAZ, Jinshajiang (Indosinian); ⓖ CMAZ, Changning-Menglian (Indosinian); ⓗ BNAZ, Bangong-Nujiang (Yanshanian); ⓘ LJCZ, Liji (Himalayan).
Strike-slip faults: Ⓐ Altun, Ⓣ Tanlu, Ⓡ Red River.

and Gantaohe aulacogens in the Wutai-Taihang region, in the Qinglong Aulacogen in north-eastern Hebei, in the Liaohe and Fenzishan aulacogens in eastern Shandong and southern Liaoning (**Figure 3**). Through the Luliangian Orogeny of 1.9–1.7 Ga age, these aulacogen rocks were intensely folded and regionally metamorphosed, sometimes to a high chlorite schist facies.

The Luliangian Orogeny brought about the formation of a protoplatform in North China, upon which was deposited the Meso- and Neoproterozoic paracover. This orogenic event is also seen in the TAP and Qaidam Massif (QD), where the Mesoproterozoic and Neoproterozoic sequences are always separated from the basement by a fragmented unconformity. In YZP, the Palaeoproterozoic basement is probably present to the south of the Chuanzhong Nucleus (Ch), and the Kangdian belt in western YZP is mainly composed of Palaeoproterozoic metamorphic rocks. In CTA, Neoarchaean and Palaeoproterozoic rocks, with Luliangian metamorphism of amphibolite facies, occur in western Zhejiang and in the Wuyi Mountains in north-western Fujian.

**The platforms and the Jinningian Orogeny** In SKP, the Mesoproterozoic and Early Neoproterozoic strata are widespread and are divisible into the Changchengian (Ch), the Jixianian (Jx), and the Qinbaikouan (Qb) 'systems'. They are generally correlatable through acritarch and stromatolite assemblages between the main platforms of China. In the Yanshan-Taihang region, aulacogen (**Figure 3**) deposits consisting of the Lower Changchengian fluvial clastics, carbonates, and high potassium volcanics, are followed by the widely transgressive Gaoyuzhuang carbonates bearing the macro-alga *Grepania* in the upper part (1.5–1.4 Ga). The Jixianian is also widespread and contains stromatolite carbonates of great thickness. The Qingbaikouan represents a platform cover and is limited in distribution. The macro-algal assemblages of the Jixianian *Grepania* and the Qingbaikouan *Tawuia-Longfengshania* Assemblages (900–800 Ma) are almost identical to those found in the Greyson Shales in Montana and in the Little Dal Formation in the McKenzie Mountains in the Belt Supergroup of western North America. The *Tawuia* beds are also found in Hainan Island, which was probably a part of the Cathaysia palaeocontinent. These indicate that Laurentia and the Chinese palaeocontinents may have been near each other during Meso- and Neoproterozoic times.

Most platforms and massifs in China were dominated by an extensional tectonic regime in the Meso- and Neoproterozoic, as shown by the Xionger Aulacogen, with bimodal eruptives in Henan and southeastern Shanxi, and the Bayan Obo Aulacogen near the northern margin of SKP. Aulacogens of similar age are also developed in the Quruktagh region of northern TAP and in the Kunming region of southwestern YZP. All these aulacogens ended without any diastrophism. On the southern margin of SKP, there may have developed an island-arc system with the Qinling Group as the arc and the Kuanping Group as the back-arc basin. Both groups yield Mesoprotrozoic isotopic ages.

The Jinningian Orogeny of ca. 1000–830 Ma age has left clear records in many parts of China. The Qinling region, mainly South Qinling, was characterized by a complicated rift system probably formed in Late Mesoproterozoic to Early Neoproterozoic, which consisted of discrete massifs (Douling, Fuping) and trough deposits with bimodal volcanicism (e.g., the Xixiang group). The Jinningian Orogeny is represented by island arcs and marginal seas in the southern margin of North Qinling by the Songshugou ophiolite (983 Ma) and by an Early Neoproterozoic granite zone (e.g., the Dehe granite of *ca.* 950 Ma), denoting a northward subduction and accretion through an arc-continent collision. A similar southward accretion, with subduction-type granites is also found in the Hanzhong area of YZP. The common cover of the Sinian over the rifted elements implies the presence of a united South Qinling Belt. Therefore, the Jinningian Orogeny brought about the consolidation of the rifted region and the mutual approach of SKP and YZP.

Along the south-western margin of YZP in central Yunnan, to the east of the Early Mesoproterozoic Dahongshan metamorphic volcanisedimentary Belt, was developed a broad aulacogen trending north–south for hundreds of kilometres, composed mainly of the Mesoproterozoic Kunyang Group and equivalent strata of huge thickness. These rocks were intensely folded and intruded by the Jinningian granites (850–750 Ma), and may be interpreted as a wide back-arc basin lying to the east of the Dahongshan arc belt in central Yunnan.

In the Jiangnan uplift along the south-eastern margin of YZP, deformed arc-type turbidites with ophiolite mélange and Adakite zones dated at *ca.* 970 Ma represent an arc-continent collision of Early Jinningian age. In fact, the Jinningian orogenic zone was charaterized by volcanisedimentaries and granites extending from northern Zhejiang right to the border area between Hunan and Guangxi. At the close of the Jinningian Orogeny, a partial collision between YZP and CTA occurred along the Jiangshan-Shaoxing convergent zone (JSCZ), which closed at around 830 Ma. The ocean basin between YZP and CTA was probably separated by an archipelagic belt composed of small massifs near the Hunan-Jiangxi border. The north-western margin of CTA in the Neoproterozoic was also active and extends from Wuyi in North-western Fujian to Hainan Island.

Jinningian granites are distributed along the southern margin of TAP and the Qaidam Massif in the northern Kunlun Mountains and are also found in the northern margin of Qaidam. These granite zones represent active continental margins or continent-arc collision zones (**Figure 3**). The Jinningian granites are seldom found to the east of East Qinling, except in eastern Shandong along the Sulu Belt, although a Jinningian metamorphic event is suspected to be represented by the Dabie UHP belt. In summary, the main platforms and massifs in China seemed to have converged to form a loosely united Cathaysiana

Supercontinent at around 850 Ma, concurrently with the formation of the Neoproterozoic Rodinia.

### China in the Post-Jinningian to the Indosinian (Nanhuan to Triassic)

The time span from the Nanhuan to the Triassic (800–208 Ma) is traditionally subdivided into Caledonian, Hercynian, and Indosinian stages. The traditional Sinian (800–543 Ma) is subdivided into the Nanhuan (800–620 Ma) and the revised Sinian (620–543 Ma) according to the new Regional Chronostratigraphical Chart of China published by the Third All-China Stratigraphical Commission in 1999. The Hercynian and Indosinian are here integrated and called the Hercynian-Indosinian Stage (**Figure 2**).

**The Caledonian stage** The Caledonian in China is usually subdivided into a lower division of Xingkaian (Salairian) covering the Nanhuan and the Sinian, and an upper division covering the Middle Cambrian to the Silurian. The Caledonides are mainly distributed in the Qilian Mountains region between SKP and the Qaidam Massif and in South China between YZP and CTA. They are also found to the north of SKP and TAP (**Figure 3**).

The Nanhuan in YZP is characterized by clastic, glaciogenic, and cold-water deposits roughly corresponding to the Cryogenian in the International Stratigraphic Chart. Continental ice-sheets of *ca.* 740–620 Ma age seem to have been confined to western YZP and the Quruktagh region of TAP, but mountain and maritime glaciations are more widespread. In the type region of the Yangtze Gorges in the YZP, the Sinian includes the Doushantuo cap carbonates, phosphate and black shales bearing the Miaohe or Wengan biota (Pb-Pb isotopic age *ca.* 600 Ma), and the Dengying Formation bearing the Ediacara biota (*ca.* 590 Ma). In SKP, only the Sinian is developed in the peripheral parts and bears a rich metazoan fauna comparable with the Ediacara in northern Anhui. A higher Luoquan glacial horizon was reported all along the south-western border of SKP, the northern border of Qaidam Massif, the northern TAP and the Tianshan regions: it is probably younger than 600 Ma in age.

In most parts of eastern China, the Lower Palaeozoic began with an Early Cambrian transgression from southern YZP to SKP, with well-established trilobite zones for correlation. This indicates that the two platforms were not far from each other, although rifting seems to have begun in the Sinian, as is shown by entirely different Nanhua-Sinian sequences between the two platforms. The Early Cambrian Chengjiang Lagerstatte (*ca.* 525 Ma), of great significance in life evolution, contains arthropods and chordates, especially *Haikouichthys*, which might be the 'first fish' on Earth. No obvious break is known between the Sinian and the Cambrian, especially in South China, where bathyal carbonaceous silicolites are continuous in central Hunan and western Zhejiang. The continuous Lower Palaeozoic passive margin bathyal deposits have provided seldom seen complete sequences of Cambrian agnostid trilobite zones and Ordovician to Silurian graptolite zones, which are ideal for the designation of chronostratigraphical boundaries. In the residual sea between YZP and CTA, rifted uplifts and troughs were developed in which varied kinds of sediments were laid down. The marine realm began to shrink in the Late Ordovician, and the Caledonian Front, which started in the Wuyi Mountains, seems to have shifted westward, until the thick foreland basin deposits of Early Silurian age were formed within the YZP in westernmost Hunan. Caledonian granites and metamorphism are scattered, and no clear collision zone is found in the broad Caledonides in South China, which were 'filled up', rather than folded orogenic zones. In southern Hainan Island, Cambrian trilobites of Australian affinity are found, which may represent the northern margin of SCS (**Figure 2**), probably still far away to the south at that time.

The Caledonides between SKP and Qaidam on both sides of the Central Qilian Massif (QL) are represented by distinct collision belts and associated granite zones. In North Qilian a complete Caledonian orogenic sequence developed composed of Cambrian and Ordovician island arc volcanics, Middle and Upper Ordovician and Lower Silurian ophiolite suites, and Devonian molasse deposits. The QQCZ (**Figure 4**) is prominent and continues south-eastwards into North Qinling. Late Cambrian ophiolite mélanges are also found in Lajishan in South Qilian, which may connected to the SQCZ. The SKP generally lacks the Upper Ordovician and Silurian except in the western margins, and is bordered to the north by a narrow Caledonide strip (south of OSAZ), composed of Cambro-Ordovician metavolcanics unconformably overlain by fossiliferous Silurian sediments. Caledonides are also known around the Junggar and Yining massifs in northern Xinjiang. In TAP, the Cambrian and Ordovician are of platform cover type, but display a slope and deep-sea facies in the Manjiar depression in the eastern part. The Silurian is incomplete and limited in distribution. Stable Lower Palaeozoic deposits reported from Himalaya, northern Gangdise, and north-western Qiangtang indicate the existence of Precambrian basement in these massifs. However, the age of the basement may be as young as *ca.* 600 Ma, comparable with the Pan-African of Gondwanaland.

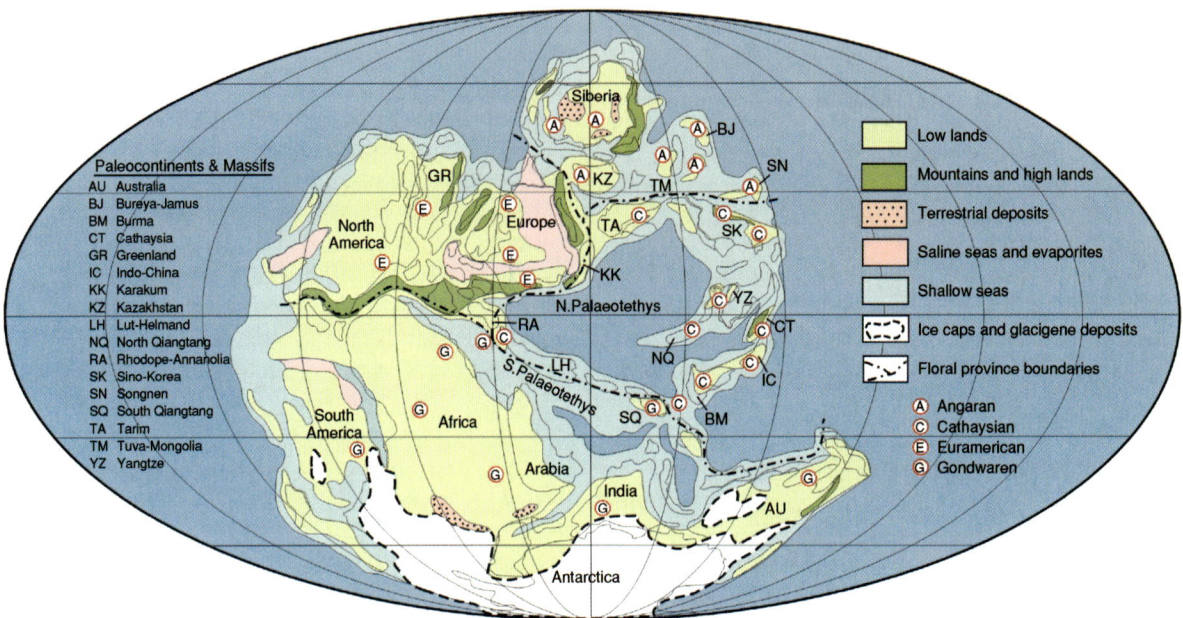

**Figure 4** A mid-Permian world palaeocontinental reconstruction.

**Laurasia Supercontinent and the Hercynian-Indosinian stage** In the Late Palaeozoic, the major part of SKP remained exposed after its upheaval in mid-Ordovician, until the Middle Carboniferous epicontinental seas began to inundate the platform. Marine facies prevailed up to the mid-Permian and are followed by Permo-Triassic paralic and terrestrial deposits. A Devonian oceanic basin existed along the China-Mongolia border, which was consumed and formed the main Hercynian HGCZ. The orogenic collision is manifested by Devonian to Early Carboniferous ophiolites and Permian A-type granites which mark the boundary between the Siberian continent marginal tract in the north and the Sino-Korean marginal tract in the south. In northern Xinjiang, the Hercynian EACZ is composed of *en echelon* sutures and represents the main boundary between the Altai-Hingan and the Tianshan-Inner Mongolia belts, and is continuous across southern Mongolia with the HGCZ. Two Hercynian collision zones documented by ophiolites occur to the south of the Junggar-Turpan and Yining massifs.

The Late Hercynian-Early Indosinian belts south of Beishan in the west, and extending via Linxi to Changchun in the east, are composed mostly of shallow marine and paralic sediments with occasional volcanics. The eastern segment is a clear boundary between the Angaran and the Cathaysian floras, but no ophiolite mélanges have been found. These marine troughs were probably filled at the end of the Palaeozoic, and are not genuine orogenic zones. Late Permian to Early Triassic terrestrial basins, formed after the Hercynian Orogeny, are widespread in northern Xinjiang, Gansu, and Inner Mongolia, and bear the tetrapod remains of *Dicynodon*, *Lystrosaurus*, and *Sinokanemeyeria*.

In the Late Palaeozoic, YZP was characterized in its northern marginal tract, the South Qinling, by thick Devonian continental slope flysch and marine to paralic Carboniferous to Triassic clastics and carbonates. To the south, both YZP and CTA witnessed Devonian transgressions from the residual Youjiang marine basin, and Late Carboniferous and Permian carbonate platforms were well developed in both regions. Extensional conditions prevailed in YZP, as is shown by the Late Permian Emei continental flood basalt. Devonian to Triassic aulacogens with different trends are known in south-western YZP, and isolated carbonate platforms were formed in the Youjiang oceanic basin, where Upper Palaeozoic deep-sea deposits with pelagic radiolarians are formed. The Youjiang marine basin was not closed until the Late Triassic, when the Indosinian Orogeny was predominant in most parts of southern China. The eastern part of YZP, now partly submerged under the sea, collided with SKP along the Dabie-Sulu HP and UHP zone through the Indosinian Orogeny, and was probably wedged in eastern SKP, with its northeastern promontary along the Imqingang Belt in Korea. Orogenic and post-orogenic Indosinian granite zones are widely distributed in southern China and in the Kunlun-Qinling Belt. Post-orogenic muscovite/two mica granites are known to the north of SKP as a result of intracontinental northward compression of SKP toward Inner Mongolia (**Figure 3**).

In western China, the main Muztagh-Maqen convergent zone (MMCZ) in the central orogenic belt closed in Late Hercynian to Indosinian times. The Indosinian Jinshajiang accretion zone, extending from West Kunlun southward to the Changning-Menglian zone (CMAZ) in western Yunnan, marks the boundary between the Yangtze-affiliated massifs in the east and the Gondwana-affiliated massifs in the west. In northern Tibet, an Indosinian suture is suspected to exist between the North and South Qiangtang massifs, mainly based on the occurrence of the Late Triassic Qiangtang ophiolite complex, which marks the southern margin of the Late Triassic flysch complex underthrust beneath South Qiangtang, and on the boundary between the Cathaysian and the Gondwanan floras (**Figure 3**). The wide Indosinides and their southern marginal massifs, North Qiangtang and Qamdo, may therefore represent the southern boundary of the newly amalgamated Laurasia Supercontinent which formed the northern half of Pangaea (**Figure 4**). The boundaries of the floral provinces are arbitrary, since mixed flora of different provinces are known; for example, the mixed Cathaysian and Angaran flora in northern Tarim. The most important world mass extinction at the end of the Permian (*ca.* 250 Ma), especially that of the marine organisms, is well represented and studied in China.

**China in Post-Indosinian Times**

The Indosinian Orogeny had caused a radical change of tectonic pattern of China from a north–south to an east–west demarcation, and the post-Indosinian marks a megastage of mainly intracontinental development. Jurassic and later seas retreated from China except in the Qinghai-Tibet region, and only sporadic marine ingressions occurred in eastern Heilongjiang, and in the border parts of TAP and CTA. The new tectonic framework and dynamics of China were chiefly controlled by interactions between the Siberia Plate in the north, the Pacific Plate in the east, and the India Plate in the south-west. The Indosinian Orogeny brought about the closure of the Late Palaeozoic Palaeotethys and the formation of the extensive Indosinides, including Kunlun-Qinling, Garze-Hon Xil, and down to Indochina and Malaysia, which formed the southern margin of the Eurasian palaeocontinent. Thus, an extensional system prevailed in East China and a successive northward accretion of the Gondwana-affiliated massifs onto Eurasia in the Tethys domain has dominated West China since the Late Mesozoic.

**Post-Indosinian tectono-magmatism and basin development in eastern China** In eastern China, the Indosinian Orogeny was followed by intracontinental collision and further welding of platforms and massifs, as is shown by the widespread Jurassic A-type granites and by the southerly-imbricated thrust zones in the Yanshan region and the westerly thrust zones in north-eastern Anhui and western Hunan. The large Tanlu fault, with a lateral shift of hundreds of kilometres in length, was probably mainly formed in the pre-Cretaceous. The newly-formed Circum-Pacific domain comprised a western belt of continental and maritime East China and an eastern belt of island arc-basin systems in the inner western Pacific. The Yanshanian Orogeny in eastern China is characterized by inner continent marginal type magmatism in the coastal belt (**Figure 3**), which originated by subduction of the Izanaqi Plate under East Asia. Consequently, inland eastern China was characterized by a combination of subduction and intra-continental collision types of magmatism, manifested by muscovite/two mica granites and high potassium calc-alkalic and shonshonitic volcanism. From a comparison between the crust and lithosphere thickness data of pre-Jurassic and Jurassic-Cretaceous in North China, based on petrogenic studies, it is found that a crustal thickening of *ca.* 15 km (40 against 50–60 km) and a lithospheric thinning of *ca.* 120 km (200–250 against 50–60 km) occurred in the Late Mesozoic. This proves that eastern China changed from a compressional to a tensional regime.

Mesozoic and Cenozoic basins in eastern China are distributed in three belts. The Ordos Basin and the Sichuan Basin in the western belt became terrestrial inland basins in the Late Permian and Late Triassic, respectively. Both were influenced by the Indosinian Orogeny that formed Late Triassic foreland basin deposits, in the southern border of Ordos north of the Qinling Orogen, and in the Longmenshan thrust belt of Sichuan Basin, respectively. The central belt comprises the Cretaceous rift basins, including the Songliao and the Liaoxi basins (bearing the Lower-Cretaceous Jehol biota (130–120 Ma), with the well preserved feathered dinosaur *Sinosauropterix* and the earliest flowering plant, *Sinocarpus*), the Cenozoic rift basins in SKP and Inner Mongolia, and the KE rifted and volcanic basins in southern China. The eastern belt consists of the offshore Cenozoic rift basins developed on the submerged palaeocontinents, which were mainly terrestrial in the Palaeogene and marine in the Neogene. Rifting and opening of the South China Sea occurred in the Oligocene and Miocene, when the SCS Massif was split into two fragments (**Figure 3**, inset map).

**The northward accretion of the Gondwana-affiliated massifs to Eurasia and the formation of Qinghai-Tibet Plateau** The Gondwana-affiliated massifs with

Precambrian basement include the Himalaya, the Gangise, the South Qiangtang, and the North Qiangtang. The Tianshuihai area of Karakorum, stable in the Lower Palaeozoic, may be a part of the North Qiangtang Massif. The Cathaysian *Gigantopteris*-flora found in the Shuanghu area of North Qiangtang, in contrast to the Gondwanan flora found in northern Gangdise, has led to a suspected biogeographic boundary between the North and South Qiangtang massifs (**Figure 3**).

The main Tethys oceanic basin to the north of the North Himalaya is represented by the main suture YZCZ. The recent discovery of Cambrian to Ordovician metamorphics in the Lhagoi Kangri zone may indicate the splitting of Gangdise from Himalaya-India in the Caledonian Stage. Continental marginal thick deposits appeared in the Permian to Late Triassic, and the marine basin may have been at its widest in the Jurassic. The northward subduction of the Himalaya oceanic plate beneath Gangdise probably began in the Late Cretaceous (*ca.* 70 Ma), and the collision of the two shown occurred in the mid-Eocene (*ca.* 45 Ma), as shown by the lower part of Linzizong volcanics in Gangdise, and the final disappearance of the residual seas. On the north side of Gangdise, a southerly subducted island-arc system of short-duration (Early Jurassic to Late Cretaceous) formed the Bangong-Nujiang suture (BNAZ). In the Cenozoic, intracontinental collisions resulting in vertical crustal thickening and shortening were prevalent in northern Tibet. Crustal thickening by subduction of the Himalaya beneath the Gangdise and subsequent collision produced muscovite/two mica granite of Early Miocene (*ca.* 20–17 Ma) age in the North Himalaya. In contrast, collision between adjacent massifs without subduction were more frequent and formed shonshonitic volcanism, as in the northern volcanic belt of North Qiangtang in the Oligocene. The Himalayan Orogeny and uplift of the Plateau occurred in two stages; Oligocene and Pliocene to Pleistocene (**Figure 2**).

It has been estimated that no less than 1000 km of north–south crustal shortening has occurred since the Early Palaeogene collision between Himalaya-India and the northern massifs, which has led to the eventual uplift of the Qinghai-Tibet Plateau. Except for the northward subduction of the Himalaya beneath Gangdise, the shortening was mostly accommodated by distributed vertical thickening of the crust, and the main strike-slip faults and thrust belts. The eastward extrusion of the Plateau in the Sanjiang belt of eastern Tibet is significant, but may not be vital in the Plateau construction. The northward indentation from Himalaya-India has been prevalent, but the southward indentation from the rigid Tarim craton and the Beishan massif is equally important in the overall dynamics of western China.

## Geology of Mongolia

### Tectonic Units and Tectonic Stages of Mongolia

Mongolia is situated in the central part of the wide and complicated orogenic belts between the Siberian Platform in the north and the North Chinese platforms SKP and TAP in the south. Mongolia is subdivided into a northern domain and a southern domain, the boundary between which runs roughly along the southern margin of the Gobi-Altai-Mandaloovo (GAB) Belt of Caledonian age (**Figure 5**).

Three megastages may be recognised in the crustal evolution of Mongolia, approximately corresponding to the megastages of China. They are: (i) Neoarchaean (An) to Early Neoproterozoic (*ca.* 850 Ma); (ii) Late Neoproterozoic to Triassic, including the Salairian, Caledonian and Hercynian Orogenies; and (iii) Indosinian stages, the Mesozoic to Cenozoic, characterized by intracontinental development (**Figure 2**).

### Mongolia in the Neoarchaean to the Early Neoproterozoic

The oldest rocks of Mongolia are found in the Neoarchaen Baidrag Complex in the southern part of the Tuva-Mongolia Massif (TMM), where a tonalite-gneiss has yielded a U-Pb zircon isotopic age of $2646 \pm 45$ Ma. In the same region, the Bumbuger Complex of granulite facies has a metamorphic zircon age of $1839.8 \pm 0.6$ Ma, which is coeval with the Luliangian Orogeny of China. Palaeoproterozoic massifs with reliable isotopic age dating are distributed in the TMM (**Figure 5**). A metamorphic age of *ca.* 500 Ma was obtained from the northern part of TMM (Songelin block), which may be attributed to the influence of the Salairian Orogeny in the region. In southern Mongolia, Palaeoproterozoic rocks may exist in the east-west trending Hutag Uul Massif (HUM) and Tzagan Uul Massif (TUM) near the southern border part of Mongolia.

Mesoproterozoic to Early Neoproterozoic rocks are widespread and form the main Precambrian basement in Mongolia. In northern TMM, the Hugiyngol Group, composed of metabasalts and metasediments including blue schists, have a metamorphic age of $829 \pm 23$ Ma. In the Mongol Altai Massif (ATM) and adjacent part of China, the thick Mongol-Altai Group, composed of highly mature terrigenous deposits, is unconformably overlain by fossil-bearing Ordovician beds. Sm-Nd model ages of *ca.* 1400–1000 Ma, obtained in westernmost Mongolia and adjacent parts of China, indicate the Precambrian age

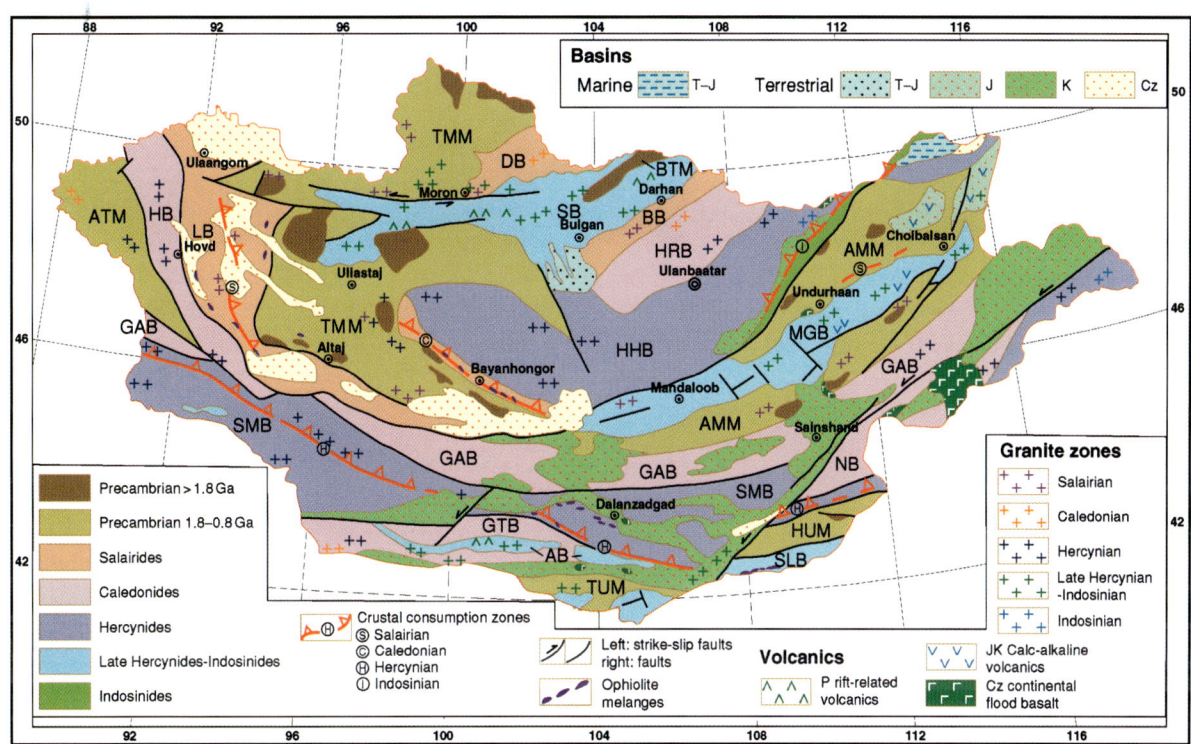

**Figure 5** Outline tectonic map of Mongolia. Compiled and integrated from the Geological Map of Mongolia, 1:5 000 000, published by Institute of Geology and Mineral Resources, 1998, Ulaanbaatar, Mongolia.
Tectonic units. Northern Mongolia: TMM Tuva-Mongolia Massif, AMM Arguna-Mongolia Massif, ATM Altai-Mongolia Massif, BTM Buteel Massif, LB Lake Salairian Belt, BB Bayangol Salairian Belt DB Dzhida Salairian Belt, HB Hovd Caledonian Belt, HRB Haraa Caledonian Belt, GAB Gobi Altai-Mandalovoo Caledonian Belt. HHB Hangay-Hentey Hercynian Belt, SB Selenge Late Hercynian-Indosinian Belt, MGB Middle Gobi Late Hercynian-Indosinian Belt. Southern Mongolia: TUM Tsagan Uul Massif, HUM Hutag Uul Massif, NB Nuketdavaa Caledonian Belt, GTB Gobi Tienshan Caledonian Belt, SMB South Mongolian Hercynian Belt. AB Atasbogd Late Hercynian-Indosinian Belt, SLB Sulinheer Late Hercynian-Indosinian Belt.

of the Mongol-Altai Group Meso- and Neoproterozoic sequences, including stromatolitic carbonates that occur in the southern belts of TMM and in AMM (Figure 5) in north-eastern Mongolia, which may be partly regarded as paracover strata on the ancient basement, as is the case in TAP and SKP of China.

## Mongolia in the Late Neoproterozoic to the Triassic

**The Salairian Stage (Late Neoproterozoic to Early Cambrian)** The Salairian Stage, ranging from Late Neoproterozoic to Early Cambrian in age, is important in Mongolia. Well-developed Salairian orogenic belts are distributed in western Mongolia and are composed of Vendian to Cambrian siliciclastics, carbonates, and volcanics that were probably partly formed in an archipelagic ocean basin with island arcs and sea-mounts. Recent studies of the accompanying ophiolites have yielded U-Pb zircon ages of 568–573 Ma and Sm-Nd ages of *ca.* 520 Ma in many places, which are in conformity with the ages formerly determined by fossils. Synorogenic granites in the Lake area bear isotopic ages of Middle to Upper Cambrian. In northern Mongolia, there are two northeast-trending Salairian orogenic belts (DB and BB) extending eastwards into Russia, in which post-orogenic granites of Ordovician age are found. Research on the well-known Bayanhongor ophiolite zone, which is situated in a Caledonian Belt, has revealed that the ophiolites was emplaced by obduction in the time interval 540–450 Ma of Salairian age. A discontinuous belt, indicating the Salairian Orogeny, is known near Choibalsan in north-eastern Mongolia, which may have some connection with the Salairian Belt in the Okhotsk region (Figure 1). In view of the discovery of granites of Salairian age on the western border of the Jamus-Xingkai Massif of China (Figure 3), it seems that the Salairian Orogeny was active in both eastern Mongolia and north-eastern China.

**The Caledonian stage (Middle Cambrian to Silurian)** In the Hovd Caledonides of western Mongolia, the Lower Palaeozoic sequence, consisting of thick flysch of Tremadocian and older ages, and unconformably overlying Ordovician and Silurian carbonates and clastics, with mafic and intermediate volcanics, are well developed. In central and eastern Mongolia, the Ordovician–Silurian sequence in the Gobi-Altai-Mandalovoo (GAB) Caledonian Belt includes a lower and an upper part separated by an unconformity, and both parts were much reworked in the Hercynian Orogeny. This extensive east-west trending belt (GAB) may mark the southern margin of the Early Palaeozoic northern Mongolian palaeocontinent. In southern Mongolia, in the Caledonian Gobi-Tianshan Belt (GTB), Ordovician to Silurian metasediments, with intercalated metavolcanics, are unconformably overlain by Devonian clastic desosits. Similar sequences are observed in the adjacent Junggar region of China. Thus the Caledonian Orogeny played an important role both in southern Mongolia and north-western China.

**The Hercynian and Indosinian stage (Devonian to Triassic)** The Early Hercynides of mainly Devonian to Early Carboniferous age are predominant in southern Mongolia. The South Mongolian Hercynian Belt (SMB) represents the main belt of Late Palaeozoic crustal increase in southern Mongolia and extends on both ends into China. To the west, it is continuous with the Ertix-Almantai zone in China and further to the west with the Zaysan fold zone in Russia (**Figures 1** and **3**). The Upper Palaeozoic in the central part of SMB is composed mainly of Devonian and Lower Carboniferous arc-related volcanics and clastics, including some Upper Silurian beds in the basal part. The ophiolite mélange zones within SMB are not continuous, and were evidently later dismembered. The lower part of the sequence includes the Berch Uul Formation consisting of a thick sequence of uppermost Silurian to Devonian tholeiitic pillow lavas, andesites, and tuffs, and Upper Devonian intermediate to felsic volcanics and olistostromes with coral limestone clastics. Frasnian conodonts and intrusive rocks dated at *ca*. 370 Ma are found in the arc-related volcanics. The upper part of the sequence comprises Lower Carboniferous fine-grained sandstones and mudstones with rich shallow marine fossils, which may have been formed after collision, although pyroclastic beds denoting volcanic activities are known to occur. In the Precambrian ATM, the Caledonian Hovd Belt (HB) and other parts of northern Mongolia, Devonian carbonates and clastics with felsic to intermediate volcanic beds are widespread as cover strata on the basement.

Thick Devonian flysch-like deposits are also developed in the broad central Mongolian Hangay-Hentey (HHB) Belt. The main part of HHB were folded after the Early Carboniferous, but a residual sea trough seems to have lingered on the eastern side of its north-eastern segment, which was essentially closed in Indosinian time. The resultant narrow Indosinide (**Figure 5**) sea extended to Russia and was probably continuous with the Mongol-Okhotsk seaway that finally closed in the Jurassic.

The Late Hercynian and Indosinian are usually inseparable in Mongolia. They are the Selenge Belt (SB) in northern Mongolia, the Middle Gobi Belt (MGB) in eastern Mongolia, the Sulinheer Belt (SLB), and the Atasbogd Belt (AB) in southern Mongolia. The latter belt extends both eastwards and westwards into China. All the belts are characterized by thick sequences of bimodal aulacogen volcanics, volcaniclastics, and pyroclastics of Late Carboniferous to Early Permian age dated by plant remains. To a certain extent, they are comparable with the sequence in the Bogda Mountains, which is a Carboniferous aulacogen that separated the Junggar and Tuha massifs in northern Xinjiang, China. There is, however, an alternative suggestion that a Carboniferous to Permian arc-basin system may have developed in the SLB on the southern border of Mongolia, which is continuous with the Late Hercynides–Early Indosinides belt in Inner Mongolia (**Figure 3**). The superimposed Triassic–Jurassic terrestrial basins developed on SB and HHB, in northern Mongolia, consist of Triassic molasse-like and coal-bearing deposits and trachyandesites and Jurassic clastic sediments, which are well dated by plant fossils. In the Noyon Basin in southern Mongolia, the Early Triassic *Lystrosaurus hedini* is found, which is almost identical to that known from the Ordos Basin.

### Mongolia in the Post-Indosinian

After the Indosinian Orogeny, Mongolia, like the main parts of China, entered a new stage of intracontinental development. No marine deposits are known in Mongolia after the Triassic. Jurassic volcani-sedimentary basins with calc-alkaline volcanics occur in north-eastern Mongolia, which are the same as in the adjacent Xingan Mountains region of China. Cretaceous basins are widely distributed in southern Mongolia, for example the Zuunbayan Basin near Sainshand, and contain the renowned Jehol biota of Early Cretaceous age as in north-eastern China. Cenozoic basins are widespread in western Mongolia. In the Transaltai Basin situated to the south-east of Bayanhongor, abundant Palaeogene mammal remains have been discovered, which may be compared with those found in Inner Mongolia within China.

## Conclusions

The history of China and Mongolia is discussed in terms of tectonic units and tectonic stages. The crustal evolution of China included three megastages in the Precambrian, marked respectively by the aggregation of continental nuclei (2.8 Ga), the lateral growth and consolidation of proto-platforms through the Luliangian Orogeny (1.8 Ga), and the cratonization and coalescence of platforms into the Cathaysiana Supercontient through the Jinningian Orogeny (830 Ma). Until the Jinningian, the crustal evolution of China seems to have been dominated by continental growth, consolidation, and convergence to form a part of the Neoproterozoic Rodinia. In Mongolia, only the last megastage, ending at 830 Ma, marked by the formation of the main massifs, is recognized.

After the Jinningian, China and Mongolia entered a megastage characterized by a tectonic pattern consisting of discrete continents and ocean basins, until their reassembly at the close of the Indosinian Orogeny (210 Ma). The Cathaysiana Supercontinent began to dissociate in the Cambrian, and ocean basins were formed between Sino-Korea and Qaidam, which was entirely closed through the Caledonian Orogeny, with marked collision zones. The wide Caledonide between Yangtze and Cathaysia was, however, folded and uplifted without clear collision. To the north of Tarim and Sino-Korea, the narrow Caledonides represent continent-arc accretion. In Mongolia, the northern Mongolian massifs were successively accreted to the Siberia Platform, and the Mongolian massifs, the Salairides and Caledonides, together formed the northern Mongolian palaeocontinent, with the Gobi-Altai Caledonian Belt as its southern margin. Two main branches of Late Palaeozoic oceans, the Zaysan-South Mongolia-Hingan in the north, and the Ural-Tianshan in the south, were consumed mainly after the Early Carboniferous, and are represented respectively by the main Hercynian sutures (**Figure 1**). The Late Carboniferous to Early Triassic marine basins in southern Mongolia and Inner Mongolia of China probably formed an ocean with scattered islands that were filled up without appreciable collision. Furthermore, the Late Hercynides-Indosinides within northern Mongolia were actually intracontinental residual seas.

To the south of the Kunlun-Qinling central orogenic belt of China, an open sea had persisted since Early Palaeozoic, and the wide Indosinides are marked by the main Indosinian (Muztagh-Maqen) convergent zone in the north and the Jinshajiang zone in the south. The main collision zones usually coincide with older collision zones; in other words, they are polyphased or superimposed collision zones. It was at the close of the Indosinian Stage that the Laurasia Supercontinent took its final shape as the northern half of the Permian-Triassic Pangaea.

The post-Indosinian megastage of China and Mongolia witnessed an entirely new tectonic regime in East Asia, due to the appearance of the Circum-Pacific domain as a result of Pangaea disintegration and the opening of the Atlantic. The subduction of the western Pacific beneath East Asia in the Jurassic caused a continent marginal magmatism along eastern China, including the Hingan belt and eastern Mongolia. This new pattern brought about an apparent change of contrast between northern and southern China to that between eastern and western China. In eastern China, and to a certain extent in eastern Mongolia, there occurred a combination of continental margin type and intracontinental type of volcanism, which was followed by the Late Cretaceous to Cenozoic tensional regime of rifted basins and consequent crustal and lithospheric thinning. In western China, the tectonic process in the Qinghai-Tibet Plateau consisted of the northward accretion of the Gondwanan massifs to Eurasia, characterized by the northward subduction of the Himalaya beneath Gangise in the south, the distributed crustal thickening and shortening in the middle, and the southward indentation from Tarim and Mongol-Siberia in the northern part. The contrast between the compressional versus extensional, and between the crustal and lithospheric thickening versus thinning regimes between western China and eastern China are evident. These features may have reflected and induced the deeper process of an eastward flow of the asthenosphere from under western China, which might have, in turn, caused mantle upwelling and crustal and lithospheric thinning in eastern China.

## See Also

**Asia:** Central; South-East. **Gondwanaland and Gondwana. Indian Subcontinent. Japan. Pangaea. Russia.**

## Further Reading

Badarch G, Cunningham WD, and Windley BF (2002) A new terrane subdivision for Monglia: implications for the Phanerozoic crustal growth of Central Asia. *Journal of Asian Earth Sciences* 21: 87–110.

Deng JF, Zhao Hailing, Mo Xuanxue, Wu Zongxu, and Luo Zhaohua (1996) *Continental roots-plume tectonics of China: key to the continental dynamics*. Beijing: Geological Publishing House. (In Chinese with English abstract.)

Dewey JF, Shackelton RM, Chang C, and Sun W (1994) The tectonic evolution of the Tibetan Plateau. *Philosophical Transactions of the Royal Society of London, Ser. A* 327: 379–413.

He Guoqi, Li Maosong, Liu Dequan, Tang Yanling, and Zhou Ruhong (1988) *Palaeozoic Crustal Evolution and*

*Mineralization in Xinjiang of China.* Urumqi: Xinjiang People's Publishing House and Educational and Cultural Press Ltd. (In Chinese with English abstract.)

Huang TK (1978) An outline of the tectonic characteristics of China. *Eclogae Geologicae Helvetiae* 71(3): 811–635.

Li Chunyu, Wang Quan, Liu Xueya, and Tang Yaoqijng (1982) *Explanatory Notes to the Tectonic Map of Asia.* Beijing: Cartographic Publishing House.

Liu Baojun and Xu Xiaosong (eds.) (1994) *Atlas of the Lithofacies and Palaeogeography of South China (Sinian-Triassic).* Beijing, New York: Science Press.

Ma Lifang, Qiao Xiufu, Min Longrui, Fan Benxian, and Ding Xiaozhong (2002) *Geological Atlas of China.* Beijing: Geological Publishing House. (61 maps and explanations.)

Ren Jishun, Wang Zuoxun, Chen Bingwei, Jiang Chunfa, Niu Baogui, Li Jingyi, Xie Guanglian, He Zhengchun, and Liu Zhigang (1999) *The Tectonics of China from a Global View-A Guide to the Tectonic Map of China and Adjacent Regions.* Beijing: Geological Publishing House.

Shi Xiaoying, Yin Jiaren, and Jia Caiping (1996) Mesozoic and Cenozoic sequence stratigraphy and sea level change cycles in the Northern Himalayas, South Tibet, China. *Newsletters on Stratigraphy* 33(1): 15–61.

Tomurtogoo O (ed.) (1998–1999) *Geological Map of Mongolia (1:1 000 000) and Attached Summary.* Mongolia: Ulanbaatar.

Wang Hongzhen (Chief Compiler) (1985) *Atlas of the Palaeogeography of China.* Beijing: Cartographic Publishing House. (143 maps, explanations in Chinese and English.)

Wang Hongzhen and Mo Xuanxue (1995) An outline of the tectonic evolution of China. *Episodes* 18(1–2): 6–16.

Wang Hongzhen and Zhang Shihong (2002) Tectonic pattern of the world Precambrian basement and problem of paleocontinent reconstruction. *Earth Science – Journal of China University of Geosciences* 27(5): 467–481. (In Chinese with English abstract.)

Xiao Xuchang, Li Tingdong, Li Guangcen, Chang Chengfa, and Yuan Xuecheng (1988) *Tectonic Evolution of the Lithosphere of the Himalayas.* Beijing: Geological publishing House. (In Chinese with English abstract.)

Yin A and Harrison TM (2000) Geologic evolution of the Himalaya-Tibetan Orogen. *Annual Review of Earth and Planetary Sciences* 28: 211–280.

Zhang Guowei, Meng Qingren, and Lai Shaocong (1995) Tectonics and structure of Qinling orogenic belt. *Science in China, Ser. B* 38(11): 1379–1394.

Zhong Dalai, et al. (2000) *Paleotethysides in West Yunnan and Sichuan, China.* Beijing, New York: Science Press.

Zonenshain LP, Kuzmin ML, and Natapov LM (1990) *Geology of USSR: A Plate-Tectonic Synthesis.* Geodynamic Series 21. Washington, DC: American Geophysical Union.

# CLAY MINERALS

**J M Huggett**, Petroclays, Ashtead, UK and The Natural History Museum, London, UK

© 2005, Elsevier Ltd. All Rights Reserved.

## Introduction

Clay minerals are a diverse group of hydrous layer aluminosilicates that constitute the greater part of the phyllosilicate family of minerals. They are commonly defined by geologists as hydrous layer aluminosilicates with a particle size <2 μm, while engineers and soil scientists define clay as any mineral particle <4 μm (*see* **Soils: Modern**). However, clay minerals are commonly >2 μm, or even 4 μm in at least one dimension. Their small size and large ratio of surface area to volume gives clay minerals a set of unique properties, including high cation exchange capacities, catalytic properties, and plastic behaviour when moist (*see* **Analytical Methods: Mineral Analysis**).

Clay minerals are the major constituent of fine-grained sediments and rocks (mudrocks, shales, claystones, clayey siltstones, clayey oozes, and argillites). They are an important constituent of soils, lake, estuarine, delta, and the ocean sediments that cover most of the Earth's surface. They are also present in almost all sedimentary rocks, the outcrops of which cover approximately 75% of the Earth's land surface. Clays which form in soils or through weathering principally reflect climate, drainage, and rock type (*see* **Weathering; Palaeoclimates**). It is now recognized that re-deposition as mudrock only infrequently preserves these assemblages, and clay assemblages in ocean sediments should not be interpreted in terms of climate alone, as has been done in the past. Most clay in sediments and sedimentary rocks is, in fact, reworked from older clay-bearing sediments, and many are metastable at the Earth's surface. This does not preclude the use of clays in stratigraphic correlation, indeed it can be a used in provenance studies. A few clays, notable the iron-rich clays form at the Earth's surface either by transformation of pre-existing clays or from solution. These clays are useful environmental indicators, so long as they are not reworked.

## Clay Structure and Chemistry

Clays can be envisaged as comprising sheets of tetrahedra and sheets of octahedra (**Figure 1**). The general formula for the tetrahedra is $T_2O_5$, where T is mainly $Si^4$, but $Al^{3+}$ frequently (and $Fe^{3+}$ less frequently) substitutes for it. The tetrahedra have a hexagonal arrangement, if not distorted by substituting cations (**Figure 2**). The octahedral sheet comprises two planes of close-packed oxygen ions with cations occupying the resulting octahedral sites between the two planes (**Figure 1B**). The cations are most commonly $Al^{3+}$, $Fe^{3+}$, and $Mg^{2+}$, but the cations of other transition elements can occur. The composite layer formed by linking one tetrahedral and one octahedral sheet is known as a 1:1 layer. In such layers, the upper-most unshared plane of anions in the octahedral sheet consists entirely of OH groups. A composite layer of one octahedral layer sandwiched between two tetrahedral layers (both with the tetrahedra pointing towards the octahedral layer) is known as a 2:1 layer. 2:1 clays of the mica and chlorite families have multiple polytypes defined by differences in stacking parallel to the c axis. If the 1:1 or 2:1 layers are not electrostatically neutral (due to substitution of trivalent cations for $Si^{4c}$ or of divalent for trivalent cations) the layer charge is neutralized by interlayer materials. These can be cations (most commonly $K^+$, $Na^+$, or $NH_4^+$), hydrated cations (most commonly $Mg^{2+}$, $Ca^{2+}$, or $Na^+$) or single sheets of hydroxide octahedral groups ($Al(OH)_3$ or $Mg(OH)_2$ (**Figure 3**). These categories approximately coincide with the illite, smectite, and chlorite-vermiculite families of clays. It is evident that the different types of interlayer cation will have a direct affect upon the thickness of the clay unit cell in the 001 dimension. This property, together with the ease with which the interlayer cations are hydrated or will interact with organic compounds, is much used in X-ray diffraction to identify clay mineral.

The principal clay physical properties of interest to the geologist are cation exchange capacity and interaction with water. Clays have charges on (001) layer surfaces and at unsatisfied bond edge sites. An important consequence of these charges is that ions and molecules, most commonly water, are attracted to and weakly bonded to clay mineral particles. In most cases cations are attracted to layer surfaces and anions to edge sites. If the interlayer charge is low, cations between 1:1 layers can be exchanged for other cations and these cations can be hydrated by up to two water layers. Water in the interlayer site is controlled by several factors including the cation size and charge. In 'normal' pore fluid, one layer of water is associated with monovalent cations, two layers with divalent cations. Water can also weakly bond to the outer surface of clay particles. The relative ease with which one cation will replace another is usually:

$$Na^+ < K^+ < Ca^{2+} < Mg^{2+} < NH_4^+$$

i.e., $NH_4^+$ is usually more strongly fixed in the interlayer sites than is $Na^+$. The exchangeability of cations is measured as the cation exchange capacity (CEC). This technique is used to characterise clay reactivity, and each clay mineral has a characteristic range of CEC values (**Table 1**).

## Classification

Clays are normally classified according to their layer type, with layer charge used to define subdivisions

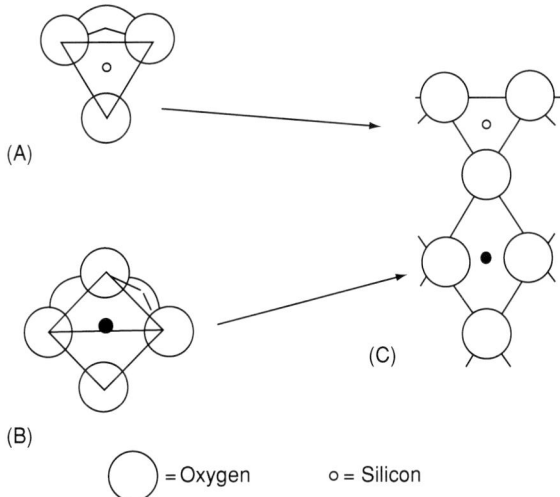

**Figure 1** (A) Tetrahedrally co-ordinated cation polyhedrons; (B) octahedrally co-ordinated cation polyhedrons; (C) linked octahedral and tetrahedral polyhedrons.

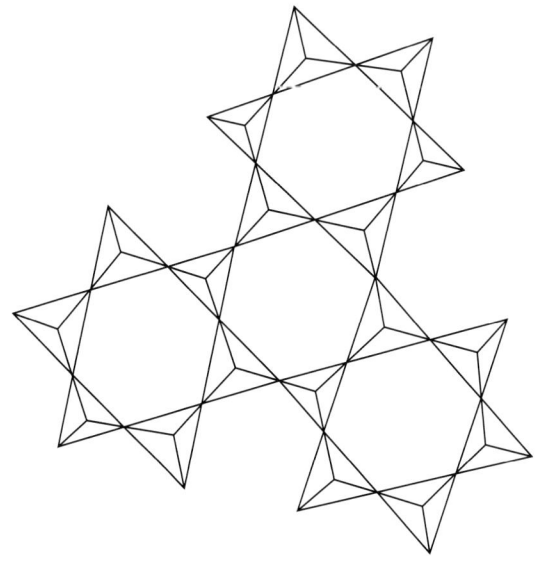

**Figure 2** Hexagonal arrangement of edge-linked tetrahedra.

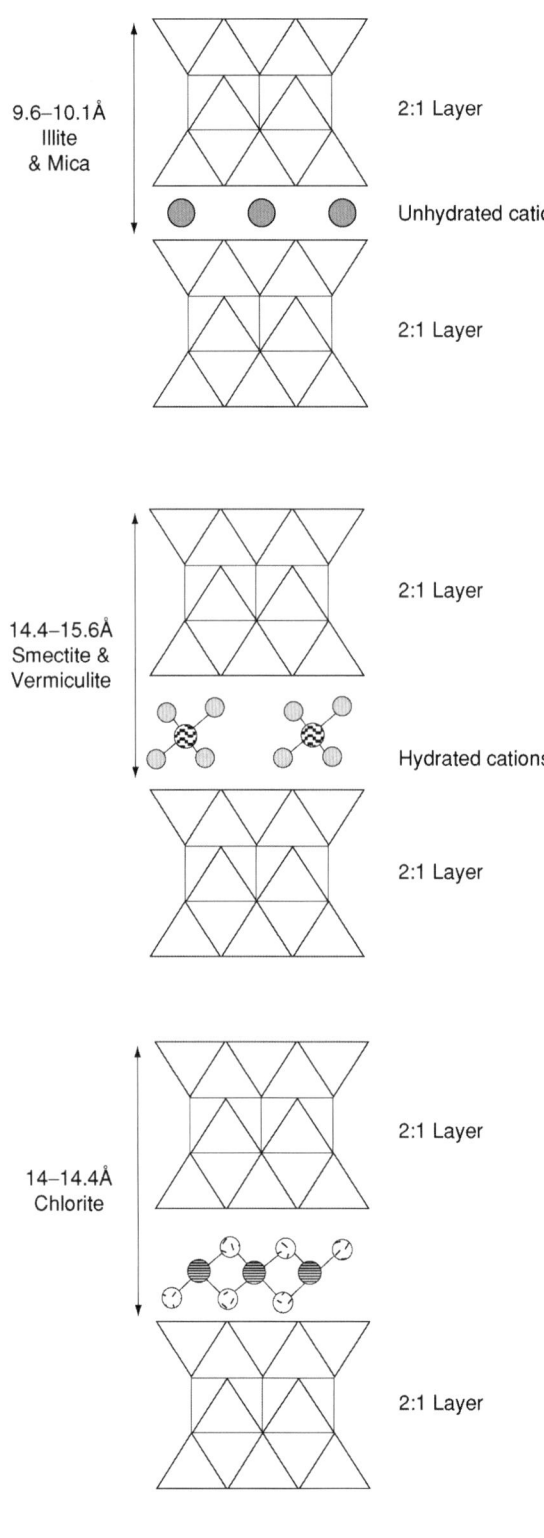

**Figure 3** 2:1 Clay structures.

**Table 1** Typical values for cation exchange capacities. Cation exchange capacities (CEC) in milliequivalents/gram. (Data from multiple sources.)

|  | meq/100 g |
|---|---|
| Kaolinite | 3 to 18 |
| Halloysite | 5 to 40 |
| Chlorite | 10 to 40 |
| Illite | 10 to 40 |
| Montmorillonite | 60 to 150 |
| Vermiculite | 100 to 215 |

(Table 2). Because of their fine particle size, clay minerals are not easily identified by optical methods, though the distinctive chemistry (and sometimes habit or morphology) of most allows identification by X-ray analysis in electron microscope studies. The most reliable method of clay identification, particularly in very fine grained rocks is X-ray diffraction, either of the bulk sample, or more reliably, of the fine fraction (usually $<2\,\mu$m). The response of the clays to glycol or glycerol solvation, cation saturation, and heat treatment is used to determine which clays are present and the extent of any interlayering (*see* **Analytical Methods: Mineral Analysis**).

### Clays (Serpentine and Kaolin) (1:1)

Berthierine, $(FeMg)_{3-x}(Fe_3Al)_x(Si_{2-x}Al_x)O_5(OH)_4$, is the FeII-rich member of the serpentine subgroup most commonly encountered in unmetamoprhosed sedimentary rocks and odinite is its FeIII-rich counterpart found (so far) in Eocene and younger sediments. Chemically, the kandite minerals are alumina octahedra and silica tetrahedra with occasional substitution of $Fe^{3+}$ for $Al^{3+}$. Of the kandite group, kaolinite is by far the most abundant clay. Kaolinite and both halloysites are single-layer structures whereas dickite and nacrite are double layer polytypes, i.e., the repeat distance along the direction perpendicular to (001) is 14 Å, not 7 Å. Dickite and nacrite were originally distinguished from kaolinite on the basis of XRD data, however this is seldom easy, especially in the presence of feldspar and quartz. Infra-red and differential thermal analysis can be successfully used to make the distinction between these two clays.

### Clays (Talc and Pyrophyllite) (2:1)

Ideal talc $(Mg_3Si_4O_{10}(OH)_2)$ and pyrophyllite $(Al_2Si_4O_{10}(OH)_2)$ are 2:1 clays with no substitution in either sheet and hence no layer charge or interlayer cations. However, minor substitution is common.

**Table 2** Clay classification by layer type Tr = trioctahedral, Di = dioctahedral, x = layer charge, Note the list of species is not exhaustive, and interstratified mixed-layer clays abound (see below). (Adapted from Brindley and Brown, 1980.)

| Layer type | Group | Sub-group | Species (clays only) |
|---|---|---|---|
| 1:1 | Serpentine-kaolin ($x \sim 0$) | Serpentines (Tr) | Berthierine, odinite |
|  |  | Kandites (di) | Kaolinite, dickite, 7 Å & 10 Å halloysite, nacrite |
| 2:1 | Talc-pyrophyllite ($x \sim 0$) | Talc (Tr) pyrophyllite (Di) |  |
| 2:1 | Smectite ($x \sim 0.2$–$0.6$) | Tr smectites | Montmorillonite, hectorite, saponite beidellite, nontronite |
|  |  | Di smectites |  |
| 2:1 | Vermiculite ($x \sim 0.6$–$0.9$) | Tr vermiculites |  |
|  |  | Di vermiculites |  |
| 2:1 | Mica-illite ($x \sim 0.8$) | Tr illite? |  |
|  |  | Di illite | Illite, glauconite |
| 2:1 | Chlorite (x variable) | Tr-Tr chlorites | Chamosite, clinochlore, ripidolite etc donbassite |
|  |  | Di-Di chlorites |  |
|  |  | Tr-Di chlorites |  |
|  |  | Di-Tr chlorites | Sudoite, cookeite |
| 2:1 | Sepiolite-palygorskite (x variable) | Sepiolites | Sepiolite |
|  |  | palygorskites | palygorskite (syn. attapulgite) |

## Clays (Smectite) (2:1)

The 2:1 clays with the lowest interlayer charge are the smectites. This group have the capacity to expand and contract with the addition or loss (through heating) of water and some organic molecules. It is this property that is used to identify smectites by glycol or glycerol solvation and heat treatments in XRD studies. This swelling is believed to be due to the greater attraction of the interlayer cations to water than to the weakly charged layer. Montmorillonite is a predominantly dioctahedral smectite with the charge primarily in the octahedral sheet ($R^+(Al_3Mg_{.33})Si_4O_{10}(OH)_2$), while beidellite ($R^+Al_2(Si_{3.67}Al_{.33})O_{10}(OH)_2$) and nontronite ($R^+ Fe(III)_2(Si_{3.67}Al_{.33})O_{10}(OH)_2$) are dioctahedral with the charge mainly in the tetrahedral sheet (where R = mono or divalent interlayer cations). Hectorite is a rare clay, similar to montmorillonite but it is predominantly trioctahedral with $Mg^{2+}$ and $Li^{2+}$ in the octahedral layer. Saponite is another uncommon smectite with a positive charge on the octahedral sheet and a negative one on the tetrahedral sheet. In smectite the interlayer cations are typically $Ca^{2+}$, $Mg^{2+}$, or $Na^+$; in high charge smectites $K^+$ may be present. When smectite expands, the interlayer cation can be replaced by some other cation. Hence the cation exchange capacities of smectite is high compared with nonexpanding clay minerals. CEC can therefore be used to identify smectite.

## Clays (Vermiculite) (2:1)

The second group of clays with exchangeable cations is vermiculite. Vermiculite has a talc-like structure in which some $Fe^{3+}$ has been substituted for $Mg^{2+}$ and some $Al^{3+}$ for $Si^{4+}$, with the resulting charge balanced by hydrated interlayer cations, most commonly $Mg^{2+}$. The layer charge typically ranges from 0.6 to 0.9. Vermiculite is distinguished from smectite by XRD after saturating with $MgCl_2$ and solvation with glycerol. This results in expansion of the interlayer to 14.5 Å, rather than the 18 Å characteristic of smectite (though there may be exceptions to this rule). Vermiculite is much less often encountered in sedimentary rocks than is smectite, probably because it is most commonly a soil-formed clay, while coarsely crystalline vermiculite deposits are formed from alteration of igneous rocks.

## Clays (Mica and Illite) (2:1)

Substitution of one $Al^{3+}$ for one $Si^{4+}$ results in a layer charge of 1, which in true mica is balanced by one monovalent interlayer cation (denoted R). In mica the cation is usually $K^+$, less often $Na^+$ or $Ca^{2+}$ and rarely $NH^+$. The term clay grade mica is sometimes used to describe mica which has been weathered resulting in loss of interlayer cations or formation of expandable smectite layers. 2:1 clays with layer charge $\sim 0.8$ are illite ($R^+(Al_{2-x}Mg, Fe II, Fe III)_x Si_{4-y}Al_y O_{10}(OH)_2$) and glauconite (the ferric iron-rich equivalent of illite). Like mica, illite and glauconite are characterised by a basal lattice spacing of $d(001) = 10$ Å which is unaffected by glycol or glycerol solvation, nor by heating. Illite is used as both a specific mineral term and as a term for a group of similar clays, including some with a small degree of mixed layering. Illite has been described as detrital clay-grade muscovite of the 2 M polytype, plus true 1 Md and 1 M illite polytypes (with or without some smectite or chlorite mixed layers). Illite is, however, chemically distinct from muscovite in having less octahedral Al, less interlayer K and more Si, Mg, and $H_2O$.

### Clays (Chlorite) (2:1)

Chlorite consists of a 2:1 layer with a negative charge $[(R^{2+}\cdot R^{2+})_3(_xSi_{4x}R_y^{2+})O_{10}OH_2]^-$ that is balanced by a positively charged interlayer octahedral sheet $[(R^{2+}\cdot R^{3+})_3(OH)_6]^+$. R is most commonly Mg, Fe II, and Fe III, with Mg-rich chlorite (clinochlore) generally being metamorphic (high temperature) or associated with aeolian and sabkha sediments, while Fe-rich chlorite (chamosite) is typically diagenetic (low temperature). Ni-rich chlorite (nimite) and Mn-rich chlorite (pennantite) are the other two less common principal varieties of chlorite. Most chlorites are trioctahedral in both sheets, i.e., the ferric iron content is low. Chlorites with dioctahedral 2:1 layers and trioctahedral interlayer sheets are called ditrioctahedral chlorite (the reverse, tri-dioctahedral chlorite is unknown). Chlorite classification is further complicated by the existence of different stacking polytypes.

Mixed layer clays are those which consist of discrete crystals of interlayered clays, usually just two clay species are present, though three is known, mainly from soils. The stacking can be random, partially regular, or regular. Different types of ordering are described using the Reichweite terminology (Reichweite means 'the reach back'), denoted as R0 for random mixed layering, R1 for regular alternating layers of two clay types (also called rectorite), and R3 for ABAA for regularly stacked sequences. R2 (ABA) has not been positively identified. The most frequently encountered mixed layer clay is illite-smectite because smectite is progressively replaced by illite during deep burial diagenesis, mostly in mudrock. This clay is probably more abundant than either illite or smectite. Interstratified chlorite-smectite is associated with alteration of basic igneous rock or rock fragments in sandstone. Regularly interstratified chlorite-smectite and chlorite-vermiculite with 50% of each component are both known as corrensite. 14Å chlorite interstratified with 7Å chlorite is becoming a more widely recognised clay, particularly in sandstones which have undergone some diagenesis.

### Fibrous Clays (Sepiolite and Palygorskite) (2:1)

Sepiolite and palygorskite (also known as attapulgite) are structurally different from other clays in two ways. Firstly, the tetrahedral sheets are divided into ribbons by inversion though they are still bonded in sheet form, and secondly the octahedral sheets are continuous in two dimensions only. They are consequently fibrous, though their macroscopic form may be flakes or fibres. An ideal sepiolite formula is approximately $Mg_8Si_{12}O_{30}OH_4(OH_2)_4 \cdot X[R^{2+}\cdot(H_2O)_8]$, and palygorskite is $MgAlS_8 O_{20}OH_3(OH_2)_4 \cdot X[R^{2+}\cdot(H_2O)_4]$. Where X = octahedral sites in sepiolite which may contain Al, Fe, Mn, or Ni, while in palygorskite may be present Na, Fe, and Mn. Cation exchange capacities are intermediate between nonexpanding and expanding 2:1 clays, and the fibrous clay structure does not swell with addition of either water or organic solvents.

## Clay Formation Through Weathering and Neoformation in Soils

Most of the clay in sedimentary rocks is probably formed by weathering or in soils. However, it is important to realize that firstly the amount of neoformed clay in soils is, at any one time, small relative to the total clay in sediments and sedimentary rocks and secondly that most clay in sediments is derived through reworking of older clay-rich sediment. Clay assemblages resulting from weathering reflect the pedoclimatic conditions (temperature, precipitation, drainage, vegetation), the composition of the rock being weathered and the length of time during which weathering occurs. This applies to palaeosols as well as recent soils (*see* **Soils: Palaeosols**), if they can be shown to be unaffected by clay diagenesis. principal process involved is hydrolysis, hence the degree of weathering increases with temperature and extent of exposure to water (precipitation and drainage), though plants and micro-organisms may also be important weathering agents (*see* **Weathering**). In tectonically unstable areas, rapid weathering and erosion may prevent formation of a stable soil clay assemblage. In general, very cold or hot and dry climate results in illite and chlorite formation. Temperate climates are characterised by illite, irregular mixed layers, vermiculite, and smectite (generally beidellite in soils, montmorillonite in altered tuff beds). Fe smectite and fibrous clays (usually palygorskite) prevail in subarid climates and hot wet climates are characterized by kaolinite (and the iron oxyhydroxide goethite). If soil-formed clays are preserved without further modification, either in the soil, during transport, or diagenesis, they may be preserved as palaeoclimatic indicators. Such preservation is, however, less common and less easily distinguished from diagenetic clays or clays reworked from older sedimentary rocks, than was formerly thought. These three modes of origin may be identified by careful provenance and petrographic studies.

Micaceous clay minerals in soils are mostly rock-derived and comprise both clay-grade micas and true illite. However, illite-rich illite-smectite can form through K fixation in smectite (especially high charge smectite) in soils subjected to repeated wetting and drying. Most soil illites, whatever their origin, are dioctahedral aluminium-rich illite. Smectite formation (**Figure 4**) is favoured by high pH, high silica activity, and an abundance of basic cations, consequently smectite is associated with poorly drained soils on base-rich parent material. Smectite formed in this way may be mostly ferruginous beidellites. However, smectites can form through transformation and neoformation in a wide range of soil types. The situation is further complicated by the frequent alteration of volcanic ash to form smectite (usually montmorillonite), which can occur dispersed through sediment or as discrete bentonite layers (*see* **Sedimentary Rocks:** Clays and Their Diagenesis).

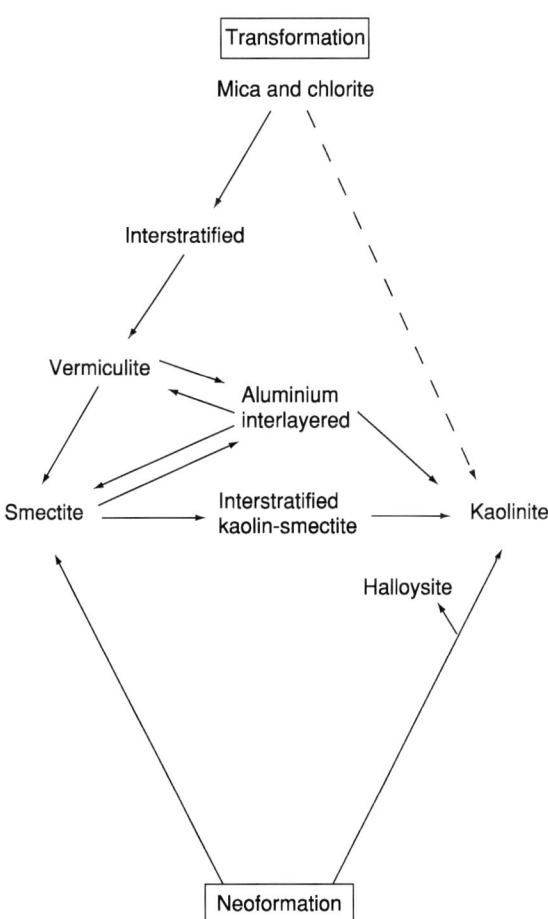

**Figure 4** Pathways for the formation of clay minerals in soils. Most smectite and kaolinite is formed through neoformation while formation of kaolinite from mica and chlorite is not a true transformation due to the structural differences. (Adapted from Wilson (1999).)

Most soil vermiculite is trioctahedral and forms through weathering of biotite, hence many soil vermiculites are interstratified illite-vermiculite (called hydrobiotite when regularly interstratified). Vermiculite can also form from chlorite through breakdown of the interlayer hydroxide sheet, though again, the mixed layer clay is most frequently encountered in soils (vermiculite-chlorite is also known as swelling chlorite). Dioctahedral vermiculite is also present in soils but is less well known.

Most trioctahedral chlorite in soils is inherited. Primary chlorites are easily weathered to chlorite-vermiculite, or if weathering is intense, dissolved altogether. This is why chlorite is characteristic of high latitude soils where surface weathering is less intense than anywhere else on the Earth. So-called pedogenic chlorite and chlorite-rich interlayered clays apparently form entirely through transformation reactions involving the introduction of nonexchangeable interlayer hydroxy-Al polymers into the interlamellar sites of pre-existing smectite, and vermiculite or interstratified expandable clays. This transformation is favoured by pH 4–5.8, and it appears that in podzols at least, the reaction may neutralise acid rain.

Neoformed kaolinite is associated with tropical soils, particularly on stable cratonic areas. The reason for this association is that the intensity of weathering to form kaolinite from rock is greater than for other clays. Kaolinite formed through transformation reactions is less common but the existence of kaolinite interstratified with other clays suggests that these reactions do occur. Soil-formed kaolinite differs from diagenetic kaolinite in having some Fe substitution for Al, being less well ordered, and having a smaller particle size. Neoformed kaolinite (**Figure 4**) is also associated with limestone karst formation (*see* **Sedimentary Processes:** Karst and Palaeokarst). Commercial deposits of kaolinite (china clay) have formed through metasomatic alteration of granite (*see* **Clays, Economic Uses**). These are almost pure kaolinite and contain only minor quartz impurities. Kaolinite-rich beds known as underclays, tonsteins, or seat-earths occur beneath coal deposits, where they form as a consequence of intense plant-assisted weathering. Dickite is typically a high temperature diagenetic clay, while both dickite and nacrite are associated with hydrothermal deposits. Halloysite is mostly associated with young, volcanic-derived soils, in which it has a characteristic tubular or spherical morphology.

Interstratified clays are very commonly neoformed in soils. This is because the represent intermediate transformation products in reactions which proceed less fast at the Earth's surface than in during buried diagenesis. Various mica-vermiculites and illite-vermiculite are common in podzols, while chlorite-vermiculite is found in a wide range of soil types on chlorite-bearing rock. Kaolinite-smectite (or more probably the kaolin mineral is halloysite, though this has only recently been recognized) is particularly abundant in soils formed on basic igneous rock.

Globally sepiolite and palygorskite are fairly rare though locally abundant. Recent and Quaternary deposits occur in North Africa and the Middle East, while there are well-known older Tertiary deposits in southern Spain and France. Neoformed sepiolite and palygorskite are commonly found in desert soils, ephemeral lakes, and playas, and occasionally in hydrothermal veins in basic and ultrabasic rocks. Formation of palygorskite requires confined conditions, either in shallow, enclosed water bodies at times of low lake level or in palaeosols and slow deposition. It occurs in mature, slow-formed calcrete, and in association with smectite and dolomite. Palygorskite does occur in deep water sediments. For example, the origin of palygorskite in deepwater sediments of the North Atlantic has been much debated. However, using SEM and TEM analysis, this clay has been demonstrated to be wind blown, reworked material from shallow water saline basins in Morocco.

## Marine Clays

The clay mineral suite of marine environments is controlled by provenance (closely linked to climate), salinity, particle size, and neoformation. Detrital clay minerals distributions in estuaries and deltas reflect the different settling rates of the clays present and their tendency to flocculate. Where flocculation does not occur, the finer clay particles (which roughly equates with smectite) remain in suspension longest and are transported further offshore. However, no such depth profile will occur if mixing of freshwater suspended clay with saline water results in flocculation, or where marine currents are strong. The effective grain-size of clay particles is also dramatically increased through the formation of mucous-bound faecal pellets by marine invertebrates. Due to the incorporation of reactive organic matter into faecal pellets, they are prone to allogenic and early diagenetic mineralisation. An example is odinite, a ferric iron-rich clay that forms exclusively within the tropics, offshore from major river systems (which introduce large amounts of iron into the marine environment). Odinite replaces faecal pellets and also forms ooidal coatings on grains. In ancient sediments, odinite is believed to transform into ferrous iron-rich 7Å (berthierine) and 14Å (chamosite) clays. Further offshore in marine shelf sediments, the better known iron-rich clay, glauconite (and intermediate clays which range from glauconitic smectite to smectitic glauconite) forms just below the sediment water interface, again, largely through replacement of faecal pellets. Glauconite forms principally on the outer shelf, in a water depth of 100–300 m, at a temperature of 7–15°C and between 50°N and 50°S. It is also favoured by a low sedimentation rate because it is slow to form and has an He close to the redox boundary as it contains both ferrous and ferric iron. Hence glauconite is associated with transgressive sediments, and can be a useful environmental indicator, though there are many instances of reworked glauconite in turbidites, shallow marine, and nonmarine sediments.

Deep sea clays cover by far the greatest area, but are deposited very slowly. Much of the clay in these sediments is a combination of detrital and neoformed Fe-rich smectite, often associated with palygorskite (which may or may not be authigenic) and authigenic clinoptilolite. In the central Pacific, where there is perhaps the least detrital input, smectite makes up >70% of the clay fraction. Nontronite has been reported from deepsea areas of volcanic activity. Apart from neoformation of smectite, the deep ocean clays largely reflect the pedogenic processes occurring on adjacent landmasses.

## Nonmarine Clays

Most continental, freshwater, and saline lacustrine clays are detrital. A few recent saline lakes contain unequivocal neoformed Fe and Mg-rich smectite. However, studies of lake clays do not always include sufficient investigation of the drainage basin to be certain that just because a lake clay is dominated by one clay type that it is neoformed. Some Tertiary lake sediments are almost monomineralic ferric rich illite, but the same provision regarding the drainage basin applies to these. There appear to be two distinct processes whereby this unusual clay forms, firstly in volcanic lakes with a high dissolved cation content, and secondly in ephemeral lakes through repeated wetting and drying of smectite. Laboratory experiments have demonstrated that such a process is possible.

## Clay Stratigraphy

As a consequence of diagenetic replacement by illite, smectite is virtually unknown from pre-Mesozoic sediments, while throughout the Mesozoic and Early Tertiary, smectite is the principal clay mineral found in sediments of most types. Glauconite is most abundant at the end of the Cambrian and the end of the Cretaceous, while berthierine/chamosite are mostly found from the Ordovician to the end of the Devonian and from the earliest Jurassic until the end of the Cretaceous. The reasons for these particular distributions through time may be related to the availability of suitable formation sites, which may in turn be linked to global climate and the formation/break up of the supercontinents. Global climatic shifts, especially in Tertiary and more recent sediments are clearly reflected in the clay minerals formed at the Earth's surface and consequently deposited in the oceans. The global cooling that began at the Eocene–Oligocene boundary and continued intermittently is reflected in the shift from a predominance of smectite to illite, plus minor chlorite and interstratified clays. At the same time, falling sea-level associated with global cooling and the build-up in polar ice resulted in the spread of nonmarine basins, hence the widespread freshwater clay deposits of the Oligocene. Recent research into clay stratigraphy is beginning to show that where post-formation modification is absent, clay mineral suites can accurately reflect Milankovitch cyclicity in sediments, and hence provide a geologically fine-scale dating tool.

Changes across stratigraphic boundaries, particularly where they are nonconformable, are not uncommon, and can be useful for dating purposes in nonmarine sediments, while within a single sedimentary unit, clear vertical changes in clay mineral suites in marine shales can be demonstrated to be linked to changes in ocean circulation and faunal changes. Note that lateral facies variations can result in lateral clay mineral suite variation, even within shales, and care should be taken when attempting to use clay stratigraphy in this way, especially if the intensity of diagenetic alteration varies between locations (*see* **Sequence Stratigraphy**).

## See Also

**Analytical Methods:** Mineral Analysis. **Clays, Economic Uses**. **Palaeoclimates**. **Sedimentary Rocks:** Clays and Their Diagenesis. **Sequence Stratigraphy**. **Soils:** Modern; Palaeosols. **Weathering**.

## Further Reading

Aplin AC, Fleet AJ, and MacQuaker JHS (1999) Muds and mudstones: physical and fluid-flow properties. In: Aplin AC, Fleet AJ, and MacQuaker JHS (eds.) *Muds and Mudstones: Physical and Fluid-Flow Properties*, Special Publications: 158, pp. 1–8. London: Geological Society.

Bailey SW (1980) Structures of Layer Silicates. In: Brindley GW and Brown G (eds.) *Crystal Structures of Clay Minerals and their X-ray Identification*, Mineralogical Society Monograph No. 5, pp. 1–124. London: Mineralogical Society.

Chamley H (1989) *Clay Sedimentology*. Berlin: Springer-Verlag.

Colson J, Cojan I, and Thiry M (1998) A hydrogeological model for palygorskite formation in the Danian continental facies of the Provence Basin (France). *Clay Minerals* 33: 333–347.

Huggett JM (1992) Petrography, mineralogy and diagenesis of overpressured Tertiary and Late Cretaceous mudrocks from the East Shetland Basin. *Clay Minerals* 27: 487–506.

Huggett JM, Gale AS, and Clauer N (2001) The nature and origin of non-marine 10Å green clays from the late Eocene and Oligocene of the Isle of Wight (Hampshire basin), UK. *Clay Minerals* 36: 447–464.

Moore DM and Reynolds RC (1997) *X-ray Diffraction and the Identification and Analysis of Clay Minerals*. Oxford: Oxford University Press.

Merriman RJ (2002) Contrasting clay mineral assemblages in British Lower Palaeozoic slate belts: the influence of geotectonic setting. *Clay Minerals* 37: 207–219.

Robert C and Chamley H (1992) Late Eocene-Early Oligocene evolution of climate and marine circulation: deep-sea clay mineral evidence. *The Antarctic Paleoenvironment: A Perspective of Global Change Antarctic Research Series* 56: 97–117.

Rupert JP, Granquist WT, and Pinnavaia TJ (1987) Catalytic properties of Clay Minerals. In: Newman ACD (ed.) *Chemistry of Clays and Clay Minerals*, Mineralogical Society Monograph No. 6: pp. 275–318. London: Mineralogical Society.

Srodon J (2002) Quantitative mineralogy of sedimentary rocks with emphasis on clays and with applications to K-Ar dating. *Mineralogical Magazine* 66: 677–688.

Velde B (1992) *Introduction to Clay Minerals*. London: Chapman & Hall.

Vogt C, Lauterjung J, and Fischer RX (2002) Investigation of the clay fraction ($<2\,\mu m$) of the Clay Minerals Society reference clays. *Clays and Clay Minerals* 50: 388–400.

Wilson MJ (1999) The origin and formation of clay minerals in soils: past, present and future perspectives. *Clay Minerals* 34: 7–26.

# CLAYS, ECONOMIC USES

**Y Fuchs**, Université Marne la Vallée, Marne la Vallée, France

© 2005, Elsevier Ltd. All Rights Reserved.

## Introduction

In common parlance, 'clays' are aggregates of minerals, soils, or rocks that commonly show plasticity. These materials contain not only minerals belonging to the clay family (*see* **Clay Minerals**) but also very fine particles with various proportions of quartz, feldspar, mica, and organic material. Clays are estimated to represent about 15% of the volume of Earth's crust as soils and as sedimentary rocks (mudstones, shales). In soils, clays play a major role in moisture retention, permeability, and adsorption of inorganic (metals) and organic solutes, as well as in the propensity for soils to shrink and swell. Thus the presence of clay minerals in soils is important for agriculture and construction. Clay has many other uses, however, and it is a particularly important commodity for industry. According to the United States Geological Survey, in 1998 there were approximately 770 active open-pit clay mines in 44 states of the USA. These mines, operated by ∼240 companies that employed ∼13 700 people in clay mining and milling, sold 43 million metric tons of clay products, with sales averaging \$2.14 billion year$^{-1}$.

## History

Clays have long been and are still used in different processes and products, from traditional house building (adobe in Latin America and toub in Africa, when mixed with straw) to personal-use products (shampoo, or 'rassoul', in North Africa); in AD 77, Pliny the Elder mentioned clays used as 'soap'. Advanced techniques now allow use of clays in the production of resistant, high-temperature ceramics and in remediation material and water-circulation barriers for waste repositories. Clay has been used as the building material of choice for millennia. Sun-dried bricks (adobe) and kiln-fired bricks were employed by the Sumerian and Babylonian civilizations of the Euphrates region for constructing ziggurats, city walls, palaces, and temples.

Clays have played an important role in pottery and vessel manufacture and in the trade in such goods. The earliest forms of clay pottery are about 10 000 years old. Stoneware items incorporating clays dating from 3400 years ago have been discovered in China. The oldest porcelain known was created in China during the Tang dynasty (AD 618–907). The import and trade of Chinese porcelain in Europe were important for centuries, and porcelain remained a very expensive product until it became more widely available in Florence at the end of sixteenth century.

## Definitions

The term 'clay' has somewhat different meanings in different contexts (**Figure 1**). To the mineralogist, 'clay' is the term describing a mineral family characterized by its crystal structure. To the geologist, 'clay' designates a type of rock in which clay minerals are particularly abundant but in which other minerals, such as calcite and quartz, also contribute to the rock composition. For industrial purposes, the definition of 'clay' is based on grain size (typically <2 μm in diameter). The importance of clays and clay minerals in economic geology derives from their fine grain size, their chemical composition, their layered structure, and the characteristics related to theses physical attributes, i.e., their role in ion exchange and their affinity for water. The importance of clays in economic geology is large and varied.

Industrial names of clay types do not correspond to the names as they are defined by the International Mineralogical Association (IMA) and as they are defined by researchers. Industrial names are generally based on the six categories established by the US Bureau of Mines: kaolin, ball clay, fire clay, bentonite, fuller's earth, and so-called common clay and shales. Kaolin, also called china clay, is composed mainly of kaolinite and other kaolinite group minerals (nacrite, halloysite, and dickite) and must be white for industrial uses (e.g., paper coating and filling, refractories, rubber, and paint). Ball clay is plastic, white-firing clay containing mainly kaolinite. It is used for manufacturing tiles, dinnerware, and bath and kitchen sinks and tubs. Fire clays withstand temperatures up to 1500°C and are used for refractories or to raise the vitrification limit. They are also composed mainly of kaolinite but may also include diaspore, an aluminium hydroxide (α-AlO(OH)), burley, ball clay, and common clay. Bentonite is an industrial clay that is composed mainly of clay minerals from the smectite group and principally of montmorillonite; it is used in foundry sands and as drilling muds. Fuller's earth is a non-plastic clay material that is characterized by its particular purifying and decolourizing properties. Fuller's earth is not very different from bentonite

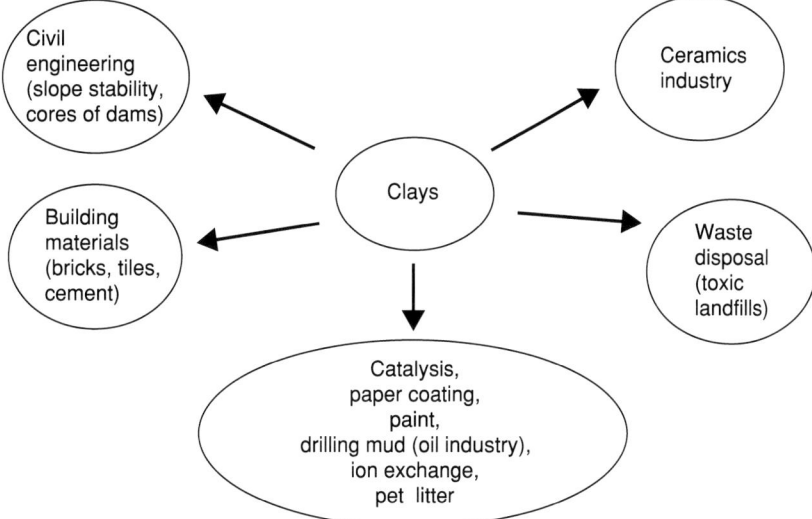

**Figure 1** Uses of clays.

and is used mainly as pet litter and as an absorbent for oil and grease. Common clay is defined industrially as material that has a vitrification point below 1100°C. It consists of mixtures of different clay minerals, mainly illite and chlorite, but may also contain kaolinite and montmorillonite.

## Civil Engineering

Volume changes in clay cause swelling and/or shrinking of soils, a major geological hazard for buildings and road pavement. Shrinkable clays are usually deposited as sediments in marine or brackish environments. They show little tendency to expand unless they have previously undergone a strong drying event. The shrinkage problems associated with this type of clay in soils are related to water uptake by deep tree roots, particularly in areas where the clay content exceeds 25%, the plasticity index is higher than 30%, and the water table is relatively deep seated. Expandable clays are characterized by their relatively high content of the clay mineral montmorillonite (and some other structurally related clay minerals of less importance). From an initial dry state, because of their montmorillonite content, these clays react to increasing moisture conditions by swelling. The same clays can show a reverse shrink–swell capacity with changed environment.

The presence of clays in soils is a problem for foundations of even small buildings. In all countries, structural foundations have to meet minimum requirements of building regulations. Under a building, the mass of soil that is appreciably affected by moisture content change is termed the 'active zone'. This zone represents the volume of soil that is influenced by the presence of an engineered structure. The impact can be related directly to the building (heat flow) or to other factors, such as changes caused by vegetation (effects of tree roots). In some countries, the depth of frost action is an important concern, particularly for structures that do not lose appreciable heat to the subsoil.

## Building Material

Clay is a vital component in the manufacture of cements and bricks. To obtain a valuable clinker (a name for the material produced in the first stage of the cement manufacturing (after furnace)), the chemical composition of the raw material must be within defined limits. CaO is the most important component (from 60 to 69% in weight content), but the crude mix must also contain $SiO_2$ (18–24%), $Al_2O_3$ (4–8%), $Fe_2O_3$ (1–8%), and minor components such as MgO, $K_2O_3$, and $Na_2O$. The four principal oxides (CaO, $SiO_2$, $Al_2O_3$, and $Fe_2O_3$) are absolutely necessary. Clays extracted from a quarry have variable concentrations of these oxides and therefore it may be necessary to add material to obtain the required composition. Limestone is the main source of CaO. Clay or shale will provide most of the $SiO_2$, $Al_2O_3$, and $Fe_2O_3$.

A typical cement raw mix comprises around 80–90% limestone and ~10–15% clay or shale. According to the United States Geological Survey, the production of cement in the United States in the year 2000 reached 90.6 million metric tons. These data indicate that the cement industry was the most important consumer of clay in the United States, requiring about 25% of the domestic clay production (total, 40.8 million metric tons). Transport of huge quantities of low-price raw material is not economical, thus the cement industry is best located near regions where geological conditions support mining of limestone and clay deposits.

## Waste Disposal

Clay is used in containment systems for landfill waste deposits to control the water fluxes in and out of the system. For this purpose, industrial designers have developed a composite material that imposes a hydraulic constraint on the clay liner. The liner is composed of a layer of montmorillonitic clay (normally sodium 'bentonite') sandwiched between two nonwoven geotextiles or bonded to a geomembrane. The thickness of the clay liner is typically about 6 mm. It has the ability to self-repair if damaged; it is equivalent to many metres of compacted clay and is not associated with problems of moisture, drying, or tree roots that normally affect compacted clay barriers.

## Ceramics Industry

The ceramics industry uses clay in the manufacture of bricks, ceramics (generic), glass, heavy clay products, aggregates, tiles, and refractory products; these accounted for two-thirds of the total clay production in the United States in 1985 (total, 45 million short tons). The most important clay mineral for ceramics is kaolinite, $Al_2Si_2O_5(OH)_4$. The basic chemical principle utilized by the ceramics industry is based on the structural and chemical transformation of clay minerals into minerals of the spinel type. In these reactions, kaolinite, heated at $\sim 550°C$, loses its (OH) groups (dehydroxylation process) and is transformed into a product called metakaolinite, $Al_2Si_2O_7$, with formation of two molecules of $H_2O$. At approximately $980°C$, metakaolinite transforms into a silicon spinel, $2Al_2O_3 \cdot 3SiO_2 + SiO_2$. This phase is rather complex and probably depends on the nature and ordering of the original clay mineral. At $1100°C$, mullite, $2(Al_2O_3 \cdot SiO_2)$, begins to form in association with quartz ($SiO_2$). The reaction continues, and at about $1400°C$ all the material has been converted to mullite and cristobalite (a high-temperature form of $SiO_2$). Pure mullite would be stable up to $1850°C$, but the presence of $SiO_2$ generates a liquid phase at about $1590°C$ that can induce silica exsolution in the material, and therefore the process is stopped at about $1650°C$. At $1850°C$, mullite would transform into corundum and a liquid-phase $SiO_2$. Ceramic mullite composition varies from $3Al_2O_3 \cdot 2SiO_2$ to $2Al_2O_3 \cdot SiO_2$, with possible partial substitution of Al by Fe: $3(Al_{0.9}Fe_{0.1})_2O_3 \cdot 2SiO_2$. The general process of ceramics formation at increasing temperature consists of going from an exsolution of silica from a primary raw material containing approximately 40% $Al_2O_3$, to mullite with an $Al_2O_3$ content up to 73%.

Refractory ceramics must have a particularly high alumina content. The phase line diagram of $SiO_2$–$Al_2O_3$ has its highest point theoretically at 100% $Al_2O_3$ and $2020°C$, thus minerals such as bauxite, with higher $Al_2O_3$ content than kaolinite has, are added to natural kaolin to obtain the most refractory products. Earthenware, wall tile, sanitary ware, stoneware porcelain, and electrical porcelain are manufactured using kaolinite or ball clay, quartz, and fluxes. One of the most important requirements is that the product must be white or ivory in colour. Temperatures between 1150 and $1250°C$ are used for vitrification except for electrical porcelain, for which it is necessary to reach $1400°C$. Because of its colours and its workability, kaolin is the main clay used, and the composition of various products ranges from 100% kaolinite for porcelain to 20–30% kaolinite for sanitary ware.

New applications for ceramics are under development, particularly as monolithic catalyst carriers for standard emission control equipment on vehicles. The ceramic for this purpose has a cordierite composition ($2MgO \cdot 2Al_2O_3 \cdot 5SiO_2$), obtained by firing a mixture of 70% talc and 30% kaolinite at $1400°C$. Ceramic membranes are another new technology. These membranes are high-temperature (up to $1000°C$) and corrosion-resistant materials that withstand the effects of corrosive products. They are used in many different processes, such as beer filtration and hot metal separation. The market for these products is steadily growing. Other applications for ceramics in modern industrial processes include permanent foundry moulds and heat exchangers.

## Physicochemical Properties of Clay Minerals

The physicochemical properties of clay minerals are strongly related to their structures and compositions, thus it is possible to organize them into three basic categories: the kaolinite group, the smectites group, and palygorskite (attapulgite) and sepiolite.

Kaolinite is a two-layer clay. Substitution of Al for Si in the $SiO_4$ tetrahedron (Tschermak substitution) is limited, i.e., the charge on the kaolinite layer is limited. Because of this limited charge, kaolinite has a low exchange capacity, particularly regarding other clay minerals (smectites, bentonite, etc.). Related to the low surface charge, kaolinite has low absorption and adsorption capacities. Kaolinite is a hydrated aluminous silicate with very limited substitutions (particularly by Fe) and is therefore white in colour. Particle size is very fine. Kaolinite is normally well crystallized, with broad dimensional distribution and low surface charge. These characteristics induce good flow properties and low viscosity even at high solid contents in water. Kaolinite is hydrophilic and will disperse easily in water with addition of a small quantity of dispersant. This quality makes kaolin an essential component in paper coating and in water-based interior paints. The paper industry is by far the largest user of kaolin. Coating paper with kaolin improves the printability of paper by making the paper surface smoother, more homogeneous, and more finely porous. To coat paper, a high-concentration solid/liquid suspension of pigment and binder is applied to the paper surface with a blade. Differently processed kaolinitic products must be further prepared for rotogravure and offset printing processes. Kaolin is also used as a filler during paper manufacture to enhance paper opacity and strength.

Smectites are a group of clay minerals characterized by their three-layered structure and their chemistry, consisting of sodium, calcium, magnesium, or lithium aluminium silicates. Chemical variations of smectite minerals include sodium montmorillonite, calcium montmorillonite, saponite, nontronite, and hectorite. Bentonites are types of sodium and calcium montmorillonites derived from alterations of volcanic glasses (usually volcanic ashes) and are of great industrial and economic importance. The three-layer structure of the smectites is very important; a single octahedral sheet is sandwiched between two tetrahedral sheets. In the octahedral sheet, $Fe^{2+}$ (ferrous iron) and/or $Mg^{2+}$ substitutes for $Al^{3+}$, creating charge deficiencies in the layers. The charge balance is also affected by Al substituting for Si in the tetrahedral sites. These deficiencies are charge balanced by exchangeable cations ($Na^+$, $Ca^{2+}$) located in interlayer positions. Two or four OH groups are associated with the exchangeable cations in the interlayer sites. The cations and OH groups located in the interlayer can be exchanged with other cations and also with polar organic molecules such as ethylene glycol, quaternary amines, and (poly)alcohols. Smectites are generally present in nature as very small particles and therefore they have a high specific surface. In association with a high charge, this induces a high degree of viscosity when smectites are mixed with water. High cationic exchange capacity and a high degree of viscosity when associated with water are the properties that determine the industrial uses of smectites.

The most important industrial user of smectites is the oil industry. Bentonite is used as drilling mud. Its excellent swelling capacity and high viscosity enhance the stabilizing effect of freshwater drilling mud. Bentonite mixed with sand of high silica content is also used in foundries. Adding a small quantity of water yields a plastic material that can be moulded. Bentonites are also used to pelletize iron ore and as sealants for water ponds, ditches, and household basement walls. Sodium bentonite can be produced from calcium bentonite by the addition of soda ash. In the presence of a soluble sodium salt, Na substitutes for Ca in the structure. These artificial bentonites, called sodium-activated bentonites, have better swelling and gelling properties than the natural ones have. Industrial products are frequently a mixture of natural and activated Na-bentonites.

Palygorskite (attapulgite) and sepiolite are hydrated magnesium aluminium silicates. Structurally, double silica tetrahedron chains are linked together through chain-like structures formed of oxygen and OH groups, forming octahedra in which the cationic site is occupied by aluminium or magnesium ions. Palygorskite and sepiolite are seen as fibres in scanning electron microscopy. This morphology is the result of the predominantly chain-like structure. Substitutions in the octahedral layer induce a moderate charge in the structure, and the exchange capacity is intermediate between that of kaolinite and smectite ($\sim 40$ meq per 100 g for palygorskite–sepiolite and 60–100 meq per 100 g for smectites). The combination in palygorskite–sepiolite of particle charge, channel structure, and large surface area results in a high absorption and adsorption capacity that has many industrial applications. The high and very stable viscosity makes these clay minerals useful in saltwater drilling mud or high-electrolyte-content drilling mud. They are currently used in drilling in oilfields where brines are normally associated with oil.

In agriculture, palygorskite and sepiolite are used as retardants in fertilizers, pesticides, and herbicides. When mixed with these products, the high adsorption and absorption properties of the silicates allow slow and progressive product release; this is both economical and environmentally beneficial. Palygorskite and sepiolite, due to their great stability in suspension, are also used as suspending agents in paints, fertilizers, pharmaceuticals, and cosmetics. They are also used for industrial cleaning of factory floors and service station bays because of their capacity to absorb grease and oil. These clay minerals also find a widespread use as pet litter.

## See Also

**Clay Minerals**. **Sedimentary Rocks:** Clays and Their Diagenesis.

## Further Reading

Boormans P (2004) *Ceramics Are More Than Clay Alone.* Cambridge: Cambridge International Science Publishers.

Bundy WM (1991) Kaolin in paper filling and coating. *Applied Clay Science* 5: 397–420.

Burst JF (1991) The application of clay minerals in ceramics. *Applied Clay Science* 5: 421–423.

Murray HH (1991) Overview – clay mineral application. *Applied Clay Science* 5: 379–395.

Murray HH, Bundy WM, and Harvey CC (eds.) (1993) *Kaolin Genesis and Utilization.* Aurora, CO: Clay Minerals Society.

Vaccari A (1999) Clay and catalysis: a promising future. *Applied Clay Science* 14: 161–198.

# COCCOLITHS

*See* **FOSSIL PLANTS: Calcareous Algae**

# COLONIAL SURVEYS

**A J Reedman**, Mapperley, UK

© 2005, Elsevier Ltd. All Rights Reserved.

## Introduction

The organization of the systematic surveying of the geology of the vast territories that formerly comprised the British Empire, mainly through the establishment of country specific geological surveys (*see* **Geological Surveys**), has been a long process and one that is not yet entirely complete. This article outlines some of the major features of that process with special attention being given to the role of the organisation known as the Directorate of Colonial Geological Surveys. Though the latter organization was only in existence for less than twenty years, in the middle of the last century, its influence on geological surveying in many parts of the world is still felt to this day.

## Early Days

In view of the fact that the Geological Survey of Great Britain was established in 1835, it may seem surprising that a systematic approach by the British Government to establishing and nurturing Geological Surveys in the many territories that constituted the British Empire took a further one hundred and twelve years to come to fruition, with the founding of the Directorate of Colonial Geological Surveys in 1947. A considerable number of Geological Surveys had been established in various British overseas territories well before 1947 but largely through the lobbying efforts of enthusiastic individuals, and usually through the local colonial administration, rather than as a result of any overarching policy of the centralized colonial power vested in the British Crown. The first of these local initiatives was taken just seven years after the Geological Survey of Britain came into being. In 1840, the former provinces of Upper and Lower Canada were joined by an Act of Union into a single province and a country called Canada came into being. In 1842, the legislature of the new country decided to set up a geological survey and William Logan was appointed as its first director. Shortly afterwards, in 1853, a geological survey for India was established for India, based in Calcutta.

## The Imperial Institute

The foundations for official British involvement in overseas geology, administered from London, the centre of the Empire, were laid through the opening of the Imperial Institute in 1893. This organization was established as the National Memorial of the Jubilee of Queen Victoria and was initially funded by grants from the Commissioners of the Great Exhibition held in London in 1851. It was reorganized on a number occasions during its relatively short life before its final demise in the 1950s, but during this time it contributed a great deal to geological knowledge of the vast territories that constituted the 'Empire'.

The prime purpose for the existence of the Imperial Institute was to promote trade in natural raw materials in the colonies and dependencies. Amongst these commodities, minerals (*see* **Mining Geology: Mineral Reserves**) were seen to be of great importance and a Mineral Resources Department, later to become a Division, was an important component of the establishment. In its early years, it carried out a number of pioneering reconnaissance surveys in colonial territories, mainly involving just one or two geologists, often traversing whole countries on foot, without the aid of any but the most rudimentary of topographic maps. As interest in the Empire's resources grew, the Division increasingly concentrated on providing services, such as mineral identification, assaying, and economic assessment, to those fledgling geological surveys that were beginning to be established.

A report published in the Bulletin of The Imperial Institute in 1943 summarised the work of the Mineral Resources Department during the fifty years it had so far been in existence and also reviewed the likely requirements for the future. Noting that some half a million pounds had already been earmarked to extend scientific investigation into 'Colonial problems', excluding welfare and development projects, the report drew attention to "the wisdom of allocating a fair share of any available grants to the purpose of expanding the work of Geological Surveying in the Colonial Empire". It noted that by 1943, local (overseas) geological departments existed for territories with a total area of one and a quarter million square miles (excluding the Dominions) but that the total number of trained staff responsible for geologically surveying this vast area amounted to only forty six.

The early work of the Institute, mainly carrying out reconnaissance surveys, was seen as a necessary preliminary to the formation of a Geological Survey and in a number of colonies this was achieved. Such colonies included British Malaya (now Malaysia, 1903), Ceylon (now Sri Lanka, a Mineral Survey under various subsequent names was established in 1903), the Gold Coast (now Ghana, 1913), Uganda (1918), Nigeria (1919), Tanganyika (Tanzania, 1926), Sierra Leone (1927), Kenya (1933), and British Guiana (Guyana, 1933) all established Surveys, often attached to Mines Departments and with very few geological specialists.

## Directorate of Colonial Geological Surveys

Though it was widely acknowledged that the Mineral Resources Division of the Imperial Institute had achieved a great deal in advancing knowledge of the mineral resources of the colonies since its inception, by 1943 it was also recognized that there was still a serious lack of geological knowledge of much of the extensive terrain that constituted the colonies and that a new organization was needed to help rectify this situation. A joint meeting of the Geological Society of London and the Institution of Mining and Metallurgy, after considering how geological surveys in the British colonies might be strengthened in the post-war era, made a number of recommendations. The Secretary of State for the Colonies accepted the advice of a committee set up to advise him on the need for an expansion of geological work and how this might be achieved, and on the first of January 1947, a central organization for overseas Geological Surveys was inaugurated in London with the appointment of Dr Frank Dixey as its Director and as the Geological Adviser to the Secretary of State. At the age of 54, Dr Dixey had considerable experience of geological surveying in the colonies, having successfully carried out a reconnaissance survey of Sierra Leone followed by almost eighteen years in Nyasaland, first as Government Geologist and then Director of the Geological Survey where he was required to spend much of his time and geological knowledge on the practical problems of establishing, and improving, rural water supplies. In 1939, the latter experience led him to be appointed to set up a Water Department in Northern Rhodesia (Zambia) and in 1944 he moved to Nigeria as Director of the Geological Survey.

The Directorate of Colonial Geological Surveys was initially housed in offices in the ornate Victorian building originally constructed for the Imperial Institute in South Kensington, London. Work commenced in a modest way; Dixey being joined by a geologist and then, in 1949, by a geophysicist. Two Deputy Director's were appointed and the task of establishing a photogeology section was commenced. In 1949, the Colonial Office took over responsibility for the Mineral Resources Division of the Imperial Institute which, as the Mineral Resources Division of the Directorate of Colonial Geological Surveys, continued its wide-ranging work on the mineral resources of the Colonial territories and compiling statistics on mineral production throughout the Empire.

It soon became apparent that an increasing workload demanded an increase in the establishment and investment on more modern equipment, with the cost being met from Colonial Development and Welfare funds. As well as continuing to service the specialist needs of the eleven Geological Surveys that had been previously set up in various the colonies with the assistance of the Imperial Institute, it became clear that some thirteen other territories could benefit from the establishment of their own Geological Surveys, and that the greatest need of both new and pre-existing Surveys would be for an increase in their staff complement of geologists, geophysicists, and geochemists together with buildings and equipment. While the costs were to be covered, during their first few years, from Colonial Development and Welfare Funds, most territories eventually assumed responsibility for the cost of their own Geological Surveys.

By 1957, ten years after its foundation, the organization had grown very considerably and was able to offer a broad range of services to the colonies, including the recruitment of staff to serve in their Geological Surveys. Apart from Dr Dixey, the Director, the organization now comprised a Directorate including Geophysical and Photogeology Sections (total staff 9), a Mineral Resources Division (36 staff) and a Joint Services Division (total staff 14) which included a library and serviced both the Colonial Products Laboratory and Colonial Geological Surveys.

The Mineral Resources Division not only continued to give advice to Colonial Administrations on all aspects of mineral development but was also responsible for a new publication, 'Colonial Geology and Mineral Resources', a successor to the Mineral Resources section of the former 'Bulletin of the Imperial Institute', and also continued to publish the 'Statistical Summary of the Minerals Industry'. Perhaps most significant of all, at least as far as geological mapping of the Colonial territories was concerned, was the establishment of the Photogeological Section, shortly to be housed in the Headquarters of Colonial (Geodetic and Topographic) Surveys where it was able to access the considerable stock of air photographs being built up by the latter organization.

Before 1947, geologists working in the Geological Surveys of the colonies usually had to make their own topographic maps by conventional surveying methods and frequently in the absence of a reliable trigonometrical network. The advent of air photographs not only allowed the greatly increased production of topographic maps of many remote areas at a scale adequate for geological mapping but also made photographs available for geological interpretation. Dixey reviewed the first nine years of Colonial Geological Surveys, and was able to report that, together with the increase in geological staff, it was the topographic mapping programme of Colonial (Geodetic and Topographic) Surveys that was more and more responsible for the increased rate of geological mapping in the colonial territories. In the East African territories of Kenya and Uganda for example, only a total area of about 25 000 sq miles had been geologically mapped at various scales prior to 1947, but by 1993 that figure had increased by almost 75 000 square miles, all mapped at scales of either 1:50 000 or 1:125 000.

With respect to its establishment of a Geophysics Section and, more particularly, a Photogeological Section, Colonial Geological Surveys was ahead of its much older sister organization, the Geological Survey of Great Britain, with its nearby headquarters in Exhibition Road in South Kensington. More innovation was to follow under Dr Dixey's leadership. With the help of grants obtained from the Department of Scientific and Industrial Research and the Colonial Welfare and Development Fund, a Geochemical Prospecting Research Centre was set up at the Imperial College of Science and Technology in London and this laid the foundations for many subsequent geochemical prospecting programmes both in the then Colonial territories and, later, throughout many other parts of the world. In 1957, grants were also made available to set up an Isotope Research Group in the Department of Geology and Mineralogy of Oxford University and this group, expanding in future years, was devoted primarily to problems in overseas geology.

The task of promoting and assisting the formation of Geological Surveys in the Colonial Territories and Dependencies, started by the Imperial Institute, gained momentum after the inauguration of Colonial Geological Surveys. In the first nine years of its existence, Geological Surveys had been established in a further nine countries, as shown below:

The increase in the number of overseas 'colonial' Geological Surveys saw a marked increase in the number of geologists, geochemists, and geophysicists amongst their staff. In the first ten years of the Directorate, the total rose almost fourfold from 58 to 212, and to 258 by 1961. The Directorate acted as advisers to the Colonial Office on matters of staff recruitment and also ran a variety of training courses for both individual and groups of newly recruited or established staff of the Surveys. Secondment of specialists to posts in colonial administrations or overseas Surveys was also arranged. The demand for, and supply of qualified professionals in the geological field varied from year to year but the bulk of the new recruits were young men just completing either their first or second degrees and it is estimated that each year, between 1947 and 1955, Colonial Geological Surveys absorbed between one quarter and one half of the UK universities output of suitably qualified geologists.

| | |
|---|---|
| Bechuanaland Protectorate (now Botswana) | 1948 |
| British Territories in Borneo (now mainly part of Malaysia) | 1949 |
| Jamaica | 1949 |
| Swaziland | 1949 |
| British Solomon Islands Protectorate | 1950 |
| Cyprus | 1950 |
| Northern Rhodesia (now Zambia) | 1950 |
| Fiji | 1951 |
| Cyprus | 1952 |
| New Hebrides | 1959 |

When the grand old Victorian building that had once housed the Imperial Institute was demolished to make way for the new buildings of Imperial College, the Directorate of Colonial Geological Surveys moved, in 1960, to new purpose-built laboratories and offices in Greys Inn Road, London. By this time, inclusion of the term 'Colonial' in the organizations name was becoming politically unacceptable, and it was changed to the Directorate of Overseas Geological Surveys. Furthermore, an increasing number of Britain's overseas possessions were gaining independence and this led the British Government to set up a committee to consider, amongst other matters, the type of technical assistance which the United Kingdom should be in a position to provide in the geological and mining fields and, in this context, the future organization, structure, and functions of the Directorate of Overseas Geological Surveys.

## An End and a Beginning

The report of the 'Brundrett' Committee was presented to the British Parliament in May 1964. Most of it recommendations were accepted, including the recommendation that "the functions of the Geological Survey of Great Britain should be expanded to cover overseas work in the geological and mineral (*see* **Mining Geology:** Exploration) assessment fields" and that "the Overseas Geological Surveys and the Atomic Energy Division of the Geological Survey of Great Britain should be amalgamated within the expanded Geological Survey of Great Britain". In June 1965, the amalgamation was completed with the establishment of the Institute of Geological Sciences. Overseas Geological Surveys, an integral part of the new organization, eventually evolved into the Overseas Division and then International Division of the British Geological Survey, later to be renamed the British Geological Survey where, building on the legacy of 'Colonial Geological Surveys', the work of geological surveying and assistance has continued in many countries of the former British Empire and in others worldwide.

## See Also

**Geological Field Mapping**. **Geological Surveys**. **Mining Geology:** Exploration; Mineral Reserves.

## Further Reading

Dixey F (1957) *Colonial Geological Surveys 1947–56*, Colonial Geology and Mineral Resources Supplement Series, Bulletin Supplement no 2, p. 129. London: HMSO.

Dunham KC (1983) Frank Dixey 1892–1982. *Bibliographical Memoirs of Fellows of the Royal Society* 159–176.

Intelligence Staff of the Mineral Resources Division, Imperial Institute (1943) *A Review of Geological Survey Work in the Colonies*. Bulletin of the Imperial Institute, vol. XLI, no 4. London: HMSO.

Pallister JW (1972) *British Overseas Aid in the Field of the Earth Sciences*. 24th IGC, Symposium 2.

*Report of the Committee on Technical Assistance for Overseas Geology and Mining.* (1964) London: HMSO.

Walshaw RD (1994) British Government Geologists Overseas – A Brief History. *Geoscientist* 2: 10–12.

# COMETS

See **SOLAR SYSTEM: Asteroids, Comets and Space Dust**

# CONSERVATION OF GEOLOGICAL SPECIMENS

**L Cornish and G Comerford**, The Natural History Museum, London, UK

Copyright 2005, Natural History Museum. All Rights Reserved.

## Introduction

This article covers the general ethics and methodologies of conservation, which can be applied to all geological material. It outlines best practice when carrying out both preventive and remedial conservation. Examples are given of unstable material and how to approach treatment. The further reading list will allow the reader to explore specific conservation issues.

There is no international standard for the care and conservation of fossils. Natural history specimens in general do not automatically fall under the protection of the UNESCO treaty. A great deal depends on the enabling legislation passed by individual countries. A single collection may contain materials as diverse as unaltered organic material, bone, shell, amber, and a very wide range of rock, mineral, and sediment materials. Fossils by their very nature can be preserved in isolation or in a rock matrix, and as such the compositions of the specimen and the surrounding matrix have to be considered as one unit for conservation purposes. The term fossil has therefore been used to encompass a range of geological material.

## Preventive Conservation

Preventive conservation is an important part of prolonging the life of an individual specimen or collection. There are two distinct aspects to preventive conservation: the organizational and the technical. This article concentrates on the technical aspect, whilst appreciating that in large organizations the management of human resources and the physical environment are required to produce good preventive conservation practice. Preventive conservation is everyone's responsibility and not just the preserve of an individual conservator. Preventive conservation involves indirect action to slow deterioration and prevent damage by creating conditions that are optimal for the preservation of the object.

### Handling

All specimens should be handled with care; even those that appear robust can be damaged by inappropriate handling. The basic principles of handling include:

- cleanliness – use clean bare hands or disposable gloves;
- avoid unnecessary handling;
- assess the weight and condition of the object;
- check for any breakages, cracks, or old repairs;
- handle specimens one at a time;
- handle associated pieces separately;
- support the specimen fully when picking it up;
- use supports to carry specimens that are fragile or cannot carry their own weight;
- use a trolley or similar device to move objects that are heavy;
- do not drag or push a specimen across a surface; and
- provide protection against environmental changes for specimens that are being moved to an area with different environmental conditions.

The first step in preventive conservation is to ensure that specimens are handled correctly. Specimens should be fully supported during transportation from the storage area to the study area. Large specimens are in danger of failing in weak areas if they are held at only one point. Before picking up a specimen, a brief assessment should be made, noting any vulnerable areas. Strain should not be placed on cracks or joins, and surfaces that are flaking or friable should be avoided. Scratching of highly polished surfaces should be avoided. Absorption of dirt, oil, and salts from the hands can occur if the specimen has a porous surface (e.g. limestone). Large specimens require thick soft padding to support them (e.g. high-density polyethylene). Specimens that are being moved from one set of environmental conditions to another should be packed to ensure a slow acclimatization.

### Storage

The correct storage of specimens will prevent unnecessary damage arising through abrasion and stacking. Ensuring that each specimen is housed correctly and fully supported in storage will avoid physical damage.

- Small objects stored in drawers may be placed in boxes or trays. These trays should be of adequate size and lined with a supporting material.
- Small specimens may be stored in drawers lined with foam and placed in cut-outs.
- Boxes or trays should be of such a size that they tessellate within a drawer, preventing trays moving when the drawer is opened or closed.
- Larger specimens may require bespoke mounts to ensure that they are fully supported.
- Mounts should be lined with inert foam to ensure that there is no adverse interaction with the specimen.
- Very heavy specimens should be stored on foam-lined boards that incorporate supports for vulnerable parts.

### Packaging Materials

All materials that come into direct contact with a specimen should be inert, in that they should not damage the specimen through abrasion or chemical interaction. The materials used to store and display specimens should be made from conservation-grade substances. Foams that are used to support specimens and absorb vibration should not be made from materials that degrade rapidly. A closed-cell inert-nitrogen-blown polyethylene foam that is highly chemical resistant and stable in the presence of ultraviolet radiation is a suitable foam for use in storage. This type of foam can be purchased in varying densities, and the density should be chosen according to the weight of the specimen. Larger specimens should be stored on foam-lined shelves or pallets. Other materials such as tissue used to pack specimens for travel should be acid free, as should the trays and boxes in which specimens are stored within collections.

### Environmental Conditions

Many rocks and minerals are environmentally sensitive. Environmental factors affecting specimens include light, heat, dust, relative humidity, and pollutants. Some mineral specimens may change colour owing to their sensitivity to light. Crystals may fracture if they are exposed to heat or to cycles of hot and cold temperatures. Dust can be potentially damaging to specimens as it disfigures surfaces and encourages corrosion by providing nucleation sites for the absorption of water and other pollutants. Where specimens have been remedially treated with adhesives or consolidants, dust can sink into the polymer surface and subsequently be very difficult to remove. Pyrite, clay, shale, and subfossil-bone specimens may split and crack if they are stored at an inappropriate or fluctuating relative humidity. Pollutants can occur in materials associated with the specimen, in other objects in the collection, and in the surrounding air. Materials that are known to release harmful vapours at room temperature include wood – particularly oak, birch, and chipboard. These materials should be avoided for use in storage as the vapours given off may react with some mineral specimens. The best environment in which to store rock and mineral specimens is dust free and clean, with low light and moderate temperature and humidity levels.

### Light

Light-sensitive specimens or specimens whose sensitivity to light is unknown should be housed in light-proof storage. Where specimens are to be displayed, the effects of light should be minimized by one or more of the following methods:

- excluding daylight;
- using low-ultraviolet light sources;
- using ultraviolet filters;
- using display methods that ensure that the specimens are lit only whilst they are being viewed; and
- rotating the specimens between display and low-light storage.

### Relative Humidity and Temperature

In purpose-built collection storage it is possible to specify particular values of relative humidity and temperature. A stable environment is paramount to providing a stable collection. Heat can damage specimens through simple excess or indirectly when it is introduced suddenly into an unheated space, destabilizing the relative humidity. Fluctuating relative humidity can lead to physical damage of some specimens. Increased moisture in the air can lead to the dissolution of specimens through the absorption of water from the environment. Conversely, loss of water to the environment can lead to changes in the chemical composition and properties of some minerals. Suitable environments for mineral, rock, and fossil collections are those with stable relative humidity and temperature. A temperature between $15°C$ and $21°C$, fluctuating by no more than $\pm1°C$, and a relative humidity of $45\% \pm 5\%$ would be suitable for most geological collections.

Where the specimen storage environment is unsuitable, microclimates may be created using humidity buffers. The buffering material, commonly a silica-based product available from conservation suppliers, can be purchased in bead, sheet, or cassette form. The buffers work by absorbing and releasing moisture from the enclosed air. Individual specimens can be placed in a sealed polyethylene container with an appropriate quantity of this material, and a constant humidity will be maintained for a period. The more often the container is opened, the more quickly the buffer will become spent and require recharging. Where specimens are too large to fit into polyethylene containers, microclimates can be produced by encapsulating the specimen in a buffer in a moisture-barrier film sealed with a heat sealer. These methods work very well for individual specimens but alternatives are required for larger collections of humidity-sensitive material.

### Environmental Monitoring

The environment of the collection or specimen requires monitoring. This can be done in a number of ways, depending on the size of the task and the funds available. Simple humidity-indicator cards can be placed with the specimen to give a general indication of the amount of moisture in the atmosphere. Dial hair hygrometers will give a more specific reading, and recording thermohygrographs will give a weekly or monthly trace. These methods all have their uses, albeit limited. Electronic data loggers are more useful; they are programmed and the data downloaded using a computer. The data are then more easily analysed. Radio telemetric systems and building management systems give real-time data. This type of monitoring allows the environment to be measured remotely and is very useful inside microclimates such as display cases where buffering agents have been used or in places that are remote or difficult to access (**Figure 1**). Sensors are commonly used to control the environments of buildings housing collections. In the past, data from building management systems have been of little value to the preventive conservator for

**Figure 1** Moa bird on display monitored by telemetry.

recording the environmental conditions. Generally, these sensors were part of a complex control system, and expensive specialist software was needed to view the values. They also recorded data in formats that were difficult to export. Recording of building management system data can now be easily configured via an internet browser, and, importantly, the data can be retrieved simply and analysed using standard graphing software. High-quality building management system sensors have now been developed that allow simple checking and re-certification by conservators using low-cost certified handheld monitors.

### Collection Surveys

The collection survey is an essential component of an environmental strategy and is used to determine the physical state and future needs of the collection (**Figure 2**). For large collections, statistical sampling methods may be needed to reduce the time and money spent on the survey. A formal documented inspection or survey will highlight problems in a collection. Typically information documented would include damage, surface pollutants, decay, environmental sensitivity, previous conservation treatment, style of storage, and environment. Once an overall picture of the collection has been built, a procedure for bringing the collection up to an acceptable standard can be determined.

### Integrated Pest Management

Insect pests can cause major problems for a natural history collection. Although pests rarely damage geological material, the accompanying documentation may be affected. The pesticides that were generally used to stop pest infestations are no longer used

**Figure 2** Survey being carried out on sub fossil bone mammal collection.

because of the risk they pose to health. It is advisable to monitor collections through trapping so that potential infestations can be prevented. In large organizations the approach to preventing pest damage is known as integrated pest management.

### Mould

Preventing contamination by fungal spores within any building is impossible. However, mould growth can be prevented by ensuring that the conditions for germination do not arise in the stores or display areas. Uncontrolled indoor environments may experience extreme seasonal changes, allowing humidity levels to rise to the point where germination of spores occurs. If relative humidity is maintained at sufficiently low levels, outbreaks will not occur. Levels of between 50% and 60% relative humidity are considered safe and will not allow such growth. When an infestation is encountered the steps to take are:

- confine it;
- stop its growth;
- eradicate it; and
- prevent it from reoccurring.

### Reduced Oxygen Environments

Reduced oxygen environments are cost-effective low-impact methods of controlling the deterioration of specimens that are sensitive to oxygen, water vapour, or pollutants. Any rock or mineral specimen that is sensitive to oxygen, water vapour, or pollutants (such as rocks containing pyrite that is likely to oxidize or has begun oxidizing) can be stored in reduced oxygen environments to prolong their life. There are three components to a reduced oxygen environment: an oxygen scavenger or oxygen-purging system; an oxygen monitor; and a barrier film. Oxygen scavengers are composed of either iron filings or molecular sieves such as zeolite or mordenite. The optimum composition of a barrier film for this purpose is currently under research. Commercially available films are currently composed of materials chosen to suit either the food or the electronics industries, and, whilst these films may be useful for preventive conservation, they may not be the optimum. Barrier films are layered polymers combined to produce a film with good tear strength, low water migration, and low oxygen migration. The enclosure is made by wrapping the specimen in the barrier film and using a heat sealer to close the film and finish the encapsulation. The specimen should be supported in conservation grade materials to ensure that no physical stresses are imposed on the specimen and no damage is created by the abrasion of the film on its surface (**Figure 3**).

## Remedial Conservation

Remedial conservation consists mainly of direct action carried out on an object with the aim of retarding further deterioration.

One of the most important and sometimes controversial stages of conservation treatment is surface cleaning. Irreversible damage can occur if inappropriate treatments are used. However, the removal from the surface of the object of contaminants or old consolidants that may otherwise cause harm is highly advantageous. For example dust can contain acidic particles that can cause surface damage. In some cases cleaning may also clarify an object's detail or reveal unseen damage that can be treated subsequently. Joins that require repair also have to be surface cleaned to improve the final bond. There are many types of surface-cleaning techniques available to the conservator, and they tend to be divided into mechanical cleaning and chemical cleaning.

### Surface Cleaning: Mechanical

**Abrasive** Abrasive cleaning methods include all techniques that physically abrade the fossil surface to remove contaminants or coatings. Such techniques involve the use of materials that impact or abrade the surface under pressure, or abrasive tools and equipment. The use of water in combination with abrasive powder may also be classified as an abrasive cleaning method. Depending on the manner in which

**Figure 3** Pyritised ammonite encapsulated in barrier film in an oxygen free environment.

it is applied, water may soften the impact of the powder, but water that is too highly pressurized can be very abrasive.

**Steam cleaning** Steam, heat, and pressure provide the means for the immediate removal of particles from a given surface, cleaning it thoroughly. This technique is usually carried out using of a handheld steam pencil and is especially useful for removing old adhesives and labels.

**Ultrasonic** Ultrasound is used in the cleaning of material because of its vibration rates. Large acoustic forces break off particles and contaminants from surfaces. This cleaning method is generally carried out either by immersing the specimen in a tank or by using a handheld pen. When the latter is used, its vibrating tip contacts the surface and fragmentation occurs. When fossils are immersed in a tank of liquid (usually water and detergent), sound waves from the transducer radiate through the tank, causing alternating high and low pressures in the solution. During the low-pressure stage, millions of microscopic bubbles form and grow. This process is called cavitation, meaning 'formation of cavities'. During the high-pressure stage, the bubbles collapse or implode, releasing enormous amounts of energy. These implosions work in all directions, on all surfaces.

**Laser cleaning** Surface cleaning by laser is a recent innovation in the treatment of geological material and presents several advantages when compared with standard cleaning methods. The laser offers the potential for selective and controlled removal of surface contaminants with a limited risk of damaging the underlying substrate. When the laser beam meets a boundary between two media, e.g. dirt and the sample surface, a proportion of the beam's energy is reflected, part is absorbed, and the rest is transmitted (**Figure 4**). The proportion of energy absorbed is determined by the wavelength of the radiation (primary source) and the chemical structure of the material. In order to remove dirt or another unwanted surface coating from an object, it is important that the dirt or coating absorbs energy much more strongly than the underlying object at the selected laser wavelength. If this is the case then, once the dirt layer has been removed, further pulses from the laser are reflected and the cleaned surface is left undamaged. The technique, therefore, is – to a certain degree – self limiting. Cleaning efficiency can also be enhanced by brushing a thin film of water onto the surface immediately prior to laser irradiation. The laser also effectively removes conducting coatings, for example gold or palladium, from microfossils; these have previously been almost impossible to remove.

### Surface Cleaning: Chemical

Surface cleaning by using chemicals should be considered carefully so that the integrity of the object is not compromised. A disadvantage of using chemicals is the inability to control the chemical movement within an object, which can lead to over cleaning and the unwanted removal of fossil material. Usually a chemical-cleaning testing regime should be carried out, commencing with the least aggressive solvent. A solvent is a liquid capable of dissolving other substances.

Care should be taken to ensure that the physical and chemical properties of the fossil and surrounding matrix (where present) are taken into account so that

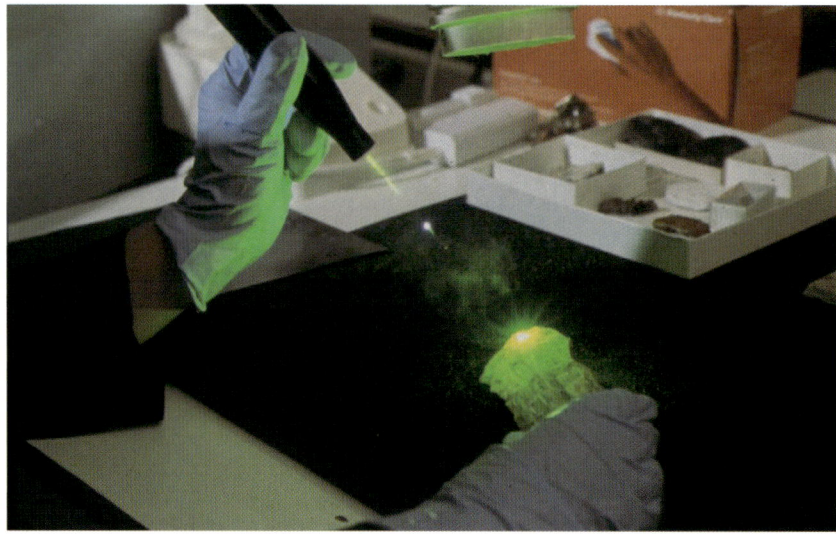

**Figure 4** Laser cleaning of fossil reptile (indet) and surrounding matrix. Visible green wavelength (533 nm) being used.

inappropriate use is avoided. For example, water is non-hazardous to the conservator and the strong polarity of its molecules means that it is effective in removing a wide range of pollutants. However, this apparently harmless material would have severe detrimental effects on humidity-sensitive pyritized fossils, causing them to deteriorate.

Poulticing is used to avoid deep penetration and to limit the action of chemicals on the object. Chemicals are mixed with absorbent powders to form a paste or poultice, which is applied to the surface. As the poultice dries, dirt is drawn out and has to be physically removed. A development of this technique is the use of solvent gels. These cleaning systems have an aqueous gel base composed of a polymer resin that thickens on the addition of water, and a surfactant – also a thickening agent – which improves the gel's contact with the surface to be cleaned. Any number of cleaning agents can be added to this gel base. Of particular concern to the conservator are the possible long-term effects on surfaces. The most pressing concern has been whether any residue of the gels is left on the treated surface that may cause future damage.

**Consolidants, Adhesives, and Gap Fillers**

The consolidants, adhesives, and gap fillers used at any given time reflect the existing knowledge of chemistry and technology; so, naturally, their diversity and quality change over time. From the viewpoint of the conservator, all should be reversible. Geological material, whether it is in scientific collections or in a private collection, must be considered valuable and irreplaceable. As such, conservationally sound and approved materials should be used wherever possible.

**Consolidants** When considering whether to use a consolidant on a specimen, it is important to remember that not all specimens require consolidation. The most important axiom of conservation is: minimal intervention is best.

A consolidant is a liquid solution of a resin (normally a synthetic polymer) that is used to impregnate a fragile object in order to strengthen its structure. Common solvents for the resins are water, acetone, alcohol, and toluene. Consolidants are generally available in two forms: pure resins and emulsions. Pure resins are mixed with their solvents to form a very thin watery solution, which is then applied to the specimen (or the specimen is immersed in the solution). The aim is to get the solution to penetrate the specimen's surface and carry resin down into the interior of the fossil bone: the consolidant must be thin otherwise it will be deposited on the surface of the bone only, like the shellac or varnish used in the past. Those treatments may have protected the surface, but did little to strengthen the whole bone.

The second class of consolidants, the emulsions, are mainly used to treat wet or moist specimens. Emulsions are suspensions, in water, of a resin and solvent solution, popularly polyvinyl acetate emulsion, and are generally white milky mixtures. Emulsions are not as desirable as pure resins. It is hard to reverse emulsions once they have dried and virtually impossible once they have crosslinked on exposure to ultraviolet light from the sun or from fluorescent bulbs. Emulsions also tend to turn yellow with age and increasing crosslinking.

**Adhesives and gap fillers** Historically, the most common adhesives have been animal glues made from bones, fishes, and hides, and these were used extensively for fossil repair and can still be purchased today. Owing to their inherent problems, such as yellowing, brittleness, and instability, they are no longer used on fossils. Consisting mainly of collagen–protein slurries, animal glues are also quite attractive to a variety of pests. This class of adhesives is mentioned here because of its long period of use. In museums and at fossil auctions, it is not uncommon to find specimens that have been repaired with these glues.

The twentieth century has given us many new classes of adhesives, all of which are organic polymers – large complex molecules formed from chains of simpler molecules called monomers. Manufacturers may extol the virtues of the newest adhesive from their laboratories, but only time can judge the effectiveness and longevity of an adhesive. Many turn yellow with age or are prone to brittle breakage, where even a slight jar or shock will cause the glued joint to break. Other polymers may crosslink with time or upon prolonged exposure to ultraviolet light, causing shrinkage that can seriously damage a fossil that has been repaired with them. Most of these polymers have unique properties and characteristics, which make some better for certain uses than others. Another class of recently developed adhesives used to repair fossils are the 'superglues' or cyanoacrylates. Their characteristics, which include rapid setting and strength, have made these adhesives increasingly popular for fossil repair; however, since they are so new (dating only from the 1980s), our knowledge of their long-term efficacy is limited. A major drawback is the difficulty of reversing bonds made with cyanoacrylates.

Adhesives are sometimes mixed with materials, e.g. glass micro balloons, to form a gap filler. Gap filling is a process whereby a sympathetic replacement material is used to fill or bridge small or large gaps. The gap filler should not conceal original material and should be easy to remove (**Figure 5**).

## Conservation of Sensitive Geological Material

### Subfossil Bone

Subfossilized bone retains a fair amount of organic material (collagen) and original mineralized bone (hydroxyapatite). To maintain its integrity, subfossil bone needs to be kept in a stable environment. Once it is removed from the matrix in which it is found, it is likely to deteriorate rapidly, especially if it is wet and the bone is subjected to extreme fluctuations of temperature and humidity.

Recently excavated material that may still be wet can be treated in two ways. The first method is the use of water-based consolidants, which can be applied by brushing, immersing, spraying, or vacuum impregnation. The second method is slow controlled drying to prevent cracking and delamination. The aim of controlling the drying procedure is to reduce the high relative humidity of the wet specimen slowly until it matches that of the storage area. Even if a water-based consolidant is used, it is advisable to apply controlled drying procedures until the specimen is stabilized.

Smaller specimens can be placed in plastic containers, and for larger specimens a chamber or tent can be constructed, for example by placing a piece of clear plastic sheeting over laboratory scaffolding. Barrier films can also be used to control drying stages. There needs to be a controlled exchange of the moist air inside with the lower-relative-humidity ambient air outside the box. To monitor changes in relative humidity, a humidity gauge can be placed in the containment area along with the specimen. Potential mould growth needs to be monitored, and placing a fungicide inside the containment area should control the problem. The optimum storage environment for subfossil bone has a relative humidity of 45–55% and a temperature between 15°C and 21°C. Low relative humidity can lead to cracking and shrinking as the specimen dries out, and high relative humidity (above 70%) encourages damaging mould and fungal growth.

Once the specimen has been dried and stabilized, an organic-solvent-based consolidant can be applied.

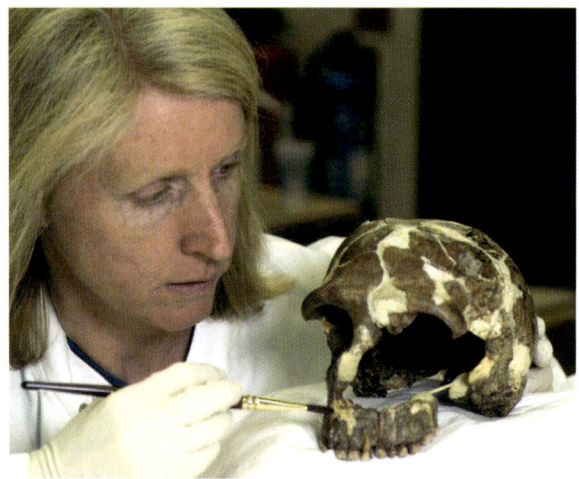

**Figure 5** Neanderthal skull (Tabun) showing gap filler which aids in its reconstruction.

**Figure 6** *Dacosuarus maximums* showing severe cracking in response to fluctuating humidity conditions.

It is not advisable to apply water-based consolidants to specimens that have thoroughly dried because the high water content of the consolidant will cause the dry specimen to swell and crack multidimensionally.

If gap fillers are to be applied to repair the bone, particular attention should be given to the choice of filler. Materials that shrink or expand upon curing should be avoided as either action can damage the specimen.

### Shale and Other Fine-Grained Sediments

Deterioration in the form of delamination, cracking, and shrinkage is common for shale and other fine-grained sediments, especially if they are stored in the wrong environment. The best approach is therefore to ensure that storage conditions are optimum, with a temperature of below 20°C and humidity level of around 50%.

### Pyrite

The deterioration and even complete decomposition of pyritized fossils through oxidation is a common problem (**Figure 6**). Deterioration is best prevented by storing at less than 45% relative humidity. Once deterioration has occurred various conservation treatments are available, which have varying levels of success. For example, ethanolamine thioglycollate or ammonia gas treatment may be used. The latter is considered more conservationally sound.

## Documentation

Existing documentation associated with individual geological specimens should be preserved as a matter of priority; without it the material is scientifically useless. Any treatments should be recorded accurately (with an image if possible) so that a history of treatment can be built up over time. This information will influence future conservation considerations.

## See Also

**Fake Fossils**. **Minerals:** Definition and Classification; Sulphides; Zeolites. **Palaeontology**. **Rocks and Their Classification**.

## Further Reading

Casaar M (1999) *Environmental Management Guidelines for Museums and Galleries*. London and New York: Routledge.

Cornish L and Doyle AM (1984) Use of ethanolamine thioglycollate in the conservation of pyritized fossils. *Palaeontology* 27: 421–424.

Cornish L and Jones CG (2003) Laser cleaning natural history specimens and subsequent SEM examination. Chapter 16 In: Townsend J, Eremin K, and Adriaens A (eds.) *Conservation Science 2002*, pp. 101–106. London: Archetype.

Cornish L, Doyle AM, and Swannell J (1995) The Gallery 30 Project: conservation of a collection of fossil marine reptiles. *The Conservator* 19: 20–28.

Croucher R and Woolley AR (1982) *Fossils, Minerals and Rocks – Collection and Preservation*. London: Cambridge University Press.

Gilroy D and Godfrey I (1998) *A Practical Guide to the Conservation and Care of Collections*. Perth: Western Australian Museum.

Howie FM (ed.) (1992) *The Care and Conservation of Geological Material: Minerals, Rock, Meteorite and Lunar Finds*. Oxford: Butterworth-Heinemann.

Institute of Paper Conservation. www.ipc.org.uk.

Larkin NR, Makridou E, and Comerford GM (1998) Plastic storage containers: a comparison. *The Conservator* 22: 81–87.

Museums, libraries and archives council. http://www.mla.gov.uk/index.asp.

Resource UK Council for Museums, Archives, Libraries (1993) *Standards in the Museum Care of Geological Collections*. London: Resource UK Council for Museums, Archives and Libraries.

Waller R (1987) *An Experimental Ammonia Gas Treatment Method for Oxidised Pyritic Mineral Specimens*, pp. 625–630. Triennial Report. Rome: ICOM Committee for Conservation.

# CREATIONISM

**E Scott**, National Center for Science Education, Berkeley, CA, USA

© 2005, Elsevier Ltd. All Rights Reserved.

## Definitions

Although evolution is a component of many scientific fields, it generally has the same meaning across disciplines: cumulative change through time. The topic of this article, 'creationism', is a term with many definitions. In theology, creationism is the doctrine that God creates new souls for each person. In anthropology, creationism is the well-confirmed thesis that almost all human societies have origin stories about the acts of gods, or a God, or powerful spirits of some kind.

Viewed socially and politically, creationism refers to a number of twentieth-century religiously-based anti-evolution movements originating in the USA, but now spreading to many countries. The most familiar of these (and the movement to which the term is most frequently applied) is 'creation science', an attempt to demonstrate with scientific data and theory the theological view known as special creationism. According to special creationism, God created the universe – stars, galaxies, Earth, and living things – in essentially their present forms. Living things were created as 'kinds' that do not have a genealogical (evolutionary) relationship to one another. This biblical literalist theology views Genesis narratives, such as the creation of Adam and Eve, their sin and expulsion from Eden, and the Flood of Noah, as historical events. In creation science and its ancestor, 'Flood Geology', the Flood of Noah has shaped most of the Earth's geology. A recent creationist movement, Intelligent Design Theory, pays little attention to the Flood of Noah or to geology, or to fact claims of any sort, contenting itself with proclaiming God's intermittent creation of supposedly 'irreducibly complex' biochemical structures, such as the bacterial flagellum or the blood clotting cascade, rather than presenting a scientific alternative to evolution. A non-Christian creationism is promoted by the Krishna Consciousness movement, whose members agree with geologists about the age of the geological column and how it was shaped, but who argue that human artefacts are found from the Precambrian on, thus supporting a literal interpretation of the Vedas that humans have existed for billions of years.

Christians who opposed evolution during Darwin's time rarely referred to themselves as 'creationists'; they used the term 'creationism' generically to refer to the idea that God purposefully creates living things, in contrast with Darwin's naturalistic explanation for the appearance of humans and other creatures (*see* **Famous Geologists:** Darwin). Nineteenth-century clergy and scientists could choose from many models of creation beyond the Biblical literalist six 24-h days. Charles Lyell (*see* **Famous Geologists:** Lyell) proposed that God had created animals adapted to 'centres of creation' around the world; Cuvier (*see* **Famous Geologists:** Cuvier) proposed a series of geological catastrophes followed by a series of creations. Theologically, the 6 days of Genesis Creation could be interpreted as very long periods of time (the 'day-age' theory), or Genesis could be read as permitting a long period of time between the first and second verses (the 'gap theory'). Some doubtless clung to a literal Genesis of six 24-h days and a historical, universal Flood, but this view was not common amongst university-educated scientists or clergy.

The evolution of creationism and its relationship to geology are the subjects of this article, and thus, befitting an evolutionary approach, we begin with a historical perspective.

## Static versus Dynamic Views of the Earth

Throughout much of the early European scientific period (1600–1700), two perspectives of the world competed: either it had remained unchanged since the special Creation described in Genesis, or it was changing now and had changed in the past. The shift from a static to a dynamic view of nature was stimulated by European exploration during the 1500–1700s. During these expeditions, vast amounts of natural history, including geology, were learned by travellers and settlers, and the new information proved to be difficult to fit into a biblical literalist framework.

The remains of molluscs and other sea creatures on mountaintops, found in the same groupings as living shellfish, encouraged da Vinci to question a literal Flood; he argued that the Flood would have mixed up the shells, not deposited them in life-like settings. Biological data also did not fit into the view of a static world: new species were discovered in the new lands that were not mentioned in the Bible, and geologists found remains of extinct species, troubling for a theology assuming a perfect Creation. Biogeography also made a literal Flood story problematic: how did marsupials in Australia and South America get there after the Ark landed on Ararat? Old views of a static Creation, unchanged since God rested on the seventh day, gradually gave way to an appreciation of an evolving world and, eventually, of the evolution of living things.

Geology is an evolutionary science, dealing as it does with cumulative change in the history of the planet. Geology came into its own as a scientific discipline during the 1700s, as more was learned about the geological characteristics of the planet, prompting speculations about the processes and mechanisms that produced them. The fruits of fieldwork and careful mapping illuminated such processes as sedimentation, erosion, volcanoes and earthquakes, mountain building, and the like, and it made sense that these processes had also operated in the past, changing the contours of the planet. The increased understanding of the Earth and the forces that produced its landforms led to the inevitable conclusion that the Earth was ancient.

An ancient Earth, however, conflicted with traditional scriptural interpretation that the Earth was young; Archbishop Ussher's calculations of a 6000-year-old Earth were endorsed in the margins of commonly used Bibles. However, believing that the Book of Nature and the Book of God must be telling the same story, Christian geologists worked to find ways to interpret the geological data in a framework that allowed them to retain at least some of the Genesis story of Creation, whilst accepting the empirical evidence of their new science.

One accommodation was that of the English clergyman, Thomas Chalmers, who attempted to harmonize geology with the Bible through the 'gap theory'. In his 1815 book, *Evidence and Authority of the Christian Revelation*, Chalmers argued that there was a temporal gap between Chapters 1 and 2 of Genesis. This preserved a literal 6-day Creation, but placed it after a long pre-Adamite creation. The evidence of the rocks that the Earth was ancient was thus acknowledged without the wholesale abandonment of Genesis. The gap or 'ruin and restoration' compromise also had the advantage of allowing for a relatively recent Creation, which pleased religious conservatives. The gap theory remained popular through the nineteenth century and well into the twentieth.

Another compromise between the Bible and science was offered by Scottish stonemason Hugh Miller. In his popular 1847 book, *Footprints of the Creator*, he proposed the 'day-age theory', in which the 6 days of Genesis were not 24-h days, but long periods of time. This compromise was even more scientifically flexible than Chalmers' gap theory: it allowed for the acceptance of virtually all the geological data by requiring that Genesis be taken more figuratively.

Other religious views were more concerned with Darwinian evolution than geology. 'Progressive creationism' accepted the sequence of fossils in the geological column: God was believed to have created increasingly more advanced forms through time. The doctrine of 'theistic evolution', in which God was thought to use evolution and natural selection to bring about the current variety of living things, similarly had little impact on geology. During the early- and mid-nineteenth century, Christian geologists worked to harmonize the 'two books' and, working with professional clergy, eased the worries of Christians that modifications in Ussher's view of a 6000-year-old Earth would create irreparable rents in the fabric of Christianity.

Not all Christians agreed, however. Some felt that these compromises were unacceptable because they required the Bible to be 'interpreted' rather than being read at face value. What came to be called 'Scriptural Geology' took the position that, when the 'two books' were in apparent conflict, the book of God's word was to be given priority. The proponents of Scriptural Geology tended to be neither trained geologists nor university-educated clergy; rather, they were self-taught, educated laymen. Most of them promoted a young Earth and a historical Flood. At best Scriptural Geology was a rearguard movement that had little effect on the views of professional science. As will be discussed later, however, some of the same challenges to evolution promoted by the Scriptural Geologists reappeared in the twentieth century in the guise of 'creation science'.

By the end of the nineteenth century, practising scientists and professional clergy in both the USA and on the continent accepted an ancient age of the Earth and rejected the Flood of Noah as a universal historical event that shaped the planet's landforms. Virtually all scientists likewise accepted biological evolution, although not necessarily Darwin's mechanism of natural selection. In the USA, however, evolution was about to be attacked with the emergence of a religious position called 'Fundamentalism' and, even more importantly, by the efforts of a dedicated amateur geologist and, later, a hydraulic engineer.

## Young Earth Creationism and Flood Geology

The Fundamentalist movement in American Protestantism began with a series of small booklets, collectively called *The Fundamentals*, published between 1910 and 1915. Fundamentalism was partly a reaction to a theological movement called Modernism that began in Germany in the 1880s. Modernism included viewing the Bible in its cultural, historical, and even literary contexts; biblical Creation and Flood stories, for example, were shown by comparison of ancient texts to have been influenced by similar stories from earlier non-Hebrew religions. With such interpretations, the Bible could be viewed as a product of human agency – with all that that suggests for the possibilities of error, misunderstanding, and contradiction – as well as a product of divine inspiration. Fundamentalists in response stressed divine inspiration and absolute accuracy of scripture, including Biblical miracles such as Noah's Flood.

Millions of copies of *The Fundamentals* booklets were printed and distributed. Most of the authors of *The Fundamentals* were day-age creationists, allowing for an old Earth, but insisting on a recent appearance of humans. Although not all *The Fundamentals* booklets were anti-evolutionary, the Fundamentalist position hardened against evolution fairly quickly. Fundamentalists were motivated by religious sentiments – if evolution were true, then what of the accuracy of the Bible? – and also a concern that evolution was the source of many corrosive social practices, such as child labour, 'laissez-faire' capitalism, and exploitation of workers and immigrants. The erroneous association of evolution with World War I German militarism further tainted evolution in the eyes of many.

Fundamentalism remained an almost exclusively American religious movement, not attracting much interest in Great Britain or Europe. However, the Bible-based religious view proved to be a fertile ground for anti-evolutionism, and science was pressed into service by a number of Fundamentalist evangelists to support the cause. Harry Rimmer, Arthur I. Brown, and others hammered evolution before large public audiences, and anti-evolution tracts and books sold well. The target audience of these evangelists was the general public, not the scientific community; the latter remained staunchly evolutionist, and professional scientists were scarce indeed among the anti-evolutionists associated with the growing Fundamentalist movement.

Into this environment came a Seventh-Day Adventist and self-trained geologist named George McCready Price, who was to develop the first version of the most influential creationist perspective of the twentieth century. Seventh-Day Adventist prophetess, Ellen G. White, had claimed visions of a 6-day Creation and a universal Flood; Price sought to scientifically validate her views. He recognized that geology was the key to disproving evolution because, if it could be demonstrated that the Earth was young, then biological evolution would be impossible. He therefore concentrated on attacking conventional interpretations of the geological record. For decades, he vigorously argued that fossils could not be used to determine the order of geological strata, because there were instances where fossils were 'in the wrong order'. He accused geologists of trying to 'shore up their theory' of evolution by proposing 'highly unlikely' theories, such as overthrusting. A favourite example was the Lewis Overthrust. Extending from central Alberta to the eastern edge of Glacier National Park in Montana, this massive structure is composed of Precambrian limestone overlying Cretaceous shales. Price argued that a limestone slab of such immense size could not have been positioned by natural forces. In general, Price ascribed all geological features to the 6 days of Creation, the period between Creation and the Flood, or the Flood itself, and called his view Flood Geology.

Price tried to distinguish his 'new catastrophism' from earlier catastrophic geological views, such as those proposed by Cuvier. He worked tirelessly to promote Flood Geology, and was given a boost when William Jennings Bryan referred to him as a noted scientist during the Scopes Trial. By the 1940s, Price's views had a wide influence on evangelical Christian views of creation and evolution. His major rivals within that community were the gap and day-age theories, but some evangelicals also believed in progressive creationism or theistic evolution. Price rejected all of these as being insufficiently biblical, and clung to the literalism of a young Earth and a global Flood.

Seventh-Day Adventists were considered to be theologically suspect by many conservative Christians. As a result, Price did not always receive the credit his pioneering work deserved. By the 1950s, Flood Geology had begun to slip out of favour with evangelicals, especially with the 1954 publication of Bernard Ramm's *The Christian View of Science and Scripture*, which promoted progressive creationism rather than the 6-day Creation and a universal Flood. The popularity of Ramm's book generated a backlash from more conservative evangelicals. Theologian John C. Whitcomb and hydraulic engineer Henry M. Morris presented the basics of Price's Flood Geology – without mentioning Price – in their 1961 book, *The Genesis Flood*, which proposed a universal Flood, Flood Geology, biblical literalist Genesis theology, and a young Earth as the only acceptable scientific and religious position for a true Christian.

In some respects, *The Genesis Flood* was even more conservative in its presentation of a purportedly scientific foundation for Genesis than were Price's publications. Whitcomb and Morris argued that the entire universe was created between 5000 and 7000 years before the present, whereas Price argued that only the solar system, the Earth, and living things were created during the Adam and Eve creation. Whitcomb and Morris revived the vapour canopy theory, an earlier view that, on the second day of creation, the Earth was shrouded in a canopy of water vapour. This provided the source of the 'waters above the firmament' that produced the forty days and nights of rain at the beginning of the Flood, and also produced a greenhouse effect enhancing the 'garden' of Eden. By shielding the surface of the planet from ionizing radiation, the vapour canopy also reduced cellular damage, allowing Methuselah and his contemporaries to have lifespans of hundreds of years. Consonant with creationists' objections to radiometric dating, the vapour canopy was also claimed to alter the ratios of $^{14}C$ in the atmosphere, making age estimations based on $^{14}C$ inaccurate. *The Genesis Flood* also presented the familiar young Earth model with the Flood as the source of sedimentary features around the planet, and reprised Price's idea of the fossil record being the result of 'hydrodynamic sorting'. Regular and streamlined animals would be naturally sorted by the Flood waters into lower strata; irregular (and more intelligent) fauna would either temporarily escape to higher ground or would float there to be interred at higher elevations.

The book seized the attention of the evangelical community in a way that the Adventist Price's books did not. It sold tens of thousands of copies during its first decade, and remains in print today. The nineteenth century effort to support the literal 6-day Genesis Creation and a young Earth through scientific data and theory was finally institutionalized when Morris and his colleagues organized the Creation Research Society in 1963 as an association of conservative Christian scientists promoting young Earth creationism. In 1972, Morris organized the California-based Institute for Creation Research (ICR) to promote 'creation science'. Its staff and publications expanded; within a decade Morris could claim an ICR publication list of 55 books. The current book list numbers in the hundreds, and these and other creationist books have been translated into Afrikaans, Chinese, Czech, French, German, Hungarian, Italian, Japanese, Polish, Portuguese, Romanian, Russian, and Spanish. The ICR and other creation science ministries, such as Answers in Genesis, have been very effective in shaping the misunderstanding – and rejection – of evolution by millions of conservative Christians. As a result, the word 'creationism' itself is now usually understood to mean young Earth creation science.

## Creation Science and Geology

Proponents of creation science argue that there are only two possible views: creationism and evolution; thus arguments against evolution are arguments in favour of creationism. Creation science literature thus centres on alleged examples of 'evidence against evolution', such as anomalies in scientific literature, that are then used to argue that the evidence for evolution is deficient. Echoing Price's contention that the geological column is not reliable, creation science proponents deny that missing strata at paraconformities ever existed: instead, they are evidence for rapid deposition by the Flood. Proponents also point to fossils that are out of order, such as an alleged human footprint occurring on top of a trilobite, or 'Moab man', a human skeleton found in the Morrison Formation (Late Jurassic) in Utah. Despite the report of the physical anthropologist who excavated the Moab skeleton that it was an intrusive burial and a $^{14}C$ date of 300 years, creationists continue to promote the burial as evidence of a distorted and misleading geological column.

The most famous 'out of place' fossils, however, are alleged human footprints occurring with dinosaur footprints, claimed to have been found in Pennsylvania, New Mexico, and Texas, the last being the best known. The Cretaceous limestones of the Paluxy River near Glen Rose, Texas, present many well-preserved tridactyl dinosaur footprints. Creationists have claimed since the 1930s that human footprints are found in the same strata, proving that dinosaurs and humans existed at the same time. Analysis of these footprints by palaeontologists and geologists has shown that none of the claimed 'mantracks' is valid: all can be explained as natural erosional features, eroded dinosaur tracks, or carvings. Although the ICR has retreated slightly from promoting the Paluxy River 'mantracks', other creationists continue to do so, as can be seen by even a cursory web search.

Like Price, modern creation science proponents know that the age of the Earth is a critical issue for evolution; without an old Earth, biological evolution could not have occurred. A great deal of creationist literature, therefore, focuses on efforts to demonstrate that the Earth is young rather than ancient. For example, when presented with evidence from ice cores, varves, or even tree rings, creationists argue that these annual indicators of age are flawed, because more than one layer, varve, or ring can be formed in a year, thus throwing off the counts.

Many of these young Earth arguments rely on extrapolations of various rates of change. When a measured rate is extrapolated backwards (or forwards), a result is produced that is incompatible with the accepted age of the Earth. Engineer Walter T. Brown's book, *In the Beginning*, includes numbered 'evidences' against evolution, including this one on the decay of the Earth's magnetic field:

> 84. Over the past 140 years, direct measurements of Earth's magnetic field show its steady and rapid decline in strength. This decay pattern is consistent with the theoretical view that a decaying electrical current inside Earth produces the magnetic field. If this is correct, then just 20 000 years ago the electrical current would have been so vast that Earth's structure could not have survived the heat produced. This implies Earth could not be older than 20 000 years.

Of course, the rate of decay of the Earth's magnetic field is not linear, but periodically reverses, and thus the fear of a too-powerful field 20 000 years ago is groundless. Other rate arguments similarly rely on taking a known rate and extrapolating it in a linear fashion, resulting in conclusions at variance with modern geology:

> 80. The continents are eroding at a rate that would level them in much less than 25 million years. However, evolutionists believe fossils of animals and plants at high elevations have somehow avoided this erosion for more than 300 million years. Something is wrong.

83. Meteoritic dust is accumulating on Earth so fast that, after four billion years, the equivalent of more than 16 feet of this dust should have accumulated. Because this dust is high in nickel, Earth's crust should have an abundance of nickel. No such concentration has been found on land or in the oceans. Therefore, Earth appears to be young.

Creation science proponents argue that, after the Flood, a strict uniformitarianism has held, but, during the Flood, God used processes that are different from those now in effect. This selective catastrophism is necessary to avoid the implications of standard uniformitarian interpretations, which are overwhelming evidence for a slowly changing planet and thus an ancient Earth. The young Earth model demands accelerated rates of change well beyond any observed today. Coral reefs would have needed to accumulate at 40 000 times their current observed rate to form as quickly as a Flood Geology model requires. The interior of the USA has sedimentary deposits of great thickness, averaging a kilometre in depth, some of which, such as calcareous muds, are the results of animal activity; these too would have needed to form at mind-boggling rates. In the Grand Canyon, the Kaibab and Redwall limestones together comprise about 300 m of sediment; to produce deposits of this thickness during the year of the Flood would have required deposition of 80 cm per day – clearly more than has ever been observed or even imagined.

As part of the effort to attack the age of the Earth, creation science proponents hold that radiometric dating methods are unreliable. The Grand Canyon Dating Project, a project of the ICR, recently attempted to demonstrate this by showing alleged discrepancies between dates of Grand Canyon lava flows collected by ICR staff and by others. In the ICR study, Pleistocene lava flows were dated earlier than the Proterozoic Cardenas basalts, but other geologists have pointed out that the technique used by the ICR (Rb/Sr) may give false isochrons by reflecting the age of the mantle source from which the lava flows were derived, rather than the eruption time of the lava itself.

A more recent ICR research project critiquing radiometric dating is the Radioisotopes and the Age of the Earth (RATE) project, begun in 1997. In 2000, Vardiman *et al.* published a 675-page book describing the project and its expectations for research. A final volume of conclusions from the project is expected in 2005.

The fact that ICR scientists have undertaken research distinguishes them from their nineteenth-century Scriptural Geology predecessors. The amount of research performed to date, however, is very small, and its quality has been questioned by professional geologists. Most of the literature consists of the selective search of scientific data for anomalies cited to disprove evolution, and the little actual research that has been performed, such as efforts to $^{14}$C date dinosaur bones or the Grand Canyon Dating Project, has not inspired confidence in the quality of creationist research efforts. As a result, the professional geological community has largely ignored creation science, except as a social movement that has had remarkable success in popularizing poor science to the general public.

## See Also

**Analytical Methods:** Geochronological Techniques. **Evolution. Famous Geologists:** Cuvier; Darwin; Lyell. **History of Geology From 1780 To 1835. History of Geology From 1835 To 1900. History of Geology From 1900 To 1962. Trace Fossils.**

## Further Reading

Cole JR, Godfrey LR, Hastings RJ, and Schafersman SD (eds.) (1985) The Paluxy River Footprint Mystery – Solved! *Creation/Evolution*, 5(1), pp 1–56.

Forrest B and Gross PR (2004) *Creationism's Trojan Horse, The Wedge of Intelligent Design*. Oxford: Oxford University Press.

Larson EJ (1997) *Summer for the Gods: The Scopes Trial and America's Continuing Debate Over Science and Religion*. New York: Basic Books.

Lindberg DL and Numbers RL (eds.) (1986) *God and Nature: Historical Essays on the Encounter Between Christianity and Science*. Berkeley, CA: University of California Press.

Livingstone DN, Hart DG, and Noll MA (eds.) (1999) *Evangelicals and Science in Historical Perspective*. New York, NY: Oxford University Press.

Lynch JM (ed.) (2002) *Creationism and Scriptural Geology, 1817–1857*. London: Bristol Thoemmes Press.

Numbers R (1992) *The Creationists*. New York: Knopf.

Scott EC (2004) *Creationism vs Evolution: An Introduction*. Westport, CT: Greenwood Publishing Group.

Stassen C (2003) *A Criticism of the ICR's Grand Canyon Dating Project*. Retrieved December 29, 2003, from http://www.talkorigins.org/faqs/icr-science.html.

Strahler A (1999) *Science and Earth History, The Evolution/Creation Controversy* (revised edn.). Buffalo, NY: Prometheus Books.

Vardiman L, Snelling AA, and Chaffin EF (eds.) (2000) *Radioisotopes and the Age of the Earth: A Young-Earth Creationist Research Initiative*. El Cajon, CA, and St. Joseph, MO: Institute for Creation Research and Creation Research Society.

Wise DU (1998) Creationism's geological time scale. *American Scientist* 86: 160–173.

# DELTAS

See **SEDIMENTARY ENVIRONMENTS: Deltas**

# DENDROCHRONOLOGY

**M Bridge**, University College London, London, UK

© 2005, Elsevier Ltd. All Rights Reserved.

## Introduction

In recent decades tree-ring studies have opened up a whole new area of climatic and other environmental reconstructions at various geographical and temporal scales. These generally go under the heading of dendrochronology – although some people reserve this term only for the use of tree rings in dating studies. Dendrochronological dating relies on the fact that whilst each tree-ring series reflects peculiarities of the life history of the individual tree, trees of the same species growing at the same time over wide areas respond similarly to the weather experienced both during an individual growth period, and perhaps also to conditioning throughout previous seasons.

Not all species have clearly distinct ring boundaries, and those without them are of little use to dendrochronology. Some species not only do have clear rings, but may have clear divisions within the rings, such that the ring can be divided into the 'earlywood' which forms during the spring and early summer, and the 'latewood' which forms through the summer and early autumn. Whilst many studies make use of the whole ring, in recent years many have focused on part of the ring, usually the latewood, which is usually more independent of the influences of previous growth in its variations. Most dendrochronological work has been done in temperate zones, although some tropical trees have been used.

With historical timber, samples of wood from naturally preserved timbers e.g. in river gravels, peat bogs and similar environments, usually take the form of cross-sections cut from trunk remains. Historical wood in situations such as standing buildings is generally cored down the radius of the trunk. Many of the studies outlined below, however, depend on samples from living, or recently felled, trees. Living trees can be sampled by extracting radial cores, usually of the order of 5 mm in diameter, taken using especially designed manually operated borers. The tree 'compartmentalizes' the wound, sealing it off from the rest of the living tissue, and any damage to the tree is generally minimal.

During the late 1960s and the 1970s, the development of radiocarbon dating and the realization that the formation of radiocarbon in the atmosphere had not been constant through time were the stimuli for several laboratories to attempt to build long (multimillennial) tree-ring chronologies. These provided wood samples of known calendar age with which to investigate changes in radiocarbon levels through time, and then to calibrate the radiocarbon time-scale. This has made possible 'wiggle matching', which in suitable circumstances can provide more accurate radiocarbon dates.

As the chronologies were being constructed it was noticed that some periods had scarce wood remains which was thought to reflect environmental changes, either in the vigour of tree growth, or in the conditions that favour the preservation of the wood. Once several long chronologies became available, it was noticed that certain periods showed common growth responses over wide geographical areas. For example, a downturn in oak growth lasting about a decade after 1628 BC was reported in Irish bog-oaks. This was subsequently found to occur at the same time in oaks growing in northern England and northern Germany.

This decade showed the lowest growth rates for many centuries in each locality, and must therefore reflect a large climatic forcing agent. It was suggested that the most likely candidate for this particular period of growth reduction was the Santorini (Thera) volcanic eruption, often linked to changes in the Minoan civilization. Accepting this as dating that particular event was controversial, but even if some were unable to accept this hypothesis, some major factor had to be responsible for the observed response.

Throughout the rapid growth of dendrochronology in the late twentieth century, many studies focused on dating discrete events such as volcanic eruptions, earthquakes, landslides, avalanches, forest fires, floods, severe frosts etc. through history, giving important background information about the frequency and intensity of these phenomena, and in many instances giving the first calendar dates for them. These events may be represented in the tree-ring series by sudden growth rate changes, scars, or abnormal cells.

Longer-term variations and rates of change for such phenomena as river-flow changes, lake-level changes, saltwater ingression, changes in ocean currents through time (mapped by dating driftwood), glacial advance and retreat (reflected in subfossil trunks from valley sides), erosion rates, human forest clearance and the like have all been subjects for dendrochronological study. Periods of building of prehistoric lakeside settlements have been found to be coincident over areas of central Europe, perhaps reflecting migration patterns in response to climatic changes.

## Dendroclimatology

It was not long before people realized that well replicated chronologies were themselves a proxy dataset of the major climatic influences on growth, and people looked for the best methods of extracting this information at annual, decadal, century and millennial time-scales. In some areas the relationships between climatic factors and growth are relatively simple, and the dominant limiting factor to growth exhibits a clear correlation with the width of the annual ring. In semi-arid areas for example, the ring width is generally a reflection of rainfall levels.

Some basic environmental information can be gained simply by looking at long-term trends in the ring width itself. For example, different centuries show quite different growth rates for similarly biologically aged oaks preserved in river gravels in Germany, used to construct a multimillennial chronology. Even without any complex calibration of the climate–growth relationship, it is possible to draw broad inferences about changing conditions at particular times, with narrow mean ring widths representing poorer growth conditions and wider mean ring widths more favourable growth conditions.

The sensitivity of trees to external changes in their environment changes with the site conditions. In more extreme environments it is possible that rings may be missing, or partially missing around the circumference, or if unfavourable conditions are experienced during the normal growth period, a slowing down and then resurgence of growth may produce apparent 'false' rings. These problems are generally detected when 'cross-matching' the samples during chronology construction. Unlike many scientific studies, sampling for dendroclimatological reconstruction relies on careful selection of the trees, not random selection. The reasons for this become clear when one considers the nature of the ring width itself. The ring width in any given growth season is the result of a number of factors: the biological age of the tree, environmental factors unique to that tree, stand-wide influences atypical of growth elsewhere, a large number of other factors, and of course the regional climatic signal that is of interest in these studies.

One can minimize the influence of non-climatic factors and maximize the climatic signal by careful site selection. Trees growing at the margins of the population, whether that be altitudinally or latitudinally are generally most sensitive to climatic factors. By choosing the dominant trees with no obvious signs of damage or disease, one enhances the relationship between ring width and regional climatic influence. By increasing the number of samples one reduces the individual tree responses and enhances the 'climatic signal'. The 'signal to noise ratio' has been studied in detail and will vary from location to location and species to species, but as a general guide a minimum of 15 to 20 trees seems to produce a representative sample.

The biological age of the tree generally has an effect on ring width, with a natural tendency to put on narrower rings as the tree increases in girth, even though overall productivity levels may be very similar. This so called 'age trend' can be readily removed by fitting curves to the overall series and then taking account of the difference of the observed ring width from the theoretical mean value at a given time. This process, known as 'standardization', results in the production of a series of indices of growth, and makes direct comparison between trees of different biological age more readily achievable.

Just how best to derive this series of indices for subsequent analysis has been the subject of a vast literature within dendroclimatology, and each new study really needs to justify the particular methods employed to remove age-related trends. The problem of course is that the more closely one fits curves to the original data, the more climatic information one may be subtracting, and whilst the great advantage of dendroclimatology is that it has the potential to produce annual resolution in the results, longer-term trends in growth may also reflect climatic information, and may be lost in the process of standardization. Even so, the resulting series is often highly autoregressive, that is to say that growth in any one year often reflects the overall vigour of the tree for the few prior years, and this autoregression needs to be removed to enhance the year-to-year variation. Autoregressive Moving Average (ARMA) models have been perhaps the most frequently used standardizing models, though the literature is full of alternative methodologies, each of which may be more suitable in particular conditions.

Having satisfactorily arrived at a well replicated index series representing growth, the next stage is generally to model how the climatic information influences the observed growth. One of the great problems here is the availability of reliable meteorological records, and the length of those records for regions where the most environmentally sensitive trees are growing. The most useful tree-ring information is often from remote areas, and even where there are local records, these are generally only available over recent decades. Not surprisingly, an early study of oak growth in Britain showed that the variation in ring width attributable to mean monthly rainfall and temperature data varied according to distance from the meteorological station used.

Nevertheless, a process of stepwise principal component regressions generally allows a response function to be derived in which individual monthly rainfall and temperatures prior to, and during, the growth season can be recognized in order of their importance. For example, in a typical temperate northern-hemisphere model, April, May and June rainfall, and May, June and July temperature may be found to have the greatest influence on growth. A transfer function can then be derived so that these major elements are reconstructed throughout the length of well replicated tree-ring chronology. Typically, in the more sensitive trees growing at the margins of distribution, one factor can be derived as the major influence on growth, and the reconstructions are concerned with aspects of either rainfall or temperature.

Given that biological responses are often not linear, and that relatively crude data in the form of monthly means are the normal tools employed, it is surprising how reproducible the results have been. In 1982 a study of response functions for oaks growing at 16 sites in the British Isles was published. This showed many similarities in responses, including the surprising result that temperature in the December prior to growth was significantly negatively correlated with ring width in the following growth season. This was mathematically derived, without reference to the biology of the trees, and a tentative explanation was put forward. This suggested that a warm December resulted in the trees metabolizing the food reserves which were subsequently not available to the tree the following spring, resulting in less growth. In a cold December, the tree had 'shut down' allowing the reserves to be utilized for growth when conditions improved. This highlighted a need for better understanding of tree physiology which is still a requirement for better modelling of the climate–growth relationship.

Some early reconstructions were tempted to use very short periods of meteorological data to calibrate the model – often this was all that was available, although the ring series may go back several centuries. There is also a need to verify the model, so the short data series was divided into two parts, with one short series being used to calibrate the model and the other part to verify the reconstructed values over an equally short period. Hundreds of years were then reconstructed on this unsatisfactory model. Whilst this is tempting, and no doubt may highlight very different periods of climate within the tree-ring time-series, it is clearly open to inappropriate interpretation. Guidelines on methodology also suggest that in order to remove any effects of changes in response over time, the calibration and verification of the model should be done both ways around, i.e. using the outer years to calibrate and verify using the inner years, and then reversing the periods used for calibration and verification.

Critics of this response function approach to dendroclimatology point out that not only are responses seldom linear, and that dramatic changes can result from crossing threshold values, but that many influences on tree growth can occur at widely varying time-scales, perhaps in a matter of hours. For example, a severe frost on a single night in the late spring, when growth has not been going on for long and leaves are tender, may damage a significant proportion of the photosynthetic capacity of the tree and have a profound effect on growth for the remainder of that growth season. Such an event is unlikely to show up in a crude measure such as mean monthly temperature. Critics also point to the observable fact that whilst generally quite representative values are derived for much of the series, reconstructed values rarely recreate the more extreme years well, tending to underestimate the meteorological data (see **Figure 1**).

A number of studies, particularly based in Europe, have looked at growth in 'pointer' years – that is, years where a large proportion of the trees show a marked growth change in the same direction – and have managed to look in greater detail, often at daily weather records, to see what factors are responsible for these growth changes.

These two approaches both yield very valuable information, and are not mutually exclusive. It is by combining several approaches that the maximum information is likely to be gained in the future, although

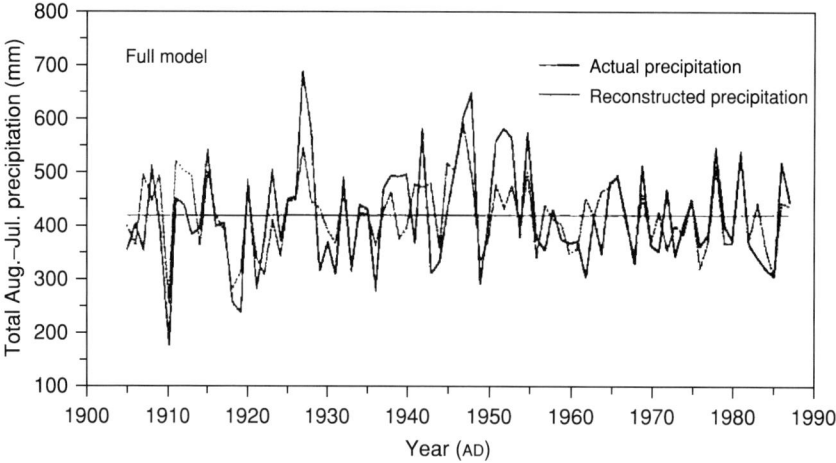

**Figure 1** A comparison of the tree-ring reconstructed and instrumentally recorded precipitation for the 1905–1987 calibration period. (Reproduced with permission from Case RA and MacDonald GM (1995) A dendroclimatic reconstruction of annual precipitation on the western Canadian prairies since AD 1505 from *Pinus flexilis* James. *Quaternary Research* 44: 267–275, copyright Elsevier 1995.)

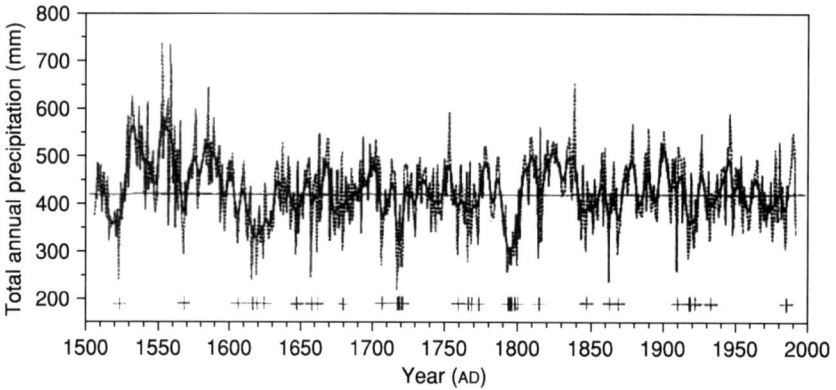

**Figure 2** A 487-year (AD 1505–1992) reconstruction of August–July annual precipitation. Drought years are indicated on the lower portion of the graph (+) and are defined as years when precipitation is more than one standard deviation less than the average of the complete series. The heavy line represents an 11-year low-pass filter. (Reproduced with permission from Case RA and MacDonald GM (1995) A dendroclimatic reconstruction of annual precipitation on the Western Canadian prairies since AD 1505 from *Pinus flexilis* James. *Quaternary Research* 44: 267–275, copyright Elsevier 1995.)

of course the latter approach is only possible where the data are available.

Many dendroclimatic studies have concerned themselves with reconstructing one particular factor in a limited geographical area, for example summer temperature in a particular country or state. Another approach, which has been invaluable, is to produce a network of chronologies over national, continental or hemispherical ranges and to use these as the basis of more wide-scale reconstructions. This approach enables an assessment of the reconstruction values at the intermediate points to be made, and can identify areas within the network where further sampling is desirable. More ecologically based studies have shown changes in tree response that are elevation dependent (**Figure 2**).

### Micro-Anatomical Variations

Whole ring, or earlywood/latewood width studies are the commonest form of dendroclimatological studies. In parallel with these studies however, there is a long history of studies in the relationship of intraring features with external growing conditions. Cell size, cell differentiation, cell distribution and other characteristics (e.g. wall thickness) have all been studied.

### Stable Isotope Studies

This discussion has so far concentrated on the use of ring-width studies, but it is important to remember that dendrochronological studies (or tree-ring studies if one prefers) can and have used many other aspects of the properties of the rings. Most notable amongst these is the use of X-ray densitometry. In this, uniform thickness cross-sections of the tree (usually from cores) are pre-treated to remove resins and other chemicals, and then subjected to an X-ray source, with film below. The resulting photomicrograph can be scanned to reveal not only ring-width information, but also the density of the cellulose across the ring. It has been found that the maximum latewood density in conifers is generally more highly correlated with meteorological factors than is simple ring width, and another advantage is that there is much less autocorrelation in density measurements. The disadvantages of this method are that is it far more costly and technologically demanding than pure ring-width studies.

Another area of tree-ring studies with a long history is the consideration of changing chemical isotopes in the wood itself. Both temporal and spatial variation can be found in the isotopic records of trees. Studies on variations in the deuterium, $^{13}C$ and $^{18}O$ isotopes have been used to look at long-term changes in temperature and precipitation.

Deuterium levels in tree rings contain information about the isotopic composition of the water taken up by the tree, and this is generally a reflection of the isotopic composition of the precipitation in the area where the tree grew. This isotope signal is closely associated with condensation temperatures, and deuterium levels in tree rings have therefore been used as records of palaeotemperature. Because of isotopic fractionation processes within the tree, deuterium levels have also been associated with humidity levels. It has also been shown that as a result of the covariation between temperature and relative humidity, that when this is taken into consideration, it is possible to produce a more sensitive temperature record from deuterium levels.

Carbon isotopes have been found to be linked with temperature, relative humidity, light intensity and water availability. Some of these associations, such as that with temperature, can vary considerably in different site conditions.

Stable carbon and hydrogen isotopes originate from different sources, the air and groundwater respectively, and hence a combination of isotope studies is likely to give a broader picture of environmental changes. Differences have been found between the isotopic composition of earlywood and latewood, and it is generally assumed that latewood values are more representative of the conditions at the time of wood formation, since there appears to be some utilization of stored food reserves during the formation of earlywood. Differences have also been found with the biological age of the tree, and therefore growth trends need to be removed as in many tree-ring studies.

In southern Germany, periods with high values in both isotope records occur around 5270 BC, 2990 BC and 2180 BC and are thought to represent times with high temperatures and low water availability. Similarly, low values around 6230 BC, 5600 BC and AD 390 probably represent unusually wet, cool periods. The 6230 BC corresponds with an event reported in North America in high-resolution palaeoclimate records, and it has been suggested that this may represent a time of rapid melting of the Laurentide ice-shield into the North Atlantic.

Other periods of coincidence in the hydrogen and carbon isotope records do not appear to be associated with information from other palaeoclimate records, and so some caution must be employed in their interpretation. It is hoped that the development of other proxy records – such as long-term varve chronologies – may help resolve these issues.

The ratio of $^{18}O$ to $^{16}O$ in the cell walls of plants results from changes in groundwater and evaporation. Levels vary with altitude and latitude, and oxygen isotope levels have long been regarded as records of palaeotemperature. It has been suggested from a study in India that relative humidity levels may be recorded in the tree rings, and it is likely that, as in other isotope studies, site conditions play a large role in determining the usefulness of particular studies.

Further research into tree physiology will undoubtedly unravel some of the more complex relationships between external environmental conditions and the incorporation of different stable isotopes into the wood. These will enhance the use of this specialist, and relatively expensive, area of tree-ring research.

## Conclusions

Although very long chronologies have been constructed for many parts of the world, the replication of tree-ring data needed limits the time-frame for which these reconstructions are considered reliable, and the majority of those so far produced cover little more than five or six centuries, with a few notable longer contributions. The tree-ring-derived temperature reconstructions generally show greater century time-scale variability than is found from other proxy datasets, and suggest a cooler period in the late fifteenth and during the sixteenth centuries than is suggested by non-tree-ring data.

In recent years it has been suggested that the climate–growth relationship has fundamentally changed in character, perhaps as a result of increasing carbon dioxide levels or decreasing ozone levels. There seems to be a decoupling of the relationships observed prior to the 1950s. Whilst this does not totally discredit the numerous reconstructions that have been carried out in recent decades, it does mean that their conclusions need to be used cautiously, and it poses serious problems for the calibration of future models used for climate reconstruction. Wide-scale ecologically based studies are highlighting variations in tree response at different site types in recent decades. It is perhaps more useful to extract climatic data from several different species growing in a variety of sites, but for which there is a limited historical dataset, rather than to attempt much longer reconstructions based on relatively few trees, in order to best understand the complexities of our changing climate.

## See Also

**Europe:** Holocene. **Palaeoclimates**. **Tertiary To Present:** Pleistocene and The Ice Age.

## Further Reading

Baillie MGL (1982) *Tree-Ring Dating and Archaeology*. London: Croom-Helm.

Baillie MGL (1995) *A Slice through Time; Dendrochronology and Precision Dating*. London: Batsford Ltd.

Baillie M and Munro M (1988) Irish tree-rings, Santorini and Volcanic dust veils. *Nature* 322: 344–346.

Becker B (1993) An 11,000-year German oak and pine dendrochronology for radiocarbon calibration. *Radiocarbon* 35: 201–213.

Becker B and Schirmer W (1977) Palaeoecological study on the Holocene valley development of the R. Main, southern Germany. *Boreas* 6: 303–321.

Briffa K, Jones PD, Wigley TML, Pilcher JR, and Baillie MGL (1983) Climate reconstruction from tree rings: part 1, Basic methodology and preliminary results for England. *Journal of Climatology* 3: 233–242.

Briffa K, Jones PD, Wigley TML, Pilcher JR, and Baillie MGL (1986) Climate reconstruction from tree rings: part 2, Spatial reconstructions of summer mean sea level pressure patterns over Great Britain. *Journal of Climatology* 6: 1–15.

Briffa K, Schweingruber FH, Jones PD, Osborn TJ, Harris IC, Shiyatov SG, Vaganov EA, and Grudd H (1998) Trees tell of past climates: but are they speaking less clearly today? *Philosophical Transactions of the Royal Society of London* 353B: 65–73.

Briffa KR, Osborn TJ, and Schweingruber FH (2004) Large-scale temperature inferences from tree rings: a review. *Global and Planetary Change* 40(1+2): 11–26.

Case RA and MacDonald GM (1995) A dendroclimatic reconstruction of annual precipitation on the Western Canadian prairies since A.D. 1505 from *Pinus flexilis* James. *Quaternary Research* 44: 267–275.

Cook ER and Kariukstis LA (eds.) (1990) *Methods of Dendrochronology: Applications in the Environmental Sciences*. Dordrecht: Kluwer.

Cook ER, Bird T, Peterson M, Barbetti M, Buckley B, D'Arrigo R, and Francis R (1992) Climatic change over the last millennium in Tasmania reconstructed from tree-rings. *The Holocene* 2(3): 205–217.

Dittmar C, Zech W, and Elling W (2003) Growth variations of common beech (*Fagus sylvatica* L.) under different climatic and environmental conditions in Europe – a dendroecological study. *Forest Ecology and Management* 173: 63–78.

Fritts H (1976) *Tree Rings and Climate*. London: Academic Press.

Fritts HC (1991) *Reconstructing Large-Scale Climatic Patterns from Tree-Ring Data*. Tucson: University of Arizona Press.

Hughes MK, Kelly PM, Pilcher JR, and LaMarche VC (1982) *Climate from Tree Rings*. Cambridge: Cambridge University Press.

Jones PD, Bradley RS, and Jouzel J (eds.) (1996) *Climate Variations and Forcing Mechanisms of the Last 2000 Years*. Berlin: Springer-Verlag.

Mayr C, Frenzel B, Friedrich M, Spurk M, Stichler W, and Trimborn P (2003) Stable carbon- and hydrogen-isotope ratios of subfossil oaks in southern Germany: methodology and application to composite record for the Holocene. *The Holocene* 13: 393–402.

McCarroll D, Jalkenen R, Hicks S, Tuovinen M, Pawellek F, Gagen M, Eckstein D, Schmitt U, Autio J, and Heikkenen O (2003) Multi-proxy dendroclimatology: a pilot study in northern Finland. *The Holocene* 13: 829–838.

McCarroll D and Loader NJ (2004) Stable isotopes in tree-rings. *Quaternary Science Reviews* 23: 771–801.

Pilcher J and Gray B (1982) The relationships between oak tree growth and climate in Britain. *Journal of Ecology* 70: 297–304.

Schweingruber FH (1988) *Tree Rings*. Dordrecht: Reidel.

Schweingruber FH (1996) *Tree Rings and Environment. Dendroecology*. Birmensdorf: Swiss Federal Institute for Forest, Snow and Landscape Research.

# DESERTS

See **SEDIMENTARY ENVIRONMENTS: Deserts**

# DIAGENESIS, OVERVIEW

**R C Selley**, Imperial College London, London, UK

© 2005, Elsevier Ltd. All Rights Reserved.

## Introduction

Lithification is the process by which unconsolidated sediment is turned into solid rock. In 1888, von Gumbel proposed the word 'diagenesis' to describe the processes of lithification. Subsequently, in 1957, Pettijohn wrote:

> Diagenesis refers primarily to the reactions which take place within a sediment between one mineral and another, or between several minerals and the interstitial or supernatant fluids.

In 1968, Dunoyer de Segonzac gave a detailed account of the history of the concept of diagenesis, and pointed out that, in addition to the chemical processes mentioned by Pettijohn, some physical processes, such as compaction, are important in diagenesis. Thus, two types of diagenesis are now recognized: physical and chemical. These will be described in turn, but first it is appropriate to define the boundaries of diagenesis.

## The Boundaries of Diagenesis

Diagenesis begins as soon as sediment is deposited. Fossilized beer bottles and other anthropogenic detritus found in modern 'beach rock' limestone picturesquely illustrate this. Ancient evidence of early diagenesis is further confirmed by the occurrence of intraformational conglomerates containing clasts, not only of contemporaneous limestone, but also of siderite-cemented sandstones and shales. As sediment is buried more deeply, temperature and pressure increase and, ultimately, diagenesis merges into metamorphism, with shale becoming slate, sandstone becoming quartzite, and limestone becoming marble. Field observation and laboratory experiments demonstrate that the boundary between these rock types, and hence diagenesis and metamorphism, is gradational. The sequence, deposition → diagenesis → metamorphism, is not a 'one-way street', however.

At any time while sediment is on its way to metamorphism, it may be uplifted to the surface again. Rocks returned to the surface show a reversal of the trend of porosity decreasing with burial, and its enhancement by both physical and chemical processes. The term 'epidiagenesis' was applied by Fairbridge in 1967 to the diagenesis resulting from uplift and weathering. Epidiagenesis is of little significance in shales. It is, however, of great importance in sandstones and carbonates, because of the way in which it restores porosity and permeability to rocks that had previously lost these features. When buried beneath an unconformity, these epidiagenetically enhanced zones may provide excellent petroleum reservoirs. Epidiagenesis is also well known to mining geologists, being responsible for the 'gossan' sulphide ore bodies, such as those of Rio Tinto, Spain. **Figure 1** delineates the boundaries of diagenesis.

## Physical Diagenesis

The main processes of physical diagenesis are the expulsion of connate waters concomitant with the compaction of sediment. There is abundant evidence of compaction in sedimentary rocks, notably observations in thin sections of warped mica flakes and buckled fragmented shells. Arguably much of this compaction occurs during shallow burial, aided by seismic (including microseismic) shocks. Vibration causes the packing of sediment to tighten as the fabric adjusts and grains snuggle up close to each other.

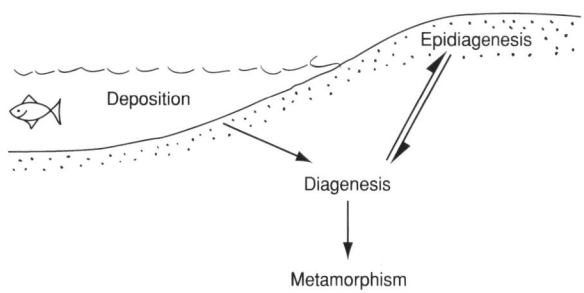

**Figure 1** Diagram to show the interfaces and pathways of diagenesis.

Theoretical studies show that loosely packed sand may have a porosity of nearly 50%, but this is reduced to 27% for the tightest packing possible, a loss of 23%. Experiments reveal that poorly sorted muddy sediments compact more than clean well-sorted sands. This is because of the ease with which clay and finer particles may be squashed into the pores between the larger framework grains. Compaction gradually increases with burial depth, resulting in a concomitant loss of porosity. The exception to this general statement is the condition termed overpressure, in which the fluid pressure within pores diminishes the pressure at grain contacts and supports the sediment load. Once sediment is lithified, however, compaction ceases, although rocks have a degree of elasticity.

Graphs of porosity against burial depth for clay and sand are different. Clay burial graphs are curved, with the rate of porosity loss declining to become linear at depth. Sand burial curves, on the other hand, tend to be linear throughout. Intuitively, this suggests that clay loses porosity quickly by dewatering and compaction, whereas sands lose porosity more by cementation. Not all physical diagenesis results in porosity loss. Fracturing is an important process that takes place only in lithified rock: sedimentary, igneous, and metamorphic. Fracturing may occur in several ways, most commonly tectonic, but also, of course, anthropogenically around a borehole. Fracturing also occurs due to the release of stress when rocks are uplifted and the overburden pressure is diminished. Thus, rocks subjected to epidiagenesis become fractured, and fracturing is commonly well developed in truncated strata beneath an unconformity.

Fractures are extremely important, not so much because of the way in which they increase the porosity of a rock, which may be minimal, but because they cause a dramatic increase in the permeability of a previously impermeable rock. The resultant fractures may permit the flow of petroleum, water, and mineralizing fluids, which may, in turn, re-cement the very fractures that permitted their invasion.

## Chemical Diagenesis

Chemical diagenesis includes mineral transformation, recrystallization, cementation, and dissolution. A wide range of minerals occur as cements in sedimentary rocks, precipitating in the pore spaces between the framework grains. The principal pore-filling cements are quartz, carbonate (calcite, siderite, and dolomite), and clay. Cement is obviously a major destroyer of porosity and permeability.

Chemical diagenesis, however, also includes the dissolution of grains, matrix, and cement by corrosive fluids. The grains most commonly dissolved are bioclasts, peloids, and ooids composed variously of aragonite (during shallow burial of limestone) and calcite. The matrix that is most commonly removed is lime mud, micrite. Similarly, the cement that is most commonly leached out is calcite, and other carbonates, rather than quartz. These observations demonstrate that dissolution results from the invasion of a sedimentary rock by acidic rather than alkaline pore fluids. Two sources of acidic fluid have been suggested. It is known that carbonic acid is expelled from shales prior to the emission of petroleum. It has been suggested that the carbonic acid serendipitously enhances the porosity and permeability of the potential reservoir, ahead of petroleum invasion. It is also noted that there is nothing new in acid rain. Sedimentary rocks exposed to epidiagenesis are subjected to flushing by acidic meteoric water. This leaches out carbonate grains, matrix, and cements. In limestones, it will enlarge existing fractures, and generate extensive zones of mouldic, vuggy, and cavernous porosity. The caves may synchronously fall in to form collapse breccias.

Dissolution, of whatever origin, will, of course, increase the porosity and permeability of sedimentary rock. Renewed burial will lead to renewed cementation, and the sedimentary rock will continue on its downward path to metamorphism.

## Summary

Diagenesis is the term that describes the physical and chemical processes that take place in sediment after deposition and before it reaches the threshold of

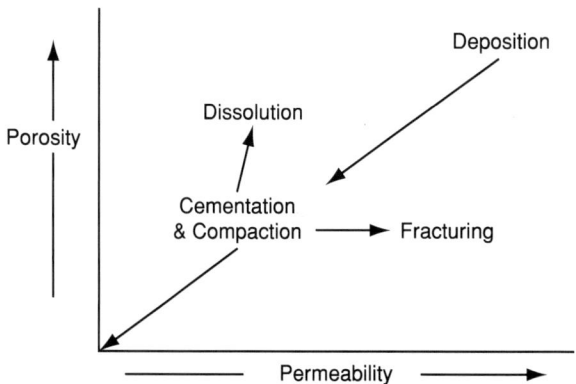

**Figure 2** Diagram to illustrate diagenetic pathways and their petrophysical responses.

metamorphism. Physical processes include the loss of contained fluid and compaction, which destroys porosity, and fracturing, which enhances permeability. Chemical processes include porosity-destroying cementation and porosity-enhancing dissolution. The sequence, deposition → diagenesis → metamorphism, is not a 'one-way street', however. At any point along this route, the sediment may be uplifted to the surface again. Rocks returned to the surface show a reversal of the trend of porosity decreasing with burial, and its enhancement by the physical and chemical processes of fracturing and dissolution, respectively. **Figure 2** summarizes the diagenetic pathways of sedimentary rocks.

Diagenesis is of great economic importance. Diagenetic studies form an integral part of the analysis of the evolution of porosity and permeability in aquifers and petroleum reservoirs. Diagenetic studies are also an important aid to understanding the formation of many mineral deposits.

## See Also

**Petroleum Geology:** Production. **Sedimentary Rocks:** Chalk; Clays and Their Diagenesis; Dolomites; Limestones; Sandstones, Diagenesis and Porosity Evolution. **Weathering**.

## Further Reading

Burley SD and Worden RH (2003) *Sandstone Diagenesis Recent and Ancient*. Oxford: Blackwell Science.

Dunoyer de Segonzac G (1968) The birth and development of the concept of diagenesis. *Earth Science Reviews* 4: 153–201.

Fairbridge RW (1967) Phases of diagenesis and authigenesis. In: Larsen G and Chilingar GV (eds.) *Developments in Sediments*, pp. 19–28. Amsterdam: Elsevier.

Giles MR (1997) *Diagenesis: A Quantitative Perspective*. Dordrecht: Kluwer Academic Publishers.

Leeder MR (1999) *Sedimentology and Sedimentary Basins: From Turbulence to Tectonics*. Oxford: Blackwell Science.

Pettijohn FJ (1957) *Sedimentary Rocks*, 2nd edn. New York: Harper Geoscience.

Selley RC (2000) *Applied Sedimentology*, 2nd edn. San Diego: Academic Press.

Tucker ME (1991) *Sedimentary Petrology*, 2nd edn. Oxford: Blackwell Scientific Publications.

# DINOSAURS

*See* **FOSSIL VERTEBRATES: Dinosaurs**

# EARTH

Contents

**Mantle**
**Crust**
**Orbital Variation (Including Milankovitch Cycles)**

## Mantle

G J H McCall, Cirencester, Gloucester, UK

© 2005, Elsevier Ltd. All Rights Reserved.

### Introduction

The mantle is the middle of the three primary concentric zones of the Earth, making up together with the core more than 99% of the planet's volume. It extends from the Mohorovicic Discontinuity (the base of the crust, which varies in depth from 25–35 km below the continents to 6–11 km beneath the ocean floor) (**Figure 1A**) to its boundary with the outer core at a depth of 2890 km.

The mantle itself is subdivided into an upper mantle, extending from the Mohorovicic Discontinuity to a depth of 670 km, and a much thicker lower mantle beneath. The upper mantle is further divided into a rigid upper zone, extending to a maximum depth of about 100 km, which together with the crust comprises the lithosphere, and a weaker ductile lower zone, which forms the asthenosphere. The ductile lower zone is commonly delineated at its base by a transition zone, extending from 400 to 670 km; the change to the lower mantle is transitional rather than abrupt. The lower mantle is delineated by a more rapid increase in density, amounting to 20%, as shown by an increase in the velocities of seismic waves. It is customary to consider the thickness of the lithosphere as varying from a maximum beneath the continents to zero beneath the mid-ocean ridges, the sites of eruptivity leading to sea-floor spreading (**Figure 1A**).

### Direct Sampling of the Mantle is not Possible

The mantle, like the core beneath it, cannot be sampled directly. We shall never have revealed to us the actual rock material of which it is composed. Thus, all our knowledge of the mantle is derived from indirect methods, and this means that there remains a considerable degree of uncertainty as to its exact composition. The indirect methods that have been used to study the mantle are discussed below.

#### Seismology

Earthquake waves and similar seismic waves generated by artificial explosions (e.g. nuclear) provide a major line of evidence. Their velocities depend on the physical properties of the materials through which they pass. Experiments are carried out in the laboratory to establish the behaviour of such waves when passing through a known range of likely materials. Radial and lateral velocities can then be compared. Solutions, however, are seldom unique; there is also the difficulty of replicating the very high pressures of the deep mantle in the laboratory, so the method is less satisfactory when considering increasing depths in the mantle.

#### Chemical-equilibrium Studies

Experiments at high pressures and temperatures can examine the exchange reactions between different minerals and between minerals and magmas, and the stability fields of minerals at different pressures and temperatures, replicating the conditions at various levels in the mantle (**Figures 2 and 3**). From such experiments the possible or probable mineral and elemental make-up of the mantle can be deduced, but again there is no certainty in these results.

#### Peridotites and Oceanic Basalts

Materials erupting at the spreading centres in the oceans are considered to be virtually uncontaminated derivatives, ascending rapidly from layers in the mantle, and valuable conclusions can be drawn from the study of these. This method is believed to give reliable information about the mantle to a depth of about 300 km. Peridotite may well exist in the deeper mantle, and some authorities entertain a wholly peridotitic mantle.

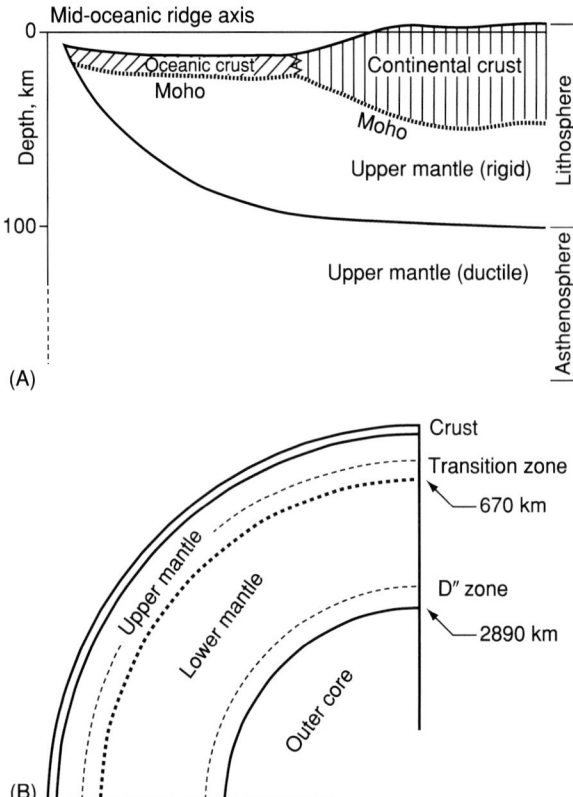

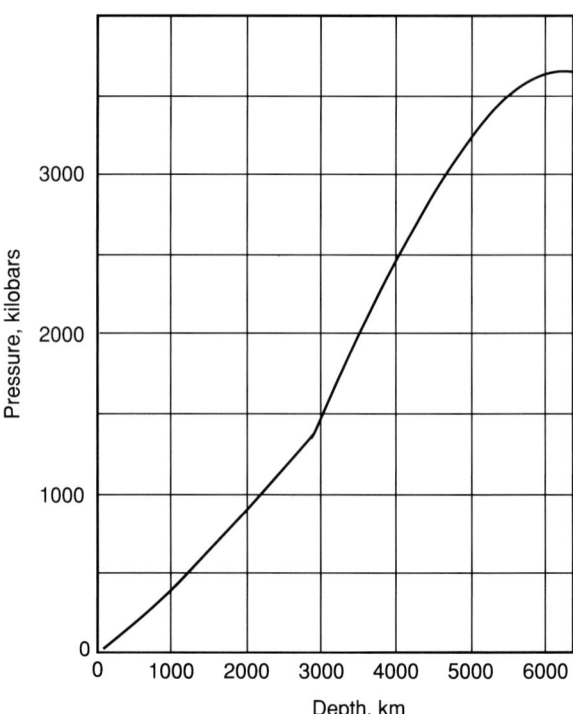

**Figure 2** Pressure variation curve for the interior of the Earth (reproduced with permission from Mason B (1966) *Principles of Geochemistry*. New York, London, Sydney: John Wiley and Sons, Inc.).

**Figure 1** (A) Cross-section from a mid-ocean ridge to the interior of a continent, showing the relationship between the crust, lithosphere, Moho, and zones in the upper mantle. (B) The relationships between the upper mantle, lower mantle, and core (after Hancock PL and Skinner BJ (eds.) (2000) *The Oxford Companion to the Earth*. Oxford: Oxford University Press).

### Kimberlites

Kimberlites (*see* **Igneous Rocks:** Kimberlite) are mica peridotites, which, unlike oceanic peridotites, are volatile-rich. They carry mica and carbonate minerals. There are two ways in which these rocks, which are erupted very rapidly and explosively through continental crust in diatremes or pipes, sample the mantle. They contain diamonds with melt or mineral inclusions that come from great depths in the mantle, possibly as deep as the base at 2890 km. The diamonds in kimberlites are older than the host rocks, as shown by isotopic evidence, and have clearly been picked up during passage of the Kimberlite through the mantle to the surface. There may also be mineral or rock xenoliths, accidental fragments also picked up by the turbulent host magma on its ascent. These may range from comminuted debris to actual rock fragments. Kimberlites occur only within the continents, so they provide information only about the composition of the subcontinental mantle. Such accidental inclusions probably sample the mantle to a depth of at least 150 km.

### Meteoritic Analogy

Geochemical modelling has been based on analogies with chondritic meteorites (*see* **Solar System:** Meteorites) because

- they are the most abundant extraterrestrial objects that strike the Earth (as measured by fall rates and excluding finds);
- they are essentially similar to terrestrial ultrabasic rocks, consisting of olivine-pyroxene assemblages with similar Fe/Mg ratios; and
- the stable-isotope data is in agreement with such modelling.

However, there are difficulties in accepting the 'chondrite Earth model', a major one being the difference in alkaline-element contents, and there is another problem involving the $^{13}C/^{12}C$ isotopic ratio, which is low in common chondrite meteorites and terrestrial basalts, but high in carbonaceous chondrites, terrestrial carbonates, and diamonds from kimberlite pipes (the latter must come from the mantle). Thus, a number of authorities believe that the mantle is derived from the accreted material of carbonaceous chondrites, rather than common chondrites. This conveniently provides a source for the water and other volatiles that are known to reside in the mantle (e.g. according to the Rubey model). The fact that

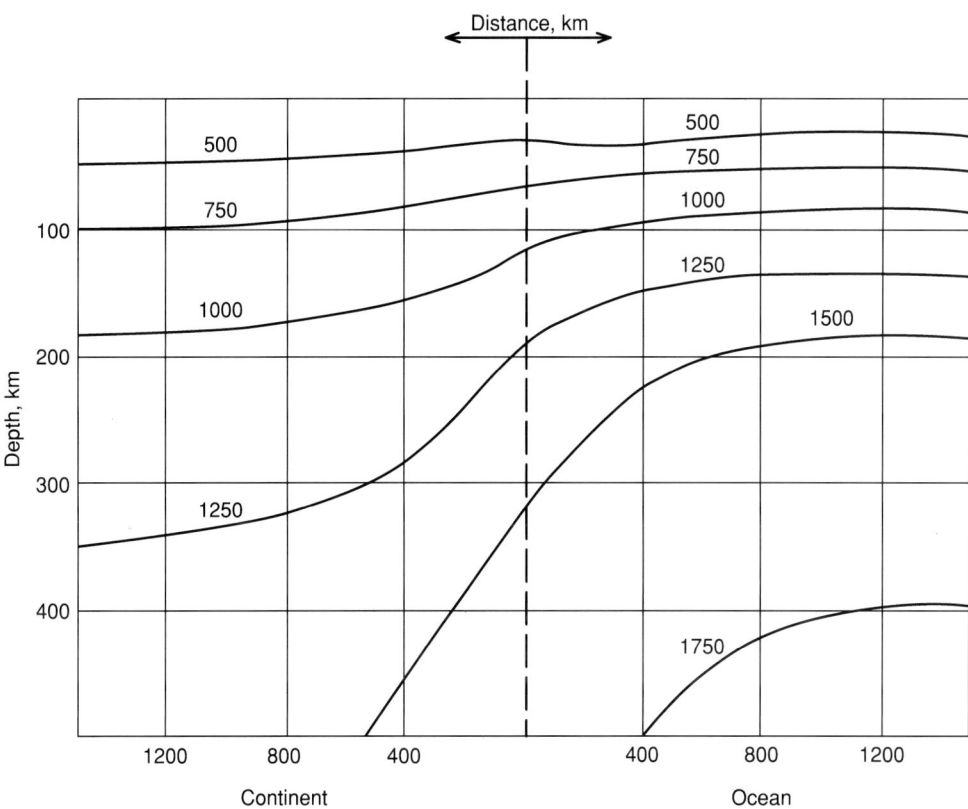

**Figure 3** Temperature (degrees C) variation curve for the interior of the Earth. Lines are isotherms (reproduced with permission from Mason B (1966) *Principles of Geochemistry*. New York, London, Sydney: John Wiley and Sons, Inc.).

carbonaceous chondrites fall to Earth extremely rarely compared with common chondrites can be explained both by their fragility and by the fact that common chondrites, according to theory, are derived from them by metamorphic processes. However, doubts must remain about whether the material of chondrites, which accreted as asteroids in the interval between Mars and Jupiter at about 4500 Ma, is really representative of the material from which the Earth's mantle accreted at about the same time. There is an assumption of a homogeneous dust cloud from the Sun outwards.

The metal forming the core is supposed to have melted and gravitated to the centre of the planet (**Figure 4**). Whether accretion of the mantle from the dust cloud was homogeneous or heterogeneous (layered) remains undecided – both possibilities are shown in **Figure 4**.

## Mantle Composition

Concepts concerning the composition of the mantle have inevitably changed since the acceptance of the plate-tectonic paradigm. It is relevant here to go back to the earlier statements. Those of Arthur Holmes and Brian Mason are concise summaries of what was then believed.

Holmes, in a 1965 reissue of his famous book *Physical Geology*, wrote

> It has long been thought that meteorites, stony and iron, might be direct clues to the nature of the Earth's mantle and core. Stony meteorites are like our terrestrial peridotites in many ways and, moreover a few varieties have compositions not unlike some of our basalts. For these and other reasons a similar range of composition for the mantle, with ultrabasic materials predominating, has seemed to be a plausible guess. The sudden change of seismic velocities at the Moho indicates either a change in composition (e.g. from basic rocks above to ultrabasic below) or a change in state (from, say, gabbro or amphibolite to a high pressure modification, which, for convenience, we may refer to as eclogite), without much change in composition. Olivine-rich peridotites and garnet-bearing varieties, at the appropriate temperatures and pressures, give seismic velocities that are about right for most parts of the upper mantle. Eclogite would do equally well, and various associations of the rocks would match seismic and density requirements. That eclogite does occur in the mantle is proved by the fact that it occurs in diamond pipes as inclusions.... Moreover, basaltic magma ascends from the mantle and

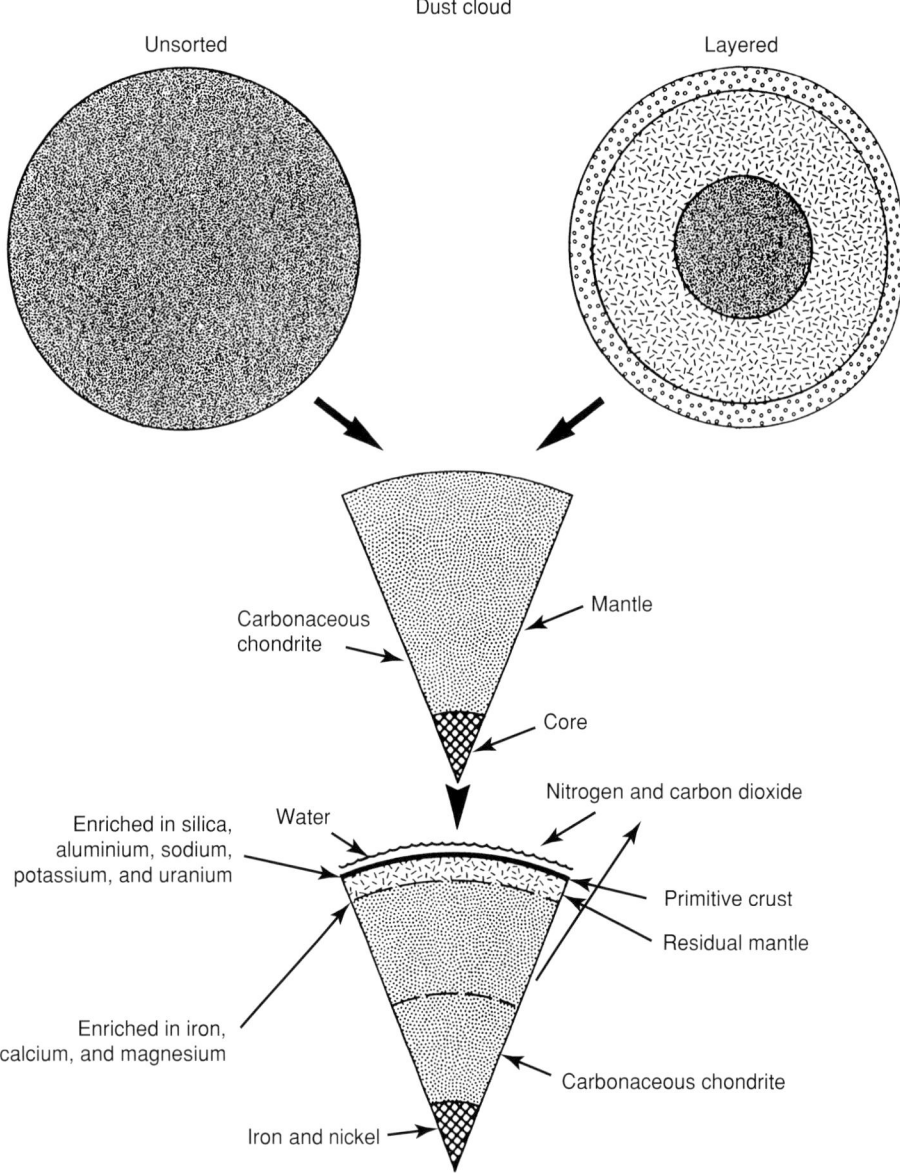

**Figure 4** Two models for the accretion of the mantle and core (reproduced from Van Andel 1994).

one very probable source of basalt supply would be the melting of eclogite.

Although eclogites might represent primordial mantle or subducted oceanic crust, one problem with an eclogitic upper mantle is that eclogite xenoliths are scarce in kimberlites, whereas peridotite xenoliths are common. Mason, in a textbook reissued in 1966, stated that the alternative of a chemical explanation for the Moho was favoured by Clark and Ringwood in 1964; they preferred an ultrabasic model for the upper mantle, with the overall composition being one part basalt to three parts dunite (an olivine rock). This composition would equate to that of a pyroxene-olivine rock, which they called 'pyrolite'. This is a hypothetical composition, not a rock type. Fractional melting of this material would yield basaltic magma, injected into the crust over geological time, leaving residual dunite or peridotite. The pyrolite could crystallize in four ways, depending on the pressure and temperature in the mantle: olivine and amphibole;. olivine, aluminium-poor pyroxene, and plagioclase; olivine, aluminium-rich pyroxene, and spinel; or olivine, aluminium-poor pyroxene and garnet. Because of pressure and temperature differences, the compositions of the upper mantle beneath continents and oceans would differ. The proposed compositions under the continents and oceans are shown in **Figure 5**. Polymorphism of minerals was considered to be

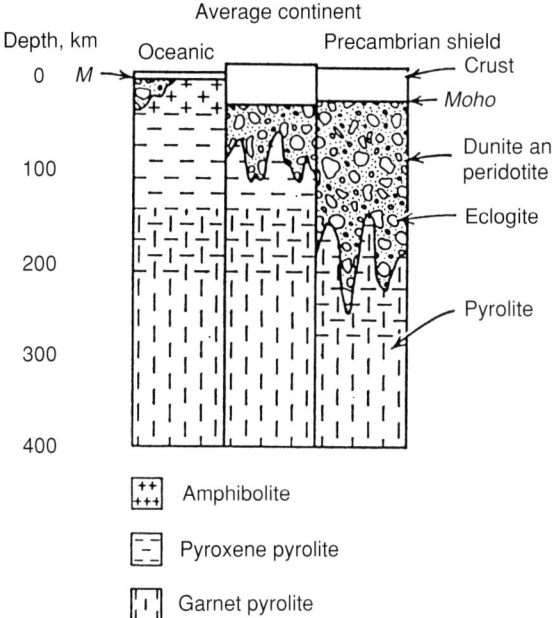

**Figure 5** Suggested compositions of the mantle according to the original pyrolite hypothesis (reproduced with permission from Mason B (1966) *Principles of Geochemistry*. New York, London, Sydney: John Wiley and Sons, Inc.).

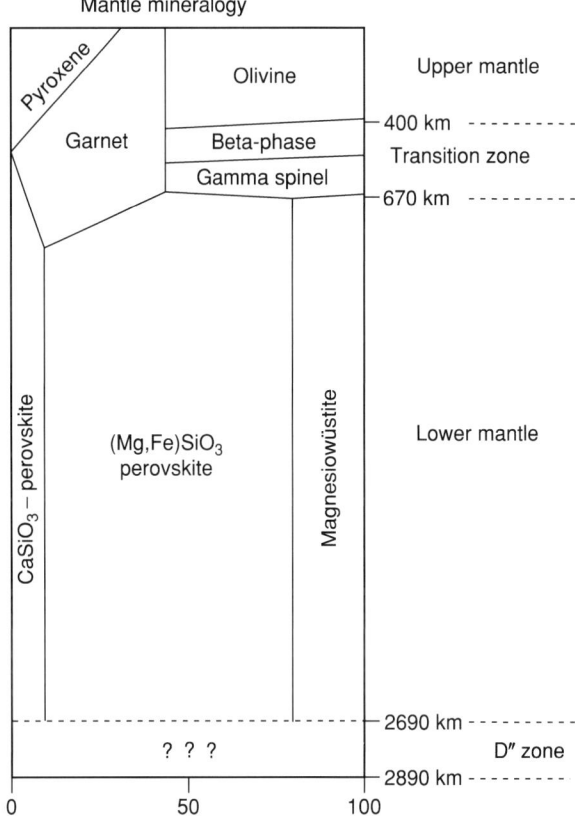

**Figure 6** Suggested composition of the mantle at various depths. The verticle dimension is depth, the horizontal dimension is minerals (modified from Agee CE (2000) Mantle and core composition. In: Hancock PL and Skinner BJ (eds.) *The Oxford Companion to the Earth*, pp. 654–657. Oxford: Oxford University Press).

an important mantle process: olivine converting to an isometric spinel structure at higher density, pyroxene to an ilmenite structure, and quartz to the high-pressure polymorph that occurs naturally in impact structures – stishovite with rutile, stable at 130 kb and 1600 °C. Transformations into closer-packed forms were suggested to occur in the transition zone. The lower mantle was believed to consist of a mixture of $(Mg,Fe)SiO_3$ with an ilmenite structure and periclase (MgO), but the possibility of the ilmenite structure converting to that of perovskite (even denser) was suggested, as was the possibility of that then converting to a CaCl structure (though experimental evidence for these changes was lacking).

The pyrolite hypothesis, which was first stated by Clark and Ringwood in 1964, has survived the appearance of the plate-tectonic paradigm and was brought up to date in a summary by Agee published in 2000. He noted that a major point at issue is whether the mantle is homogenous or heterogeneous. The abrupt jumps or discontinuities in seismic properties within the transition zone are attributed to changes of minerals to denser, more closely-spaced phases; olivine is converted at a depth of 400 km to wadsleyite, an orthorhombic form found in meteorites, and at 520 km to ringwoodite (a gamma spinel form). At approximately 670 km olivine breaks down into perovskite ($[Mg,Fe]SiO_3$) and magnesiowustite ($[Mg,Fe]O$). Pyrolite has been given a hypothetical composition of 57% forsteritic olivine, 17% enstatitic pyroxene, 12% diopsidic pyroxene, and 14% pyrope garnet.

Pyrolite is a hypothetical analogue of the primordial composition of the mantle that existed after planetary accretion and core formation (the rigid part of the upper mantle, above the asthenosphere, is thus termed the 'residual mantle') (*see* **Earth Structure and Origins**). The shallowest part of the mantle is believed to have been continuously depleted in basaltic melt throughout geological time.

An up-to-date diagram of the likely composition of the mantle at various depths under the continents and oceans is given in **Figure 6**.

It has been noted above that there are two possibilities for the original accretion of the mantle from the dust cloud – homogeneous and layered. D L Anderson and others favour a compositionally layered mantle rather than the model of progressive downwards change under increasing conditions of pressure and temperature. The plate-tectonic paradigm requires the solid upper mantle to be rheid and slowly convecting as shown in **Figure 7**. Advocates of

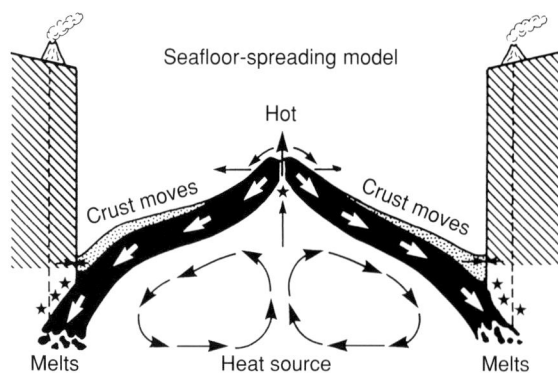

**Figure 7** Mantle convection under a spreading ocean (reproduced from Van Andel 1994).

the compositionally layered mantle believe that the discontinuity at the base of the upper mantle, at approximately 670 km, is not only a phase-change boundary but also a compositional boundary. The layered model requires layered convection with little or no mass transport across this discontinuity. The pyrolite hypothesis is retained, but, whereas the upper mantle is taken to be composed of olivine-rich peridotite similar to pyrolite, the lower mantle is believed to be enriched in perovskite. This requires a lower mantle that is enriched in silica compared with the upper mantle, something that is in agreement with the chondritic Earth model. The compositional layering of the mantle could stem from the accretion phase, but could also be a product of an early melting stage, with the mantle crystallizing like a giant layered igneous intrusion.

## The Problem of Subducted Slabs

It is a requirement of the plate-tectonic paradigm that lithospheric slabs descend into the mantle in subduction zones. One hypothesis is that they sink right through the mantle to a 'slab graveyard' at the core-mantle boundary. There is some seismological evidence for this in the basal 200 km of the mantle (the $D''$ zone). However, the seismological anomaly here may be explained in a number of ways, and advocates of the compositionally layered mantle believe that there is a high enough contrast of density and viscosity across the 670 km boundary to prevent any passage of material through it and that the slabs must remain in the upper mantle, the 'graveyard' being situated in the transition zone and affecting its composition. The sedimentary cover in these slabs contributes to the calc-alkaline eruptives in the crust above the descending plate (e.g. the Peruvian granitoid batholiths). There is no real certainty about what happens to the sediments descending into the mantle.

## Tomography

Tomography delineates in three-dimensions regions of fast and slow seismic-wave in the Earth's interior, both radially and laterally. These may simply represent colder downwellings and hotter upwellings. They may alternatively represent compositional differences in a primordially layered mantle being complicated and eradicated by convection. Tomographical irregularities are invoked by advocates of slab descent to the base of the mantle and of plumes ascending through the 670 km discontinuity from that region (see **Earth Structure and Origins**), whereas advocates of 'closed upper mantle circulation of plate tectonics', such as Warren Hamilton, believe that plumes could not penetrate the 670 km discontinuity and that convection cells are restricted to the upper mantle.

## What Drives Plates and Initiates the Pattern?

The plate-tectonic paradigm has been with us for four decades, but there is no certainty at all about what drives the plate motion. What actually starts the process? Is the rise of magma from the already convecting mantle at the site of the future mid-ocean ridge the splitting initiator? Or is initiation controlled by lithospheric irregularities above the mantle before convection begins? That is, is the primary drive exerted by the mantle or by the lithosphere? Though there has been much discussion of the driving forces of plate movement, none of the theories suggested seems adequate. However, we do have a clue in the Rift Valley of East Africa (see **Africa**: Rift Valley). This is manifestly an aborted ocean, a site of magmatism and rifting, which has, nevertheless, not developed into a spreading ocean – it has been 'halted in its tracks'. This surely is evidence that the process starts at the spreading centre, before any subduction takes place. It would seem to invalidate slab pull or the passive sinking of oceanic lithosphere, as favoured by Warren Hamilton, as the prime driving mechanism; rather, either already commenced mantle convection forces a split or the whole process is started by a lithospheric split along a favourable line of weakness in the crust– both mechanisms seem to be consistent with this evidence.

## Conclusion

The mantle is of critical importance in understanding geological processes, yet remains elusive. Numerous models have been erected, and no doubt many more will be derived in the coming decades. An aspect that has not been touched on here, the variations in isotopic composition in basalts derived from the mantle, illustrates the complexity of the problems. It is not possible to explain all the isotopic complexities by simple mixing of one enriched and one depleted reservoir, and the answer would seem to be that, even taking into account crustal contamination on ascent, there are a number of different reservoirs, a fact that argues for a heterogeneous mantle. Yet the exact petrological, spatial, and physical nature of the source regions remains as elusive as ever.

## See Also

**Africa:** Rift Valley. **Earth:** Crust. **Earth Structure and Origins**. **Igneous Rocks:** Kimberlite. **Mantle Plumes and Hot Spots**. **Plate Tectonics**. **Solar System:** Meteorites. **Tectonics:** Earthquakes.

## Further Reading

Agee CE (2000) Mantle and core composition. In: Hancock PL and Skinner BJ (eds.) *The Oxford Companion to the Earth*, pp. 654–657. Oxford: Oxford University Press.

Anderson DL (1989) *Theory of the Earth*. Oxford: Blackwell Scientific Publications.

Anderson DL (1999) A theory of the Earth. In: Craig GV and Hull JH (eds.) *James Hutton – Present and Future*, pp. 13–35. Special Publications 150. London: Geological Society.

Clark SP and Ringwood AE (1964) Density distribution and constitution of the mantle. *Reviews of Geophysics* 2: 35–88.

Hamilton WB (2002) The closed upper mantle circulation of plate tectonics. In: Stein S and Freymueller JT (eds.) *Plate Boundary Zones*, pp. 359–410. Geodynamics Series 30. American Geophysical Union.

Holmes A (1965) *Physical Geology*. Edinburgh, London: Nelson.

Mason B (1966) *Principles of Geochemistry*. New York, London, and Sydney: John Wiley and Sons.

Price NJ (2001) *Major Impacts and Plate Tectonics*. London and New York: Routledge.

Ringwood AE (1975) *Composition of the Earth*. New York: McGraw-Hill.

Rubey WW (1955) Development of the hydrosphere and atmosphere with special reference to the probable composition of the early atmosphere. In: Poldervaart A (ed.) *Crust of the Earth*, pp. 631–650. Colorado Geological Society of America Special paper 62 Part IV. Geological Society of America.

Van Andel TH (1994) *New Views on an Old Planet*. Cambridge: Cambridge University Press.

Wylie PJ (1967) *Ultramafic and Related Rocks*. New York, London, Sydney: John Wiley and Sons.

# Crust

**G J H McCall**, Cirencester, Gloucester, UK

© 2005, Elsevier Ltd. All Rights Reserved.

## Introduction

Holmes, in 1965, defined the 'crust' or 'lithosphere' as the outer shell of the solid Earth, the two terms being synonymous at that time. He described it as being composed of a great variety of rocks, with a blanket of loose soil or superficial deposits (e.g., alluvium, desert sands). The term dates back to Descartes (1596–1650), who saw the crust as a shell of heavy rocks which was covered by lighter sands and clays and rested on a metallic interior. Leibnitz (1646–1716) believed that the Earth had cooled from an incandescent state and that the crust was the first consolidated rocky part covering a still molten interior. However, it became clear from the work of Kelvin (1862) that the Earth's tidal behaviour precluded a still molten interior beneath a thin solid crust.

The plate tectonics paradigm appeared in the late 1960s and required a rigid lithosphere extending to ca. 100 km (see **Plate Tectonics**), including part of the upper mantle, above the convecting asthenosphere. This redefinition made the terms 'crust' and 'lithosphere' no longer synonymous. The crust is the rocky upper layer of the Earth with a loose superficial cover, and extends to the Mohorovičić discontinuity (Moho) at 5–35 km below the surface, whereas the lithosphere extends down to 100 km and includes the crust as its upper layer and also the uppermost part of the mantle beneath. The crust is thus the top compositional layer of the lithospheric plates, which move about the Earth's surface, the remainder of the plates being formed of the uppermost, rigid part of the mantle.

The oceanic crust and continental crust are of different character (**Figure 1**). The oceanic crust, at its thinnest, only extends down to 5 km minimum, but locally may be thicker, up to 15 km. It forms 59% of the total crust by area. The continental crust extends down to 30–80 km and forms 79% of the crust by volume. Under islands, continental margins, and island arcs, the crust is transitional and is 15–30 km thick.

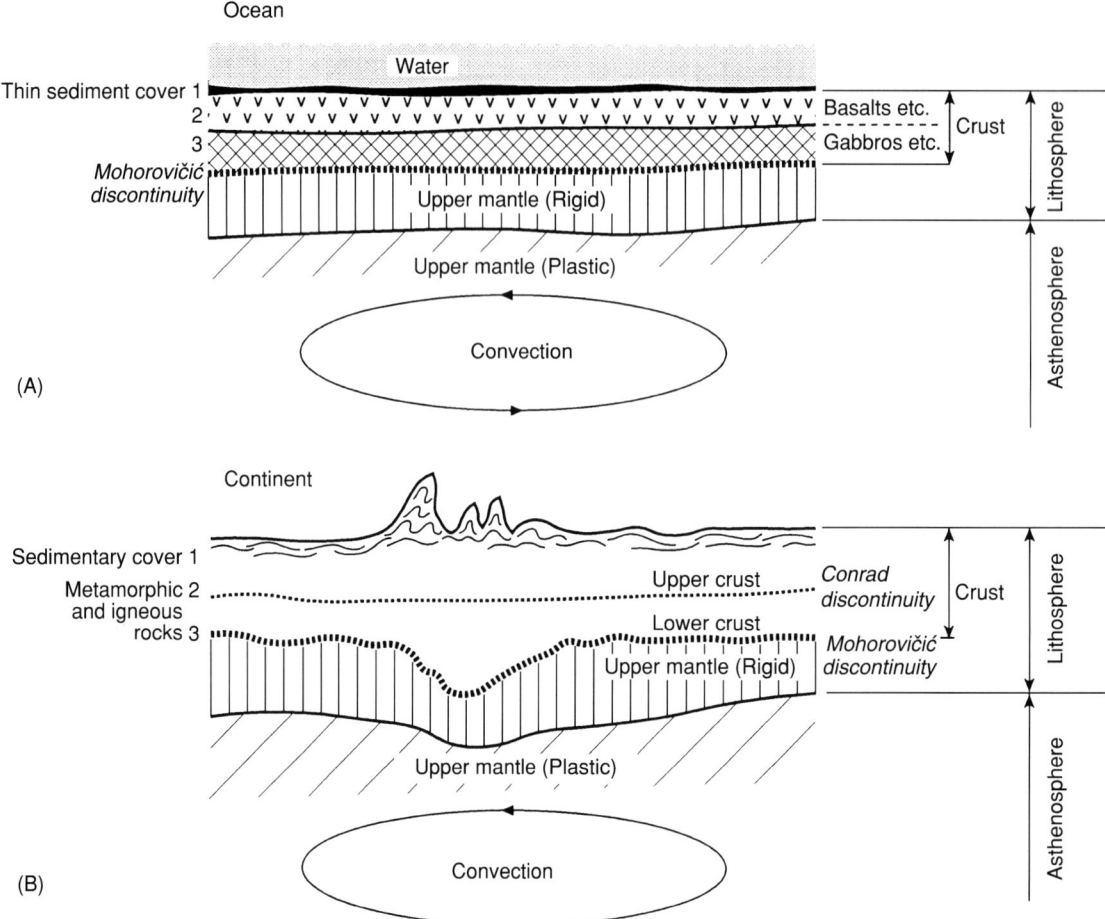

**Figure 1** Diagrams showing the threefold layering of the oceanic (A) and continental (B) crust. The continental crust is much thicker than the oceanic crust. The two diagrams are not to scale.

## Recycling of the Crust

Plate tectonics requires that crustal slabs are recycled in subduction zones and taken down to the mantle (**Figure 2**). However, this is by no means an even process, with oceanic crust being subducted very rapidly, but continental crust resisting recycling. The result of this is that the oceans are all products of the last plate dispersion cycle, which commenced about the beginning of the Mesozoic, and no oceanic crust is older than about 200 million years. In contrast, continental crust has been preserved almost indefinitely, and there is some continental crust in the Slave Province of Canada which is more than 4000 million years old. Minerals (zircons) in continental rocks in Western Australia are also up to 4300 million years old.

## Physical Regions of the Crust

Continental crust is subdivided into a number of physical regions.

- Shields: deeply eroded expanses of low relief, which have been stable since Precambrian times.
- Platforms: similar to the above, but mantled by thick sedimentary cover, which may be entirely or in part Phanerozoic in age.
- Orogens: long, curved belts of folded rocks, usually forming mountain chains, mostly formed by continental collisions.
- Rifts: linear, fault-bounded depressions, traversing continents; these are the structures which originate crustal splitting and dispersion, and lead to mid-ocean ridge formation, but they may, as in the case of the East African Rifts, be aborted, i.e., never developed into oceans.

Oceanic crust is also subdivided into a number of physical regions.

- Volcanic islands: scatterings or chains of islands which mark mantle hotspots.
- Volcanic arcs: above subduction zones (e.g., Aleutian Islands), these may be represented by their

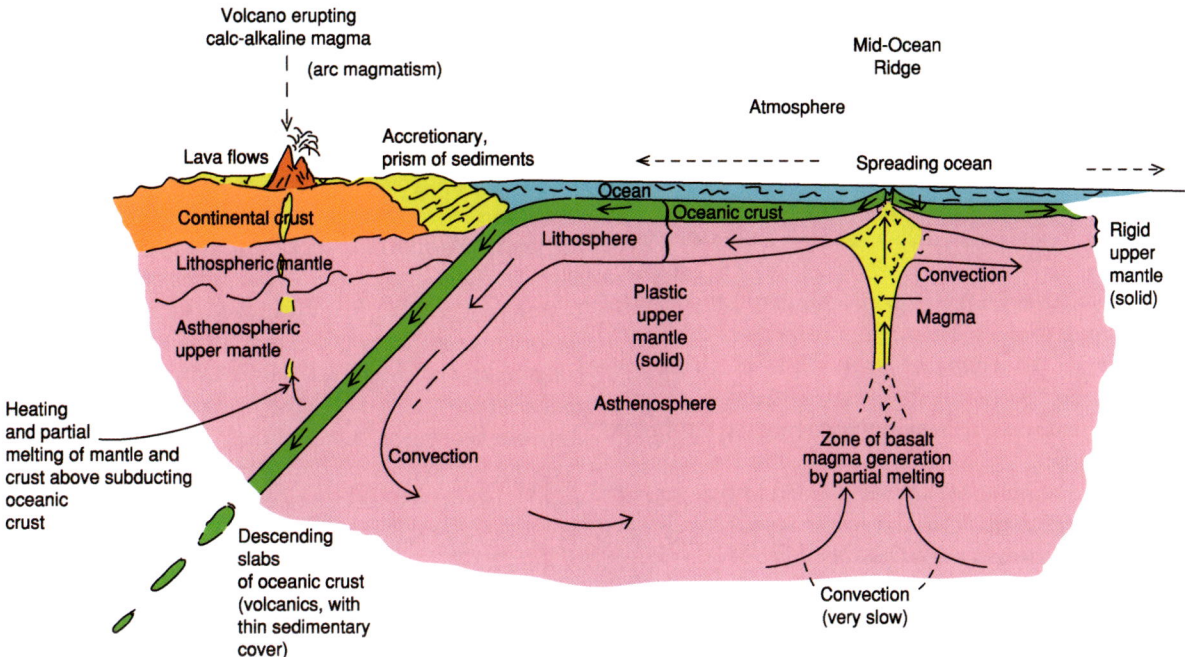

**Figure 2** Diagram showing the recycling of the crust by subduction. After Edwards K and Rosen B (2000) *From the Beginning*. London: Natural History Museum.

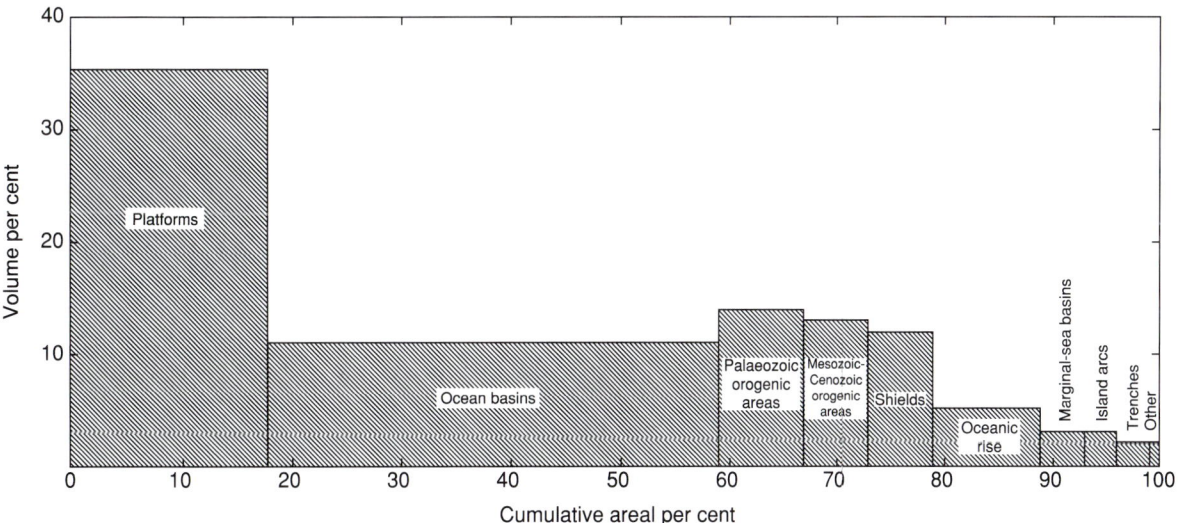

**Figure 3** Area and volume proportions of the major crustal types. From *Plate Tectonics and Crustal Evolution* by K. C. Condie (1976). Reprinted by permission of Elsevier Ltd.

eroded underworks – chains of calc-alkaline batholiths (e.g., Peru).
- Trenches: the outer margin of subduction zones; the deepest parts of the oceans.
- Ocean basins/abyssal plains: the extensive flat, deep areas of oceans, beyond continental slopes.
- Marginal basins: small basins separating arcs, or landward from arcs (back-arc basins).
- Inland seas: seas within continents (e.g., Caspian).

The relative area and volume of the major crustal types are shown in **Figure 3**.

## Crustal Structure

We can broadly determine the contrasting layers in the crustal structure by means of seismic wave velocities, determined from earthquake waves and artificial (e.g., nuclear) explosions. This shows us that there are three contrasting layers in the oceanic crust and also three in the continental crust, but they are different. In the oceanic crust, we can use analogies with 'ophiolites' (*see* **Tectonics: Convergent Plate Boundaries and Accretionary Wedges**), which have been recognized as old oceanic crust, subducted and later obducted onto the continental surface, where they form bodies of surface rocks. Ophiolites are present in surface rock assemblages of all ages, as far back as the Archaean upper boundary (2500 Ma). The three layers of the oceanic crust are believed to comprise a sediment layer (0–1 km thick), a basement layer (mainly basalt) (0.7–2.0 km thick), and an oceanic layer (sheeted basaltic dykes, gabbros) (3–7 km thick).

The continental crust can be divided into an upper layer (10–20 km thick) and a lower layer (15–25 km thick). Sedimentary cover, possibly up to several kilometres thick, may or may not be present above the upper layer. The boundary between the upper and lower layer may be well defined by a change in seismic velocities, although it may not be detectable. It is called the 'Conrad discontinuity'.

## Chemical Composition of the Crust

Much of the crust is buried, and we have no direct access to the material from which it is composed; therefore, much reliance must be placed on indirect methods of determination.

### Seismic Wave Velocities

An important method is based on the measurement of seismic wave velocities, which indicate what materials are possible for any particular layer according to density (*see* **Seismic Surveys**).

### Chemical Analyses

The many different rock materials that make up the continental shields can be analysed directly and weighted according to observations of relative abundance to give an average for the upper continental crust. Fine sediments, such as clays, can be analysed, being assumed to match closely the average composition of the upper continental crust. This assumption is based on the fact that clays give consistent values for thorium and rare-earth elements, which are relatively insoluble in natural water; amounts of such elements vary widely in concentration between different rock types of the upper continental crust, and this suggests that mixing in such fine-grained sediments of the contributory source rocks is very thorough.

### Studies of Sections of Deep Crustal Rocks Exposed by Tectonic Processes: Deep-Sourced Xenoliths in Volcanic Rocks

There are some regions in which tectonic processes have revealed sections of the crust, on end, so that a geologist can traverse across the surface terrain, encountering progressively deeper crustal levels, including deep metamorphic rocks, such as granulites, at the surface. One such case is the Kapuskasing Belt in Ontario, Canada. Such exposed sequences of the deeper crustal rocks indicate that, in the deeper continental crust, below the Conrad discontinuity, there is much variation in both the horizontal and vertical dimensions, and that models invoking uniform layering of, for example, a granulite layer are not valid. The evidence from both exposed sequences and deep-sourced xenoliths in volcanic rocks suggests, besides heterogeneity, that the lower continental crust is of more basic, less siliceous, composition than the upper continental crust – that is, more rocks of basaltic composition are represented and metamorphic granulites are characteristic.

### Oceanic Crust

The composition of the oceanic crust can be derived from the study of ophiolitic sequences, which have a systematic layering from lavas at the top, through sheeted basaltic dykes, to gabbros or layered differentiates of gabbroic magma below. Ophiolites can commonly be studied in traverses on the ground, through sections exposed on end, and even, as in Oman and western Newfoundland, from exposures of the oceanic crust–mantle boundary. Deep ocean sediments are very thin and can be ignored in making a weighted assessment according to the observed abundance of these three components. Such an estimate yields a basaltic composition for the oceanic crust, but poor in $K_2O$.

The values derived from such methods and given in **Table 1** seem to be reasonable.

**Table 1** Chemical composition (wt.%) of the crust

| Component | Continental upper layer | Continental lower layer | Ocean crust |
|---|---|---|---|
| $SiO_2$ | 65.5 | 49.2 | 49.6 |
| $Al_2O_3$ | 15.0 | 15.0 | 16.8 |
| $TiO_2$ | 0.5 | 1.5 | 1.5 |
| FeO | 4.3 | 13.0 | 8.8 |
| MgO | 2.2 | 7.8 | 7.2 |
| CaO | 4.2 | 10.4 | 11.8 |
| $Na_2O$ | 3.6 | 2.2 | 2.7 |
| $K_2O$ | 3.3 | 0.5 | 0.2 |
| Total | 98.6 | 99.6 | 98.6 |

### Partial Melting within the Crust

At the mid-ocean ridges, where the oceanic crust is thinnest, partial melting of the upper mantle material leaves behind a depleted mantle. Some crustal material may also suffer partial melting, modifying the residual oceanic crust, but this is probably not a major process, because the ascent of magma is rapid here. However, in the island arc situations in the overriding plate above the subducting plate, the eruptives that form the volcanic arcs and batholiths beneath them, mainly of calc-alkaline igneous rocks, pass quite slowly through the quite thick crustal rocks of the overriding plate; therefore, here there is much mixture of partial melting products from both mantle and crust, and the crust is always suffering modification and displacement. This process makes for a very heterogeneous continental crust at all levels.

## Crustal Growth and Loss

The fact that some isotopically derived ages of continental crustal rocks are more commonly obtained than others, e.g., 2700 Ma and 2000 Ma being particularly common, suggests that the rate of new continental crust growth compared with loss by subduction has not been constant through geological time. This may be explained in terms of the plate tectonics paradigm, but may also be explained, at least partly, if plate tectonics processes did not operate or did not operate significantly in Archaean times (something for which there is considerable evidence). This is, however, an area of considerable uncertainty. In any case, there are a number of processes by which continents grow: (1) by continental plate collision (which does not affect the overall total of continental crust, but can cause crustal loss by metamorphism of the deep roots of collision zones to denser *PT*-compatible mineral phases, gravitational instability, and sinking to the mantle); (2) by underplating by basaltic magma from mantle hotspots or plumes (assuming that plumes do exist), or surface eruption therefrom to form lava plateaux (e.g., the Deccan); (3) by the addition of ophiolite material of old oceans; (4) by obduction onto continental edges; or (5) by collisions with oceanic terranes (islands, island arcs, submarine plateaux). The return process may be the direct subduction of continental slabs, or a three-stage process of erosion of continental crustal material, deposition of it as sediments, and their subduction.

Oceanic crust is either rapidly subducted or lost to continents by obduction onto their edges or by collisions of terrains with continents as mentioned above. It does not have a protracted geological history as does continental crust. It can, however, be added to during its relatively ephemeral history by the addition of magma from mantle hotspots or plumes beneath (e.g., Ontong-Java Plateau).

## The Primitive Crust

It is widely believed that the early Earth possessed a primitive globe-encircling basaltic crust. The present surfaces of Mercury, the Moon, or Mars may provide an analogue as they have apparently never experienced plate tectonics and there is evidence suggesting a basaltic surficial shell still covering the whole of Mars (*see* **Solar System: Mars**). We now know that there must have been some continental crust very early in the Earth's history, the evidence being in the form of zircons of continental crust provenance up to 4300 Ma at Mt Narryer in western Australia. The process of continental crust formation has been compared to a scum or slag of light material forming on the surface of a basaltic melt, repeated again and again, until extensive continents emerged after a repetition of cycles. Thick continental blocks appear to have been in place by 3500 Ma, being represented by the oldest cratons of South Africa and Australia, but the earlier protocontinents were probably small and thin. This is as far as we can go at the present state of knowledge.

## The Crust and Isostasy

The concept that gravitational equilibrium controls the heights of continents and ocean floors was proposed by Dutton in America in 1889. The analogy with wooden blocks floating in water was elegantly illustrated by Holmes in 1965 (**Figure 4**). The concept is one of hydrostatic balance, the Earth's major relief being compensated by the underlying differences in density, whereas minor relief features are compensated by the strength of the underlying rock. The compensation depth was traditionally placed at the Moho, but the plate tectonics paradigm necessitates a rethink of this. Isostasy is, however, a real crustal process and can be seen in action in the central sag of Greenland under the weight of the enormous

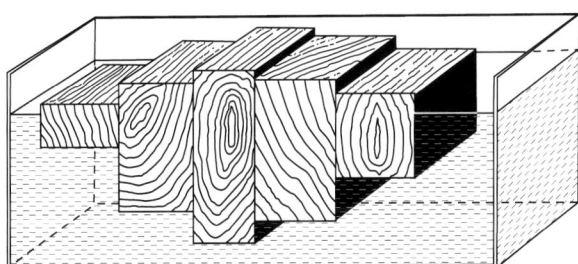

**Figure 4** Representation of isostasy using floating wooden blocks of unequal dimensions. Reproduced with permission from Holmes A (1965) *Principles of Physical Geology*. London: Nelson.

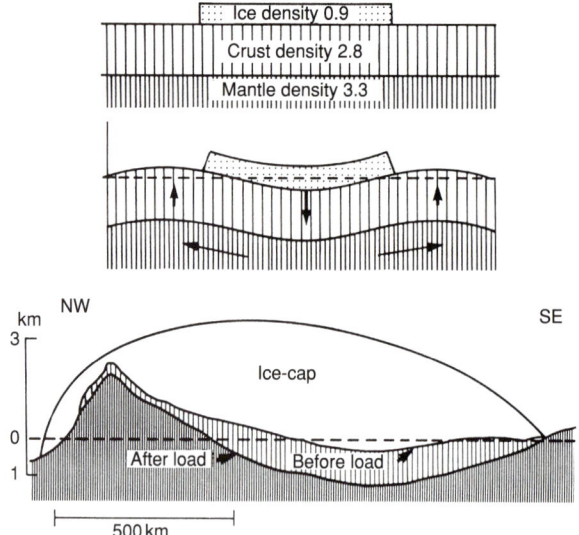

**Figure 5** The Earth's crust floats on a dense mantle that behaves as a viscous liquid. If we place an ice-cap on it, enough mantle material is displaced sideways to equal the additional crustal load, and a low broad welt is raised peripheral to the ice-cap. The mantle adjustment is very slow, so that depression and rebound after removal of the ice-cap are delayed. The situation at present in Greenland is shown in the lower part of the figure, and the delayed rebound is being experienced in the Baltic. From Van Andel TJ (1994) *New Views on an Old Planet*. Cambridge: Cambridge University Press.

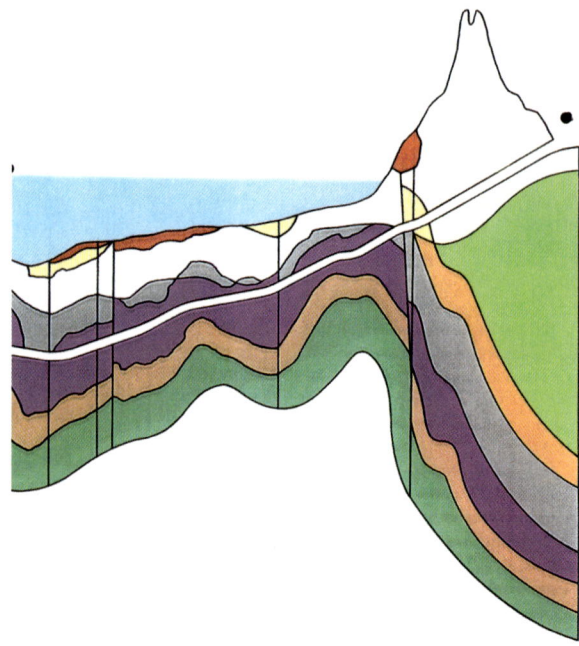

**Figure 6** A sectional diagram of the French entry section to the Channel Tunnel showing tectonic deformation (folding, faulting) of the Cretaceous and Quaternary strata of the crust. Extract from BREAKTHROUGH by Derek Wilson published by Century Hutchinson. Reproduced from The Random House Group Limited.

ice-sheet which has sagged below sea-level in places (**Figure 5**), and also in the rebound occurring now in the Baltic after the ice-sheet has gone – the renowned Swedish geologist, Harry von Eckerman, used to go to the Alnö Island carbonatite complex yearly to see what new rocks had surfaced due to this rebound.

## Heat Flow to the Crustal Surface

The continental surface heat flow is linearly related to the heat productivity of the near-surface granitic rocks. Variability from region to region is mainly related to the distribution of near-surface radioactivity, derived from certain minerals in the granitic rocks. Most models favour an exponential decrease in radioactive heat source with depth in the continental crust and a residual amount of heat rising from the upper mantle. The weighted average heat flow from both continental and oceanic crust is 1.5 heat flow units (HFU). This equivalence is explained by the heat flow to the oceanic crust surface coming mainly from the mantle, whereas it comes mainly from radioactive minerals in the crust at the continental surface. This requires increased radioactive sources in the mantle under the oceans, or that convective heat from the mantle under the oceans happens to equate to the radioactive-sourced heat from the continental crust, or that the mantle-derived convective heat is much the same under continental and oceanic lithosphere, but the thicker continental lithosphere is more depleted in radioactive heat sources than the oceanic lithosphere. The latter is preferred because it allows equal movements of both continental and oceanic crust-dominated plates.

## Crustal Deformation

The crustal rocks of the Earth are subject to many and diverse deformation processes. The most significant are 'tectonic' processes (folding and faulting; **Figure 6**), which act very slowly through long periods of geological time and are mostly related to the movement of tectonic plates (especially collision and subduction). Stress on faults is subject to long build-up, but may be relieved by abrupt, almost instantaneous dislocation (which may form a linear earthquake scarp on the surface; **Figure 7**), or prolonged dissipation by creep movements without any earthquake. Earthquake foci are mostly situated within the crust, but some, especially those in margins of continental areas of plates, may have foci several hundred kilometres deep, within the mantle (*see* **Tectonics**: Earthquakes).

In contrast with tectonic deformation, 'superficial' deformation of the crust, such as landsliding, sink-hole formation, and submarine slumping, occurs abruptly, within a matter of seconds, minutes, hours, or days (**Figure 8**) (*see* **Sedimentary Processes**: Landslides).

**Figure 7** A low fresh scarp of the Nojima Fault, Awaji Island, caused by the Kobe Earthquake, Japan, 1995. This was the almost instantaneous product of relief of one of many successive prolonged build-ups of stress on an active fault. Reproduced with permission from Esper P and Tachibana E (1998) The Kobe Earthquake. In: JG Maurel and M Eddleston (eds.) *Geohazards in Engineering Geology, Geological Society of London Engineering Geology Special Publication 15*, pp. 105–116. London: Geological Society.

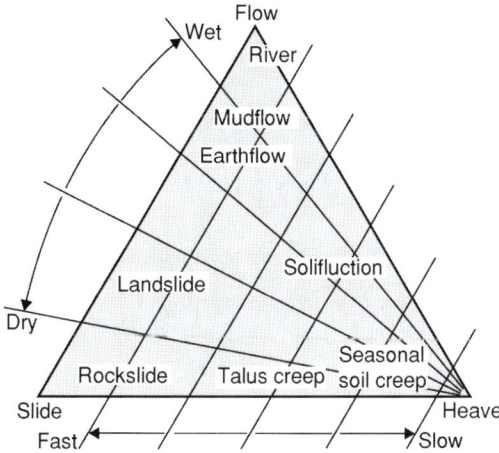

**Figure 8** Classification of superficial mass movements on slopes and examples of some major types. Reproduced with permission from Goudie A (1989) *The Nature of the Environment*. Oxford: Blackwell.

## Conclusion

The Earth's crust is the site of the great majority of processes studied by geologists, and thus is better known and understood than the mantle or core. However, there remain aspects of uncertainty, particularly the mechanisms of plate movement and earthquakes, and the exact nature of the lower parts of the crust.

## See Also

**Earth:** Mantle. **Earth Structure and Origins**. **Mantle Plumes and Hot Spots**. **Plate Tectonics**. **Seismic Surveys**. **Sedimentary Processes:** Landslides. **Solar System:** Mars. **Shields**. **Tectonics:** Convergent Plate Boundaries and Accretionary Wedges; Earthquakes; Mountain Building and Orogeny; Ocean Trenches.

## Further Reading

Condie KC (1976) *Plate Tectonics and Crustal Evolution*. New York, Toronto, Oxford, Sydney, Braunschweig, Paris: Pergamon Press.

Edwards K and Rosen B (2000) *From the Beginning*. London: Natural History Museum.

Goudie A (1989) *The Nature of the Environment*. Oxford: Blackwell.

Hancock PL and Skinner BJ (2000) *Oxford Companion to the Earth*. Oxford: Oxford University Press.

Holmes A (1965) *Principles of Physical Geology*. London: Nelson.

Stamp LD (1951) *The Earth's Crust*. London, Toronto, Wellington, Sydney: GGHarrap.

Van Andel TJ (1994) *New Views on an Old Planet*. Cambridge: Cambridge University Press.

# Orbital Variation (Including Milankovitch Cycles)

**H Pälike**, Stockholm University, Stockholm, Sweden

© 2005, Elsevier Ltd. All Rights Reserved.

## Introduction

Earth's orbital variations, caused by the mutual gravitational interaction between the sun, the planets, and their satellites, have been invoked as a mechanism for long-term variations (time-scales of $10^4$–$10^6$ years) of Earth's climate system. The orbital variations are known as 'Milankovitch cycles', after the Serbian mathematician Milutin Milankovitch. In this article, the relationships between the different parameters that affect orbital variations are explored, with a further view towards long-term patterns.

## Celestial Mechanics

The time-varying motion of the planets and other satellites around the sun and around each other is controlled by mutual gravitational interactions. This relationship can be described by Newton's laws, corrected by Einstein's principles of general relativity. The overall behaviour of the entire solar system is even more complex; it is posed as an N-body problem (which is solved numerically), and is further complicated by physical parameters such as the tidal dissipation of energy and the detailed and changing distribution of mass within each body. Given this complexity, the planets do not follow stationary orbits around the Sun, but rather undergo perturbed quasiperiodic variations that, from a climatic point of view, this affects the amount, distribution, and timing of solar radiation received at the top of Earth's atmosphere. Milutin Milankovitch explained glaciations and de-glaciations (on time-scales of $10^4$–$10^6$ years) by variations of solar radiation distribution at the top of Earth's atmosphere, coupled with the latitudinal distribution of land masses on Earth (*see* **Palaeoclimates**).

At any given time, the orbit of a body can be described by six parameters. These parameters define the position, shape, and orientation of an orbit and the location of a body in the orbit, with respect to a frame of reference. The trajectory of the orbiting body follows an ellipse; in the case of the Solar System, the sun is located at one focal point. Due to gravitational interactions between the different planets, the orientation and dimension of the ellipse change over time. **Figure 1** illustrates how the relationships between the six 'Keplerian orbital elements' ($a$, $e$, $i$, $\lambda$, $\tilde{\omega}$, and $\Omega$) apply to Earth in its orbit. A reference plane is fixed with respect to the stars; it is typically chosen as the orbital plane of Earth at a particular time (say, AD 1950 or 2000) and is called the 'ecliptic of date'. Alternatively one can choose a plane perpendicular to the total angular momentum of the Solar System (the 'invariant' plane). The 'invariant' plane almost coincides with the orbital plane of Jupiter due to its large mass. The reference plane is defined by two axes, both originating from the position of the Sun ($S$; **Figure 1**) on the reference plane. One of these, $\gamma$, is typically the position of the mean vernal (spring) equinox on the reference plane at a given time; the second axis, $Z$, is perpendicular to $\gamma$ as well as to the reference plane. The parameter $a$, the semi-major axis of the orbital ellipse, corresponds to the average radius. The eccentricity $e$ of the ellipse is defined as $e = (a^2 - b^2)^{1/2}/a$, where $b$ is the semi-minor axis of the ellipse. The inclination of the orbit with respect to the reference plane is given by the angle $i$. The position of the ascending node $N$ is specified by the angle $\Omega$ ('longitude of the node'), measured from the fixed direction in the reference plane ($\gamma$). The parameter $\tilde{\omega}$ specifies the position of the moving perihelion $P$ (the closest approach to the Sun) and is defined as $\tilde{\omega} = \Omega + \omega$ ('longitude of the perihelion'). Finally, the sixth Keplerian element, $\lambda$, specifies the position of the orbiting body ($J$) on its elliptical orbit, and is defined as $\lambda = \omega + M$, where the mean anomaly $M$ is an angle that is proportional to the area $SPJ$ (Kepler's third law).

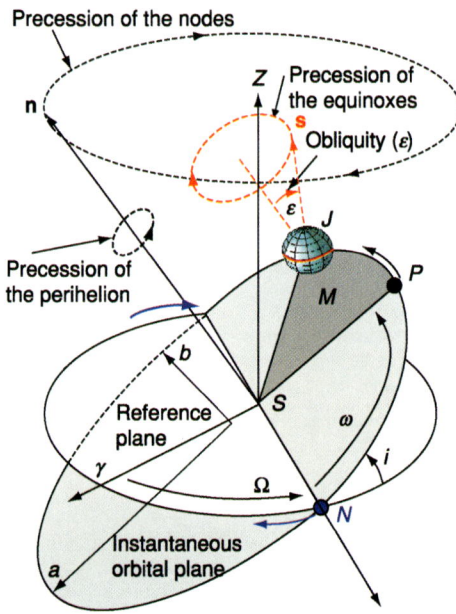

**Figure 1** Orbital elements of the Earth's movement around the Sun.

## Origin of Orbital Frequencies

If Earth were the only planet orbiting the sun, and in the absence of dissipative effects and other physical processes, the position and orientation of its orbit would remain fixed for all times. In this situation, the only Keplerian element that would change over time is $\lambda$, which describes the position of a body in its orbit. In this case, the position and velocity of the orbiting body would vary according to the Keplerian laws of motion around a fixed ellipse. However, gravitational interactions between all bodies of the Solar System cause changes in the shape and orientation of the elliptical orbit on various time-scales, which are typically of the order of $10^4$–$10^6$ years. From a long-term climatic point of view, the relevant variations are those obtained after averaging the planetary orbits over their long-term orbital periods. These are termed 'secular variations', and they can be related to a set of fundamental frequencies. Variations in the orbital elements that characterize the secular variations can be separated into two categories that are related to different types of precession movements. The first category, variation within an orbital plane, is described by the variation of the eccentricity $e$ and the rotation of the location of the perihelion (described by $\omega$). The second category, variation of the orientation of an orbital plane, is described by the inclination angle $i$ and the location of the ascending node $N$ (described by $\Omega$). These oscillations are coupled such that it is possible to investigate the behaviour of these parameters as pairs: $(e, \omega)$ and $(i, \Omega)$.

## Fundamental Frequencies of the Solar System

Computing the orbital elements for the eight main planets (Pluto can be excluded due to its small mass) obtains eight characteristic modal frequencies for each of the paired elements $(e, \omega)$ and $(i, \Omega)$. Table 1 illustrates these fundamental frequencies, estimated over the past 20 My. Individual frequencies $g_i$ are related to variations in the pair $(e, \omega)$, whereas frequencies $s_i$ are related to variations in the pair $(i, \Omega)$. The individual $g_i$ and $s_i$ frequencies arise as eigenvalues from a matrix of linear differential Lagrange/Laplace equations that are used to expand the orbital elements for the eight planets. As eigenvalues of a matrix, they are not strictly associated with a particular planet. However, because the matrix from which they are obtained has a rear-diagonal structure, suppressing a planet removes one set of frequencies but does not change the other frequencies significantly. Thus, the indices $g_i, s_i$ can be used to indicate which planet provides the strongest contribution to a particular frequency (indices $g_1, s_1$ correspond to Mercury; indices $g_3, s_3$ correspond to Earth; and so on). Note that all of the $g_i$ terms are positive, indicating that the perihelia advance counterclockwise if viewed from the 'north' of the orbital axis shown in **Figure 1**. In contrast, seven out of the eight $s_i$ terms are negative, indicating that the positions of the nodes, which mark the intersection of the orbital plane with the reference plane, regress (rotate clockwise). Due to considerations of angular momentum the eighth frequency is set to zero, and by conversion is chosen to be $s_5$ because the invariant plane is close to the orbital plane of Jupiter, due to Jupiter's large mass (*see* **Solar System:** The Sun; Asteroids, Comets and Space Dust; Meteorites; Mercury; Venus; Moon; Mars; Jupiter, Saturn and Their Moons; Neptune, Pluto and Uranus).

## General Precession of Earth

In addition to the fundamental orbital frequencies, which apply to the Solar System as a whole, two additional fundamental frequencies are necessary to

**Table 1** Fundamental orbital frequencies[a]

| Related to (e, ω) | | | Related to (i, Ω) | | | |
| --- | --- | --- | --- | --- | --- | --- |
| Term | Frequency ('' year$^{-1}$) | Period (ky) | Term | Frequency ('' year$^{-1}$) | Period (ky) | Associated planet |
| $g_1$ | 5.596 | 231.0 | $s_1$ | −5.618 | 230.0 | Mercury |
| $g_2$ | 7.456 | 174.0 | $s_2$ | −7.080 | 183.0 | Venus |
| $g_3$ | 17.365 | 74.6 | $s_3$ | −18.851 | 68.7 | Earth |
| $g_4$ | 17.916 | 72.3 | $s_4$ | −17.748 | 73.0 | Mars |
| $g_5$ | 4.249 | 305.0 | $s_5$ | 0.000 | | Jupiter |
| $g_6$ | 28.221 | 45.9 | $s_6$ | −26.330 | 49.2 | Saturn |
| $g_7$ | 3.089 | 419.0 | $s_7$ | −3.005 | 431.0 | Uranus |
| $g_8$ | 0.667 | 1940.0 | $s_8$ | −0.692 | 1870.0 | Neptune |

[a]Fundamental orbital frequencies of the precession motions in the Solar System, computed as mean values over 20 million years. The $g_i$ and $s_i$ are eigenvalues that characterize the evolution of the orbital elements $(e, \omega)$ and $(i, \Omega)$, respectively, and are loosely associated with the eight planets considered; i.e., the indices $g_1, s_1$ correspond to Mercury, and the indices $g_8, s_8$ correspond to Neptune. The period $a$, in years, can be calculated from the frequency, in arcseconds ('') per year, as $a = (360 \times 60 \times 60)/''\,\text{year}^{-1}$. Data from Laskar J (1990) The chaotic motion of the Solar System – a numerical estimate of the size of the chaotic zones. *Icarus* 88(2): 266–291.

describe the orbital motion of Earth. The formation of an equatorial bulge is caused by the rotation of Earth, and other processes (e.g., the formation of ice-caps at high latitudes and mantle convection) redistribute mass on Earth. These processes result in a torque that is applied to Earth by the Sun, the Moon, and the other planets. Approximately two thirds of the torque is caused by the Moon, and one third by the Sun. Similar to a spinning top, this applied torque results in the nutation and precession of Earth's spin axis. Nutation is the short-term period portion (periods ≤18.6 years) of these variations; precession is the long-period portion. The nutational component leads to a 'nodding' motion of Earth's spin axis, with main periods of ~13 days, 6 months, 1 year, 9.3 years, and ~18.6 years, whereas the precessional component makes Earth's spin axis trace out a cone shape. The short-term nutation component that is superimposed on top of the longer term precession component is illustrated in **Figure 2**, and the precession component is illustrated in **Figures 1** and **3**. From a climatic perspective on geological time-scales, only the precession component has a significant effect; nutational variations result in small, mainly atmospheric effects.

With respect to the fixed stars, the frequency $p$ of the precessional cycle has a period of approximately 25.8 ky. The precession of Earth's spin axis has several effects on Earth's climate system, one of which is that the position of the seasons with respect to Earth's orbit, defined by the solstices and equinoxes with respect to the perihelion and aphelion of the orbit, changes over time. For this reason, the precession of Earth's spin axis is also called the 'precession of the equinoxes'. As shown in **Figure 1**, the precession of Earth's spin axis traces out a cone shape that forms an angle with Earth's orbital plane. This angle, denoted by $\varepsilon$, is the obliquity (tilt) of Earth. The angle $\varepsilon$ changes due to the combined effect of the precession of Earth's spin axis and the changing orientation of Earth's orbital plane (this will be discussed in more detail in a following section).

As an approximation, the fundamental frequencies $g_i$ and $s_i$ can be used together with the precession constant $p$ to explain the origin of almost all periodicities that affect the climate system, which arise from 'beats' between the fundamental frequencies. However, Jacques Laskar discovered that additional resonance terms are present in the Solar System, and these lead to the presence of chaos. The presence of chaos in the solar system has important consequences (see later). How the three orbital parameters, eccentricity, obliquity, and climatic precession, which are involved in the calculation of the solar radiation, are related to the fundamental frequencies of the Solar System is discussed in the following sections. The main parameters, known as 'Milankovitch cycles', are illustrated in **Figure 3**.

### Eccentricity

Earth's orbital eccentricity $e$ quantifies the deviation of Earth's orbital path from the shape of a circle. It is the only orbital parameter that controls the total amount of solar radiation received by Earth, averaged over the course of 1 year. The present eccentricity of Earth is $e \approx 0.01671$. In the past, it has varied between 0 and ~0.06. The eccentricity value can be used to compute the difference in the distance from Earth to the Sun between their closest and furthest approaches (perihelion and aphelion); presently, this amounts to $2e \approx 3.3\%$. At maximum eccentricity, the annual variation of solar insolation due to eccentricity is thus 24%. Although the exact values of orbital parameters should be computed by numerical integration, it is possible to approximate the calculation as a series of quasiperiodic terms, some of which are listed in **Table 2**. It is important to point out that the eccentricity frequencies are completely independent of the precession frequency $p$. Earth's eccentricity frequency component with the largest amplitude has a period of approximately 400 ky, which arises mainly from the interactions of the planets Venus and Jupiter, due to their close approach and large mass, respectively. This component is called the 'long' eccentricity cycle, and of all of Earth's orbital frequencies, it is considered to be the most stable. Additional terms can be found with periods clustered around ~96 and ~127 ky. These are called 'short' eccentricity cycles.

An important feature of all orbital components is the presence of 'beats'. These arise from the interaction between different frequency components and they produce a modulation in amplitude. This results, for example, in an amplitude modulation of the short eccentricity cycle, because the difference between the

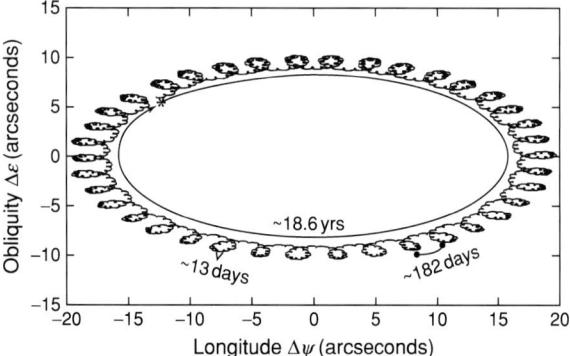

**Figure 2** Short-term nutation motion of Earth's spin axis, based on nutation and precession model IAU2000A of the International Astronomical Union. The nutation components in longitude and obliquity are with respect to the equinox and ecliptic of date. The figure is plotted for an ~18.5-year period, beginning on 1 January 2004.

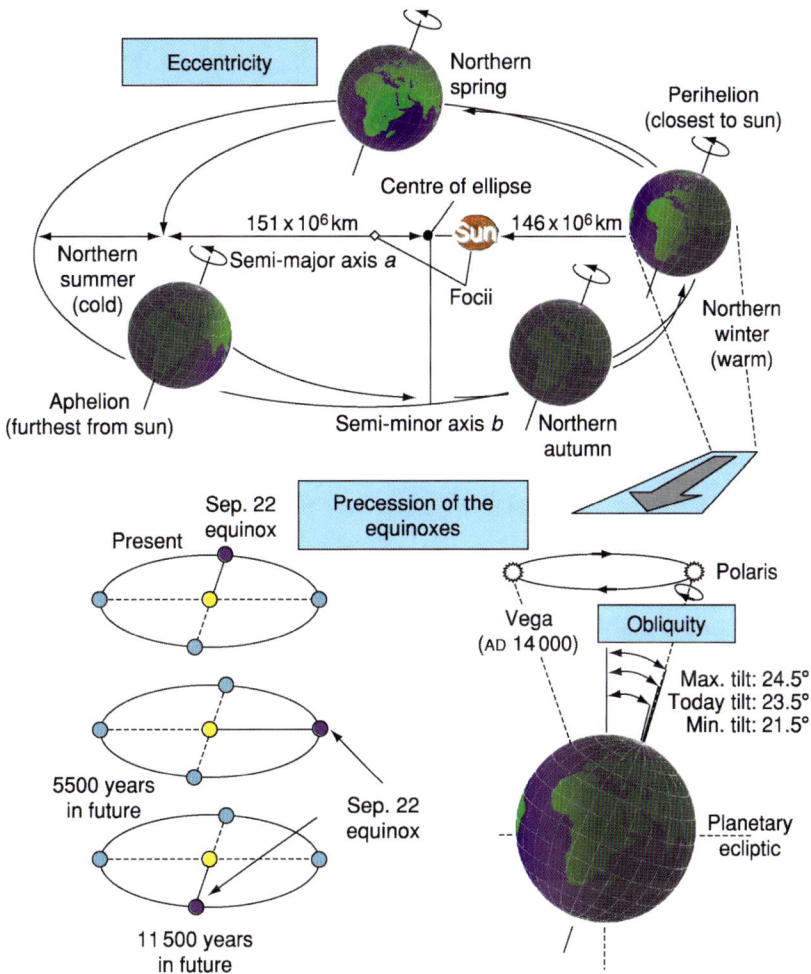

**Figure 3** Orbital motions of Earth, showing the main Earth orbital parameters, eccentricity, obliquity, and precession of the equinoxes.

**Table 2** Five leading terms for Earth's eccentricity[a]

| Term | Frequency (" year$^{-1}$) | Period (ky) | Amplitude |
| --- | --- | --- | --- |
| $g_2 - g_5$ | 3.1996 | 406.182 | 0.0109 |
| $g_4 - g_5$ | 13.6665 | 94.830 | 0.0092 |
| $g_4 - g_2$ | 10.4615 | 123.882 | 0.0071 |
| $g_3 - g_5$ | 13.1430 | 98.607 | 0.0059 |
| $g_3 - g_2$ | 9.9677 | 130.019 | 0.0053 |

[a]Principal eccentricity frequency components in an astronomical solution analysed over the past 4 My. The frequency terms $g_i$ refer to those given in **Table 1**.
Data from Laskar J (1999) The limits of Earth orbital calculations for geological time-scale use. *Philosophical Transactions of the Royal Society of London, Series A, Mathematical, Physical and Engineering Sciences* 357(1757): 1735–1759.

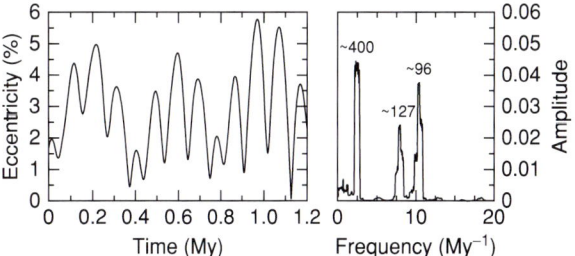

**Figure 4** Earth's orbital eccentricity over time (1.2 million years) and frequency analysis for a 10-My time-span. The peaks in the frequency analysis correspond to the frequencies given in **Table 2**; the numbers over the peaks represent the periods (in thousands of years).

second and third strongest eccentricity components is $(g_4 - g_5) - (g_4 - g_2) = (g_2 - g_5)$, which corresponds to the ~400-ky eccentricity cycle. The same modulation is observed for the fourth and fifth strongest terms. This type of amplitude modulation can be found in all orbital components of Earth. The nature of eccentricity variations is illustrated in **Figure 4**. The superposition of the long and short eccentricity cycles, and their variation in amplitude, are clearly visible. The right-hand side of the plot shows the results of a frequency analysis, which was run over a 10-My-long interval to better resolve the position of individual peaks. The peaks correspond to the frequencies given in **Table 2**.

## Obliquity

The obliquity (tilt) $\varepsilon$ of Earth's axis with respect to the orbital plane (see **Figure 1**) is defined by the angle between Earth's spin vector **s** and that of the orbital plane **n**, and can be computed as $\cos \varepsilon = \mathbf{n} \diamond \mathbf{s}$, using unit vectors. As the inclination and orientation of the orbital plane vary, the obliquity is not constant, but oscillates due to the interference of the precession frequency $p$ and the orbital elements $s_i$. As shown in **Table 3**, if the variation in obliquity is approximated by quasiperiodic terms, the result is a strong oscillation with a period of approximately 41 ky, with additional periods around 54 and 29 ky. The ~41-ky period arises from the simultaneous variation in Earth's orbital inclination, given by $s_3$, and the precession of Earth's spin direction, expressed by $p$. Table 3 also shows that the obliquity signal contains contributions from the $g_i$ as well as the $s_i$ fundamental frequencies, due to their combined effect on the change of the orbital plane normal. The present day obliquity of approximately 23.45° has varied between ~22.25° and ~24.5° during the past 1 million years. The main climatic effect of variations in Earth's obliquity is in control of the seasonal contrast. The total annual energy received on Earth is not affected, but the obliquity controls the distribution of heat as a function of latitude, and is strongest at high latitudes.

It is important to note that all of the obliquity frequency components contain the precession constant $p$. Due to tidal dissipation, the frequency of the precession constant $p$ has been higher in the past, a fact that can be shown from geological observations, such as ancient growth rings in corals, and from tidal laminations. **Figure 5** illustrates the variation in obliquity over 1.2 million years. The oscillation is dominated by a ~41-ky period cycle, and a variation in amplitude is also observed. This variation is due to beats arising from the presence of additional ~29- and ~54-ky periods, which are just visible in the frequency analysis shown on the right-hand side of **Figure 5**.

## Climatic Precession

The precession of Earth's spin axis has a profound effect on Earth's climate, because it controls the timing of the approach of perihelion (the closest approach to the Sun) with respect to Earth's seasons. At present, perihelion occurs on the 4 January, close to the winter solstice. With respect to the stars, the precessional movement of Earth's spin axis traces out a cone shape with a period of ~25.8 ky. However, due to the precession of the perihelion within the orbital plane, the period of precession, measured with respect to the Sun and the seasons, is shorter. The motion of the perihelion is not steady but is caused by a superposition of the different $g_i$ frequencies. For this reason, the precession of the equinoxes with respect to the orbital plane lurches with a superposition of several periods around ~19, 22, and 24 ky. The effect of the precession of the equinoxes on the amount of solar radiation received by Earth also depends on the eccentricity. If the eccentricity is zero (i.e., the orbit of Earth follows a circle), the effect of the precession of the equinoxes is also zero. From a climatic point of view, the eccentricity and longitude of the perihelion combine to create what is termed the 'climatic precession', defined as $e\sin(\tilde{\omega})$, where $e\sin(\tilde{\omega})$ is the longitude of perihelion from the moving equinox. This means that the climatic precession index is modulated in amplitude by variations in Earth's eccentricity.

A quasiperiodic approximation of the climatic precession time series reveals the contribution from different frequency components, as shown in **Table 4**.

**Table 3** Six leading terms for Earth's obliquity[a]

| Term | Frequency ($''$ year$^{-1}$) | Period (ky) | Amplitude |
|---|---|---|---|
| $p + s_3$ | 31.613 | 40.996 | 0.0112 |
| $p + s_4$ | 32.680 | 39.657 | 0.0044 |
| $p + s_3 + g_4 - g_3$ | 32.183 | 40.270 | 0.0030 |
| $p + s_6$ | 24.128 | 53.714 | 0.0029 |
| $p + s_3 - g_4 + g_3$ | 31.098 | 41.674 | 0.0026 |
| $p + s_1$ | 44.861 | 28.889 | 0.0015 |

[a]Principal obliquity frequency components analysed over the past 4 My. The frequency terms $g_i$ and $s_i$ refer to those given in **Table 1**.
Data from Laskar J (1999) The limits of Earth orbital calculations for geological time-scale use. *Philosophical Transactions of the Royal Society of London, Series A, Mathematical, Physical and Engineering Sciences* 357(1757): 1735–1759.

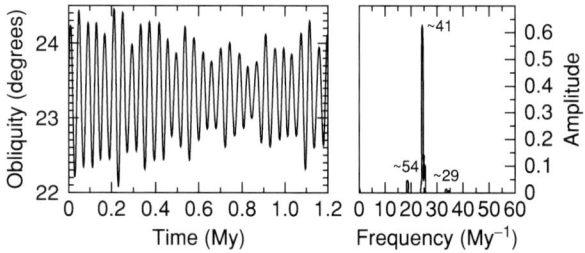

**Figure 5** Earth's obliquity over time (1.2 million years) and frequency analysis for a 10-My time-span. The peaks in the frequency analysis correspond to the frequencies given in **Table 3**; the numbers over the peaks represent the periods (in thousands of years).

Note that the components of climatic precession can be constructed from the precession constant $p$ and the fundamental frequencies $g_i$. **Figure 6** illustrates the variation in the climatic precession index, and its modulation in amplitude by eccentricity, over 1.2 million years. The frequency analysis, shown on the right-hand side of the plot, reveals three peaks corresponding to frequencies given in **Table 4**.

### Insolation

Conceptually, the actual forcing of Earth's climate by orbital variations is applied through the radiative flux received at the top of the atmosphere at a particular latitude and time, which is then transferred through oceanic, atmospheric, and biological processes into the geological record. The integral of the radiative flux over a specified interval of time, termed 'insolation' (French from the Latin *insolare*), can be computed from the eccentricity $e$, the obliquity $\varepsilon$, and the climatic precession $e\sin(\tilde{\omega})$. The amount of solar radiation received at a particular location depends on the orientation towards the Sun of that location. Calculation of insolation becomes complex if it is to be calculated over a particular time interval. Averaged over 1 year and the whole Earth, the only factor that controls the total amount of insolation received, apart from the solar constant, is the changing distance from Earth to the Sun, which is determined by Earth's semi-major axis $a$ and its eccentricity $e$.

If insolation variations are computed for a particular latitude, and over a particular length of time, the main contribution arises from the climatic precession signal, with additional contributions from the variation in obliquity. The exact nature of the insolation signal is complicated. Certain general statements can be made, though. The signal arising from the climatic precession is always present in insolation time series. Also, if the obliquity signal is present, it typically shows a larger amplitude towards high latitudes. In addition, the climatic precession signal in the insolation calculation depends on the latitude at which it is calculated, such that the signal in the southern-hemisphere summer shows opposite polarity to that in the northern-hemisphere summer (see **Figure 7**). If the mean monthly insolation is computed for a particular latitude, each month corresponds to a difference in phase (i.e., a difference in time of a particular insolation maximum or minimum) of approximately 2 ky (12 months approximately correspond to the (on average) ~24-ky-long climatic precession cycle).

It is unlikely that geological processes that encode the insolation signal are driven by variations at the same latitudes and times of the year throughout geological time. Depending on the latitude and the time interval over which insolation quantities are computed, the calculation can be very complex, and the question of time lags and forcing can be resolved only through the application of climate models. A very revealing study to this effect was reported by Short and colleagues in 1991. At the present level of understanding, it is probably appropriate to avoid a strict

**Table 4** Five leading terms for Earth's climatic precession[a]

| Term | Frequency ('' year$^{-1}$) | Period (ky) | Amplitude |
|---|---|---|---|
| $p + g_5$ | 54.7064 | 23.680 | 0.0188 |
| $p + g_2$ | 57.8949 | 22.385 | 0.0170 |
| $p + g_4$ | 68.3691 | 18.956 | 0.0148 |
| $p + g_3$ | 67.8626 | 19.097 | 0.0101 |
| $p + g_1$ | 56.0707 | 23.114 | 0.0042 |

[a]Principal climatic precession components analysed over the past 4 My. The frequency terms $g_i$ refer to those given in **Table 1**. Data from Laskar J (1999) The limits of Earth orbital calculations for geological time-scale use. *Philosophical Transactions of the Royal Society of London, Series A, Mathematical, Physical and Engineering Sciences* 357(1757): 1735–1759.

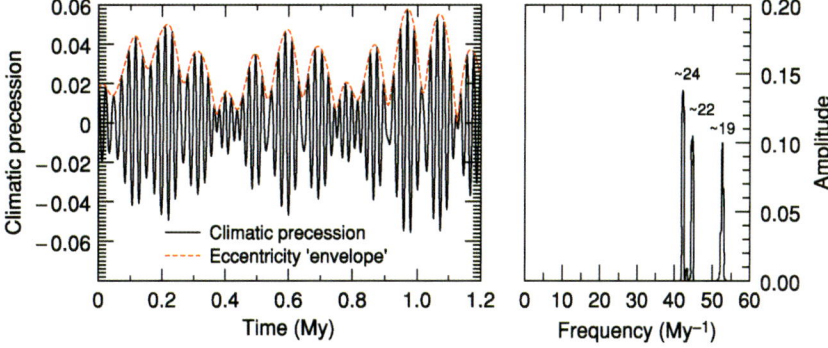

**Figure 6** Climatic precession index and eccentricity envelope over time (1.2 million years) and frequency analysis for a 10-My time-span. The peaks in the frequency analysis correspond to the frequencies given in **Table 4**; the numbers over the peaks represent the periods (in thousands of years).

interpretation of Milankovitch's theory, which would imply that the ice-ages are best explained by the summer insolation curve computed at 65° N. Instead, a better understanding of the complex mechanisms of the climate system will have to be achieved through the use of geological data providing boundary conditions for climate models.

## Amplitude Modulation Patterns: The 'Fingerprint' of Orbital Cycles

A very important feature of Earth's eccentricity, obliquity, and climatic precession variations is that they display modulations in amplitude and frequency. These modulations provide a 'fingerprint' of a particular astronomical calculation at a given time. The modulation terms arise through the interference of individual cycles to produce 'beats', with periods ranging from hundreds of thousands to millions of years. The most prominent amplitude modulation cycles are listed in **Table 5**. An excellent visual representation of amplitude modulation cycles in astronomical calculations can be obtained by computing evolutionary or wavelet spectra, which show the variation in amplitude at different frequencies over time. This is shown in **Figure 8** for the past 10 million years.

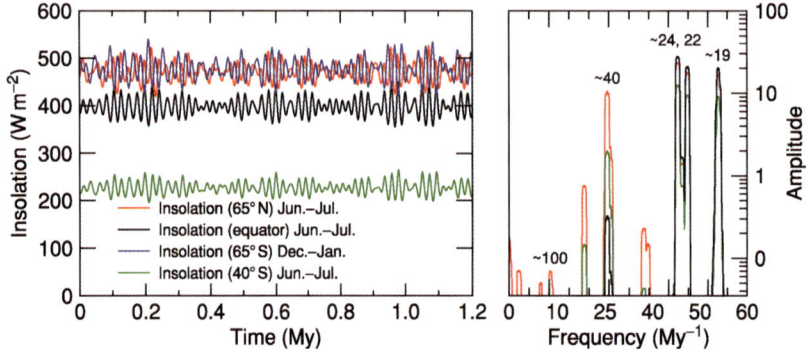

**Figure 7** Earth's insolation over time (1.2 million years) at various latitudes and frequency analysis for a 10-My time-span (on a logarithmic amplitude scale); the numbers over the peaks represent the periods (in thousands of years). The precession component is out of phase between the northern-hemisphere summer and the southern-hemisphere summer. Note that the eccentricity component in insolation computations is always very weak compared to the obliquity and precession components. Also note the latitudinal dependence of the relative strength of the obliquity cycle in the insolation curves.

**Table 5** Modulation terms[a]

| Type | Interfering terms | 'Beat' term | Period |
| --- | --- | --- | --- |
| Short eccentricity amplitude modulation terms | $(g_4 - g_5) - (g_4 - g_2)$ <br> $(g_3 - g_5) - (g_3 - g_2)$ | $= (g_2 - g_5)$ | $\approx$ 400 ky |
|  | ... | ... | ... |
| Short and long eccentricity amplitude modulation terms | $(g_4 - g_5) - (g_3 - g_5)$ <br> $(g_4 - g_2) - (g_3 - g_2)$ | $= (g_4 - g_3)$ | $\approx$ 2.4 My |
|  | ... | ... | ... |
| Climatic precession amplitude modulation terms |  |  |  |
| Identical to eccentricity frequencies and amplitude modulation terms |  |  |  |
| Obliquity amplitude modulation terms | $(p + s_3) - (p + s_4)$ | $= (s_3 - s_4)$ | $\approx$ 1.2 My |
|  | $(p + s_3 + g_4 - g_3) - (p + s_3 - g_4 + g_3)$ | $= (2g_4 - 2g_3)$ | $\approx$ 1.2 My |
|  | $(p + s_3) - (p + s_3 + g_4 - g_3)$ <br> $(p + s_3) - (p + s_3 + g_4 - g_3)$ | $= (g_4 - g_3)$ | $\approx$ 2.4 My |
|  | $(p + s_3) - (p + s_6)$ | $= (s_3 - s_6)$ | $\approx$ 173 ky |
|  | ... | ... | ... |

[a]Origin of amplitude modulation terms that affect Earth's eccentricity, obliquity, and climatic precession. Individual $g_i$ and $s_i$ terms refer to those given in **Table 1**. The $\sim$100-ky (short) eccentricity cycles are modulated with a period of $\sim$400 ky by the long eccentricity cycle. Both short and long eccentricity cycles are modulated with a period of 2.4 My, but with a phase difference of 180° (i.e., an amplitude maximum of the $\sim$100-ky eccentricity coincides with an amplitude minimum of the $\sim$400-ky eccentricity). Because eccentricity directly modulates the climatic precession, all eccentricity amplitude modulation terms are also present in the climatic precession signal. The obliquity signal is weakly amplitude modulated with a period of $\sim$170 ky, and more strongly with a period of 1.2 My. This amplitude modulation cycle is dynamically linked with the 2.4-My cycle present in the eccentricity modulation. All period cycles listed here are shown in **Figure 8**.

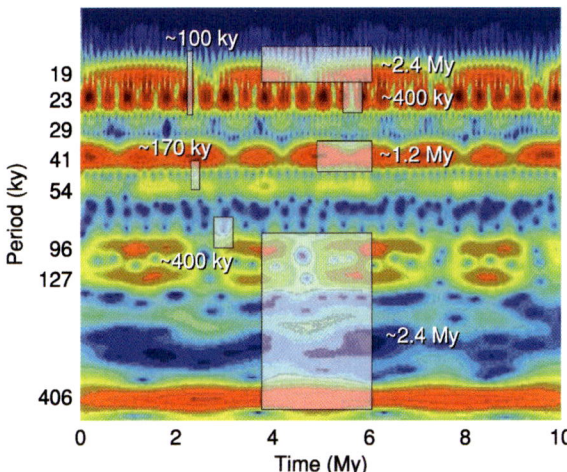

**Figure 8** Joint time–frequency analysis of an arbitrary mixture of the eccentricity, obliquity, and climatic precession signal to allow a better visual representation of the main climatically important orbital variations. The amplitude at a particular frequency and time is colour coded, red corresponding to a high relative amplitude. A selection of amplitude modulation terms is highlighted (boxes inside figure); these represent the 'fingerprint' of an astronomical model. The numbers near the fingerprints represent periods (see **Table 5**).

The significance of amplitude modulation cycles is twofold. First, if these cycles can be detected in the geological record, they allow the placement of geological data into a consistent framework within these amplitude modulation envelopes, even in the absence of individual cycles and in the presence of gaps. The extraction of long amplitude modulation cycles typically requires high-fidelity geological records that are millions of years long. Of particular value for the generation of geological time-scales beyond the Neogene is the ~400-ky-long eccentricity cycle, because it is considered to be very stable. In addition, if the eccentricity signal could be found in the geological data directly, as well as through its modulation of the climatic precession amplitude, it might be possible to evaluate phase lags between the astronomical forcing and the geological record. The second significant use of amplitude modulation cycles is that they are related to specific dynamical properties of a given astronomical model. These properties are related to the chaotic nature of the Solar System, and potentially allow the use of geological data to provide constraints on the dynamical evolution of the Solar System and astronomical models.

### Tidal Dissipation and Dynamical Ellipticity

The mean fundamental orbital frequencies $g_i$ and $s_i$, as well as the precession constant $p$, are likely to have changed throughout geological time. However, whereas the changes in $g_i$ and $s_i$ have probably been small, the precession constant $p$ is likely to have changed significantly. Changes are caused by the effects of energy (tidal) dissipation as well as by redistribution of mass (due, e.g., to waxing and waning ice-caps and mantle convection). Changes in $p$ are related to changes of the length of day, which is reflected in geological and palaeontological records. In particular, Earth's tidal response to the gravitational pull from the Sun and the Moon is not instantaneous. This means that the tidal bulge that develops on Earth (and on the Moon), is not aligned with the direction of the Moon's gravitational pull. This pull exerts a torque on Earth, which leads to a gradual decrease in its rotational velocity. In addition, conservation of angular momentum leads to an increase in the distance from Earth to the Moon over time, and a change in Earth's rotational velocity leads to a redistribution of mass on Earth ('dynamical ellipticity'). The dynamical ellipticity of Earth can also be affected by mantle convection and ice loading.

These processes modify Earth's precession constant $p$, the frequency of which has decreased over geological time. Because $p$ is contained in the expressions for obliquity and climatic precession (see **Tables 3** and **4**), the periods of obliquity and climatic precession also change. In 1994, Berger and Loutre estimated possible values for changes of astronomical periods, based on astronomical and geological observations. Their results are illustrated in **Figure 9**. The effects of changing tidal dissipation and dynamical ellipticity values have a large impact on astronomical calculations, and have to be obtained from observation. Strictly speaking, astronomical calculations cannot be performed independently of the chosen Earth model, and particularly, there cannot be separate treatment of the Earth–Moon system. This is why numerical computations are invaluable.

## Chaos in the Solar System

Probably the most significant development of astronomical theory in recent times has been the discovery of the chaotic behaviour of the Solar System by Jacques Laskar in 1990. Laskar established that the dynamics of the orbital elements in the Solar System are not fixed for all times, but rather are unpredictable over tens of millions of years. This is due to the non-linear gravitational interaction of the different bodies in the Solar System, which makes it theoretically impossible to calculate the exact movements of celestial bodies from their present-day masses, velocities, and positions over long periods of time. This feature poses limits on the use of astronomical theory for the purposes of creating astronomically calibrated

418 **EARTH/Orbital Variation (Including Milankovitch Cycles)**

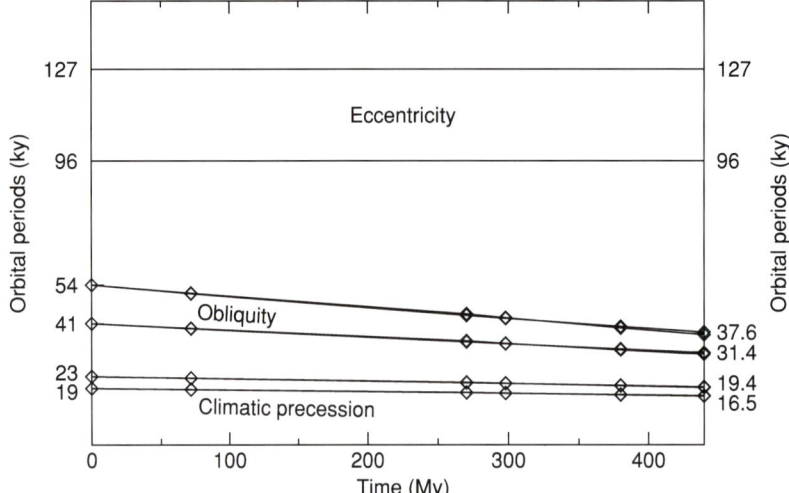

**Figure 9** Estimated changes in the obliquity and climatic precession periods related to a changing distance between Earth and the Moon over the past 440 My, according to calculations by Berger and Loutre in 1994. Note that the periods of obliquity and climatic precession change due to a change of the precession constant $p$, whereas the eccentricity periods remain unaffected. Note that this diagram is for illustrative purposes only, and the exact variation of orbital frequencies over tens of millions of years is not yet known in detail.

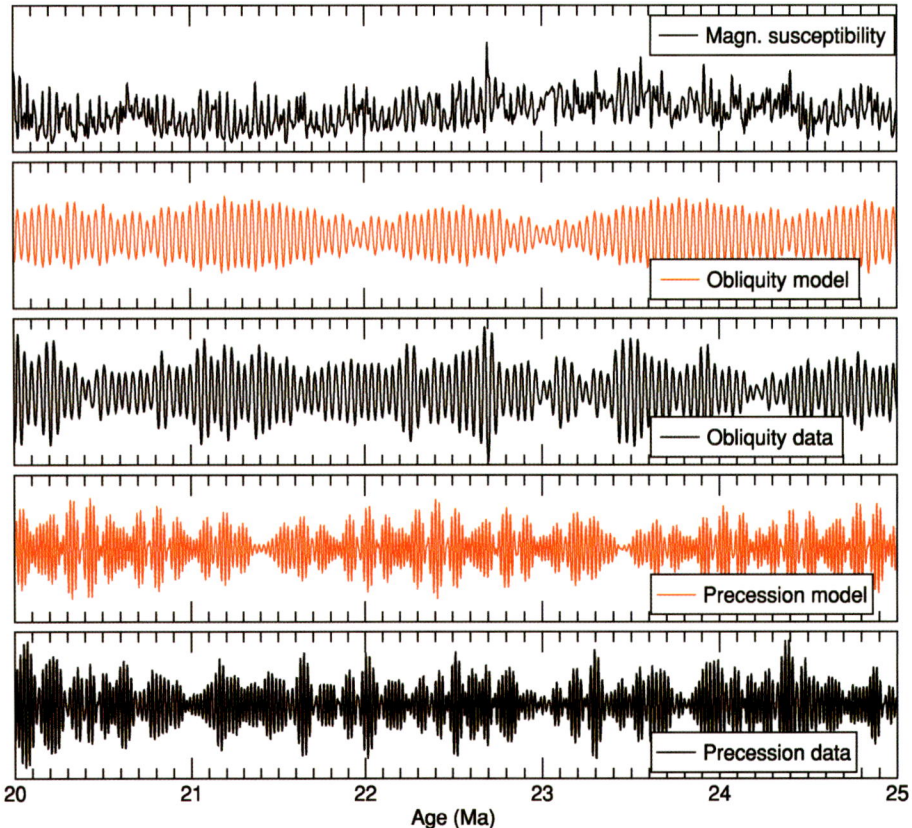

**Figure 10** Lithological magnetic susceptibility data from Leg 154 of the Ocean Drilling Program. These data, analysed by Shackleton and colleagues in 1999, demonstrate the fidelity with which orbital variations can be encoded in the geological record. The magnetic susceptibility parameter, plotted together with filters that extract the visible cycles, chiefly reflects variations between terrestrial (clay-rich) and marine (carbonate-rich) sediments, being forced by a dominant obliquity signal.

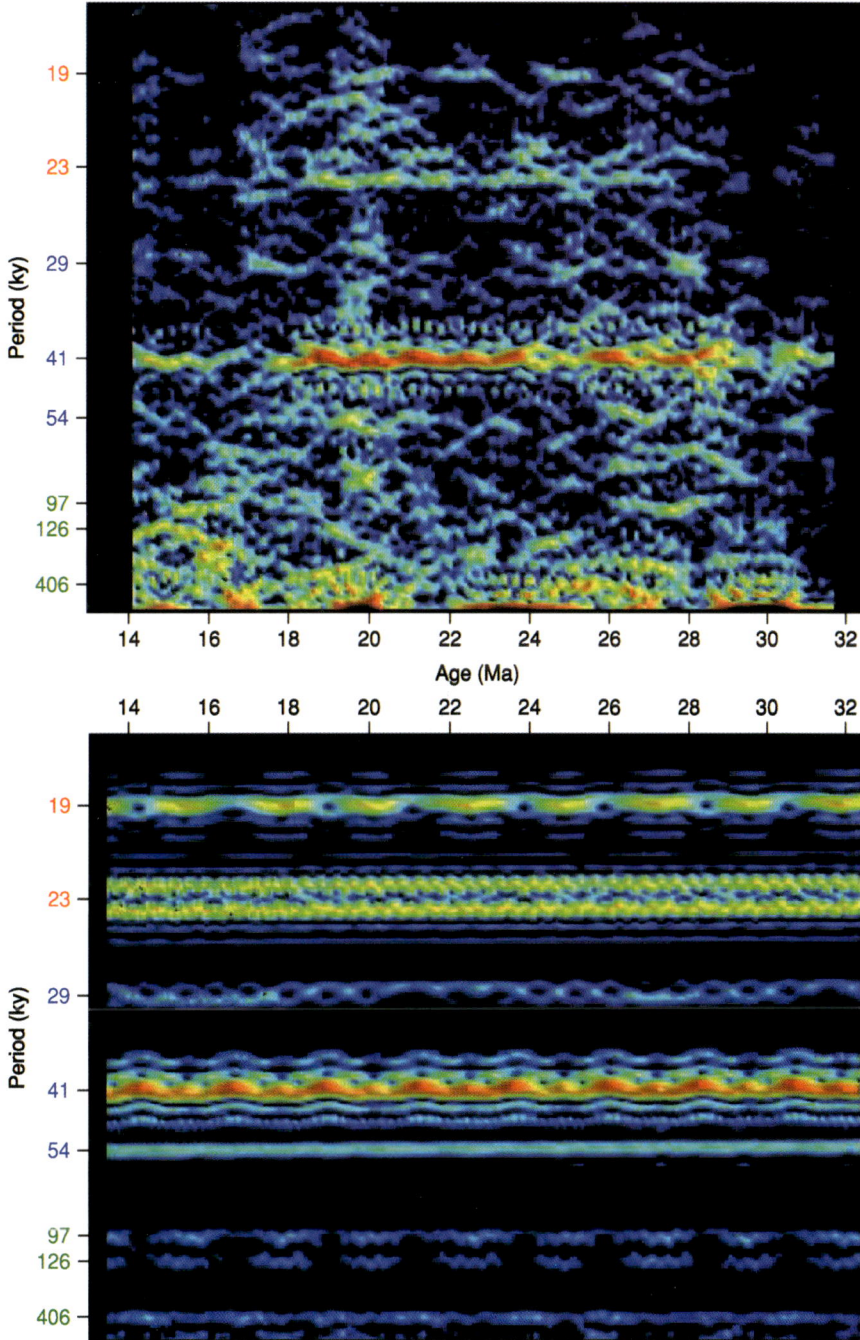

**Figure 11** (Top) Evolutionary spectrum (frequency vs age, amplitude colour coded) of Ocean Drilling Program Leg 154 lithological data (combined magnetic susceptibility and colour reflectance) (cf. **Figure 10**). These data, analysed by Shackleton and colleagues in 1999, and contrasted to astronomical calculations of Earth's orbital variations, demonstrate that the geological data contain an imprint of long-term amplitude modulations of astronomical forcing components. (Bottom) The astronomical solution provided by J Laskar.

geological time-scales. Laskar found that the calculations of orbits diverge exponentially with time for a given set of initial conditions. This implies, for example, that if the present position of a planet is known with a relative error of $\sim 10^{-10}$, this error increases to $\sim 10^{-9}$ after 10 My, and reaches the order of 1 after 100 My. Despite these limitations, the determination of astronomical parameters, such as the planetary masses, velocities, and positions, is improving due to additional satellite measurements.

Together with improved and more detailed astronomical models, the limits of accuracy of astronomical calculations are also likely to extend further back in time.

A clearer representation of chaos in the Solar System is provided by amplitude modulation terms. The ~2.4- and 1.2-My-long amplitude modulation terms that occur in the calculations of eccentricity and obliquity, respectively, are in resonance, and the expression $(s_4 - s_3) - 2(g_4 - g_3) = 0$ can evolve into a new state, where $(s_4 - s_3) - (g_4 - g_3) = 0$. This implies a change from a 1:2 resonance to a 1:1 resonance. Laskar found this behaviour to be the main source and representation of chaos in the Solar System. As shown in **Table 5**, these terms are present in several astronomical frequencies, and should be possible to detect in the geological record.

## Earth's Orbital Variation Encoded in Geological Data

The imprint of Earth's orbital variation in geological records, first statistically demonstrated in 1976 for the recent geological past, in the seminal paper by Hays and colleagues, has now been found throughout parts of the Cenozoic and beyond, through variations of stable isotope measurements that act as a proxy for Earth's climate system, as well as in a large number of lithological parameters. A review of this body of data is beyond the scope of the present discussion, but it is illustrative to show at least one example of very high-quality data that demonstrate the imprint of Earth's orbital variations in the rock record. **Figure 10** shows part of a record used to correlate geological data for the past ~30 My with astronomical calculations. The record shows exceptionally well-encoded obliquity and climatic precession cycles. Most importantly, this record also demonstrates the consistent variation in amplitude of the obliquity cycles. This can be illustrated with the help of evolutionary spectral analysis, whereby the relative amplitude at individual frequencies is evaluated at different times. This is shown in **Figure 11**.

Linking Earth's orbital variations (Milankovitch cycles) with geological records has caused controversies as to whether the theory that orbital variations driving Earth's climate conditions can be correct. The controversies stem from observations of palaeoclimatic proxies from the recent past that reveal an imprint that does not correspond to the expected strength of orbital variations in insolation calculations, and instead suggest a nonlinear response of the climate system at different orbital frequencies. In particular, during the past 800 ky, the imprint of eccentricity in stable isotope records has been much stronger than expected. It is now becoming clear that geological data show a wide variety of responses to individual orbital frequencies, depending on factors such as palaeolatitude, the prevailing oceanographic system at the study site, global ice volume, etc., with records showing much more variation than would be expected from a simple insolation calculation. A better understanding of the interaction between Earth's orbital variations and their imprint on the climate system and geological records is likely going to be gained from integrated Earth system modelling studies, making use of the growing body of observations that has been provided recently by ocean drilling. As a final note, orbital variations also affect the other planets of the Solar System, and recent attempts have been made to link these to climatic variations on Venus and Mars.

## See Also

**Analytical Methods:** Geochronological Techniques. **Carbon Cycle. Earth Structure and Origins. Famous Geologists:** Agassiz. **Gaia. Magnetostratigraphy. Microfossils:** Foraminifera. **Palaeoclimates. Solar System:** The Sun; Asteroids, Comets and Space Dust; Meteorites; Mercury; Venus; Moon; Mars; Jupiter, Saturn and Their Moons; Neptune, Pluto and Uranus. **Tektites. Tertiary To Present:** Pleistocene and The Ice Age. **Time Scale**.

## Further Reading

Berger A, Imbrie J, Hays J, Kukla G, and Saltzman B. (eds.) (1984) *Milankovitch and Climate: Understanding the Response to Astronomical Forcing*. Dordrecht and Boston: D. Reidel Publishing Company.

Berger A and Loutre MF (1994) Astronomical forcing through geological time. In: de Boer PL and Smith DG (eds.) *Orbital Forcing and Cyclic Sequences (IAS Special Publication)*, vol. 19, pp. 15–24. Oxford: Blackwell Scientific.

Berger A, Loutre MF, and Tricot C (1993) Insolation and Earth's orbital periods. *Journal of Geophysical Research* 98(D6): 10341–10362.

de Boer PL and Smith DG (eds.) (1994) *Orbital Forcing and Cyclic Sequences (IAS Special Publication)*, vol. 19. Oxford: Blackwell Scientific Publications.

Croll J (1875) *Climate and Time in their Geological Relations: A Theory of Secular Changes of the Earth's Climate*. London: Daldy, Tsbister and Co.

Einsele G, Ricken W, and Seilacher A (eds.) (1991) *Cycles and Events in Stratigraphy*. Berlin: Springer Verlag.

Gilbert GK (1895) Sedimentary measurement of Cretaceous time. *Journal of Geology* III: 121–127.

Hays JD, Imbrie J, and Shackleton NJ (1976) Variations in the Earth's orbit: pacemaker of the Ice Ages. *Science* 194(4270): 1121–1131.

Hinnov LA (2000) New perspectives on orbitally forced stratigraphy. *Annual Review of Earth and Planetary Science* 28: 419–475.

House MR and Gale AS (eds.) (1995) *Orbital Forcing Timescales and Cyclostratigraphy, Geological Society Special Publication*, vol. 85. London: The Geological Society.

Imbrie J and Palmer-Imbrie K (1979) *Ice Ages; Solving the Mystery*. Short Hills, NJ/Cambridge, MA: Enslow Publishers/Harvard University Press.

Lambeck K (1980) *The Earth's Variable Rotation: Geophysical Causes and Consequences*. Cambridge: Cambridge University Press.

Laskar J (1990) The chaotic motion of the solar system – a numerical estimate of the size of the chaotic zones. *Icarus* 88(2): 266-291.

Laskar J (1999) The limits of Earth orbital calculations for geological time-scale use. *Philosophical Transactions of the Royal Society of London, Series A, Mathematical, Physical and Engineering Sciences* 357(1757): 1735–1759.

Milankovitch M (1941) *Kanon der Erdbestrahlung und seine Anwendung auf das Eiszeitenproblem, Special Publication 132*, vol. 33. Belgrade: Royal Serbian Sciences. ('Canon of Insolation and Ice Age Problem', English translation by Israël Program for Scientific Translation and published for the US Department of Commerce and the National Science Foundation, Washington, DC, 1969).

Schwarzacher W (1993) *Cyclostratigraphy and the Milankovitch Theory*. Amsterdam: Elsevier.

Shackleton NJ, Crowhurst SJ, Weedon GP, and Laskar J (1999) Astronomical calibration of Oligocene–Miocene time. *Philosophical Transactions of the Royal Society of London, Series A, Mathematical Physical and Engineering Sciences* 357(1757): 1907–1929.

Shackleton NJ, McCave IN, and Weedon GP (eds.) (1999) Astronomical (Milankovitch) calibration of the geological timescale (9–10 December 1998, London) *Philosophical Transactions of the Royal Society of London, Series A, Mathematical Physical and Engineering Sciences* 357(1757).

Short DA, Mengel JG, Crowley TJ, Hyde WT, and North GR (1991) Filtering of Milankovitch cycles by Earth's geography. *Quaternary Research* 35(2): 157–173.

# EARTH STRUCTURE AND ORIGINS

**G J H McCall**, Cirencester, Gloucester, UK

© 2005, Elsevier Ltd. All Rights Reserved.

## Introduction

The Earth is a planet of the inner zone of the solar system, which is occupied by four 'terrestrial' planets – Mercury, Venus, Earth, and Mars from the Sun outwards. These planets are composed mainly of rock and metal, whereas the outer planets – the so called 'gas-giants' and 'ice-giants' – have immense gaseous envelopes. The four differ greatly in character. Mercury has a dead cratered surface closely resembling that of the Moon, though there is evidence that it is compositionally very different. Venus is shrouded by a dense greenhouse (carbon dioxide-rich) atmosphere that is not visually penetrable, but radar images reveal a largely volcanically moulded surface. Mars, like Venus and Earth, has a very complex surface, moulded by volcanism, impacts, and wind action, and has polar ice caps. The Earth is unique in possessing oceans that cover the majority of the surface and form part of the hydrosphere, which includes surface and underground water bodies. The Earth is also unique in possessing an oxygen-containing atmosphere dominated by nitrogen; Mars has a thin carbon dioxide-rich atmosphere with negligible oxygen, and Venus has a thick carbon dioxide-rich atmosphere. Mercury and the Moon have negligible atmospheres dominated by inert gases. The unique character of the Earth is well seen from Space (**Figure 1**): the oceans appear blue in reflected sunlight, and the single satellite is uniquely

**Figure 1** The Earth and Moon as seen from Space (NASA image kindly supplied by Sir Patrick Moore).

large compared with the mother planet, and uniquely orbits it at the same rate as it rotates so that the same lunar face is always Earthwards. Of the other inner planets, only Mars has moons (two of them), and they may be captured asteroids. The origin of the Earth's Moon remains highly controversial despite vigorous championing of various models.

## The Unique Biosphere

As far as we know, the Earth is unique in having a biosphere on and below its surface, in its waters, and in the air above it – a quite thin zone compared with the planetary interior. Minute living organisms may be carried naturally high in the atmosphere, and the exact upper boundary of the biosphere is not closely defined. In both the oceanic and continental realms, bacteria can be found deep beneath the surface of the rocks: 92% of the Earth's bacteria occupy the deep biosphere, and they can live at depths of up to 4 km and at temperatures of up to 90°C.

## Properties

The essential properties of the Earth are given in **Table 1**, where they are compared with those of the other inner planets and the Moon (*see* **Solar System: Moon**).

The Earth is not perfectly spherical but is an oblate spheroid; this was recognized by Newton to explain the precession of the equinoxes, the shift of the equinoctial point eastwards each year. The figure of 1/298.25 is widely accepted for the degree of flattening of the polar diameter compared with the equatorial diameter. The Earth is subject to various secular changes that affect the length of the day; these include secular orbital changes and rotational changes, amounting to several parts per $10^8$ over time-scales ranging from days to millennia. Besides the rate of rotation, there are secular changes related to the celestial frame. Polar motion is a combination of the effect of the Chandler Wobble (departure from pure spin amounting to an axial deviation of $4''$ of arc with a 428 day cycle) and a contributing effect of variation of the distribution of mass within the Earth and in the oceans and atmosphere. The precession of the equinoxes mentioned above is due to lunar and solar gravity acting on an oblate spheroid with an axis of rotation inclined at $23.5°$ to the perpendicular. There is also planetary precession, which reduces the obliquity of the axis to the ecliptic (the mean plane of the orbit around the sun) by $0.5''$ per year. The combined precession has a period of 25870 years. The Milankovich model of climate change is generally accepted to be related to orbital eccentricity, obliquity of the ecliptic, and the precession of the equinoxes (*see* **Earth: Orbital Variation (Including Milankovitch Cycles)**), and this has a periodicities of 90 000–110 000, approximately 42 000 and approximately 26 000 years. This model is believed to control glaciations in the geological record. The world's temperate zone is advancing into the tropics, and this is the Milankovich cycle in operation before our eyes. All these secular changes complicate extremely accurate calculations of time and geodetic surveying, and there is the further complication of magnetism, in particular the Earth's magnetic field, which is at present subject to increasing variation (see below).

## Earth Tides

The gravitational attraction within the Earth–Moon system keeps them in orbit about the common centre of mass, a point about 4670 km from the centre of the Earth. The pull on the side of the Earth facing the Moon at any time is strongest, and the result is a tendency to pull the Earth into a prolate ellipsoid aligned with the Earth–Moon axis (**Figure 2**). The

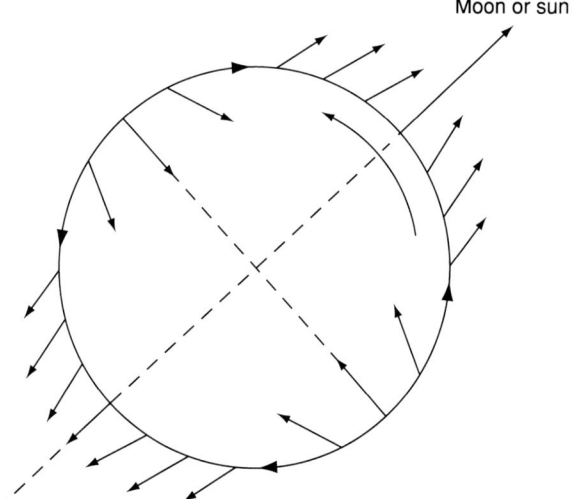

**Figure 2** The pattern of tidal forces at the surface of the Earth resulting from the gravitational pull of the Moon and Sun (reproduced with permission from F D Stacey (2000) Earth Tides. In: Hancock PL and Skinner BJ (eds.) *Oxford Companion to the Earth*, pp. 280–281. Oxford: Oxford University Press.).

**Table 1** The essential properties of the Earth, the other inner planets, and the Moon

|  | Radius (km) | Mass ($10^{20}$ kg) | Density ($10^3$ kg m$^{-3}$) |
| --- | --- | --- | --- |
| Mercury | 2439 | 3300 | 5.4 |
| Venus | 6051 | 48700 | 5.3 |
| Earth | 6371 | 59970 | 5.517 |
| Moon | 1738 | 734.9 | 3.34 |
| Mars | 3394 | 6420 | 3.9 |

variation in gravitational pull imposed by the rotation of the Earth causes Earth tides. There is also a smaller contribution to Earth tides by the Sun.

## Internal Configuration

What we know of the internal configuration of the Earth comes from the following sources:

- study of surface rocks and rocks brought up by eruptive processes (kimberlite diatremes are particularly important with respect to the mantle as they sample regions within it that would otherwise be unknown to us directly. (*see* **Volcanoes; Igneous Rocks: Kimberlite**);
- analogies with meteorites (in particular between nickel–iron meteorites and the core, and between carbonaceous chondrites and the mantle);
- instrumental measurements (geophysical), especially seismology (**Figure 3**); and
- geochemical research (relating to the Earth, other planetary bodies, and meteorites).

As a result of these approaches it is accepted that the Earth is made up of a series of concentric shells, each with its own physical and chemical properties (**Figure 4**). The outermost is the crust, beneath it is the mantle (divided into an upper mantle and lower mantle), and at the centre is the core (divided into the inner core and outer core). The plate-tectonics paradigm also accepts the existence of a lithosphere, which is not synonymous with the crust, but is a solid and rigid zone extending into the upper part of the mantle and bounded below by a transition zone. This separates the strong elastic layer from the weak and ductile asthenosphere, which is solid but very slowly convecting. The crust is bounded below by a discontinuity at which the velocity of seismic P waves increases abruptly; this is the Mohorovicic discontinuity and forms the lower boundary of the zone of isostatic adjustment (*see* **Earth: Mantle; Crust**).

The properties of the various concentric zones of the Earth are summarized in **Table 2**.

## Magnetic Field

The Earth's magnetic field is environmentally critical because it forms a protective shield that blocks out solar radiation and harmful incoming particles. It has the character of a dipole, but has, throughout geological time, been subject to reversals, with the poles

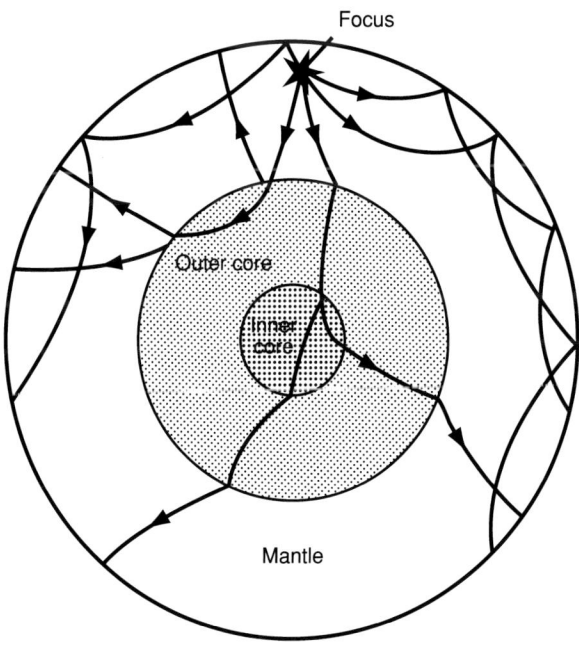

**Figure 3** Earthquake energy travels down and out from the focus as shown, but at each internal boundary part of it is reflected or refracted and returns to the surface. Times of travel of the energy waves and their behaviour have given us much of the information that we have about the depths of the boundaries and the nature of the material forming the concentric shells of the Earth (reproduced from Van Amdel TJ (1994) (eds.) *New Views on an Old Planet*, 2nd edn. Cambridge: Cambridge University Press, with permission from the author and publisher).

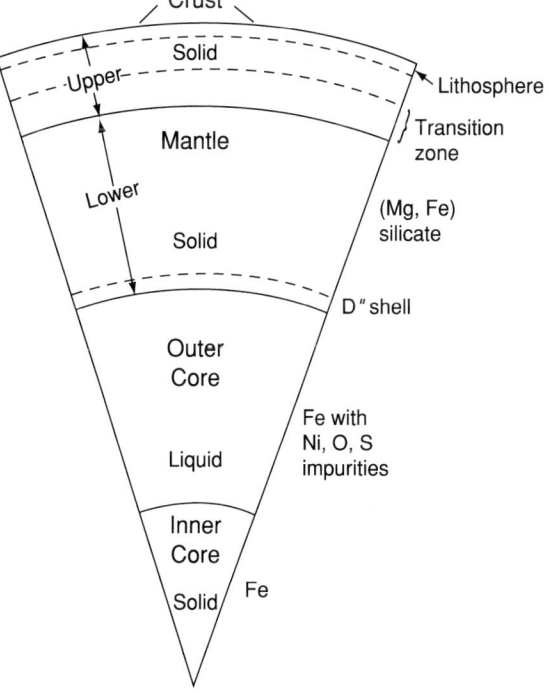

**Figure 4** The configuration of the Earth's internal zones as deduced from the four indirect sources listed in the text (reproduced with permission from C M R Fowler (2000) Mantle convection, plumes, viscosity and dynamics. In: Hancock PL and Skinner BJ (eds.) *Oxford Companion to the Earth*, pp. 649–652. Oxford: Oxford University Press).

**Table 2** Volumes, masses, and densities of the various concentric shells of the Earth (from C M R Fowler (2000) Earth Structure. In: Hancock PL and Skinner BJ (eds.) *Oxford Companion to the Earth*, pp. 276–280. Oxford: Oxford University Press)

| | Depth (km) | Volume ($10^{18} m^3$) | Volume (% of total) | Mass ($10^{21}$ kg) | Mass (% of total) | Density ($10^3$ kg m$^{-3}$) |
| --- | --- | --- | --- | --- | --- | --- |
| Crust | 0–Moho* | 10 | 0.9 | 28 | 0.5 | 2.60–2.90 |
| Upper mantle | Moho–670 | 297 | 27.4 | 1064 | 17.8 | 3.35–3.99 |
| Lower mantle | 670–2891 | 600 | 55.4 | 2940 | 49.2 | 4.38–5.56 |
| Outer core | 2891–5150 | 169 | 15.6 | 1841 | 30.8 | 9.90–12.16 |
| Inner core | 5150–6371 | 8 | 0.7 | 102 | 1.7 | 12.76–13.8 |
| Whole Earth | | 1083 | 100 | 5975 | 100 | |

*The Moho is 25–35 km deep under the continents and 6–7 km deep under the oceans.

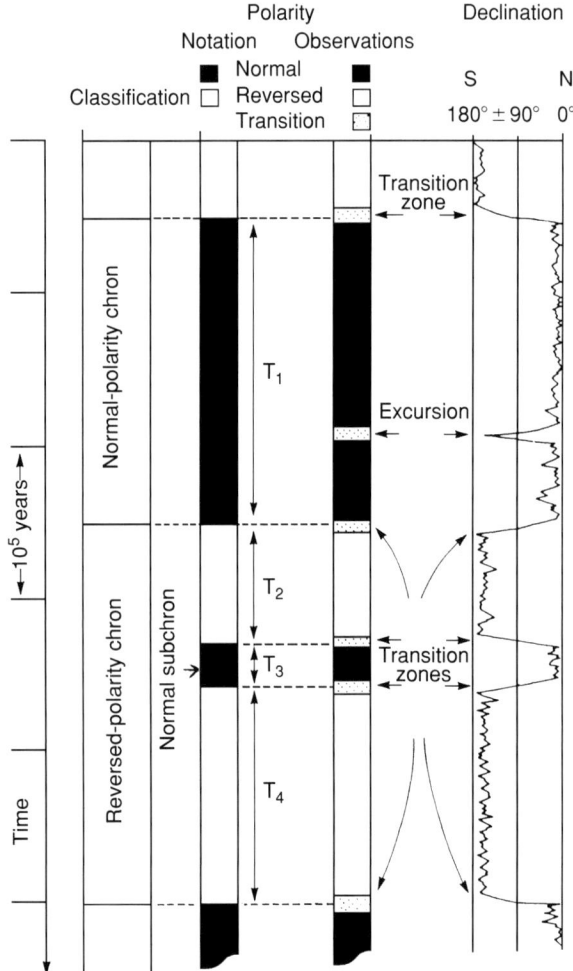

**Figure 5** Magnetic-polarity divisions of geological time (chrons, subchrons, transitional zones, and excursions) (reproduced from Harland WB, Armstrong RL, Cox AV, et al. (1990) *A Geologic Timescale 1989*. Cambridge: Cambridge University Press).

switching positions (**Figure 5**). Such reversals may occur over periods of a few hundred thousand years, but there have been longer periods without reversal of around 10 Ma. The intensity of the dipole field varies, being particularly strong during long stable periods and weakening just before reversal. The position of the north magnetic pole, first located by James Clark Ross in 1831 in the Boothia Peninsula, north-east Canada, shows secular displacement and is at present accelerating northwards towards Siberia (**Figures 6A and 6B**). The magnetic field is believed to be due to convection in the liquid outer core (**Figure 7**) (*see* **Magnetostratigraphy**).

## Ozone Layer

There is a layer of ozone ($O_3$), a blue-green poisonous gas, in the stratosphere at an altitude of between 15 and 40 km (**Figure 8**). This is important because it absorbs the carcinogenic part of the solar spectrum (ultraviolet B). It is produced by photochemical reactions, and its concentration displays natural variations, being more abundant at the poles and less so in equatorial regions, though it is created at the equator and destroyed at the poles, where the concentration shows seasonal variation. Chlorofluorocarbon compounds, used as refrigerants and aerosol sprays, are believed to threaten the ozone layer; stratospheric jet planes would also do so, but relatively few are in use.

## Plate Tectonic Movement and Mantle Convection

Space exploration has revealed no other body (planet, satellite, or asteroid) with evidence of lateral plate-tectonic displacement attributable to mantle convection, with accompanying ridge eruptivity, spreading away from ridges, subduction at plate margins, and continental collision (**Figure 9**). The upper mantle, though solid, is capable of very slow thermal convection over the vast periods of geological time. Mantle convection and plate tectonics probably did not operate in the Archaean and earliest Proterozoic (before 2000 Ma), though some minor shuffling of numerous small plates may have occurred. There is evidence that it operated through most of the Proterozoic and Phanerozoic, and the Earth's surface at present consists of relatively few

EARTH STRUCTURE AND ORIGINS 425

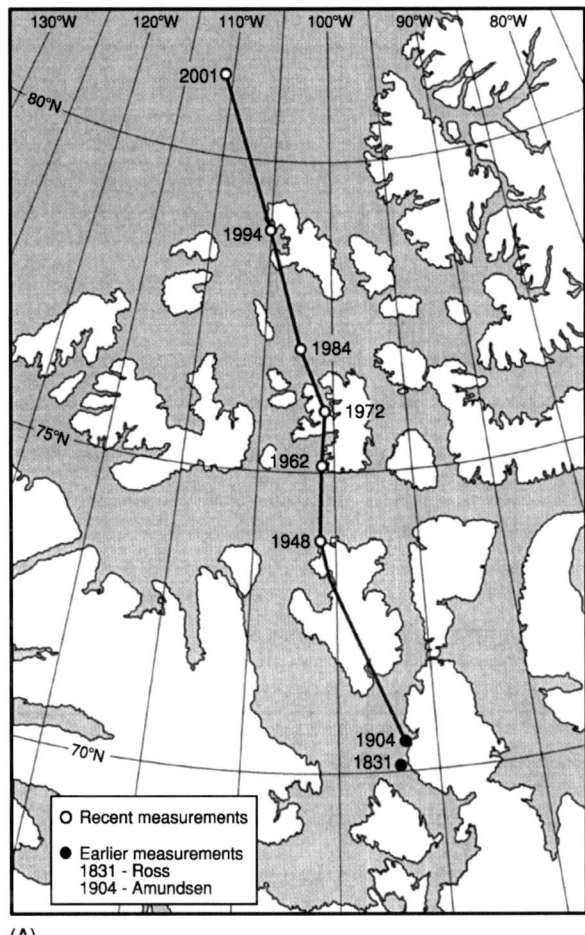

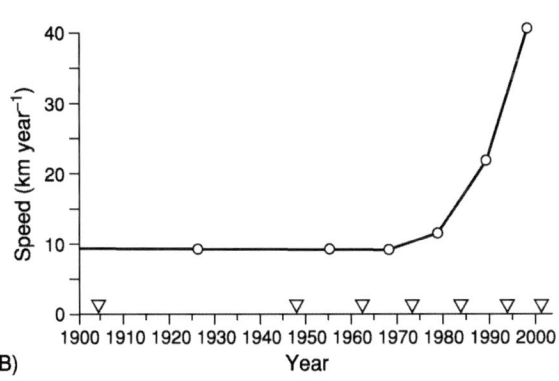

**Figure 6** (A) The movement of the north magnetic pole since James Clark Ross located it in 1831. (B) The accelerating movement of the pole at the present time, which could herald a reversal (reproduced from McCall GJH (2003) Pole up the pole. *Geoscientist* 13: 9, with the permission of the Geological Society Publishing House).

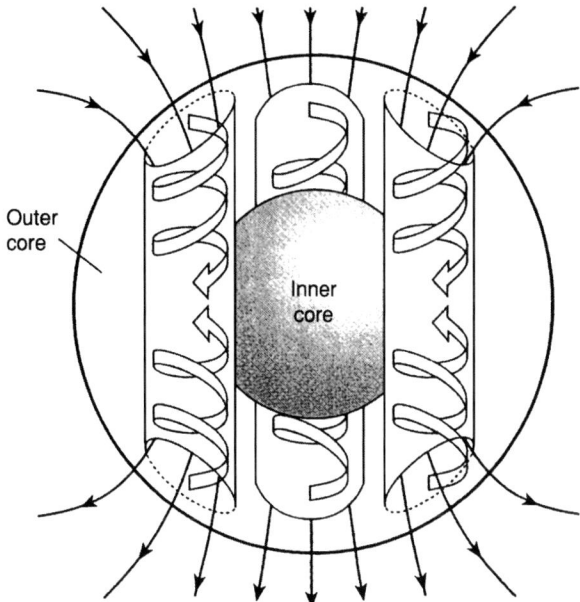

**Figure 7** The magnetic field of the Earth is due to convection in the outer core, for which the configuration is not determined (reproduced from D Ravat (2000) Magnetic field, origin of the Earth's internal field. In: Hancock PL and Skinner BJ (eds.) *Oxford Companion to the Earth*, pp. 630–631. Oxford: Oxford University Press).

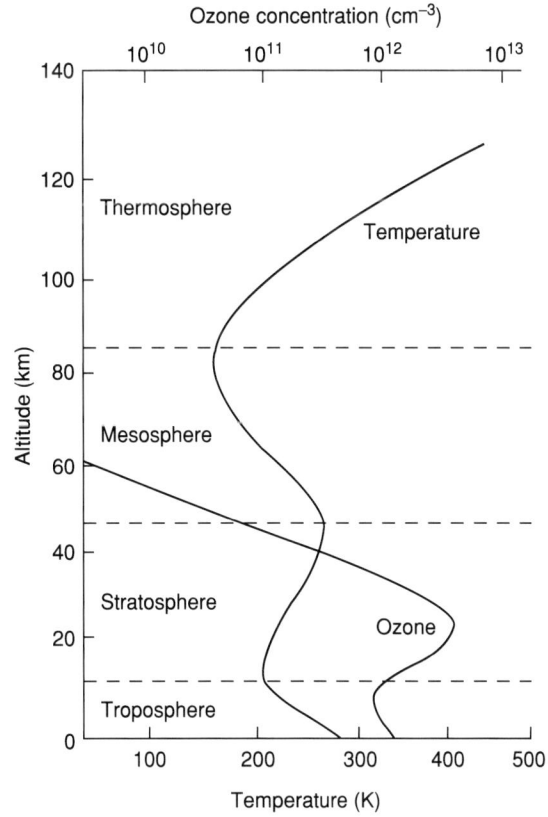

**Figure 8** The configuration of the ozone layer (reproduced from F Drake (2000) Ozone-layer chemistry. In: Hancock PL and Skinner BJ *Oxford Companion to the Earth*, pp. 772–773. Oxford: Oxford University Press).

plates, some of them immense (for example the one includes India, the Indian Ocean, Australia, and the Tasman Sea) (**Figure 10**). That plate movement has occurred and is occurring cannot be seriously doubted, though attempts at instrumental measurements have

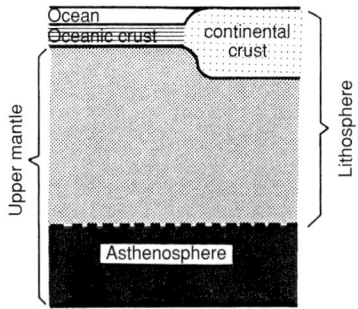
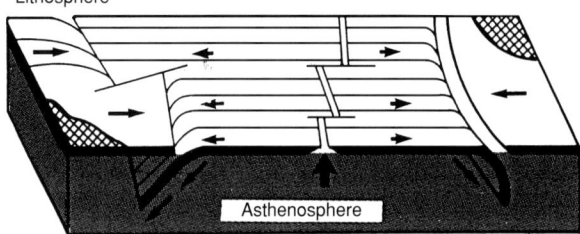

**Figure 9** The widely accepted configuration of plate movement away from a mid-ocean ridge and down subduction zones beneath continents (reproduced from Van Andel TJ (1994) *New Views on an Old Planet*, 2nd edn. Cambridge: Cambridge University Press, with permission from the author and publisher).

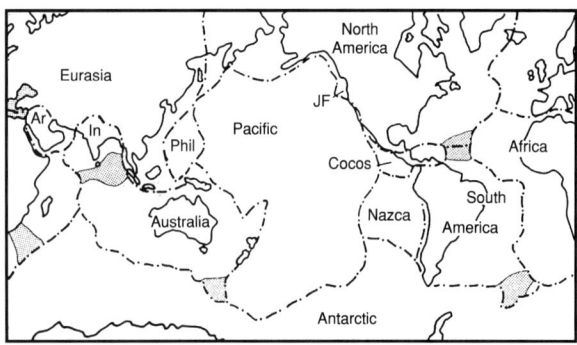

**Figure 10** The tectonic plates of the Earth, Ar = Arabia; Phil = Philippines (reproduced from Van Andel TJ (1994) *New Views on an Old Planet*, 2nd edn. Cambridge: Cambridge University Press, with permission from the author and publisher).

yielded conflicting results and suggested differences in movements at margins and within plates. There are conflicting opinions as to whether convection operates down into the lower mantle (the more orthodox view) or only above the 670 km boundary, which forms a barrier. The latter view ('closed upper-mantle circulation') is held by D L Anderson and W B Hamilton and considers the reality or otherwise of deep plumes, rising from the base of the mantle. Plumes were originally invoked to explain chains of seamounts and/or islands such as the Emperor Seamount and Hawaiian Islands in the Pacific Ocean, where the ages of the individual islands become progressively younger towards the south-east. Oceanic lithosphere is hypothesized to move over a static hotspot. Sceptics believe that the 670 km barrier cannot be penetrated by a rising plume (*see* **Mantle Plumes and Hot Spots**). There is also no agreement on the mechanism of plate movement. Ridge push, subducting-slab pull, and a combination of both have all been favoured, as has a mechanism of gravity gliding. The plate movement must start with spreading away from the ridge before subduction begins to operate (no slab yet to pull), so slab pull can never provide the entire answer. It seems fair to say that, whereas the plate-tectonics paradigm must reasonably be accepted as fact, the mechanism remains unexplained, and mathematical calculations do not really explain how the forces needed to move the plates are derived (*see* **Plate Tectonics**).

## Geochronological Comparisons – Earth and Other Solar System Bodies

It is early days yet in Space exploration, so, whereas we can derive a very detailed and accurate chronology for our own planet, based on the radioactive decay of isotopes, palaeontology, palaeomagnetism etc., we have very little chronological information about other bodies. There is no palaeontological data for other bodies (there may have been no life there). We have isotope-based data for the Moon, but there is really no geochronological data from there for the last 3000 Ma or so. Palaeomagnetism cannot be applied to bodies other than the Earth. There is a set of dates for the formation of the rocks comprising the SNC meteorites (shergottites, Nakhlites and chassignites), which are supposed to have come from Mars, and these have contributed to a Martian geochronology (which may yet prove to be wildly incorrect). Isotopic dates of around 4500 Ma for the formation of the asteroids have been derived from chondritic meteorites, and slightly younger dates, recording a slightly later melting process in the asteroidal parent bodies, have been derived for achondritic meteorites (*see* **Solar System: Meteorites**). For Venus and Mercury we have no yardsticks at all, although surmises have been made from the radar-derived images of Venus. Mercury resembles the Moon in its surface cratering and appears, like the Moon, to be a 'dead' multicratered body with no or negligible eruptivity since the deduced bombardment that formed the craters (*see* **Solar System: Mercury; Venus**). Sequential relationships without any benchmarks have been derived from the superposition orders of crater populations and volcanic structures. The comparison diagram (**Figure 11**) utilizes the very limited evidence from outside the Earth to make this comparison.

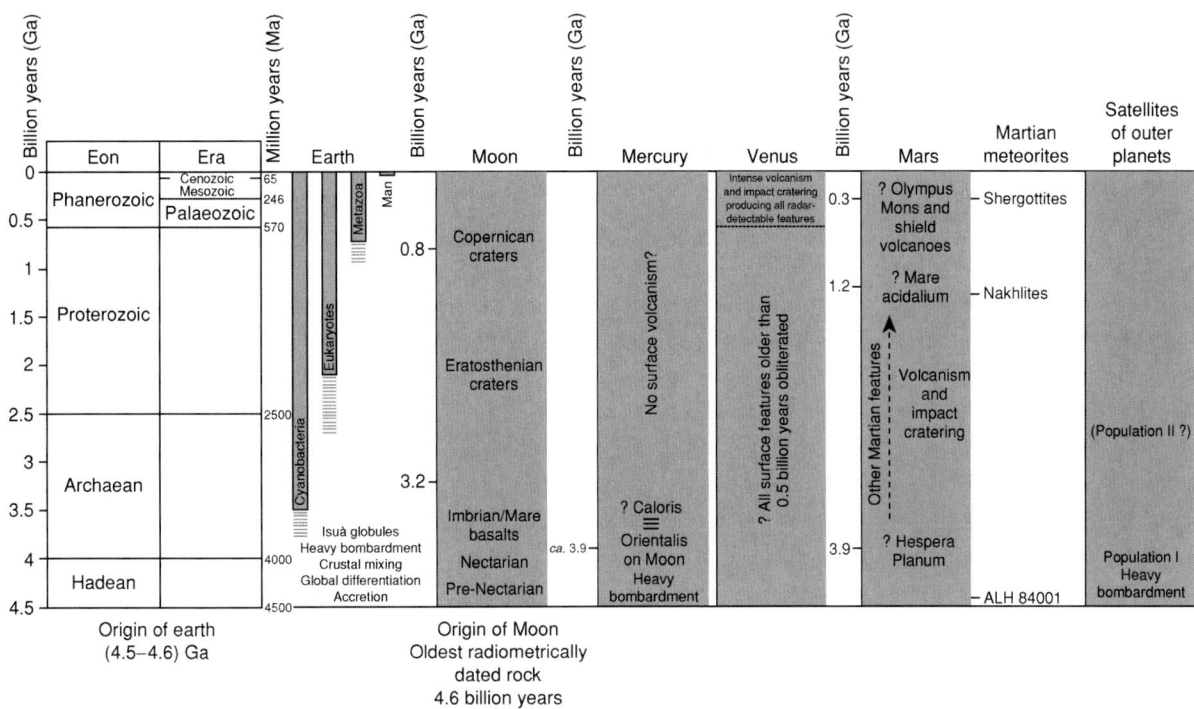

**Figure 11** A comparison between the eons and eras of the geological column and sequences of events that have been deduced for the other inner planets, the Moon and satellites of the outer planets on the basis of the very limited evidence available from Space and meteoritical research (modified from G J H McCall (2000) Age and early evolution of the Earth and Solar System. In: Hancock PL and Skinner BJ (eds.) *Oxford Companion to the Earth*, pp. 8–11. Oxford: Oxford University Press).

## The Origin of the Earth

The Earth was formed from the nebula that produced the Solar System. It is almost universally accepted that the Sun, the planets and their satellites, the asteroids, and the comets of the Oort 'cloud' grew from a cloud of gas and dust that contracted under its own gravity. The cloud of gas and dust had some degree of rotation (**Figure 12**) and, as the centre contracted, the angular momentum forced the remains into a flattened disk, in the plume perpendicular to the axis of rotation of the proto-Sun. This was apparently a very rapid process, perhaps taking place over 10 000 years. The dust particles accreted heterogeneously to produce lumps that formed planetesmals, the first small solid bodies, and cooling occurred.

We can estimate the composition of the original nebula from the solar abundances of elements, obtained spectrographically, and analysis of primitive meteorites (e.g. carbonaceous chondrites) (*see* **Solar System**: Meteorites); however, these are only estimates, and a second method is based on an assumption that may not be entirely accurate. Radioactivity-based dating methods (relying on the decay of radiogenic isotopes), as applied to terrestrial igneous rocks, allow us to date the condensation to 4500 Ma ago (the so-called 'age of the Earth'), and it is believed that condensation was rapid, taking about 100 000 years.

Accretion of the Earth and the other planets may have been slow and heterogeneous or rapid and homogeneous. Which of these two models is correct remains uncertain, but the first model, if correct, could have produced a layered mantle (*see* **Earth**: Mantle).

After condensation and accretion, the Earth evolved rapidly into a planet with most of its present properties. This could have taken no more than 500 Ma and was probably effected more rapidly: the oldest known minerals are zoned zircons from the Mount Narryer gneiss terrain, Western Australia, which give spot SHRIMP ages of 4400 Ma and are believed to have been formed by partial melting of pre-existing granitic crust (*see* **Minerals**: Zircons). This evidence indicates that there was solid rock at the Earth's surface astonishingly soon after condensation and accretion.

The oldest actual rock dated is a component of the Acasta Gneiss in the Slave Province of Canada, which has an age of 4000 Ma. Despite the existence of the Mount Narryer minerals, there is no other really significant evidence from geology of the Earth pertaining to the 500 Ma prior to the formation of the Acasta Gneiss, and this is called the 'Hadean Eon',

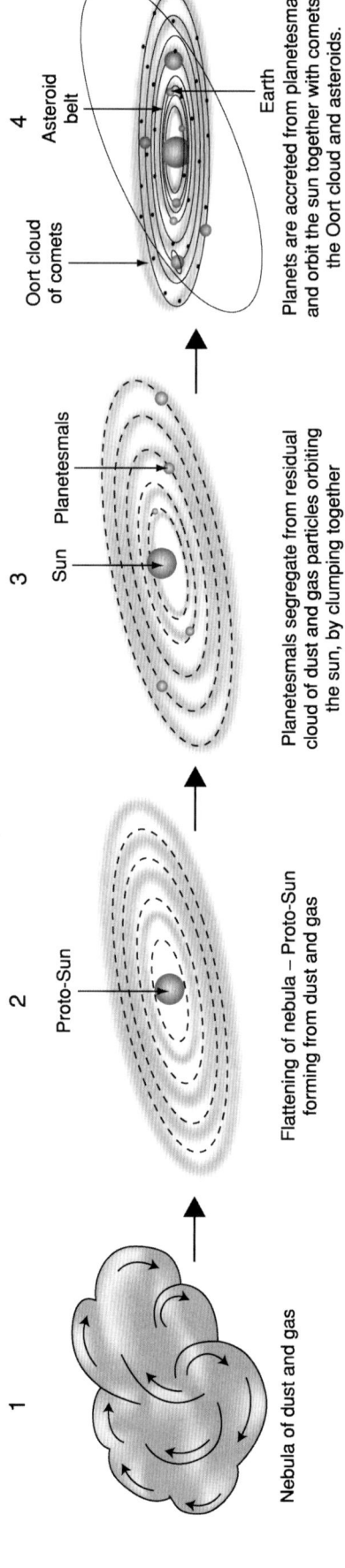

**Figure 12** The development of the Earth as a planet in the Solar System from the nebula.

geology's dark ages. We do know something of this period from the Moon – the lunar rocks brought back by Apollo and Lunik date from this period – and it is reasonable to invoke a bombardment of the Earth, like the Moon, during this period. However, there is no trace of this in the geological record of the Earth. The lack of evidence of ejecta from the Moon in Space means that it is difficult to attribute all the heavy early cratering of the Moon to bombardment by impacting objects from Space (see **Solar System: Meteorites**). Nevertheless, the early Earth must have swept up asteroids, comets, and any other debris in its path for a few million years after it attained its present size.

The Earth, after accretion, must have heated up internally and melted in order to separate, gravitationally, the core (part liquid and part solid), mantle, and crust. This heating would have been caused by the radioactive decay of elements and by gravitational energy – the latter as the dust, or dust and planetesmals, condensed and accreted into a ball. It has also been suggested that heat could have been produced by impacts on the surface, though much of this heat would have dissipated into Space. The heavy elements, mainly iron, would have gravitated to the core, and the lighter silicate material would have ended up in the mantle and crust. A primitive and highly mobile surface layer, perhaps a magma ocean, would have initiated the crust, and from this small-scale proto-continents separated. More 'way-out' models for the formation of the crust invoke early impacts, the subsidence of large impact-generated basins in the basaltic proto-crust, and partial melting of the basalt to form the proto-continents. The early proto-continents were certainly small, and the extent of the continents increased until around 3200 Ma, when the Kaapvaal (South Africa) and Pilbara (Western Australia) cratons were formed.

The very lightest volatile elements would have separated at the same time to form the proto-ocean and proto-atmosphere. This volatile component would have had a bulk composition of 10–20% water and contents of other volatiles in line with those of carbonaceous chondrite meteorites, according to the chondritic Earth model. The first atmosphere was almost certainly reducing rather than oxygenating. Our present oxygen-containing atmosphere was derived later, though how much later is debatable, and relates, in its origin, to the onset of biological activity. How and when life originated is still a mystery – there are models that suggest that it originated within the early developing Earth, but there are also models that suggest that it arrived on the planet from comets or primitive meteorites (see **Origin of Life**).

## See Also

**Earth:** Mantle; Crust; Orbital Variation (Including Milankovitch Cycles). **Igneous Rocks:** Kimberlite. **Magnetostratigraphy. Mantle Plumes and Hot Spots. Minerals:** Zircons. **Origin of Life. Plate Tectonics. Solar System:** Meteorites; Mercury; Moon; Venus. **Volcanoes.**

## Further Reading

Anderson DL (1999) A theory of the Earth: Hutton, Humpty Dumpty and Holmes. In: Craig GY and Hull JH (eds.) *James Hutton: Present and Future*, pp. 13–35. Special Publication 150. London: Geological Society.

Edwards K and Rosen B (2000) *From the Beginning*. London: Natural History Museum.

Hamilton WB (2002) The closed upper mantle circulation in plate tectonics. In: *Plate Boundary Zones*, pp. 359–410. Geodynamics Series 30. American Geophysical Union.

Hancock PL and Skinner BJ (2000) *Oxford Companion to the Earth*. Oxford: Oxford University Press.

Harland WB, Armstrong RL, Cox AV, et al. (1990) *A Geologic Timescale 1989*. Cambridge: Cambridge University Press.

Holmes A (1965) *Principles of Geology*, revised edn. London and Edinburgh: Nelson.

McCall GJH (2000) Age and early evolution of the Earth and Solar System. In: Hancock PL and Skinner BJ (eds.) *Oxford Companion to the Earth*, pp. 8–11. Oxford: Oxford University Press.

McCall GJH (2003) Pole up the pole. *Geoscientist* 13: 9.

Price NJ (2001) *Major Impacts and Plate Tectonics*. London and New York: Routledge.

Rothery DA (1992) *Satellites of the Outer Planets*. Oxford: Clarendon.

Van Andel TJ (1994) *New Views on an Old Planet*, 2nd edn. Cambridge: Cambridge University Press.

Vita-Finzi C (2002) *Monitoring the Earth*. Harpenden: Terra.

Wilde SA, Valley JW, Peck WH, and Graham CM (2001) Evidence for the early growth of continents and oceans from <4 Ga detrital zircons. *AGSO – Geoscience Australia Record* 2002/3: 6–8.

# EARTH SYSTEM SCIENCE

**R C Selley**, Imperial College London, London, UK

© 2005, Elsevier Ltd. All Rights Reserved.

## Introduction

### What on Earth is Earth System Science?

Earth system science is founded on the precept that the Earth is a dynamic system that is essentially closed materially, but open with respect to energy. This statement needs to be qualified by noting that the Earth continues to accrete matter from space in the form of meteorites, asteroids, and comets. Incoming energy is principally derived from solar radiation at a rate that may fluctuate with time. Earth processes in the lithosphere, the biosphere, and the atmosphere are linked, with a change in one impacting on one or more of the others. Earth system science has grown out of an appreciation of the need to integrate geology with other scientific disciplines, not only to understand the planet on which we live, but also most particularly to predict its future in general and climate change in particular. Earth system science requires a holistic approach to education in which students learn geoscience, bioscience, climatic science, astroscience, and space science synchronously and seamlessly.

### The Genesis of Earth System Science

In the fourteenth century Richard de Bury (Bishop of Durham 1333–45) divided all research into 'Geologia' (geology), the study of earthly things, and 'Theologia' (theology), the study of heavenly things. Masons, miners, and engineers were the first investigators of rocks. The Church taught them that the Earth was an inert mass of rock formed in 7 days. Subsequently natural philosophers pondered the meaning of fossiliferous strata and how it was that they were intruded by crystalline rocks and truncated by unconformities. Gradually it dawned on these natural philosophers that the rocks were not formed in an instant, but resulted from processes, many of which could be observed on the modern surface of the Earth. This was formulated in Hutton's principle of uniformitarianism, (*see* **Famous Geologists: Hutton**) and was epitomized in the dictum that 'the present is the key to the past.'

As Lyell (*see* **Famous Geologists: Lyell**) wrote in his *Principles of Geology* in 1834:

The entire mass of stratified deposits in the Earth's crust is at once the monument and measure of the denudation which has taken place.

By the beginning of the twentieth century it was realized that the history of the Earth could be interpreted in terms of cycles. Davies (1850–1934) recognized the landscape cycle, commencing with uplift, followed by youthful, mature, and senile landforms, followed by rejuvenation. Stratigraphers recognized cycles of weathering (*see* **Weathering**), erosion, transportation, deposition, and diagenesis. Sedimentologists discovered that all sedimentation is cyclic, although some is more cyclic than others. The hydrologic cycle was revealed, in which water fell as rain on land, flowed into rivers, was discharged into the world's oceans, evaporated and reprecipitated. Geochemists identified the cycles of carbon (*see* **Carbon Cycle**) and other key elements, such as nitrogen, oxygen, and sulphur.

Agassiz (1807–73) (*see* **Famous Geologists: Agassiz**) was the first geologist to establish the existence of ancient glaciation. Subsequently evidence accumulated for past climatic cycles of alternating 'greenhouse' and 'icehouse' phases, as they became picturesquely termed.

Thus cyclicity was revealed in rocks, water, and air – or the lithosphere, the hydrosphere, and the atmosphere. The extent to which these cycles interrelated with one another was little understood. Initially palaeontologists took the view that the evolution of the biosphere responded to external changes, and had little inter-reaction with them. Subsequently it was realized that this is far from the case. The stromatolitic limestones (*see* **Fossil Plants: Calcareous Algae**) of the Late Precambrian, some 3400 My BP, are a dramatic example (**Figure 1**). These limestones are the relics of primitive algae and cyanobacteria. They are found worldwide in Late Precambrian and Phanerozoic strata, and form today in tidal-flat environments.

Stromatolitic limestones provide the earliest case preserved in the stratigraphical record of the interaction of organic and inorganic processes to form rock. Their creators, the first abundant photosynthesizers, took carbon dioxide from the atmosphere, replaced it to some degree with oxygen, extracted calcium from sea water, and precipitated the vast limestone rock formations that are preserved all over the world to this day (*see* **Atmosphere Evolution**).

Analysis of ice cores from modern polar regions shows a strong positive correlation between carbon

**Figure 1** A bedding surface showing the characteristic colloform brassicamorphic structure of a stromatolitic limestone. Late Precambrian, Ella Island, East Greenland. Extensive Late Precambrian stromatolitic limestones provide early evidence of the interrelationship between the biosphere, the hydrosphere, and the atmosphere. Earth system science is all about these interrelationships. For modern example see **Minerals: Carbonates Figure 2G**.

dioxide concentrations in the atmosphere and temperature. It is reasonable to assume such linkage in the past, and that the precipitation of stromatolitic and other limestones may be one of several causes of global cooling. In the history of the Earth extensive episodes of limestone formation have been followed by global cooling several times. The Late Precambrian stromatolitic limestones demonstrate the linkage between biosphere, lithosphere, and atmosphere, and provide a good starting point to examine what is now called Earth system science.

## Biogeochemical Cycles

The appreciation of the interplay of atmosphere, biosphere, and hydrosphere, demonstrated by stromatolites, coupled with the realization that many Earth processes are cyclic, has lead to the concept of the biogeochemical cycle. This term is applied to the flux of material in and out of the lithosphere, hydrosphere, biosphere, pedosphere, and atmosphere (which collectively constitute the geosphere), and the chemical and physical changes that occur therein. The material Earth, or geosphere, is a closed dynamic system, within which there is a constant recycling of its components. It is not, however, a closed system with respect to energy, because this constantly reaches the Earth in the form of solar radiation (**Figure 2**).

A variation in solar energy will impact on the dynamics of biogeochemical cycles. As a generalization, an increase in solar radiation may result in global warming, and a decrease may result in global cooling.

The situation is more complex than this, however, because of the interconnected nature of the cycles or systems.

### Some Definitions Defined

Biogeochemical cycles are described and modelled in terms of reservoirs, material, fluxes, sources, sinks, and budgets. The reservoir is the amount of material in a given Earth system, such as oxygen in the atmosphere, or water in the ocean. The flux is the amount of material moved from one reservoir to another – for example, the amount of water lost from the ocean to the atmosphere by evaporation. The source is the flux of material into a reservoir, and the sink is the amount of material removed from it. The budget is the balance sheet of the amount of material lost or gained in a system. If source and flux are in equilibrium, such that the amount of material in the reservoir is constant, it is in a steady state. A system of two or more reservoirs, in which material is transferred cyclically without an external flux, is termed closed. Commonly Earth systems are interconnected or coupled: a variation in the flux of one affects the dynamics of the other (**Figure 3**).

Awareness that the geosphere consists of a multitude of coupled systems, where a variation in the flux of one will have a 'knock-on' effect on others, poses the question 'why is not the Earth in a permanent state of dynamic chaos?' Why, for example, does the Earth's climate alternate between glacial and interglacial episodes, without fluctuating wildly all the time? The answer lies in feedback, a concept familiar to physiologists and engineers. For example, the temperature of the human body remains constant, thanks to the feedback effect of the hypothalamus, and the temperature of a house remains constant owing to regulation of the central heating system by a thermostat. A feedback is when a process in one system causes changes in another one that in turn influences the first system. A positive feedback accelerates the original process; a negative feedback retards it. A positive feedback may be illustrated by considering the effect of increasing the $CO_2$ content of the atmosphere (by burning fossil fuels, by volcanic eruption, or whatever). This greenhouse gas will result in an increase in temperature, which will result in evaporation increasing from the oceans to the atmosphere and an increase in water vapour, another greenhouse gas, which will accelerate the increase in temperature. Negative feedbacks are rarer than positive feedbacks in the geosphere, but include heat transfer towards the poles and the cooling effect of volcanic ash clouds.

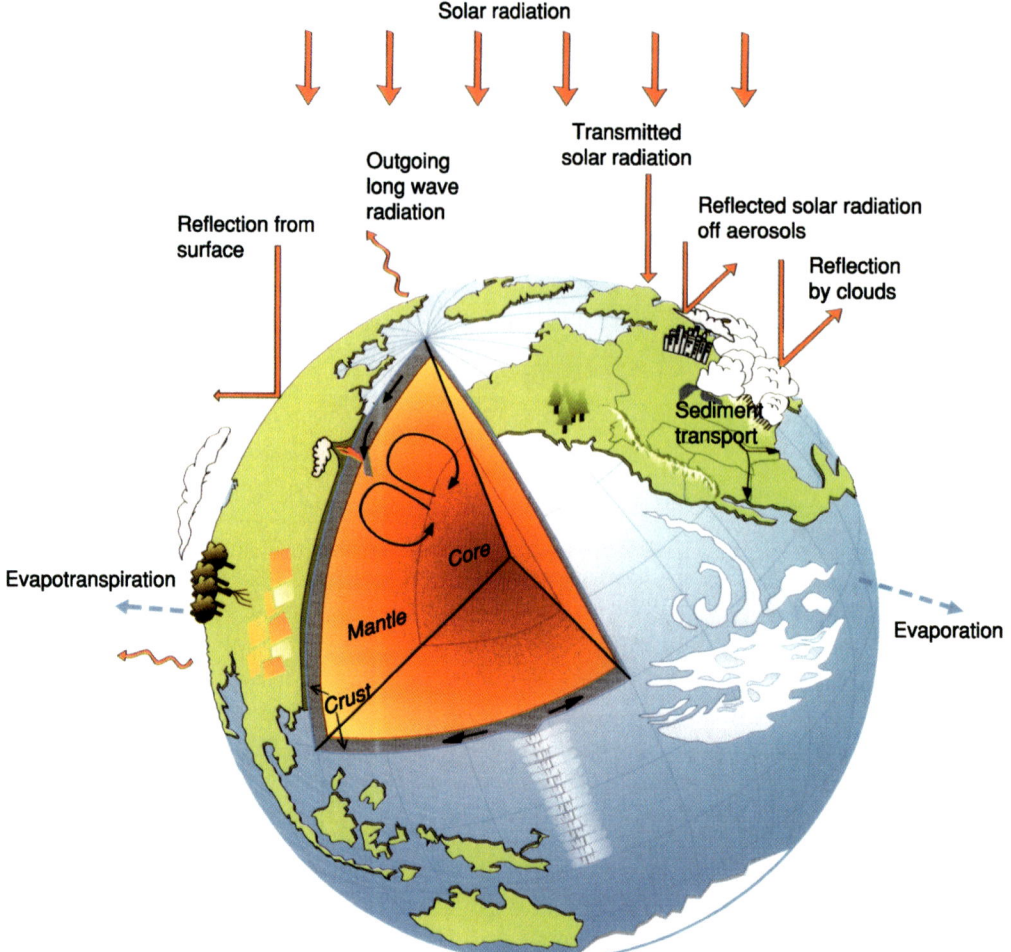

**Figure 2** Overview of the cyclic processes on Earth. The placements of the geological and geographical features are not meant to represent their true positions on the Earth, but to provide an overview of some of the Earth systems. Reproduced with permission from Jacobson MC, Charlson RJ, and Rodhe H (2000) Introduction: Biogeochemical cycles as fundamental constructs for studying Earth system science and global change. In: Jacobson MC, Charlson RJ, Rodhe H, and Orians H. (eds.) *Earth System Science*, pp. 3–13. San Diego: Academic Press.

## Earth System Science and the 'Gaia' Hypothesis

Since the 1970s James Lovelock developed the Gaia hypothesis, named after the ancient Greek goddess of the Earth (*See* **Gaia**). As originally conceived the 'Gaia' concept envisages the Earth as a super-organism that operates to regulate its own environment, principally temperature, to keep it habitable for the biosphere. Lovelock has never argued that the biosphere consciously anticipates environmental change, but only that it automatically responds to it. Nonetheless some sections of the public have construed it that way, and in the popular mind Gaia gained a quasi-mystical connotation, enhanced by its name. The great value of the Gaia hypothesis is that it presents the interdependence of the constituents of the geosphere in a media-friendly way. Earth system science also involves a holistic approach to the geosphere, but without the 'ghost in the machine'. Nonetheless Amazon, the internet book shop, still classifies books on Earth system science under 'Religion and Spirituality > New Age > Earth-Based Religions > Gaia'.

## Impact of Earth System Science on Geology

When the dust of history has settled over the present period it may be argued that Earth system science has had as large an effect on geology as did plate tectonics some 40 years previously (*See* **History of Geology Since 1962**). Its import may be wider still however. This is because Earth system science is an important

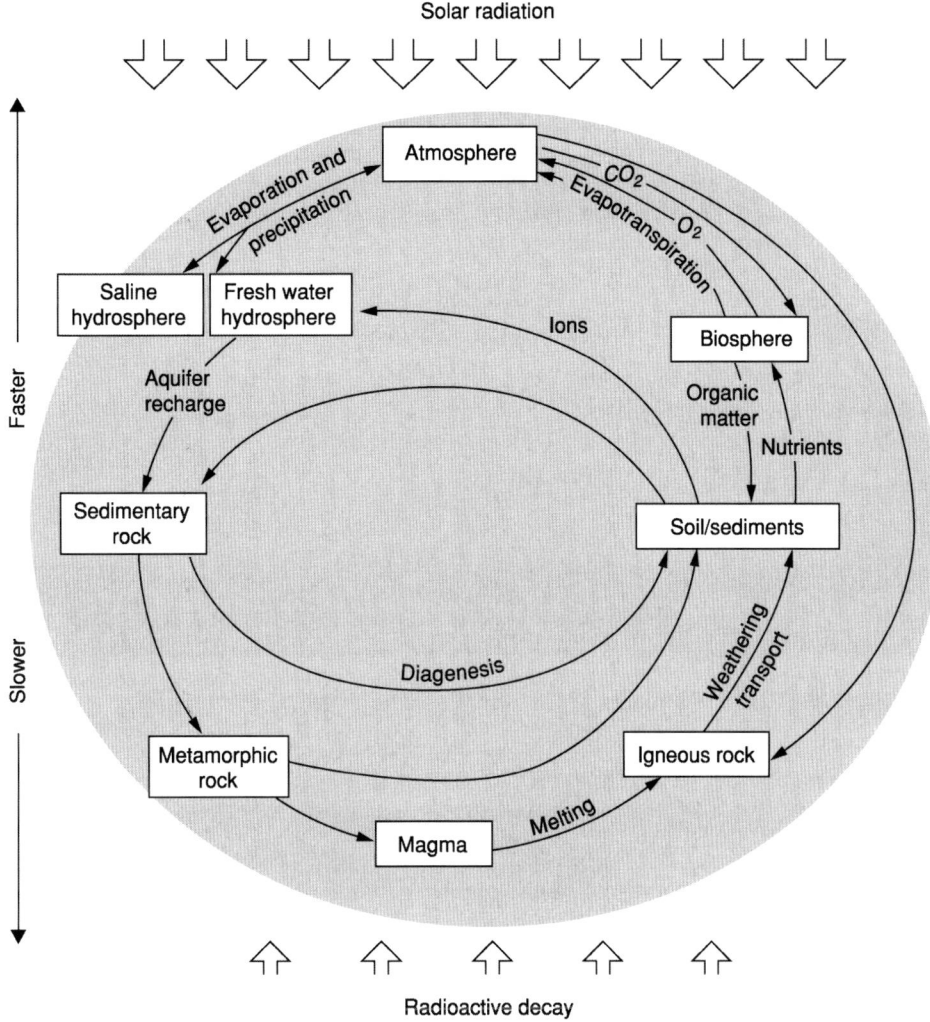

**Figure 3** The cyclic processes and fluxes between major reservoirs on the Earth. Reproduced with permission from Jacobson MC, Charlson RJ, and Rodhe H (2000) Introduction: Biogeochemical cycles as fundamental constructs for studying Earth system science and global change. In: Jacobson MC, Charlson RJ, Rodhe H, and Orians H. (eds.) *Earth System Science*, pp. 3–13. San Diego: Academic Press.

tool in the study of climate change, a topic of great current concern. In particular it focuses attention on how anthropomorphic activities, such as the combustion of fossil fuels and deforestation, may have fed back into atmospheric systems, and hence affected global climate (*see* **Palaeoclimates**).

Moving from the practical to the theoretical, Earth system science has brought about a decline in the reductionist approach to science in general and to geology in particular. (Reductionism is the process of knowing more and more about less and less.) In its infancy geology was advanced by natural philosophers (the term 'scientist' was not popularized until 1858 by Huxley) who were polymaths. It was relatively easy to be a polymath in the nineteenth century, because the pool of scientific knowledge was limited. As the pool of knowledge expanded into a lake, then a sea, and finally an ocean, scientists have had to focus their attention on progressively smaller and smaller areas of knowledge, thus losing sight of the wood for the trees. Distinct disciplines of chemistry, physics, life science, and geology have evolved, all with their own specialized subsets. Earth system science, by taking a holistic view of the Earth, has had a beneficial effect on the development of interdisciplinary scientific research. Concomitant with this, however, has been a decline in the training of geologists who can identify rocks, minerals, and fossils, and map their distribution over the Earth. Geology has become disseminated into Earth science. University Geology Departments have metamorphosed into Departments of Earth Science, coupled with geography, environmental science, etc. These departments no longer produce graduates with focused geological knowledge, but Earth scientists who are Jacks-of-all-trades and masters of none.

Bishop Richard of Durham would have been saddened to see his word 'Geologia' fall into desuetude after 700 years.

## See Also

**Atmosphere Evolution. Carbon Cycle. Famous Geologists:** Agassiz; Hutton; Lyell. **Fossil Plants:** Calcareous Algae. **Gaia. History of Geology Since 1962. Minerals:** Carbonates. **Palaeoclimates. Weathering.**

## Further Reading

Ernst WGF (ed.) (2000) *Earth Systems: Processes and Issues*. Cambridge: Cambridge University Press.

Hamblin K (2002) *The Earth's Dynamic Systems: A Textbook in Physical Geology*. New York: Prentice Hall.

Jacobson MC, Charlson RJ, Rodhe H, and Orians H (2000) *Earth System Science*. San Diego: Academic Press.

Kump LR, Keating JF, Crane RG, and Kasting JF (2003) *The Earth System: An Introduction to Earth Systems Science*. New York: Prentice Hall.

Lovelock J (2000) *The Ages of Gaia: A Biography of Our Living Earth*. Oxford: Oxford University Press.

Nisbet E (2002) The influence of life on the face of the Earth. In: Fowler CMR, Ebinger CJ, and Hawkesworth CJ (eds.) *The Early Earth: Physical, Chemical and Biological Development*, pp. 275–307. Special Publication 199. London: Geological Society.


# EARTHQUAKES

*See* ENGINEERING GEOLOGY: Aspects of Earthquakes; TECTONICS: Earthquakes


# ECONOMIC GEOLOGY

**G R Davis**, Imperial College London, London, UK

© 2005, Elsevier Ltd. All Rights Reserved.

## Introduction

If you visit the strangely dimpled landscape known from Saxon times as Grimes Graves in Norfolk, England, you will see not an ancient burial ground, but hundreds of backfilled shafts in the white Cretaceous Chalk. Here Neolithic man discovered and exploited a thin subsurface layer of dark flint (**Figure 1**) so extensive and of such superior quality that, after due process of mining and treatment, the finished product could profitably be traded and fashioned into the finest flint tools and weapons at the cutting edge of technology. If you visit the flat forested landscape known as Weipa in the Yorke Peninsula, Australia, you will see where twentieth century man discovered and now exploits a surface layer of red bauxite (**Figure 2**) so extensive and of such superior quality that, after due process of mining and treatment, the finished product can profitably be traded and fashioned into the finest aluminium utensils and machines at the cutting edge of technology.

In his long march from hunter-gatherer to moonwalker, *Homo* has used his *sapiens* to exploit the Earth's bountiful mineral resources that, apart from food and clothing, have provided the materials for his advancement. At about 2000 BC flint from Grimes Graves and many other flint workings in the Cretaceous Chalk of Europe was an economic industrial mineral because insufficient copper and bronze was available to meet demand. The occurrence in nature of native copper and gold had already led to their use

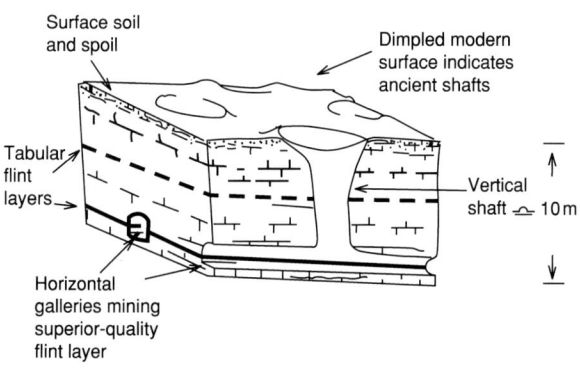

**Figure 1** A sketch to illustrate how high-quality flint was mined in the chalk of Britain about 4000 years ago.

**Figure 2** The cliffs facing the Gulf of Carpentaria in Queensland, Australia, and a working face in the vast expanse of pisolitic bauxite. Scale shown by white penknife near the centre. Photos: GR Davis.

for over 2000 years as tools, ornaments and weapons. Supplies of metal increased when smelting was discovered to reduce the natural ores of Cu, Pb, Sn and later Fe into metal. The ownership of valuable metals and other minerals, such as salt, begat wealth and power. The rise and fall of ancient civilizations is intimately bound up with the riches derived from control of mineral resources, and the historical theme continues to this day. For example the shift of power and control of trade routes from the Persians to the Greeks was greatly influenced by wealth from the newly discovered Laurium Ag-Pb deposits near Athens, and then gold in Macedonia. Finance was available to build a powerful navy that enabled Themistocles to conquer Xerxes at the decisive battle of Salamis in 480 BC. The Greeks enjoyed a golden age of philosophy and art; the mining revenues assisted Philip of Macedonia to establish his dominance, and for his son Alexander the Great to finance the first campaign in his great Middle Eastern conquests about 330 BC. In that same region, the 20th century has witnessed the global economic and political impact of the discovery and development of major oilfields upon the countries of the Middle East.

**Figure 3** (with logarithmic scale on the vertical axis) illustrates how world demand for mineral products over the last 150 years of the industrial age has been met by production growth rates that exceed even the 'explosive' growth rate of world population. Annual production rates for the basic industrial materials iron and petroleum have grown dramatically, and the production rate of copper, a primary base metal, has increased a hundredfold. During the

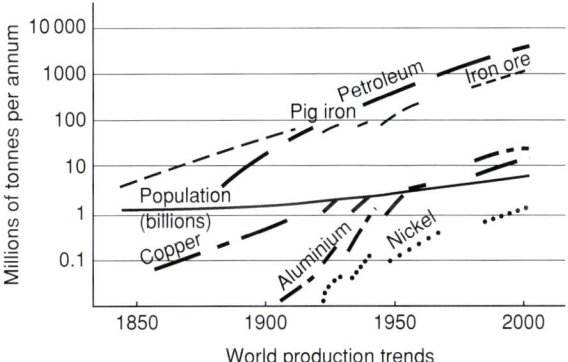

**Figure 3** Over the past 150 years, world demand for mineral products has exceeded the 'explosive' growth in population, shown on a logarithmic scale.

twentieth century newer metals such as nickel became industrially important, and aluminium, once rare and precious, reached production rates exceeding that of copper, and at a lower cost per tonne. Many other 'new' elements continue to emerge in response to the growth of modern high-technology industry. For instance, demand is now growing rapidly for indium (mine production about 340 tonnes in 2002), which is needed for making flat-panel electronic displays as used in mobile telephones and plasma-screen televisions. World population is increasing, and the average per capita demand for mineral products, while heavily skewed by industrialized Western world usage, is also increasing. Economic geology plays an integral part in meeting this challenge to the mineral resource base of the world.

# Characteristics of Economic Mineral Deposits

The AGI Glossary of Geology defines economic geology as... "the application of geologic knowledge and theory to the search for and the understanding of mineral deposits; study and analysis of geologic bodies and materials that can be utilized profitably by man, including fuels, metals, non-metallic minerals, and water". In this work (Encyclopedia) water and hydrocarbon fuels are treated separately in view of their major industrial and economic importance. The solid mineral resources also fall into specialist groups, as indicated in the definition quoted above. Collectively, they comprise the non-renewable resource base of the extractive industries. To qualify for "profitable utilisation by man" they must all demonstrate certain characteristics that are, appropriately, both economic and geological.

## Geological Attributes of Economic Deposits

- The size (tonnage) must be sufficient to sustain exploitation for a period long enough (mine life) to justify development.
- The valuable content (grade) must be high enough to repay all costs. Taken together, the tonnage and average grade express the comparative geological potential of the deposit.
- The shape, attitude, depth and physical properties of the deposit must be amenable to extraction by existing mining technology.
- The valuable constituent(s) must be amenable to separation from the unwanted portions (waste), and beneficiation to marketable product by existing technologies.

## Economic Attributes of Economic Deposits

- The price received for the product(s) must cover all production costs and assure competitive advantage.
- The geographical situation must be amenable to mine development, including water and power supplies; and within economic transport distance to market, especially for bulky products.
- Socio-political conditions must be favourable. These include planning permission and mineral rights, various taxes and royalties, labour and safety laws and conditions, regulations for waste disposal, and rehabilitation when the mine is worked out.

A new mining development is typically capital intensive with a lead-time of years between discovery and profitable production. Brief consideration of the seven attributes listed reveals that the risk of financial failure is shared between measurable geological, engineering and logistical natural factors on the one hand and indefinite state-imposed constraints or incentives on the other. Economic geologists need a sound appreciation of mineral economics, mining engineering and mineral processing technology, but their core expertise rests upon their knowledge of mineral deposits.

Figure 4 illustrates the hierarchical progression from mineral occurrence to mineral prospect to mineral deposit, and then through mineral resources and proved mineral reserves to profitable mining. This progression rests for its success, above all else, on the quality of geological observation and interpretation, and its application in industry.

It is important also to appreciate that the mines of today are exploiting only those selected deposits that have met all the criteria for current economic development. They represent only the tip of the iceberg of the Earth's mineral resources. Large deposits of many mineral commodities, currently identified but sub-economic, may in future become payable because of improvements in mining or mineral extraction technology, commodity demands and prices, and similar conditions. Those deposits may one day join the stock of future discoveries that will provide the economic deposits and mines of tomorrow.

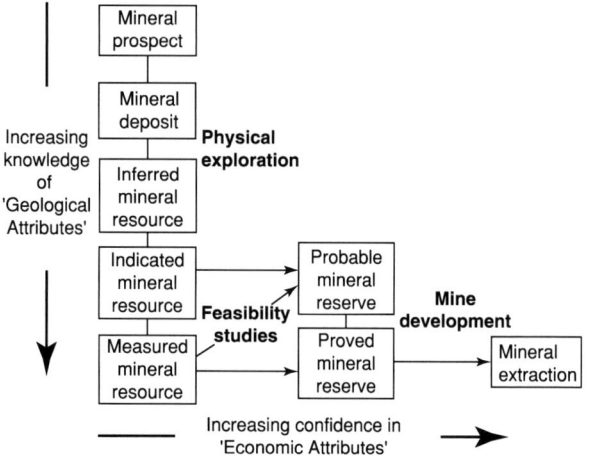

**Figure 4** The progression from a mineral prospect to mineral extraction. Expenses increase as work becomes more detailed. Mine development requires capital expenditure before cash flow starts from mineral product sales. The categories of resources and reserves, and the technical terms proved, probable, measured, and indicated are strictly defined in codes of practice formulated by professional mining institutions in some leading mining countries (Australia, Canada, South Africa, UK/Europe, and USA) and accepted by financial organizations such as stock exchanges.

## Variety and Use of Mineral Deposits

The range of 'geologic bodies and materials' that is utilized by man has grown in scope and volume over the course of history, but falls into a few practical groups.

### Construction Materials

These include building stones, sand, aggregates, clay, and cement raw materials. The high cost of transporting these bulk products requires source rocks to be as close as possible to the place of utilization, but typically the costs of extraction and treatment are low. A long history of use (the stone walls of Jericho were packed with a clay mortar circa 8000 BC) has built a body of experience concerning desirable source materials and their properties. In developed countries today, geological knowledge is applied to not only ensure that technical specifications are met, but also to discover and delineate those particular sources that best suit land development planning and minimize environmental problems. Some industrial rock products are valuable enough for export, such as clays for ceramics and ornamental stone for decorative use in buildings.

### Fuel or Energy Mineral Deposits

These include hydrocarbon fluids, coal and uranium. Coal is a sedimentary rock derived from plant remains, the fossil fuel on which modern industrial development was built. Together with petroleum, this versatile material supplies most of the worlds' energy needs. It is also a major industrial raw material for the manufacture of chemicals, and coke for iron and steel production. Economic deposits of uranium-bearing minerals, the base on which the atomic age is built, occur at a scale of magnitude nowhere near that of coal and oilfields.

### Industrial Minerals

Sometimes termed non-metallic minerals, these are valued for their chemical and/or physical properties and the fact that they are not of widespread occurrence. In general, prices are sensitive to market demand and product specifications (with premium prices for premium grade products), and quality is a major factor in the economic geological evaluation of mineral reserves and productive life of industrial mineral deposits. A vast range of industrial minerals is produced in an equally impressive range of tonnages. Minerals with valuable chemical properties, used mainly in the chemical and fertilizer industries, include rock phosphate, potash and mixed chloride salts, sulphur, nitrates and borates. Fluorspar and limestone are prominent as fluxes in the metallurgical industry and ceramics, and other process industries consume silica sand, feldspars, and kyanite. Physical attributes useful in filler and extender applications make talc, limestone and kaolin competitive in paints, paper and plastics. Other minerals with useful physical properties include asbestos, barytes, diatomaceous earth, and the lightweight aggregates perlite, pumice and vermiculite. Hardness is utilized in abrasives such as corundum, garnet and industrial diamond. The extensive list of industrially useful minerals makes it clear that a wide range of geological knowledge finds application in the search for industrial minerals and in ensuring products that conform to specifications set by the industrial user.

### Metallic Mineral Deposits

The metallic ore minerals are commonly metal compounds such as sulphides and oxides in which the metal content is high compared with rock-forming minerals. Natural concentrations of ore minerals form discrete ore bodies that may typically contain only a few percent of the valuable metal. Unlike many industrial minerals that find direct use after mining and limited beneficiation, the ore minerals, in general, must be reduced to metal by complex processing. Modern mining and mineral extraction procedures are tending towards greater use of chemical and bacterial leaching methods for suitable ores, such as in-situ extraction of some uranium deposits, and heap leaching of some gold ores. The great bulk of metalliferous ore, however, is mined by surface or underground rock-breaking methods. Run of mine ore is first crushed and milled to a fine pulp, from which the desired ore minerals are separated from the gangue minerals by various methods to produce a concentrate. Metal is then extracted from the mineral concentrate by further treatment, which may include smelting or various chemical methods such as solvent extraction and electrowinning, followed by refining to market standards. The expensive multi-stage process of extraction results in complex engineering works at the site of large ore bodies, often in remote locations. The commonly used metals are sometimes grouped for convenience by their geochemical or industrial affinities. Base metals include Pb, Zn, Sn and Cu. Iron and ferroalloy metals include Cr, Co, Ni, Mn and V. Light metals include Al, Ti, Mg, Li and Be. The precious metals comprise Ag, Au and PGM (platinum group metals) (**Table 1, Figure 5**).

The attached statistical data illustrate some world production rates and the relative importance of the broad mineral groups in the world economy. It should be noted that most metal production tonnages are less by one or two orders of magnitude

**Table 1** Annual world production levels of selected metals and industrial minerals

| | 1982–1983 | 2000–2001 | Main producing countries | |
|---|---|---|---|---|
| | Tonnes per annum (millions) | | | |
| Coal | 4000 | 4500 | China | USA |
| Asbestos | 4.1 | 2.0 | Russia | Canada |
| Bauxite | 78 | 139 | Australia | Guinea |
| Chromite | 8.2 | 13 | South Africa | Kazakhstan |
| Copper | 8.0 | 13.6 | Chile | USA |
| Diatomite | 1.9 | 1.7 | USA | China |
| Fluorspar | 4.4 | 4.2 | China | Mexico |
| Iron ore | 750 | 1150 | China | Brazil |
| Kaolin | (21) | 23 | USA | UK |
| Lead | 3.5 | 3.0 | Australia | China |
| Manganese ore | 24 | 22 | China | South Africa |
| Phosphates | 130 | 130 | USA | China |
| Potash | 25 | 27 | Canada | Belarus |
| Talc | 7.4 | 8.4 | China | USA |
| Zinc | 6.5 | 9.0 | China | Australia |
| | Tonnes per annum (thousands) | | | |
| Antimony | ? | 123 | China | Guatamala |
| Gold | 1.4 | 2.5 | South Africa | USA |
| Mercury | 6.0 | 1.8 | Kyrgyzstan | Spain |
| Platinum group metals | 0.3 | 0.5 | South Africa | Russia |
| Silver | 12 | 19 | Mexico | Peru |
| Tin | 205 | 245 | China | Indonesia |
| Uranium | 50 | 34 | Canada | Australia |
| Vermiculite | 540 | 376 | South Africa | USA |

Data compiled from World Mineral Statistics (2003) by permission of British Geological Survey. © NERC. All rights reserved. IPR/46–28cw.

than the tonnages of rock mined to produce them. The relative value of metallic mineral and industrial mineral production varies considerably from one country to another. For instance, for the mean of the years 2001 and 2002, industrial minerals accounted for 78% of the total value of non-fuel minerals in the USA whereas in neighbouring Canada the metallic minerals accounted for 57% of the non-fuel group. (Compiled from various sources).

## World Distribution of Economic Mineral Deposits

Like all other geological bodies, mineral resources are unevenly distributed in the Earth's crust. Many nations owe a major part of their wealth to discovery

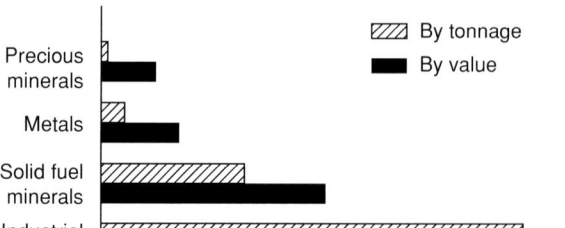

**Figure 5** The importance of industrial minerals in the modern world is not always appreciated. The relative values of the four commodity groups is a better measure than tonnages, because the amount of rock mined to produce metals and precious metals from their ores is at least an order of magnitude higher. Compiled from data quoted in Evans (1995), after Noetstaller.

and development of economic mineral deposits, while others may be either very poorly endowed by nature, or ignorant of their undiscovered mineral resources. The people of Nauru are richly blessed because their small Pacific island homeland contains 40 million tonnes of high grade phosphate rock. In developed countries even with a large mining industry, the lack of domestic sources of certain vital commodities leads to the concept of 'strategic minerals' and defensive stockpiling against national emergencies. Before entry into World War II the USA included Cr, Mn and Sn in a list of 14 strategic commodities. Statistics of world mineral production illustrate the imbalances that cause geopolitical concern about national vulnerability to imported supplies (**Figure 6**), or to damaging price changes in key commodities. For example South Africa, a country about twice the size of France, dominates Pt production, supplies almost half the worlds Cr and major amounts of Au, Mn and V, but has inadequate domestic resources of Mo and Al, oil and gas, potash, sulphur and kaolin. China, a large country with a rapidly expanding economy, dominates world production of the rare earth elements (REE), produces over 80% of world Sb, over one-third of world V, Sn and W, and is a key nation in world minerals trade.

The observed major world patterns of distribution of economic mineral deposits become less arbitrary and more understandable when viewed in the light of metallogenic and geological maps. Relationships appear between mineral resources and the geological environments and rock systems in which they occur. For instance, oilfields and coalfields are associated with large sedimentary basins of upper Phanerozoic age, and one would not search hopefully for these

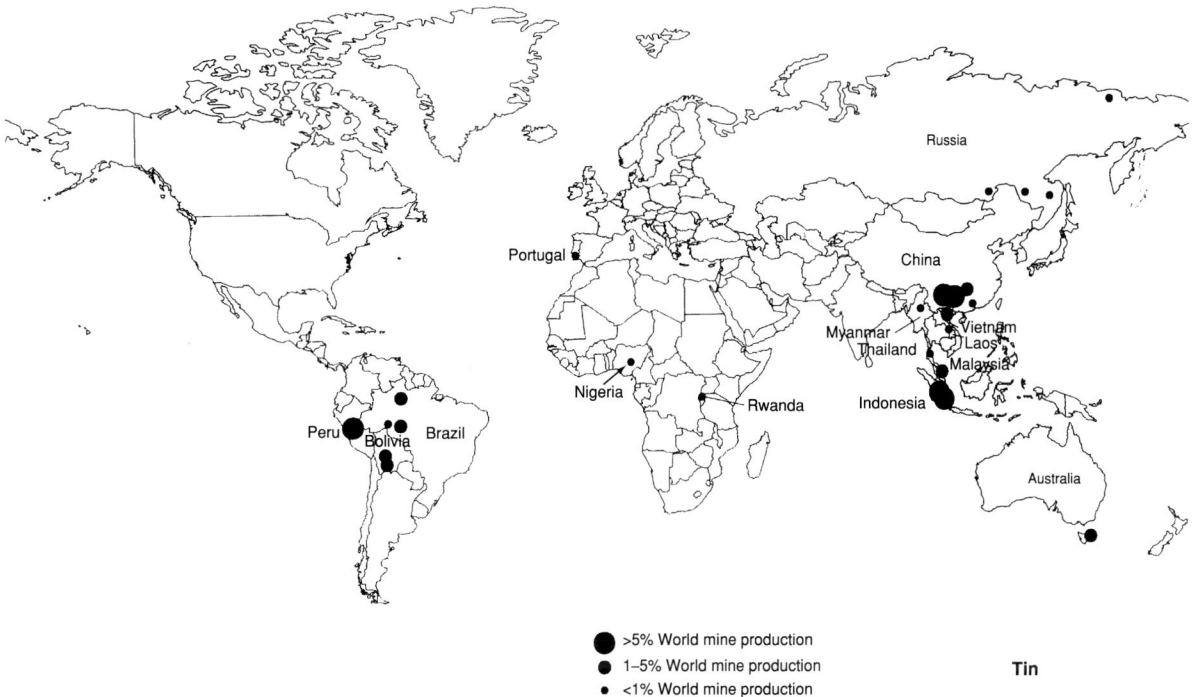

**Figure 6** This map, showing the sites of tin production, is an example of the uneven global distribution of several mineral commodities. Modified from World Mineral Statistics (2003) by permission of the British Geological Survey. © NERC. All rights reserved. IPR/46–28CW.

great energy resources in sedimentary basins within ancient shield areas of Archaean rocks. The converse applies for the great banded iron formations (BIF) in Precambrian sedimentary basins that are host to most of the world's high-grade Fe ore deposits (*see* **Sedimentary Rocks: Banded Iron Formations**). Similarly, large and highly productive porphyry copper deposits are strongly grouped in the mountain chains made up of volcanic rocks and granitic intrusions along the western margins of the Americas. This pattern has been more readily understood over the past 40 years in the light of geological mapping and plate tectonic theory. Empirical associations between mineral deposits and their host rocks have long been noted and put to use by miners, and the list is still growing as geological knowledge of Earth's physico-chemical systems advances. Concepts have developed of mineral provinces and mineral epochs, and in recent decades regional patterns of mineralization have been related to the various kinds of tectonic plate boundaries (**Figure 7**) (*see* **Plate Tectonics**).

## Economic Geology and the Extractive Minerals Industries

Every producing mineral deposit is a non-renewable resource with a finite life. In order to continue to meet world demand, the extractive industries are geared to a life cycle of activities as shown in **Table 2**.

These four activities are very briefly outlined below, and the reader is referred to specific topics for more comprehensive information.

### The State of Relevant Geological Knowledge

Economic geologists are widely employed in the many specialized areas of the extractive industries. All, however, rely on the quality and appropriate application of their geological knowledge of mineral deposits (*see* **Mineral Deposits and Their Genesis**). Great advances have been made in the geological understanding of the 4500 million year history of the earth, and of the igneous, sedimentary and metamorphic processes in which the genesis of mineral deposits is an integral part. Concentrations of valuable minerals into economic deposits are no longer seen as special, isolated events, but as the result of processes, often sequential and superimposed, that operate in geological environments of every kind and in every age from the Archaean to the present. Advancing views on regional patterns of mineral deposition related to space and time (i.e., in mineral provinces and mineral epochs) were dramatically boosted through definition of crustal tectonic regimes by the theory of plate tectonics. The scientific overview of mineral deposits has moved away from worthy attempts at genetic classification, notably that of Lindgren, towards development of 'mineral

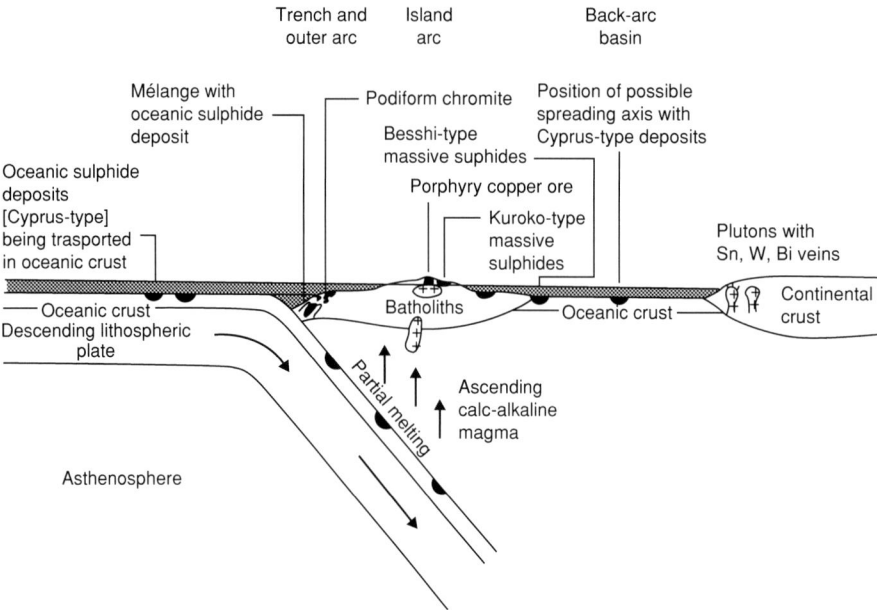

**Figure 7** Development of the theory of plate tectonics in the 1960s led rapidly to a new view of the genesis of some ore deposits as shown on this early diagram in 1972. Reproduced with permission from Evans AM (1995) *Introduction to Mineral Exploration*. Oxford: Blackwell, p. 40 (modified from Sillitoe 1972).

**Table 2** Life cycle of extractive industry activities

| Activity | Applied geology input |
| --- | --- |
| Discovery of new deposits | Dominant role |
| Feasibility studies and mine development | Major role in defining reserves and critical technical parameters |
| Mineral extraction | Essential engineering role in operational efficiency |
| End of life of deposit | Long-term environmental assessment |

deposit models'. These models marshal all the pertinent facts concerning related deposits that are seen as 'types'. Modern advances in nearly all fields of geology are of relevance and application to some aspect of economic geology. For example structural geology from the scale of electron microscopy in rock deformation studies to megastructures on satellite imagery; evolution and structure of sedimentary basins; the geochemistry of sediments as source rocks for hydrocarbons and metalliferous brines; the volcanic realm and its diverse products; magma generation and crystallization; metamorphic effects on rock structures and mineral chemistry; and even the profound effects of major meteorite impacts on the Earth's crust.

Specialist fields of knowledge are required for each phase of the industrial cycle of activity (**Table 2**) in each branch of the extractive industries, and economic geologists have formed several specialist groups such as petroleum geologists, mining geologists and exploration geologists.

### Discovery of New Deposits

To maintain a stable extractive industry long term, the huge tonnages of minerals extracted each year must be replaced, on average, by discovery and/or development of an equivalent amount of new materials. The difficulty imposed by this high demand is compounded by the fact that new mineral fields and individual deposits become progressively harder to discover as the easier finds are made. Scientifically based mineral exploration has developed into a sub-industry, in which the participants include government agencies, mineral resource companies of all sizes, and specialist groups offering contract and consulting services.

Exploration objectives are targeted on either certain commodities or certain regions of interest. The search area for bulk materials such as aggregates or low-grade coal is restricted by transportation costs and environmental factors; some mines concentrate their search close to existing infrastructure; and state organizations are of course interested in their own territories. Most mineral commodities are internationally traded, and exploration money and effort tend to favour countries that offer the most attractive combination of geology, political stability, mining taxation and operational infrastructure. Commodity-targeted exploration is based solidly on favourable geological environments, the oil and gas sector being a prime example. Views on the factors that have controlled or generated mineral deposits (e.g., where in the world to look for kimberlite pipes?) are of paramount importance to cost-effective exploration. Many small specialized groups undertake exploration on this basis. Their successes may be sold to large companies, which have the policy options of either buying their future mineral resources or organizing their own exploration teams on a long-term basis.

Exploration programmes generally follow phases of target definition, reconnaissance, selective follow-up, and detailed exploration of prospects. A great and growing number of exploration technologies, including heavy mineral, geochemical and geophysical surveys, are available to supplement the essential art and science of sound geological mapping based on wide-awake observation. At the detailed stage of exploration, physical methods are used, such as pitting, trenching and especially the various types of drillholes now available.

Mineral exploration is a high-risk enterprise justified by potentially high reward, but a historically low success rate overall. Organizations with a relatively high success rate tend to be well managed, exploring in carefully researched and targeted areas, employing teams of well-qualified and motivated people to use the most cost-effective sequence of search technologies, with financial backing stable enough to ensure long-term effort. The constant application of sound geological judgement to ensure that exploration data at all stages make geological sense, and especially when deciding whether to abandon a prospect or press on at higher cost and effort, makes disciplined mineral exploration a professionally rewarding activity (**Figure 8**).

### Feasibility Studies and Mine Development

At some point enough is known about a prospect through physical exploration to either abandon it or proceed to further detailed exploration work. A decision must then be made whether the prospect is worth the high cost risk of development into an operating mine. This is the period of feasibility study, when a multidisciplinary team will rigorously test the prospect against all the criteria for full economic status. The professional contribution from economic geology is the foundation stone upon which the entire edifice will be built, and confidence in the reliability of all geological data is essential. The main geological information will include the following:

- Knowledge of the general characteristics of the type of deposit involved.
- Complete mineralogical and chemical information, including grain size and distribution, and useful by-products or deleterious constituents, that affect amenability of the deposit to processing.
- The three-dimensional geometry and depth of the deposit, the physical condition of the ore and its wall rocks, and other factors affecting amenability to mining methods.
- A set of clear records including reports, geological and other survey maps, drillhole data and logs, and assay data.
- Mineral resource and reserve estimates, including estimates of the degree of error attached to such basic data as sampling methods, sample preparation and assays, bulk density of ore and gangue types, the range and distribution of values within the deposit, and the continuity of mineralization between drillholes or other exposures.

The process is illustrated by some geological aspects of the case history of the Lihir gold mine, situated in a volcanic caldera on the island of Lihir, Papua New Guinea (**Figure 9**). After 3 years of encouraging physical exploration, in 1985, the epithermal pyritic gold ore failed the test of amenability to conventional processing. Tests on an alternative advanced technology worked well, but meant that

**Figure 8** McArthur River, Queensland, Australia. A large stratiform pyritic Zn–Pb sulphide deposit. (A) The gossanous discovery outcrop, broken open to reveal white secondary Pb and Zn minerals. (B) Polished surface (about 20 mm across) of very fine-grained stratiform sulphide ore. The 2–5 micron grain size of the base metal sulphides in this large and easily mined deposit made it too difficult and expensive to process by conventional technology. Photographs: GR Davis.

**Figure 9** The open pit gold mine at Lihir Island, Papua New Guinea, under development in 1998, twenty months after inception of mining operations. The two pits, then about 40 m deep, and centred on ore bodies 1.5 km apart, will merge into a pit about 2 km across, with a planned depth of 300 m. The background and foreground hills are the walls of the volcanic caldera in which the epithermal gold-pyrite mineralization occurs. Photograph by courtesy of Rio Tinto PLC. Reproduced with permission from Evans AM (1995) Introduction to Mineral Exploration. Oxford: Blackwell p. 34.

only large-scale mining would provide an economic return. A further detailed drilling campaign was followed in 1992 by a feasibility study. This was based on a data bank that included 56 000 samples from 83 000 m of core from 343 drillholes, and 4479 specific gravity tests on the 12 ore types recognized. Four adjacent and partly overlapping ore zones amenable to open-pit mining were defined within an area of 2.0 km by 1.5 km. Using three computerized geostatistical methods to optimize confidence in ore continuity and distribution of gold values, an overall mineral resource of 478 million tonnes was estimated. Mining engineers then designed an open pit to mine 445 million tonnes in 15 years. Of this total, 341 million tonnes will be waste rock, 42 million tonnes will be high-grade ore reserves averaging 6.9 g/t Au to be processed through the plant, and the remaining 62 million tonnes will be stockpiled as low-grade ore averaging 2.9g/t Au to be processed during years 15 to 36. Gold production commenced in 1998, some 16 years after initial work on the discovery. As usual in a new mining area, exploration will continue, with a view to extending the life of the mining infrastructure beyond the presently planned project.

## Mineral Extraction

Geological services are now widely employed in all forms of mineral extraction operations, the mainstay of the minerals industries. The proven usefulness of applied geology lies in its contribution towards optimizing the efficiency of the overall production engineering process, and the geologist is an integral member of the engineering team and its objectives. The nature of the work extends the quantitative aspects of exploration at the feasibility study stage into an even more detailed and practical realm where geological predictions are at once put to the test. Geological services fulfill the following main functions, differing only in detail across the spectrum from bulk material quarries through industrial minerals to metalliferous mines.

**Maintaining mineral reserves** A new mine is brought into production with only enough proved reserve to provide reasonable assurance that the invested capital will be returned. The greater part of the detailed exploration of the deposit must continue over the whole life of the operation. The planning and execution of this vital work is the special responsibility of the geologist, as part of long-term management planning. It calls for skills in detailed geological mapping at scales up to 1/250, knowledge of sampling theory and practice, and competence in computerized data manipulation.

**Services to the mining department** Clean mining and low production costs depend on physically defining in the greatest detail the three kinds of material in a deposit – waste that must be dumped, payable ore or feed that goes to the processing plant, and marginal material that may be stockpiled for future processing. This daily task of the geologist at the working faces finalizes the preliminary outlines predicted from drillholes, sampling and assaying, and contributes also to quality control at the processing plant and towards planned production targets. This definitive work is as important to product quality and cost in a cement raw materials quarry as it is in an underground vein gold mine. Mining procedures rely also on information from rock mechanics, a specialized subject in which knowledge of the detailed structural geology of the deposit is linked with engineering parameters. This overlap with engineering geology occurs also in hydrogeology, as control of groundwater is important in nearly all mining operations.

**Services to mineral processing plants** Improvements in mineral recovery and efficiency over the life of a deposit may be brought about not only by processing technology, but by awareness of changes in the geological conditions to which the plant is tuned. The geologist is in the best position to forecast and make known changes in run-of-mine deliveries, such as grade and mineral composition of the ore, or clay content of the gangue, that are expected in long-life operations, and which upset the routine best performance of the processing plant.

**Other engineering services** Geological advice is often called for, especially in remote mining locations, in selecting sites for buildings, works, tailings disposal, stockpiles, etc., to select the best foundation conditions and especially to ensure that permanent works are not built in places that would sterilize future mineral resources. Geologists are also well placed to advise concerning natural hazards such as earthquakes, and landslides in mountainous terrain.

Even in so curtailed a summary, it must be mentioned that the overall usefulness of applied geology rests heavily on effective communication by geologists with the working industrial community. Good personal relationships at all levels are helpful, but clearly written reports that can be readily understood by non-geologists are essential.

### End of Life of Deposit

In the environmentally conscious world of the twenty-first century, few mineral deposits are developed to the stage of extraction without a searching environmental impact investigation and statement. Similarly, when mineral workings reach the inevitable end of their productive lives, the impact of the closure may be planned so as to reduce deleterious effects and optimize beneficial effects upon the environment and the community that has grown around the deposit. For example, the Lihir gold mining project (mentioned in the feasibility study above) has from inception of operations provided finance and facilities for periodic review and action on environmental and community effects. Applied geology has entered the new professional field of environmental science in an age of positive and proactive approach.

## Conclusion

From time to time ill-founded alarms have arisen that world resources of some minerals will be insufficient to meet mankind's growing demands. An influential, but ill-advised report of the Club of Rome in 1972, for instance, concluded that the limiting factor to world economic growth would be mineral resource deficiencies. Led by applied economic geology, exploration within the minerals sector has in practice demonstrated an abundant resource base open to ever-improving mining and processing technologies. There is no indication that future needs for traditional and new materials will not be met. In a crowded world, the concept of transition to sustainable development has become a serious global issue that geological science is well placed to appreciate. In the spirit of this new drive forward, the world's leading mining companies have proactively organized the Global Mining Initiative, a programme to ensure that the issues relevant to the extractive industries are positively identified and addressed. Both the theoretical and practical aspects of economic geology are involved in this broadening framework to its contribution in discovering and extracting the precious non-renewable mineral resources of our world.

## See Also

**Aggregates**. **Geochemical Exploration**. **Mineral Deposits and Their Genesis**. **Mining Geology:** Exploration; Mineral Reserves. **Petroleum Geology:** Exploration; Reserves. **Plate Tectonics**. **Sedimentary Rocks:** Mineralogy and Classification; Banded Iron Formations; Ironstones; Phosphates.

## Further Reading

British Geological Survey (2003) *World Mineral Statistics 1997–2001*. Minerals Programme Publication 13, Keyworth.

Davis GR (1978) Geology in the minerals industry. In: Knill JL (ed.) *Industrial Geology*, pp. 78–110. Oxford: Oxford University Press.

Dixon CJ (1979) *Atlas of Economic Mineral Deposits*. London: Chapman & Hall.

Evans AM (1997) *An Introduction to Economic Geology and its Environmental Impact*. Oxford: Blackwell Scientific.

Evans AM (ed.) (1995) *Introduction to Mineral Exploration*. Oxford: Blackwell Science.

Hartman L (1987) *Introductory Mining Engineering*. New York: Wiley.

Holland HD and Petersen U (1995) *Living Dangerously: The Earth, its Resources and the Environment*. New York: Princeton University Press.

Jones MP (1978) *Applied Mineralogy – A Quantitative Approach*. London: Graham & Trotman.

Manning DAC (1995) *Introduction to Industrial Minerals*. London: Chapman & Hall.

Wills BA (1997) *Minerals Processing Technology*, 6th edn. Oxford: Pergamon Press.

# ENGINEERING GEOLOGY

**Contents**

**Overview**
**Codes of Practice**
**Aspects of Earthquakes**
**Geological Maps**
**Geomorphology**
**Geophysics**
**Seismology**
**Natural and Anthropogenic Geohazards**
**Liquefaction**
**Made Ground**
**Problematic Rocks**
**Problematic Soils**
**Rock Properties and Their Assessment**
**Site and Ground Investigation**
**Site Classification**
**Subsidence**
**Ground Water Monitoring at Solid Waste Landfills**

## Overview

**M S Rosenbaum**, Twickenham, UK

© 2005, Elsevier Ltd. All Rights Reserved.

Engineering geology embraces the whole of geoscience, gathering information pertinent to the (infra-)structure of society, preservation and enhancement of the environment, and sustainable development. The engineering geologist interprets this information with the relevant application of knowledge and judgement to support the engineering profession, and to protect the public and the environment.

An engineering geologist in essence reads the ground like a book, anticipating where adverse conditions might arise. The primary concern is with defining the likelihood of a geological hazard ('geohazard') occurring, whose existence may be predicted from an appropriate consideration of the ground. This requires defining the state of the ground conditions: the geometry of the various soil and rock units, locating their boundaries; determining their structure, composition and (geotechnical) properties; identifying displacements or distortions (actual or potential), for instance those arising from neotectonic forces (stress), mining or excavation; and identifying the presence of fluids within and flow through the units comprising the ground profile.

Three disciplines are particularly important for the synthesis of engineering geology: hydrogeology, soil mechanics, and rock mechanics. Engineering geology facilitates their integration with the science of geology, particularly by bringing to bear the skills of observation, balancing idealization with reality, and injecting both experience and sound judgement. Hydrogeology concerns the character of fluids (water and gas) within the ground (as distinct from hydrology which concerns fluid flow, generally at the surface), whereas soil mechanics and rock mechanics are engineering disciplines, together forming the profession of geotechnical engineering. Each activity has its own distinct methodology and its own rigour. Each is interlinked and so there is a need to strive towards coherence and integration.

The term 'environmental geology' has evolved relatively recently, but as a discipline is really a component of engineering geology, concentrating on issues related to our industrial legacy, in particular the chemical character of the ground and groundwater, with focus on contaminated land.

## The Profession

Another term requires consideration: 'geological engineering'. The distinction between this and 'engineering geology' lies in the requirements of the engineering institutions. Accreditation of education and training falls within the remit of these professional institutions. They set levels of training if direct registration of the individual is to be approved, such as the minimum time devoted to subjects like mathematics, mechanics, and design. In the United States, this is undertaken by the Accreditation Board for Engineering and Technology (ABET), leading to appointment as a Professional Engineer (PE); in the United Kingdom this is done by the Engineering Council (EC), leading to appointment as a Chartered Engineer (CEng). Such accredited courses are commonly entitled 'Geological Engineering', especially in the USA. The training of a 'Geological Engineer' in essence follows that of an engineer with additional geological knowledge, whereas the 'Engineering Geologist' remains a scientist; this difference has ramifications for professional registration and professional indemnity.

A relatively new route is now opening up for those working primarily within geoscience, accredited by the Geological Society of London, which leads to appointment as a Chartered Geologist (CGeol). This accreditation is also accepted by the European Federation of Geologists, qualifying the registrant as a Euro Geologist (Eur Geol), equivalent to a Euro Engineer (Eur Ing), but only the latter currently has legal standing.

## Development

Engineering geology addresses the issues resulting from adverse ground behaviour. Records of early applications may be traced back five millennia, to the Ancient Egyptians mitigating the impact of flooding by the River Nile and constructing the Pyramids. There follow records of the Romans and Medieval Europeans utilising their understanding of engineering geology in a wide range of public works as well as for the extraction of mineral resources and support of military campaigns, notably sapping and mining. In the early nineteenth century, the creator of the first comprehensive geological map of England and Wales, the engineer William Smith, expressed the essence of an engineering geologist, applying geological principles (largely stratigraphic) to the design and construction of the British canal system as well as to groundwater control.

However, the development of engineering geology as a recognizable discipline in its own right arguably had to wait until the early twentieth century and realization of the important impact of geology on civil engineering. This was exemplified by failure of slopes along the Panama Canal in the first decade and the devastation brought about by the 1906 San Francisco earthquake, focusing attention to the wider potential adverse impact of the ground on the built environment. This encouraged the pioneering work by Karl Terzaghi, transcending the disciplinary boundaries between the science of geology and the art of engineering, seeding the growth of the geotechnical profession worldwide.

Publication lagged behind the increased understanding of ground behaviour that was being developed within the construction industry, from research underway in government-funded establishments, and from the lectures being delivered to engineering institutions and university students. Books containing 'engineering geology' in the title had been published by the 1880s, for instance the volume by Henry Penning, but were not widely distributed.

The compendium of case histories published in 1939 by Robert Legget has become an important milestone, helping establish the subject's profile, soon to be followed by 'A Geology for Engineers' by Blyth (in 1943) and the Berkey Volume (in 1950). The latter volume, entitled 'Application of Geology to Engineering Practice', was published by the Geological Society of America (GSA) in memory of the stimulating contributions by Charles Berkey, whose main professional work had been the provision of consultancy advice for dam construction but had also published his important paper, 'Responsibilities of the geologist in engineering projects' (AIME Tech. Pub. 215).

Collations of published works then followed in the early 1960s, notably the 'Reviews on Engineering Geology' series published by the Geological Society of America and the 'Engineering Geology Special Publications' produced by the Geological Society of London. The first Congress (in 1970) of the International Association of Engineering Geology (IAEG) established the discipline at an international level, since when it has provided a framework for promotion and development of the subject. The IAEG added 'and the Environment' to its title in 1997, thereby acknowledging the wider remit of the organisation.

## Engineering Context

Ground behaviour of concern arises from the adverse combination of geological processes and ground conditions precipitated by human activity with the potential to cause harm. Examples may be drawn from a wide range of situations selected from many parts of the world. These include slope instability, subsidence, volume change, hazards relating to water, erosion, seismicity, volcanism, glacial and periglacial phenomena, and pollution. Knowledge concerning several processes, notably slope instability and subsidence, have reached a level of maturity but others, such as natural pollution, are still underrepresented and are in need of further research and development.

The causes of adverse ground behaviour are of scientific interest, but it is the consequences which are of greatest concern to the engineering profession. The properties of the geological profile and the processes operating within it determine where and when an adverse combination of circumstances might become linked together, quite possibly induced by human action, so bringing about an imbalance that triggers failure. Taking into account the likelihood of a geologically-related hazard occurring, together with an assessment of the vulnerability of people, property, and environment in the vicinity, provides support for decisions as to what should be done to mitigate the consequences.

Engineering geologists essentially interpret geological information for use by others, applying judgement as appropriate to provide the knowledge necessary for effective decision-making concerning resource abstraction and identification of constraints on development and regeneration. In essence, the aim is to identify the most probable geological conditions together with the most unfavourable conditions that might be plausible, and an estimate of their likelihood. This informs the broader hazard assessment and risk analysis procedures undertaken for most major engineering projects.

The general approach adopted is to build a three-dimensional model in advance of the site investigation, using existing knowledge of the distribution and properties of the geological materials and the geological processes acting upon them. Geological materials comprise three components: solids, liquids, and gases. Of equal relevance to the natural geological materials classically studied by geologists (bedrock, or 'Solid', geology) are the Quaternary deposits (superficial, or 'Drift' geology), artificial deposits ('Made Ground') and the weathered profile.

The conceptual ground model can then be used to plan the site investigation in an interactive way to reduce uncertainty without adopting unnecessarily expensive approaches (for example, close-spread fixed separations between boreholes). The potential exists for financially-based decision-making in terms of value, for example by linking reduction in geostatistical error of estimation with improvement in uncertainty (that is, risk reduction), and so develop an optimal site investigation design.

## Social Context

The social and economic significance of ground failure is still largely underestimated, particularly in urban and suburban areas. Essentially the impact of each of a series of possible strategies which could be pursued needs to be measured, including the option of doing nothing. The aim should be to establish cost effective and socially acceptable management of the built environment. The scale of the effect brought about by the occurrence of adverse ground behaviour determines who might be affected and therefore who needs to know. This could range from the owner of a property damaged by a local landslide up to the national government for a major catastrophe such as the devastation brought about by a high magnitude earthquake.

The overview and focus for advice is undertaken by the planning profession. The most important step for planners and the public alike is to become more aware of the importance of the nature of the ground and of the need to take this into account. The method of presenting relevant geological information is crucially important if it is to have effect. Users often have little or no knowledge of geology, the likely behaviour of the ground, or know what effect unexpected ground conditions and the causal processes might have on their use or enjoyment of that ground.

The most widely used tool for communicating ground conditions is the geological map, indicating the general (geological) character of the ground. However, this requires skilled interpretation to derive an indication of any potential ground-related problems. What the user requires is an indication of when and what advice is needed, and where it can be obtained, linked to a database of relevant information concerning each location.

The development of digital techniques for the manipulation and presentation of three-dimensional spatially defined data (notably Geographical Information Systems – GIS) is revolutionizing the way spatial data can be portrayed. It is now possible to interactively alter a geological model as new data becomes available from the site investigation, and soon it will be possible to overlay the proposed engineering works and iterate the design to reflect a variety of alternatives, observing the effects as the ground model changes. As important is the need to enhance public perception, increasing education, awareness, and information.

## Further information

Further information on engineering geology as a profession may be obtained from the Engineering Group at the Geological Society of London (http://www.geolsoc.org.uk/template.cfm?name=geogroup10), from the Secretary of the Association of Engineering Geologists (http://www.aeg.com), and from the General Secretary of the International Association of Engineering and the Environment (http://cgi.ensmp.fr:88/iaeg/). General information on what is covered by engineering geology is summarized in Masters course introductions, for instance those at Leeds University (http://earth.leeds.ac.uk/msc/eng1.htm) and at Imperial College, London (http://www.cv.imperial.ac.uk/research/soils/engeo/enggeol%20home%20page1.html). General information on what is encompassed by the term 'geohazard' is summarized by the British Geological Survey (http://www.bgs.ac.uk/enquiries/hazards.html) and the Australian Geological Survey (http://www.ga.gov.au/urban/factsheets/geo_index.jsp). The professional requirements to register as a Chartered Geologist are set out by the Geological Society of London (http://www.geolsoc.org.uk/template.cfm?name=chartered) and the requirements to register as a Euro Geologist are set out by the European Federation of Geologists (http://www.eurogeologists. de/).

## See Also

**Engineering Geology:** Codes of Practice; Aspects of Earthquakes; Geological Maps; Geomorphology; Geophysics; Seismology; Natural and Anthropogenic Geohazards; Liquefaction; Made Ground; Problematic Rocks; Problematic Soils; Rock Properties and Their Assessment; Site and Ground Investigation; Site Classification; Subsidence; Ground Water Monitoring at Solid Waste Landfills. **Environmental Geology. Geological Engineering. Soil Mechanics.**

## Further Reading

Anon (1993) *Without Site Investigation Ground is a Hazard.* Report of the Site Investigation Steering Group of the Institution of Civil Engineer. London: Thomas Telford.

Blyth FGH (1943) *A Geology for Engineers.* London: Edward Arnold.

Clayton CRI (2001) *Managing Geotechnical Risk*, p. 80. London: Thomas Telford.

Fookes PG (1997) Geology for Engineers: the Geological Model, Prediction and Performance. *Quarterly Journal of Engineering Geology* 30: 293–424.

Goodman RE (1999) *Karl Terzaghi. The Engineer as an Artist.* Reston; Virginia: American Society of Civil Engineers.

Kiersch GA (ed.) (1991) *The Heritage of Engineering Geology; The First Hundred Years.* Boulder, Colorado: Geological Society of America, Centennial Special Volume 3.

Knill JL (2001) Environmental change and engineering geology: our global challenge. In: Marinos PG, Koukis GC, Tsiambaos GC, and Stournaras GC (eds.) *Engineering Geology and the Environment*, 4, pp. 3355–3361.

Knill JL (2003) Core values: the first Hans-Cloos lecture. *Engineering Geology* 62: 1–34.

Legget RF (1939) *Geology and Engineering.* New York: McGraw-Hill.

McCall GJH, de Mulder EFJ, and Marker BR (eds.) (1996) *Urban Geoscience.* Rotterdam: Balkema.

Müller-Salzburg L (1976) Geology and engineering geology. Reflections on the occasion of the 25th anniversary of the death of Hans Cloos. *Bulletin of the International Association of Engineering Geology* 13: 35–36.

Paige S (ed.) (1950) *Application of Geology to Engineering Practice.* Berkey Volume. New York: Geological Society of America.

Rosenbaum MS and Culshaw MG (2003) Communicating the risks arising from geohazards. *Journal of the Royal Statistical Society, Series A* 166(2): 261–270.

Terzaghi K (1925) *Erdbaumechanik auf Bodenphysikalischer. Grundlage.* Vienna: Franz Deuticke.


# Codes of Practice

**D Norbury**, CL Associates, Wokingham, UK

© 2005, Elsevier Ltd. All Rights Reserved.

## Introduction

The practice of engineering geology requires effective communication of observations, test results, and a conceptual model of the ground. This communication has to be unambiguous and clearly understood if the works are to proceed smoothly. Engineering projects have become increasingly international, increasing the importance of clear communication. National codification of descriptive terminology and field and laboratory test procedures has been appearing over the last 30 years; the next step is for these national standards to be subsumed within international standards.

## The History of Codification

Engineering geology as an established professional practice has been in existence for some 70 years, although some may argue that the practice has been around for as long as man has been carrying out engineering works in and on the ground. As the industry grew it became increasingly clear that the meanings of words, observations, and results were too often being misunderstood, making effective work increasingly difficult.

Initially there was no published guidance, but a range of publications aiming to standardize practice have emerged in two distinct areas. The procedures to be used in the field and in the laboratory have become standardized: guidelines have been prepared at a national level but with limited coordination between countries. However, the description of soils and rocks, which is arguably the basis of all engineering geological studies, has not seen the same rapid progress. Most of the guidance in this area has been advisory rather than compulsory, possibly because geologists tend to be independently minded practitioners.

As construction projects and engineering geology have become increasingly international, the need for common procedures and practices has increased. One of the primary aspirations of the International Organization for Standardization (ISO) is to provide such commonality, leading to better communication and fairer competitive tendering for work.

The development of standards essentially takes place in committee and is coordinated by the Comité Européen de Normalisation (CEN) and the ISO.

This article outlines the history of the development of codes in the practice of engineering geology, largely by reference to publications in the UK, which is the author's base of experience. Developments in other countries have been along similar lines at similar times, so the example of the UK situation is a useful illustration of the general picture. Examples of standards from other countries are included in the list of Further Reading. We are looking at a profession where the guidance is moving from national and advisory to international and normative.

## What is the Role of Engineering Geology?

Engineering geology is a core component of the profession of ground engineering, which concerns engineering practice with, on, or in geological materials. Ground engineering is concerned with the well-being and advancement of society, including

- the safety of residential, commercial, and industrial structures,
- the essential supply of energy and mineral resources,
- the mitigation of geological hazards,
- the alleviation of human-induced hazards,
- the efficient functioning of the engineering infrastructure, and
- contributing towards a sustainable environment.

Ground engineering is based on the professional input of geologists and engineers, and specifically includes the scientific disciplines of engineering geology, soil mechanics, rock mechanics, hydrogeology, and mining geomechanics. Examples of work activities are as follows.

- Geotechnics is concerned with the foundations of any type of building or structure, such as dams and bridges, and with excavations, slopes, embankments, tunnels, and other underground openings.
- The exploitation of natural resources involves surface and underground mining, the extraction and protection of groundwater, and the extraction of natural materials for construction, hydrocarbons, and geothermal energy.
- Geo-environmental considerations include protection and conservation of the geological environment, rehabilitation of contaminated land (soil and groundwater) and of mining areas, waste disposal (domestic and toxic), and the subsurface emplacement of chemical and radioactive waste.
- Geo-risk is the process of mitigating geological hazards (e.g. earthquakes, slope instabilities, collapsible ground, gas) in land-use planning.

Ground engineering is of considerable economic importance and benefit to society because it provides a means of building efficient structures and facilitating the sustainable use of resources and space. This is often not fully appreciated by the general public. In stark contrast to other engineered structures, most geoengineered solutions are hidden in the ground and so are not visible. Nevertheless, ground-engineered structures can present a major challenge to engineering design and construction and, if successfully completed, are testament to substantial technological and intellectual achievements.

The execution of such projects requires input from a range of scientific and engineering specialties, and the relevant specialists must be able to communicate with each other in order to agree on conceptual models and parameters to apply to the design and must leave an audit trail to ensure quality and safety. In addition, and perhaps even more importantly, there is a need to communicate with other interested parties, not least the owner of the project and the general public. Subjects on which efficient communication is required include observations of the condition of the ground in and around the works and the quality of that ground as revealed by physical records, the logging of cores or exposures, and parameters measured in field and laboratory tests. In order for such communication to be possible, an internationally agreed library of linguistic and scientific terminology, test procedures, and overall investigation processes needs to be available.

## What are Codes in Engineering Geology?

Before considering the trends and requirements in the codification of the practice of engineering geology, it is important to remind ourselves of the role of the practitioner in this field. The fundamental role of an engineering geologist is to observe and record evidence of geological conditions at the site of proposed or current engineering works and to communicate these observations to other (non-geological) members of the team. The evidence for the ground conditions may be in the form of exposures, such as cliff or quarry faces, or may be in the form of cores or samples recovered from boreholes. It is almost universally the case that this geological information comes from the proximity, but not the actual location, of the proposed works. There may also be indirect information, such as the results of geological mapping or geophysical surveys or evidence from previous engineering works in the same area or geological setting. The engineering geologist therefore has to develop an understanding of the geology of the area and make predictions about the geology that will be encountered by, or will affect or be affected by, the engineering works. It is rare for the geologist to have sufficient information to understand the ground conditions fully, and there is always a point beyond which further investigation cannot be justified by a further reduction in uncertainty. It is therefore not uncommon for the geologist to have less information than might be obtained from a small number of boreholes. For instance, road and rail tunnels driven at low level through mountains cannot sensibly be investigated: borehole locations may not be available, and the cost of drilling hundreds of metres before reaching the zone of interest can be prohibitive.

Notwithstanding the source and detail of the information available, the engineering geologist has to collate and interpret the geological information in order to produce a realistic geological model that includes realistic assessments of the degree of uncertainty. The first stage is to create an essentially factual model, before moving on to the interpretation phase. The key aspect of the engineering geologist's role then comes into play: the communication of all aspects of this conceptual model to other members of the design team, the project owner or client, and, increasingly, the public.

To some extent this communication of information can be carried out using existing geological nomenclature in a qualitative sense. However, such an approach by geologists has often left listeners confused. Usually, standard geological nomenclature is qualitatively, rather than quantitatively, defined, so even other geologists can be left uncertain as to the exact meaning intended. Over the years, a language of better-defined terms has developed, which should better enable geologists to communicate, not least because there is now a core of standard terminology with which the listeners will be familiar. It is the derivation and definition of this standard terminology that is one of the main reasons for recent advances in the drafting and implementation of codes in engineering geology.

The intention of the ISO and CEN in preparing international standards is to help raise levels of quality, safety, reliability, efficiency, compatibility, and interchangeability, and to provide these benefits economically. Standards contribute to making the development, manufacturing, and supply of products and services more efficient, safer, and cleaner. They make trade between countries easier and fairer. ISO and CEN standards also safeguard consumers, and users in general, of products and services and make their lives simpler.

## The History of Codification in Description

Prior to 1970, there was no international standard terminology to allow the communication of descriptions of geological materials or their properties, although some of the larger contracting companies had developed internal guides. The first British Code of Practice (CP 2001) was published in the UK in 1957. This code laid down key underlying precepts for the description of soils, in particular that soils should be described in terms of their likely engineering behaviour. This basic need is often lost in today's welter of published guidance. However, CP 2001 did not cover the description of rocks. Geological sciences continued to develop an increasingly variable use of terminology, but the increasing size of the ground engineering industry made the terminology increasingly irrelevant and no longer tenable. The need for a defined and wide-ranging terminology had arrived.

The use of undefined and narrow terminologies caused confusion and ambiguity in communication and, as a result, frequent contractual arguments arose, often leading to claims based on 'unforeseen ground conditions'. This was hardly surprising as, if the terminology is variable and undefined, there will always be someone who could misread the ground conditions being predicted. For instance, terms such as 'highly fissured' and 'moderately jointed' were not defined and therefore meant different things to different readers. This was addressed when the Engineering Group of the Geological Society of London published, in the early 1970s, a series of Working Party Reports for guidance, for example on core logging (in 1970) and the preparation of maps and plans (in 1972). These reports formed the basis of UK practice and, as it turned out, international practice in many respects. Similar activities on the international scene resulted, by 1981, in publications from the International Association of Engineering Geologists and the International Society of Rock Mechanics on field investigation, geological mapping, and soil and rock description. In the UK, this decade of guidance culminated with BS 5930 (published in 1981), the seminal National Standard for engineering geological activities. It is important to note, however, that even at this stage BS 5930 was designated as a code of practice, meaning that the guidance was advisory rather than normative (i.e. mandatory). This designation was maintained in the updated version of BS 5930, published in 1999. The reasons for this relate to the preference of many geologists to not be given edicts on geological terminology. However, the designation of codes as advisory has little practical relevance. The codes are referenced in contract specification documents and thus become binding. In legal arguments about claims or failures, the courts will expect the national guidance to have been followed. Therefore, at least by default, the codes of practice have become *de facto* Standards.

The publication of international guidance has continued, albeit at a slower pace, with individual nations publishing national guidance documents. The designation of such guidance as mandatory or advisory varies, and it is this variance that is now being resolved in the name of international normalization.

## Particular Problem Areas in Combining National Codes

Despite the codifications in various countries proceeding independently, there are many cases where the practices of one country have been adopted by another. For this reason, the preparation of international codes by the ISO has not been as difficult as might have been anticipated, at least as far as the description of soils and rocks is concerned. There have nevertheless been a few difficulties, as outlined below. However, the historical development of local codes has tended to reflect and emphasize local geological conditions, and the classifications were rather more difficult to bring together into an all-embracing international standard. This proved particularly difficult for the classification of soils and resulted in the need for a simple and separate ISO Standard on this topic.

The Scandinavian countries have different types of soil (coarse glacial deposits and fine 'quick' clays) from Italy (volcanic sands) and Japan (silts and loose sands that are liable to liquefaction). National practice has, for sound technical reasons, tended to reflect the regional geological source materials, active geological processes, and impacts of climate and time of exposure, all of which vary across the world. The work carried out by the ISO Technical Committees has needed to incorporate these variations into practice.

### Fine-Grained Materials

The definition of the grain-size boundaries used to classify soils and rocks for engineering purposes has, by and large, developed along similar lines throughout the world, as shown in **Table 1**. Exceptions to this are the USA, where a range of definitions are available, and some Pacific Rim countries. The agreement is wide ranging and complete except for the finer-grained materials, specifically rocks, where terms such as claystone, siltstone, mudstone, shale, and slate are all variously used. The simplification of the terminology to mirror the descriptive terms for soils is proposed in the latest codes.

### Weathering Classifications

Weathering classifications are provided in a number of the existing national and international guidance documents, but there is little commonality between publications. The provision of such classifications within a descriptive framework has proved unhelpful, as weathering profiles are rarely amenable to global classification (because they depend strongly on climate and local topography) and fail to embrace the actual description of the weathering features that are present. For this reason, UK practice, since the 1999 publication of BS 5930, requires the description of weathering profiles. Further classification schemes are permitted only if relevant, available, useful, and unambiguous. However, this practice is not yet generally accepted at an international level.

### Core Indices

Core indices such as Rock Quality Designation (RQD) have been proposed for the logging of rock not only from borehole cores but also from the mapping of exposures, in order to provide a ready indicator of rock quality. A number of indices and a range of definitions have been proposed, without much commonality. Recent International Standards have addressed this and provide unambiguous and clear definitions of which fractures should and should not be included in the quality assessment and of how the index should be measured. Although the RQD is only a crude indicator of rock quality, it is very widely used, on its own and correlated with other properties,

**Table 1** Particle size definitions

| Soil name | Soil fractions | Dimensions | Sedimentary rock name |
| --- | --- | --- | --- |
| Boulders |  | over 200 mm |  |
| Cobbles |  | 200–60 mm |  |
| Gravel | Coarse | 60–20 mm |  |
|  | Medium | 20–6 mm | Conglomerate (boulders, cobbles and gravel) |
|  | Fine | 6–2 mm |  |
| Sand | Coarse | 2–0.6 mm | Coarse-grained sandstone |
|  | Medium | 0.6–0.2 mm | Medium-grained sandstone |
|  | Fine | 0.2–0.06 mm | Fine-grained sandstone |
| Silt | Coarse | 0.06–0.02 mm |  |
|  | Medium | 0.02–0.006 mm | Siltstone (coarse, medium and fine) |
|  | Fine | 0.006–0.002 mm |  |
| Clay |  | less than 0.002 mm | Mudstone or claystone |

Few soils comprise merely a single principal size fraction. Soils usually include secondary constituents of a different size fraction that could significantly influence their engineering behaviour. There are also minor constituents that can help in geological identification but do not influence characteristic behaviour. Claystone is used in some countries to match the soil terms, but this can be confused with nodular concretions. In these cases the term Mudstone is preferred.

in rock-mass classification systems (e.g. compressibility, diggability, tunnel stand-up time). The lack of consistency in the measurement of the index is not widely appreciated, and so the international and normative standardization of the index is long overdue. This neatly illustrates the need for standardization wherever a property, however simple, is used to communicate a measure of ground quality. It will be valid only if the measurement process is consistently applied and understood.

The ongoing evolution of the codification of terminology and test procedures means that users need to be aware of the date of data collection when considering the descriptive and measurement standards that are likely to have been incorporated in borehole logs, maps, and records. In particular, archival records are likely to have been prepared according to different, outdated, Standards and definitions.

## History of the Codification of Field and Laboratory Testing

The standardization of field and laboratory procedures is as important as the standardization of the description and includes forming the borehole or describing the exposure, executing field tests, recovering samples, and carrying out laboratory tests. If the results of any of these activities are to be applicable and relevant in the minds of others, the procedures used need to be clearly identifiable and standardized.

The codification of laboratory test procedures for soils commenced in the 1940s, when the first machines were developed. The procedures covered everything from how to drill a borehole, conduct the basic field tests, and take and describe samples, to their storage, transport, and laboratory testing. In fact, even the design of the testing machines had to be specified as test procedures evolved. Work by the ISO and CEN that is underway at the time of writing will describe the procedures and concentrate on the preparation of Technical Specifications (**Table 2**), building on earlier test procedures drafted by the International Society of Soil Mechanics.

The testing of rocks in commercial practice started somewhat later, by which time the need for international cooperation was better appreciated. It has therefore been possible for the rock-testing procedures to be better established, under the auspices of the International Society of Rock Mechanics, who have published a series of suggested methods. As there were no precedent procedures in place, these suggested methods have been rapidly adopted by the professional community and have become internationally recognized without the involvement of national standards bodies. It would therefore be comparatively straightforward, but not necessarily easy, to prepare normative international standards for most rock tests.

## Professional Qualifications

A further aspect of codification is the identification of relevant qualifications and experience necessary for those who plan, execute, and interpret ground investigations. The guidance documents and codes prepared up to the end of the 1980s did not try to lay down rules about the qualifications and experience required by those working as specialists in engineering geology. In the very early days, this was felt to be unnecessary; the number of practitioners was small, and their capabilities and limitations were known through reputation. This has become increasingly less reliable, with even academic programmes of training changing significantly from previously accepted standards of coverage and achievement. It is now necessary to define the roles and who may be permitted to practice. It is interesting to consider who benefits most from such codification. Is it the client, who can feel better protected by the knowledge that they have properly trained and experienced professional advisors, is it the insurers, who feel they have lower exposure to risk, is it the individual practitioners, who feel this improves their status in society, is it the employers, who can recognize a qualified practitioner, or is it the companies, who can see a market with fair competition? The truth is probably a combination of all of these. The position taken by the Standards Institutions is based on the latter view, and ISO/CEN documents currently in preparation include definitions of specialist practitioners. These definitions will therefore become requirements in the practice of ground engineering.

The development of such defined roles is closely linked to the development of Directives in the European Union. In order to facilitate the mobility of workers, the availability of internationally recognized qualifications is essential. The Directive on this subject is likely to be enacted by about 2005. The requirement for practitioners to be able to practice, at least for limited periods, in any European Union country is the holding of a recognized qualification. This qualification is likely to be represented by the common platform of the European Federation titles of EurIng, awarded by the European Federation of Engineers, and EurGeol, awarded by the

**Table 2** Codes in preparation by CEN/TC 341 and ISO/TC 182/SC 1 concerning geotechnical investigation and testing

| ISO reference number | Title | Committee draft | Draft International Standard | Final draft International Standard | Publication as EN/ISO standard | Remarks |
|---|---|---|---|---|---|---|
| 14688-1 | Identification of soil | | 2001–June | 2002–March | 2002–June | ISO |
| 14688-2 | Classification of soil | | 2001–September | 2002–September | 2002–December | ISO |
| 14689 | Identification of rock | | 2001–September | 2002–September | 2002–December | ISO |
| 22475-1 | Sampling methods | 2003–September | 2004–September | 2006–March | 2006–June | |
| 22475-2 | Sampling – Qualification criteria | 2003–September | 2004–September | 2006–March | 2006–June | |
| 22476-1 | Cone penetration tests | 2003–September | 2004–September | 2006–March | 2006–June | |
| 22476-2 | Dynamic probing | 2001–December | 2002–September | 2004–September | 2004–December | |
| 22476-3 | Standard penetration test | 2001–December | 2002–September | 2004–September | 2004–December | |
| 22476-4 | Menard pressuremeter test | 2003–September | 2004–September | 2006–March | 2006–June | |
| 22476-5 | Flexible dilatometer test | 2003–September | 2004–September | 2006–March | 2006–June | |
| 22476-6 | Self-boring pressuremeter test | 2003–September | 2004–September | 2006–March | 2006–June | |
| 22476-7 | Borehole jack test | 2003–September | 2004–September | 2006–March | 2006–June | |
| 22476-8 | Full displacement pressuremeter | 2003–September | 2004–September | 2006–March | 2006–June | |
| 22476-9 | Field vane test | 2003–September | 2004–September | 2006–March | 2006–June | |
| 22476-10 | Weight sounding test | 2002–June | | | 2002–December | TS |
| 22476-11 | Flat dilatometer test | 2002–June | | | 2002–December | TS |
| 22477-1 | Testing of piles | 2003–September | 2004–September | 2006–March | 2006–June | |
| 22477-2 | Testing of anchorages | 2003–September | 2004–September | 2006–March | 2006–June | |
| 22477-3 | Testing of shallow foundations | 2003–September | 2004–September | 2006–March | 2006–June | |
| 22477-4 | Testing of nailing | 2003–September | 2004–September | 2006–March | 2006–June | |
| 22477-5 | Testing of reinforced fill | 2003–September | 2004–September | 2006–March | 2006–June | |
| 17892-1 | Water content | 2002–December | | | 2003–June | TS |
| 17892-2 | Density of fine-grained soils | 2002–December | | | 2003–June | TS |
| 17892-3 | Density of solid particles | 2002–December | | | 2003–June | TS |
| 17892-4 | Particle size distribution | 2002–December | | | 2003–June | TS |
| 17892-5 | Oedometer test | 2002–December | | | 2003–June | TS |
| 17892-6 | Fall cone test | 2002–December | | | 2003–June | TS |
| 17892-7 | Compression test | 2002–December | | | 2003–June | TS |
| 17892-8 | Unconsolidated triaxial test | 2002–December | | | 2003–June | TS |
| 17892-9 | Consolidated triaxial test | 2002–December | | | 2003–June | TS |
| 17892-10 | Direct shear test | 2002–December | | | 2003–June | TS |
| 17892-11 | Permeability test | 2002–December | | | 2003–June | TS |
| 17892-12 | Atterberg limits | 2002–December | | | 2003–June | TS |

TS: technical specification, not initially normative.
All document-drafting committees are led by CEN except where noted to the contrary.

European Federation of Geologists. These titles show that the bearer has undertaken an appropriate course of study, carried out appropriate training, and gained sufficient experience to be able to act as a professional engineer or geologist, and that this record has been submitted to the bearer's peers for validation. The holder of such a title agrees to work within the code of conduct operated by the awarding Federation and will be able to work in any European country, at least for limited periods, without needing to qualify separately in that country. These are major developments in providing commonality of professional standards and represent development exactly as hoped for by the ISO and CEN, but driven forward by the European Commission.

## Codification into the Twenty-First Century

After 25 years in preparation, the suite of Eurocodes is, in the early 2000's, becoming a reality. These Eurocodes bring together codes of practice for building and civil engineering structures, and provide a world-class standard for all aspects of construction. Included in Eurocode 7 (geotechnical design) are elements of codes on the description and classification of soils and rocks, field investigation methods, field and laboratory testing, assessment of engineering parameters, and design procedures. For the first time, engineering geologists throughout Europe will be talking in a common language when reporting the findings of their field observations. This progress extends beyond Europe, as, in accordance with the Vienna agreement, the standards drawn up by CEN and ISO undergo parallel voting procedures for common adoption. Thus, for example, the proposals for the description of soils and rocks prepared by the ISO in 2003 will also be incorporated into Eurocode 7. Thus, engineering geologists around the world will be able to pass on their geological information without misunderstanding and ambiguity. The Japanese 'brown sandy clay' will be the same as the Swedish 'brown sandy clay'. Similarly, the results of field or laboratory testing will be transferable around the world. Major exceptions to this rule are China and the USA, who are not members of either of these standards bodies and who have had no input into the drafting of the codes.

A schedule of the codes and specifications being prepared by the ISO and CEN in this area is given in **Table 2**. The codes for the description and classification of soils and rocks were prepared by a Technical Committee of the ISO. The reason for this work item being proposed initially was, in accordance with the ISO mission, to encourage and allow international communication in applied science and therefore to allow more and fairer competition in international trade. This enshrines the concept of engineering geologists the world over sharing a single interpretation of a conceptual ground model. They will therefore be able to compete equally for contracts and cooperate more effectively on design briefs.

The primary benefit of such codes is that they will standardize ground investigation, reporting, and design approaches that are common over much of the world. The market for investigation and design, and thus the market for engineering geology, thus becomes both much larger and much more competitive, but with a level playing field.

However, the Eurocodes do not completely subsume the national practices, which have been built up over the years and which are pertinent to local situations. This is right and proper given that variations in engineering geological practice have their base in the different geological conditions in different countries. For instance the Scandinavian extensive crystalline basement crust and soft 'quick' clays varies significantly from the deep weathering profiles of the tropics. These different conditions require different approaches to investigation and testing. The description of the materials can nevertheless be based on a single standardized approach.

The national differences in the approach required can be incorporated into National Annexes, which allow key safety and technical issues to remain a national rather than a European responsibility and allow geological and climatic variations to be taken into account. However, these National Annexes are enhancements of, rather than local rewrites of, the overarching international codes.

## Concluding Remarks

Over a remarkably short time-span, engineering geology has developed from a situation where the early professionals acted as individuals and managed without codes to today's world where teamwork predominates and there is an increasing need for accountability. The codes coming into place in the 2000s define how we should drill holes, take and test samples, and describe soils and rocks. Perhaps the biggest change will be that we will have to carry internationally recognized qualifications if we want to be able to practice all over the world.

## See Also

**Engineering Geology:** Geological Maps; Site and Ground Investigation; Site Classification. **Geological Engineering. Geology, The Profession. Geotechnical Engineering. Sedimentary Rocks:** Mineralogy and Classification. **Tectonics:** Fractures (Including Joints). **Weathering**.

## Further Reading

ASTM D 4879-89 (1989) *Standard Guide for Geotechnical Mapping of Large Underground Openings in Rock*. West Conshohocken: ASTM International.

ASTM D2487-93 (1993) *Classification of Soils for Engineering Purposes (Unified Soil Classification System)*. West Conshohocken: ASTM International.

ASTM D3282-93 (1993) *Standard Classifications of Soils and Soil Aggregate Mixtures for Highway Construction Purposes*. West Conshohocken: ASTM International.

ASTM D5878-95 (1995) *Standard Guide for using Rock Mass Classification Systems for Civil Engineering Purposes*. West Conshohocken: ASTM International.

ASTM D653-96 (1996) *Standard Terminology Relating to Soil, Rock, and Contained Fluids*. West Conshohocken: ASTM International.

Brown ET (ed.) (1978) *Rock Characterisation, Testing and Monitoring. ISRM Suggested Methods*. Oxford: Pergamon Press.

BS 5930 (1981) *Code of Practice for Site Investigations*. London: BSI.

BS 1377 (1990) *Methods of Test for Soils for Civil Engineering Purposes: Parts 1 to 9*. London: BSI.

BS 5930 (1999) *Code of Practice for Site Investigations*. London: BSI.

Deere DU and Deere DW (1988) The rock quality designation (RQD) index in practice. In: *Rock Classification Systems for Engineering Purposes*. ASTM STP 984. West Conshohocken: ASTM International.

DIN 4022-1 (1987) *Classification and Description of Soil and Rock. Borehole Logging of Rock and Soil not Involving Continuous Core Sample Recovery*. Berlin: Deutsche Norm.

DIN 18196 (1988) *Soil Classification for Civil Engineering Purposes*. Berlin: Deutsche Norm.

Engineering Group of the Geological Society of London (1970) The logging of rock cores for engineering purposes. Working Party Report of the Engineering Group of the Geological Society of London. *Quarterly Journal of Engineering Geology* 3: 1–25.

Engineering Group of the Geological Society of London (1972) The preparation of maps and plans in terms of engineering geology. Working Party Report of the Engineering Group of the Geological Society of London. *Quarterly Journal of Engineering Geology* 5: 297–367.

Engineering Group of the Geological Society of London (1977) The description of rock masses for engineering purposes. Working Party Report of the Engineering Group of the Geological Society of London. *Quarterly Journal of Engineering Geology* 10: 355–388.

Engineering Group of the Geological Society of London (1995) Description and classification of weathered rocks for engineering purposes. Working Party Report of the Engineering Group of the Geological Society of London. *Quarterly Journal of Engineering Geology* 28: 207–242.

IAEG (1981) Rock and soil description and classification for engineering geological mapping. Report by IAEG Commission on engineering geological mapping *Bulletin of the Engineering Geology and the Environmental* 24: 235–274.

ISO (2003) *Geotechnical Engineering – Identification and Classification of Soil – Part 1 Identification and Description*. EN 14688 – 1. ISO/TC 182/SC 1. Geneva: International Organization for Standardization.

ISO (2003) *Geotechnical Engineering – Identification and Classification of Soil – Part 2 Classification*. EN 14688 – 2. ISO/TC 182/SC 1. Geneva: International Organization for Standardization.

ISO (2003) *Geotechnical Engineering – Identification and Classification of Rock*. EN 14689. ISO/TC 182/SC 1. Geneva: International Organization for Standardization.

Japanese Geotechnical Society (2000) *Method of Classification of Geomaterials for Engineering Purposes*. 0051-2000. Tokyo: Japanese Geotechnical Society.

Kulander BR, Dean SL, and Ward BJ (1990) *Fractured Core Analysis: Interpretation, Logging and Use of Natural and Induced Fractures in Core*. AAPG Methods in Exploration Series 8. Tulsa: American Association of Petroleum Geologists.

Swedish Geotechnical Society (1981) *Soil Classification and Identification D8:81*. SGF Laboratory Manual, Part 2. Byggforskringsrådet. Linköping: Laboratory Committee of the Swedish Geotechnical Society.

# Aspects of Earthquakes

**A W Hatheway**, Rolla, MO and Big Arm, MT, USA

© 2005, Elsevier Ltd. All Rights Reserved.

## Introduction

The main product arising from the work of engineering geologists is their site characterizations for engineered works. In terms of the impact of earthquakes and their mitigation, design engineers, architects, and planners need to both accommodate people and provide functionality. With these goals in mind, engineering geologists develop three-dimensional descriptions of the ground to the depth that will be affected by the static loads resulting from the engineered works and by the expected levels of earthquake-induced ground motion.

The damaging effects of earthquakes can be mitigated, but the earthquakes themselves cannot yet be controlled or stopped. Engineering geology can be applied in a variety of ways to assist in mitigating the effects of earthquakes; for instance by

- recommending the avoidance of sites that are likely to experience ground displacement as a result of fault movement;
- grading the site to a configuration that promotes stability of the ground mass;
- improving the engineering characteristics of the ground so as to resist damaging deformation;
- controlling groundwater that is likely to experience excess pore pressure leading to ground failure under the involved structural loads; and
- creating green space to isolate active faults and possible seismically induced slope failures.

Engineering geologists also record details of and analyse the damage done by earthquakes (**Figure 1**). In this context, the role of the geologist is expanded from considering the known structural-loading conditions imposed by the designers to considering the unknown, but anticipated, ground motion during a hypothetical earthquake felt to be appropriate to the role and function of the project. Worst-case-earthquake design predictions are appropriate for critical facilities or for projects housing dependent populations. Critical facilities are those projects for which the consequences of failure are intolerable.

Worst-case-earthquake design is based on the concept of the maximum credible earthquake (MCE), which is the greatest magnitude and duration of strong motion that can reasonably be expected to occur.

Ancillary to this concept is risk analysis to define magnitudes and durations of strong motion for which there are cost benefits and for which damage can be tolerated.

**Figure 1** Engineering geological mapping records evidence of earthquake motion such as this railroad deformation indicating north-to-south shortening of the ground along the rail right of way. Most evidence of earthquake damage is fragile and likely to be destroyed by human activity within hours or days. (Photographed by Richard J. Proctor, then Chief Engineering Geologist, Metropolitan Water District of Southern California, at San Fernando Valley, California, February 1971.)

## Why Ground Fails During an Earthquake

The behaviour of the ground during an earthquake depends not only on the character of the incidental ground motion but also on the properties of the site subsurface. Engineering geology is used in this context to detect and delimit bodies of ground that are most susceptible to earthquake damage and to identify, where possible, the interfaces between geological units where shaking displacements may be concentrated or otherwise focused.

Engineering geological investigation looks for horizons and pockets of damage-susceptible soil and potential surfaces of failure (displacement, created where differential stress is greatest and exceeds *in situ* shear strength). For rock masses, the failures representing displacements generally occur along pre-existing discontinuities.

Geometry plays a strong role in the development of earthquake-induced failure conditions. In the worst case, the gross motion vector generally enters a hillside mass and leaves at a hillside or some form of cut slope where there is no restraint to the passing, outward-bound ground motion, hence leading to failure. River banks and incised stream valleys are very susceptible to such conditions. Added to this basic geometry, if rainfall has been heavy for a week or more, the earth media have been infiltrated by rain or snowmelt, or, for reservoirs, drawdown has been recent and of some magnitude (a few metres or more) then a significantly unstable situation will result.

Where soil masses serve as the foundation and have little or no lateral ground support, new soil failure surfaces can be created leading to a condition of lateral spreading, where the foundation shifts in the direction of least lateral support, generally causing considerable damage to the affected component of the engineered structure.

Another poor geological situation is where major open rock joints strike at an oblique angle to the face of a hillside or cut, dipping towards the face at a dip angle greater than that of the internal friction of the failure surfaces, which themselves form a wedge of rock pointing out of the hillside.

Little warning of impending failure is given (perhaps just a couple of seconds in which audible grinding noises are heard) before the accelerated mass of debris, separated from the hill by the surfaces of the geometric wedge, begins to move.

Earthquake strong motion typically lasts between 10 and 30 seconds, which, given the presence of the instability factors listed above, is long enough to create a failure motion.

## Linking Earthquakes to Ground Effects

As already noted, engineering geologists assess the likelihood of earthquakes occurring, based on data from observed and historic earthquakes in the same seismotectonic zone. Geological observation becomes the basis for predicting the potential occurrence of various ground effects.

Near-region geological evidence indicates the general seismogenic character of the site. Relevant factors include seismic history (magnitude and frequency of recurrence), palaeoseismic patterns, geomorphological evidence, fault sources, and the delimitation of seismogenic zones within which individual capable faults occur.

Displacement effects occur when bodies of earth materials are displaced along curvilinear failure surfaces (in soil) or along pre-existing discontinuities (in weak rock). Normally the displacement occurs in the direction of least lateral constraint.

Slope failures may affect unstable masses, defined by hillside geometry, valley walls, river banks, and cut slopes, when the arriving earthquake ground motion acts momentarily within the up-gradient mass to reactivate pre-existing rock discontinuities or newly formed curvilinear failure surfaces. Gravitational acceleration moves the disequilibrated earth mass down the slope.

Liquefaction is a dynamic process that mainly affects cohesionless unconsolidated soil in the presence of near-surface groundwater. Earthquake strong motion raises pore-water pressure in such strata, mobilizing the soil, often upward along linear or pipe-like channels, which may reach the surface to generate a fountain, leaving geomorphological evidence in the form of sand blows.

## Geologically Based Mitigation

One of the most successful methods of mitigating potential earthquake damage is to employ engineering geological studies to devise methods by which human development can be governed or designed so as to be less susceptible to seismic damage.

### Avoiding Damage-Prone Areas

Development projects should avoid damage-prone areas, and site characterization should be employed to discover and delimit ground at an unacceptably high risk. This is naturally difficult to accomplish where land development is governed by commercial interests and imperatives. There are two basic means by which development can be limited; both have been successfully employed in the State of California using local laws (generally enacted as city or county ordinances) and State laws.

The first measure is the creation of a greenbelt, whereby ground felt to be prone to damage is assigned to park or promenade use, thereby remaining landscaped but available for use by residents rather than being given over to the construction of residential housing or commercial buildings.

The second measure is the zonation of capable (active) faults. This was devised by the California legislature in 1969. Named after its politician promoters, State Senator Alquist and State Assemblyman Priolo, the Alquist–Priolo Act requires the California Geological Survey to define exclusion zones (440 m wide) bounding the most likely central-axis location of the subject fault and also designed in band-width to meet the usual California alluvial fault-rupture situation in which repetitions of historic ground ruptures often occur with positional variance. The maps are updated regularly and are available for public inspection.

**Improving Ground Conditions**

For engineered works that must be built in seismically active regions, but not within or adjacent to active faults, there remains a variety of geotechnical methods that can improve the ability of the foundation or the surrounding ground to resist the expected MCE ground shaking. Such geotechnical methods generally fall into one of the following categories, all of which require engineering geological input in order to be successfully designed and emplaced.

- Dewatering avoids the pore-pressure build-up that may cause liquefaction or other forms of ground failure. Measures can be passive (e.g. French drains and vertical gravel-packed columns, wick drains and subhorizontal drain pipes) or active (e.g. dewatering points or wells).
- Retention generally uses iron or steel tiebacks and ground anchors to counter vectoral earthquake motion acting outwards from the foundation mass towards unconfined ground (such as hillsides and cuts) and upward acceleration towards the ground surface itself.
- Densification of the ground is usually achieved by dynamic compaction using falling weights (e.g. a demolition ball dropped from a crane).
- Less stable layers, pockets, or zones may be excavated and replaced. Selection of this option is generally controlled on the basis of (low) cost.

**Predicting Collateral Damage**

Earthquake-induced ground motion can dislodge or otherwise set in motion boulders, joint blocks, and slope masses. These situations generally can be identified from photogeological interpretation and confirmed by geological mapping. Each situation will have a slightly different nature, but is likely to fit into one of the following categories.

- Boulder dislodgment is the displacement of boulders by long-term geological processes (mainly slope erosion) or earthquake excitation. Much has been done by the Utah Geological and Mineralogical Survey to mitigate this hazard by educating land developers and home buyers. The situation is perhaps worst along the front of the Wasatch Mountains, at Salt Lake, and in Utah cities between Provo and Ogden.
- Rockfall is perhaps the most prevalent and dangerous of the slope-motion earthquake threats, recognizable mainly along major transportation routes through rugged mountain terrain, especially in areas and at elevations not supporting dense vegetation that would obstruct the passage of debris. The leader in the mitigation of individual-event rockfall is the Colorado Geological Survey.
- Slope failure has been noted primarily in the South American Andes and in the Coast Range immediately south of San Francisco, California. The South American slope failures have been particularly spectacular and perhaps the most devastating in terms of loss of life. Here, the dominant geological factors have been extreme topographical relief, narrow mountain-front valleys, concentrated human occupation in valley-sited towns, and, of course, the general position on the actively subducting Nazca tectonic plate. It is probable that each large earthquake in the greater Andean region will kill hundreds or thousands of people in these mountain-front towns, because huge masses of ground, often supported on naturally compressed air masses, will be dislodged so rapidly that the use of alarm sounds and evacuation principles will be impractical.
- Engineering geological evidence relies on the recognition of the topographical circumstances that will lead to the occurrence of tsunamis in coastal regions. Palaeoseismic evidence of tsunami inundation has been observed along the coast of Oregon. In continental interiors, vertical components of earthquake-induced ground rupture can create a similar inundation effect alongside lakes. All that is required is instantaneous displacement of the down-thrown side of a fault or an associated above-lake slope failure. Either case would displace masses of soil or rock into the body of the lake. This situation is very difficult to mitigate in terms of public rights and awareness.

## Seismotectonic Zonation

Zonation is a technique whereby known and predicted earthquake damage can be assigned to areas for planning, zoning, and regulatory purposes. A number of concepts apply to zoning. Each determination should be made on site-specific needs.

### Active Faults

Where ground rupture or the associated effects of surficial damage have been noted it is appropriate to assign a status of 'active' faulting. State-of-the-art determination is represented by the Alquist–Priolo Active Fault Zones, legislatively mandated in 1969 by the California Legislature. These zones are named, delineated, and maintained by the California Geological Survey and constitute 400 m wide strips centred along the best-identified trace or centre line of the subject fault. Zones are plotted on 1 : 24 000 US Geological Survey topographical quadrangle bases and are available for reference on the world wide web.

### Credible Faults

The definition of credible faults was an innovation of the nuclear power plant safety programme of the US Nuclear Regulatory Commission (USNRC); credible faults are known faults or areas where there is strong geomorphological or structural geological evidence of fault displacement at least once during the Holocene and more than once in the past 500 000 years. The half-million-year bracket is an artefact of the fusilinid age determination resolution affinity of California marine seashore terraces, and was applied to the Diablo Canyon Nuclear Power Station (NPS) in the early 1960s. The concept, however, is not applicable elsewhere, though it is frequently specified. Credible faults are considered capable of experiencing ground rupture and propagation of ground motion along their entire known length. Bonilla was the first to associate ground-rupture length with the magnitude of the causative event.

### Deterministic Design Input

Design-level input by engineering geologists forms the basis for carrying actual earth science evidence into 'seismic withstand' design by civil and other engineers. This geological input consists of the following process of analysis and evaluation. First, the area around the project is seismically zoned, and credible faults are identified or a most likely epicentre for a candidate earthquake is assigned for each seismogenic zone identified. Then, where active faults are defined, the design-input ground motion is considered to occur at the location closest (measured perpendicular to the fault trace) to the site. Where an active fault has not been identified, a most likely location is selected for each seismogenic zone; the expected ground motion is attenuated from the hypocentre to the site. This becomes the MCE for each of the candidate source faults or zones. Finally, the style of earthquake to be represented for each of the candidates is evaluated, in terms of vector displacement, duration of strong motion, and peak ground acceleration.

### Attenuation

This assessment looks at the ground motion from the hypocentre to the project. Waves of ground motion move through the earth media, are concentrated in the bedrock, and there suffer the physical effects of trying to displace the surfaces of innumerable rock discontinuities encountered during the passage to the site. This infinite number of surface-to-surface grinding processes consumes much of the energy of the earthquake.

Attenuation curves have been developed empirically for a variety of geological terranes and are published in the literature, particularly in the reports of the former US Army Waterways Experiment Station (now Engineer Research and Development Command, Vicksburg, MS, USA).

Each geological terrane has its own directional effect on attenuation. The greatest effect is generally exerted perpendicular to strongly foliated older metamorphic terranes, such as the north-east-foliated Lower Palaeozoic sequence of New England, USA.

### Comparison of Candidate Design Earthquakes

A selection of potentially impacting design earthquakes are defined in terms of the parameters listed above. Each candidate is moved through a defined scenario and attenuated from each stipulated source to the site. The worst cases are then declared the MCEs, and their characteristics are used as input in seismic-withstand design.

## Engineering Geological Mapping

A primary method for addressing and meeting seismic threats uses engineering geological mapping, which records the evidence of past earthquake damage and uses this alongside geological conditions.

Earthquake damage was generally not mapped in detail until the Alaskan 'Good Friday' earthquake of 1964. The former (regrettably now defunct) Branch of Engineering Geology responded to this event under the constructive leadership of its Director, the late Edwin B. Eckel, also later the Executive Director of the Geological Society of America. Eckel's lobbying in Washington DC was immediate, and the Survey mounted a field effort to observe and map the damage of the magnitude 8.3 event as a long series of Professional Papers, the content of which has served as the ongoing model for the application of engineering geology to earthquake mitigation learning.

Subsequent American earthquakes have been given similar, yet lesser overall, attention, for instance the San Fernando (California) earthquake of 9 February 1971 (magnitude 6.6) and the Loma Prieta event of 17 October 1989 (magnitude 7.1) located just south of San Francisco (California) on a San Andreas splinter fault.

### Purpose

Strictly speaking, the purpose of engineering geological mapping in this context is to record the physical nature of ground-rupture earthquakes and to elucidate the related types of motion-induced damage (**Table 1**). Once defined, these can be applied on a worldwide basis, wherever similar conditions of tectonics and near-field geology and topography occur.

Engineering geological mapping for earthquake mitigation (**Table 2**) determines how engineered works and human safety can be protected by judicious design considerations, most of which are governed by geological conditions related to the site characterization.

### Geological Profile (or Ground Profile)

This term geological profile has taken over in engineering geology from 'geological section' and considers only the depth of influence of engineering works (generally less than 15 m; **Table 3**).

### Exploration Trenches and Trench Logging

During the peak of nuclear power-plant siting and construction in the 1970s considerable effort was made to confirm the absence of active faults within the footprint of the power block (i.e. the location of the reactor and the critical cooling linkage). The concept of avoiding active fault traces was promoted so that the risk of damage to the reactor and its containment as a result of rupture-type earthquake ground motion could be minimized. As a result, the technique of exploratory trenching advanced from the original method of observation from hand-dug pits (**Table 4**).

## Site Characterization

As an integral part of site characterization in seismogenic regions, attention should be given to identifying geological conditions that may make the site susceptible to physical damage from strong motion (**Table 5**).

## Post-Event Surveys

Certain elements of the earth media are particularly susceptible to being lifted, shifted, toppled, or cracked by earthquake strong motion. The patterns of damage reveal much about the frequency characteristics of the incident ground motion and the relative duration of the strong motion. Particularly affected are fine soils, boulders on slopes, blocks of rock defined by joints, overly steep stream and shore banks and cliffs, and hillside masses saturated with groundwater.

Geologists have but hours to locate, photograph, and map these features before they are destroyed, first by human visitors and soon after by rainfall and other natural erosive agents (**Table 6**).

Cultural features within the built environment (including engineered works) offer additional potential

**Table 1** Engineering geological mapping of earthquake effects

| Elements | Purpose | Important considerations |
| --- | --- | --- |
| Stratigraphy | Identify and describe the geological formational units to be expected in design and construction | Individual engineering geological units |
| Groundwater regime | Define character of groundwater, as it is affected by ground motion and diminishes the shear strength of earth media to resist dynamic deformation | Perched water<br>Vadose zone and fluctuations<br>Peizometric surface<br>Potentiometric surface |
| Rock-mass characterization | Delimit observable or likely subsurface bounds of each detectable hard rock unit | Identify bodies of discontinuity-bounded rock masses that may become unstable from shaking |
| Presence of weak rock | Basis of definition, including why the rock is determined to be weak | Recommendations as to how and why such weak rock may pose problems to design and/or construction, operation, and maintenance |
| Potential problems related to sedimentological, structural, or geomorphological conditions | Portions of surface or subsurface that appear to be related to mapped patterns of earthquake damage | Use special map symbols to portray these features; the Geological Society of London Engineering Geomorphological map symbols are ideal for this purpose |

**Table 2** Engineering geological mapping for earthquake mitigation

| Elements | Purpose | Important considerations |
|---|---|---|
| **Before earthquake** | | |
| Conceptual site geological model | Forms the basis for expectation and detection of geological features critical to the mission of defining geological controls over site-specific earthquake threats | Define the presence and interrelationships between site geology and topographical configuration, to include hydrogeological conditions, as well as the character of existing or planned development |
| Existing evidence of presence of active or credible faults in near proximity to the project | Provide maximum means for avoidance of ground-rupture potential to the project works | Regional geological maps<br>Known active or credible faults<br>Site explorations to preclude the presence of active or credible faults |
| Geomorphological evidence of past displacements or earthquake-induced ground failure | Generally high potential for recurrence of failure or displacement in the future | Image interpretation of aerial photographs, especially those executed at low sun angle with shadows |
| **After earthquake** | | |
| Ground deformational evidence | basis for reconstruction, lessons learned for basic knowledge<br>Locate and plot evidence of ground rupture from activated faults | Map evidence immediately before it is destroyed<br>Provides felt-intensity information useful in meisoseismal mapping for quantification of attenuation |
| Groundwater perturbation | Establish fluctuation in groundwater surface from existing water wells and observation wells (if such are present) | Instructive as basis for pore-pressure effects in deformation as well as liquefactions |
| Evidence of dislodgement | Helps to establish ground acceleration magnitude and vector of ground motion | Sometimes possible to locate fresh slope debris and to establish provenance of dislodgment |
| Displacements | Traces of offset pavement, curbs and traffic-control lines<br>Fracture pattern and offsets in sidewalks and retaining and structural walls | Establishes ground-motion vectors and possible ground rupture |
| Distortion of infrastructure | Collapsed power lines, broken water and sewer pipes, deformed railroad rails | Facility may need significant improvement in replacement, if design or location is faulty in terms of future earthquake motion |
| Water displacement | Wave-motion characteristics and indication of tsunami or seiche character | Evidence will persist for only a few days, even without human intervention |
| Survey triangulation | Establishes bounds of displaced ground | May indicate tilting of ground masses from ground shaking |

**Table 3** Ground profiles for earthquake mitigation

| Elements | Purpose | Important Considerations |
|---|---|---|
| Primary profile | Placed perpendicular to known or most likely trace (strike) of active or credible fault | The profile integrates the Site Conceptual Model with actual site conditions and sets the stage for exploration trenching and logging |
| Display geological conditions related to expected worst-case site response to and damage from the MCE | Promotes rational consideration of all geological site conditions that could conceivably affect performance of the planned works, in the context of earthquake threat | Generally means that one or more exploratory trenches will be placed and logged perpendicular to secondary geotechnical profiles |
| Secondary profile(s) | Examination of potential failure scenarios other than perpendicular to nearby active or credible fault | Prime example would be to construct a geotechnical profile perpendicular to the face of a slope or other face thought to be susceptible to earthquake damage |

evidence to the discerning eye of the engineering geologist. Relatively tall and thin objects (such as gravestones, lamp standards, free-standing walls, plate window glass, roadway signboards, and utility poles and towers) tend to register a deformational response to ground motion. In addition to revealing much about the character of the underlying ground, such damage also gives an insight into the vector, acceleration, and duration of local ground motion.

**Table 4** Exploration trenches and trench logging for earthquake mitigation

| Elements | Purpose | Important considerations |
|---|---|---|
| Orientation of trench | Perpendicular to suspected fault trace (strike) or related geomorphological feature | Maximum exposure of displacement and age-determination |
| Trench log | True-scale representation as a hand-drawn graphic image | Show running horizontal scale and repetitive vertical scale |
| Observation and recording | Capture dimensional relationships in true scale and position | Establish vertical and horizontal control. Locate sample and photographic locations |
| Graphic scale | Faithful representation | Generally at 1:3 to 1:5, for detail |
| Sampling | Mainly for purposes of age determination of horizons and faulting displacements | Show location of samples on the graphical log |
| Labelling | Viewer identification of key geological features related to the nature and age determination of faulting | Mark also on the archival photograph and/or video image record of both faces of the trench |
| Faces | The surface portrays interrelationships of stratigraphy and relative ages of elements; normally only one of the two opposing faces is logged | Carbon-14 samples give minimum ages of offset strata; other techniques (e.g. fluid-inclusion analysis) refine minimum age of displacement of trench-intercepted faults |

**Table 5** Site characterization for earthquake mitigation

| Elements | Purpose | Important considerations |
|---|---|---|
| Stratigraphy | Define horizons, zones, or pockets that may be improved by engineering to withstand earthquake induced failures or movement | Engineers to consider methods of improvement of the geotechnical characteristics of the critical slope-failure mass |
| Displacement surfaces | Defines source and magnitude of ground rupture as input to design accommodation of future displacement on known faults | Generally applied only to earth-fill dam embankments where it is deemed worth the effort, cost, and risk to attempt withstand design |
| Palaeoseismic evidence | Sand blows, as well as geomorphological evidence of ground displacements as seen in exploratory trenches, sand and gravel pits, and stream banks | Evidence mainly occurs as stratigraphical displacement or cross-cutting relationships |

**Table 6** Post-event geological mapping

| Elements | Purpose | Important considerations |
|---|---|---|
| Ground rupture | Define nature of displacements from activation of a causative active fault | Displacement on active fault; displacement from ground heaving; tensile failures of lateral spreading |
| Liquefaction | Locates ground below which to excavate exploratory trenches | Need to define structural and stratigraphical locus of liquefaction |
| Lateral deformation | Define how cultural features have been affected | Transportation features, towers, pipelines, walls, gravestones, and many other thin and/or spindly objects that are moved in a bodily fashion |
| Ground failure | Delimit the failed ground then excavate, observe, sample, and define causes of damage | Forms more bases for lessons learned and then applied elsewhere |

## Summary

The engineering geologist's expertise in characterizing ground for all manner of engineered works can be extended to seismogenic zones. Engineering geologists can detect and describe geological, hydrogeological, and topographical conditions, defining the potential hazard at sites that are at risk of damage from earthquakes.

The description of fresh earthquake damage improves the ground model, helping to mitigate the adverse effects of future earthquakes. These analyses can be applied more widely to assist civil engineers in the development of effective measures to withstand future events.

## See Also

**Engineering Geology:** Seismology; Natural and Anthropogenic Geohazards; Liquefaction. **Sedimentary Processes:** Particle-Driven Subaqueous Gravity Processes. **Seismic Surveys. Tectonics:** Earthquakes.

## Further Reading

Bonilla MG (1967) Historic surface faulting in continental United States and adjacent parts of Mexico. Unnumbered open file report to the US Atomic Energy Commission. US Geological Survey. Menlo Park, California.

Bonilla MG (1973) Trench exposures across surface fault ruptures associated with San Fernando Earthquake. In: Murphy ML (ed.) *San Fernando, California Earthquake of February 9, 1971*, pp. 173–182. Washington DC: US Department of Commerce.

Bonilla MG, Mark RK, and Lienkaemper JJ (1984) Statistical relations among earthquake magnitude, surface rupture length, and surface fault displacement. *Bulletin of the Seismological Society of America* 74: 2379–2412.

Brown RD Jr, Ward PL, and Plafker G (1973) *Geologic and Seismologic Aspects of the Managua, Nicaragua, Earthquakes of December 23, 1972*. USGS Professional Paper 838. Washington DC: US Geological Survey.

Eckel EB (1964) *The Alaska Earthquake, March 27, 1964: Lessons and Conclusions*. USGS Professional Paper 546. Washington DC: US Geological Survey.

Hays WW (1981) *Facing Geologic and Hydrologic Hazards: Earth Science Considerations*. USGS Professional Paper 1240-B. Washington DC: US Geological Survey.

Krinitzsky EL (2003) How to combine deterministic and probabilistic methods for assessing earthquake hazards. *Engineering Geology* 70: 157–163.

Obermeier SF, Jacobson RB, Smoot JP, et al. (1990) *Earthquake-Induced Liquefaction Features in the Coastal Setting of South Carolina and in the Fluvial Setting of the New Madrid Seismic Zones*. USGS Professional Paper 1504. Washington DC: US Geological Survey.

Scott GR (1970) *Quaternary Faulting and Potential Earthquakes in East-Central Colorado*. USGS Professional Paper 700-C. Washington DC: US Geological Survey.

Youd TL and Hoose SN (1978) *Historic Ground Failures in Northern California Associated with Earthquakes*. USGS Professional Paper 993. Washington DC: US Geological Survey.

Wells DL and Coppersmith KJ (1994) New empirical relationships among magnitude, rupture length, rupture width, rupture area, and surface displacement. *Bulletin of the Seismological Society of America* 84: 974–1002.

# Geological Maps

**J S Griffiths**, University of Plymouth, Plymouth, UK

© 2005, Elsevier Ltd. All Rights Reserved.

## Introduction

Engineering geological maps provide ground information of relevance to civil engineering planning, design, and construction. Given that geology and civil engineering deal with three-dimensional structures and how they behave through time, albeit usually on different time-scales (geology $10^1$–$10^9$ years; civil engineering $10^1$–$10^2$ years), it is to be expected that the preparation of maps and plans is a essential component of both disciplines. Hence engineering geology maps have been regarded as an effective means of conveying information between geologists and engineers from the earliest days of the emergence of engineering geology as an identifiable subject. Indeed, the first classic stratigraphic maps and sections of William Smith (*see* **Famous Geologists:** Smith) in the late eighteenth and early nineteenth centuries arose out of his work on canal construction and the need to anticipate ground conditions.

An engineering geological map differs from the standard geological map by not being limited to scientific categories of bedrock and/or superficial geology. The map may group materials according to similar engineering characteristics and also present relevant data on topography, hydrology, hydrogeology, geomorphology (*see* **Engineering Geology:** Geomorphology), and geotechnics, plus information on man-made structures such as landfill sites or earthworks. This additional ground information may be shown on the map itself, or by using tables, charts, and diagrams accompanying the map, as exemplified by **Table 1** from the Hong Kong Geotechnical Area Studies Programme (GASP), and **Figure 1** from the Vía Inter-Oceánica, Quito, Ecuador. Whilst most maps in the past were cartographically drafted and produced as hardcopy, there is increasing use of computer-based Geographical and Geoscience Information Systems (GIS and GSIS) for compiling the data in formats that can be readily updated and analyzed.

The data for an engineering geological map will be compiled from a wide range of data sources and will

**Table 1** Example of an extended engineering geology legend based on bedrock data from the Hong Kong GASP programme

| Material description | | | | Evaluation of material | | |
|---|---|---|---|---|---|---|
| Map unit | Lithology | Topography | Weathering | Material properties | Engineering comment | Uses/excavation |
| Lower Cretaceous Dolerite (dyke rock) | Black to very dark grey, fine to medium-grained rock. Smooth joints normal to boundaries result of cooling | Generally occurs as linear structural features transecting the volcanic and granite units. May be of slightly depressed or elevated topographic form due to variable resistance of the country rocks. This geological structure often controls local surface runoff and may act as a loci for subsurface water concentrations | Weathers deeply to a dark red silty clay | Weathered mantle will contain a high proportion of clay and iron oxides leading to low Ø values. Intact rock strength will be very high, >100 Mpa when fresh | Restricted extent precludes detailed comment. Weathered mantle will have relative low permeability and will affect groundwater hydrology by forming barriers, and variable boundary conditions. Sub-vertical dykes may dam groundwater leading to unexpectedly high groundwater levels. | Restricted extent precludes deliberate borrow or quarry activities. Weathered material would make poor fill but fresh rock would make suitable high density aggregate or railway ballast. |
| Upper Jurassic Hong Kong Granite | Pink to grey medium-grained equigranular, non-porphyritic rock Mineral include quartz, potassium feldspar, plagioclase, biotite and muscovite. Rough sheeting joints and widely spaced tectonic joints widespread | Forms extensive areas of moderate to steep convexo-concave slopes. High-level infilled valeys are common. Drainage pattern is often dendritic in nature and is commonly dislocated by major tectonic discontinuities. These units are characterised by moderate to severe gully and sheet erosion associated with hillcrest and upper sidelong terrain | Shallow to deep residual soils over weathered granite. Local development of less weathered outcrops in stream beds and occasional cliff faces. Residual core boulders common on surface of sidelong ground and gullies. Weathering depths >20 m | Material properties vary with depth within the weathering profile. For completely weathered granite typical values are $c'\approx$ 0–25 kPa, $\emptyset \approx$ 31–43°; permeablity $\approx 10^{-6}$–$10^{-8}$ m/s; dry density 1500 kg m$^3$. Moisture content 15% near surface, 30% at depth. Fresh rock UCS 80–150 MPa. Rock mass strength dependent on joint characteristics. Roughness angles for tectonic joints 5–10°; for sheet joints 10–15°; basic friction angle $\approx 39°$ | Weathered mantle subject to sheet and gully erosion with landslides in steep slopes or if severely undercut. Perched water tables conform with highly permeable upper weathered zones. Rock is prone to discontinuity controlled failures in fresh to moderately weathered state. Stream and drainage lines align with geological weakness. Large structures may require deep foundations. Cut slope design may be governed by the large depths of weathered material | Extensively quarried and used as concrete aggregate. Weathered material widely used as fill as it is easily excavated by machinery. Core boulders can cause problems during excavation |

| | | | | |
|---|---|---|---|---|
| Middle and Lower Jurassic Lok Ma Chau Formation | Metamorphosed sedimentary and volcanic rocks, including schist, phyllite, quarzite, metasediments and marble. | Forms hills of moderate to low relief due to its low resistance to erosion. Occurs extensively beneath colluvial and alluvial cover. Local areas of surface boulders and occasional rock outcrops on sidelong ground and in gullies | Metasediments generally weather to produce moderately deep (1–2 m), uniform or gradational red-brown clay | Near-surface completely weathered residual soil acts as a silt with a void ratio 0.25–0.33. Gradings show 5–15% clay, 40–60% silt, 20–30% fine sand. PL 25–35%; LL 34–40%. Typical shear strength $c' \approx 0$–15 kPa, $\varnothing \approx 35°$. Weathered materials dry density 1600–1800 kg m$^{-3}$. Fresh rock UCS 40–90 MPa. Discontinuity strength parameters $c' \approx 0$–5 kPa, $\varnothing \approx 25$–30° | Considerable care is required during investigation, design, and construction. Bearing capacity reasonable for low and moderate loads. Stability is dependent on the very closely spaced discontinuities. Discontinuity surveys are essential for cut slope design. Material prone to failure along discontinuities when weathered and saturated | Material can be used as a source of bulk fill but may break down to silt if over-compacted. Excavation by machine is relatively easy |

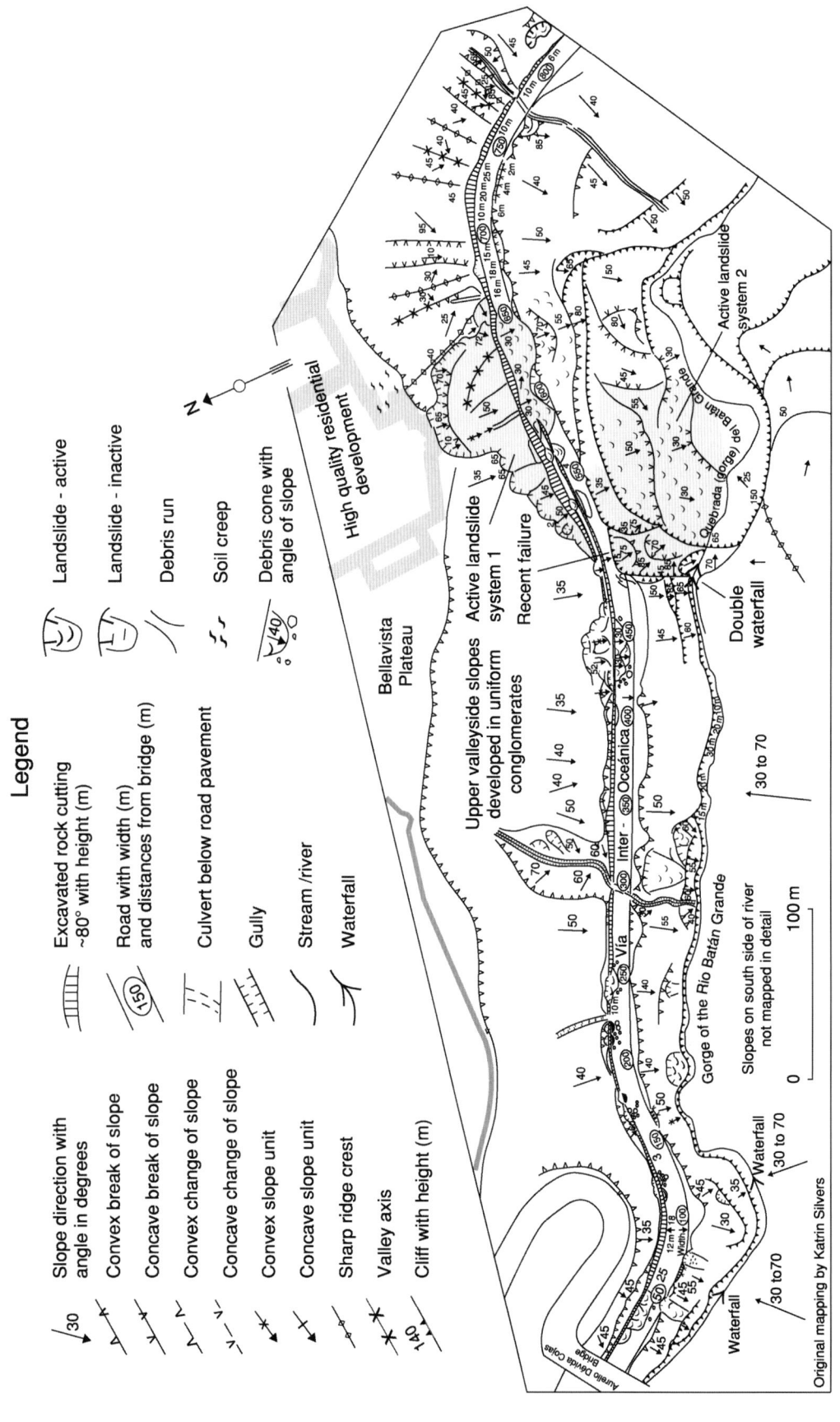

**Figure 1** Large-scale engineering geology map of a section of the Vía Inter-Oceánica, Quito, Ecuador.

**Table 2** Scale of Engineering Geology Maps suitable for different purposes

Large-scale
- 1 : 10 000 or greater. Plans and sections up to 1 : 500 scale can be used in particular circumstances
- Based on detailed field mapping and ground investigation data, supplemented by interpretation of large-scale aerial photographs or high resolution airborne and satellite imagery
- Uses: designing and interpreting ground investigations; background data for detailed design of foundations and structures; claims and litigation
- End-users: civil engineers; engineering geologists; geotechnical engineers; quantity surveyors

Medium-scale
- 1:10 000 to 1:100 000 scale
- Based on terrain evaluation techniques, remote sensing interpretation (small-scale aerial photographs and satellite imagery) and ground-truth mapping of selected areas
- Uses: geohazard evaluations; locating construction sites and route alignment planning; resource evaluation; hydrogeological studies; development planning at local and regional level; disaster relief planning
- End-users: civil engineers; engineering geologists; local and regional planners; local and regional government; insurers; water authorities; emergency services; military

Small-scale
- 1:100 000 or smaller
- Based on terrain evaluation techniques, remote sensing interpretation (primarily satellite imagery) and limited reconnaissance of field area
- Uses: generalized geohazard evaluation; general development planning at the regional or national level
- End-users: local and regional planners; regional and central government; insurers; military

not be collected solely through field mapping. However, as with any map, the value of the engineering geological map is dependent on the accuracy of the information used in its compilation. Thus, engineering geologists responsible for producing engineering geology maps must have a broad range of skills and be able to recognize and correctly compile data from a wide spectrum of Earth science disciplines.

Engineering geological maps are not just academic interpretations of the ground; they are produced to meet the specific requirements of a project. Therefore, the material presented and the scale used on the map will vary to meet the requirements of the end-user. Consequently, there is no unique format or content for an engineering geology map and throughout the mapping programme, the engineering geologist will need to be aware of the end-use to ensure that collection and presentation of the data are appropriate to the project requirements. **Table 2** provides an indication of the application of engineering geological maps at different scales to suit different purposes.

## The Type of Data to be Recorded

Although the specific content of any one engineering geological map will depend on the application, the main aim of an engineering geological mapping programme is to produce a map on which the mapped units are defined by engineering properties or behaviour. The limits of the units are determined by changes in the physical and mechanical properties of the materials. The boundaries of the mapped units may not correlate or coincide with the underlying geological structure or the chrono-stratigraphic units as depicted on conventional geological maps. However, experience has shown that the lithology of engineering soils and rocks can often be effective in defining the engineering geological map units. Apart from these map units, **Table 3** presents the type of additional data that are relevant and could be recorded on the engineering geological map through observations made in the field, supplemented by desk studies and ground investigations using exploratory pits and geophysics.

Whilst **Table 3** provides an indication of the range of data that might be compiled, the requirements for an individual engineering geological map will be tailored to suit the specific issues to be investigated. For example, in an area of earthquake risk there is likely to be more emphasis on the location of active faults, extent of soils liable to liquefy under dynamic loading, unconsolidated deposits that can amplify ground shaking, and zones potentially liable to the affects of tsunami or seiche. Similar specific details will be appropriate for different types of geohazard evaluation (see **Engineering Geology:** Natural and Anthropogenic Geohazards). For example, **Figure 1** is an example of a large-scale engineering geological map/plan of a road (Vía Inter-Oceánica) east of Quito, Ecuador, which has been affected by landsliding. Because the concern of the engineering geologist in this study is slope instability, the map emphasises the landslides and man-made features but contains relatively little geological detail.

Some of the engineering geological data may be used to identify 'zones'. These areas on the map

**Table 3** Data to be recorded on an Engineering Geology Map

Geological data
- Map units (chrono-and/or lithostratigraphy)
- Geological boundaries (with accuracy indicated)
- Description of soils and rocks (using standard engineering codes of practice)
- Description of exposures (cross referenced to field notebooks)
- Description of state of weathering and alteration (note depth and degree of weathering)
- Description of discontinuities (as much detail as possible on the nature, frequency, inclination and orientation of all joints, bedding, cleavage, etc.)
- Structural geological data (folding, faulting, etc.)
- Tectonic activity (notably neotectonics, including rates of uplift)

Engineering geology data
- Engineering soil and rock units (based on their engineering geological properties)
- Subsurface conditions (provision of subsurface information if possible, e.g., rockhead isopachytes)
- Geotechnical data of the engineering soil and rock units
- Location of previous site investigations (i.e., the sites of boreholes, trial pits, and geophysical surveys)
- Location of mines and quarries, including whether active or abandoned, dates of working, materials extracted, and whether or not mine plans are available
- Contaminated ground (waste tips, landfill sites, old industrial sites)
- Man-made features, such as earthworks (with measurements of design slope angles, drainage provision, etc.), bridges and culverts (including data on waterway areas), tunnels and dams

Hydrological and hydrogeological data
- Availability of Information (reference to existing maps, well logs, abstraction data)
- General hydrogeological conditions (notes on: groundwater flow lines; piezometric conditions; water quality; artesian conditions; potability)
- Hydrogeological properties of rocks and soils (aquifers, aquicludes, and aquitards; permeabilities; perched watertables);
- Springs and Seepages (flows to be quantified wherever possible)
- Streams, rivers, lakes, and estuaries (with data on flows, stage heights, and tidal limits)
- Man-made features (canals, leats, drainage ditches, reservoirs)

Geomorphological data
- General geomorphological features (ground morphology; landforms; processes; Quaternary deposits)
- Ground Movement Features (e.g., landslides; subsidence; solifluction lobes; cambering)

Geohazards
- Mass movement (extent and nature of landslides, type and frequency of landsliding, possible estimates of runout hazard, snow avalanche tracks)
- Swelling and shrinking, or collapsible, soils (soil properties)
- Areas of natural and man-made subsidence (karst, areas of mining, over-extraction of groundwater)
- Flooding (areas at risk, flood magnitude and frequency, coastal or river flooding)
- Coastal erosion (cliff form, rate of coastal retreat, coastal processes, types of coastal protection)
- Seismicity (seismic hazard assessment)
- Vulcanicity (volcanic hazard assessment)

have approximately homogenous engineering geological conditions, usually defined by more than just the lithostratigraphy. The zoning system would be derived from the factual data contained on the base map. It would not, therefore, normally form part of the original mapping programme and represents a derivative or interpretative engineering geological map. Zoning maps can be particularly effective in geohazard studies where the magnitude of a particular hazard can be represented by an interpretative map containing data on probability of occurrence or frequency. However, interpretative zoning maps can be used in many ways relevant to engineering and examples can be found of maps showing: foundation conditions, excavatability of materials, sources of construction material, bearing capacity of soils and rocks, and general constraints to development.

## Map Scale

For engineering work, mapping may be carried out during feasibility studies, before any site work is started, just prior to or during the site investigation phase (see **Engineering Geology:** Site and Ground Investigation), whilst construction is underway, or as a means of compiling data if there have been problems with a structure after it has been built. To meet this range of applications, three broad-scale categories are recognized (large, medium, and small). These categories are presented in **Table 2**, and whilst the boundaries to the categories should be regarded as flexible, there are differences in the way the maps are compiled, the techniques employed in their construction, and their intended end-use. A wide range of examples of all the categories of engineering geology map can be found in the published literature, and **Figure 1** is presented as an illustration of a typical large-scale engineering geological map. This map was produced specifically for the investigation of a single stretch of road and provides data of direct relevance to the design of remedial measures.

Medium-scale engineering geological maps are probably the most widely used and excellent examples can be found in the UK, where a national programme of applied geological mapping, predominantly by the British Geological Survey, has been underway for over 20 years. This programme has resulted in over 35 studies and produced a wide range of maps for use in engineering construction and planning development (**Table 4**). These maps have mainly been produced at a scale of 1:25 000 and represent compilations of engineering geological data in map form that can used as the basis for engineering feasibility studies. Similar applied geological mapping programmes were carried out in a number of other countries. In France, the ZERMOS programme has produced a number of 1:25 000 scale maps of selected area. Originally under the auspices of the Geotechnical Control Office, the former territory of Hong Kong was mapped at 1:20 000 during the GASP programme. The USA also has a long commitment to engineering geological mapping at a wide range of scales which is illustrated in the compilation of maps provided by the US Geological Survey Profession Report 950, entitled 'Nature to be Commanded.' However, despite their widespread production, the actual use of engineering geological maps in planning and development is variable.

Small-scale maps are best exemplified by the PUCE (Pattern-Unit-Component-Evaluation) system developed by the Commonwealth Scientific Research Organisation (CSIRO) in Australia, but also used in Papua New Guinea. This programme produced maps at a scale of 1:250 000 that define broad terrain patterns, within which particular assemblage of landforms, pedological soils, and vegetation sequences occur. These have been predominantly used in regional planning but there are examples of their use for aggregate resource surveys; route corridor alignment; water resources; military and off-road mobility; flood hazard; land capability assessment; and aesthetic landscape appreciation.

## Data Collection

Primary mapping for engineering geology follows the same basic rules and uses the same techniques established for conventional geological mapping. However, a number of additional decisions need to be made when undertaking engineering geological mapping. These are to identify the types of data that are to be collected to meet the survey requirements (**Table 3**); the scale of the mapping (**Table 2**); the methods to be used for data collection; and the intended final map products (**Table 4**).

**Table 4** Applied Geological Maps compiled in the United Kingdom 1983–96 utilizing Geological and Engineering Geological Data

Data points
- Location of exploratory holes and wells
- Distribution of geotechnical data test results
- Point rockhead information

Disturbed ground (human activity)
- Distribution (general)
- Distribution of mines and mine workings (all types, including surface and sub-surface)
- Distribution of made-ground

Superficial geology
- Soil types, extent, lithology, and thickness
- Drift thickness/rockhead contours
- Geotechnical properties of soils

Bedrock geology
- Rock types, extent, lithology, lithostratigraphy
- Structure contours
- Geological structure
- Geotechnical properties of rocks

Engineering geology
- Foundation conditions
- Hydrogeological conditions
- Ground conditions in relation to groundwater
- Nature and distribution of geohazards (subsidence, instability, flooding, earthquakes, etc.)
- Engineering geological zones (i.e., areas of homogenous engineering geological conditions)
- Aggregate and borrow material sources

Geomorphology
- Geomorphological landforms and process
- Drainage
- Areas of slope instability
- Flood frequency limits

Derived construction constraints maps
- Slope steepness
- Ground instability (e.g., subsidence, cambering, landslides, soft ground, etc.)
- Landslide hazard and risk maps
- Previous industrial usage (brownfield sites and contaminated land)

Derived resources maps
- Nature, extent, and properties of mineral resources (superficial and bedrock)
- Groundwater resources
- Distribution of aggregates
- Sites of Special Scientific Interest (SSSI)

Summary maps
- Development potential
- Summary of construction constraints
- Statutory protected land

**Table 5** Example of an extended engineering geology legend based on the superficial geology map data in the Applied Geology Map of Stoke-On-Trent

| | Description | Characteristics | Planning & Engineering Considerations | | | |
|---|---|---|---|---|---|---|
| | | | Slope stability | Excavation | Foundations | Engineered fill |
| Alluvium | Silts & clays 0.6–9 m thick. Occurs mainly in the Trent valley and tributaries | Very soft to firm, low to high plasticity, medium to high compressibility. May be desiccated near top | not applicable as occurs in flat areas | Diggable by excavator. Heave may occur at base of excavations. Trench support required | Low acceptable bearing capacity (<75 kPa) | Generally unsuitable |
| | Organic in places, with peat lenses | Very soft to soft, intermediate to extremely high plasticity. Highly compressible | | | Sulphate protection usually required for concrete | |
| | Sands and gravels often occur at base | Loose to very dense. Water bearing | | Running ground conditions will require cut-offs or dewatering | | |
| Periglacial head | Variable soils derived from bedrock other superficial deposits. Composition varies according to the parent material. Generally consists of sandy, silty clays with gravel and cobbles. Forms a thin veneer on slopes and may thicken downslope. Perched water tables may occur within coarser horizons | Variable. Usually cohesive, soft to stiff, with low to high plasticity. Compressibility usually intermediate, but may be high. Pre-existing shear surfaces may be present, with low residual friction angles | Natural slopes often marginally stable. The variability of this deposit necessitates careful site investigation, preferably by trial pitting, to determine the site characteristics in relation to any proposed development. Clayey slopes may require drainage and other stabilisation works prior to construction | Diggable by excavator | Consolidation settlement usually small. Differential settlement likely where soft compressible zones present | Generally unsuitable |
| Glacial sand and gravel | Coarse sand and subangular to subrounded gravel Occasional subrounded cobbles | Loose to dense granular deposit. Water bearing | not applicable as occurs in flat areas | Diggable by excavator Support and groundwater control required | Consolidation settlements small. Pile driving may be difficult in cobbles. Sulphate protection may be required for concrete | Suitable for use in embankments if the soft clay zones are removed |
| | Some horizons of laminated clay/silt occur | Clay/silt horizons are usually soft to stiff, of low to intermediate plasticity | | | | |

| | | | | |
|---|---|---|---|---|
| Glacial till (boulder clay) | Variable deposit, generally sandy, silty clays with gravel, cobbles and occasional boulders<br><br>Water bearing lenses of sand and gravel may occur | Generally firm to stiff, with low to intermediate plasticity and intermediate compressibility | Cut slope of 1V:2.5H generally adequate for long term stability | May be difficult to dig and can require ripping. Excavations generally stable in the short term but deteriorate on exposure and wetting. Support required for deep excavations, and where sand lenses occur | Usually forms a good founding medium with acceptable bearing capacities typically 150 to 600+ kPa) | Suitable if placed in dry weather when moisture content is low |
| Landslide debris | In the field area all known occurrences occur in weathered mudstones of the Etruria Formation | Deposits contain shear surfaces with low residual strengths. Remoulded clay debris is generally poorly drained with possible perched water tables | | Areas of landslide debris should be avoided if possible. Constructional activity is likely to reactivate slope movement unless appropriate remedial measures are taken<br><br>Detailed site investigation is essential with extensive use of exploratory holes and geophysics. If construction is unavoidable groundwater and ground movement monitoring is essential | | Unsuitable |

In most engineering situations there will be four phases in the preparation of an engineering geological map: desk study; field mapping (*see* **Geological Field Mapping**); interpretation; and reporting. During the desk-study phase all existing data are compiled, remote sensing interpretation is carried out, a preliminary field reconnaissance may be undertaken, and the field programme is planned. Field mapping requires the collection of primary data in the field. Even if the available data is quite comprehensive and it is only intended to produce small-scale maps, some primary field mapping will be necessary. Interpretation of the data involves bringing together the field and desk study data and preparing the suite of maps that meet the project requirements. Finally, the maps will need to be supplemented by a written report for the end-user that expands on the details shown on the map and, in engineering situations, may provide some design guidance or recommendations.

## Map Presentation

The presentation of engineering geological maps follows normal cartographic rules over scale, north arrow, and locational data, but the information displayed will be based on end-user requirements. Because the information on the map is variable, it is usually necessary to create a bespoke legend for the map, as exemplified by the key to **Figure 1**. However, general guidance on the typical symbols to use can be found in the standard literature on engineering geological maps listed below.

Often, with engineering geological maps, it is necessary to include quite comprehensive data in the map legend. These data will not have just been compiled from field observations but will include data from the desk studies and any detailed ground investigations carried out in the area. An example of a comprehensive, or extended, legend is provided in **Table 5**, based on the UK Applied Geological Map for Stoke-on-Trent. This uses the superficial geological map as the basis for identifying the engineering geological units. **Table 1** provides an example of a similar compilation of data based on bedrock properties of the type used in the tropical weathering environment investigated during the Hong Kong GASP programme. In both examples, additional data on the geotechnical and engineering characteristics of the various materials are included in the tables as well as comments on

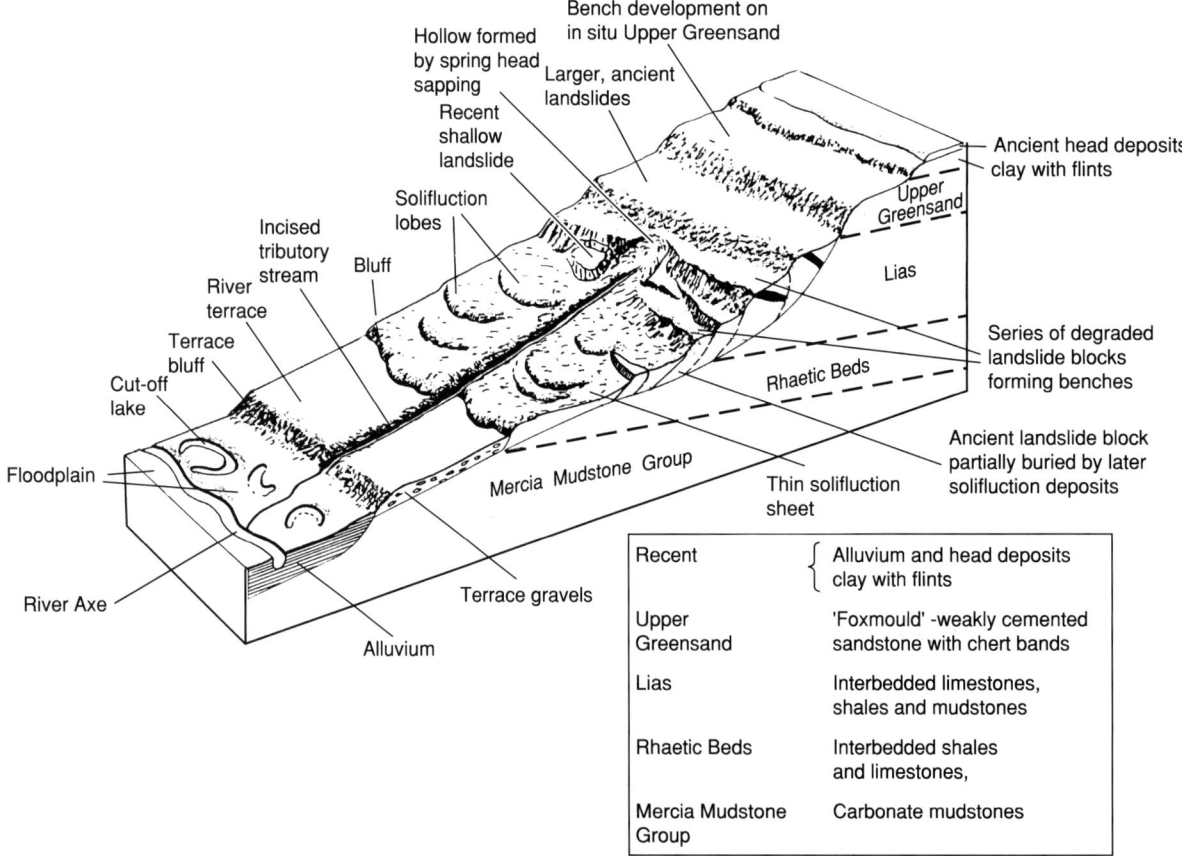

**Figure 2** Three-dimensional engineering geology ground model developed for the Axminster By-pass, Devon, England. Reproduced with permission from Croot D and Griffiths JS (2001). Engineering geological significance of relict periglacial activity in South and East Devon. *Quarterly Journal of Engineering Geology and Hydrogeology* 34: 269–282.

engineering issues, such as slope stability, excavatability, and groundwater conditions. These data may be shown in summary on the actual map or in an accompanying report.

Another method of conveying engineering geological data that can be included on the map or in an accompanying report is shown in **Figure 2**. This three-dimensional block diagram, compiled during mapping for the Axminster by-pass in Devon, provides a synopsis of the ground conditions that illustrates how the bedrock geology, superficial geology, and geomorphology create a landscape that contains a number of technical problems for the engineering geologist. This type of figure is generally referred to as a 'ground model'.

These various techniques of data presentation are very effective in conveying detailed information in a way that can be readily understood by the nonspecialist. However, there is always a concern that a report can become separated from its maps and, as a general recommendation, the map itself should be able to stand alone and be understood by all potential users, without having to refer to a separate report.

## Integration with Site Investigation

Site investigation for engineering is the process by which data appropriate for the design and construction of structures is collected. Whilst this primarily involves the exploration of the ground using invasive techniques such as drilling and trial pitting, it is recommended that engineering geological mapping be integral to the process. Mapping has proved itself to be extremely cost effective and can be used to design a more efficient ground investigation by defining the engineering geological units that will represented by the exploratory holes. This is illustrated by the mapping carried out for the UK portal and terminal areas of the channel tunnel. The UK channel tunnel portal is located in a late-glacial multiple-rotational landslide that was subject to detailed mapping in order to plan the ground investigations. The mapping provided the basis for the development of a ground model against which additional data were checked as they were acquired. This demonstrates how mapping can ensure that any geohazards that might affect a project will be identified early on and thereby allowed for in the design.

## See Also

**Engineering Geology:** Geomorphology; Natural and Anthropogenic Geohazards; Site and Ground Investigation. **Famous Geologists:** Smith. **Geological Field Mapping**. **Geological Maps and Their Interpretation**.

## Further Reading

Anon (1976) *Engineering Geology Maps: A Guide to their Preparation*. Paris: The UNESCO Press.

Barnes J and Lisle RJ (2004) *Basic Geological Mapping*, 4th edn. Chichester: Wiley.

Brunsden D (2002) Geomorphological roulette for engineers and planners: some insights into an old game. *Quarterly Journal of Engineering Geology and Hydrogeology* 35: 101–142.

Culshaw MG, Bell FG, Cripps JC, and O'Hara M (eds.) (1987) *Planning and Engineering Geology*. Geological Society Engineering Special Publication No. 4.

Dearman WR (1991) *Engineering Geological Mapping*. Oxford: Butterworth-Heinemann.

Doornkamp JC, Brunsden D, Cooke RU, Jones DKC, and Griffiths JS (1987) Environmental geology mapping: an international review. In: Culshaw MG, Bell FG, Cripps JC, and O'Hara M (eds.) *Planning and Engineering Geology*. Geological Society Engineering Special Publication, No. 4, 215–219.

Eddleston M, Walthall S, Cripps JC, and Culshaw MG (1995) *Engineering Geology of Construction*. Geological Society Engineering Special Publication No. 10.

Finlayson AA (1984) Land surface evaluation for engineering practice: applications of the Australian PUCE system for terrain analysis. *Quarterly Journal of Engineering Geology* 17: 149–158.

Fookes PG (1997) Geology for engineers: the geological model, prediction and performance. *Quarterly Journal of Engineering Geology* 30: 293–424.

Fookes PG, Baynes FJ, and Hutchinson JN (2000) Total geological history: a model approach to the anticipation, observation and understanding of site conditions. *GeoEng 2000, an International Conference on Geotechnical & Geological Engineering*, Melbourne, 1: 370–460.

Griffiths JS (ed.) (2001) *Land Surface Evaluation for Engineering Practice*. Geological Society Engineering Geology Special Publication No. 18.

Griffiths JS (2002) *Mapping in Engineering Geology*. The Geological Society, Key Issues in Earth Scienes, 1.

Griffiths JS, Brunsden D, Lee EM, and Jones DKC (1995) Geomorphological investigations for the Channel Tunnel terminal and portal. *The Geographical Journal* 161(3): 275–284.

Hutchinson JN (2001) Reading the ground: morphology and geology in site appraisal. *Quarterly Journal of Engineering Geology and Hydrogeology* 34: 7–50.

Kiersch GA (ed.) (1991) *The Heritage of Engineering Geology; the First Hundred Years*. Geological Society of America Centennial Special Volume 3.

Lawrence CJ, Byard RJ, and Beaven PJ (1993) *Terrain Evaluation*. Transportation Research Laboratory Report SR 378, TRRL, Crowthorne.

Maund JG and Eddleston M (1998) *Geohazards in Engineering Geology*. Geological Society Engineering Geology Special Publication No. 15.

Porcher M and Guillope P (1979) Cartography des risques ZERMOS appliqués a des plans d'occupation des sols en

Normandie. *Bulletin Mason Laboratoire des Ponts et Chaussees* 99.

Robinson GD and Speiker AM (eds.) (1978) *Nature to be Commanded*. US Geological Survey, Professional Paper 950.

Rosenbaum MS and Turner AK (eds.) (2003) *Characterisation of Shallow Subsurface: Implications for Urban Infrastructure and Environmental Assessment*. Dusseldorf: Springer-Verlag.

Smith A and Ellison RA (1999) Applied geological maps for planning and development: a review of examples from England and Wales, 1983 to 1996. *Quarterly Journal of Engineering Geology* 32: S1–S44.

Styles KA and Hansen A (1989) *Geotechnical Area Studies Area Programme: Territory of Hong Kong*. GASP Report XII, Geotechnical Control Office, Civil Engineering Services Department.

# Geomorphology

**E M Lee**, York, UK
**J S Griffiths**, University of Plymouth, Plymouth, UK
**P G Fookes**, Winchester, UK

© 2005, Elsevier Ltd. All Rights Reserved.

## Introduction

Geomorphology is the study of landforms and landform change. Engineering geomorphology is concerned with evaluation of the implications of landform change for society. The focus of the engineering geomorphologist is primarily on the risks from current surface processes (i.e., the impact of geohazards), the characteristics of near-surface materials (i.e., the products of processes and changes, including landslide debris, river terrace gravels, duricrusts, and metastable soils), and the effects of development on the environment, notably on surface processes and any resulting changes to landforms or increased level of risk (i.e., environmental impacts).

Engineering geomorphology has developed in the past few decades to support a number of distinct areas of engineering, including river, coastal, geotechnical, and agricultural engineering. River engineering involves studies of the nature and causes of alluvial river channel change (e.g., channel migration, bank erosion, and bed scour); coastal engineering studies emphasize the understanding of the occurrence and significance of shoreline changes, especially in response to changes in sea-level and sediment supply. Geotechnical engineering has complemented engineering geology and has been proved to be valuable for rapid site reconnaissance. Geomorphology provides a spatial context for developing site models and for explaining the distribution and characteristics of particular ground-related problems (e.g. landslides, permafrost, or the presence of aggressive soils) and resources (e.g., sand and gravel deposits). In agricultural engineering geomorphology has contributed notably to the investigation and management of soil erosion problems. To a large degree, each of these applications of geomorphology has developed separately in response to specific engineering needs. In recent years, however, all have started to become integrated as a coherent discipline.

## A Framework for Evaluating Change: Physical Systems

Earth's surface is dynamic, and landforms change through time in response to weathering and surface processes (e.g., erosion, mass movement, and deposition). Most of the changes occur in response to variations in the energy inputs into physical systems, including variations in rainfall intensity or total, in temperature, in river flows (discharge and sediment load), and in wave/tidal energy arriving at the coast, over a range of time-scales. Physical systems are a means of describing the interrelationships between different landforms. They form a useful spatial framework for evaluating how hazards and risks to a particular site can arise as a result of processes operating elsewhere. For example, changes in land use in the catchment headwaters can lead to changes in flood frequency and river channel change elsewhere. In addition, systems can be used to evaluate the potential impacts of a project on landforms at sites distant from the project; for example, reclamation of an intertidal wetland can have significant effects on the whole estuary, through the resulting changes to the tidal prism and mean water depth. Systems can be defined at a range of scales, from river drainage basins (watersheds or catchments) and coastal cells (sediment transport cells) to individual hillslopes, dunes, or cliffs. Irrespective of the scale, each system comprises an assemblage of individual components (i.e., the landforms) and transfers of energy and sediment (**Figure 1**).

Engineering geomorphology is directed towards understanding the way systems respond to relatively short- to medium-term changes in energy inputs

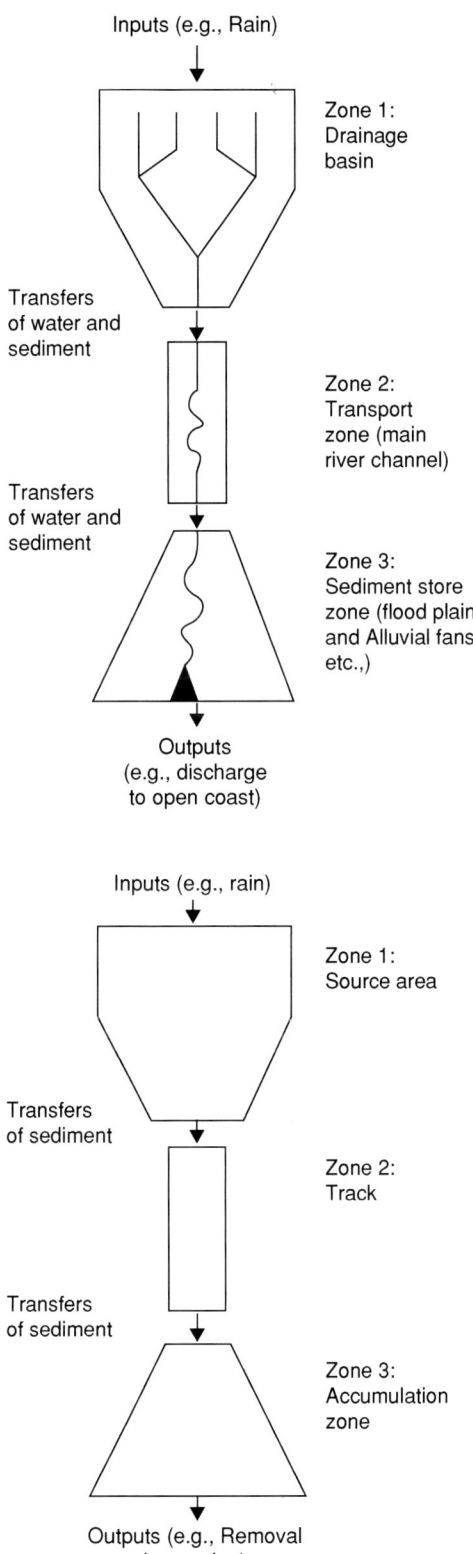

**Figure 1** Examples of physical systems. (Top) A river catchment; (bottom) a mudslide. Each system comprises an assemblage of landforms and transfers of energy and sediment. Reproduced from Fookes PG, Lee EM, and Griffiths JS (2004) *Foundations of Engineering Geomorphology*. Latheronwheel, Caithness: Whittles Publishing.

(e.g., resulting from climatic variability, changes in sediment supply, sea-level rise, or the effects of humans), rather than to long-term landscape denudation and evolution. However, there is also a need to be aware of the significance of longer term trends (e.g., the Holocene decline in sediment availability experienced on many temperate coastlines) and the presence of potential geohazards inherited from the past (e.g., ancient landslides, periglacial solifluction sheets, karst features). Engineering project cycles are generally in the order of 10–100 years (occasionally longer), and this relatively limited duration of 'engineering time' imposes a constraint on the types of landscape changes that are relevant to engineering geomorphology. Abrupt and dramatic landscape changes are likely to be significant over a time-scale of 10 to ~100 years. Relevant examples include establishment of gully systems, migration of sand dunes, river planform changes, coastal cliff recession, and the growth and breakdown of shingle barriers. High-probability to relatively high-probability events are important features and include wind-blown sand, soil erosion, shallow hillside failures, flooding, scour and river bank erosion, and coastal erosion and deposition. Low-probability events that could have a major impact on an engineering project or development are often a key issue and include flash floods, major first-time landslides, and tsunamis.

## Investigation Methods

Engineering geomorphology provides practical support for engineering decision making (i.e., project planning, design, and construction). Engineering geomorphologists need to work as part of an integrated team and must provide information at many levels, ranging from crude prefeasibility-stage qualitative approximations to sophisticated quantitative analyses in support of detailed design and construction. The level of precision and sophistication required needs to be sufficient for a particular problem or context, so as to allow adequately informed decisions to be made.

An engineering geomorphologist is typically required to ask questions that address a number of situations:

- Will the project be at risk from instability, erosion, or deposition processes over its design lifetime (e.g., is the development set back sufficiently from a retreating cliff top)?
- How could problems arise (e.g., will removal of support during unregulated excavations at the base of a slope triggering a landslide)?
- What is the likelihood of the project being affected by instability, erosion, or deposition processes over

its design lifetime (e.g., is the chance of being affected by a channelized debris flow, or being engulfed by blown sand, very low)?
- What will be the effect of climate change or sea-level rise on the project risks (e.g., will there be accelerated cliff recession or a higher chance of landslide reactivation)?
- Why has a problem arisen (e.g., are there leaking water pipes or is there disruption of sediment transport along a shoreline)?
- What effects will the project have elsewhere (e.g., will there be reduction in floodplain storage or changes in surface run off and erosion potential)?
- What magnitude event should be designed for to provide a particular standard of defence (e.g., what is the expected volume/depth of the 1- in 50-year debris flow event)?

The answers to these questions can be both qualitative and quantitative; for example, recognition of pre-existing landslides with potential for reactivation can be related to rates of change and the magnitude/frequency of events.

Most projects will involve a combination of approaches, including application of historical data, taking measurements, and making maps and models. Historical records can be used to define the timing and frequency of past events, and historical maps, charts, and aerial photographs may aid in determining past rates of change. Measurement and monitoring of change are accomplished using a variety of field-based or remote-survey techniques (e.g., aerial photography and satellite imagery). Mapping and characterization of the landscape (terrain evaluation and geomorphological mapping) can be applied in developing geomorphological models to provide a framework for the prediction of hazards and future changes. Geographical information systems (GIS) are important tools for the storage, management, and analysis of the geomorphological information collected throughout a project.

### Historical Records and Maps

In many countries, there exists a wide range of sources that can provide useful information on the past occurrence of events, including aerial photographs, topographic maps, satellite imagery, public records, local newspapers, consultants' reports, scientific papers, journals, diaries, and oral histories. For example, a

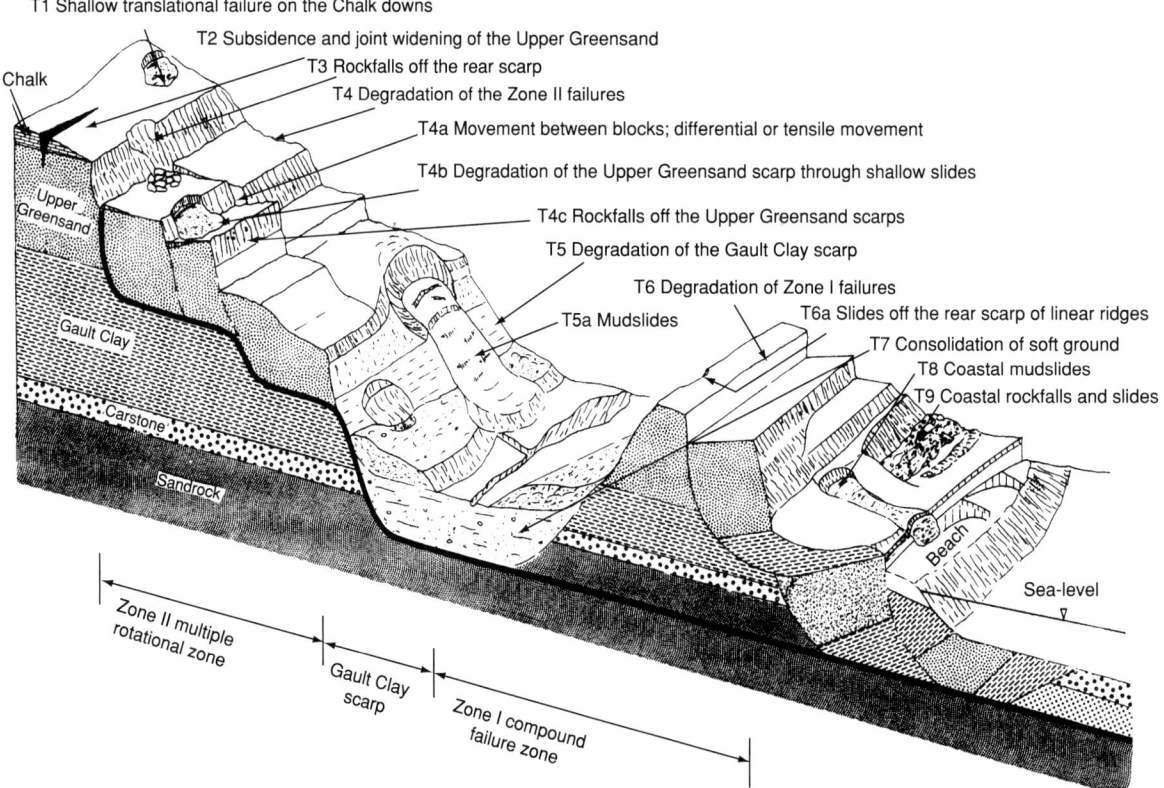

**Figure 2** Landslide model of the Ventnor Undercliff, UK, showing different landslide processes (T1–T9). Systematic study of local records revealed that the contemporary movements are largely confined to the superficial degradation of the ancient, deep-seated landslide complex. Reproduced from Lee EM and Moore R (1991) *Coastal Landslip Potential Assessment: Isle of Wight Undercliff, Ventnor*. London: Department of the Environment, UK.

systematic survey of local newspapers from 1855 to the present day was used to establish the pattern of contemporary ground movement in the Ventnor landslide complex, Isle of Wight, UK. The search identified over 200 individual incidents of ground movement and allowed a detailed understanding of the relationship between landslide activity and rainfall to be developed (**Figures 2** and **3**). Historical topographical maps, charts, and aerial photographs provide a record of the former positions of various

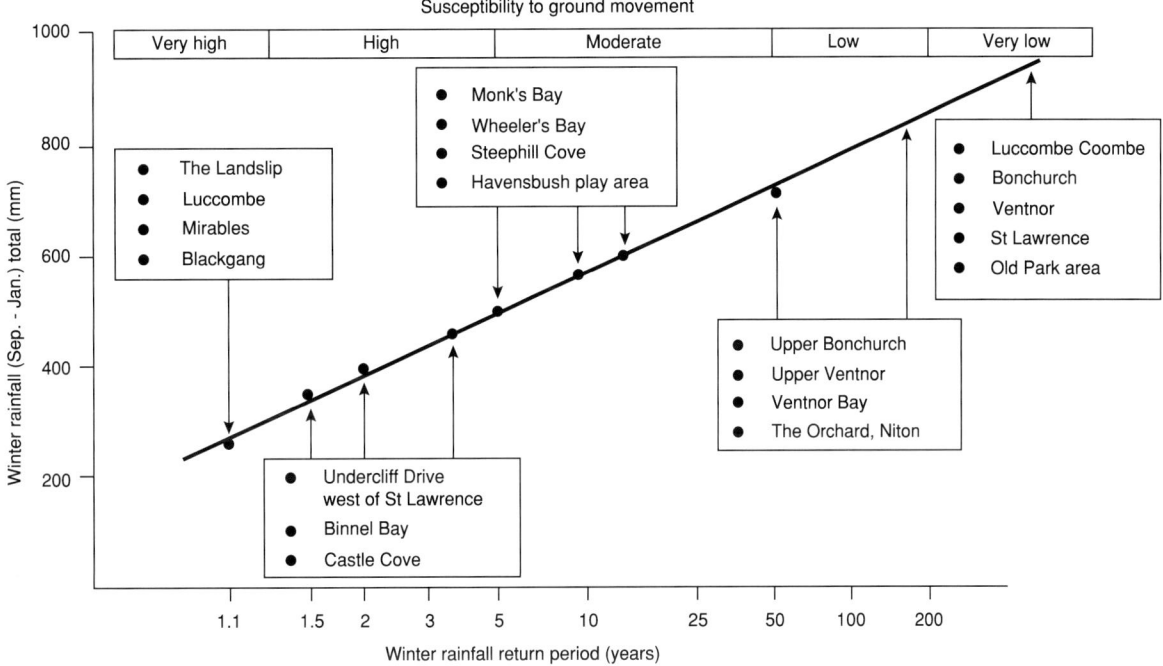

**Figure 3** The relationship between rainfall and landslide activity in different parts of the Ventnor Undercliff, UK. This relationship was established by defining the minimum winter rainfall total associated with triggering ground movement over the past 150 years. Reproduced from Lee EM, Moore R, and McInnes RG (1998) Assessment of the probability of landslide reactivation: Isle of Wight Undercliff, UK. In: Moore D, and Hungr O (eds.) *Engineering Geology: The View from the Pacific Rim*, pp. 1315–1321. Rotterdam: Balkema.

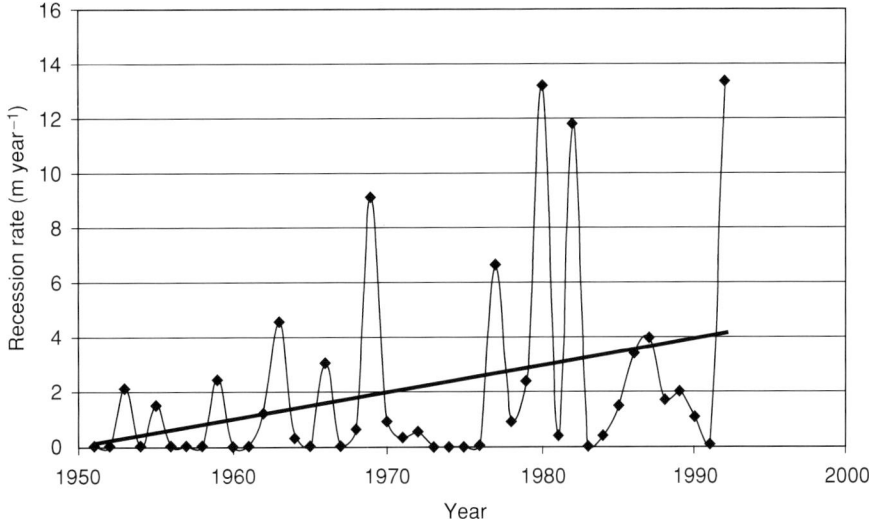

**Figure 4** Annual measurements of cliff recession on the Holderness coast, UK. The measurements are for a single erosion marker post along the cliffline. Note that there has been a trend of increasing annual cliff recession over the 40-year period; this may reflect a combination of sea-level rise and the impact of coastal defences on the adjacent coastline.

features, such as clifflines and river channel banks. In many cases, historical maps and charts may provide the only evidence of evolution over the past 100 years or more. When compared with recent surveys or photographs, these sources can provide the basis for estimating cumulative land loss and the average annual erosion rate between survey dates. However, great care is needed in using historical data because of the potential problems of accuracy and reliability.

**Measurement and Monitoring**

Rates of change can be determined through direct measurement of the positions of features at fixed points and at regular intervals. Channel cross-sections, beach profiles, or cliff tops, for example, can be assessed on an annual or biannual basis. In the early 1950s, on the Holderness coast, UK, the local authority initiated a programme of cliff recession measurement that has been continued on an annual basis ever since. A series of 71 marker posts were installed at 500-m intervals along the coastline, each post located at a distance of between 50 and 100 m normal to the coast. Annual measurements from each post to the cliff top commenced in 1953 (**Figure 4**).

Analytical photogrammetry can be used to quantify the nature and extent of landform changes over time, using aerial photographs taken on different dates, by comparing the three-dimensional coordinates of the same points. This approach was used to establish the rate of contemporary building movement in the Ventnor landslide complex, Isle of Wight. A total of 129 points distributed throughout the town were selected for measurement. The coordinates of each point were determined for each photograph date (1949, 1968, and 1988) and then compared to produce 'discrepancy' vectors. Where the coordinate discrepancy was greater than the standard deviation of coordinates, significant movement was assumed to have occurred.

**Terrain Evaluation**

Terrain evaluation has its origins in the need to organize and communicate specific earth science information or intelligence in a way that is of direct relevance

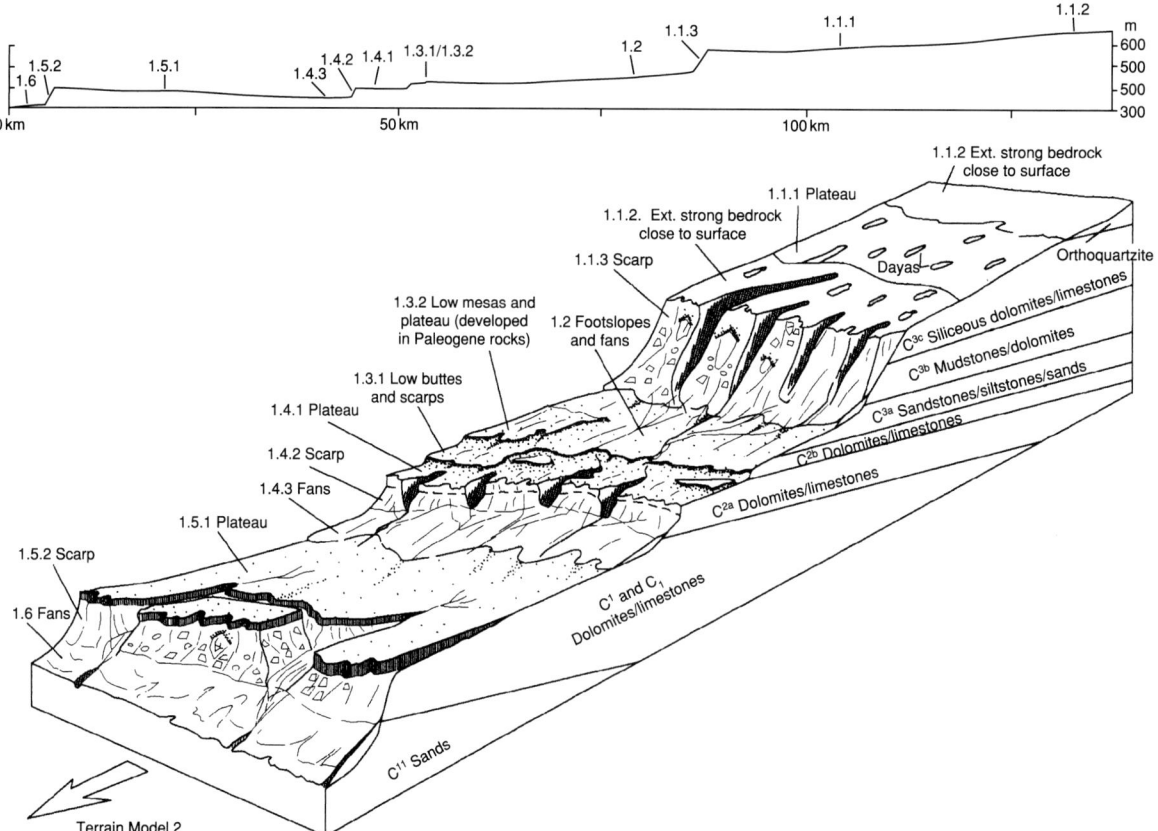

**Figure 5** Terrain model 1 of the Tademait Plateau, site of the In Salah Gas pipeline project, Algeria. The numbers (e.g., 1.4.1) represent terrain units (see **Table 1**); geological codes ($C^{3c}$ etc.) refer to notation on original geological maps of the region. Reproduced with permission from Fookes PG, Lee EM, and Sweeney M (2001) Terrain evaluation for pipeline route selection and characterisation, Algeria. In: Griffiths JS (ed.) *Land Surface Evaluation for Engineering Practice, Geological Society Special Publication 18*, pp. 115–121. Bath: Geological Society Publishing House.

**Table 1** The In Salah gas pipeline, Algeria: characteristics for terrain unit 1.1.1[a]

| Terrain type | Plateau |
| --- | --- |
| Surface form | Almost flat, very gently sloping ($<1°$) surface; extensive stone pavement surface, locally disrupted by vehicle traffic; frequent circular and linear enclosed silt/sand-filled depressions (dayas) with no stone pavement |
| Geomorphological processes | Water: surface drainage is primarily into enclosed depressions (dayas); locally, channel flow occurs in straight, single-thread ephemeral streams (wadis), less than 50 m wide, with sheet sand and nebkha floors |
| | Solution: dayas are likely to be formed by localized solution weathering of limestones |
| Hazards | Flooding: dayas will become flooded during and immediately after rain events; wadi flows are likely to be localized, with rapid transmission losses with limited potential for serious flooding and scouring; estimated $Q_{max} = <50\,m^3\,s^{-1}$; $V_{max} = <1\,m\,s^{-1}$ |
| | Subsidence: dayas may overlie infilled solution holes, with potential for subsidence |
| Superficial materials | Stone pavement over loose to dense calcareous silty fine sands (up to 3 m thick), with calcrete cobbles and gravels |
| Duricrust | Duricrust occurs as draped sequences on floors and flanks of broad wadi channels; strong to very strong siliceous calcrete within 0.5 m of surface |
| Bedrock | Massive crystalline limestones (very strong to strong), generally $>3$ m below surface; bedrock is partly karstified |
| Water table | Not observed |
| Rock mass structure | Not observed |
| Fracture spacing | Duricrust: 0.2 m (estimated) |
| Rock strength | Siliceous calcrete (strong–very strong); limestone (strong–very strong) |
| Excavatability | Easy to hard digging; hard to very hard ripping |
| Borrow | Random and select fill |
| Trafficability | No significant constraints |
| Principal constraints | Potential for subsidence within dayas needs to be considered; very strong materials within 3 m of the surface, especially along floors and flanks of wadis |

[a]See **Figure 5**.

to the end user (civil and military engineers, land use planners, agriculturalists, and foresters). The principles of terrain evaluation involve defining areas of terrain that have similar physical characteristics, i.e., a typical range of topographic, geohazard, and engineering factors. The objectives of terrain evaluation are to identify clear associations between surface forms (the terrain units), near-surface materials, and processes; to simplify the complexity of ground conditions and surface processes within a particular area, highlighting those of significance to the planning of projects; and to provide a tool for predicting terrain characteristics within a particular area or region. This is based on the assumption that the terrain units are sufficiently homogeneous and mutually distinctive to allow valid prediction.

At the broadest scale, landscape types (terrain models or land systems) can be defined (e.g., mountains, desert plateaus, or coastal plains); this level of subdivision may be suitable for prefeasibility overviews of very large areas (see **Figure 5** and **Table 1**). Within a landscape type, it will be possible to identify a variety of landform assemblages (terrain units), such as river floodplains, escarpment faces, extensive areas of unstable hillslopes, and ridge crests; this level of detail may be sufficient for corridor assessment. Within a terrain unit, there will be numerous individual landforms (terrain sub-units) that will each present slightly different levels of challenge to a project; an escarpment face, for example, may contain a variety of sub-units, including bare rock faces and discrete landslide systems separated by stable ridges and spurs.

### Geomorphological Mapping

The production of some form of map underpins many engineering geomorphological studies. The map might be the product of an intensive fieldwork programme, a sketch map of part of an area based on a walkover survey, or interpretation of remote imagery (e.g., aerial photographs or satellite images). Most map making follows three stages: recognition of landforms or landform elements (units) that provide a practical framework for addressing the problem in hand, characterization of these units in terms of the significant surface processes (these may be active or relict) and the near-surface materials, and interpretation of the significance of these forms, processes, and materials to the problem facing the project. This may involve producing some form of derivative map (e.g., a landslide hazard map or aggregate resource map). A good example is the extensive programme of geomorphological mapping that was carried out as part of the investigations at the Channel Tunnel portal and terminal areas near

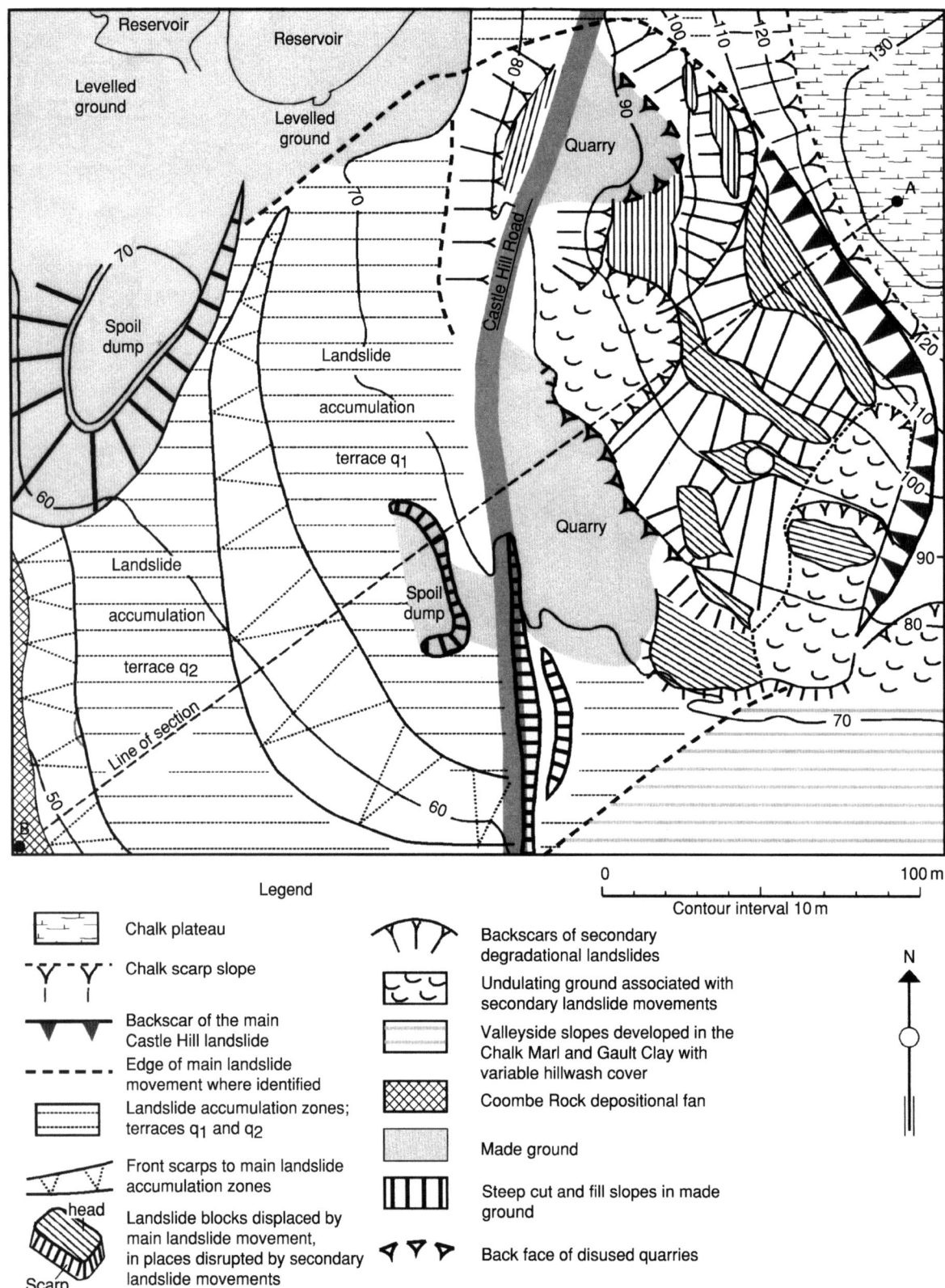

**Figure 6** Engineering geomorphological map of part of the Channel Tunnel terminal area, Folkestone, UK. The map defines the extent of the Castle Hill landslide at the site of one of the portals for the tunnel. The line of section relates to another map in the source paper. Reproduced with permission from Griffiths JS, Brunsden D, Lee EM, and Jones DKC (1995) Geomorphological investigations for the Channel Tunnel Terminal and Portal. *The Geographical Journal* 161(3): 275–284.

Folkestone, UK (**Figure 6**). The Channel Tunnel terminal and portal on the UK side is located immediately below a Lower Chalk escarpment (the North Downs) that was known to be mantled by a series of large, ancient landslides. The objective of the mapping exercise was to delimit and define the nature of past and contemporary landslide activity within the general area of the construction site. Field mapping was undertaken at 1:500 scale. The large-scale geomorphological maps refined the boundaries of the landslides shown on earlier geological maps. The mapping was also of value in showing the form and complexity of the landslide units. This complexity could then be allowed for during the interpretation of borehole data, and for the engineering design.

### Geomorphological Models

One of the main uncertainties in engineering is the risk of unexpected conditions (such as previously unrecognized hazards) or impacts and unanticipated changes. Failure to anticipate hazards and risks generally results from an inadequate understanding of a site and its context within a broader system. These problems can be minimized by the development of a model (a 'total geomorphological model') of the expected conditions at an early stage in the project.

Initial (conceptual) models can be based on desk study of available sources, concentrating on identifying the relevant systems and subsystems, along with the main environmental controls. The model should develop progressively to be site specific as the understanding of local conditions improves during the project, leading to a site model. Typical models include hazard models, which identify the nature and distribution of geohazards that might occur within an area, and ground models, which characterize the landforms and near-surface materials within an area. These models can be used to develop a checklist of questions to be answered during subsequent stages of investigations. Another commonly used model, the landform change model, identifies past changes within a system, including the interrelationships between system components. This type of model can provide the framework for developing future scenarios (e.g., the impact of climate change on the risks to a project).

## See Also

**Engineering Geology:** Overview; Natural and Anthropogenic Geohazards; Made Ground; Site and Ground Investigation; Subsidence. **Remote Sensing:** GIS. **Sedimentary Processes:** Landslides; Fluxes and Budgets.

## Further Reading

Fookes PG, Baynes FJ, and Hutchinson JN (2000) Total geological history: a model approach to the anticipation, observation and understanding of site conditions. In: *Proceedings, GeoEng 2000—International Conference on Geotechnical and Geological Engineering, 1, Melbourne*, pp. 370–460. Basel: Technomic Publishing.

Fookes PG, Lee EM, and Griffiths JS (2005) *Foundations of Engineering Geomorphology*. Latheronwheel, Caithness: Whittles Publishing.

Fookes PG, Lee EM, and Milligan G (eds.) (2004) *Geomorphology for Engineers*. Latheronwheel, Caithness: Whittles Publishing.

Fookes PG, Lee EM, and Sweeney M (2001) Terrain evaluation for pipeline route selection and characterisation, Algeria. In: Griffiths JS (ed.) *Land Surface Evaluation for Engineering Practice, Geological Society Special Publication 18*, pp. 115–121. Bath: Geological Society Publishing House.

Griffiths JS (ed.) (2001) *Land Surface Evaluation for Engineering Practice, Geological Society Special Publication 18*. Bath: Geological Society Publishing House.

Griffiths JS, Brunsden D, Lee EM, and Jones DKC (1995) Geomorphological investigations for the Channel Tunnel Terminal and Portal. *The Geographical Journal* 161(3): 275–284.

Lee EM and Clark AR (2002) *Investigation and Management of Soft Rock Cliffs*. London: Thomas Telford.

Lee EM and Moore R (1991) *Coastal Landslip Potential Assessment: Isle of Wight Undercliff, Ventnor*. London: Department of the Environment, UK.

Lee EM, Moore R, and McInnes RG (1998) Assessment of the probability of landslide reactivation: Isle of Wight Undercliff, UK. In: Moore D and Hungr O (eds.) *Engineering Geology: The View from the Pacific Rim*, pp. 1315–1321. Rotterdam: Balkema.

Morgan RPC (1986) *Soil Erosion and Conservation*. Harlow: Longman.

Thorne CR, Hey RD, and Newson MD (eds.) (1997) *Applied Fluvial Geomorphology for River Engineering and Management*. Chichester: John Wiley and Sons Ltd.

# Geophysics

**J K Gascoyne and A S Eriksen**, Zetica, Witney, UK

© 2005, Elsevier Ltd. All Rights Reserved.

## Introduction

Geophysics can be defined as the study of the Earth through the measurement of its physical properties. Use of the discipline dates back to ancient times, but only since the advent of modern-day instrumentation has its application become widespread. The development of modern geophysical techniques and equipment was initially driven by oil and mineral exploration during the early to middle parts of the twentieth century, and many of the instruments used today in engineering geophysics owe their evolution to the field of exploration geophysics.

Engineering geophysics involves using geophysical techniques to investigate subsurface structures and materials that may be of significance to the design and safety of an engineered structure. Unlike the deeper investigations associated with exploration geophysics (up to 2–3 km), engineering surveys are usually concerned with investigation of the near-surface, at depths in the range of 1–100 m.

The key advantages of geophysics over intrusive site-investigation techniques, such as digging trial pits or drilling boreholes, are that geophysical methods are comprehensive and non-invasive. Large areas can be evaluated rapidly without direct access to the subsurface. One class of engineering geophysics, borehole geophysics, is an exception in that it makes use of boreholes already drilled to sample the local area around the borehole.

When combined with intrusive methods, geophysics provides a cost-effective means of analysing the undisturbed subsurface to aid selection of, and interpolation between, widely spaced sampling locations.

Engineering geophysics can be applied throughout the life cycle of an engineered structure, starting with the initial ground investigation to determine the suitability of a particular site and provide design-sensitive and critical safety information. This may be followed by materials testing during the various stages of construction, monitoring the impact of construction on surrounding structures, on-going monitoring of the integrity of structures after completion, and helping to determine when to schedule essential maintenance tasks, such as pavement or ballast renewal on a road or railway, respectively.

The success of all geophysical methods relies on there being a measurable contrast between the physical properties of the target and those of the surrounding medium. The properties used are typically density, elasticity, magnetic susceptibility, electrical conductivity, and radioactivity. Knowledge of the material properties likely to be associated with a target is thus essential to guide the selection of the correct method to be used and to interpret the results obtained. Often a combination of methods provides the best means of solving complex problems. It is sometimes the case that, if a target does not provide a measurable physical contrast, the association of the target with other measurable conditions may indirectly lead to detection.

## Methods

Engineering geophysical methods can be split into two main categories – passive and active.

With passive methods, naturally occurring sources, such as the Earth's magnetic field, over which the observer has no control, are used to detect abnormal variations in background caused by the presence of the target. Interpretation of this data is non-unique and relies heavily on the knowledge of the interpreter.

Active methods involve generating signals in order to induce a measurable response associated with a target. The observer can control the level of energy input to the ground and measure variations in energy transmissibility over distance and time. Interpretation of this data can be more quantitative with improved depth control compared with passive methods, but ease of interpretation is not guaranteed.

**Table 1** lists of some of the techniques most commonly used in engineering geophysics.

Measurements are commonly taken at the surface and from boreholes, underground mineworkings, over or under water, or from aircraft platforms. The advent of powerful computer-aided modelling has led to the development of a number of sophisticated imaging techniques, such as cross-hole seismic and resistivity tomography and reflective tomography, which are capable of imaging the properties of the ground in three dimensions between the surface and two or more boreholes or beyond the face of a tunnel.

Armed with a knowledge of the physical properties of a target (see **Table 2**), its burial setting, and the requirements of the survey, a feasibility assessment is carried out by a geophysicist to determine the likely deliverables of a geophysical survey. Based on the results of this assessment, an appropriate geophysical

**Table 1** List of techniques commonly used in engineering geophysics

| Technique | Passsive/active | Physical property used | Source/signal |
|---|---|---|---|
| Magnetics | Passive | Magnetic susceptibility | Earth's magnetic field |
| Microgravity | Passive | Density | Earth's gravitational field |
| Continuous wave and time-domain electromagnetics | Active/passive | Electrical conductivity/resistivity | Radio-frequency electromagnetics (Hz/kHz band) |
| Resistivity imaging/sounding | Active | Electrical resistivity | DC electrical (<1000 V) |
| Induced polarization | Active | Electrical resistivity/complex resistivity, and chargeability | DC electrical |
| Self potential | Passive | Electrokinetic | Streaming and diffusion potentials |
| Seismic refraction and reflection/sonic NDT | Active/passive | Density/elasticity | Explosives, weight drop, vibration, earthquakes |
| Radiometrics | Active/passive | Radioactivity | Natural or controlled radioactive source |
| Ground penetrating radar | Active | Dielectric (permittivity) | Pulsed or stepped frequency microwave electromagnetics (50–2000 MHz) |
| Wireline logging | Active/passive | Variety | Variety |

**Table 2** Typical range of physical property values for selected soils, geology, and man-made materials

| | Physical property and relevant example method(s) | | | |
|---|---|---|---|---|
| | Seismic P-wave velocity range ($m\,s^{-1}$) | Electrical resistivity range ($\Omega m$) | Density range ($kg\,m^{-3}$) | Relative dielectric range |
| Rock/soil/material type | Seismic refraction and reflection, crosshole seismic tomography, borehole seismics | Two-dimensional resistivity imaging, resistivity sounding, electromagnetics | Microgravity, seismic reflection, gamma–gamma wireline logging | Ground penetrating radar |
| Air | 330 | — | 0 | 1 |
| Water (fresh) | 1450–1500 | 20–100 | 1000 | 81 |
| Water (saline) | 1450–1500 | <10 | 1000 | 81 |
| Ice | 3100–4000 | $>5 \times 10^4$ | 880–920 | 3–4 |
| Granite | 4000–6000 | $300-3 \times 10^6$ | 2500–2800 | 5–10 |
| Basalt | 5500–6500 | $10-1.3 \times 10^7$ | 2700–3300 | 8–12 |
| Limestone | 1700–7000 | $50-1 \times 10^7$ | 1900–2900 | 6–9 |
| Chalk | 1800–3500 | 40–200 | 1500–2600 | 8–15 |
| Sandstone | 1400–4500 | $1-7.4 \times 10^8$ | 1600–2800 | 4–6 |
| Sand (dry) | 200–1100 | 80–1000 | 1700–2300 | 3–6 |
| Sand (saturated) | 1500–2000 | 20–200 | — | >20 |
| Gravel | 1000–2500 | 100–1400 | 1700–2500 | 5–20 |
| Sand and gravel | 400–2300 | 30–250 | 1700–2400 | 6–30 |
| Glacial till | 800–2300 | 15–50 | 1800–2100 | 8–30 |
| Clay | 1000–2500 | 4–150 | 1600–2600 | 3–15 |
| Alluvium | 1800–2000 | 10–100 | 1950–2000 | >10 |
| Concrete | 3000–4000 | $10^1-10^6$ | 1750–2400 | 5–30 |
| Ballast | — | — | — | 3–8 |
| Made ground | 160–600 | 20–2000 | 1400–2000 | 8–15 |
| Tarmac | >2000 | — | — | 3.5–6 |

methodology can be designed to meet the survey objectives.

Once acquired, geophysical data needs to be processed and interpreted to provide meaningful information to the engineer. Data analysis plays a vital role in the successful application of engineering geophysics and is just as important as the selection of the survey method, survey design, and the practical skills needed to collect good data.

A key aspect of processing geophysical data ready for interpretation is the recognition and removal or separation of effects unrelated to the target being investigated. As an example, consider raw data collected during a microgravity survey. These data incorporate significant effects, such as the elevation of the site relative to sea-level, change in the Earth's radius with latitude, tides, local topography, and density of intervening materials, in addition to any mass variation caused by the target itself. Successful use of microgravity data relies on measuring and removing the known influences related to position and elevation and contributions to the observed field caused by non-target sources. The importance of this is emphasized by the fact that these extraneous effects can be orders of magnitude greater than the signal related to the target itself.

Interpretation may be either qualitative or quantitative depending on the type of method employed, the end use of the data, and available budgets. Qualitative analysis simply involves delineating zones of property change (so-called anomalous areas) within the subsurface, with little or no attempt being made to determine the physical parameters of the target. This type of analysis is common in brownfield site characterization surveys where the objective of the survey is to locate buried objects such as utilities and underground storage tanks (**Figure 1**). Quantitative interpretation of the data, on the other hand, may include modelled estimates of physical properties, including target size and depth, and a model of the site ground conditions (**Figure 2**).

As evidenced above, the display of the data and final interpretation can take a variety of forms, depending on the method deployed and the objectives of the survey.

The detection of buried hazards, such as cavities or ordnance, generally involves the acquisition of data over a wide area, using techniques such as magnetic profiling or microgravity, and consequently results are normally presented as contoured or colour-coded plots.

In the case of linear surveys, such as bedrock profiling using seismic refraction or resistivity imaging, the results are generally presented as profiles or cross-sections. Three-dimensional visualization of processed geophysical data and geophysical models is also an effective tool to simplify communication to non-geophysicists. **Figure 3** illustrates some typical data-presentation formats.

## Survey Design

As we have seen, modern engineering geophysics encompasses the use of a wide suite of geophysical methods. The first job of the geophysicist when faced with a particular engineering problem is to design a suitable survey based on the survey objectives and the nature of the target and its surroundings (both above and below ground). The survey proposed must also be cost-effective and offer cost benefits over 'traditional' site-investigation methods (*see* **Engineering Geology: Site and Ground Investigation**).

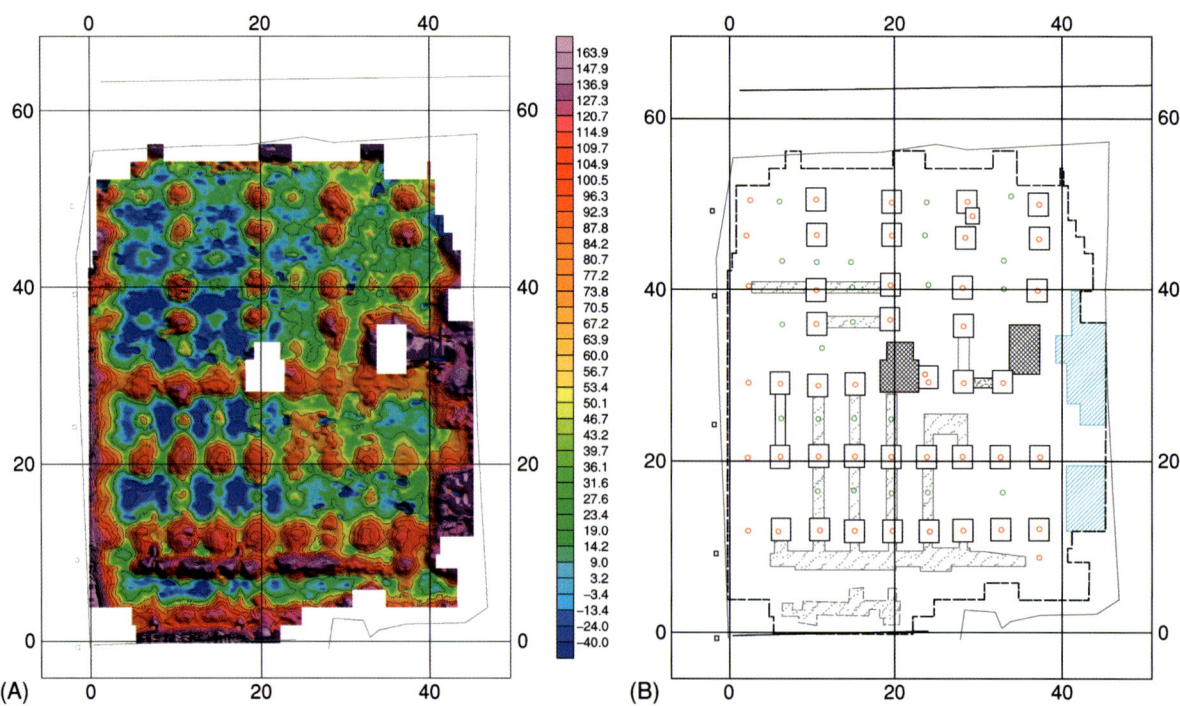

**Figure 1** Example of qualitative interpretation of geophysical data to locate buried piles. (A) Time domain electromagnetic data collected on a 1 m × 0.5 m grid and interpolated onto a grid. (B) Interpretation showing the location of all in-ground piles and other structures. The survey was 100% effective, facilitating the installation of new services without hitting any obstructions.

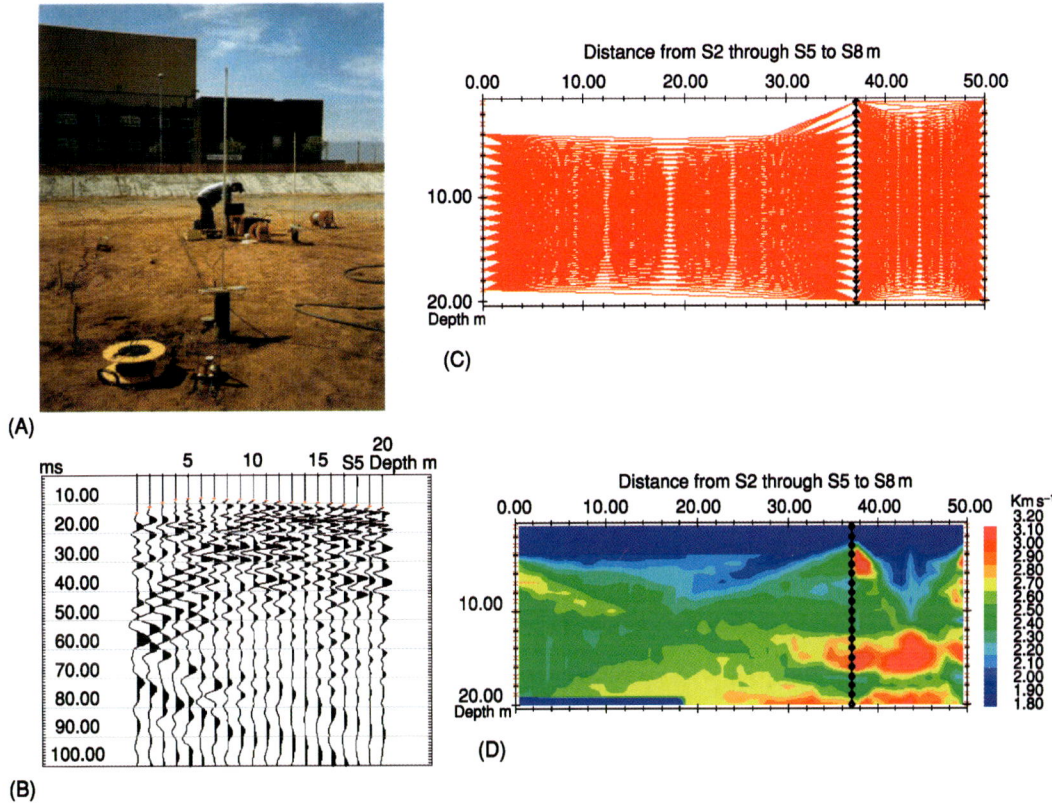

**Figure 2** Example of quantitative interpretation of cross-hole seismic data using two-dimensional computer modelling. (A) Data collection on site using a pair of boreholes, one containing receivers and the other used for locating the seismic source, which is lowered at regular intervals. (B) Raw data showing first arrivals and later events. (C) Modelled ray trace based on matching travel times to measured first arrivals from the raw data. (D) Modelled velocities showing localized high-velocity features (red).

It is worth re-emphasizing the fact that geophysical techniques cannot be applied indiscriminately. Knowledge of the material properties likely to be associated with a target is essential to choosing the correct method(s) and interpreting the results obtained.

The successful design and implementation of a survey requires careful consideration of the following main factors.

- Target discrimination. The nature and degree of contrast in physical properties between the target and its surroundings is a primary influence on the feasibility assessment of geophysics and on the choice of techniques. Information regarding the target and the expected surrounding materials may be limited or non-existent, and in these cases the geophysicist should recommend a trial survey or the application of multiple techniques on the site.
- Detection distance. All geophysical methods are sensitive to the relationship between target size and detection distance in addition to the material composition of the target and its surrounding materials. In general, the greater the detection distance, the larger a target's volumetric size and/or cross-sectional area must be for it to be detectable and the greater the required spacing between sample points (**Figure 4**).
- Survey resolution. The correct choice of sampling interval (frequency or spacing of sampling points) is critical to the success of the survey and its cost-effectiveness. For surveys conducted from the surface the sampling interval is dictated by the geophysical 'footprint' of the target, which may be tens of centimetres for small-diameter shallow pipes, a few metres for narrow fault zones, or tens of metres for large-diameter voids at depth. The optimum sampling interval will be the one that samples the anomaly adequately (commensurate with the perceived risk of non-detection) to meet the survey objectives with minimal loss of information. It is also important to note that undersampling can result in so-called 'aliasing' of the anomaly (**Figure 5**).
- Site conditions. The suitability of a site for collecting good-quality geophysical data is often overlooked in the design of works. The issues affecting data quality that could be of concern depend on the

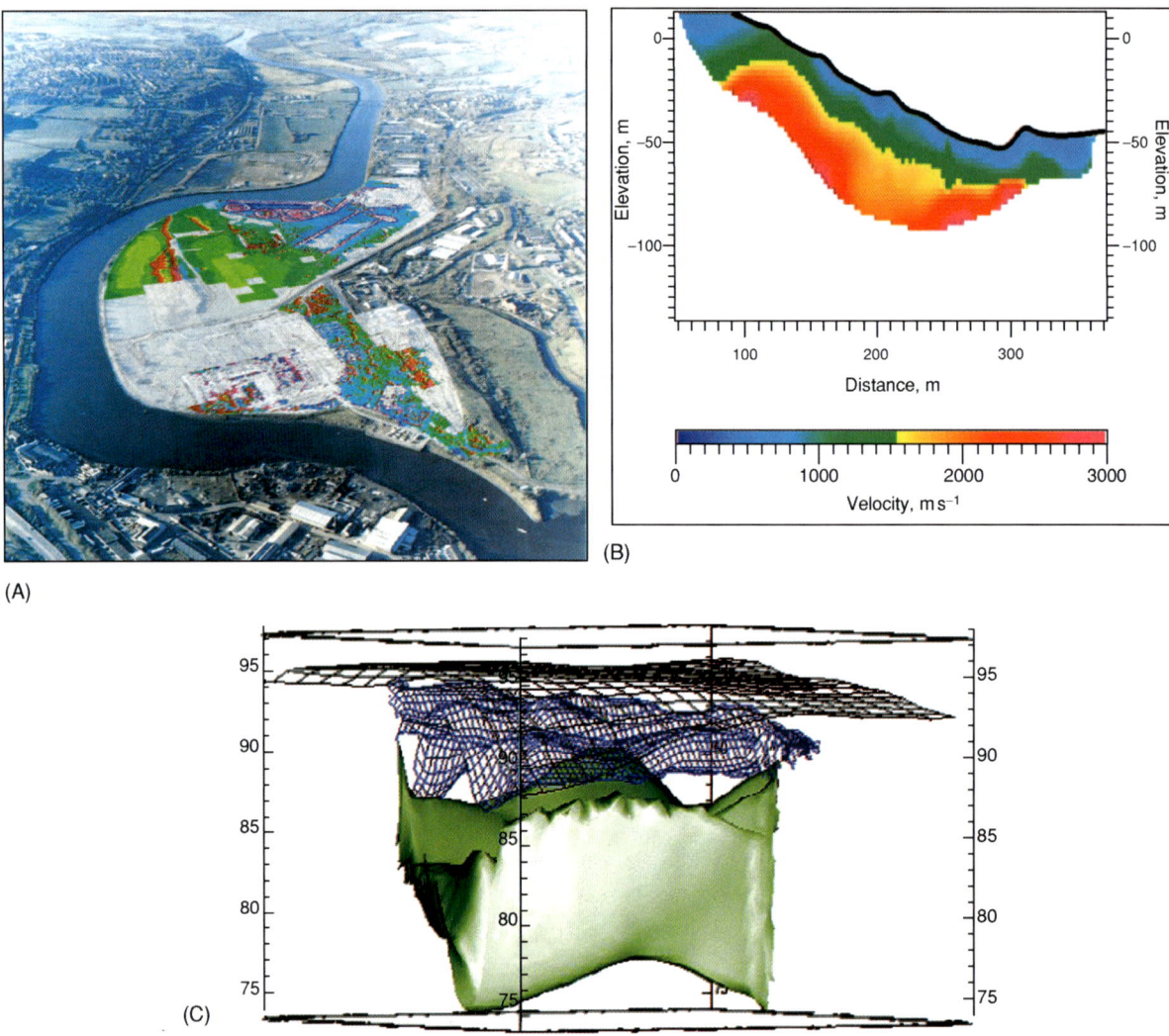

**Figure 3** Examples of different display formats. (A) Plot of residual magnetics data showing ferrous targets, draped on an aerial photograph of the 80 ha brownfield site. (B) Modelled seismic velocity cross-section for proposed pipeline route. (C) Three-dimensional visualization of the geometry of the base of a landfill cell (green) and the top of leachate (blue mesh) derived from a combined two-dimensional resistivity imaging and time-domain electromagnetic survey.

method or methods being proposed. For example, in the case of electromagnetic and magnetic methods, signal degradation or geophysical 'noise' may be introduced by the presence of surface metallic structures and overhead power lines. For microgravity or seismic surveys, noise may result from traffic movements or wind and waves. Where the level of noise exceeds the amplitude of the anomaly due to the target and where this cannot be successfully removed, the target will not be detectable. The best way to assess the likely influence of site conditions is to visit the site at the design stage and/or carry out a trial survey.

The design of a survey can be aided by the use of powerful two-dimensional and three-dimensional forward-modelling geophysical software. With information on the expected target size, depth, and composition, a geophysicist can evaluate the feasibility of a particular method, including likely error bars on modelled size or depth, and can determine optimum survey design parameters such as the sampling interval and the optimum configuration for the method (e.g. transmission frequency for ground-penetrating radar and electromagnetic surveys or the geophone spacing in a seismic survey).

**Figures 6** and **7** illustrate the results of a feasibility study for a tunnelling project in London, UK. **Figures 6A** and **7A** are the client's conceptual model of a Victorian well thought to be associated with a risk of subsidence caused by tunnelling at 20 m depth. The feasibility analysis facilitated an early appreciation of

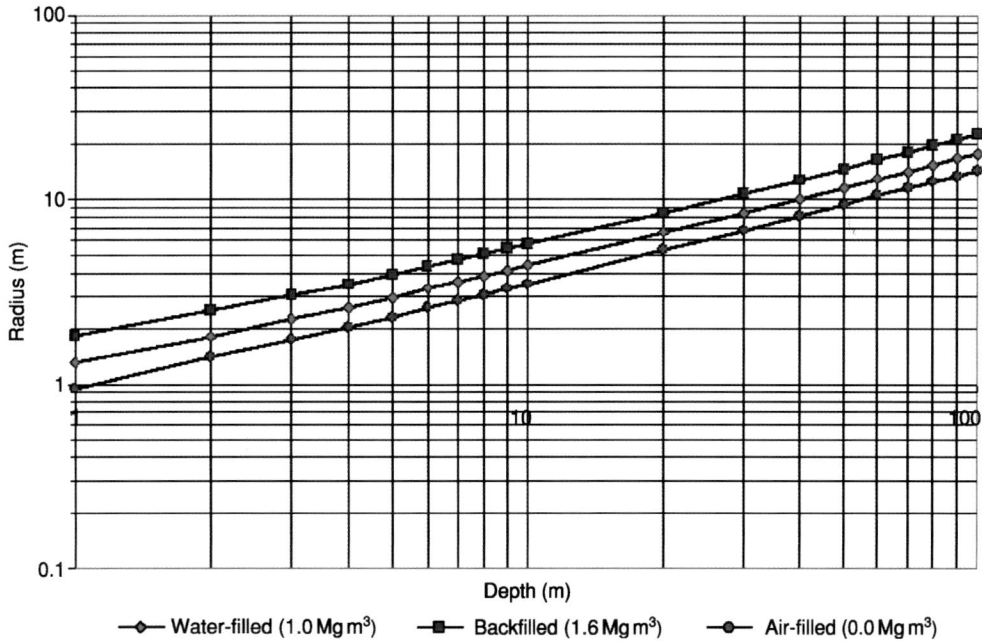

**Figure 4** Approximate minimum radius of cavity that will produce a 15 μgal microgravity anomaly in host rock of density 2.3 Mg/m³.

the limitations of the selected methodology and enabled a go/no-go decision to be made on carrying out a full-scale geophysical survey.

Working closely with the engineer, a geophysical advisor can clarify the likely value of using geophysics in a site investigation. Assuming that geophysics is viable for solving a particular problem, then the next step is to help the client decide what combination of geophysical screening and direct sampling would give the optimum information on the subsurface for the budget available (**Figure 8**).

## Applications

Common applications of engineering geophysics can be broadly separated into seven main classes, each of which encompasses a range of different target types:

1. transport infrastructure – the application of geophysics in the design and monitoring of structures such as roads, railways, bridges, tunnels, and canals;
2. foundation design – the measurement of the engineering properties of soils and bedrock, for example in determining the depth to a pile-bearing layer;
3. pipeline route evaluation – the selection of appropriate route and design criteria, for example measuring soil corrosivity for cathodic protection;
4. hazards identification – the identification and monitoring of natural hazards such as cavities and landslides and man-made structures including mineshafts and unexploded ordnance;
5. non-destructive testing of structures;
6. containment structures – integrity testing and monitoring of structures such as dams and landfills; and
7. buried assets – mapping the location of utilities and underground storage tanks.

Targets tend to be either isolated structures – such as a mineshaft, tunnel, foundations, or underground tank – or laterally extensive – such as bedrock layers, overburden, or groundwater. **Table 3** presents some examples of the use of geophysics to solve common engineering problems.

### Transport Infrastructure

Geophysical methods have a valuable role to play in both the design and monitoring of linear engineering structures, such as roads, railways, and canals, and related structures, such as bridges and tunnels.

The near-continuous lateral coverage afforded by a technique such as continuous-wave electromagnetic profiling is particularly useful during route selection, as it provides a rapid and cost-effective means of assessing near-surface ground conditions. Reconnaissance surveys of this type may be followed up by more detailed studies using quantitative techniques such as seismic refraction profiling and two-dimensional resistivity imaging within areas identified as being of particular risk or that display anomalous ground conditions (**Figure 9**).

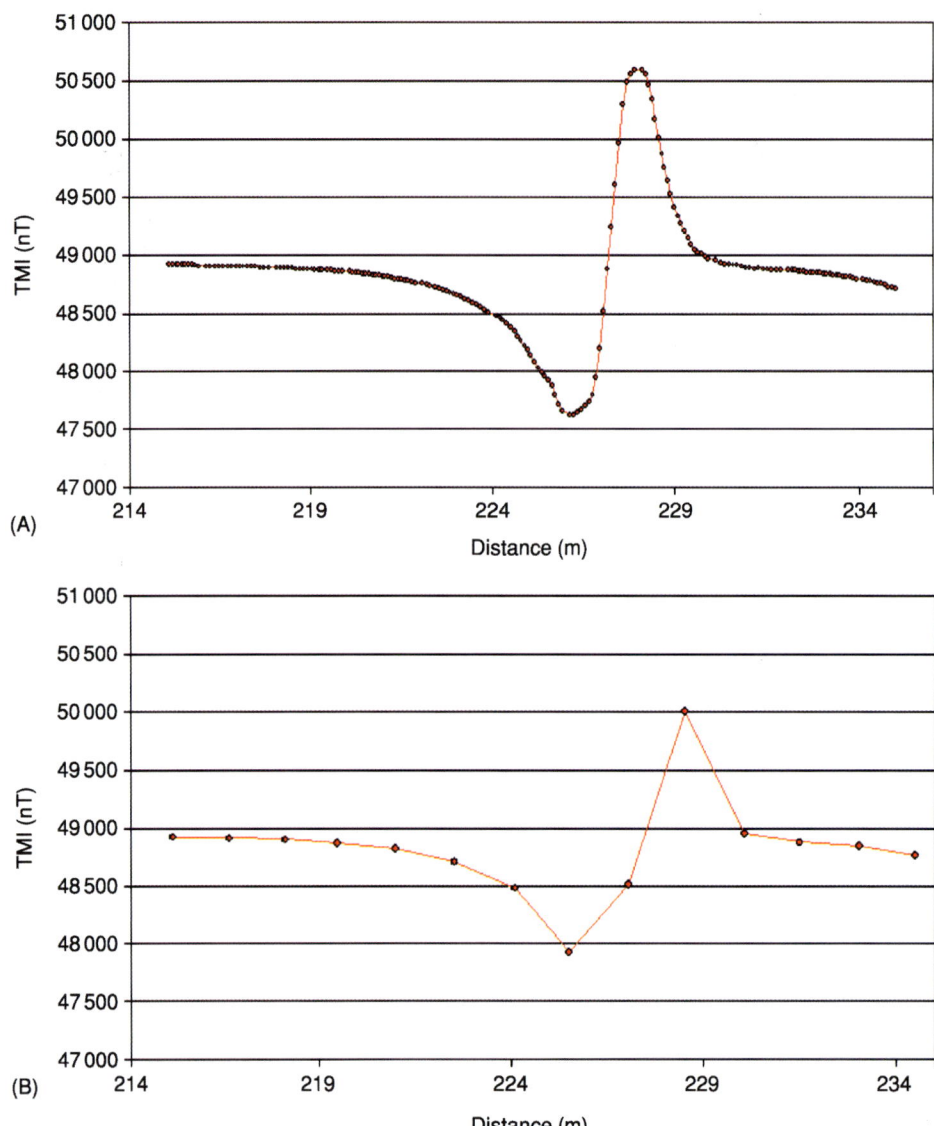

**Figure 5** Example of the effects of aliasing on the detection of a pipe using magnetic profiling. The profile shown in **Figure 5A** illustrates correct sampling of the anomaly, whilst the profile shown in **Figure 5B** illustrates the smoothing effects of under-sampling.

The main application of seismic refraction profiling is in the determination of bedrock depth and depth to groundwater (**Figure 10**). However, analysis of the P-wave velocity values obtained during such surveys can also be employed to map the variability in the strength of the bedrock along a route. The evaluation of so-called bedrock rippability is particularly useful in areas where excess material may need to be removed, such as in cuttings and the formation of embankments, in helping to determine the type of plant that will be required.

The heavy weight of traffic on roads and railways means that monitoring the condition of road pavements and track ballast is essential for proactive maintenance and planning of renewals. Ground penetrating radar (GPR) is particularly suited to this task. Modern digital GPR acquisition systems enable detailed mapping of changes in road-pavement construction, layer thicknesses, and ballast thickness and condition at speeds in excess of $100 \, \text{km} \, \text{h}^{-1}$ (**Figure 11**).

GPR can also play an important role in the evaluation of structures such as bridges and tunnels. Common applications include the identification of voiding behind brick or concrete tunnel linings and mapping the location and condition of bridge-deck reinforcing.

In order to determine layer thicknesses accurately the velocity of the GPR signal within the materials

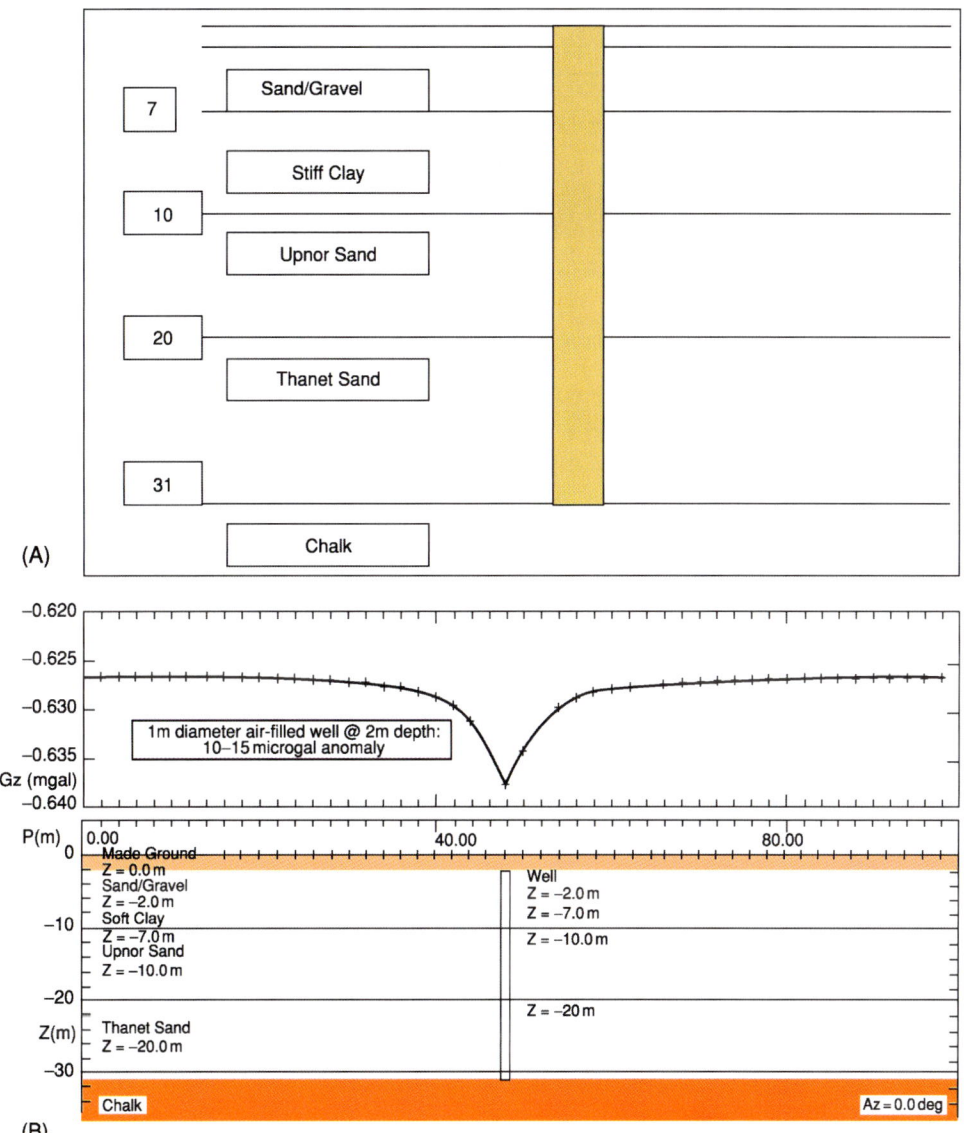

**Figure 6** (B) The use of two-dimensional forward modelling to determine the feasibility of using the microgravity method to detect (A) Victorian water wells. The size of the anomaly is 10 μGals, which, in the conditions prevalent on site (busy streets in London), would have been below the resolution of modern digital microgravity meters.

overlying an interface needs to be determined. This is normally done directly through the use of core samples but can also be derived indirectly using geometric techniques such as curve-fitting on hyperbolic anomalies and common mid-point analysis. The effective use of core samples requires that the data be accurately located.

### Foundation Design

Geophysical methods are routinely used to measure the engineering properties of soils and bedrock as an input to the design of foundation structures including piles. These properties include but are not limited to soil resistivity, shear modulus, Poisson's ratio, and percentage moisture content. These properties are usually derived from measurements taken from the surface or from boreholes. Methods for deriving low-strain elastic moduli include surface-wave seismics, seismic refraction, cross-hole seismics, seismic tomography (**Figure 2**), and sonic logging.

Wireline logging has been used to correlate stratigraphy between boreholes and provide a clearer definition of stratal boundaries and type. In the

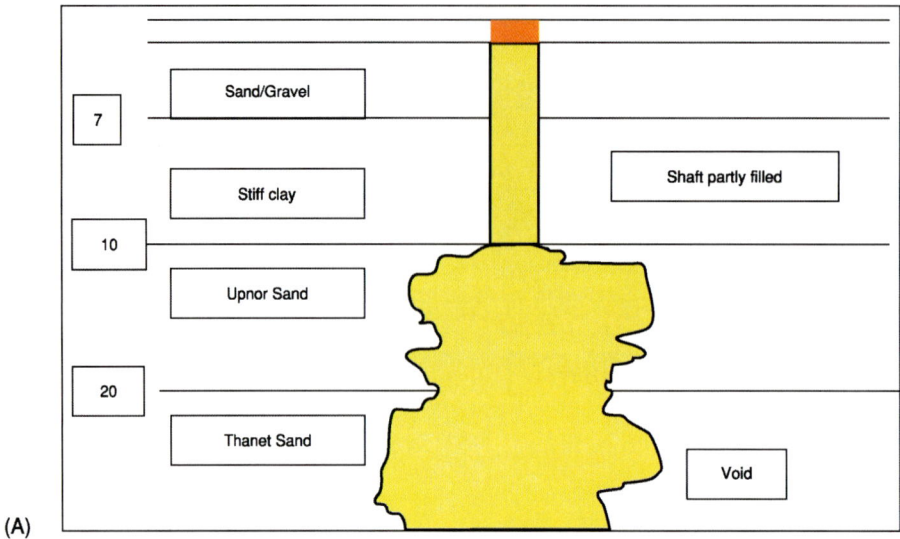

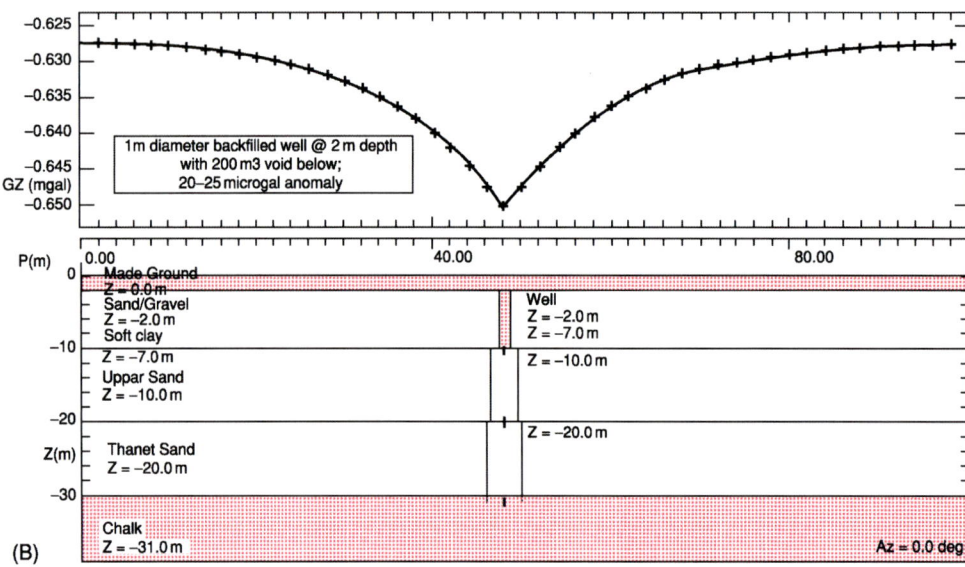

**Figure 7** (B) The use of two-dimensional forward modelling to determine the feasibility of using the microgravity method to detect (A) collapse structures associated with Victorian water wells. The size of the anomaly is 25 μGals, which is readily detectable using modern digital microgravity meters.

example shown in **Figure 12**, taken from a log from the Isle of Dogs in London, UK, the electrical log demonstrated a difference between the upper and lower parts of the Thanet Sand that was not obvious when looking at standard soils data such as gradings. The electrical log was matched to measurements of limit pressure in pressuremeter tests, and as a lower bound to SPT (Standard Penetration Test) data. The geophysical tool proved to be a powerful and cost-effective alternative for delineating the pile-bearing horizon in the Thanet Sand.

**Pipeline Investigations**

Determination of the most suitable route for a pipeline and the type of materials or protective measures required to prevent its corrosion in the ground can both be aided by the use of geophysical techniques.

As in the case of other linear structures such as roads, route evaluation may require the determination of bedrock depth, mapping of lateral variations in the nature of overburden materials, and the

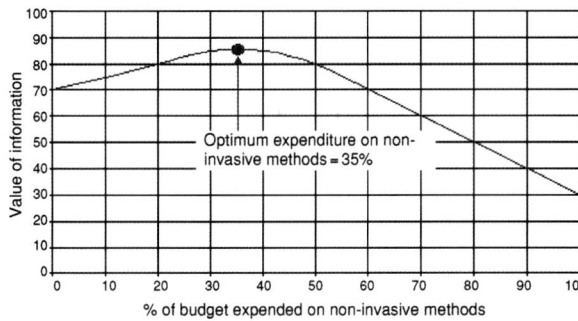

**Figure 8** Example of a cost–benefit analysis to determine the highest-value spend (i.e. most information derived) for a fixed site-investigation budget; in this brownfield case 35% of the budget was allocated for site-wide geophysical screening, to target trial pits and boreholes that might otherwise miss important features.

location of subsurface hazards or anomalous ground conditions that may affect the construction or long-term integrity of the structure. Methods commonly employed include seismic refraction and electrical resistivity imaging for profiling bedrock depth, together with resistivity soundings and electromagnetic profiling for overburden characterization. The linear nature of pipelines means that they often traverse a diverse range of terrains, and, consequently, the type of hazard that may be encountered will be varied and not always obvious. **Figure 13** illustrates data from a detailed microgravity survey undertaken along the proposed route of a pipeline in the UK with the objective of locating buried mineshafts. Forward and inverse modelling of the data allowed the likely size and depth of the shaft to be determined.

**Table 3** Examples of the application of geophysical methods in engineering investigations

| Application | Example | Typical methods |
| --- | --- | --- |
| *Transport infrastructure* | | |
| Route evaluation | Determination of thickness of overburden | Seismic refraction, 2D resistivity imaging depending on nature of expected lithologies and water table |
| | Location of buried channel | Seismic refraction or 2D resistivity imaging for large-scale structures |
| | Identification of karstic structures (swallow holes, cavities) | Continuous-wave electromagnetic profiling (reconnaissance), microgravity and seismic refraction for detailed modelling |
| Pavement/ballast assessment | Measurement of thickness of pavement or track layers and mapping location of construction or condition changes | High-speed GPR (200–1200 MHz) |
| Structural assessment | Mapping location, depth, and condition of reinforcing within concrete | High-frequency (900–1500 MHz) GPR, continuous-wave electromagnetic profiling |
| | Identification of voiding, moisture retention behind tunnel wall lining | High-frequency (900–1200 MHz) GPR, infrared thermography |
| *Foundation design* | | |
| Stratigraphic | Determination of depth to pile-bearing layer | Seismic refraction/reflection, cross-hole seismics, seismic tomography, surface-wave seismics, wireline geophysics |
| | Mapping boundaries of layers | Seismic refraction, seismic reflection, cross-hole seismics, seismic tomography, surface-wave seismics, wireline geophysics |
| | Measuring the stiffness of soils | Seismic refraction, seismic reflection, P- and S-wave cross-hole, up-hole, and down-hole seismics, cross-hole seismic tomography, surface-wave seismics, wireline geophysics |
| | Rock rippability | Seismic refraction |
| | Groundwater table | Seismic refraction, 2D resistivity imaging, time-domain electromagnetic sounding, 1D resistivity soundings, wireline logging |
| Structural | Mapping fault zones | Very-low-frequency continuous-wave electromagnetic profiling, seismic reflection |
| *Pipeline route evaluation* | | |
| Route evaluation – geological | Determination of thickness of overburden | Seismic refraction, seismic reflection (P- and S-wave), 2D resistivity imaging, microgravity, continuous-wave electromagnetic, time-domain electromagnetic sounding |
| | Rock rippability | Seismic refraction |
| | Corrosivity of soils | Resistivity, continuous-wave electromagnetic, redox potential |

*Continued*

**Table 3** Continued

| Application | Example | Typical methods |
|---|---|---|
| Archaeological assessment | Map anthropogenic features such as buried pits, walls, foundations, crypts | Magnetics, resistivity profiling, GPR, 2D resistivity imaging, magnetic susceptibility |
| Hazards identification | See below | See below |
| *Hazards identification* | | |
| Geological | Natural cavities and sinkholes | Microgravity, GPR, 2D resistivity imaging, continuous wave electromagnetic |
| Brownfield and current industrial | In-ground obstructions | Continuous-wave electromagnetic, magnetics, time-domain electromagnetic, GPR, microgravity |
| | Mineshafts | Microgravity, continuous-wave electromagnetic, magnetics, 2D resistivity imaging, GPR, infrared thermography |
| | Detection of live cables | Passive electromagnetic |
| | Detection of unexploded ordnance | Magnetics, time-domain electromagnetic, GPR |
| | Mapping contaminant sources and pathways | Continuous-wave electromagnetic, 2D resistivity imaging, spectral induced polarization, self potential, GPR, microgravity |
| | Mapping existence and number of underground storage tanks | Continuous-wave electromagnetic, magnetics, time-domain electromagnetic, GPR, microgravity |
| *Containment structures* | | |
| Construction | Quality control of construction process | Electric leak location, GPR, resistivity imaging |
| | Integrity of engineered structure over time | Self potential, 2D/3D resistivity imaging, continuous-wave electromagnetic, GPR, borehole geophysics |
| *Buried assets* | | |
| Utilities | Verifying existing maps and producing 3D plans | Passive electromagnetic, continuous-wave electromagnetic, magnetics, time-domain electromagnetic, GPR |
| Underground storage tanks | Pre-purchase audit | Magnetics and/or time-domain electromagnetic, GPR, microgravity (for non-metallic tanks) |

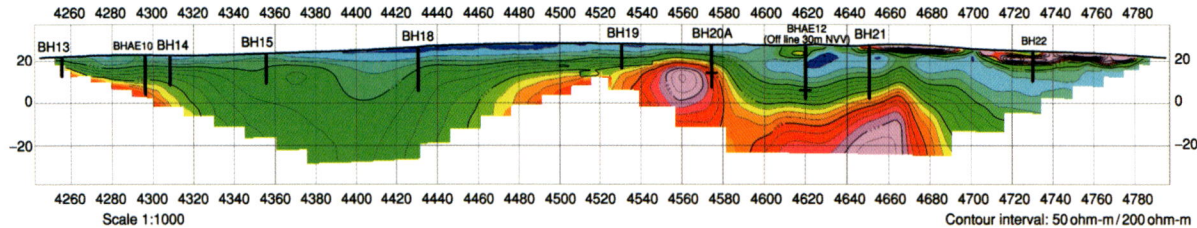

**Figure 9** Cross-section illustrating the results of a two-dimensional resistivity imaging survey along a proposed road route in southern Ireland. The objective of the survey was to determine the thickness of the glacial overburden overlying the limestone bedrock. Areas of high resistivity, relating to the bedrock, appear orange and red, whilst areas of lower resistivity, due to the overlying glacial deposits, appear green and blue. The deep area of overburden on the left of the image relates to infilling of a solution feature.

The prevention of corrosion of buried pipelines requires knowledge of the aggressivity of the soils along the proposed route. A soil's aggressiveness depends on three main factors: its pH, redox potential, and electrical conductivity. Once these factors have been determined suitable preventative measures can be put in place. These can include cathodic protection, zinc coating, or the use of plastic rather than steel pipes.

The soil electrical conductivity (or resistivity) might be determined at selected positions along the route using techniques such as one-dimensional resistivity or time-domain electromagnetic sounding or in a more continuous manner using methods such as electromagnetic profiling and capacitively coupled towed resistivity imaging.

Resistivity soundings have the advantage of providing a profile of the variation in soil resistivity with depth, thereby enabling aggressive layers within the soil to be avoided by altering the depth of the pipe. However, as with other point-sampling investigation tools such as boreholes, areas of importance along a

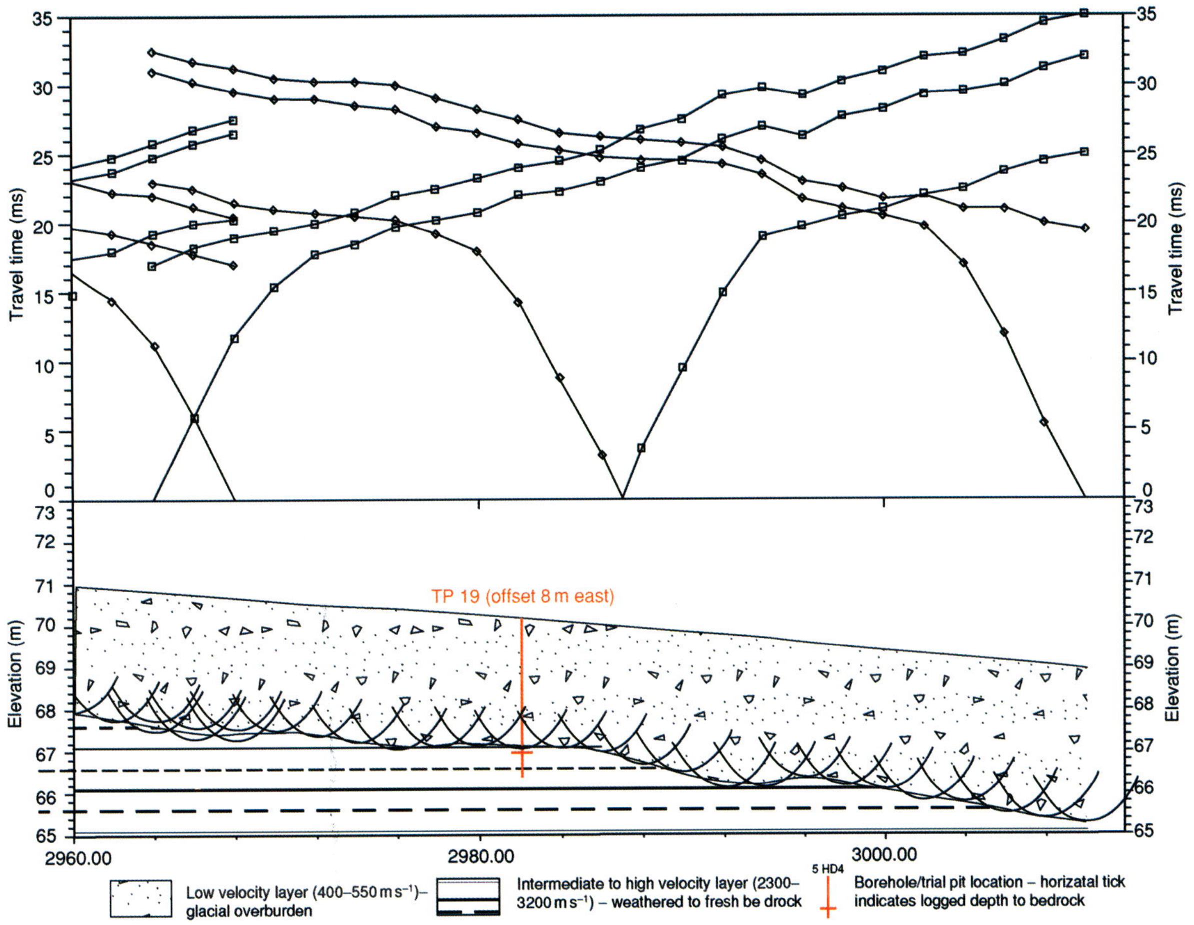

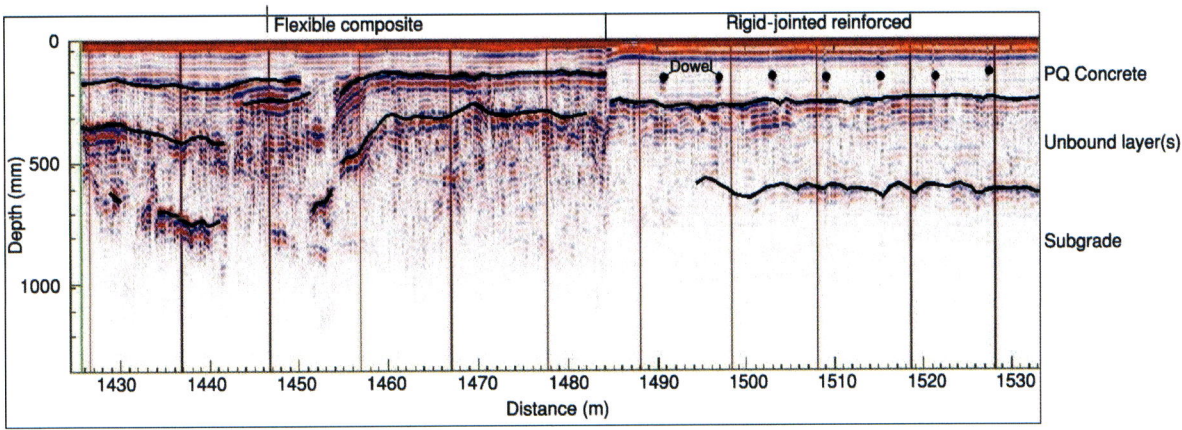

**Figure 10** Seismic refraction survey to map depth to bedrock along a proposed road route.

**Figure 11** GPR pavement survey, illustrating two types of pavement construction.

route may be missed. Techniques such as continuous wave electromagnetic profiling enable more continuous coverage but require multiple passes with different instrument configurations to determine variations in resistivity with depth. As with other route investigations, it is often beneficial to undertake an initial reconnaissance survey using a profiling method and then target areas of interest with soundings.

### Hazard Identification

Knowledge of potential subsurface hazards is critical to the safe development and maintenance of

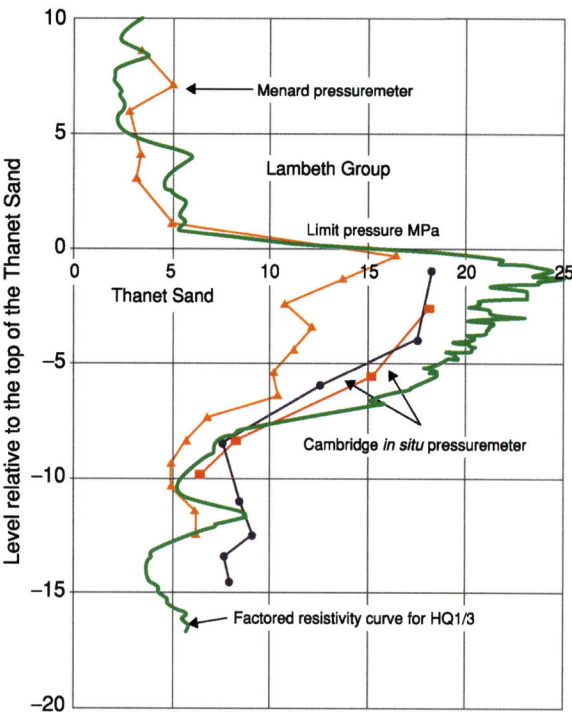

**Figure 12** Comparison of geophysical (resistivity in green) and geotechnical (Menard pressuremeter and Cambridge *in situ* pressuremeter) results for a borehole in London, UK. (Published with permission of Arup 2003.)

an engineering structure. A lack of information on the location and nature of both natural and man-made hazards, such as caves, mineshafts, buried landfills, and unexploded ordnance, could potentially be catastrophic and could lead to costly delays.

An example of how geophysics can be used to mitigate the risk posed by natural sinkholes is illustrated in **Figure 14**. A previous site investigation, incorporating four boreholes located one in each corner of the site, had determined that the depth to chalk bedrock and the condition of the chalk were essentially similar in each borehole. The regulator was informed and permission was given to construct the landfill cell. Not long after commencement, the earthworks contractor encountered large sinkhole features in the chalk, and the works were halted. The regulator requested that a geophysical survey be carried out to clarify the extent of the problem. A ground-conductivity map of the 1 ha area highlighted numerous sinkhole features as ground-conductivity highs. These features were then further investigated using resistivity imaging and cone penetrometer testing to determine the depth of the sinkholes. A hypothetical exercise was carried out to determine how many boreholes would have been required to match the detail provided by the geophysical survey.

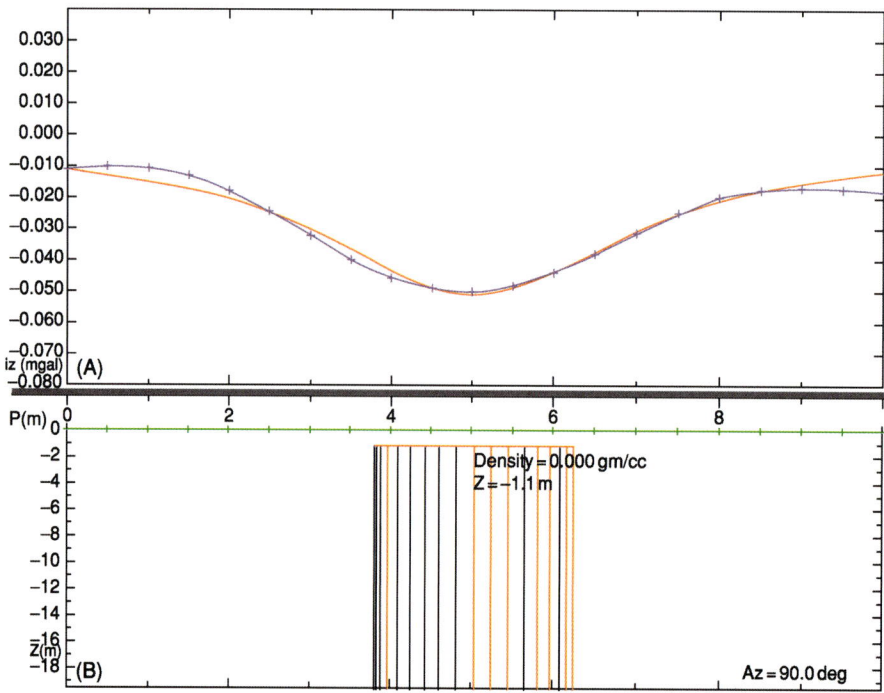

**Figure 13** Modelling the microgravity response of a buried mineshaft. (A) The blue line represents the observed gravity data, whilst the red line represents the calculated response due to (B) a cylindrical source body.

**Figure 14** shows that for a three-by-three (total of nine) or even a five-by-five (total of 25) grid of boreholes the information on the extent of sinkhole development would have been patchy at best. The most cost-effective sequence of events would have been to carry out the ground-conductivity survey first and then drill three or four boreholes in targeted areas that were revealed by geophysics to be representative of background and anomalies. **Figure 15** presents an example of the use of ground penetrating radar to locate potentially hazardous voids within karstic limestone. The high resolution obtained from GPR surveys is ideal for the detection of small-scale near-surface cavities but under good conditions, such as in unweathered hard rock, the technique is also capable of providing information to significant depth. In these circumstances it is often used to complement data obtained from more traditional techniques such as microgravity profiling.

Projects on land associated with, or affected by, military activities may have an increased risk of encountering unexploded ordnance. Examples include

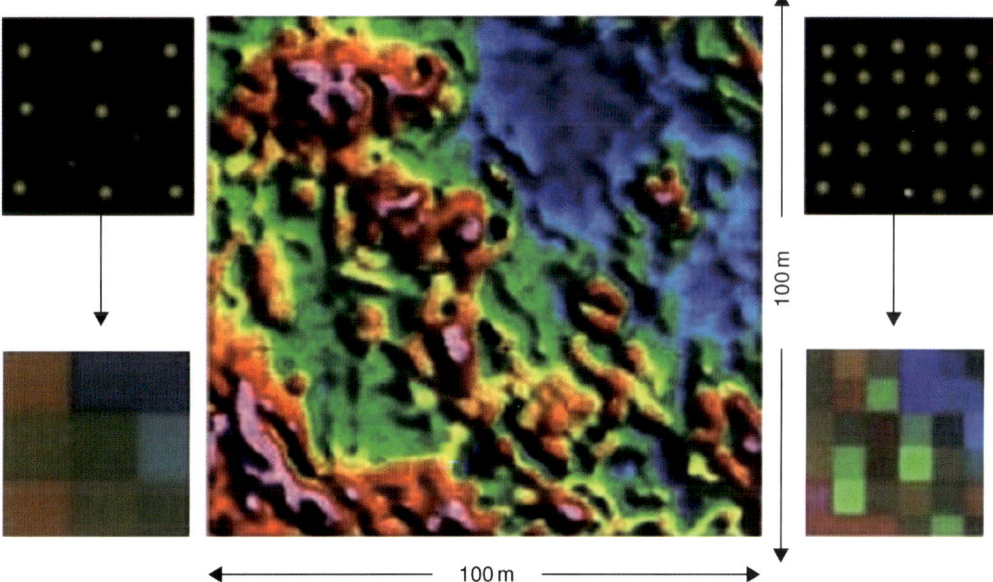

**Figure 14** Comparison of the results of undertaking an electromagnetic (apparent conductivity) survey (centre) across a site to map solution features in chalk and a more traditional site-investigation approach using just boreholes (left and right). Red and magenta areas on the apparent-conductivity map represent sinkholes filled with moist sand. The blue and green areas correspond to competent chalk.

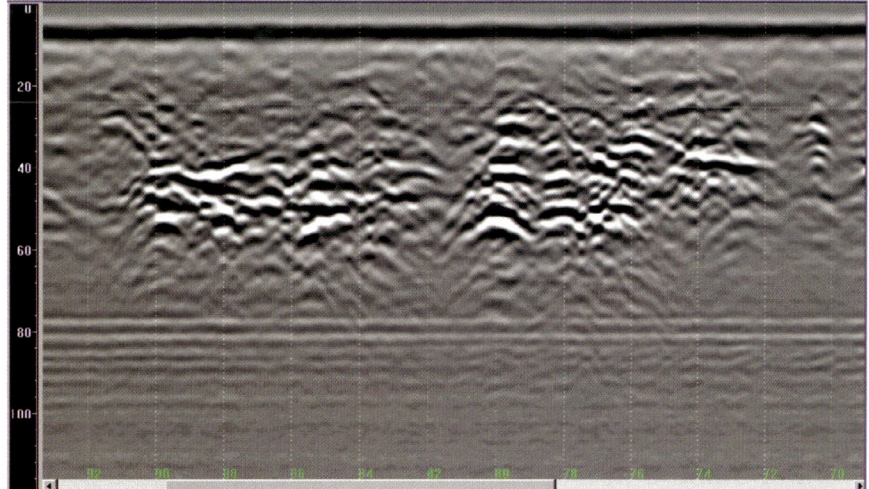

**Figure 15** The use of GPR to locate voiding within karstic limestone for a port development in the Dominican Republic. The areas of white and black on this B-scan plot (or 'radargram') are indicative of clustered small-scale (less than 0.5 m in diameter) voids within otherwise competent limestone. The vertical axis on the plot represents the two-way travel time of the GPR signal through the rock measured in nanoseconds.

munitions-factory sites, military bases, military training grounds, and wartime target areas. Apart from low-metal landmines, most ordnance is composed of metal casing, which can be targeted by geophysical methods (depending on the size and depth of burial). In London, UK, which was heavily bombed during the Second World War, there may be a risk of the presence of unexploded bombs to depths of 10 m or more, depending on the size of bomb dropped and soil conditions. Deep bombs such as these are undetectable using surface geophysical methods but can be detected using sensors lowered down boreholes or mounted in push rods. An example of how geophysics can be used to mitigate the risk posed by unexploded ordnance is shown in **Figure 16**. A sitewide gradient magnetometer survey was carried out on a close grid spacing to detect small ferrous ordnance items. These were then flagged in the field for identification and disposal. Following removal of identified targets a second survey was carried out to confirm that all located ordnance had been removed.

### Non-Destructive Testing

Engineering geophysics forms a significant component of the field of non-destructive testing, which encompasses all methods used to detect and evaluate the integrity of materials. Non-destructive testing is used for in-service inspection and condition monitoring of structures and plant including the measurement of physical properties such as hardness and internal stress.

Examples of geophysical methods used in non-destructive testing are tabulated in **Table 4**.

### Containment Structures

As well as playing a role in the assessment of ground conditions prior to construction of a containment structure such as a lagoon or dam, geophysics can also be used to effect in the assessment and monitoring of the integrity of a structure prior to and during use and in the investigation of breaches in containment during its lifetime. In the latter case, where the storage of hazardous materials such as chemical and nuclear wastes is involved, a geophysical survey is often a cost-effective means of investigating the problem as it avoids the need to remove the stored material.

The most common requirement for geophysics at the pre-commissioning stage is in the assessment of the integrity of lined structures such as domestic landfills and leachate lagoons using the electrical leak location method. This technique originated in the USA during the mid-1980s and has since been adopted widely. It now forms a recommended part of the Construction Quality Assurance testing of new

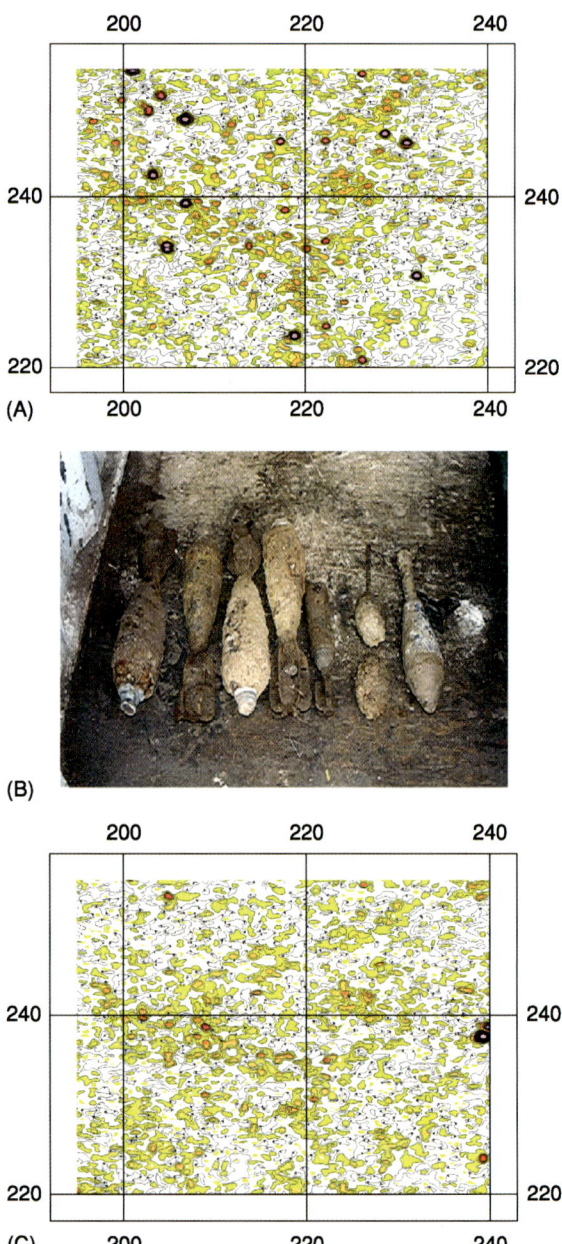

**Figure 16** (A) The results of scanning an area using gradient magnetic profiling to detect ferrous ordnance items. (B) The located items were removed by trained EOC (Explosive Ordnance Clearance) personnel, and (C) the area was rescanned to prove that all items had been removed.

landfill cells in the UK. The technique comprises a novel form of resistivity profiling, which takes advantage of the insulating properties of geomembrane materials, such as HDPE (high density polyethylene) and PVC (polyvinyl chloride), in order to pinpoint holes or tears as small as $0.5\ mm^2$. Variants of the basic method have been developed for testing soil-covered liners, cell side slopes (with no cover material), and water-filled lagoons and ponds.

**Table 4** Geophysical methods used in non-destructive testing (after McDowell et al., 2002)

| Material | Method | Application |
| --- | --- | --- |
| Concrete | Ultrasonic pulse velocity, GPR | Measure strength, voidage, reinforcing, cracking, cover, and fire damage |
| Metal | Ultrasonic pulse velocity | Measure thickness and cracking |
| Walls and roofs | Thermography, GPR, electromagnetic scanners, resistance/capacitance | Cavity insulation, wall ties, cladding fixings, moisture penetration, roof leaks |
| Buried objects | GPR, Cable Avoidance tool (CAT), magnetometer | Location of services and foundations, archaeological remains, buried objects |
| Machinery | Vibration metres, thermography, sound metres | Worn bearings, overheating (electrical) |

Permanent geomembrane electrical leak location systems, comprising a grid of electrodes below or on either side of the geomembrane, allow continuous monitoring of the integrity of the liner throughout its lifetime. These systems can also be employed to create two- or three-dimensional resistivity models for the strata beneath the containment in order to monitor the spread of pollutants in the event of a breach.

Where access to a structure is unavailable or impossible owing to the hazardous nature of the contents, the structure can be monitored remotely using techniques such as electrical resistance tomography and excitation of the mass. Unlike electrical leak location, these methods also enable the detection of leaks from conductive containment structures such as steel underground storage tanks. Both methods involve the use of an array of electrodes placed around the structure, with excitation of the mass additionally employing an electrode placed within the containment.

Investigations into breaches in containment structures such as dams, locks, and canals can involve a wide variety of geophysical techniques, depending on the nature of the problem. Methods such as GPR and sonic testing can be used to identify voiding within and behind concrete and brick structures, whilst resistivity imaging might be used to map changes in electrical resistivity resulting from the outflow of water from a dam or canal. The self-potential method involves measuring low-amplitude voltages that result from the movement of fluids through soil and rock and can be used to locate seepage pathways within embankments and retaining walls.

### Buried Assets

The inaccuracy and incompleteness of site plans showing services is infamous. Failing to confirm the location of underground services and other assets, such as underground storage tanks, can have serious cost and safety implications for a project. In the UK, the requirement to locate services prior to the commencement of any intrusive work has now been included in Health and Safety legislation. Confirmation of the position and depth of existing utility lines during the site-investigation phase allows for the planned diversion of essential services and ensures that exploratory boreholes and trial pits can be safely positioned.

The detection and tracing of services presents a particular problem for geophysics owing to the range of materials and target sizes that can be encountered. For this reason it is common to approach services detection with a broad suite of both passive and active techniques. Those most commonly used techniques include continuous-wave electromagnetic profiling (**Figure 17**), and passive electromagnetic methods such as the cable avoidance tool, magnetic profiling, and GPR (**Figure 18**). The latter is the only method capable of detecting non-metallic utility lines such as high-pressure water and gas mains.

In order to ensure optimum detection of linear targets such as services it may be necessary to undertake the survey in two orthogonal directions unless the majority of the targets are known to trend in a particular direction. This is particularly important for electromagnetic profiling and GPR, where the target response is highly dependent on the relative orientation of the instrument and target.

Where access to a particular pipe is available tracing may be simplified by using a pipe sonde. This comprises an electromagnetic transmitter mounted in a cylindrical housing that can be fed into the pipeline through an access cover or outfall. The signal from the transmitter is detected on the surface using a suitable hand-held receiver.

It is important to note that the detectability of services is dependant on burial setting and on the material type of the services being mapped. For example, an inaccessible concrete waste pipe 30 cm in diameter buried more than 2 m deep in clayey soils could be undetectable with any geophysical method. Services running beneath reinforced concrete slabs may be detectable only within 0.5 m of the surface with GPR as the presence of reinforcing will preclude

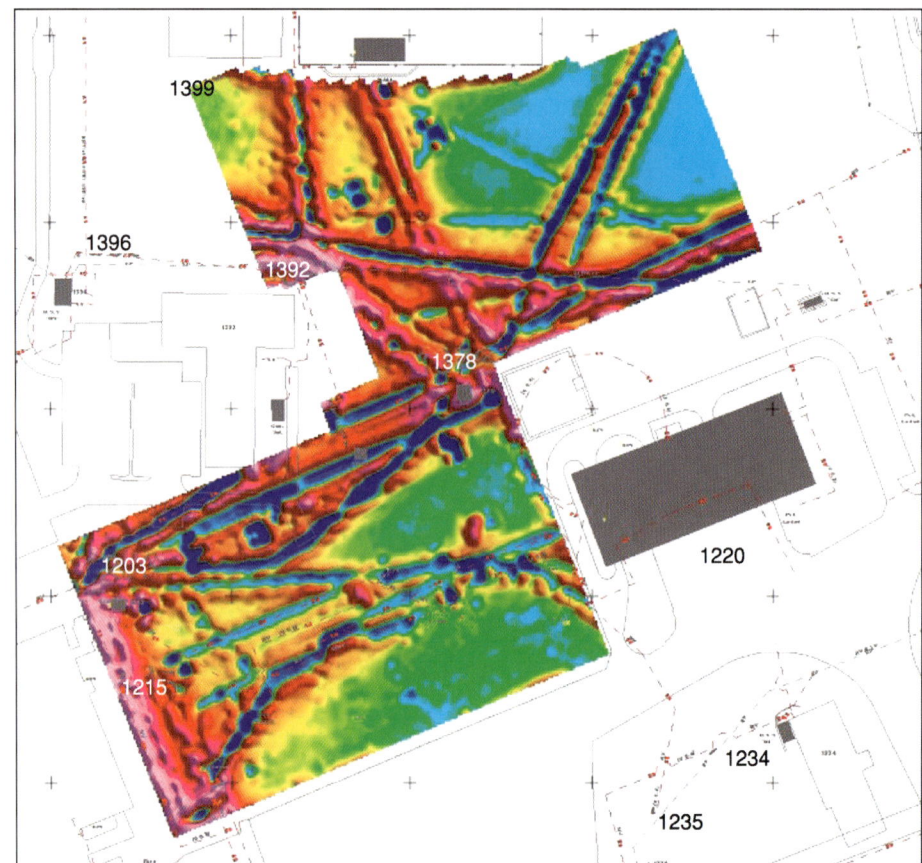

**Figure 17** Electromagnetic profiling survey to map the location of services at an air base in the UK.

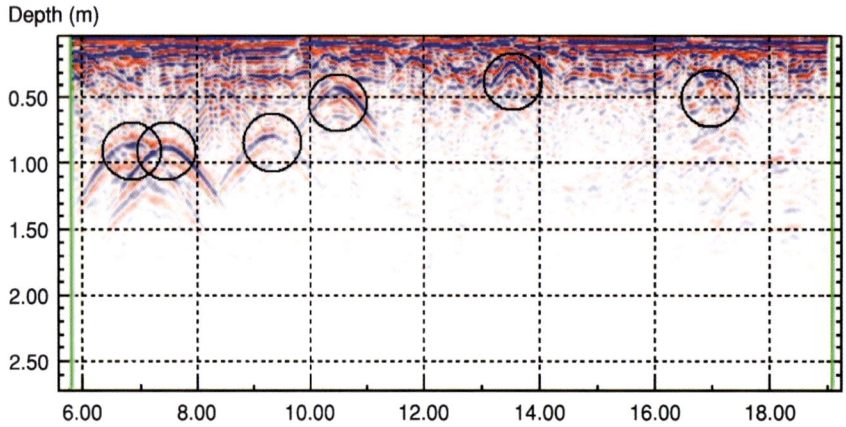

**Figure 18** The location of buried utilities using GPR.

the use of electromagnetic methods. Exaggerated claims that '95% of all buried services will be mapped' should be weighed against the limitations of geophysical methods in general.

Estimating the depth to the top of a target is possible, although the degree of accuracy varies widely. Cable-avoidance tools provide a quick method of determining approximate depths on site, although these are generally accurate only to within ±20%. Advanced processing of magnetic data using techniques such as Euler Deconvolution or 2.5D inverse and forward modelling enables estimates of depths to within ±15–20%. In the case of GPR, calculation of target depths is dependent on determining the average velocity of the signal through the overlying materials.

## See Also

**Analytical Methods:** Gravity. **Engineering Geology:** Seismology; Natural and Anthropogenic Geohazards; Site and Ground Investigation; Site Classification. **Military Geology. Rock Mechanics. Seismic Surveys. Soil Mechanics.**

## Further Reading

Environmental and Engineering Geophysical Society. http://www.eegs.org/whatis/

Health and Safety Executive (UK) (2000) *Avoiding Danger from Underground Services*. Health and Safety Executive, London.

McCann DM, Eddleston M, Fenning PJ, and Reeves GM (eds.) (1997) *Modern Geophysics in Engineering Geology*. Engineering Geology Special Publication 12. London: Geological Society.

McDowell PW, Barker RD, and Butcher AP (eds.) (2002) *Geophysics in Engineering Investigations*. Engineering Geology Special Publication 19. London: CIRIA.

Milsom J (2003) *Field Geophysics*, 3rd edn., The Geological Field Guide Series. Chichester: John Wiley & Sons.

Reynolds JM (1997) *An Introduction to Applied and Environmental Geophysics*. Chichester: John Wiley & Sons.

Telford WM, Geldart LP, and Sheriff RE (1990) *Applied Geophysics*, 2nd edn. Cambridge: Cambridge University Press.

Zeltica. Geophysical Method Descriptons. http://www.geophysics.co.uk/methods.html

# Seismology

**J J Bommer**, Imperial College London, London, UK
**D M Boore**, United States Geological Survey, Menlo Park, CA, USA

© 2005, Elsevier Ltd. All Rights Reserved.

## Introduction

Engineering seismology is an integral part of earthquake engineering, a specialized branch of civil engineering concerned with the protection of the built environment against the potentially destructive effects of earthquakes. The objective of earthquake engineering can be stated as the reduction or mitigation of seismic risk, which is understood as the possibility of losses – human, social or economic – being caused by earthquakes. Seismic risk exists because of the convolution of three factors: seismic hazard, exposure, and vulnerability. Seismic hazard refers to the effects of earthquakes that can cause damage in the built environment, such as the primary effects of ground shaking or ground rupture or secondary effects such as soil liquefaction or landslides. Exposure refers to the population, buildings, installations, and infrastructure encountered at the location where earthquake effects could occur. Vulnerability represents the likelihood of damage being sustained by a structure when it is exposed to a particular earthquake effect.

## Earthquake Hazards and Seismic Risk

Risk can be reduced in two main ways, the first being to avoid exposure where seismic hazard is high. However, human settlement, the construction of industrial facilities, and the routing of lifelines such as roads, bridges, pipelines, telecommunications, and energy distribution systems are often governed by other factors that are of greater importance. Indeed, for lifelines, which may have total lengths of hundreds of kilometres, it will often be impossible to avoid seismic hazards by relocation. Millions of people are living in areas of the world where there is appreciable seismic hazard. The key to mitigating seismic risk therefore lies in control of vulnerability in the built environment, by designing and building structures and facilities with sufficient resistance to withstand the effects of earthquakes; this is the essence of earthquake engineering.

This is not to say, however, that the aim is to construct an earthquake-proof built environment that will suffer no damage in the case of a strong earthquake. The cost of such levels of protection would be extremely high, and, moreover, it might mean protecting the built environment against events that may not occur within the useful life of a particular building. It is generally not possible to justify such investments, especially when there are many competing demands on resources. Objectives of earthquake-resistant design are often stated in terms that relate different performance objectives to different levels of earthquake motion, such as no damage being sustained due to mild levels of shaking that may occur frequently, damage being limited to non-structural elements or to easily repairable levels in structural elements in moderate shaking that occurs occasionally, and collapse being avoided under severe ground shaking that is only expected to occur rarely. These objectives are adjusted according to the consequences of damage to the structure; for rare occurrences of intense shaking, the performance target for a single-family dwelling will be to avoid collapse of structural elements, so that the occupants may escape from the building without injury. For a hospital or fire station, the performance target will be to remain fully operational under the same level of shaking, because the services provided will be particularly important in the aftermath of an earthquake. For a nuclear power plant or radioactive waste repository, the performance objective will be to maintain structural integrity even under extreme levels of shaking that may be expected to occur very rarely.

Once planners and developers have taken decisions regarding the location of civil engineering projects, or once people have begun to settle in an area, the level of exposure is determined. In general, seismic hazard cannot be altered, hence the key to mitigating seismic risk levels lies in the reduction of vulnerability or, stated another way, depends on the provision of earthquake resistance. In order to provide effective earthquake protection, the civil engineer requires quantitative information on the nature and likelihood of the expected earthquake hazards. As already indicated, this may mean defining the hazard, not in terms of the effects that a single earthquake event may produce at the site, but as a synthesis of the potential effects of many possible earthquake scenarios and quantitative definitions of the particular effects that may be expected to occur with different specified frequencies. This is the essence of engineering seismology: to provide quantitative assessments of earthquake hazards.

Earthquake generation creates a number of effects that are potentially threatening to the built environment (**Figure 1**). These effects are earthquake hazards, and engineering seismology, in the broadest sense, is concerned with assessing the likelihood and characteristics of each of these hazards and their possible impact in a given region or at a given site of interest. Earthquakes are caused by sudden rupture on geological faults, the slip on the fault rupture ranging from a few to tens of centimetres for moderate earthquakes (magnitudes from 5 to about 6.5) to many metres for large events (*see* **Tectonics: Earthquakes**). In those cases where the fault extends to the ground surface (which is often not the case), the relative displacement of the two sides of the fault presents an obvious hazard to any structure crossing the fault trace. In the 17 August 1999 Kocaeli earthquake in Turkey, several hundred houses and at least one industrial facility were severely damaged by the surface deformations associated with the slip on the

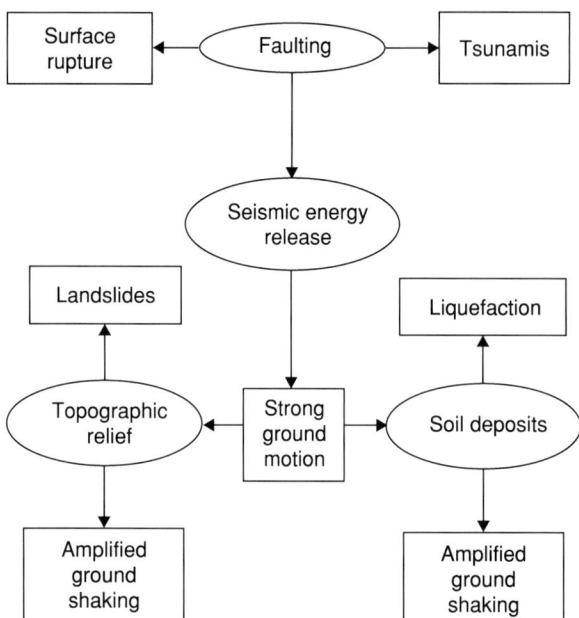

**Figure 1** Potentially destructive effects of earthquakes, showing the elements of the earthquake generation process and the natural environment (ovals) and the resulting seismic hazards (rectangles).

North Anatolian fault. However, buildings must be situated within a few metres of the fault trace for surface rupture to be a hazard, whereas ground shaking effects can present a serious hazard even at tens of kilometres from the fault trace; hence, even if the fault rupture is tens or even hundreds of kilometres in length, the relative importance of surface rupture hazard is low. An exception to this is where lifelines, particularly pipelines, cross fault traces, although identification and quantification of the hazard allows the design to take into account the expected slip in the event of an earthquake: in the vicinity of the Denali Fault, the trans-Alaskan oil pipeline is mounted on sleepers that allow lateral movement, which permitted the pipeline to remain functional despite more than 4 m of lateral displacement on the fault in the magnitude 7.9 earthquake of 3 November 2002.

Surface fault ruptures in the ocean floor can give rise to tsunamis, seismic sea waves that are generated by the sudden displacement of the surface of the sea and travel with very high speeds. As the waves approach the shore and the water depth decreases, the amplitude of the waves increases to maintain the momentum, reaching heights of up to 30 m. When the waves impact on low-lying coastal areas, the destruction can be almost total.

All of the other hazards generated by earthquakes are directly related to the shaking of the ground caused by the passage of seismic waves. The rapid movement of building foundations during an earthquake generates inertial loads that can lead to damage and collapse, which is the cause of the vast majority of fatalities due to earthquakes. For this reason, the main focus of engineering seismology, and also of this article, is the assessment of the hazard of ground shaking. Earthquake ground motion can be amplified by features of the natural environment, increasing the hazard to the built environment. Topographic features such as ridges can cause amplification of the shaking, and soft soil deposits also tend to increase the amplitude of the shaking with respect to rock sites. At the same time, the shaking can induce secondary geotechnical hazards by causing failure of the ground. In mountainous or hilly areas, earthquakes frequently trigger landslides, which can significantly compound the losses: the 6 March 1987 earthquake in Ecuador triggered landslides that interrupted a 40-km segment of the pipeline carrying oil from the production fields in the Amazon basin to the coast, thereby cutting one of the major exports of the country; the earthquake that struck El Salvador on 13 January 2001 killed about 850 people, and nearly all of them were buried by landslides. In areas where saturated sandy soils are encountered, the ground shaking can induce liquefaction (*see* **Engineering Geology:** Liquefaction) through the generation of high pore-water pressures, leading to reduced effective stress and a significant loss of shear strength, which in turns leads to the sinking of buildings into the ground and lateral spreading on river banks and along coasts. Extensive damage in the 17 January 1994 Kobe earthquake was caused by liquefaction of reclaimed land, leaving Japan's second port out of operation for 3 years.

The assessment of landslide and liquefaction hazard involves evaluating the susceptibility of slopes and soil deposits, and determining the expected level of earthquake ground motion. The basis for earthquake-resistant design of buildings and bridges also requires quantitative assessment of the ground motion that may be expected at the location of the project during its design life. Seismic hazard assessment in terms of strong ground motion is the activity that defines engineering seismology.

## Measuring Earthquake Ground Motion

The measurement of seismic waves is fundamental to seismology. Earthquake locations and magnitudes are determined from recordings on sensitive instruments (called seismographs) installed throughout the world, detecting imperceptible motions of waves generated by events occurring hundreds or even thousands of kilometres away. Engineering seismology deals with ground motions sufficiently close to the causative rupture to be strong enough to present a threat to engineering structures. There are cases in which destructive motions have occurred at significant distances from the earthquake source, generally as the result of amplification of the motions by very soft soil deposits, such as in the San Francisco Marina District during the 18 October 1989 Loma Prieta earthquake, and even more spectacularly in Mexico City during the 19 September 1985 Michoacan earthquake, almost 400 km from the earthquake source. In general, however, the realm of interest of engineering seismology is limited to a few tens of kilometres from the earthquake source, perhaps extending to 100 km or a little more for the largest magnitude events.

Seismographs specifically designed for measuring the strong ground motion near the source of an earthquake are called accelerographs, and the records that they produce are accelerograms. The first accelerographs were installed in California in 1932, almost four decades after the first seismographs, the delay being caused by the challenge of constructing instruments that were simultaneously sensitive enough to produce accurate records of the ground acceleration while being of sufficient robustness to withstand the shaking without damage.

Prior to the development of the first accelerographs, the only way to quantify earthquake shaking was through the use of intensity scales, which provide an index reflecting the strength of ground shaking at a particular location during an earthquake. The index is evaluated on the basis of observations of how people, objects, and buildings respond to the shaking (**Table 1**). Some intensity scales also include the response of the ground with indicators such as slumping, ground cracking, and landslides, but these phenomena are generally considered to be dependent on too many variables to be reliable indicators of the strength of ground shaking. At the lower intensity degrees, the most important indicators are related to human perception of the shaking, whereas at the higher levels, the assessment is based primarily on the damage sustained by different classes of buildings. A common misconception is that intensity is a measure of damage, whereas it is in fact a measure of the strength of ground motion inferred from building damage, whence a single intensity degree can correspond to severe damage in vulnerable rural dwellings and minor damage in engineered constructions. The most widely used intensity scales, both of which have 12 degrees and which are broadly equivalent, are the Modified Mercalli (MM), used in the Americas, and the 1998 European Macroseismic Scale (EMS-98), which has replaced the Medvedev–Sponheuer–Karnik (MSK) scale.

**Table 1** Summary of the 1998 European Macroseismic Scale

| Intensity | Definition | Effects on people and buildings[a] |
|---|---|---|
| I | Not felt | Not felt; detected only by sensitive instruments |
| II | Scarcely felt | Felt by very few people, mainly those who are in particularly favourable conditions at rest and indoors |
| III | Weak | Felt by a few people, mostly those at rest |
| IV | Largely observed | Felt by many people indoors and very few outdoors; a few people are awakened by the shaking |
| V | Strong | Felt by most people indoors and a few outdoors; a few people are frightened and many who are sleeping are awakened by the shaking |
| | | Fine cracks in a few of the most vulnerable types of buildings, such as adobe and unreinforced masonry |
| VI | Slightly damaging | Shaking felt by nearly everyone indoors and by many outdoors; a few people lose their balance |
| | | Minor cracks in many vulnerable buildings and in a few poor-quality RC structures; a few of the most vulnerable buildings have cracks in walls and spalling of fairly large pieces of plaster |
| VII | Damaging | Most people are frightened by the shaking, and many have difficulty standing, especially those on upper storeys of buildings |
| | | Many adobe and unreinforced masonry buildings sustain large cracks and damage to roofs and chimneys; some will have serious failures in walls and partial collapse of roofs and floors; minor cracks in RC structures with some earthquake-resistant design, more significant cracks in poor-quality RC structures |
| VIII | Heavily damaged | Many find it difficult to stand |
| | | Many vulnerable buildings experience partial collapse and some collapse completely; large and extensive cracks in poor-quality RC buildings and many cracks in RC buildings with some degree of earthquake-resistant design |
| IX | Destructive | General panic (this is the highest degree of intensity that can be assessed from human response) |
| | | General and extensive damage in vulnerable buildings; cracks in columns, beams, and partition walls of RC buildings, with some of those of poor quality suffering heavy structural damage and partial collapse; non-structural damage in RC structures with high level of earthquake-resistant design |
| X | Very destructive | Partial or total collapse in nearly all vulnerable buildings; extensive structural damage in RC and steel structures |
| XI | Devastating | Extensive damage and widespread collapse in nearly all building types; very rarely observed, if ever |
| XII | Completely devastating | Cataclysmic damage; has never been observed and is probably not physically realizable |

[a] RC, Reinforced concrete.

A very useful picture of the strength and distribution of ground shaking in an earthquake can be obtained by mapping intensity observations. The modal value is assigned to each given location, such as a village, from all the individual point observations gathered, and then lines called isoseismals are drawn to enclose areas where the intensity reached the same degree (**Figure 2**). Such isoseismal maps have many applications, one of the most important of which is to establish correlations between earthquake size (magnitude) and the area enclosed by isoseismals. These empirical relations can then be used to estimate the magnitude of historical earthquakes that occurred before the advent of the global seismograph network at the end of nineteenth century, but for which it is possible to compile isoseismal maps from written accounts of the earthquake effects. For engineering design and analysis, however, intensity values are of limited use because they cannot be reliably translated into numerical values related to the acceleration or displacement of the ground. Indeed, intensities are usually expressed in Roman numerals precisely to reinforce the idea that they are broad indices rather than numerical measurements. This is why accelerograms are invaluable to earthquake engineering, providing detailed measurements of the actual movement of the ground during strong earthquake-induced shaking.

Accelerograms generally consist of three mutually perpendicular components of motion, two horizontal and one vertical, registering the ground acceleration against time (**Figure 3**). The first generation of accelerographs produced analogue records on paper or film, which had to be digitized in order to be able to perform numerical analyses. The analogue recording and the digitization process both introduce noise into the signal; this noise then requires processing of the time-series to improve the signal-to-noise ratio, particularly at long periods. This processing generally involves the application of digital filters. The current models of accelerographs record digitally, thus bypassing the time-consuming and troublesome process of digitization, and producing records with much lower noise contamination. Another advantage of

ENGINEERING GEOLOGY/Seismology 503

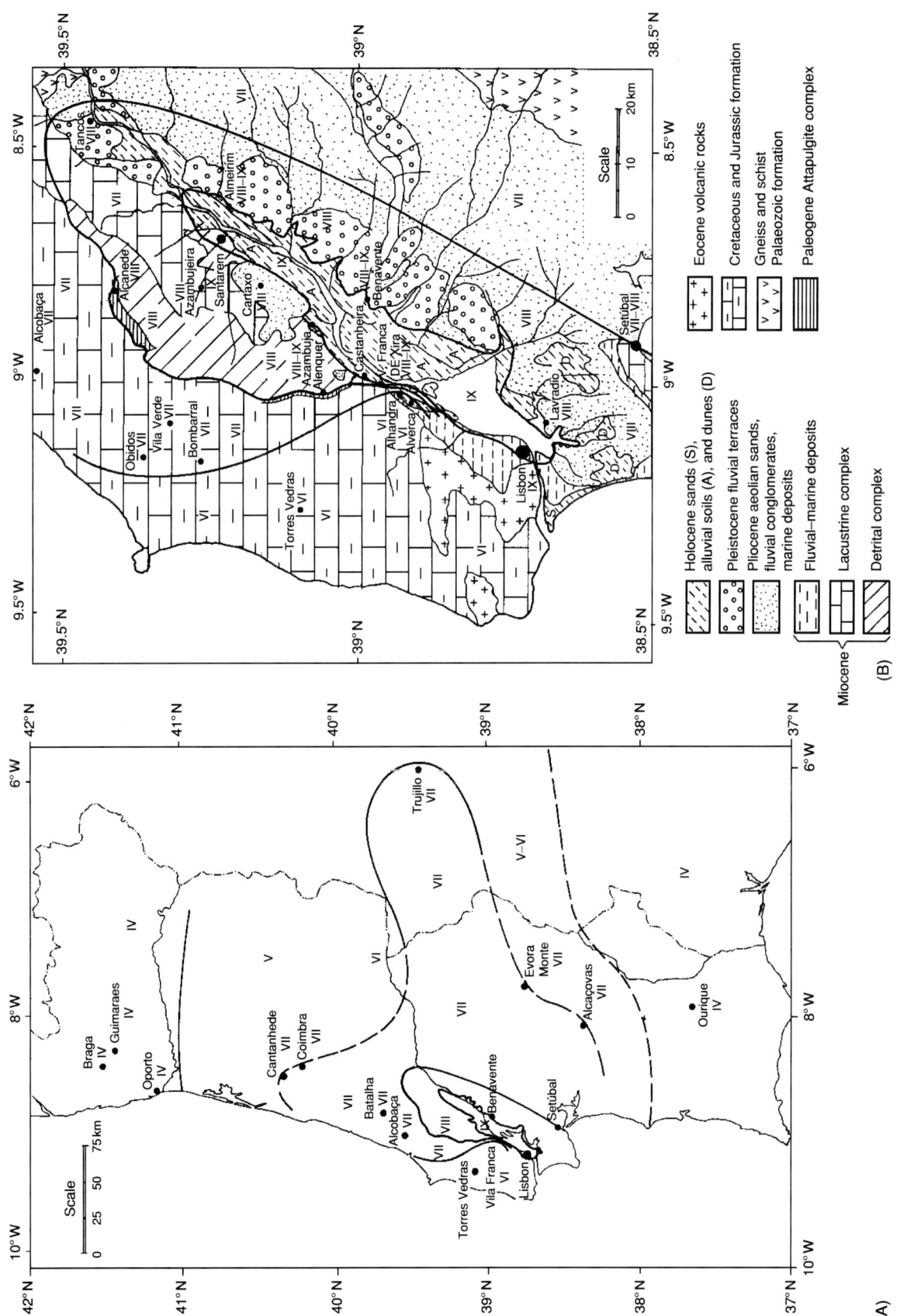

**Figure 2** (A) Regional isoseismal map of the Lisbon earthquake of January 1531. (B) Isoseismal map for the epicentral area of the Lisbon earthquake of January 1531, superimposed on a map of the local geology. Reprinted with permission from Justo JL and Salwa C (1998) The 1531 Lisbon earthquake. *Bulletin of the Seismological Society of America* 88(2): 319–328. © Seismological Society of America.

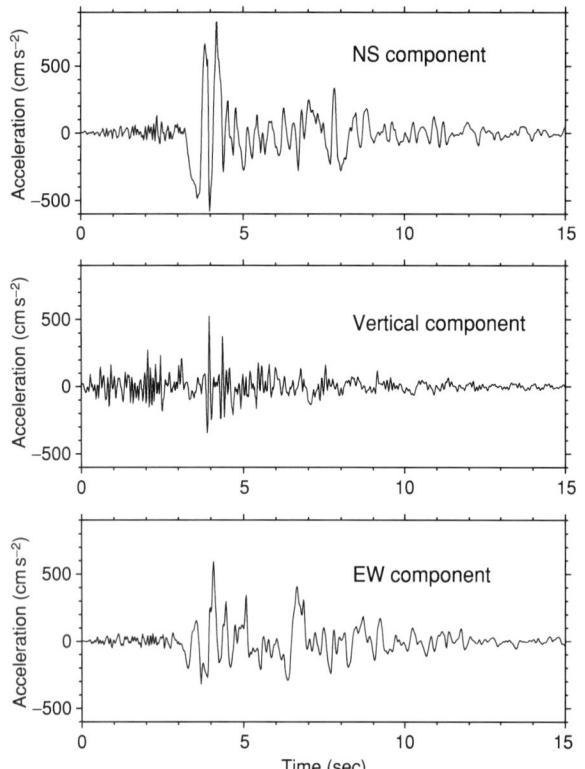

**Figure 3** Three-component accelerogram (NS, north–south; EW, east–west) digitized from an analogue recording at the Sylmar Hospital free-field site during the Northridge, California, earthquake of January 1994.

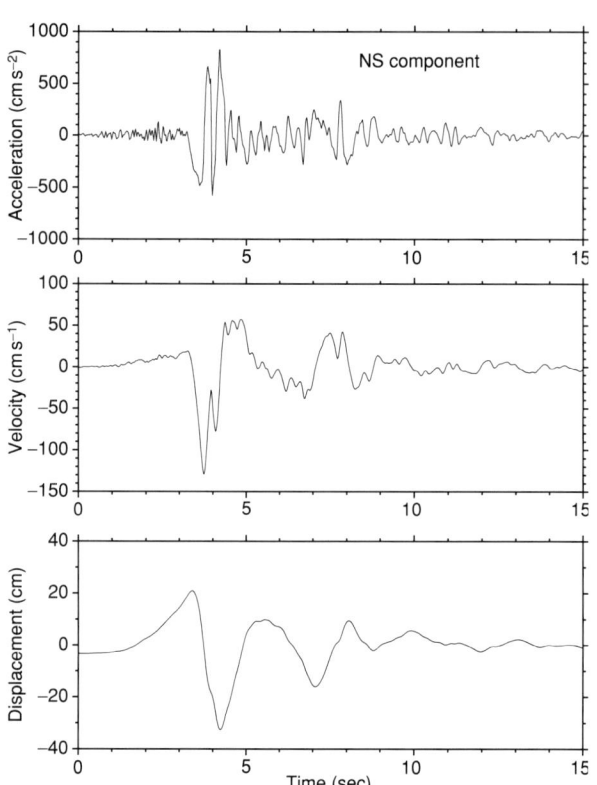

**Figure 4** Filtered acceleration, velocity, and displacement time-series from the north–south (NS) horizontal component of the Sylmar Hospital free-field site record (1994 earthquake, Northridge, California).

digital accelerographs is that they record continuously on reusable media, whereas optical–mechanical instruments remain on standby until triggered by a minimum-threshold acceleration, thus missing the very first-wave arrivals. Although the motions missed by an analogue instrument are generally very small, the advantage with a digital record is that the boundary conditions of initial velocity and displacement are known with greater confidence; the time-series of velocity and displacement are obtained by simple integration of the acceleration time-series, modified as needed by filtering and/or baseline correction to account for long-period noise (**Figure 4**).

## Characterizing Strong Ground Motion

A number of parameters are used to characterize the nature of the earthquake ground motion captured on an accelerogram, although in isolation, no single parameter is able to represent fully all of the important features. The simplest and most widely used parameter is the peak ground acceleration (PGA), which is simply the largest absolute value of acceleration in the time-series; the horizontal PGA is generally treated separately from the peak vertical motion. Similarly the values of the peak ground velocity (PGV) and peak ground displacement (PGD) are also used to characterize the motion, although the latter is difficult to determine reliably from an accelerogram because of the influence of the unknown baseline on the records and the double integration of the long-period noise in the record. These parameters are particularly poor for characterizing the overall nature of the motion because they reflect only the amplitude of a single isolated peak (**Figure 5**). In engineering seismology, unlike geophysics, accelerations are generally expressed in units of $g$, the acceleration due to gravity ($9.81 \, \mathrm{m\,s^{-2}}$).

Another important characteristic of the ground motion is the duration of shaking, particularly of the portion of the accelerogram where the motion is intense. There are many different ways in which the duration of the motion can be measured from an accelerogram, one of the more commonly used definitions being the total interval between the first and last excursions of a specified threshold, such as $0.05 \, g$. Duration of shaking is particularly important in

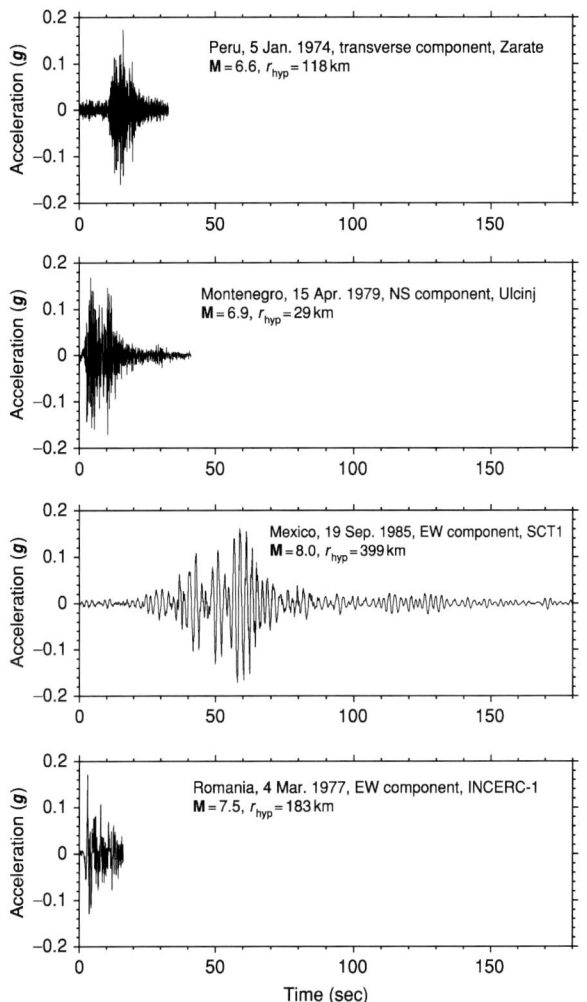

**Figure 5** Four horizontal accelerogram components with nearly the same peak ground acceleration. **M**, Moment magnitude; $r_{hyp}$, distance from the hypocentre.

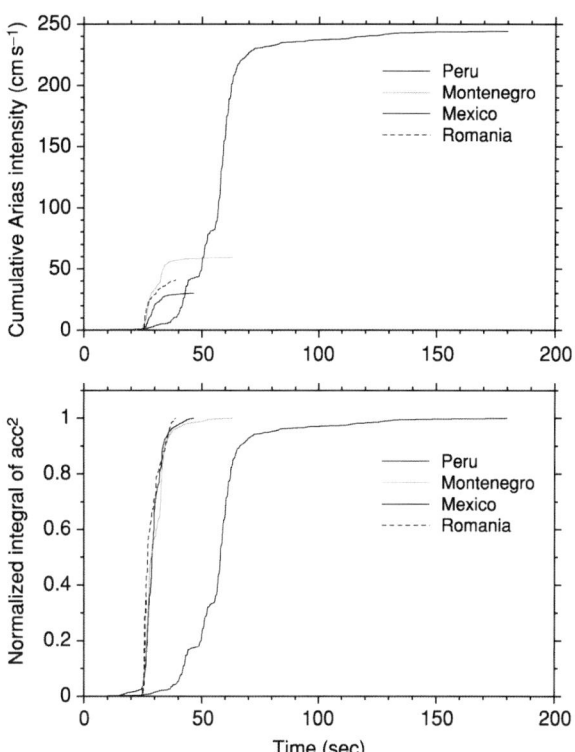

**Figure 6** Husid plots for the four acceleration time-series shown in **Figure 5**, showing the build-up of Arias intensity with time. The upper plot shows the absolute values of Arias intensity; in the lower plot, the curves are normalized to the maximum value attained.

the assessment of liquefaction hazard because the build-up of pore water pressures is controlled by the number of cycles of motion as well as by the amplitude of the motion.

A parameter that measures the energy content of the ground shaking is the Arias intensity, which is the integral over time of the square of the acceleration. A plot of the build-up of Arias intensity with time is known as a Husid plot (**Figure 6**) and it serves to identify the interval over which the majority of the energy is imparted. The root-mean-square acceleration ($a_{rms}$) is the equivalent constant level of acceleration over any specified interval of the accelerogram; the $a_{rms}$ is also the square root of the gradient of the Husid plot over the same interval. Arias intensity has been found to be a useful parameter to define thresholds of shaking that trigger landslides.

For engineering purposes, the most important representation of earthquake ground motion is the response spectrum, which is a graph of the maximum response experienced by a series of single-degree-of-freedom (SDOF) oscillators when subjected to the acceleration time-series at their bases (**Figure 7**). The dynamic characteristics of a SDOF system, such as an idealized inverted pendulum, are fully described by its natural period of free vibration and its damping, usually modelled as an equivalent viscous damping and expressed as a proportion of the critical damping that returns a displaced SDOF system to rest without vibrations. In earthquake engineering, the default value generally used is 5% of critical damping, the nominal level of damping in a reinforced concrete structure. The response can be measured in terms of the absolute acceleration of the mass of the system, or in terms of its velocity or displacement relative to the base. The most widely used spectrum currently is the acceleration response, because the acceleration at the natural period of the structure can be multiplied by the mass of the building to estimate the lateral force exerted on the structure by the earthquake shaking. The natural period of vibration of a building can be very approximately estimated, in seconds, as the number of storeys

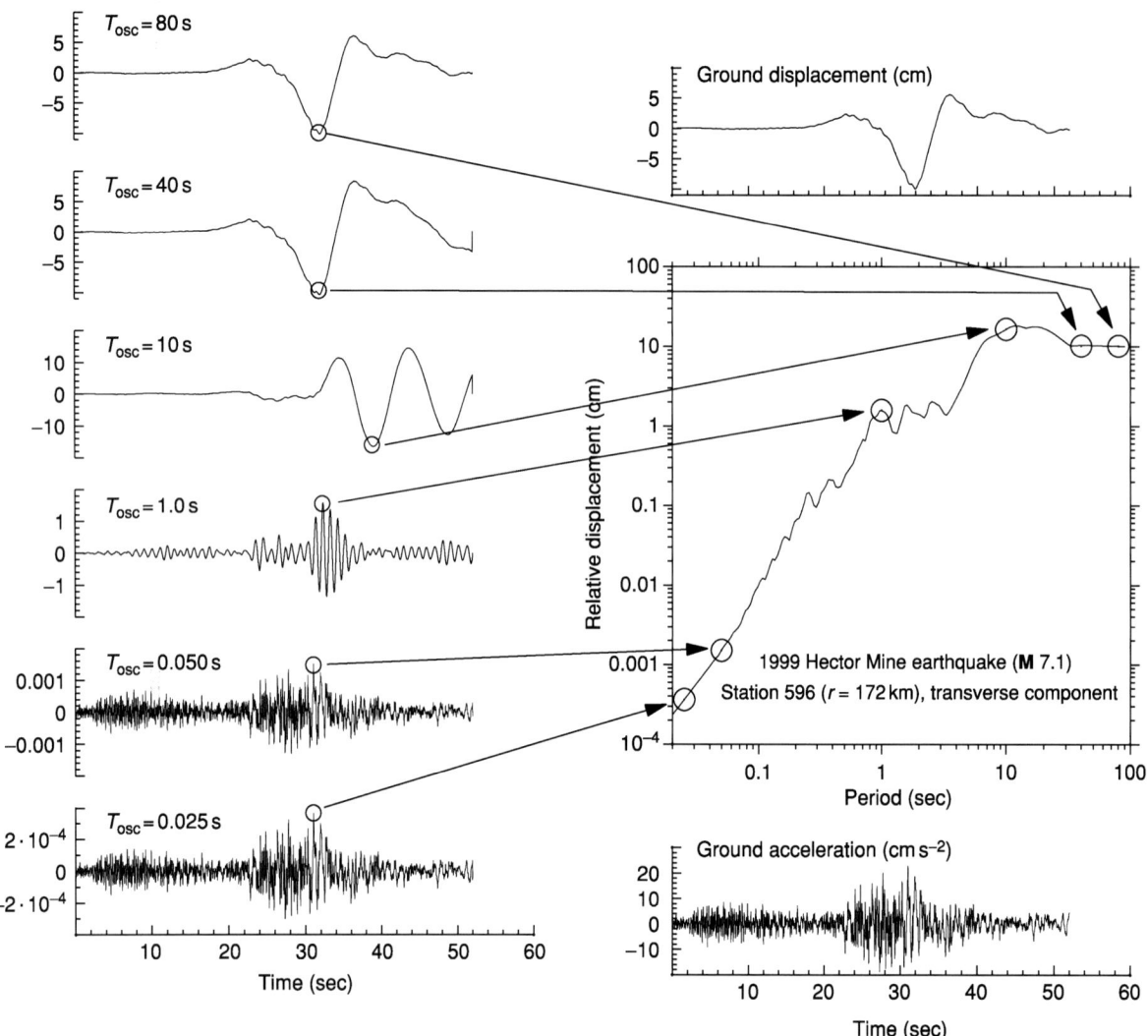

**Figure 7** Illustration of the concept of the response spectrum. The trace in the lower right-hand corner is the ground acceleration, and the traces on the left-hand side are the displacement response time-series for simple oscillators with different natural periods of vibration ($T_{osc}$). The maximum displacement of each oscillator is plotted against its natural period to construct the response spectrum of relative displacement. The lowest two traces in the figure illustrate that for short-period oscillators, the response mimics the ground acceleration, whereas for long-period oscillators, the response imitates the ground displacement (shown in the upper right-hand corner). **M**, Moment magnitude; r, distance to fault rupture.

divided by 10. The acceleration response spectrum intersects the vertical axis at PGA (**Figure 8**). PGA as a ground-motion parameter has many shortcomings, including a very poor correlation with structural damage, but one of the main reasons that it is persistently used is the fact that it essentially defines the anchor point for the acceleration response spectrum. The relationship between the natural period of vibration of the building and the frequency content of the ground motion is a critical factor in determining the impact of earthquake shaking on buildings.

In recent years, there has been a gradual tendency to move away from the use of acceleration response spectra in force-based seismic design towards displacement-based approaches, because structural damage is much more closely related to displacements than to the forces that are imposed on buildings very briefly during ground shaking. Techniques for assessing the seismic capacity of existing buildings have already adopted displacement-based approaches, giving rise to greater interest in displacement response spectra (**Figure 9**).

## Prediction of Earthquake Ground Motion

Recordings of ground motions in previous earthquakes are used to derive empirical equations that may be used to estimate values of particular ground-motion parameters for future earthquake

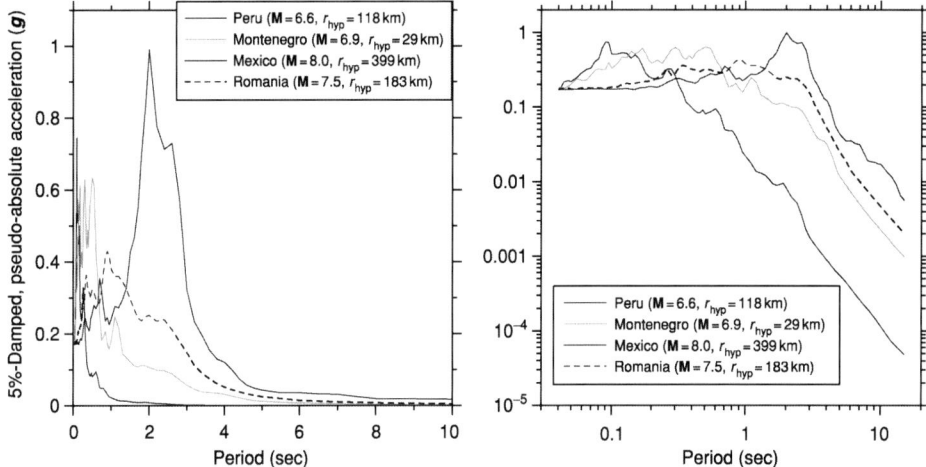

**Figure 8** Acceleration response spectra (5% damped) of the four accelerograms shown in **Figure 5**; the plots are identical except that the one on the left uses linear axes and the one on the right uses logarithmic axes. Each type of presentation is useful for viewing particular aspects of the motion, depending on whether the interest is primarily at short or long periods of response. **M**, Moment magnitude; $r_{hyp}$, distance from the hypocentre.

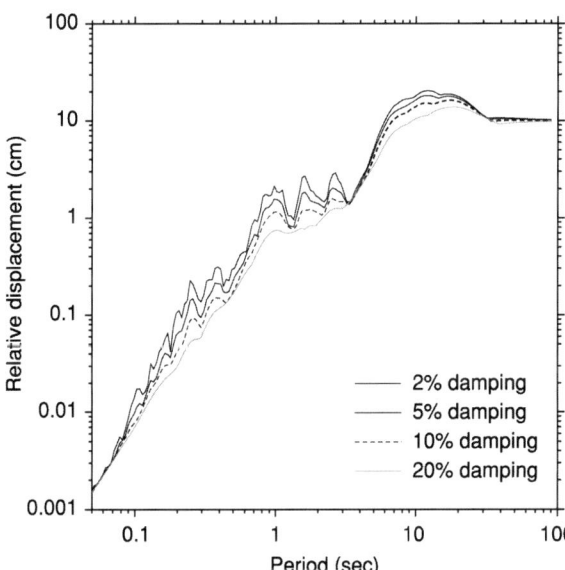

**Figure 9** Displacement response spectra of a single accelerogram from the 1999 California Hector Mine earthquake (moment magnitude 7.1; station 596; $r$ = 172 km, r is the distance from the fault rupture in km, transverse component), plotted for different levels of damping. The spectra all converge to the value of peak ground displacement at very long periods.

scenarios. Such empirical equations have been derived for a variety of parameters, but the most abundant are those for predicting PGA and ordinates of response spectra. The equations are often referred to as attenuation relationships, but this is a misnomer because the equations describe both the attenuation (decay) of the amplitudes with distance from the earthquake source and the scaling (increase) of the amplitudes with earthquake magnitude. These two parameters, earthquake magnitude and source-to-site distance, are always included in ground-motion prediction equations (**Figure 10**). The equations are simple models for a very complex phenomenon and as such there is generally a large amount of scatter about the fitted curve (**Figure 11**). The residuals of the logarithmic values of the observed data points are generally found to follow a normal or Gaussian distribution about the mean, and hence the scatter can be measured by the standard deviation. Predictions of PGA at the 84-percentile level (i.e., one standard deviation above the mean value) will generally be as much as 80% higher than the median predictions (**Figure 12**).

The nature of the surface geology can also exert a pronounced effect on the recorded ground motion. The presence of soft soil deposits of more than a few metres thickness will tend to amplify the ground motion as the waves propagate from the stiffer materials below to the surface (**Figures 2B** and **13**). To account for this effect, site classification is generally included as a third explanatory variable in prediction equations for response spectrum ordinates, allowing spectra to be predicted for different sites (**Figure 14**). The modelling of site amplification effects in ground-motion prediction equations is generally crude, using simple site classification schemes based on the average shear-wave velocity of the upper 30 m at the site and assuming that the degree of amplification is independent of the amplitude of the input motion, whereas it is generally observed that weak motion is amplified more than strong motion; this non-linear response of soils is reflected only in a few predictive equations. For this

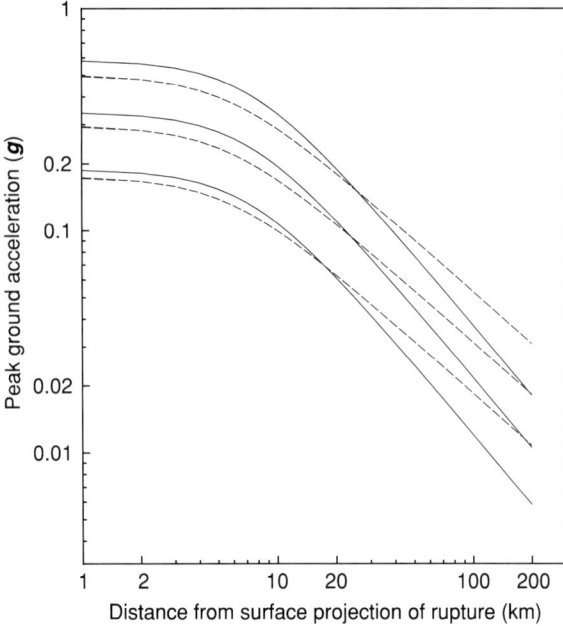

**Figure 10** Predicted median values of peak ground acceleration as a function of distance for earthquakes of moment magnitude 5.5, 6.5, and 7.5, using equations derived from western North American (solid lines) and European (dashed lines) data sets. The European equation is based on surface-wave magnitude, so an appropriate empirical conversion to moment magnitude has been used. The equations are based on different definitions of the horizontal component, the European equation using the larger of the two horizontal accelerations, the North American equation using their geometric mean; the former definition, on average, yields values about 1.17 times larger than the latter. Different criteria were used in the two studies for selecting records, especially at greater distances, for regression. In light of these various observations, it is not possible to draw conclusions about differences or similarities in strong ground motions between the two regions. The North American equation is from Boore DM, Joyner WB, and Fumal TE (1997) *Seismological Research Letters* 68(1): 128–153; the European equation is from Tromans IJ and Bommer JJ (2002) *Proceedings of the 12th European Conference on Earthquake Engineering*, Paper no. 394, London.

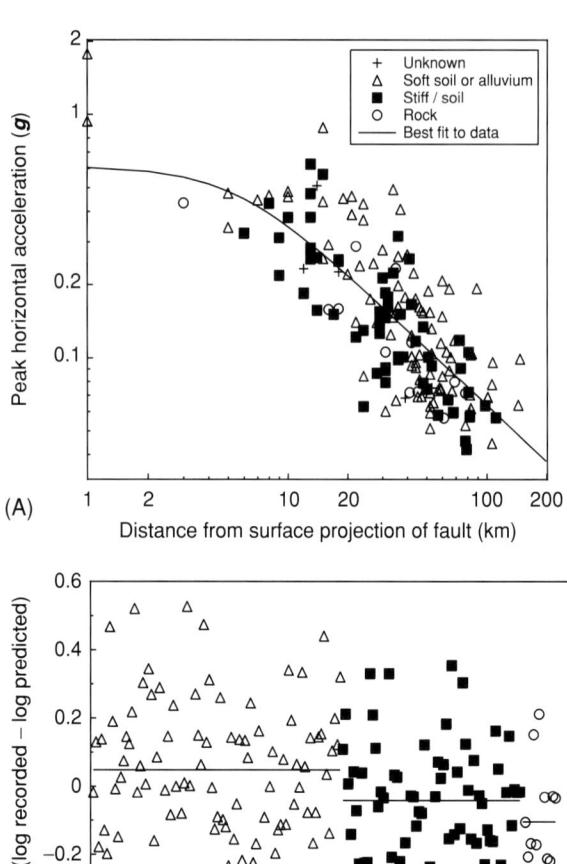

**Figure 11** (A) Values of peak horizontal ground acceleration recorded in the 1994 Northridge earthquake in California, plotted as a function of distance from the earthquake source, with an indication of the surface geology at the recording site. The solid line is the result of a regression on the data using a typical model employed in ground-motion prediction equations. (B) Residuals group by site class; the horizontal lines show the mean of the residuals in each class, indicating that there is a tendency for stronger motions on softer ground. However, the differences between the mean lines are small compared with the overall dispersion, thus inclusion of site classification in the predictive equation would result in only a modest reduction of the aleatory variability.

reason, for site-specific predictions, it is often preferred to predict the bedrock motions first and then model the dynamic response of the site separately (**Figure 15**).

The nature of ground shaking at a particular site during an earthquake is influenced by many factors, including the distribution and velocity of the slip on the fault rupture, the depth at which the fault rupture is located, the orientation of the fault rupture with respect to the travel path to the site, and the geological structure along the travel path and for several kilometres below the site. A situation that produces particularly destructive motions is the propagation of the fault rupture towards the site, which produces large-amplitude, high-energy pulses of velocity (**Figure 16**). The variable most often included in predictive equations after magnitude, distance, and site classification is the style of faulting; equations that include the rupture mechanism as an explanatory variable all predict higher amplitudes of motion from reverse-faulting earthquakes than from strike–slip events. The addition of this fourth explanatory variable, however, has an almost negligible impact on the scatter in the equations, with no appreciable reduction of the standard deviation.

Dense networks of accelerographs now exist in many countries around the world, but for many decades these were limited to a few regions such as California, Japan, Italy, Greece, and Yugoslavia. In

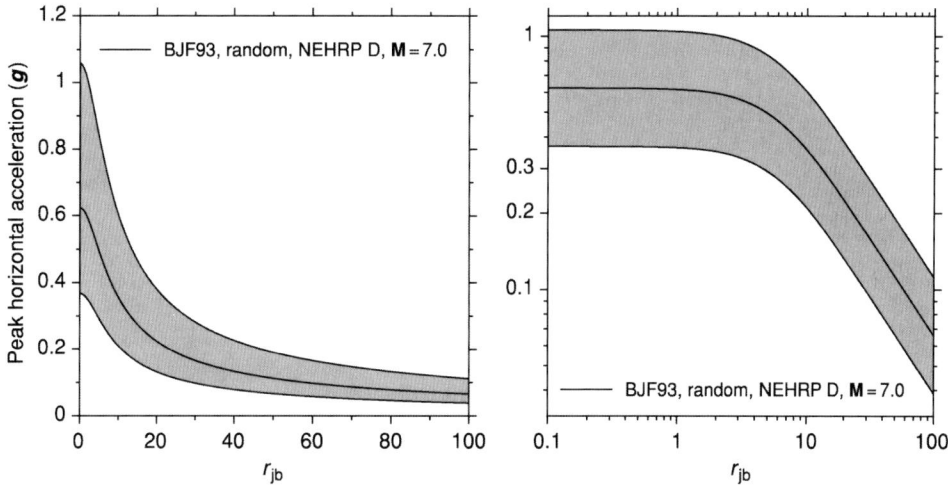

**Figure 12** Predicted values of horizontal peak ground acceleration on soft soil (National Earthquake Hazards Reduction Program (NEHRP) class D) sites for an earthquake of magnitude 7 as a function of distance from the surface projection of the fault rupture ($r_{jb}$). The predictions are made using an equation derived from earthquakes recorded in western North America, by DM Boore, WB Joyner, and TE Fumal. The thick black line of Boore, Joyner, and Fumal (BJF93) shows the median predicted values and the grey bands indicate the range of the median multiplied and divided by $10^\sigma$, where $\sigma$ is the standard deviation of the logarithmic residual; on average, 68% of observations would be expected to fall within the grey area.

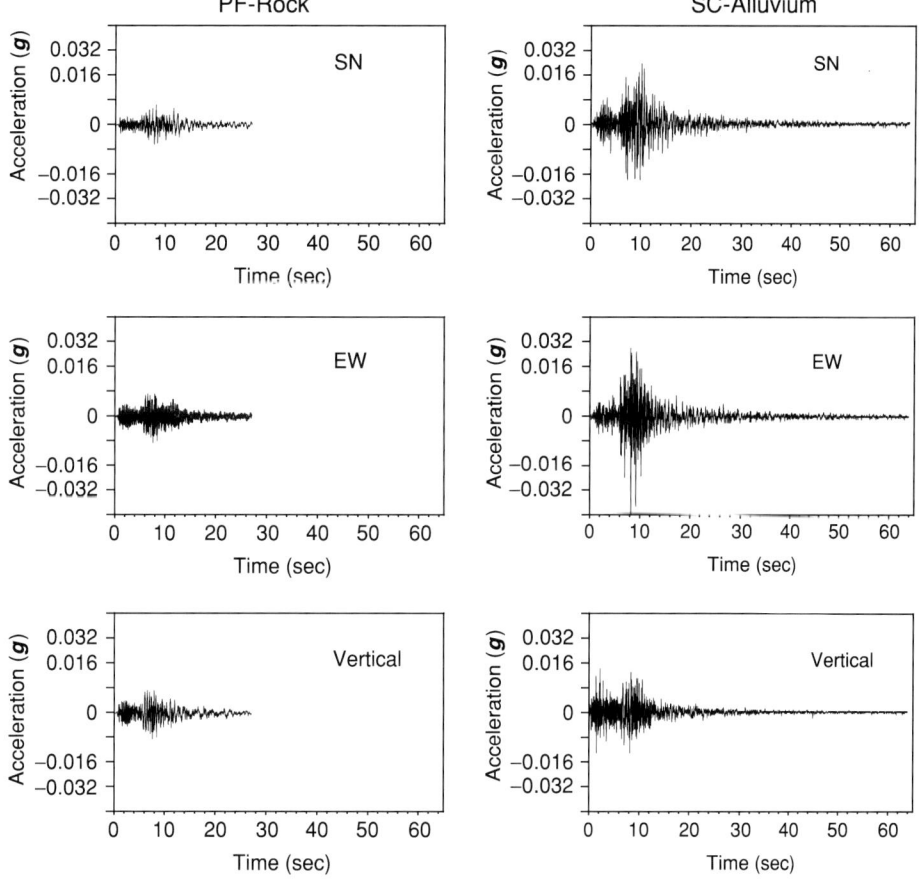

**Figure 13** Accelerograms recorded on two instruments in the Gemona (Italy) array during the **M** 5.6 Bovak (Slovenia) earthquake of 12 April 1998 (**M** = moment magnitude). Both instruments are 38 km from the earthquake epicentre, and are separated by about 700 m. The Piazza del Ferro station (PF) is on rock with a shear-wave velocity reported as 2500 m s$^{-1}$, whereas the Scugelars station (SC) is on alluvium with a shear-wave velocity of 500 m s$^{-1}$. Because the distance and azimuth of the two stations with respect to the earthquake source are almost identical, the difference between the amplitude of the two recordings is primarily due to the different surface geology at the two recording locations (SN, south–north; EW, east–west).

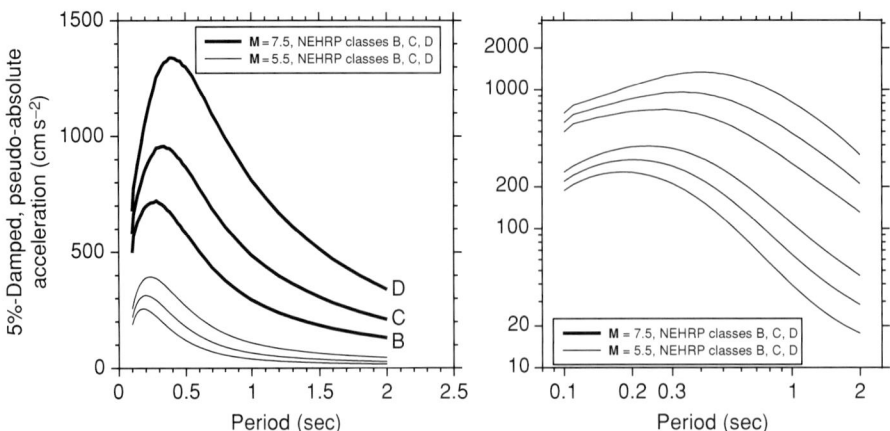

**Figure 14** Predicted median ordinates of spectral acceleration (5% damped) for earthquakes of magnitudes 5.5 and 7.5, recorded at 10 km on sites classified as B, C, and D in the National Earthquake Hazards Reduction Program (NEHRP) scheme, corresponding to soft rock, stiff soil, and soft soil, respectively. The two graphs are identical except for the left-hand plot being on linear axes and the right-hand plot being on logarithmic axes.

other regions, the number of available accelerograms is further reduced by the relatively low frequency of felt earthquakes. Seismic hazard assessment in such regions is hampered by the absence of indigenous recordings, and a solution that is often employed is to use stochastic simulations to generate motions based on representing the ground shaking as random motion with frequency content and duration that are controlled by theoretical and empirical descriptions of seismic radiation and propagation (**Figure 17**).

Complete acceleration time-histories can also be generated using more sophisticated seismological models that represent the source as a finite fault rupture. These simulations usually take the form of a kinematic model of the fault rupture, with rupture velocities and slip amplitudes specified across the fault plane (although these can be specified using statistical distribution in order to be able to generate a suite of motions for a given rupture). Dynamic fault modelling, in which the stress conditions are specified and subsequent rupture is controlled by friction laws on the fault surface as well as the heterogeneous three-dimensional stress distribution, is less commonly used for the simulation of ground motions for engineering purposes because the necessary physical quantities are not known with sufficient precision and because the computational effort is much greater than with the kinematic models (which can be quite demanding if lateral variations in geology are taken into account). In the future, as computers become more powerful and the details of the geologic structure and velocity distributions are better determined, it is likely that the kinematic models will continue to be used, but the simulations will better represent the effects of wave propagation.

## Seismic Hazard Assessment

Engineering seismology is often described as the link between Earth sciences and engineering, and this is most evident in seismic hazard assessment, the aim of which is to estimate the earthquake ground motions that can be expected at a particular location. Seismic hazard assessments may be performed for a number of reasons, the most common being to determine seismic loads to be considered in earthquake-resistant design of buildings. Another common application is the evaluation of the ground motions required as input to the assessment of landslide or liquefaction hazard. A rapidly expanding market for seismic hazard assessments is being created by the demand for earthquake loss models by local governments, emergency services, seismic code developers, and particularly the insurance and reinsurance industries.

There are many different ways to approach seismic hazard assessment, but all methods and approaches include two essential elements: a model for earthquake occurrence and a model for predicting ground motions from each earthquake. The starting point for building a model for earthquake occurrence, known as a seismicity model, is to first identify the locations where earthquakes will be expected to occur in the future. Data used to define seismic source zones include previous earthquake activity, the existence of geological faults, and crustal deformations. Ideally, all seismic sources should be identified as active geological faults, but this is very often not possible, hence zones are defined as areas within which future earthquakes are expected to occur (**Figure 18**). Some recent approaches to seismic hazard assessment dispense with source zones altogether, basing the sources of

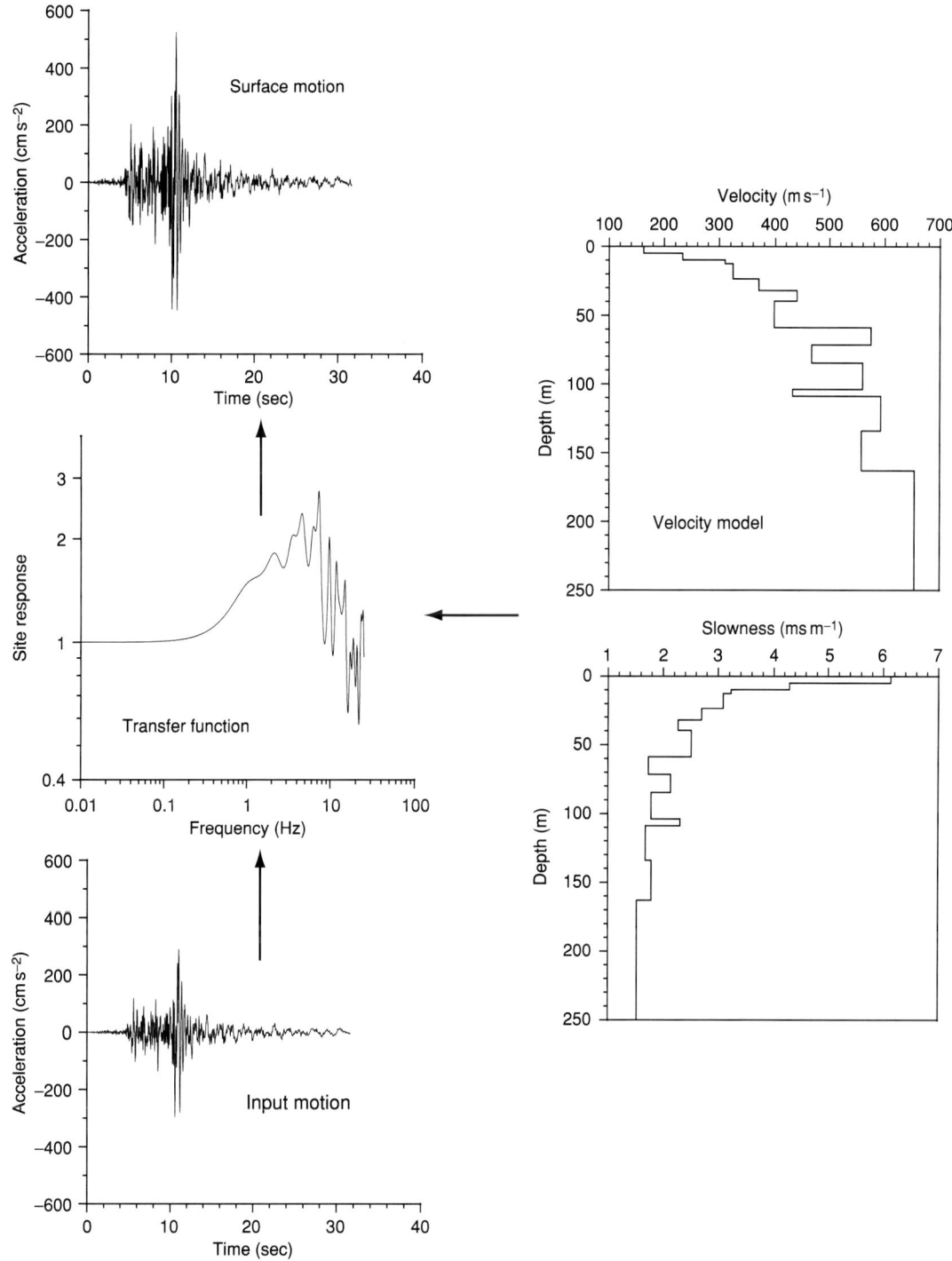

**Figure 15** Illustration of site response analysis: The input motion in the underlying rock (bottom left) passes through a model of soil profile to produce the surface motions (top left). The soil characteristics are represented either by the shear-wave velocity or the shear-wave slowness (right); the slowness is simply the reciprocal of the shear-wave velocity, but there are many reasons for using it in place of the velocity. The theoretical response of layered systems involves travel times across the layers, which are directly proportional to the slowness; the proportionality to velocity is inverse. Furthermore, to obtain a representative profile for a soil class using several boreholes, individual measures of slowness can be directly averaged, whereas it is not correct to do this for velocities. Finally, the largest influence on the surface motion is due to the differences in velocities near the surface, and these differences become more clearly apparent when the slowness profile is shown; larger velocity differences in the stiffer layers at greater depths, which can dominate a velocity profile, are less important.

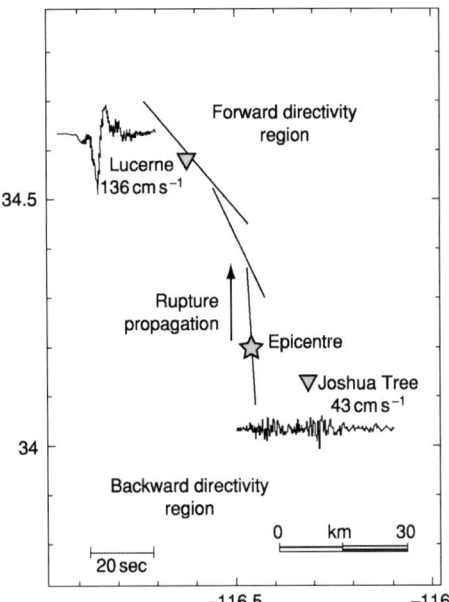

**Figure 16** Velocity time-series obtained from integration of horizontal accelerograms recorded at Lucerne and Joshua Tree stations during the 1992 Landers earthquake in California. The fault rupture propagated towards Lucerne and away from Joshua Tree, creating forward directivity effects (short-duration shaking consisting of a concentrated high-energy pulse of motion) at the former, and backward directivity effects (long-duration shaking of several small pulses of motion) at the latter. Reprinted with permission from: Somerville PG, Smith NF, Graves RW, and Abrahamson NA (1997) Modification of empirical strong ground motion attenuation relations to include the amplitude and duration effects of rupture directivity. *Seismological Research Letters* 68(1): 199–222. © Seismological Society of America.

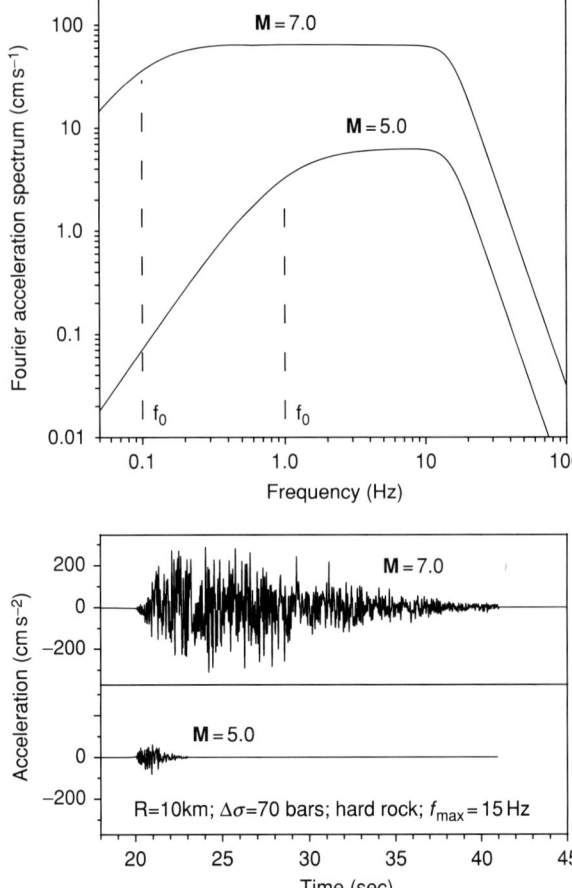

**Figure 17** Theoretical Fourier spectra for earthquakes of magnitudes 5 and 7 at 10 km from a rock site, assuming a stress parameter of 70 bars; $f_0$ is the corner frequency, which is inversely proportional to the duration of rupture. The lower part of the figure shows stochastic acceleration time-series generated from these spectra, from which the influence of magnitude on amplitude and duration (or number of cycles) can be clearly observed. From the spectra in the upper part of the figure, it can be appreciated that scaling with magnitude is frequency-dependent, with larger magnitude events generating proportionally more long-period radiation.

postulated future events on past activity. The catalogue of instrumentally recorded earthquakes extends back at the very most to 1898, which is a very short period of observation for events for which recurrence intervals can extend to hundreds or even thousands of years. For this reason, great value is to be obtained from extending the catalogue through the careful interpretation of historical records, making use of empirical relationships between magnitude and intensity referred to previously. The seismic record can also be extended through palaeoseismology, by quantifying and dating coseismic displacements on geological faults.

Although the distinction masks many equally marked differences among the various methods that fall within each camp, a basic division exists between deterministic and probabilistic approaches to seismic hazard assessment. In the deterministic approach, only a few earthquake scenarios are considered, and sometimes just one is selected to represent an approximation to the worst case. The controlling earthquake will generally correspond to an event with the nominal maximum credible magnitude, located at the location closest to the site within the seismogenic source; the ground motion is generally calculated as the 50- or 84-percentile value from the prediction equation. In the probabilistic approach, all possible earthquake scenarios are considered, including events of every magnitude, from the minimum considered to be of engineering significance (∼4) up to the maximum credible, occurring at every possible location within the source zones, and for each magnitude–distance combination various percentiles of the motion are considered to reflect the scatter in the ground-motion prediction equations. Alternative options for the input parameters may be considered by using a logic-tree formulation, in which weights are assigned to different options that reflect the

ENGINEERING GEOLOGY/Seismology 513

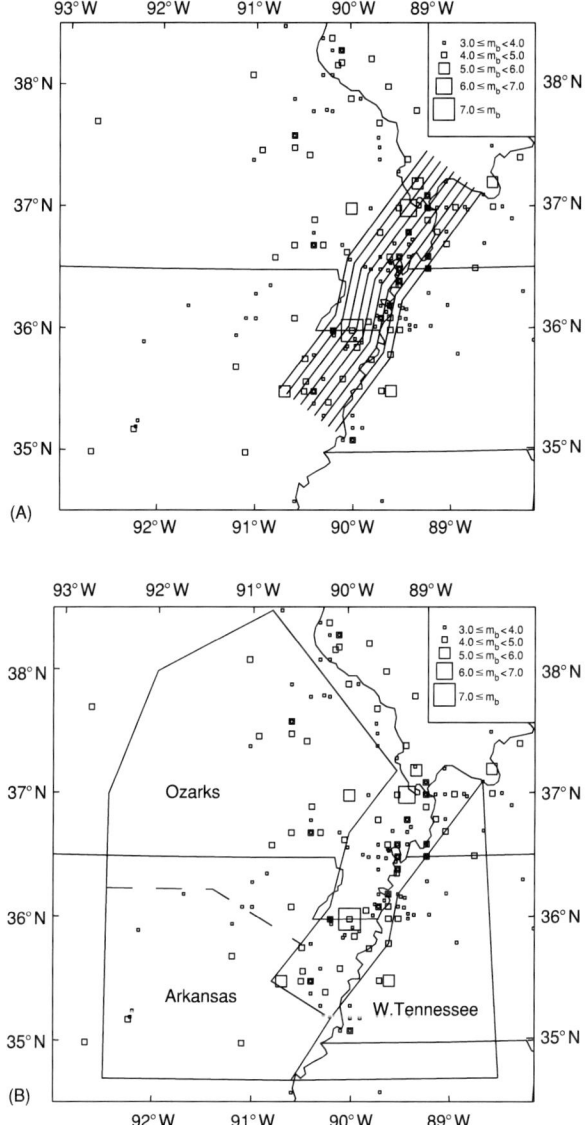

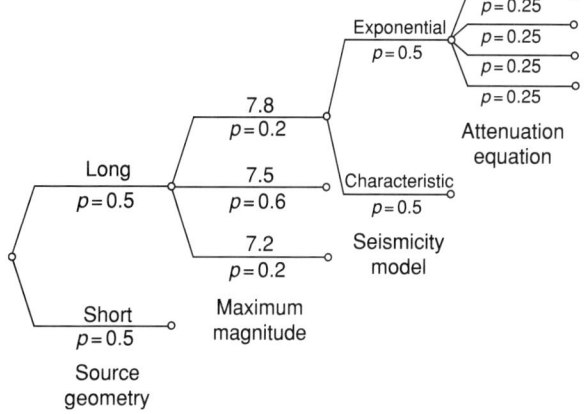

**Figure 19** Logic-tree formulation for probabilistic seismic hazard study of the Mississippi embayment. For different input parameters, different options are considered at each node, which are then assigned weights (p) to reflect the relative confidence in each option; the values of p at each node always sum to unity. The source geometry options are the faults shown in **Figure 18**, and an alternative model in which the faults are longer. Reprinted with permission from Toro GR, Silva WJ, McGuire RK, and Herrmann RB (1992) Probabilistic seismic hazard assessment of the Mississippi Embayment. *Seismological Research Letters* 63(3): 449–475. © Seismological Society of America.

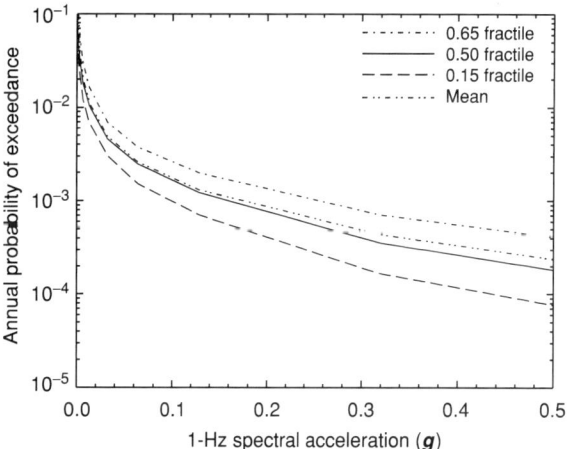

**Figure 18** Seismic sources used for probabilistic hazard study of the Mississippi embayment. Small squares show the epicentres of earthquakes with the size corresponding to the body-wave magnitude $M_b$. (A) The New Madrid seismic source zone represented as a number of parallel hypothetical faults; (B) the area sources used to represent seismicity not associated with the New Madrid zone. Reprinted with permission from Toro GR, Silva WJ, McGuire RK, and Herrmann RB (1992) Probabilistic seismic hazard assessment of the Mississippi Embayment. *Seismological Research Letters* 63(3): 449–475. © Seismological Society of America.

**Figure 20** Seismic summary hazard curves for spectral acceleration at 1-s response period for Memphis (including site response), obtained using the logic-tree formulation in **Figure 19**. Reprinted with permission from Toro GR, Silva WJ, McGuire RK, and Herrmann RB (1992) Probabilistic seismic hazard assessment of the Mississippi Embayment. *Seismological Research Letters* 63(3): 449–475. © Seismological Society of America.

relative confidence in their being the best representation (**Figure 19**). The output is a ranking of these earthquake scenarios in terms of the ground-motion amplitudes that they generate at the site and their frequency of occurrence (**Figure 20**).

Seismic hazard assessment can be carried out for an individual site, the output including a response spectrum calculated ordinate by ordinate, or even a suite of accelerograms. Hazard assessment is often performed for regions or countries, deriving hazard curves for a large number of locations and then drawing contours of a ground-motion parameter (most often PGA) for a given annual frequency of occurrence, which is often expressed in terms of its inverse, known as a return period (**Figure 21**). Such hazard maps in terms of PGA are the basis of

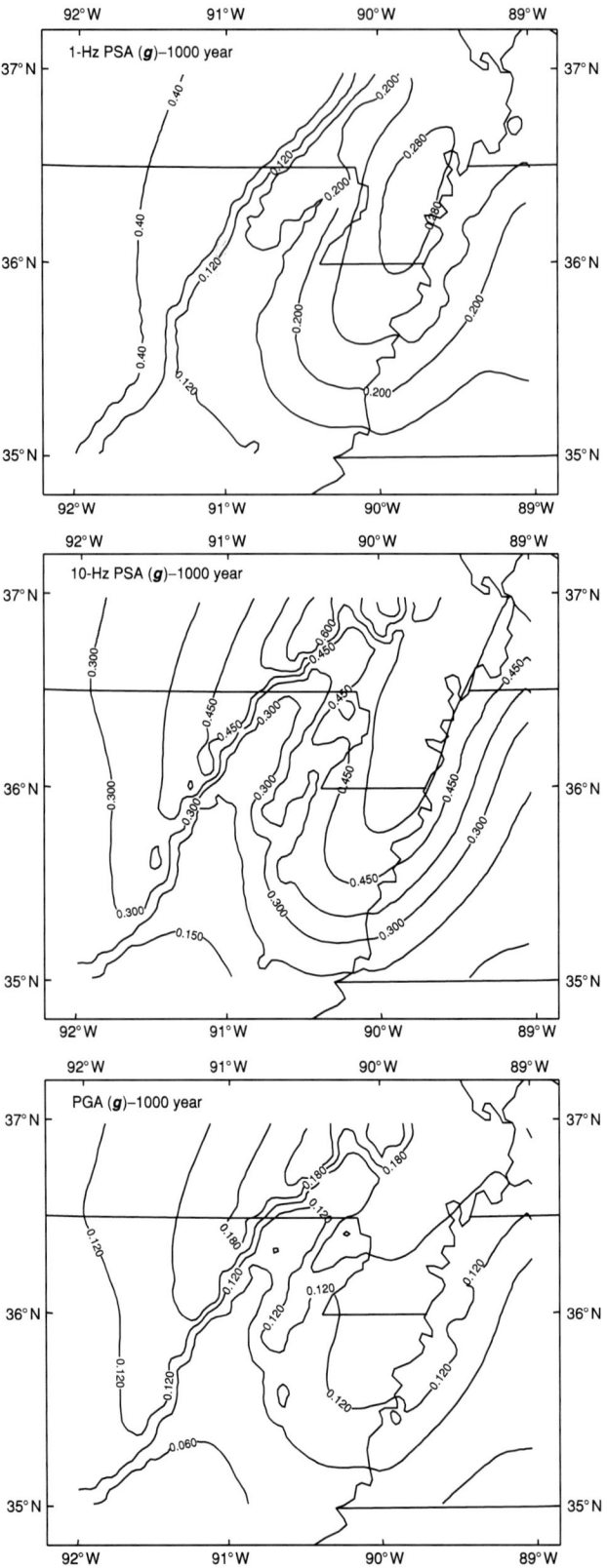

**Figure 21** Contours of spectral accelerations in the Mississippi embayment with a 1000-year return period, for different response frequencies corresponding to the response periods of 1.0s (top) and 0.1s (middle) and to PGA (botom). The maps show the spectral accelerations read from the mean hazard curve. Reprinted with permission from Toro GR, Silva WJ, McGuire RK, and Herrmann RB (1992) Probabilistic seismic hazard assessment of the Mississippi Embayment. *Seismological Research Letters* 63(3): 449–475. © Seismological Society of America.

zonation maps included in seismic design codes, from which design response spectra are constructed by anchoring a spectral shape selected for the appropriate site class to the PGA value read from the zonation map.

## See Also

**Engineering Geology:** Aspects of Earthquakes; Natural and Anthropogenic Geohazards; Liquefaction. **Sedimentary Processes:** Particle-Driven Subaqueous Gravity Processes. **Tectonics:** Earthquakes; Faults; Neotectonics.

## Further Reading

Abrahamson NA (2000) *State of the Practice of Seismic Hazard Evaluation*. GeoEng 2000, 19–24 November, Melbourne, Australia.

Ambraseys NN (1988) Engineering seismology. *Earthquake Engineering & Structural Dynamics* 17: 1–105.

Beskos DE and Anagnostopoulos SA (1997) *Computer Analysis and Design of Earthquake Resistant Structures: A Handbook*, chs 3–5. Southampton: Computational Mechanics Publications.

Bommer JJ (2003) Uncertainty about the uncertainty in seismic hazard analysis. *Engineering Geology* 70: 165–168.

Boore DM (1977) The motion of the ground in earthquakes. *Scientific American* 237: 68–78.

Boore DM (2003) Simulation of ground motion using the stochastic method. *Pure and Applied Geophysics* 160: 635–676.

Chen W-F and Scawthorn C (2003) *Earthquake Engineering Handbook*, chs 1–10. Boca Raton, FL: CRC Press.

Douglas J (2003) Earthquake ground motion estimation using strong-motion records: a review of equations for the estimation of peak ground acceleration and response spectral ordinates. *Earth Science Reviews* 61: 43–104.

Dowrick D (2003) *Earthquake Risk Reduction*. Chichester, England: John Wiley.

Giardini D (1999) The global seismic hazard assessment program (GSHAP) 1992–1999: summary volume. *Annali di Geofisica* 42(6): 957–1230.

Giardini D and Basham P (1993) Technical planning volume of the ILP's global seismic hazard assessment program for the UN/IDNDR. *Annali di Geofisica* 36(3/4): 3–257.

Hudson DL (1979) *Reading and Interpreting Strong Motion Accelerograms*. EERI Monograph, Earthquake Engineering Research Institute, Oakland, California.

Jackson JA (2001) Living with earthquakes: know your faults. *Journal of Earthquake Engineering* 5(special issue 1): 5–123.

Kramer SL (1996) *Geotechnical Earthquake Engineering*. Upper Saddle River, NJ: Prentice-Hall.

Lee WHK, Kanamori H, Jennings PC, and Kisslinger C (2003) *International Handbook of Earthquake and Engineering Seismology, Part B*. chs. 57–64. Amsterdam: Academic Press.

McGuire RK (2004) *Seismic Hazard and Risk Analysis*. EERI Monograph MNO-10. Earthquake Engineering Research Institute, Oakland, California.

Naeim F (2001) *The Seismic Design Handbook*, 2nd edn. chs. 1–3. Boston: Kluwer Academic Publ.

Reiter L (1990) *Earthquake Hazard Analysis: Issues and Insights*. New York: Columbia University Press.

# Natural and Anthropogenic Geohazards

**G J H McCall**, Cirencester, Gloucester, UK

© 2005, Elsevier Ltd. All Rights Reserved.

## Introduction

The topic of geohazards became popular with scientists and the media in the early 1990s at the time of the International Decade for Natural Disaster Reduction aimed, by the United Nations, specifically at developing nations in the Third World. It became popular with the media in the belief that we were moving into an age of disaster. In particular, the appreciation of the reality of global climate change and humankind's contribution in the Industrial Age to global warming has led to an awareness of the vulnerability of the world in which we live. Geohazards operate at local scales, e.g., in villages and towns, at a regional scale, and at the largest scale of all, global. Urban geohazards represent a specialized and increasingly important type of hazard.

An increasing incidence and scale of risk and disaster have recently occurred due to a number of factors:

- increased population concentrations;
- increased technological development;
- over-intensive agriculture and increased industrialization;
- excessive use of the internal combustion engine and other noxious fume emitters, and wasteful transport systems;
- poor technological practices in construction, water management, and waste disposal;
- excessive emphasis on commercial development;
- increased scientific tinkering with Nature without due concern for long-term effects.

Some hazards are natural, others are purely anthropogenic, and many have both natural and anthropogenic causal elements.

There is a need to consider geohazards from a sociological as well as a scientific viewpoint, as the perception and response of the local populace and all tiers of government are as important as technological understanding. Geohazards can also only be successfully evaluated with a full interdisciplinary approach: an international, possibly even a global, approach is also called for.

## Definitions

There are two published definitions of geohazards.

> A geohazard is a hazard of geological, hydrological nature which poses a threat to Man and his activities. (McCall & Marker 1989)

and

> A geohazard is one that involves the interaction of man and any natural process on the planet. (McCall and others 1992)

The second is wider, but the first brings in the essential element of human vulnerability, which is well expressed by MR Degg in his diagram for the disaster equation (**Figure 1**) – the relationship between the hazard, disaster, and vulnerability. The larger and more severe the event and the more vulnerable the population, the greater the disaster; there is a threshold below which there is no disaster. If there are no human's present, strictly speaking there is no hazard.

The words 'hazard' and 'risk' are often used interchangeably, but the risk is a quantification. RJ Blong defined 'risk' as the product of 'hazard' and 'vulnerability'. The word 'disaster' refers rather loosely to the consequence of hazards.

## Types of Natural Geohazard

RJ Blong lists the following as the principal natural geohazards:

- earthquakes (r);
- volcanic eruptions (r);
- landslides (r);
- tsunamis (r);
- subsidence (s);
- coastal erosion (s);
- coastal progradation (s);
- soil erosion (s);
- expansive soils (s).

Already, we see two distinct classes of geohazard: the rapid-onset, intensive hazards (r) and the slow-onset, pervasive hazards (s). Attention has been traditionally focused by scientists, the media, and the populace on the former, but the latter can be equally damaging.

SA Thompson separated 'natural geohazards' from 'meteorological hazards':

- cyclones;
- tornadoes;
- floods;
- heatwaves;
- thunderstorms and gales.

All of these have some geological consequences, especially the first three.

GJH McCall later added to the list and separated those that are potentially catastrophic and cannot at present be reliably predicted, unless in general terms or not at all. These can be treated mainly by emergency planning, warning systems, education and communication, and evacuation. Mitigation and prevention procedures are of limited application. The list is:

- earthquake hazards (e.g., Kobe 1995);
- volcanic hazards (lava, pyroclastic flows, nuées ardentes, ash falls, lahars, blasts, and gas releases) (e.g., Etna, Martinique, Montserrat, Nevado del Ruiz 1985, Lake Nyos 1986);
- tsunamis (related to either of the above) (e.g., Hilo, Hawaii, 1960);
- extraterrestrial impacts (e.g., Tunguska 1908).

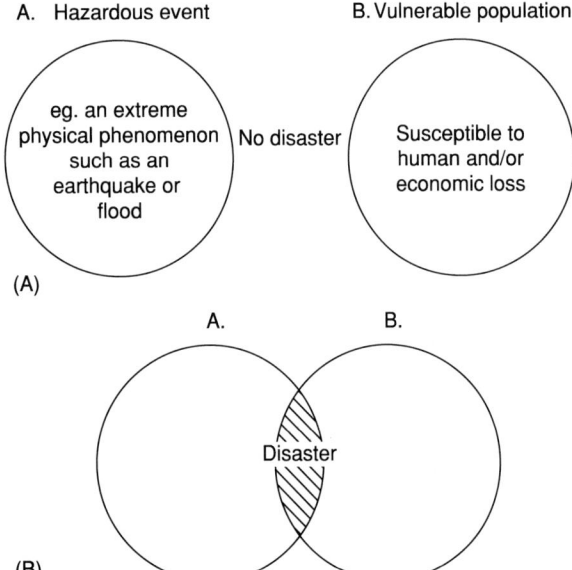

**Figure 1** The 'disaster equation'. After Degg MR, in McCall GJH and Marker BR (1989) *Earth Science Mapping for Planning, Development and Conservation, Engineering Geology Special Publication No. 4*. London: Kluwer.

The remainder can be reasonably counteracted by engineering procedures at a cost. The list is:

- mass movements (landslides, avalanches, debris flows, ice-related movements) (e.g., Ventnor, Swiss Alps, Cotopaxi, Cordillera Blanca);
- subsidence (swallow holes, karstic processes, gypsum-related sink holes, sinking cities) (e.g., Pennines, Wuzhan, Ripon, Venice);
- flooding after abnormal rainfall and cyclones (Bangladesh);
- coastal erosion (sea-level rise) and coastal progradation (China, Burma);
- riverbank failure and silting up of rivers (Mississippi River);
- expansive and collapsing soils, thixotropic sands (e.g., Anchorage in Alaska);
- permafrost (e.g., Canada, Siberia);
- hazardous gas emission (radon) (e.g. Cornwall).

A third category was listed of those that fall between the two:

- combustion and wildfire (e.g., Australia);
- neotectonic deformation and fissuring (e.g., Xian);
- desertification (e.g., Sahel).

K Hewitt and I Burton, in 1975, published a list of parameters for 'selecting' hazard events:

- property damage extending to more than 20 families, or economic loss (including loss of income, a halt to production, costs of emergency action) in excess of US$50 000;
- major disruption of social services, including communications failure and closure of essential facilities of establishments of economic importance;
- a sudden, unexpected or unscheduled event, or series of events, which puts excessive strain on essential services (police, fire service, hospitals, public utilities) and/or requires the calling in of men, equipment, or funds from other jurisdictions;
- an event in which 10 or more persons are killed or 50 or more injured.

Such quantification of thresholds is applicable to rapid-onset, intensive hazards, but of little use for the slow-onset, pervasive hazards, such as karstic processes or coastal erosion. Also, it appears to be applicable to developed countries, but of little relevance to less developed countries, such as islands in the Pacific Ocean. Monetary loss and body count alone are thus poor indicators of the magnitude of an effect of a hazard on an afflicted community. Vulnerability should take into account not merely the risk, but also the endemic conditions inherent in the society. Deaths due to starvation consequent on hazards are not considered in this scheme.

**Table 1** Frequency of hazards and deaths per event. From Thompson SA (1982)

| Hazard | Hazard frequency | % frequency | Deaths per event |
|---|---|---|---|
| *Geological hazards* | | | |
| Landslides | 29 | 2.7 | 190 |
| Tsunamis | 10 | 1.0 | 856 |
| Volcanoes | 18 | 1.7 | 525 |
| Earthquakes | 161 | 15.2 | 2652 |
| *Meteorological hazards* | | | |
| Cyclones | 211 | 19.9 | 2373 |
| Tornadoes | 127 | 12.0 | 66 |
| Floods | 343 | 32.3 | 571 |
| Heatwaves | 22 | 2.1 | 315 |
| Thunderstorms and gales | 36 | 3.4 | 587 |

SA Thompson, in 1982, produced a table combining frequency with deaths per event (**Table 1**). This table shows that earthquakes and cyclones are the most lethal per event on a global scale and both are of high frequency; however, flooding, because of its very high frequency, is nearly as lethal. The low frequency and number of deaths per event for landslides can be misleading – for example, in Basilicata, Italy, virtually all of the dense cluster of hilltowns are threatened by landslides, and it is unquestionably a major hazard, even if not a great killer. Likewise, in Nepal and the Pamirs (**Figure 2**), the steep topography means that landslides repeatedly wipe out villages and communications and present a major and intractable hazard problem.

SA Thompson also produced a table of fatalities covering Asia and Australasia between 1947 and 1981, showing how regional statistics can reveal extraordinary contrasts (**Table 2**). Asia had 85.8% of the global count of 1 208 044 global deaths from the hazards listed in **Table 2** during this period. However, this figure is a combination of both geological and meteorological disasters, and, for the geological disasters, the figure is slightly below the global average. Australasia accounted for only 0.4% of the global deaths. The death counts and magnitude of the event cannot be correlated, because of factors such as variations in populations at risk, ground conditions, and building construction quality and type. In the case of volcanic eruptions, there is a wide variation in the expected magnitude and type of eruption according to the classification of the volcano and its petrological products.

Volcanoes and earthquakes are covered by separate entries in this encyclopedia (*see* **Tectonics: Earthquakes, Volcanoes**).

**Figure 2** Beneath this peaceful scene in the Pamirs, Tadzhikistan, is buried Xait, one of 33 villages buried by landslides triggered by earthquakes in 1948, with 50 000 lives lost which geological knowledge could have saved. de Mulder EFJ, Holland, International Union of Geological Sciences 'COGEOENVIRONMENT' photograph.

**Table 2** Fatalities per event in Asia and Australasia. From Thompson SA (1982)

| Event | Asia | Australasia |
|---|---|---|
| Landslides | 3576 | 0 |
| Tsunamis | 7864 | 44 |
| Volcanoes | 2806 | 4000 |
| Earthquakes | 33 3623 | 133 |

## Procedures for the Mitigation of Natural Hazards

These include:

- the study of existing conditions and evidence of past activity, including mapping;
- prediction in time and space (when?, where?, and on what scale? risk assessment);
- prevention or mitigation by engineering or other methods;
- planning of land use;
- monitoring and installation of warning systems;
- emergency planning; proactive, including planning for evacuation; post-event relief.

Each of the above will have variable relevance to a particular hazard and subject site/area/region, and may have none at all.

## Types of Anthropogenic Geohazard

JD Mather and others included:

- surface movements – subsidence (**Figure 3**) related to mineral or fluid (hydrocarbon, water) extraction and surface collapse into voids left by mineral workings;
- contaminated land – left by mineral workings, chemical plants, gas works, etc. ('brownfield sites');
- rising groundwater levels;
- modification of groundwater quality (unsewered sanitation, organic solvent pollution);
- waste disposal.

GJH McCall and others also included:

- loss of soil and agricultural land (including the effects of urbanization on the soil resource);
- reduction in biodiversity.

The anthropogenic hazards are mostly of long build up – even surface collapses due to old mine workings (**Figure 4**), which appear abruptly without warning, as the roof of the mined-out cavity gradually moves towards the surface due to the fall-out of blocks.

Three of these quiet, pervasive, slow build-up hazards are mentioned below to illustrate this type of geohazard.

**Figure 3** Leaning church tower due to differential settlement in soft Holocene sediments, a foundation problem near Lake Ijssel, The Netherlands. de Mulder EFJ, Holland, International Union of Geological Sciences 'COGEOENVIRONMENT' photograph.

### Rising Groundwater

Rising groundwater was described in 1992 by DJ George – a problem that has affected many cities in the Middle East (Kuwait, Doha, Cairo, Riyadh, Jeddah, Madinah), and is a major hazard in terms of cost, and loss of and injury to human life. A typical cause is the installation of new water supply systems piped in from outside a city area, which previously relied on numerous small wells within it. The water table under the city is no longer utilized and the water table rises. This produces:

- damage to buildings and structures;
- damage to services and roads;
- overloading of sewer systems and treatment plants;
- salting and waterlogging of soils;
- public health hazards.

This is essentially anthropogenic but, like all such hazards, is due to interference with the natural system.

### Sea-Level Change on the China Coast

S Wang and X Zhao covered this hazard, which is likely to become more significant with global climate change – warming related to the 'greenhouse effect' – which has now been shown to have been unquestionably influenced since the beginning of the Industrial Age (see summary report by McCall GJH (2003); **Solar System: The Sun**). On the long, sinuous, and island-dotted coast of China, there are many large, industrialized cities and regions – e.g., Shanghai, Tianjin, Guangzhou – located on coastal plains and deltas. The evidence reveals a complex picture of overall oscillations of the order of a thousand years and different effects of Quaternary crustal movements, up and down in different regions; however, subtracting the latter, a general rise of 0.65 mm per year has been recognized. A future rise of several centimetres, which is predicted (also allowing for the effect of global warming), will have disastrous effects on the coastal and coastal hinterland environment and the industry and agriculture there. The wetlands will be inundated. Engineering constructions proposed in the Yangtze delta will be under the following main headings:

- control of land subsidence due to human activity;
- prevention of seawater seepage and windstorm tides;
- drainage of low-lying land and land improvement;
- taking into account sea-level rise in planning new urban areas.

### Global Soil Loss: Biodiversity Loss

WS Fyfe has emphasized the fact that the explosive growth of the human population (a taboo subject as WI Stanton has emphasized) (**Figure 5**) is at the root of most of the quiet, pervasive hazards. He indicates that topsoil loss is occurring at a global rate of 0.7% per year. The global climate bioproductivity is vital to the functioning of the global thermostat, and species continue to be wiped out wholesale due to the activities of humankind.

As described by GJH McCall *et al.*, in an original definitive account, urban geohazards are becoming more and more significant as urbanization mushrooms in the early twenty-first century (**Figure 6**). These include natural and anthropogenic hazards.

## Hazard and Risk Mapping

JG Doornkamp, in 1989, published a concise account of hazard mapping. He covered:

- landslips and avalanches;
- natural ground subsidence;

**Figure 4** The result of a void migrating to the surface in old mine workings beneath a suburb of Glasgow. British Geological Survey photograph, in McCall GJH, de Mulder EFJ, and Marker BR (1996) *Urban Geoscience*. Rotterdam: Balkema.

- hazards of quarrying and mining;
- erosion and deposition;
- flooding;
- saline soils;
- permafrost;
- seismological hazards;
- volcanic hazards.

The third in this list is an anthropogenic hazard – the general principles apply to both natural and anthropogenic hazards. JG Doornkamp noted that the distinction is ambiguous.

Every hazard has three key properties:

- magnitude;
- frequency;
- location.

The last of these is most amenable to mapping. Mapping may be taken for different reasons:

- to define the nature and extent of a historical event;
- to define the nature and extent of a recent event;
- to define present conditions in order to assess the likelihood of a recurrent event.

The social and economic content of the affected subject must be included in any such programme. A hazard cannot in itself be mapped, as it is a process. Rather, the results of a hazard are mapped. The resultant landforms, the effects on soil and rocks, and the effects on constructions are mapped.

JG Doornkamp devised a useful diagram showing the context of hazard mapping (**Figure 7**).

Aids to mapping include:

- satellite imagery;
- airborne multispectral scanning;
- photogrammetry;
- shallow geophysics;
- global positional systems.

DKC Jones has illustrated well the procedures in hazard mapping in an account of the application to landslides. The graphical methods are only part of the exercise, which includes:

- graphical methods, mainly involving mapping;
- the analysis of the empirical relationships between landslides and individual causal factors, such as slope steepness, material type, vegetation cover, rainfall intensity, and human constructions;
- multivariate analysis of the causes of landslides.

True hazard maps display the extent of past hazardous events (zoned), together with an internal division reflecting the magnitude frequency distribution

ENGINEERING GEOLOGY/Natural and Anthropogenic Geohazards  521

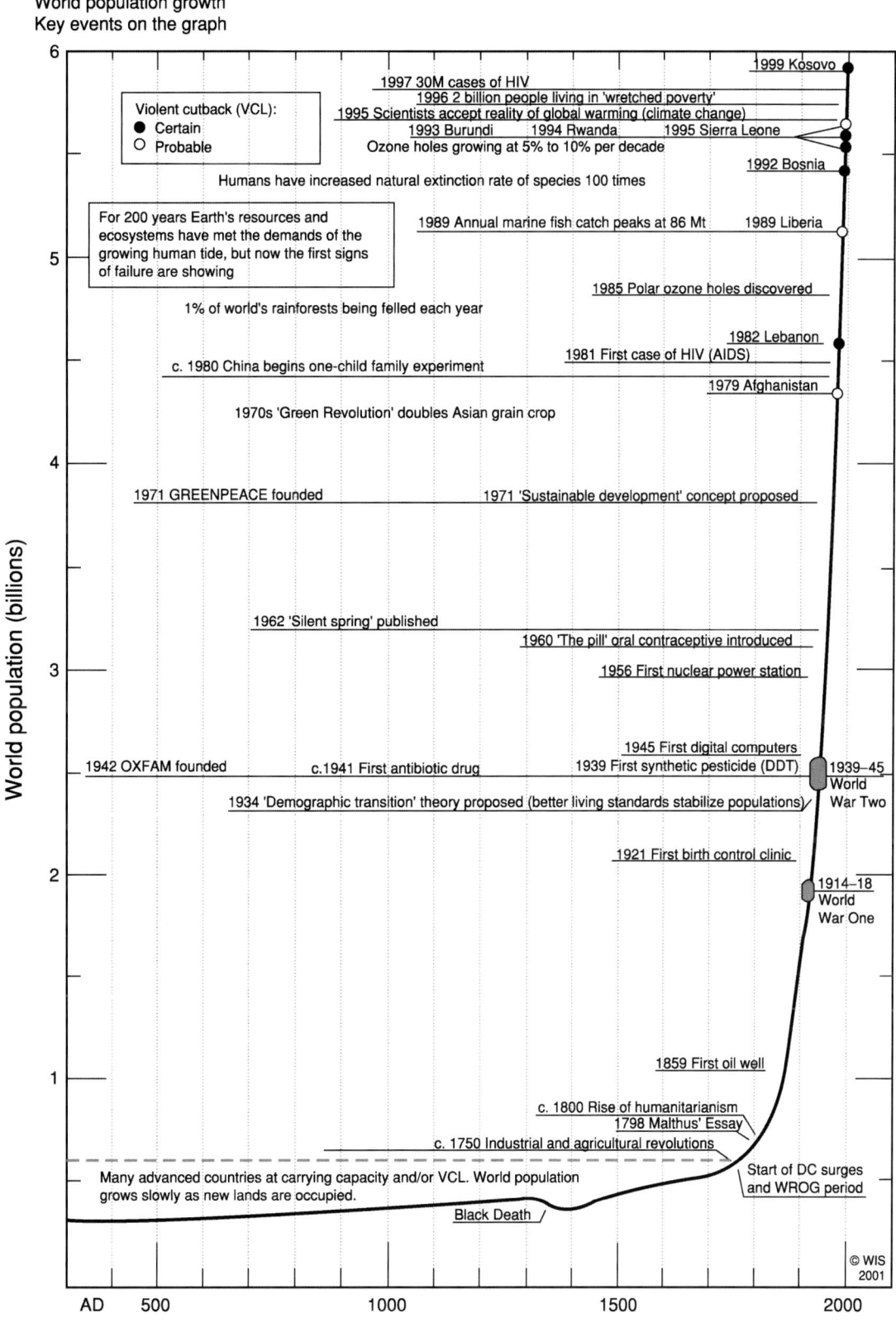

**Figure 5** The escalation of the world population. After Stanton WI (2003) *The Rapid Growth of Human Populations: 1750–2000*. Brentwood: Multi-science Publishers.

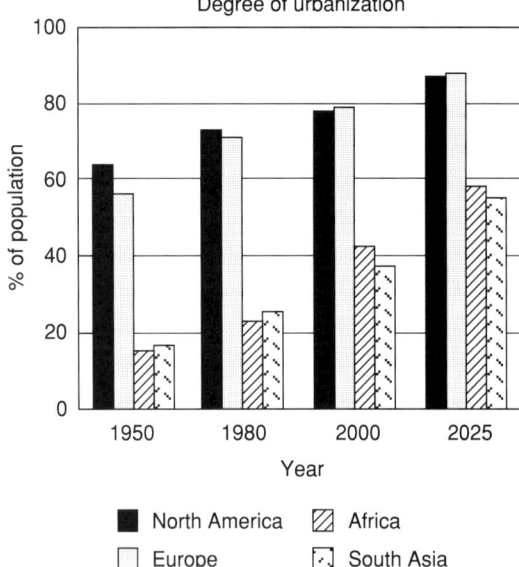

**Figure 6** The escalation of urbanization. After de Mulder EFJ, in McCall GJH, de Mulder EFJ, and Marker BR (1996) *Urban Geoscience*. Rotterdam: Balkema.

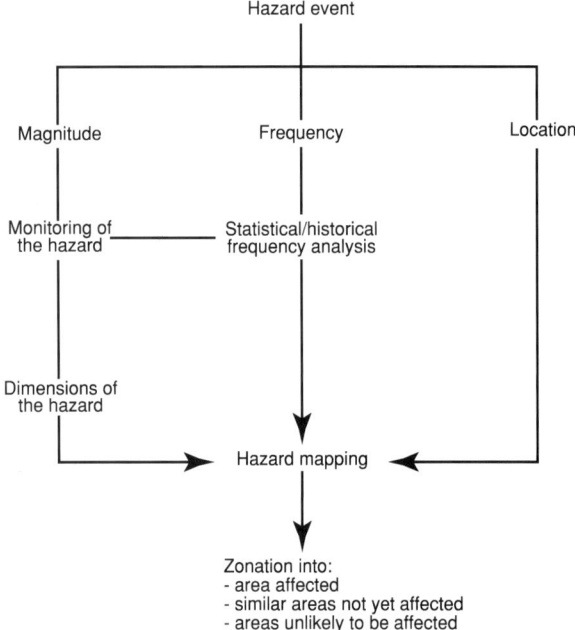

**Figure 7** The context of hazard mapping. After Doornkamp JG, in McCall GJH and Marker BR (1989) *Earth Science Mapping for Planning, Development and Conservation, Engineering Geology Special Publication No. 4*. London: Graham and Trotman.

of past events (hazard rating). Evidence of past failures must be used to predict the likelihood of future events, their scale, and frequency. Risk maps are quite different in that they assess the potential losses that may be incurred by society. They may involve numerical quantification (for example, the table for the Central Pacific by Blong RJ (1988)) (**Table 3**). They may be followed by maps showing geographical divisions of recommended action (for example, the map of Basilicata; **Figure 8**). Hazard mapping may be very complex and involve numerous maps of different type. For example, DKC Jones, in 1992, listed the following as all used in landslide mapping:

- geological structure;
- chronostratigraphy of rocks;
- lithostratigraphy of rocks;
- lithostratigraphy of soils;
- rockhead contours;
- geotechnical properties of soils;
- geotechnical properties of rocks;
- hydrological conditions;
- geomorphological processes;
- geomorphological history, palaeodeposits, surfaces, or residual conditions;
- seismic activity;
- climate (including precipitation);
- ground morphology (especially slope height, length, and angle);
- land use;
- vegetation type, cover, root density, and strength;
- pedological soils, type and thickness of regolith;
- past landslide deposits;
- past landslide morphometry.

## Wider Responses

JG Doornkamp listed four wider responses beyond mapping considerations that can be adopted to mitigate hazards:

- land use planning: place development in less hazardous places;
- economic planning: investment in the right places and maintenance of financial reserves to cover disasters; adequate insurance schemes;
- development control: designing built structures to cope with known hazards (e.g., earthquake-resistant designs);
- emergency planning: evacuation procedures to be planned and rehearsed in all areas at risk from major hazards; rescue teams trained and tested with modern equipment, able to move quickly to the site of a disaster.

## Engineering and Geohazards

Experience has shown that geologists, on the one hand, and engineers and developers, on the other,

**Table 3** Geological risk rating for selected towns in Papua New Guinea. After Blong RJ (1988)

| Town | Earthquake rating | Volcano rating | Tsunami rating | Risk rating |
|---|---|---|---|---|
| Arawa | 6 | 2 | 0.1 | 8 |
| Daru | 2 | 0.1 | 0.03 | 2 |
| Goroka | 5 | 0.1 | – | 5 |
| Kavieng | 3 | 0.1 | 0.1 | 3 |
| Kiet | 6 | 2 | 0.1 | 8 |
| Kimbe | 10 | 4 | 0.1 | 14 |
| Kokopo | 10 | 4 | 0.2 | 14 |
| Lae | 5 | 0.1 | 0.2 | 5 |
| Madang | 5 | 1 | 0.2 | 6 |
| Mt Hagen | 3 | 0.1 | – | 3 |
| Port Moresby | 3 | 1 | 0.01 | 3 |
| Rabaul | 10 | 5 | 0.2 | 15 |
| Wewak | 6 | 2 | 0.2 | 8 |

tend to have rather different viewpoints concerning geohazards. The differences concern site investigation, which is the keystone of any engineering project. There is a tendency amongst engineers and developers to omit consideration of the hazards pertaining to a particular site from previous site investigations, with the view that the problems will only be addressed if and when they arise. For the past three decades, environmental geologists have stressed that these factors must be evaluated as part of the previous site investigation, and this is becoming, certainly in the UK, a regulatory condition for construction and development.

Engineers are greatly involved in investigation, mitigation, and repair programmes related to both natural and anthropogenic hazards. For example,

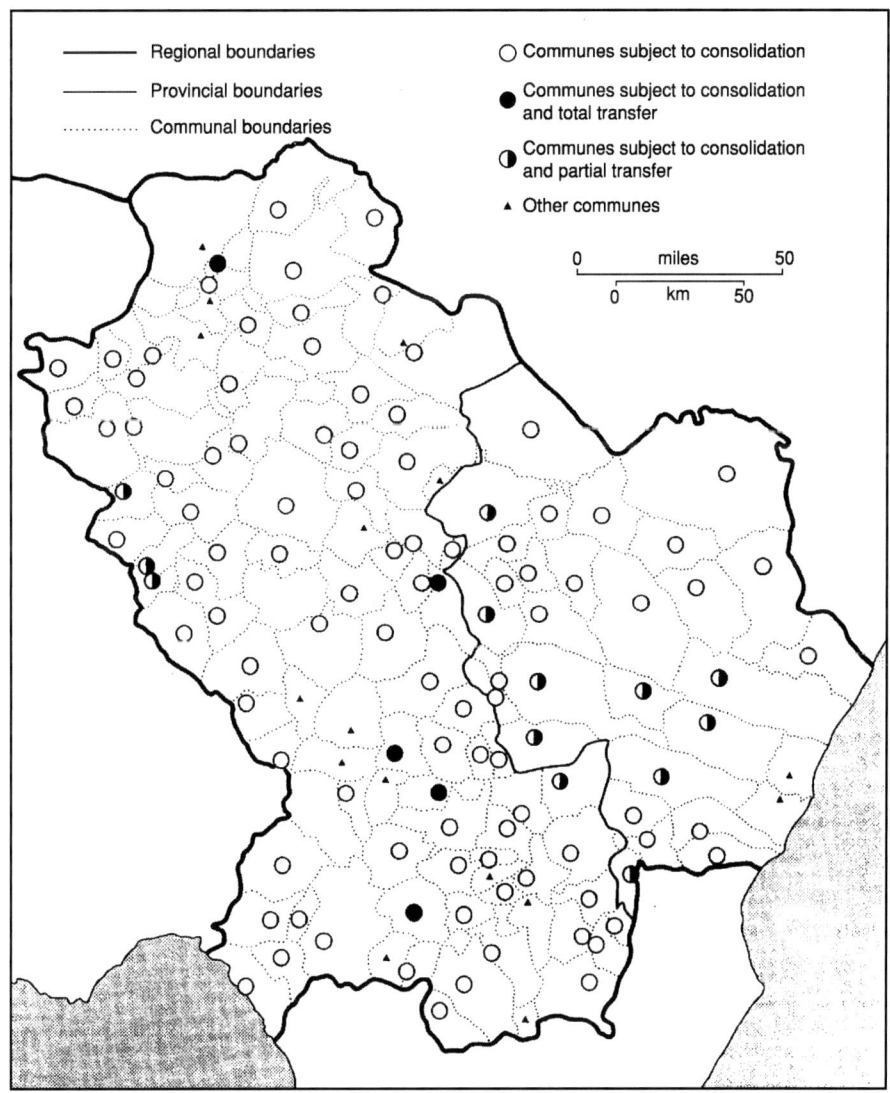

**Figure 8** Landslide hazard assessment map, Basilicata, Italy. After Jones DKC, in McCall GJH, Laming DJC, and Scott SR (1992) *Geohazards – Natural and Man Made*. London: Chapman and Hall.

they design earthquake-proof buildings in Japan and deviation works to prevent lava flows or lahars from reaching villages and towns (as in Sicily). To illustrate the variety of engineering applications to geohazards, a list is given below of the types of work in which engineers have been involved, and which are represented by reports to Special Publications of the Engineering Group of the Geological Society:

- liquefaction of sediments in the Fraser River Delta (Canada);
- flooding at Ladysmith (South Africa);
- fluvial hazards (Bihar, India);
- ice- and snow-related high-altitude problems in glacial lakes (Himalayas, Peru);
- cliff erosion (Isle of Wight, UK);
- lessons from the Kobe earthquake (Japan);
- risks from low-seismicity earthquakes (Holland);
- gypsum-related subsidence at Ripon, Yorkshire (UK);
- subsidence related to solution in chalk (south-east England);
- groundwater recharge under rapid urbanization (Mexico, Thailand);
- organic solvent pollution of groundwater (Coventry, UK);
- landfill disposal of urban wastes (Tanzania, Gambia, Mauritius);
- acid mine drainage (Transvaal);
- heavy metal contamination (Cornwall, UK);
- foundation conditions and site investigations (UK);
- land restoration, brownfield sites;
- detection of karst features by remote sensing (England).

## Conclusion

There have been great advances in hazard-related earth science and engineering in recent decades, but these hazards are far from understood and there is an ongoing need for more data and statistics. Countries such as China, with its immense population, many of whom live in sites of high risk, and Colombia or Nepal, where the climate and physiography militate to give a situation in which natural geohazards are quite unavoidable, but can be mitigated, if only by emergency planning, are extreme examples; however, no country is free from these problems. With the growth of megacities, urban geohazards are assuming increasing importance (**Figure 6**).

## See Also

**Engineering Geology:** Liquefaction; Made Ground; Problematic Rocks; Problematic Soils; Site and Ground Investigation; Subsidence. **Environmental Geology. Soil Mechanics. Solar System:** The Sun. **Tectonics:** Earthquakes. **Volcanoes**.

## Further Reading

Appleton JD, Fuge R, and McCall GJH (1996) *Environmental Geochemistry and Health, Special Publication No. 113*. London: Geological Society.

Blong RJ (1988) Assessment of eruption consequences. *Kagoshima International Conference on Volcanoes Proceedings*, pp. 569–572.

Bullock P and Gregory PJ (1991) *Soils in the Urban Environment*. Oxford: Blackwell Scientific Publications.

Culshaw MG, Bell FG, Cripps JC, and O'Hara M (1987) *Planning and Engineering Geology, Engineering Geology Special Publication No. 4*. London: Geological Society.

de Mulder EFJ COGEOENVIRONMENT, International Union of Geological Sciences (IUGS) (undated leaflet) *Planning and Management, the Human Environment – the Essential Role of the Geosciences*.

Maund JG and Eddleston M (1998) *Geohazards in Engineering Geology, Engineering Geology Special Publication No. 15*. London: Geological Society.

McCall GJH (2003) Global climate change – a view from the floor. *Geoscientist* 13(6): 18–20.

McCall GJH, Laming DJC, and Scott SR (1992) *Geohazards – Natural and Man Made*. London: Chapman and Hall.

McCall GJH and Marker BR (1989) *Earth Science Mapping for Planning, Development and Conservation, Engineering Geology Special Publication No. 4*. London: Graham and Trotman.

McCall GJH, de Mulder EFJ, and Marker BR (1996) *Urban Geoscience*. Rotterdam: Balkema (see, especially, papers by McCall, Simpson, and Mather).

Stanton WI (2003) *The Rapid Growth of Human Populations: 1750–2000*. Brentwood: Multi-science Publishers.

Thompson SA (1982) *Trends and Developments in Global Natural Disasters 1947–1981*. University of Colorado Institute of Behavioural Science Natural Hazards. Research Working Paper No. 45.

Wang S (1997) *Engineering Geology, Proceedings of the 30th International Geological Congress, Beijing, 1997*. Vol. 2–3. Utrecht: VSP.

Zhang Z, de Mulder EFJ, Liu T, and Zhou L (1997) *Geosciences and Human Survival, Environment, Natural Hazards, Global Change. Proceedings of the 30th International Geological Congress, Beijing, 1997*. Vol. 2–3. Utrecht: VSP.

# Liquefaction

**J F Bird**, Imperial College London, London, UK
**R W Boulanger and I M Idriss**, University of California, Davis, CA, USA

© 2005, Elsevier Ltd. All Rights Reserved.

## Introduction

Liquefaction-induced ground failure has caused widespread damage and devastation; a recent example is the closure of the Port of Kobe, Japan's busiest port, following the 17 January 1995 earthquake, largely due to the liquefaction-related failure of reclaimed land (**Figure 1**). The potential consequences of liquefaction are far reaching, ranging from settlement or tilt of individual building foundations to the spread of fire when the water supply system is damaged by permanent ground deformations.

Liquefaction is the loss of shear strength of a saturated cohesionless soil due to increased pore water pressures and the corresponding reduction in effective stress during cyclic loading. Liquefaction and its associated ground deformation are very complex phenomena, and the term 'liquefaction' has been used to describe a wide range of soil behaviour. This article focuses on the susceptibility, triggering, and consequences of earthquake-induced liquefaction; liquefaction caused by non-earthquake-related loading is not included. Although the emphasis here is mainly the liquefaction of saturated cohesionless deposits such as sands or silty sands, it is important to consider that strength loss in fine-grained soils under earthquake loading can also pose a significant hazard. The assessment of liquefaction hazard is an essential part of any engineering project in seismic regions and is needed to make informed decisions with respect to mitigation options, foundation design, and emergency response and recovery plans based on what is considered to be an acceptable level of risk.

## The Principles of Liquefaction

As a saturated cohesionless soil is cyclically loaded, its particle structure can tend to collapse to a denser arrangement. If the soil's permeability and the site stratigraphy are such that drainage cannot occur immediately, then, as the collapse occurs, stresses will be transferred from the soil grain contacts to the pore water, leading to an increase in pore water pressure. In simple terms, when the pore water pressure increases, the effective stress (total stress minus pore water pressure) on the particle structure will reduce, and as it approaches zero, the shear resistance of the soil will also approach zero. This loss of effective stress and shear resistance is known as liquefaction.

The stress–strain behaviour of liquefying sand depends strongly on its relative density. When loose sand liquefies, the gravitational static shear stresses may exceed the shear resistance of the soil and rapid deformation with very large shear strains can commence; this is referred to as flow deformation. The soil behaviour is termed 'contractive', and the shear resistance exhibited by the liquefied soil during flow deformation is termed the 'residual strength'. When moderately dense granular soils are cyclically sheared, pore pressures may similarly rise and liquefaction can be triggered. However, rather than undergoing flow deformation, the soil particle structure may try to expand as it reaches a certain level of shear strain, resulting in what is termed 'dilative' behaviour. For undrained conditions, this leads to a reduction in the pore water pressures and a corresponding increase in effective stress and shear resistance. A shear stress reversal, however, such as will occur many times during earthquake shaking, may cause the soil particle structure to be incrementally contractive and the state of zero effective stress may be temporarily reached once more. This continued cycle of zero effective stress and strength regain is termed 'cyclic mobility'. The cumulative deformations can be significant, particularly if the duration of shaking is long, but dilative soils do not exhibit very large flow deformations in the way that contractive soils do (**Figure 2**).

**Figure 1** Liquefaction of reclaimed fills at the Port of Kobe in 1995 caused complete suspension of operations. The lateral displacement of the quay walls in this picture pulled apart the crane legs, causing collapse. Photograph by Leslie F. Harder, Jr.

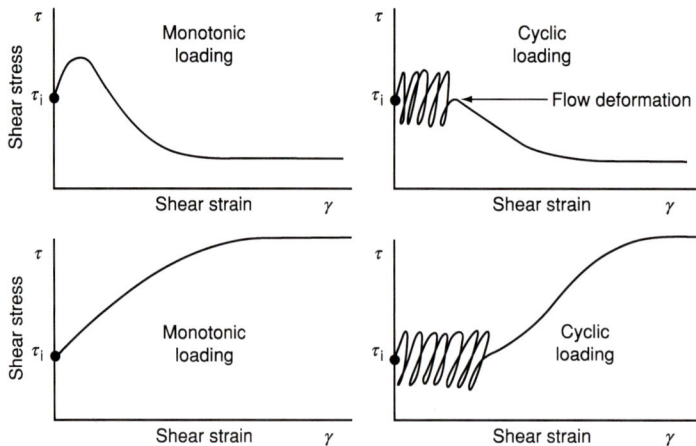

**Figure 2** Stress–strain behaviour of contractive (top) and dilative (bottom) soils under monotonic and cyclic loading. $\tau_i$ = initial (static) shear stress. In all cases, the steady-state undrained shear strength is ultimately reached. Reproduced from Castro (1976).

**Figure 3** Sand boil in liquefied soil following an earthquake; Japan, 1983.

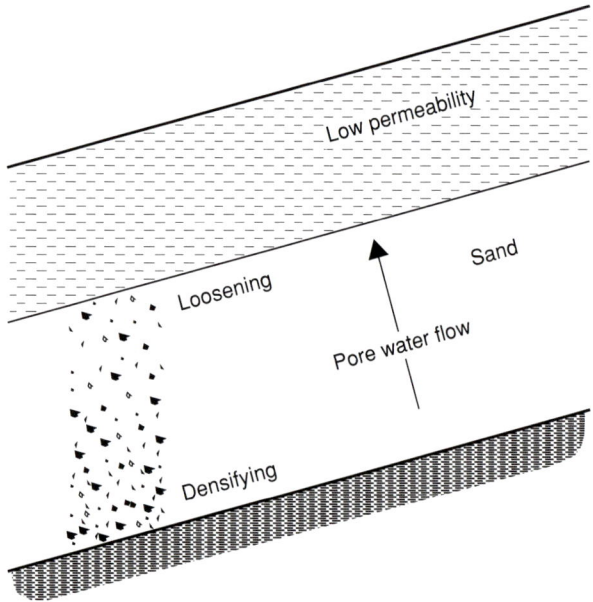

**Figure 4** Schematic illustration of void redistribution resulting from spreading of pore pressure and global volume changes. Reproduced from the National Research Council (1985) *Liquefaction of Soils during Earthquakes. Report by the Participants in the Workshop on Liquefaction, Committee on Earthquake Engineering, National Research Council, Report No. CETS-EE-001.* Washington, DC: National Academy Press.

Sand boils are caused by the tendency of the excess pore water pressures to dissipate upwards, towards the free surface, carrying soil particles up from the liquefied layer through cracks or channels in the overlying material and ejecting them at the ground surface (**Figure 3**). The upward seepage of pore water, driven by the earthquake-induced excess pore pressures, can be impeded by less permeable overlying soil layers, which can result in water accumulating near the interface between the liquefied soil and the overlying lower permeability soil. The accumulation of water can loosen the soil and possibly even result in the formation of water films, either of which greatly reduces the available shear resistance along the interface (**Figure 4**). This phenomenon is referred to as 'void redistribution'. A key consequence of void redistribution is that the residual shear strength of liquefied soil does not just depend on the pre-earthquake properties of the soil, but also depends on those factors affecting the dissipation and movement of pore water following earthquake shaking.

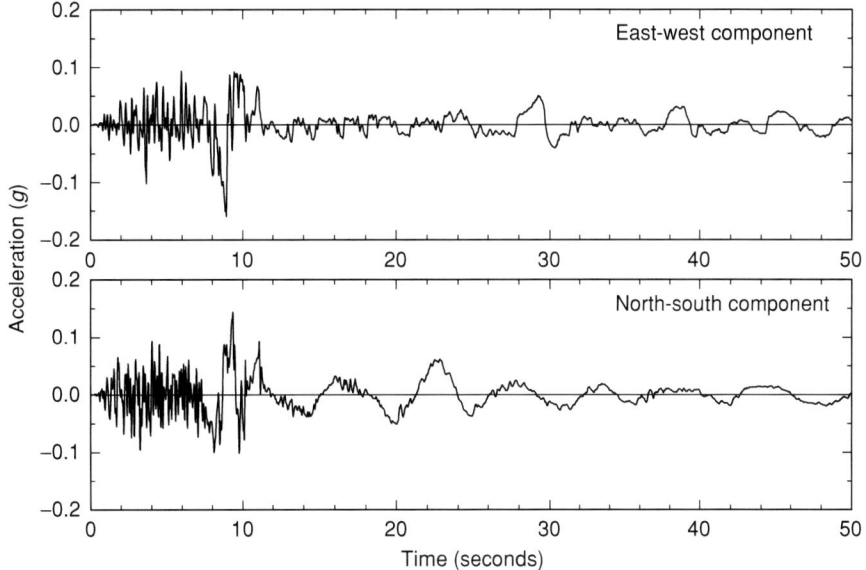

**Figure 5** Horizontal accelerograms recorded in Niigata, Japan in the 1964 earthquake. The area surrounding the recording station was heavily damaged by liquefaction. The onset of liquefaction appears to have occurred at approximately 10 s after strong ground shaking commenced.

Another by-product of triggering of liquefaction is the potential effect it has on the characteristics of the strong ground motion. A number of accelerograms have clearly shown the triggering of liquefaction beneath the instrument (e.g., **Figure 5**).

## Assessment of Liquefaction Hazard

The potential for liquefaction to cause damage is assessed in three stages: the susceptibility to liquefaction, the likelihood of liquefaction being triggered by the design earthquake scenario, and the consequences related to liquefaction (**Figure 6**). The variables that influence the onset of liquefaction include environmental factors such as the location of the water table, site stratigraphy, and depositional and seismic loading history of the soil; soil characteristics such as relative density, grain size distribution, mineralogy, and the presence of any cementing agents; and the characteristics of the earthquake under consideration, mainly the amplitude of ground shaking and its duration.

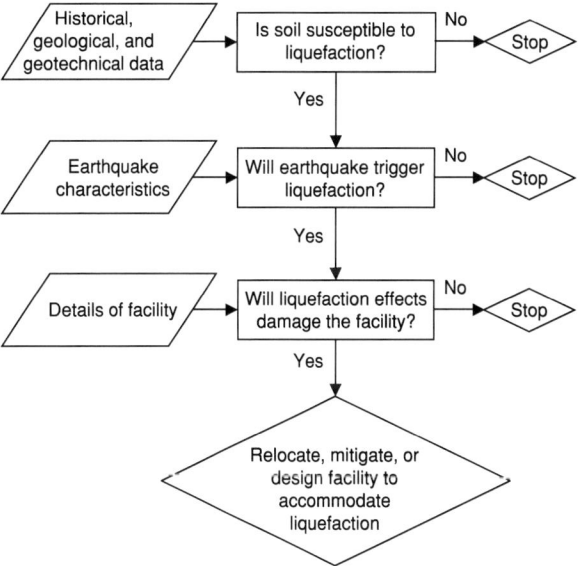

**Figure 6** Flow chart for the three stages of assessing the potential of earthquake-induced liquefaction to damage structures, lifelines, and other facilities.

### Liquefaction Susceptibility

The most obvious evidence that a particular deposit is susceptible to liquefaction is historical precedence, either at the same site or in similar conditions elsewhere; it is not unusual for liquefaction to re-occur in the same location. In regions of potential hazard where there is no historical evidence, other information must be used to make an initial assessment of the liquefaction risk. The criteria in **Table 1**, classifying different levels of liquefaction susceptibility for various types and ages of soil deposits, are widely used by engineers and engineering geologists for the purposes of either producing regional maps of liquefaction hazard for planning purposes or for preliminary assessment of potential hazards for site selection purposes. Saturated loose cohesionless soils are the most

**Table 1** Susceptibility of sedimentary deposits to liquefaction during strong seismic shaking[a]

| Type of deposit | Distribution of cohesionless sediments in deposit | Likelihood that cohesionless sediments, when saturated, would be susceptible to liquefaction | | | |
|---|---|---|---|---|---|
| | | <500 years | Holocene | Pleistocene | Pre-Pleistocene |
| **Continental** | | | | | |
| River channel | Locally variable | Very high | High | Low | Very Low |
| Flood plain | Locally variable | High | Moderate | Low | Very Low |
| Alluvial fan and plains | Widespread | Moderate | Low | Low | Very low |
| Marine terraces and plains | Widespread | — | Low | Very low | Very low |
| Delta and fan delta | Widespread | High | Moderate | Low | Very low |
| Lacustrine and playa | Variable | High | Moderate | Low | Very low |
| Colluvium | Variable | High | Moderate | Low | Very low |
| Talus | Widespread | Low | Low | Very low | Very low |
| Dunes | Widespread | High | Moderate | Low | Very low |
| Loess | Variable | High | High | High | Unknown |
| Glacial till | Variable | Low | Low | Very low | Very low |
| Tuff | Rare | Low | Low | Very low | Very low |
| Tephra | Widespread | High | High | ? | ? |
| Residual soils | Rare | Low | Low | Very low | Very low |
| Sebka | Locally variable | High | Moderate | Low | Very low |
| **Coastal zone** | | | | | |
| Delta | Widespread | Very high | High | Low | Very low |
| Estuarine beach | Locally variable | High | Moderate | Low | Very low |
| High wave energy | Widespread | Moderate | Low | Very low | Very low |
| Low wave energy | Widespread | High | Moderate | Low | Very low |
| Lagoonal | Locally variable | High | Moderate | Low | Very low |
| Fore shore | Locally variable | High | Moderate | Low | Very low |
| **Artificial fill** | | | | | |
| Uncompacted fill | Variable | Very high | — | — | — |
| Compacted fill | Variable | Low | — | — | — |

[a]Reproduced with permission from Youd TL and Perkins DM (1978) Mapping liquefaction-induced ground failure potential. *Journal of the Geotechnical Engineering Division, ASCE* 104(GT4): 443–446.

vulnerable to liquefaction. Older deposits generally have higher densities and thus greater resistance to liquefaction.

The mode of deposition is also an important factor in liquefaction susceptibility, as can be seen in **Table 1**. Alluvial, fluvial, marine, deltaic, or wind-blown deposits generally have a higher susceptibility than do either residual soils or glacial tills; materials that are more highly consolidated. There are few case histories of liquefaction occurring at depths greater than about 10–15 m. This is partly due to the fact that soils at greater depths tend to be older, and therefore more resistant to liquefaction, and also that the earthquake-induced cyclic shear stress ratio (see eqn [1]) will be lower at greater depths. In addition, liquefaction at greater depths beneath level ground surfaces would not necessarily be evident at the ground surface, and thus may go undetected. The issue of a maximum depth for liquefaction is of particular importance for the construction and evaluation of large earth dams, where the presence of significant overburden above natural deposits should not be construed as a reduction in liquefaction susceptibility.

When some site-specific information is available, compositional criteria can be used to make preliminary assessments of liquefaction susceptibility. The fines content and clay content, the Atterberg limits, and the particle size distribution are all relevant. Though clays are generally resistant to liquefaction, it is also understood that plastic silts and clays may develop significant shear strains under sufficiently strong and sustained earthquake loading, leading to strength loss and rapid deformation or even instability. Compositional characteristics are used in various criteria, such as the often-cited 'Chinese criteria,' to evaluate a soil's liquefaction susceptibility, but a classification of 'non-liquefiable' should not be equated with the absence of a problem. Thus, liquefaction evaluation procedures may not be appropriate for plastic silts and clays, but the potential for such soils to fail and induce damage under earthquake loading should not be discounted.

### Evaluation of Liquefaction Potential

The state-of-the-practice approach for evaluating liquefaction potential at a specific site, often referred to as the 'simplified procedure', was first published by H B Seed and I M Idriss in 1971. This procedure compares the shear stresses induced by an earthquake with those required to cause liquefaction in the soil profile, where the liquefaction resistance is based on empirical data. The procedure is based on the following relationship:

$$\text{CSR} = \frac{\tau_{av}}{\sigma'_v} = 0.65 \cdot \frac{a_{max}}{g} \cdot \frac{\sigma_v}{\sigma'_v} \cdot \frac{r_d}{\text{MSF}} \quad [1]$$

where CSR is the cyclic shear stress ratio, $\tau_{av}/\sigma'_v$, at the depth of the soil layer under consideration, $a_{max}$ is the peak ground acceleration at the surface, $g$ is acceleration due to gravity, $\sigma_v/\sigma'_v$ is the ratio of total to effective vertical stress in the soil layer, and $r_d$ is a stress reduction coefficient (**Figure 7**). MSF is a magnitude scaling factor, equal to unity for M = 7.5 earthquakes, and greater than unity for M < 7.5, to correct in an implicit manner for the duration of shaking. Recommended procedures for the evaluation of $r_d$ and MSF for use in conjunction with eqn [1] are described in a 2004 paper by Idriss and Boulanger.

The factor of safety (*F*) against liquefaction is then determined as follows:

$$F = \text{CRR}/\text{CSR} \quad [2]$$

where CRR is the cyclic resistance ratio of the soil layer. The evaluation of CRR can be determined directly by laboratory tests on high-quality undisturbed samples and reconstituted samples (for new fills), or indirectly by *in situ* testing. An effective method (albeit an expensive one) of evaluating the cyclic resistance involves collection of high-quality undisturbed soil samples, using techniques such as ground freezing, and undertaking cyclic testing in the laboratory (e.g., cyclic simple shear or triaxial tests). The most common approach used by engineers to determine the CRR is based on *in situ* testing, and in particular the standard penetration test (SPT) and the cone penetration test (CPT). Shear wave velocity ($V_s$) measurements and the Becker penetration test (BPT) are also used in some specific applications.

Several publications provide guidance on the estimation of CRR using *in situ* measurements. Such correlations are obtained through back analysis of case history data, and as such are related to the variables used to define the CSR (eqn [1]) for the case histories, which can vary considerably among different authors. For this reason, it is important that practitioners are consistent in the methodology they adopt, because the combination of different recommendations could lead to incorrect results. An example of a recently completed correlation relating CRR to $(N_1)_{60}$ for clean sands (i.e., fines content ≤5%) is presented in **Figure 8**. The parameter $(N_1)_{60}$ represents the SPT blowcount corrected to a hammer energy level of 60% (to compensate for variations in testing equipment and procedures) and normalized to an effective vertical stress of 1 atmosphere (to compensate for variations in the depth of the *in situ* testing between the site and the case history data). Similar curves for cohesionless soils with higher fines content are also available, based on the same data used to construct the curve in **Figure 8**.

## Permanent Ground Deformation

The extent of the permanent ground deformation as a result of liquefaction will dictate the engineering solutions in terms of mitigation or redesign. The interaction between liquefied and non-liquefied soils, foundations

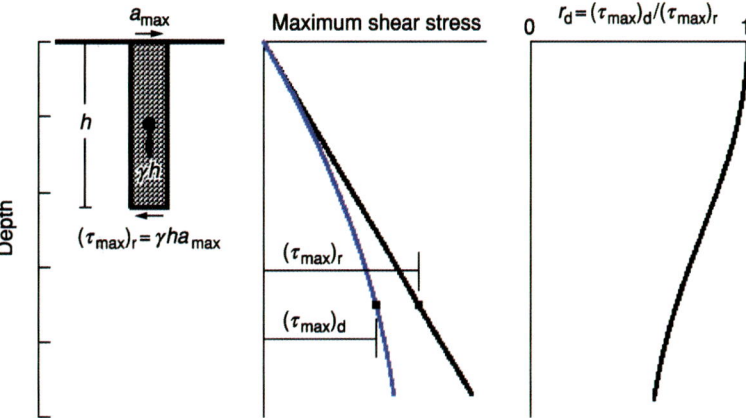

**Figure 7** Schematic for determining maximum shear stress ($\tau_{max}$) and the stress reduction coefficient ($r_d$). Reproduced with permission from Seed HB and Idriss IM (1971) Simplified procedure for evaluating soil liquefaction potential. *Journal of Soil Mechanics and Foundations Division, ASCE* 97: 1249–1273.

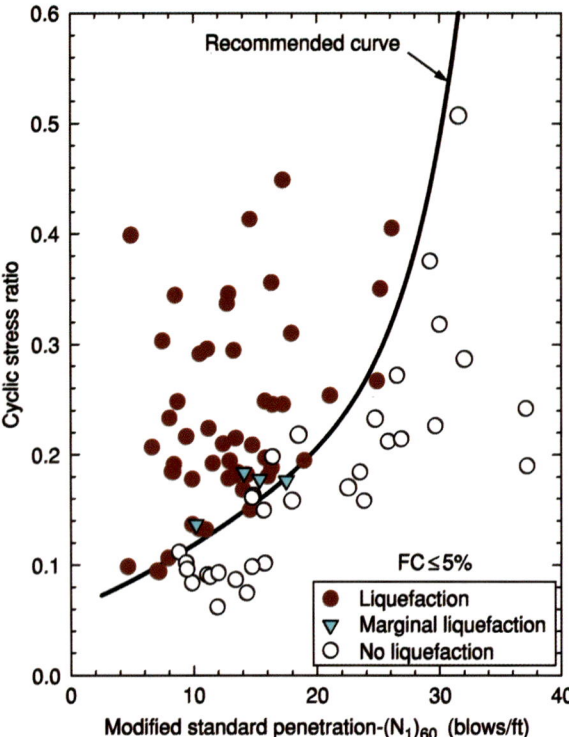

**Figure 8** Correlation between cyclic shear stress ratio (for an earthquake of magnitude 7.5) causing liquefaction and standard penetration resistance for clean sands (FC, fines content ≤5%).

**Figure 9** The failure of the upstream slope of the Lower San Fernando Dam shortly after the 1971 San Fernando earthquake, California narrowly avoided catastrophic flooding of 80 000 downstream homes. Liquefaction of the hydraulic fill caused loss of strength and slope instability, which left less than 1 m of freeboard after the earthquake. This picture shows the slide after emptying the reservoir; the paved road to the left indicates the former dam crest. Photograph by Leslie F. Harder, Jr.

and superstructures, and soils and embedded structures (e.g., pipelines, tunnels, tanks) will determine whether the potential for liquefaction-induced damage is unacceptable.

**Flow failure** Liquefaction flow failure occurs when the static shear stresses exceed the residual shear strength of a liquefied soil. Normally such failures occur on sloping ground that was stable in its static condition but became unstable due to the reduced shear resistance of the liquefied soil. Once movement is mobilized, displacements can be very large (up to tens of metres) and very rapid. The potential for flow failure can be assessed using standard slope stability analyses, substituting the residual undrained shear strength (discussed later) of the liquefied layer for its static properties, where a factor of safety below 1 indicates a flow failure hazard.

**Lateral spreading** Lateral spreading is also a downslope failure mechanism, related to cyclic mobility (**Figure 9**). The disturbing forces are a combination of the gravitational static downslope forces and the inertial loads generated by the earthquake. Lateral spreads can occur on gentle slopes or where there is a free face. The factor of safety against slope failure may remain above 1, and the ground deformation is a result of the progressive movement of surface layers as a result of the oscillation of the ground. Lateral spread-induced movements of as much as 10 m have been observed in past earthquakes.

The amount of horizontal movement can be estimated using relationships developed from empirical data, soil mechanics theory, or numerical modelling. Simplified relationships developed using empirical data are the simplest to employ, but users should consider the uncertainty associated with any simplification of this very complex phenomenon; simplifications necessarily neglect three-dimensional effects, local effects, or redistribution of pore water pressures, for example.

**Ground oscillation** Ground oscillation, another form of lateral ground deformation, can occur where the underlying soil has liquefied but there is no slope or free face for permanent lateral deformations to occur. Ground oscillation will manifest itself as large-amplitude transient ground waves with little or no resultant permanent deformation.

**Settlement** Liquefaction is a result of the tendency for saturated granular soils to densify under earthquake shaking. The eventual manifestation of this behaviour is settlement at the ground surface as excess pore pressures dissipate after the earthquake. Settlements are usually estimated using free-field, one-dimensional relationships for the volumetric strain induced in the soil as a function of both the relative density of the soil and the maximum shear

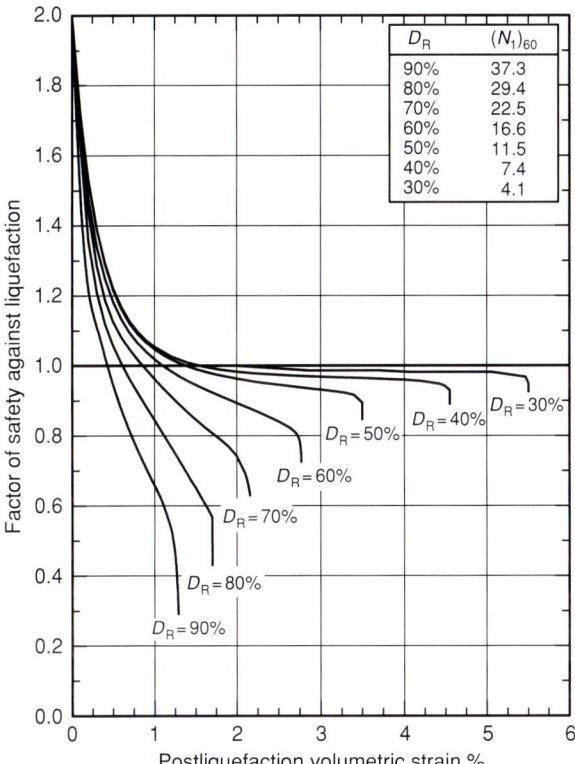

**Figure 10** Chart for the estimation of volumetric strain as a function of the factor of safety against liquefaction ($F$; eqn [2]) and the relative density of the liquefied soil layer ($D_R$). ($N_1)_{60}$ is the standard penetration test blowcount normalized to an equivalent overburden pressure of 1 atmosphere and hammer energy of 60%.

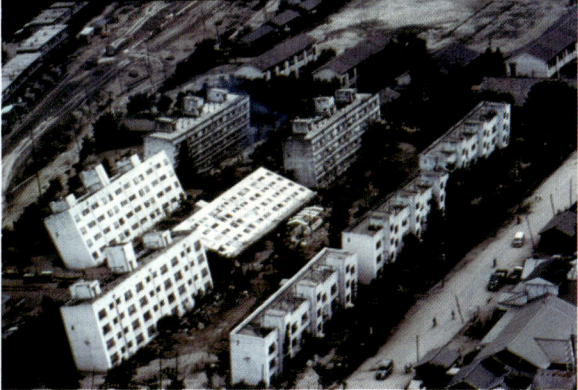

**Figure 11** Tilted buildings with shallow foundations in Niigata, Japan (1964) as a result of liquefaction-induced bearing-capacity failure.

**Figure 12** An elevated transport link in Kobe following the 17 January 1995 earthquake. The bridge piers were piled, and the structure is generally undamaged despite the settlement of the surrounding soil by up to 1 m.

strain induced by the earthquake (**Figure 10**). In practice, the degree of settlement is complicated by the heterogeneity of the soil, the interaction between vertical and lateral movements, and the presence of the structure, and most simplified methods for predicting volumetric strains have quite high associated uncertainties.

**Bearing failure** If the residual shear strength of the liquefied soil beneath is sufficiently low, then bearing capacity failure will occur, causing the structure to settle or tilt uniformly (**Figure 11**). These types of failures are generally accompanied by heaving of the ground around the foundation. The potential for a bearing-capacity failure can be determined using simple static bearing-capacity formulas, substituting the residual undrained shear strength of the liquefied layers. Failure will occur if the factor of safety is less than unity. However, structures that are safe against bearing failure can still develop excessive settlements (total and differential), depending on the strains (shear and volumetric) that develop in the underlying soils.

**Residual undrained shear strength** The residual shear strength of a liquefied soil can be determined in the laboratory or can be estimated using correlations between *in situ* test data and liquefied shear strength obtained through back analysis of field case histories. Field data, compared to laboratory tests, tend to indicate much lower shear strengths. As well as the difficulty in obtaining true, undisturbed samples, some phenomena cannot be replicated in laboratory tests. The layered nature and contrasting permeability of *in situ* soil deposits can impede the flow of water as earthquake-induced excess pore water pressures dissipate; this can lead to localized weakening of the soil at permeability interfaces (see **Figures 4** and **12**). A further *in situ* phenomenon that cannot be replicated by testing small samples is the potential intermixing of soil layers with different characteristics, due to the shear deformation, which can reduce the shear strength. Field data, however, are implicitly likely to include such phenomena. Consequently, the use of the empirical correlations to estimate postliquefaction shear strengths is recommended.

## Consequences of Earthquake-Induced Liquefaction

The importance of liquefaction from an engineering aspect is damage to the built environment, as represented by the following examples:

- Settlements, which can be of the order of 5% of the thickness of the liquefied soil, causing uniform or differential settlement of foundations, pipelines, or transportation routes.
- Damage to piled foundations as result of loss of support and flow of soil around piles that extend into or through the liquefied soil layer(s).
- Loss of bearing support to shallow or piled foundations causing settlement or tilt.
- Lateral spreads damaging building foundations, bridge piers, highways and railways, river banks, or embankment dams or pipelines.
- Lateral flow failures.
- Embankment failures due to liquefaction of underlying material or embankment fill.
- Service interruption to buildings due to connection damage caused by foundation displacements.
- Increased lateral pressures against retaining structures such as quay walls.
- Induced hazards, such as fire caused by gas pipeline rupture, spread of fire due to interrupted water supply, or floods caused by dam failure.

The severity of the damage is dependent on the strength and duration of the earthquake ground shaking, the thickness of the liquefied layer, the material properties of the liquefied soil, and the proximity of free faces such as slopes or retaining structures, as well as other variables related to the structures. The response of a building to liquefaction-induced vertical or lateral permanent ground deformation depends to a large extent on the building's foundations (**Figures 13–15**). Bridge embankments are vulnerable to lateral spreading due to the combination of susceptible river channel deposits and sloping or free faces. Bridge piers in liquefied ground can rotate, twist, or displace laterally. In extreme cases, when liquefaction is combined with poor design in the form of insufficient bearing width of the bridge deck, excessive movement can cause loss of support to the bridge deck and hence collapse (**Figure 15**). Damage to

**Figure 13** Tilted apartment block in Adapazari, Turkey, due to extensive liquefaction in the 1999 magnitude 7.4 Kocaeli earthquake. This building had thick, continuous, shallow foundations, which caused it to rotate as a rigid body. There is very little damage to load-bearing elements, walls, or ceilings; nonetheless, the building would have been demolished. Photograph by Beyza Taskin.

**Figure 14** Damage to the Marine Laboratory at Moss Landing, Monterey Bay, California, during the 1989 magnitude 6.9 Loma Prieta earthquake. Lateral spreading caused the building pad foundations to spread apart. The laboratory was subsequently demolished. Photograph by Leslie F. Harder, Jr.

roadways or railways due to liquefaction most commonly comprises settlement or lateral spreading of embankments, highway fills, or natural soil, which causes cracking and uneven surfaces. Such damage is relatively easy and inexpensive to repair but, nonetheless, the disruption and the indirect losses must be considered.

The performance of a region's lifelines networks, including water, electricity, and gas distribution systems, is an essential factor in the immediate emergency response and the subsequent recovery from the earthquake impact. Pipeline damage can be induced both by transient ground oscillations and by vertical or lateral permanent ground deformations, which can shear, compress, or pull apart pipelines at their joints. Foundation displacements are likely to damage pipeline connections to buildings, which can lead to building shutdown even in the absence of structural damage. Ruptured gas and water pipelines can trigger fires and impede the control of fires, respectively, creating a significant induced hazard related to liquefaction damage to pipelines. Buried pipelines, as well as other buried vessels such as tanks, can become buoyant in liquefied soils. Ports and harbours tend to be particularly vulnerable to liquefaction damage due to the combination of high water tables, reclaimed soils, and retained vertical faces. Dam failures due to liquefaction at the toe, or flow failure of the embankment, can cause catastrophic damage and massive loss of life as a result of the flooding induced by such a failure (see **Figure 9**). Mitigation of liquefaction hazard in the design of dams is therefore of the utmost importance.

## Mitigation

When the potential liquefaction risk is considered to be unacceptable for the performance requirements of an engineering project, the three principal options are relocation, prevention, and effective design. Relocating a structure or facility to avoid susceptible zones may be the most straightforward option in terms of mitigating the effect of liquefaction. For regionally distributed facilities such as lifeline or transportation networks, this option is unlikely to be practicable. The use of ground improvement to limit or prevent the occurrence of liquefaction has been shown to be effective in many past earthquakes. Ground improvement may involve increasing the density of the liquefiable soil through compaction, vibration, or replacement; reinforcing and densifying the soil through jet grouting or deep soil mixing; or providing additional drainage to allow excess pore water pressures to dissipate more rapidly (**Figure 16**). Often a combination of several techniques is adopted. Again, this is unlikely to be an appropriate solution for regionally distributed facilities.

Accommodation through design is the third option to manage liquefaction risk. Foundations can be designed to withstand expected ground deformations

**Figure 15** This span of the Nishinomaya Bridge collapsed following the 1999 magnitude 6.9 Kobe earthquake in Japan as a result of liquefaction-related foundation deformations. Ground cracks behind the quay walls and parallel to the water edge are indicative of the lateral ground movements that occurred. Sand boils are visible on the ground surface. Photograph by Leslie F. Harder, Jr.

**Figure 16** Installation of stone columns to mitigate liquefaction through densification and strengthening of liquefiable layers (courtesy of Hayward Baker, Inc.).

and to ensure that movements are not translated to the superstructure. Piled foundations can be designed to accommodate additional lateral loads imparted by soil movement, or to have sufficient vertical capacity even in the case of negative skin friction due to settlement. Shallow foundations can be designed to be sufficiently strong so as to behave as a rigid body when subjected to ground deformations. In the case of lifeline networks, the accommodation of the expected movements, and the implementation of appropriate response and repair measures, are frequently the only available solution.

## See Also

**Engineering Geology:** Aspects of Earthquakes; Seismology; Site and Ground Investigation; Site Classification. **Geotechnical Engineering**. **Soil Mechanics**. **Tectonics:** Earthquakes.

## Further Reading

Berrill J and Yasuda S (2002) Liquefaction and piled foundations: some issues. *Journal of Earthquake Engineering* 6(Special Issue 1): 1–41.

Boulanger RW and Idriss IM (2004) *State Normalization of Penetration Resistance and the Effect of Overburden on Liquefaction Resistance,* pp. 484–491. 11th International Conference on Soil Dynamics and Earthquake Engineering and 3rd International Conference on Earthquake Geotechnical Engineering, University of California, Berkeley.

Harder LF, Jr (1997) *Application of the Becker Penetration Test for Evaluating the Liquefaction Potential of Gravelly Soils.* Proceedings of the NCEER Workshop on Evaluation of Liquefaction Resistance of Soils, NCEER-97-0022.

Idriss IM and Boulanger RW (2004) *Semi-empirical Procedures for Evaluating Liquefaction Potential during Earthquakes,* pp. 32–56. 11th International Conference on Soil Dynamics and Earthquake Engineering and 3rd International Conference on Earthquake Geotechnical Engineering, University of California, Berkeley.

Ishihara K (1993) Liquefaction and flow failures during earthquakes. (Rankine Lecture). *Geotechnique* 43: 351–415.

Ishihara K and Yoshimine M (1992) Evaluation of settlements in sand deposits following liquefaction during earthquakes. *Soils and Foundations* 32(No. 1): 173–188.

Kramer SL (1996) Geotechnical earthquake engineering. Upper Saddle River, NJ: Prentice Hall.

National Research Council (1985) *Liquefaction of Soils during Earthquakes. Report by the Participants in the Workshop on Liquefaction, Committee on Earthquake Engineering, National Research Council, Report No. CETS-EE-001.* Washington, DC: National Academy Press.

Seed HB (1987) Design problems in soil liquefaction. *Journal of Geotechnical Engineering, ASCE* 113(8): 827–845.

Seed RB and Harder LF, Jr (1990) SPT-based analysis of cyclic pore pressure generation and undrained residual strength. In: *H. Bolton Seed – Volume 2. Memorial Symposium Proceedings,* pp. 351–376. Vancouver, BC: BiTech Publishers Ltd.

Seed HB and Idriss IM (1971) Simplified procedure for evaluating soil liquefaction potential. *Journal of Soil Mechanics and Foundations Division, ASCE* 97: 1249–1273.

Seed HB and Idriss IM (1982) *Ground Motions and Soil Liquefaction during Earthquakes. Monograph No. 5.* Oakland, CA: Earthquake Engineering Research Institute (Note: A new edition of this monograph is expected in 2005).

Tokimatsu K and Seed HB (1987) Evaluation of settlements in sand due to earthquake shaking. *Journal of Geotechnical Engineering, ASCE* 113(8): 861–878.

Youd TL, Hansen CM, and Bartlett SF (2002) Revised multilinear regression equations for prediction of lateral spread displacement. *Journal of Geotechnical and Geoenvironmental Engineering, ASCE* 128(12): 1007–1017.

Youd TL and Idriss IM (2001) Liquefaction resistance of soils: Summary report from the 1996 NCEER and 1998 NCEER and NSF workshops on evaluation of liquefaction resistance of soils. *Journal of Geotechnical and Geoenvironmental Engineering, ASCE* 127(4): 297–313.

Youd TL and Perkins DM (1978) Mapping liquefaction-induced ground failure potential. *Journal of the Geotechnical Engineering Division, ASCE* 104(GT4): 433–446.

# Made Ground

**J A Charles**, Formerly Building Research Establishment, Hertfordshire, UK

© 2005, Published by Elsevier Ltd.

## Introduction

The term 'made ground' is used to describe ground that has been formed by human activity rather than by natural geological processes. The material of which made ground is composed is described as 'fill', and in practice the terms 'made ground' and 'fill' are often used interchangeably. Fill materials include not only natural soils and rocks but also the waste products of mining and industrial processes, and commercial and domestic refuse.

Throughout history, mankind has deliberately adjusted the topography of the Earth by excavating soil and rock and placing the excavated material in more convenient locations. The casual disposal of waste materials has also changed landforms in urban areas. These processes greatly accelerated during the twentieth century, and in some localities, such as major conurbations and mining areas, a significant proportion of the surface area is now made ground. It is therefore important that the extent, depth, and nature of such deposits are reliably mapped and that the geotechnical behaviour of the fill materials is adequately understood.

A simple classification system has been developed for describing ground affected by human activity, and this classification is used in the geological mapping of the UK.

- Made ground – areas where it is known that fill material has been placed.
- Worked ground – areas where excavations have been made in natural ground.
- Disturbed ground – areas of surface and near-surface disturbance (including ground that has subsided), typically associated with mining.
- Landscaped ground – areas where the ground surface has been remodelled, but made ground and worked ground cannot be distinguished.

A broad distinction can be made between 'engineered fill' and 'non-engineered fill'. The distinction is essentially one of purpose: engineered fill is placed for a specific purpose, whereas non-engineered fill is a by-product of the disposal of waste material. Engineered fill is selected, placed, and compacted to an appropriate specification in order that it will exhibit the required engineering behaviour; engineering design focuses on the specification and control of filling. Non-engineered fill is placed with no subsequent engineering application in view. The distinction between engineered and non-engineered fill is clear in principle, but not always in practice.

## History

Early earthmoving activities were undertaken to meet practical objectives, although in some cases that objective is not obvious to the modern mind. Large-scale earthmoving has provided many durable reminders of previous ages.

Silbury Hill, situated in Wiltshire in the south of England, is the largest prehistoric manmade mound in Europe and a remarkable early civil-engineering achievement. Archaeological evidence suggests that the 40 m high mound (**Figure 1**) is more than 4000 years old. It was carefully engineered in a series of six stepped horizontal layers created from concentric rings of chalk block walls. There are radial as well as circumferential walls, and the compartments between them are filled with chalk rubble. The steps in the outer slope were infilled to give a smooth slope. It has been suggested that the hill is a burial mound, but no graves have been found.

One practical reason for placing fill is to increase the area of land suitable for building on within the confines of a town or city. This has been achieved in hill country by cut-and-fill earthworks on hillsides and in low-lying areas by raising the level of marshy ground. Some 3000 years ago rock-fill platform terracing was formed on the eastern slopes of Jerusalem, thereby substantially increasing the building area. This is thought to be King Solomon's *Millo*, the

**Figure 1** Silbury Hill, England.

construction of which is referred to in *I Kings* 9:15,24 and 11:27.

The construction of embankment dams to retain reservoirs of water has a long history. The 14 m high Sadd-el-Kafara was built with engineered fill in Egypt at the beginning of the age of the pyramids and impounded $0.5 \times 10^6 \, m^3$ of water. The dam had a central core of silty sand supported by shoulders of rock fill. Inadequate provision was made for floods, and the dam was probably overtopped and breached during a flood.

Fills were placed to provide a suitable elevation for defence or for control of the local population. The mound for Clifford's Tower was built in York in 1069 by William I during his campaign to subdue the north of England. The 15 m high mound, which was built of horizontal layers of fill comprising stones, gravel, and clay, was founded on low-lying ground to provide a suitable elevation for the construction of a stronghold. The original tower was made of timber, and the present stone structure known as Clifford's Tower dates from the middle of the thirteenth century. Soon after the erection of the stone tower, severe floods in 1315–1316 softened the fill in the mound, and in 1358 the tower was described as being cracked from top to bottom in two places. These cracks, which were repaired at great expense before 1370, are still visible.

In many parts of the world, low-lying wet ground has been reclaimed by filling, and in the last few centuries this type of land reclamation has taken place on a large scale. Reclamation of the marshes bordering the Baltic Sea, on which St Petersburg is built, began in 1703. On the opposite side of the Atlantic Ocean, much of downtown Manhattan was built on made ground created before 1900. When present-day maps of cities are superimposed on old maps, the extent of such made ground is revealed.

In urban locations where the land has been continuously occupied for centuries, there are likely to be large areas of made ground. Fills have arisen inadvertently from the rubble of demolished buildings and the slow accumulation of refuse. Made ground of this type may contain soil, rubble, and refuse and may be very old. It can be quite extensive in area but is often relatively shallow. Some towns in the Middle East provide examples of the unplanned accumulation of fills. The most common building material was mud brick, and so the walls had to be thick. New construction took place on the ruins of older buildings, and in Syria and Iraq villages stand on mounds of their own making. The ruins of an ancient city may rise 30 m above the surrounding plain.

This gradual rise of debris has been much less common in Great Britain, although in some locations deep fills have accumulated. By the third century AD the Wallbrook in the City of London was already half buried, and mosaic pavements of Roman London lie 8–9 m below the streets of the modern city.

The Royal Scottish Academy in Edinburgh was completed in 1826 on the Mound, which was formed in the late 1700s using clay spoil from the construction of the New Town. The building was founded on square timber piles that, in the course of time, rotted because they were above the water-table, leaving large voids under the stone footings. Remedial works involving compensation grouting were carried out recently.

With the coming of the Industrial Revolution mankind's capacity to generate waste materials, and to cover significant portions of the Earth's surface with them, greatly increased. Where minerals were extracted from underground workings, it was impracticable to avoid extracting quantities of other materials with the desired mineral, and the resulting spoil was brought to the surface and placed in heaps.

The need to supply unpolluted water to the rapidly expanding industrial cities in the north of England led to the construction of large numbers of embankment dams in the nineteenth century. Dale Dyke was one of the dams that was built to supply water to Sheffield. The 29 m high embankment followed the traditional British form of dam construction, with a narrow central core of puddle clay forming the watertight element. The reservoir capacity was $3.2 \times 10^6 \, m^3$. By 10 March 1864, during the first filling of the reservoir, the water level behind the newly built dam was 0.7 m below the crest of the overflow weir. In the late afternoon of 11 March 1864 a crack was observed along the downstream slope near the crest of the dam. At 23.30 the dam was breached, and the resulting flood destroyed property estimated to be worth half a million pounds sterling, and caused the loss of 244 lives. Developments in geotechnical engineering in the twentieth century have enabled safe embankment dams to be built with confidence.

## Twentieth Century

The twentieth century saw a massive expansion of made ground. Large-scale earthmoving machinery made it possible to place fill rapidly and cheaply in quantities never before experienced. This applied both to engineered fills placed to construct embankment dams, road embankments, and sites for buildings, and to non-engineered fills placed as mining, industrial, chemical, building, dredging, commercial, and domestic wastes. **Table 1** provides details of some fills placed over the last 4000 years and illustrates how the

**Table 1** Some examples of made ground

| Structure | Location | Purpose | Date built | Height or depth (m) | Volume ($10^6 m^3$) | Surface area (ha) |
|---|---|---|---|---|---|---|
| Silbury Hill | Wiltshire, England | Unknown | pre-2000 BC | 40 | 0.25 | 2 |
| Sadd-el-Kafara | Egypt | Retain water | pre-2000 BC | 14 | 0.09 | 2 |
| Clifford's Tower mound | York, England | Military | AD 1069 | 15 | 0.04 | 0.4 |
| Dale Dyke dam | Sheffield, England | Retain water | AD 1864 | 29 | 0.4 | 4 |
| Fort Peck dam | Montana, USA | Retain water | AD 1940 | 76 | 96 | 200 |
| Scammonden dam | Huddersfield, England | Retain water + motorway embankment | AD 1969 | 76 | 4.3 | 15 |
| Tarbela dam + upstream blanket | Pakistan | Retain water | AD 1976 | 148 | 118 | 700 |
| Nurek dam | Tajikistan | Retain water | AD 1980 | 300 | 58 | 60 |
| Gilow impoundment | South-western Poland | Retain copper tailings | AD 1980 | 22 | 68 | 540 |
| Lounge opencast site[a] | Ashby-de-la-Zouch, England | Engineered fill for road embankment | AD 1990 | 40 | 3.7 | 15 |
| Dixon opencast site | Chesterfield, England | Backfill | AD 1992 | 74 | 16 | 119 |
| Kansai airport (Phase 1) | Osaka, Japan | Island for airport | AD 1994 | 33 | 180 | 510 |
| Chek Lap Kok airport | Hong Kong | Platform for airport | AD 1996 | 25 | 194 | 1248 |

[a]Data refers only to that part of the backfill that was placed as an engineered fill.

scale of operations vastly increased in the twentieth century. Cross-sections of a number of the earth structures included in **Table 1** are shown in **Figure 2**.

Although urban redevelopment has continued throughout history, modern programmes of urban regeneration are carried out at a rate and on a scale not seen before. Much of this redevelopment is carried out on made ground.

The reclamation of land from the sea can be achieved either by the construction of water-retaining embankments, which prevent the sea flooding land below sea-level, as in the Netherlands, or by the placement of fill to form made ground whose surface is above sea-level. Examples of the latter approach include the massive reclamation projects carried out in Hong Kong, Japan, and Singapore for the construction of new airports. Kansai airport is located 5 km off the Japanese mainland, and the placement of a 33 m thickness of fill to form the island has caused the underlying deposits to settle by 14 m. The construction of Phase 2 of Kansai airport has commenced, and this will increase the surface area of the made ground to 1100 ha.

The number and size of embankment dams increased greatly during the twentieth century. Fort Peck dam in the USA was built of hydraulically placed earthfill and consequently has flat slopes and contains a large volume of fill. When the embankment was nearly complete, there was a slide in the upstream slope involving $4 \times 10^6 m^3$ of fill. At the time of its completion in 1940 it was the largest dam in the world, and it remained so until Tarbela dam was completed in 1976 in Pakistan. Tarbela has an impervious

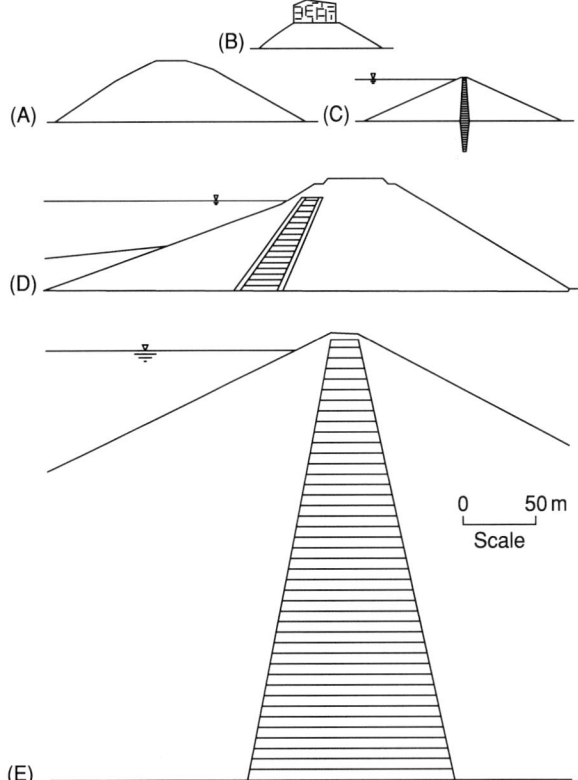

**Figure 2** Fill structures: (A) Silbury Hill, England; (B) Clifford's Tower mound, England; (C) Dale Dyke dam, England; (D) Scammonden dam, England; and (E) Nurek dam, Tajikistan.

upstream blanket, which continues for 2 km upstream of the upstream toe of the dam. The 300 m high Nurek dam in Tajikistan was completed in 1980 and is one of the world's highest dams.

The 76 m high Scammonden dam, which was completed in 1969, is the highest embankment dam in England. The embankment was designed to serve the dual purpose of impounding a reservoir with a capacity of $7.9 \times 10^6\,\text{m}^3$ and carrying the M62 trans-Pennine motorway across the Scammonden valley. A cross-section of the embankment is shown in **Figure 2D**. The shoulders were built of compacted sandstone and mudstone rock fill, and it has an upstream-sloping rolled clay core in the upstream part of the embankment.

In metal mining, rock is crushed to extract the desired mineral, leaving large amounts of crushed rock as a waste material. The fine waste material, which is known as tailings, is discharged from the wet process as a saturated slurry and commonly pumped through a pipeline from the plant to an impoundment formed by an embankment dam. Large waste-impounding embankments have been built in the twentieth century, some more than 150 m high and some impounding more than $10^8\,\text{m}^3$ of waste materials. The Gilow impoundment in Poland has a surface area of 540 ha. The principal hazard posed by such dams and their retained waste impoundments is the risk of rapid discharge of the impounded waste material in an uncontrolled manner should there be a breach of the embankment. There have been a number of failures of large tailings dams in recent years, and some waste materials pose major hazards to the natural environment; for example, the breach on 24 April 1998 of the Aznalcollar tailings dam at Seville in Spain released $7 \times 10^6\,\text{m}^3$ of mining waste and water, threatening the Coto de Donana National Park.

Where iron ore, coal, and other minerals are found in thin seams not too far below the surface, they can be won by opencast or strip mining; that is, the overlying soil or rock is excavated to reach the mineral seams without the need for subsurface tunnelling. When a mineral has been extracted by opencast mining, the overburden soils and rocks are replaced in the excavation, and these mining operations have been major producers of deep non-engineered fills in many parts of the world. By 1986 more than 1000 residential buildings and farms had been established on deep uncompacted backfills in the Rhenish brown coal area of Germany. During the 1980s opencast coal production in Great Britain was about 14 million tonnes per annum, and the extraction of one tonne of coal typically involved the excavation of $15\,\text{m}^3$ of overburden. Sites were often restored to agricultural use without compaction during backfilling of the opencast excavation, but latterly systematic compaction has become more common. Two British opencast sites are included in **Table 1**.

Waste from the deep mining of coal, which is known as colliery spoil, is derived from the rocks adjacent to the coal seams, and during mining operations quantities of these rocks, unavoidably extracted with the coal or in driving the tunnels that give access to the coalface, are brought to the surface. Towards the end of the twentieth century, world coal production approached $5 \times 10^9$ tonnes per annum.

The coarse discard from coal mining used to be dumped in heaps, which could become very large. Following the Aberfan disaster in Wales in 1966, when the failure of an unstable colliery spoil heap caused great loss of life, it became the usual practice in Great Britain to place the spoil in thin layers with compaction. Geotechnical problems can be largely overcome by adequate compaction during placement. By 1974 there were $3 \times 10^9$ tonnes of colliery spoil in Great Britain, and in 1984 the coal mining industry produced $5 \times 10^7$ tonnes of coarse spoil annually, which was placed in tips adjacent to the collieries. In April 1988 there were 4700 ha of derelict land associated with colliery spoil heaps in England alone. The rapid decline of the coal industry in the 1990s meant that the annual production of these wastes reduced, but large stocks of colliery spoil remain in the coalfields. In 1996 it was planned to reclaim 900 ha of colliery land for residential, commercial, and retail uses.

A major proportion of the domestic waste generated in the UK is disposed of in landfill. In 1986, over 90% of controlled domestic, commercial, and industrial solid wastes (excluding mining and quarrying wastes) were disposed of by landfilling. In the year 1986/1987 nearly $2 \times 10^7$ tonnes of household or domestic waste were disposed of in England and Wales, and this figure did not change substantially over the subsequent 10 years. Despite environmental initiatives, landfilling has continued on a large scale. In the USA at the end of the twentieth century $1.2 \times 10^8$ tonnes of municipal solid waste were landfilled each year.

## Functions of Made Ground

The function of non-engineered fills is to dispose of unwanted waste material. By contrast, the large quantities of fill material that are placed as engineered fills form part of carefully controlled civil-engineering works. Three major types of engineered fills are embankment dams, road embankments, and fills that support buildings.

As we have seen, embankment dams are constructed from fill and are built to retain reservoirs of water, which may be required for hydropower, water supply to towns, irrigation, or flood control. Similar embankments can be used to retain canals. An embankment dam will usually be composed of several different types of fill: a low-permeability fill to form the watertight element, stronger fill to support the watertight element, and fills to act as filters, drains, and transition materials. Possible hazards affecting the dam include internal erosion, slope instability, and overtopping during floods.

Embankment dams are also used to retain lagoons of sedimented waste material from mining and industrial activities. If the waste is not toxic, these embankments may be designed so as to allow water to drain through. The embankment may be built in stages as waste disposal progresses.

Another use for a special type of embankment is to protect harbours and the shoreline from the sea. Such embankments are often built as rubble-mound breakwaters and require high-quality quarried rock for the fill. Placement to the required profile presents obvious difficulties.

Embankments to carry roads and railways are usually built to reduce gradients. The engineered fills use material excavated from adjacent cuttings during the cut-and-fill operations, so there is limited scope for material selection and whatever is excavated has to be placed in the adjacent embankment unless it is clearly unsuitable. In England, during the construction of the M6 motorway between Lancaster and Penrith in the 1960s, much of the soil excavated along the line of the road was very wet, and a geotechnical design was developed that involved the use of drainage layers built into the embankment during construction.

Buildings may be founded on made ground. Old excavations are infilled, and sometimes embankments are built above the level of the surrounding ground. The objective is to support buildings safely while minimizing the risk of damaging settlement, and, where structures sensitive to settlement are to be built on made ground, a high-quality fill that is not vulnerable to large post-construction movement is required.

Opencast-mining sites have often been restored for agricultural use, but where the sites are close to urban areas they may subsequently be used for housing and commercial developments. Loose backfill usually has considerable settlement potential, and some existing areas of loose fill have been improved by preloading with a temporary heavy surcharge of fill. Because of the free-draining nature of the loose fill, it soon consolidates, and this type of treatment has made some sites quite suitable for normal housing. Where building is foreseen prior to backfilling an opencast mining site, the fill material should be placed in thin layers and heavily compacted as an engineered fill.

## Fill Placement

There are two basic elements in the quality management of engineered fills: placement of a fill with the required quality; and evidence that the fill has the required quality. Both an appropriate specification and rigorous quality-control procedures are required. It is not possible to prevent some variability in the made ground, as there will be a degree of heterogeneity in the source material and some segregation during placement. It is necessary to determine how the required properties can be achieved with an acceptable degree of uniformity.

Placement in thin layers with heavy compaction at an appropriate water content is the method usually adopted to obtain the required performance from an engineered fill used in dam, road, or foundation applications. **Figure 3** shows the compaction of a clay fill, and **Figure 4** shows a rock fill being watered during placement. There are three basic approaches to the specification of engineered fills.

**Figure 3** Compacting clay fill.

**Figure 4** Watering rock fill during placement.

- In a method specification, the procedure for placement and compaction is described. The type and mass of the compactor, the number of passes, and the layer thickness are specified. Reliance is placed on close inspection to ensure compliance with the specification.
- An end-product specification is based on required values for properties of the fill as it is placed. The basic measurements of the *in situ* state of compaction are density and water content, but these measurements on their own are not adequate indicators – the density needs to be interpreted in terms of the density at a specified water content under some standard type of compaction. For a clay fill, the specification could be in terms of percentage air voids or undrained shear strength. Compliance is tested as filling progresses.
- In a performance specification some facet of the post-construction behaviour of the fill is specified, such as a permissible post-construction settlement or a load-test result. With this approach the specification is directly related to one or more aspects of the performance requirements.

Non-engineered fills have usually been placed in the simplest and cheapest way feasible. The fill may have been tipped in high lifts with no systematic compaction. A non-engineered fill can be effectively converted to an engineered fill by *in situ* ground treatment subsequent to the completion of fill placement, although the limitations of such post-construction treatment should be recognized, since unsuitable fill material is still present after treatment and most forms of compaction applied to the ground surface are effective only to a limited depth.

Natural fine soils or waste materials can be mixed with enough water to enable them to be transported in suspension. Usually the suspension is pumped through a pipe and then discharged onto the surface being filled. The deposit is described as hydraulic fill. **Figure 5** shows a lagoon of pulverized fuel ash; this waste product from coal-fired power stations has been mixed with sufficient water to enable it to be pumped to the lagoon.

Where a waste material contains particles of different sizes, segregation may occur during deposition. As the suspension flows away from the discharge point, the larger soil particles settle out almost immediately and the water and fines flow away. Eventually, the fine material also settles out but at a much greater distance from the discharge point. The placement of hydraulic sand fills beneath water involves settling from a slurry, and such subaqueous hydraulic fills can be placed by bottom dumping from a barge or by pipeline placement.

**Figure 5** Lagoon filled with pulverized fuel ash.

## Fill Properties and Behaviour

The required behaviour of made ground is closely linked to the purpose of the engineered fill and the processes and hazards to which it may be exposed. There may be performance requirements that can be expressed in terms of a wide range of geotechnical properties, such as shear strength, stiffness, compressibility, and permeability.

Made ground exhibits as wide a range of engineering properties as does natural ground; both the nature of the fill material and the mode of formation have a major influence on subsequent behaviour. Needless to say, there is a vast difference between the behaviour of an engineered heavily compacted sand-and-gravel fill and that of recently placed domestic refuse. Made ground may have been formed by fill placed in thin layers and heavily compacted or by fill end-tipped in high lifts under dry conditions or into standing water. The method of placement affects the density of the fill and the homogeneity of the made ground. Many of the problems with non-engineered fills are related to their heterogeneity. Depending on the method of placement and the degree of control exercised during placement, there may be variability in materials, density, and age.

The engineering properties of hydraulic fills can be expected to be very different from the properties of fills placed at lower water contents under dry conditions. Hydraulic transport and deposition of materials generally produces fills with a high water content in a relatively loose or soft condition. Segregation of different particle sizes is likely.

The amount of densification of a clayey soil that can be achieved for a given compactive effort is a function of water content. Laboratory compaction-test results are plotted as the variation of dry density with water content. Such plots are a simple and useful way of representing the condition of a partially

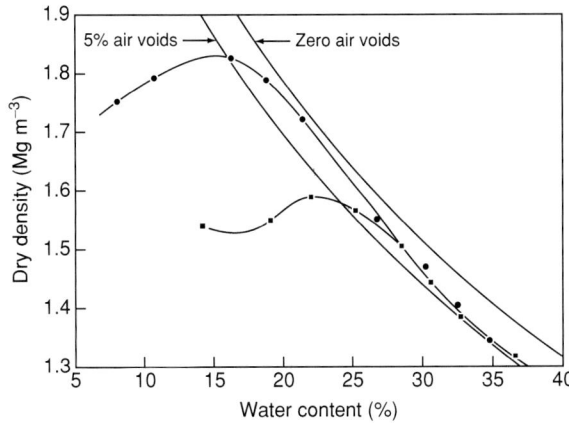

**Figure 6** Laboratory compaction of clay fill using two different compactive efforts. ●, Heavy compaction (4.5 kg rammer); ■, Standard Proctor compaction (2.5 kg rammer).

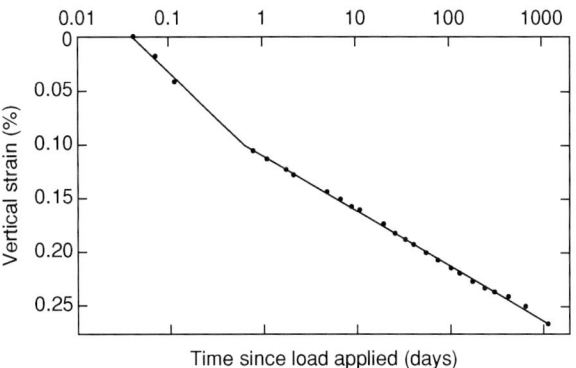

**Figure 7** Creep compression of heavily compacted sample of sandstone rock fill.

saturated fill and can be used to identify two parameters of prime interest – the maximum dry density and the corresponding optimum water content. The terms maximum dry density and optimum water content refer to a specified compaction procedure and can be misleading if taken out of the context of that procedure. **Figure 6** shows laboratory test results using two different compactive efforts. The heavier compaction produces a greater maximum dry density at a lower optimum water content. At a water content less than optimum, the specified compaction procedure may result in a fill with large air voids. At a water content significantly higher than optimum, the specified compaction procedure should produce a fill with a minimum of air voids, typically between 2% and 4%.

Where made ground is built upon, the magnitude and rate of long-term settlement is an important issue. Fill will undergo some creep compression, which occurs without changes in the load applied to the fill or the water content of the fill. For many fills there is a linear relationship between creep compression and the logarithm of the time that has elapsed since the load was applied. This type of relationship can be characterized by the parameter $\alpha$, which is the vertical compression occurring during a log cycle of time (i.e. between, say, 1 year and 10 years). For loosely placed coarse fills, $\alpha$ is typically between 0.5% and 1%. For heavily compacted fills, $\alpha$ is much smaller and strongly dependent on stress. **Figure 7** shows the rate of compression of a heavily compacted sandstone rock fill measured in a large laboratory oedometer test under a constant vertical applied stress of 0.7 MPa. From 1 day to 1000 days after the load was applied there was a linear relationship between creep compression and the logarithm of elapsed time, corresponding to $\alpha = 0.05\%$.

When first inundated, most types of partially saturated fill are susceptible to collapse compression if they have been placed in a relatively loose or dry condition. This reduction in volume can occur under a wide range of stresses and without a change in applied total stress. For buildings on fill, the risk of collapse compression is normally the main concern; where it occurs after construction has taken place, buildings may be seriously damaged. The causes of collapse compression on inundation fall into three categories: weakening of interparticle bonds; weakening of particles in coarse fills; and softening of lumps of clay in fine fills. Inundation can result from a rising groundwater table or from downward infiltration of surface water. An important objective of the specification and control procedures adopted for fills is to eliminate or at least minimize the potential for collapse compression.

## Future Trends

It may be questioned whether the massive increase in the generation of fill materials that occurred in the twentieth century can or should continue. Large-scale earthmoving, which is visually intrusive and involves major modifications to the environment, is meeting increasing opposition, particularly where large dams are built. The principal objection to new dams has centred on the flooding of large tracts of land, but all engineering projects involving major earthmoving are likely to face similar opposition. While it is not unreasonable for earthmoving projects to be subjected to close scrutiny, it is important that the benefits conferred by many types of made ground are not overlooked.

Carefully controlled engineered fill is likely to form only a small fraction of new made ground, the bulk of which will be created by the landfilling of various types of waste material. Every year, billions of tonnes of solid waste are generated by industrial and mining activities and deposited either as spoil heaps on the surface of the ground or as waste dumps within excavations. Mining has brought much prosperity to human societies, but it is now widely questioned whether these activities, which are being carried out on an unprecedented scale, are sustainable and whether such massive quantities of waste are environmentally acceptable. Despite the huge rate of consumption during the twentieth century, the known resources of many minerals have increased rather than diminished as new sources have been identified. It seems probable that, for the foreseeable future, the needs of successive generations will require the extraction of minerals from the surface of the Earth on a large scale and that there will be sufficient resources to meet this demand. These mining operations will mean that spoil heaps and waste dumps will cover ever greater portions of the Earth's surface.

While increased concern over environmental issues is not likely to halt the industrial and mining activities that generate waste, it will have a major effect on waste disposal. In most countries, plans for reclamation of the land must be approved before mining begins, and it is usually required that the land will be restored as closely as possible to its original state. Environmental factors will increasingly influence how, where, and in what form waste fills are deposited and the treatment that such deposits receive. There will be increasing pressure to reduce the amount of domestic waste that is placed as landfill.

## See Also

**Engineering Geology:** Geological Maps; Natural and Anthropogenic Geohazards; Liquefaction; Site and Ground Investigation; Subsidence. **Environmental Geology**. **Geotechnical Engineering**. **Soil Mechanics**. **Urban Geology**.

## Further Reading

Charles JA and Watts KS (2001) *Building on Fill: Geotechnical Aspects*, 2nd edn. Building Research Establishment Report BR 424. Garston: BRE Bookshop.

Charles JA and Watts KS (2002) *Treated Ground: Engineering Properties and Performance.* Report C572. London: CIRIA.

Clarke BG, Jones CJFP, and Moffat AIB (eds.) (1993) *Engineered Fills.* Proceedings of conference held in Newcastle-upon-Tyne, September 1993. London: Thomas Telford.

Grace H and Green PA (1978) The use of wet fill for the construction of embankments for motorways. In: *Clay Fills.* Proceedings of conference held in London, November 1978, pp. 113–118. London: Institution of Civil Engineers.

Johnston TA, Millmore JP, Charles JA, and Tedd P (1999) *An Engineering Guide to the Safety of Embankment Dams in the United Kingdom,* 2nd edn. Building Research Establishment Report BR 363. Garston: BRE Bookshop.

Lomborg B (2001) *The Skeptical Environmentalist: Measuring the Real State of the World.* Cambridge: Cambridge University Press.

Matsui T, Oda K, and Tabata T (2003) Structures on and within man-made deposits – Kansai airport. In: Vanicek I, Barvinek R, Bohac J, Jettmar J, Jirasko D, and Salak J (eds.) *Geotechnical Problems with Man-made and Man Influenced Ground.* Proceedings of the 13th European Conference on Soil Mechanics and Geotechnical Engineering, Prague, vol. 3, pp. 315–328. Prague: Czech Geotechnical Society.

Parsons AW (1992) *Compaction of Soils and Granular Materials: A Review of Research Performed at the Transport Research Laboratory.* London: HMSO.

Penman ADM (2002) Tailings dam incidents and new methods. In: Tedd P (ed.) *Reservoirs in a Changing World.* Proceedings of the 12th Conference of the British Dam Society, Dublin, pp. 471–483. London: Thomas Telford.

Perry J, Pedley M, and Reid M (2001) *Infrastructure Embankments – Condition Appraisal and Remedial Treatment.* CIRIA Report C550. London: CIRIA.

Proctor RR (1933) Fundamental principles of soil compaction. *Engineering News Record* 111: 245–248.

Rosenbaum MS, McMillan AA, Powell JH, *et al.* (2003) Classification of artificial (man-made) ground. *Engineering Geology* 69: 399–409.

Schnitter NJ (1994) *A History of Dams: The Useful Pyramids.* Rotterdam: Balkema.

Trenter NA and Charles JA (1998) A model specification for engineered fills for building purposes. *Proceedings of the Institution of Civil Engineers, Geotechnical Engineering* 119: 219–230.

# Problematic Rocks

F G Bell, British Geological Survey, Keyworth, UK

© 2005, Elsevier Ltd. All Rights Reserved.

## Introduction

As far as engineering behaviour is concerned, a distinction has to be made between rock as a material and the rock mass. 'Rock' usually refers to the intact rock, which may usually be considered as a continuum, that is, as a polycrystalline solid consisting of an aggregate of minerals or grains with void or pore space. The properties of intact rock are governed by the physical properties of the materials of which it is composed and by the manner in which they are bonded together. The properties that influence the engineering behaviour of rock material therefore include its mineralogical composition, texture, fabric, minor lithological characteristics, degree of weathering or alteration, density, porosity, strength, hardness, intrinsic (or primary) permeability, seismic velocity, and modulus of elasticity. Swelling and slaking are taken into account where appropriate, for example in argillaceous rocks.

On the other hand, a 'rock mass' includes the fissures and flaws as well as the rock material and may be regarded as a discontinuum of rock material transected by discontinuities. A discontinuity is a plane of weakness within the rock mass, across which the rock material is structurally discontinuous. Although discontinuities are not necessarily planes of separation, most of them are, and they possess little or no tensile strength. Discontinuities vary in size from small fissures to huge faults. The most common discontinuities in all rocks are joints and bedding planes. Other important discontinuities are planes of cleavage and schistosity, which occur in some metamorphic rock masses, and lamination, in some sedimentary rock masses. Obviously, discontinuities will have a significant influence on the engineering behaviour of rock in the ground. Indeed, the behaviour of a rock mass is, to a large extent, determined by the type, spacing, orientation, and characteristics of the discontinuities present. As a consequence, the parameters that should be used when describing a rock mass include the nature and geometry of the discontinuities, as well as overall strength, deformation modulus, secondary permeability, and seismic velocity of the rock mass.

## The Influence of Weathering on Engineering Behaviour

The process of weathering represents an adjustment of the constituent minerals of a rock to the conditions prevailing at the surface of the Earth (see Weathering). Importantly, in terms of engineering behaviour, weathering weakens the rock fabric and exaggerates any structural discontinuities, thereby further aiding the breakdown processes. Rock may become more friable as a result of the development of fractures both between and within grains. Also, some weathered material may be removed, leaving a porous framework of individual grains.

Weathering is controlled by the presence of discontinuities because these provide access for the agents of weathering. Some of the earliest effects of weathering are seen along discontinuity surfaces. Weathering then proceeds inwards, so that the rock mass may develop a marked heterogeneity, with cores of relatively unweathered material within a highly weathered matrix (Figure 1). Ultimately, the whole of the rock mass can be reduced to a residual soil.

Weathering generally leads to a decrease in density and strength and to an increase in the deformability of the rock mass. An increase in the mass permeability frequently occurs during the initial stages of weathering owing to the development of fractures, but, if clay material is produced as minerals break down, the permeability may be reduced. Widening of discontinuities and the development of karstic features in carbonate rock masses lead to a progressive increase

**Figure 1** Weathered basalt showing corestones with spheroidal weathering; basalt plateau, northern Lesotho.

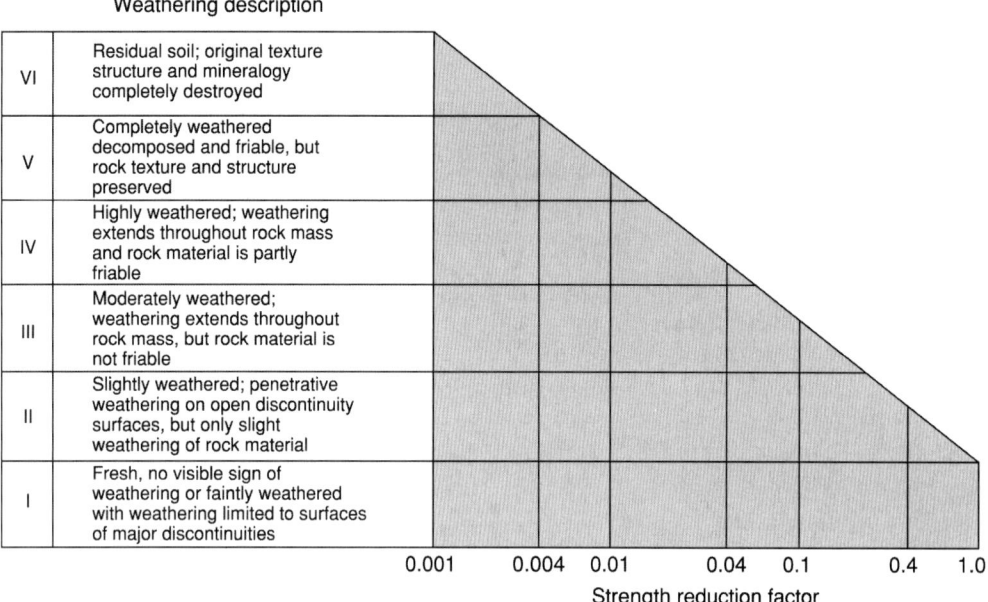

**Figure 2** Generalized strength reduction due to increasing grade of weathering.

in secondary permeability. Weathering of carbonate and evaporite rock masses takes place primarily by dissolution. The dissolution is concentrated along discontinuities, which are thereby enlarged. Continuing dissolution ultimately leads to the development of sinkholes and cavities within the rock mass.

The grade of weathering refers to the stage that the weathering has reached. This varies between fresh rock on the one hand and residual soil on the other. Generally, about six grades of weathering are recognized, which, in turn, can be related to engineering performance (**Figure 2**). One of the first classifications of the engineering grade of weathering was devised for granite found in the vicinity of the Snowy Mountains hydroelectric power scheme in Australia. Subsequently, similar classifications have been developed for different rock types. Usually, rocks of the various grades lie one above the other in a weathered profile developed from a single rock type, the highest grade being found at the surface (**Figure 3**). However, this is not necessarily the case in complex geological conditions. Such a classification can be used to produce maps, sections, or models showing the distribution of the grade of weathering at a particular site. The dramatic effect of weathering on the strength of a rock is illustrated, according to the grade of weathering, in **Figure 2**.

## Igneous and Metamorphic Rocks

Intrusive igneous rocks in their unaltered (unweathered) state are generally sound and durable, with adequate strength for any engineering requirement (**Table 1**). In some instances, however, they may be highly altered by weathering or hydrothermal processes (**Figure 4**). Furthermore, fissure zones are by no means uncommon in granites. Such granite rock masses may be highly fragmented along these zones; indeed, the granite may be reduced to sand-sized material and/or have undergone varying degrees of kaolinization (development of clay minerals). Generally, the weathered products of plutonic rocks have a large clay content. Granitic rocks can sometimes be porous and may have a permeability comparable with that of medium-grained sand. Some saprolites derived from granites that have been weathered in semi-arid climates may develop a metastable fabric and therefore be potentially collapsible. They may also be dispersive. Joints in plutonic rocks are often quite regular steeply dipping structures. Sheet joints tend to be approximately parallel to the topographical surface and develop as a result of stress relief following erosion (**Figure 5**). Consequently, they may introduce a dangerous element of weakness into valley slopes. The engineering properties and behaviour of gneisses are similar to those of granites.

Turning to extrusive igneous rocks, generally speaking, older volcanic rocks are not problematical; for instance, ancient lavas normally have strengths in excess of 200 MPa (**Table 1**). However, younger volcanic deposits have, at times, proved treacherous. This is because they often consist of markedly anisotropic sequences, in which lavas (generally strong), pyroclastics (generally weak or loose), and mudflows

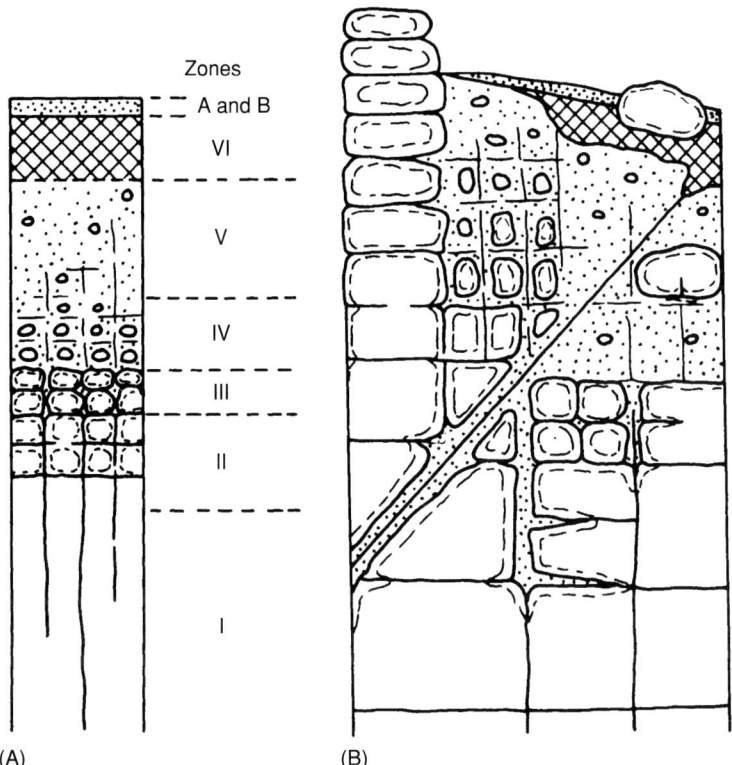

**Figure 3** Weathering profile: (A) idealized profile; and (B) more complex profile.

**Table 1** Geotechnical properties of some British igneous and metamorphic rocks

| | Specific gravity | Unconfined compressive strength (MPa)[a] | Young's modulus (GPa)[b] |
|---|---|---|---|
| Mount Sorrel Granite (Leicestershire) | 2.68 | 176.4 (VS) | 60.6 (VL) |
| Eskdale Granite (Cumbria) | 2.65 | 198.3 (VS) | 56.6 (L) |
| Dalbeattie Granite (Kirkcudbrightshire) | 2.67 | 147.8 (VS) | 41.1 (L) |
| Markfieldite (Leicestershire) | 2.68 | 185.2 (VS) | 56.2 (L) |
| Granophyre (Cumbria) | 2.65 | 204.7 (ES) | 84.3 (VL) |
| Andesite (Somerset) | 2.79 | 204.3 (ES) | 77.0 (VL) |
| Basalt (Derbyshire) | 2.91 | 321.0 (ES) | 93.6 (VL) |
| Slate[c] (North Wales) | 2.67 | 96.4 (S) | 31.2 (L) |
| Slate[d] (North Wales) | — | 72.3 (S) | — |
| Schist[c] (Aberdeenshire) | 2.66 | 82.7 (S) | 35.5 (L) |
| Schist[d] (Aberdeenshire) | — | 71.9 (S) | — |
| Gneiss (Aberdeenshire) | 2.66 | 162.0 (VS) | 46.0 (L) |
| Hornfels (Cumbria) | 2.68 | 303.1 (ES) | 109.3 (VL) |

[a]Classification of strength: ES, extremely strong, over 200 MPa; VS, very strong, 100–200 MPa; S, strong, 50–100 MPa.
[b]Classification of deformability: VL, very low, over 60 GPa; L, low, 30–60 GPa.
[c]Tested normal to cleavage or schistosity.
[d]Tested parallel to cleavage or schistosity.

are interbedded. Hence, foundation problems arise in volcanic sequences because weak beds of ash, tuff, and mudstone (formed from lahars) occur within lava piles, giving rise to problems of differential settlement and sliding. In addition, some volcanic materials weather relatively rapidly, so that weathering during periods of volcanic inactivity may readily lead to weakening and the development of soils, which are of much lower strength.

Clay minerals may be formed within newly erupted basaltic rocks by the alteration of primary minerals. Their presence can mean that the parent rhyolite, andesite, or basalt breaks down rapidly once exposed. The disintegration is exacerbated by the swelling of

**Figure 4** Weathered granite grading into soil and showing a large corestone; near Hillcrest, Natal, South Africa.

**Figure 6** Crazing developed in basalt on exposure in the Transfer Tunnel; Lesotho Highlands Water Scheme.

**Figure 5** An inselberg north of Mbabane, Swaziland, showing sheet joint up.

**Figure 7** Amygdales and pipes in basalt, exposed in the Transfer Tunnel; Lesotho Highlands Water Scheme.

expansive clay minerals as they absorb water. This breakdown process has been referred to as slaking. These expansive clay minerals form when basic volcanic glass, olivine, pyroxene, or plagioclase is subjected to deuteric alteration brought about by hot gases and fluids from a magmatic source migrating through the rock. Disintegration can also be brought about by the absorption of water by zeolites, minerals that commonly develop in groundwater and are present in gas bubbles (amygdales). However, the breakdown of basalt also depends on its texture, since water must have access to the minerals that swell as water is absorbed. The disintegration of some basalts may take the form of crazing, an extensive microfracturing that develops on exposure to the atmosphere or to moisture (**Figure 6**). Such microfractures expand with time, causing the basalt to disintegrate into gravel-sized fragments. Some dolerites – intrusive rocks with a similar mineral composition to basalt – also undergo slaking.

Individual lava flows may be thin and transected by a polygonal pattern of cooling joints. They may also be vesicular or contain pipes, cavities, or even tunnels (**Figure 7**).

Pyroclastics (*see* **Pyroclastics**) usually give rise to extremely variable ground conditions owing to wide variations in their strength, durability, and permeability. Ashes tend to be weak and are often highly permeable. Those that are metastable are likely to undergo hydrocompaction on saturation. Moreover, ashes deposited on slopes are frequently prone to sliding, yet the irregular shapes of their constituent particles, which can therefore interlock, may enable very steep slopes to be excavated that can stand up, at least in the short term. The strength and behaviour of tuffs depend on their degree of induration. However, the durability of some basaltic tuffs is poor or very poor, and they may be susceptible to frost. An ignimbrite is a pyroclastic rock consisting predominantly of pumiceous material that shows evidence of having been formed from a hot and concentrated pyroclastic flow. Once deposited, induration may be brought about by welding of viscous glassy fragments, by devitrification of glassy material, by deposition of material from escaping gases, and by compaction. Accordingly, ignimbrites have a wide variety of geotechnical characteristics, which are attributable to their modes of eruption, transportation, and deposition. At one extreme they are weak materials that behave as soils in the engineering sense; at the other extreme they are strong hard rocks in which extensive sets of essentially vertical cooling joints are developed. In fact, non-durable, intermediate, and highly durable ignimbrites have been recognized. The non-durable ignimbrites are characterized by low densities, high porosities, and low unconfined compressive strengths (5 MPa or less). A hyaloclastite consists of a mixture of rock and glass fragments. Initially, hyaloclastites are loose deposits, but as a result of weathering and diagenesis they usually become harder, particularly due to palagonitization, which is a solution–precipitation mechanism whereby the glass is hydrated and ions are leached out of it. The process slows down as the precipitation of authigenic minerals reduces the porosity and permeability of the rock mass. An increase in strength due to palagonitization can occur quickly; for example, loose pyroclastic material can be transformed into quite hard compact rocks within 20 years. Weaker varieties are probably more frequent than stronger types, often being hidden below the palagonized hyaloclastites that form the uppermost altered beds.

Slates, phyllites, and schists are characterized by textures that have a marked preferred orientation. Anisotropic structures such as cleavage and schistosity, attributable to such textures, not only adversely affect the strengths of metamorphic rocks but also make them much more susceptible to decay. Generally speaking, slates, phyllites, and schists weather relatively slowly, but the areas of regional metamorphism in which they occur have been subject to extensive folding so that, in places, such rocks may be extensively fractured and highly deformed. The quality of schists, slates, and phyllites is generally suspect. Care must be taken to detect weaker components; for instance, talc, chlorite, and sericite schists are weak rocks containing planes of schistosity only 1 mm or so apart. Some schists become slippery upon weathering and can therefore fail under a moderately light load.

## Sandstones

Sandstones (*see* **Sedimentary Rocks:** Sandstones, Diagenesis and Porosity Evolution) exhibit a wide range of strengths (from less than 5.0 MPa to over 150 MPa), depending on their porosity, the amount and type of cement and/or matrix material, and the composition and texture of the individual grains. Higher cement or matrix content and lower porosity are characteristic of the stronger sandstones. In addition, their strength, like that of other rocks, is reduced by moisture content. The strength of saturated sandstone may be half of what it is when dry. For instance, the Kidderminster Sandstone (Triassic) has an average dry unconfined compressive strength of around 2.5 MPa, but when saturated this may be reduced to as low as 0.5 MPa. Nevertheless, sandstones generally do not give rise to notable engineering problems. Indeed, sandstones usually have sufficiently high coefficients of internal friction to give them high shearing strength when restrained under load. As a foundation rock, even poorly cemented sandstone is not normally susceptible to plastic deformation. Moreover, with the exceptions of shaly sandstone and rocks where the cement is readily soluble, clastic rocks are not subject to rapid deterioration on exposure. Nonetheless, salt action can give rise to honeycomb weathering in sandstone, which can cause relatively rapid disfigurement and deterioration when sandstone is used as a building stone. The process involves the progressive development of closely spaced cavities in the rock. Individual cavities range from a few millimetres to several centimetres in diameter, although larger cavernous weathering features, termed tafoni, can develop in sandstone exposures.

The presence of discontinuities can obviously adversely affect the behaviour of sandstone, reducing its mass strength. When inclined, discontinuities may cause rock to slide into unprotected excavations. Laminations impart a notable anisotropy to sandstone, reducing its strength to a significant degree along the planes of lamination. Furthermore, in certain engineering situations additional problems can develop. Friable sandstones, for example, can introduce problems of scour within dam foundations. Sandstones are also highly vulnerable to scouring and plucking actions in the overflow from dams and consequently have to be adequately protected by suitable hydraulic structures such as stilling basins. Quartzose sandstone in a tunnel being excavated by a tunnel-boring machine can prove highly abrasive to the cutting head, and the cuttings produce an abrasive slurry when mixed with water.

Sandstones are frequently interbedded with shale. When such a formation is inclined, the layers of shale may represent potential sliding surfaces. Sometimes such interbedding accentuates the undesirable properties of the shale by concentrating water along the sandstone–shale contacts, thereby further weakening the shale.

## Mudrocks

Mudrock is the commonest sedimentary rock, the two principal types being shale and mudstone. Shale is characterized by its lamination. Mudrock composed of grains of a similar size range and composition that is not laminated is usually referred to as mudstone. There is no sharp distinction between shale and mudstone; one grades into the other.

Shale is frequently regarded as an undesirable material to work with. Certainly, there are many examples of failures of structures founded on slopes in shales. Nonetheless, shales do vary in their engineering behaviour, largely according to their degree of compaction and cementation. Cemented shales are invariably stronger and more durable than compacted shales. The degree of packing – and hence the porosity, void ratio, and density of shale – depends on the mineral composition, grain-size distribution, mode of sedimentation, subsequent depth of burial, tectonic history, and effects of diagenesis. When the natural moisture content of shales exceeds 20%, they frequently become suspect and tend to develop potentially high pore-water pressures. Generally, shales with a cohesion of less than 20 MPa and an apparent angle of friction of less than 20° are likely to present engineering problems.

The higher the degree of fissility possessed by a shale, the greater the anisotropy with regard to strength, deformation, and permeability. For instance, the compressive strength and deformation modulus at right angles to the laminations may be more than twice those parallel to the laminations.

The greatest variation in the engineering properties of mudrocks can be attributed to the effects of weathering. Weathering reduces the amount of induration or removes it completely, leading to an increase in moisture content and a decrease in density. Indeed, weathering ultimately returns mudrock to a normally consolidated remoulded condition by destroying the bonds between the grains. Initially, mudrocks degrade rapidly to form a dominantly gravel-sized aggregate; this is facilitated by the presence of polygonal fracture patterns, joints, fissures, and bedding.

Depending on the relative humidity, some shales may slake almost immediately when exposed to the atmosphere. Alternate wetting and drying causes the rapid breakdown of compaction within the shale. Low-grade compaction shales, in particular, completely disintegrate after just a few cycles of drying and wetting. On the other hand, well-cemented shales are fairly resistant to slaking. If mudrocks undergo desiccation, air is drawn into the outer pores and capillaries as high suction pressures develop. On saturation, the entrapped air is pressurized as water is drawn into the rock by capillarity. Slaking therefore stresses the fabric of the rock. Disintegration consequently takes place as a result of air breakage after a sufficient number of wetting and drying cycles.

The swelling properties of certain mudrocks have proved to be extremely detrimental to the integrity of many civil engineering structures. Swelling, especially in clay shales, is attributable to the absorption of free water by expansive clay minerals, notably montmorillonite. Highly fissured overconsolidated shales have a greater tendency to swell than poorly fissured clayey shales, because the fissures provide access for water. The failure of poorly cemented mudrocks occurs during saturation, when the swelling pressure or internal saturation swelling stress developed by capillary suction pressures exceeds the tensile strength.

Uplift is common in excavations in shales, and can be attributed to swelling and heave. Rebound on unloading of the shale during excavation is attributed to heave due to the release of stored strain energy. Shale relaxes towards the newly excavated face, and sometimes this occurs as offsets along weaker seams in the shale. The greatest amount of rebound occurs in heavily overconsolidated compaction shales.

Settlement of shales can generally be managed by reducing the unit bearing load, for instance, by widening the base of a structure or by using spread footings. In some cases, appreciable differential settlement is provided for by designing an articulated structure that is capable of accommodating differential movement of individual sections without damage. Severe settlement may take place in low-grade compaction shales. However, compaction shales contain fewer open joints and fissures, which can be compressed beneath a heavy structure, than would the equivalent cemented shales.

When a load is applied to an essentially saturated shale foundation, the pore space in the shale decreases and the pore water attempts to migrate to regions of lesser load. Owing to the relative impermeability of shale, water becomes trapped in the voids and can migrate only slowly. As the load is increased, there comes a point at which it is partly transferred to the pore water, resulting in a build-up of pore-water pressure. Depending on the permeability of the shale and the rate of loading, the pore-water pressure can increase, to the extent that it can equal the pressure imposed by the load. This greatly reduces the shear strength of the shale, and structures can fail under such conditions. By contrast, problems with pore-water pressure are generally less important in cemented shales.

Sulphur compounds are frequently present in argillaceous rocks. An expansion in volume large enough to cause structural damage can occur when sulphide minerals, such as pyrite and marcasite, oxidize, yielding products such as anhydrous and hydrous sulphates. Significant heave can occur when sulphur compounds resulting from the breakdown of pyrite combine with calcium to form gypsum and jarosite. Movements in excess of 100 mm have been recorded, with heave rates of about 2 mm per month. The reactions are exothermic, and the resulting increase in temperature can also have adverse effects, especially on the ventilation and insulation systems of basement and underground structures. The decomposition of sulphur compounds gives rise to aqueous solutions of sulphate and sulphuric acid, which react with the tricalcium aluminate in Portland cement to form calcium sulpho-aluminate or ettringite. This reaction is accompanied by significant expansion and leads to the deterioration of concrete.

## Carbonate Rocks

Carbonate rocks are those that contain more than 50% carbonate minerals (such as calcite and dolomite). The term limestone (see **Sedimentary Rocks: Limestones**) is applied to those rocks in which the carbonate fraction exceeds 50%, over half of which is calcite or aragonite. If the carbonate material consists chiefly of dolomite, the rock is named dolostone.

Some representative geotechnical properties of carbonate rocks are listed in **Table 2**. It can be seen that, in general, the densities of these rocks increase with age, whilst the porosities are reduced. Furthermore, the porosity has a highly significant influence on the unconfined compressive strength: as the porosity increases, the strength declines. As both dolomitization and dedolomitization can give rise to increased porosity, both can be responsible for lower compressive strengths.

Both lithology and age frequently influence the strength and deformation characteristics of carbonate rocks. For instance, micritic (microcrystalline calcium carbonate) limestones may have a higher strength than sparitic (coarsely crystalline calcite) types. Most Silurian, Devonian, and Carboniferous limestones in Britain have compressive strengths of over 50 MPa, whereas Jurassic and Cretaceous limestones are often moderately weak, having unconfined compressive strengths of less than 12.5 MPa.

**Table 2** Some geotechnical properties of British carbonate rocks

| Property | Limestone Carboniferous (Derbyshire) | Magnesian Limestone Permian (South Yorkshire) | Great Oolite Jurassic (Wiltshire) | Lower Chalk | Upper Chalk |
|---|---|---|---|---|---|
| Dry density (Mg m$^{-3}$) | | | | | |
| Range | 2.55–2.61 | 2.46–2.58 | 1.91–2.21 | 1.85–2.13 | 1.35–1.61 |
| Mean | 2.58 | 2.51 | 1.98 | 2.08 | 1.44 |
| Porosity (%) | | | | | |
| Range | 2.4–3.6 | 8.5–12.0 | 13.8–23.7 | 17.2–30.2 | 29.6–45.7 |
| Mean | 2.9 | 10.4 | 17.7 | 20.6 | 41.7 |
| Dry unconfined compressive strength (MPa) | | | | | |
| Range | 65.2–170.9 | 34.6–69.6 | 8.9–20.1 | 19.1–32.7 | 4.8–6.2 |
| Mean | 106.2 | 54.6 | 15.6 | 26.4 | 5.5 |
| Saturated unconfined compressive strength (MPa) | | | | | |
| Range | 56.1–131.6 | 25.6–49.4 | 7.8–10.4 | 8.6–16.2 | 1.4–2.2 |
| Mean | 83.9 | 36.6 | 9.3 | 13.7 | 1.7 |
| Young's modulus (GPa) | | | | | |
| Range | 53.9–79.7 | 22.3–53.0 | 9.7–27.8 | 7.5–18.4 | 4.2–4.6 |
| Mean | 68.9 | 41.3 | 16.1 | 12.7 | 4.4 |

Dry density: very low, less than 1.8 Mg m$^{-1}$; low, 1.8–2.2 Mg m$^{-1}$; moderate, 2.2–2.55 Mg m$^{-1}$; high 2.55–2.75, Mg m$^{-1}$.
Porosity: low, 1–5%; medium, 5–15%; high, 15–30%; very high, over 30%.
Unconfined compressive strength: moderately weak, 5–12.5 MPa; moderately strong, 12.5–50 MPa; strong, 50–100 MPa; very strong, 100–200 MPa.
Young's modulus: very highly deformable, less than 5 GPa; highly deformable, 5–15 GPa; moderate, 15–30 GPa; low, 30–60 GPa; very low, over 60 GPa.

Similarly, the oldest limestones tend to possess the highest Young's moduli.

As can be seen from **Table 2**, the unconfined compressive strengths of the four limestones are reduced by saturation. The smallest reduction, 21%, is exhibited by the limestone of Carboniferous age, which is also the strongest. The reductions in strength of the Magnesian Limestone (Permian), Lincolnshire Limestone (Jurassic), and Great Oolite (Jurassic) are 35%, 40%, and 42% respectively. In other words, the strongest material undergoes the smallest reduction in strength on saturation, and there is a progressive increase in the average percentage reduction in strength after saturation as the dry strength of the limestone decreases; the weakest material shows the greatest reduction in strength.

Thickly bedded horizontally lying limestones that are relatively free from solution cavities afford excellent foundations. On the other hand, thinly bedded, highly folded, or cavernous limestones are likely to present serious foundation problems. The possibility of sliding may exist in highly bedded folded sequences. Similarly, when beds are separated by layers of clay or shale, especially when they are inclined, these may serve as sliding planes and result in failure.

Limestones are commonly transected by joints. These have generally been subjected to various degrees of dissolution, so much so that some may gape (**Figure 8**). Rain water is usually weakly acidic, and further acids may be taken into solution from carbon dioxide or from organic or mineral matter in the soil. The aggressiveness of the water to a limestone can be assessed on the basis of the relationship between the dissolved carbonate content, the pH, and the temperature of the water. At any given pH, the cooler the water, the more aggressive it is. If dissolution continues, its rate slackens, and it eventually ceases when the water becomes saturated. Hence, solution is greatest when the carbonate concentration is low. This occurs when water is circulating such that fresh supplies with low lime saturation are continually available. Freshwater can dissolve up to $400\ \text{mg}\,\text{l}^{-1}$ of calcium carbonate. The solution of limestone, however, is a very slow process. For example, the mean rates of surface lowering of limestone areas in Britain tend to range from $0.04\ \text{mm}\ \text{year}^{-1}$ to $0.1\ \text{mm}\ \text{year}^{-1}$. Nevertheless, dissolution may be accelerated by manmade changes in the groundwater conditions or by a change in the character of the surface water that drains into the limestone.

An important effect of dissolution within limestone is the enlargement of the pores, which enhances water circulation, thereby encouraging further dissolution. This increases the stress within the remaining rock fabric, which reduces the strength of the rock mass and leads to increasing stress corrosion. On loading, the volume of the voids is reduced by fracture of the weakened cement between the particles and by the reorientation of intact aggregates of rock that have been separated by the loss of bonding. Most of the resultant settlement takes place rapidly, within a few days of the application of the load.

The progressive opening of discontinuities by dissolution leads to an increase in mass permeability. Sometimes dissolution produces a highly irregular pinnacled surface, which may become exposed as a limestone pavement following soil erosion (**Figure 8**). The size, form, abundance, and downward extent of the aforementioned features depend on the geological structure and the presence of interbedded impervious layers. Solution cavities present numerous problems for the construction of large foundations.

Sinkholes may develop where opened joints intersect, and these may lead to an interlocking system of subterranean galleries and caverns (**Figure 9A**). Such features are associated with karstic landscapes (*see* **Sedimentary Processes:** Karst and Palaeokarst). The latter are characteristic of thick massive carbonate rocks. Sinkholes may be classified on the basis of origin into dissolution, collapse, caprock, dropout, suffusion, and buried sinkholes. Dissolution sinkholes form by slow dissolutional lowering of the outcrop or rockhead when surface water drains into carbonate rocks. Collapse sinkholes are formed by the collapse of cavern roofs that have become unstable owing to removal of support. Rapid collapse of a roof is very unusual; progressive collapse of a heavily fissured zone is more common. Caprock sinkholes are formed when an insoluble rock above a cavity in a carbonate rock collapses. Dropout sinkholes are formed by subsidence of a cohesive soil into cavities beneath. For example, clay soil

**Figure 8** Gaping joints due to dissolution, forming grykes in a limestone pavement above Malham Cove, North Yorkshire, England.

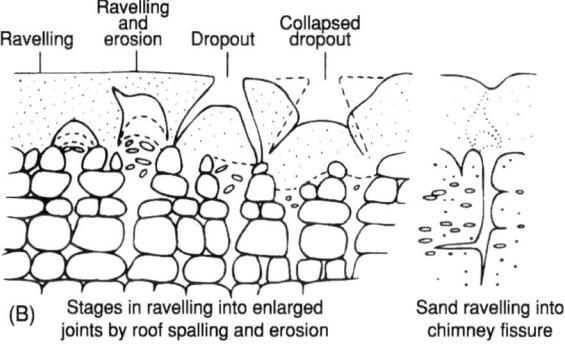

**Figure 9** (A) A sinkhole developed as a result of groundwater lowering at a miner's recreation centre at Venterspoort, South Africa. (B) Mechanisms of ravelling.

may be capable of bridging a void developed in an underlying limestone. As the void is enlarged, the clay cover ravels until it eventually fails to form a dropout sinkhole (**Figure 9B**). Ravelling (breaking away) failures are the most widespread and probably the most dangerous of all the subsidence phenomena associated with karstic carbonate rocks. Rapid changes in moisture content lead to aggravated slabbing in clays. In particular, lowering of the water table increases the downward seepage gradient and accelerates downward erosion. It also reduces capillary attraction and increases the instability of flow through narrow openings, and gives rise to shrinkage cracks in highly plastic clays, which weaken the mass in dry weather and produce concentrated seepage during rains. Increased infiltration often initiates failure, especially when it follows a period when the water table has been lowered. Suffusion sinkholes develop in non-cohesive soils where percolating water washes the soil into a cavity in the underlying limestone. Buried sinkholes occur in limestone rockhead and are filled or buried with sediment. They may be developed by subsurface solution or as normal subaerial sinkholes that are later filled with sediment.

Unfortunately, the appearance of sinkholes at the surface is commonly influenced by human activity, especially groundwater withdrawal or an increased flow of water into the ground from a point source. As an example, more than 4000 sinkholes have been catalogued in Alabama as being caused by human activities, with the great majority developing since 1950. The largest is called the 'December Giant' because it developed suddenly in December 1972. It measures 102 m in diameter and is 26 m deep. Sinkholes are particularly dangerous when they form instantaneously by collapse, and they often occur in significant numbers within a short time-span. They have resulted in costly damage to a variety of structures and are a major local source of groundwater pollution.

Areas underlain by highly cavernous carbonate rocks possess the most sinkholes; hence, sinkhole density has proven to be a useful indicator of potential subsidence. Solution voids preferentially develop along zones of high secondary permeability because these concentrate groundwater flow. Data on fracture orientation and density, fracture intersection density, and the total length of fractures have been used to model the presence of solution cavities in limestone. Accordingly, the locations of areas at high risk of cavity collapse have been estimated from the intersections of lineaments formed by fracture traces and lineated depressions. Aerial photographs have proved particularly useful in this context. Subsidence-susceptibility maps can be developed using a geographical information system that incorporates the relevant data in a spatial context.

The chalk in southern England exhibits a wide range of dry density values, ranging from as low as 1.25 Mg m$^{-3}$ up to 2.5 Mg m$^{-3}$. Generally, chalk in northern England (i.e. Yorkshire) is denser than that in the south-east (Kent). The average porosity of chalk ranges between about 25% and 40%. The dry unconfined compressive strength of chalk varies from moderately weak to moderately strong; the chalk in Yorkshire tends to be appreciably stronger than that in Kent. When saturated, chalk undergoes a notable reduction in compressive strength, frequently by over 50%. The deformation properties of chalk *in situ* depend on its strength and on the spacing, tightness, and orientation of its discontinuities. In addition, the values of Young's modulus are influenced by the amount of weathering that the chalk has undergone. Six grades of weathering have been recognized, varying from completely unweathered material to a structureless mélange consisting of unweathered and partially weathered fragments of chalk in a matrix of deeply weathered chalk. Fresh chalk has a low compressibility and compresses elastically up to a critical pressure, which has been termed the apparent preconsolidation pressure (the yield stress). Marked breakdown and substantial consolidation occur at high pressures. The yield stress of intact chalk may mark the point of pore collapse in high-porosity chalks (where the porosity exceeds 30%) as the cement bonds break down. Some interparticle slip and local grain crushing may also occur. With initial values of Young's modulus of 100 MPa and over, settlement in chalk is not normally a problem provided that the yield stress is not exceeded.

Chalk, being a relatively pure form of limestone, is subject to dissolution along discontinuities in the same way as the other carbonate rocks. However, subterranean solution features are less common in chalk, possibly because it is usually weaker than limestone and so collapses as solution occurs. Nevertheless, solution pipes and swallow holes are known to occur in chalk, commonly being found near the contact of Cretaceous chalk with the overlying Tertiary and Quaternary sediments. High concentrations of water, including run-off from roads, can lead to the reactivation of swallow holes and the formation of small pipes within a few years. Moreover, voids can gradually migrate upwards through chalk due to material collapse. Dissolution can lower the chalk surface beneath the overlying deposits, disturbing the latter and lowering their degree of packing. Hence, the chalk surface may be extremely irregular in places, causing problems of differential settlement to the foundations of buildings.

## Evaporite Rocks

Evaporite deposits are formed by precipitation from saline waters in arid areas. They include anhydrite ($CaSO_4$), gypsum ($CaSO_4 \cdot nH_2O$), rock salt or halite (NaCl), and sylvite (KCl).

Dry densities and porosities for anhydrite and gypsum are given in **Table 3**, from which it can be seen that the ranges are relatively low. Anhydrite, according to its unconfined compressive strength, tends to be very strong, whereas gypsum tends to be only moderately strong. It appears that the impurity content of gypsum affects its strength and that material containing more impurities is stronger. This may be because impurities within calcium sulphate rocks tend to reduce the crystal size and consequently increase the strength. The Young's moduli are generally significantly higher for anhydrite than for gypsum. In other words, the deformability of anhydrite is either very low or low, whereas that of gypsum varies from low to high. In both rock types the amount of creep usually increases with increasing levels of constant loading.

The unconfined compressive strength of rock salt (halite) is generally moderately weak, whereas that of sylvite is moderately strong. These two rocks are either very highly or highly deformable (**Table 3**). In rock salt the yield strength may be as little as one-tenth of the ultimate compressive strength, whereas in sylvite plastic deformation is generally initiated at somewhere between 20% and 50% of the load at failure. Creep may account for anything between 20% and 60% of the strain at failure in these two evaporitic rock types.

Gypsum is more readily soluble than calcite. Sinkholes and caverns can therefore develop in thick beds of gypsum more rapidly than they would in

**Table 3** Some geotechnical properties of British evaporite rocks

| Property | Anhydrite Cumbria | Gypsum North Yorkshire | Halite Cheshire | Sylvite Cleveland |
|---|---|---|---|---|
| Dry density (Mg m$^{-3}$) | | | | |
| Range | 2.77–2.82 | 2.16–2.32 | 1.92–2.09 | 1.86–1.99 |
| Mean | 2.79 | 2.21 | 2.03 | 1.94 |
| Porosity (%) | | | | |
| Range | 3.1–3.7 | 3.4–9.1 | 2.7–7.9 | 3.2–8.7 |
| Mean | 3.3 | 5.1 | 4.8 | 5.4 |
| Unconfined compressive strength (MPa) | | | | |
| Range | 77.9–126.8 | 19.0–40.8 | 9.4–14.9 | 18.5–31.8 |
| Mean | 102.9 | 27.5 | 11.7 | 25.8 |
| Young's modulus (GPa) ($E_{t50}$) | | | | |
| Range | 57.0–86.4 | 15.6–36.0 | 1.9–6.3 | 3.5–11.5 |
| Mean | 78.7 | 24.8 | 3.8 | 7.9 |

Dry density: very low, less than 1.8 Mg m$^{-1}$; low, 1.8 to 2.2 Mg m$^{-1}$; moderate, 2.2 to 2.55 Mg m$^{-1}$; high 2.55 to 2.75, Mg m$^{-1}$; very high, over 2.75 Mg m$^{-1}$.
Porosity: low, 1 to 5%; medium, 5 to 15%. Porosity derived by air porosimeter method.
Unconfined compressive strength: moderately weak, 5 to 12.5 MPa; moderately strong, 12.5 to 50 MPa; strong, 50 to 100 MPa; very strong, over 100 MPa.
Young's modulus: very highly deformable, less than 5 GPa; highly deformable, 5 to 15 GPa; moderately deformable, 15 to 30 GPa; low deformability, 30 to 60 GPa; very low deformability, over 60 GPa.

limestone. Indeed, in the USA, such features have been known to form within a few years, for instance where beds of gypsum are located beneath a dam. Extensive surface cracking and subsidence may be associated with the collapse of cavernous gypsum. The problem is accentuated by the fact that gypsum is weaker than limestone and therefore collapses more readily. Where beds of gypsum approach the surface, their presence is often indicated by collapse sinkholes. Such sinkholes can take only a matter of minutes to appear at the surface. However, where gypsum is effectively sealed from the ingress of water by overlying impermeable strata, such as mudstone, dissolution does not occur.

Massive deposits of gypsum are usually less dangerous than those of anhydrite, because gypsum tends to dissolve steadily, forming caverns or causing progressive settlement. In fact, the solution of massive gypsum is not likely to give rise to an accelerating deterioration beneath a foundation if precautions, such as grouting, are taken to keep seepage velocities low. On the other hand, massive anhydrite can be dissolved, leading to runaway situations in which seepage flow rates increase in a rapidly accelerating manner. Even small fissures in massive anhydrite can prove to be dangerous.

If gypsum or anhydrite occur in particulate form in the ground, their subsequent removal by dissolution can give rise to significant settlement. In such situations, the width of the solution zone and its rate of progress are obviously important as far as the location of hydraulic structures is concerned. Anhydrite is less likely than gypsum to undergo catastrophic solution in a fragmented or particulate form. Another point that should be borne in mind, and this particularly applies to conglomerates or breccias cemented with such soluble material, is that when this material is removed by solution the rock is greatly reduced in strength. In addition, when anhydrite comes into contact with water it may become hydrated to form gypsum. In so doing, there is a volume increase of between 30% and 58%, which exerts swelling pressures that are commonly between 1 MPa and 8 MPa and on rare occasions exceed 12 MPa. No great length of time is required to bring about such hydration.

Rock salt is even more soluble than gypsum, and the evidence of slumping, brecciation, and collapse structures in rocks that overlie saliferous strata bear witness to the fact that rock salt has gone into solution in past geological times. It is generally believed, however, that, in humid and semi-arid areas underlain by saliferous beds, measurable surface subsidence is unlikely to occur, except where salt is being extracted. Perhaps this is because an equilibrium has been attained between the supply of unsaturated groundwater and the salt available for solution. Exceptional cases of rapid subsidence have been recorded, such as the Meade salt sink in Kansas. Karstic features, particularly sinkholes, may develop in salt formations in arid areas. Salt extraction by some types of solution mining can give rise to serious or even catastrophic subsidence of the ground surface.

## Organic Rocks: Coal

Coal is an organic deposit composed of different types of macerated plant tissue, which occurs in association with other sedimentary rocks such as shales, mudstones, and sandstones. Many coal seams have a composite character. At the bottom the coal is frequently softer, with bright coal in the centre and dull coal predominating in the upper part of the seam, reflecting changes in the type of plant material that accumulated and in the drainage conditions. Coal seams may be split, wholly or partially, by washouts. Coal usually breaks into small blocks that have three pairs of faces approximately parallel to each other. These surfaces are referred to as cleat. The cleat direction is fairly constant and is best developed in bright coal. Cleat may be coated with films of mineral matter, commonly calcite, ankerite, and iron pyrite. The breakdown of iron pyrite frequently gives rise to acid mine drainage, which is associated with coal mining activity and has caused problems in mining areas around the world. The heavy metals in, and the acidity and sulphate and iron contents of, acid mine drainage may pollute groundwater and surface water, and contaminate soils and sediments.

The unconfined compressive strength of coal varies, but generally it is less then 20 MPa. Exceptionally, the unconfined compressive strengths of some coals, such as the Barnsley Hard Coal, may exceed 50 MPa and their Young's moduli may be greater that 25 GPa. Consequently, the presence of coal seams in foundations does not usually present a problem. However, when coal has been mined at shallow depth, the presence of abandoned bell-pit or pillar-and-stall workings frequently cause foundation problems. Longwall mining of coal, which involves the total extraction of panels of coal, especially when it occurs at shallow depth, can give rise to notable subsidence at the ground surface, which, in turn, may damage buildings and structures or lead to flooding of agricultural land.

Another problem associated with coal is that it can spontaneously combust on exposure to air. This can be regarded as an atmospheric oxidation process in which self-heating occurs (i.e. an exothermic reaction emitting between 5 kCal and 10 kCal per gram of coal). What is more, coal may be oxidized in the presence of air at temperatures below its ignition point. If heat is lost to the atmosphere, then the ignition temperature for coal is between 420°C and 480°C. However, where the heat of reaction is retained, the ignition point falls appreciably to between 35°C and 140°C. If the heat generated cannot be dissipated, the temperature rises, which increases the rate of oxidation so that the reaction becomes self-sustaining if there is a continuous supply of oxygen. Some coals ignite more easily than others; for example, high-rank coals are less prone to spontaneous combustion than are low-rank coals. It is not just coal seams that may spontaneously combust when exposed at or below the surface: colliery spoil heaps are also subject to spontaneous combustion. Obviously, the spontaneous combustion of coal, especially when it is being worked by opencast mining, and of spoil heaps when being reclaimed present appreciable problems.

## See Also

**Lava**. **Pyroclastics**. **Rock Mechanics**. **Sedimentary Processes:** Karst and Palaeokarst. **Sedimentary Rocks:** Chalk; Evaporites; Limestones; Sandstones, Diagenesis and Porosity Evolution. **Tectonics:** Fractures (Including Joints). **Weathering**.

## Further Reading

Bell FG (ed.) (1992) *Engineering in Rock Masses*. Oxford: Butterworth-Heinemann.

Bell FG (2000) *Engineering Properties of Soils and Rocks*, 4th edn. Oxford: Blackwell Scientific Publications.

Bell FG (2004) *Engineering Geology and Construction*. London: E & FN Spon.

Bell FG, Waltham AC, and Culshaw MG (2004) *Sinkholes and Subsidence*. Chichester/New York: Praxis/Springer.

Goodman RE (1993) *Engineering Geology: Rock in Engineering Construction*. New York: Wiley.

# Problematic Soils

**F G Bell**, British Geological Survey, Keyworth, UK

© 2005, Elsevier Ltd. All Rights Reserved.

## Introduction

Most civil engineering operations are founded in the uppermost layers of the ground and are therefore generally carried out in soil. Poor soil conditions in terms of engineering increase the cost of construction by necessitating special foundation structures and/or mean that some type of engineering soil treatment is required. Consequently, it important to understand the nature of the soil that is being dealt with. However, all soil types at times can be problematic, depending upon the conditions existing at a particular engineering site. For instance, if saturated gravel is to be excavated into, then it will have to be dewatered prior to the commencement of the operation in order to avoid flooding the site. Furthermore, soils can be contaminated and thereby present special engineering problems if the area where they occur is to be developed. Be that as it may, only the more troublesome soil types are dealt with here.

In the engineering sense, soil consists of an unconsolidated assemblage of particles between which are voids, which may contain air or water or both. As such, soil consists of three phases, that is, solids, water, and air. Under certain circumstances, soil can contain other gases such as methane or other liquids such as nonaqueous phase liquids. The interrelationships of the weights and volumes of the three phases are important since they help define the character of a soil. The solid phase of soil is derived from the breakdown of rock material by weathering and/or erosion, and it may have suffered a varying amount of transportation prior to deposition. It also may contain organic matter, the total organic content of soils varying from less than 1% in the case of some immature or desert soils to over 90% in the case of peats. The type of breakdown process(es) and the amount of transport undergone by sediments influence the nature of the macro- and microstructure of the soil that, in turn, influence its behaviour (**Table 1**). Furthermore, the same type of rock can give rise to different types of soils, depending on the climatic regime and the vegetative cover under which it develops. Indeed, the character of a soil frequently is influenced to a significant extent by the climatic regime in which it is formed and exists. This is especially the case with some soils formed in more extreme climates such as arid and semi-arid zones, notable examples being sabkha soils and dispersive soils, respectively, or quick clays in cold climates. Such soils possess their own peculiar characteristics that provide problems for the engineer. Time also is an important factor in the development of a mature soil.

**Table 1** Effects of transportation on sediments

|  | Gravity | Ice | Water | Air |
|---|---|---|---|---|
| Size | Various | Varies from clay to boulders | Various sizes from boulder gravel to muds | Sand size and less |
| Sorting | Unsorted | Generally unsorted | Sorting takes place both laterally and vertically. Marine deposits often uniformly sorted. River deposits may be well sorted | Uniformly sorted |
| Shape | Angular | Angular | From angular to well rounded | Well rounded |
| Surface texture | Striated surfaces | Striated surfaces | Gravel: rugose surfaces. Sand: smooth, polished surfaces. Silt: little effect | Impact produces frosted surfaces |

Any system of soil classification involves grouping soils into categories that possess similar properties, so providing a systematic method of soil description by which soils can be identified quickly. Although soils include materials of various origins, for purposes of engineering classification it is sufficient to consider their simple index properties, which can be assessed easily, such as their particle size distribution and consistency limits. For instance, coarse-grained soils are distinguished from fine on a basis of particle size, gravels and sands being the two principal types of coarse-grained soils. Plasticity also is used when classifying fine-grained soils, that is, silts and clays.

## Quicksands

As water flows through sands or silts and slows down, its energy is transferred to the particles past which it is moving that, in turn, creates a drag effect on the particles. If the drag effect is in the same direction as the force of gravity, then the effective pressure is increased and the soil is stable. Conversely, if water flows towards the surface, then the drag effect is counter to gravity, thereby reducing the effective pressure between particles. If the upward flow velocity is sufficient, it can buoy up the particles so that the effective pressure is reduced to zero. This represents a critical condition where the weight of the submerged soil is balanced by the upward acting seepage force. If the upward velocity of flow increases beyond this critical hydraulic gradient, then a quick condition develops. As the velocity of the upward seepage force increases further from the critical gradient, the soil begins to boil more and more violently (**Figure 1A**). At such a point structures fail by sinking into the quicksand (**Figure 1B**).

Quicksands, if subjected to deformation or disturbance, can undergo a spontaneous loss of strength, causing them to flow like viscous liquids. A number of conditions must exist for quick conditions to develop. Firstly, the sand or silt concerned must be saturated and loosely packed. Secondly, on disturbance the constituent grains become more closely packed, which leads to an increase in pore-water pressure, reducing the forces acting between the grains. This brings about a reduction in strength. If the pore water can escape very rapidly the loss in strength is momentary. Hence, the third condition requires that pore water cannot escape readily. This is fulfilled if the sand or silt has a low permeability and/or the seepage path is long.

Quick conditions frequently are encountered in excavations made in fine sands that are below the watertable. Liquefaction of potential quicksands also may be brought about by sudden shocks caused by the action of heavy machinery (notably pile driving), blasting, and earthquakes (**Figure 1B**). Such shocks increase the stress carried by the pore water, and give rise to a decrease in the effective stress and shear strength of the soil. There is also a possibility of a quick condition developing in a layered soil sequence containing fine sands where the individual beds have different permeabilities.

## Collapsible Soils

Soils, such as loess and brickearth, commonly are regarded as being of aeolian origin and some believe that the material of which they are composed was formed initially by glacial action. These soils often have a metastable fabric and so possess the potential to collapse. In addition, some wind-blown silts, such as those found in subtropical arid areas like Arizona and the Kalahari Desert in southern Africa, have metastable fabrics and are potentially collapsible. Certain residual soils also are prone to collapse. For example, when granites undergo notable chemical weathering in subtropical regions that involves appreciable leaching, then the resulting saprolite and

**Figure 1** (A) Liquefaction of sand due to the Tangshan earthquake (28 July 1976) near Dougtian, China. (B) Collapse of buildings because of liquefaction of fine sands and silts, Niigata, Japan, 1964.

residual soil tend to develop a metastable fabric as fine particles are removed. As a consequence, the weathered products are potentially collapsible.

Truly collapsible soils are those in which collapse occurs on saturation since the soil fabric cannot support the weight of the overburden. When the saturation collapse pressure exceeds the overburden pressure, soils are capable of supporting a certain level of stress on saturation and can be regarded as conditionally collapsible soils. The maximum load that such soils can support is the difference between the saturation collapse and overburden pressures. These soils generally consist of 50 to 90% silt particles, and sandy, silty, and clayey types have been recognized, with most falling into the silty category (**Figure 2**). The fabric of collapsible soils generally takes the form of a loose skeleton of

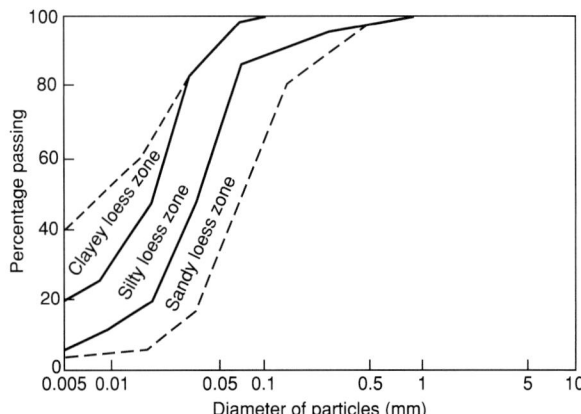

**Figure 2** Particle size distribution of loess.

grains (generally quartz) and micro-aggregates (usually assemblages of clay particles). These tend to be separate from each other, being connected by bonds and bridges, with uniformly distributed pores (**Figure 3**). The bridges are formed of clay-sized minerals. As grains are not in contact, mechanical behaviour is governed by the structure and quality of the bonds and bridges.

The structural stability of collapsible soils also is related to the amount of weathering undergone. Younger weakly weathered loess generally has a high potential for collapse whereas older weathered loess is relatively stable. In addition, highly collapsible loess tends to occur in regions where the landscape and/or the climatic conditions are not conducive to development of long-term saturated conditions within the soil. The size of the pores is all important, collapse normally occurring as a result of pore-space reduction taking place in pores greater than 1 μm in size and more especially in those exceeding 10 μm in size. Hence, collapsible soils possess porous textures with high void ratios and relatively low densities. At their natural low moisture content, these soils possess high apparent strength but they are susceptible to large reductions in void ratio upon wetting. In other words, their metastable texture collapses as the bonds between the grains break down when the soil is wetted. Collapse on saturation normally only takes a short period of time, although the more clay particles such a soil contains, the longer the period tends to be.

From the above it may be concluded that significant settlements can take place beneath structures founded on collapsible soils after they have been wetted, in some cases in the order of metres. These have led to foundation failures. A number of techniques can be used to stabilize collapsible soils; these are summarized in **Table 2**.

## Expansive Clays

An important characteristic of some clay soils is their susceptibility to slow volume change that can occur independently of loading due to swelling or shrinkage. Some of the best examples of expansive clays are provided by vertisols or black clays. Such soils typically are developed in subtropical regions with well-defined wet and dry seasons. Unfortunately, expansive clays frequently are responsible for significant national costs due to damage caused to property. For example, in the United States, the cost of repair of damage to property built on expansive clays exceeds two billion dollars annually. This frequently is twice the cost of flood damage or landslide damage and 20 times the cost of earthquake damage. Many clay soils in Britain, especially south-east England, possess the potential for large volume change. However, the mild damp climate means that significant deficits in soil moisture developed during the summer are confined to the upper 1.0 to 1.5 m of the soil and that the field capacity is re-established during the winter. Even so, deeper permanent deficits can be brought about by

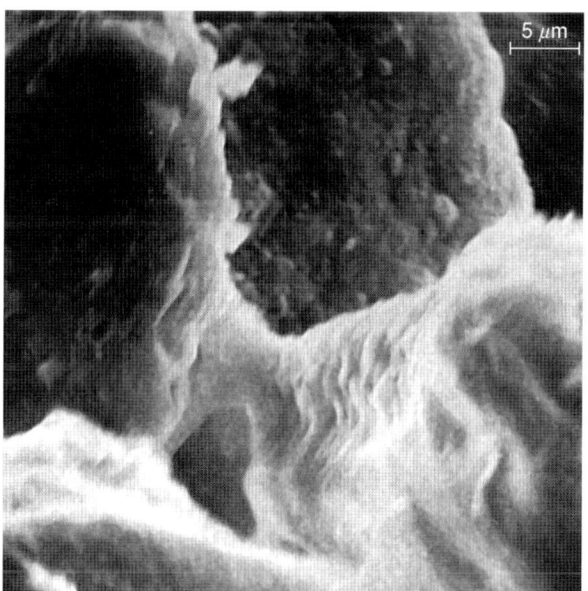

**Figure 3** Scanning electron photomicrograph of brickearth from south-east Essex, showing three silt particles, two of which are bridged by clay particles.

**Table 2** Methods of treating collapsible foundations

| Depth of subsoil treatment | |
| --- | --- |
| 0–1.4 m | Moistening and compaction (conventional extra heavy impact or vibratory rollers) |
| 1.5–10 m | Over-excavation and recompaction (earth pads with or without stabilization by additives such as cement or lime). Vibro-compaction (stone columns). Vibroreplacement. Dynamic compaction. Compaction piles. Injection of lime. Lime piles and columns. Jet grouting. Ponding or flooding (if no impervious layer exists). Heat treatment to solidify the soils in place. |
| Over 10 m | Any of the aforementioned or combinations of the aforementioned, where applicable. Ponding and infiltration wells, or ponding and infiltration wells with the use of explosive. |

transpiration from large trees, and significant damage to property on potentially expansive clay can be caused during notably dry summers.

Generally, the clay fraction of expansive clay exceeds 50%, silty material varying between 20 and 40%, and sand forming the remainder. Montmorillonite normally is present in the clay fraction and is the principal factor determining the appreciable volume changes that take place in these soils on wetting and drying.

The depth of the active zone in expansive clays (i.e., the zone in which swelling and shrinkage occurs in wet and dry seasons, respectively) can be large. For instance, the maximum seasonal changes in the moisture content of expansive clays in Romania are around 20% at 0.4 m depth, 10% at 1.2 m depth, and less than 5% at 1.8 m depth. The corresponding cyclic movements of the ground surface are between 100 and 200 mm. Surface heaves approaching 500 mm have been recorded in some expansive clays in South Africa. During the dry season, profiles in some regions can dry out to depths of 15 to 20 m.

The potential for volume change in expansive clay soils depends on the initial moisture content, initial voids ratio, the microstructure, and the vertical stress, as well as the type and amount of clay minerals present. The type of clay minerals are responsible primarily for the intrinsic expansiveness, whilst the change in moisture content or suction controls the actual amount of volume change that a soil undergoes at a given applied pressure. Changes in soil suction are brought about by moisture movement through the soil, due to evaporation from its surface in dry weather, by transpiration from plants, or alternatively by recharge consequent upon precipitation. Alternating wet and dry seasons may produce significant vertical movements as soil suction changes. The rate of expansion depends upon the rate of accumulation of moisture in the soil. In semi-arid regions, it is limited by the availability of water and this, together with the available void volume, governs the rate of penetration of the heave front in the soil. When these expansive clays occur above the water table, they can undergo a high degree of shrinkage on drying. Seasonal changes in volume also produce shrinkage cracks so that expansive clays are often heavily fissured. Sometimes the soil is so desiccated that the fissures are wide open and the soil is shattered or micro-shattered.

Transpiration from vegetative cover is a major cause of water loss from soils in subtropical semi-arid regions. Indeed, the distribution of soil suction in soil is controlled primarily by transpiration from vegetation and this represents one of the most significant changes made in loading (i.e., to the state of stress in a soil). The suction induced by the withdrawal of moisture fluctuates with the seasons, reflecting the growth of vegetation. The maximum soil suction that can be developed is governed by the ability of vegetation to extract moisture from the soil. The level at which moisture is no longer available to plants is termed the permanent wilting point. In fact, the moisture content at the wilting point exceeds that of the shrinkage limit in soils with high clay contents and is less in those possessing low clay contents. This explains why settlement resulting from the desiccating effects of trees is more notable in low to moderately expansive soils than in highly expansive ones.

These volume changes can give rise to ground movements that may result in damage to buildings. Low-rise buildings are particularly vulnerable to such ground movements since they generally do not have sufficient weight or strength to resist. Be that as it may, three methods can be adopted when choosing a design solution for building on expansive soils. Firstly, a foundation and structure can be provided that can tolerate movements without unacceptable damage; secondly, the foundation and structure can be isolated from the effects of the soil; and thirdly, the ground conditions can be altered or controlled. These soils also represent a problem when they are encountered in road construction, and shrinkage settlement of embankments composed of such clay soils can lead to cracking and breakup of the roads they support (**Figure 4**).

## Dispersive Soils

Dispersive soils occur in subtropical semi-arid regions, normally where the rainfall is less than 850 mm annually. Dispersion occurs in such soils when the repulsive forces between clay particles exceed the attractive forces, thus bringing about deflocculation so that in the presence of relatively pure water the particles repel each other to form colloidal suspensions. In non-dispersive soil there is a definite threshold velocity below which flowing water causes no erosion. By contrast, there is no threshold velocity for dispersive soil, the colloidal clay particles going into suspension even in quiet water. Therefore, these soils are highly susceptible to erosion and piping. Dispersive soils contain a moderate to high content of clay material but there are no significant differences in the clay fractions of dispersive and non-dispersive soils, except that soils with less than 10% clay particles may not have enough colloids to support dispersive piping. Dispersive soils contain a higher content of dissolved sodium (up to 12%) in their pore water than ordinary soils. The clay particles in soils with high salt contents exist as aggregates and coatings around silt and sand particles (**Figure 5**).

For a given eroding fluid the boundary between the flocculated and deflocculated states depends on the

**Figure 4** Breakup along the CapeTown Johannesburg road, South Africa, due to expansive clay.

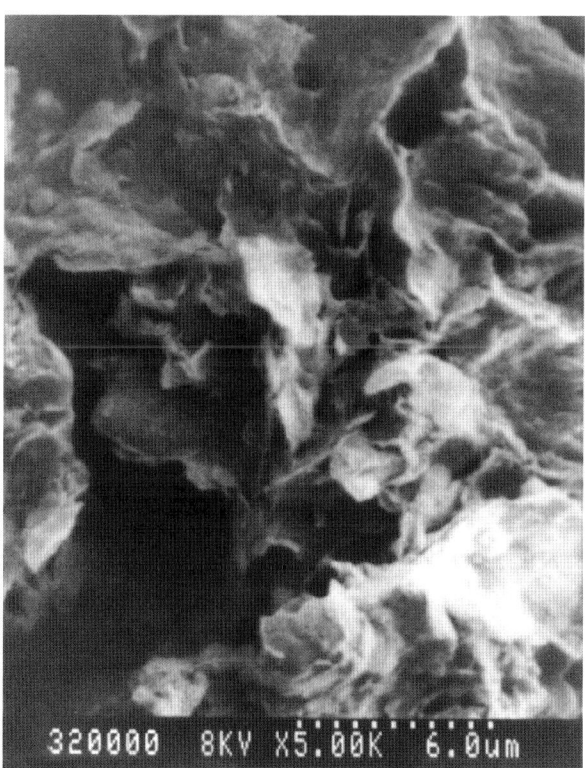

**Figure 5** Scanning electron photomicrograph illustrating the fabric of dispersed soil from Natal, South Africa.

value of the sodium adsorption ratio, the salt concentration, the pH value, and the mineralogy. The sodium adsorption ratio (SAR) is used to quantify the role of sodium where free salts are present in the pore water and is defined as:

$$\mathrm{SAR} = \frac{\mathrm{Na}}{\sqrt{0.5(\mathrm{Ca} + \mathrm{Mg})}} \qquad [1]$$

with units expressed in meq/litre of the saturated extract. It has been suggested that an SAR value greater than 10 is indicative of dispersive soils, between 6 and 10 as intermediate, and less than 6 as non-dispersive. However, dispersion has occurred in soils with values lower than 6. The presence of exchangeable sodium is the main chemical factor contributing towards dispersive behaviour in soil. This is expressed in terms of the exchangeable sodium percentage (ESP):

$$\mathrm{ESP} = \frac{\text{exchangeable sodium}}{\text{cation exchange capacity}} \times 100 \qquad [2]$$

where the units are given in meq/100 g of dry clay. A threshold value of ESP of 10% has been recommended, above which soils that have their free salts leached by seepage of relatively pure water are prone to dispersion. Soils with ESP values above 15% are highly dispersive. Those with low cation exchange values (15 meq/100 g of clay) have been found to be completely non-dispersive at ESP values of 6% or below. Similarly, soils with high cation exchange capacity values and a plasticity index greater than 35% swell to such an extent that dispersion is not significant.

Severe erosion or worse, serious piping damage to embankments and piping failures of earth dams have occurred when dispersive soils have been used in their construction. Indications of piping take the form of small leakages of muddy coloured water from an earth dam after initial filling of the reservoir. The pipes become enlarged rapidly and this can lead to failure of the dam (**Figure 6**). Experience, however, indicates

**Figure 6** Failed earth dam constructed of dispersive soil, showing piping outlets on the downstream side, near Ramsgate, South Africa.

that if an earth dam is built with careful construction control and incorporates filters, then it should be safe enough, even if it is constructed with dispersive soil. Alternatively, hydrated lime, pulverised fly ash, gypsum, or aluminium sulphate have been used to treat dispersive soils used in earth dams.

## Humid Tropical Zone Soils

In humid tropical regions, weathering of rock is more intense and extends to greater depth than in other parts of the world. Residual soils develop in place as a consequence of weathering, primarily chemical weathering. The mineralogy of residual soils is partly inherited from the parent rock from which they were derived and partly produced by the processes of weathering. Hence, the mineralogy varies widely, as does grain size and unit weight. The particles and their arrangement evolve gradually as weathering proceeds. In addition, weathering of parent rock *in situ* may leave behind relict structures that may offer weak bonding even in extremely weathered material. Low strength along relict discontinuities may be attributable to particles being coated with low-friction iron/manganese organic compounds.

Reproducible results from some standard tests may be difficult to obtain from residual tropical soils. Different results can be obtained depending upon whether the soil is pre-dried prior to testing or kept close to its natural moisture content. Also, disaggregation of the soil structure, especially in relation to particle size analysis has proved problematic. Consequently, conventional index tests frequently have been modified in an attempt to make them more applicable for use with tropical residual soils. Of course, it would be wrong to assume that all tropical soils behave differently from those found in other climatic regions. For instance, alluvial clays and sands behave in the same manner and have similar geotechnical properties, regardless of the climatic conditions of the region of deposition.

Drying brings about changes in the properties of residual clay soils in that it initiates two important effects, namely, cementation by the sesquioxides and aggregate formation on the one hand, and loss of water from hydrated clay minerals on the other. In the case of halloysite, the latter causes an irreversible transformation to metahalloysite. Drying can cause almost total aggregation of clay size particles into silt and sand size ranges, and a reduction or loss of plasticity. Cycles of wetting and drying may increase the stiffness of the soil fabric, which increases its shear strength and decreases its compressibility.

Laterite can be regarded as a highly weathered material that forms as a result of the concentration of hydrated oxides of iron and aluminium in such a way that the character of the deposit in which they occur is affected. These oxides may be present in an unhardened soil, as a hardened layer, as concretionary nodules in a soil matrix or in a cemented matrix enclosing other materials. When a hardened crust is present near the surface, then the strength of laterite beneath decreases with increasing depth.

Red clays and latosols are residual ferruginous soils formed primarily by chemical weathering of the parent rock. This results in the release of iron and aluminium sesquioxides, increasing loss of silica and increasing dominance of new clay minerals such as smectites, allophane, halloysite and, with increasing weathering, kaolinite. The microstructure also is developed by chemical weathering processes and consists of an open-bonded fabric of silt and sand size peds, which are formed mainly of clay minerals and fine disseminated iron oxides. Relatively weak bonds exist between the peds and are formed of iron oxides and/or amorphous aluminium silicate gels. Such soils differ from laterite in that they behave as clay and do not possess strong concretions. They do, however, grade into laterite.

Allophane-rich soils or andosols are developed from basic volcanic ashes in high temperature-rainfall regions. Allophane is an amorphous clay mineral. These soils have very high moisture contents, usually in the range 60% to 80% but values of up to 250% have been recorded; and corresponding high plasticity. The soils also are characterized by very low dry densities and high void ratios (sometimes as high as 6). Moisture content does affect the strength of andosols significantly as the degree of saturation can have an appreciable affect on cementation. Soils containing halloysite, or its partially dehydrated form metahalloysite, have high moisture contents (30% to 65%) and can possess high plasticity. Some of these soils are susceptible to collapse.

## Soils of Hot Arid Regions

Most soils in arid regions consist of the products of physical weathering of rock material. This breakdown process gives rise to a variety of rock and mineral fragments that may be transported and deposited under the influence of gravity, wind, or water. Many arid soils are of aeolian origin and sands frequently are uniformly sorted. Uncemented silty soil may possess a metastable fabric and hence be potentially collapsible. The precipitation of salts in the upper horizons of an arid soil, due to evaporation of moisture from the surface, commonly means that some amount of cementation has occurred, which generally has been concentrated in layers, and that the pore water is likely to be saline. High rates of evaporation in hot arid areas may lead to ground heave due to the precipitation of minerals within the capillary fringe.

Where the watertable is at a shallow depth the soils may possess a salty crust and be chemically aggressive due to the precipitation of salts from saline groundwater. Occasional wetting and subsequent evaporation frequently are responsible for a patchy development of weak, mainly carbonate, and occasionally gypsum cement, often with clay material deposited between and around the coarser particles. These soils, therefore, may undergo collapse, especially where localized changes in the soil–water regime are brought about by construction activity. Collapse is attributed to a loss of strength in the binding agent and the amount of collapse undergone depends upon the initial void ratio. Loosely packed aeolian sandy soils, with a density of less than 1.6 Mg m$^{-3}$, commonly exhibit a tendency to collapse.

Silts may have been affected by periodic desiccation and be interbedded with evaporite deposits. The latter process leads to the development of a stiffened crust or, where this has occurred successively, to a series of hardened layers within the formation. Loosely packed aeolian silty soils formed under arid conditions often undergo considerable volume reduction or collapse when wetted. Such metastability arises from the loss in strength of interparticle bonds resulting from increases in water content. Thus, infiltration of surface water, including that applied during irrigation, leakage from pipes, and rise of water table, may cause large settlements to occur.

Low-lying coastal zones and inland plains in arid regions with shallow water tables, are areas in which sabkha conditions commonly develop. Sabkhas are extensive saline flats that are underlain by sand, silt, or clay and often are encrusted with salt. Groundwater is saline, containing calcium, sodium, chloride, and sulphate ions. Evaporative pumping, whereby brine moves upward from the water table under capillary action, appears to be the most effective mechanism for the concentration of salt in groundwater and the precipitation of minerals in sabkha. Salts are precipitated at the ground surface when the capillary fringe extends from the water table to the surface.

One of the main problems with sabkha is the decrease in density and strength, and increased permeability that occur, particularly in the uppermost layers, after rainfall, flash floods, or marine inundation, due to the dissolution of soluble salts that act as cementing materials. Changes in the hydration state of minerals, such as calcium sulphate, also cause significant volume change in soils. There is a possibility of differential settlement occurring on loading due to the different compressibility characteristics resulting from differential cementation of sediments. Excessive settlement also can occur, due to the removal of soluble salts by flowing groundwater. This can cause severe disruption to structures within months or a few years. Movement of groundwater also can lead to the dissolution of minerals to an extent that small caverns, channels, and surface holes can be formed. On the other hand, heave resulting from the precipitation and growth of crystals can elevate the surface of a sabkha in places by as much as 1 m.

Because sabkha soils frequently are characterized by low strength, their bearing capacity and compressibility frequently do not meet routine design requirements. Various ground improvement techniques, therefore, have been used in relation to large construction projects such as vibro-replacement, dynamic compaction, compaction piles, and underdrainage. Soil replacement and preloading have been used when highway embankments have been constructed.

Cementation of sediments by precipitation of mineral matter from the groundwater may lead to the development of various crusts or cretes (e.g., precipitation of calcite gives rise to calcrete and of gypsum to gypcrete). These may form continuous sheet or isolated patch-like masses at the ground surface when the water table is at or near this level, or at some other position within the ground profile. Therefore, duricrusts or pedocretes are soils that have been, to a greater or lesser extent, cemented. These materials take three forms, namely, indurated (e.g., hardpans and nodules), non-indurated (soft or powder forms), and mixtures of the two (e.g., nodular pedocretes). Indurated pedocretes that occur at the surface may be underlain by loose or soft material that, at times, may be potentially expansive or collapsible. Nodular pedocrete with a smectitic clay matrix could have the potential to be expansive, whereas collapsibility may be associated with cemented soils, powder, or nodular calcretes. Small-scale karst-like features have been recorded in some weathered calcretes.

## Soils Developed in Cold Regions

Tills are deposited by ice-sheets and consist of a variable assortment of rock debris ranging from fine rock flour to boulders; they are characteristically unsorted. On the one hand, tills may consist predominantly of sand and gravel with very little binder, whereas on the other they may have an excess of clayey material. The nature of a till deposit depends on the lithology of the material from which it was derived, on the position in which it was transported in the glacier, and on the mode of deposition. The underlying bedrock usually constitutes up to about 80% of basal tills, depending on its resistance to abrasion. Till sheets can comprise one or more layers of different material, not all of which are likely to be found at any one locality. Shrinking and reconstituting of an ice-sheet can complicate the sequence.

Distinction has been made between tills derived from rock debris carried along at the base of a glacier and those deposits that were transported within or at the terminus of the ice. The former is referred to as lodgement till, whilst the latter is termed ablation till. Lodgement till is commonly stiff, dense, and relatively incompressible. Because of the overlying weight of ice, such deposits are overconsolidated. Fissures frequently are present in lodgement till and influence its shear strength and therefore its stability. Ablation till accumulates on the surface of the ice when englacial debris melts out. It, therefore, is normally consolidated and non-fissile. The proportion of silt and clay size material is relatively high in lodgement till (e.g., the clay fraction typically varies from 15 to 40%). Ablation till is characterized by abundant large angular stones, the proportion of sand and gravel is high, and clay is present only in small amounts (usually less than 10%).

The most familiar pro-glacial deposits are varved clays. These sediments accumulated on the floors of glacial lakes and are characteristically composed of alternating laminae of finer and coarser grain size, each such couplet being termed a varve. The thickness of the individual varve is frequently less than 2 mm. Generally, the coarser layer is of silt size and the finer of clay size. Clay minerals tend to show a high degree of orientation parallel to the laminae.

Varved clays have a very wide range of plasticity, although most tend to be highly plastic and many possess the potential to swell. They tend to be normally consolidated or lightly overconsolidated. Furthermore, they often undergo a notable reduction in strength when remoulded, that is, they are medium sensitive to sensitive. However, the remoulded strength increases with time (**Figure 7**). The anisotropic behaviour of varved clay is explained by the influence of their lamination.

Quick clays were formed at the end or after the last ice age, when meltwater from glaciers carried large quantities of fine-grained sediments into the sea. The material of which quick clays are composed is predominantly smaller than 0.002 mm but some deposits seem to be poor in clay minerals, containing a high proportion of ground down fine quartz. The open fabric that is characteristic of quick clays has been attributed to their initial deposition, during which time colloidal particles interacted to form loose aggregations by gelation and flocculation. This fabric may have been retained to the present day.

Quick clays often exhibit little plasticity. However, the most extraordinary property possessed by quick clays is their very high sensitivity, that is, a large proportion of their undisturbed strength is permanently lost following shear (**Figure 8**). Consequently,

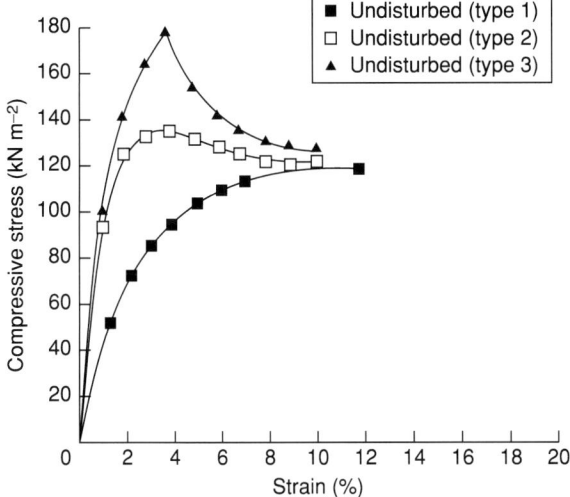

**Figure 7** Examples of stress-strain curves of undisturbed and remoulded Tees Laminated Clay tested in quick undrained triaxial conditions: (1) from Stockton, (2) from Cowpen Bewley, (3) from Middlesbrough.

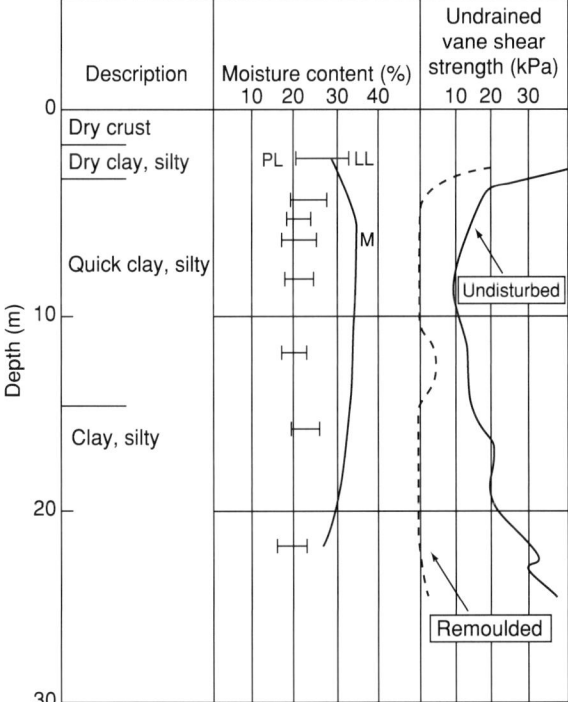

**Figure 8** Moisture content, consistency indices, undrained shear strength, and sensitivity of quick clay from near Trondheim, Norway.

quick clays are associated with several serious engineering problems. Not only is their bearing capacity low, with large settlements occurring under load, but quick clays can liquefy on sudden shock. This can result in an almost instantaneous loss in strength.

Slides, which have been referred to as mud-runs and bottleneck slides, therefore are a common feature of quick clay areas and sometimes have proved disastrous. Quick clays can be stabilized by adding salts or lime, which increase their remoulded shear strength.

Frozen ground phenomena are found in regions that experience a tundra climate, that is, in those regions where the winter temperatures rarely rise above freezing point and the summer temperatures are only warm enough to cause thawing in the upper metre or so of the soil. Beneath the upper or active zone, the subsoil is permanently frozen and so is known as permafrost. Because of the permafrost layer, summer meltwater cannot seep into the ground, the active zone then becomes waterlogged and soils on gentle slopes are liable to flow. Generally, the depth of thaw is less, the higher the latitude. It is at a minimum in peat or highly organic sediments and increases in clay, silt, and sand, to a maximum in gravel. Layers or lenses of unfrozen ground, termed taliks, may occur, often temporarily, in the permafrost.

The mechanical properties of frozen soil are influenced largely by the grain size distribution, mineral content, density, frozen and unfrozen water contents, and presence of ice lenses and layering. Ice has no long-term strength, that is, it flows under very small loads. When loaded, stresses at the point of contact between soil particles and ice bring about pressure melting of the ice. Because of differences in the surface tension of the meltwater, it tends to move into regions of lower stress, where it refreezes. The process of ice melting and the movement of unfrozen water are accompanied by breakdown of ice and bonding with grains of soil. This leads to plastic deformation of the ice in the voids and to a rearrangement of particle fabric. The net result is time-dependent deformation of the frozen soil, namely, creep. Frozen soil undergoes appreciable deformation under sustained loading, the magnitude and rate of creep being governed by the composition of the soil, especially the amount of ice present, the temperature, the stress, and the stress history.

As the soil thaws downwards the upper layers become saturated, and since water cannot drain through the frozen soil beneath, they may suffer a complete loss of strength. What is more, as ice melts, settlement occurs. Excess pore-water pressures develop when the rate of ice-melt is greater than the discharge capacity of the soil. This can lead to the failure of slopes and foundations.

Shrinkage, which gives rise to polygonal cracking of the ground, presents a problem when soil is subjected to freezing. Individual cracks may be over 1 m wide at their top and may penetrate to depths of 10 m. Water that accumulates in the cracks is frozen to form ice wedges. When the ice disappears, an ice wedge pseudomorph is formed by sediment filling the crack. The ground also may undergo notable disturbance due to mutual interference of growing bodies of ice or excess pore-water pressures developed in confined water-bearing lenses. Involutions are plugs, pockets, or tongues of highly disturbed material, generally possessing inferior geotechnical properties, which have been intruded into overlying layers. Pseudomorphs of ice wedges and involutions usually mean that one material suddenly replaces another. This can cause problems in shallow excavations. Prolonged freezing gives rise to shattering in the frozen layer. Frost shattering may extend to significant depths and is responsible for an increase in deformability and permeability of the ground affected. It should be borne in mind that periglacial conditions extended over a much greater area of the land surface during the Pleistocene epoch and as a result the features mentioned are found in many present temperate climate regions such as Britain.

There are two types of methods of construction in permafrost, namely, passive and active methods. In the former, the frozen ground is not disturbed and heat from a structure is prevented from thawing the ground below, thereby reducing its stability. In other words, the aim is to maintain the ground beneath buildings in a frozen state. This is accomplished either by ventilation or by insulation. In the former case, a building is raised on piles above the ground so that air can circulate beneath it, thereby dissipating heat generated by the building. Alternatively, a building may be constructed on a layer of gravel to insulate the ground beneath. An insulating material such as polystyrene may be incorporated into the layer of gravel. The ground is thawed prior to construction in the active method. It is either kept thawed or removed and replaced by materials not affected by frost action. The active method is used where permafrost is thin, sporadic, or discontinuous, and where thawed ground has an acceptable bearing capacity. Foundations frequently are taken through the active layer into the permafrost beneath. Hence, piles often are used as foundation structures.

Frost action in a soil obviously is not restricted to periglacial regions. If frost penetrates down to the capillary fringe in fine-grained soils, especially silts then, under certain conditions, lenses of ice may be developed. The formation of such ice lenses may, in turn, cause frost heave and frost boil that may lead to the breakup of roads and the failure of slopes. Such problems are experienced in temperate climatic zones, as well as those of tundra regions.

## Peat Soils

Peat represents an accumulation of partially decomposed and disintegrated plant remains that have been

**Figure 9** Holme Post, a cast iron pillar erected in 1851 on the south-west edge of Whittlesey Mere, near Peterborough, England, to indicate the amount of peat shrinkage and associated subsidence caused by drainage. The post was driven through 7 m of peat into clay until its top was flush with the ground. Within 10 years ground level had fallen 1.5 m through shrinkage. A second post was erected in 1957 with its top at the same level as that of the original post (right-hand side). Between 1851 and 1971 the ground subsided by some 4 m.

preserved under conditions of incomplete aeration and high water content. It accumulates in areas where there is an excess of rainfall and the ground is poorly drained. Nonetheless, peat deposits tend to be most common in those regions with a comparatively cold wet climate. The high water-holding capacity of peat maintains a surplus of water, which ensures continued plant growth and consequent peat accumulation. Drying out, groundwater fluctuations, and snow loading bring about compression in the upper layers of a peat deposit. As the water table in peat normally is near the surface, the effective overburden pressure is negligible.

The void ratio of peat is very large, ranging from 9 up to 25. It usually tends to decrease with depth within a peat deposit. Such high void ratios give rise to phenomenally high water content, varying from a few hundreds per cent dry weight to over 3000% in some coarse fibrous varieties. Put another way, the water content may range from 75 to 98% by volume of peat. Peat, therefore, undergoes significant shrinkage on drying out. The magnitude of pore-water pressure is particularly significant in determining the stability of peat. With the exception of those peats with low water contents (less than 500%) and high mineral contents, the average bulk density of peat is slightly lower than that of water. Gas is formed in peat as plant material decomposes, the volume of gas varying from around 5 to 7.5%. Most of the gas is free and so has a significant influence on the rate of consolidation, pore pressure under load and permeability.

Differential and excessive settlement are the principal problems confronting the engineer working on a peat soil. Serious shearing stresses are induced, even by moderate loads. Worse still, should the loads exceed a given minimum, then settlement may be accompanied by creep, lateral spread or, in extreme cases, by rotational slip and upheaval of adjacent ground. At any given time, the total settlement in peat, due to loading, involves settlement with and without volume change. Settlement without volume change is the more serious for it can give rise to the types of failure mentioned above. What is more, it does not enhance the strength of peat.

Because of the potential problem of settlement arising from loading peat, especially in the construction of embankments carrying roads, some method of dealing with this problem has to be employed. Bulk excavation of peat frequently is undertaken if the deposit is less than 3 m in thickness. When a deposit exceeds 3 m or peat occurs as layers within soft sediments, precompression, involving surcharge loading, commonly is used. With few exceptions, improved drainage has no beneficial effect on the rate of consolidation.

When peatlands are drained artificially for reclamation purposes, the ground level can experience significant subsidence. The subsidence is not simply due to the consolidation that occurs as a result of the loss of the buoyant force of groundwater but also is attributable to desiccation and shrinkage associated with drying out in the zone of aeration and oxidation. For instance, in some parts of the Fenlands of eastern England, the thickness of peat has been almost halved as a result of drainage (e.g., Holme Post was installed in 1848 and by 1932 the thickness of the peat had been reduced from 6.7 m to 3.4 m; **Figure 9**).

## See Also

**Engineering Geology:** Problematic Rocks; Rock Properties and Their Assessment; Site and Ground Investigation; Subsidence.

## Further Reading

Andersland OB and Ladanyi B (1994) *An Introduction to Frozen Ground Engineering*. New York: Chapman and Hall.

Bell FG (1993) *Engineering Treatment of Soils*. London: E and FN Spon.

Bell FG (2000) *Engineering Properties of Soils and Rocks*, Fourth Edition. Oxford: Blackwell Scientific Publications.

Blight GE (ed.) (1997) *Mechanics of Residual Soils*. Rotterdam: AA Balkema.

Chen FH (1988) *Foundations on Expansive Soils*. Amsterdam: Elsevier.

Charman JM (1988) *Laterite in Road Pavements*. Special Publication 47, London: Construction Industry Research and Information Association (CIRIA).

Fookes PG (ed.) (1997) *Tropical Residual Soils*. Engineering Group Working Party Revised Report, London: Geological Society.

Fookes PG and Parry RHG (eds.) (1994) *Characteristics of Arid Soils*. Rotterdam: AA Balkema.

Jefferson I, Murray EJ, Faragher E, and Fleming PR (eds.) (2001) *Problematic Soils*. London: Thomas Telford Press.

Jefferson IF, Rosenbaum MS, and Smalley IJ (eds.) (2004) *Silt and Siltation, Problems and Engineering Solutions*. Berlin: Springer-Verlag.

# Rock Properties and Their Assessment

**F G Bell**, British Geological Survey, Keyworth, UK

© 2005, Elsevier Ltd. All Rights Reserved.

## Introduction

Those properties that can be used to describe rock materials in terms of engineering classification are referred to as index properties. Index properties frequently show a good correlation one with another. In order for the test of an index property to be useful it is desirable that it should be simple to obtain, rapidly performed, and inexpensive. The test results must be reproducible and the index properties must be relevant to the engineering requirement.

## Density and Porosity

The density of a rock is one of its most fundamental properties. It is influenced principally by its mineral composition on the one hand and the amount of pore space on the other; as the proportion of pore space increases so the density decreases. Four different types of density are recognised. Firstly, grain density is the mass of the mineral aggregate per volume of solid material. The grain density is similar to the specific gravity (or relative density) of a rock except that the specific gravity is not expressed in units, it being the ratio of solid rock to that of an equal volume of water at a specified temperature. Secondly, dry density is the mass of the mineral aggregate per volume. Thirdly, bulk density is the mass of mineral aggregate and natural water content per volume. Fourthly, saturated density is the mass of mineral aggregate and saturated water content per volume. The unit weight can be used instead of density, and is expressed in terms of stress ($kN\,m^{-3}$). It can be derived from density simply by dividing by 98.8 (e.g. density of $2200\,kg\,m^{-3}$ 98.8 = $22.26\,kN\,m^{-3}$ unit weight; density is usually expressed in $Mg\,m^{-3}$).

The specific gravity can be determined by grinding a rock to powder and using a density bottle or alternatively by the immersion in water method. Determination of dry density, bulk density, and saturated density is dependent upon accurately weighing the rock specimen concerned and upon the accurate measurement of its volume. In the case of the dry density the rock specimen is dried in a ventilated oven at 105°C until a constant mass is reached and then allowed to cool in a desiccator. In order to obtain the saturated density, the rock specimen is first saturated by immersing in water under vacuum with periodic agitation to remove any trapped air. This method is not suitable for rocks that slake when immersed in water. The bulk density simply requires the mass of the specimen as obtained from the field, that is, with its natural moisture content. The volume of specimens can be determined by the caliper method or the buoyancy method. Regular shaped specimens are required for the caliper method so that accurate measurements of dimensions are obtained; the volume is obtained by immersion in water with the buoyancy method.

The porosity of a rock can be defined as the volume of the pore space divided by the total volume expressed as a percentage (**Figure 1**). Grades of dry density and porosity were suggested by the International Association of Engineering Geology (IAEG) and are provided in **Table 1**. Total or absolute porosity involves the total pore volume, that is, it includes the occluded pores (occluded pores are isolated and not interconnected with other pores). However, the effective or net porosity is a more practical measurement of porosity and it may be regarded as the pore space from which water or fluid can be removed. The

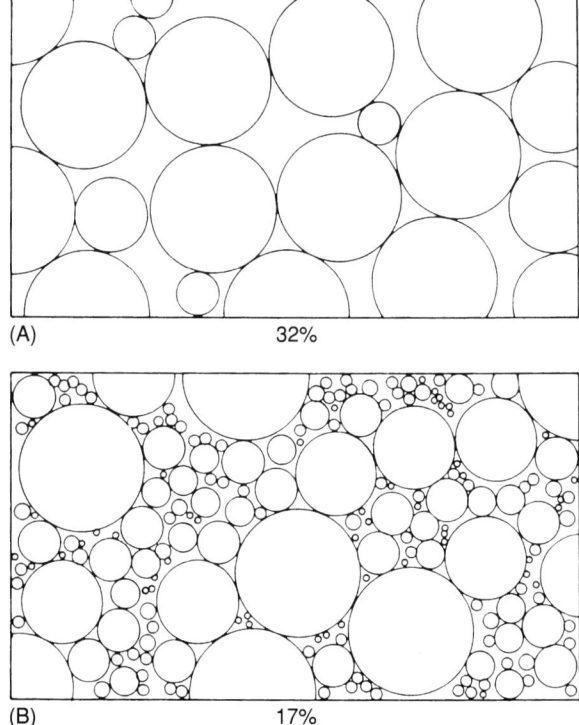

**Figure 1** Different values of porosity attributable to different degrees of sorting. Numbers beneath diagrams refer to percentage porosity.

**Table 1** Dry density and porosity

| Class | Dry density (Mg m$^{-3}$) | Description | Porosity (%) | Description |
|---|---|---|---|---|
| 1 | Less than 1.8 | Very low | Over 30 | Very high |
| 2 | 1.8–2.2 | Low | 30–15 | High |
| 3 | 2.2–2.55 | Moderate | 15–5 | Medium |
| 4 | 2.55–2.75 | High | 5–1 | Low |
| 5 | Over 2.75 | Very high | Less than 1 | Very low |

Reproduced from Bulletin International Association Engineering Geology, No. 19, 364–371, 1979.

latter generally is determined by using the standard saturation method. Alternatively, an air porosimeter can be used. In both methods, the pore volume is obtained by saturating with water or with air, the total volume being found by the caliper or bouyancy method. The porosity as determined by these two tests does not provide an indication of the way in which the pore space is distributed within a rock, or whether it consists of many fine pores or a smaller number of coarse pores. Two tests have been used to investigate the distribution of pore sizes and the microporosity of a rock specimen, namely, the suction plate test and the mercury porosimeter test. Microporosity in the suction plate method is defined as the volume of water retained (expressed as a percentage of the total available pore space) when a suction equivalent to 6.4 m of head of water is applied to the specimen. In effect, this measures the percentage of pores with an effective diameter of less than 5 $\mu$m. Such pores are able to retain water against applied suction and influence the amount of damage that can be caused by frost or by the crystallization of soluble salts that a rock used for building stone may undergo. Mercury is forced to penetrate the pores of the specimen under an applied pressure in the mercury porosimeter test. Obviously, the finer the pores, the higher the pressure that must be used to bring about penetration. In this way it is possible to derive the dimensions and pore size distribution from a graph showing the distribution of pores sizes. A line is drawn on the curve at the position where 10% of the pore space has been filled with mercury. The pores below this size limit can be regarded as the microporosity.

## Hardness

Hardness is one of the most investigated properties of materials, yet it is one of the most complex to understand. It does not lend itself to exact definition in terms of physical concepts. The numerical value of hardness is as much a function of the type of test used as a material property. The concept of hardness is usually associated with the surface of a material. For

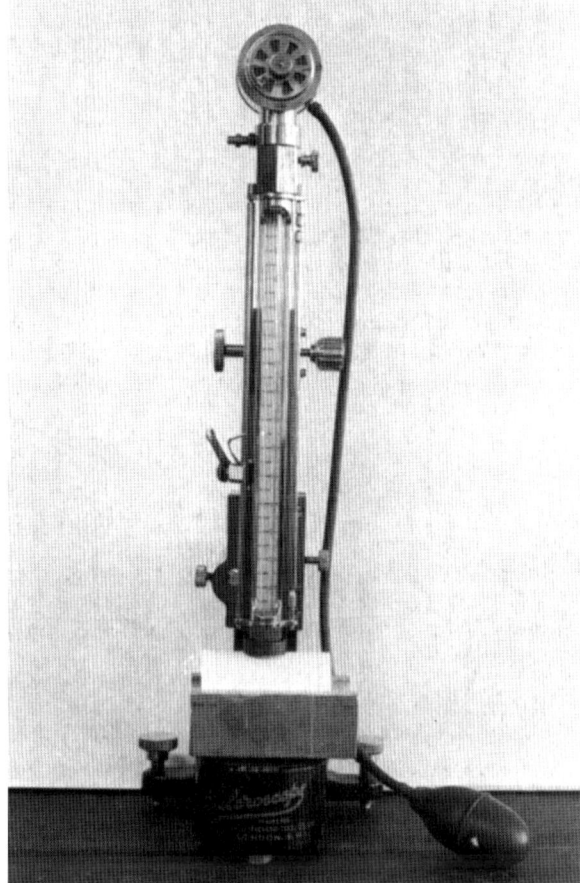

**Figure 2** The Shore scleroscope.

instance, the hardness of a rock can be considered as its resistance to a penetrating force, whether static or dynamic, or the resistance to displacement of surface particles by tangential abrasive force. As such, hardness is controlled by the efficiency of the bond between minerals or grains, as well as the strength of these two components.

A number of tests have been used to assess indentation hardness, of which the two most often used in rock mechanics testing are the Shore scleroscope and Schmidt hammer tests. The Shore scleroscope is a nondestructive hardness measuring device that indicates the relative values of hardness from the height of rebound of a small diamond-pointed hammer that is dropped vertically onto a securely clamped test surface from a height of 250 mm (**Figure 2**). Because a rock is not a homogeneous material, several hardness tests must be made over the surface of the specimen and the results averaged. Hence, at least 20 hardness determinations should be taken and each point of test should be at least 5 mm from any other. The Shore hardness value can be used to derive an approximate value of uniaxial compressive strength from **Figure 3** for rocks with strengths in excess of 35 MPa.

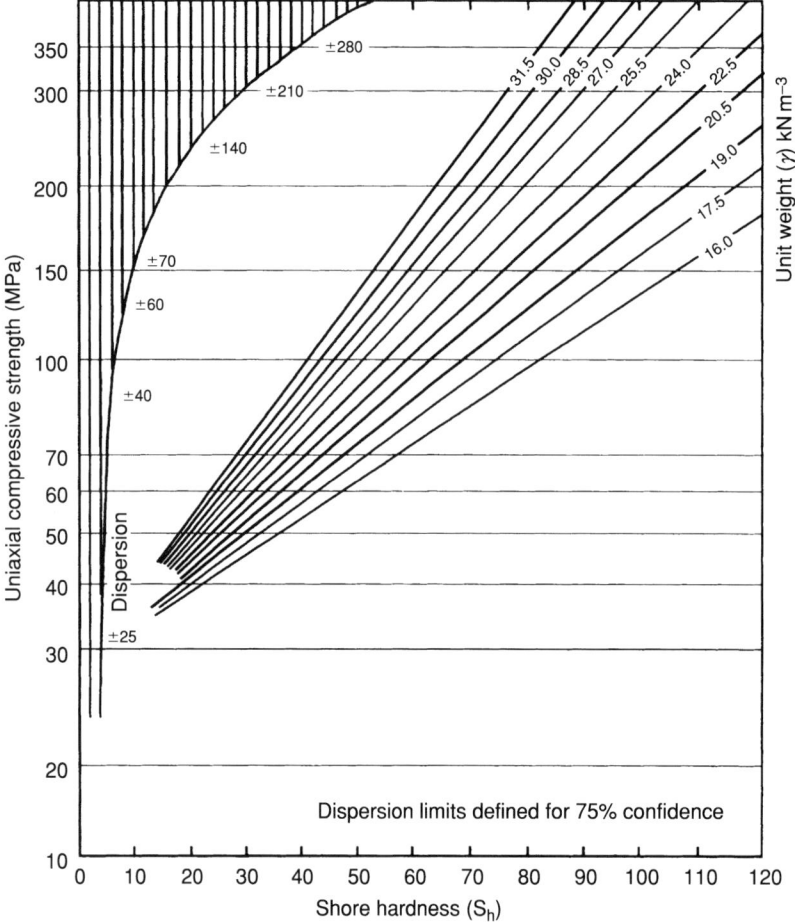

**Figure 3** Correlation chart for Shore hardness, relating unit weight of rock, unconfined compressive strength and hardness value (After Deere DU and Miller RP (1966) *Engineering Classification of Engineering Properties for Intact Rock*. Technical Report AFWL-TR-65-116, Air Force Weapons Laboratory, Kirkland Air Base, New Mexico).

The Schmidt hammer is a portable non-destructive device that expends a known amount of stored energy from a spring and indicates the degree of rebound of a hammer mass, following impact, within the instrument (**Figure 4**). Tests are made by placing the specimen in a rigid cradle and impacting the hammer at a series of points along its upper surface. The hammer is held vertically at right angles to the axis of the specimen. The specimens should have a flat smooth surface where tested and the rock beneath this area should be free from cracks. Test locations should be separated by at least the diameter of the plunger. At least 20 readings should be taken from each specimen. The lower 50% of the test values should be discarded and the average obtained from the upper 50%. This average is multiplied by the correction factor of the Schmidt hammer to obtain the hardness. However, the Schmidt hammer test is not a satisfactory method for the determination of the hardness of very soft or very hard rocks. Like the Shore sclerascope test, Schmidt hardness values can be used to derive approximate values of uniaxial compressive strength, as shown in **Figure 5**.

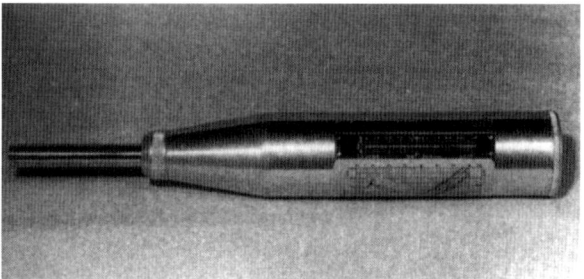

**Figure 4** The Schmidt hammer. Reproduced from Engineering Geology, vol. 4, 1979, Elsevier.

Abrasion tests measure the resistance of rocks to wear. The two abrasion tests most frequently used are the Dorry and Los Angeles tests. As both these tests are used to assess the resistance to wear of aggregate for road making, they are not dealt with here.

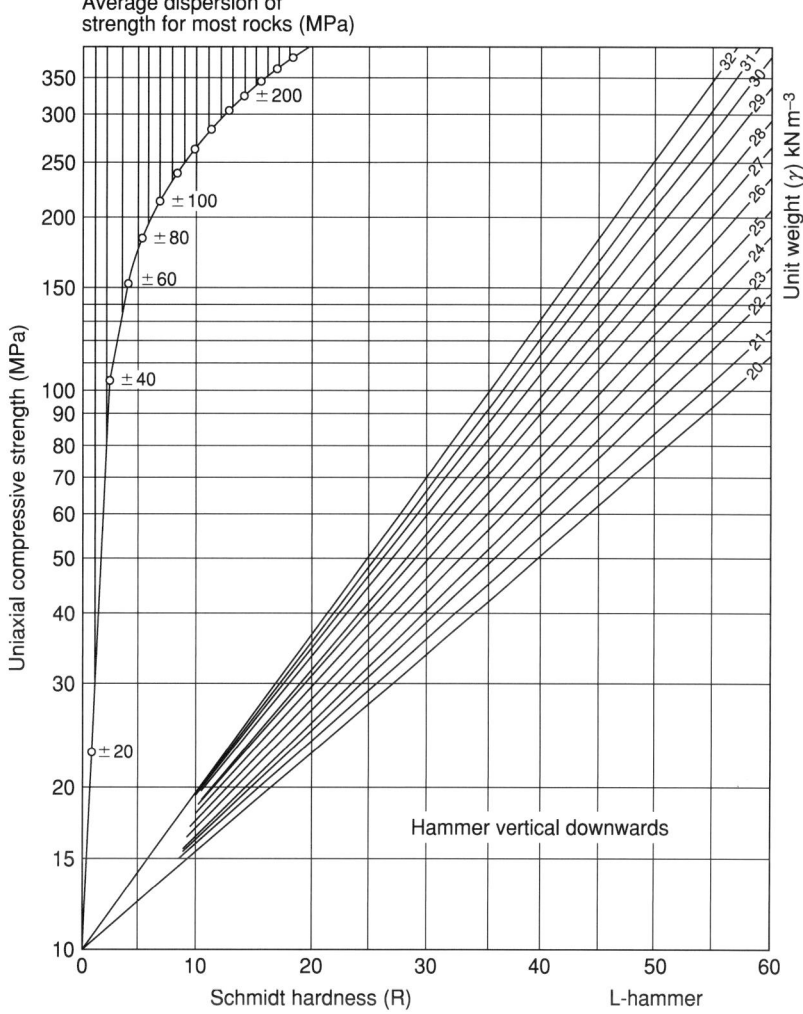

**Figure 5** Correlation chart for Schmidt hardness, relating unit weight, unconfined compressive strength and hardness value (After Deere DU and Miller RP (1966) *Engineering Classification of Engineering Properties for Intact Rock*. Technical Report AFWL-TR-65-116, Air Force Weapons Laboratory, Kirkland Air Base, New Mexico).

## Deformation of Rocks

The factors that influence the deformation characteristics and the failure of rock can be divided into two categories: internal and external. The internal category includes the inherent properties of the rock itself, whilst the external category is that of its environment at a particular point in time. As far as the internal factors are concerned, the mineralogical composition and texture are obviously important, but fractures within a rock and the degree of mineral alteration are frequently more important. The temperature-pressure conditions under which a rock exists, significantly affect its mechanical behaviour, as does its pore water content. Importantly, the length of time that a rock experiences a change in stress, and the rate at which this is imposed, significantly affects its deformation characteristics.

The composition and texture of a rock are governed by its origin and geological history. Few rocks are composed of only one mineral species and even when they are the properties of that species vary slightly from mineral to mineral. Such variations within minerals may be due to cleavage, twinning, inclusions, cracking, and alteration, as well as to slight differences in composition. This, in turn, is reflected in the mechanical behaviour of the resultant rocks. As a consequence, few rocks can be regarded as homogeneous isotropic substances. The size and shape relationships of the component minerals/grains are also significant in this respect, generally the smaller the size, the stronger the rock. One of the most important features of texture as far as mechanical behaviour, particularly strength, is concerned, is the degree of interlocking of the component minerals or grains. Breakage is more likely to take place along grain boundaries (intergranular fracture) than through grains (transgranular fracture) and therefore irregular boundaries make fracture more difficult. The bond between grains in many sedimentary rocks is provided by the cement and/or matrix, rather than by grains interlocking. The amount and, to a lesser extent, the type of cement/matrix is important, not only influencing strength and elasticity, but also density, porosity, and primary permeability. Rocks are not uniformly coherent materials, but contain defects that include microfractures, grain boundaries, mineral cleavage, twinning planes, inclusion trains, and elongated shell fragments. Obviously, such defects influence the ultimate strength of a rock and may act as surfaces of weakness that control the direction in which failure occurs. Crystal grain orientation in a particular direction facilitates breakage along that direction.

The presence of moisture in rocks adversely affects their engineering behaviour. For instance, moisture content increases the strain velocity and lowers the strength. More specifically the angle of internal friction is not affected significantly by changes in moisture content whereas the cohesion undergoes a notable reduction. It has therefore been suggested that the reduction in strength with increasing moisture content is due primarily to a lowering of the tensile strength, which is a function of the molecular cohesive strength of the material.

Although all rock types undergo a decrease in strength with increasing temperature and an increase in strength as the confining pressure is increased, the combined effect of these is notably different for different rock types. With increasing temperature there is a reduction in yield stress and strain hardening decreases. Heating enhances the ductility of rocks and their ability to deform permanently without loss of integrity. The transition from brittle to ductile deformation in porous rocks is characterized by an abrupt change from dilational behaviour at low stress to compaction during inelastic axial strain at high stress. This type of behaviour differs from that of rocks with low porosity. With the latter, dilatancy persists well into the ductile zone. The compaction that occurs during ductile deformation in porous rocks at high confining stress is due to collapse of the pore space and the rearrangement of grains to give more compact packing.

Four stages of deformation have been recognized, namely: elastic, elastico-viscous, plastic, and rupture. The stages are dependent upon the elasticity, viscosity, and rigidity of the rock, as well as on stress history, temperature, time, pore water, and anisotropy. An elastic deformation is defined as one that disappears when the stress responsible for it ceases. Ideal elasticity exists if the deformation on loading and its disappearance on unloading are both instantaneous. This is never the case with rocks since there is always some retardation, known as hysteresis, in the unloading process. With purely elastic deformation the strain is a linear function of stress, that is, the material obeys Hooke's law. Therefore, the relationship between stress and strain is constant, and is referred to as Young's modulus, $E$. Rock only approximates to an ideal Hookean solid. In fact, Young's modulus is not a simple constant but is related to the level of applied stress. Just how closely rock approximates to an ideal material depends on its homogeneity, isotropy, and continuity. Homogeneity refers to the physical continuity of a material, that is, the constituent particles are evenly distributed throughout its volume so that the elastic properties are the same at all points. Isotropy represents a measure of the directional properties of a rock. Hence, a rock is only isotropic if it is monomineralic and the crystals/grains have a random orientation. Since most rocks are composed of two or more essential minerals, which may possess preferred orientation, they are generally anisotropic. Continuity refers to the pore space and fractures within a rock. The degree of continuity affects the cohesion and so the transmission of stress throughout a rock.

The change in deformability at the elastic limit from elastic to plastic deformation is referred to as the yield point or yield strength. If the stress acting on a rock exceeds its elastic limit, then it becomes permanently strained, the latter being brought about by plastic flow. Within the zone of plastic flow there is a region where elastic stress is still important, referred to as the field of elastico-viscous flow. Plasticity may be regarded as time-independent, non-elastic, non-recoverable, stress-dependent deformation under uniform sustained load. Solids are classified as brittle or ductile according to the amount of plastic deformation they exhibit. In brittle materials the amount of plastic deformation is zero or very little, whereas it is large in ductile substances. Rupture occurs when the stress exceeds the strength of the material involved. It represents the maximum stress a rock is able to withstand prior to loss of cohesion by fracturing. The initiation of rupture is marked by an increase in strain velocity.

Most strong rocks exhibit little time-dependent strain or creep. However, creep in evaporitic rocks, notably salt, may greatly exceed the instantaneous elastic deformation. The time-strain pattern exhibited by such rocks, when subjected to a constant uniaxial stress, can be represented diagrammatically as shown in **Figure 6**. The instantaneous elastic strain,

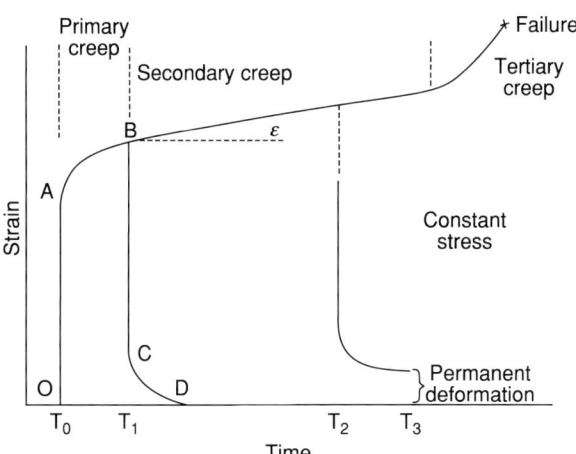

**Figure 6** Theoretical strain curve at constant stress (creep curve).

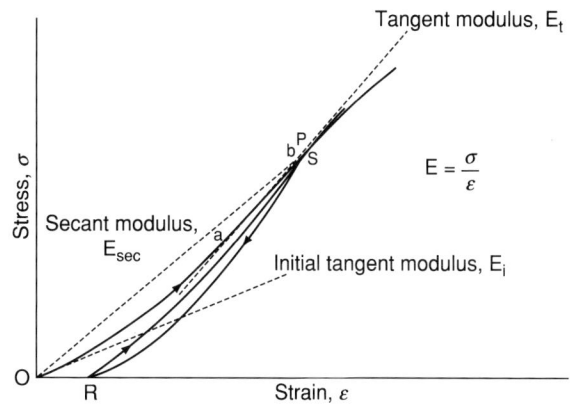

**Figure 7** Representative stress–strain curve for rock in unconfined compression, showing hysteresis.

which takes place when a load is applied, is represented by OA. There follows a period of primary or transient creep (AB) in which the rate of deformation decreases with time. If the stress is removed, the specimen recovers. At first this is instantaneous (BC), but this is followed by a time elastic recovery, illustrated by curve CD. On the other hand, if the loading continues, the specimen begins to exhibit secondary or pseudo-viscous creep. This type of creep represents a phase of deformation in which the rate of strain is constant. The deformation is permanent and is proportional to the length of time over which the stress is applied. If the loading is continued further, then the specimen suffers tertiary creep during which the strain rate accelerates with time and ultimately leads to failure.

## Elastic Properties

Young's modulus, E, is the most important of the elastic constants and can be derived from the slope of the strain–stress curve obtained when a rock specimen is subjected to unconfined compression (i.e., static loading), it being the ratio of stress to strain. The strains are measured by attaching strain gauges to the test specimens, or by displacement transducers, and recording their outputs. Strain measurements on specimens less than 50 mm in diameter, however, are high and not representative of the material behaviour. Most crystalline rocks have S-shaped stress–strain curves (**Figure 7**). At low stresses the curve is non-linear and concave upwards, that is, Young's modulus increases as the stress increases. The initial tangent modulus is given by the slope of the stress-strain curve at the origin. Gradually a level of stress is reached where the slope of the curve becomes approximately linear. In this region Young's modulus is defined as the tangent modulus or secant modulus. At this stress level the secant modulus has a lower value than the tangent modulus because it includes the initial 'plastic' history of the curve. A classification of deformability has been suggested by the IAEG and is given in **Table 2**.

**Table 2** Classification of deformability

| Class | Deformability (MPa × $10^{-3}$) | Description |
|---|---|---|
| 1 | Less than 5 | Very high |
| 2 | 5–15 | High |
| 3 | 15–30 | Moderate |
| 4 | 30–60 | Low |
| 5 | Over 60 | Very low |

Reproduced from Bulletin International Association Engineering Geology, No. 19, 364–371, 1979.

In addition to their non-elastic behaviour, most rocks exhibit hysteresis. Under uniaxial stress the slope of the stress–strain curve during unloading initially is greater than during loading for all stress values (**Figure 7**). As stress is decreased to zero a residual strain, OR, is often exhibited. On reloading the curve RS is produced that, in turn, is somewhat steeper than OP. Further cycles of unloading and reloading to the same maximum stress give rise to hysteresis loops, which are shifted slightly to the right. The non-linear elastic behaviour and elastic hysteresis of brittle rocks under uniaxial compression is due to the presence of flaws or minute cracks in the rock. At low stresses these cracks are open but they close as the stress is increased and the rock becomes elastically stiffer, that is, E increases with stress. Once the cracks are closed the stress–strain curve becomes linear.

When a specimen undergoes compression it is shortened and this generally is accompanied by an increase in its cross-sectional area. The ratio of lateral unit deformation to linear unit deformation, within the elastic range, is known as Poisson's ratio, $v$. This similarly can be obtained by monitoring strains during an unconfined compression test. The ideal geometrical value of Poisson's ratio is 0.333.

Another elastic constant is compressibility, K, which is the ratio of change in volume of an elastic solid to change in hydrostatic pressure. A further measure of elasticity is rigidity, G, which refers to the resistance of a body to shear. These four elastic constants (E, $v$, K, G) are not independent of each other and, if any two are known, it is possible to derive the other two from the following expressions:

$$G = E/2(1 + v) \quad [1]$$

and

$$K = E/3(1 - 2v) \quad [2]$$

Of the four constants, Young's modulus and Poisson's ratio are more readily determined experimentally.

Methods used to determine the dynamic values (as opposed to static values, see above) of Young's modulus and Poisson's ratio generally depend upon determining the velocities of propagation of elastic waves through a specimen of rock. These can be measured by using the high frequency ultrasonic pulse method, the low-frequency ultrasonic pulse technique, or the resonant method. For example, the high-frequency ultrasonic pulse method is used to determine the velocities of compressional, $v_p$, and shear, $v_s$, waves in rock specimens of effectively infinite extent compared to the wavelength of the pulse used. The condition of infinite extent is satisfied if the average grain size is less than the wavelength of the pulse that, in turn, is less than the minimum dimensions of the specimen. These two velocities can be substituted in the following expressions to derive the dynamic values of Young's modulus and Poisson's ratio:

$$E = \rho v_p^2 \frac{(1+v)(1-2v)}{(1-v)} \quad [3]$$

or

$$E = 2v_s^2 \rho (1 - v) \quad [4]$$

or

$$E = v_s^2 \rho \frac{[3(v_p/v_s)^2 - 4]}{[(v_p/v_s)^2 - 1]} \quad [5]$$

$$v = \frac{0.5(v_p/v_s)^2 - 1}{(v_p/v_s)^2 - 1} \quad [6]$$

where $\rho$ is density.

## Strength

### Uniaxial Compression

The uniaxial strength, also known as the unconfined compressive strength, of a rock may be regarded as the highest stress that a rock specimen can carry when a unidirectional stress is applied, normally in an axial direction to the ends of a cylindrical specimen. It represents the maximum load supported by a specimen during the test divided by the cross-sectional area of the specimen. Grades of unconfined compressive strength are shown in **Table 3**. Although its application is limited, the uniaxial compressive strength allows comparisons to be made between rocks and affords some indication of rock behaviour under more complex stress systems.

The behaviour of rock in uniaxial compression is influenced to some extent by the test conditions. The most important of these is the length-diameter or slenderness ratio of the specimen, the most satisfactory slenderness ratio being 2.5 since it provides a reasonably good distribution of stress throughout

**Table 3** Grades of unconfined compressive strength

| Geological Society (1977) | | IAEG (1979)* | | ISRM (1981)** | |
|---|---|---|---|---|---|
| Term | Strength (MPa) | Term | Strength (MPa) | Term | Strength (MPa) |
| Very weak | Less than 1.25 | Weak | Under 15 | Very low | Under 6 |
| Weak | 1.25–5.00 | Moderately strong | 15–50 | Low | 6–10 |
| Moderately weak | 5.00–12.50 | Strong | 50–120 | Moderate | 20–60 |
| Moderately strong | 12.50–50 | Very strong | 120–230 | High | 60–200 |
| Strong | 50–100 | Extremely strong | Over 230 | Very high | Over 200 |
| Very strong | 100–200 | | | | |
| Extremely strong | Over 200 | | | | |

*International Association of Engineering Geology.
**International Society for Rock Mechanics.

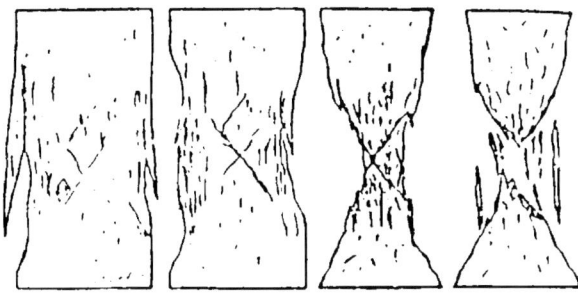

**Figure 8** Stages in the development of fracturing with increasing unconfined compressive loading.

the specimen. Secondly, the rate at which loading occurs also influences the compressive strength, a loading rate of between 0.5 and 1.0 MPa normally is recommended. Thirdly, the ends of a specimen should be lapped so that they are exactly perpendicular to the long axis.

The onset of failure in a rock specimen subjected to compressive loading is marked initially by the formation of a large number of isolated fractures, which characterizes the relief of stress concentration produced by the mechanical inhomogeneities in the rock. Most cracks are orientated parallel to the applied stress (**Figure 8**). This is quickly followed by the development of two groups of macroscopic shear failures, at the boundary and in the interior of the specimen, which suggests that most of the major sources of induced lateral tensile stresses have been eliminated. The interior macroscopic shear failures are extended and become interconnected to form a conjugate set of open shear fractures. Two central cones are formed, which either abrade during the large shear displacement or produce major fractures in the remaining rock material. If one shear failure surface becomes dominant, then cones are not developed and the sample ultimately fails in two parts along a diagonal plane of shear.

### Triaxial Compression Strength

A triaxial test is necessary if the complete nature of the failure of a rock is required. In this test a constant hydraulic pressure (the confining pressure) is applied to the cylindrical surface of the rock specimen, whilst applying an axial load to the ends of the sample. The axial load is increased up to the point where the specimen fails. Testing of the rock specimen is carried out within a special high pressure cell (**Figure 9**). A series of tests, each at higher confining pressure, are carried out on specimens from the same rock. These enable Mohr circles and their envelope to be drawn, from which the strength parameters, that is, the angle of friction ($\phi$) and value of the cohesion (c) are obtained (**Figure 10**). The shear strength ($\tau$) is then derived from the Mohr–Coulomb criterion:

$$\tau = c + \sigma \tan\phi \qquad [7]$$

where $\sigma$ is the normal stress.

### Direct Shear

A number of other tests can be used to deduce the shear strength of intact rock and of rock fracture surfaces, in addition to the triaxial test. These include the shear box test, the direct single and double shear tests, the punch shear test, and the torsion test. The most commonly used method is the shear box test, which is particularly useful for assessing the shear strength of weaker rocks and also for obtaining the shear strength along discontinuity surfaces. In this test a constant normal force is applied to the specimen, which then is sheared. A number of tests on the same material are undertaken, each at a higher normal stress, so that a shear strength versus normal stress graph can be drawn from which the value of cohesion and angle of friction are derived (**Figure 11A**). The other shear tests are used infrequently but are illustrated in **Figure 11B**.

### Tensile Strength

Rocks have a much lower tensile strength than compressive strength. Brittle failure theory predicts a ratio of compressive strength to tensile strength of about 8:1 but in practice it is generally between 15:1 and 25:1. The direct tensile strength of a rock can be obtained by attaching metal end caps with epoxy resin to the specimen, which are then pulled into tension by wires. In direct tensile tests the slenderness ratio of cylindrical specimens should be 2.5 to 3.0 and the diameter preferably should not be less than 54 mm. The ratio of the diameter of the specimen to the largest crystal/grain in the rock should be at least 10:1. Unfortunately, the determination of the direct tensile strength often proves difficult since a satisfactory method has not yet been devised to grip the specimen without introducing bending stresses. Accordingly, most tensile tests have been carried out by indirect methods.

The Brazilian test is an indirect method of assessing the tensile strength of rocks, based on the observation that most rocks in biaxial stress fields fail in tension when one principal stress is compressive. In this test a cylindrical specimen of rock is loaded in a diametrical plane along its axis. The sample usually fails by splitting along the line of diametrical loading and the indirect tensile strength ($T_b$) can be obtained from:

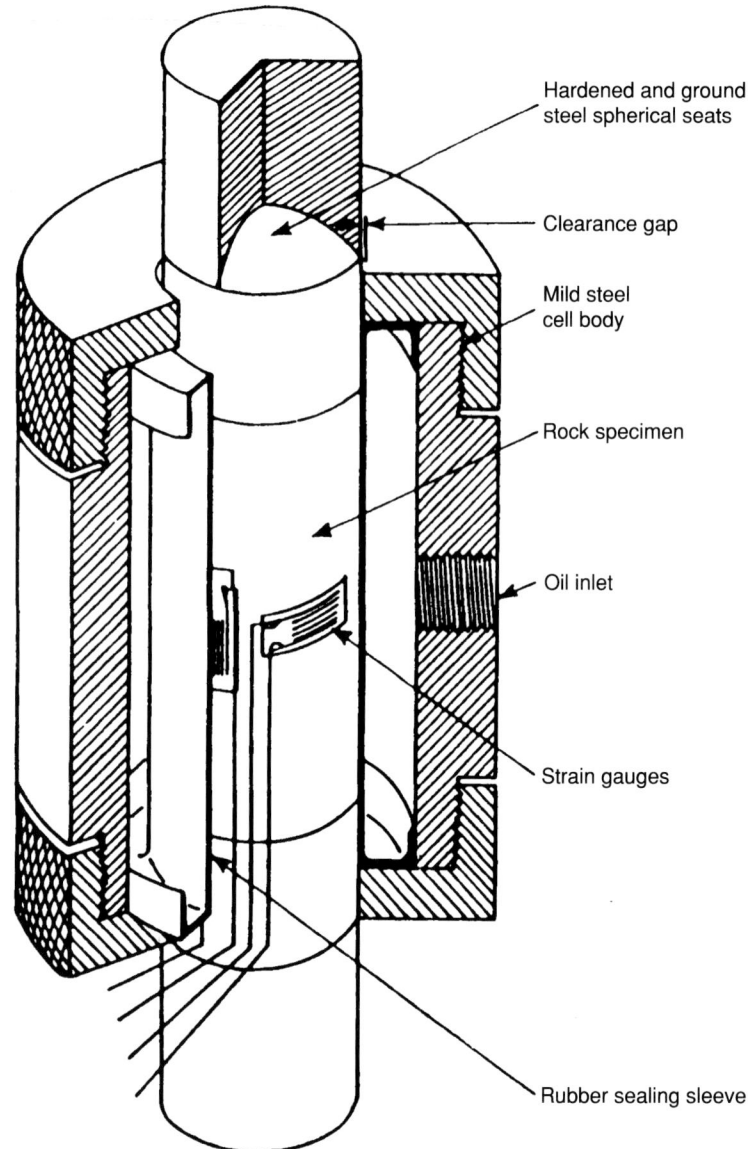

**Figure 9** Cutaway view of a triaxial cell for testing rock.

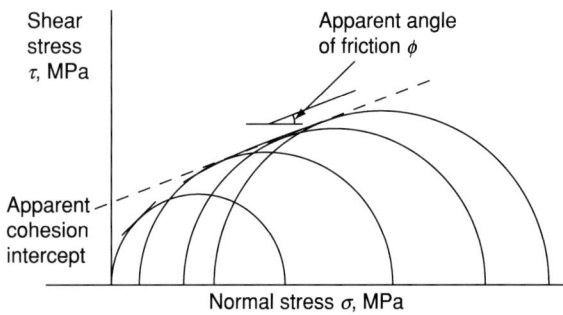

**Figure 10** Mohr envelope and circles.

$$T_b = 2P/\pi LD \qquad [8]$$

where P is the load at failure, and L and D are the length and diameter of the specimen respectively. Disc-shaped specimens are used in the Brazilian disc test. In this case curved jaw loading platens are used to improve loading conditions. Uncertainties associated with the premature development of failure can be removed by drilling a hole in the centre of the disc-shaped specimen (sometimes this has been referred to as a ring test). A disc-shaped specimen should be wrapped around its periphery with a layer of masking tape and the specimen should not be less

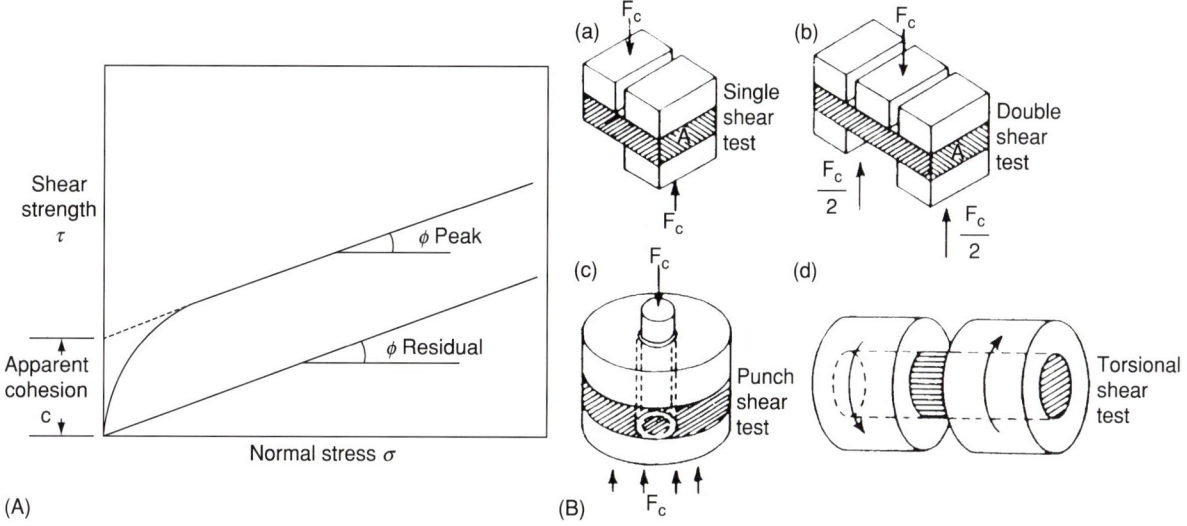

**Figure 11** (A) Graph derived from shear test, showing curves derived for both peak and residual strength. (B) Other forms of shear strength tests.

54 mm in diameter, with a thickness that is approximately equal to the radius of the specimen. The tensile strength of the specimen is obtained as follows:

$$T_b = 0.636P/DH \qquad [9]$$

where H is the thickness of the specimen. The Brazilian test is useful for brittle materials but for other materials it may give erroneous results.

In the point load test the specimen is placed between opposing cone-shaped platens and subjected to compression. This generates tensile stresses normal to the axis of loading and the indirect tensile strength ($T_p$) is then derived from:

$$T_p = P/D^2 \qquad [10]$$

Loading can take place across the diameter of the specimen, as in Figure 12, or along the axis. Rocks that are anisotropic should be tested along and parallel to the lineation. Irregular-shaped specimens can also be tested, but at least 20 tests should be made on the same sample material and the results averaged to obtain a value. The point load test is limited to rocks with uniaxial compressive strengths exceeding 25 MPa (i.e., point load index above 1 MPa).

The effect of the size of specimens is greater in tensile than compression testing because in tension, cracks open and give rise to large strength reductions, whilst in compression the cracks close and so disturbances are appreciably reduced. This is especially the case in the axial and irregular lump point load tests. Accordingly, a standard distance between the two cones of 50 mm has been recommended, to which other sizes should be corrected by reference to a correction chart. Once determined, the point load index can be used to grade the indirect tensile strength of rocks, as shown in **Table 4**.

Finally, in the flexural test a cylindrical specimen of rock is loaded between one lower and two upper supports until the sample fails. The flexural strength gives a higher value of tensile strength than that determined in direct tension.

## Durability of Rocks

Durability refers to the resistance that a rock offers to the various processes that lead to its breakdown and therefore durability tests can be used to provide a general impression of how a rock will behave in relation to weathering, especially mechanical weathering. Durability tests most frequently are used to assess the behaviour of suspect rocks, that is, those that tend to breakdown more readily such as mudrocks, some chalks, and certain basalts and dolerites. There are a large number of tests that have been used to assess the durability and many of them are used to determine the durability of rock as a material for building purposes, for aggregate, or for armourstone. The latter types of test are not dealt with here.

Some of the more simple tests include the water absorption test, the wet and dry test, the freeze-thaw test, and soak tests. The water absorption test involves oven drying a rock specimen at 105°C until it has attained a constant weight and then saturating it under vacuum. The percentage saturation is determined and reflects porosity. As rocks break down, their porosity increases and so the water absorption test has been used to indicate the degree of

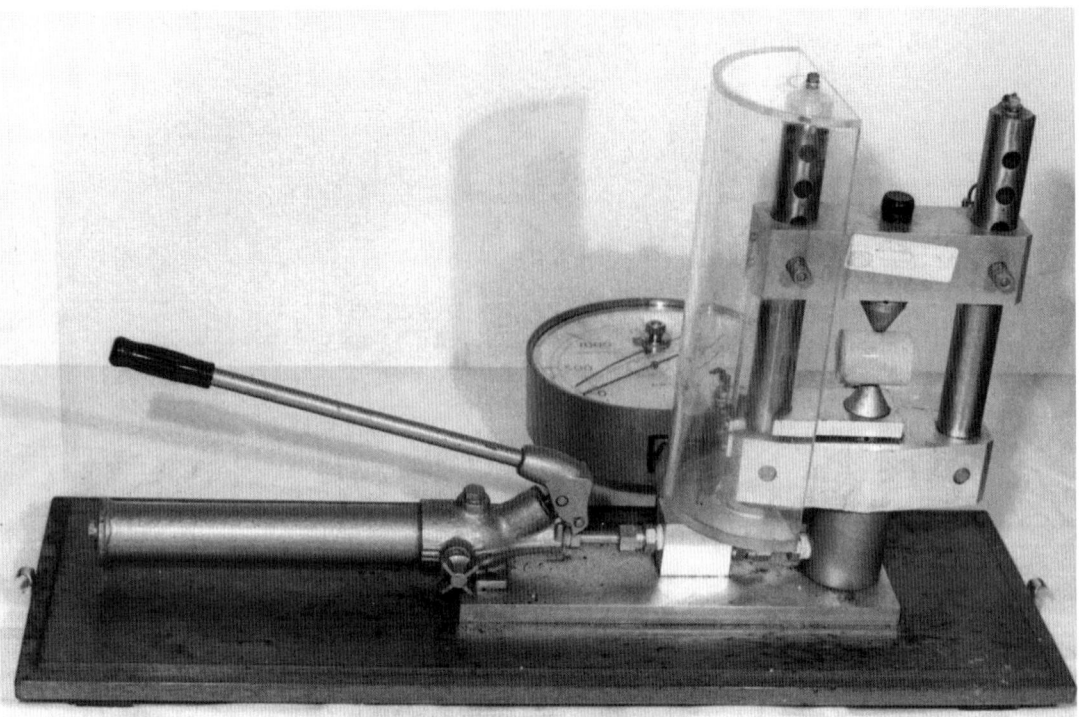

**Figure 12** Point load test apparatus.

**Table 4** Point load strength classification

| Description | Point load strength index (MPa) | Equivalent uniaxial compressive strength (MPa) |
|---|---|---|
| Extremely high strength | Over 10 | Over 160 |
| Very high strength | 3–10 | 50–160 |
| High strength | 1–3 | 15–60 |
| Medium strength | 0.3–1 | 5–16 |
| Low strength | 0.1–0.3 | 1.6–5 |
| Very low strength | 0.03–0.1 | 0.5–1.6 |
| Extremely low strength | Less than 0.03 | Less than 0.5 |

weathering a rock, especially a crystalline rock such as granite, has undergone. In other words, is it fresh, slightly weathered, moderately weathered, or highly weathered. However, some moderately and highly weathered rocks may break down before becoming saturated. Similarly, in the wet and dry test the rock specimen is first dried and then saturated, but this time for a given number of cycles. The effect that cyclic wetting and drying has on the specimen is recorded (i.e., no effect, softening, minor spalling, minor hairline cracking, severe hairline cracking, breakdown before last cycle noting cycle number). In the freeze-thaw test the specimen is saturated and then frozen for 24 h. The specimen is subjected to a given number of cycles and the effects are recorded in the same way as for the wet and dry test. The freeze-thaw test has been used to assess the frost resistance of building stone. However, it is no longer used for this purpose in Britain, it being regarded as unsatisfactory primarily because of the difficulty of interpreting the results in relation to a period of time over which a rock will perform as required. Soak tests are used to assess the breakdown of rocks as a result of swelling brought about by the absorption of water, especially of those rocks that contain swelling minerals. The rock specimen either may be soaked in water or ethylene glycol $(CH_2OH)_2$ for a given number of days. Ethylene glycol is much more effective than water as far as assessment of those rocks that contain swelling minerals are concerned. The soak test allows five classes of rock disintegration and the time when the worst condition occurs to be recognized:

Degree of disintegration:
    Class 1 : No obvious effects, or only very minor spalling of sand-sized particles.
    Class 2 : Flaking and/or swelling.
    Class 3 : Fracturing without extensive spalling.
    Class 4 : Fracturing with extensive spalling.
    Class 5 : Complete disintegration.

Time required to reach worst condition:
    Class 6 : 0–1 day
    Class 5 : 2–3 days

**Table 5** Ethylene glycol soak test index values

| Degree of disintegration class | Time class | | | | | | |
|---|---|---|---|---|---|---|---|
| | 6<br>0–1 day | 5<br>2–3 days | 4<br>4–10 days | 3<br>11–15 days | 2<br>16–20 days | 1<br>21–30 days | 0<br>Over 30 days |
| 1 | 7 | 6 | 5 | 4 | 3 | 2 | 1 |
| 2 | 8 | 7 | 6 | 5 | 4 | 3 | 2 |
| 3 | 9 | 8 | 7 | 6 | 5 | 4 | 3 |
| 4 | 10 | 9 | 8 | 7 | 6 | 5 | 4 |
| 5 | 11 | 10 | 9 | 8 | 7 | 6 | 5 |

Class 4 : 4–10 days
Class 3 : 11–15 days
Class 2 : 16–20 days
Class 1 : 21–30 days
Class 0 : Over 30 days

An index value can be derived from the integration of the degree of disintegration and time taken to reach the worst condition. In other words, a scale of relative durability values can be developed, ranging from 1 in which specimens are more or less unaffected to 11 when they are totally disintegrated (**Table 5**). The ethylene glycol soak test has been used to distinguish certain unsound basalts and dolerites, commonly described as slaking types, from sound types, particularly in relation to their use as road aggregate.

Slaking refers to the breakdown of rocks, especially mudrocks, by alternate wetting and drying. If a fragment of mudrock is allowed to dry out, air is drawn into the outer pores and high suction pressures develop. When the mudrock is next saturated the entrapped air is pressurised as water is drawn into the rock by capillary action. This slaking process causes the internal arrangement of grains to be stressed. Given enough cycles of wetting and drying, breakdown can occur as a result of air breakage, the process ultimately reducing the mudrock to gravel-sized fragments. The slake-durability test estimates the resistance to wetting and drying of a rock sample, and is particularly suitable for mudrocks and shales. The sample, which consists of 10 pieces of rock, each weighing about 40 g, is placed in a test drum, oven dried, and weighed. After this, the drum, with sample, is half immersed in a tank of water and attached to a rotor arm, which rotates the drum for a period of 10 min at 20 revolutions per minute (**Figure 13**). The cylindrical periphery of the drum is formed using a 2 mm sieve mesh, so that broken down material can be lost whilst the test is in progress. After slaking, the drum and the material retained are dried and weighed. The slake-durability index is then obtained by dividing the weight of the sample retained by its original weight, and expressing the answer as a percentage. The following scale of slake-durability is used:

under 25%, very low
25–50%, low
50–75%, medium
75–90%, high
90–95%, very high
over 95%, extremely high.

**Figure 13** Slake-durability test apparatus.

However, it has been suggested that durable mudrocks may be better distinguished from non-durable types on the basis of compressive strength and three-cycle slake-durability index (i.e. those mudrocks with a compressive strength of over 3.6 MPa and a three cycle slake-durability index in excess of 60% are regarded as durable). In fact, the value of the slake durability test as a means of assessing mudrock durability has been questioned as the results obtained frequently do not compare well with those of other durability tests. This has led to a number of adaptations being made to the test, for example, ethylene glycol has been used instead of water, the time taken to carry out the test has been extended and the number of cycles has been increased.

Failure of weak rocks occurs during saturation when the swelling pressure (or internal saturation swelling stress, $\sigma_s$) developed by capillary suction pressures, exceeds their tensile strength. An estimate of $\sigma_s$ can be obtained from the modulus of deformation (E):

$$E = \sigma_s / \varepsilon_D \quad [11]$$

where $\varepsilon_D$ is the free swelling coefficient. The latter is determined by a sensitive dial gauge recording the amount of swelling of an oven-dried core specimen per unit height along the vertical axis during saturation in water for 12 h, $\varepsilon_D$ being obtained as follows:

$$\varepsilon_D = \frac{\text{change in length after swelling}}{\text{initial length}} \quad [12]$$

A durability classification has been developed based on the free-swelling coefficient and uniaxial compressive strength (**Figure 14**).

When the free expansion of rocks liable to swell is inhibited, stresses of sufficient magnitude to cause damage to engineering structures may develop. A number of tests may be conducted that provide an indication of the swelling behaviour of such rock. These include the swelling strain index test performed under unconfined conditions, and the swelling pressure index test, carried out under conditions of zero volume change. Swelling strain measurements can also be undertaken on radially confined specimens under various conditions of axial loading. The swelling strain represents maximum expansion of an unconfined rock specimen when it is submerged in water. Test specimens may either be oven-dried or retain their natural moisture content. As a test specimen is submerged in water it is advisable to prevent rock prone to slaking from collapsing before the ultimate swelling strain has developed by wrapping the specimen in muslin. During the test the specimen is supported by a frame between a fixed point and a measuring point. The latter may consist of a dial gauge or a deformation transducer. The specimen and frame are placed in a container that is filled

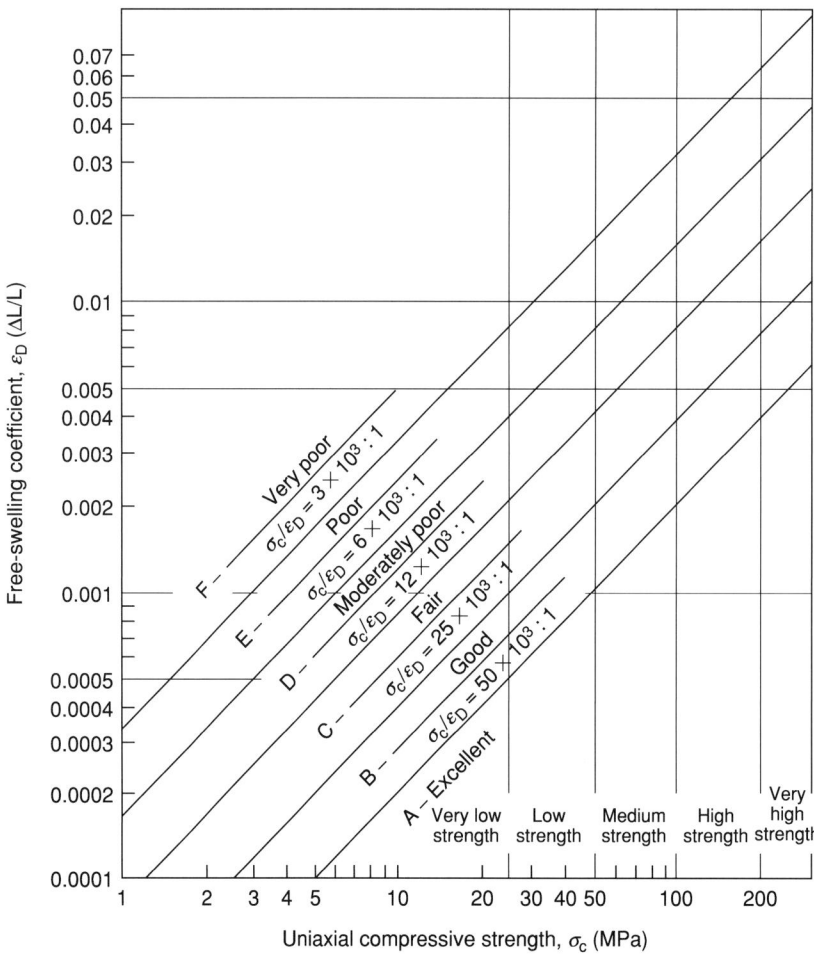

**Figure 14** Geodurability classification chart (After Olivier HJ (1979) A new engineering-geological rock durability classification. *Engineering Geology* 4: 255–279).

with water just above the level of the specimen. Strain is measured in all three perpendicular directions. No strain is allowed to develop in the test specimen during the swelling pressure test. This is accomplished by tightly mounting a cylindrical specimen inside a rigid ring that provides radial constraint. The specimen absorbs water through porous end plates. Any axial strain that develops is monitored and compensated for by increasing the axial load. The stress required to prevent expansion when equilibrium conditions are established is equal to the swelling pressure index.

## Permeability

Permeability considers the ability of a rock to allow the passage of fluids into or through it without impairing its fabric. In ordinary hydraulic usage, a substance is called 'permeable' when it permits the passage of a measurable quantity of fluid in a finite period of time and 'impermeable' when the rate at which it transmits that fluid is slow enough to be negligible under existing temperature-pressure conditions. The permeability of a particular rock is defined by its coefficient of permeability or hydraulic conductivity. Grades of permeability are given in **Table 6**.

Determination of the permeability of many rock types in the laboratory is made by using a falling-head permeameter (**Figure 15A**). The sample is placed in the permeameter, which is then filled with water to a certain height in the standpipe. The stopcock is then opened and the water allowed to infiltrate the

**Table 6** Grades of permeability (IAEG, 1979)*

| | Permeability | |
| --- | --- | --- |
| Class | $m\,s^{-1}$ | Description |
| 1 | Greater than $10^{-2}$ | Very highly |
| 2 | $10^{-2}$–$10^{-4}$ | Highly |
| 3 | $10^{-4}$–$10^{-5}$ | Moderately |
| 4 | $10^{-5}$–$10^{-7}$ | Slightly |
| 5 | $10^{-7}$–$10^{-9}$ | Very slightly |
| 6 | Less than $10^{-9}$ | Practically impermeable |

*International Association of Engineering Geology.

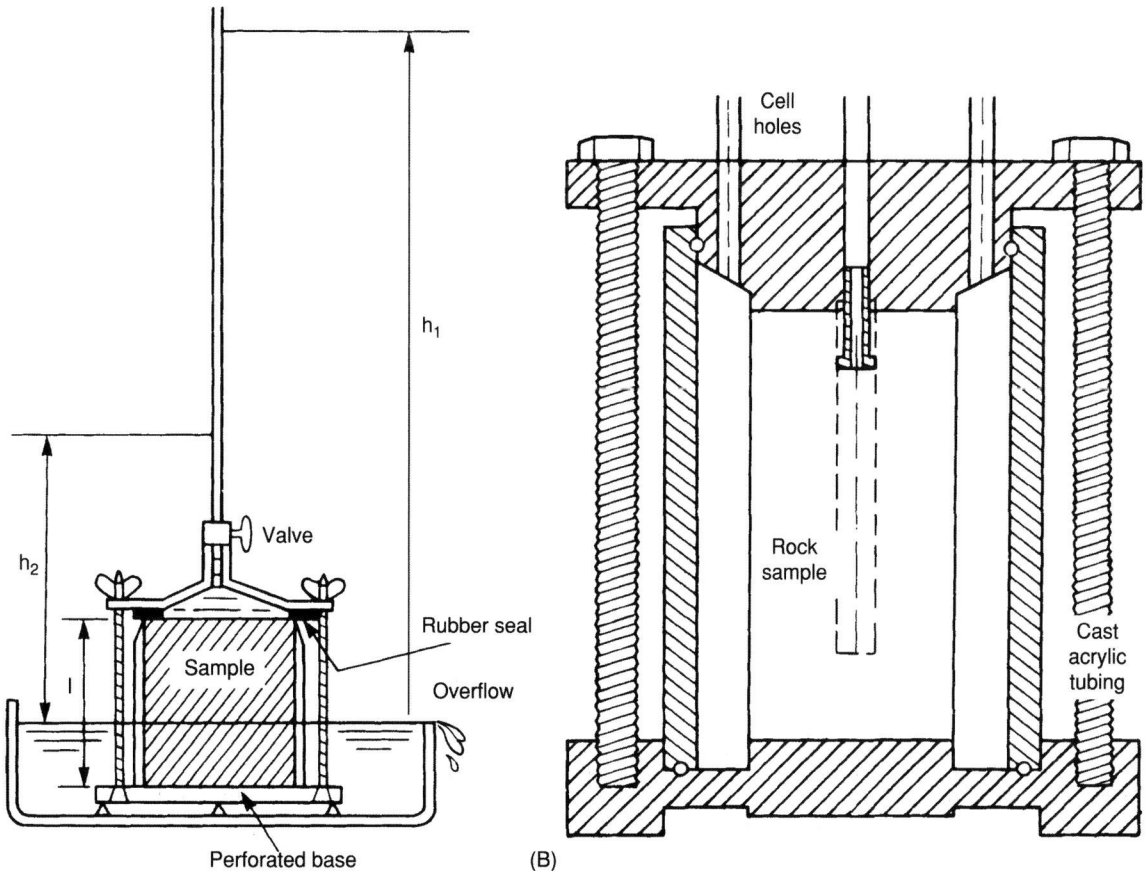

**Figure 15** (A) Falling head permeameter. (B) Radial percolation test apparatus.

sample, the height of the water in the standpipe falling. The times at the beginning, $t_1$, and end, $t_2$, of the test are recorded and these, together with the two corresponding heights, $h_1$ and $h_2$, the cross-sectional area of the standpipe, a, and cross-sectional area, A, and length, l, of specimen are substituted in the following expression to derive the coefficient of permeability, k:

$$k = \frac{2.303 a l}{A(t_2 - t_1)} \times (\log_{10} h_1 - \log_{10} h_2) \quad [13]$$

Variations of permeability in rocks under stress can be obtained by using a radial percolation test. A cylindrical specimen, in which an axial hole is drilled, is placed in the radial percolation cell. The latter can either contain water under pressure where the axial hole is in contact with atmospheric pressure; or water can be injected under pressure into the hole (**Figure 15B**). The flow is radial over almost the whole height of the sample and is convergent when the water pressure is applied to the outer face of the specimen, and divergent when the water is under pressure within the axial hole. Porous rocks remain more or less unaffected by pressure changes. On the other hand, fissured rocks exhibit far greater permeability in divergent flow than in convergent flow. Moreover, fissured rocks exhibit a continuous increase in permeability as the pressure attributable to divergent flow is increased.

## See Also

**Aggregates. Engineering Geology:** Codes of Practice; Natural and Anthropogenic Geohazards; Problematic Rocks.

## Further Reading

Anon (1975) *Methods of Sampling and Testing Mineral Aggregates, Sands and Fillers, BS 812*. London: British Standards Institution.

Anon (1982) *Standard Test Methods for Absorption and Bulk Specific Gravity of Natural Building Stone, C93–117*. Philadelphia: American Society for Testing Materials.

Bell FG (ed.) (1992) *Engineering in Rock Masses*. Oxford: Butterworth-Heinemann.

Bell FG (2000) *Engineering Properties of Soils and Rocks*. Oxford: Blackwell Science.

Brown ET (ed.) (1981) *Rock Characterization, Testing and Monitoring*. Oxford: Pergamon Press.

Farmer IW (1983) *Engineering Behaviour of Rocks*, 2nd ed. London: Chapman and Hall.

Goodman RE (1989) *An Introduction to Rock Mechanics*, 2nd ed. New York: Wiley.

Hudson JA and Harrison JP (1997) *Engineering Rock Mechanics: An Introduction to the Principles*. Oxford: Pergamon.

# Site and Ground Investigation

**J R Greenwood**, Nottingham Trent University, Nottingham, UK

© 2005, Elsevier Ltd. All Rights Reserved.

## Introduction and Terminology

The procedure of 'investigation' is fundamental to any project or activity involving the ground. The historical records need to be reviewed, current conditions need to be established, and the consequences of the proposed activity, works, or construction need to be carefully considered.

Investigation is an on-going process of establishing and reviewing the facts and processing the information to assist our future activities. With respect to construction works the following definitions are used:

- 'Site investigation' is a continuous process, as the construction project develops, involving both the site under consideration and the interaction with the surrounding areas. It is not confined to obtaining information on geotechnical aspects but may include hydrological, meteorological, geological, and environmental investigation.
- 'Ground investigation' is more site-specific and aims to investigate ground and groundwater conditions in and around the site of a proposed development or an identified post-construction problem.

The term 'site characterization' is now also used; it stems from the environmental specialist's study of contaminated sites but is equally applicable to any site. 'Characterization' perhaps implies the results of 'investigation'.

This article reviews the procedures necessary for quality site investigation to be carried out and describes some of the ground investigation techniques commonly applied in advance of construction or remedial works.

## Responsibilities

Investigation work must be entrusted to competent professionals with appropriate geotechnical engineering or engineering geology experience. Advice on the qualifications and experience required of such a professional is given in the publications of the Site Investigation Steering Group. The Geotechnical Advisor, who is an appropriately experienced Chartered Engineer or Chartered Geologist in the UK, will be a key figure in ensuring the correct geotechnical input to the project and should be appointed at an early stage and continue to work with the project team throughout the life of the project. Projects proceeding without the benefit of specialist geotechnical advice are more likely to encounter unforeseen problems, resulting in delay and budget overspend.

## The Investigation Process

### Investigation Stages

The investigation work for most projects can be divided into stages, as illustrated in **Table 1**. The Geotechnical Advisor will ensure appropriate geotechnical input at each stage (see **Engineering Geology: Codes of Practice**).

### The Procedural Statement

The key to successful investigation lies in the planning process. If all aspects of the investigation work are considered in advance, together with necessary actions relating to the probable findings, then the outcome is likely to be satisfactory for all parties involved.

A convenient way to bring together and record the proposals for each stage of site and ground investigation is by preparing a 'Procedural Statement' or 'Statement of Intent'. This approach was formally introduced for the United Kingdom by the Department of Transport/Highways Agency in the 1980s and has now become widely accepted as good practice. An example of the topics covered in a Procedural Statement is given in **Table 2**. Headings and content will change slightly for each phase of the investigation process as more information is accumulated. The Procedural Statement is usually prepared by the geotechnical engineer or specialist responsible for the work and should be agreed by all interested parties, and in particular the client, before the investigation proceeds.

The Procedural Statement encourages the designer to consider relevant aspects of the proposed investigation and to seek authority to proceed. It forms a valuable document within a quality management system, and it becomes a base reference as the investigation proceeds in case changes are needed in the light of the findings.

## The Desk Study

The desk study, sometimes referred to as the 'initial appraisal' or 'preliminary sources' study, is vital for gaining a preliminary understanding of the geology of the site and the likely ground behaviour. The term 'desk study' can be misleading because, in addition to collection and examination of existing information, it must include a walk-over survey. The study will determine what is already known about the site and how the ground should be investigated.

Before embarking on groundwork, much valuable information may be readily gleaned from existing

**Table 1** Stages of an investigation (developed from Clayton CRI, Matthews MC, and Simons NE (1995) *Site Investigation: A Handbook for Engineers*, 2nd edn. Oxford: Blackwell Scientific)

| Construction phase | Investigation work |
| --- | --- |
| Definition of project | Appointment of Geotechnical Advisor for advice on likely design issues |
| Site selection | Preliminary sources study (desk study) to provide information on the relative geotechnical merits of available sites |
| Conceptual design | Detailed preliminary sources study (desk study) and site inspections to provide expected ground conditions and recommendations for dealing with particular geotechnical design aspects and problems<br>Plan ground investigation (Procedural Statement) |
| Detailed design | Full ground investigation and geotechnical design; (additional ground investigation if necessary for design changes or for problematic ground conditions) |
| Construction | Comparison of actual and anticipated ground conditions; assessment of new risks (additional ground investigation if necessary) |
| Performance/maintenance | Monitoring, instrumentation, feedback reporting |

**Table 2** Example of a Procedural Statement's contents to be prepared before the ground investigation phase (HD 22/02)

The Procedural Statement (sometimes referred to as 'statement of intent' or the 'ground investigation brief') should be prepared by the responsible Geotechnical Advisor and agreed by the client and interested parties

*1. Scheme*
Details of scheme and any alternatives to be investigated; key location plan

*2. Objectives*
(For example) to provide information to confirm and amplify the geotechnical and geomorphological findings of the desk study as reported separately and to obtain detailed knowledge of the soils encountered and their likely behaviour and acceptability (for earthworks). To ascertain groundwater conditions and location of any underground workings (work limits to be defined)

*3. Special problems to be investigated*
Location of structures. Subsoil conditions below high embankments. Aquifers and likely water-bearing strata affecting the proposed works. Rock-stability problems. Manmade features to be encountered. Effects on adjacent properties

*4. Existing information*
List of all relevant reports and data

*5. Proposed investigation work*
*Fieldwork* – Details of exploratory work proposed for specific areas with reasons for choice of investigation methods selected. Proposed sampling to match laboratory testing
*Laboratory work* – Details of proposals with reasons for choice of tests and relevance to design

*6. Site and working restrictions*
Assessment of risk associated with proposals. Site safety, traffic management, difficult access, railway working

*7. Specialist consultation*
Details of specialist needed to support proposals

*8. Programme, cost, and contract arrangements*
Anticipated start date, work programme, contract arrangements, cost estimates, specification and conditions of contract. Arrangements for work supervision

*9. Reporting*
Responsibility for factual and interpretive reporting. Format of reports and topics to be covered

**Table 3** Some sources of information for desk study work (preliminary sources study)

| Topic | Possible sources |
|---|---|
| Site topography | Topographical maps, aerial photographs |
| Geology and soil conditions | Geological maps, regional guides and publications (British Geological Survey, sheet memoirs), learned society journals |
| Geotechnical problems and parameters | Published technical journal articles, civil engineering and geological journals, newspapers, previous ground investigation reports, local authorities |
| Groundwater conditions | Topographical maps, aerial photographs, well records, previous ground investigation reports, water authorities (flood records) |
| Meterological conditions | Meterorological Office |
| Existing site use and services | As-built drawings, utility information, mining records, construction press, land-use maps, commercial records, contamination records |
| Previous land use | Aerial photographs, old maps, archaeological records, agricultural records, mining records, local resident knowledge, local library records |

sources, such as geological and Ordnance Survey maps, aerial photographs, and archival material. Such documents can yield significant information about site conditions and, following the walk-over survey, a geotechnical plan of the site may be prepared. A checklist of the type of information to be sought in a desk study is given in **Table 3**.

The desk study is often the most cost-effective element of the entire site-investigation process, revealing facts that cannot be discovered in any other way. The preliminary engineering concepts for the site are prepared and developed at the desk-study phase, based on the acquired information. The ground investigation in the field is then designed to confirm that the conditions are as predicted and to provide ground information for the detailed design and project construction.

**Figure 1** illustrates the types of ground-related problem that might be investigated. It is important to identify the zone of influence relating to each example. Each project is unique and therefore requires a specially designed ground investigation.

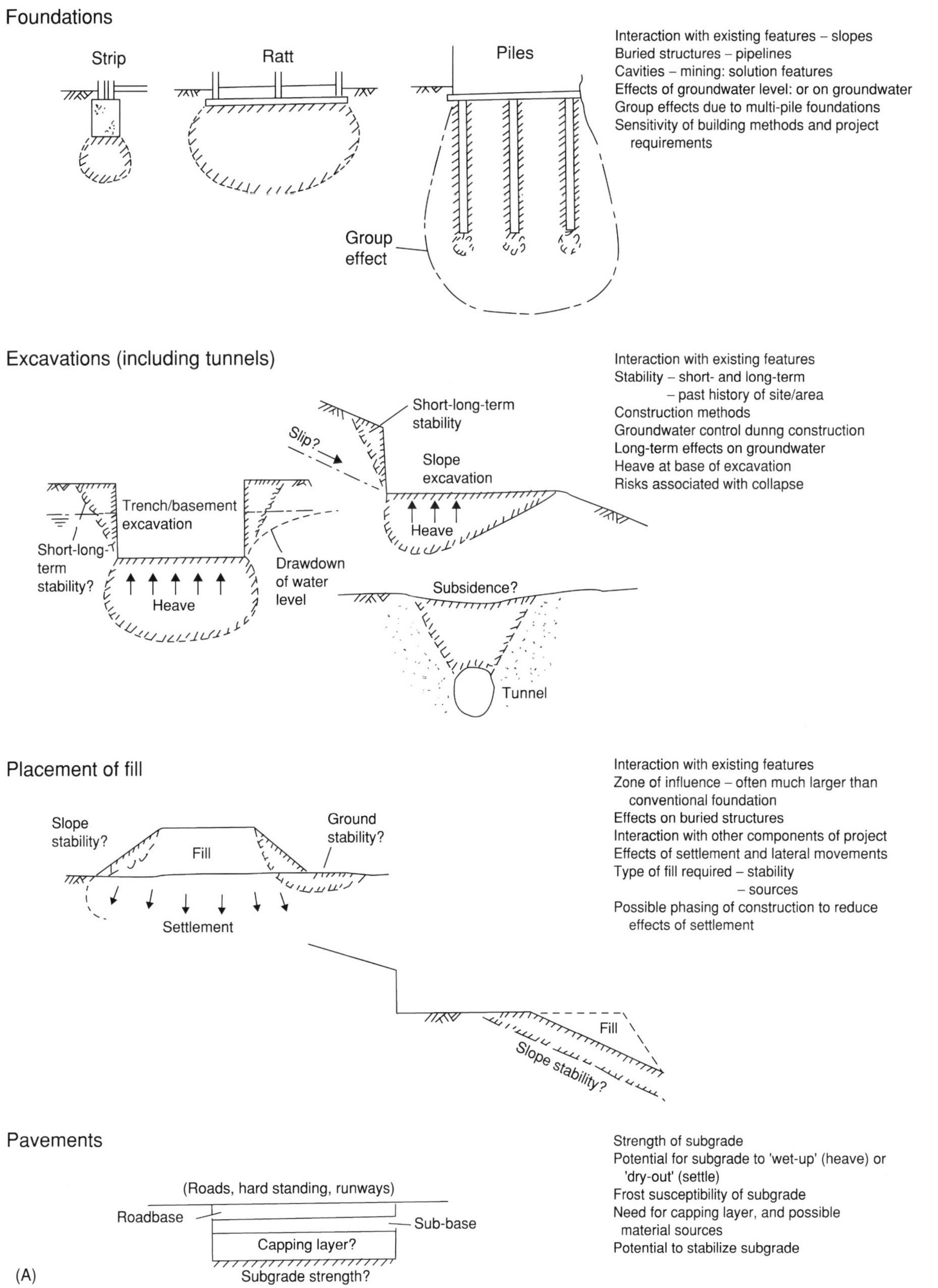

**ENGINEERING GEOLOGY/Site and Ground Investigation** 583

## Foundations

Strip  Ratt  Piles

Group effect

Interaction with existing features – slopes
Buried structures – pipelines
Cavities – mining: solution features
Effects of groundwater level: or on groundwater
Group effects due to multi-pile foundations
Sensitivity of building methods and project requirements

## Excavations (including tunnels)

Short-long-term stability
Slip?
Slope excavation
Heave
Trench/basement excavation
Short-long-term stability?
Heave
Drawdown of water level
Subsidence?
Tunnel

Interaction with existing features
Stability – short- and long-term
 – past history of site/area
Construction methods
Groundwater control during construction
Long-term effects on groundwater
Heave at base of excavation
Risks associated with collapse

## Placement of fill

Slope stability?
Fill
Ground stability?
Settlement
Fill
Slope stability?

Interaction with existing features
Zone of influence – often much larger than conventional foundation
Effects on buried structures
Interaction with other components of project
Effects of settlement and lateral movements
Type of fill required – stability
 – sources
Possible phasing of construction to reduce effects of settlement

## Pavements

(Roads, hard standing, runways)
Roadbase
Sub-base
Capping layer?
Subgrade strength?

(A)

Strength of subgrade
Potential for subgrade to 'wet-up' (heave) or 'dry-out' (settle)
Frost susceptibility of subgrade
Need for capping layer, and possible material sources
Potential to stabilize subgrade

**Figure 1** Construction zones to consider within site investigation. (Reproduced from Site Investigation Steering Group (1993) *Part 2 Planning, Procurement and Quality Management*. London: Thomas Telford.)

*Continued*

### Drainage/groundwater effects

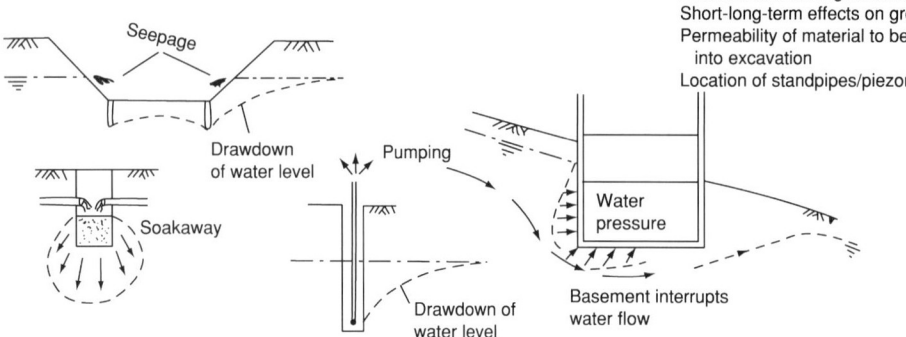

Type of drainage – soakaways, fin drains, slope drainage
Need for shor-term groundwater control
Short-long-term effects on groundwater levels
Permeability of material to be drained and/or rate of inflow into excavation
Location of standpipes/piezometers

### Vegetation: removal and planting

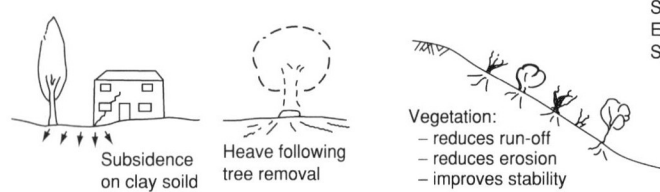

Effects of removel
Effects of planting
Slopes – depth/stability of topsoil
Erosion if topsoil omitted, e.g. chalk slopes
Special feature/requirements – liaise with landscape architect/horticulturalist

### Environmental factors, e.g. landfills, contaminated land

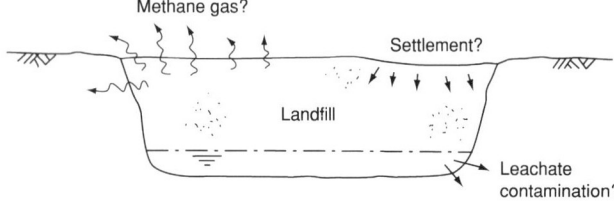

Nature and condition of existing material
Need to install monitoring equipment
Possible treatments prior to construction
Special factors e.g. isolation of vibration
Planning permission constraints
Designated areas – Sites of Special Scientific interest areas of outstanding natural beauty, protected flora/launa
Health and safety matters on contaminated sites

### Ground improvement

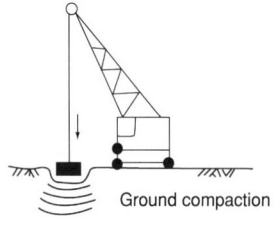
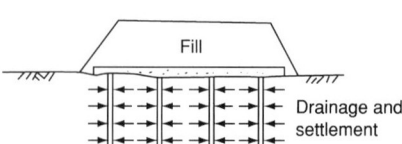

Parameter to be improved. e.g. strength, compaction, construction speed
Methods available and information required to assess their suitability
Dewatering – effects on groundwater levels
Consult specialist contractors

(B)

**Figure 1** Continued

### The Walk-Over Survey

The walk-over survey is a detailed inspection of the site. It is often done in stages, with an initial visit for familiarization, photography, and checking of the current site conditions, and subsequent visits to check out features noted on historical maps and photographs. Features should be sketched at an appropriate scale on a base plan for inclusion in the desk-study report.

### Derelict, Reused, and Contaminated Land

The desk study may reveal that a site has a history of previous use. Industrial use is likely to have left a legacy of pollution, which will require identification and possibly treatment depending on the intended reuse of the site. Guidance on the investigation of contaminated ground on the basis of 'source-pathway-receptor' is given in the key reference documents listed. The desk study will form an important element of the 'phase 1 risk assessment' that is required by contaminated-land legislation. An environmental specialist will need to be consulted whenever contamination is suspected, and appropriate safety precautions should be taken at all stages of the investigation work.

### Reporting the Desk Study

On completion of the desk study, the results are formally presented in a report that brings together details of site topography, geology, geotechnical problems and parameters, groundwater conditions, existing construction and services, previous land use, expected construction risk, and proposed ground-investigation methods. A hazard plan is a good way of presenting a summary of the data accumulated during the desk study and walk-over survey.

## Properties to be Determined

The prime purpose of a site investigation is generally to determine the ground profile and the continuity and properties of the various strata. This will include ranges of values and assessments of probability and risk, so that construction may proceed within a defined framework of knowledge.

## Ground Investigation

### Ground Investigation Design

The objective of a ground investigation is to ensure economical design and construction by reducing to an acceptable level the uncertainties and risks that the ground poses to the project.

The initial appraisal (desk study) will identify what is already known and will facilitate a preliminary understanding of the ground and its behaviour. This provides a basis for assessing the nature, location, extent, and duration of subsequent fieldwork and for preparing a programme of laboratory tests of samples obtained during the fieldwork. It is the geotechnical model against which every piece of acquired data can be checked. As the field programme progresses, so the model will be either confirmed or amended.

The design of the ground investigation can commence only once the following information has been obtained and client approval given:

1. a clearly defined purpose for the investigation;
2. an assessment of what information is required and when;
3. the areas and depths of ground to be investigated;
4. the time required for the investigation; and
5. an estimate of the cost.

These items can be conveniently covered by the Procedural Statement, as discussed previously.

### Ground Investigation Methods

The ground investigation can include many different activities such as:

- trial pits with descriptions of the material exposed;
- boreholes with sampling for later laboratory testing;
- tests in boreholes using simple or sophisticated instruments;
- probing from the ground surface;
- loading tests at the surface or in excavations;
- geophysical testing; and
- geochemical sampling.

Ground-contamination investigations are often required, especially on derelict sites; appropriate expertise is essential as there are important health and safety considerations.

### How Many? How Deep?

The scope and size of the ground investigation will depend both on what is known about the site and on the nature of the project. The ground investigation may, therefore, vary from a few trial pits dug by an excavator in one day, for a small housing project, to a major undertaking lasting many months, for a large earth dam. There are often alternative approaches that are equally acceptable technically, but sometimes one method is preferable to the other because of plant availability or access restraints. Boreholes may to some extent be replaced with trial pits, cone penetrometer probes, or geophysical surveys to provide similar or complimentary information. Cost is always important, but the designer should prepare the 'right' investigation for the project rather than working to a fixed budget provided by the client.

Teamwork is important and involves pooling ideas and suggestions and taking account of restraints imposed by the client, structural designers, auditors, or environmental specialists, who may have valuable input that can be used to support the geotechnical engineer in preparing the investigation. Again, the Procedural Statement is a useful vehicle for conveying proposals to all interested parties.

Time should be allowed for innovative design and review of proposals. 'Sleeping' on ideas for a day or two can often lead to the development of better alternatives. Checklists are useful to remind the designer of aspects that should be covered.

Guidance on spacing and depth of and sampling in exploratory holes may be found in the publications of the Site Investigation Steering Group and other key texts.

There may be a temptation to believe that everything about a site should be discovered at the investigation stage. It may be more realistic to accept that certain local features (for example swallow holes in chalk, mine workings beneath a highway cutting, or precise founding levels in variable strata) are best picked up and reviewed during the construction or remediation phase. This could be more cost-effective than attempting to detect every void and strata variation during the investigation, but it does require an element of flexibility and the application of appropriate expertise during the construction work. Such flexibility is more readily available in the 'design and construct' type of construction contract and in the use of 'observation methods', which are gaining popularity with clients.

## Ground Investigation Fieldwork

The various approaches that might be considered for gaining information on ground conditions are briefly reviewed in this section.

### Trial Pitting

Trial pitting is a relatively cheap and efficient method of exploring the ground. Techniques vary from digging by hand – useful where services may be present or access by machine is difficult – and using mini excavators – where access is restricted – to using large-tracked back-hoe type excavators, which can reach depths of 6 m or more (**Figure 2A**).

Trial pits enable the stratification and the true nature of each soil horizon to be logged and representative samples selected. Block samples can be cut for subsequent laboratory testing, and localized samples can be taken from shear surfaces. However, access to the pit for close visual inspection is permitted only where temporary side support (shoring) is provided and a careful assessment of the risk of instability and the presence of gas has been made.

When inspecting the ground, a small wedge of soil is frequently sheared away from the side of the pit to reveal the true nature of the soil behind the zone of smear resulting from excavation. A moisture-content profile using close centres (typically 75 mm) can provide much information about the consistency of the soil. This is easily and accurately undertaken using the 'moisture in the bag' technique. The moisture

**Figure 2** Two common methods of exploration. (A) The trial pit, with shoring and ladder for safe access and egress. (B) The percussion boring rig.

samples are sealed in a lightweight polythene bag (**Figure 3**). They are then weighed and dried in the bag, avoiding the need for separate moisture-content containers and reducing the error due to dehydration during sample storage.

Observation of the sides of the pit will provide an indication of the stability of the ground. The nature of the excavated material will provide information on possible reuse during earthworks. If a trial pit can be safely protected and left overnight, drying of the soil surface often reveals fissures, discontinuities, and possible existing slip surfaces within the strata.

Groundwater conditions and their effect on ground stability may be closely observed in the pit. Instability due to groundwater flow may prevent safe advancement of the pit in finer-grained granular soils. Permeability testing (by pumping or by water addition) may be carried out to give an indication of the ground permeability.

On completion of the excavation, strata should be replaced and compacted back to the depth excavated. The location (grid coordinates) of the pit must be carefully recorded as the ground is now disturbed and could present a hazard to future construction. For this reason, pits are generally excavated outside the planned area of the proposed foundations. The positions of exploratory holes are readily established by modern handheld global positioning systems.

### Boring Techniques – Soft Ground

The light cable tool percussion ('shell and auger') technique of soft-ground boring at a diameter of 150 mm is perhaps the best-established, simplest, and most flexible method of boring vertical holes, particularly in the UK (**Figure 2B**). It generally allows data to be obtained for strata other than rock. A tubular cutter (for cohesive soil) or a shell with a flap valve (for granular soil) is repeatedly lifted and dropped using a winch and rope operated from a tripod frame. The soil that enters these tools is regularly removed and laid to one side for backfilling. Steel casing is used where necessary to prevent collapse of the sides of the borehole.

The technique can determine conditions to depths in excess of 30 m under suitable circumstances and usually causes less surface disturbance than trial pitting.

Small disturbed samples are taken on encountering each new stratum and at regular intervals within it. Larger samples of the materials are taken typically at 1 m intervals in the top 5 m of the borehole and subsequently at 1.5 m intervals. In cohesive or 'fine grained' soils, sampling is by nominal 100 mm diameter 0.45 m long open-tube drive sampling (U100), and the sample is often referred to as a (relatively) 'undisturbed' sample. These are sealed with wax to preserve moisture and capped at each end prior to transport to the laboratory for testing.

In granular soils a standard penetration test (SPT) is typically undertaken at 1 m to 1.5 m intervals, depending on depth, and a bulk sample is removed from the tested length for subsequent description and classification. The SPT involves driving a 50 mm diameter tube or cone into the ground using a 65 kg weight falling a distance of 760 mm and controlled using a trip mechanism. Blows are counted for successive penetrations of 75 mm for a distance of 450 mm. The SPT value is the number of blows required to penetrate the final 300 mm. This test is sometimes also used in firm clays. The results of the SPT are correlated empirically with the density and angle of shear resistance in granular soils and with shear strength in cohesive soils.

Alternative techniques that might be considered for advancing boreholes include the less reliable wash-boring method, where the soil is washed out of the cased hole, and flight auger boring, where disturbed samples are brought to the surface by the rotation of the auger and, where hole stability permits, the auger is removed for undisturbed sampling.

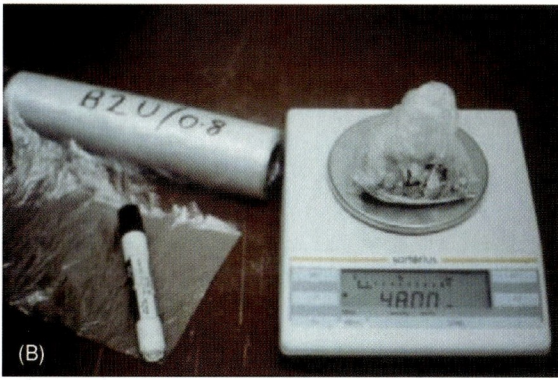

**Figure 3** 'Moisture in the bag'; an efficient method of determining moisture contents.

The hollow-stem flight auger is a more reliable system, which supports the hole whilst undisturbed samples are recovered through the central stem.

For shallow holes where access is difficult, hand augering may be employed. The use of a lightweight mechanical hammer associated with dynamic probing has led to the recent development of the 'window' sampling technique. This involves driving a tube with a section of the side wall removed to allow the soil sequence to be logged and samples to be selected from the side for classification and, increasingly, for contamination testing. Sample-tube lengths are typically 1 m or 1.5 m. Deeper sampling is possible by progressively reducing the tube diameter to reduce side friction from the strata above the sampling level.

### Boring Techniques – Hard Ground

Rotary drilling produces rock cores by rotating an annular diamond-impregnated tube or barrel into the ground. This is the technique that is most appropriate for making investigation holes in rock or other hard strata. It may be used vertically or at any angle.

Core diameters of less than 100 mm are most common for site-investigation purposes. A flushing fluid such as air, water, mist, or foam is used to cool the bit and carry cuttings to the surface.

Examination of rock cores allows detailed description and generally enables angled discontinuity surfaces to be observed. However, it does not necessarily reveal the presence of near-vertical fissures or joint discontinuities. The core can be tested in the field or in the laboratory. Core recovery depends on rock type and the techniques employed.

Where open-hole rotary drilling is employed, descriptions of the strata depend on the examination at the surface of small particles ejected from the borehole in the flushing medium. Consequently, no indication of fissuring, bedding, consistency, or degree of weathering can be obtained. Depths in excess of 60 m can be investigated using rotary techniques with minimal surface disturbance.

### Dynamic and Static Probing

Dynamic probing involves driving a rod with a fixed or detachable cone into the ground by allowing a falling weight to impact on an anvil attached at the top of the rod and counting the number of blows required to achieve successive penetrations, typically of 100 mm. The apparatus can be handheld (for example the Mackintosh probe), or a machine-mounted weight and guide can be used for larger-diameter rods (for example the Pennine probe) (**Figure 4**).

(A) Lightweight drilling, dynamic probe rig (Courtesy Exploration Associates)

(B) Lightweight dynamic (Mackintosh) hand probing for indication of soil density

**Figure 4** Examples of dynamic probing (A) by lightweight drilling and (B) by hand. (A) A dynamic probe rig (photograph courtesy of Soil Mechanics). (B) Lightweight dynamic (Mackintosh) hand probing to obtain an indication of soil density.

Whilst the number of blows may be correlated empirically with the shear strength or density of granular soils, these are generally profiling tools best used for interpolating between boreholes or trial pits. Samples are not normally taken, but the technique has been extended by using the hammer apparatus to take 'window' samples, as described above.

The static cone technique (**Figure 5**) involves pushing an instrumented cone and sleeve into the ground at a defined rate and recording the point resistance and skin friction using electronic means. More sophisticated instruments can now monitor the pressure of pore water just behind the cone tip. By considering the ratio of end resistance to shaft friction resistance, an indication of the soil type and its density may be deduced.

The apparatus can measure changes in soil type over very short distances and avoids the problems of disturbance, particularly of granular soils, below the water table. It is a very rapid technique, often used as a profiling tool between boreholes, and is capable of electronic interpretation. It has been used extensively in North Sea oil and gas structure foundations and is now used widely for land-based investigations.

**Figure 5** The static cone apparatus. (A) 'Dutch' cone penetrometer testing truck with (B) control console and (C) electric static cone (photographs courtesy of Fugro Ltd).

As with all probing techniques, the presence of large cobbles and boulders can prevent penetration in certain soils and in made ground. Although linked with boring techniques, the static and dynamic cones are regarded as a form of *in situ* test.

## *In Situ* Testing

The main objectives of a ground investigation are generally to identify and classify the soil types into groups of materials that exhibit broadly similar engineering behaviour and to determine the parameters that are required for engineering design calculations.

Some soils, such as certain clays, may be readily sampled and transported to the laboratory for quality testing under controlled conditions. Other soils, such as very soft or sensitive clays, stony soils, sands, and gravels, and weak, fissile, or fractured rock are not easily sampled in an intact 'undisturbed' state, and therefore *in situ* testing is required to obtain the necessary engineering parameters. *In situ* tests may take the form of geophysical tests (as described in the following section), *in situ* soil-testing techniques, or links with field instrumentation.

*In situ* tests fall into three typical groups.

1. Empirical tests, where no fundamental analysis is possible and stress paths, drainage conditions, and rate of loading are either uncontrolled or inappropriate (example is SPT).
2. Semi-empirical tests, where a limited relationship between parameters and measurements may be developed (examples are cone penetrometer tests, California Bearing Ratio (CBR) and borehole permeability tests).
3. Analytical tests, where stress paths are controlled (but strain levels and drainage often are not) (examples are vane shear strength, pressuremeter tests, pumping tests and packer permeability).

Details of the tests and their application are given in British Standards 1377 and 5930 and other references listed in the further reading section. The main point to note with *in situ* testing is that the drainage conditions are almost impossible to control, and, therefore, there is uncertainty as to the drained/undrained nature of the test. However, the very act of testing *in situ* provides a good indication of the actual ground response provided that scale factors are taken into account. For example, a plate-load test having a plate width of $B$ will affect the ground to a depth of approximately $1.5B$. The actual foundation may be many times the width of the test plate, and the depth of influence will therefore be much deeper.

The SPT is a simple cost-effective means of assessing granular soils. More analytically correct tools such as the pressuremeter are expensive and their results are not always straightforward to interpret. Judgement must be used to determine the level of sophistication required for a particular project.

## Geophysical Investigation

Measurement of the geophysical properties of the ground may provide an indication of the location of strata boundaries and anomalous ground conditions. Such surveys may include measurements of ground conductivity, magnetic and gravity fields, electrical resistivity, and seismic responses (*see* **Analytical Methods:** Gravity, **Seismic Surveys**).

The use of geophysical techniques is generally non-intrusive, with little site disturbance, and is therefore sometimes carried out at the desk-study stage. However, the main benefit comes when some exploratory hole data are available to permit more accurate calibration, correlation, and processing of the geophysical data. The ability of the current generation of computers to process large amounts of data rapidly has led to a resurgence of interest in the potential value of geophysical investigation techniques.

## Groundwater and Instrumentation

The presence of groundwater leads to many engineering problems, and it is vital that all water observations are carefully recorded. Whenever water is struck in a borehole or trial pit the point of entry must be noted. It is normal practice to cease the advancement of the hole for 20 min whilst any rise in the water level is observed. Longer-term observations of groundwater fluctuations may be made by installing a standpipe or standpipe piezometer. This typically comprises a porous filter at the base of a 19 mm diameter plastic tube inserted to an appropriate depth in a borehole (**Figure 6**). Water enters through the porous filter, and its level in the plastic tube can be monitored over a period of time using a dip meter that emits an audible signal when an electrical circuit is completed as it meets the water in the borehole.

Water-level records taken over an appropriate period of time will reveal the likely range of water conditions to be allowed for in the design and to be encountered during the project construction.

More sophisticated equipment such as hydraulic, electrical, or pneumatic piezometers with transducer systems and automatic recording can be used where the information obtained is critical to the project.

Other instrumentation, such as settlement gauges, plates, and extensometers to measure vertical movement, and slip indicators and inclinometers to measure lateral movement at depth, are available for use in conjunction with the ground investigation or with advance trial constructions and main works constructed by 'observational' methods.

## Laboratory Testing

It should be remembered that soil is an assemblage of solid particles, which may contain organic matter. The voids between the particles are filled with gas – usually air – and water. The geotechnical engineer will need to determine the composition of the ground *in situ* and also explore the changes that will occur to the soil as a result of the proposed construction.

Laboratory tests tend to be divided into two main classes.

Classification tests involve grouping soil types into categories of soils that possess similar properties. For coarse-grained soils the particle size distribution is often the key property, whereas for fine-grained soils the moisture content and mineralogy, as reflected in the Atterberg limits (index properties), most often hold the key to grouping soils of similar characteristics. For general classification purposes, the moisture content and index properties (liquid limit, plastic limit, and sometimes shrinkage limit) are used together with the particle size distribution, density of the soil, and, most importantly, an accurate description carried out in accordance with the standards in the code of practice.

Engineering tests assess the engineering properties of the soil, such as routine shear strength, compressibility (see **Figure** 7), compaction characteristics, and permeability. Where more specific soil properties are required for complex design functions, 'advanced' laboratory tests may be used, in which high-quality undisturbed samples are carefully prepared and tested under precise laboratory control and stress and small strains are measured.

Full details of laboratory tests and their relevance to engineering design are given in the appropriate British Standards and the key references in the further reading section.

It is important that the testing regime is planned in advance of the fieldwork in order that suitable samples are recovered from the ground. The Procedural Statement will help with this planning process, as the reasons for the sampling and subsequent

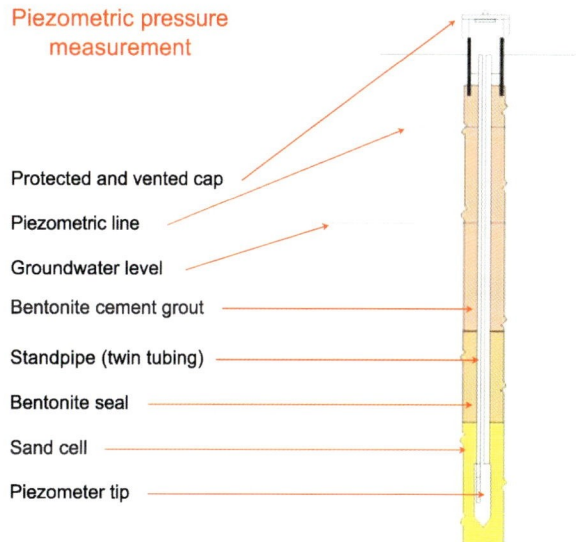

**Figure 6** Schematic of standpipe piezometer (courtesy of Geotechnical Instruments UK Ltd).

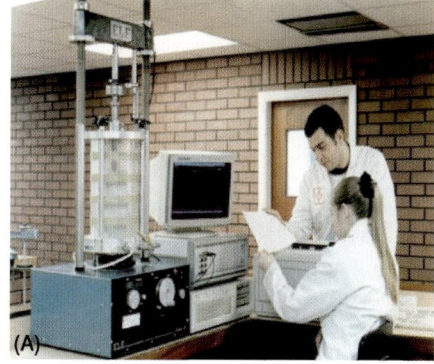

**Figure 7** Commercial laboratory testing for shear strength and consolidation. (A) A 100 mm triaxial test cell. (B) A bank of consolidation (oedometer) test machines. (Photographs courtesy of CSL Ltd.)

laboratory testing can be logically considered together with the significance of the anticipated results.

## Reporting

The reporting of site and ground investigation work is most important. The factual report is generally the only tangible output from the investigation and represents a considerable financial investment. The factual report should include:

- a statement as to the purpose and rationale of the investigation;
- a description of the work carried out, including references to the specification and standards adopted and any deviations from them;
- exploratory hole logs (including location and ground level);
- *in situ* test records;
- laboratory test results; and
- a site location plan.

The development by the Association of Geotechnical and Environmental Specialists of a standard format for geotechnical data files in digital form has led to the efficient collection, transfer, processing, presentation, and storage of the data for all parties involved in the investigation process. Data are ideally input manually to the computer only once or are collected directly in electronic form from monitoring and testing devices. Specialist software houses have developed programs to help the investigator collect, store, and process the data and to present it in the form most suited to the requirements of a particular project (**Figure 8**).

The exploratory hole log can perhaps be regarded as the heart of the factual report because it records all the information obtained at that particular location. An example log prepared from digital data is presented in **Figure 9**.

Interpretation of the geotechnical data to relate them to the proposed design and construction will be required for each project and is normally the responsibility of the Geotechnical Advisor. Formal interpretive reports (also known as assessment reports or appraisal reports) may be prepared as required by the design team. It is good practice to prepare a

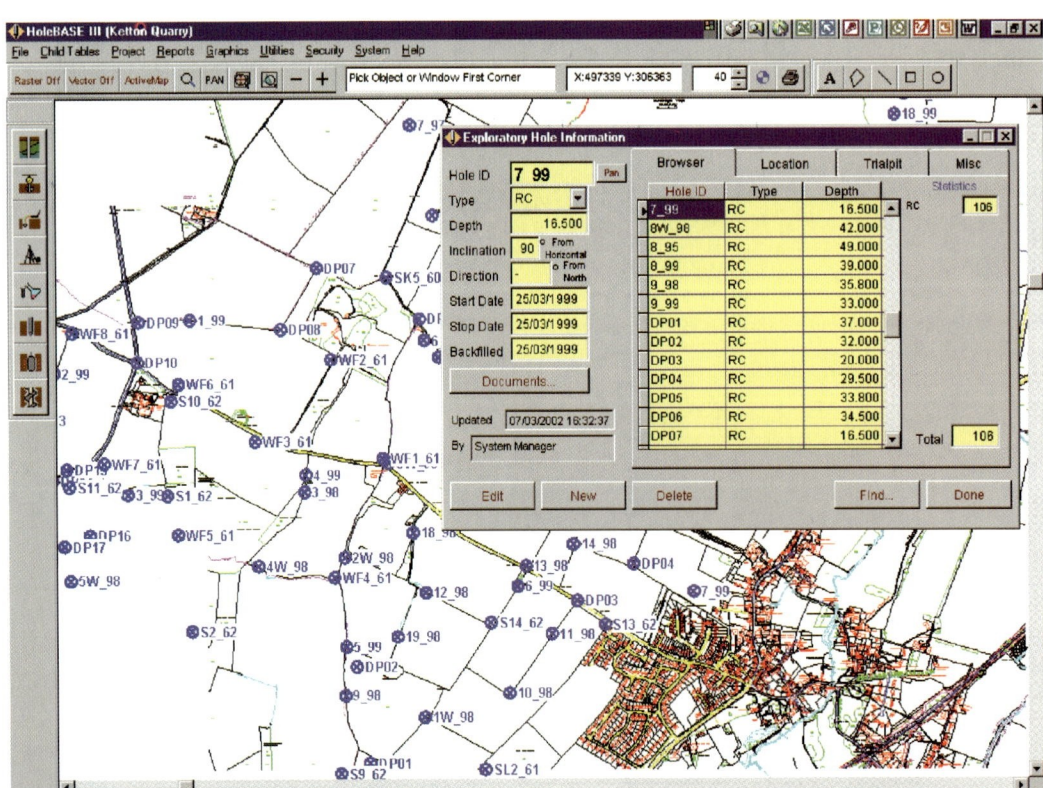

**Figure 8** Use of digital techniques for efficient handling of site investigation data (image courtesy of Key Systems and Geotechnical Developments).

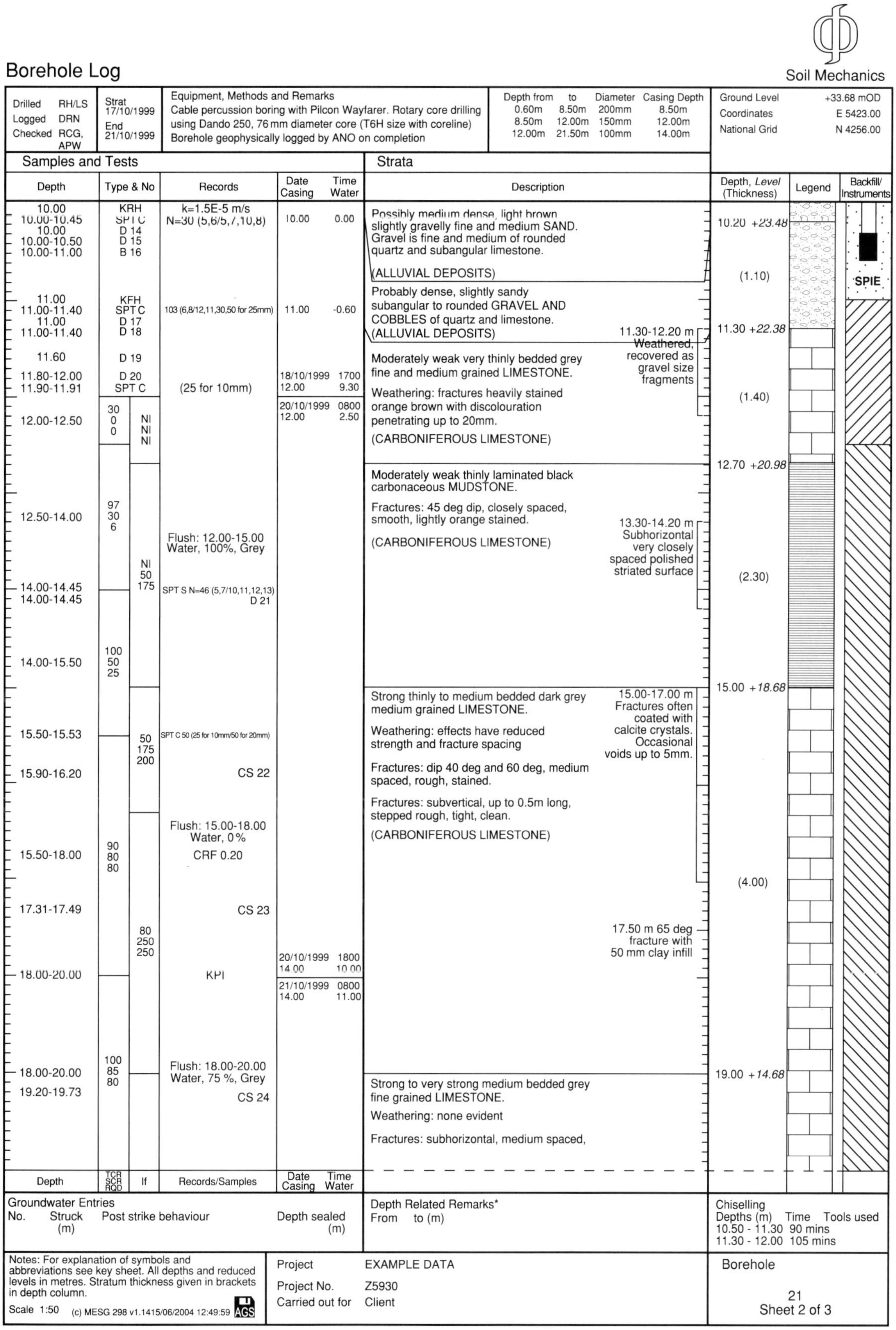

Figure 9  Example exploratory hole log (Courtesy Soil Mechanics).

geotechnical design brief to assist the construction team on site and, at the end of the project, to prepare a feedback (maintenance) report on any issues from which lessons can be learned and on which future scheme development can be based.

The various stages of quality control and reporting that help to ensure maintenance of the standards and accuracy of the output from a site and ground investigation are described in the publications of the Site Investigation Steering Group.

## Concluding Remarks

Site investigation involves establishing facts concerning the ground; the excitement of discovery is tempered by the professional responsibility for much of the financial risk and safety associated with each engineering project.

The combination of appropriately experienced investigators working to agreed team procedures within a sensible permitted time frame with limited competition is likely to produce the best solutions to engineering challenges presented by particular ground conditions. This article has touched on some of the important issues relevant to site investigation but is in no way intended to be a complete guide. A good investigator will set out to define the investigation but to expect the unexpected.

## See Also

**Analytical Methods:** Gravity. **Engineering Geology:** Codes of Practice; Site Classification; Ground Water Monitoring at Solid Waste Landfills. **Geotechnical Engineering. Seismic Surveys. Soil Mechanics.**

## Further Reading

BS 5930 (1999) *Code of Practice for Site Investigation*. London: BSI.

BS 1377 (1990) *British Standard Methods of Test for Soils for Civil Engineering Parts 1–9*. London: BSI.

BS 10 (2000) *Contaminated Land*. London: BSI.

BS 10175 (2001) Investigation of potentially contaminated sites. London: BSI.

Clayton CRI, Matthews MC, and Simons NE (1995) *Site Investigation: A Handbook for Engineers*, 2nd edn. Oxford: Blackwell Scientific.

HD22/02 Managing geotechnical risk. DMRB. London: Highways Agency.

Institution of Civil Engineers (2003) *Conditions of Contract Ground Investigation*, 2nd edn. London: Thomas Telford.

McCann DM, Eddlestone M, Fenning PJ, and Reeves GM (eds.) (1997) *Modern Geophysics in Engineering*. Special Publication 12. London: Geological Society.

Perry J (1996) *Sources of Information for Site Investigations in Britain*. TRL Report 192. Crowthorne: Transport Research Laboratory.

Simons N, Menzies B, and Matthews M (2002) *A Short Course in Geotechnical Site Investigation*. London: Thomas Telford.

Site Investigation Steering Group (1993) *Part 1: Without Site Investigation Ground is Hazard*. London: Thomas Telford.

Site Investigation Steering Group (1993) *Part 2 Planning, Procurement and Quality Management*. London: Thomas Telford.

Site Investigation Steering Group (1993) *Part 3 Specification for Ground Investigation*. London: Thomas Telford.

Site Investigation Steering Group (1993) *Part 4 Guidelines for Safe Investigation by Drilling of Landfills and Contaminated Land*. London: Thomas Telford.